Edited by
Sushil K. Misra

**Multifrequency Electron
Paramagnetic Resonance**

Details of the figures on the book cover

Left: The black traces are a set of model 1.6 GHz second-harmonic experimental EPR spectra corresponding to mixtures of 100–0% Cu(II)-imidazole and 0–100% Cu(II)-KTSM [3-ethoxy-2-oxobutyraldehyde-bis(N4-methylthiosemicarbazonato) copper(II)], respectively, over the perpendicular region and the m(I) = +1/2 parallel line. The red traces are an automated fit to the model experimental spectra using simulations of the individual components as basis spectra. The simulations provided anisotropic copper and nitrogen hyperfine terms. The fits predicted the fraction of the major component to within 5% for each mixture. The feature at 580 G is a free radical contaminant in the Cu(II)-KTSM sample.

Right, from top to bottom: (i) SECSY: Two-Dimensional EPR spectrum of a spin labeled peptide: Gramicidin, in an aligned lipid membrane near room temperature taken at 95 GHz and showing its g-tensor resolution; (ii) An illustration of the model used to analyze multifrequency EPR spectra of spin-labeled proteins, in terms of internal and overall modes of motion. (iii) EPR micro-images of a LiPc crystal: (upper) 2D microimages; (middle) a series of 2D images for different Z-slice sections, lower right image is the optical image of the LiPc crystal; (bottom) a 3D stacking of the 2D images. (iv) The time domain Pulse-Dipolar EPR signal taken at 17 GHz from an RNA/DNA duplex corresponding to a distance between spin labels of 80Å. The time evolution shown extends to 20 micro-secs.

Edited by Sushil K. Misra

Multifrequency Electron Paramagnetic Resonance

Theory and Applications

WILEY-VCH Verlag GmbH & Co. KGaA

The Editor

Prof. Dr. Sushil K. Misra
Concordia University
Physics Department
1455 de Maisonneuve Blvd. West
Montreal, QC H3G 1M8
Canada

■ All books published by Wiley-VCH are carefully produced. Nevertheless, authors, editors, and publisher do not warrant the information contained in these books, including this book, to be free of errors. Readers are advised to keep in mind that statements, data, illustrations, procedural details or other items may inadvertently be inaccurate.

Library of Congress Card No.: applied for

British Library Cataloguing-in-Publication Data
A catalogue record for this book is available from the British Library.

Bibliographic information published by the Deutsche Nationalbibliothek
The Deutsche Nationalbibliothek lists this publication in the Deutsche Nationalbibliografie; detailed bibliographic data are available on the Internet at http://dnb.d-nb.de.

© 2011 Wiley-VCH Verlag & Co. KGaA, Boschstr. 12, 69469 Weinheim, Germany

All rights reserved (including those of translation into other languages). No part of this book may be reproduced in any form – by photoprinting, microfilm, or any other means – nor transmitted or translated into a machine language without written permission from the publishers. Registered names, trademarks, etc. used in this book, even when not specifically marked as such, are not to be considered unprotected by law.

Cover   Formgeber, Eppelheim
Typesetting   Toppan Best-set Premedia Limited
Printing and Binding   Strauss GmbH, Mörlenbach

Printed in the Federal Republic of Germany
Printed on acid-free paper

ISBN: 978-3-527-40779-8
ePDF ISBN: 978-3-527-63354-8
ePub ISBN: 978-3-527-63355-5
oBook ISBN: 978-3-527-63353-1

Contents

Preface *XXIX*
List of Contributors *XXXI*

1	**Introduction** *1*	
	Sushil K. Misra	
1.1	Introduction to EPR *1*	
1.1.1	Continuous-Wave EPR *1*	
1.1.2	Pulsed EPR *2*	
1.1.3	EPR Imaging *2*	
1.2	Historical Background of EPR *2*	
1.2.1	Literature Pertinent to the Early History of EPR *3*	
1.3	Typical X-Band, Low-, and High-Frequency Spectrometers *3*	
1.3.1	EPR Spectrometer Design *3*	
1.3.2	X-Band Spectrometer *4*	
1.3.2.1	Source of Microwave Radiation *4*	
1.3.2.2	Transmission of Microwaves *6*	
1.3.2.3	The Cavity (Resonator) System *6*	
1.3.2.4	Magnetic Field System *7*	
1.3.2.5	Modulation and Detection System *8*	
1.3.3	EPR Line Shapes and Determination of Signal Intensity *9*	
1.3.4	Low-Frequency Spectrometers *9*	
1.3.5	High-Frequency Spectrometers *10*	
1.3.5.1	Sources of Radiation *10*	
1.3.5.2	Transmission of Submillimeter Waves *12*	
1.3.5.3	Resonators and Sensitivity *13*	
1.3.5.4	Magnetic Field *14*	
1.3.5.5	Detectors *14*	
1.3.6	Pertinent Literature *15*	
1.4	Applications of EPR *15*	
1.4.1	Pertinent Literature *20*	

1.5	Scope of This Book 20
	Acknowledgments 21
	Further Reading 21

2 Multifrequency Aspects of EPR 23
Sushil K. Misra

2.1	Frequency Bands 23
2.2	X-Band EPR 23
2.3	EPR at Higher Frequencies (HF) 24
2.3.1	Advantages 24
2.3.2	Disadvantages 32
2.4	Low-Frequency EPR 34
2.4.1	Advantages 34
2.4.2	Disadvantages 38
2.5	Multifrequency EPR 39
2.5.1	Advantages of Using Multifrequency EPR 39
2.5.2	Limitations of Using Multifrequency EPR 51
2.5.3	Size of Resonant Cavity at Different Frequencies 51
2.5.4	Signal-to-Noise Ratios at Different Frequencies 53
2.5.5	Multifrequency Aspects of Using Home-Built versus Commercial Spectrometers 53
2.5.6	Multifrequency Aspects of Sample-Related Problems 53
	Acknowledgments 53
	Pertinent Literature 53
	References 54

3 Basic Theory of Electron Paramagnetic Resonance 57
Sushil K. Misra

3.1	Introduction 57
3.2	Crystal-Field Theory 58
3.2.1	Introduction to CFT 58
3.2.2	Free Atoms and Ions 59
3.2.3	The Crystal-Field Description of Transition Group Ions in Crystals 62
3.2.3.1	p-Orbitals 63
3.2.3.2	d-Orbitals 63
3.2.4	Crystal Field Potential 65
3.2.5	Point Charge Model 67
3.2.5.1	Potentials for Cubic and Lower Symmetry 68
3.2.6	Equivalent Operators and the Wigner–Eckart Theorem 69
3.2.7	Properties of d-Electrons in Crystal Fields 71
3.2.7.1	Ions with Several d-Electrons: Strong- and Weak-Field Cases 71
3.2.7.2	Energies and Wave-Functions for d-Electrons 74
3.2.7.3	Crystal-Field Parameters for d-Electrons 75
3.2.7.4	Crystal-Field Splittings for $3d^1$ and $3d^9$ Configurations 77

3.2.7.5	The Ground State and its Relationship to EPR: Quenching of Orbital Angular Momentum and Calculation of g-Factors	*78*
3.2.8	The Rare-Earth Ions	*80*
3.2.8.1	Crystal Fields for Rare-Earth Ions: Dominant Spin–Orbit Coupling	*81*
3.2.9	Irreducible Representations for CF Energy Levels	*81*
3.2.10	Critique of Crystal-Field Theory	*82*
3.2.11	Kramers' Theorem	*82*
3.3	Superposition Model (SPM)	*83*
3.4	Molecular Orbital (MO) Approach	*85*
3.4.1	Linear Combination of Atomic Orbitals (LCAO)	*85*
3.4.2	Extended Hückel Molecular Orbital Theory (EHMO)	*88*
3.4.3	Ligand Field Theory: The Angular Overlap Model (AOM)	*88*
3.5	The Jahn–Teller (JT) Effect	*92*
3.5.1	Theory of the JT Effect	*95*
3.5.1.1	General Theory of the JT Effect	*96*
3.5.2	Perturbation within the Vibronic Ground State	*98*
3.5.3	Three-State Model	*100*
3.5.4	Transition from Dynamic to Static JT Effect	*101*
3.6	The Spin Hamiltonian	*102*
3.6.1	The Abragam and Pryce Spin Hamiltonian for the Iron Group	*102*
3.6.1.1	Incorporation of Covalency	*105*
3.6.2	Zero-Field Splitting (ZFS)	*105*
3.6.2.1	Cubic Zero-Field Splitting ($S > 3/2$)	*106*
3.6.3	The Phenomenological Spin Hamiltonian	*106*
3.6.3.1	Triclinic Symmetry	*107*
3.6.3.2	Monoclinic Symmetry (C_{2h}, C_2, C_{2s})	*108*
3.6.3.3	Orthorhombic Symmetry (D_{2h}, D_2, D_{2v})	*108*
3.6.3.4	Tetragonal (D_{4h}, D_4, C_{4v}, D_{2d}, C_{4h}, S_4, and C_4)	*108*
3.6.3.5	Cubic (O_h, O, T_d, T_h, and T) and Spherical Symmetry	*108*
3.6.3.6	Additional Spin-Hamiltonian Terms with Higher Powers of Components of S	*108*
3.6.4	The Generalized Spin Hamiltonian	*110*
3.6.5	The Effective Spin Hamiltonian for EPR	*110*
3.7	Concluding Remarks	*111*
	Acknowledgments	*111*
	Pertinent Literature	*111*
	References	*111*

Part One   Experimental   *115*

4	**Spectrometers** *117*	
4.1	Zero-Field EPR *117*	
	Sushil K. Misra	

4.1.1	Introduction 117
4.1.2	Preliminary Theory of ZFR 118
4.1.3	The ZFR Spectrometer 119
4.1.3.1	Examples of ZFR Spectra 119
4.1.4	Advantages of Using Resonant Systems 121
4.1.5	Examples of ZFR 121
4.1.5.1	The Case of the Mn^{2+} Ion 122
4.1.6	Concluding Remarks 125
	Pertinent Literature 126
	References 128
4.2	Low-Frequency CW-EPR Spectrometers: 10 MHz to 100 GHz 128
	Harvey A. Buckmaster
4.2.1	Introduction 128
4.2.2	CW-EPR Spectrometer Configurations 132
4.2.3	Theoretical Sensitivity 146
4.2.4	EPR Lineshapes and Modulation Broadening 148
4.2.5	Microwave Power Sources 149
4.2.6	Reflex Klystrons 151
4.2.7	Solid-State Devices 151
4.2.8	Frequency Synthesizers 153
4.2.9	Microwave CW-EPR Sample Cavity Designs 153
4.2.10	Transmission Cavities 156
4.2.11	Reflection Cavities 157
4.2.12	Re-Entrant Cavities 159
4.2.13	Loop–Gap Cavities 160
4.2.14	Other Resonant Structures 163
4.2.15	Microwave Detectors or Demodulators 164
4.2.15.1	Point Contact Diodes 164
4.2.15.2	Schottky Barrier Diodes 165
4.2.15.3	Backward Diodes 165
4.2.15.4	Bolometers 165
4.2.16	Electromagnets 166
4.2.17	Zero-Field CW-EPR 167
4.2.18	Support Instrumentation 168
4.2.19	Concluding Remarks 168
4.2.20	Pertinent Literature 169
	References 169
	Appendix 4.2.I 171
	Appendix 4.2.II 173
	Appendix 4.2.III 174
4.3	High-Frequency EPR Spectrometers 175
	Edward Reijerse
4.3.1	Introduction 175
4.3.2	High-Frequency EPR Spectrometer Configurations 176
4.3.3	Sensitivity Considerations 182

4.3.3.1 Cavity and Sample Holder *183*
4.3.3.2 Reflection Cavity with Square-Law Detector *184*
4.3.3.3 Reflection Cavity with Linear Detector *184*
4.3.3.4 Spectrometer Bridge and Detector *185*
4.3.4 Conclusions and Future Perspectives *188*
Pertinent Literature *188*
References *188*
4.4 Pulsed Techniques in EPR *190*
Sankaran Subramanian and Murali C. Krishna
4.4.1 Introduction *190*
4.4.2 Components of a Pulsed EPR Spectrometer *193*
4.4.2.1 K_a-Band (26.5–40 GHz) Pulsed EPR Spectrometer *194*
4.4.2.2 Radiofrequency Pulsed EPR Spectrometers Operating at 300, 500, and 750 MHz *197*
4.4.3 Resonators *199*
4.4.4 Pulsed Excitation and Relaxation *202*
4.4.5 Fourier Transform in Magnetic Resonance *202*
4.4.6 Simple Pulsed EPR Experiments *203*
4.4.6.1 Inversion Recovery and Hahn Echo Pulse Sequences, T_1 and T_2 *204*
4.4.7 Pulsed ENDOR, ESEEM, and HYSCORE *208*
4.4.7.1 Nuclear Modulation Effects Leading to ENDOR and ESEEM *209*
4.4.7.2 Mims and Davis Pulsed ENDOR Sequences *211*
4.4.8 Electron Spin Echo Envelope Modulation (ESEEM) and Hyperfine Sublevel Correlation Spectroscopy (HYSCORE) *214*
4.4.9 Electron–Electron Double Resonance (ELDOR), Double Electron–Electron Resonance (DEER), or Pulsed ELDOR (PELDOR) *218*
4.4.10 Double-Quantum EPR *220*
4.4.11 Concluding Remarks *222*
Pertinent Literature *224*
References *225*

5 Multifrequency EPR: Experimental Considerations *229*
5.1 Multiarm EPR Spectroscopy at Multiple Microwave Frequencies: Multiquantum (MQ) EPR, MQ-ELDOR, Saturation Recovery (SR) EPR, and SR-ELDOR *229*
James S. Hyde, Robert A. Strangeway, and Theodore G. Camenisch
5.1.1 Introduction *229*
5.1.2 Review of Frequency-Translation Techniques *231*
5.1.3 Review of Multiarm Bridges *233*
5.1.4 Multiarm Bridges at Higher Millimeter-Wave Frequencies *236*
5.1.5 Resonator Considerations for Multiarm Experiments *238*
5.1.6 Reference Arm and Receiver Design Considerations for Multiarm Experiments *239*
5.1.7 Discussion *241*
Pertinent Literature *243*

	Acknowledgments *243*
	References *243*
5.2	Resonators for Multifrequency EPR of Spin Labels *244*
	James S. Hyde, Jason W. Sidabras, Richard R. Mett
5.2.1	Introduction *244*
5.2.2	Methods *247*
5.2.2.1	Computer-Based Simulations *247*
5.2.2.2	Fabrication and Testing *251*
5.2.3	Aqueous Samples *252*
5.2.3.1	The Complex Dielectric Constant as a Function of Frequency and Temperature *252*
5.2.3.2	Dielectric Loss Types and Parallel and Perpendicular E-Field Geometries *253*
5.2.3.3	Results in Commercial Resonators at X-Band Using Extruded Sample Tubes *255*
5.2.3.4	Multichannel Design *256*
5.2.4	Uniform Field Cavities and Loop–Gap Resonators *258*
5.2.4.1	Intrinsic Uniformity *258*
5.2.4.2	Uniform Field Cavities *258*
5.2.4.3	Uniformity in Two Dimensions *258*
5.2.4.4	Loop–Gap Resonators *259*
5.2.5	Coupling *261*
5.2.5.1	Coupling at Low Frequencies *262*
5.2.5.2	Coupling at High Frequencies *262*
5.2.6	Field Modulation Penetration *263*
5.2.7	Sample Access Stacks *265*
5.2.8	Conclusions *268*
	Pertinent Literature *269*
	Acknowledgments *269*
	References *269*
5.3	Multifrequency EPR Sensitivity *270*
	George A. Rinard, Richard W. Quine, Sandra S. Eaton, and Gareth R. Eaton
5.3.1	Introduction *270*
5.3.1.1	Nomenclature *271*
5.3.2	Frequency Dependence of Sensitivity for an Ideal Spectrometer, at the Thermal Noise Limit *272*
5.3.2.1	General Expression for SNR *272*
5.3.2.2	Explanation of Table 5.3.2 *275*
5.3.2.3	On Beyond the Predictions of Table 5.3.2 *276*
5.3.2.4	Dependence of SNR on g-Anisotropy *277*
5.3.2.5	Source Noise *277*
5.3.3	Experimental Validation of Predicted Dependence of Sensitivity on Frequency *279*
5.3.3.1	CW Spectrometers at Frequencies <10 GHz *279*

5.3.3.2 Pulsed EPR Spectrometers in the Frequency Range 250 MHz to 9.5 GHz *279*
5.3.3.3 Summary of Experimental Validation of SNR of CW and Pulsed Spectrometers at Frequencies of <10 GHz *280*
5.3.4 Reference Samples for SNR: Weak Pitch *281*
5.3.5 Performance of High-Frequency (≥94 GHz)/High-Field EPR Spectrometers *282*
5.3.5.1 CW Spectrometers *282*
5.3.5.2 Pulsed EPR Spectrometers *282*
5.3.6 Reported Sensitivities of CW and Pulsed Spectrometers at Various Frequencies *285*
5.3.6.1 Further Details on CW EPR Sensitivity *286*
5.3.7 Sensitivity Aspects Beyond the Minimum Detectable Number of Spins: Frequency Dependence of Pulse and CW Measurements Related to Distances Between Spins *288*
5.3.7.1 Electron–Electron Coupling *288*
5.3.7.2 Electron–Nuclear Coupling *289*
5.3.7.3 Summary *289*
5.3.8 Limitations of Sensitivity Considerations *289*
5.3.8.1 CW Spectrometers *289*
5.3.8.2 Resonators *290*
5.3.8.3 Samples *190*
5.3.8.4 Pulse Spectrometers *290*
5.3.9 Conclusions *290*
Acknowledgments *291*
Pertinent Literature *291*
References *292*

Part Two   Theoretical *295*

6   First Principles Approach to Spin-Hamiltonian Parameters *297*
Frank Neese
6.1 Introduction *297*
6.2 The Spin Hamiltonian *298*
6.3 Electronic Structure Theory of Spin-Hamiltonian Parameters *300*
6.3.1 Electronic Structure Methods *300*
6.3.2 Additional Terms in the Hamiltonian *305*
6.3.3 Sum-Over States Theory of Spin Hamiltonian Parameters *307*
6.3.4 Linear Response Theory *310*
6.3.5 Expression for Spin-Hamiltonian Parameters for Self-Consistent Field Methods *314*
6.3.6 Practical Aspects *320*
6.3.6.1 Choice of Molecular Model *320*
6.3.6.2 Choice of Geometry *320*

6.3.6.3	Choice of Theoretical Method *321*
6.3.6.4	Choice of Basis Set *322*
6.3.6.5	Summary and Recommendations *323*
6.4	Concluding Remarks *323*
	Acknowledgments *324*
	Pertinent Literature *325*
	References *325*

7 Spin Hamiltonians and Site Symmetries for Transition Ions *327*
Sushil K. Misra

7.1	Introduction *327*
7.2	Spin Hamiltonians *328*
7.3	Spin-Hamiltonian Terms for Various Site Symmetries *332*
7.4	Transition Ions *333*
7.4.1	Introduction to Transition-Metal Ions *333*
7.4.2	First-Transition Series Ions ($3d^n$, Iron-Group Ions) *333*
7.4.3	Second and Third Transition Series (The 4d, Palladium and 5d, Platinum Groups) *345*
7.4.4	Rare-Earth Ions *347*
7.4.4.1	Odd Number of 4f Electrons *350*
7.4.4.2	Even Number of 4f Electrons *350*
7.4.5	Actinide Ions ($5f^n$) *354*
7.4.5.1	$5f^1$ Configuration *354*
7.4.5.2	$5f^2$ Configuration *356*
7.4.5.3	$5f^3$ Configuration ($^4I_{9/2}$; U^{3+}, Np^{4+}) *357*
7.4.6	S-State Ions *358*
7.4.6.1	Introduction *358*
7.4.6.2	Spin Hamiltonian *358*
7.4.6.3	Theoretical Considerations *359*
7.5	Concluding Remarks *363*
	Acknowledgments *363*
	Pertinent Literature *363*
	References *363*
	Appendix 7.I Spin Operators and Their Matrix Elements *365*
	Appendix 7.II Descent of Symmetry *381*
	Appendix 7.III Site Symmetries of Host Crystals *382*

8 Evaluation of Spin-Hamiltonian Parameters from Multifrequency EPR Data *385*
Sushil K. Misra

8.1	Introduction *385*
8.2	Perturbation Approach *386*
8.2.1	Spin Hamiltonian *387*
8.2.1.1	$S = 7/2$ *390*
8.2.1.2	$S = 5/2$ (Fe^{3+}) *391*

8.3	Brute-Force Methods to Evaluate SHP	394
8.3.1	Variation of One Parameter at a Time	394
8.3.2	Variation of Parameters in Subgroups	395
8.4	Least-Squares Fitting (LSF) Method	395
8.4.1	Introduction	395
8.4.2	Details of the LSF Method as Applied to EPR	397
8.4.3	Determination of Parameter Errors	399
8.4.4	General Strategies for Achieving Convergence	400
8.4.4.1	Use of Interpolated Fields: Calculation of Resonant Field Values	400
8.4.4.2	Use of Interpolated Frequencies	401
8.4.4.3	Use of Binary Chop (Misra, 1976)	401
8.5	Other Applications of the LSF Method	401
8.5.1	Electron–Nuclear Spin-Coupled Systems (Misra, 1983)	402
8.5.1.1	Estimation of Initial Values of FS SHPs	402
8.5.1.2	Estimation of HFS Parameters	403
8.5.1.3	Identification of Energy Levels Participating in Resonance	403
8.5.1.4	Construction of the SH Matrix for ENSC Systems	403
8.5.1.5	Absolute Signs of SH Parameters	404
8.5.2	Fitting of ENDOR Data	404
8.5.3	Calculation and Fitting of Line Intensities to SHP	404
8.5.3.1	The Intensity Operator	405
8.5.3.2	Fitting of Line Intensities and Line Positions to SHP	405
8.5.3.3	Normalized Intensity and its Derivatives	406
8.5.3.4	Limits of Applicability of the Method	408
8.6	Concluding Remarks	408
	Acknowledgments	410
	Pertinent Literature	410
	References	410
	Appendix 8.I Historical Review	411
9	**Simulation of EPR Spectra**	**417**
	Sushil K. Misra	
9.1	Introduction	417
9.2	Simulation of Single-Crystal Spectrum	417
9.2.1	Transition Probability	418
9.2.2	Single-Crystal Lineshape Function $F(B_{ri}, B_k)$	420
9.3	Simulation of a Polycrystalline Spectrum	421
9.3.1	Angular Variation of EPR Spectra: Homotopy Technique	421
9.3.1.1	Computation of the Initial Resonant Fields $B_r(\theta, \phi)$	422
9.3.1.2	Computation of the First and Second Derivatives of χ^2 with Respect to B	422
9.3.1.3	Problems Encountered in the Application of Homotopy Method, and their Solutions	423
9.3.2	Lineshapes	424
9.3.3	Transition Probabilities	424

9.3.4 Resonance Eigenpairs 424
9.3.5 Integrals 425
9.3.6 (θ_j, φ_j) Grid 426
9.3.7 Steps Required in Simulation of Powder Spectrum 426
9.3.7.1 Calculation of First-Derivative EPR Spectrum 428
9.3.8 Illustrative Example 429
9.3.9 Additional remarks 429
9.4 Evaluation of Spin-Hamiltonian (SH) Parameters and the Linewidth from a Polycrystalline EPR Spectrum 429
9.4.1 Estimation of Spin-Hamiltonian Parameters from a Polycrystalline Spectrum 430
9.4.1.1 Estimation of D and E Parameters for the Mn^{2+} Ion from Forbidden Hyperfine Doublet Separations in Polycrystalline Samples in the Central Sextet 430
9.4.1.2 Rigorous Evaluation of SH Parameters from a Polycrystalline Spectrum by Using Matrix Diagonalization and Least-Squares Fitting 432
9.4.1.3 Evaluation of SH Parameters and Linewidths for the Case of Two Magnetically Inequivalent Species 436
9.4.1.4 Illustrative Example 436
9.4.1.5 General Remarks 437
9.5 Simulation of EPR Spectra in Disordered Materials: Application to Glassy Materials 437
9.5.1 Introduction 437
9.5.2 Computer Simulation of EPR Spectra in Glasses 437
9.5.3 Computer-Simulated Spectra and Comparison with Experiment 440
9.5.4 Shape of EPR Spectra in Glasses: Effect of SH Parameters 442
9.5.4.1 Distribution of the Fine-Structure Parameters D and E 442
9.5.4.2 Sharp Features in Spectra 447
9.5.4.3 Broad Resonances in Spectra 448
9.6 Simulation of EPR Spectra in Disordered Random Network Materials 448
9.6.1 Introduction 448
9.6.2 CW-EPR Spectrum for Random Distribution of SH Parameters at Various Sites in Glasses 449
9.6.2.1 Calculation of Eigenvectors $|i'\rangle$ and $|i''\rangle$ Required in Equation 9.57 450
9.6.3 Limitations of the Original Implementation and its Assumptions 450
Acknowledgments 451
Pertinent Literature 451
References 451
Appendix 9.I The Eigenfield Equation 453

10 Relaxation of Paramagnetic Spins 455
Sushil K. Misra
10.1 Introduction 455

10.2	Equilibrium Magnetization of a Paramagnetic Spin System 457
10.3	Relaxation Phenomena: Spin–Lattice and Spin–Spin Relaxation Times 458
10.3.1	Bloch's Equations 459
10.4	Rotating Frame 459
10.5	Experimental Techniques to Measure Relaxation Times 460
10.5.1	CW-EPR Techniques (Bertini, Martini, and Luchinat, 1994) 461
10.5.1.1	CW Saturation 461
10.5.2	Longitudinally Detected Paramagnetic Resonance (LODEPR) to Measure Short Relaxation times (10^{-8} s) (Giordano et al., 1981) 463
10.5.3	Amplitude Modulation Technique to Measure Very Short Relaxation Times ($10^{-6} - 10^{-9}$ s) (Misra, 2005) 463
10.5.4	Pulsed EPR Techniques to Measure Relaxation Times 463
10.5.5	Long Pulse Saturation Recovery Using CW Detection (Huisjen and Hyde, 1974; Percival and Hyde, 1975; Eaton and Eaton, 2000) 464
10.5.6	Inversion Recovery (Eaton and Eaton, 2000) 464
10.5.7	Electron Spin Echo (ESE) Technique (Schweiger and Jeschke, 2001) 464
10.5.8	Long-Pulse Saturation with Spin-Echo Detection 465
10.5.9	Picket-Fence Excitation (Eaton and Eaton, 2000) 465
10.5.10	Echo Repetition Rate (Eaton and Eaton, 2000) 465
10.5.11	Three-Pulse-Stimulated Echo (Eaton and Eaton, 2000) 466
10.5.12	Longitudinally Detected Pulsed EPR (LODPEPR) (Schweiger, 1991; Schweiger and Ernst, 1988) 466
10.5.13	Other Pulse Techniques 466
10.5.14	Measurements of Relaxation Time by Line-Shape Analysis: Linewidth and Spin–Spin Relaxation Time 466
10.5.15	Temperature-Dependent Contribution to EPR Linewidth (Poole and Farach, 1971) 467
10.5.16	Non-EPR Techniques to Measure Relaxation Times 467
10.6	Relaxation Mechanisms 468
10.6.1	Spin-Lattice Relaxation in Diluted Ionic Solids in the Crystalline State 468
10.6.1.1	General Background 468
10.6.1.2	The Direct Process 469
10.6.1.3	The Orbach Process (Orbach and Stapleton, 1972; Orbach, 1961a, 1961c) 471
10.6.1.4	Two-Phonon Raman Process 471
10.6.1.5	SLR due to Exchange Interaction 473
10.6.2	Relaxation in Amorphous Systems 474
10.6.2.1	Relaxation via TLS Centers 475
10.6.2.2	SLR Effected by Electron–Nuclear Dipolar Coupling to a TLS Center 477
10.6.2.3	SLR due to Fermi-Contact Hyperfine Interaction with a TLS Center 478

10.6.2.4	Temperature Dependence of Relaxation Rate in Amorphous Materials due to Exchange Interaction *478*	
10.6.2.5	Relaxation for the Case of Strong Cross-Relaxation and Weak Spin-Lattice Relaxation of Single Ions in Amorphous Materials (Al'tshuler, 1956) *478*	
10.6.3	Relaxation in Diluted Liquid Solutions *479*	
10.6.4	Effect of Intramolecular Dynamics of Molecular Species on Relaxation *483*	
10.6.4.1	Dephasing by Methyl Groups in Solvent or Surroundings *483*	
10.6.4.2	Shape of the Echo-Decay Curve *484*	
10.6.4.3	Averaging of Electron-Nuclear Couplings due to Rotation of Methyl Groups *484*	
10.6.4.4	Effect of a Rapidly Relaxing Partner on Electron–Electron Spin–Spin Coupling *484*	
10.6.4.5	Librational Motion *484*	
10.6.4.6	Molecular Tumbling *484*	
10.6.4.7	Biomolecules *485*	
10.6.4.8	Macromolecules *485*	
10.6.5	Relaxation among Different Paramagnetic Centers in Concentrated Solution *485*	
10.6.6	Spin-Fracton Relaxation *485*	
10.6.6.1	One-Fracton Emission *486*	
10.6.6.2	Two-Fracton Inelastic Scattering (Localized Electronic State) (Alexander, Entin-Wohlman, and Orbach, 1985a) *487*	
10.6.7	Frequency/Field Dependence of Paramagnetic Relaxation *489*	
	Pertinent literature *490*	
	Acknowledgments *491*	
	References *491*	
	Appendix 10.I Early History of Paramagnetic Spin-Lattice Relaxation *494*	
11	**Molecular Motions** *497*	
	Sushil K. Misra and Jack H. Freed	
11.1	Introduction *497*	
11.2	Historical Background *498*	
11.3	High-Field Multifrequency CW-EPR Experiments to Unravel Molecular Motion *500*	
11.3.1	Determination of the Axes of Motion from High-Field, High-Frequency (HFHF) EPR Spectra: Orientational Resolution *503*	
11.3.2	Observation of Motion as a Function of Frequency *504*	
11.3.3	Virtues of Multifrequency EPR in Studying Molecular Motion *504*	
11.3.4	Stochastic Liouville Equation (SLE) to Describe Slow-Motional EPR Spectra *505*	
11.3.4.1	Calculation of Slow-Motion Spectrum *506*	
11.3.4.2	MOMD and SRLS Models *511*	

11.4	Pulsed EPR Study of Molecular Motion	*514*
11.4.1	T_2-Type Field-Swept 2D ESE	*515*
11.4.2	Magnetization Transfer by Field-Swept 2-D-ESE	*515*
11.4.3	Stepped-Field Spin-Echo ELDOR	*517*
11.4.4	2-D Fourier Transform EPR	*517*
11.4.4.1	Lineshapes of the Auto and Cross-Peaks: Homogeneous (HB) and Inhomogeneous Broadening (IB)	*518*
11.4.5	MOMD and SRLS Models and 2-D-ELDOR	*520*
11.4.6	Extension of 2-D-ELDOR to Higher Frequencies	*522*
11.5	Simulation of Multifrequency EPR Spectra Using More Atomistic Detail Including Molecular Dynamics and Stochastic Trajectories	*522*
11.5.1	Augmented SLE	*522*
11.5.2	MD Simulations Using Trajectories	*524*
11.5.3	Use of Dynamic Trajectories to Simulate Multifrequency EPR Spectra	*525*
11.5.4	Numerical Integrators	*526*
11.5.4.1	Integration of the Quantal Spin Dynamics	*526*
11.5.4.2	Generation of Stochastic Trajectories for Rotational Diffusion	*531*
11.5.4.3	Testing the Integrators: Generation of Trajectories for Typical Stochastic Models of Spin-Label Dynamics	*535*
11.6	Concluding Remarks	*541*
	Acknowledgments	*541*
	Pertinent Literature	*541*
	References	*542*
12	**Distance Measurements: Continuous-Wave (CW)- and Pulsed Dipolar EPR**	*545*
	Sushil K. Misra and Jack H. Freed	
12.1	Introduction	*545*
12.2	The Dipolar Interaction and Distance Measurements	*547*
12.2.1	Unlike Spins	*547*
12.2.2	Like Spins	*548*
12.2.3	Intermediate Case	*548*
12.3	CW EPR Method to Measure Distances	*548*
12.4	Pulsed Dipolar EPR Spectroscopy (PDS)	*549*
12.5	Double Electron–Electron Resonance (DEER)	*550*
12.5.1	Orientation-Selection Considerations in DEER	*552*
12.5.2	Three-Pulse DEER	*553*
12.5.3	Four-Pulse DEER	*555*
12.5.4	Merits and Limitations of DEER as Compared to CW-EPR and FRET	*557*
12.6	Six-Pulse DQC	*559*
12.6.1	Theoretical Background and Computation of Six-Pulse DQC Signal	*562*
12.6.2	Illustrative Examples	*566*

12.6.3		Conclusions and Future Prospects of Six-Pulse DQC Echo Signal Simulation *566*
12.7		Sensitivity Considerations: Multifrequency Aspects *570*
12.7.1		Frequency Dependence of Sensitivity of PDS *572*
12.8		Distance Distributions: Tikhonov Regularization *573*
12.9		Additional Technical Aspects of DEER and DQC *574*
12.10		Concluding Remarks *576*
		Acknowledgments *576*
		Pertinent Literature *576*
		References *576*
	Appendix 12.I	Density-Matrix Derivation of Echo Signal for Three-Pulse DEER *578*
	Appendix 12.II	Density-Matrix Derivation of the Echo Signal for Four-Pulse DEER *582*
	Appendix 12.III	Spin Hamiltonian for Coupled Nitroxides Used in Six-Pulse DQC Calculation *584*
	Appendix 12.IV	Algorithm to Calculate Six-Pulse DQC Signal *586*
	Appendix 12.V	Approximate Analytic Expressions for 1-D DQC Signal *587*

Part Three   Applications *589*

13 Determination of Large Zero-Field Splitting *591*
Sushil K. Misra
13.1   Introduction *591*
13.2   ZFS of Kramers and Non-Kramers Ions in Different Environments *592*
13.3   Concluding Remarks *596*
   Acknowledgments *597*
   Pertinent Literature *597*
   References *597*

14 Determination of Non-Coincident Anisotropic $\tilde{g}^2$, $\tilde{A}^2$, $\tilde{D}$, and $\tilde{P}$ Tensors: Low-Symmetry Considerations *599*
Sushil K. Misra
14.1   Introduction *599*
14.2   Spin Hamiltonian *599*
14.3   Eigenvalues *601*
14.3.1   Perturbation Approach *601*
14.3.1.1   Complexities Associated with the Use of Second-Order-Perturbed Eigenvalues in the Application of Least-Squares Fitting (LSF) Procedure *604*
14.3.2   Exact Matrix Diagonalization *605*
14.4   Evaluation of SHPs by the LSF Technique *606*
14.4.1   First-Order Perturbation *606*

14.4.2	Second-Order Perturbation *607*
14.4.3	Use of Special Coordinate Axes *609*
14.4.3.1	"Allowed" Line Positions *609*
14.4.3.2	"Forbidden" Line Positions *611*
14.4.4	Use of Arbitrary Coordinate Axes *612*
14.4.5	Simultaneous LSF Fitting of Both the "Allowed" and "Forbidden" Line Positions *613*
14.5	Numerical Evaluation of the Derivatives Required in the LSF Procedure *614*
14.6	General Remarks *616*
	Acknowledgments *618*
	Pertinent Literature *618*
	References *618*

15 Biological Systems *619*
Boris Dzikovski

15.1	Introduction *619*
15.2	VHF EPR as the g-Resolved EPR Spectroscopy *620*
15.2.1	Spectral Resolution of g-Factor Differences *620*
15.2.2	Precise Determination of the g-Tensor Principal Values *621*
15.2.3	Resolution of g-Factors of Different Paramagnetic Centers *622*
15.3	Effect of Polarity of the Environment on the g-Factor *623*
15.3.1	Examples *623*
15.3.1.1	Derivatives of 2,2,6,6-tetramethylpiperidine-1-oxyl (TEMPO) *623*
15.3.1.2	Spin-Labeled Phospholipid Membranes: 1,2-Dipalmitoyl-*sn*-Glycero-3-Phosphocholine (DPPC) and 1-Palmitoyl-2-Oleoyl-*sn*-Glycero-3-Phosphocholine (POPC) *625*
15.3.1.3	Bacteriorhodopsin (BR) *625*
15.3.1.4	Azurin *626*
15.3.1.5	Tyrosyl and Tryptophan Radicals *626*
15.3.1.6	Flavin *626*
15.3.1.7	Biliverdin Radical *627*
15.3.2	Polarity Measurements Outside of Rigid Limit Conditions *627*
15.4	Improvement in Orientational Resolution for Spin Labels *628*
15.5	Simulation of EPR Spectra at Various Frequencies: Simple Limiting Cases *630*
15.6	Macroscopically Aligned Phospholipid Membranes *631*
15.6.1	A "Shunt" Fabry–Pérot Resonator. The study of DMPC and DMPS (1,2-dimyristoyl-sn-glycero-3-phospho-L-serine) Membranes with 3-doxyl-5(-cholestane) (CSL) Spin Label *632*
15.6.2	Microtome Technique on Isopotential Spin-Dry Ultracentrifugation (ISDU)-Aligned Membranes *633*
15.6.3	Other Membrane-Alignment Techniques *635*
15.7	Metalloproteins *636*
15.7.1	Fe^{3+} Systems *638*

15.7.2	Mn^{2+} Systems	638
15.7.3	Cu^{2+} Systems	639
15.8	Concluding Remarks	641
	Acknowledgments	642
	Pertinent Literature	642
	References	643

16   Copper Coordination Environments   647
William E. Antholine, Brian Bennett, and Graeme R. Hanson

16.1	Introduction	647
16.2	Multifrequency EPR Toolkit	649
16.2.1	g-Value Resolution and Orientation Selection	649
16.2.2	Magnitude of the Microwave Frequency	650
16.2.3	State Mixing	650
16.2.4	Angular Anomalies	652
16.2.5	Distribution of Spin Hamiltonian Parameters	653
16.2.6	Numerical Differentiation and Fourier Filtering	655
16.2.7	High-Resolution EPR Techniques	655
16.2.8	Computer Simulation	656
16.2.9	Computational Chemistry	658
16.3	Multifrequency EPR Simulation of Square–Planar-Based Cu(II)	660
16.3.1	EPR of Square–Planar-Based Cu(II)	660
16.3.2	Multifrequency EPR of Square–Planar-Based Cu(II): S- and L-Band EPR	660
16.3.3	Multifrequency EPR of Square–Planar-Based Cu(II): Very Low-Frequency EPR	661
16.3.4	Multifrequency EPR of Square–Planar-Based Cu(II): Experimental Considerations for Low-Frequency EPR	663
16.3.5	Introduction to Multifrequency EPR Simulations of Square–Planar Cu(II)	664
16.3.6	Optimum Frequency Selection	665
16.3.7	Sensitivity Analysis	668
16.3.8	Global Fitting	669
16.3.8.1	Mo(V) Complexes	671
16.3.8.2	Low-Spin Co(II) Crossover Complexes	673
16.3.8.3	Future Developments	675
16.3.9	Heterogeneity	675
16.4	Copper-Coordination Environments: Multifrequency EPR of Three-Coordinate Copper and Mixed-Valence Dinuclear Copper [Cu(1.5$^+$) ... Cu(1.5$^+$)]	677
16.4.1	Introduction: Spectrum and Structure	677
16.4.1.1	X-Band EPR Spectrum for Mononuclear, Light Blue Cu^{2+}	677
16.4.1.2	Peisach–Blumberg-Like Table (EPR Parameters Assembled by the Author)	677
16.4.1.3	Type 1 (Blue) Copper Centers, Three-Coordinate Cu	678

16.4.2   EPR for New Three-Coordinate Copper Complexes   681
16.4.2.1   Three-Coordinate CuL(SCPh$_3$) and Copper(II)Phenolate Complexes   681
16.4.2.2   CuPPN, Three-Coordinate Copper Amido and Aminyl Complexes (More Like a Free Radical)   681
16.4.2.3   Simulation of Spectra for CuPPN (Quenched EPR Parameters Expected for a Radical)   682
16.4.2.4   EPR Parameters for CuPPN (Unpaired Electron Density Delocalized as Expected for a Radical)   685
16.4.3   Spectra for Mixed-Valence Dinuclear Copper Complexes   686
16.4.3.1   Nitrous Oxide Reductase, N$_2$OR (^{15}N Example)   686
16.4.3.2   Perturbation of the EPR Spectrum of Cu$_A$, H120X   689
16.4.3.3   Cytochrome c Oxidase (CcO): Best Demonstration of the Use of Low-Frequency for Mixed-Valence Sites   691
16.4.3.4   Model Diamond Core Complexes, $\{Cu(LXL)\}_2^+$ 695
16.4.3.5   X-Band EPR Spectra of $\{Cu(PPP)\}_2^+$, $\{Cu(PNP)\}_2^+$, and $\{Cu(SNS)\}_2^+$ 695
16.4.3.6   Q-Band EPR Spectra of $\{Cu(PPP)\}_2^+$, $\{Cu(PNP)\}_2^+$, and $\{Cu(SNS)\}_2^+$ 695
16.4.3.7   S-Band Spectra of $\{Cu(PPP)\}_2^+$, $\{Cu(PNP)\}_2^+$, and $\{Cu(SNS)\}_2^+$ 697
16.4.3.8   EPR Parameters and Simulations for $\{Cu(SNS)\}_2^+$ 697
16.4.3.9   First-Harmonic S-Band Spectrum for $\{Cu(PPP)\}_2^+$ 697
16.5   Structural Characterization of Copper(II) Cyclic Peptide Complexes Employing Multifrequency EPR and Computational Chemistry   699
16.5.1   Copper(II) Complexes with Marine Cyclic Peptides   701
16.5.2   Copper(II) Complexes with Westiellamide and Synthetic Analogs   707
16.6   Summary   711
           Acknowledgments   711
           Pertinent Literature   712
           Section 16.3   712
           Section 16.4   713
           References   714

17 Multifrequency Electron Spin-Relaxation Times   719
Gareth R. Eaton and Sandra S. Eaton

17.1   Introduction and Scope of the Chapter   719
17.2   Spin–Spin Relaxation, T_2 and T_m 720
17.2.1 T_m for Fremy's Salt in Glassy Solvents   723
17.2.2   Exchange-Narrowed Species and the 10/3 Effect   724
17.2.3   Conducting Systems   725
17.2.4   Metal Ions in Solution   726
17.2.5   Pb^{3+} in Calcite   726
17.3   Spin–lattice Relaxation, T_1 726
17.3.1   Phonon Densities   727
17.3.2   Practical Interpretation of Relaxation Time Data as a Function of Temperature   729

17.3.3 Glasses versus Crystals *729*
17.3.4 Spectral Diffusion and Cross-Relaxation *731*
17.3.5 Effect of Pairs and Clusters *732*
17.3.6 Magnetic Field Dependence of Relaxation *732*
17.3.6.1 The Direct Process *732*
17.3.6.2 The Raman Process *734*
17.3.6.3 The Orbach Process *734*
17.3.6.4 The Thermally Activated Process *735*
17.3.6.5 Local Modes *735*
17.3.7 Dependence of Relaxation on Magnetic Field Position in a CW-EPR Spectrum *736*
17.3.8 Case Studies of Experimental Data *737*
17.3.8.1 Nitroxyl Spin Labels *737*
17.3.8.2 Semiquinones *741*
17.3.8.3 Triarylmethyl (Trityl) Radicals *742*
17.3.8.4 DPPH *742*
17.3.8.5 Conducting Spin Systems *742*
17.3.8.6 Metal ions in Fluid Solution *743*
17.3.8.7 Relaxation at 2 mm Wavelength (150 GHz) *746*
17.3.9 Fullerenes *747*
17.3.10 Summary *747*
Acknowledgments *748*
Pertinent Literature *748*
References *748*

18 EPR Imaging: Theory and Instrumentation *755*
Rizwan Ahmad and Periannan Kuppusamy
18.1 Introduction *755*
18.2 EPR Principle: Zeeman Effect *756*
18.2.1 Hyperfine Splitting *757*
18.2.2 Spin Relaxation *759*
18.2.3 Comparison to NMR *759*
18.2.4 EPR Probes *759*
18.3 CW-EPR Imager *760*
18.3.1 Magnets and Magnetic Field Control *761*
18.3.2 Gradient Coil Assembly *762*
18.3.3 RF Bridge *764*
18.3.4 EPR Resonator *765*
18.3.5 Signal Channel *768*
18.4 Data Acquisition for CW-EPR and EPRI *769*
18.4.1 Spectroscopy *769*
18.4.2 Spatial EPRI *770*
18.4.3 Spectral–Spatial EPRI *772*
18.5 Important Imaging Parameters *774*

18.5.1	Time Constant of Lock-In Amplifier	774
18.5.2	Modulation Amplitude	775
18.5.3	Gradient Strength	775
18.6	Image Reconstruction	776
18.6.1	Direct Methods	777
18.6.1.1	Filtered Backprojection (FBP) Method	777
18.6.1.2	Fourier-Based Reconstruction	778
18.6.2	Iterative Methods	779
18.6.3	Spectral–Spatial Reconstructions	781
18.6.4	Image Quality and Resolution	782
18.7	Other Data Collection Modalities	783
18.7.1	Pulsed-EPR	783
18.7.2	Single Point Imaging	784
18.7.3	Rapid Scan	784
18.7.4	Spinning Gradient	784
18.8	Constraints for Biological Applications	785
18.9	Special Imaging Applications	786
18.9.1	EPR Oximetry Mapping	786
18.9.2	Imaging Redox Metabolism in Tissues	788
18.9.2.1	Differential Distribution of Nitroxide Probes in Normal versus Tumor Tissue	788
18.9.2.2	Differential Metabolism of Nitroxide Probes in Normal versus Tumor Tissue	789
18.10	Scope and Limitations	790
	Acknowledgments	791
	Pertinent Literature	791
	References	791
19	**Multifrequency EPR Microscopy: Experimental and Theoretical Aspects**	**795**
	Aharon Blank	
19.1	General	795
19.2	Introduction	795
19.2.1	Definition	795
19.2.2	Historical Overview	796
19.2.3	"Induction Detection" versus Other Detection Methods	797
19.3	General Experimental Aspects of EPR Microscopy	798
19.3.1	CW-EPR Microscopy	798
19.3.1.1	System Configuration	798
19.3.1.2	Signal-to-Noise Ratio	803
19.3.1.3	Resolution	805
19.3.2	Pulsed-EPR Microscopy	805
19.3.2.1	System Configuration	805
19.3.2.2	SNR	808

19.3.2.3 Resolution *810*
19.4 Specific Aspects of Multifrequency EPR Microscopy at Various Temperatures *811*
19.4.1 SNR in a Multifrequency Context *812*
19.4.2 Resolution in a Multifrequency Context *814*
19.5 Illustrative Examples *815*
19.5.1 Pulsed-EPR Microscopy of Solid Samples at Room Temperature *816*
19.5.2 Pulsed-EPR Microscopy of Liquid Samples at Room Temperature *816*
19.5.3 CW-EPR Microscopy of Solid and Liquid Samples at Room Temperature *819*
19.6 Conclusions and Future Prospects *821*
Acknowledgments *821*
Pertinent Literature *821*
References *822*

20 EPR Studies of Nanomaterials *825*
Alex Smirnov
20.1 Introduction *825*
20.2 EPR Studies of Magnetic Nanostructures *827*
20.3 Characterization of Nanostructured Oxide Semiconductors for Photoactivated Catalysis and Solar Energy Conversion *832*
20.4 Surface Radicals, Catalytic Activity, Cytotoxicity, and Radical-Scavenging Properties of Nanomaterials *833*
20.4.1 Cayalytic Activity *833*
20.4.2 Cytotoxicity *834*
20.4.3 Radical-Scavenging Properties *835*
20.5 Spin-Labeling EPR Studies of Ligand-Protected Nanoparticles and Hybrid Nanostructures *835*
20.6 Summary and Future Perspectives *841*
Acknowledgments *842*
Pertinent Literature *842*
References *842*

21 Single-Molecule Magnets and Magnetic Quantum Tunneling *845*
Sushil K. Misra
21.1 Introduction *845*
21.1.1 Intramolecular Coupling *846*
21.1.2 Examples of SMMs Reported in the Literature *847*
21.1.3 Applications *851*
21.2 Multifrequency EPR of SMMs: Magnetic Hysteresis and MQT *852*
21.2.1 The Effective Spin Hamiltonian *853*
21.2.2 Magnetic Quantum-Mechanical Tunneling (MQT) and MF-EPR *854*
21.2.3 Zero-Field EPR with Variable Frequency *854*

21.2.4	Low-Field (X-band) EPR	854
21.2.5	MF High-Frequency EPR	855
21.2.5.1	EPR Spectrometers with MF Cavity (40–350 and Extended Range 18–350 GHz), and up to 650 GHz Without a Cavity	855
21.2.5.2	Polycrystalline Powder EPR Spectrum	855
21.2.5.3	The Virtues of Single-Crystal Measurements	855
21.2.5.4	A Typical SMM Spectrum	857
21.2.5.5	EPR Linewidth Measurements: Effect of D-Strain, g-Strain, Dipolar and Exchange Interactions	857
21.2.5.6	Study of Intermolecular Exchange Interactions and Dipolar Interactions	860
21.2.5.7	EPR Spectra for Mn_4 Family	861
21.2.6	Effect of Molecular Site Symmetry on Tunneling Phenomenon (MQT) as Revealed by EPR	863
21.3	Magnetic Quantum Tunneling (MQT): Pure and Thermally Assisted Tunneling	867
21.3.1	Relaxation of Magnetization for SMMs	867
21.3.2	Magnetic Hysteresis, Resonant Magnetization Tunneling in High-Spin Molecules and Thermally Assisted Resonant Tunneling Between Quantum States	868
21.4	Concluding Remarks	872
	Acknowledgments	872
	Pertinent Literature	872
	References	872
22	**Multifrequency EPR on Photosynthetic Systems**	**875**
	Sushil K. Misra, Klaus Möbius, and Anton Savitsky	
22.1	Introduction	875
22.2	Nonoxygenic Photosynthesis	880
22.3	Multifrequency EPR on Bacterial Photosynthetic Reaction Centers (RCs)	882
22.3.1	X-band EPR Experiments	882
22.3.2	95-GHz EPR on Primary Donor Cations $P^{\cdot+}$ in Single-Crystal RCs	883
22.3.3	360-GHz EPR on Primary Donor Cations $P^{\cdot+}$ in Mutant RCs	884
22.3.4	Results of g-tensor Computations of $P^{\cdot+}$	885
22.3.5	95-GHz EPR and ENDOR on the Acceptors $Q_A^{\cdot-}$ and $Q_B^{\cdot-}$	885
22.3.6	95-GHz ESE-Detected EPR on the Spin-correlated Radical Pair $P^{\cdot+}Q_A^{\cdot-}$	892
22.3.7	95-GHz RIDME and PELDOR on the Spin-Correlated Radical Pair $P^{\cdot+}Q_A^{\cdot-}$	893
22.3.8	Multifrequency EPR on Primary Donor Triplet States in RCs	895
22.4	Oxygenic Photosynthesis	897
22.4.1	Multifrequency EPR on Doublet States in Photosystem I (PS I)	897
22.4.2	Multifrequency EPR on Doublet States in Photosystem II (PS II)	900

22.5	Concluding Remarks 902
	Acknowledgments 904
	Pertinent Literature 904
	References 905

23	**Measurement of Superconducting Gaps** 913
	Sushil K. Misra
23.1	Introduction 913
23.2	The Superconducting Gap 913
23.3	Measurement of SCG 914
23.4	Concluding Remarks 917
	Acknowledgments 918
	References 919

24	**Dynamic Nuclear Polarization (DNP) at High Magnetic Fields** 921
	Thomas Prisner and Mark J. Prandolini
24.1	Introduction 921
24.2	Historical Aspects (Metals, Solids and Liquids) at Lower Magnetic Fields 922
24.3	Theory 924
24.3.1	The Overhauser Effect (OE) 924
24.3.2	Two-Spin Cross-Polarization: Solid Effect (SE) 930
24.3.3	Many-Spin Cross-Polarization: Thermal Mixing (TM) 931
24.3.4	Three-Spin Cross-Polarization: Cross Effect (CE) 933
24.3.5	Beyond Classical DNP Methods: Coherent Polarization Transfer 935
24.4	Hardware (High-Frequency Microwave Equipment, SS-MAS DNP, HF-Liquid DNP, Dissolution DNP, Shuttle-DNP) 936
24.4.1	High-Frequency Microwave Sources 936
24.4.2	Transmission Lines 937
24.4.3	Spectrometer Types 938
24.4.3.1	Solid-State Magic Angle Spinning (MAS) DNP 938
24.4.3.2	Low-Temperature Dissolution Polarizer 939
24.4.3.3	*In-Situ* Temperature-Jump DNP (Laser Melting) 940
24.4.3.4	High-Field (HF) Liquid-DNP Spectrometers 940
24.4.3.5	Shuttle DNP 941
24.5	First Applications and Outlook 942
24.5.1	Application Areas of High-Field DNP 942
24.5.2	Outlook 943
	Acknowledgments 943
	Pertinent Literature 943
	References 944

25	**Chemically Induced Electron and Nuclear Polarization** 947
	Lawrence J. Berliner and Elena Bagryanskaya
25.1	Introduction 947

25.2	History of the CIDNP Phenomenon 948
25.3	The Radical Pair Mechanism 948
25.3.1	The Mechanism of Singlet–Triplet Conversion in RPs 949
25.4	Chemically Induced Dynamic Nuclear Polarization 952
25.4.1	The CIDNP Experiment 955
25.4.2	Time-Resolved CIDNP 956
25.4.3	Low Magnetic Field CIDNP 958
25.4.4	The Application of CIDNP to Biological Systems 960
25.4.5	Photo-CIDNP in the Study of Protein Folding 961
25.4.6	CIDNP Application to Study Primary Processes in the Bacterial Photosynthetic Center 963
25.4.7	CIDNP Applications to Electron Transfer in Peptide and Amino Acids 966
25.5	Chemically Induced Dynamic Electron Polarization 967
25.5.1	Triplet Mechanism of CIDEP 967
25.5.2	Radical-Pair Mechanism of CIDEP 969
25.5.2.1	CIDEP Due to $S–T_0$ Transitions 969
25.5.3	CIDEP Due to $S–T_-$ and $S–T_+$ Transitions 970
25.5.4	CIDEP Due to the Radical-Triplet Pair Mechanism 971
25.5.5	CIDEP Due to the SCRP Mechanism 972
25.5.6	CIDEP Kinetics 974
25.5.6.1	Modified Bloch Equations 974
25.5.7	Time-Resolved EPR Spectroscopy 974
25.5.8	CIDEP Applications 976
25.5.9	Applications of CIDEP to Biological Systems 980
25.5.10	Applications of CIDEP to Study Photochemical Reaction Centers 981
25.5.11	RTPM CIDEP in Spin-Labeled Peptides 981
25.5.12	Applications of CIDEP to Studies of Biological Function: Protein Dynamics and Protein–Surface Interactions 982
25.5.13	CIDEP Study of Amino Acid Photooxidation 983
25.6	Conclusion 984
	Pertinent Literature 986
	References 988

Part Four Future Perspectives 993

26 Future Perspectives 995
Sushil K. Misra

26.1	Spectroscopic Techniques Currently Available in EPR 995
26.1.1	Future Perspectives in EPR Instrumentation 997
26.1.2	Desirable Advancements in EPR Instrumentation 998
26.2	Cutting-Edge Topics 999
26.2.1	Topics Related to the Theoretical Interpretation of EPR Data 1002

26.3 Desirable Applications of EPR *1003*
26.4 Future of EPR *1003*
Acknowledgments *1004*

Appendix A1   Fundamental Constants and Conversion Factors used in EPR *1005*
Index *1009*

Preface

This book started as a joint effort with Charlie Poole in 2003, when I was invited to celebrate his 70th anniversary at the University of South Carolina, Columbia, USA. Charlie was very excited with the recent developments in very-high-frequency spectrometers, opening up new dimensions in EPR research, and together we prepared a list of appropriate chapters, and potential authors, on multifrequency (MF) EPR. Charlie had tremendous experience in writing research monographs, in particular his classic treatise "*Electron Spin Resonance*" (Wiley-Interscience, New York, 1967) on experimental techniques in EPR. Life is not always as easy as it sounds at times. Soon after his wife passed away (in 2004), Charlie was unable to continue as a coauthor of the book, but encouraged me to write it solo, while promising his continued availability for advice and support. Fortunately, I was able to enlist several distinguished researchers to participate in this project. Then, Wiley-VCH publishers honored us by unconditionally approving the tentative contents. Although we signed the contract in early 2007, an excellent monograph is like a fine bottle of wine – it needs time to evolve and mature. In the meantime – my emergency quadruple bypass surgery (no heart attack) notwithstanding – steady, consistent work by all of the excellent authors has resulted in the completion of the current monograph, updating the EPR community in the areas of MF-EPR in a timely fashion.

I am grateful to the various authors, who have written the book's chapters, all of which include pertinent literature on the topics covered for a more complete access to current research. A landscape of the contents of the book is as follows. In addition to a general coverage of the building blocks of the theme of the book, for example, MF-EPR, among others, by myself, there are chapters on experimental (ZF-EPR, Misra; low-frequency EPR spectrometers, Buckmaster; pulsed-EPR spectrometers, Subramanian (Subu) and Murali Krishna; HF-EPR spectrometers, Reijerse) and theoretical (manipulating spin Hamiltonians to evaluate parameters used in MF-EPR and to simulate spectra, Misra) techniques, along with applications and future perspectives (Misra). Particularly noteworthy chapters in the context of applications of MF-EPR are: multiarm EPR spectroscopy at multiple microwave frequencies (Hyde), resonators (Hyde), EPR sensitivity (Rinard, Quine, S. Eaton, G. Eaton), first-principles approach to spin-Hamiltonian parameters (Neese), paramagnetic relaxation (Misra), molecular motions and distance measurements-pulsed dipolar EPR (Misra and Freed), biological systems (Dzikovsky), copper-coordination compounds (Antholine, Bennett, and Hanson), relaxation times (G. Eaton and S. Eaton), EPR imaging (Ahmad and Kuppusamy), EPR microscopy (Blank), nanomaterials (Smirnov), single-molecule magnets and macroscopic quantum tunneling (Misra), photosynthetic systems (Misra, Möbius, and Savitsky), superconducting gaps (Misra), DNP at high magnetic fields (Prisner and Prandolini), and CIDNP and CIDEP (Berliner and Bagryanskaya).

I am grateful to Larry Berliner, who has edited many volumes on current topics in biological magnetic resonance spectroscopy, for his valuable advice throughout the course of writing this book. I also express my gratitude to Jack Freed for introducing me to many cutting-edge topics in MF-EPR research. Once again, I appreciate very profoundly all the inspiration and help that Charlie Poole has provided me to complete this book.

I dedicate this book to my parents, Mr Rajendra Misra and (late) Mrs Prakash Wati Misra, and my three children—Manjula and Shivali (daughters) and Paraish (son)—all of whom have always been proud of my accomplishments, and have been a source of my inspiration.

I can now spend more time on my various hobbies, such as sports, dance, music, and theater, for which have I have made a gross sacrifice to complete the writing of this book.

Montreal, Canada
May 2010

List of Contributors

Rizwan Ahmad
The Ohio State University
Davis Heart and Lung Research Institute
Columbus, OH 43210
USA

William E. Antholine
Medical College of Wisconsin
Department of Biophysics
Milwaukee, WI 53226
USA

Elena Bagryanskaya
International Tomography Center
SB RAS
Institutskaya 3a
Novosibirsk 630090
Russia

Brian Bennett
Medical College of Wisconsin
Department of Biophysics
Milwaukee, WI 53226
USA

Lawrence J. Berliner
University of Denver
Department of Chemistry and Biochemistry
Denver, CO 80208
USA

Aharon Blank
Technion
Schulich Faculty of Chemistry
Haifa, 32000
Israel

Harvey A. Buckmaster
University of Victoria
Department of Physics and Astronomy
Victoria
British Columbia, V8W 3P6
Canada

Theodore G. Camenisch
Medical College of Wisconsin
Department of Biophysics
8701 Watertown Plank Road
Milwaukee, WI 53226-0509
USA

Boris Dzikovski
Cornell University
Department of Chemistry and Chemical Biology
Baker Laboratory
Advanced Center for ESR Technology (ACERT)
Ithaca, NY 14853
USA

List of Contributors

Gareth R. Eaton
University of Denver
Department of Chemistry and
Biochemistry
2101 E. Wesley Ave
Denver, CO 80208
USA

Sandra S. Eaton
University of Denver
Department of Chemistry and
Biochemistry
2101 E. Wesley Ave
Denver, CO 80208
USA

Jack H. Freed
Cornell University
ACERT
Department of Chemistry and
Chemical Biology
Ithaca, NY 14853-1301
USA

Graeme R. Hanson
The University of Queensland
Centre for Advanced Imaging
St. Lucia
Qld 4072
Australia

James S. Hyde
National Biomedical EPR Center
Medical College of Wisconsin
Department of Biophysics
8701 Watertown Plank Road
Milwaukee, WI 53226-0509
USA

Murali C. Krishna
National Institutes of Health
National Cancer Institute
Bethesda, MD 200892
USA

Periannan Kuppusamy
The Ohio State University
Davis Heart and Lung Research
Institute
Columbus, OH 43210
USA

Richard R. Mett
National Biomedical EPR Center
Medical College of Wisconsin
Department of Biophysics
8701 Watertown Plank Road
Milwaukee, WI 53226-0509
USA

Sushil K. Misra
Concordia University
Physics Department
1455 de Maisonneuve Boulevard
West
Montreal
QC H3G 1M8
Canada

Klaus Möbius
Free University Berlin
Department of Physics
Arnimallee 14
14195 Berlin
Germany
and
Max Planck Institute for
Bioinorganic Chemistry
Stiftstr. 34-36
45470 Mülheim (Ruhr)
Germany

Frank Neese
Universität Bonn
Institut für Physikalische und
Theoretische Chemie
Lehrstuhl für Theoretische
Chemie
53115 Bonn
Germany

Mark J. Prandolini
Goethe-Universität Frankfurt
Institute for Physical und
Theoretical Chemistry and Center
for Biomolecular Magnetic
Resonance
Max-von-Laue-Str. 7
60438 Frankfurt am Main
Germany

Thomas Prisner
Goethe-Universität Frankfurt
Institute for Physical und
Theoretical Chemistry and Center
for Biomolecular Magnetic
Resonance
Max-von-Laue-Str. 7
60438 Frankfurt am Main
Germany

Richard W. Quine
University of Denver
Department of Engineering
2135 E. Wesley Ave
Denver, CO 80208
USA

Edward Reijerse
Max-Planck Institüt für
bioanorganische Chemie
Stiftstrasse 34-36
45470 Mülheim
Germany

George A. Rinard
University of Denver
Department of Engineering
2135 E. Wesley Ave
Denver, CO 80208
USA

Anton Savitsky
Max Planck Institute for
Bioinorganic Chemistry
Stiftstr. 34-36
45470 Mülheim (Ruhr)
Germany

Jason W. Sidabras
National Biomedical EPR Center
Medical College of Wisconsin
Department of Biophysics
8701 Watertown Plank Road
Milwaukee, WI 53226-0509
USA

Alex Smirnov
North Carolina State University
Department of Chemistry
Raleigh, NC 27695-8204
USA

Robert A. Strangeway
Medical College of Wisconsin
Department of Biophysics
8701 Watertown Plank Road
Milwaukee, WI 53226-0509
USA

Milwaukee School of Engineering
Department of Electrical
Engineering and Computer
Science
1025 North Broadway
Milwaukee, WI 53202-3109
USA

Sankaran Subramanian
National Institutes of Health
National Cancer Institute
Bethesda, MD 200892
USA

1
Introduction
Sushil K. Misra

1.1
Introduction to EPR

In earlier days, electron paramagnetic resonance (EPR) was referred to as paramagnetic resonance (PMR), but today it is also referred to as electron spin resonance (ESR) and, more recently – in analogy with nuclear magnetic resonance (NMR) – as electron magnetic resonance (EMR). For simplicity and consistency, however, the term EPR will be used throughout this book.

With the advent of high-frequency spectrometers, EPR has today become a very sensitive and unique technique. Historically, the most commonly used frequency band for EPR has been X-band (~9.5 GHz; 1 GHz = 10^9 Hertz]), requiring a waveguide resonator of the size ~3.0 cm. The processing and analysis of EPR spectra have become much more easy and accurate with the development of the personal computer (PC), which today is becoming equipped with increasingly faster processors as time marches on. As a result, computational times have been reduced by many orders of magnitude over the past three decades, and this is expected to be reduced even further in future. In addition, faster and more accurate EPR data acquisition softwares have been developed which no longer require the use of chart recorders. When combining these developments with parallel processing in computation, the acquisition, simulation, and analysis of EPR data have become extremely rapid and efficient.

1.1.1
Continuous-Wave EPR

In continuous-wave EPR (CW-EPR) spectroscopy, the microwave field is applied to the sample over the entire period of time, while the EPR signal is recorded for a sweep of the external magnetic field over a chosen range of time, normally over 0.0–0.9 Tesla (1 Tesla (T) = 10 000 Gauss) at X-band. Although the use of X-band spectrometers has provided certain advantages over the years, spectrometers that function at microwave bands of higher and lower frequencies than that of X-band have now been developed, and these may be preferable over X-band for some types

Multifrequency Electron Paramagnetic Resonance, First Edition. Edited by Sushil K. Misra.
© 2011 Wiley-VCH Verlag GmbH & Co. KGaA. Published 2011 by Wiley-VCH Verlag GmbH & Co. KGaA.

of investigation. Finally, the variety of experimental data acquired at various frequencies indicates that a multifrequency approach in EPR is preferable in order to achieve a complete and unambiguous description of the interaction of a paramagnetic ion with its environment.

1.1.2
Pulsed EPR

In contrast to CW-EPR, and following NMR for which pulsed techniques were first developed, pulsed EPR spectroscopy is now commonly used. In this technique, microwave pulses are applied to the sample over selected finite intervals of time, so that the orientation of the precessing magnetic moment about the external magnetic field (Larmor precession) can be turned by chosen angles about the x-, y-, or z-axes (conventionally, the z-axis is chosen to be parallel to the external magnetic field). To this end, it is possible to use a single pulse or more pulses with chosen intervals of time in between, recording a signal (referred to as an "echo signal") that is proportional to the magnetization of the sample at appropriate intervals following application of the last pulse. As a consequence, several varieties of pulsed EPR techniques have been developed; these include electron spin echo (ESE), ESE envelope modulation (ESEEM), electron–electron double resonance (ELDOR), double quantum coherence (DQC), double electron–electron resonance (DEER) – which is also referred to as pulse ELDOR (PELDOR) and ELDOR in ESE – two-dimensional Fourier transform EPR (2-D-FTEPR), spin-echo correlated spectroscopy (SECSY), correlation spectroscopy (COSY), and hyperfine spin-correlation spectroscopy (HYSCORE). Pulsed EPR has certain advantages over CW-EPR, as it provides information that cannot be acquired by using CW-EPR. However, the instrumentation required for pulse EPR is much more costly, and analysis of data much more complex, compared to CW-EPR.

1.1.3
EPR Imaging

In analogy with NMR imaging, the technique of EPR imaging – which is based on the same principle as its NMR counterpart – has undergone a rather rapid development. EPR imaging provides information that is complementary to NMR imaging. In this book, attention is focused on the theory and applications of multifrequency EPR carried out at frequencies in the ~0.1 to 1200 GHz range, covering the latest developments.

1.2
Historical Background of EPR

The historical background of EPR has occurred in chronological order: (i) the developments in paramagnetism up to 1939, which greatly influenced the achievement of magnetic resonance (Bleaney, 1997); (ii) subsequent historical developments in EPR research, which include background research leading to EPR (Poole and Farach, 1997), the early history of EPR spectrometers (Eaton and Eaton, 2004; Grinberg and Dubinskii, 2004; Freed, 2004), and early commercial spectrometers (Varian: Hyde, 2004; Bruker: Schmallbein, 2004). The early history of paramagnetic spin-lattice relaxation was described by Manenkov and Orbach (1976). Details of these references, along with the respective titles, are provided below.

1.2.1
Literature Pertinent to the Early History of EPR

The details of the early history of EPR can be found, for example, in the following sources:

- B. Bleaney (1997) Paramagnetism, Before Magnetic Resonance, in *Foundations of Modern EPR* (eds G.R. Eaton, S.S. Eaton, and K.M. Salikhov), World Scientific, New Jersey, 1997 (hereafter Ref. 1), pp. 22–36;

- C.P. Poole, Jr and H.A. Farach, Preparing the Way for Paramagnetic Resonance, Ref. 1, pp. 13–24;

- C.P. Poole, Jr and H. A. Farach, The First Sesquidecade of Paramagnetic Resonance, Ref. 1, pp. 63–83;

- G.R. Eaton and S.S. Eaton (2004) EPR Spectrometers at Frequencies Below X-band, in *Biological Magnetic Resonance*, vol. 21 (eds L.J. Berliner and C.J. Bender), Kluwer Academic/Plenum Publishers, New York, pp. 59–114;

- O.Y. Grinberg and A.A. Dubinskii (2004) The Early Years, in *Biological Magnetic Resonance*, vol. 22, Kluwer Academic/Plenum Publishers, New York;

- O. Grinberg and L.J. Berliner (eds) (hereafter Ref. 2), pp. 1–18, and J.H. Freed, The Development of High Field/High Frequency ESR. Historical Review, Ref. 2, pp. 19–43;

- J.S. Hyde, EPR at VARIAN: 1954–1974, Ref. 1, pp. 695–716

- D. Schmalbein, A Bruker History, Ref. 1, pp. 717–730;

- A.A. Manenkov and R. Orbach (1976) *Spin-Lattice Relaxation in Ionic Solids* (eds A.A. Manenkov and R. Orbach), Harper and Row Publishers, New York, pp. ix–x.

1.3
Typical X-Band, Low-, and High-Frequency Spectrometers

1.3.1
EPR Spectrometer Design

The basic details of the designs of CW-EPR spectrometers operating at various frequencies are detailed in this section. Typically, in a CW-EPR spectrometer the microwave frequency is kept constant and the magnetic field is varied linearly with time. This changes the spacing of the energy levels, so that the condition of resonance is fulfilled when the spacing of energy levels participating in resonance becomes equal to $h\nu$, where ν is the frequency of the microwave radiation and h is Planck's constant. Such a spectrometer consists of four components:

- A *source of microwave radiation*, with components that control and measure the frequency and intensity of that radiation.
- A *magnetic-field system*, which provides a stable, linearly varying, and homogeneous magnetic field of arbitrary magnitude.
- A *cavity*, or resonator, system characterized by a fixed resonant frequency, which holds the sample, and directs and controls the microwave beam to and from the sample.
- A *signal-detection system*, which is accomplished by modulating the magnetic field by an alternating magnetic field, providing a sinusoidal voltage at the modulation frequency that is amplified as the useful EPR absorption signal for its recording.

In the rudimentary system, the cavity can be avoided by simply placing the sample in the waveguide. The next stage of sophistication in design used for "old-fashioned" spectrometers was to place the sample in the cavity and to use the transmission mode. Finally, in the commonly used modern design, the cavity is used in reflection mode in conjunction with a microwave circulator to prevent the reflected microwaves from passing back to the source of radiation. Today, most EPR spectrometers are of the reflection type, whereby changes in the amount of radiation reflected back from the microwave cavity containing the sample are measured. Such EPR spectrometers all have the same design over the range of frequencies from 8 to 70 GHz.

1.3.2
X-Band Spectrometer

This is the most common spectrometer of those used at various frequencies. The schematic of a typical X-band spectrometer is shown in Figure 1.1, and the details of the various components are as follows.

1.3.2.1 Source of Microwave Radiation

While klystrons (a type of vacuum tube) were used almost exclusively in old spectrometers as the source of microwave, in virtually all new spectrometers a variety of different sources are used at all frequencies. The radiation emitted from the source is monochromatic (the following description is given in terms of klystrons). In a klystron, microwave oscillations are produced that are centered over a small range of frequencies. This output, when expressed as a function of frequency, is referred to as the *klystron mode*. Among the several modes available, the mode with the highest power output is usually selected. A mode can be displayed on an oscil-

1.3 Typical X-Band, Low-, and High-Frequency Spectrometers

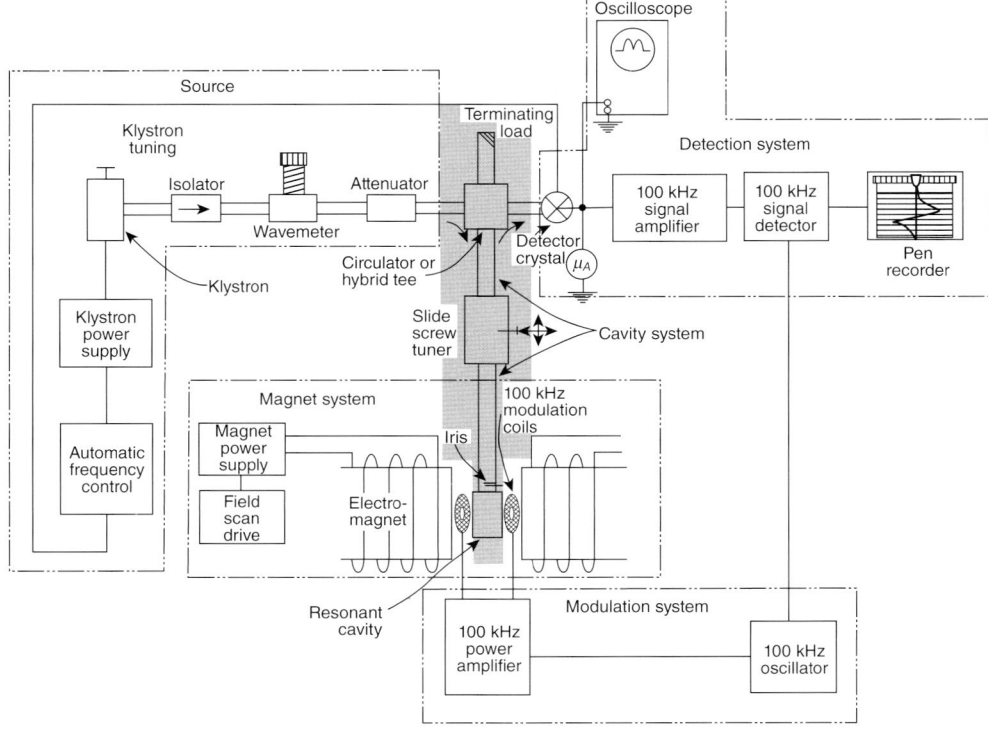

Figure 1.1 Schematic of a typical X-band EPR spectrometer equipped with 100 kHz phase-sensitive detector (Wertz and Bolton, 1972).

loscope by varying the klystron frequency over the range of the mode. The voltage applied to the klystron will then determine the frequency of the microwave radiation produced by the klystron, which can be tuned over the range of the mode by adjusting a tuning stub on the klystron. Subsequently, a sharp dip is seen over the frequency region that corresponds to the cavity resonance. The klystron is tuned in such a way that the cavity dip is set at the center of the mode, where the power and stability are maximized. The klystron frequency must remain very stable, because the energy density in the resonant cavity is very sensitive to the frequency of the incident radiation. This requires that frequency variations should be small in comparison with the true EPR linewidth. An automatic frequency control system is used to stabilize the frequency of the klystron. Normally, the klystron frequency is locked to the resonant frequency of the cavity in which the sample is placed. Any significant backward reflections of the microwave energy from the system fed to the klystron may result in serious perturbations of the microwave frequency. However, this can be avoided by the use of an *isolator* – a nonreciprocal device that passes the microwave radiation in the forward direction readily, but strongly attenuates any reflections. In this manner, any variations in the microwave frequency due to backward reflections in the region between the circulator and the klystron are minimized. The frequency of the radiation is measured with a *wavemeter*. This is a cylindrical resonant cavity, the length of which is adjustable to an integral number of half-wavelengths by using a micrometer. Any decrease in the power will be detected by a silicon crystal when the resonant frequency of the wavemeter matches the frequency of the incident microwaves, because of the resonant absorption of the microwaves by the wavemeter. The wavemeter is usually read to a precision of ±1 MHz, having been calibrated in frequency units, although typically its accuracy will be ±9 MHz. For greater accuracy in frequency measurements it is possible to use a frequency meter by coupling it to the microwave system. It should be noted that wavemeters most likely retain some use at very high frequencies, but at Q-band and lower frequencies a frequency counter would be used rather than a wavemeter. The microwave power incident on the sample can be adjusted with an *attenuator*, which consists of an absorptive element.

1.3.2.2 Transmission of Microwaves

Single-mode waveguides are used for the transmission of microwaves. The size of a waveguide increases with wavelength, or decreases with frequency in a linear fashion. For example, the dimension (cross-section) of an X-band (9.8 GHz) waveguide are: 0.9×0.4 in $(2.3 \times 1.0\,cm)$, whereas they are 6.50×3.25 in $(16.5 \times 8.3\,cm)$ at L-band (1 GHz). At lower frequencies – including that of X-band – higher-order modes are not excited, so that simple bends can be made and the directional couplers and other passive components twisted in simple manner. As the wavelengths at low frequencies are rather large, the resistive losses in the waveguides at X-band are minimal, being proportional to $v^{3/2}$ for a single-mode waveguide, and of the order of $0.7\,dB\,m^{-1}$ for TE_{10} rectangular mode. It should be noted that semi-rigid coaxial cables are frequently used in the range extending from L- to X-band frequencies; however, at frequencies above 10 GHz their propagation losses become rather large. Nonetheless, they provide a compact size and thus are convenient in usage.

1.3.2.3 The Cavity (Resonator) System

The sensitivity of a spectrometer is enhanced by a resonator, which stores the microwave energy, and concentrates the microwave power at the sample. Cavities are used as the most common type of resonator, with the most frequently used being rectangular-parallelepiped and cylindrical in nature. The cavities are constructed as a short section, which is an integral number of half-wavelengths, of a rectangular or circular wave guide in which a standing wave is produced. Cavities become progressively larger as the frequency decreases. In this context, the *fundamental resonance frequency* is defined as that which corresponds to a cavity dimension. It is possible to excite more than one type of standing wave pattern, called a *mode*. These are referred to as transverse electric (TE) modes, where the subscript denotes the number of half-wavelengths along the various dimensions (e.g., TE_{102} for rectangular-parallelepiped and TE_{011} for cylindrical cavities). A cylindrical cavity can be tuned by using a piston fitted at one end; this property is also used in a wavemeter (as described above), which is a cylindrical cavity used to measure the wavelength of the microwave radiation. Since, for microwaves, the wavelength is typically of the order of centimeters, the size of the resonant cavity is convenient to handle reasonably large samples. For microwaves, it is important to consider both the electric (E_1) and magnetic (B_1) fields in the cavity. The lines of forces for the electric and magnetic fields are calculated using Maxwell's equations, with the maxima of the electric and magnetic fields occurring at different places, depending on the mode. The requirements for EPR resonance are: (i) a cavity mode should permit a high-energy density; (ii) the sample can be placed where the density of B_1 is maximum; and (iii) the B_1 field is perpendicular to the static field B to observe the allowed transitions, for which $\Delta M = \pm 1$, where M is the electron spin magnetic quantum number. The Q-factor of a cavity represents the sharpness, or figure of merit, of the response of the cavity. This is defined as the ratio of 2π times the maximum microwave energy stored in the cavity to the energy dissipated per cycle, and implies that Q will increase with the cavity volume for a fixed frequency. It also means that Q will be increased by reducing the energy losses from currents flowing in the cavity walls or in the sample. Thus, Q can be maximized by heavy silver plating, the further deterioration of which can be prevented by an additional thin gold plating. Q is also reduced if a sample with a high dielectric constant (e.g., an aqueous sample) extends into the regions of appreciable electric field. In order to reduce the fractional reflection of microwave energy from the cavity, and at the same time to increase the fractional change in the reflected power, an *iris* – a small hole at the entrance port to the cavity – is used. This is accomplished with an adjustable screw, the setting of which depends on the size and nature of the sample in the cavity. In this way, standing waves are formed due to discontinuities in the waveguide, or to an imperfect matching of microwave elements, from areas other than the cavity, and this causes a decrease in the sensitivity of the spectrometer. In order to minimize such effects, a small metallic probe known as a *slide-screw tuner* is used to produce standing waves of such amplitude and phase as to minimize the existing standing waves, by varying its depth of insertion and position along the waveguide. In order to attenuate any microwaves traveling in the reverse direction, a *circulator* (a nonreciprocal device) is used. This passes the microwaves in the forward direction with little loss, but strongly attenuates any microwaves traveling in the reverse direction. Thus, the microwaves are directed to the cavity and the signal is reflected from the cavity to the detector. If a reduction in the size of the resonator is desired, a dielectric resonator can be used in which the wavelength is considerably shorter than that in the free space. Alternatively, a different type of structure can be used, such as a split-ring or loop-gap resonator. Size reduction is important both at lower frequencies and for pulsed EPR at X-band, where an enhanced microwave magnetic field (B_1) is required.

1.3.2.4 Magnetic Field System

The magnet used at X-band is an electromagnet, consisting of two Helmholtz coils with sufficient space between the pole pieces to house the resonator. The magnetic field should be stable and uniform over the sample, with a stability of better than ±10 mG for organic free radicals in liquid solution and ~1 G for most inorganic samples. Stability of the magnetic field is achieved by the use of a highly regulated power supply to drive the magnet, which uses a Hall-effect probe to detect any fluctuations in the magnetic field directly. These are then corrected by a feedback system in the power supply to maintain the magnetic field value. A highly reproducible and linear scanning system, connected to the power supply, regulates the current in the magnet to accomplish the variation of the magnetic field in a linear fashion.

1.3.2.5 Modulation and Detection System

A phase-sensitive detection technique is used in which a small amplitude modulation of the static magnetic field is carried out to limit the noise-contributing components to frequencies very close to the modulating frequency. Commonly, the modulation frequency used is 100 kHz, achieved by placing Helmholtz coils on the two sides of the cavity along the axis of the static magnetic field. In order that the 100 kHz signal penetrates through to the sample, the cavity walls must be very thin. With this arrangement, the rectified signal arriving at the detector is amplitude-modulated at 100 kHz. For the detector, a silicon crystal, serving as a Schottky barrier diode, is most often used; this acts as a microwave rectifier, so that the detector current will vary as the square-root of the microwave power if the average incident power is greater than 1 mW. In this mode, the detector is termed a "linear detector." For optimal sensitivity, the diode is operated in the linear region with the diode current being approximately 200 μA. When the power is less than 1 μW, the electric current is proportional to the microwave power, and the detector is termed a "square-law detector." In order to supply the detector with extra power, or *bias*, to ensure that the detector is functioning at the correct level, a reference arm is used in conjunction with a *phase shifter* to bring the microwaves in phase with the reference arm. With 100 kHz frequency, the "$1/f$ detector noise," which is inversely proportional to the frequency, is appreciably less than that due to other sources. However, with the use of a "backward diode" it is possible to operate at modulation frequencies as low as 10 kHz, with the same sensitivity as that achieved with a silicon crystal operating at 100 kHz. The sensitivity of detection at low microwave power can be further increased by using the principle of *superheterodyne detection*, in which the signal is mixed with the output of a local oscillator to produce an intermediate frequency, which is then amplified and detected. Typically, in EPR, this local oscillator operates at a frequency of 30 MHz above or below that of the klystron. At the output of the mixing stage, the difference frequency of 30 MHz contains all the relevant information, with a negligible $1/f$ noise due to high value of this frequency. In addition, low-frequency modulation can be used without any loss of sensitivity, since this adds negligible noise at the detection frequency of 30 MHz. Further, when a sample requires a very small microwave power, the superheterodyne detection method has proved to be of unparalleled sensitivity.

Returning to the case of modulation at 100 kHz, the signal after detection is subjected to a narrow-band amplification. A further reduction in noise can then be achieved by using phase-sensitive detection, which rejects all noise components except those in a very narrow band (~±1 Hz), about 100 kHz. Basically, in this technique, the amplified signal is mixed with the output of the modulating 100 kHz oscillator. Then, if the two are exactly in phase the output from the unit will be a maximum, but if they are opposite in phase then the output will be a minimum. The output of the detected 100 kHz signal is approximately proportional to the slope of the absorption curve at the mid-point of the modulating field, provided that the amplitude of the modulation is small compared to the linewidth. The sign of the slope determines the output polarity of the phase-sensitive detector. As a consequence, when the amplitude of modulation is small, the output signal resembles the first derivative of the absorption signal, with the line shape becoming distorted for large modulation amplitudes approaching the linewidth. The time constant of the circuit filtering the output affects the noise of level of the signal; notably, the noise becomes less for larger time constants.

1.3.3
EPR Line Shapes and Determination of Signal Intensity

Two types of line shape are usually observed in EPR – *Lorentzian* and *Gaussian* – in which the line shapes are expressed as $y = a/(1 + bx^2)$ and $y = a\exp(-bx^2)$, respectively, where y and x represent, respectively, the EPR signal and the magnetic field. The intensity of the first-derivative absorption signal is proportional to $Y'_{max}(\Delta B_{pp})^2$, where $2Y'_{max}$ is the peak-to-peak derivative amplitude and ΔB_{pp} is the peak-to-peak width. Gaussian line shapes are produced when there occurs superposition of many components, known as *inhomogeneous broadening*. On the other hand, if there is no variation of the parameters determining the line shape from one spin to another spin, one observes a *Lorentzian* line shape, referred to as *homogeneous broadening*. The half of the Lorentzian linewidth at half height, Γ, is related to the relaxation time T_2, as follows (Wertz and Bolton, 1792, pp. 196–197): $1/T_2 = \kappa \gamma_e \Gamma$, where κ is a constant that depends on the line shape, and γ_e is the electron magnetogyric ratio.

1.3.4
Low-Frequency Spectrometers

The differences in CW-EPR spectrometer design at frequencies lower than that at X-band, described above, are listed as follows. The frequency range of these spectrometers is anywhere from ~250 MHz to ~4 GHz.

- *Source of radiation*: The selection of a microwave, or a radiofrequency, source is made based on consideration of the phase noise and the ability to control the source frequency with an automatic frequency control system. No EPR spectrometer recently described has used a klystron in the L (1 GHz)- and S (4 GHz)-bands, as these cavity-stabilized oscillators used earlier are no longer available. Some sources currently in use are: super-regenerative field-effect transistor (FET) oscillator, signal generator, microwave oscillator, phase-locked crystal oscillator.

- *Transmission*: In most cases, a coaxial cable is used for the transmission of waves at low frequencies.

- *Resonators and sensitivity*: At low frequencies, more compact devices – such as coaxial cavities, helices, and coaxial cavity containing a helix – are used rather than cylindrical or rectangular cavities, as their sizes become very large. These include: (i) a strip-line resonator, consisting of a center conductor and ground plates; (ii) a lumped-circuit resonator; (iii) a loop-gap resonator (LGR); (iv) a single-turn solenoid; (v) a tuned one-loop–two-gap bridged LGR; and (vi) a surface-coil resonator. Finally, the size of the resonator can be significantly reduced by using a dielectric resonator, in which the wavelength is considerably shorter than that in the free space. As for the sensitivity, it is noted that if the sample is unlimited, and if the lossiness is not important, then the signal-to-noise ratio (SNR) is enhanced at lower frequencies (Eaton and Eaton, 1988a).

- *Magnetic field*: In the frequency range 1–9 GHz, standard EPR iron-core magnets are used. Field control is used in iron-core and/or iron-yoke magnets, whereas current control can be used only for air-core magnets. Other sources of magnetic field that have been used at low frequencies include solenoid and Helmholtz-type coils.

- *Detectors*: At low frequencies, the detectors used are normally crystals or double-balanced mixers.

1.3.5
High-Frequency Spectrometers

The various frequency ranges considered here – which often are referred to as "submillimeter waves" – are greater than or equal to 95 GHz, or the wavelengths are less than or equal to 2.5 mm. The block diagram of a 1 mm EPR spectrometer is shown in Figure 1.2 (Lynch *et al.*, 1988). As compared to X-band, the following differences are noted for the various components of an EPR spectrometer at high frequencies.

1.3.5.1 Sources of Radiation
The sources currently in use are described as follows:

- *Gunn diodes and multipliers*: These are semiconductor devices, which produce microwave radiation, maintained by a resonant cavity coupled to the device. They are usually mounted on a single-mode waveguide, generating the output in a well-defined mode. The maximum frequency that can be produced is about 140 GHz, with the output power varying from about 500 mW at 10 GHz to 60 mW at 140 GHz. In order to produce higher frequencies, frequency multipliers can be used; this is achieved with Schottky diodes. However, the power falls rather rapidly with multiplication in accordance with the power loss being of the order of 25 log(N) dB, where N is the order of the harmonic to be generated. This reduces the power typically by 7.5 dB (16%) and 17.5 dB (1.8%) for

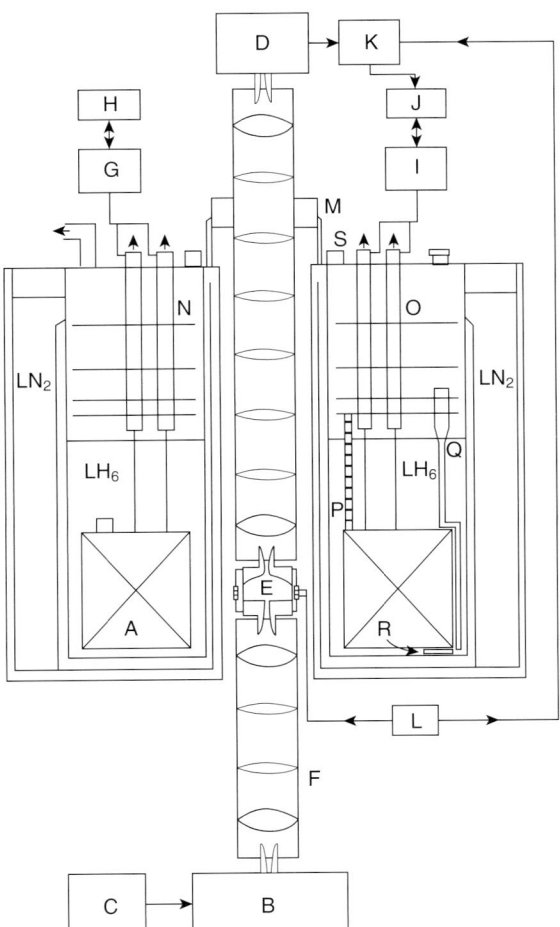

Figure 1.2 Block diagram of a 1 mm EPR spectrometer (Lynch et al., 1988) A, 9 T superconducting solenoid and 500 G sweep coils; B, phase-locked 250 GHz source; C, 100 MHz reference oscillator for 250 GHz source, D, Schottky diode detector; E, Fébry–Perot semiconfocal cavity and field-modulation coils; F, 250 GHz quasi-optical waveguide; G, power supply for main coil (100 A); H, current-ramp control for main magnet; I, power supply for sweep coil (50 A); J, PC to control field sweep, data acquisition, and data manipulation; K, lock-in amplifier for final signal amplification and manipulation; L, field modulation and lock-in reference source; M, Fébry–Perot cavity tuning screw; N, vapor-cooled leads for sweep coil (non-tractable); O, vapor-cooled leads for sweep coil (non-tractable); P, ^{4}He bath level indicator; Q, ^{4}He transfer tube; R, bath temperature/bath heater resistance pod; S, ^{4}He blow-off valves.

the second and fifth harmonics, respectively. The noise of these harmonics is quite acceptable, when used as sources of radiation in EPR. Related to the Gunn diode is the IMPATT (Impact Ionization Avalanche Transit Time Mode) device, which produces a higher power but accompanied by a higher noise. The IMPATT is commonly used as an injection-locked amplifier.

- *Molecular gas lasers*: These have also been used as sources for millimeter waves in EPR. They consist of a laser cavity with a length of the order of 1.5 m, filled with molecular gas at a medium pressure (ca. 0.1–1.0 mbar). The cavity is a metallic or a dielectric tube that is fitted with mirrors at the two ends, contains the gas, and serves as an oversized waveguide. The gas is excited by a tunable single-mode CO_2 laser with a wavelength of about 10 μm and a power of the order of 50 W. Coupling out of the laser is achieved through a hole in one of the mirrors, but the mode is not always well defined. The actual laser action takes place between the rotational levels of the gas; the frequencies produced here are discrete, and therefore cannot be changed or locked to an external frequency source. The efficiency increases with frequency, and good power levels can be obtained in the range of 200 to 2000 GHz. A major concern here is that of stability, as the system requires rather long warm-up times. Although the $1/f$ noise is quite high, it is better or equal to that of a Gunn diode at frequencies higher than 10–20 kHz.

- *Backward-wave oscillators (carcinotrons)*: These are related to traveling-wave tubes. The microwaves are produced here as a result of the bunching of electrons generated in an electron gun, as in a klystron. The power available here is higher than that produced by a Gunn diode, but the noise is higher. The frequencies available range from about 30 GHz to 1000 GHz, but the lifetime is rather limited to about 2000 h. It should be noted that similar devices, such as the extended interaction oscillator and the Orotron, operate on the principle of bunching of electrons to produce microwaves.

- *Gyrotrons*: Here, the bunching of electrons is accomplished by cyclotron resonance. This results in a very high efficiency and very high power levels of the order of 200 W. The magnetic fields required for the operation of gyrotrons are of the same order of magnitude as those required for EPR, which makes the devices bulky and expensive. They can be injection-locked, but are noisy in operation.

1.3.5.2 Transmission of Submillimeter Waves

It is possible to use either oversized (or corrugated) waveguides, or free space for the transmission of high-frequency microwaves.

- *Waveguides*: As with the single-mode waveguides, the resistive losses at W-band (95 GHz) are 16 dB m^{-1}, whereas they are 21 dB m^{-1} at 300 GHz. Thus, taking into account the fact that the power levels at frequencies of 95 GHz and higher are rather low, such losses are unacceptable; this makes the use of single-mode waveguides impractical at these frequencies over large distances. A solution is to use oversized, rather than single-mode, waveguides, for which the losses are much smaller. Thus, X-band waveguides can be used for which the attenuation will still increase with frequency, but at a much smaller rate, proportional to $v^{1/2}$. The use of all elements susceptible of exciting higher modes, and not just a straight waveguide, should be avoided; however, single-mode waveguides are still required for bends and turns. In some applications, it is not necessary to have a well-defined mode, in which case shelf brass pipes can be used. These are practical, cheap, and capable of accommodating a wide range of frequencies, with losses of the order of 2–3 dB m^{-1}. Recently, corrugated waveguides have been used in EPR (Smith *et al.*, 1998); these are a special form of oversized waveguide, that consist generally of a circular metallic waveguide having corrugations of the order of $\lambda/4$ width and depth. Such waveguides behave like dielectric waveguides, with the principal mode EH_{11}. As a result, the losses may be as small as about 0.1 dB m^{-1}.

- *Free space*: Attenuation is zero in vacuum, and very small in free space. As the divergence of radiation in this case is appreciable, it is necessary to keep the beam contained; this can be accomplished by using conventional optics with lenses and elliptical mirrors, referred to as *"quasi-optics."* Horns can be used for conversion from single-mode or corrugated waveguides to free-space propagation, using optical elements such as mirrors, lenses, and polarizers to act on the beam. The optical equations used in this context are described by Gaussian beam-mode optics. The schematics of a quasi-optical induction-mode spectrometer are shown in Figure 1.3 (Smith *et al.*, 1998).

1.3.5.3 Resonators and Sensitivity

The resonator used preferentially in VHF (~250–360 GHz) EPR is the Fébry–Perot resonator. This has the advantage of having an open structure, which makes it accessible for optical excitation, and its *Q*-factor is typically of the order of 2000–3000. The Fébry–Perot resonator can support two orthogonal modes, reflection and induction; thus, it serves as a *bimodal* cavity. Its main advantages are a small filling factor and critical sample loading. Although this type of resonator has been initially used at 95 GHz, it has now been largely replaced by a cylindrical TE_{011} cavity, which is difficult to fabricate at this frequency, as it is very small. Finally, it may be possible not to use any cavity at all; whilst this option reduces the absolute sensitivity, it still provides an acceptable sensitivity when an appropriately concentrated sample is used. The absolute sensitivity depends on frequency as $v^{11/4}$, assuming constant B_1, a constant sample size, taking into account the fact that the resonator size is inversely proportional to frequency, and considering only resistive losses in the cavity. Thus, as the sample size for the same resonator type decreases with frequency, the concentration sensitivity remains almost independent of frequency, and actually decreases at a higher frequency. In fact, in some cases, the use of a cavity at high frequencies does not have any significant advantage over a system without a cavity. [It should be noted that, whilst this information was obtained concisely from *Biological Magnetic Resonance*, vol. **22** (2004), many important details have been omitted; the same volume should be consulted to obtain precise data.]

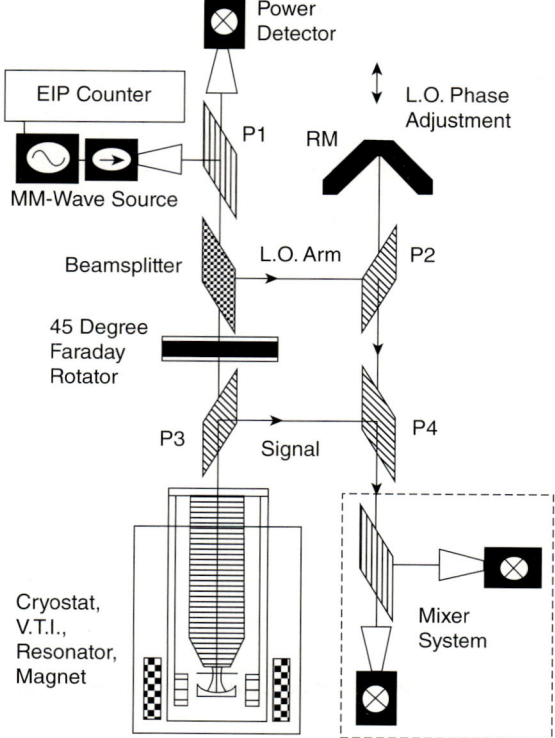

Figure 1.3 Quasi-optical induction-mode spectrometer (G. M. Smith et al., 1998). Power is transmitted to the resonator in the variable temperature insert (V.T.I.), by reflecting of polarizer P1 and passing through the bean splitter, Faraday rotator, and polarizer P3. The EPR signal is polarization (and amplitude) -encoded into the RF via the magnetic field modulation, and the induction signal is stripped off using the polarizer P3 and combined with the local oscillator signal using the polarizer P4. The local oscillator signal is stripped off using the beam splitter, and may have its phase or amplitude changed by rotating or translating the roof mirror (RM) after reflection from the polarizer P2.

1.3.5.4 Magnetic Field

At high frequencies, where magnetic fields extending to 2–14 T are required, superconducting solenoids are commonly used, but these require liquid helium for their operation. It is difficult to sweep the field with superconducting magnets quickly, due to their high inductance; consequently, they may also be fitted with room-temperature resistive coils so that the magnet can be quickly swept about its persistent field, in convenient manner. Cryogen-free magnets have also been developed, and these are available commercially.

1.3.5.5 Detectors

Two main types of detection system are currently used for submillimeter waves, namely bolometric or heterodyne with Schottky diodes.

- *Bolometric detectors*: Although these operate over a large frequency range, they are not suitable for pulsed EPR, as they possess a small bandwidth. The various bolometers in use are: (i) Ge/Ga, Si bolometers, which have a high sensitivity but possess a small bandwidth of only 300 Hz; and (ii) InSb hot-electron bolometers, which operate over the range 50 to 1500 GHz, with a bandwidth of 1 MHz. These can also be used as homodyne mixers.
- *Schottky diodes*: These are suitable for heterodyne detection, and possess a bandwidth of greater than 1 GHz.

1.3.6
Pertinent Literature

Details on EPR spectrometers are found in the following references:

X-band spectrometers

- J. Pilbrow, *Transition Ion Electron Paramagnetic Resonance*, Clarendon Press, Oxford (1990);
- J.E. Wertz and J.R. Bolton, *Electron Spin Resonance: Elementary Theory and Practical Applications*, Mc-Graw Hill, New York (1972);
- C.P. Poole, Jr, *Electron Spin Resonance: A Comprehensive Treatise on Experimental Techniques*, Wiley-Interscience, New York (1983);
- R.S. Alger, *Electron Paramagnetic Resonance: Techniques and Applications*, Wiley-Interscience, New York (1966).

Low-frequency spectrometers

A review with detailed list of references is provided by G.R. Eaton and S.S. Eaton, EPR Spectrometers at frequencies below X-band, in *Biological Magnetic Resonance*, vol. **21**, *EPR Instrumental Methods* (eds C.J. Bender and L.J. Berliner), Kluwer Academic/Plenum Publishers, New York (2004), p. 59.

High-frequency spectrometers

A good outline is provided by J. van Tol, in *Multi-frequency EPR Workshop*, 29 July 2001 (unpublished); see also Very High Frequency (VHF) ESR/EPR, in *Biological Magnetic Resonance*, vol. **22** (eds O. Grinberg and L.J. Berliner), Kluwer Academic/Plenum Publishers, New York (2004), which contains a number of review articles.

1.4
Applications of EPR

EPR is used to determine the electronic structure of free radical, transition-metal and rare-earth ions, to study the interaction between molecules, and to measure nuclear and electronic magnetic moments. It has some key advantages over NMR, because the electron has a much greater magnetic moment than a nucleus, and is thus more sensitive per spin. In time-domain experiments, such as free-induction decays, spin-echoes, or 2-D spectroscopy, the time scale in EPR is nanoseconds as compared to milliseconds for NMR. Being sensitive to materials containing even trace amounts of paramagnetic ions, EPR finds applications in the fields of physics, chemistry, biology, medicine, archeology, geology, mineralogy, radiation damage, and radiation dosimetry, among others. Apart from the applications described below, particular mention is made here of the topics included in Part Three of this book, which describes some applications of EPR. These are:

- The precise determination of fine- and hyperfine-structure spin-Hamiltonian parameters, especially large zero-field splitting (ZFS) parameters and elements of $\tilde{g}^2$-tensors using high-frequency EPR.
- The study of copper-coordination environments.
- Measurements of relaxation times.
- Chemically induced dynamic nuclear polarization, dynamic electron polarization, and dynamic nuclear polarization.
- The study of nanomaterials.
- Microscopic quantum tunneling and single-molecule magnets.

Some features of EPR relevant to its applications are briefly described as follows. EPR spectra in single crystals, containing small concentrations of paramagnetic ions substituting for regular diamagnetic ions, provide the maximum amount of information. Such spectra, which consist of numerous lines, change with the orientation of the magnetic field with respect to the crystal axes, and are thus *anisotropic*. The presence of numerous lines in the EPR spectra is due to the interactions of the orbital moments of electrons with the electric potential due to the local atoms surrounding the electrons, as well as to hyperfine interactions between the paramagnetic electrons and nuclear magnetic moments of the paramagnetic ions and surrounding atoms. On the other hand, organic molecules, containing small fractions of free radicals produced by ionizing radiation, also provide detailed information from their spectra. Symmetric (or nearly symmetric) characteristic hyperfine patterns are observed in the case of free radicals. The electron distribution throughout a molecule may be determined from a knowledge of hyperfine interactions with nuclei, the spins and magnetic moments of which are known. In addition, structural information may be deduced from EPR spectra, since hyperfine interactions vary as the inverse of the cube of the distance between the center of the free radical and the nucleus. The changes in magnitude of the hyperfine splitting and g-shift have been used to monitor features of local surroundings, such as its polarity. Furthermore, unpaired electron spins from different nitroxides, either on the same or on different molecules, interact weakly through long-range magnetic dipolar interactions, or strongly through short-range Heisenberg spin-exchange interactions. These latter interactions have been used in fluid media to monitor microscopic translational dynamics. Dipolar interactions have been used in frozen or very viscous media to measure distances, either with CW-EPR or pulsed EPR, especially in two dimensions, such as 2-D ELDOR, which is very sensitive to translational and rotational motions. This is because 2-D ELDOR also supplies off-diagonal (cross) peaks that directly report on the translational and rotational motions of labeled biomolecules. On the other hand, DQC is a new and powerful application of pulsed EPR for distance measurements between interacting spins in frozen samples. Thus, distances can be measured to 8 nm, comparable to the capabilities of the FET technique. The EPR spectra of transition-metal ions and rare-earth ions exhibit a much greater anisotropy than that exhibited by free radicals, because the former possess intrinsic anisotropy of the electron magnetic moments, and experience additional interactions when there are present more than one unpaired electron in the unfilled electronic shells. In solutions, the EPR spectra of free radicals can become quite simple due to motional averaging; only in the extreme case in which all anisotropies are fully averaged will the high-field limit be satisfied and no overlap of transitions occur. At temperatures intermediate to room temperature, however, where rapid motional averaging of the EPR spectrum takes place, and/or at the freezing point of the solution, where a powder-like spectrum is observed, information can be obtained from EPR spectra regarding slow molecular motions. Moreover, the spectra change dramatically as the tumbling motion of the probe slows, which in turn results in a great sensitivity to fluidity in the neighborhood of the spin probe. This is especially important for nitroxide spin labels that have been attached selectively to different parts of macromolecules, such as the components of natural and synthetic phospholipid membranes, liquid crystals and proteins. Important structural and functional information can be obtained from such measurements in this way.

In physics, EPR has been exploited to provide a theoretical basis for studying the modification of electronic structures by the surrounding atoms. To this end, the diamagnetic host crystals can be doped with paramagnetic impurities, such as transition-metal or rare-earth ions. In particular, it is possible to study phase transitions in solids, and the interactions that exist in pairs and triads of paramagnetic ions. Studies of other topics in physics to which EPR has been applied include superparamagnetism, ferro- and antiferro-magnetism, susceptibility, semiconductors, quantum dots, quantum wires, and defect centers.

In chemistry, EPR has been applied to characterize free radicals, to study organic reactions, and to investigate the electronic properties of paramagnetic inorganic molecules. All of this information can be exploited to understand molecular structure. Studies of some other topics in chemistry to which EPR has been applied include reaction kinetics, organometallic compounds, and molecular magnets.

In biology, EPR has been used widely to study transition-metal ions (e.g., Cu, Mn, Ni, Co, Mo, Fe), as well as complexes in proteins (enzymes), nitroxide spin labels attached to cysteine or nucleic acids, enzyme reactions, protein folding, amino acid radicals of the protein backbone (e.g., tyrosine, tryptophan, and glycine), hemes and FeS clusters in electron-transfer reactions in protein, and the dynamics of proteins, protein-bound cofactor radicals (e.g., semiquinones and flavins), and radicals produced during reaction processes in proteins and other biomolecules. Special reagents have been developed to study spin-label molecules of interest, that are of special interest in biology. Specifically designed nonreactive radical molecules can be attached to specific sites in a biological cell; this is termed *site-directed spin labeling*, whereby the EPR spectra of these spin labels, or spin probes, can provide information on their environments. Currently, more than 20 proteins that function in the mitochondrial respiratory chains of mammals have been identified using EPR, and this has been invaluable in elucidating the details of their electron-transfer processes. In order to increase the natural concentration of free radicals in cells to levels detectable with EPR, the cells can be subjected to irradiation or ionizing radiation, or rendered unstable by lowering their temperature. The nitroxide radical (NO·), in which an electron is delocalized between N and O atoms, produces EPR spectra the shapes of which are sensitive to the viscosity, polarity, and structure of the surrounding media. Thus, nitroxide radicals have been used extensively as molecular probes; in this case, they are referred to as *spin probes* or *spin labels*, because they can be bound to proteins or lipids, so as to provide information concerning the local mobility of the polypeptide chain, liquid-crystal structure, and the mobility of lipids in biomembranes. EPR has also been applied in conjunction with site-directed spin labeling to study the structure and conformational dynamics of a wide range of systems, including high-molecular-weight soluble proteins, membrane proteins, nucleic acids, and nucleic acid–protein complexes (Hubbell *et al.*, 2000). The technique of *spin trapping* enables the detection of oxygen-generated active radicals, which is extremely important in biology. For this, a nitrone compound is allowed to interact with a free short-lived radical; the latter serves as a trap and is transformed into a spin adduct – a long-lived nitroxyl-like radical – the EPR spectrum of which has been shown to be unique for a given active radical or family of radicals. Based on their chemical nature, spin traps form two basic classes, namely nitrons compounds (e.g., *C*-phenyl-*N-tert*-butyl; PBN) and nitroso compounds (e.g., 5,5-dimethyl-pirrolin-1-oxyl; DMPO). The rate of reaction with radicals, the stability of the formed spin adducts, and the variability in the spectra of spin adducts for the separate radicals each differ for the different spin traps. This, in turn, allows the precise identification and determination of free radical types based on the EPR spectra of spin adducts, by using data banks that contain the parameters of EPR spectra of spin adducts. Recently, multifrequency EPR at high magnetic fields has been exploited to obtain detailed information regarding the structure and dynamics of transient radicals and radical pairs that occur in biological electron and ion (electron) -transfer processes (Mobius *et al.*, 2005). In this manner, it is possible to acquire a much better understanding of the relationship between structure dynamics, and the function of molecular switches. Advanced EPR techniques are also able to provide the essential ingredients of structure–function relationships for the electron-transfer intermediates of the photocycle. In addition to information obtained by EPR on photosynthetic reaction centers, that obtained for site-specifically, spin-labeled bacteriorhodospin proton pump and the colicin ion-channel forming toxin, represent the potential of high-frequency EPR (HF-EPR) in characterizing the key factors that control vectorial transfer process in proteins. An understanding of these factors is especially important for the molecular engineering community, to modulate such transfer processes by the desired point mutations. Interdisciplinary applications of EPR have also been made to investigate photosynthetic systems and transient paramagnetic chromophores in light-driven processes.

EPR imaging (EPRI) is the spatial discrimination, in one to three spatial dimensions, of EPR signals due to paramagnetic species. Thus, it can be used to image the distribution of paramagnetic centers in solids, the spatial degradation of polymers, and spin-trapped endogenous free radicals either *in vitro* or *in vivo*. Further, by using stable paramagnetic molecules or particles as probes, such parameters as oxidation status, pH and oxygen concentration in biological samples and living animals can be determined by using EPRI. The biomedical applications of EPRI can be listed as: (i) EPRI of skin; (ii) whole-body *in-vivo* imaging of exogenous nitroxide probes; (iii) assessment of oxidative stress or redox status; (iv) imaging of nitroxide generation *in vivo*; (v) imaging of oxygen concentrations; and (vi) imaging of free radicals in plants.

The nonbiological applications of EPR may be listed as follows: (i) the study of polymer degradation; (ii) imaging of the distribution of the paramagnetic center diamond, including the defects formed by radiation damage; and (iii) the spatial study of radiation dosimetry.

It should be noted that the oxygen molecule, in being paramagnetic, interacts with the nitroxide radical, making its EPR spectrum broader; this forms the basis of biological *oximetry*, which can be carried out *in vivo*. Thus, it is possible to record three-dimensional (3-D) images of biological tissue, and to characterize changes in the transport and distribution of oxygen in liver, brain, and other organs of laboratory animals as a result of different pathology, or after drug treatment and/or surgery.

During recent years, new instrumentation has greatly enhanced the capability of EPR applications to study basic molecular mechanisms in membranes and proteins by using nitroxide spin labels, in particular with high-field/HF-EPR and 2-D Fourier transform EPR. Specific applications of this development have been made to accurately determine distances in biomolecules, to unravel details of the complex dynamics in proteins, to characterize the dynamic structure of membrane domains, and to discriminate between bulk lipids and boundary lipids that coat transmembrane peptides or proteins. Other applications have provided time resolution to studies of the functional dynamics of proteins.

EPR is very sensitive for the detection of free radicals and paramagnetic metal ions, which are intermediates or participants in many catalytic processes. Thus, EPR has been applied to the detection of paramagnetic intermediates on the surface of palladium catalysts during heterolytic hydrogenation reactions. In conjunction with spin trapping, EPR provides a convenient simple method for detecting hydrogen atoms generated by the dissociative chemiadsorption of hydrogen onto alumina-supported palladium catalysts (<0.4% Pd) at room temperature. The occurrence of deuterium/hydrogen spillover onto the alumina surface by H/D atom transfer to surface hydroxy groups has been demonstrated directly by using deuterium (D_2). Both, alkyl and aromatic free-radical intermediates formed at a catalyst surface have also been observed using EPR (for more details, see Carley *et al.*, 1994, as cited below).

Other applications of EPR have included the studies of: metal clusters; colloids and interfaces; single-molecule magnets; coal, graphite, and fossil fuels; electrochemically generated radical ions and complexes; ferromagnets, antiferromagnets, and superparamagnets; glasses and spin glasses; geological and mineralogical systems, as well as precious stones such as amethyst, beryl and chrisoberyl, diamond, emerald, opal, ruby, rock crystal (α-quartz), sapphire, topaz, tourmaline, turquoise, and zircon; liquid crystals; point defects in semiconductors, such as alkali halides, oxides, and semiconductors; polymers; radiation damage and dating, and the irradiation of food; spin traps; and trapped atoms and molecules (a detailed listing is provided by Weil, 2007).

1.4.1
Pertinent Literature

The references cited below (each of which has the relevant applications listed in brackets) represent a typical selection of articles on the applications of EPR discussed in this section. Whilst, by any means, this is not an exhaustive list, the references will provide access to much more information on this topic.

1.5
Scope of This Book

In accordance with the title of the book, the central theme here is multifrequency EPR. Thus, all chapters in the book are written in such a manner that they cover all aspects of multifrequency; this is highlighted particularly in Chapters 1 and 2. Although a large majority of the material in this book is based on CW-EPR, a section on pulsed spectrometers is also included. Likewise, a chapter on pulsed EPR is also included in Part Two, which covers molecular motions and distance measurements, using the techniques of ELDOR, DQC, and DEER.

With regards to data tabulations, Part Three incorporates a review of multifrequency relaxation data, while experimental multifrequency EPR considerations are discussed in Part One. One important topic of the book is the analysis of EPR data to evaluate spin-Hamiltonian parameters, and the simulation of EPR spectra for single crystals, powder, and amorphous samples in Part Two. To this end, spin-Hamiltonians appropriate to various point-group symmetries, and a listing of the matrix elements of spin operators is included. Some typical applications of EPR are described in detail in Part Three. Finally, cutting-edge topics in EPR and the future of EPR – which today is of particular importance – form the discussion in Part Four. In preparing this volume, every effort has been made to include listings of pertinent literature that is relevant to each topic, to provide the reader with access to the EPR literature at large. Notably, the book should provide a reasonably up-to-date coverage of the necessary multifrequency EPR background, pertinent literature, and research topics of current interest not only to graduate and post-doctoral students, but also to advanced research workers.

Acknowledgments

I am grateful to Professors C.P. Poole, Jr and Gareth Eaton for helpful comments to improve the presentation of the chapter, and to the Natural Sciences and Engineering Research Council (NSERC) of Canada for partial financial support.

Further Reading

Beerton, K. and Stesmans, A. (2005) *Q. Sci. Rev.*, **24**, 223. (EPR dating)

Berliner, L.J. (ed.) (1976) *Spin Labeling – Theory and Applications*, Academic Press, New York. (spin labels)

Berliner, L.J. (ed.) (1979) *Spin Labeling II – Theory and Applications*, Academic Press, New York. (spin labels)

Blank, A., Dunnum, C.R., Borbat, P.B., and Freed, J.H. (2004) A three-dimensional electron spin resonance microscope. *Rev. Sci. Instrum.*, **75**, 3050. (EPR microscopy)

Borbat, P.B., Costa-Filho, A.J., Earle, K.A., Moscicki, J.K., and Freed, J.H. (2001) *Science*, **291**, 206. (membranes and proteins)

Bourgoin, J. and Lannoo, M. (1983) *Point Defects in Semiconductors*, Springer, Berlin. (semiconductors)

Buettner, G.R. (1987) *Free Radical. Biol. Med.*, **3**, 259. (spin traps)

Bulthuis, J., Hilbers, C.W., and MacLean, C.M. (1972) NMR and ESR in liquid crystals, in *Magnetic Resonance*, vol. 4 (ed. C.A. McDowell), MTP International Review of Science, Series One, Butterworths, London, p. 201. (liquid crystals)

Calas, G. (1988) *Rev. Mineral. (Spectrosc. Methods Miner. Geol.)*, **18**, 513. (geological systems)

Carley, A.F., Edwards, H.A., Mile, B., Roberts, M.W., Rowlands, C.C., Hancock, F.E., and Jackson, S.D. (1994) *J. Chem. Sci. Faraday Trans.*, **90**, 3341. (free radicals on the surface of a catalyst)

Chumak, V.V., Sholom, S.V., Bakhnova, E.V., Pasalskaya, L.F., and Musijachenko, A.V. (2005) *Appl. Radiat. Isotopes*, **62**, 141. (high-precision EPR dosimetry)

Clarkson, R.B., Ceroke, P., Norby, S.-W., and Odinstov, B.M. (2003) *Biol. Magn. Reson.*, **18**, 233. (use of particulate paramagnetic materials in EPR oximetry: coals, lithium phthalocynine, and carbon chars) (coal and charcoal)

Crawford, J.H., Jr and Slifkin, L.M. (1972) *Point Defects in Solids*, vol. 1, Plenum Press, New York. (point defects in alkali halides and oxides)

Dinse, K.-P. (2000) *Electron Paramagn. Reson.*, **17**, 78. (trapped atoms and molecules)

Eaton, G.R., Eaton, S.S., and Salikhov, K.M. (eds) (1997) *Foundations of Modern EPR*, World Scientific, Singapore. (this is a collection of articles, many of which deal with applications)

Foster, M.A. (1979) *Magnetic Resonance in Medicine and Biology*, Academic Press, New York. (biological applications)

Fraissard, P. and Resing, H.A. (eds) (1980) *Magnetic Resonance in Colloid and Interface Science*, Reidel, Hingham, MA, USA. (colloids)

Griscom, D.L. (1980) *J. Non-Cryst. Solids*, **40**, 211. (glasses)

Halliwell, B. and Gooderidge, B. (1999) *Free Radicals in Biology and Medicine*, 3rd edn, Oxford University, Oxford. (nitroxides)

Hubbell, W.L., Cafisco, D.S., and Altenbach, C. (2000) *Nat. Struct. Biol.*, **7**, 735. (identifying conformational changes of proteins with site-directed spin-labeling)

Ikeya, M. (1989) *Ann. Sci.*, **5**, 5. (dating and dosimetry)

Janzen, E.G. and Haire, D.L. (1990) Two decades of spin-trapping, in *Advances in Free Radical Chemistry*, vol. 1 (ed. D.D. Tanner), JAI Press, Greenwich, CT, USA, pp. 253–295. (spin traps)

Kennedy, T.A. (1981) *Magn. Reson. Rev.*, **7**, 41. (semiconductors)

Kochirngsky, N. and Swartz, H. (1995) *Nitroxide Spin Labels. Reactions in Biology and Chemistry*, CRC Press. (nitroxides)

Layadi, A. and Artman, J.O. (1997) *J. Phys. D Appl. Phys.*, **30**, 3312. (ferromagnets)

Li, L., Shi, Q., Mino, M., Yamazaki, Y., and Imada, I. (2005) *J. Phys. Condens. Matter*, **17**, 2749. (antiferromagnets)

Lurie, D.J. (2002) Techniques and applications of EPR imaging. *Electron Paramagn. Reson.*, **18**, 137. (EPR imaging)

Marfunin, A.S. (1994) *Advanced Mineralogy*, Springer-Verlag, Germany. (mineralogy)

Mile, B., Howard, J., Histed, M., Morris, H., and Hampson, C.A. (1991) *Faraday Discuss.*, **92**, 129. (clusters)

Möbius, K., Savitsky, A., Schnegg, A., Plato, M., and Fuchs, M. (2005) *Phys. Chem. Chem. Phys.*, **7**, 19. (biological systems: characterization of molecular switches for electron ion transfer)

Oshio, H. and Nakano, M. (2005) *Chem. Eur. J.*, **11**, 5178. (high-spin single-molecule magnets)

Petrakis, L. and Fraissard, J.P. (1984) *Magnetic Resonance: Introduction, Advanced Topics and Applications to Fossil Energy*, NATO ASI series C124, Reidel, Dordrecht, Netherlands. (clusters)

Poole, C.P., Jr, Farach, H.A., and Bishop, T.P. (1977) *Magn. Reson. Rev.*, **4**, 137; **4**, 225 (1977). (minerals)

Rhodes, C.J. (2000) *Toxicology of the Human Environment – The Critical Role of Free Radicals*, Taylor and Francis, London. (free radicals in biology and use of EPR)

Schlick, S. and Jeschke, G. (2004) Electron Spin Resonance, in *Encyclopedia of Polymer Science and Engineering* (ed. J.I. Kroschwitz), Wiley-Interscience, New York, pp. 5–37. (polymers)

Sharma, V.K. and Baiker, A. (1981) *J. Chem. Phys.*, **75**, 5596. (super-paramagnets)

Symons, M. (1978) *Chemical and Biological Aspects of Electron Spin Resonance Spectroscopy*, John Wiley & Sons, Inc., New York. (chemical and biochemical applications)

Wadhawan, J.D. and Compton, R.G. (2002) EPR Spectroscopy in Electrochemistry, in *Encyclopedia of Electrochemistry*, vol. 2, Interfacial Kinetics and Mass Transport (eds A.J. Bard, M. Strassman, and E.J. Calvo), Wiley-VCH Verlag GmbH, Weinheim, Germany, Sec. 3.2, pp. 170–220. (electrochemical EPR)

Weil, J.A. (2000) A demi-century of magnetic defects in α-quartz, in *Defects in SiO2 and Related Dielectrics: Science and Technology* (eds G. Pacchioni, L. Skuja, and D.L. Griscom), Kluwer Academic, Dordrecht, Netherlands, p. 197 ff. (point defects)

Weil, J.A. and Bolton, J.R. (2007) *Electron Paramagnetic Resonance: Elementary Theory and Practical Applications*, 2nd edn, John Wiley & Sons, Inc., New York, pp. 414–421. (a collection of references on various applications of EPR; some, but not all, of which are included here)

Wieser, A. and Regulla, D.F. (1990) *Radiat. Prot. Dosim.*, **34**, 291. (EPR dosimetry)

Zitdinov, A.M. and Kainara, V.V. (2002) The nature of conduction ESR linewidth temperature dependence in Graphite, in *EPR in the 21st Century* (eds A. Kawamori, J. Yamauchi, and H. Ohta), Elsevier, Amsterdam, Netherlands, pp. 293–297. (graphite)

2
Multifrequency Aspects of EPR

Sushil K. Misra

This chapter provides a review of the advantages and disadvantages of EPR at various frequencies.

2.1
Frequency Bands

Historically, X-band (~9.5 GHz) has been the most commonly used frequency bands for EPR studies. Despite all of the conveniences offered by X-band, EPR at lower and higher frequencies is required to study more precisely many other phenomena than can be accomplished using only X-band. A multifrequency approach has been found to be most appropriate to acquire a more complete knowledge of the interactions of a paramagnetic ion with its environment. The various frequency bands being used currently in EPR, along with the wavelength, corresponding waveguide dimensions, the outer diameter (OD) of the sample tube, and the value of the magnetic field required to observe resonance at $g = 2.0$, are listed in Table 2.1. It should be noted that the frequencies of these bands are such that they are between maxima in the atmospheric absorption of microwaves by H_2O (at about 25, 500, 350 GHz) and by O_2 molecules (at about 60 and 120 GHz). It should also be noted that there are absorptions over a wide range of frequencies; for example, the EPR of gas-phase O_2 is used as a sensitivity standard in the Varian Q-band spectrometer.

2.2
X-Band EPR

The advantages of using EPR at X-band are as follows:

- The required magnetic field can be produced with high homogeneity and stability by iron-core electromagnets. At much higher frequencies, liquid helium-cooled superconducting magnets are required.
- X-band microwave components are highly engineered and readily available.

2 Multifrequency Aspects of EPR

Table 2.1 The various frequency bands commonly used in EPR spectrometers.

Frequency band	Microwave frequency (GHz)	Approx. wavelength (mm)	Waveguide dimensions (cm)	Sample tube OD (mm)	Magnetic field (mT) for $g = 2.0$ (calculation)
L	~1 (1–2)	300	~16.5 × 8.3	30	35.682 48
S	~4 (2–4)	75	~8.6 × 4.3		142.729 93
C	~6 (4–8)		~4.3 × 2.2		
X	~9.8 (8.2–12.4)	30	~2.3 × 1.0	4	349.688 34
Ku	~18 (12.4–18)		~1.4 × 0.6		
K	~24 (18–26.5)	10	~1.07 × 0.43		
Q	~34 (26.5–40)		~0.71 × 0.36	2	1213.204 43
V	~55 (40–75)				
W	~94 (75–110)	3	~0.25 × 0.13	0.9	3354.153 43
D	~140 (110–170)				
Far infra-red	170, 240, 360 (140–1120)				

The wavelength ($= c/\nu$), outer diameter (OD) of the commonly used sample tube, and the magnetic field required for $g = 2.0$, along with their waveguide cross-sectional dimensions (cm) are given for frequencies up to 95 GHz. The range of frequency is given in brackets after the central frequency. (For another table, which includes alternative labeling of frequency bands, see Eaton and Eaton (1993).)

- The dimension of the resonant cavity required for the wavelength of ~3 cm characterizing X-band, is very convenient to accommodate the sizes of easily prepared samples.

2.3
EPR at Higher Frequencies (HF)

EPR at microwave frequencies higher than X-band, up to ~1200 GHz, will now be discussed. In this context, the term very high-frequency (VHF) refers to frequency bands equal to or greater than 140 GHz (D band).

2.3.1
Advantages

The main advantages of HF-EPR may be considered as follows:

Study of faster motions of EPR probes: When the motion of a spin label is in the fast-motion limit for a frequency band (e.g., X-band), a better observation of

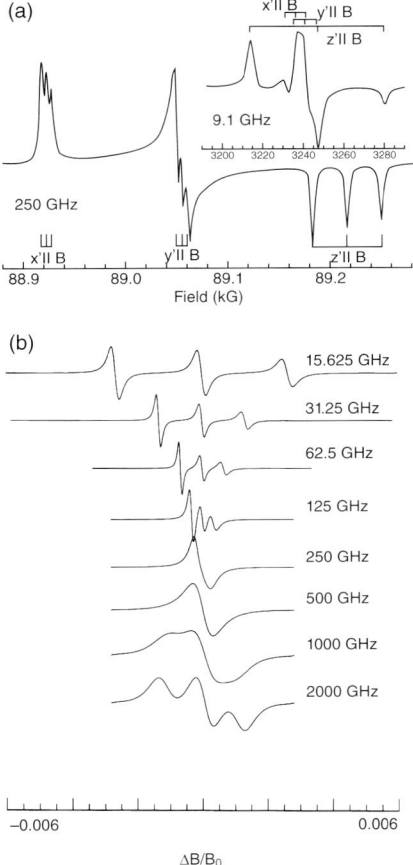

Figure 2.1 (a) Simulated derivative EPR spectra at 9.1 and 250 GHz for a dilute powder containing cholesterol-like nitroxide (CSL). The short vertical lines represent fields where CSL absorbs when its x′, y′, z′ axes are parallel to **B** (Barnes and Freed, 1998); (b) Simulated derivative EPR spectra for a nitroxide, reorienting with a rotational diffusion coefficient $R = 10^8 \, \text{s}^{-1}$ (corresponding to rotational correlation time $\tau_R = 1.67 \, \text{ns}$) for a wide range of frequencies. Adapted from Freed, 2000).

molecular motion is made possible by increasing the frequency to much higher than 9.5 GHz, as the molecular motion is not completely averaged out at the faster time scale. This is shown clearly in Figures 2.1 and 2.2, where EPR spectra were recorded at various frequencies for a nitroxyl radical and for a spin-labeled copolymer, as reported by Freed (2000) and Pilar et al. (2000), respectively.

Enhanced resolution due to field-dependent interactions: The Zeeman interaction, which is proportional to the applied magnetic field, is enhanced at higher frequencies due to the higher magnetic fields required (see next subsection). Examples of the resolution of small g-value differences at high frequencies are shown in Figure 2.3. In this case, with two different nitroxyl radicals in the mixture of phenyl-PBN, and a difference in the respective g-values of

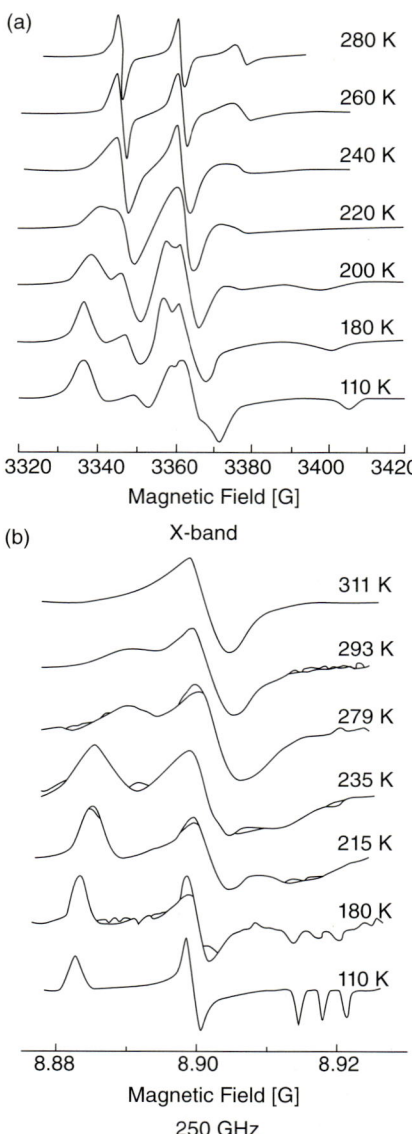

Figure 2.2 EPR spectra of the copolymer SL–ST–AA in toluene at the given temperatures (experimental spectra shown as full lines, simulated spectra as dotted lines). (a) X-band; (b) Far-IR. Reproduced from Pilar et al. (2000).

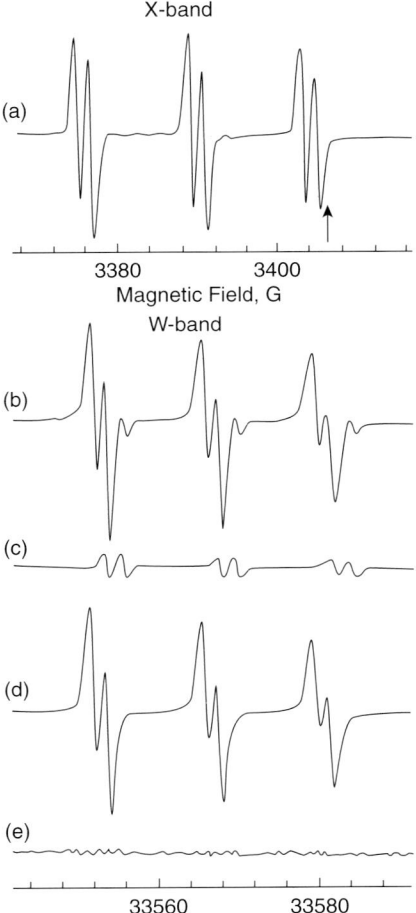

Figure 2.3 Spectrum a: X-band (9.5 GHz) EPR spectrum from a mixture of phenyl-PBN and trichloromethyl-PBN adducts in benzene at 297 K. The arrow indicates some extra broadening on the line shoulder, which is caused by the presence of two species. Spectrum b: Experimental W-band (95 GHz) spectrum of the mixture and the corresponding least-squares simulated spectra of phenyl-PBN (spectrum c) and trichloromethyl-PBN (spectrum d) adducts, respectively. Spectrum e: Residual: difference between experimental (spectrum b) and simulated spectra (c,d). Adapted from Smirnova et al. (1997).

$\Delta g \sim 1.2 \times 10^{-4}$, the EPR spectra overlap considerably at X-band but are resolved at W-band (95 GHz) (Smirnova et al., 1997). In Figure 2.4 it is clear that, from the EPR spectrum of the semiquinone radical at W-band, it is possible to read off the three principal values of the g-matrix, but that this is not the case for X-band (Burghaus et al., 1992). It should be noted, however, that there would be an optimum resolution for a set of g, A, and their distributions and anisotropies, and this makes EPR at low frequencies especially advantageous, for example,

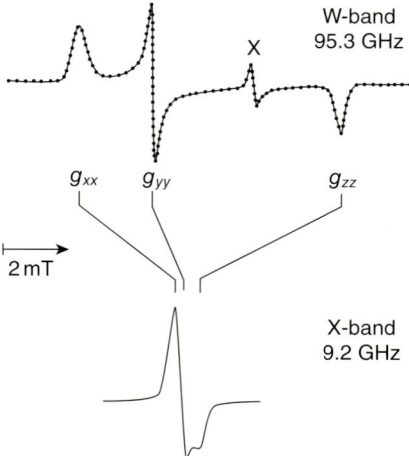

Figure 2.4 EPR spectra of perdeuterated 1,4-benzoquinone anion radical in frozen perdeuterated isopropyl alcohol at 130 K. Upper: The W-band spectrum (experiment shown as full line, simulation as dotted line) is taken at 95.3 GHz. The W-band spectrum is recorded simultaneously with a g-standard sample of powdered Mn^{2+} in MgO from which one h.f. component is seen (X). Lower: X-band spectrum of the same benzosemiquinone sample, taken at 9.2 GHz. Adapted from Burghaus *et al.* (1992).

at L-band (1 GHz) for Cu^{2+} EPR in proteins. The impact of distributions of g and A is more important for $S = ½$ than for second-order effects (for further details, see Chapter 16).

Enhanced orientation resolution at higher frequencies: Another benefit of HF-EPR over that at low frequencies is the excellent orientational resolution that it provides for studies involving the nitroxyl spin label, as seen clearly in Figure 2.1 (upper). This shows the positions of the resonant lines for the canonical orientations in a powder simulation at 9.1 and 250 GHz. In this figure, at 250 GHz, the regions corresponding to molecules with their x-, y-, and z-axes parallel to the external magnetic field are well separated because of the dominant role of the g-tensor, which is certainly not the case at 9.1 GHz

Suppression of second-order perturbation effects at higher frequencies: Due to the increased strength of the Zeeman interaction, the second-order perturbation effects—and, in particular, those due to anisotropy of the $\tilde{g}$-matrix—are decreased with increasing frequency. Thus, the EPR lines in powders are narrower at higher frequency as a result of the suppression of complex features in the EPR spectrum due to anisotropy effects. This is in contrast to lower frequencies, where zero-field splitting (ZFS) can lead to very broad lines, caused by anisotropy.

Observation of EPR transitions for large ZFS: Estimation of the ZFS parameter D: One important advantage of HF-EPR is the ability to measure very large ZFS

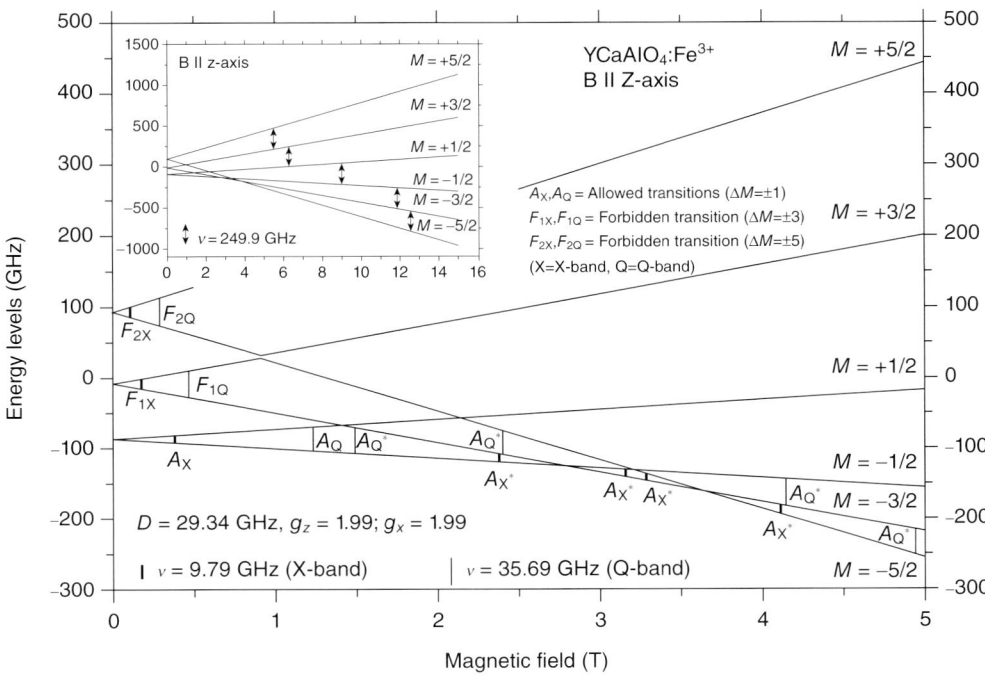

Figure 2.5 A plot of the eigenvalues of the Fe^{3+} ion in $YCaAlO_4$, calculated using the parameters listed in Misra et al. (2001), versus the intensity of the external magnetic field (**B**) for values of **B** up to 5 T, showing all expected $\Delta M = \pm 1$ transitions at X (9.79 GHz)- and Q (35.69 GHz)-bands. The inset shows a plot of eigenvalues for values of **B** up to 15 T, showing all expected $\Delta M = \pm 1$ transitions at 249.9 GHz. In order to determine the absolute sign of the ZFS parameter D, it is necessary to compare the intensities of the extreme lines, $+5/2 \leftrightarrow +3/2$ and $-3/2 \leftrightarrow -5/2$ transitions. Reproduced from Misra et al. (2001).

parameters ($D > 20$ GHz), which is typical for metal ions in biological systems, such as metalloproteins. It should be noted that the ability to observe *all* the allowed transitions ($\Delta M = \pm 1$) is important when estimating the value of the ZFS parameter (D) accurately, since for a Kramers' ion, with half odd-integral spin, the allowed central transition: $M = 1/2 \leftrightarrow M = -1/2$, is the only one that can be observed at a lower frequency (LF) for large ZFS. This is because the quantum of microwave energy has insufficient energy to match the energy differences between all pairs of levels participating in resonance for an allowed transition at LF. However, the central transition depends on D only in second order in perturbation, and hence it does not provide a precise estimate of D. Furthermore, the observation of all noncentral allowed transitions is necessary to determine the absolute sign of D, and thus of all other fine-structure parameters, which requires comparison of the intensities of the two extreme lines of the whole set of allowed transitions. This is seen clearly in Figure 2.5 (Misra et al., 2001), for the energy levels and transitions for Fe^{3+} in a $YCaAlO_4$ single crystal both at 9.79, 35.69, and 249.9 GHz. Here, it is seen that some of the same $\Delta M = \pm 1$ transition can be observed at more than one magnetic-field value.

Simplification of spectral shape for S > 1/2: Being strongly influenced by the relative values of the ZFS and hyperfine (h.f.) interactions, and of the g anisotropy, the EPR spectral shapes become simpler at higher frequencies, and therefore easier to analyze. This is because, at higher frequencies, the Zeeman interaction dominates both the ZFS and h.f. interactions. This is illustrated with the help of three examples of EPR spectra of ions with different spins, with and without hyperfine interaction.

- *Example 1:* The angular variation of EPR line positions for the non-Kramers' Ni^{2+} ion with electron spin $S = 1$, occupying two magnetically inequivalent sites in $NiCdCl_6 \cdot 6H_2O$ single crystal at 249.9 GHz as compared to that at X-band (9.5 GHz) are shown in Figure 2.6. From this figure, the magnitude of the ZFS parameter D can be read directly from the plot of angular-variation of the EPR line positions at 249.9 GHz, where the splitting of the two lines is $2D$ for the orientation of **B** along the magnetic z-axis. This is not possible at X-band, where the angular variation of line positions is rather complex.

- *Example 2:* Figure 2.7 shows the angular variation of EPR line positions at 249.9, 35.69, and 9.79 GHz for the Fe^{3+} ion in $YCaAlO_4$, which shows that the angular variation pattern is rather simple at the high frequency of 249.9 GHz.

- *Example 3:* Figure 2.8 shows the angular variation of the EPR line positions of Mn^{2+} in ZnV_2O_7 at 249.9 and 9.6 GHz. The figure shows that there is greater simplicity in the simulated angular variations of the line positions at 249.9 GHz shown in Figure 2.8 (top) as compared to those at 9.6 GHz in Figure 2.8 (bottom). This is due to the fact that the former, being in the high-field limit, yield a simple fine structure pattern from which the ZFS parameter (D) can be read directly from the spectrum for the orientation of **B** along the magnetic z-axis with a very good first approximation, as in Example 1.

More precise determination of the $\tilde{g}$-matrix: Due to increased magnitude of the Zeeman term at higher frequencies, it is possible to determine the $\tilde{g}$-matrix with greater accuracy, for example, by an additional significant figure at 249.9 GHz from that at X-band (Misra *et al.*, 2001).

Determination of the absolute signs of Spin-Hamiltonian parameters (SHP) at relatively higher temperatures: At high frequencies, the absolute sign of SHP can be determined from the relative intensities of the high-field lines and the low-field lines at temperatures much higher than those required at lower frequencies. This is because the Boltzmann factor, $\exp(-h\nu/k_B T)$, where ν is the microwave frequency and k_B is Boltzmann's constant, which governs the ratios of the population between the levels participating in resonance, increases linearly with frequency. Therefore, the same intensity of the low-field lines relative

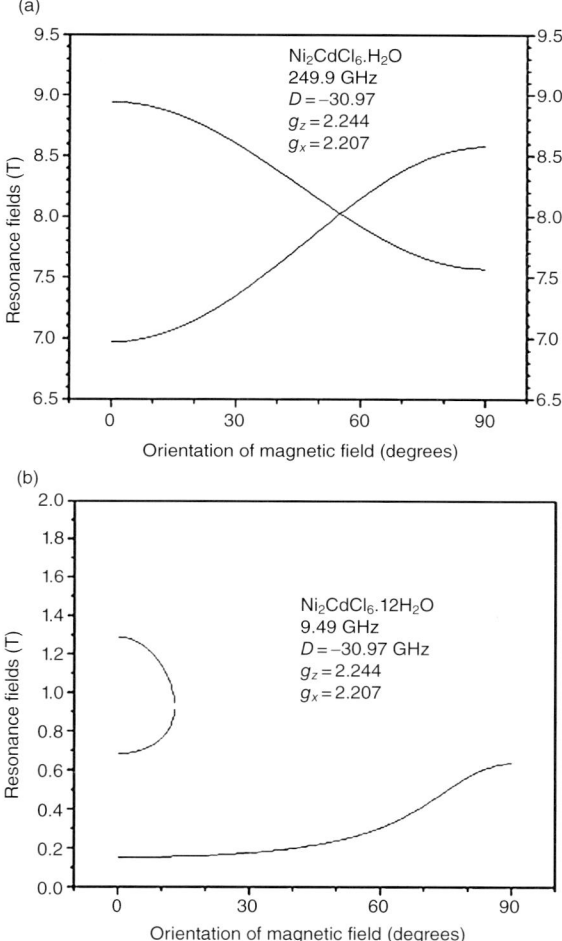

Figure 2.6 Plot of the angular variations of the simulated line positions for the two allowed transitions of the Ni^{2+} ion (S = 1) in a Ni$_2$CdCl$_6$·6H$_2$O single crystal at 249.9 GHz (a) and at 9.49 GHz (b) using the spin-Hamiltonian parameters as indicated. Reproduced from Misra et al. (2001).

to the higher-field lines is achieved at higher temperatures at higher frequencies as compared to that at lower frequencies. As shown in the formula, the effect is linearly proportional to the frequency. For example, the same relative intensity is achievable at ~100 K at 250 GHz as at ~4 K at 10 GHz. It is noted that, in order to determine the absolute sign of the largest ZFS parameter, D, if the intensities of the lower-field lines relative to those of the higher field lines increase/decrease at lower temperatures compared to that at room temperature, the absolute sign of the largest ZFS parameter, D, is positive/negative (Abragam and Bleaney, 1970, p. 161). This can be seen clearly in Figure 2.9, which shows

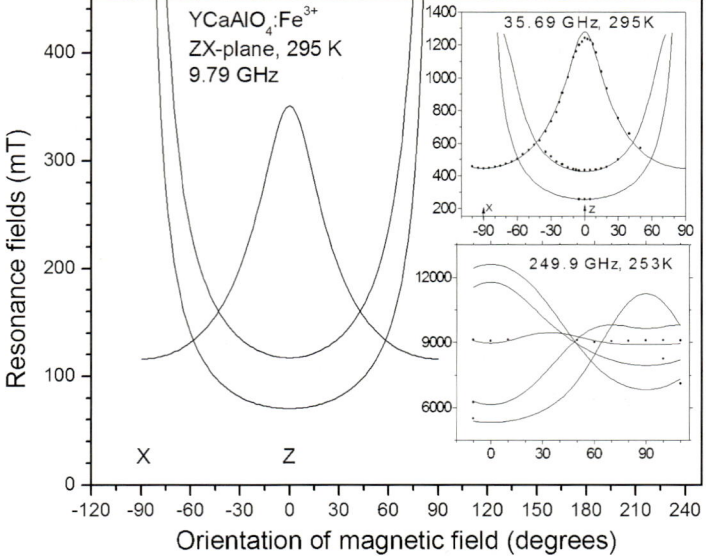

Figure 2.7 Simulated and observed angular variation of Fe^{3+} EPR line positions in YCaAlO$_4$ for the orientation of **B** in the magnetic ZX plane at 9.79 GHz at 295 K. The insets show corresponding spectra at 249.9 GHz at 253 K and at 35.69 GHz at 295 K. Because of the large value of D (~29 GHz), no lines were experimentally observed at X-band. The continuous lines represent simulation, while the points represent experimental values. Reproduced from Misra and Andronenko (2002); Misra et al. (2001).

the EPR spectra of the Mn^{2+} ion in NH4I$_{0.9}$Cl$_{0.1}$ at 170 GHz at various temperatures (Misra et al., 2003), where the absolute sign of D was determined to be negative, since the intensity of highest-field sextet relative to the lowest-field sextet decreased significantly as the temperature was lowered from 312 K to 30 K, the highest-field sextet almost disappeared at 10 K! In this context, it should be noted that the least-squares fitting method used to evaluate the SHPs yields the correct relative signs of all the fine-structure SHPs. Thus, knowing the absolute sign of D enables determination of the absolute signs of all the other fine-structure SHPs (further details are provided in Chapter 8.)

2.3.2
Disadvantages

The disadvantages of HF EPR are as follows:

- The EPR linewidth increases at higher frequencies due to $\tilde{A}$-, $\tilde{g}$-, and $\tilde{D}$-strain broadenings, caused by the variations of $\tilde{g}$-, $\tilde{A}$- and $\tilde{D}$-matrices from site to site,

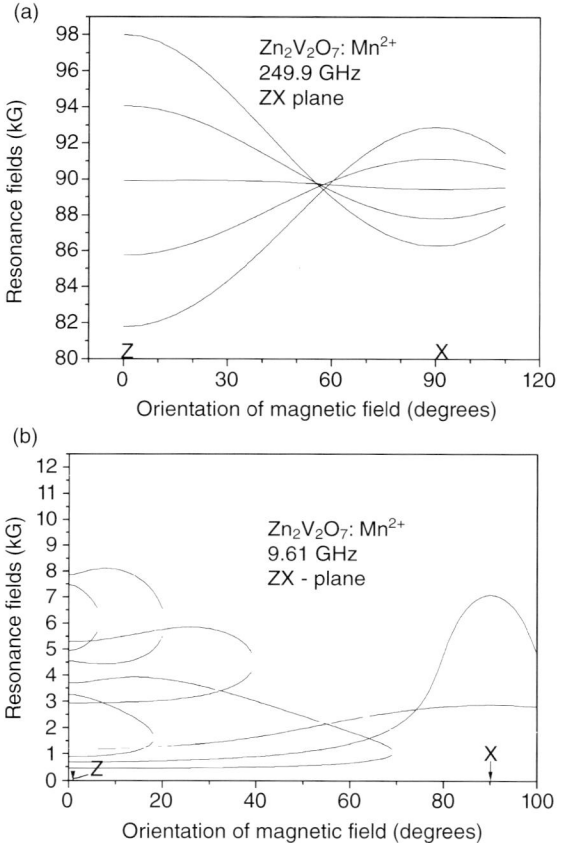

Figure 2.8 Plot of the angular variations of the simulated line positions for the allowed transitions of the Mn^{2+} ion ($S = 5/2$) in a Zn$_2$V$_2$O$_7$ single crystal at 249.9 GHz (a) and at 9.61 GHz (b) using the published spin-Hamiltonian parameters. Reproduced from Misra et al. (2001).

which mask small splittings between lines, such as those due to the hyperfine interaction.

- Measurement at higher frequencies of samples with high dielectric losses, such as aqueous samples, is not convenient because of problems related to the absorption and penetration of microwaves, as well as reflections due to impedance mismatch. For these reasons, it is necessary to use a very thin sample in a flat cell for recording EPR spectra at high frequencies.

- A third limitation of using HF-EPR is that, except for spectrometers built for special types of sample, it is often more difficult to control the sample environment, such as the oxygen concentration and/or time after preparation, in HF spectrometers than at X-band and lower frequencies.

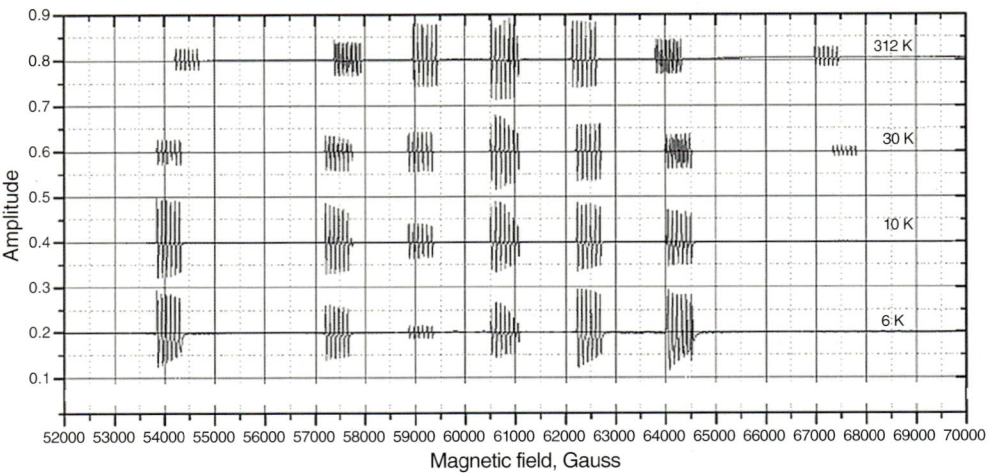

Figure 2.9 The first-derivative EPR spectra as observed for Mn^{2+} in $NH_4Cl_{0.9}I_{0.1}$ single crystal at 170 GHz at 312, 30, 10, and 6 K. The two sets of spectra for B ∥ Z- and B ∥ X/Y-axes are also obtained for the orientation of the cubic crystal such that one of its edges is coincident with **B**. They are identified as follows, referring to the spectrum at 312 K: The five sextets corresponding to B ∥ Z are in the first, second, fourth, sixth, and seventh multitudes of lines, whereas for B ∥ X/Y they are in the second, third, fourth, fifth, and sixth multitudes of lines. The sextets for X and Y directions of B overlap each other. The highest-field sextet clearly disappears at 10 K and below, due to depletion of populations of the levels participating in resonance. Reproduced from Misra et al. (2003).

2.4
Low-Frequency EPR

The low frequencies in this context are those equal to or below that of X-band, specifically the C-, S-, and L-bands.

2.4.1
Advantages

The main advantages of low-frequency EPR include:

The study of slower motion of molecules in fluids: The motion of Cu^{2+} complexes has been monitored successfully by employing EPR at lower frequencies, which provides a better resolution of room-temperature aqueous solution spectra of various Cu^{2+} complexes, for example, the complexes $Cu(DOPA)_2$ and $Cu(carnosine)_4$. This is seen in Figure 2.10 (Hyde and Froncisz, 1982), which shows a better resolution in the EPR spectra of two Cu^{2+} complexes at the lower frequency of 2.62 GHz, as compared to those at 9.30 and 3.79 GHz (see Antho-

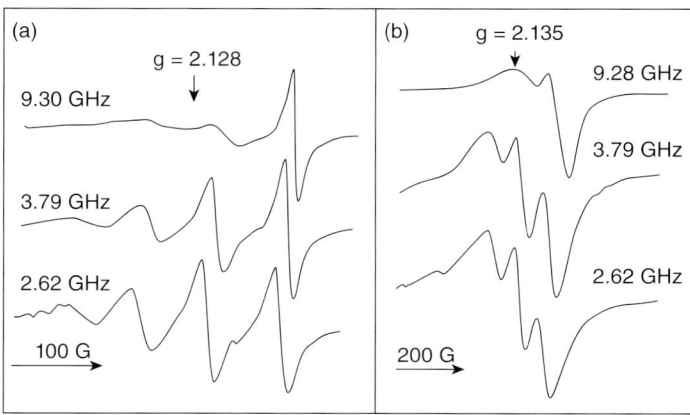

Figure 2.10 (a) Room-temperature EPR spectra of aqueous solution of Cu(DOPA)$_2$ (molecular weight = 489); (b) Room-temperature EPR spectra of aqueous solution of Cu(carnosine)$_4$ (molecular weight = 968). Adapted from Hyde and Froncisz (1982).

line, 2005). The analysis of the data led to the calculation of rotational correlation times using the linewidth model of Kivelson (Kivelson, 1960; Wilson and Kivelson, 1966a, 1966b; Atkins and Kivelson, 1966). They were found to be in the range ~30–150 ps. In this context, it should be noted that, when simulating Cu^{2+} spectra in fluid solutions at low frequencies, account should be taken of the nonsecular terms in the spin Hamiltonian (Paskenkiewicz-Gierula, Subczynski, and Antholine, 1997), which contain the operators $S_{\pm}I_{iz}, S_{\pm}I_{i\pm}, S_{\pm}I_{i\mp}$.

Study of copper-copper sites in proteins: EPR data obtained at several low-frequency bands, specifically at L-, S-, C-, along with those at X-, and Q-bands, have been exploited to unravel the details of copper sites in cytochrome-c oxidase (CcO or COX) and nitrous oxide reductase (N$_2$OR); Figure 2.11 shows such spectra for N$_2$OR at X-, C-, and S-bands (Antholine *et al.*, 1992). The first well-resolved hyperfine structure in these compounds was observed at S-band by Froncisz *et al.* (1979). This led ultimately to the proposal of the occurrence of a mixed-valence [Cu(1.5).Cu(1.5)] state for the CuA center in CcO and N$_2$OR (Kroneck *et al.*, 1988). Similar studies on the Cu^{2+} ion in a particulate methane mono-oxygenase have been reported by Yuan, Collins, and Antholine (1997).

Resolution of hyperfine interactions in copper and the effect of g- and A-strains: Because the Zeeman interaction increases with increasing magnetic field, the three main themes in the context of usage of multifrequency EPR of Cu^{2+} proteins, as identified by Hyde and Froncisz (1982), are: (i) state mixing as determined by the relative values of electron Zeeman, nuclear Zeeman, nuclear hyperfine, and nuclear quadrupole interactions; (ii) EPR line broadening due to g-, A- and D-strains; and (iii) dependence of the shape of EPR spectra

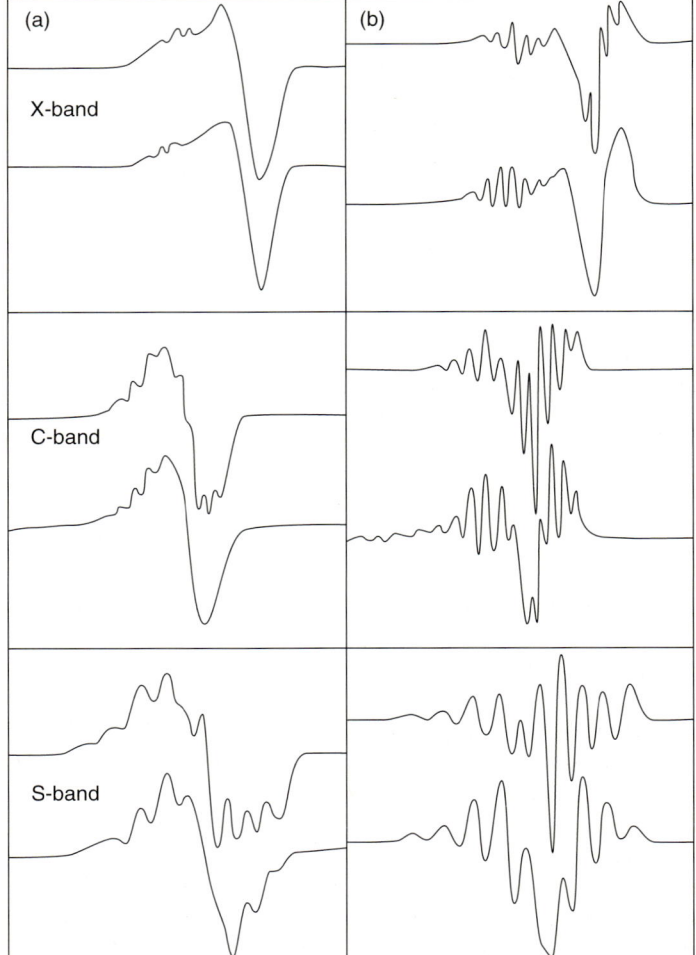

Figure 2.11 (a) First-derivative and (b) second-derivative X-band, C-band, and S-band EPR signal and their computer simulations, placed above the respective experimental spectra for N$_2$OR of *Pseudomonas stutzeri*, N$_2$OR I at 15 K; 25 spectra were averaged for the second derivative. Adapted from Antholine *et al.* (1992).

in fluid solutions on rotational correlation time and g-anisotropy. There exists a correlation between $A_\|$ and $g_\|$ values, such that the g- and A-strains partly cancel out each other's contributions to the linewidth to the h.f. lines situated at low-field at S-band. This results in a decreased linewidth, which enhances the resolution of ^{14}N h.f. splitting, enabling the determination of the number of N atoms coordinated to Cu^{2+} ions in some proteins. On the other hand, the linewidths of the h.f. lines situated at high-field increase resulting in decreased

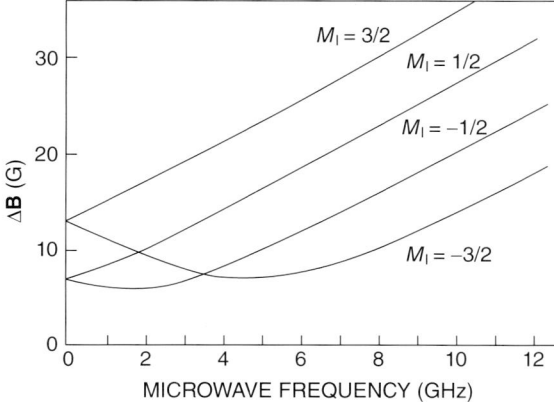

Figure 2.12 Width of the $g_{\parallel}$ turning points as calculated by theory, using the data of Cu(catechol)$_2$ to fix the parameters in the theory. Adapted from Hyde and Froncisz (1982).

resolution (Hyde and Froncisz, 1982). This is shown in Figure 2.12. Another example of advantage of Cu^{2+} EPR at low frequencies is the accurate estimation of h.f. parameters, for example, in azurin (Antholine, Hanna, and McMillin, 1993; Froncisz and Hyde, 1980) and transferrin (Hyde and Froncisz, 1982). The transferrin spectrum is shown in Figure 2.13; this figure shows that the ligand h.f. structure was not resolved at X-band, but it became indeed resolved at the lower frequencies of 1.1 and 2.4 GHz. Further, at the low frequencies, ligand h.f. coupling was observed in the $g_{\perp}$ region, but not in the $g_{\parallel}$ region. In order to achieve self-consistent estimates of the $\tilde{g}$ and $\tilde{A}$ matrices, it was necessary to simulate CW spectra at both L-band and S-band. The allowed Cu^{2+} h.f. lines characterized by the nuclear spin magnetic quantum numbers $m = -3/2$ and $-\frac{1}{2}$ exhibit minima at ~6 GHz and ~2 GHz, respectively, when plotted as functions of the microwave frequency, as shown in Figure 2.12. As for the other two h.f. lines corresponding to $m = \frac{1}{2}$ and $3/2$, it should be noted that their widths increase monotonically with increasing microwave frequency, without exhibiting any minima.

Observation of superhyperfine interaction: The h.f. interaction of an ion with the ligands is known as the "superhyperfine interaction." As discussed in Section 2.5.1 below, a narrower EPR linewidth is observed in a solution, or a frozen solution, at lower microwave frequencies. This allows the observation of a superhyperfine interaction, providing a simpler interpretation of spectra; for example, that of Mo^{5+} dimer spectra (Hanson *et al.*, 1987).

Study of large and/or lossy samples: In general, these are better studied at lower frequencies, where there are no problems associated with the absorption and penetration of microwaves.

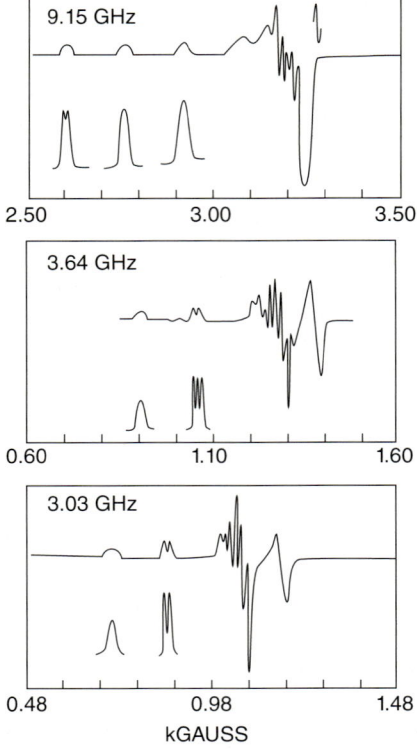

Figure 2.13 EPR spectra of a frozen solution of monocupric transferring at three frequencies. Coupling to a single nitrogen ligand is well-resolved at low frequency. Adapted from Hyde and Froncisz (1982).

2.4.2
Disadvantages

The following disadvantages of EPR studies at low frequencies are noted:

- The effects due to second-order terms in perturbation in the energies of the levels become significant at lower frequencies, because the Zeeman and non-Zeeman interactions can be of the same order of magnitude. This results in the appearance of complex features in the spectrum at lower frequencies. Further complexities to this arise from the $\tilde{g}$-anisotropy; this is shown clearly in Figures 2.6–2.8.

- At low microwave frequencies, with the typical available magnetic fields, for example, <2.0 T at X- and Q-bands, in many cases only a few – and certainly not all – of the allowed transitions can be observed, other than the central $1/2 \leftrightarrow -1/2$ transition, when the value of the ZFS parameter D is large. This is shown in Figure 2.5 for the Fe^{3+} ion in $YCaAlO_4$.

2.5
Multifrequency EPR

Today, multifrequency (MF) EPR, covering the range of frequencies from ~0.1 ~1200 GHz, is the preferred technique to obtain a more complete and unambiguous picture of the various processes that affect a paramagnetic ion in a given environment. In this context, it was noted by both Husted *et al.* (1997) and Hyde and Froncisz (1982) that, at the minimum, three frequencies are required to obtain complete information about the interaction of a paramagnetic ion with the environment by EPR, although a reasonable amount of information can be acquired by using just one low-frequency and one high-frequency spectrometer. The role played by different frequencies in determining these processes effectively is summarized in Table 2.2 (Eaton and Eaton, 2001).

2.5.1
Advantages of Using Multifrequency EPR

The advantages associated with MF-EPR are as follows:

Distinction between field-dependent and field-independent processes: The Zeeman interaction, described by the $\tilde{g}$-matrix, is a field-dependent interaction, which dominates at higher frequencies. On the other hand, the hyperfine interaction, described by the $\tilde{A}$ matrix, is a field-independent process, which dominates at lower frequencies. By choosing a particular frequency, it is possible to study a particular process in the EPR spectrum. The factors that govern the EPR linewidth, such as unresolved h.f. structure, relaxation, and g-strain, should be taken into account when choosing a particular frequency band. For example, the optimum frequency for resolution of h.f. splitting in immobilized Cu^{2+} spectra lies at lower frequencies, such as S-band. On the other hand, a greater resolution of the EPR spectrum can be achieved by going to higher frequencies for moderate fields to observe the same transition, for example, from X- to Q-band, since the effect of g-strain will not then be sufficiently predominant as the magnetic field is not sufficiently high. In this context, it should be noted that any accidental overlap of different spectral features occurring at a particular microwave frequency will render that frequency disadvantageous (Hyde and Froncisz, 1982).

Spectral resolution: The frequency dependence of the linewidth in a solution is expressed as (Kivelson, 1960; Wilson and Kivelson, 1966a, 1966b; Atkins and Kivelson, 1966):

$$\sigma_v = a + bm + cm^2 + dm^3, \tag{2.1}$$

where the coefficients a, b, c, and d are related to the solvent viscosity, correlation time, molecular hydrodynamic radius, and the anisotropy of the $\tilde{g}$, $\tilde{A}$ matrices of the system under study. Originally, Froncisz and Hyde (1980) and Hyde and Froncisz (1982) developed a correlated $\tilde{g}$, $\tilde{A}$ model, which has been

Table 2.2 EPR phenomena monitored at various frequencies.

EPR phenomenon to be studied	General conditions required	Special cases or conditions	Comments on field/frequency required
Measure g-values	Better accuracy at higher frequencies		Accurate field and frequency required
Multiple species	Δg > linewidth	Free radicals	X, Q, W
To resolve nuclear neighbors	(i) CW: $A > \sigma(A)$, ΔB $M_I \sigma(A_\parallel) = -\sigma(g_\parallel)(h\nu_0/g_\parallel^2 \mu_B)$ (Hyde and Froncisz, 1982) (ii) ESEEM: $A = 2\nu_N$	(i) L, S for CW Cu^{2+}	(ii) Select for exact cancellation in ^{14}N ESEEM
To measure intermediate motion	Tumbling correlation time comparable to motion-averaged anisotropies of $\tilde{g}$, $\tilde{A}$ matrices	Wide range of frequencies required	Each transition and correlation time characterized by an optimal frequency
To measure distance between electron spins	Increases at lower frequencies for half-field ($\Delta M = \pm 2$) transition; increases at higher frequencies for the transition at $g = 2$		S, X, Q, W, and higher; fit data at multiple frequencies
In vivo or large samples	Depth penetration for large resonators	Physiological motion	VHF or L-band
$S = 1/2, 3/2, 5/2$ (Kramers ions)	The central transition $+\frac{1}{2} \leftrightarrow -\frac{1}{2}$ observable at all frequencies	$h\nu$ > ZFS	Multiple frequencies
$S = 1, 2$ (Non-Kramers ions)	The transitions are sometimes observable in parallel mode, and sometimes at X-band, but for many transitions it is necessary to use frequencies above X-band.	$h\nu$ > ZFS	Multiple frequencies; for integer spins one needs frequencies higher than X-band
$S = 3$ (Non-Kramers ions)	See *J. Chem. Phys.* **129**, 174510 (2008) for details.		X-band

Here, $\sigma(A)$ and $\sigma(g)$ are the widths of the Gaussian distributions of A and g, respectively, in the sample; ΔB is the EPR linewidth. From Eaton and Eaton (2001).

successfully used to account for the linewidth variations encountered in $S = \frac{1}{2}$ systems, particularly in Cu^{2+} and low-spin Co^{2+} complexes. Expressed in the frequency-domain, the linewidth in this model is expressed as:

$$\sigma_\nu^2 = \left(\sum_{i=x,y,z} \left\{ \sigma_{R_i}^2 + \left[\frac{\Delta g_i}{g_i} \nu_0(B) + \Delta A_i m \right]^2 \right\} g_i^2 l_i^2 \right) \Big/ g^2, \qquad (2.2)$$

where $\sigma_{R_i}(i = x, y, z)$ are the residual linewidths due to unresolved metal and/or ligand h.f. splittings, homogeneous broadening, and other sources, $v_0(B)$ is the microwave frequency, ΔA_i, Δg_i represent the Gaussian-distribution widths of the A_i, g_i values, and l_i are the direction cosines of B with respect to the principal axes of the g-matrix. The g–A strain model involves nine parameters for a rhombically distorted metal-ion site. As far as the choice of a microwave frequency is concerned, Equations 2.1 and 2.2 imply that: (i) for spin systems with no h.f. coupling, the lower the frequency the narrower is the linewidth; and (ii) for spin systems which contain nuclei, the optimum frequency for the best resolution of a given h.f. resonance depends on a quadratic in m and the frequency. Generally, lower frequencies, such as L-, or S-band, are better. The fact that the lines are narrower at lower frequencies, allows observation of superhyperfine interactions (Hanson et al., 1987), where interpretation of spectra becomes simpler due to a better resolution of h.f. resonances.

A more precise estimation of spin-Hamiltonian parameters: This is accomplished by combining the LF- and HF-EPR line positions in a simultaneous fit, using the least-squares fitting technique (Misra, 1976, 1983, 1986, 1999). The fine-structure ZFS parameters (D and E), when not too large, and the h.f. structure parameters (A, B), are determined more precisely from the line positions at lower frequencies, whereas the $\tilde{g}$-matrix is determined more accurately from the line positions at high-field/high-frequency (Misra et al., 2001).

Differentiation between the spectra of multiple species: This is accomplished by using the fact that the position and relative spacing of EPR lines, and shapes for different species change differently with varying frequency, v. It should be noted that the difference in resonant fields for resonance, ΔB, for two species characterized by the isotropic g-factors g and $g + \Delta g$, is approximately proportional to the microwave frequency:

$$\Delta B \approx (hv\, \Delta g/\mu_B g[g + \Delta g]). \tag{2.3}$$

Thus, if ΔB is made distinctly larger by increasing the microwave frequency, the two species can be unambiguously identified. This is clearly seen in Figures 2.3 and 2.4, as discussed in Section 2.3.1. Specifically, Figure 2.3 shows two different nitroxyl radicals in the mixture of phenyl-PBN, with the difference in the respective g-values being $\Delta g \sim 1.2 \times 10^{-4}$, whose EPR spectra overlap considerably at X-band but become resolved at 95 GHz (Smirnova et al., 1997), whereas from the EPR spectrum of the semiquinone radical at W-band shown in Figure 2.4 it is possible to read off the three principal values of the g-matrix, which is not possible at X-band (Burghaus et al., 1992).

Identification of nuclei in the vicinity of electron spins using electron spin echo envelope modulation (ESEEM): The number of nitrogen (N) ions bound to nuclei can be determined using MF-EPR. For example, ESEEM at X-band identifies the individually bound nitrogen (N) ions in imidazole, but not the directly bound N, which experiences too large a hyperfine field from Cu^{2+} unpaired electron to experience sufficient electron-nuclear state mixing to produce a significant EPR signal at X-band. The ESEEM signal for directly bound nuclei depends on mixing of the nuclear and electron states, which is enhanced at higher frequencies as the size of Zeeman interaction increases. An exception to this is N^{14}, for which there is an optimum frequency, the important consideration being the relative values of the electron Zeeman field and the hyperfine field due to the electron at the nucleus (see Flanagan and Singel, 1987; Gerfen and Singel, 1990; Hyde and Froncisz, 1982).

Study of paramagnetic ions with spins greater than ½: These ions are all characterized by nonzero ZFS parameters, since $S \geq 1$:

- *Non-Kramers' ions*: these possess an even number of electrons, characterized by integral spin values. The resonance condition for these ions can be expressed as: $h\nu = (g_{eff} \mu_B B)^2 + \Delta^2$, where Δ is the ZFS. Consequently, the condition $h\nu > \Delta$ should be fulfilled in order to observe resonance. In addition, EPR data are required at two frequencies in order to determine the two unknowns, Δ and g_{eff}, in the resonance condition. It is possible, however, to observe resonance at X-band, even though the condition $h\nu > \Delta$ is not satisfied. This is accomplished by observing the forbidden, parallel-mode transitions, $\Delta M = 0$, induced by the configuration $\mathbf{B}_1 \parallel \mathbf{B}$, when there exist non-axial ZFS terms in the spin Hamiltonian to mix the wavefunctions, so that the probability of these transitions becomes nonzero. Some observed MF spectra for non-Kramers' ions are described as follows: $S = 1$: The relevant ions are V^{3+}, Ni^2. Their EPR spectra of interest are shown in Figure 2.14a, along with the corresponding energy-level diagram in Figure 2.14b for V^{3+} in $CsGa(SO_4)_2 \cdot 12H_2O$, in Figure 2.15a and b for V^{3+} in corundum, in Figure 2.16a and b at 150 and 10 GHz, respectively, for V^{3+} in VCl_2, and in Figure 2.17a, b, and c for Ni^{2+} in tris(ethylenediamine) zinc dinitrate at X-, Q-, and D-bands, respectively; $S = 2$: The relevant ions are Fe^{2+}, Mn^{3+}, and Cr^{2+}; their spectra are shown in Figures 2.18, 2.19a,b, and 2.20, respectively, in the hosts deoxyHr azide (Q-band), polycrystalline [Mn(dbm)$_3$] (349.3 and 245.0 GHz), and frozen solution of $[Cr(H_2O)_6]^{2+}$.

- *Kramers' (odd electron) ions*: the relevant ions for which MF spectra have been observed are: (i) Cr^{3+} ($S = 3/2$) at X- and Q-bands in $LiScGeO_4$ by Galeev et al. (2000) reporting observation of more lines at Q-band than at X-band as shown in Figure 2.21. Some lines were also observed at 70 GHz; (ii) Mn^{2+} ($S = 5/2$) by Wood et al. (1999) in a complex with Mn(γ-picoline)$_4$I$_2$ at 249.9 GHz in metamyoglobin, as shown in Figure 2.22; and (iii) multifrequency (1–285 GHz) EPR spectra of high-spin Fe^{3+} ($S = 5/2$), as displayed in Figure 2.23.

Measurement of distances based on dipolar coupling: The dipolar coupling between two electron spins is proportional to the inverse cube of the distance between them. Thus, the position of the additional line in the EPR spectrum due to the dipolar interaction of two coupled nitroxides, with electron spin ½ each, can be exploited to estimate the distance between the two nitroxides. It was shown by Husted et al. (1997), that a global analysis of data obtained at

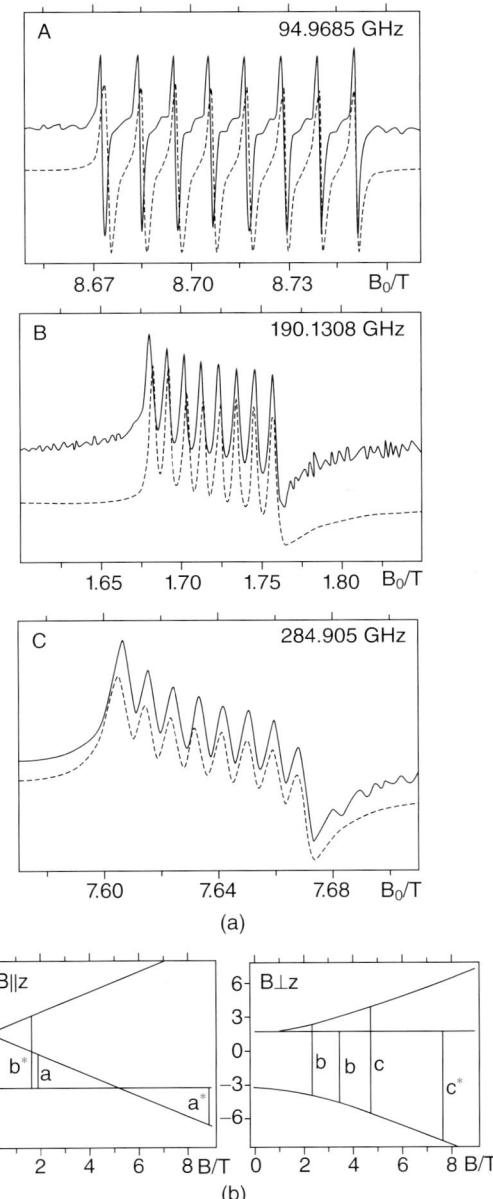

Figure 2.14 (a) Representative single-crystal (A) and powder (B, C) EPR spectra (solid lines) of Cs[Ga:V]SD and Cs[Ga:V]SH. Spectral simulations are shown by the broken lines; (b) Energy-level diagram showing the observed EPR transitions (vertical lines) between states of the ^3A_g (S_6) ground term: a, b, and c correspond to frequencies of 95, 190, and 250 GHz, respectively. The EPR spectra corresponding to the transitions marked * are displayed in (a) above. Adapted from Tregenna-Piagott et al. (1999).

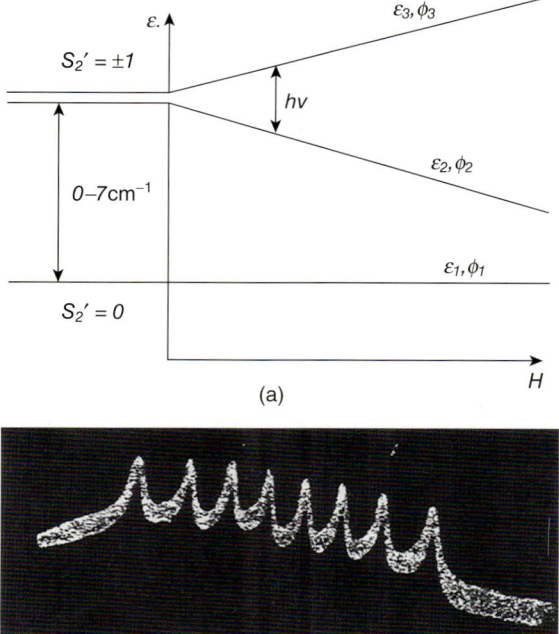

Figure 2.15 (a) Splitting of the lowest triplet for V^{3+} ($S = 1$) in corundum (Al_2O_3) by spin-orbit coupling into a ground-state singlet and an excited doublet. The forbidden $|\Delta M| = 2$ trantion can be observed at normal X- and K-band frequencies; (b) EPR line of V^{3+} in corundum at 37.45 GHz recorded at 4.2 K for orientation of the magnetic field along the magnetic z-axis. Reproduced from Zverev and Prokhorov (1960).

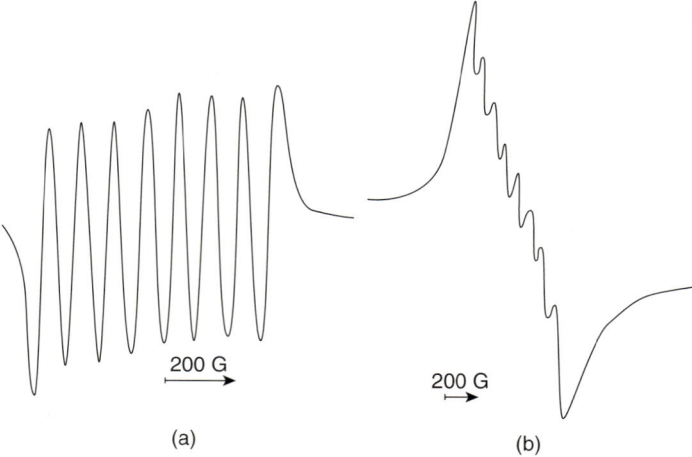

Figure 2.16 EPR spectra of a frozen methanol solution of VCl_2 (3×10^{-2} M) at 150 GHz (a) and 10 GHz (b). The high-frequency spectrum becomes resolved due to suppression of forbidden transitions. Adapted from Lebedev (1990).

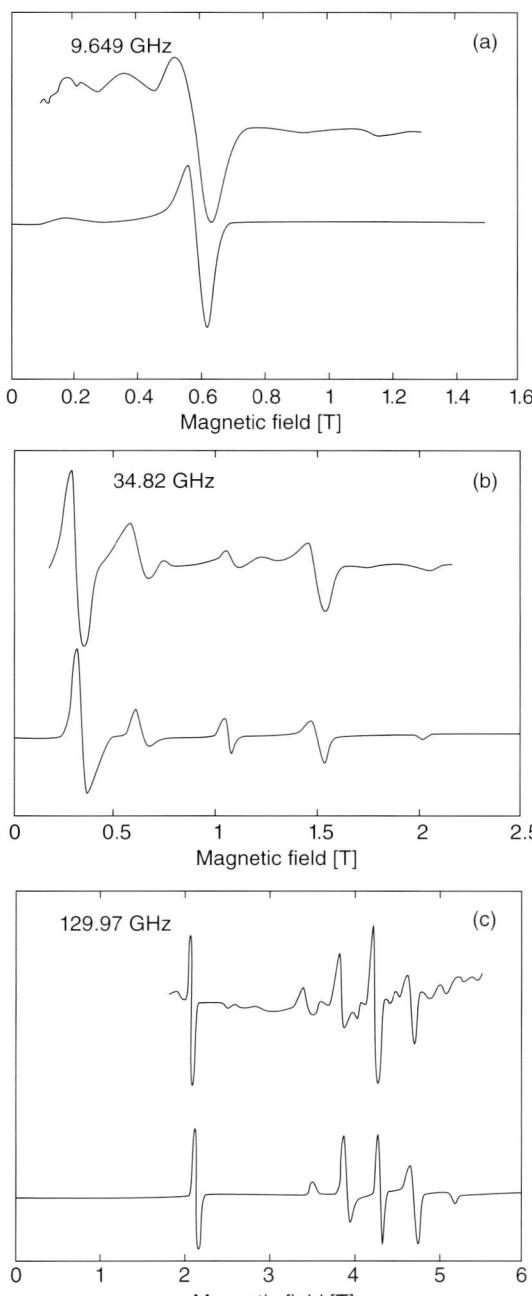

Figure 2.17 (a) X-band, (b) Q-band, and (c) D-band room-temperature EPR spectra of Ni^{2+} in [Zn(ethylenediamine)$_3$](NO$_3$)$_2$. The lower plot in each figure is the simulation using the parameters: $D = 0.832$ cm^{-1}, $E = 0.0$ cm^{-1}, $g_x = 2.156$, $g_y = 2.156$, $g_z = 2.181$. Reproduced from van Dam et al. (1998).

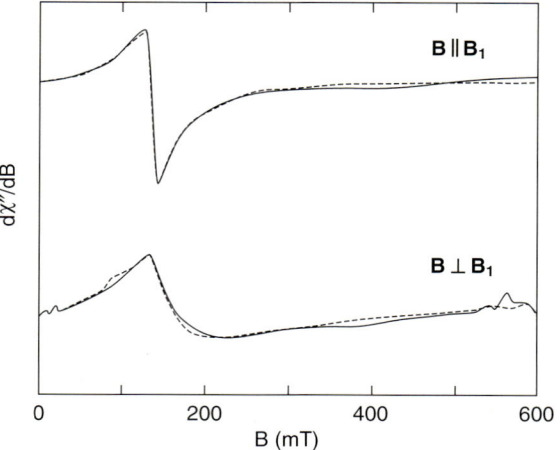

Figure 2.18 Q-band (34.071 GHz) EPR spectra of Fe^{2+} ($S = 2$) in a frozen solution of deoxyHr azide (solid lines) along with simulated spectra (dotted lines) with the parameters: $D = 2.92\,cm^{-1}$, $E = 0.45\,cm^{-1}$, $\sigma_E = 0.21\,cm^{-1}$, $g = 2.23$. Adapted from Petasis and Hendrich (1999).

three frequencies, for example, X-, Q-, and W-bands, would provide a unique fit to estimate inter-spin distance. However, there are additional parameters to be taken into account to simulate the EPR spectrum, for example, the relative orientations of the magnetic axes of the two spins in spin-labeled tetrameric glyceraldehydes 3-phosphate dehydrogenase. These two aspects are illustrated in Figures 2.24 and 2.25 (Husted et al., 1997), which show EPR spectra, respectively, for the cases of dilute (showing no dipolar lines) and concentrated (showing dipolar lines) samples, noting that the dipolar interaction increases in concentrated samples due to decreased inter-ion distances. Similar studies were reported by Calvo et al. (2000) on a semiquinone biradical formed in the photosynthetic reaction center at X-, Q-, and W-bands. The distance between the spins thus found (17.2 ± 0.2 Å) was in good agreement with that determined using X-ray crystallography (17.4 ± 0.4 Å).

Identification of relaxation mechanisms: Normally, the temperature-dependence of relaxation data is analyzed to determine the relaxation mechanism. However, this does not always lead to an unambiguous answer. With the use of MF-EPR, it is possible to identify the dominant relaxation mechanisms from a knowledge of both the temperature and frequency dependences of relaxation times. For example, the spin-lattice relaxation times (T_1), as governed by Raman, or local-mode processes, are frequency independent, whereas T_1 times for processes, such as direct and thermally activated ones, are frequency-dependent. MF data revealed that T_1 for TEMPOL in 4-OH-2,2,6,6 tetramethyl 1-piperidinol at temperatures higher than 160 K was frequency-dependent; this indicated that it was actually a thermally activated process that was contributing to T_1, as the direct process is not effective at higher temperatures (Weber, 2001; Eaton et al., 2001).

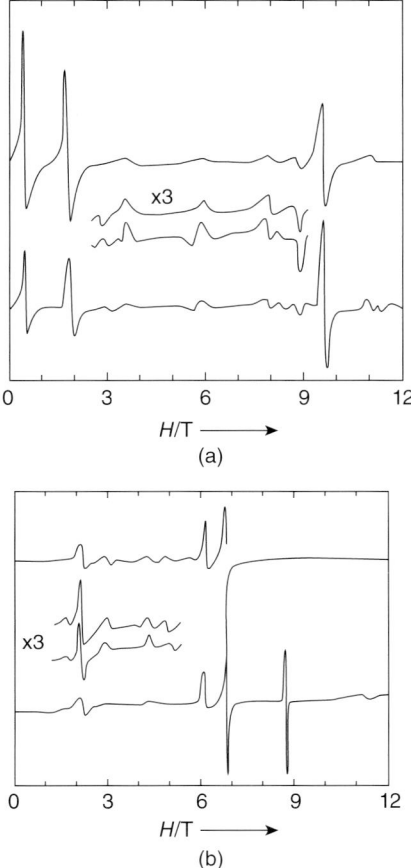

Figure 2.19 Experimental (lower plot) and simulated (upper plot) of polycrystalline [Mn(dbm)$_3$] at 15 K recorded at (a) 349.3 GHz, at which the free electron resonates at about 12.46 T, and (b) 245.0 GHz, where the narrow line at about 8.75 T arises due to DPPH (α,α-diphenyl-β-picrylhydrazyl). Adapted from Barra et al. (1997).

If a paramagnetic ion is rapidly tumbling, the anisotropies are averaged out, causing the spin–spin relaxation time (T_2) to become longer, resulting in narrower lines. However, if the anisotropy is rather large, the anisotropy is not completely averaged out. The T_2 times then become shorter, resulting in broader EPR lines, which depend on nuclear spin magnetic quantum number, m, of the h.f. lines causing dramatic changes in the linewidths at higher microwave frequencies due to an enhanced effect of g-anisotropy combined with A-anisotropy, since incomplete motional averaging depends on both g- and A-anisotropy.

Study of motion: Low-frequency (e.g., X-band) EPR is more sensitive to study slower motions, such as overall motion of a macromolecule, whereas high-frequency (e.g., 250 GHz) EPR is more sensitive to study faster motions, such

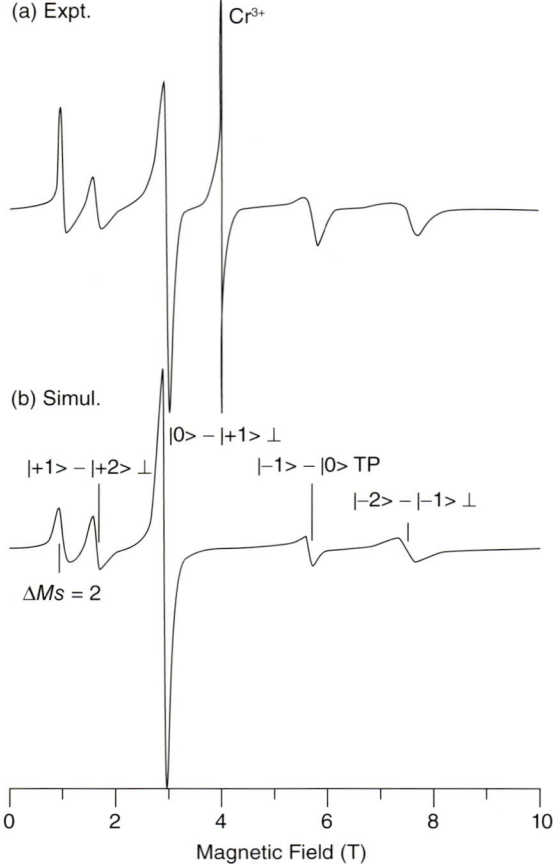

Figure 2.20 High-frequency (109.564 GHz) EPR spectrum of aqueous Cr^{2+} (S = 2) in $[Cr(H_2O)_6]^{2+}$ (~0.1–0.2 M) with a sulfate counterion at 10 K. (a) Experimental spectrum. The sharp signal from an aqueous Cr^{3+} impurity at g = 2.00 is indicated; (b) Simulated spectrum with the parameters: $D = -2.2\,\text{cm}^{-1}$, $E = 0.00\,\text{cm}^{-1}$, g (isotropic) = 1.98, single-crystal linewidth = 100 mT. The allowed EPR transitions with $\Delta M = \pm 1$ are assigned as to |M> ground and excited states and magnetic-field orientation (|| stands for **B** along the z-axis, ⊥ stands for **B** along the x-axis, as indicated on the simulated spectrum). The resonance corresponding to |−1>→ |0> occurs at an off-axis extremum or "turning point" when **B** is ~60° from the z-axis, and is so identified by the label TP. A partially allowed transition with $\Delta M = \pm 2$ is also indicated. Adapted from Telser et al. (1998).

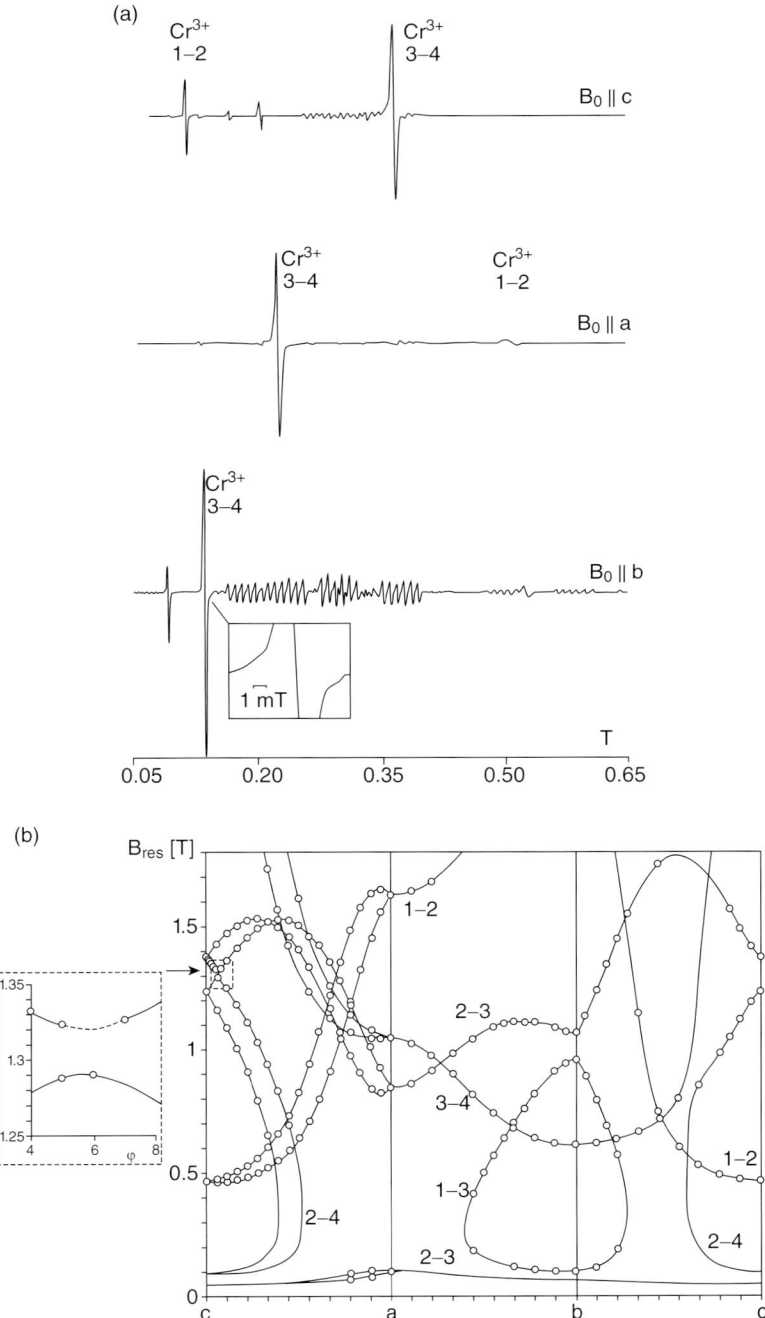

Figure 2.21 (a) X-band EPR spectra of LiScGeO$_4$ crystal for B along the crystallographic a-, b-, and c-axes at room temperature. The transition lines of Cr^{3+} are indicated by two numbers; (b) Angular dependence of the Cr^{3+} Q-band EPR transitions at room temperature in the three crystallographic planes. Circles denote experimental data, whereas the continuous lines are calculated data. The inset illustrates the vanishing of the transition 2–3 (broken curve) in the vicinity of the principal axis of the ZFS tensor. Adapted from Galeev et al. (2000).

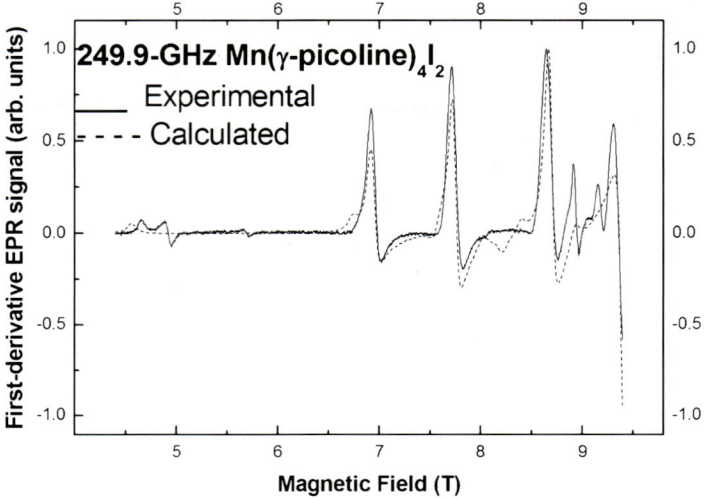

Figure 2.22 EPR spectrum at 249.9 GHz of Mn(γ-picoline)$_4$I$_2$, as represented by a continuous line. The dotted lines show the simulated spectrum as calculated using matrix diagonalization. Adapted from Wood et al. (1999).

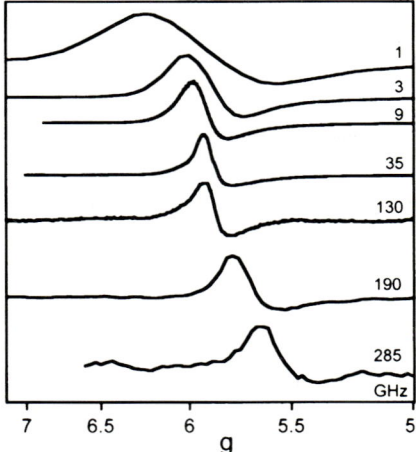

Figure 2.23 Multifrequency (1–285 GHz) EPR spectra of high-spin Fe^{3+} in metamyoglobin around g_{eff} = 6. The x-axis is reciprocal to increasing **B**, thus showing the variation in resolution at different frequencies. The changes in the spectra are attributed to distributions in g, referred to as g-strain, and in the ZFS parameter D. The value of $g_\perp$ decreases as the Zeeman energy approaches D. Adapted from van Kan et al. (1998).

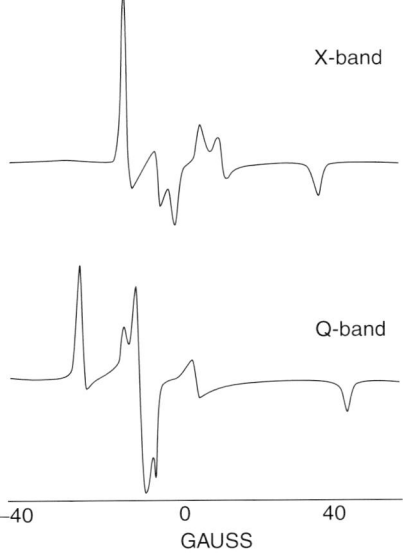

Figure 2.24 CW-EPR spectra (dots) at X- and Q-bands of N⁰-SL-NAD⁻ bound to GAPDH at a ratio of 0.3 spin label per GAPDH tetramer. The X- and Q-band spectra have been fitted (solid lines) simultaneously to obtain the following values of the parameters: $g_{xx} = 2.008588$, $g_{yy} = 2.005640$, $g_{zz} = 2.002302$, $A_{xx} = 10.804$ Gauss, $A_{yy} = 12.191$ Gauss, $A_{zz} = 49.008$ Gauss, Lorentzian linewidth $\Gamma = 0.463$ Gauss, Gaussian convolute width: $\sigma = 0.470$ Gauss. Adapted from Husted et al. (1997).

as local motion of a spin label on a macromolecule (Liang and Freed, 1999; Pilar et al., 2000).

2.5.2
Limitations of Using Multifrequency EPR

Despite the above-described advantages, the multifrequency requirement poses one serious limitation, namely that HF spectrometers are not readily available to individuals working in research, on the basis that they are too costly. On the other hand, the great virtue of multiple frequencies, and especially of high frequencies for the types of study described in this chapter, represents a trade-off with the cost of custom-built spectrometers and the operating costs of scanning superconducting magnets. In many countries (e.g., US, France, Japan), national laboratories have been funded which make HF-EPR facilities available to research groups unable to acquire such spectrometers in their own laboratories.

2.5.3
Size of Resonant Cavity at Different Frequencies

It should be noted that different frequencies require different sizes of resonant cavities, thus imposing restrictions on the size of sample that can be investigated.

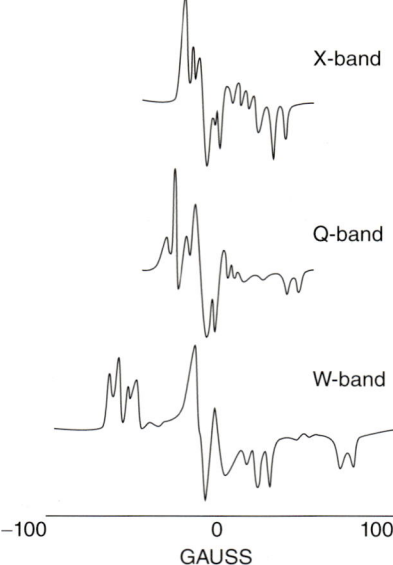

Figure 2.25 CW-EPR spectra (dots) at X-, Q- and W-bands of N^0-SL-NAD$^-$ bound to GAPDH at a ratio of 3.4 spin labels per GAPDH tetramer. The X-, Q-, and W-band spectra have been fitted (solid lines) simultaneously and overlaid on the data. The following best-fit values of the parameters were obtained: $\xi = 75.4°$, $\eta = 43.5°$, $R = 12.84$ Å, $\alpha = 16.1°$, $\beta = 53.5°$, and $\gamma = 19.0°$. The calculated labeling stoichiometry was 3.7 SL-NAD$^-$ per GAPDH tetramer at X-band, 3.7 at Q-band, and 3.4 at W-band; at X-band $\Gamma = 0.46$ Gauss (fixed) for the spatially isolated component and $\Gamma = 0.83$ Gauss for the dipolar-coupled component; at Q-band $\Gamma = 0.46$ Gauss (fixed) for the spatially isolated component and $\Gamma = 0.68$ Gauss for the dipolar-coupled component; at W-band $\Gamma = 1.27$ Gauss for both components; σ was fixed at 0.47 Gauss for all spectra. Adapted from Husted et al. (1997).

This is particularly the case at higher frequencies, where the size of the resonant cavity may become quite small as it decreases in proportion to the microwave frequency. Samples of different sizes and properties are optimally studied at different frequencies due to the varying size of the resonant cavity with frequency (see Table 2.1). The size of the cavity decreases with increasing frequency in a waveguide arrangement. However, by using a Fabry–Pérot type of arrangement, samples of a larger size can be investigated at higher frequencies. In contrast, the density of the microwave field (B_1) decreases with decreasing frequency, but can be enhanced by the use of a loop-gap resonator to concentrate the B_1 field in selected regions of the cavity. It is further noted that the B_1 per square root incident Watt can be high at low frequency if a resonator is used which is as small as that used at the higher frequency. An advantage of the loop-gap and other lumped-circuit-type resonators is that the size can be selected for the sample. Further, an enhanced signal-to-noise ratio (SNR) is achieved at higher frequencies when a sample of only very small size is available, due to difficulties in isolating, synthesizing, or crystallizing a sample, or when the sample is a single cell (as described in the following section).

2.5.4
Signal-to-Noise Ratios at Different Frequencies

If the sample is limited in size and is not lossy, a higher SNR is obtained at higher frequencies. On the other hand, if the sample is not limited in size, or it *is* lossy, then a better SNR is obtained at lower frequencies. A detailed discussion on multifrequency EPR sensitivity is provided in Chapter 5, Section 5.3.

2.5.5
Multifrequency Aspects of Using Home-Built versus Commercial Spectrometers

Special requirements may warrant opting for a home-built apparatus at VHF, for which commercial spectrometers are not available. However, the qualities of the following three factors are compromised in noncommercial spectrometers, when built with limited funds and limited engineering skills: (i) a high SNR, even at higher frequencies; (ii) magnetic-field homogeneity and accuracy; and (iii) the elimination of background signals from the resonator.

2.5.6
Multifrequency Aspects of Sample-Related Problems

The following considerations should be taken into account when deciding which frequency would be advantageous to study a given sample. These are, specifically: (i) g-strain; (ii) increased validity of first-order approximations at higher frequency; (iii) the requirement that smaller samples should be studied at higher frequency; (iv) the occurrence of field-dependent relaxation mechanisms, (v) the resonance requirement of equality of energy level separations and $h\nu$ at any frequency, and (vi) the need to control sample conditions, including deoxygenation.

Acknowledgments

I am grateful to Professors C.P. Poole, Jr, Gareth Eaton, and Sandra Eaton for helpful comments to improve the presentation of this chapter, and to the Natural Sciences and Engineering Research Council (NSERC) of Canada for partial financial support.

Pertinent Literature

Details of various aspects of MF EPR can be found in the handouts of the two multifrequency EPR workshops:

- Multi-Frequency EPR Workshop, 24th International Symposium, 29 July 2001, held by the University of Denver and Bruker Biospin.
- Electron Magnetic Resonance Developments and Applications in Chemistry, Biology and Materials Science, International EMR Workshop, December 13–14, 2002 at the National High magnetic Field Laboratory, Tallahassee, Florida. Many considerations included in this chapter were presented at the 2005 Euromar international conference (Misra, 2005).

References

Abragam, A. and Bleaney, B. (1970) *Electron Paramagnetic Resonance of Transition Metal Ions*, Oxford University Press, Oxford.

Antholine, W.E. (1997) *Adv. Biophys. Chem.*, **6**, 217.

Antholine, W.E. (2005) Low frequency EPR of Cu^{2+} in proteins, in *Biological Magnetic Resonance*, vol. 23 (eds S.S. Eaton, G.R. Eaton, and L.J. Berliner), Kluwer/Plenum Publishers, p. 417.

Antholine, W.E., Kastrau, D.H.W., Steffens, G.C.M., Buse, G., Zumft, W.G., and Kroneck, P.M.H. (1992) *Eur. J. Biochem.*, **209**, 875.

Antholine, W.E., Hanna, P.M., and McMillin, D.R. (1993) *Biophys. J.*, **64**, 267.

Atkins, P.W. and Kivelson, D. (1966) *J. Chem. Phys.*, **44**, 169.

Barnes, J.P. and Freed, J.H. (1998) *Biophys. J.*, **95**, 2532.

Barra, A.-L., Gatteschi, D., Sassoli, R., Abbati, G.L., Cornia, A., Fabretti, A.C., and Uytterhoeven, M.G. (1997) *Agnew. Chem. Int. Ed. Engl.*, **36**, 2329.

Burghaus, O., Rohrer, M., Götzinger, T., Plato, M., and Möbius, K. (1992) *Meas. Sci. Technol.*, **3**, 765.

Calvo, R., Abresch, E.C., Bittl, R., Feher, G., Hofbauer, W., Issacson, R.A., Lubitz, W., Okamura, M.Y., and Paddock, M.L. (2000) *J. Am. Chem. Soc.*, **122**, 7327.

Eaton, S.S. and Eaton, G.R. (1993) *Magn. Reson. Rev.*, **16**, 157–181.

Eaton, G.R. and Eaton, S.S. (2001) Case studies of questions addressed by EPR at various frequencies in multi-frequency EPR workshops. 24th International EPR Symposium, 29 July 2001 (University of Denver and Bruker Biospin).

Eaton, S.S., Harbridge, J., Rinard, G.A., Eaton, G.R., and Weber, R.T. (2001) *Appl. Magn. Reson.*, **20**, 151–157.

Flanagan, H.L. and Singel, D.J. (1987) *J. Chem. Phys.*, **87**, 5606.

Freed, J.H. (2000) *Annu. Rev. Phys. Chem.*, **51**, 655.

Froncisz, W. and Hyde, J.S. (1980) *J. Chem. Phys.*, **73**, 3123.

Froncisz, W., Scholes, C.P., Hyde, J.S., Wei, Y.-H., King, T.E., Shaw, R.W., and Beinert, H. (1979) *J. Biol. Chem.*, **254**, 7482.

Galeev, A.A., Khasanova, N.M., Rudowicz, C., Shakurov, G.S., Bykov, A.B., Bulka, G.R., Nizatmutinov, N.M., and Vinukarov, V.M. (2000) *J. Phys. Condens. Matter*, **12**, 4465.

Gerfen, G.J. and Singel, D.J. (1990) *J. Chem. Phys.*, **93**, 4571.

Hanson, G.R., Wilson, G.R., Bailey, T.D., Pilbrow, J.R., and Wedd, A.G. (1987) *J. Am. Chem. Soc.*, **109**, 2609.

Husted, E.J., Smirnov, A.I., Laub, C.F., Cobb, C.E., and Beth, A.H. (1997) *Biophys. J.*, **74**, 1861.

Hyde, J.S. and Froncisz, W. (1982) *Annu. Rev. Biophys. Bioeng.*, **11**, 391.

Kivelson, D. (1960) *J. Chem. Phys.*, **33**, 1094.

Kroneck, P.M.H., Antholine, W.A., Riester, J., and Zumft, W.G. (1988) *FEBS Lett.*, **242**, 70.

Lebedev, Y.S. (1990) High-frequency continuous-wave electron spin resonance, in *Modern Pulsed and Continuous-Wave Electron Spin Resonance* (eds L. Kevan and M.K. Bowman), Wiley Interscience, New York, Chapter 8, pp. 365–404.

Liang, Z. and Freed, J.H. (1999) *J. Phys. Chem. B*, **103**, 6384.

Misra, S.K. (1976) *J. Magn. Reson.*, **23**, 403.
Misra, S.K. (1983) *Physica*, **121B**, 193.
Misra, S.K. (1986) *Mag. Reson. Rev.*, **10**, 285.
Misra, S.K. (1999) *J. Magn. Reson.*, **140**, 179.
Misra, S.K. (2005) Virtues of high-field/high-frequency EPR: advantages of multi-frequency approach in EPR & single-crystal versus powder EPR. Euromar Conference, held in Eindhoven, Holland, July-August, 2005.
Misra, S.K. and Andronenko, S.I. (2002) *Phys. Rev. B*, **65**, 104435.
Misra, S.K., Andronenko, S.I., Earle, K.A., and Freed, J.H. (2001) *Appl. Magn. Reson.*, **21**, 549–561.
Misra, S.K., Andronenko, S.I., Rinaldi, G., Chand, P., Earle, K.A., and Freed, J.H. (2003) *J. Magn. Reson.*, **160**, 131–138.
Paskenkiewicz-Gierula, M., Subczynski, W.K., and Antholine, W.E. (1997) *J. Phys. Chem. B*, **101**, 5596.
Petasis, D.T. and Hendrich, M.P. (1999) *J. Magn. Reson.*, **136**, 200.
Pilar, J., Labsky, J., Marek, A., Budil, D.E., Earle, K.A., and Freed, J.H. (2000) *Macromolecules*, **33**, 4438.
Smirnova, T.I., Smirnov, A.I., Clarkson, R.B., Belf, R.L., Kotake, Y., and Janzen, E.G. (1997) *J. Phys. Chem. B*, **101**, 3877.
Telser, J., Pardi, L.A., Krzystek, J., and Brunel, L.C. (1998) *Inorg. Chem.*, **37**, 5769.
Tregenna-Piagott, P.L.W., Wiehle, H., Bendix, J., Barra, A.-L., and Güdel, H.-U. (1999) *Inorg. Chem.*, **38**, 5928.
van Dam, P.J., Klassen, A.A.K., Reijerse, E.J., and Hagen, W.R. (1998) *J. Magn. Reson.*, **130**, 140.
van Kan, P.J.M., van der Horst, E., Reijerse, E.J., van Bentum, P.J.M., and Hagen, R. (1998) *J. Chem. Soc. Faraday Trans.*, **94**, 2975.
Weber, R. (2001) Overview of multi-frequency EPR in multi-frequency EPR workshops. 24th International EPR Symposium, 29 July 2001 (University of Denver and Bruker Biospin).
Wilson, R. and Kivelson, D. (1966a) *J. Chem. Phys.*, **44**, 15.
Wilson, R. and Kivelson, D. (1966b) *J. Chem. Phys.*, **44**, 4440.
Wood, R.M., Stucker, D.M., Jones, L.M., Lynch, W.B., Misra, S.K., and Freed, J.H. (1999) *Inorg. Chem.*, **38**, 5384.
Yuan, H., Collins, M.L.P., and Antholine, W.E. (1997) *Biophys. J. Am. Chem. Soc.*, **119**, 5073.
Zverev, G.M. and Prokhorov, A.M. (1960) *Sov. Phys. JETP*, **11**, 330.

3
Basic Theory of Electron Paramagnetic Resonance
Sushil K. Misra

3.1
Introduction

The crystal and ligand field theories attempt to calculate splittings of the d-electron energy levels to explain the optical absorption data, to calculate the magnetic and EPR properties using the wavefunctions of the ground state, and to understand effects such as covalency. The various models so used represent approximations. Crystal-field theory (CFT) was most likely among the first of the electrostatic models to be used for calculating the energy levels of transition metal ions. Indeed, van Vleck's CFT calculations were quite successful during the early days (the 1930s), owing to the small number of parameters used, and taking into account the symmetry of the configuration. More recently, Abragam and Bleaney (1970) have provided a good review on this subject.

Subsequently, the ligand-field theory (LFT) and its derivatives were developed to accommodate covalency, using the molecular orbital (MO) theory of Mulliken, notably by Van Vleck during the 1930s. The role of LFT was to investigate the origin and consequences of the effect of the surrounding atoms on the orbitals of the transition metal ions. It appears that CFT and LFT are complementary, and of equal importance (Sealey, Hyde, and Antholine, 1985). Additional phenomenological models have been developed to parameterize the splittings of the transition-metal ion energies; most notable among these are the "angular overlap model" (AOM) (Gerloch, 1983) and a similar, but less-sophisticated ligand superposition model (SPM) (Bradbury and Newman, 1968). Finally, the Xα method has been developed for carrying out many-electron calculations; this is based on an approximation for band-structure calculations, as introduced by Slater (1951a, 1951b).

Before describing the details of the various methods, it would be worth globally reviewing the similarities between the CFT, SPM, and AOM models, which have been used to explain – with varying degrees of success – the interpretation of optical, magnetic, and EPR data obtained mainly for the first-row transition ions (Pilbrow, 1990). As with CFT, in the point-charge model (PCM) the basis of the theory is the spectroscopic term of the free ion in the weak-field approximation. It also includes the Slater determinants, based on the products of one-electron wavefunctions, and possibly including the Racah B and C parameters directly for the correlation term in the strong-field approximation. In SPM, the basis of the theory is the spectroscopic term of the free ion. As for the extension of molecular orbital (MO) theories, the following are noted. In the LCAO-MO model, a symmetry-adapted LCAO (linear combination of atomic orbitals) is used, on a non-orthogonal basis, starting from the spectroscopic terms of the free ion. The Hückel theory also uses a non-orthogonal basis, making certain assumptions about the off-diagonal elements. The original AOM is based on the Wolfsberg–Helmholz analysis, as developed from the Hückel theory, and focuses on angular factors in overlap between the metal and ligand orbitals. In the AOM, when used in context with LFT, the AOM parameters are not tied to the original association with MO theory; the parameters used are not "free", but are consistent with the electron-donating/accepting character of the ligands, which are consistently assumed for the same ligand when bound to different metal ions. The X_α method represents a many-electron approximation, and is based on a simple model of electron–electron correlation. Further comments describing relationships between various models are as follows. There exists a relationship between the CFT and AOM models via multipole expansions used; the SPM and CFT are related by expressing the crystal-field parameters and crystalline potential, respectively, as an expansion in terms of products of radial- and angular-dependent parts. Local parameters are used in the SPM and AOM models to describe the local splittings of a paramagnetic ion in a crystal field. The SPM assumes the interactions to be axially symmetric, whereas in the AOM nonaxial bonding is permitted. CFT breaks down for low-symmetry complexes, as it is a global model. It is easier to use the SPM than the AOM, the former being an extension of the CFT. The actual calculations carried out in the SPM and AOM are different from each other. In the SPM, which is essentially a crystal-field model, expressions equivalent to the elementary crystal-field operators are used, on a basis which is often solely determined by a single spectroscopic term. In contrast, in AOM the full energy matrix involves the Racah B parameter, as the AOM calculations are normally carried out for all the terms from a particular d^n configuration.

In this chapter, attention will be focused on the simplest models, including ions of the first-row transition metals, based on treatments in books by Griffith (1961), Gerloch (1983), Abragam and Bleaney (1970), and as described by Poole and Farach (1972), Pake and Estle (1973), and Pilbrow (1990).

3.2
Crystal-Field Theory

3.2.1
Introduction to CFT

The crystalline electric fields of surrounding ions influence the energies of a paramagnetic ion in a significant manner. Typically, a very strong Stark effect is produced on the $3d^n$ partly filled shells of first-series transition metal ions, comparable to the main electronic energies of the ion, and far exceeding the spin–orbit interaction. In the case of the rare-earth ions, the situation is different, where $5s^2p^6$ filled shells shield the inner paramagnetic $4f^n$ electrons. As a result, the crystalline electric field (CEF) Stark effect splitting of the rare-earth ions is of the order of a few hundreds of cm^{-1}, which is less than the spin–orbit coupling energy. For $3d^n$ ions, these CEF energy splittings are much larger than the electronic Zeeman interaction (~0.3 cm^{-1}), and they are detected directly in optical spectroscopy. In any case, the magnitude and anisotropy of the spin-Hamiltonian parameters are strongly affected by them.

The energy levels and wavefunctions for some typical paramagnetic ions will now be described using the CFT. It transpires that the inner d- or f-shells of atoms are more impervious to their environment in solids, unlike the s and p electrons, which are likely to be lost from the shell to other ions, or the shell requires to be filled in the solid. Accordingly, the d or f shells are likely to be more incomplete in crystalline surroundings. Therefore, the transition-group ions, consisting of unfilled d and f shells, will be examined here. These are most commonly responsible for magnetism, and represent roughly one-half of all the elements. The five transition groups are listed in Chapter 7 of this book, along with some of their properties. The most commonly studied transition groups are the iron group, which consists of scandium, titanium, vanadium, chromium, manganese, iron, cobalt, nickel, and copper, and the rare-earth group.

3.2.2
Free Atoms and Ions

The crystalline environment does not possess spherical symmetry. Rather, it is considered as an additional electrostatic potential, as seen by the electrons of the ion. The effect of the CEF will be considered here as a perturbation on a free transition group ion with an unfilled shell, for which the Hamiltonian is:

$$H = -\sum_i \frac{\hbar^2}{2m}\nabla_i^2 - \sum_i \frac{Ze^2}{r_j} + \sum_{i>j} \frac{e^2}{r_{i,j}} + \sum_i \xi(r_i) l_i \cdot s_i, \qquad (3.1)$$

where the first three terms together, constituting the electronic Hamiltonian, H_{el}, have the dominant effect. They are, respectively, the kinetic energy, an effective spherical potential representing the nuclear attraction, and an average screening term due to the electron–electron interactions. The order of magnitude of H_{el} is 10^4–10^5 cm^{-1}, which is many orders of magnitude larger than the Zeeman energy at X-band (~9.5 GHz). It belongs to the optical region of the spectrum. As these dominant terms possess spherical symmetry, the angular momentum, l_i, of each electron is a good quantum number, along with the principal quantum number, n_i, and the magnetic quantum number, m_i. The term *orbital* refers to the individual electron configuration, which can be occupied only by two electrons with opposite spins, as a consequence of *Pauli's exclusion principle*. By proceeding in this manner, it is possible to calculate the configuration that describes the total energy and states of all electrons. Thus, the wavefunctions are seen to be Slater determinants, while the energy is the sum of all the individual energies. In this approximation, a large number of states of the ion possess the same energy, with all corresponding to the same configuration. The energies of the ion under the action of the various interactions are such that, the lowest energy configuration is usually separated from the next highest energy configuration by about $10\,000\,cm^{-1}$, or more. The magnitude of this separation, and the configuration with the lowest energy, depend on whether the ion is free, or is in a crystal field. In EPR, the states of interest are those which arise from the splitting of the lowest-energy configuration due to the remaining terms in the Hamiltonian. It will be assumed hereafter that these splittings are small as compared to the spacing between the two lowest-energy configurations. For light elements, the electron–electron interaction, and its spherical average, which produce bigger splittings than the spin–orbit interaction, should be considered first. This is termed Russell–Saunders (or L–S), coupling, and is larger than the spin–orbit interaction. In the case of heavy nuclei, the reverse constitutes a good approximation, termed j–j coupling.

The orbital angular momenta for the individual electrons are no longer good quantum numbers, because the electron–electron interaction has nonvanishing matrix elements between the states, for which the orbitals of the two electrons differ. In the absence of spin-dependent operators, the total orbital angular momentum, $L = \sum_i \ell_i$, is not coupled to the spin. Thus, its eigenvalues, L, as well its projections, M_L^i, are good quantum numbers. Similarly, since individual orbitals can no longer be defined at this point, individual spin functions do not exist, but the projections, M_S, of the total spin, $S = \sum_i s_i$, are good quantum numbers.

Taking into account the relevant interactions, it emerges that each term, when defined as a group of states specified by L and S, and arising out of the lowest configuration, will – in general – have a different energy. The farther the electrons stay from each other, the lower will be the energy of the term. Finally, by applying Pauli's exclusion principle, it emerges that the lower energy terms will be those characterized by higher spin. This is a slight generalization of what is known as the *first Hund's rule* (Pake and Estle, 1973). According to the *second Hund's rule*, the lowest term will be characterized by the maximum L consistent with the maximum value of S; this arises from the Coulomb repulsion term in Equation 3.1. The resulting terms are denoted by the spectroscopic notation $^{2S+1}\Gamma_J$, where $\Gamma = S, P, D, F, G$, etc. represent $L = 0, 1, 2, 3, 4$, etc. The configuration of the lower energy levels of an atom or ion with two d-electrons in addition to the closed shells (for example, Ti^{2+} or V^{3+}) is shown in Figure 3.1. Here, the closed shell is referred to as $[Ar]3d^2$ configuration, which means that there are two $3d$ electrons added to the closed-shell configuration [Ar] of Argon. The 45-fold degenerate lowest energy configuration for this case, considering that for the iron-group ions in solids the $3d$ electrons are lower in energy than the $4s$ electrons (as shown on the left), breaks up into five terms by electron–electron Coulomb interactions. When the L–S coupling is taken into account as a perturbation to all that has been considered so far, the spin–orbit interaction splits these terms further into energy levels, characterized by J. This can be explained as follows. The operator $\sum_i \xi(r_i) l_i \cdot s_i$ in Equation 3.1 can be replaced

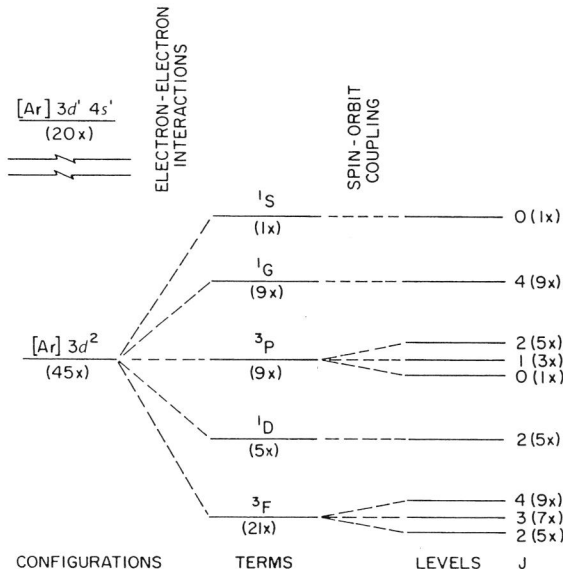

Figure 3.1 Energy level scheme for the lowest energy states of an ion with two 3d electrons outside the closed-shell argon core to various degrees of accuracy. The two lowest levels for the orbital approximation are shown on the left. Since, in crystals, the 3d level is lower than the 4s, the lowest configuration is taken to be [Ar]3d², where the symbol [Ar] stands for $1s^2 2s^2 2p^6 3s^2 3p^6$, representing the ground configuration for argon. There is a break-up into five terms of the 45-fold degenerate ground configuration as a result of electron–electron Coulomb interactions. The spin–orbit interaction splits the terms into energy levels characterized by J due to L–S coupling as shown on the right. Adapted from Pake and Estle (1973).

by $\lambda \mathbf{L} \cdot \mathbf{S}$, where $\lambda = \pm \dfrac{\xi_{3d}}{2S}$, with $\xi_{3d} = \langle 3d_{radial\ fn.} | \xi(r_i) | 3d_{radial\ fn.} \rangle$. (Here the positive sign should be chosen, as the shell is less than half full for a $3d^1$ ion; otherwise, the minus sign should be used here.) In this context, a useful relation is $\lambda \mathbf{L} \cdot \mathbf{S} = \dfrac{\lambda}{2}[J(J+1) - L(L+1) - S(S+1)]$, which determines the splitting of the energy levels belonging to a term. One has $j = \ell \pm 1/2$ for a single electron outside a closed-shell core; then, the spin–orbit splitting between the levels is given by the *Lande interval rule*:

$$2\xi[\ell + 1/2] = 2\xi j_{\max}.$$

The term $\lambda \mathbf{L} \cdot \mathbf{S}$ has the effect of characterizing the resulting split levels by the total angular momentum, J, shown in the second column on the right of Figure 3.1 for the [Ar]3d² configuration.

When, instead of L–S coupling, j–j coupling is taken into account, the spin–orbit interaction is considered first. The quantum numbers $j_i (= \ell_i + s_i)$ are good, since

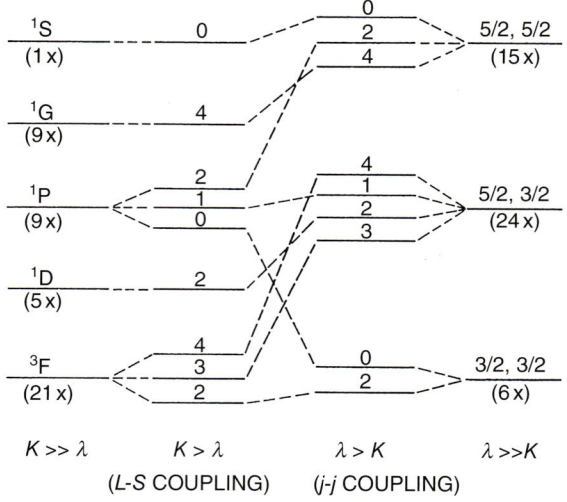

Figure 3.2 Energy-level scheme for an ion with [Ar]$3d^2$ configuration. The extreme L–S coupling limit is shown on the left, whereas the extreme j–j coupling limit is shown on the right. The levels for the two intermediate cases are labeled by their j-values. The levels on the right are labeled by the two j-values of the two electrons. Here, λ is the spin–orbit coupling constant, whereas K, being of the order of the splitting between terms, is the characteristic of the strength of the electron–electron interactions. Adapted from Pake and Estle (1973).

$\sum_i \xi(r_i) \mathbf{l}_i \cdot \mathbf{s}_i$ is the sum of one-electron operators, but l_i and s_i are not good quantum numbers. Now, the addition of electron–electron interactions couples the $\{j_i\}$, which results in the total J and M_J as the only good quantum numbers. Figure 3.2 shows, schematically, the relationship to the Russell–Saunders limit, where K and λ are quantities which characterize the strength of the electron–electron and spin–orbit interactions, respectively.

Finally, having understood the two limiting cases of L–S and j–j coupling, it is possible to calculate the energies using the lowest order of perturbation theory, as well as the wavefunctions by the use of coupling of angular momenta.

3.2.3
The Crystal-Field Description of Transition Group Ions in Crystals

The crystal field produced at the site of a paramagnetic ion should possess the same point symmetry as exists at the ionic site, as this field is due to the charged particles that constitute the crystal. Accordingly, an additional term is introduced into the atomic Hamiltonian due to the crystal field. The dominant symmetry is, in general, cubic symmetry. In addition, there are present smaller terms due to lower symmetries present in the crystal. Hereafter, the behavior of a single electron, plus a closed-shell ion core, will be explored in a cubic crystal, using the

3.2.3.1 *p*-Orbitals

An electron is now considered in a central potential acted upon by a perturbing field, without taking into account the spin–orbit coupling. Although, the crystal field cannot remove any degeneracy of the *s*-orbital, it will lift the threefold degeneracy of a *p*-orbital. The free-ion term is ^2P for a single electron in the absence of spin–orbit coupling. The real wavefunctions for the *p*-orbitals are conveniently described as

$$|1, x\rangle = \sqrt{\frac{3}{4\pi}} \frac{x}{r}; |1, y\rangle = \sqrt{\frac{3}{4\pi}} \frac{y}{r}; |1, z\rangle = \sqrt{\frac{3}{4\pi}} \frac{z}{r}, \quad (3.2)$$

where 1 in the kets denotes a *p*-orbital ($l = 1$). The functions given by Equation 3.2 determine the probability densities as functions of (x, y, z), which results in different interaction energies due to the charges present in the vicinity of a paramagnetic ion. For example, there will be a splitting of the threefold degenerate *p*-orbitals by two negative charges on the positive and negative z-axis. This is because an electron in a $|1, z\rangle$ orbital will approach the charges more closely and, as a consequence, its energy will be higher than those of the $|1, x\rangle$ and $|1, y\rangle$ orbitals, which are equivalent, and extend farther from the charges on average. On the other hand, if the negative charges were to exist at all three axes, x, y, z, and also at their negative sides, there would emerge three states with different energies.

3.2.3.2 *d*-Orbitals

For understanding transition-group paramagnetism, when crystal-field effects dominate spin–orbit effects, the *d*-electrons are most interesting. As for the five degenerate *d*-orbitals for a single *d*-electron, for which the free-ion term is ^2D, the real wavefunctions are:

$$|2, xy\rangle = \frac{1}{2}\sqrt{\frac{15}{\pi}} \frac{xy}{r^2}; |2, yz\rangle = \frac{1}{2}\sqrt{\frac{15}{\pi}} \frac{yz}{r^2}; |2, zx\rangle = \frac{1}{2}\sqrt{\frac{15}{\pi}} \frac{zx}{r^2};$$
$$|2, x^2 - y^2\rangle = \frac{1}{4}\sqrt{\frac{15}{\pi}} \frac{x^2 - y^2}{r^2}; |2, 3z^2 - r^2\rangle = \frac{1}{4}\sqrt{\frac{15}{\pi}} \frac{3z^2 - r^2}{r^2}, \quad (3.3)$$

where 2 in the kets denotes a *d*-orbital ($l = 2$). It emerges that, under cubic symmetry, with the x-, y-, and z-axes directed along the cube edges, the first three of the wavefunctions in Equation 3.3 are equivalent. Thus, they are degenerate in energy, and are referred to as t_2 orbitals (they are also referred to as *t*-orbitals in *T* symmetry, t_g orbitals in T_h symmetry, and t_2 orbitals in *O* and T_d symmetry; occasionally, the symbols γ_5, or *d*ε, are also used). In contrast, the last two wavefunctions are also equivalent to each other in cubic symmetry, are thus degenerate in energy, and are referred to as *e*-orbitals (they are also referred to as e-orbitals in *T*, *O*, and T_d symmetry, and as e_g-orbitals in T_h and O_h symmetry; the notations γ_3 or *d*γ are also used in this case).

In order to determine the order of energies of the two orbitals, t_2 and e, the surrounding charges should also be taken into account. When the transition-group ion is surrounded by six negative point charges, each placed equidistant from the ion along the positive and negative cubic axes, it corresponds to the octahedral symmetry, O_h, with the ion being six-coordinated. It emerges that the two e-orbitals, $|2, x^2 - y^2\rangle$ and $|2, 3z^2 - r^2\rangle$, possess large electron densities in the direction of the negative ions, whereas the three t_2 orbitals, $|2, xy\rangle$, $|2, yz\rangle$, and $|2, zx\rangle$, have their maxima along the directions midway between the directions to two of the negative charges. These considerations imply that the e-orbitals are higher in energy than the t_2 orbitals. The magnitude of this splitting is denoted as Δ, and also referred to as $10Dq$. If, however, one considers eight-coordinated point negative charges showing cubic symmetry, being situated at the corners of the cube, while the paramagnetic d-ion is at the body center of the cube, it emerges that e-orbitals avoid the negative charges more efficiently than do the t_2 orbitals. Then, the energy of the t_2 orbitals will be higher than those of the e-orbitals, by Δ. This same ordering is true for four-coordinated cubic symmetry, since in this case the two interpenetrating tetrahedra, with negative charges located on their vertices, create eight-coordinated, as well as 12-coordinated cubic symmetry, wherein the charges lie along the $\langle 110 \rangle$, and equivalent, directions, as do the lobes of the t_2 orbitals.

Consider now the effect of a lower symmetry superimposed over the cubic symmetry, which is characterized by only one parameter Δ. As an example, consider the case of six-coordination, but with an additional crystal field of tetragonal symmetry. This can be considered equivalently as a compression, or elongation, of the two negative charges on the z-axis. In the case of compression, the energies of the states will be in decreasing order of energies: $|2, 3z^2 - r^2\rangle > |2, x^2 - y^2\rangle > (|2, yz\rangle = |2, zx\rangle) > |2, xy\rangle$. In a strong tetragonal field, the four energy levels as obtained for the weak tetragonal field will be modified such that the relative order of the energies of the $|2, x^2 - y^2\rangle$ and that of the pair ($|2, yz\rangle$, $|2, zx\rangle$) states, will be interchanged. Finally, for a pure axial field, there will result splitting into three distinct energy levels, corresponding to the states, in increasing order of energy, ($|2, xy\rangle$, $|2, x^2 - y^2\rangle$), ($|2, yz\rangle$, $|2, zx\rangle$), and $|2, 3z^2 - r^2\rangle$. The tetragonal field requires two parameters to describe its effect within an incomplete d-shell configuration, including the cubic parameter, Δ. The tetragonal field is weak for the majority of transition-group impurities, frequently not being more than the spin–orbit interaction even when the cubic field is greater in energy than the spin–orbit coupling. Similar considerations apply to other symmetries lower than cubic.

So far, the sources of the potential – that is, charges – have been taken into account to consider the energies of the paramagnetic ion. One can, alternatively, calculate the potential about the d-electron itself inside the region of negative charge. One can then expand the potential, V, in a series of spherical harmonics. It emerges, in view of the characteristics of coupled angular momenta and the Wigner–Eckart theorem, that very few terms in the expansion are needed, of which many are eliminated by symmetry. Finally, for cubic symmetry, there is just one term for d-electrons: $V \propto Y_4^0(\theta,\phi) + \sqrt{\frac{5}{14}}\left[Y_4^4(\theta,\phi) + Y_4^{-4}(\theta,\phi)\right]$, to which are added the terms $Y_2^0(\theta,\phi)$ and $Y_4^0(\theta,\phi)$ for tetragonal symmetry. Here $Y_l^m(\theta,\phi)$ are spherical harmonics. This approach is frequently adopted for actual computations using crystal-field theory (Ballhausen, 1962; Griffith, 1961).

3.2.4
Crystal Field Potential

The dominant component of CEF in most crystals arises from a basically octahedral, tetrahedral, or cubic array of the surrounding negative charges, or ligands, as illustrated in Figure 3.3. The nearest-neighbor ions dominantly determine the crystal-field potential. Normally, the distortions of the surrounding charges are axial, with tetragonal or trigonal axis of symmetry. However, distortions which are lower in symmetry than these also occur frequently.

For an array of N negative charges q_i, the electric potential near its center can be expressed as:

$$V = \sum_{i=1}^{N} q_i/r_i, \tag{3.4}$$

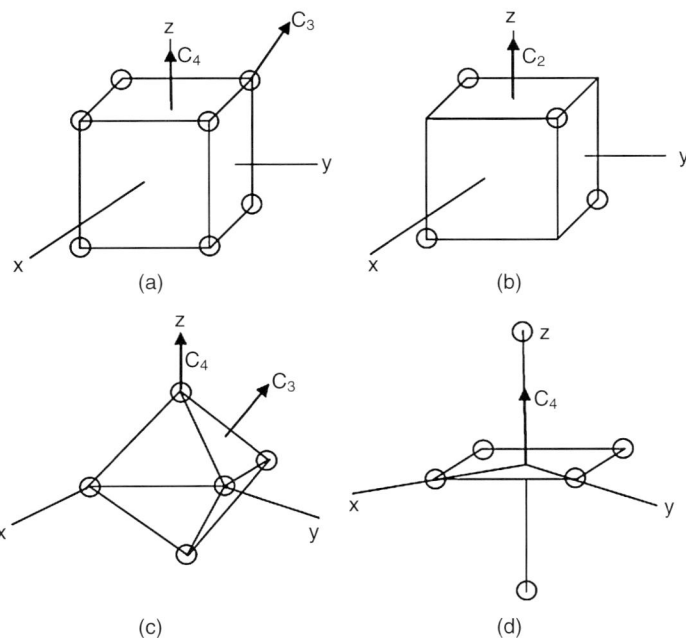

Figure 3.3 Charge distributions. (a) Cubic; (b) Tetrahedral; (c) Octahedral; (d) Square planar plus axial (tetragonally distorted octahedral). Adapted from Poole and Farach (1972.)

where r_i is the distance of the ith ion from the center. Conveniently, one may select the center of the array of charges as the reference point. Then, the potential at an arbitrary point in terms of its radius vector from the center, r, from the origin, as follows:

$$V = \sum_{i=1}^{N} (q_i/r_i) \left[1 + \left(\frac{r}{R_i}\right)^2 - 2\frac{r}{R_i}\cos\alpha_i \right]^{-1/2}, \tag{3.5}$$

where α_i denotes the angle between the vectors $\mathbf{R}_i$ and $\mathbf{r}$ for each ion. Since, in CEF, one is interested in expressing the potential at points near the center at the distance r, which are closer to the origin than the ligand distance, R_i, one can make a power-series expansion in terms of the ratio r/R_i, as follows:

$$V = \sum_{i=1}^{N} (q_i/R_i) \sum_{i=0}^{\infty} \left(\frac{r}{R_i}\right)^{\ell} P_\ell(\cos\alpha_i), \tag{3.6}$$

where $P_\ell(\cos\alpha_i)$ is the lth order Legendre polynomial. Using the spherical-harmonics addition theorem, one may express this in polar coordinates, as follows:

$$P_\ell(\cos\alpha_i) = \frac{4\pi}{2\ell+1} \sum_{m=-\ell}^{\ell} (-1)^m Y_l^{-m}(\theta_i, \phi_i) Y_l^m(\theta, \phi).$$

The above expansion can be expressed in terms of the real spherical harmonics, $Z_{\ell m}(\theta, \phi)$, called Tesseral harmonics, related to the usual spherical harmonics as follows:

$$Z_{l0} = Y_{l o}; \quad Z_{lm}^c = \frac{1}{\sqrt{2}}[Y_l^{-m} + (-1)^m Y_l^m]; \quad Z_{lm}^s = \frac{1}{\sqrt{2}}[Y_l^{-m} - (-1)^m Y_l^m],$$

where $m > 0$, and the superscripts c refers to cosine and s to sine, respectively. The above expressions can be rewritten as follows:

$$Z_{l0} = \sqrt{\frac{(2l+1)}{4\pi}} P_l(\cos\theta); \quad Z_{lm}^c = \sqrt{\frac{(2l+1)(l-m)}{2\pi(l+m)}} P_l^m(\cos\theta)\cos m\varphi;$$

$$Z_{lm}^s = \sqrt{\frac{(2l+1)(l-m)}{2\pi(l+m)}} P_l^m(\cos\theta) \sin m\varphi,$$

where the associated Legendre polynomials, P_l^m, are functions of θ. The commonly occurring Tesseral harmonics, both in Cartesian and spherical coordinates, are listed by Poole and Farach (1972).

Using the above, the crystal-field potential can be expressed as

$$\mathbf{V}(r, \theta, \varphi) = \sum_{l=0}^{\infty} \sum_{m=-l}^{+l} r^l A_{lm} Z_{lm}(\theta, \varphi). \tag{3.7}$$

For a discrete charge distribution of N ligands, the explicit expressions for the coefficients A_{lm} are:

$$A_{l0} = \frac{4\pi}{2l+1} \sum_{i=1}^{N} q_i \frac{Z_{l0}(\theta_i, \varphi_i)}{R_i^{l+1}} \tag{3.8}$$

$$A_{lm}^c = \frac{4\pi}{2l+1}\sum_{i=1}^{N} q_i \frac{Z_{lm}^c(\theta_i, \varphi_i)}{R_i^{l+1}} \quad (3.9)$$

$$A_{lm}^s = \frac{4\pi}{2l+1}\sum_{i=1}^{N} q_i \frac{Z_{lm}^s(\theta_i, \varphi_i)}{R_i^{l+1}} \quad (3.10)$$

There are, in most cases, N identical nearest-neighbor ligand charges, q, located at the same distance R from the transition metal ion site, so that

$$A_{lm}^c = \frac{4\pi}{2l+1}\frac{q_i}{R^{l+1}}\sum_{i=1}^{N} Z_{lm}^c(\theta_i, \varphi_i).$$

In the above equation, the summation amounts to a geometrical factor. Similar expressions exist for $A_{\ell m}^s$ and $A_{\ell 0}$. The summation in the above equation is over infinite values of ℓ; however, for high symmetry most terms vanish. Furthermore, one need consider only the terms $\ell \leq \ell'$, where $\ell' = 2$ for d-electrons, and $\ell' = 3$ for f-electrons.

3.2.5
Point Charge Model

For a particular array of charges, the crystal-field potential is obtained by substituting the ligand coordinates into Equation 3.7; this is known as the PCM, and it provides a good first approximation to the potential. Hereafter, the $\ell = 0$ term is omitted, as it shifts all of the energies levels equally.

Some specific examples, addressing higher symmetries, cubic, tetrahedral, and octahedral, are now considered.

Cubic array of charges (Figure 3.3a): In this case, the nonzero coefficients are expressed as:

$$A_{00} = \frac{8Ze\sqrt{4\pi}}{R},\ A_{40} = \frac{-28}{9}\frac{Ze}{R^5}\sqrt{\frac{4\pi}{9}},\ A_{60} = \frac{16}{9}\frac{Ze}{R^7}\sqrt{\frac{4\pi}{13}} \quad (3.11)$$

From Equation 3.11 one obtains A_{44}^c and A_{64}^c using the following relationships:

$$A_{44}^c = \sqrt{\frac{5}{7}}A_{40},\ A_{64}^c = -7A_{60} \quad (3.12)$$

Tetrahedral array of charges (Figure 3.3b): The corresponding coefficients $A_{\ell m}$ are one half those for the cubic array as given by Equation 3.11.

Octahedral array of charges (Figure 3.3c): For this case, the coefficients are:

$$A_{00} = \frac{6Ze\sqrt{4\pi}}{R},\ A_{40} = \frac{7}{2}\frac{Ze}{R^5}\sqrt{\frac{4\pi}{9}},\ A_{60} = \frac{3}{4}\frac{Ze}{R^7}\sqrt{\frac{4\pi}{13}}.$$

In addition, the A_{44}^c and A_{64}^c coefficients are obtained by using the same relations as those given by Equation 3.12 for this symmetry.

It is noted that, for d-electrons, only the A_{40} and A_{44} terms contribute to the energy splitting, and for cubic symmetry, the coefficient A_{4m} are eight to nine times as large as those for the octahedral coefficients. As a consequence, for the same distances and charges, the octahedral configuration is slightly more effective, by a factor of 9/8, than the eight cubic charges in splitting the energy levels.

The above formulas are valid for the case when the point, at which the potentials are being calculated, is at a distance r which is less than the nearest-neighbor ligand. When this is not the case, the roles of r and R are reversed in Equation 3.6, so that the expansions should be carried out in powers of $(R_i/r)^\ell$, thereby interchanging r and R, which should be done in subsequent formulas.

3.2.5.1 Potentials for Cubic and Lower Symmetry

The general formulas given in the previous section will now be specialized to write explicit expressions for the configurations of charges (as shown in Figure 3.3), by substituting for the coordinates of the charges. The following apply to the first-transition series ions ($3d^n$), for which $\ell = 2$.

Cubic: Here, the eight charges are at the vertices of a cube as shown in Figure 3.3a:

$$V_c = \frac{8Ze}{R} - \frac{56\sqrt{\pi}}{27}\left(\frac{Ze}{R^5}\right)r^4\left(Z_{40} + \sqrt{\frac{5}{7}}Z^c_{44}\right).$$

Tetrahedral: This configuration is obtained by removing half of the charges from the cube to form a regular tetrahedron, as shown in Figure 3.3b:

$$V_t = \frac{4Ze}{R} - \frac{28\sqrt{\pi}}{27}\left(\frac{Ze}{R^5}\right)r^4\left(Z_{40} + \sqrt{\frac{5}{7}}Z^c_{44}\right).$$

Octahedral: For this case, as shown in Figure 3.3c, where the charges are located along the coordinate axes, the potential is:

$$V_o = \frac{6Ze}{R} + \frac{7\sqrt{\pi}}{3}\left(\frac{Ze}{R^5}\right)r^4\left(Z_{40} + \sqrt{\frac{5}{7}}Z^c_{43}\right).$$

Two charges along the Z-axis: For this configuration, as shown in Figure 3.3d, one obtains:

$$V_Z = \frac{2Ze}{R} + 4\sqrt{\frac{\pi}{5}}\left(\frac{Ze}{R^3}\right)r^2 Z_{20} + \frac{4\sqrt{\pi}}{3}\left(\frac{Ze}{R^5}\right)r^4 Z_{40}$$

Square-planar array of four charges: For this configuration, as shown in Figure 3.3d, one obtains

$$V_{xy} = \frac{4Ze}{R} - 4\sqrt{\frac{\pi}{5}}\left(\frac{Ze}{R^3}\right)r^2 Z_{20} + \sqrt{\pi}\left(\frac{Ze}{R^5}\right)r^4\left(Z_{40} + \frac{\sqrt{35}}{3}Z^c_{44}\right)$$

Octahedral potential: This is the sum of the axial and square-planar potentials: $V_o = V_z + V_{xy}$. It emerges that the angular-dependent part of the cubic potential, apart from the A_{00} term, is minus eight to nine times the octahedral potential (the minus sign inverts the crystal-field energy levels).

Tetragonally distorted octahedral configuration: The potential for this arrangement is obtained by adding the potentials V_z and V_{xy} for the case when the two axial ions are at a different distance from the origin than the square-planar ones. Accordingly, $V = PV_z + QV_{xy}$, where the normalization of the dimensionless coefficients $P + Q = 2$. For a perfect octahedron, $P + Q = 1$. As for the tetragonally distorted potential, one has the polynomial in Cartesian coordinates of the form: $K(P-Q)(3z^2 - r^2) + x^4 + y^4 + z^4 - \frac{3}{5}r^4$, where K is a constant.

As for tetragonal distortion, $(P-Q)$ is positive for stretching, whereas $(P-Q)$ is negative for compression along the z-axis. The term $\pm\left(x^4 + y^4 + z^4 - \frac{3}{5}r^4\right)$ occurs in the expressions for the potential for cubic, tetrahedral, or octahedral symmetry, wherein the positive sign applies to octahedral symmetry, while the negative sign applies to cubic and tetrahedral cases.

The tetragonal distortion occurs along the Z-axis, and is therefore easy to handle.

For trigonal distortion, with the Z-axis directed along the threefold axis of the cube, the octahedral potential is given as follows:

$$V = \frac{6Ze}{R} - \frac{14\sqrt{\pi}}{9}\left(\frac{Ze}{R^5}\right)r^4\left(Z_{40} + 2\sqrt{\frac{5}{7}}Z_{43}^s\right). \tag{3.13}$$

Other special cases can be similarly treated.

3.2.6
Equivalent Operators and the Wigner–Eckart Theorem

The equivalent-operator method, which is used to transform the potentials to enable use of the angular momentum operators, is based on the *Wigner–Eckart theorem*, which relates the matrix element $\langle \alpha jm|T_k^q|\alpha' j'm'\rangle$ of the qth component of a kth-order irreducible tensor operator to a Clebsch–Gordan coefficient $\langle j'km'q| jm\rangle$ as follows:

$$\langle \alpha jm|T_k^q|\alpha' j'm'\rangle = \frac{(\alpha j\|T_k\|\alpha' j')}{\sqrt{2J+1}}\langle j'km'q| jm\rangle, \tag{3.14}$$

where $(\alpha j\|T_k\|\alpha' j')$ denotes the reduced matrix element, which is independent of m and the particular component of T. It is an intrinsic property of the individual tensor operator T_k. The components of the irreducible tensor operator T_k^q obey the following commutation relations with respect to the components of the total angular momentum operator J:

$$\left[J_\pm, T_k^q\right] = \sqrt{k(k+1) - q(q\pm 1)}T_k^{q\pm 1}; \left[J_z, T_k^q\right] = qT_k^q$$

This leads to the property that the qth component of an irreducible tensor operator transforms under rotation into a sum of one or more of the $2k + 1$ components of T_k^q: $P_R T_k^q P_R^{-1} = \sum_{q'=-k}^{k} T_k^{q'} D_{k(R)qq'}$, where P_k is a $(2k+1)\times(2k+1)$ general rotation matrix, and $D_{k(R)qq'}$ are coefficients characteristic of the particular rotation and k.

A detailed discussion of irreducible tensor operators will not be given here, as it is a vast topic. A list of spherical tensor operators sufficient for use in context with the spin Hamiltonian, expressed in terms of the angular-momentum operators $J_x, J_y, J_z, J_\pm = J_x \pm iJ_y$, which transform as the spherical harmonics Y_ℓ^m under rotation of coordinate axes is given in Chapter 7 of this book.

In order to calculate the matrix elements of the potential given by Equation 3.7, it is helpful to express it in terms of polynomials in (x,y,z), and then to convert it into equivalent angular-momentum operators. This is necessary in order to calculate the crystal-field energy levels, required for an optical spectrum, among others.

The evaluation of equivalent operators is facilitated by using *operator equivalents*, since the matrix elements of r, x, y, z are proportional to those of J, J_x, J_y, J_z, respectively, in terms of the eigenvectors of J_z, that is, $|J,M\rangle$, so that $\langle J_{M\pm 1}|J_\pm|J_M\rangle = \sqrt{(J \mp M)(J \pm M + 1)}$ and $\langle J_M|J_z|J_{M'}\rangle = M\delta_{MM'}$. Accordingly, one replaces $x \to CJ_x, y \to CJ_y, z \to CJ_z, z^2 \to C'J_z^2, r^2 \to C'J(J+1)$. For the other multiplicative terms, for example, one replaces xy by symmetrical combinations: $xy \to \dfrac{C'}{2}(J_xJ_y + J_yJ_x)$, where the factor ½ is the normalization constant, since the angular-momentum operators J_x, J_y, J_z do not commute with each other. In general, $\{AB\}_s = (1/2)(AB+BA)$, where $\{AB\}_s$ stands for the symmetrized product. Therefore, products such as $x^m y^n z^p$ are replaced by an average of the symmetrized product of $J_x^m J_y^n J_z^p$, where the operators J_x, J_y, J_z appear m, n, and p times. Further simplification of this average is obtained by using the commutation relations between angular momentum operators, such as

$$[J_x, J_y] = J_xJ_y - J_yJ_x = iJ_z (\hbar = 1).$$

The particular combinations of angular momentum operators so obtained are referred to as *equivalent operators* O_k^q. Abragam and Bleaney (1970) and Misra (1999) list the most commonly used equivalent operators; these are also listed in Chapter 7 of this book.

The crystal-field Hamiltonian due to a charge distribution is expressed in terms of operators O_k^q as:

$$H_{CF} = e\sum_j V(r_j, \theta_j, \varphi_j) = \sum_{k=0}^{\infty}\sum_{q=-k}^{+k} B_k^q O_k^q, \tag{3.15}$$

where the angular momentum involved is the orbital angular momentum, **L**. For symmetrical charge configurations, many coefficients B_k^q will be zero, depending on the symmetry. The higher the symmetry, the smaller is the number of nonzero coefficients. Furthermore, if there exists a center of inversion at the ionic site, no terms will occur with odd k in the above expansion. In addition, it should be noted that the number of terms required in the expansion is restricted by the electronic configuration of the particular transition metal ion interacting with the crystal field. Thus, the maximum required value of $k = 2l$, where l is the individual electron's orbital quantum number. One has, for d electrons $l = 2$, whereas for f electrons $l = 3$. It should be noted that it is not the total $L = \sum_i l_i$ that restricts the summation. As a consequence, a $3d^3$ configuration with ^4F ground state ($L = 3$) is characterized by the cutoff of $2L = 4$, and not restricted by the limit of $2L = 6$.

For typical symmetric charge configurations, one obtains

$$\mathbf{H}_{CF} = \sum_{k=2,4} \sum_{q=0}^{k} B_k^q O_k^q \; (d\text{-electrons}),$$

$$\mathbf{H}_{CF} = \sum_{k=2,4} \sum_{q=0}^{k} B_k^q O_k^q \; (f\text{-electrons}),$$

where the term $B_0^0 O_0^0$ is omitted, as it only gives a constant shift of energy.

Octahedral or cubic coordination:

$$\mathbf{H}_{CF} = B_4(O_4^0 + 5O_4^4) \text{ for } d \text{ electrons} \quad (3.16)$$

$$\mathbf{H}_{CF} = B_4(O_4^0 + 5O_4^4) + B_6(O_6^0 - 21O_6^4) \text{ for } f\text{-electrons,} \quad (3.17)$$

where the coefficients have been simplified by not using any superscripts in these special cases.

Predominantly cubic or octahedral with a tetragonal or trigonal distortion: For this case, one has

$$\mathbf{H}_{CF} = \sum_{k=2,4} \sum_{q=0}^{k} [B_{k\,cube}^q + B_{k\,D}^q] O_k^q \; (d\text{-electrons}),$$

where $B_{k\,cube}^q$ represents the cubic or octahedral terms, and $B_{k\,D}^q$ the distortion terms. The $B_{k\,cube}^q$ terms appear in Equations 3.16 and 3.17 as B_4 and B_6 instead of B_4^0 and B_6^0, respectively.

Tetragonal distortion: For this case, one has

$$\mathbf{H}_{CF} = B_4(O_4^0 + 5O_4^4) + B_{2\,D}^0 O_2^0 + B_{4\,D}^0 O_4^0 \; (d\text{-electrons}) \quad (3.18)$$

Predominantly cubic with trigonal distortion: Here, the potential is expressed as

$$\mathbf{H}_{CF} = -\frac{2}{3} B_4(O_4^0 + 20\sqrt{2}\, O_4^3) + B_{2\,D}^0 O_2^0 + B_{4\,D}^0 O_4^0 \; (d\text{-electrons}),$$

where the coordinate system described in context with Equation 3.10 is used.

As for f-electrons, additional terms with $k = 6$ are required, but are not considered here.

3.2.7
Properties of *d*-Electrons in Crystal Fields

3.2.7.1 Ions with Several *d*-Electrons: Strong- and Weak-Field Cases
The presence of more than one electron requires consideration of the electron–electron Coulomb interaction. In this context, one must consider the relative order

Table 3.1 Ground cubic crystal-field states for d-electrons in six-coordination.

Number of d-electrons	Weak field			Strong field		
	Configuration	Spin	Orbital degeneracy	Configuration	Spin	Orbital degeneracy
1	t_2^1	1/2	3×	t_2^1	1/2	3×
2	t_2^2	1	3×	t_2^2	1	3×
3	t_2^3	3/2	1×	t_2^3	3/2	1×
4	$t_2^3 e^1$	2	2×	t_2^4	1/2	3×
5	$t_2^3 e^2$	5/2	1×	t_2^5	0	3×
6	$t_2^4 e^2$	2	3×	t_2^6	1/2	1×
7	$t_2^5 e^2$	3/2	3×	$t_2^6 e^1$	1	2×
8	$t_2^6 e^2$	1	1×	$t_2^6 e^2$	1	1×
9	$t_2^6 e^3$	1/2	2×	$t_2^6 e^3$	1/2	2×

of λ (the spin–orbit coupling constant), Δ (the cubic crystal-field splitting parameter), and K (a parameter that represents the splittings due to electron–electron interactions). In what follows, an approximate description of the *weak-field* ($K \gg \Delta \gg \lambda$) and *strong-field* ($\Delta \gg K \gg \lambda$) cases will be considered. Although this treatment is simple, it possesses many of the features of more elaborate treatments (Orgel, 1960), and it also provides information on the important ground states in EPR. Some useful ideas are as follows. The spin function corresponding to the maximum possible spin for a configuration is an even function under interchanges of two particles. Then, the associated spatial function is odd under such interchanges, as the total wavefunction is antisymmetric under particle exchange. In that case, the electrons avoid each other, so that the higher the spin of a term, the lower is its energy. This applies well to the ground state, but not so well to the excited states. When speaking of the t_2- and e-orbitals into which d-orbitals are split in a cubic field, a six-coordinated ion will be lower in energy by Δ for each electron that can be removed from an e-orbital and put into a t_2-orbital. Thus, there are two competing effects to be considered when calculating the energy of the ground state. The results for ions with one to nine d-electrons in a six-coordinated environment are listed in Table 3.1. In the weak-field case ($\Delta \ll K$), the electrons are put into orbitals until maximum spin is allowed by Pauli's exclusion principle, after which they are put in the next lowest-energy orbital to maximize spin, as seen in Table 3.1. In contrast, for the strong-field case ($\Delta \gg K$), the electrons are placed in the lowest-energy orbitals; first, the spin is maximized consistent with this orbital placement (see Table 3.2). The indicated orbital degeneracy in Table 3.1 represents the number of equivalent ways of obtaining the configuration with the indicated spin. In Table 3.2 are listed the results for four-, eight-, and twelve-coordination, where the order of t_2- and e-orbitals is inverted. Because the electron and hole behaviors are similar, Tables 3.1 and 3.2 are obtained from each other by inverting their order; that is, $d^n \leftrightarrow d^{10-n}$. Many of the most

Table 3.2 Ground cubic crystal-field states for d-electrons in 4-, 8-, and 12-coordination.

Number of d-electrons	Weak Field			Strong field		
	Configuration	Spin	Orbital degeneracy	Configuration	Spin	Orbital degeneracy
1	e^1	1/2	2×	e^1	1/2	2×
2	e^2	1	1×	e^2	1	1×
3	$e^2 t_2^1$	3/2	3×	e^3	1/2	2×
4	$e^2 t_2^2$	2	3×	e^4	0	1×
5	$e^2 t_2^3$	5/2	1×	$e^4 t_2^1$	1/2	3×
6	$e^3 t_2^3$	2	2×	$e^4 t_2^2$	1	3×
7	$e^4 t_2^3$	3/2	1×	$e^4 t_2^3$	3/2	1×
8	$e^4 t_2^4$	1	3×	$e^4 t_2^4$	1	3×
9	$e^4 t_2^5$	1/2	3×	$e^4 t_2^5$	1/2	3×

useful qualitative features of the transition-group ions for EPR are found in these two tables. The ground states given in the tables in terms of crystal-field configurations are strictly correct only in the extreme strong-field limit. In *weak fields*, other crystal-field configurations arising out of the d^n free-ion configuration can be mixed in. Such a configuration mixture appears only for $t_2^2, t_2^5 e^2, e^2 t_2^1$, and $e^4 t_2^4$ among the configurations employed in a weak field; all ground configurations that occur only for strong fields can mix with any other of these crystal-field configurations.

In place of Tables 3.1 and 3.2, a more detailed – but merely qualitative – description is provided by schematic energy-level diagrams (correlation diagrams), such as that in Figure 3.4. Quantitative versions of Figure 3.4, which are referred to as *Tanabe–Sugano diagrams*, in which energy is plotted against the crystal-field parameter, Δ, have been calculated for certain values of the ratio, C/B, of the two Racah parameters specifying the interelectronic Coulomb interaction (Figgis, 1966). In generating diagrams such as Figure 3.4, and also Tanabe–Sugano diagrams, the first step is to calculate how the terms split in a cubic field. This is analogous to determining how the energy levels of a single d-electron split, noting that one d-electron produces a ^2D term. It emerges (without taking into account the spin) that the S and P terms do not split, whereas the D-term splits into a doublet and a triplet, and an F-term splits into two triplets and a singlet. The levels are ordered in such a way that the ground state always turns out to be that predicted from Tables 3.1 and 3.2 for weak fields.

So far, considerations have been made only for incomplete multielectron configurations in cubic symmetry. When a lower symmetry is considered, it is necessary to take into account more independent crystal-field parameters, which in turn makes the results more complicated. Frequently, departures from cubic field have been treated as perturbations, requiring the use of the solutions already obtained in the cubic-field case.

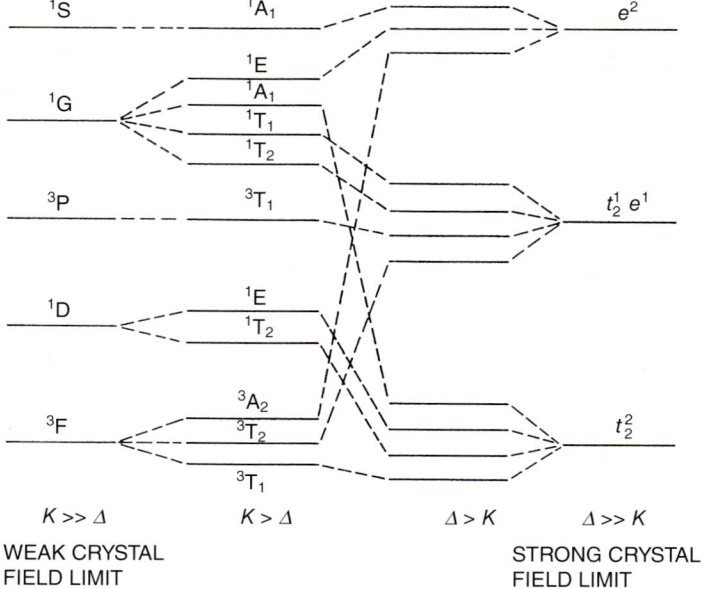

Figure 3.4 Energy level scheme for the configuration [Ar]$3d^2$ for various relative magnitudes of the term splitting K and the cubic crystal-field splitting (Δ) for six-coordination, neglecting the spin–orbit coupling. It should be noted that the states formed from the ground strong-field configuration with spin aligned is the lowest state throughout; hence, in Table 3.2 the $3d^2$ configuration is the same in both strong and weak field cases. If the symmetry were fully cubic (O_h), a subscript g would appear conventionally on all state designations, except those for terms, that is, $^1A_{1g}$, $^3T_{2g}$, e_g, etc. Adapted from Pake and Estle (1973).

The CFT can be applied to calculate the energies and wavefunctions, as well as the CF parameters, for d-electrons. The details of calculations can be found, for example, in Abragam and Bleaney (1970) and Poole and Farach (1972). Here, only the results will be described in an outline form as extracted from the latter book.

3.2.7.2 Energies and Wave-Functions for d-Electrons

The d^n electronic configurations are D and F states. As the spin–orbit constant for these ions (which belong to the first transition series) is much less than the CF splittings, this point is ignored in the following discussion, illustrating the D-state. One takes into account the 5 × 5 matrix for the CF Hamiltonian given by Equation 3.16 for the octahedral case. Its five eigenvalues turn out to be such that there is an orbital doublet (twofold degeneracy) with the energy $72B_4$ and an orbital triplet (threefold degeneracy) with the energy $-48B_4$. The corresponding wavefunctions are, respectively, $|2^s\rangle = \frac{1}{\sqrt{2}}[|2\rangle + |-2\rangle]$ and $|2^a\rangle = \frac{1}{\sqrt{2}}[|2\rangle - |-2\rangle]$, where the superscripts s and a denote the symmetric and anti-symmetric combina-

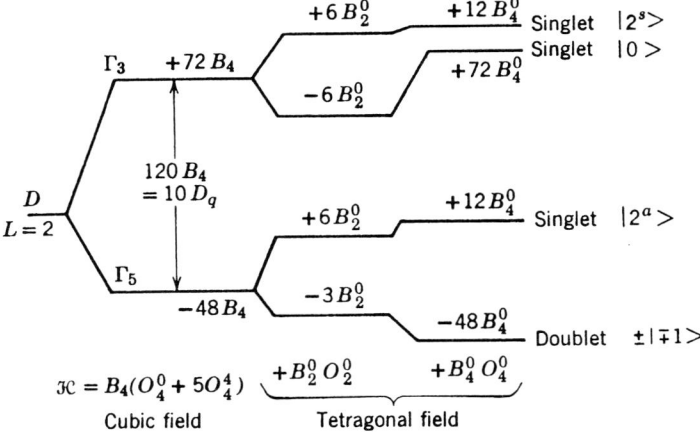

Figure 3.5 Splitting of a $3d^9$, D-state in cubic and tetragonal crystalline electric fields. The corresponding wavefunctions are shown on the right in $|M_L\rangle$ representation; the superscripts s and a indicate symmetric and antisymmetric combinations of $|\pm 2\rangle$ basis states. Adapted from Poole and Farach (1972.)

tions, respectively, of the angular-momentum states $|m_L = \pm 2\rangle$). The center of gravity of the original D level remains invariant. The separation between these two energies is $\Delta = 10\ Dq = 120\ B_4$ in the usual notation. For tetragonal distortion of the cube or octahedron, one takes into account Equation 3.18, finds its eigenvalues, and then adds them to those of Equation 3.16. In addition, the eigenfunctions of the sum of the two Hamiltonians, given by Equations 3.16 and 3.18, are calculated. The results for the eigenvalues, with the corresponding eigenfunctions in brackets, are listed as follows in decreasing order of energy: $72B_4 + 6B_2^0 + 12B_4^0$ (singlet $|2^s\rangle$); $72B_4 - 6B_2^0 + 72B_4^0$ (singlet $|0\rangle$); $-48B_4 + 6B_2^0 + 12B_4^0$ (singlet $|2^a\rangle$); $-48B_4 - 3B_2^0 - 48B_4^0$ (doublet $\pm|\mp 1\rangle$). These are shown in Figure 3.5, while Figures 3.6–3.8 illustrate, respectively, the results for trigonal distortion for a D-state, tetragonal distortion for an F-state, and trigonal distortion for an F-state.

3.2.7.3 Crystal-Field Parameters for d-Electrons

The energy levels produced under the action of crystalline electric field should be calculated from the CF potential, expressed by Equation 3.7. In order to accomplish this, one needs to compare the two expressions for the Hamiltonians, given by Equations 3.7 and 3.15. Setting now the matrix elements of these Hamiltonians equal to each other, one obtains:

$$eA_{kq}\left\langle Lm_L \left| \sum_j r_j^k Z_{kq}(\theta_j, j_j) \right| Lm_L' \right\rangle = B_k^q \left\langle Lm_L \left| O_k^q \right| Lm_L' \right\rangle,$$

where the summation is over all of the electrons. The summation over the ligands, as indicated in Equations 3.8–3.10, is already contained in A_{kq}. One obtains the identity:

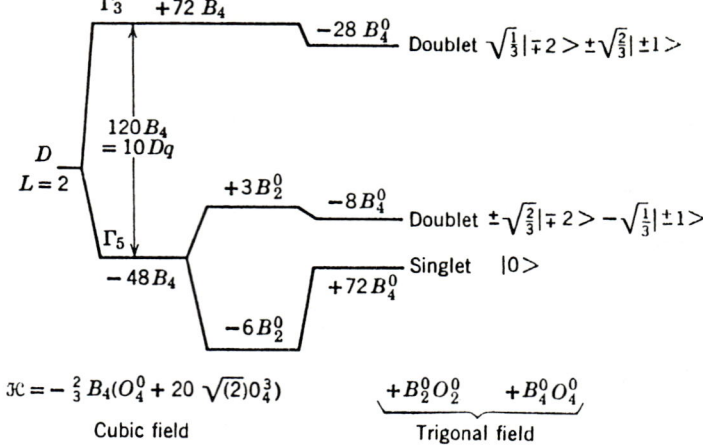

Figure 3.6 Splitting of a $3d^9$, D-state in cubic and trigonal crystalline electric fields. The notation of Figure 3.5 has been used. Adapted from Poole and Farach (1972).

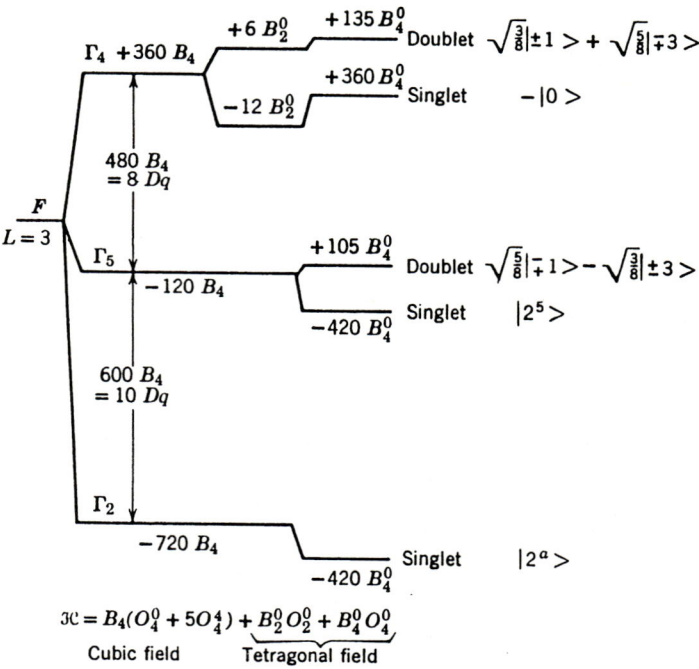

Figure 3.7 Splitting of an F-state in cubic and tetragonal crystalline electric fields. The notation of Figure 3.5 has been used. Adapted from Poole and Farach (1972).

3.2 Crystal-Field Theory

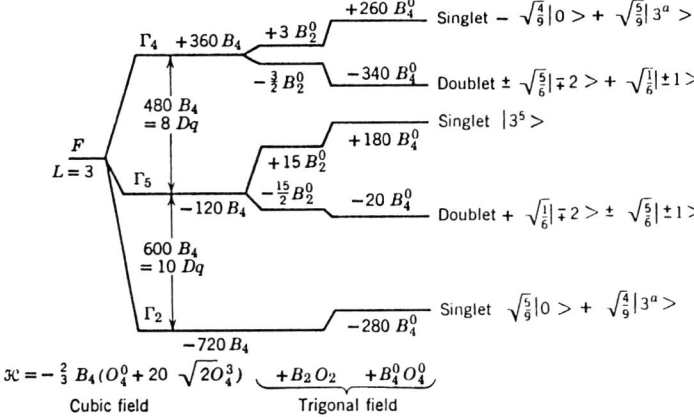

Figure 3.8 Splitting of an F-state in cubic and trigonal crystalline electric fields. The notation of Figure 3.5 has been used. Adapted from Poole and Farach (1972).

$$B_k^q = e a_k N_{kq} A_{kq} \langle r^k \rangle, \tag{3.19}$$

where the Stevens multiplicative coefficients are defined by Tinkham (1964):

$$a_k = \frac{1}{N_{ka} \langle r^k \rangle} \left[\frac{\langle LM_L | \sum_i r_i^k Z_{kq} | LM_L' \rangle}{\langle LM_L | O_k^q | LM_L' \rangle} \right], \tag{3.20}$$

and the coefficient N_{kq} is the normalization constant for the corresponding Tesseral harmonic, for example, $N_{40} = 3/(16\sqrt{\pi})$ and $N_{60} = (13/\pi)^{1/2}/32$. It is noted that the ratio in square brackets in Equation 3.20 involves the integration of a polynomial in Cartesian coordinates, and another integration of angular-momentum operators. This ratio is independent of q in the light of Wigner–Eckart theorem, expressed by Equation 3.14.

3.2.7.4 Crystal-Field Splittings for $3d^1$ and $3d^9$ Configurations

Using the details described above, the crystal field splittings for $3d^n$ ions can be calculated. In particular, for $3d^1$ and $3d^9$ configurations, with constant ligand distance R, the energies of the orbital t_{2g} relative to the orbital e_g in the various symmetries are as follows: $E(t_{2g}) > E(e_g)$ for $3d^1$ in cubic and tetrahedral symmetries and for $3d^9$ in octahedral symmetry; whereas $E(e_g) > E(t_{2g})$ for $3d^1$ in octahedral symmetry and for $3d^9$ in cubic and tetrahedral symmetries. It is noted that the levels become inverted for the same ion in going from an octahedral to a cubic or tetrahedral field. Furthermore, the $3d^1$ and $3d^9$ configurations are inverted with respect to each other, since $3d^9$ configuration may be treated as a $3d^1$ positive hole in a filled ($3d^{10}$) shell. As for the F-state energy levels, they also invert under the same conditions.

3.2.7.5 The Ground State and its Relationship to EPR: Quenching of Orbital Angular Momentum and Calculation of g-Factors

Consider an unfilled shell, the Hamiltonian of which is expressed above by Equation 3.1. By including the Zeeman interaction, $H_{Ze} = \mu_B(\mathbf{L} + g_e\mathbf{S}) \cdot \mathbf{B}$, one can determine the energy levels arising out of the ground state. As an illustration, the case of a $3d^1$ configuration in addition to the closed shells is considered in tetragonal symmetry, with $\lambda < \Delta, \delta$. Here, $L = 2$, and the energy levels resulting from the ground state are shown in the following figure, where Δ and δ are the energies of the levels which are mixed into the ground state.

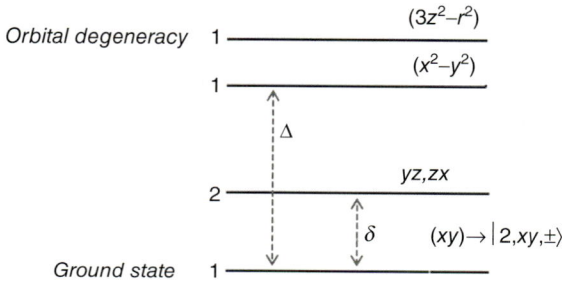

The effect of the spin–orbit coupling, $\lambda \mathbf{L} \cdot \mathbf{S}$, is now considered. Noting that

$$\lambda \mathbf{L} \cdot \mathbf{S} = \lambda L_z S_z + \frac{1}{2}\lambda L_x(S_+ + S_-) + \frac{1}{2i}\lambda L_y(S_+ - S_-);$$
$$S_\pm = S_x \pm iS_y;$$
(3.21)

The ground state is a spin doublet, characterized by twofold spin degeneracy. It is expressed approximately as $|2, xy, \pm\rangle$ (as discussed next); here, 2 represents the spin degeneracy, xy the orbital wavefunction, and $\pm$ the spin projections.

Calculation of the Effect of First-Order Perturbation Due to $\lambda \mathbf{L} \cdot \mathbf{S}$ One has for the eigenfunctions calculated by first-order perturbation:

$$|\pm\rangle = |2, xy, \pm\rangle + \sum_{i,\alpha} \frac{\langle 2, i, \alpha | \lambda \mathbf{L} \cdot \mathbf{S} | 2, xy, \pm\rangle}{E_{xy} - E_i} |2, i, \alpha\rangle,$$

where the second term is summed over the spin states α and i, the excited orbital states. Now, by using Equation 3.21 and the matrix elements of the various components of the angular momentum operator between the d-orbitals, listed in Pake and Estle (1973), one obtains:

$$|\pm\rangle = |2, xy, \pm\rangle \pm \frac{i\lambda}{\Delta}|2, x^2-y^2, \pm\rangle - \frac{i\lambda}{2\delta}|2, zx, \mp\rangle \mp \frac{\lambda}{2\delta}|2, yz, \mp\rangle$$
(3.22)

It is seen that the matrix elements of the components of the orbital angular momentum between the zero-order wavefunction of the ground state are zero:

$\langle 2, xy, \pm | L | 2, xy, \pm \rangle = 0$. This is known as *quenching of the orbital angular momentum*, and is a common occurrence among the ions of the iron transition group. However, not all the matrix elements of L and S between the perturbed levels $|\pm\rangle$ are zero. The nonzero matrix elements, as calculated using Equation 3.22, are:

$$\langle \pm | L_z | \pm \rangle = \mp \frac{4\lambda}{\Delta}; \langle \pm | L_x | \pm \rangle = \mp \frac{\lambda}{\Delta}; \langle \pm | L_y | \pm \rangle = \mp i \frac{\lambda}{z};$$
$$\langle \pm | S_z | \pm \rangle = \pm \frac{1}{2}; \langle \pm | S_x | \pm \rangle = \frac{1}{2}; \langle \pm | S_y | \pm \rangle = \mp i \frac{1}{2}. \quad (3.23)$$

All other matrix elements of L and S are equal to zero, such that now, $H_{Ze} = \mu_B (L + g_e S) \cdot B$. Using Equation 3.23, one obtains for the matrix elements of the Zeeman term as

$$\langle \pm | H_{Ze} | \pm \rangle = \left(\pm \frac{1}{2} g_e \mp 4 \frac{\lambda}{\Delta} \right) \mu_B B_z;$$

$$\langle \pm | H_{Ze} | \mp \rangle = \left(\frac{1}{2} g_e - \frac{\lambda}{\delta} \right) \mu_B B_x \pm i \left(\frac{1}{2} g_e - \frac{\lambda}{\delta} \right) \mu_B B_y$$

Now constructing the 2×2 matrix for H_{Ze} between the states $|\pm\rangle$, and solving the secular determinant $|H_{Ze} - E\mathbf{1}| = 0$, where $\mathbf{1}$ is the unit matrix, one obtains

$$E = \pm \frac{1}{2} \left(\sqrt{\left[g_e - 8\left(\frac{\lambda}{\Delta}\right) \right]^2 \cos^2 \theta + \left[g_e - 2\left(\frac{\lambda}{\delta}\right) \right]^2 \sin^2 \theta} \right) \mu_B B \equiv \pm \frac{1}{2} g \mu_B B, \quad (3.24)$$

where θ is the angle between B and the tetragonal z axis. It is noted that g, as defined by Equation 3.24, $\neq g_e$, and is anisotropic, due to a small contribution by the orbital angular-momentum in perturbation. Consequently, the parallel and perpendicular values of the g-factor are:

$$g_\parallel = g_e - 8\frac{\lambda}{\Delta} \text{ for } \theta = 0°; \quad g_\perp = g_e - 2\frac{\lambda}{\Delta} \text{ for } \theta = 90°,$$

so that $g_\parallel, g_\perp < 2.0023$ (the free electron value, g_e), since $\lambda > 0$. For $\lambda < 0$, which is the case when the electronic shell is more than half full, one obtains $g_\parallel, g_\perp > 2.0023$ (the free electron value).

The g-value of a transition-metal ion differs from the g-factor of the free electron due to the admixture of the higher lying crystal-field states into the ground state, which depends directly on the spin–orbit coupling constant and inversely on the crystal field splitting, Δ. An illustrative example, the $3d^9$ Cu^{2+} ion in a tetragonal crystalline field, is now considered, for which calculation shows (Polder, 1942):

$$g_\parallel = 2 - \frac{8\lambda}{E_3 - E_1}; \quad g_\perp = 2 - \frac{8\lambda}{E_4 - E_1},$$

where E_1, E_2, E_3 and E_4 are the four levels into which the crystal field splits the $3d^9$, ^2D configuration, E_1 being the ground state of the group. For Cu^{2+}, $\lambda = -852 \text{ cm}^{-1}$. Since the energy denominators in the above equations are an order of magnitude higher in energy than λ, the g-factors for the Cu^{2+} ion lie in the range 2.15–2.4.

Similar formulas for g-factors for other transition metal ions are listed by Misra (1999); and also in Chapter 7 of this book.

3.2.8
The Rare-Earth Ions

The rare-earth ions, with $4f$ electrons, should be treated using an approach that is different from that used for $3d$ ions, as described above. Here, the crystal field interaction is a perturbation to the free-ion interactions, which in turn results in Russell–Saunders coupling. In particular, for these ions, $K \gg \lambda \gg \Delta$, which should be taken into consideration for successive approximations.

As an illustration, a simple example will be considered, that of a $4f^1$ configuration, for example, that of the Ce^{3+} ion, in a cubic environment at a six-coordinated site. Coupling of the orbital angular momentum ($L = 3$) and the spin angular momentum ($S = ½$) by the spin–orbit interaction results in two states for this ion with the total angular momentum $J = 5/2$ and $J = 7/2$, wherein the $J = 5/2$ state lies lower in energy by $2000\,cm^{-1}$. The cubic crystal field partially lifts the sixfold degeneracy of this state, splitting it into two states. Of these, the excited state possesses fourfold degeneracy, whereas the lower one is twofold degenerate. Using the notation $|J, M\rangle$ for basis wavefunctions, the lower state is described by the eigenvectors (Lea, Leask, and Wolf, 1962),:

$$\psi_\pm = \frac{1}{\sqrt{6}}\left[|5/2, \pm 5/2\rangle - \sqrt{5/6}|5/2, \mp 3/2\rangle\right] \tag{3.25}$$

Using now the expression for the g-factor:

$$g_L = 1 + \frac{J(J+1) + S(S+1) - L(L+1)}{2J(J+1)}(g_e - 1),$$

and the fact that the magnetic moment of the ion is $\mu = -g\mu_B J$, it emerges that the Hamiltonian of this lower state with $J = 5/2$ in the magnetic field is:

$$H = \frac{6}{7}\mu_B \mathbf{B} \cdot \mathbf{J}.$$

The matrix elements of this Hamiltonian between the states given by Equation 3.25, considering the magnetic field to be oriented along a cubic axis, taken to be the z-axis, are as follows:

$$\langle \psi_\pm | H | \psi_\pm \rangle = \mp \frac{5}{7}\mu_B B.$$

All other matrix elements are zero. The above equation implies, taking into account the fact that the energies of the sublevels given by $g\mu_B BM$, with $M = \pm 1/2$ in the present case, that the energy levels in this case are the same as that for a two-level system with an effective spin of ½, characterized by an isotropic g-value of $10/7$. In other words, the ground state of the $4f^1$ ion under consideration has a g-value of $10/7 = 1.43$. Effectively, it is equivalent to a spin-only angular momentum of ½.

3.2 Crystal-Field Theory

3.2.8.1 Crystal Fields for Rare-Earth Ions: Dominant Spin–Orbit Coupling

In this case, the spin–orbit coupling is much greater than the CF splitting. The calculations will be illustrated here again for the $4f^1$ Kramers ion, Ce^{3+}, considered above. The spin Hamiltonian will first be constructed for the smaller CF terms, and thereafter it will be combined with the spin–orbit coupling interaction; it will then be solved after making certain approximations. The simplified CF Hamiltonian (Abragam and Bleaney, 1970) is:

$$H_{CF} = I_2 \times \left[B_2 O_{2(L)}^0 + B_4 O_{4(L)}^0 + B_6 O_{6(L)}^0 \right], \quad (3.26)$$

where I_2 represents the 2×2 unit matrix ($S = \frac{1}{2}$). Using the 7×7 matrices for the spin operators ($L = 3$) in the above equation (as given in Chapter 7), the matrix representing the Hamiltonian given by Equation 3.26 can be constructed. The matrix for the Hamiltonian describing the spin–orbit interaction $H_{SO} = \lambda \left(S_x L_x + S_y L_y + S_z L_z \right)$ can be likewise constructed. The resulting sum of the two Hamiltonians, $H_{CF} + H_{SO}$, is now expressed in the $|m_L m_S\rangle$ representation, after which the matrix elements are regrouped into submatrices of the type $|m_L m_S = 1/2\rangle$ and $|m_L m_S = -1/2\rangle$. If the term $B_6 O_{6(L)}^0$ is neglected, which is rather small, since $B_6 \ll |\lambda|$, then the resulting matrix so obtained will be of the form $\begin{pmatrix} H' & 0 \\ 0 & H''(=H') \end{pmatrix}$. The matrix for H' can now be diagonalized using a matrix, C, consisting of Clebsch–Gordan coefficients: $H'_{diag} = C^{-1} H' C$, to transform to the coupled representation spanned by the kets $|JM_J\rangle$, where $JM_J = 7/2\ 7/2;\ 7/2\ 5/2;\ 5/2\ 5/2;\ 7/2\ 3/2;\ 5/2\ 3/2;\ 7/2\ 1/2;\ 5/2\ 1/2$; which turns out to be a diagonal matrix, with the diagonal elements corresponding to these eigenkets being $\frac{3}{2}\lambda, \frac{3}{2}\lambda, -2\lambda, \frac{3}{2}\lambda, -2\lambda, \frac{3}{2}\lambda, -2\lambda$, respectively.

3.2.9 Irreducible Representations for CF Energy Levels

For the orbital electrons in octahedral, cubic, or tetrahedral environments the energy levels are described by five symmetry types called *irreducible representations*. These are listed in the following table.

Value of J	Multiplicity	Notation (1)	Notation (2)
Integral	Singlet	A_1	Γ_1
Integral	Singlet	A_2	Γ_2
Integral	Doublet	E	Γ_3
Integral	Triplet	T_1	Γ_4
Integral	Triplet	T_2	Γ_5
Half-integral	Doublet	E'	Γ_6
Half-integral	Doublet	E''	Γ_7
Half-integral	Quartet	U	Γ_8

The notation often used for d-electrons is A_{1g}, T_{2g}, Γ_{2g}, etc., where g stands for *gerade*, meaning that the wavefunction is symmetric under inversion. The wavefunctions that change sign under inversion are denoted by u (*ungerade*). Levels with half-integral angular momenta are split by octahedral, cubic, and tetrahedral crystal fields into combinations of irreducible representations which are doublets (Γ_6 and Γ_7) and quartets (Γ_8). This splitting of levels is easily calculated by the application of group theory. Integral spins have five irreducible representations in tetragonal environments: four singlets (A_1, A_2, B_1, B_2) and one doublet (E), whereas half-integral spins have two doublet (E', E'') representations. Trigonal symmetry is characterized by two singlet (A_1, A_2) and one doublet (E) representations for integral spin, whereas two doublets (E', E'') are present for half-integral J. Poole and Farach (1972) list the irreducible representations for spectroscopic states J in cubic (O_h), tetrahedral (T_d), octahedral (O_h), tetragonal (D_4), and trigonal (D_3) crystalline electric fields.

3.2.10
Critique of Crystal-Field Theory

Although CFT is successful in writing down the crystal potentials of the correct symmetry, it does have some drawbacks (Pilbrow, 1990). These can be summarized as follows:

- Covalency is not taken into account specifically, except for considering its effect in an empirical manner by using reduced values of the Racah B and C parameters and those of the spin–orbit coupling constants.

- CFT is based only on Coulomb repulsion between the d-electrons of the metal ion and ligand electrons.

- There is no provision here to take into account the effect of changing just one ligand, as it is a global model.

3.2.11
Kramers' Theorem

Due to the time-reversal invariance of the spin Hamiltonian in the absence of an external magnetic field, there exist certain degeneracies. As a result, Kramers' theorem states that there must exist even degeneracies for all systems with an odd number of electrons in zero magnetic field. In other words, no nondegenerate states can occur for an odd-electron system in the absence of a magnetic field. The proof of this theorem can be found, for example, in Pake and Estle (1973). It is noted that the two wavefunctions that are degenerate for an odd-electron system are connected to each other by the time-reversal operator. The Kramers' conjugate functions are obtained, in general, by reversing the signs of all M values, and taking the complex conjugate of the coefficients, for example, the two eigenvectors given by Equation 3.25.

3.3
Superposition Model (SPM)

This method was originally derived from the equations of the PCM developed in context with the CFT. Although the equations of SPM imply nothing about the PCM, it bears a certain resemblance to the angular overlap model. Originally, it was described by Bradbury and Newman (1968) in a context with rare-earth ions; subsequently, an application was made to Cr^{3+} by Stedman (1969). The objectives of the SPM are to: (i) provide crystal-field parameters for new systems; (ii) determine the signs of these parameters; (iii) predict the effects of local strains and distortions; and (iv) provide a connection between experiment and *ab initio* calculations. A review on SPM was produced by Newman and Urban (1975); the details of SPM in this chapter are based on that, as described by Pilbrow (1990).

The procedure for SPM calculations is as follows. The individual contributions to the crystal field from each of the surrounding ions are taken into account individually in order to correlate experimental crystal-field parameters on different systems. Inter-atomic interactions are neglected here, and only single-ion contributions are taken into account. There are two important features of SPM:

- The dominant contribution is not electrostatic, but rather due to covalency and overlap.
- The coordinated ions, or ligands, provide the dominant contribution to the crystal field. An assumption is made that the total crystal-field is made up of a sum of axially symmetric contributions from the ligands.

Basically, in the SPM model a single ligand at (R, θ, ϕ) relative to a coordinate system centered on the transition-metal ion makes the contributions to the crystal field of the following type:

$$K_k^q(\theta, \phi)\overline{B}_k(R),$$

where $\overline{B}_k(R)$ is the intrinsic parameter, representing the axial crystal field due to a single ligand distant R from the center of the magnetic ion, the precise dependence of which on R is not known previously. As for the correlation factors, $K_k^q(\theta, \phi); q \geq 0$, they are explicit functions of the angular position of the ligand, as listed below. The corresponding factors for $q < 0$ are obtained by substituting $\sin \theta$ for $\cos \theta$.

Angular correction factors for the superposition model as listed by Newman and Urban (1975)

$$K_2^0 = (3\cos^2\theta - 1)/2$$
$$K_2^1 = 3\sin 2\theta \cos\phi$$
$$K_2^2 = (3\sin^2\theta \cos 2\phi)/2$$
$$K_4^0 = (35\cos^4\theta - 30\sin^2\theta + 3)/8$$
$$K_4^2 = \frac{5}{2}(7\cos^2\theta - 1)\sin^2\theta \cos 2\phi$$
$$K_4^3 = 35\cos\theta \sin^3\theta \cos 3\phi$$
$$K_4^4 = \frac{35}{8}\sin^4\theta \cos 4\phi$$

The observed crystal-field parameters are written in the SPM as a simple sum over the individual ligands i:

$$\tilde{B}_k^q \langle r^k \rangle = \sum_i K_k^q(\theta_i, \phi_i) \bar{B}_k(R_i).$$

In the presence of several ligands at the same distance, R_r, the above sum can be simplified to:

$$\tilde{B}_k^q \langle r^k \rangle = \sum_r \bar{B}_k(R_r) \sum_{j(r)} K_k^q(\theta_{j(r)}, \phi_{j(r)}),$$

where $j(r)$ labels all ligands at a distance R_r. By grouping together the coordination factors:

$$K_q^r(r) = \sum_{j(r)} K_k^q(\theta_{j(r)}, \phi_{j(r)}),$$

one obtains a simple form:

$$B_k^q \langle r^k \rangle = \sum_r \bar{B}_k(R_r) K_k^q(r).$$

In order to determine the values of K_q^r and R_r, X-ray data are required. By using these values and the experimental values of the $B_k^q \langle r^k \rangle$, as determined from optical spectra, it is possible to determine the magnitudes and signs of the various $\bar{B}_k(R_r)$ values. The distance dependence of the parameters is usually assumed to be:

$$\bar{B}_q(R) = \bar{B}_q(R_0)(R_0/R)^{t_q},$$

over a limited range of R values, where t_q is an empirical parameter.

A modified form of the superposition model to apply to the case of determination of fine-structure terms in the spin Hamiltonian of S-state ions was given by Newman and Urban (1975), because some of the mechanisms that give rise to the splitting in the iron-group ion spin-Hamiltonian parameters do not satisfy the required conditions. The essence of this approach is that the contribution due to each neighboring ligand is parameterized. The following criteria in context with $L = 0$, or S-states, is delineated, taking into account the theorem that restricts the number of terms in the spin Hamiltonian due to time-reversal invariance as being fundamental (see Section 3.6.3). A spin Hamiltonian is then constructed that is invariant with respect to the symmetry group of the open-shell ion. Thereafter, the terms which represent all the physical processes applicable to the system in question must be included in the spin Hamiltonian. Finally, the spin Hamiltonian should be constructed with as few free parameters as possible; in particular, the number of free parameters should be less than the independent data points available experimentally. Among the reports which have considered the S-state ions in context with the SPM are included: (i) Murrietta et al. (1979) for Fe^{3+} in fluoroperovskite compounds ($KMgF_3$); (ii) Seth, Yadav, and Bansal (1985) for Mn^{2+} in double nitrate crystals; (iii) Newman (1982) for Cr^{3+} in ruby; (iv) Clare and Devine (1984), who used the modified SPM to determine spin–strain coupling tensors for Cr^{3+} in ruby; (v) and Elbers, Remme, and Lehman (1986), who determined the spin Hamiltonian for Cr^{3+} in tris(acetylacetonato) gallium (III) single crystals.

A few examples of the application of SPM worthy of mention include: (i) Co^{2+} in sites of trigonal symmetry in $CdCl_2$, $CdBr_2$, and K_4CdCl_4 by Edgar (1976); (ii) Fe^{2+} in pyrope-almandine garnets by Newman, Price, and Runciment (1978), which showed a reasonable agreement; (iii) Fe^{2+} in quasi-tetrahedral surroundings in quartz by Mombourquette, Tenant, and Weil (1986); and (iv) Gd^{3+} in rare-earth trifluorides by Misra, Mikolajczak, and Lewis (1981).

3.4
Molecular Orbital (MO) Approach

The history of ligand-field theory has been summarized by Gerloch (1983), along with the empirical results for the spectrochemical series, as well as the basis of simple MO descriptions of electronic structure for metal ions in ligand fields. MO models are based on the fact that the magnetic and electronic properties of transition-metal ions complexes, including the d–d optical transitions, are due to electrons in antibonding orbitals formed from metal–ligand combinations.

3.4.1
Linear Combination of Atomic Orbitals (LCAO)

As a result of interactions with the atoms and ions near the transition-group ion, its electronic states are affected, so that its electrons in unfilled shells occupy molecular, rather than atomic, orbitals (Ballhausen and Gray, 1965). These molecular orbitals are spread over a molecular complex, which consists of its nearest neighbors (known as ligands), and possibly other close neighbors. When the overlap of atoms is not significant, the wavefunction near any atom will appear atomic. As a reasonable approximation, the molecular orbital may be written as a linear combination of atomic orbitals (LCAO), which are centered on the various atoms, that constitute the complex. These provide the basis for a plausible discussion of the electronic properties of ionic to moderately covalently bound transition ions. Although such a description may appear to be quite different from that of crystal-field theory, the two approaches can be considered to be equivalent in every respect (Griffith, 1964).

The LCAO method can be illustrated by considering a p-electron on an ion midway between two other ions, with the assumption that only one s-orbital is present on each of the side ions and the three p-orbitals on the central ion are significant. Figure 3.9 illustrates the geometry of the atoms and the angular dependences. The molecular orbitals are assumed to be of the form

$$\psi = a\psi_{s1} + b\psi_{s2} + \sum_{i=x,y,z} c_i \psi_{pi}.$$

The next step is to solve the eigenvalues problem using the basis of $\psi_{s1}, \psi_{s2}, \{\psi_{pi}\}$ states, with respect to the Hamiltonian, which is the analog of the central potential

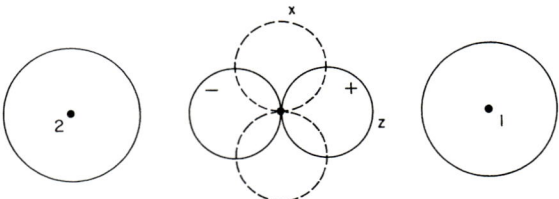

Figure 3.9 Configuration used to describe the molecular orbital treatment of the splitting of p-orbitals resulting from two negative ions along the positive and the negative z-axis. It should be noted that the p_z function can mix the two s-orbitals, but the nodal plane of p_x or p_y blocks any such mixing. Adapted from Pake and Estle (1973).

in the atomic case. The Hamiltonian under consideration excludes the electron–electron interaction, except as an effective potential, as well as the spin–orbit interaction. In addition, this Hamiltonian is characterized by axial symmetry as dictated by the geometry of the configuration, which implies that it can only have nonzero matrix elements if the product of the two basis functions is also axial. This is only possible when one uses the two basis functions $\frac{\psi_{s1} - \psi_{s2}}{\sqrt{2}}$ and ψ_{pz} out of all the possible functions. Therefore, only a 2 × 2 Hamiltonian matrix needs to be solved to calculate the energies and resulting wavefunctions of LCAO states. To determine the eigenvalues, E, taking into account the matrix elements of H between the two states, one has the secular determinant:

$$\begin{vmatrix} H_{11} - E & \sqrt{2}H_{1z} \\ \sqrt{2}H_{1z} & H_{zz} - E \end{vmatrix} = 0, \tag{3.27}$$

where

$$H_{11} \equiv \langle \psi_{s1}|H|\psi_{s1}\rangle = \langle \psi_{s2}|H|\psi_{s2}\rangle,$$
$$H_{1z} \equiv \langle \psi_{s1}|H|\psi_{pz}\rangle = \langle \psi_{s2}|H|\psi_{pz}\rangle,$$
$$H_{zz} \equiv \langle \psi_{pz}|H|\psi_{pz}\rangle.$$

The solutions of Equation 3.27 for the eigenvalues are:

$$E = \frac{1}{2}(H_{zz} + H_{11}) \pm \frac{1}{2}\left[(H_{zz} - H_{11})^2 + 8H_{1z}^2\right]^{1/2}, \tag{3.28}$$

with the eigenfunctions

$$\psi_i = \frac{a_i}{\sqrt{2}}(\psi_{s1} - \psi_{s2}) + c_i \psi_{pz}; i = 1, 2,$$

The corresponding ratios, c_i/a_i ($i = 1, 2$), of the coefficients are:

$$\frac{c_i}{a_i} = -\frac{H_{11} - E_i}{\sqrt{2}H_{1z}}; i = 1, 2. \tag{3.29}$$

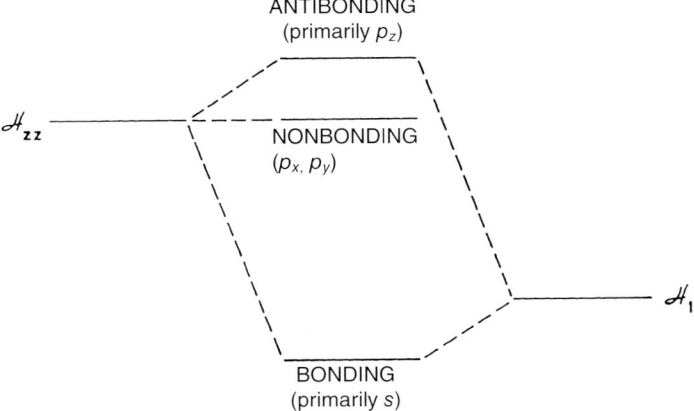

Figure 3.10 Splitting of the p-orbitals as viewed in the molecular-orbital treatment. It should be noted that the p_z orbital lies above the p_x and p_y orbitals, because it is an antibonding orbital. Adapted from Pake and Estle (1973).

In reality, $H_{zz} > H_{11}$, since H_{zz} is approximately the central ion p-orbital energy in the free state, whereas H_{11} is approximately the s-electron energy for the outer atoms. According to Equation 3.28, the molecular energy levels are somewhat more than half the atomic energy difference above and below the average atomic energy. The splitting of the p-orbitals in the molecular orbital approximation is shown in Figure 3.10. For the higher eigenvalue with the positive sign in Equation 3.28, $|c_i/a_i| \gg 1$, and $c_i/a_i < 0$, as seen from Equation 3.29, since $H_{1z} \ll H_{11} - E$. As a consequence, the wavefunction is primarily ψ_{pz} with nodes between its lobes and the two s functions, and is primarily an *anti-bonding* orbital, because its energy is above the average atomic energy (Wertz and Bolton, 1972, p. 95). The orbital with the lower energy, which is below the average atomic energy, is a *bonding* orbital, with the corresponding wavefunction being a primarily s wavefunction. Further, the p_x and p_y orbitals have nearly the same energy as that in the free ion; they are *nonbonding* orbitals. The anti-bonding character of p_z causes it to have a higher energy than the p_x and p_y orbitals, as well as to have an admixture of neighboring s-orbitals. A similar result was obtained earlier in this chapter (see Section 3.2.7), using the crystal-field theory.

The MO method can be applied to an ion with a single d-electron surrounded by six atoms (ligands) along the cubic axes. Considering only bonding with tightly held s-electrons on the six neighbors, the two e-orbitals will form bonding and antibonding orbitals with the s-orbitals, as shown in Figure 3.11. On the other hand, the nodal planes of the t_2-orbitals will prevent bonding, resulting only in nonbonding orbitals (Figure 3.11). The electrons from the neighbor fill the bonding orbitals, whereas the d-electrons are accommodated in the nonbonding t_2- and antibonding e-orbitals of higher energy, so that one obtains the same result as that obtained using the crystal-field theory for six-coordination.

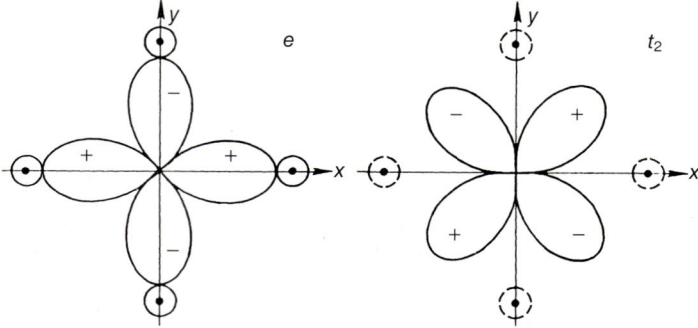

Figure 3.11 Plots of e and t_2 orbitals for the geometry used for the molecular orbital treatment of an ion in a cubic and six-coordinated environment. This shows that the e-orbital can mix the ligand s-orbitals, whereas the t_2 orbitals cannot. As a consequence, an antibonding, primarily e-orbital lies above the nonbonding t_2-orbital in energy. Adapted from Pake and Estle (1973).

3.4.2
Extended Hückel Molecular Orbital Theory (EHMO)

The extended EHMO, as outlined by Keijzers, DeVries, and Van Der Avoird (1972), was applicable to transition-metal ions in a ligand field. This involves two adjustable parameters: (i) the Wolfsberg–Helmholz parameter; and (ii) the charge dependence of the Hamiltonian matrix elements (Keijzers, DeVries, and Van Der Avoird, 1972; Keijzers and De Boer, 1972). The basic concepts were as follows. A suitable effective one-electron Hamiltonian is used in a nonorthogonal basis of atomic orbitals, denoted as ϕ_i. One needs to solve the secular equations:

$$\sum_j (H_{ij} - ES_{ij})C_{ij} = 0;\; H_{ij} = \langle j_i | H_{\mathit{eff}} | j_j \rangle,\; S_{ij} = \langle j_i | j_j \rangle. \tag{3.30}$$

The orbital energies and the LCAO coefficients, C_{ij}, are obtained as solutions of Equation 3.30.

The extended EHMO was later exploited by Keijzers and De Boer (1972; 1975a, 1975b) in calculations of spin-Hamiltonian parameters.

3.4.3
Ligand Field Theory: The Angular Overlap Model (AOM)

This theory is especially aimed to take into account covalency. The most notable of these methods, as outlined here, is the angular overlap model (AOM), which is conceptually similar to the SPM (Steenkamp, 1984), wherein the contributions from each ligand are simply added after transforming them from the various ligands' frames of reference. The AOM, as originally developed by Schaffer (1968), is a reasonable numerical method, which takes into account the electronic properties of transition-metal complexes as exhibited in their physical and chemical properties. It has also been exploited from a different perspective by Gerloch (1983). In this theory, d-electron orbitals, acted on by an effective potential at each ligand, are used; each ligand (or a lone electron where appropriate) is associated with a volume cell. In the AOM, low symmetries are as easy to handle as higher symmetries, and consequently it not only demonstrates an enormous advantage over other methods, but is also less restrictive than the SPM model. With the AOM, it is very easy to introduce the chemical properties of individual ligands, such as O^{2-}, CN^-, SCN^-, and F^-. According to its name, the AOM is based on the idea that when ligand and metal orbitals are in any arbitrary relationship where overlap occurs, a suitable overlap integral can be set up, the value of which will depend on the relative orientation of the simple metal and ligand orbitals. Accordingly, one begins with two LCAOs based on a nonorthogonal basis set, χ_M and χ_L belonging to the metal and ligand ions, respectively, as follows:

$$\psi_b = C_{bM}\chi_M + C_{bL}\chi_L; \psi_a = C_{aM}\chi_M + C_{aL}\chi_L,$$

and diagonalizes the secular equation:

$$\begin{vmatrix} H_M - E & H_{ML} - ES_{ML} \\ H_{ML} - ES_{ML} & H_L - E \end{vmatrix} = 0,$$

subjected to the Helmholz–Wolfsberg approximation to approximate the off-diagonal elements:

$$H_{ij} = \frac{1}{2}KS_{ij}(H_{ii} + H_{jj}),$$

where K lies in the range of 1.5 to 3.0. Then, the MO energies become

$$E_a = H_M + \Delta E_a;$$

$$\Delta E_a = \frac{x\, S_{ML}^2}{H_M - H_L}, x = \left[\frac{1}{2}K(H_M + H_L) - H_M\right]^2;$$

$$E_b = H_L - \Delta E_b;$$

$$\Delta E_b = \frac{y\, S_{ML}^2}{H_M - H_L}, y = \left[\frac{1}{2}K(H_M + H_L) - H_L\right]^2.$$

These results are depicted in Figure 3.12. Further, one defines,

$$S_{ML} = F_\lambda^\ell S_{ML}^*; \lambda = \sigma, \pi, \delta,$$

which are illustrated in Figures 3.13 and 3.14 for single σ- and π-bonds, respectively.

In these considerations, according to Gerloch (1983), emphasis should be placed on local interactions and the concept of the functional group. Further, the angular part of S_{ML} transforms in exactly the same way as the metal-basis functions. Gerloch's approach to the AOM can be summarized as follows. The ligand-field

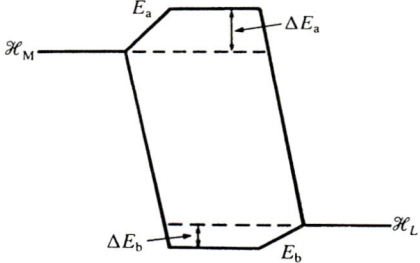

Figure 3.12 The interaction between metal and ligand orbitals in a simple molecular-orbital model. Adapted from Pilbrow (1990).

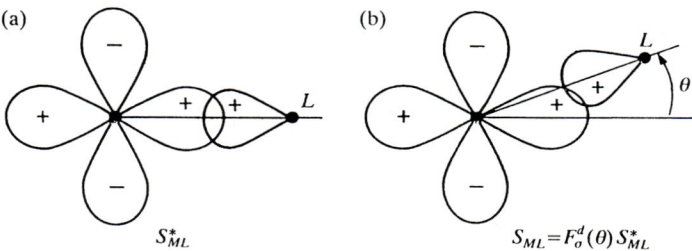

Figure 3.13 Definition of the radial and overlap integrals in a single metal–ligand σ-bond. Adapted from Pilbrow (1990).

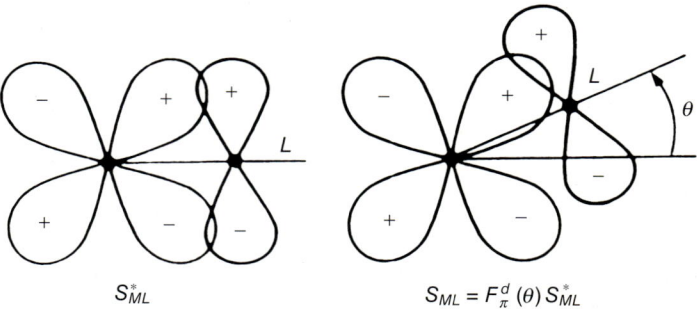

Figure 3.14 Definition of the radial and overlap integrals in a single metal–ligand π-bond. Adapted from Pilbrow (1990).

Hamiltonian is

$$H_{LF} = \sum_{i<j} U(i,j) + \sum_{i=1}^{N} V_{LF}(r_i) + \xi \sum_{i=1}^{N} l_i \cdot s_i, \qquad (3.31)$$

which acts on a restricted set of $N (= 5)$ d-orbitals.

3.4 Molecular Orbital (MO) Approach

The multipole expansion of the matrix element of $\boldsymbol{H}_{LF}$: $\langle i|H_{LF}|j\rangle$ depends on the product of an angular and a radial part, the latter depending on the ligand coordinates:

$$\langle \psi|V_{LF}(r)|\psi'\rangle = \sum_{k=0}^{\infty}\sum_{q=-k}^{k} a_{kq}\langle \psi|\rho_k(r)Y_k^q(\theta,\phi)|\psi'\rangle,$$

where $\rho_k(r)$ is the radial part of V_{LF}, $|\psi\rangle = |lR(r)A(r)\rangle$ (R = radial, A = angular),

$$\langle lR(\hat{r})A(\hat{r})|V_{LF}(\hat{r})|lR(\hat{r})A'(\hat{r})\rangle$$
$$= \sum_k \sum_q a_{kq}\langle R(r)|\rho_k(\hat{r})|R(r)\rangle \langle A(\hat{r})|Y_q^k(\theta,j)|A'(\hat{r})\rangle$$
$$= \sum_k \sum_q c_{kq}\langle A(\hat{r})|Y_q^k(\theta,j)|A'(\hat{r})\rangle$$

In the last equality the c-terms include the radial integrals, and have been tabulated in some cases. For example, it can be shown that

$$c_{21} = \left\{\frac{4\pi}{5}\right\}^{1/2}\left[\sqrt{6}\langle 2|V|1\rangle + \langle 1|V|0\rangle\right]; c_{41} = \left\{\frac{4\pi}{5}\right\}^{1/2}\left[-\langle 2|V|1\rangle + \langle 1|V|0\rangle\right],$$

where the c-terms are expressed on the M_L basis, e.g. $|0\rangle = |M_L = 0\rangle$.

In order to calculate the ligand-field potential, it is necessary to sum over the local potentials for each ligand, and then to further sum it over N cells, excluding those cells that are associated with a lone pair on the metal ions, and are empty. Therefore,

$$V_{LF} = \sum_{i=1}^{N} v^i.$$

The relationship between the global matrix elements to the local (*l*) elements is as follows:

$$V_{ij} = \langle d_i|V_{LF}|d_j\rangle = \sum_{i=1}^{N}\langle d_i|v^l|d_j\rangle = \sum_{i=1}^{N} v_{ij}^l.$$

The AOM parameters (*e*) are obtained from the *v*-terms by a 5 × 5 transformation matrix, $\boldsymbol{R}$, so that, for the *l*th ligand:

$$R^l v^l R^{l\dagger} = e^l.$$

The transformed orbitals are obtained as linear combinations of the global orbitals transformed by R:

$$\langle d_k^l| = \sum_i R_{ki}^l \langle d_i|,$$

and the exact energy shift of the local orbitals produced by the local potential v^l is

$$e_k^l = \langle d_k^l|v^l|d_k^l\rangle.$$

By proceeding in this manner, one obtains the basic equation of AOM:

$$V_{ij} = \sum_{i=1}^{N} \sum_{k} R_{ik}^{l\dagger} R_{kj}^{l} e_{k}^{l}, 1 \leq i, k \leq 5 \text{ and } 1 \leq l \leq N.$$

The original form given by Schaffer (1968) is:

$$\langle d_{z^2}^l | v^l | d_{z^2}^l \rangle = e_\sigma^1 \quad \langle d_{yz}^l | v^l | d_{yz}^l \rangle = e_{\pi y}^1$$
$$\langle d_{xz}^l | v^l | d_{xz}^l \rangle = e_{\pi x}^1 \quad \langle d_{x^2-y^2}^l | v^l | d_{x^2-y^2}^l \rangle = e_{\delta x^2-y^2}^1$$
$$\langle d_{xy}^l | v^l | d_{xy}^l \rangle = e_{\delta \pi x}^1$$

In the procedure for AOM, one first chooses the values for e, and then transforms the matrix elements to the global frame of reference, finally diagonalizing the full Hamiltonian, as given by Equation 3.31, which includes the Racah parameter, B, which accounts for the electron correlation term $U(i, j)$. The advantage of using the AOM is that each ligand is characterized by a local potential by choosing the e terms appropriately, and is not tied down to the MO framework. Typical values for Ni^{2+} and Co^{2+} coordinated to nitrogen are $e_\sigma \sim 3500\ cm^{-1}$ and $e_\pi \sim 500\ cm^{-1}$ (Gerloch, 1983). The $\tilde{g}$-matrix and paramagnetic susceptibility can be calculated by using the wavefunctions obtained by diagonalization of the Hamiltonian given by Equation 3.31 (Gerloch, 1983). Expressions have been derived by Bencini, Benelli, and Gatteschi (1979) to determine the hyperfine interaction by extending the use of the AOM method. [Apart from the references already cited, see Banci et al. (1982) and Deeth and Gerloch (1985) for some other applications of the AOM.]

3.5
The Jahn–Teller (JT) Effect

In the above discussion of the ground state of ions with several d-electrons, some of the ground states listed had two- or threefold degeneracies. However, in practice, these degeneracies do not persist due to the Jahn–Teller (JT) effect. The JT theorem states that a nonlinear molecule or molecular complex having electronic degeneracy will distort spontaneously to a nuclear configuration of lower symmetry, leaving only the twofold Kramers' degeneracy. Only orbital degeneracy needs to be taken into account, as the interaction between the spin angular momentum and nuclear coordinates is very weak. The behavior exhibited in the JT effect arises because of different interactions of one of the several degenerate states with its various neighbors. This can be illustrated by considering the example of a single electron in an e-orbital of a six-coordinated ion; it can occupy either of the degenerate states, $|2, x^2 - y^2\rangle$ or $|2, 3z^2 - r^2\rangle$, or any linear combinations of these. The distortion shown in Figure 3.15, denoted as $-Q_3$, will reduce the energy of the electron in the state $|2, x^2 - y^2\rangle$ as a consequence of reduced Coulomb interaction with the four negative point charges in the x–y plane, whereas it will raise

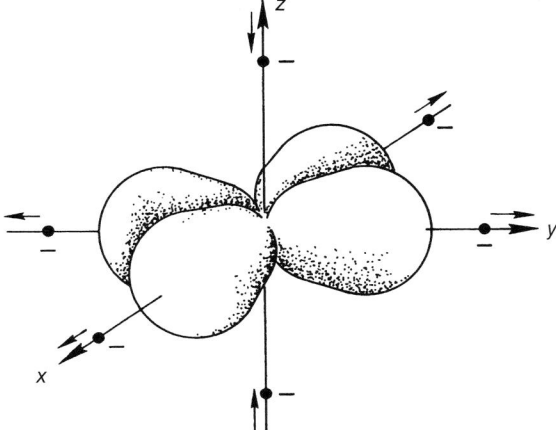

Figure 3.15 The distortion responsible for lowering of the energy of the x^2-y^2 state. The same distortion will raise the energy of the $3z^2-r^2$ state. These types of electrostatic force are responsible for the Jahn–Teller effect. Adapted from Pilbrow (1990).

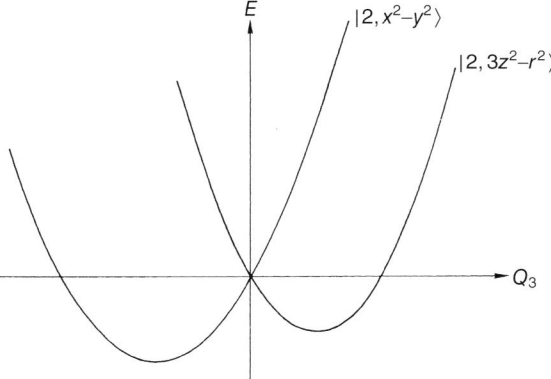

Figure 3.16 The energies of the x^2-y^2 and $3z^2-r^2$ states as a function of the distortion shown in Figure 3.15, assuming a negative distortion. Eventually, the elastic restoring forces cause the energy to increase, producing minima for a nonzero distortion. Adapted from Pilbrow (1990).

the energy of the electron in the $|2, 3z^2 - r^2\rangle$ state. (The opposite distortion will affect the energies of the two states in the opposite manner.) The energy levels for the distortion $-Q_3$ are shown in Figure 3.16. The presence of other interactions within the complex change the initial linear dependence on $-Q_3$, leading to minima in the energy which correspond to a significant distortion from the original high symmetry configuration to a tetragonal one. This type of distortion is experienced, for example, by the Ag^{2+} ($4d^9$) ion, for which several equivalent distorted ions are observed (Sierro, 1967).

More detailed reviews of the JT effect are provided by Sturge (1967), Ham (1972), and Müller (1967), Bersuker (1984a, 1984b), Perlin and Wagner (1984), and Bill (1984). The following discussion is extracted from the book by Pilbrow (1990). A schematic illustration of the two main types of specific adiabatic potential behavior due to vibronic interactions, leading to the JT effect and pseudo-JT effect (Bersuker, 1984a), is shown in Figure 3.17. In Figure 3.18 is shown the familiar "Mexican hat" potential for linear vibronic coupling, which provides an appropriate insight into the temperature-dependent behavior (Perlin and Wagner, 1984). The ground state is mixed with the first excited vibronic state by a weak linear vibronic interaction. It is necessary to consider the three-cornered hat potential when quadratic

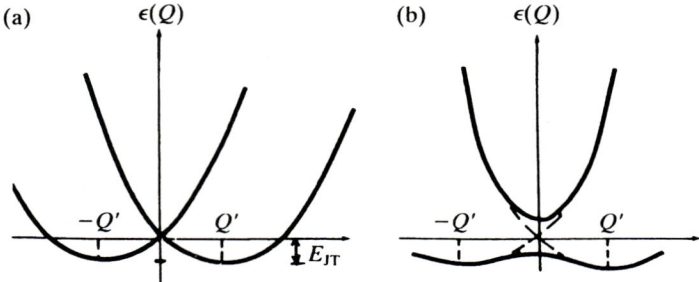

Figure 3.17 The two main types of specific adiabatic potential behavior due to vibronic interactions. (a) Electronic degeneracy – the Jahn–Teller effect; (b) Electronic pseudo-degeneracy. Adapted from Pilbrow (1990).

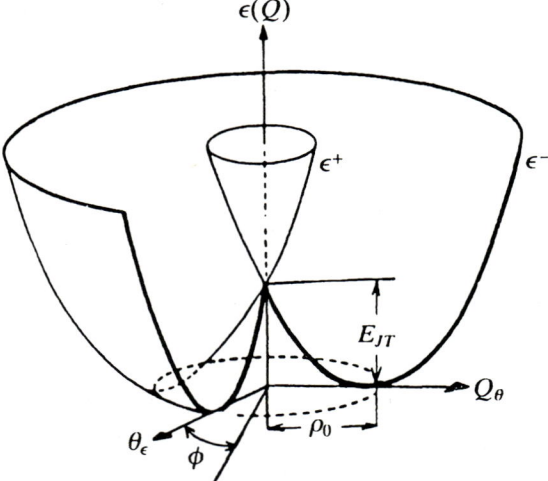

Figure 3.18 Mexican-hat-type adiabatic potential for twofold degenerate E-term interacting linearly with the twofold degenerate e-type vibrations, Q_θ and Q_ϵ. Adapted from Pilbrow (1990).

coupling is included since, in such cases, tunneling splitting (e.g., inversion splitting for ammonia) can occur. In addition, the pinning of distortions by random strain has the effect of rendering one of the three minima lower in energy than the other two, thereby providing information on the orientation of the principal axes of the spin Hamiltonian at each site.

The theories of the JT effect rely on the adiabatic approximation, applicable when $M/m \gg 1$, where M is the typical nuclear mass and m is the mass of an electron (Bersuker, 1984b). Further, it is assumed that each instantaneous nuclear configuration is associated with a stationary electronic state. An extreme form of the adiabatic approximation—the Born–Oppenheimer approximation—when used, permits the separation of electronic and vibrational wavefunctions. However, this often breaks down, so that vibronic mixing of the wavefunctions occurs, leading to new phenomena. In the simplest case—which is referred to as the $E \otimes e$ case—the twofold electronic degeneracy in the ideal configuration becomes a twofold degenerate vibronic state when vibronic effects are taken into account. When the electron spin is considered in addition to this, there result four electronic–vibronic states, treated as an effective spin $S = 3/2$ system. When the weak JT effect is considered, wherein an isotropic spectrum is observed at low temperatures, it becomes necessary to include an excited vibronic state; this is referred to as the "three-state" model.

Several different types of JT effect have been observed, the most classic of which was due to Cu^{2+} in $ZnSiF_6 \cdot 6H_2O$ (Bleaney and Bowers, 1952a; Dang, Buisson, and Williams, 1974). In this case, a static JT effect was exhibited at low temperatures, but this turned into a form of dynamic JT effect at ~77 K as a result of reorientation between the static distortions. The systems of Cu^{2+}-doped MgO (Coffman, 1965a, 1965b; Coffman et al., 1968) and Sc^{2+}-doped CaF_2 and SrF_2 (Hochli and Estle, 1967), exhibited a different type of JT effect, wherein isotropic spectra were found to coexist with anisotropic spectra up to ~10 K. This was attributed to manifestations of a weak JT effect for degenerate ^2E electronic states with moderate vibronic coupling. Thermal excitation to an excited vibronic state leads to the low-temperature dynamic JT effect (Ham, 1965, 1968). Subsequently, two reduction factors, p and q, were introduced by Ham, the values of which depend on the strength of the linear vibronic coupling. For weak coupling, p and q both approach 1, whereas $p \to 0$ and $q \to \frac{1}{2}$ for strong coupling. This situation, which is referred to as the "Ham effect," occurs for a doubly degenerate electronic state, when at least one of the states is coupled to the nuclear system. Large random strains are responsible for unusual lineshapes. The pseudo-JT effect occurs when near-degeneracy exists (Figure 3.17b).

3.5.1
Theory of the JT Effect

The general theory of the JT effect will be discussed first, followed by details of: (i) static JT effect; (ii) linear JT effect; (iii) perturbation within the vibronic ground state; (iv) three-state model; (v) and transition from dynamic to static JT effect.

3.5.1.1 General Theory of the JT Effect

First, it is necessary to solve the Schrödinger equation for the combined electronic and vibrational system for a degenerate electronic state of a paramagnetic ion in a crystal, where the energy difference between the partners of the degenerate system is smaller than the typical vibrational energies. Vibrational modes are taken into account in terms of the coordinates of the normal modes of vibration, denoted as Q_θ, Q_ε, for the doubly degenerate electronic E state for an ion in octahedral cubic symmetry (O_h). The corresponding electronic wavefunctions are written as $|\varepsilon\rangle, |\theta\rangle$, transforming as E_ε, E_θ of the irreducible representation of O_h. Within the context of the adiabatic approximation, motion of the nuclei is ignored when solving the purely electronic part of Schrödinger's equation. The potential for nuclear motion is chosen to be the mean electronic energy, and the initial nuclear configuration is chosen for the situation at which the electronic state is degenerate, or nearly degenerate. Neglecting the vibronic mixing of different electronic states: $\hbar\omega \ll |\varepsilon'_m - \varepsilon'_k|$, the regime of adiabatic approximation becomes applicable. The effective JT Hamiltonian can be expressed as:

$$H = \left\{E_0 + \frac{1}{2M}(p_\theta^2 + p_\varepsilon^2) + \frac{K}{2}(Q_\theta^2 + Q_\varepsilon^2)\right\}u_1 + V_E(Q_\theta u_\theta + Q_\varepsilon u_\varepsilon)$$
$$+ V_2\left[(Q_\varepsilon^2 - Q_\theta^2) + 2Q_\theta Q_\varepsilon u_\varepsilon\right] + V_3 Q_\theta(Q_\theta^2 - 3Q_\varepsilon^2)u_1 \quad (3.32)$$
$$= H_0 + H_{JT} + H_2$$

In the above equation, H_0 describes the uncoupled system, H_{JT} the linear JT coupling, and H_2 the quadratic coupling and third-order anharmonicity of the potential energy. The perturbations must transform as singlet states (A_{1g}, A_{2g}) and/or doublets (E_g) in the electronic part, where the subscript g refers to even representations. The matrix elements of any operator acting within the irreducible representation reduce, according to the Wigner–Eckart theorem (Messiah, 1965; as discussed in Section 3.2, Equation 3.14), to the product of two factors, a reduced matrix element and a Clebsch–Gordan coefficient. The matrices u in Equation 3.32 are nothing else but a set of 2×2 Pauli matrices with symmetries shown here for the cubic case:

$$u_1(A_1) = \begin{pmatrix} 1 & 0 \\ 0 & 1 \end{pmatrix}; \ u_2(A_2) = \begin{pmatrix} 0 & -i \\ i & 0 \end{pmatrix};$$
$$u_\theta(E_\theta) = \begin{pmatrix} -1 & 0 \\ 0 & 1 \end{pmatrix}; \ u_\varepsilon(E_\varepsilon) = \begin{pmatrix} 0 & 1 \\ 1 & 0 \end{pmatrix} \quad (3.33)$$

where the letters in the brackets in front of the u terms indicate their transformation properties.

The various specific cases of the JT effect are described as follows.

Static JT effect: Here, the nuclear kinetic energy in the Schrödinger equation is taken to be zero, and the normal coordinates are expressed as: $Q_\theta = \rho\cos\theta; Q_\varepsilon = \rho\sin\theta$. Hence, the energy levels of this particular Schrödinger equation are:

$$E_\pm = E(\rho, \theta) = E_0 + \frac{1}{2}K\rho^2 + V_3\rho^3 \cos 3\theta \pm \left[V_E^2 + V_2^2\rho^4 - 2V_E V_2\rho^3 \cos 3\theta\right]^{\frac{1}{2}}$$

The Mexican hat potential surface, as shown in Figure 3.18, is obtained by considering only the linear term in the above equation, by substituting $V_2 = V_3 = 0$. The minimum energy is then obtained when $\rho_{min} = |V_E|/K$, and the energy decrease by this substitution, referred to as E_{JT}, the "JT energy," is independent of θ, and is given by: $E_{JT} = V_E^2/2K = \rho_{min}^2/2K$. As a consequence, the upper potential sheet has an energy $\Delta E = 2V_E^2/K = 4E_{JT}$ at $\rho = \rho_{min}$. The corresponding wavefunctions are:

$$|\psi +\rangle = \sin\frac{1}{2}\theta |E_\theta\rangle + \cos\frac{1}{2}\theta |E_\varepsilon\rangle;$$

$$|\psi -\rangle = \cos\frac{1}{2}\theta |E_\theta\rangle - \sin\frac{1}{2}\theta |E_\varepsilon\rangle$$

When the second-order and harmonic terms (H_2) are included, the Mexican hat potential becomes warped, wherein the three energy minima occur at $\theta_0 = 0, 2\pi/3$, and $4\pi/3$ radians when $V_E(V_2 + V_3) < 0$, and at $\theta_0 = \pi/3, \pi$, and $5\pi/3$ radians when the inequality is reversed. No exact solution of this problem exists.

Linear JT effect: For this case, $V_2 = V_3 = 0$, the vibronic energy levels are all degenerate, and their wavefunctions are of the form:

$$|j, n_1 n_2\rangle = \chi_{n_1}(\rho, \theta)|\psi -\rangle + \chi_{n_2}(\rho, \theta)|\psi -\rangle; i = 1, 2, \theta, \varepsilon$$

In the above equation n_1 is the first vibrational state of level n, and so on. As these states cannot be simply factorized into a product of electronic and vibrational states, they are not Born–Oppenheimer states. Further,

$$E = E_0 + \hbar\omega - E_{JT},$$

where ω is the angular frequency of the doubly degenerate vibration; $E_{JT} = V_E^2/2M\omega^2$, where M is the mass of a ligand atom, and the force constant is $M\omega^2$.

The first excited vibrational state turns out to be a doublet, separated in energy by the following amounts:

$$\Delta E = \hbar\omega[1 - (2E_{JT}/\hbar\omega)], \text{ when } E_{JT}/\hbar\omega \ll 1;$$

$$\Delta E = \hbar\omega[\hbar\omega/2E_{JT}], \text{ when } E_{JT}/\hbar\omega \gg 1$$

The second of the above two equations must be modified if the JT effect is strong, because in that case H_2 should be included in the calculation.

Strong JT effect: Here, one should take into account the warping terms mentioned above. This requires transforming the Hamiltonian, given by Equation 3.32, by the operator represented by the matrix:

$$S = \begin{pmatrix} \sin\theta/2 & \cos\theta/2 \\ \cos\theta/2 & -\sin\theta/2 \end{pmatrix},$$

which is equivalent to a rotation of the axes in the 2-D subspace. Thereafter, only the diagonal part of the warping term is retained, which is:

$$H_2(diag) = -V_3\rho^3 \cos 3\theta \pm V_2\rho^2 \cos 3\theta.$$

As for the quadratic coupling, it is negative for $\langle\psi+|H_2|\psi+\rangle$, and positive for $\langle\psi-|H_2|\psi-\rangle$. The wavefunctions of the lowest state are localized near the minima of the warped potential, consisting of approximate Born–Oppenheimer states. The ground state turns out to be approximately a radial harmonic oscillator function at $\rho = \rho_0$. Further details can be found in the article by O'Brien, (1964). Accordingly, the ground state is a vibronic doublet (E), whereas the excited states are partly nondegenerate. The first excited state turns out to be either of A_1 or of A_2 symmetry.

3.5.2
Perturbation within the Vibronic Ground State

The perturbations to be considered here can be listed in decreasing order of importance as follows: (i) strain; (ii) Zeeman; (iii) hyperfine; (iv) quadrupole interaction; and (v) ligand hyperfine interaction. The relevant Hamiltonian, within an orbital electronic doublet of a transition ion in octahedral symmetry, is:

$$H_P = H_s + H_Z + H_{hfs} + H_Q + \sum_i H_{\ell i}. \qquad (3.34)$$

Further, for tetragonal and isotropic strain, one has

$$H_s = V_s(e_\theta + e_\varepsilon) + V_0 e_s$$

As for the symmetry-adapted components of strain, one has

$$e_\theta = \frac{1}{2}(2\varepsilon_{zz} - \varepsilon_{xx} - \varepsilon_{yy}); \; e_\varepsilon = \frac{1}{2}\sqrt{3}(\varepsilon_{xx} - \varepsilon_{yy});$$

$$e_\rho = \frac{1}{3}(\varepsilon_{xx} + \varepsilon_{yy} + \varepsilon_{zz}).$$

According to Ham (1972), the internal strain in crystals that determines the shape of most ground-state systems is random. Further, the unperturbed vibronic ground state has at least the same symmetry as the electronic orbital state which is active in exhibiting the JT effect. It remains degenerate until acted upon by the strains with strengths of a few cm^{-1}. The local fields at different centers are different. In order to calculate the lineshape, one needs the strain distribution (Stoneham, 1969). Accordingly, one has, explicitly, for the various terms in the Hamiltonian:

$$H_z = g_1\mu_B \mathbf{B} \cdot \mathbf{S} u_1 + \frac{1}{2}g_2\mu_B(3B_z S_z - \mathbf{B}\cdot\mathbf{S})u_\theta + \sqrt{3}(B_x S_x - B_y S_y)u_\varepsilon;$$

$$H_{HFS} = A_1 \mathbf{S}\cdot\mathbf{I} u_1 + \frac{1}{2}A_2(3S_z I_z - \mathbf{S}\cdot\mathbf{I})u_\theta + \sqrt{3}(S_x I_x - S_y I_y)u_\varepsilon;$$

$$H_Q = C\{[3I_z^2 - I(I+1)]u_\theta + \sqrt{3}(I_x^2 - I_y^2)u_\varepsilon\};$$

where
$$C = \pm\frac{3e^2Q}{I(2I-1)}\langle r^{-3}\rangle\xi;$$

$$\xi = \frac{2\ell + 1 - 4S}{S(2\ell-1)(2\ell+3)(2L-1)} \quad (\ell = 2 \text{ for } d\text{-electrons})$$

The vibronic Hamiltonian needs to be solved in full, unless the static JT effect is being considered. Further, the numerical values of the matrix elements of the electronic operators that transform as E_θ, E_ε, and A_2 are reduced in value, because the overlap integrals between the vibrational part of two different vibronic wavefunctions enter as multiplying factors.

When considering the two-state model, one requires two reduction factors, p and q, to modify the values of the elements of the matrices for the u-matrices, as given by Equation 3.33, so that the matrices are now: $u'_1 = u_2$; $u'_2 = pu_2$; $u'_\theta = qu_\theta$; $u'_\theta = qu_\theta$. These modified u' matrices are formally equivalent to the u matrices of Equation 3.33, except that they act only within the vibronic doublet. The parameters p and q are adjustable, with the limiting values (Ham, 1968):

$$p = 1 - 4E_{JT}/\hbar\omega, q = 1 - 2E_{JT}/\hbar\omega; \text{ when } E_{JT}/\hbar\omega \ll 1;$$

$$p \gg 0, q \geq \frac{1}{2}; \text{when } E_{JT}/\hbar\omega \gg 1;$$

subject to the condition $2q - p - 1 = 0$.

The complete behavior of the reduction factors p and q as functions of the dimensionless parameter $\lambda_E = E_{JT}/\hbar\omega$ can be found in Ham (1968). Some relevant discussion is provided in Bersuker (1984a). Within the two-state model, $0.5 \leq q \leq 1$. The upper limit applies to weak vibronic coupling, whereas the lower limit applies to strong linear coupling with no warping.

When the calculations are made, H_s, in Equation 3.34, is considered first, keeping only the anisotropic part. The strains are expressed in polar coordinates as

$$e_\theta = \xi_s \cos\phi, e_\varepsilon = \xi_s \sin\phi,$$

and the corresponding transformed vibronic states are

$$|+v\rangle = \cos\frac{1}{2}\phi|E_\varepsilon v\rangle + \sin\frac{1}{2}\phi|E_\theta v\rangle,$$

$$|-v\rangle = -\sin\frac{1}{2}\phi|E_\varepsilon v\rangle + \cos\frac{1}{2}\phi|E_\theta v\rangle$$

For which the resulting eigenvalues are

$$E_\pm = \pm qV_s\xi_s.$$

One can now use the perturbation calculation, assuming the strain splittings to be larger than the magnitude of the other terms, to determine their effect the EPR spectra.

The resonance fields for $S = \frac{1}{2}$, applicable to d^1 and d^9 states, taking into account the hyperfine interaction, have been calculated by Bill (1984) and Herrington, Estle, and Boatner (1971). These reduce to the following expressions, as calculated by Ham (1972), when there is no hyperfine interaction:

$$B_\pm = B_0[1 \mp (qg_2/g_1) f \cos(\phi - \alpha)],$$

where $f = [1 - 3(\ell_1^2\ell_2^2 + \ell_2^2\ell_3^2 + \ell_3^2\ell_1^2)]^{1/2}$; $\tan\alpha = \sqrt{3}(\ell_1^2 - \ell_2^2)/(3\ell_2^2 - 1)$; $g_1 = g_s - (4\lambda/\Delta E)$, $g_2 = -(4\lambda/\Delta E)$, and ℓ_1, ℓ_2, ℓ_3 are the direction cosines of **B** with respect to the three cube axes.

The dependence of EPR transitions on the angle $(\phi - \alpha)$ of the strain orientation with respect to the direction of the magnetic field has been discussed in detail by Bersuker (1984a, pp. 113ff). The spectrum is isotropic when $(\phi - \alpha) = \pm\pi/2$ and $g = g_1$. When $(\phi - \alpha) = 0$ or π, the spectrum is the same as that in the absence of strain. The anisotropic part of the g-factor varies from 0 to $\pm qg_2 f$ at intermediate values of $(\phi - \alpha)$ between 0 and π. In addition, the transition probabilities are independent of strain and the direction of **B**. This leads to an inhomogeneous broadening of the spectrum, resulting in a spectrum with powder-like features. Furthermore, relaxation transitions can be significant, and an additional feature appears at the center of the absorption spectrum.

Further complications arise due to the effect of tunneling splitting when the vibronic singlet Kramers' doublets, either A_1 or A_2, approaches the ground state. A shift in one of the transitions and a skewing of the resulting spectrum occurs due to the combined effect of tunneling splitting and random strain. Further details are available elsewhere (Bersuker, 1984a).

3.5.3
Three-State Model

This is used for strong JT coupling when either A_1 or A_2 approaches the ground state (Ham, 1972; Englman, 1972; Sester, Barksdale, and Estle, 1975; Reynolds and Boatner, 1975). The 3-D subspace is a useful approximation to understand the EPR spectrum, taking into account an additional reduction factor r, if the inversion splitting 3Γ is much smaller than the energy of the excited vibronic levels, and is comparable to the perturbation H_p (as referred to above; see Equation 3.34). The energy matrix in this case is a function of the reduction factors p, q, r, in addition to 3Γ, and the cubic strain parameters e_θ, e_ε, and V_s.

The solution approaches the static JT effect result, when V_E is large, so that the warping of the potential is important. The wavefunctions are of the Born–Oppenheimer type, when the three minima are deep with respect to the zero-point energy. The tetragonal g-factors are $g_\parallel = g_1 \mp 2qg_2$; $g_\perp = g_1 \pm qg_2$. The comparison between theory and experiment for the two limiting cases of angular dependence of the g-factor for the linear $E \otimes e$ problem with strong coupling and without vibronic coupling is given in Bersuker (1984a).

3.5.4
Transition from Dynamic to Static JT Effect

This occurs when there is a low-lying excited vibronic state which interacts with the ground vibronic level. The inversion or tunneling splitting, δ, is here equal to 3Γ. Although, details of the dependence of the JT EPR spectra on tunneling splitting are given in Bersuker (1984a), brief details are as follows. There are three regions to be considered: Region I, where $g_2\mu_B B \ll 3\Gamma$; Region III, where $g_2\mu_B B \gg 3\Gamma$; and Region II, which is intermediate between Regions I and III. The mixing of the ground state and the excited, A_2, state is weak in Region I. Here, three EPR lines are expected to be observed, two of which arise from the ground state and are weak, and one more (isotropic) line, which arises from the orbital singlet excited state. As for Region III, three allowed lines are also expected here. Within the limit of strong vibronic coupling, for which $q = \frac{1}{2}$ and $3\Gamma = 0$, the g-factors are $g_\| = g_1 + g_2$; $g_\perp = g_1 - g_2/2$ for A_2, the active excited level, whereas for the level A_1 they are $g_\| = g_1 - g_2$; $g_\perp = g_1 + g_2/2$, with similar relationships for the hyperfine parameters. The configuration here is such that the system lies in a static distortion at the minima of the adiabatic potential. Region I corresponds to free rotations in A_1- or A_2-states and hindered rotations in the E-doublet. In order to observe EPR transitions in Region III, the lifetime of the distorted configuration in pulsating motion must be greater than the equivalent time represented by the anisotropy in the Zeeman interaction; that is, the tunneling frequency must be less than one nanosecond.

A useful summary of the JT discussion covered in this section is depicted in Figure 3.19 (Reynold and Boatner, 1975). This illustrates how electronic degeneracy

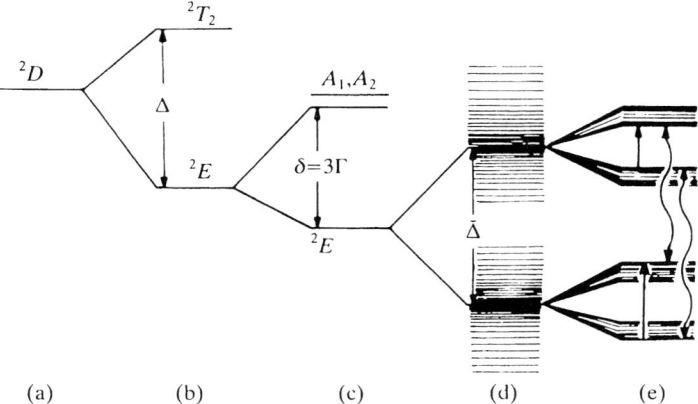

Figure 3.19 Scheme of the lowest energy levels for an octahedral system in the ^2E-state. (a) Free-ion term; (b) Crystal-field splitting; (c) Tunneling splitting; (d) splitting of the ground vibronic level into two Kramers' doublets by random strain; (e) Zeeman splitting by the magnetic field. The allowed EPR transitions are shown by straight arrows, whereas the relaxation transitions without spin reversal are shown by wavy arrows.

is replaced by vibrational degeneracy, and also how the combination of tunneling splitting and random strain leads to the current model for understanding the simplest form of the JT cases; specifically, this relates to the $E \otimes e$ problem.

3.6
The Spin Hamiltonian

In order to describe the theory of EPR, one approach has been to use the spin-Hamiltonian, the matrix of which is diagonalized to obtain the required eigenvalues and eigenvectors for calculating the positions and intensities of EPR lines. Rudowicz and Misra (2001) have reviewed the various types of spin Hamiltonian permitted by time-reversal invariance. Among these are: (i) *microscopic*, including the Abragam and Pryce version; (ii) *phenomenological*, where the required terms are guessed, and (iii) *generalized*, or group theoretical, which includes all terms allowed by symmetry considerations. The effective spin, S, describes the low close-lying levels (within a few cm^{-1}), widely separated from all the upper levels. Then, the total number of levels in the group is equal to $(2S + 1)$, among which EPR transitions are induced by microwave fields of appropriate frequencies. A simple approach would be to write down the polynomial operators to describe this effective spin, including the various interactions allowed for by symmetry and time-reversal invariance. This leads to the phenomenological spin Hamiltonian, without restricting the magnitude of any of the coefficients.

3.6.1
The Abragam and Pryce Spin Hamiltonian for the Iron Group

It was originally Pryce who developed a modified form of perturbation theory to derive a number of perturbation calculations of comparable magnitude (Pryce, 1950). The general formulae so obtained can be readily used to calculate estimates of measurable parameters. The main disadvantage of this approach is that the spin operators are taken for the true spin for ions, for which the crystal field produces an orbital singlet state. The scheme was subsequently extended by Abragam and Pryce (1951a) to treat those cases for which the orbital angular momentum was not quenched. A modified version used an effective orbital angular momentum, leading to an effective spin of ½ for the ground state, for example, for d^7 (Abragam and Pryce, 1951b). In fact, effective spins could be calculated for all the iron-group ions in strong and weak crystal fields. Furthermore, the spin Hamiltonian as formulated originally was based on crystal-field theory of the electronic energy levels. The necessary modifications to the theory to allow for covalency are described below (see Section 3.6.1.1). The following description is extracted from Pilbrow (1990).

In the approach of Pryce (1950), all calculations involving an orbital angular momentum operator were carried out explicitly, while leaving spin variables as operators, when applied to the iron-group ions possessing an orbital singlet ground state. This resulted in the concepts of spin Hamiltonian and effective spin, taking into account the Zeeman effect, fine and hyperfine structures. These calculations were then extended by Abragam and Pryce (1951a) to include electron–nuclear hyperfine structure, quadrupole, and nuclear terms, in addition to the cases where quenching of the orbital angular momentum was not complete. Their spin Hamiltonian is as follows:

$$H = H_{FS} + \mu_B \mathbf{B} \cdot \tilde{g} \cdot \mathbf{S} + \mathbf{I} \cdot \tilde{A} \cdot \mathbf{S} + \mathbf{I} \cdot \tilde{P} \cdot \mathbf{I} - \mu_N g_N \mathbf{B} \cdot \mathbf{I};$$
$$H_{FS} = \mathbf{S} \cdot \tilde{D} \cdot \mathbf{S},$$
(3.35)

where H_{FS} is the second-order fine-structure term for $S \geq 1$; $\tilde{g}$, $\tilde{A}$ are 3×3 matrices, the principal values of which are the square roots of the diagonal matrices $g^\dagger g$ and $A^\dagger A$, respectively; and $\tilde{P}$ is a 3×3 tensor. The above spin Hamiltonian is obtained by calculating the perturbation of the ground-state levels caused by crystal-field mixing of a limited set of low-lying electronic states, which are usually (but not always) derivable from the lowest spectroscopic term. The various interactions considered, restricting the discussion to a set of energy levels from a single spectroscopic term, including some others for completeness, are expressed as follows.

Spin-orbit coupling: $\lambda \mathbf{L} \cdot \mathbf{S}$. The order of magnitude of this interaction is ~$10^2 \mathrm{cm}^{-1}$. This is much smaller than typical crystal-field splittings for the first transition series, but it is larger than those for the rare earths.

Spin-spin coupling: $-\rho \sum_{p,q} \left\{ \frac{1}{2}[L_p L_q + L_q L_p] - \frac{1}{3} L(L+1)\delta_{pq} \right\} S_p S_q;$

Electron Zeeman interaction: $\mu_B [2.0023\mathbf{S} + \mathbf{L}] \cdot \mathbf{B} = \mu_B \mathbf{S} \cdot \tilde{g} \cdot \mathbf{B}$. The order of magnitude of this interaction is $0.3 \mathrm{cm}^{-1}$ at X-band (10 GHz) for which $\mathbf{B} \sim 3500 \mathrm{G}$.

Electron nuclear hyperfine interaction:

$$P\left[\{\xi L(L+1) - \kappa\} \mathbf{S} \cdot \mathbf{I} - \frac{3}{2}\xi(\mathbf{L}\cdot\mathbf{S})(\mathbf{L}\cdot\mathbf{I}) - \frac{3}{2}\xi(\mathbf{L}\cdot\mathbf{I})(\mathbf{L}\cdot\mathbf{S}) + \mathbf{L}\cdot\mathbf{I} \right]$$

where $P = 2g_N \mu_B \mu_N \langle r^{-3} \rangle$, and ξ is defined by Equation 3.36 below.

Further explanation of the above expressions is as follows. The electron–nuclear hyperfine structure includes a sum over all d-electrons of the electron–nuclear dipole–dipole interaction; κ is an empirical parameter which accounts for the core polarization of the inner s-electron shells and their isotropic contribution to the hyperfine interaction, and is equal to $-\kappa P$. The core polarization occurs due to a spin unpairing of inner 1s, 2s, and 3s electrons caused by the d-shell electrons; this results in a finite s-electron spin density at the nucleus and different radial wavefunctions for spin-up and spin-down electrons.

Super-hyperfine interaction (H_{SHF}): This was not considered by Abragam and Pryce (1951a), but is included here for completeness. The magnetic moment of an electron in an atom also interacts with the local magnetic fields originating

$$b_2^0 O_2^0(S) + \frac{1}{3} b_2^2 O_2^2(S) = 3 B_2^0 O_2^0(S) + B_2^2 O_2^2(S),$$

where $D = 3B_2^0 = b_2^0$, $O_2^0(S) = [3S_z^2 - S(S+1)]$, and $E = B_2^2 = \frac{1}{3} b_2^2$, $O_2^2(S) = [S_x^2 - S_y^2]$.

Over and above the D and E ZFS terms, there are present other zero-field interactions in different symmetries are described below.

3.6.2.1 Cubic Zero-Field Splitting ($S > 3/2$)

This is described by the ZFS parameter a, which is defined by the spin Hamiltonian

$$\frac{1}{6} a [S_x^4 + S_y^4 + S_z^4 - \frac{1}{5} S(S+1)(3S^2 + 3S - 1)] = B_4^0 O_4^0(S) + B_4^4 O_4^4(S)$$

In the case of a fourfold coordinate-axis system, the following relations exist:

$$\frac{1}{6} a = 20 B_4^0 = 4 B_4^4,$$

whereas for the threefold coordinate-axis system, one has $\frac{1}{6} a = -\frac{40}{3} B_4^0 = -\frac{800\sqrt{2}}{3} B_4^3$.

The zero-field F term, which is applicable to distortions with axial symmetry for $S > 3/2$, is expressed as:

$$\frac{F}{180} \{35 S_z^4 - 30 S(S+1) S_z^2 + 25 S_z^2 - 6 S(S+1)[3 S^2 (S+1)^2 + S(S+1) - 2]\} = B_4^0 O_4^0(S)$$

where $F = 180 B_4^0$, so that $b_4^0 = 60 B_4^0 = F/3$.

3.6.3
The Phenomenological Spin Hamiltonian

It should be noted that the ground manifold does not, in general, exhibit the isotropic behavior of a free spin. Thus, the spin Hamiltonian must contain electron spin operators consistent with the point symmetry of the environment, as well as those for the nuclear spin, if there is a nonzero nuclear spin present. The time-reversal invariance of energy, $H(-t) = H(t)$, dictates existence of the terms in the spin Hamiltonian of the type $B^p S^q I^r$ with $p + q + r = 2n$ ($n > 0$, each p, q, r is an integer), where, $p \leq 1$, $q \leq 2S$, $r \leq 2I$; in what follows the terms, which do not contain **B**, the magnetic field, are considered when $p = 0$, whereas the terms that depend on **B** (Zeeman terms) are considered when $p = 1$.

The various independent terms required for the various symmetries are listed below, where A, B, and C represent B, S, or I.

ground state. This resulted in the concepts of spin Hamiltonian and effective spin, taking into account the Zeeman effect, fine and hyperfine structures. These calculations were then extended by Abragam and Pryce (1951a) to include electron–nuclear hyperfine structure, quadrupole, and nuclear terms, in addition to the cases where quenching of the orbital angular momentum was not complete. Their spin Hamiltonian is as follows:

$$H = H_{FS} + \mu_B B \cdot \tilde{g} \cdot S + I \cdot \tilde{A} \cdot S + I \cdot \tilde{P} \cdot I - \mu_N g_N B \cdot I;$$
$$H_{FS} = S \cdot \tilde{D} \cdot S, \tag{3.35}$$

where H_{FS} is the second-order fine-structure term for $S \geq 1$; $\tilde{g}, \tilde{A}$ are 3×3 matrices, the principal values of which are the square roots of the diagonal matrices $g^\dagger g$ and $A^\dagger A$, respectively; and $\tilde{P}$ is a 3×3 tensor. The above spin Hamiltonian is obtained by calculating the perturbation of the ground-state levels caused by crystal-field mixing of a limited set of low-lying electronic states, which are usually (but not always) derivable from the lowest spectroscopic term. The various interactions considered, restricting the discussion to a set of energy levels from a single spectroscopic term, including some others for completeness, are expressed as follows.

Spin-orbit coupling: $\lambda L \cdot S$. The order of magnitude of this interaction is $\sim 10^2 \, \text{cm}^{-1}$. This is much smaller than typical crystal-field splittings for the first transition series, but it is larger than those for the rare earths.

Spin-spin coupling: $-\rho \sum_{p,q} \left\{ \frac{1}{2}[L_p L_q + L_q L_p] - \frac{1}{3} L(L+1)\delta_{pq} \right\} S_p S_q;$

Electron Zeeman interaction: $\mu_B [2.0023 S + L] \cdot B = \mu_B S \cdot \tilde{g} \cdot B$. The order of magnitude of this interaction is $0.3 \, \text{cm}^{-1}$ at X-band (10 GHz) for which $B \sim 3500 \, \text{G}$.

Electron nuclear hyperfine interaction:

$$P\left[\{\xi L(L+1) - \kappa\} S \cdot I - \frac{3}{2}\xi(L \cdot S)(L \cdot I) - \frac{3}{2}\xi(L \cdot I)(L \cdot S) + L \cdot I \right]$$

where $P = 2g_N \mu_B \mu_N \langle r^{-3} \rangle$, and ξ is defined by Equation 3.36 below.

Further explanation of the above expressions is as follows. The electron–nuclear hyperfine structure includes a sum over all d-electrons of the electron–nuclear dipole–dipole interaction; κ is an empirical parameter which accounts for the core polarization of the inner s-electron shells and their isotropic contribution to the hyperfine interaction, and is equal to $-\kappa P$. The core polarization occurs due to a spin unpairing of inner $1s$, $2s$, and $3s$ electrons caused by the d-shell electrons; this results in a finite s-electron spin density at the nucleus and different radial wavefunctions for spin-up and spin-down electrons.

Super-hyperfine interaction (H_{SHF}): This was not considered by Abragam and Pryce (1951a), but is included here for completeness. The magnetic moment of an electron in an atom also interacts with the local magnetic fields originating from the magnetic moment of nonzero nuclear spins of the ions in the first coordination sphere around the atom, expressed as $H_{SHF} = \mathbf{S} \cdot \sum_i \tilde{A}_i \cdot \mathbf{I}_i$, where $\tilde{A}_i$ is the hyperfine tensor, for the interaction of the electron spin $\mathbf{S}$ with the nuclear spin I_i; this is termed *super-hyperfine* interaction. The splittings due to these are mostly weak, and usually unresolved.

Nuclear electric quadrupole interaction:

$$\sum_{p,q} q' \left\{ \frac{1}{2}(L_p L_q + L_q L_p) - \frac{1}{3}\delta_{pq}L(L+1) \right\} \left\{ \frac{1}{2}(I_p I_q + I_q I_p) - \frac{1}{3}\delta_{pq}I(I+1) \right\},$$

where $q' = \dfrac{3e^2 Q}{2I(I+1)} \langle r^{-3} \rangle \langle L \| \alpha \| L \rangle$ and Q is the quadrupole interaction,

$$\xi = \frac{2l+1-4S}{S(2l-1)(2l+1)(2L-1)}, \text{ with } \ell = 2 \text{ for d-electrons;} \tag{3.36}$$

and $\langle L \| \alpha \| L \rangle$, related to ξ, is defined by Abragam and Bleaney (1970, p. 671). A nucleus with spin $I \geq 1$ possesses an electric quadrupole moment Q because of its nonspherical charge distribution, that interacts with the electric field gradient, due to the surrounding charges, and which arises from uneven distributions of electric charges around the nucleus. The Hamiltonian of the nuclear quadruple interaction is given by

$$\mathbf{H}_Q = \frac{e^2 Q}{4I(2I-1)} \left(\frac{\partial^2 V}{\partial z^2} \right) [\{3I_z^2 - I(I+1)\} + \eta(I_x^2 - I_y^2)] = C_2^0 O_2^0(I) + C_2^2 O_2^2(I),$$

where $\eta = \left(\dfrac{\partial^2 V}{\partial x^2} - \dfrac{\partial^2 V}{\partial y^2} \right) \Big/ \dfrac{\partial^2 V}{\partial z^2}$ is the asymmetry parameter, which is equal to zero in the case of axial symmetry; the spin operators are $O_2^0(I) = [3I_z^2 - I(I+1)]$ and $O_2^2(I) = [I_x^2 - I_y^2]$.

Nuclear Zeeman interaction: $-g_N \mu_N \mathbf{B} \cdot \mathbf{I}$, where μ_N is the nuclear magneton. For the presence of more than one nucleus, the Hamiltonian is $-\sum_i \mu_N \mathbf{B} \cdot \tilde{g}_{Ni} \cdot \mathbf{I}_i$, and the sum is over the various nuclei. As the mass of the nucleus is much larger – approximately 2000-fold the mass of the electron – and since the magnetic moment is inversely proportional to the mass, the nuclear magnetic moment is much smaller than the electronic moment. Thus, the nuclear Zeeman interaction is three orders of magnitude smaller than the electronic Zeeman interaction.

The above parameters, as described by Equation 3.35, and used to describe the spin Hamiltonian, using the convention that repeated indices imply summation, are as follows:

$$g_{ij} = 2.0023(\delta_{ij} - \lambda \Lambda_{ij}); \quad A_{ij} = -P(\kappa \delta_{ij} + 3\xi \ell_{ij} + 2\lambda \Lambda_{ij} - 3\xi u_{ij}); \quad D_{ij} = \lambda^2 \Lambda_{ij} - \rho \ell_{ij};$$
$$\text{and } P_{ij} = q' \ell_{ij};$$

where $\Lambda_{ij} = \Lambda_{ji} = \sum_{n \neq 0} \dfrac{\langle 0 | L_i | n \rangle \langle n | L_j | 0 \rangle}{E_n - E_0}; \quad u_{ij} = -\dfrac{1}{2} \varepsilon_{jkl} \sum_{n \neq 0} \dfrac{\langle 0 | L_i | n \rangle \langle n | L_j L_k + L_k L_j | 0 \rangle}{E_n - E_0};$

$u_{ii} = 0; \quad \dfrac{1}{2} \langle 0 | L_i L_j + L_j L_i | 0 \rangle = \dfrac{1}{3} L(L+1)\delta_{ij} + \ell_{ij}; \quad \ell_{ii} = 0.$

Here, the ground state is denoted by $|0\rangle$, with energy E_0, and the excited states by $|n\rangle$, with energies E_n. One uses $L = 2$ for D terms and $L = 3$ for F terms, in order to calculate ξ from Equation 3.36.

The quadrupole interaction, characterized by the tensor $\tilde{P}$, depends on the product of the electric field gradient at the nucleus from both the unfilled shell of magnetic electrons and ligands. This should be corrected by enhancing it with the Sternheimer anti-shielding factor, γ_∞, which can be approximately 100 for atoms towards the middle of the Periodic Table (Slichter, 1963). In addition, there may be a lattice contribution to the electric field gradient.

It should be noted that, in contrast to the tensors $\tilde{D}, \tilde{P}$, which are symmetric and are bilinear in the same vectors S and I, respectively, $\tilde{g}, \tilde{A}$ are coupling matrices between different vectors. For more details, the reader is referred to Abragam and Pryce (1951a), Kneubuhl (1963), and Ham (1965), as well as to Chapter 7 of this book.

3.6.1.1 Incorporation of Covalency

This can be accomplished in the Abragam and Pryce spin Hamiltonian by means of LCAO-MO theory, described as follows. The states $|0\rangle, |n\rangle$ are now replaced by LCAOs as $\alpha|0\rangle + \beta|0\rangle_L; \alpha'_n|0\rangle + \beta'_n|n\rangle_L$, respectively, where the eigenvectors with subscript L are symmetry-adapted linear combinations of ligand orbitals. This leads to a modification of the g_{ij}, A_{ij} values as follows:

$$g_{ij} = 2.0023(\delta_{ij} - \lambda \Lambda_{ij} - \lambda_L \Lambda^L_{ij})$$

$$A_{ij} = -P(\kappa \delta_{ij} + 3\xi l_{ij} + 3\xi_L l^L_{ij} + 2\lambda \Lambda_{ij} + 2\lambda_L \Lambda^L_{ij} - 3\xi u_{ij} - 3\xi_L u^L_{ij}),$$

The above discussion in this section is limited to the case when an orbital-singlet lies lowest. In order to consider more general cases, for example an orbital-triplet state, the theory must be modified.

3.6.2 Zero-Field Splitting (ZFS)

This second-order fine-structure term, H_{FS}, expressed by Equation 3.35, applies to $S \geq 1$, and is expressed in traceless form as:

$$\mathbf{H}_{SS} = \mathbf{S} \cdot \tilde{D} \cdot \mathbf{S} = D\left[S_z^2 - \frac{1}{3}S(S+1)\right] + E(S_x^2 - S_x^2), \tag{3.37}$$

where D and E are electron–electron coupling constants and, for axial symmetry, $E = 0$. The order of magnitude of this interaction is 0.1 to 10 cm^{-1}, being frequently of the same order of magnitude as the electronic Zeeman energy. This results in a strong angular dependence of the EPR spectrum with respect to the orientation of the external magnetic field.

The ZFS describes the various interactions affecting the energy levels of an electron spin ($S > \frac{1}{2}$) in the absence of an applied magnetic field. The second-order interaction can be expressed in terms of two independent constants D and E, as follows:

$$b_2^0 O_2^0(S) + \frac{1}{3} b_2^2 O_2^2(S) = 3B_2^0 O_2^0(S) + B_2^2 O_2^2(S),$$

where $D = 3B_2^0 = b_2^0$, $O_2^0(S) = [3S_z^2 - S(S+1)]$, and $E = B_2^2 = \frac{1}{3} b_2^2$, $O_2^2(S) = [S_x^2 - S_y^2]$.

Over and above the D and E ZFS terms, there are present other zero-field interactions in different symmetries are described below.

3.6.2.1 Cubic Zero-Field Splitting (S > 3/2)
This is described by the ZFS parameter a, which is defined by the spin Hamiltonian

$$\frac{1}{6} a[S_x^4 + S_y^4 + S_z^4 - \frac{1}{5} S(S+1)(3S^2 + 3S - 1)] = B_4^0 O_4^0(S) + B_4^4 O_4^4(S)$$

In the case of a fourfold coordinate-axis system, the following relations exist:

$$\frac{1}{6} a = 20 B_4^0 = 4 B_4^4,$$

whereas for the threefold coordinate-axis system, one has $\frac{1}{6} a = -\frac{40}{3} B_4^0 = -\frac{800\sqrt{2}}{3} B_4^3$.

The zero-field F term, which is applicable to distortions with axial symmetry for $S > 3/2$, is expressed as:

$$\frac{F}{180} \{35 S_z^4 - 30 S(S+1) S_z^2 + 25 S_z^2 - 6 S(S+1) [3 S^2 (S+1)^2 + S(S+1) - 2]\} = B_4^0 O_4^0(S)$$

where $F = 180 B_4^0$, so that $b_4^0 = 60 B_4^0 = F/3$.

3.6.3
The Phenomenological Spin Hamiltonian

It should be noted that the ground manifold does not, in general, exhibit the isotropic behavior of a free spin. Thus, the spin Hamiltonian must contain electron spin operators consistent with the point symmetry of the environment, as well as those for the nuclear spin, if there is a nonzero nuclear spin present. The time-reversal invariance of energy, $H(-t) = H(t)$, dictates existence of the terms in the spin Hamiltonian of the type $B^p S^q I^r$ with $p + q + r = 2n$ ($n > 0$, each p, q, r is an integer), where, $p \leq 1$, $q \leq 2S$, $r \leq 2I$; in what follows the terms, which do not contain **B**, the magnetic field, are considered when $p = 0$, whereas the terms that depend on **B** (Zeeman terms) are considered when $p = 1$.

The various independent terms required for the various symmetries are listed below, where A, B, and C represent B, S, or I.

List of terms required in spin Hamiltonian for various symmetries

Point group	Symbolic operator type	A^2	A^4	A^6	AB	AB^3	AB^5	AB^7	A^2B^2	A^2BC
Spherical		0	0	0	1	0	0	0	1	1
Cubic	O_h, O, T_d	0	1	1	1	1	2	2	2	2
	T_h, T	0	1	2	1	2	3	4	3	4
Hexagonal	$D_{6h}, D_6, D_{3h}, C_{6v}$	1	1	2	2	2	3	5	3	5
	C_{6h}, C_6, C_{3h}	1	1	3	3	3	5	9	5	9
Tetragonal	$D_{4h}, D_4, C_{4v}, D_{2d}$	1	2	2	2	3	5	6	4	6
	C_{4h}, S_4, C_4	1	3	3	3	5	9	11	7	11
Trigonal	D_{3d}, D_3, C_{3v}	1	2	3	2	4	6	8	5	8
	S_6, C_3	1	3	5	3	7	11	15	9	15
Orthorhombic	D_{2h}, D_2, C_{2v}	2	3	4	3	6	9	12	7	12
Monoclinic	C_{2h}, C_2, C_s	3	5	7	5	11	17	23	13	23
Triclinic	C_i, C_1	5	9	13	9	21	33	45	25	45
Axial	$D_{\infty h}, C_{\infty v}$	1	1	1	2	2	2	2	3	5
	$C_{\infty h}$	1	1	1	3	3	3	3	5	9

The following spin-Hamiltonian forms can be written, using the nonzero BS, SI, S^2 terms as listed in the above table. The quadratic and bilinear forms ($p + q + r = 2$) are considered first, for which the spin Hamiltonian, containing all but three terms required for S-values ½ and 1, is as follows.

$$H = \mathbf{S} \cdot \tilde{D} \cdot \mathbf{S} + \mu_B \mathbf{B} \cdot \tilde{g} \cdot \mathbf{S} + \mathbf{S} \cdot \tilde{A} \cdot \mathbf{I} - \mu_n \mathbf{B} \cdot \tilde{g}_n \cdot \mathbf{I} + \mathbf{I} \cdot \tilde{Q} \cdot \mathbf{I}, \tag{3.38}$$

which can be explicitly written for various symmetries, as follows.

3.6.3.1 Triclinic Symmetry

The properties of the unit cell for this case are: (i) all three axes are of unequal length: $a \neq b \neq c$; and (ii) no two axes are at right-angles (90°): $\alpha \neq \beta \neq \gamma \neq 90°$. There is no symmetry here, only rotation through 360° will bring the unit cell back to itself. Here, the maximum possible number of coefficients in Equation 3.38 occur; that is, all of the $\tilde{D}, \tilde{g}, \tilde{A}, \tilde{g}_n, \tilde{Q}$ tensor components. (Each of the matrices $\tilde{g}, \tilde{A}, \tilde{g}_n$ has nine components, and are independent of each other.) However, there occur only five components of $\tilde{D}$ and $\tilde{Q}$ tensors. This is illustrated for $\tilde{D}$, for which three bilinear combinations out of nine $S_\alpha S_\beta; \alpha, \beta = x, y, z$ can be eliminated using the commutation relationship: $[S_x, S_y] = iS_z$; and cyclic permutations, that is, $S_x S_y$ in terms S_x, S_y and S_z; and one more term can be eliminated by using $S_x^2 + S_y^2 + S_z^2 = S(S + 1)$. Similar considerations apply to the $\tilde{Q}$ tensor in terms of the bilinear combinations of the nuclear spin components I_x, I_y and I_z. Further, by a rotation of the coordinate system, requiring three Euler angles, the coupling tensors $\tilde{D}$ and $\tilde{Q}$ can be brought to diagonal forms. Since the trace of a matrix does not change in a transformation, and using $S_x^2 + S_y^2 + S_z^2 = S(S+1)$, one obtains two independent diagonal components of each of $\tilde{D}$ tensor. This constitutes a total of five independent components, along with the three Euler angles. Similar considerations apply to the $\tilde{Q}$ tensor.

3.6.3.2 Monoclinic Symmetry (C_{2h}, C_2, C_{2s})

Here, $a \neq b \neq c$ in the unit cell, and $\alpha = \beta = 90°$; $\gamma \neq 90°$. Only $3A^2$, $5AB$-type quadratic and bilinear terms occur here, where the letters A and B denote B, S, and I terms in the spin Hamiltonian. The principal axis is along the c-axis, which is a twofold axis; this is either normal to the reflection plane or parallel to the twofold rotation axis. This choice eliminates two of the off-diagonal elements for the quadratic and bilinear coupling tensors and two skew-symmetric components of the bilinear form.

There are also present single skew-symmetric components for each of the symmetries $C_{\infty h}$, S_6, C_3, C_{4h}, S_4, C_4, C_{6h}, C_6, and C_{3h} as dictated by the above table listing the independent terms for various symmetries, in addition to the skew-symmetric combinations for triclinic and monoclinic symmetries. However, in practice these play only minor roles. Following these considerations the following forms are obtained for the various point-group symmetries.

3.6.3.3 Orthorhombic Symmetry (D_{2h}, D_2, C_{2v})

$$H = \mu_B(g_x S_x B_x + g_y S_y B_y + g_z S_z B_z) + A_x S_x I_x + A_y S_y I_y$$
$$+ A_z S_z I_z + D\left[S_z^2 - \frac{1}{3}S(S+1)\right] + E\left(S_x^2 - S_y^2\right)$$

3.6.3.4 Tetragonal (D_{4h}, D_4, C_{4v}, D_{2d}, C_{4h}, S_4, and C_4)

$$H = g_\parallel \mu_B B_z S_z + g_\perp \mu_B (B_x S_x + B_y S_y) + A S_z I_z + B(S_x I_x + S_y I_y) + D\left[S_z^2 - \frac{1}{3}I(I+1)\right]$$

3.6.3.5 Cubic (O_h, O, T_d, T_h, and T) and Spherical Symmetry

For this case, there do not exist any quadratic terms, only one bilinear form remains:

$$H = g\mu_B \mathbf{B} \cdot \mathbf{S} + A\mathbf{S} \cdot \mathbf{I}$$

As for the terms containing nuclear spin in the above spin Hamiltonians, the nuclear Zeeman interaction is always considered to be isotropic: $-g_n \mu_n \mathbf{B} \cdot \mathbf{I}$, where g_n is the nuclear g-factor.

3.6.3.6 Additional Spin-Hamiltonian Terms with Higher Powers of Components of S

The following fourth-order terms can be written.

Cubic symmetry:

$$H_{Cubic}^s = \frac{1}{6}a\{S_x^4 + S_y^4 + S_z^4 - \frac{1}{5}S(S+1)[3S(S+1)-1]\} \tag{3.39}$$

Axial or hexagonal symmetry with the rotation axis parallel to the z-axis:

$$H_{Axial,hexagonal} = \frac{1}{180} F[35S_z^4 - 30S(S+1)S_z^2 + 25S_z^2 - 6S(S+1) + 3S^2(S+1)^2] \quad (3.40)$$

For D_{4h}, D_4, D_{4v}, and D_{2d} symmetries, it is necessary to sum Equations 3.39 and 3.40 to include the quartic spin operators, where the tetragonal axis is the z-axis, and the x- and y-axes are determined by twofold axes or mirror planes. As for D_{3d}, D_3, and C_{3v} symmetries, one again has to sum Equations 3.39 and 3.40), but now the z-axis in Equation 3.40 is to be taken as the threefold axis, and the axes of the three Cartesian coordinates in Equation 3.39 all make an angle of $\cos^{-1}(1/\sqrt{3})$ with the z-axis. Their orientations are dictated by the twofold axes or mirror planes. Accordingly, one can write for tetragonal and trigonal distortions, the following spin Hamiltonians.

Tetragonal distortion:
$H = H_{Cubic} + H_{Axial}$, as given by Equations 3.39 and 3.40 above.

Trigonal distortion: By transforming Equation 3.39 to a coordinate system with the z-axis along the trigonal axis, and x and y perpendicular to it, one obtains:

$$H = \frac{1}{180}(F-a)[35S_z^4 - 30S(S+1)S_z^2 + 25S_z^2 - 6S(S+1)$$
$$+ 3S^2(S+1)^2] + \frac{a\sqrt{2}}{36}[S_z(S_+^3 + S_-^3) + (S_+^3 + S_-^3)S_z]$$

where $S_\pm = S_x \pm iS_y$

For f^7 configuration: $S = \frac{7}{2}$ (Gd^{3+}, Eu^{2+}), one requires additional sixth-order terms, as follows, for the various symmetries.

Cubic:

$$H = B_6(O_6^0 - 21O_6^4)$$

AB³ terms in the spin Hamiltonian

These appear in the following cases:

O_h, O, T_d symmetries.:

$$H = u\mu_B \left\{ B_x S_x^3 + B_y S_y^3 + B_z S_z^3 - \frac{1}{5}[3S(S+1)-1]\mathbf{B}\cdot\mathbf{S} \right\}$$
$$+ U \left\{ I_x S_x^3 + I_y S_y^3 + I_z S_z^3 - \frac{1}{5}[3S(S+1)-1]\mathbf{I}\cdot\mathbf{S} \right\} \quad (3.41)$$

T_h and T symmetries:

$$H \propto B_x[S_x(S_y^2 - S_z^2) + (S_y^2 - S_z^2)S_x]$$
$$+ B_y[S_y(S_z^2 - S_x^2) + (S_z^2 - S_x^2)S_y]$$
$$+ B_z[S_z(S_x^2 - S_y^2) + (S_x^2 - S_y^2)S_z]$$

For lower symmetries, the expressions are more complicated. Furthermore, one can write expressions similar to Equation 3.40 for BS^5 and IS^5 terms.

Spherical symmetry

The A^2B^2 term here has the form similar to that of the quadrupole interaction in free ions and ions:

$$H \propto 3(S \cdot I)^2 + \frac{3}{2} S \cdot I - S^2 I^2.$$

The terms of the type I^2 are usually referred to as the quadrupole interaction terms. The terms I^2 and $S^2 I^2$ terms may be proportional to the quadrupole moment of the nucleus. In lower symmetry, many more complicated terms of the type A^2B^2 occur.

3.6.4
The Generalized Spin Hamiltonian

When the real local symmetry at the site of a paramagnetic ion is taken into account, the concept of *generalized spin Hamiltonian* is arrived at. Koster and Statz (1959a, 1959b) were the first to propose such a Hamiltonian for cubic symmetry when they found, for example, the existence of the term $S_i^3 B_i$. On the other hand, Ray (1964) – who considered the most general spin Hamiltonian in cubic symmetry – found the occurrence of both $S_i^3 B_i$ and $S_i^5 B_i$ terms. The second-order fine-structure term, Equation 3.37, requires additional cross terms $S_x S_y + S_y S_x$ in C_2 symmetry, equivalent to the fact that the original Cartesian system does not coincide with the principal-axis frame of the ZFS tensor $\tilde{D}$. In a single crystal, the local symmetry dictates consideration of 32 point-groups, of which 11 are axially symmetric, and there exist 11 distinguishable classes of symmetry to be considered for describing EPR spectra, as explained by Roitsin (1981).

3.6.5
The Effective Spin Hamiltonian for EPR

The effective spin S describes the close-lying levels (within a few cm^{-1}), widely separated from all the upper levels. The effective spin S is such that $2S + 1$ = total number of levels in the group (further details are provided by Rudowicz and Misra, 2001). It is helpful to simplify the use of complex interactions encountered by transition-series ions by using this effective spin, and a Hamiltonian expressed in terms of either the actual, or an effective, spin. The following general spin Hamiltonian is responsible for EPR transitions:

$$H = \mu_B S \cdot \tilde{g} \cdot B - \mu_N B \cdot \tilde{g}_N \cdot I + S \cdot \tilde{A} \cdot I + \sum_{\substack{k=0 \\ \text{even}}}^{\leq S} \sum_{q=-k}^{+k} B_k^q O_k^q(S) + \sum_{\substack{n=0 \\ \text{even}}}^{\leq 2I} \sum_{m=-n}^{n} C_n^m O_n^m(I)$$

where $O_k^q(S)$, $O_n^m(I)$ and C_n^m represent the electron and nuclear Stevens spin operators and the crystal field parameters, respectively. The $O_k^q(S)$ and $O_n^m(I)$ operators are referred to as equivalent operators, as discussed in Section 3.2; these can be expressed in terms of the spin operators S_i, I_i; $i = x, y, z$. Tables of matrix elements of $O_k^q(S)$ are available; see for example, Misra (1999), and listed in Chapter 7 of this book; from these, the nuclear spin operators $O_n^m(I)$ can be obtained by replacing S with I. The specific values of the coefficients B_k^q and C_n^m characterizing a transition-metal ion in a crystal or ligand field depend on the interactions and the symmetry site of the ion in a lattice, and are determined from EPR spectra – in particular, from line positions.

3.7
Concluding Remarks

This chapter has provided an overview of the various types of interaction experienced by a paramagnetic ion in an environment that exists, most frequently, in a crystal, and which is constituted by charged and/or paramagnetic particles, that are ions or molecules. The totality of these interactions is represented by various types of Hamiltonian. An outline is provided of what these types are, and how they are manipulated to calculate the energy levels and corresponding eigenfunctions of transition-metal ions, required for interpreting EPR spectra at various frequencies. Additional details on these subjects are provided in Chapters 7, 8, and 9 of this book.

Acknowledgments

I am grateful to Professor C.P. Poole, Jr, for helpful comments in improving the presentation of this chapter, and to the Natural Sciences and Engineering Research Council (NSERC) of Canada for partial financial support.

Pertinent Literature

The main sources of information for this chapter were: (i) Abragam and Bleaney (1970); this is a classic on the subject, though now slightly dated, having been written at a time when high-frequency spectrometers were not available; (ii) Poole and Farach (1972); (iii) Pake and Estle (1973); and (iv) Pilbrow (1990).

References

Abragam, A. and Bleaney, B. (1970) *Electron Paramagnetic Resonance of Transition Ions*, Clarendon Press.

Abragam, A. and Pryce, M.H.L. (1951a) *Proc. R. Soc. A*, **205**, 135.

Abragam, A. and Pryce, M.H.L. (1951b) *Proc. R. Soc. A*, **206**, 173.

Ballhausen, C.J. (1962) *Introduction to Ligand Field Theory*, McGraw-Hill, New York.

Ballhausen, C.J. and Gray, H.B. (1965) *Molecular Orbital Theory*, Benjamin, New York.

Banci, L., Bencini, A., Benelli, C., Gatteschi, D., and Zanchini, C. (1982) *Struct. Bond.*, **52**, 37.

Bencini, A., Benelli, C., and Gatteschi, D. (1979) *J. Magn. Reson.*, **34**, 653.

Bersuker, I.B. (1984a) *The Jahn-Teller Effect and Vibronic Interactions in Modern Chemistry*, Plenum Press, New York.

Bersuker, I.B. (1984b) *The Jahn-Teller Effect: A Bibliographic Review*, IFI/Plenum Press, New York.

Bill, H. (1984) Observation of the Jahn-Teller effect with EPR, in *The Dynamical Jahn-Teller Effect in Localised Systems* (eds Yu.E. Perlin and M. Wagner), Elsevier, Amsterdam, p. 709.

Bleaney, B. and Bowers, K.D. (1952) *Proc. Roy. Soc. Ser A*, **214**, 451–465.

Bradbury, J.H. and Newman, D.J. (1968) *Chem. Phys. Lett.*, **1**, 44.

Clare, J.F. and Devine, S.D. (1984) *J. Phys. C*, **17**, L581.

Coffman, R.E. (1965a) *Phys. Lett.*, **19**, 475.

Coffman, R.E. (1965b) *Phys. Lett.*, **21**, 381.

Coffman, R.E., Lyle, D.L., and Mattison, D.R. (1968) *J. Phys. Chem.*, **72**, 1392.

Dang, L.S., Buisson, R., and Williams, F.I.B. (1974) *Journal de physique*, **35**, 49.

Deeth, R.J. and Gerloch, M. (1985) *Inorg. Chem.*, **24**, 4490.

Edgar, A. (1976) *J. Phys. C*, **9**, 4303.

Elbers, G., Remme, S., and Lehman, G. (1986) *Inorg. Chem.*, **25**, 896.

Englman, R. (1972) *The Jahn-Teller Effect in Molecules and Crystals*, Wiley-Interscience, New York.

Figgis, B.N. (1966) *Introduction to Ligand Fields*, John Wiley & Sons, Inc., New York.

Gerloch, M. (1983) *Magnetism and Ligand Field Analysis*, Cambridge University Press.

Griffith, J.S. (1961) *The Theory of Transition-Metal Ions*, Cambridge University Press.

Griffith, J.S. (1964) *J. Chem. Phys.*, **41**, 576.

Ham, F.S. (1965) *Phys. Rev.*, **A138**, 1727.

Ham, F.S. (1968) *Phys. Rev.*, **166**, 307.

Ham, F.S. (1972) Jahn-Teller effects in electron paramagnetic resonance spectra, in *Electron Paramagnetic Resonance* (ed. S. Geschwind), Plenum, New York, pp. 1–119.

Herrington, J.R., Estle, T.L., and Boatner, L.A. (1971) *Phys. Rev.*, **B3**, 2933.

Herrington, J.R., Estle, T.L., and Boatner, L.A. (1973) *Phys. Rev.*, **B7**, 3003.

Hochli, U.T. and Estle, T.L. (1967) *Phys. Rev. Lett.*, **18**, 128.

Keijzers, C.P. and De Boer, E. (1972) *J. Chem. Phys.*, **57**, 1277.

Keijzers, C.P. and De Boer, E. (1975a) *Molec. Phys.*, **29**, 1007.

Keijzers, C.P. and De Boer, E. (1975b) *Molec. Phys.*, **29**, 1743.

Keijzers, C.P., DeVries H.J.M., and Van Der Avoird, A. (1972) *Inorg. Chem.*, **11**, 1388.

Kneubuhl, F.K. (1963) *Phys. Kondens. Mater.*, **1**, 410.

Koster, G.F. and Statz, H. (1959a) *Phys. Rev.*, **113**, 445.

Koster, G.F. and Statz, H. (1959b) *Phys. Rev.*, **115**, 1568.

Lea, K.R., Leask, J.M., and Wolf, W.P. (1962) *J. Phys. Chem. Solids*, **23**, 1381, p. 1.

Messiah, A. (1965) *Quantum Mechanics*, vol. I, Eq. [A18], North-Holland, Amsterdam.

Misra, S.K. (1999) Transition Ion Hamiltonians, in *Handbook of Electron Spin Resonance* (eds C.P. Poole Jr and H.A. Farach), A. I. P. Press, Springer-Verlag, New York, pp. 115–152.

Misra, S.K., Mikolajczak, P., and Lewis, N.R. (1981) *Phys. Rev. B*, **24**, 3279.

Mombourquette, M.J., Tenant, W.J., and Weil, J.A. (1986) *J. Chem. Phys.*, **85**, 68.

Müller, A. (1967) Jahn-Teller effects in magnetic resonance, in *Magnetic Resonance and Relaxation* (ed. R. Blinc), North Holland, Amsterdam, p. 192.

Murrietta, S.H., Rubio, O.J., and Aguilar, S.G. (1979) *Phys. Rev. B*, **19**, 5516.

Newman, D.J. (1982) *J. Phys. C*, **15**, 6627.

Newman, D.J. and Urban, W. (1975) *Adv. Phys.*, **24**, 793.

Newman, D.J., Price, D.C., and Runcimen, W.A. (1978) *Ann. Mineral.*, **63**, 1278.

O'Brien, M.C.M. (1964) *Proc. R. Soc. A*, **281**, 340.

Orgel, L.E. (1960) *Introduction to Transition Chemistry*, John Wiley & Sons, Inc., New York.

Pake, G.E. and Estle, T.E. (1973) *The Physical Principles of Electron Paramagnetic Resonance*, 2nd edn, W. A. Benjamin, Inc., Reading, MA.

Perlin, Yu.E. and Wagner, M. (1984) *The Dynamical Jahn-Teller Effect in Localized Systems*, Elsevier, Amsterdam.

Pilbrow, J.R. (1990) *Transition Ion Electron Paramagnetic Resonance*, Clarendon Press, Oxford.

Polder, D. (1942) *Physica*, **9**, 709.

Poole, C.P., Jr and Farach, H.A. (1972) *The Theory of Magnetic Resonance*, Wiley-Interscience, New York.

Pryce, M.H.L. (1950) *Phys. Rev.*, **80**, 1107.

Ray, T. (1964) *Proc. R. Soc. A*, **277**, 76.

Reynolds, R.W. and Boatner, L.A. (1975) *Phys. Rev.*, **B12**, 4735.

Roitsin, A.B. (1981) *Phys. Status Solidi B*, **104**, 11.

Rudowicz, C. and Misra, S.K. (2001) *Appl. Spectrom. Rev.*, **36**, 11.

Schaffer, C.E. (1968) *Struct. Bond.*, **5**, 68.

Sealey, R., Hyde, J.S., and Antholine, W.E. (1985) Electron Spin Resonance, in *Modern Physical Methods in Biochemistry* (eds A. Heuberger and L.L.M. Van Deenan), Elsevier, New York, Chap. 2, p. 69.

Sester, G.G., Barksdale, A.O., and Estle, T.L. (1975) *Phys. Rev.*, **B12**, 4720.

Seth, V.P., Yadav, A., and Bansal, R.S. (1985) *Phys. Status Solidi B*, **131**, 255.

Sierro, J. (1967) *J. Phys. Chem. Solids*, **28**, 417.

Slater, J.C. (1951a) *Phys. Rev.*, **81**, 385.

Slater, J.C. (1951b) *The Self-Consistent Field for Molecules and Solids*, vol. 4 *of Quantum Theory of Molecules and Solids*, McGraw-Hill, New York.

Slichter, C.P. (1963) *Principles of Magnetic Resonance*, Harper and Row, New York.

Stedman, G.E. (1969) *J. Chem. Phys.*, **50**, 1461.

Steenkamp, P. (1984) *Aust. J. Chem.*, **37**, 679.

Stoneham, A.M. (1969) *Rev. Mod. Phys.*, **41**, 82.

Sturge, M.D. (1967) The Jahn-Teller effect in solids, in *Solid State Physics*, vol. 20 (eds F. Seitz, D. Turnbull, and H. Ehrenreich), Academic Press, New York, p. 90.

Tinkham, M. (1964) *Group Theory and Quantum Mechanics.* McGraw-Hill, New York.

Wertz, J.E. and Bolton, J.R. (1972) *Electron Spin Resonance: Elementary Theory and Practical Applications*, McGraw-Hill, New York.

**Part One
Experimental**

4
Spectrometers

4.1
Zero-Field EPR
Sushil K. Misra

4.1.1
Introduction

This section, which serves as an introduction to Section 4.2, covers low-frequency spectrometers, and includes a discussion of frequency-swept EPR in low magnetic fields. In zero-field resonance (ZFR), the frequency is swept in the absence of a static magnetic field, with only the microwave magnetic field inducing transitions being present. Frequency-swept EPR is not always ZFR, however, as the frequency may be swept in a finite magnetic field. ZFR and normal fixed frequency, magnetic field-swept EPR are complementary techniques. The short discussion presented here is based on that of Pilbrow (1990) which, in turn, was based on the earlier reports of Bramley and Strach (1983, 1985). ZFR spectrometer design and practice have been reviewed by Newman and Urban (1975).

The main advantages of ZFR, and low-field frequency-swept EPR over normal CW-EPR are that:

- The field-dependent line broadening is considerably reduced, or even completely removed, so that zero-field lines are generally narrower than those in the corresponding CW-EPR spectrum. Furthermore, orientational errors do not occur in ZFR.

- Only one spectrum is observed for magnetically inequivalent, but otherwise identical, EPR centers.

- Quadrupole effects appear in first order in ZFR.

- Hyperfine coupling also contributes to the zero-field splitting (ZFS).

- As the magnetic field is absent, there is no competition for quantization of the spin S, so that powders, glasses, and polycrystalline materials are amenable to

ZFR thereby avoiding the difficult overlap problems that occur in EPR when studying these systems.

- Theoretical treatment of ZFR is simpler, as it requires only one diagonalization of the spin Hamiltonian. This is unlike CW-EPR, where many diagonalizations are required (as discussed in Chapter 8).
- It is possible to measure large ZFS over $1\,\text{cm}^{-1}$ (~30 GHz), as the frequency can be swept over a wide range, including the near infra-red (1–40 GHz). Presently, within the context of multifrequency EPR, frequencies can be accessed up to about 1000 GHz.

In contrast, the main disadvantages of ZFR are: (i) the frequency (v); (ii) the sample filling factor (η); and (iii) the unloaded quality factor (Q_0), which change across a spectral line, unlike those in CW-EPR. As a consequence, it is necessary to take into account the frequency dependences of η and Q_0, in addition to that of the Boltzmann factor. This renders the quantitative evaluation of spectra more difficult in the frequency-swept case.

4.1.2
Preliminary Theory of ZFR

As an illustrative example, the case of $S = 1$ is discussed. The zero-field spin Hamiltonian is expressed as

$$H = D\left\{S_z^2 - \frac{1}{3}S(S+1)\right\} + E\left(S_x^2 - S_y^2\right), \tag{4.1.1}$$

where D and E are ZFS parameters. The interaction of the excitation field is expressed as

$$H_1 = \mu_B \sum_{i=x,y,z} g_i l_i S_i B_1, \tag{4.1.2}$$

where B_1 represents the excitation field that induces transitions among energy levels, μ_B is the Bohr magneton, and l_i ($i = x, y, z$) are the direction cosines of B_1 with respect to the crystal axes. The eigenvalues of the 3 × 3 matrix corresponding to Equation 4.1.1 are: $E_0 = -\frac{2}{3}D$, $E_\pm = \frac{1}{3}D \pm E$, with the corresponding eigenvectors being $|0\rangle, [|1\rangle \pm |-1\rangle]/\sqrt{2}$, respectively. The interaction with the excitation field, given by Equation 4.1.2, induces transitions between the levels m and n, the transition probabilities of which are given by $P_{mn} = |\langle m|\mu_B B_1 g_i l_i S_i|n\rangle|^2$. It emerges that not all transition probabilities are nonzero. For example, for the orientation of B_1 along the x-axis, it is easily seen that the only nonzero probability is $P_{+0} = g_x^2$ of the three possible transitions for this polarization of B_1, with the resonance condition being $hv = D + E$. Likewise, for B_1 polarized along the y-axis, only $P_{-0} = g_y^2$ is nonzero. However, for B_1 polarized along the z-axis, all transition probabilities are zero. For an arbitrary orientation of B_1, both the transitions expected for polarizations along x and y occur, with the relative intensities depending on the x and y components of the microwave field. In any case, the ZFR experiment does not provide the relative signs of D and E, a situation similar to that in nuclear quadrupole resonance (NQR).

As for the case of $S = 3/2$, it is easily seen that the ZFS is $2\sqrt{(D^2 + 3E^2)}$, which represents the spacing between the sets of doubly degenerate levels $\left|\pm\frac{1}{2}\right\rangle$ and $\left|\pm\frac{3}{2}\right\rangle$, and only a single transition is observed. Furthermore, so one cannot determine D and E separately.

Finally, it is noted that in zero magnetic fields, the polarization of the transitions is determined solely by the polarization of the B_1-field.

4.1.3
The ZFR Spectrometer

Currently, there are several very sound spectroscopic and technical reasons which dictate the use of the microwave range of frequencies of 1 to 40 GHz for ZFR (Pilbrow, 1990). In principle, all that is needed is a matched system, consisting of a tunable frequency source, a transmission line containing the sample, and a detector to observe ZFR. However, the sensitivity of such a simple system is far from optimum. It is possible to use modulation methods with lock-in phase-sensitive detection (PSD) as in normal EPR. Sine-wave modulation is not appropriate, as the Zeeman effect is not linear at low fields for $S \geq 1$ ions. The use of bipolar Zeeman modulation, instead of ordinary square-wave modulation, and detection at twice the modulation frequency ($2\omega_m$) was described by Bramley and Strach (1983). This results in an output signal, with half the signal grounded, which is the oscillating difference between the field-on and field-off spectra. A block diagram of a field-modulated nonresonant transmission spectrometer is shown in Figure 4.1.1. Sweep oscillators in octave bands up to 8 GHz, in waveguide bands up to 40 GHz, are the most versatile. There is lower noise if reflex klystrons are used instead of sweep oscillators, but their narrow frequency band presents a limitation. The sample can be inserted through a sample port in the waveguide above 8 GHz, whereas in coaxial transmission lines it is simply packed into the air space between the inner and outer conductors. Although, there are problems due to reflections, these methods can be used up to 26 GHz. Coaxial lines are particularly suitable for ZFR because of their frequency independence. About 10^{17} Fe^{3+} ions were required in order to obtain satisfactory results (Bogle et al., 1961).

4.1.3.1 Examples of ZFR Spectra
ZFR spectra due to Cr^{3+} ions in ruby (Al_2O_3), recorded using bipolar field modulation and phase-sensitive detection at $2\omega_m$, are shown in Figure 4.1.2; the relevant energy levels, along with the field-on, field-off transitions are shown in Figure

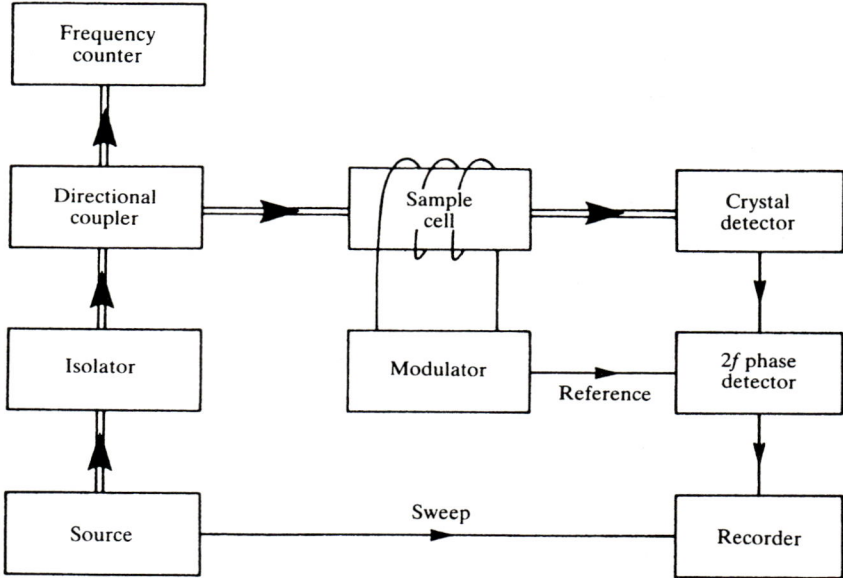

Figure 4.1.1 Block diagram of a ZFR spectrometer operating in a nonresonant transmission mode. Adapted from Pilbrow (1990).

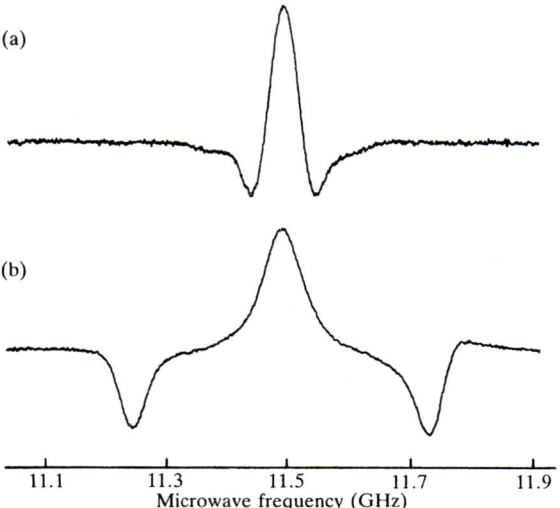

Figure 4.1.2 Cr^{3+} ZFR spectrum of synthetic ruby (α-Al_2O_3), with the center of the transition located at 11.493 GHz at 295 K. The experimental set-up was such that the microwave polarization was perpendicular to the threefold symmetry (z) axis of the crystal, and the square-wave bidirectional on-off magnetic field modulation had a frequency of 700 Hz with the direction being parallel to the z-axis. The field-modulation strength, B_m, used was ±2 mT (a) and 9 mT (b). The small distortions of the line shapes situated in the wings are most likely due to the 10% abundant isotope ^{53}Cr ($I = 3/2$). The Teflon tube that supported the sample passed through the waveguide in a transmission-mode spectrometer. Adapted from Pilbrow (1990).

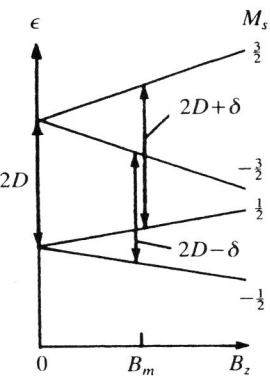

Figure 4.1.3 Diagram showing the energy-levels and transitions occurring at zero field, and at the field B_m for Cr^{3+} ($S = 3/2$) in ruby. Here, $\delta = g\mu_B B_m$. The modulation field B_m is directed along the symmetry axis, and the B_1 field is perpendicular to this direction. Adapted from Pilbrow (1990).

4.1.3. The phase relationships between the modulation and the spectral responses, at their line centers, of the two field-on transitions and the single field-off transition (which is really two superimposed transitions) are illustrated in Figure 4.1.4. A three-dimensional (3-D) representation of ZFR using the bipolar field modulation is shown in Figure 4.1.5, wherein the magnitude of δ is adjusted in Bramley and Strach's experiments to give a resulting signal with a line shape which has the appearance of a second-derivative (as observed in Figure 4.1.2a), although it is not really a true second derivative.

4.1.4
Advantages of Using Resonant Systems

The signal is enhanced by the Q-factor when the sample is placed in a resonant system. If a reflection bridge is used, the operation is simplified, since in that case only a single adjustable coupling is required as compared with two adjustments needed for a transmission system (Bramley and Strach, 1983). A cavity was described by Urban (1966), tunable from 8 to 13 GHz, wherein a plunger or dielectric load had to be moved smoothly. Both, Shing (1972) and Shing and Buckmaster (1976) described the use of a re-entrant cavity which was tunable from 1 to 4 GHz and could be used at both 77 and 290 K (see Chapter 4.2). Newman and Urban (1975) have described helix, or slow-wave structures, in considerable detail. In contrast, Bramley and Strach (1985) used a tunable loop-gap resonator (LGR) with a fine rutile rod which was moved up and down the center of the capacitive gap over more than one octave.

4.1.5
Examples of ZFR

Several examples of ZFR are available, of which the following are discussed by Pilbrow (1990), which cover the cases of: (i) $S = \frac{1}{2}$, $I = 7/2$ [VO(IV)]; (ii) $S = 1$ (Ni^{2+}); (iii) $S = 3/2$ [intermediate spin Fe(III)]; (iv) $S = 3/2$, $I = 7/2$ [$^{53}Cr^{3+}$];

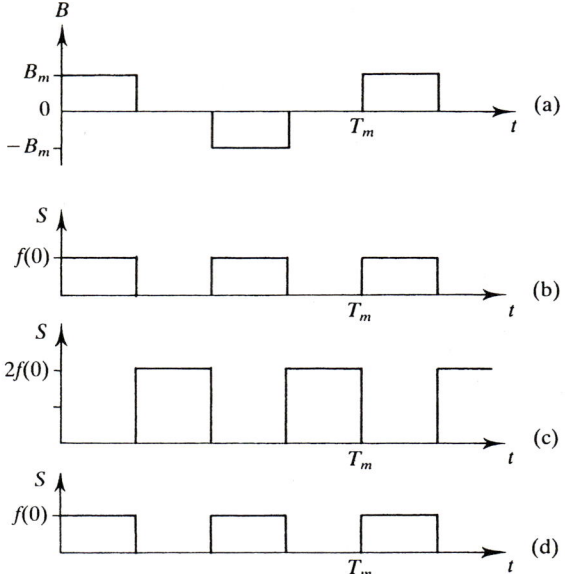

Figure 4.1.4 Schematic of the form of detected signal in Figure 4.1.2. (a) Bipolar modulation of the amplitude B_m; (b) The spectral response at the center of the line at $2D + \delta$ ($\delta = g\mu_B B_m$); (c) The spectral response at the center of the line at $2D$, which is of twice the intensity as the two signals are superimposed on each other; (d) The spectral response at the center of line at $2D - \delta$. It is noted that the responses in (b) and (d) are out of phase with that in (c). This results in a 'second-derivative'-like appearance when B_m (or δ) is adjusted to obtain the spectrum shown in Figure 4.1.2a. Adapted from Pilbrow (1990).

(v) $S = 5/2$ [Fe^{3+}]; and (vi) $S = 5/2$, $I = 5/2$ [Mn^{2+}]. For further detail, the reader is referred to Pilbrow (1990), and the references cited therein. The extensive literature concerning ZFR and low magnetic field measurements on $S = 7/2$ [Gd^{3+}] in the lanthanum ethyl sulfate lattice has been summarized by Shing and Buckmaster (1976). These authors discuss the discrepancies between the values of the ZFS parameters obtained at high and weak magnetic fields by various research groups. At this point, only the case of Mn^{2+} will be discussed in some detail, as it exhibits many important aspects of ZFR.

4.1.5.1 The Case of the Mn^{2+} Ion

The zero-field energy levels, spin states, transition energies and intensities for high-spin d^5 ions ($S = 5/2$) in a cubic field with trigonal and tetragonal distortions, as listed by Pilbrow (1990; see Tables 12.2 and 12.3 of Pilbrow's book), are reproduced here in Figures 4.1.6 and 4.1.7, respectively. There result three Kramers' doublets in zero field (see also Chapter 3). The effect of different values of spin

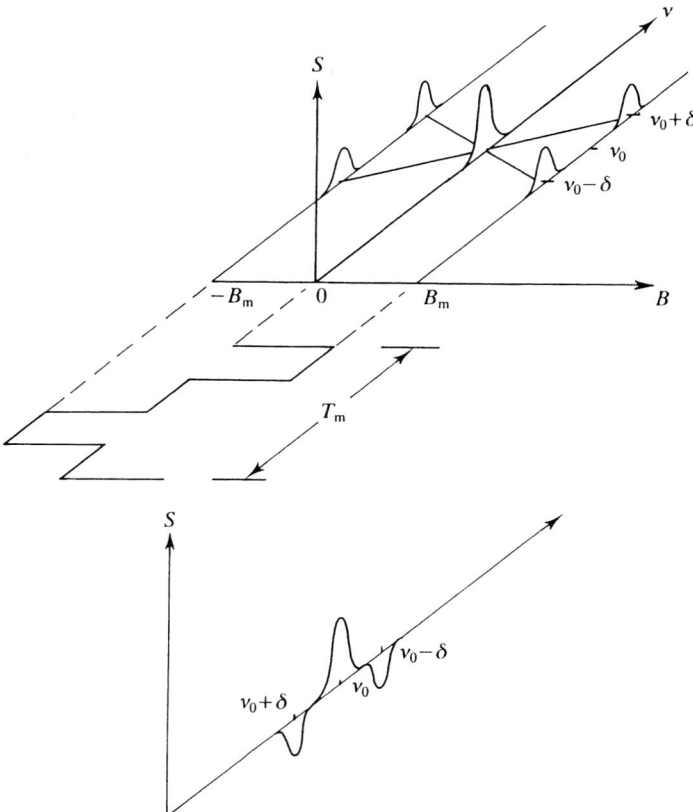

Figure 4.1.5 A three-dimensional representation of the spectra obtained for the schematic shown in Figure 4.1.4, which also shows the phase relationships. The resulting spectrum is that shown in Figure 4.1.2. It is clearly seen that the in- and out-of-phase responses differ in amplitude by a factor of two. Adapted from Pilbrow (1990).

Hamiltonian (SH) parameters on the expected ZFR resonant frequencies has been discussed by Bramley and Strach (1983). Figure 4.1.8 exhibits the energy-level diagram for $S = 5/2$ as a function of the magnetic field parallel to a threefold symmetry axis showing the field-dependent and ZFR transitions (further details are provided in the caption of Figure 4.1.8.) Due to the presence of hyperfine structure ($I = 5/2$), Mn^{2+} represents a more complicated case than Fe^{3+}, which shows only fine structure, both possessing electron spin $S = 5/2$. As compared to three transitions for Fe^{3+}, one expects a total of 630 transitions for Mn^{2+}. Bramley and Strach (1981) studied the ZFR of Mn^{2+} in $MgSO_4 \cdot 7H_2O$, and found significant differences between the ZFR SH parameters and those reported earlier as determined by EPR; such differences were ascribed to crystal misalignment and field errors. The same

TABLE 12.2 Zero-field energy levels, spin states, transition energies and intensities for high-spin d^5 ions in a cubic field with trigonal distortion (after Bramley and Strach 1983)

Energy

$$\varepsilon_1 = \tfrac{1}{3}D - \tfrac{1}{2}(a - F) + \tfrac{1}{6}[(18D + a - F)^2 + 80a^2]^{1/2}$$

$$\varepsilon_2 = -\tfrac{2}{3}D + (a - F)$$

$$\varepsilon_3 = \tfrac{1}{3}D - \tfrac{1}{2}(a - F) - \tfrac{1}{6}[(18D + a - F)^2 + 80a^2]^{1/2}$$

Spin state

$$|\phi_1\rangle = \cos\alpha|\pm\tfrac{5}{2}\rangle \mp \sin\alpha|\mp\tfrac{1}{2}\rangle$$

$$|\phi_2\rangle = |\pm\tfrac{3}{2}\rangle$$

$$|\phi_3\rangle = \sin\alpha|\pm\tfrac{5}{2}\rangle \pm \cos\alpha|\mp\tfrac{1}{2}\rangle$$

Transition energies

$$\Delta\varepsilon_1 = \varepsilon_2 - \varepsilon_3 \approx 2D + \tfrac{5}{3}(a - F) + \cdots$$

$$\Delta\varepsilon_2 = \varepsilon_1 - \varepsilon_2 \approx 4D - \tfrac{4}{3}(a - F) + \cdots$$

$$\Delta\varepsilon_3 = \varepsilon_1 - \varepsilon_3 \approx 6D + \tfrac{1}{3}(a - F) + \cdots$$

Intensities

x,y-polarization	z-polarization
$\tfrac{5}{2} + \tfrac{3}{2}\cos^2\alpha$	0
$\tfrac{5}{2} + \tfrac{3}{2}\sin^2\alpha$	0
$\tfrac{9}{8}\sin^2 2\alpha$	$\tfrac{9}{2}\sin^2 2\alpha$

where $\tan 2\alpha = \sqrt{(80)}a/(a - F + 18D)$. The approximate transition energies correspond to small values of α or $\tan 2\alpha$.

Figure 4.1.6 The information noted in Section 4.1.4 to be included for d^5 ions for the case of a cubic field with trigonal distortion. Adapted from Pilbrow (1990); Table 12.2 from Bramley and Strach (1983).

group also studied the ZFR of Mn^{2+} in four Tutton salts (Bramley and Strach, 1984), and found more complex spectra than that in $MgSO_4 \cdot 7H_2O$; subsequently, they were able to determine the sign of the parameter D from the second-order shifts in the hyperfine structure (see Chapter 8). As for Mn^{2+} in NH_4Cl (Bramley and Strach, 1981), a better agreement was found between ZFR and EPR than that for the Tutton salts. A schematic energy-level diagram for the hyperfine case of $S = I = 5/2$ is shown in Figure 4.1.9, including the stick spectra for the expected 11-line and six-line ZFR spectra (further details are provided in the caption of Figure 4.1.9.)

TABLE 12.3 Zero-field energy levels, spin states, transition energies and intensities for high-spin d^5 ions in a cubic field with tetragonal distortion. (After Bramley and Strach 1983.)

Energy

$$\varepsilon_1 = \tfrac{4}{3}D - \tfrac{3}{2}(a - \tfrac{2}{3}F) + [(2D + a - \tfrac{2}{3}F)^2 + \tfrac{5}{4}a^2]^{1/2}$$

$$\varepsilon_1 = \tfrac{4}{3}D - \tfrac{3}{2}(a - \tfrac{2}{3}F) - [(2D + a - \tfrac{2}{3}F)^2 + \tfrac{5}{4}a^2]^{1/2}$$

$$\varepsilon_3 = -\tfrac{8}{3}D + a + \tfrac{2}{3}F$$

Spin state

$$|\phi_1\rangle = \cos\alpha\,|\pm\tfrac{5}{2}\rangle + \sin\alpha\,|\mp\tfrac{3}{2}\rangle$$

$$|\phi_2\rangle = \sin\alpha\,|\pm\tfrac{5}{2}\rangle - \cos\alpha\,|\mp\tfrac{3}{2}\rangle$$

$$|\phi_3\rangle = |\pm\tfrac{1}{2}\rangle$$

Transition energies

$$\Delta\varepsilon_1 = \varepsilon_2 - \varepsilon_3 \approx 2D - \tfrac{5}{2}(a + \tfrac{2}{3}F) + \cdots$$

$$\Delta\varepsilon_2 = \varepsilon_1 - \varepsilon_2 \approx 4D + 2(a + \tfrac{2}{3}F) + \cdots$$

$$\Delta\varepsilon_3 = \varepsilon_1 - \varepsilon_3 \approx 6D - \tfrac{1}{2}(a + \tfrac{2}{3}F) + \cdots$$

Intensities

x,y-polarization	z-polarization
$\cos^2\alpha$	0
$\tfrac{3}{2}\cos^2 2\alpha$	$8\sin^2 2\alpha$
$4\sin^2\alpha$	0

where $\tan 2\alpha = \sqrt{5}a/[2(a + \tfrac{2}{3}F + 4D)]$. The approximate transition energies correspond to small values of α or $\tan 2\alpha$.

Figure 4.1.7 The information noted in Section 4.1.4 to be included for d^5 ions for the case of a cubic field with tetragonal distortion. Adapted from Pilbrow (1990); Table 12.3 from Bramley and Strach (1983).

4.1.6
Concluding Remarks

The above discussion on ZFR studies on the Mn^{2+} ion clearly shows that the spin-Hamiltonian parameters determined by ZFR differ from those determined by EPR, in that significant errors arise in ZFR. These errors are due mainly to the inherent complexity caused by structural and magnetic inequivalence, as well as to the assumptions used in EPR about the Zeeman axis being parallel to those for the other interactions. The advantages and disadvantages of ZFR have been discussed in Section 4.1.1. Consequently it would appear that, in order to obtain a complete picture of a paramagnetic system, both EPR and ZFR should ideally be used and the two sets of data correlated.

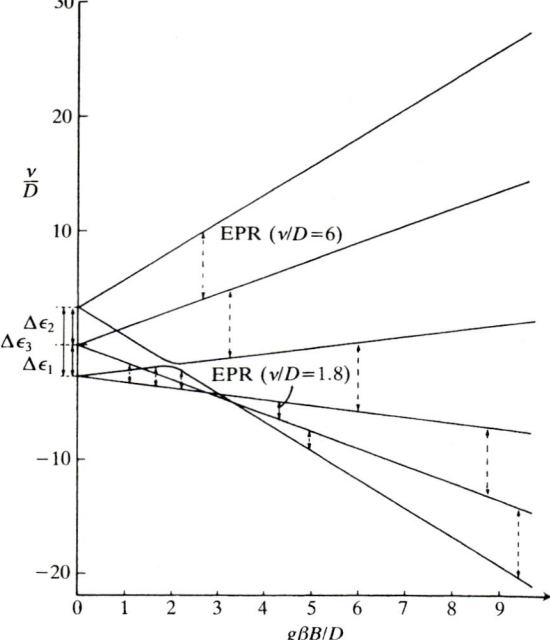

Figure 4.1.8 Energy levels for $S = 5/2$ as functions of the magnetic field, which is parallel to a threefold symmetry axis for Fe^{3+} in sapphire. Here, β is the Bohr magneton. The unit of energy is the ZFS parameter, D. The expected three ZFR, and five strong-field transitions are shown for a constant microwave frequency $\nu(=30\,\text{GHz}) \gg D$. The three X-band ($\nu \sim 9.3\,\text{GHz}$) transitions used to determine the minimum energy gap between the two repelling levels are also shown. Adapted from Pilbrow (1990).

Pertinent Literature

Bramley and Strach (1984) provide a summary of the literature on ZFR up to 1981, whereas Shing (1972) and Newman and Urban (1975) have reviewed aspects of ZFR spectrometer design and practice (see also Chapter 4.2). The chapter on ZFR in the book by Pilbrow (1990) provides a good, concise coverage of this subject.

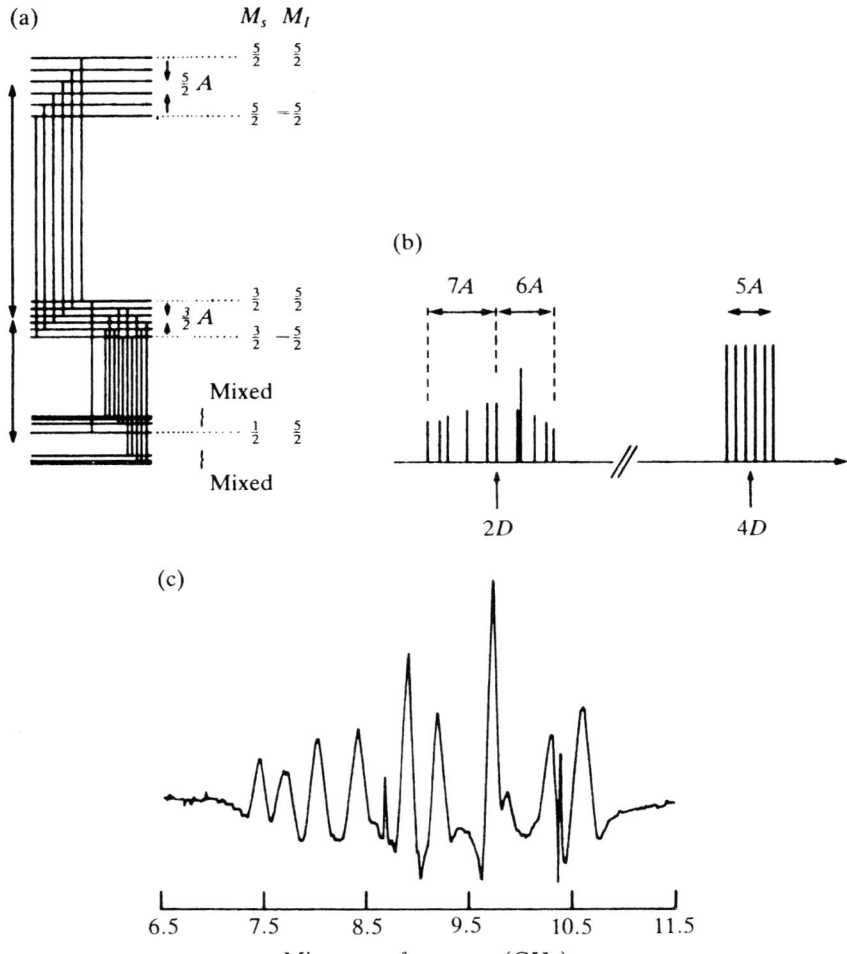

Figure 4.1.9 (a) The energy-level diagram for $S = I = 5/2$ for the case when $D \gg A$ (isotropic hyperfine interaction); (b) The resulting ZFR transitions, excluding those in the $v = 5A$ region, as indicated in panel (a); (c) The observed ZFR spectrum for 0.1% Mn^{2+}-doped NH_4Cl at 295 K, which can be simulated with the parameters $D = -4.508$ GHz, $a + 2F/3 = 21$, $A_\parallel = -247$ and $A_\perp = -254$ MHz. The experimental conditions were as follows. A coaxial sample cell in a reflection microwave apparatus and magnetic field modulation ($\mathbf{B}_m \perp \mathbf{B}_1$) of ± 3.3 mT were used. The observed 2-D set of lines were slightly different from the spaced pattern in panel (b) due to hyperfine anisotropy. The sharp features result from spurious effects of the reflection spectrometer. Adapted from Pilbrow (1990).

References

Bogle, G.S., Symmons, H.F., Burgess, V.R., and Sierins, J.V. (1961) *Proc. Phys. Soc. B*, **77**, 561.

Bramley, R. and Strach, S.J. (1981) *Phys. Lett.*, **79**, 183.

Bramley, R. and Strach, S.J. (1983) *Chem. Rev.*, **83**, 49.

Bramley, R. and Strach, S.J. (1984) *J. Magn. Reson.*, **56**, 10.

Bramley, R. and Strach, S.J. (1985) *J. Magn. Reson.*, **61**, 245.

Newman, D.J. and Urban, W. (1975) *Adv. Phys.*, **24**, 793.

Pilbrow, J.R. (1990) *Transition Ion Electron Paramagnetic Resonance*, Clarendon Press, Oxford.

Shing, Y.H. (1972) Theoretical and Experimental Studies in Electron Paramagnetic Resonance II – An Investigation of the Ground State Splitting of the Gd^{3+} [$4f^7$, $^8S_{7/2}$] Ion in Lanthanum Ethyl Sulphates and the Design of 1–8 GHz Coaxial Electron Paramagnetic Resonance Spectrometers. Unpublished PhD Thesis. Department of Physics, University of Calgary, Calgary, AB, Canada.

Shing, Y.H. and Buckmaster, H.A. (1976) *J. Magn. Reson.*, **21**, 295.

Urban, W. (1966) *Z. Angew. Phys.*, **20**, 215.

4.2
Low-Frequency CW-EPR Spectrometers: 10 MHz to 100 GHz
Harvey A. Buckmaster

4.2.1
Introduction

The design of continuous-wave (CW) electron paramagnetic resonance spectrometer (EPR) over a very large frequency range, from 10 MHz to 100 GHz, will be discussed in this chapter. This frequency range corresponds to $g = 2$ magnetic fields, ranging from 0 to 3 T or zero-field EPR to millimeter EPR. The aim of the chapter is to provide contemporary research workers who use CW-EPR spectrometers with an understanding of the basic electronic engineering principles that are essential in the design of such instrumentation, and how it functions. Whilst these basic principles are independent of the measurement frequency, the method of their realization will vary, since at frequencies below 500 MHz lumped parameter circuits and components are used, whereas at higher frequencies coaxial components are the norm up to above 26 GHz. Yet, coaxial components can be used at all lower frequencies. Waveguide components predominate above 26 GHz and, eventually, are replaced by quasi-optical components. Although some 60 years ago, waveguide components were the norm above about 1 GHz, rapidly changing technological innovation has enabled coaxial and quasi-optical components to replace waveguide components. Indeed, today the technology exists to create a CW-EPR spectrometer using an integrated circuit. Currently, millimeter-wave monolithic integrated circuits are used routinely by the military and in radio astronomy, while commercial applications are also being developed (Wang *et al.*, 2009 and references therein). The chapter also provides some specific examples of spectrometer design, to illustrate how certain features are incorporated. It is not the intention that the chapter is of direct assistance in the design and construction of an EPR spectrometer, as most present-day research groups will use commercial spectrometers. For further detail, the reader is referred to the classic EPR instrumentation books by Poole (1967, 1983) and Alger (1968), as well as more recent books, including those of Pilbrow (1990), Weil, Bolton, and Wertz (1994) and Weil and Bolton (2007), and to various journals regarding the specific instrument designs employed. The details of pulsed EPR, ENDOR and ELDOR spectrometers, as well those EPR spectrometers operating at frequencies above 100 GHz, are discussed in Chapters 4.3 and 4.4. Appendix 4.2.I describes the instrument systems for a EPR spectrometer; Appendix 4.2.II provides the names and concise descriptions for the coaxial/microwave components used in EPR spectrometers; and Appendix 4.2.III provides useful hints concerning problems that may arise in EPR spectrometer operation.

A *spectrometer* is an instrument which is used to measure a physical process that involves the absorption or emission of a quantum of energy $h\upsilon$ between two energy levels, E_i and E_f in accordance with Planck's law $h\upsilon = E_i - E_f$. The instrumentation required to make spectroscopic measurements has three common factors that are independent of the measurement frequency: (i) a source of radiation; (ii) a sample irradiation region; and (iii) a radiation detector. In this respect, EPR spectrometers are no exception, and their design and realization has been evolving since the first successful measurements were made over 60 years ago.

The essential components for a MHz or GHz CW-EPR spectrometer are: (i) a source of radiofrequency or microwave power; (ii) a resonant structure where the sample under study is located in a magnetic field; (iii) a radiofrequency or microwave detector; (iv) an electromagnet; and (v) a data acquisition system (Figure 4.2.1). MHz versions of CW-EPR spectrometers are used at lower frequencies,

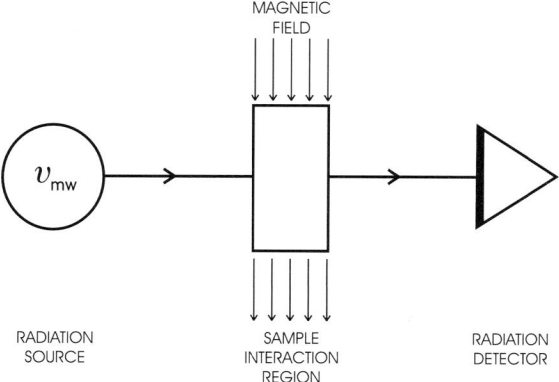

Figure 4.2.1 Block diagram for the basic components for an EPR spectrometer.

while further refinements are necessary to control the temperature of the sample when measurements are made at temperatures either above or below that of the laboratory.

The earliest CW-EPR spectrometers were very simple in their construction. The pioneer measurements at MHz frequencies used vacuum tube oscillators, and were made in or near-zero magnetic fields. As the paramagnetic species in the samples were not diluted with diamagnetic species, this resulted in large resonance linewidths so that only a portion of the resonance could be detected. These early developments were summarized by Bramley and Strach (1983). Following World War II, most CW-EPR measurements were made primarily at microwave frequencies above 1 GHz. These spectrometers used reflex klystrons to generate the microwave power, a tunable transmission resonant cavity operating in the TE_{111} cylindrical mode, and a microwave detector diode. The coupling of the microwave power into and out of the cavity was adjusted to ensure that the detector diode operated in its square law region by changing the size of the coupling holes. This design was pioneered by Bleaney and coworkers at the Clarendon Laboratory (Bleaney and Stevens, 1953), who made measurements primarily at 24 GHz and at liquid hydrogen (20 K) temperatures. The magnetic field was adjusted until a resonance could be observed and centered on the oscilloscope screen, after which the magnetic field was measured using a proton magnetometer to about $1:10^4$. The amplitude of the magnetic field modulation at the 50 Hz power line frequency was adjusted to optimize the resonance displayed, and the microwave frequency was determined using a tunable cavity wavemeter to about $1:10^{3-4}$. The process was simple but crude and slow, and the accuracy was determined by the frequency measurement. The sensitivity was limited primarily by the 50 Hz magnetic field modulation frequency. A typical early EPR spectrometer, using a 115 kHz magnetic field modulation frequency (Llewellyn, 1957) is shown in block diagram form in Figure 4.2.2.

The development of radar systems during World War II occurred in a number of stages as technological developments enabled the microwave frequency of the power sources to be increased from, initially, L-band (1 GHz or 30 cm), to S-band (3 GHz or 10 cm), then to X-band (10 GHz or 3 cm), and finally to K-band (24 GHz or 1.25 cm). On approaching the end of World War II, the Q-band (35 GHz or 0.8 cm) radar became operational. Whilst the pioneering EPR measurements were made at frequencies as low as 100 MHz, the majority of post-war CW-EPR measurements were made near 9 and 24 GHz, because klystrons and waveguide components at these frequencies were readily available from surplus military radar instrumentation, and the waveguide was of manageable dimensions. Many of the early research groups had gained experience while working at these frequencies, because they had been involved with the development of radar during World War II. The very large body of information regarding the design of microwave transmitters and receivers and their components, developed at the Massachusetts Institute of Technology (MIT) was published between 1947 and 1951 in a monumental, 28-volume *Radiation Laboratory Series*; these volumes are listed in Poole (1983, p. 67/8). In fact, the early EPR research workers relied very heavily on these texts

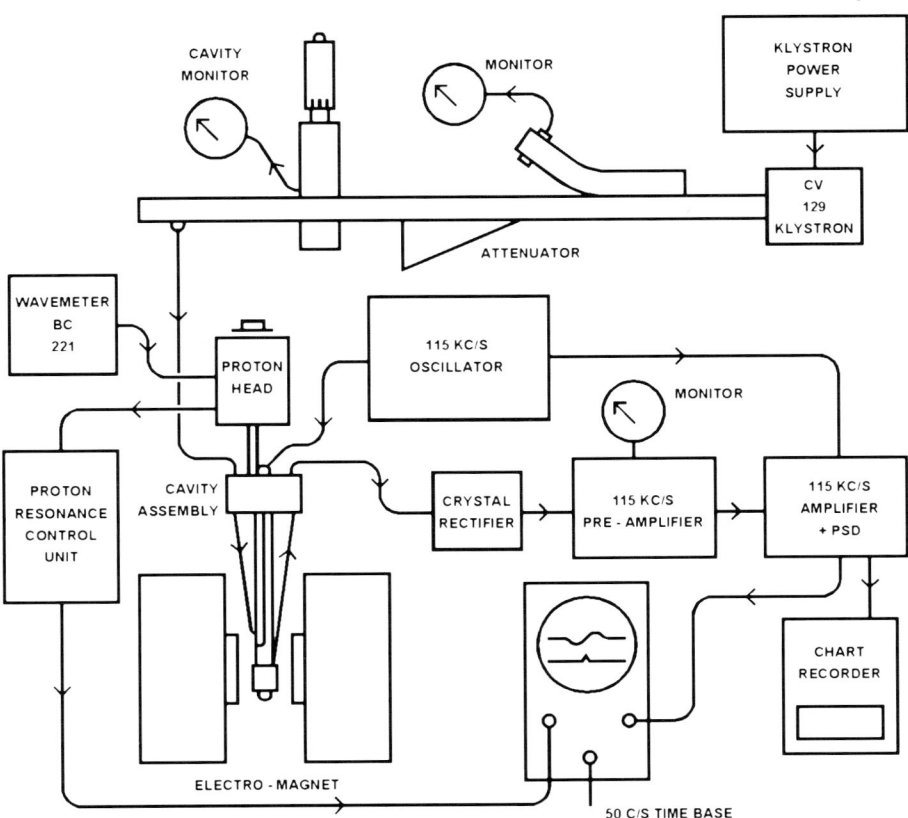

Figure 4.2.2 Typical early video EPR spectrometer using 115 kHz magnetic field modulation. From Llewellyn (1957).

when designing components, as no commercial sources for microwave components existed at that point in time.

Since the post-war period, the design of CW-EPR spectrometers has undergone enormous change, due mainly to the development of solid-state devices to replace vacuum tube technology, to improved production techniques for microwave detector diodes, and to the use of phase lock loops, analog and later digital coherent signal processing, digital instrumentation, digital data acquisition, computer instrumentation control. In addition, waveguide components at frequencies up to at least 26 GHz have been replaced with the miniaturized coaxial components described above. Notably, the physical dimensions of an EPR spectrometer have decreased, while the accuracy, reliability and repeatability of measurements have undergone significant improvements. Whilst the physical dimensions and weight of the electromagnet and its power supply remain a limiting factor, commercially available desktop EPR spectrometers are now available and suitable for routine testing and analysis. Interestingly, it would be highly instructive to compare the details of the earliest commercial EPR spectrometers with those available today.

In theory, CW-EPR measurements can be made at either a fixed frequency or at a fixed magnetic field. CW-NMR measurements can be realized either way because the frequencies used are below about 800 MHz (~2 T). In practice, they are made at a fixed magnetic field because the magnetic field homogeneity, which is very important in NMR because the required sample size and resonance linewidths are easier to optimize when the magnetic field is constant. At the microwave frequencies used in most EPR spectroscopy, it is extremely difficult to lock the microwave power source output frequency and to track the resonant frequency of the sample cavity when this frequency is slowly varied to display a spectrum when the magnetic field is fixed. This tracking process is even more difficult to realize if the temperature of the sample cavity is also varied. Consequently, most CW-EPR measurements are made at fixed frequency and the magnetic field is varied over the interval required to display the EPR spectrum being studied. The exception is swept frequency zero and near-zero magnetic field CW-EPR, using a broadband nonresonant sample holder.

4.2.2
CW-EPR Spectrometer Configurations

Some early research groups attempted to improve the sensitivity achievable using audio frequency magnetic field modulation by constructing superheterodyne EPR spectrometers. However, it proved difficult to obtain a stable and repeatable improvement in sensitivity because the available microwave power and local oscillator sources were not designed to be stable. Unlike a radar system, which is operated at a fixed frequency, the operating frequency of an EPR spectrometer must be varied to take account of changes in the resonance frequency due to changes in the temperature of the sample cavity, and the presence of a sample. Consequently, locking two microwave oscillators at a fixed intermediate frequency difference, and also keeping the sample resonant cavity tuned, was extremely difficult. Although these spectrometers were complex and time-consuming to operate, they demonstrated that the available microwave diodes were much more sensitive at 30 MHz (the intermediate frequency used in World War II radar superheterodyne receivers) than when demodulation was at frequencies below 10 kHz. The noise performance of the available microwave diodes as a function of frequency had not been studied, as most radar systems used 30 MHz superheterodyne detection systems. However, Buckmaster and Scovil (1956) showed that the EPR sensitivity could be improved by a factor of about 100 by using a magnetic field modulation frequency of 455 kHz rather than 60 Hz. This frequency was chosen arbitrarily because it was used in AM (amplitude modulation) radios, so that the required tuned transformers were available. The technical problem was to produce a magnetic field inside the sample resonant cavity. Consequently, a 24 GHz TE_{111} mode was slotted into a cylindrical resonant cavity so as to form a half-turn loop for the modulation field current. The slot was oriented so that it did not cut any of the microwave current loops, and hence did not cause any significant degradation in Q. The realization that the sensitivity could be increased by using high-frequency modulation opened a flood gate of innovative proposals to avoid the obvious disadvantages of a split cavity. Subsequently, 9 GHz rectangular and cylindrical resonant cavities made from silver- or gold-plated glass, machined epoxy resins and sintered lava or "Wonderstone," as well as inserting a coil around the sample, showed that many solutions were feasible without degrading the cavity Q-factor. In contemporary designs for 9 GHz TE_{011} mode cylindrical cavities, the modulation coils were embedded in the epoxy resin, while the orientation of the modulation magnetic field was parallel to the static external magnetic field (these fields must be perpendicular to the microwave magnetic field that induces the resonance). The inner polished surface of the dielectric was then silver- or gold-plated to a thickness of about four to five skin depths at the 100 kHz modulation frequency to maximize the Q-factor. This frequency—which was chosen simply because the required tuned transformers were readily available—represented a significant advance in sensitivity, as it avoided operating the detector (demodulator) diodes at frequencies where the flicker noise component would predominate. Consequently, most EPR spectrometers have for many years used 100 kHz magnetic field modulation with silver- or gold-plated sample resonant cavities made from various nonconducting materials, so as to optimize the sensitivity of the microwave diode demodulation system. The Q-factor of the sample cavity is not compromised, provided that the thickness of the plated surface exceeds the skin depth at microwave frequencies by a factor of about five, but is much less than the skin depth at the modulation frequency. This constraint arises because eddy or "Foucault" currents induced in a conductor are the limiting factor; these create a magnetic field at the same frequency as, but in the opposite direction to, the applied field. Smythe (1950, p. 417) described the relationship relating the fraction of an external AC magnetic field applied perpendicular to the axis of a conducting cylindrical conductor that penetrates this cylinder as a function of the skin depth of the conductor, the modulation frequency and the thickness and radius of the cylinder. This problem has been discussed in detail by Talpe (1971).

This innovation was followed by the use of the so-called "phase-sensitive" or "lock-in" detector, as originally proposed by Dicke (1946), to narrow the detection bandwidth and eliminate signals at extraneous frequencies that are not phase-related to the input signal. Whilst it may be more precise and correct to describe this as a synchronous demodulator, the name given by Dicke (1946) persists in the literature. This is an early example of a phase-locked loop. A block diagram of the essential components for a lock-in or synchronous demodulator is provided in Figure 4.2.3. This measurement technique has been used extensively in NMR measurements (where sensitivity was very important) and was an early example of using information theory, as the measurement sensitivity could be increased by decreasing the bandwidth and increasing the measurement time. The peak-to-peak modulation amplitude B_m^{pp} was limited to be less than the resonance linewidth at half-maximum amplitude (Figure 4.2.4). If the modulation amplitude was

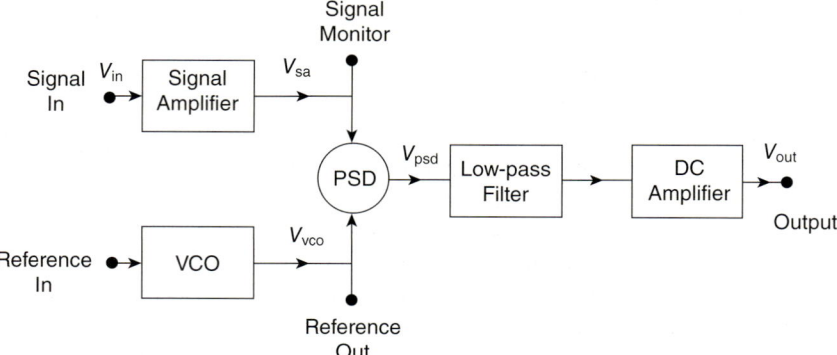

Figure 4.2.3 Block diagram for a lock-in or synchronous demodulator. PSD, phase-sensitive detector; VCO, voltage-controlled oscillator; DC, direct current. From Schofield (1994).

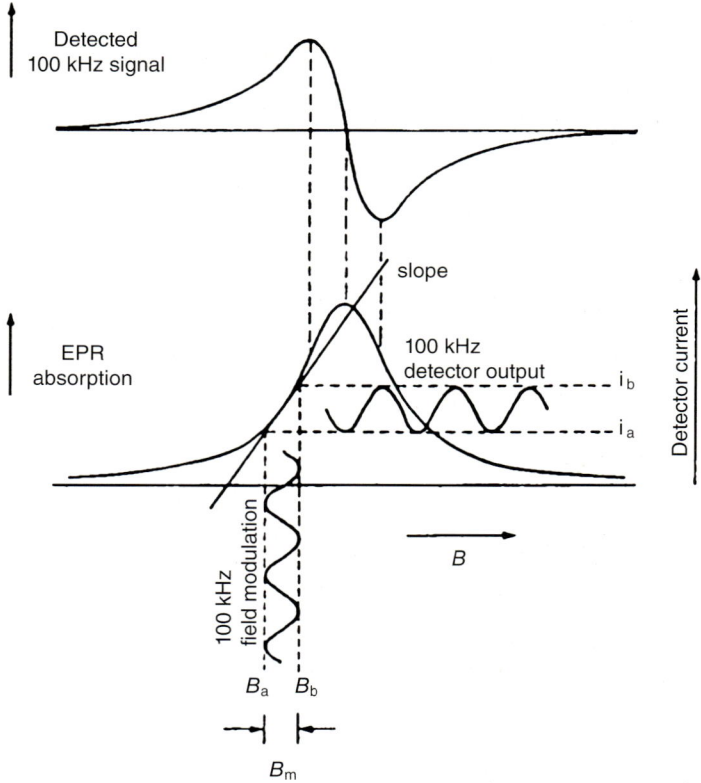

Figure 4.2.4 Diagram showing the resultant output signal a_1 (first derivative of the absorption) at the magnetic field modulation frequency f_m when the modulation has a peak-to-peak amplitude B_m that is less than the magnetic resonance linewidth. From Weil and Bolton (2007).

large compared to the linewidth, then the output of the detector would contain the Fourier components of the resonance lineshape. However, if it were less than the linewidth, then the so-called "first derivative" (more correctly the first Fourier component of the absorption lineshape) a_1 is observed at the modulation frequency (this topic is discussed in greater detail below).

Strandberg et al. (1956) discussed the design of CW-EPR spectrometers operating at 20, 60, and 225 MHz and 3, 9, and 24 GHz. As a specific example, the design of a 9 GHz spectrometer (see Figure 4.2.5) was described which used a magic-T sample cavity bridge, while the microwave oscillator frequency was locked to the resonant frequency of the sample cavity by modulating the balanced arm opposite to the sample cavity at 30 MHz. This frequency locking system was an improved version of that originally developed by Pound (1946) and later reported by Zaffarano and Galloway (1947). These authors emphasized that this approach would ensure that the EPR spectrum consisted of only the absorption component, and had the advantage that microwave oscillator frequency modulation noise was minimized. Their analysis of both reflection and transmission cavity designs revealed that the use of a magic-T and a reflection cavity did not lead to a factor of 2 lost in signal, because there was a $2^{1/2}$ loss using a transmission cavity. The MHz versions produced by these research groups used two pairs of crossed coils as the lumped parameter equivalent of a magic-T. The use of a magic-T bridge with the sample cavity connected to one of the balanced arms proved to be the next significant advance in EPR spectrometer design. Feher (1957) provided the first comprehensive discussion regarding the sensitivity of CW-EPR spectrome-

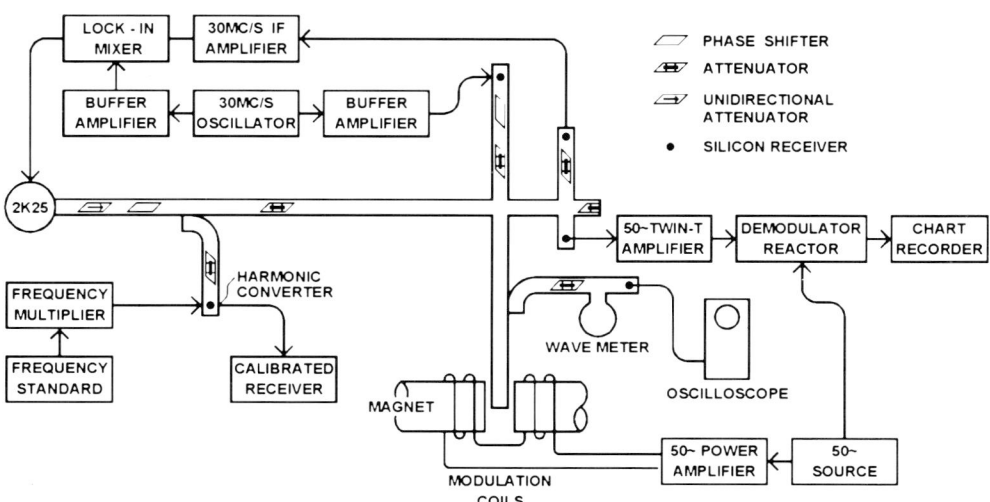

Figure 4.2.5 Block diagram of the first CW-EPR spectrometer to incorporate a magic-T sample cavity bridge, frequency locking of the microwave oscillator to the sample cavity resonant frequency, synchronous demodulation at the magnetic field modulation frequency and ferrite isolators operating at 9 GHz CW-EPR. From Strandberg et al. (1956).

ters, and also described an experimental study of the sensitivity of various spectrometer configurations. Feher also examined the sensitivity of bolometers and crystal diodes as detectors, and showed superheterodyne detection to be more sensitive by a factor of ten. Feher also described a 9 GHz superheterodyne CW-EPR spectrometer (Figure 4.2.6) that included a magic-T sample cavity bridge, a magic-T bridge superheterodyne mixer, and lock-in detection at a magnetic field modulation frequency that could be varied from 100 to 1000 Hz, and reported the details of its sensitivity together with a wealth of valuable experimental information. Subsequently, the details provided by Feher (1957) and Strandberg et al. (1956) provided the basis for the design of all later-developed CW-EPR spectrometers. Two years later, Misra (1958) described an EPR spectrometer that used a sample cavity magic-T bridge to achieve near-balance with another cavity in the other balanced arm, in addition to a magic-T bridge that provided the local oscillator power to bias the microwave detectors. These design developments highlighted the fact that instrumentation advances followed a similar pattern to those for many seminal conceptual advances in science. Most design concepts are rapidly absorbed and become the norm shortly after having been shown as feasible, having already been pregnant concepts within the scientific community. However, some design concepts may lay fallow for many years.

A magic-T balanced mixer is shown in Figure 4.2.7. Here, the microwave power enters on the H-plane arm (one of the unbalanced arms), while the microwave signal exits on the E-plane arm (the other unbalanced arm). It should be noted that the mixer diodes are reversed in this configuration; however, if the local oscillator bias enters on the E-arm then the mixer diodes are not reversed. Either configuration ensures that noise due to the local oscillator is canceled to first approximation by ~40 dB because the bridge is symmetric. The other balanced arm is terminated with a matched load. A magic-T can be used to form a sample cavity bridge in which the cavity coupling is adjusted to near-critical coupling to provide sufficient microwave power to correctly bias the detector diodes to operate a near-optimum noise figure. These characteristics arise because the E-arm and H-arm are orthogonal, so that the two asymmetric arms have different cut-off frequencies because the waveguide is a high-pass filter. The cut-off frequency f_c for a rectangular waveguide is related to the waveguide width a by $f_c = c/2a$ in air, where c is the velocity of light. The choice of sample resonant structure, and how it can be coupled to the microwave power, are discussed later in the chapter. If the characteristic impedance of the E-plane unbalanced arm is Z_o, then the combined characteristic impedance of the balanced arms is $2Z_o$ whereas if the characteristic impedance of the H-plane unbalanced arm is Z_o then the combined impedance of the balanced arms is $Z_o/2$. Matching ensures that the four arms each have a characteristic impedance, Z_o. A magic-T is a matched hybrid-T, but the bandwidth over which the match is achieved is less than the operating bandwidth of the waveguide. The characteristic impedance of a rectangular waveguide of width a and thickness b at frequency f is

$$Z_o = 754b/a(1-f^2)^{1/2} \qquad (4.2.2.1)$$

in air.

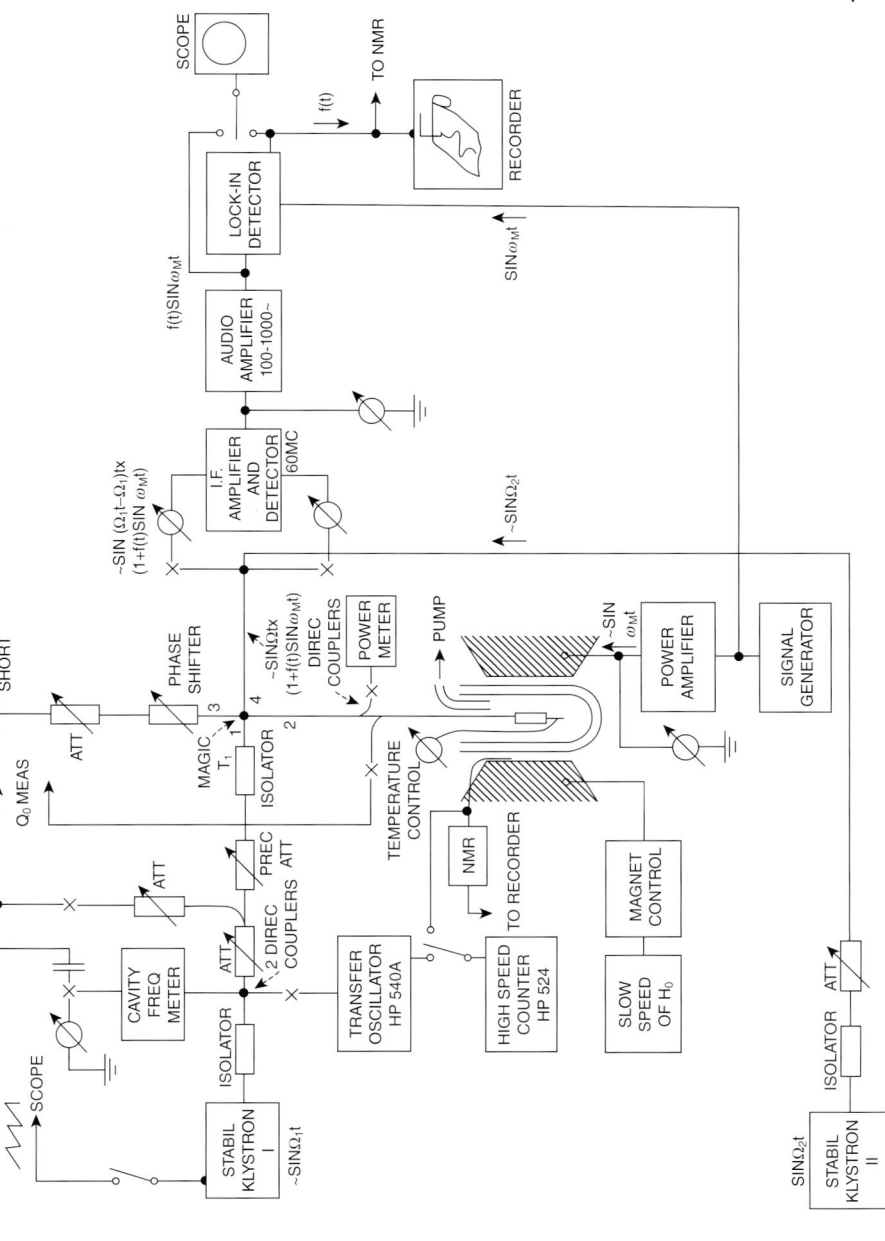

Figure 4.2.6 Block diagram of a 9 GHz CW-EPR superheterodyne spectrometer, incorporating many innovative features. From Feher (1957).

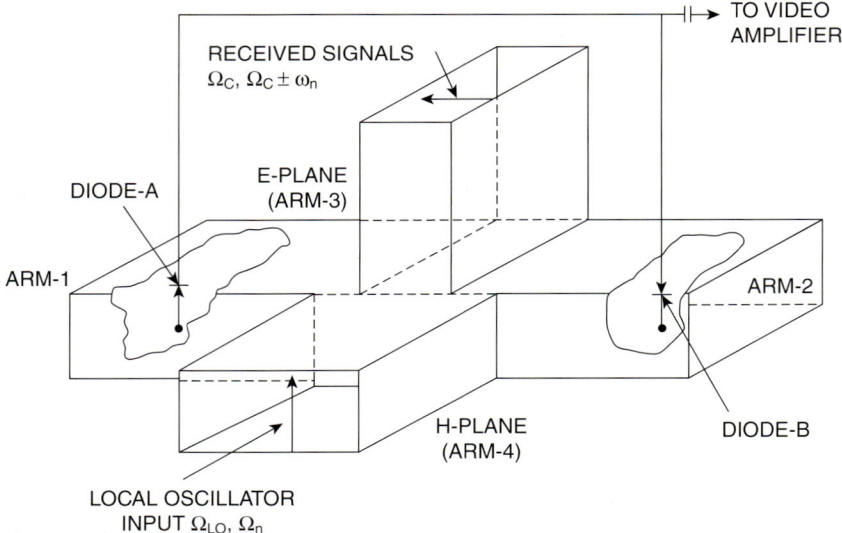

Figure 4.2.7 Diagram of a magic-T bridge mixer showing the local oscillator input in the H-plane arm and the signal input in the E-plane arm with the mixer diodes reversed to cancel local oscillator noise. From Ondria (1968).

A *three-port circulator* is another microwave device that can be used instead of a magic-T. This has the advantage that all of the input power is coupled into the second port to which a microwave sample cavity can be connected, and all of the signal information from this port is coupled into the third port to which a balanced synchronous demodulator can be connected. The isolation between the three ports is never better than about 40 dB, however, and there are also small losses between the ports. This isolation is achieved by the introduction of ferrite material into the structure, and this will vary if there are stray magnetic fields present that change. At MHz frequencies, two pairs of crossed coils have similar properties to a magic-T. Such coils have been used extensively in CW-NMR as well as in zero- and low-magnetic field CW-EPR (see Strandberg et al., 1956; Bramley and Strach, 1983 and Eaton and Eaton, 2004).

Three factors have played a major role in improving the sensitivity and stability of CW-EPR spectrometers. The introduction of microwave telephone relay systems to replace cable links has meant that the frequency stability and spectral purity of the low power klystrons used in these links, and also the relay receivers, became very important. The development of semiconductor technology has led to significant advances in both the quality and purity of the semiconductor materials used in microwave diodes, their design, and the modes in which they operated. As a consequence, the flicker or "$1/f$" noise component knee frequency has been lowered from $\sim 10^5$ kHz to below 10^2 Hz. The third development was the realization that the optimum approach to the improvement in the stability, as well as the sensitivity and repeatability of EPR measurements, was to employ phase-coherent or synchronous signal-processing techniques. The Dicke phase-sensitive or lock-in detector mentioned above presaged the evolution of these synchronous demodulation techniques. *Synchronous demodulation* is a very efficient method for both narrowing the signal bandwidth to a fraction of a cycle and eliminating other sources of noise and signals at the same frequency that are not phase-related to, or synchronous with, the signal being demodulated. The output of a detector is a DC signal, whereas the output of a demodulator is a DC signal that varies at the sideband frequency. The term "demodulator" will be used throughout this chapter because the microwave carrier is AC-modulated, whereas the output of a "detector" is a DC voltage.

Buckmaster and Dering (1965) reported the first study of the operating conditions that minimize the noise temperature conversion gain ratio for the microwave point contact diodes used in an EPR spectrometer. These authors measured the noise temperature and conversion gain parameters for selected 1N23WE silicon point-contact crystal mixer diodes as a function of the modulation frequency, from 10 to 10^4 Hz, while the biasing power was varied (Figure 4.2.8a). The results obtained were then verified by measuring the EPR spectral signal-to-noise ratio (SNR) of a diphenyl picryl hydrazil (DPPH) sample as the parameters were varied (Figure 4.2.8b). The measurements showed that the SNR obtained when using these microwave mixer diodes did not improve when the modulation frequency was above ~0.2 MHz. This showed that the knee of the "$1/f$" or flicker noise component in 1N23WE microwave diodes had been reached by this frequency. The thus-developed microwave synchrodyne EPR spectrometer used a magic-T sample cavity bridge, critical cavity coupling, and a balanced mixer synchronous demodulator with a separate reference arm for the microwave power. A later version of this spectrometer configuration is shown in block diagram form in Figure 4.2.12. In early reports, the term "homodyne" was used when the local oscillator had the same frequency as the input; this was in order to distinguish it from the term "superheterodyne," when the local oscillator had a different frequency.

The study of noise sources in electronic devices and systems is a specialized area of study, and is beyond the scope of this chapter; however, those readers interested in this topic should consult other reports. For example, Mumford and Schelbe (1968) have provided a concise and readable summary of the subject, as well as an excellent list of other sources of information. In general, the EPR spectroscopist need know few facts concerning the sensitivity of EPR spectrometers.

Although the first circuit for a microwave phase-lock oscillator synchronizer was first described by Peter and Strandberg (1955) during the 1950s, this device was not used in an EPR spectrometer. Previously, Bosch and Gambling (1962) had shown that frequency-modulated (FM) noise in reflex klystrons was the dominant factor at sideband frequencies below 10 MHz. The commercial availability of instrumentation capable of phase-locking a microwave power source to a harmonic of a high-stability, MHz frequency crystal oscillator, led to the provision of a method for minimizing FM noise in reflex klystrons. Buckmaster and Dering

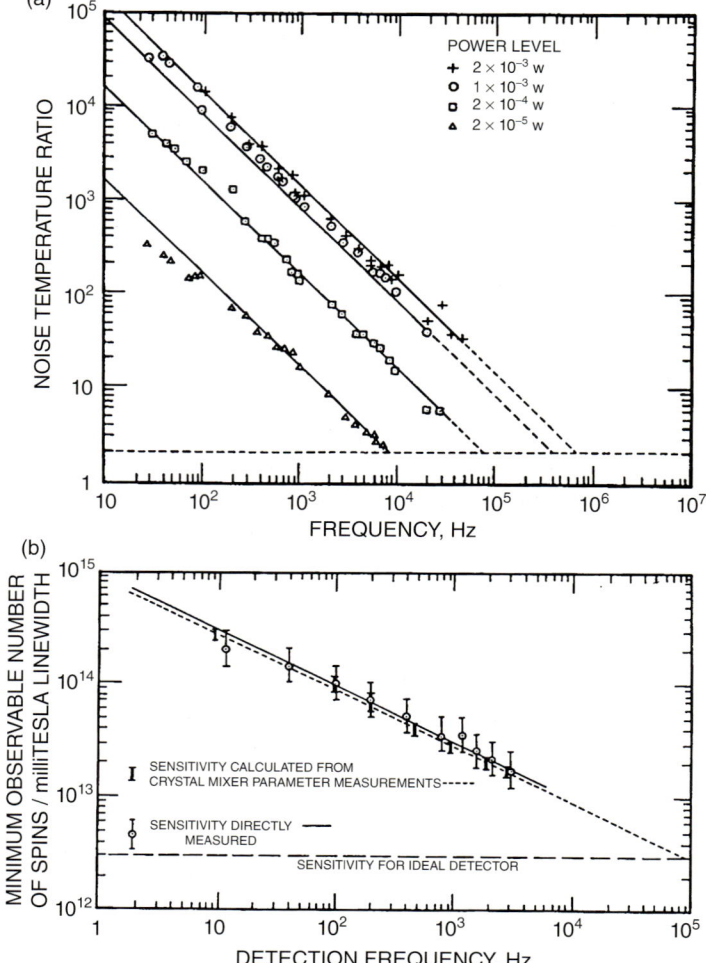

Figure 4.2.8 (a) Graph showing the frequency dependence of the noise temperature–conversion gain ratio (t/G) of a 1N23WE silicon point-contact diode for various incident microwave power levels. From Buckmaster and Dering (1965); (b) Graph showing the EPR signal-to-noise ratio for a DPPH sample as a function of the modulation frequency when a matched pair of 1N23WE silicon point-contact diodes were used in the 9 GHz CW-EPR spectrometer with a magic-T microwave synchronous demodulator as a function of the modulation frequency. From Buckmaster and Dering (1965).

(1966) showed that this instrumentation would enable the klystron output frequency deviations to be minimized to less than $1:10^8 \text{s}^{-1}$ and to $1:10^7 \text{h}^{-1}$ in a CW-EPR spectrometer. A block diagram of a microwave oscillator frequency synchronizer is shown in Figure 4.2.9 (Shing, 1972).

Teaney, Klein, and Portis (1961) had shown, using a reflection bridge superheterodyne EPR spectrometer, that the sensitivity did not increase above a cavity power level of about 10 mW. These authors suggested that the stability and sensitivity of their spectrometer was limited by FM oscillator noise above cavity power levels of about 20 mW. Subsequently, Buckmaster and Dering (1967a) re-examined this problem when the microwave oscillator was phase-locked to a high-stability MHz crystal oscillator and frequency-locked to the sample cavity. It was shown that the observed SNR of an EPR resonance would increase linearly up to at least 1 W. Moreover, the results of Teaney, Klein, and Portis (1961) could be reproduced by introducing FM white noise modulation into the reference oscillator (Figure 4.2.10). The results of this study also verified the expression that Buckmaster and Dering (1967b) had derived for the role that discriminated FM microwave oscillator noise plays in limiting the theoretical spectrometer sensitivity. Buckmaster and Dering (1967c) also examined the possible role of parametric and maser preamplifiers in EPR spectrometers, and showed that any sensitivity improvement could be more easily obtained by increasing the sample cavity power level in the absence of saturation effects. Strandberg (1972) later provided a similar, but more sophisticated, analysis of the factors that determined the sensitivity of an EPR spectrometer. Strandberg's commercial 9 GHz EPR spectrometer (Strand Labs) used a magnetic field modulation frequency of 6 kHz, based on the realization that it was unnecessary to use a higher frequency because the sensitivity would not increase. This design was also innovative as it used a circulator rather than a magic-T to couple the microwave power into, and the signal power out of, the sample cavity. A separate arm was also used to couple microwave power from the power source to bias the microwave diodes in the signal demodulation bridge. Unfortunately, the introduction of a circulator to replace a magic-T in the resonant cavity bridge had one disadvantage, in that the bridge balance was affected by the stray magnetic field from the electromagnet – an effect which was changed when the field was swept through the resonance under study.

A further refinement was introduced by Buckmaster and Dering (1967d), who showed that a theoretical sensitivity could be achieved and maintained over time intervals of many hours by using a single klystron CW-EPR spectrometer. For this, superheterodyne demodulation was used, if the local oscillator frequency and the microwave power oscillator frequency were phase-related or quasi-coherent, and both were phase-locked to high-stability MHz crystal oscillator synchronizers. Synchronous demodulation was used at the microwave, intermediate, and magnetic field modulation frequencies in this spectrometer. A block diagram of the microwave configuration for the spectrometer is shown in Figure 4.2.11. Quasi-coherence between two frequency sources arises when two frequencies are generated by the same oscillator power source, after which one of the frequencies is shifted to another frequency. These frequencies are not clones of each other but

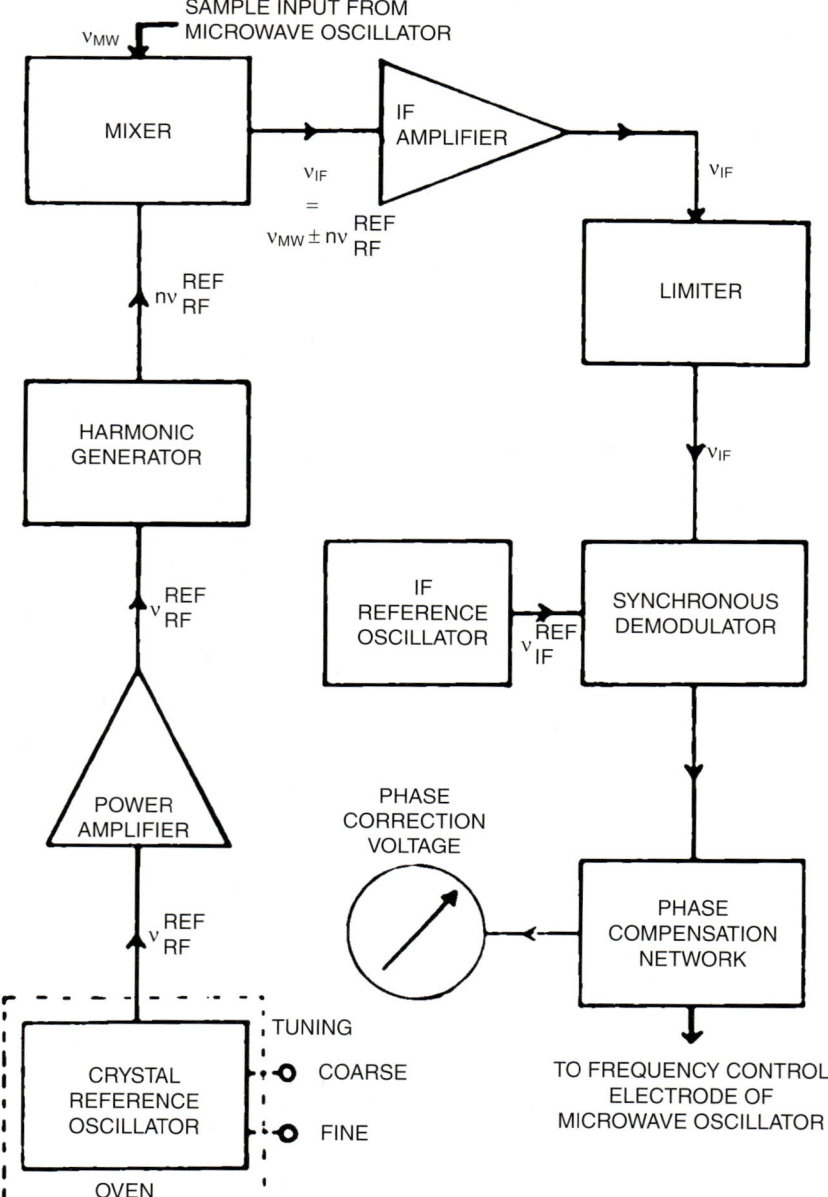

Figure 4.2.9 Block diagram of a microwave oscillator frequency synchronizer. From Shing (1972).

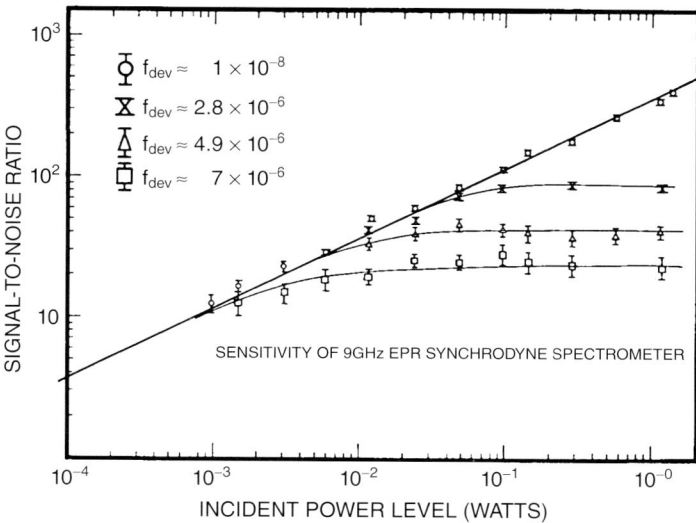

Figure 4.2.10 Graph showing the measured signal-to-noise ratio as a function of the microwave sample cavity power for various microwave spectral purities. From Buckmaster and Dering (1967a).

rather are genetically related, which enables synchronous demodulation to be employed at the superheterodyne intermediate frequency. Buckmaster and Gray (1971) showed that a single klystron superheterodyne spectrometer could also be realized using a serrodyne single sideband generator with a ferrite phase modulator driven by a sawtooth waveform to generate the local oscillator frequency. Whilst this alternative simplifies the design particularly at frequencies above 26 GHz, it has the disadvantage that the carrier and the other sidebands are suppressed by only about 17 dB.

Current CW-EPR spectrometer designs do not use superheterodyne demodulation, because advances in the fabrication of microwave demodulator diodes have reduced the frequency of the knee of the flicker noise to below 10 kHz, and thus have improved their conversion gain. This means that modulation frequencies of about 500 Hz can be used without any significant loss in sensitivity whenever higher sample cavity power levels can be employed. Nevertheless, most commercial spectrometers continue to use 100 kHz magnetic field modulation. Strandberg (1972) has shown that the degree of bridge balance required, and the spectral purity of the microwave power source, must be greater when 100 kHz rather than a 6 kHz magnetic field modulation being used because the microwave power source AM noise becomes the limiting factor. The experience of the present author is that it is easy to obtain a high degree of bridge balance and of microwave power source spectral purity; consequently, even lower magnetic field modulation frequencies can be used, and the AM microwave power source noise plays no detectable role.

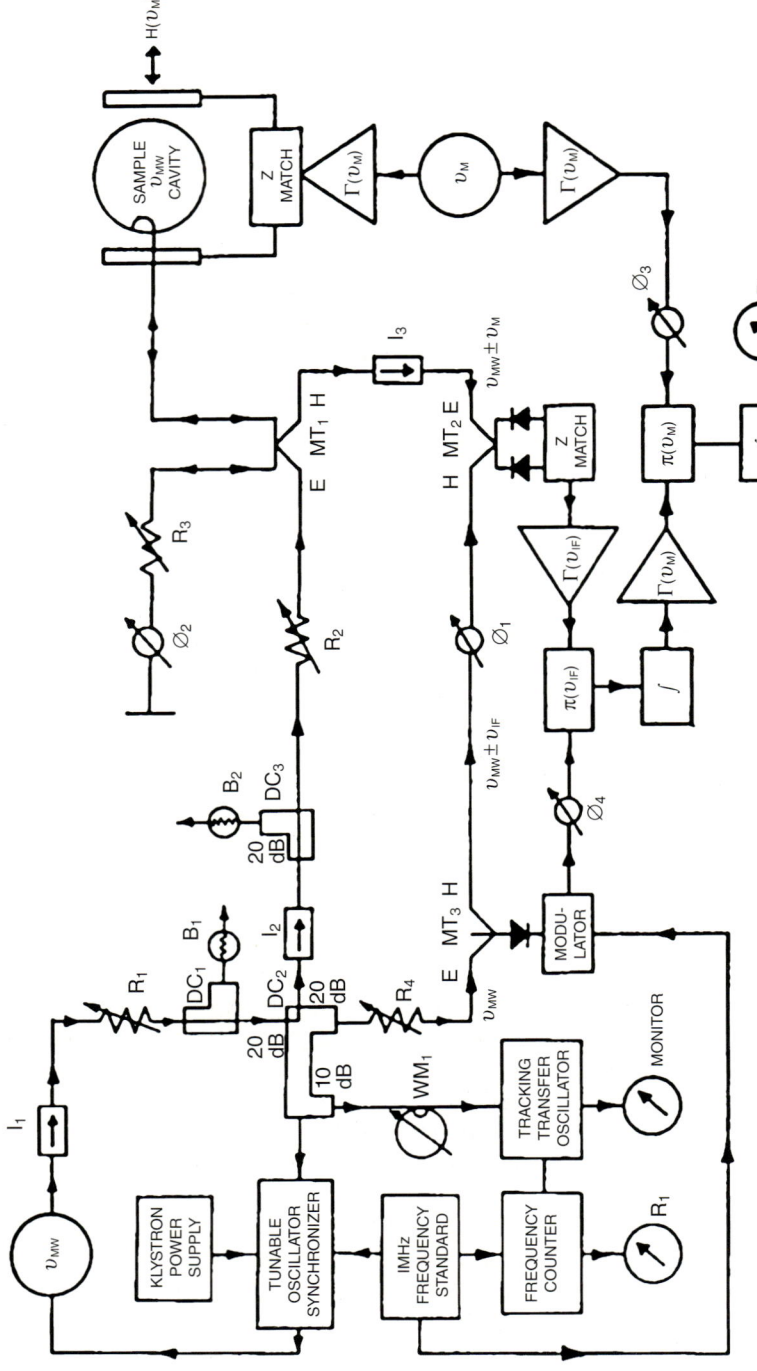

Figure 4.2.11 Block diagram for a CW-EPR single klystron superheterodyne spectrometer using synchronous demodulation at the microwave, intermediate, and modulation frequencies. From Buckmaster and Dering (1967d).

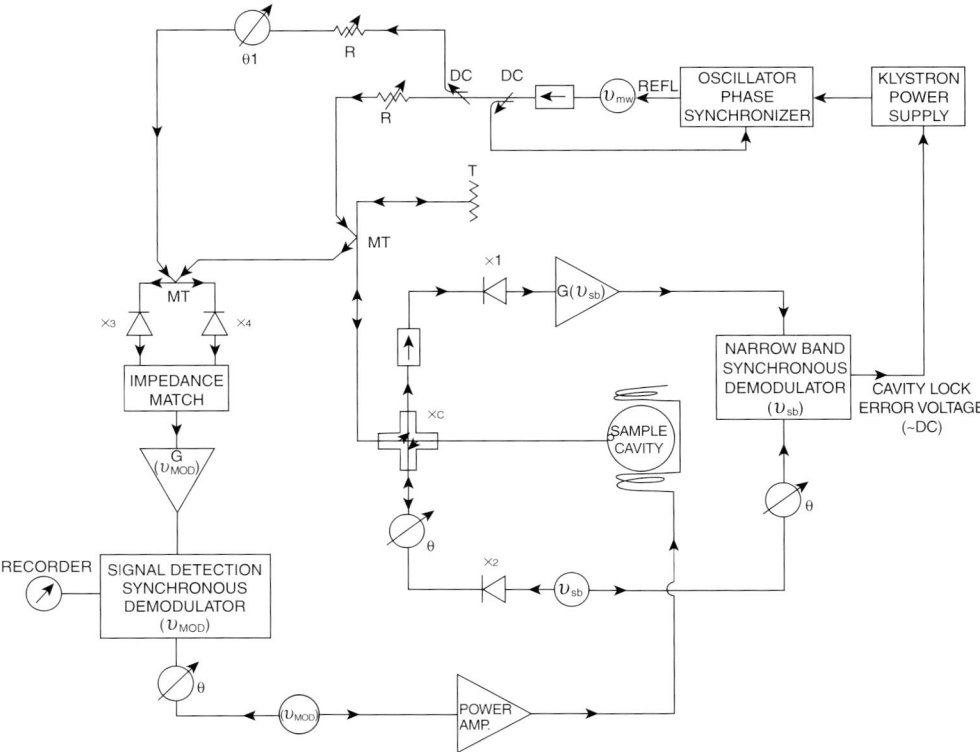

Figure 4.2.12 Block diagram for a typical CW-EPR spectrometer incorporating a microwave phase-locked oscillator synchronizer, sample cavity frequency locking, and synchronous demodulation at the microwave and magnetic field modulation frequencies. From Buckmaster and Skirrow (1969).

A block diagram of a CW-EPR spectrometer designed and operated for many years by the present author is shown in Figure 4.2.12. This system incorporates all of the concepts discussed above, the aim being to assist the reader in appreciating the overall configuration of such a spectrometer. The system includes cavity frequency locking circuitry, as experience has shown that metal sample cavities are much less stable than those constructed from dielectric materials, unless they are carefully temperature-stabilized. Dielectric cavities can also benefit from temperature stabilization.

Although, commercial spectrometers will differ in many details, they are designed to take account of the factors discussed here, that determine the operational sensitivity and stability of a CW-EPR spectrometer. Such systems are designed to operate at microwave cavity power levels less than ~100 mW, since digital signal averaging techniques can be used to increase the sensitivity when is required. Cost considerations play a major role in determining how these concepts are realized in commercial EPR spectrometers; operational simplicity is also an important objective for the research worker without microwave and electronics expertise. The fundamental principle of good instrumentation design is to use phase-locked loops wherever possible, as this results in maximum stability and minimum noise, thus maximizing the achievable sensitivity.

4.2.3
Theoretical Sensitivity

The first expression for the sensitivity of an EPR spectrometer was given by Bleaney and Stevens (1953). A more comprehensive examination of this subject was provided by Feher (1957), concerning the design of EPR spectrometers. Poole (1967, 1983) has summarized the various EPR spectrometer sensitivity calculations made previously, and has shown the minimum detectable susceptibility χ''_{min} to be given by:

$$\chi''_{min} = (2/\eta Q)\sqrt{FkT_0B/P} \qquad (4.2.2.2)$$

where Q is the unloaded sample cavity quality factor, η is the sample cavity filling factor, F is the EPR spectrometer detection system noise factor, P is the microwave cavity power, B is the final spectrometer bandwidth, k is Boltzmann's constant, and T_0 is the reference temperature. Equation 4.2.2.1 can be expressed in terms of the minimum detectable number of spins obeying Curie's law and the Bloch equation (Bloch, 1946):

$$N_{min} = (2\ \eta Q) \cdot [FkT_0B/P]^{1/2} \cdot [3kT/\mu_B^2 S(S+1)] \cdot (\Delta B/B) \qquad (4.2.2.3)$$

where μ_B is the Bohr magneton, S is the electron spin, T is the sample temperature, and $(\Delta B/B)$ is the fractional resonance linewidth.

The expression given Equation 4.2.2.1 for the minimum detectable susceptibility assumes that the microwave power that irradiates the EPR sample is stable in amplitude, has very high spectral purity, and that there is a perfect microwave radiation detector. These assumptions are not realized in practice, which means that the actual number of spins that can be detected is greater than that given by Equation 4.2.2.2. These two limitations will now be discussed in detail, following a brief discussion of the electronic noise encountered in various spectrometer components.

The three sources of noise in electronic circuits are Johnson or Nyquist noise, shot noise, and flicker or "$1/f$" noise. Johnson noise arises from the equilibrium fluctuations of an electric current in a conductor due to the random thermal motion of the charge carriers. Shot noise arises because of fluctuations in the current due to the discrete nature of the charge carriers (electrons).

Flicker or "$1/f$" noise has a power spectrum which varies inversely with the frequency and arises because all conducting and semi-conducting materials contain impurities that cause the charge carriers to be scattered. The noise power $P_n = kT\Delta f$, where k is the Boltzmann constant since the noise voltage, $v_n = v_n = (4kTR\Delta f)^{1/2}$, is measured in a resistance R at temperature T in a bandwidth Δf in the Rayleigh–Jeans approximation. $v_n = 4.07\,nV/(Hz)^{1/2}$ for a $1\,k\Omega$ resistor at $300\,K$. A perfect transducer does not degrade the noise performance of a system by its introduction. The noise factor F is defined $F \equiv P_{out}/GP_{in}$, where P_{in} is the available noise power into the transducer and P_{out} is the available noise power out of the transducer. It follows that $F = 1 + (P_{trans}/P_{in})$, where P_{trans} is the equivalent input noise power introduced by the transducer and $F(T_{ref}) = 1 + T_{trans}/T_{ref}$. It can be shown that, for a cascade of transducers $(F - 1) = (F_1 - 1) = (F_2 - 1)/G_1 + (F_3 - 1)/G_1 G_2 + \ldots$, where F_i and G_i are the noise figure and gain of the ith transducer. This expression is known as the *Friis formula*. In EPR spectroscopy, mixer diodes, amplifiers and oscillators are transducers. The value of T_{ref} is usually $290\,K$, since this is the IEEE standard for noise measurements.

Buckmaster and Skirrow (1972) re-examined Equation 4.2.2.3 and drew attention to the large number of arbitrary factors that must be specified when either Equation 4.2.2.1 or Equation 4.2.2.2 are used. They showed that these factors can significantly alter the value of the sensitivity calculated, although the noise factor of the microwave demodulation system remains unaltered. They also extended the equation derived by Buckmaster and Dering (1967c), to take account of the possibility of introducing a microwave preamplifier to amplify the very weak microwave signal from the sample prior to its detection. Strandberg (1972) has also provided a comprehensive analysis of the factors involved in determining the sensitivity of a CW-EPR spectrometer. (This analysis will be referred to later in the chapter, as it duplicates work by the present author and coworkers regarding determination of the sensitivity of an CW-EPR spectrometer.)

The expression derived by Buckmaster and Dering (1967c), as corrected by Buckmaster and Skirrow (1972), and assuming that there is no microwave preamplifier, is given by:

$$(F_{syst} - 1) = (F_{mo} - 1) \cdot (1 - |\Gamma|^2)^2 [Q_0^2 (\Delta f/f)^2 / 2^{3/2}] + [\{(t + F_{rec} - 1)/G_{md}\} - 1] \quad (4.2.2.4)$$

where F_{syst} is the noise figure of the overall system, F_{mo} is either the AM or FM noise figure of the microwave oscillator, Γ is the reflection coefficient of the microwave cavity, Q_0 is the unloaded Q of the microwave sample cavity, Δf is the spectral purity of the microwave power source, f is the resonant frequency of the microwave sample cavity, t is the noise temperature ratio of the microwave demodulator diodes, F_{rec} is the noise figure of the post-demodulation amplification system, and G_{md} is the power conversion gain of the microwave diodes. The first term in Equation 4.2.2.4 gives the contribution to the overall system noise figure due to the degree of balance of the microwave bridge with the microwave sample cavity. This expression was first derived by Buckmaster and Dering (1967c) and generalized by Buckmaster and Kloza (1969).

Although Equation 4.2.2.4 may appear formidable, it is not complicated provided that the microwave sample cavity is adjusted to critical coupling so that $|\Gamma|$ is zero in the absence of a resonance and a high spectral purity microwave power source is used where $(\Delta f/f)$ is better than 10^{-8} and stable over periods in excess of an hour. Strandberg (1972) has estimated that it should be possible to detect less than 5×10^{10} $S = 1/2$ spins with a resonance linewidth of $0.1\,T$ using a well-designed

10 GHz CW-EPR spectrometer and a sample cavity power level of less than 100 mW.

4.2.4
EPR Lineshapes and Modulation Broadening

Most resonances observed in EPR spectroscopy have a Lorentzian lineshape, because the electron spin system is constrained by interactions with the lattice in which it is embedded. Gaussian lineshapes arise when a system consists of random uncoupled oscillators. Tsallis (1997) has shown that a generalized Levy distribution function exists which includes the Gaussian, Lorentzian, and Dysonian lineshapes as special cases. Howarth, Weil, and Zemprel (2003) have shown how the Tsallis distribution function can be applied to the analysis of magnetic resonance lineshapes. The discussion of the effect of magnetic field modulation on the shape of an observed EPR resonance will be confined to the Lorentzian lineshape, because it is the most common encountered in EPR spectroscopy.

The fact that the magnetic susceptibility of a paramagnetic sample $\chi = \chi' - i\chi''$ is complex where χ' is the absorption and χ'' is the dispersion component implies that the representation of a modulated magnetic resonance by a Fourier series consists of Fourier absorption a_n and dispersion d_n coefficients that are connected by the Kramers–Kronig relations. These relations are valid for any complex function which is analytic in the upper half plane, and hold for the response functions in physical systems. The existence of these relations implies that the same information about a resonance lineshape is available from either the dispersion or the absorption coefficients, since knowing one set enables the other set to be calculated. The real part χ' is an even function, and the imaginary part χ'' is an odd function.

An extensive literature exists concerning modulation broadened unsaturated Lorentzian and Gaussian lineshapes. Wahlquist (1961) was the first to obtain a closed expression for the Fourier absorption coefficients for a Lorentzian lineshape, while Berger and Gunthart (1962) obtained them by taking the Fourier transform of the Fourier series for the lineshape function. Buckmaster (1968) used their methodology for a Gaussian lineshape and obtained closed expressions in the form of a double infinite series that could not be summed. Buckmaster and Dering (1968) derived expressions for the location and amplitude of the extrema and the slope extrema, the ratio of the slope extrema and the location of the zeros of a_n^L for $n = 0, I, II$ and Buckmaster and Skirrow (1971) added a_{III}^L. Buckmaster (1968) showed that an expression relating the absorption coefficients a_n^L with the dispersion coefficients d_n^L existed as the Kramers–Kronig relation requires. Buckmaster and Skirrow (1971) verified the expression for d_I^L obtained by Evans and Brey (1968), and extended the previous analysis for a_n^L to d_n^L for $n = 0, I,$ and II.

The importance of these findings was to show that it was unnecessary to make measurements with the magnetic field modulation amplitude very much less than the resonance lineshape, in order to estimate the number of paramagnetic ions in a sample with the attendant loss of sensitivity. It also enabled the linewidth to be determined from measurements of the location of the amplitude extrema and zeros of two or more a_n^L and d_n^L using a minimum of two EPR spectra.

Buckmaster and Skirrow (1969) showed that the dispersion components d_n (the nth Fourier dispersion coefficient) of an EPR resonance could be observed with the same SNR as the absorption components a_n of this resonance, using an EPR spectrometer of the design developed in the present author's laboratory. They studied four different methods to observe these coefficients:

- Unbalancing of the microwave sample cavity bridge.

- An appropriate adjustment of the phase of the microwave synchrodyne reference power to yield d_n.

- A synchronous demodulation of the sample cavity frequency-lock error signal to yield d_n.

- The direct measurement of d_0, by recording the microwave oscillator frequency as a function of the magnetic field when this oscillator is frequency-locked to the instantaneous sample cavity resonant frequency.

This capability was not widely appreciated, and its potential application has never been exploited; neither has the fact that dispersion measurements can be made without loss in sensitivity in a correctly designed CW-EPR spectrometer.

4.2.5
Microwave Power Sources

The characteristics required for a suitable CW-EPR spectrometer power source are: (i) a low AM noise; (ii) a low FM noise; (iii) a high spectral purity; (iv) low thermal drift; and (v) a low modulation index.

The *AM noise* in an oscillator is characterized by the random fluctuations in the amplitude of the output power, while the *FM noise* in an oscillator is characterized by the random fluctuations in the frequency of the output power. The *spectral purity* of an oscillator describes the spread in the frequencies that the oscillator generates, and should not be confused with the FM noise, which causes the entire output spectral to fluctuate. *Thermal drift* is characterized by the fact that both the output frequency and power will vary as the temperature of the oscillator drifts. Usually, a microwave oscillator can be modulated in either amplitude or frequency by applying an AC voltage to one of its electrodes. If the oscillator is not modulated, then these electrodes can introduce AM or FM noise into the output power. This noise can be minimized by ensuring that voltages applied to these electrodes are very stable. Ondria (1968) has discussed the role of AM and FM noise in an oscillator. Figure 4.2.13a shows, diagrammatically, the noise spectrum of a carrier Ω_c into sinusoidal noise voltages ω_n, while Figure 4.2.13b is a vector diagram of a carrier that has simultaneously phase or FM and AM noise components. Figure

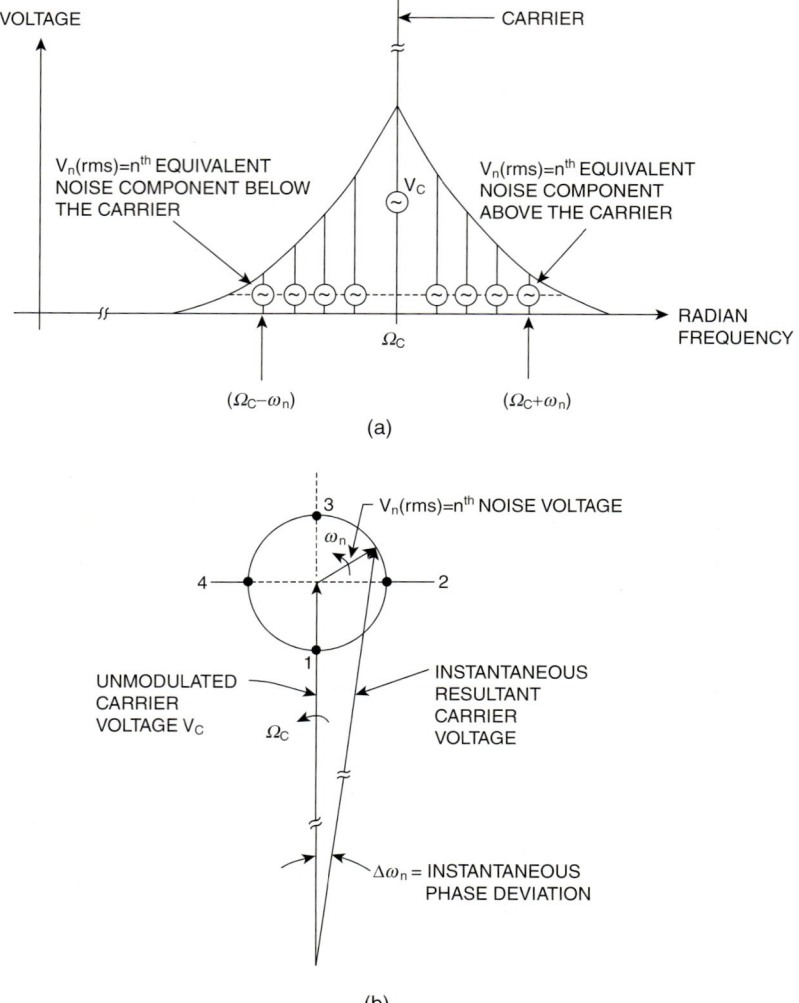

Figure 4.2.13 (a) Decomposition of a noise spectrum of a carrier Ω_c into sinusoidal noise voltages ω_n; (b) A vector representation of a carrier Ω_c that has simultaneously both phase or FM and AM noise ω_n components. From Ondria (1968).

4.2.14 shows, diagrammatically, the fact that the effect of a pair of noise sidebands ω_n above and below the carrier frequency Ω_c is to create a FM vector perpendicular to the carrier vector. Figure 4.2.15a and b are diagrammatic representations of the conversion of FM noise ω_n into AM noise ω_n, by suppressing the carrier Ω_c and reintroducing it phase-advanced by 90°, and how AM noise is simultaneously converted into FM noise due to the presence of a dispersive circuit element. Strandberg (1972) includes a comprehensive mathematical analysis of these con-

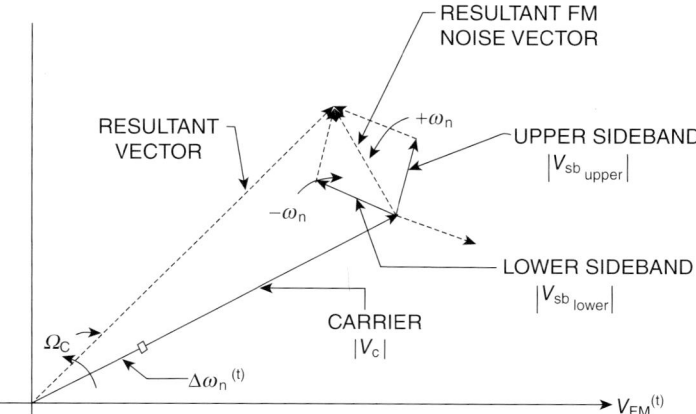

Figure 4.2.14 Diagrammatic representation of a carrier (signal) frequency modulated by two noise components located, respectively, above and below the carrier at frequency Ω_c by a radian frequency ω_n. The resultant FM noise vector is perpendicular to the carrier vector. From Ondria (1968).

version processes for FM and AM noise as applied to EPR spectrometers. Whilst many excellent references are available on this subject, the tutorial by Lee (2000) and the book by Robbins (1982) provide concise and comprehensive discussions, respectively, of this subject.

4.2.6
Reflex Klystrons

Low-power microwave reflex klystrons were used in World War II radar as the local oscillator for the superheterodyne detection system. They were locked in frequency offset by the intermediate frequency (usually 30 MHz) to that of the high-power pulsed magnetron signal power source; consequently, little effort was put into ensuring that they generated a stable frequency. Their design evolved rapidly as many other applications were found. For example, the Varian V-153 and other low-noise, frequency-stable reflex klystrons were required for various microwave repeater systems. In general, these reflex klystrons generated microwave power at levels of 200–400 mW. The Varian VA-232 reflex klystron is an example of a 9 GHz microwave power source that is very suitable for use in EPR spectrometers. Commercial EPR spectrometers have used both Varian V-153 and V-279 klystrons.

4.2.7
Solid-State Devices

Solid-state diodes capable of generating microwave power first became available during the 1990s, when they were used as the active component in a resonant

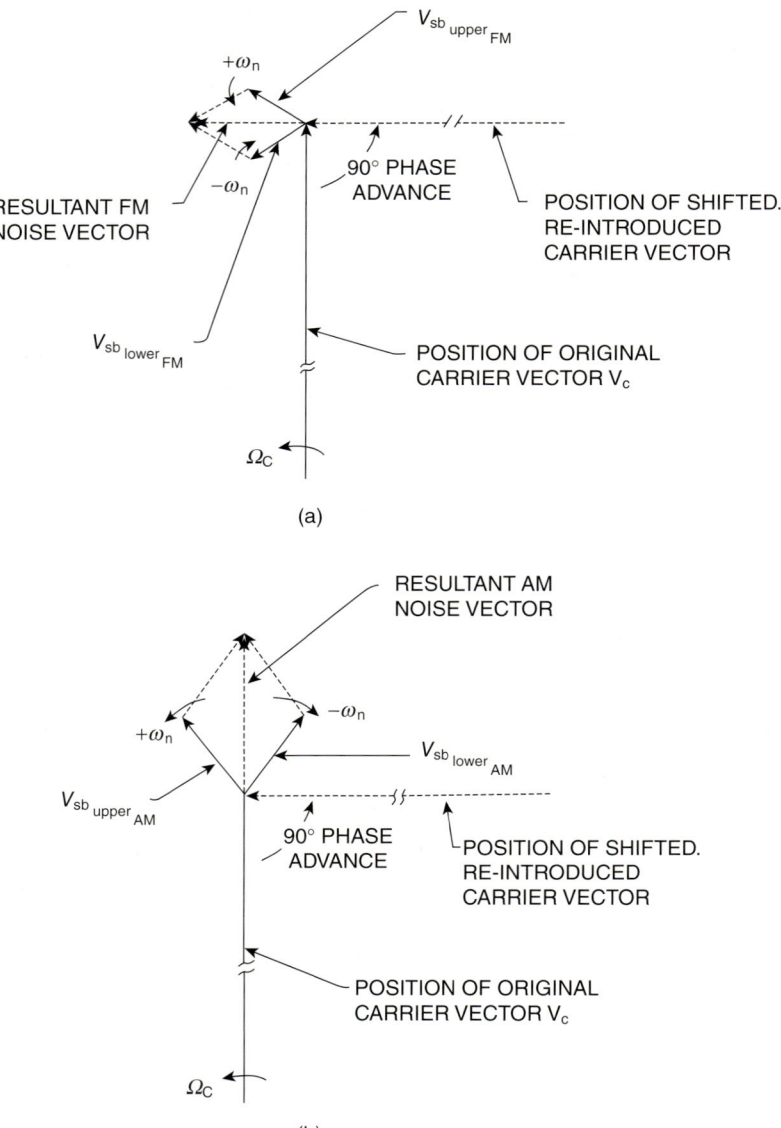

Figure 4.2.15 Diagrammatic representation of (a) the conversion of FM noise ω_n into AM noise ω_n by suppressing the carrier Ω_c and (b) AM noise ω_n into FM noise ω_n by suppressing the carrier Ω_c and reintroducing it phase-advanced by 90° by a dispersive circuit element. From Ondria (1968).

cavity that could be tuned. Whilst the early designs were capable of generating only low levels of microwave power, similar devices of contemporary design are capable of much higher levels. The early versions did not find much application in EPR spectroscopy because they could not be tuned over a very large frequency range, and they also had stability and spectral purity limitations. However, the Gunn diode oscillator design subsequently underwent immense improvements to a point where they are now widely used as microwave power sources for CW-EPR spectrometers. For example, Shing (1972) used a solid-state oscillator and power amplifier in a CW-EPR spectrometer to make 1–4 GHz measurements. It should be noted that the Q-factor for the cavity in which a semiconductor oscillator diode is located is generally less than that in a reflex klystron, so that the ability to phase-lock a solid-state oscillator to a high-stability/high-purity MHz crystal oscillator is very important. More basic solid-state designs can be used for the power oscillators required at frequencies below 500 MHz, because the Q-factor of the sample resonant circuit is lower than at microwave frequencies. Frequency locking and power oscillator spectral purity assume diminished roles at MHz frequencies. Photonic technologies now provide another source of power at frequencies from above 30 GHz to 10 THz (Nagatsuma, 2009).

4.2.8
Frequency Synthesizers

To date, the present author is unaware of any attempt to use a microwave frequency synthesizer coupled to a low-noise power amplifier as a source of power for a CW-EPR spectrometer. Nevertheless, this approach has considerable potential, as they are tunable electronically, the spectral purity has improved and the output power of commercial instrumentation has also increased. Today, as the output power is low (<10 mW), it would be necessary to use a narrow-band microwave power amplifier. In general, however, the major limitation is due to the higher noise floor near the output frequency rather than to the spectral purity, which is higher than in semiconductor oscillators and reflex klystrons.

4.2.9
Microwave CW-EPR Sample Cavity Designs

Foe CW-EPR, an important problem is to determine the most appropriate sample resonant cavity for use in any particular experimental study, in order to maximize the EPR signal amplitude. The choice depends not only on whether the sample is monocrystalline, polycrystalline or a liquid, but also on the measurement frequency and the required temperature range. Liquid samples or samples with large dielectric loss factors at the measurement frequency must not be located in a region of the microwave sample cavity, where the electric field is nonzero, because the Q-factor for the sample cavity will be significantly degraded. If the size of the available sample is a limitation, then the sample should be located where the sample filling factor is maximized. Monocrystal studies frequently involve rotating either the sample or the applied magnetic field in order to study their angular variation properties. Moreover, measurements taken at temperatures above or below the laboratory temperature impose restrictions on the design of the sample resonant cavity. A comprehensive discussion of the design of resonant cavities, and the various versions developed, is available (Poole, 1983, Poole, 1967), while Alger (1968) and Talpe (1971) have also provided valuable information on the subject.

At frequencies below about 500 MHz, a resonant circuit consists of an inductance and a capacitance in parallel, which are lumped parameter circuit components. As the frequency increases, these lumped parameter circuit elements are replaced by distributed parameter circuits. The sample resonant cavities used in EPR experiments at microwave frequencies are the distributed parameter versions of the low-frequency LC resonant circuits.

The four most important factors in the design of a resonant cavity are: (i) the resonant frequency; (ii) the Q-factor; (iii) the method of coupling; and (iv) the filling factor η. The waveguide cavities may be either rectangular or cylindrical, whilst in addition there are loop–gap-resonant and strip line-resonant cavities which are a crossover between a lumped and a distributed parameter circuit. A coupling structure is used to provide a match between the transmission line and the resonant structure. The design depends on the type of transmission line and resonant structure, and may be a loop or probe for a coaxial line or a shaped hole in the conducting plate inserted between the waveguide and the resonant cavity. Above 75 GHz, quasi-optical resonators such as the Fébry–Perot cells are used.

The unloaded or intrinsic Q_i-factor of a resonant cavity is defined as:

$$Q_i \equiv 2\eta \text{ (maximum energy stored per cycle)}/$$
$$\text{(energy dissipated per cycle).}$$

where the energy dissipated is due to the finite resistance of the cavity walls.

The effect introducing a coupling hole into the cavity results in a coupling Q_c-factor, defined as:

$$Q_c \equiv 2\pi \text{ (maximum energy stored per cycle)}/$$
$$\text{(energy dissipated by the coupling iris per cycle).}$$

The effect of dielectric losses inside the cavity results in another Q_ε-factor defined as:

$$Q_\varepsilon \equiv 2\pi \text{ (maximum energy stored per cycle)}/$$
$$\text{(energy dissipated in dielectric per cycle).}$$

The external or effective or loaded Q_L-factor is then given by:

$$(1/Q_L) = (1/Q_i) + (1/Q_c) + (1/Q_\varepsilon) \qquad (4.2.2.5)$$

A cavity is critically coupled if the internal cavity losses are equal to the coupling losses. It is undercoupled if $\Gamma < 1$ and overcoupled is $1 < \Gamma$, where Γ is the coupling factor. Note that beta is commonly used as the symbol for the coupling factor in engineering literature.

$$\Gamma \equiv Q_i/Q_c \tag{4.2.2.6}$$

The coupling factor can be determined using a standing-wave measurement, and adjusted by modulating the frequency source to display the cavity resonance and adjusting the microwave resonant cavity bridge balance by varying the cavity coupling with the Gordon coupler or the depth of the conducting iris post to minimize the power reflected at the resonant frequency. A Gordon coupler (Gordon, 1961) (see Figure 4.2.16) consists of a section of waveguide beyond cut-off inserted between the resonant cavity that is tapered to the dimensions of the connecting waveguide. A sliding dielectric (usually Teflon) plunger matches the two transmission lines. The coupling is varied by moving the plunger away from the cavity coupling iris. Yariv and Clapp (1959) have outlined a procedure for determining critical coupling in a reflection-type cavity.

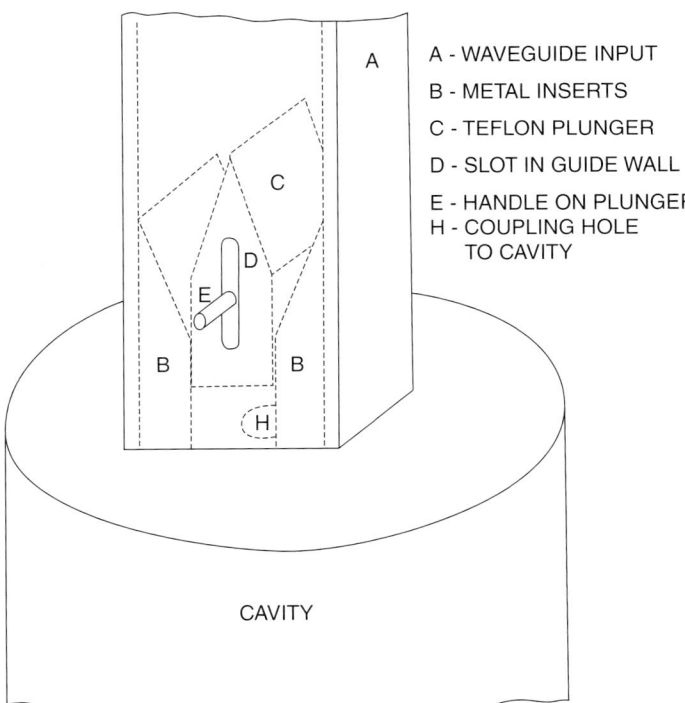

Figure 4.2.16 Schematic diagram showing the design of a Gordon coupler (not to scale). From Gordon (1961).

The Q of a cavity can be measured in various ways. In a microwave measurements laboratory, the standard method is to use a standing-wave meter approach. It is easier and more convenient to triangular wave modulate the frequency source and to display the cavity resonance curve on an oscilloscope when the cavity forms part of the EPR spectrometer. The screen can be calibrated by decreasing the microwave input power by 3 dB; the −3 dB points on the Q-curve can then be determined and the superposition of a wavemeter resonance used to measure the three frequencies. Ginzton (1957) has shown that $Q_L = f_0/(f_u - f_l)$, where f_0 is the resonant frequency, and f_u and f_l are the higher- and lower-frequency 3 dB points. Most resonant cavities used in CW-EPR spectroscopy have Q-factors of the order of 10^3 or 10^4 at 293 K.

The filling factor η is related to a Q_χ-factor, which is defined as the ratio of the magnetic energy stored in the cavity to the magnetic energy absorbed by the paramagnetic sample:

$$Q_\chi = 1/(\chi''\eta) \qquad (4.2.2.7)$$

where χ'' is the magnetic loss component of the complex paramagnetic susceptibility of the sample. Expressions for the filling factors for various modes of rectangular and cylindrical resonant cavities are given in Poole (1967, 1983). The choice of cavity configuration depends on the factors discussed above. The role that the dielectric loss ε'' of the sample plays an important role in this choice, since it is large in liquid samples and water is very lossy at microwave frequencies.

The sample cavities of primary interest in CW-EPR spectroscopy at microwave frequencies are the rectangular TE_{omn} cavity, the cylindrical TE_{lmn} cavity, the loop-gap cavity, and the re-entrant cavity. Among other cavity designs developed for special EPR applications are a bimodal cylindrical TE_{111} mode cavity as described by Teaney, Klein, and Portis (1961), in which the input and output waveguide were orthogonal and isolated the ports by 60 dB. This was the EPR analog of a nuclear induction NMR spectrometer (Bloch, 1946), and has the advantage that higher cavity power levels can be used, without sacrificing sensitivity. However, there is a complex trade-off with its complexity and the loss of flexibility Strip-line cavities have not been used in CW-EPR spectroscopy because of their low Q-factor, but this feature is advantageous in pulsed EPR spectroscopy (Davis and Mims, 1978). They also have potential application in the EPR studies of thin films. Quasi-optical cavities have been used above 75 GHz, whilst at MHz frequencies the sample is located inside a lumped parameter coil.

4.2.10
Transmission Cavities

Bleaney and coworkers used cylindrical cavities operating in the TE_{111} mode. The sample was located on the tuning screw base of the cavity, where the magnetic field was a maximum and the electric field was a minimum. The adjustable base was sealed using a tapered metal cap that was greased to prevent liquid refrigerants from entering the cavity.

Although transmission cavities were used in the earliest EPR spectrometers, they are rarely used today because two waveguides, rather than one waveguide or coaxial coupling to the sample cavity, are required. Reflection sample cavities are easier to use because critical coupling can be easily achieved using either an adjustable post located near the coupling iris, or a Gordon coupler which enables critical coupling to be achieved very accurately (Gordon, 1961) (see Figure 4.2.16). As the former are less stable mechanically than the latter, they are a source of both AM and FM noise due to coupling instability. Both, Poole (1983) and Alger (1968) have provided detailed information regarding the design of Gordon couplers, though the details of many other cavity designs have also been reported.

4.2.11
Reflection Cavities

Following the introduction of the magic-T microwave sample cavity bridge into the design of EPR spectrometers, a variety of rectangular and cylindrical cavities has been used. The simplest rectangular cavity was designed to operate in the TE_{101} mode (Figure 4.2.17a). In this case, the sample was located on the base of the cavity where the electric field was zero, while the sealed cavity could be immersed in various refrigerants for low-temperature measurements. Whilst this was a good design for solid samples, liquid samples usually required larger filling factors; consequently, the TE_{102} (Figure 4.2.17a) rectangular cavity was more suited to 293 K measurements, because the filling factor for the same sample was twice as large. Today, cylindrical-mode TE_{011} cavities represent the "workhorse" cavity used in EPR spectrometers, because of the high Q-value, large filling factor, and flexibility. Indeed, a cold finger Dewar can be inserted into the cavity's axis without causing any significant degradation in Q-value (Figure 4.2.17b). The cylindrical TE_{111} mode does have some advantages for solid samples, however, as these can be mounted on the lower end of the cavity that is removable (because it is threaded) and is, therefore, also tunable.

The resonant frequency of a rectangular or cylindrical cavity can be calculated using Maxwell's equations. The resonant frequency for a rectangular TE_{mnp} mode cavity of cross-section a and b, and length c is given by:

$$k = (k_x^2 + k_y^2 + k_z^2)^{1/2} = \pi[(m/a)^2 + (n/b)^2 + (p/c)^2]^{1/2} \tag{4.2.2.8}$$

or

$$f_r = (1/2c)[(m/a)^2 + (n/b)^2 + (p/c)^2]^{1/2} \tag{4.2.2.9}$$

where the symbols are defined in Figure 4.2.17b (see Poole (1967, 1983) for further details).

The resonant frequency for a cylindrical TE_{mnp} of diameter a and length d is given by:

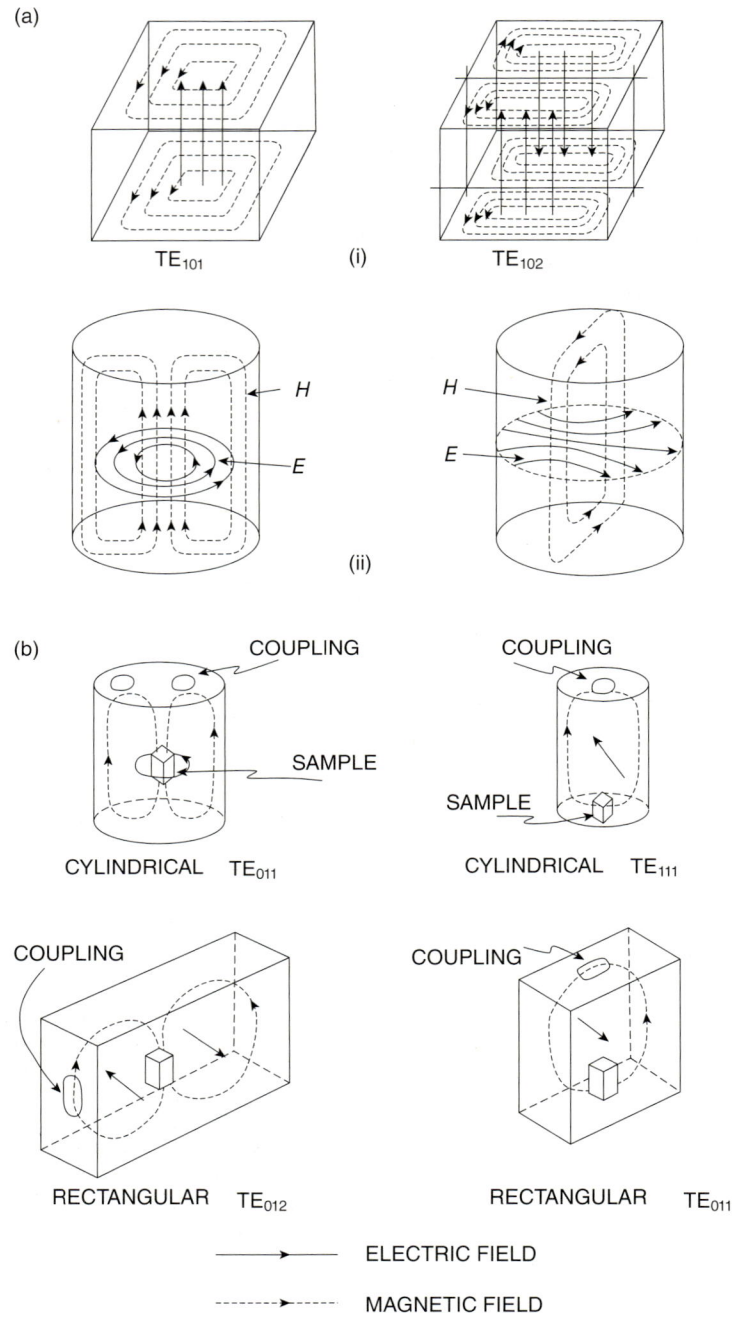

Figure 4.2.17 (a) Diagrams for (i) rectangular TE_{011} and TE_{012} and (ii) cylindrical TE_{011} and TE_{111} mode resonant cavities showing the electric **E** (– lines) and magnetic **H** (---- lines) field distributions and directions. from Ishii (1966); (b) Diagrams showing the location of the sample in the cylindrical TE_{011} and TE_{111} and rectangular TE_{012} and TE_{011} cavities and the location of the coupling hole. From Dering (1964).

$$k = [(k_ca)_{mn}^2/a^2 + (p\pi/d)^2]^{1/2} \qquad (4.2.2.10)$$

or

$$f_r = (1/2c)[(k_ca)_{mn}^2/a^2 + (p\pi/d)^2]^{1/2} \qquad (4.2.2.11)$$

where $(k_ca)_{mn}$ is the nth root of the mth order Bessel function derivative $[J_m(k_ca)]'$. These roots are tabulated in Poole (1967, 1983) as are the corresponding roots of $J_m(k_ca)$ required for the TM$_{nmp}$ modes.

4.2.12
Re-Entrant Cavities

Re-entrant cavities are very useful as they can be designed for operation as low as 1 GHz, without their physical dimensions becoming excessive. Erickson (1966) reported the first application of a re-entrant sample cavity in EPR spectroscopy, and Shing (1972) has discussed this application in detail. Shing emphasized that such cavities have the distinct advantage of being tunable over an octave, so that two cavities are required to tune from 1–2 and 2–4 GHz. This feature meshed well with the fact that early coaxial microwave components were designed to cover octave bands – that is, 1–2, 2–4, and 4–8 GHz. Figure 4.2.18a shows a diagram for the physical configuration of a re-entrant cavity, while Figures 4.2.18b and c show (diagrammatically) the microwave electric E and magnetic H field configurations in a re-entrant cavity. Re-entrant cavities can be used down to about 500 MHz before their physical size becomes a significant limitation.

Moreno (1958) has shown that the approximate resonant frequency f_r of a re-entrant cavity is given by:

$$f_r = (c/2\pi)\{(Z_0\rho_1/2\delta)\ln(\rho_2/\rho_1)]^{1/2} \qquad (4.2.2.12)$$

where the symbols are defined in Figure 4.2.18a. Ishii (1966, p. 100) provides a more exact expression for this resonant frequency.

Figure 4.2.19 is a diagram showing the design of the low-temperature, tunable re-entrant microwave resonant cavity system used by Shing (1972). The paramagnetic sample is located where the electric field is zero, as shown in Figure 4.2.18. The loaded Q-factor of the cavity was ~1500 at 290 K, but this was degraded significantly to ~500 at 77 K. It is suspected that contraction of the flexible beryllium fingers used to provide the electrical contact when the cavity was cooled were responsible for this problem. The introduction of a dielectric rod using Teflon might represent a possible solution if the Q-factor is not degraded too much. Figure 4.2.20 is a block diagram of a 1–8 GHz coaxial CW-EPR spectrometer using synchronous demodulation at the microwave and magnetic field modulation frequencies (Shing, 1972). Buckmaster, Fry, and Skirrow (1968) showed that coaxial components could be used to construct a MHz frequency CW-NMR spectrometer, and that this design could also be used for zero-field CW-EPR.

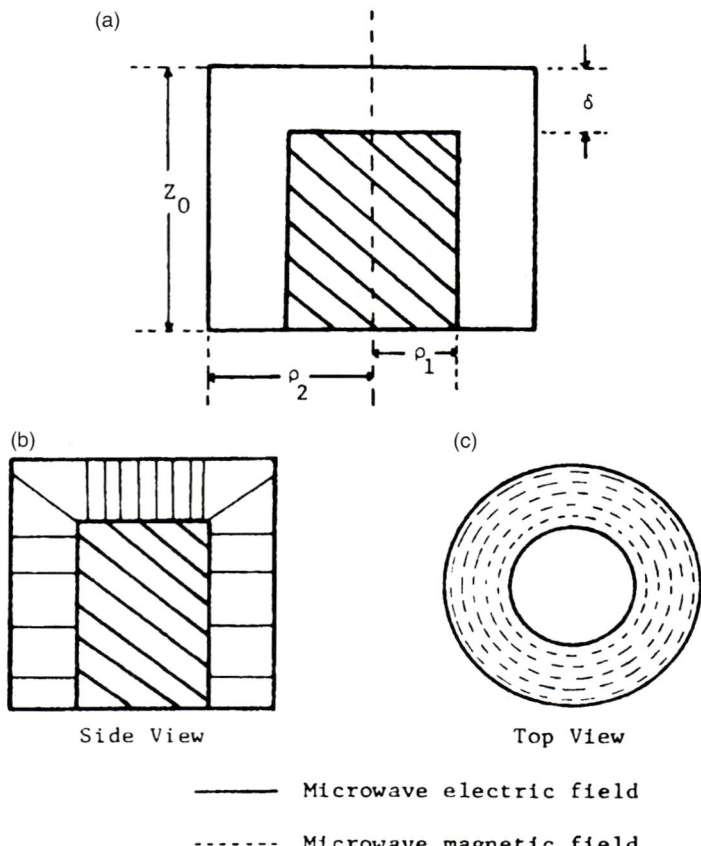

Figure 4.2.18 Schematic diagrams of a re-entrant resonant cavity. (a) Definitions of symbols used in Equation 4.2.2.12 (Z_o, resonator length, δ, gap width, ρ_1, resonator inner radius, ρ_2, resonator outer radius); (b) Side view of cavity showing the microwave electric and magnetic field distribution; (c) Top view showing the magnetic field distribution. From Shing (1972).

4.2.13
Loop–Gap Cavities

The loop–gap cavity is a very useful design as it can be used over a wide frequency range because of its small size. This type of cavity was first used in EPR spectroscopy by Froncisz and Hyde (1982), who not only demonstrated its use with more than one gap, but also provided complete design information for the cavities that they had used. The loop–gap cavity is shown schematically in Figure 4.2.21; subsequently, Hardy and Whitehead (1981) used this design for an NMR cavity and conducted a comprehensive analysis of it properties. The resonant frequency f_r is given by:

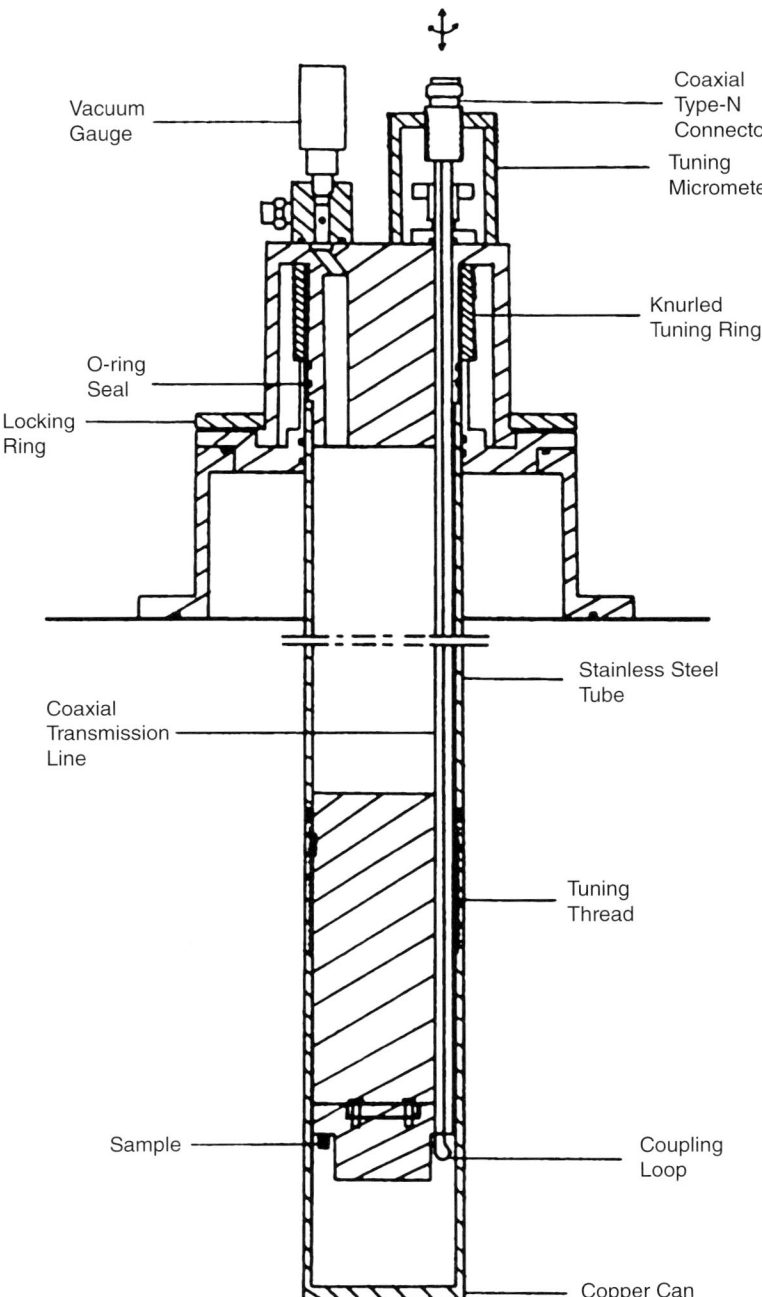

Figure 4.2.19 Diagram showing the design of a low-temperature, tunable, re-entrant microwave resonant cavity system with micrometer adjustable coupling loop, and the sample location. From Shing (1972).

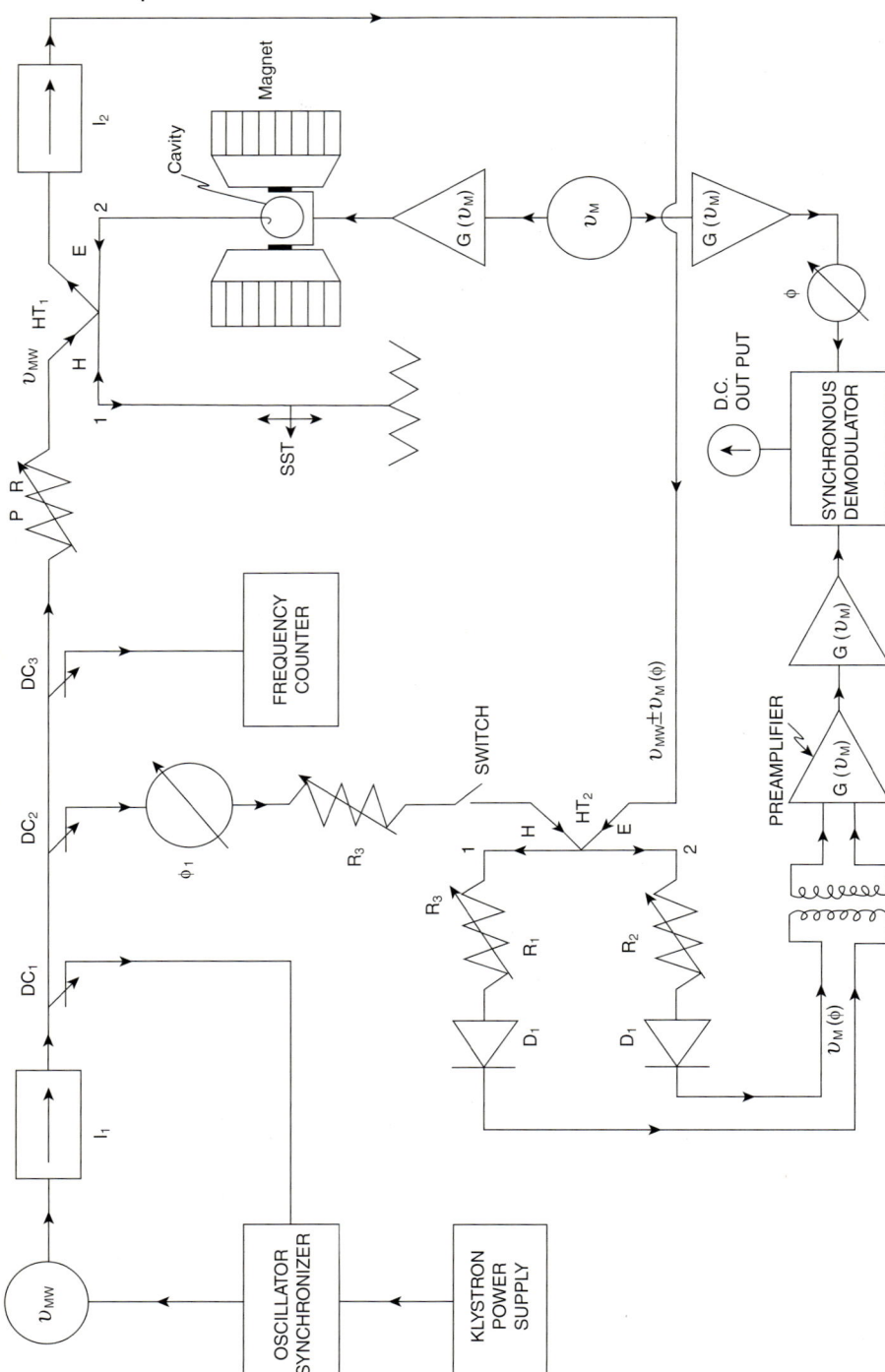

Figure 4.2.20 Block diagram of 1–8 GHz coaxial component CW-EPR spectrometer using synchronous demodulation at the microwave and magnetic field modulation frequencies. I, isolator; DC, directional coupler; R, variable attenuator; f, phase shifter; D, microwave diode; SST, slide screw tuner; HT, hybrid Tee; $G(v_M)$, bandpass amplifier gain G at magnetic field modulation frequency v_M. From Shing (1972).

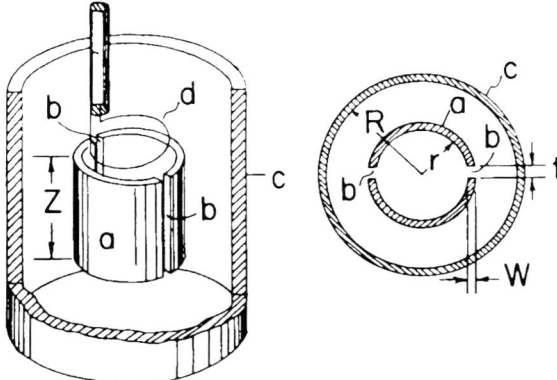

Figure 4.2.21 Diagram of a typical loop–gap resonant cavity, showing the principal components. a, loop; b, gaps; c, shield; d, inductive coupler) and the critical dimensions (Z, resonator length; r, resonator radius; R, shield radius; t, gap separation; W, gap width) (see Equation 4.2.2.14). From Froncisz and Hyde (1982).

$$f_r = (1/2\pi r)n^{1/2}(t/W)^{1/2}(1/\pi\varepsilon\mu_o)^{1/2} \qquad (4.2.2.13)$$

where t is the gap in tube, W is the thickness of the tube, and n is the number of tubes.

It is desirable to shield the radiation from this cavity which can occur if the dimensions approach $\lambda/4$. This shielding and the fringing fields from the gap cause the resonant frequency to shift, such that the resonant frequency f_r is now given by the semi-empirical equation:

$$f_r = (1/2\pi)[1+\{r^2/[R^2-(r^2+W^2)]\}]^{1/2}[nt/\pi W\varepsilon\mu_o]^{1/2}(1/r)[1/\{1+2.5(t/W)\}]^{1/2} \qquad (4.2.2.14)$$

where the symbols are defined in Figure 4.2.21.

The Q of the loop–gap resonator is given by

$$Q \sim (r/\delta) \propto r(f_r)^{1/2} \qquad (4.2.2.15)$$

if the effects of the shield and electric field losses in the gaps are neglected. Hardy and Whitehead (1981) derived an expression for Q which takes account of the effect of the shield.

4.2.14
Other Resonant Structures

When research on *in vivo* objects became an important area of study using MHz frequency EPR, it became necessary to develop specialized resonant structures in which human limbs and small animals could be located. These measurements were made at frequencies below 1 GHz because of the dielectric loss in the samples, allowing the application of many new structures to be explored. These include lumped parameter coils, pairs of crossed coils (as used in NMR spectrometers), helical, strip line, coaxial, lumped parameter delay line, birdcage, and flat loop surface coil resonators. A comprehensive summary of the literature describing the many design configurations that have been used successfully is provided by Eaton and Eaton (2004; see their Table 6). The Q-values of these low-frequency structures are usually of the order of a few 100, the effect of FM oscillator noise is less likely to be a limiting factor for the sensitivity of low-frequency CW-EPR spectrometers. Moreover, the decrease in Q-value is compensated by an increase in the filling factor, so that sensitivity is not necessarily reduced at MHz frequencies. However, this does mean that locking the frequency of the power source to the sample cavity resonant structure will introduce much more noise, and these two factors must therefore be carefully evaluated and controlled. For this reason, it is probably preferable to improve the spectral purity of the MHz frequency source rather than to lock its output frequency to the resonant frequency of the sample resonator.

4.2.15
Microwave Detectors or Demodulators

The first microwave detectors, as developed for use with the radar systems of World War II, were silicon point-contact diodes, the design of which evolved as the operating frequency increased. These included 1N21 unshielded diodes (L- and C-bands), 1N23 unshielded diodes (X-band), 1N26 shielded diodes (K-band), and 1N53 shielded diodes (Q-band) (Torrey and Whitmer, 1948). Later, these designs were refined as the technology for their manufacture improved and, indeed, some of these designs are still in use. The 1N23H is the most recent version of the 1N23 for use at 9 GHz, and the 1N26C at 24 GHz.

A number of different devices exist that can be used to either detect or demodulate microwave radiation; these include Si point-contact diodes, Si or GaAs Schottky barrier diodes, backward diodes, and bolometers. At frequencies below 500 MHz, many more demodulator choices are available because the noise considerations are much less severe. In general, the noise figure is lower and conversion gain greater in the majority of semiconductors that operate at these frequencies.

4.2.15.1 Point Contact Diodes
These diodes, which are the modern equivalent of the early radio so-called "cats whisker" detector, consist of a pointed metal wire in contact with a block of N-type semiconductor (usually Si for microwave frequencies). Until recently, these diodes had a large $1/f$ or flicker noise component with a knee frequency higher than 100 kHz. The 1N23H point contact diodes have a noise figure of 6.0 dB and a conversion gain (G) of about −5.0 dB when the microwave bias is 1.0 mW. Their application to CW-EPR spectrometers was first investigated by Buckmaster and Dering (1965), and later by Rathie (1970), Buckmaster and Rathie (1971) and

Zaghoul et al. (1990), who measured their noise temperature (t), G, and sideband impedance as a function of their biasing power, and compared these results to those obtained by EPR measurements as a function of the magnetic field modulation frequency. The value of G for 1N23WE diodes was found to be about -7 dB, while the knee of the $1/f$ noise spectrum was about 100 kHz when the microwave bias power was about 0.3 mW. An excellent agreement was obtained between the observed EPR spectral amplitude and the value of $(t/G)^{1/2}$ obtained from point-contact diode measurements.

4.2.15.2 Schottky Barrier Diodes

In these diodes, which are also known as Schottky or "hot carrier" diodes, the metal–semiconductor junction has a nonlinear rectifying characteristic. Such diodes are normally manufactured from N-type silicon or gallium arsenide (GaAs). The latter type have a higher cut-off frequency, and so can operate up into the THz frequency range because the electrons have a higher mobility. They are made in various packages, and have been found to have good noise performance for use in EPR spectrometers. When Buckmaster and Rathie (1971) characterized a matched pair of HP 2627 Schottky barrier diodes, they showed the $(t/G)^{1/2}$ ratio to be in good agreement with that calculated from EPR measurements. The flicker noise knee of these diodes (ca. 7 kHz) was higher than that expected from selected diodes. The optimum sideband frequency matching resistance of less than 10 Ω was a disadvantage, because most matching transformers have a higher winding resistance. Further details are provided by Rathie (1970).

4.2.15.3 Backward Diodes

These diodes are also known as tunnel diodes because they are heavily doped P–N diodes in which the electron tunneling from the N-type conduction band to the P-type valence band produces a negative resistance region. Buckmaster and Rathie (1971) characterized two matched Philco 4154 backward diodes by measuring t and G after determining the optimum sideband frequency matching resistance to be about 7 Ω. The $(t/G)^{1/2}$ ratio, as verified by EPR measurements, was found to become independent of the magnetic field modulation frequency above about 2 kHz. More recently, Zaghoul et al. (1990) showed, using a new microwave evaluation methodology, that the noise performance of tunnel diodes was essentially independent of the microwave power level, and that the $1/f$ noise knee was about 4 kHz with near-optimum microwave diode bias.

4.2.15.4 Bolometers

Early bolometers had long time-constants, and so were of limited use in EPR spectrometers (although they are frequently referred to in the literature). The original bolometers, which used Wollaston wire on a mica substrate, are still used in microwave power meters because they have a linear calibration curve. Contemporary designs of semiconductor bolometers are used at frequencies above 100 GHz.

4.2.16
Electromagnets

Most commercial electromagnets that are manufactured for use in CW-EPR spectrometers have 15 cm (6"), 23 cm (9"), or 30 cm (12")-diameter pole faces and air gaps that range from 3 to 5 cm. Small desktop CW-EPR spectrometers, which are designed for use in routine 293 K measurements where high sensitivity is not required, are available from several manufacturers. In this case, the pole faces are shimmed to improve the magnetic field homogeneity, and tapered when magnetic fields in excess of 2 T are required when $g = 2$ are studied above 50 GHz.

The homogeneity of the magnetic field is very important in EPR studies because the magnetic linewidths may be a narrow as ~10 µT. However, the requirements are less stringent than in NMR, where the linewidths are ~10^{-1} µT. Inhomogeneity manifests itself in a form of magnetic resonance lineshape broadening. Most paramagnetic impurities in monocrystalline samples have linewidths of ~1–10 mT, while liquid samples with free radicals may have ~0.1 mT. These observed linewidths require that the magnetic field homogeneity approximates 1 ppm. Fortunately, this requirement is not difficult to achieve with well-designed electromagnet pole faces and careful shimming.

The following table lists the typical magnetic field values required for $g = 2$ EPR measurements at various frequencies:

Frequency (GHz)	0.01–0.1	0.8–1.2	3.4–3.8	9–10	22–26	33–36	95
Band	ZF	L	S	X	K	Q	W
Magnetic field (T)	~0.000	0.035	0.13	0.34	0.86	1.25	3.4

Before the advent of power transistors, electromagnets were designed with high-impedance coils to match the high-output impedance of the vacuum tubes used in a highly regulated power supply. One consequence was that the cooling coils had to be well insulated from the field coils, so thermal stability was a serious problem. As power transistor-regulated power supplies were low impedance, so too were the electromagnet coils low impedance, which meant that the coolant could be circulated through these coils. Subsequently, thermal stability ceased to be a serious limiting problem when it was found possible to circulate the coolant through the magnetic field-generation coils. In this way, with careful design, it became possible to attain a current stability of $1:10^6$.

The magnetic field is normally regulated using a temperature-controlled Hall effect probe, and swept by using a triangular wave probe voltage. Digital sweep ramps have improved the linearity of the magnetic field sweep. The magnetic field is measured separately, using a proton resonance magnetometer; the magnetic field values are then obtained by converting the resonant frequency. The accuracy of magnetic field measurements made in this way is determined by the gyromagnetic ratio of the protons in the probe material. This ratio is rarely known absolutely to be better than $1:10^5$, except in free space and in water, although magnetic field differences can be determined by a factor of ten to a hundred more accurately depending on the linewidth of the sample used in the magnetic field measurement probe. The ratio of the frequency to the magnetic field (γ) was 42.576 388 1(12) MHz/T for protons in water, and 6.535 906(1) MHz/T for deuterons in deuterium. The magnetic fields required for $g = 2$ EPR at frequencies above 100 GHz can be measured using a deuterium magnetic resonance magnetometer. These measurements are discussed in Section 4.3.

Air core electromagnets using a pair of Helmholtz coils can generate magnetic fields of up to ~0.10 T, which is sufficient to observe $g = 2$ EPR resonances up to ~3 GHz. The Earth's magnetic field in the laboratory is a source of error, as it is ~0.06 mT in many locations and has a horizontal component of ~0.01 mT. Thus, the air core electromagnet should be mounted on a rotary base so that EPR measurements can be made when the magnetic field it is oriented both parallel and anti-parallel to the horizontal component of the Earth's magnetic field in the laboratory. This should eliminate any error introduced by this horizontal component, while the vertical component can be canceled or nulled by using another pair of small Helmholtz coils. A well-designed power supply can provide the current necessary to generate magnetic fields with a stability of better than $1:10^5$ if the magnet is correctly cooled. The sample size may be larger than at higher magnetic (frequencies) fields, as the Helmholtz configuration optimizes the magnetic field homogeneity along and near the magnet axis. Zero-field splitting (ZFS) EPR measurements are frequently made in the 1–8 GHz frequency interval; these may then be plotted as a function of the magnetic field for the $\Delta M = -1$ and $+1$ transitions, to determine the ZFS values. These measurements do not require the Helmholtz coil current to be calibrated, nor the magnetic field that it creates to be calculated (Shing, 1972).

Near zero-field measurements are made at frequencies below about 100 MHz for $g \sim 2$ CW-EPR. Although no magnetic field is required, precision measurements require the Earth's magnetic field in the laboratory to be nulled. This can be accomplished by using two pairs of small Helmholtz coils in the horizontal and vertical planes. The nulling procedure outlined above should be used to ensure that the horizontal component of the Earth's magnetic field in the laboratory has been accurately taken into account. Shing and Buckmaster (1976) provide a useful discussion of an application of this procedure.

4.2.17
Zero-Field CW-EPR

Zero-field CW-EPR measurements are made in two distinct ways, depending on the information available. If ZFS measurements are made, and the values have been estimated from CW-EPR spectral data obtained at 3 GHz and higher frequencies, then the ZFS transitions can be observed in low magnetic fields at a sequence of frequencies and extrapolated to zero magnetic field. The sample under study is located in a resonant structure, so the sample size can be very small. Alternatively, CW-EPR measurements can be made in a zero magnetic field if the sample is located in a nonresonant structure and the frequency varied linearly. The sample size must then be much greater, because the product of the Q-factor of the structure and the filling factor determines the spectrometer sensitivity, as shown in Equation 4.2.2.2. Surveys conducted by both Bramley and Strach (1983) and Eaton and Eaton (2004) have provided a comprehensive summary of the numerous reports on zero and near-zero magnetic field CW-EPR spectroscopy. It is realistic to perform such spectroscopy using a regenerative oscillator of the type used in proton resonance magnetometers because of its high sensitivity and simplicity, although this design appears not to have been mentioned elsewhere. This marginal oscillator may occasionally be referred to as a "Q-multiplier", although in fact any magnetometer with a search capability could be used to carry out zero-field CW-EPR. The coaxial CW-NMR spectrometer design, as created by Buckmaster, Fry, and Skirrow (1968), and the design of Shing (1972), as shown in Figure 4.2.20, can be used at MHz frequencies and possesses all of the advantages outlined earlier in this section. It is notable that the design of both CW-NMR and zero-field CW-EPR spectrometers did not pass through the same evolutionary stages as did CW-EPR spectrometers at microwave frequencies, mainly because there was little scope or need to increase the overall sensitivity of the spectrometers. However, the introduction of solid-state electronic circuitry and digital signal processing followed the same time scale trajectory at all frequencies.

4.2.18
Support Instrumentation

It is useful to have a stand-alone low-frequency oscilloscope to conduct supplementary measurements, including cavity Q; moreover, this addition may also be used to determine when the cavity is critically coupled. Whilst some commercial spectrometers do not include any provision for microwave power measurements, the addition of a 20 dB directional coupler and a power head and matching power meter in the sample cavity arm may be useful, if only on a temporary basis. In particular, it is important to monitor the microwave power when EPR measurements are made on samples that can be saturated if the microwave sample cavity power is too high. Exactly how this instrumentation can be incorporated into a CW-EPR spectrometer design is shown in Figures 4.2.5 and 4.2.6 show.

4.2.19
Concluding Remarks

The aim of this chapter has been to provide a concise and brief survey of the major engineering design features of all CW-EPR spectrometers and, in particular, of those that operate at frequencies from below 10 MHz to 100 GHz. An historical approach has been used to explain why, and when, these features were introduced, and how they have evolved over time, with attempts having been made to dispel the many erroneous ideas and folklore that has evolved over the years with regards to CW-EPR spectrometer design. Examples of various designs that satisfy different design criteria have been provided to assist the reader in appreciating when certain design principles are not essential for satisfactory spectrometer performance.

4.2.20
Pertinent Literature

The reader is referred to the classic EPR instrumentation books by Poole (1967, 1983), Alger (1968), and Talpe (1971), while more recent books include Pilbrow (1990), Weil, Bolton and Wertz (1994) and Weil and Bolton (2007), as well as the journal literature concerning specific designs referred to in these works. The internet is a rich source of information and definitions and Wikipedia is a good starting point for such searches on most of the subjects discussed in this chapter.

References

Alger, R.S. (1968) *Electron Paramagnetic Resonance – Techniques and Applications*, Interscience, New York, p. xvii + 580.

Bender, C.J. (2004) EPR Instrumental Methods, in *Biological Magnetic Resonance* (eds C.J. Bender and L.J. Berliner), Kluwer Academic/Plenum Publishing Co., New York, vol. 21, Chapter 5, p. 212.

Berger, P.A. and Gunthart, H.H. (1962) *Z. Angew. Math. Phys.*, **13**, 310.

Bleaney, B. and Stevens, K.W.H. (1953) *Rep. Prog. Phys.*, **16**, 108.

Bloch, F. (1946) *Phys. Rev.*, **70**, 460.

Bosch, B.G. and Gambling, W.A. (1962) *Br. Inst. Radio Engrs*, **24**, 389.

Bramley, R. and Strach, S.J. (1983) *Chem. Rev.*, **83**, 49.

Buckmaster, H.A. (1968) *J. Appl. Phys.*, **39**, 1986.

Buckmaster, H.A. and Dering, J.C. (1965) *Can. J. Phys.*, **43**, 1088.

Buckmaster, H.A. and Dering, J.C. (1966) *J. Sci. Instrum.*, **43**, 554.

Buckmaster, H.A. and Dering, J.C. (1967a) *Proceedings, XIVth Colloque Ampere, Ljubljana 6–10 September 1966*, North-Holland Publishing Co., Amsterdam, pp. 1017–1022.

Buckmaster, H.A. and Dering, J.C. (1967b) *IEEE Trans. Instrum. Meas.*, **IM-16**, 13.

Buckmaster, H.A. and Dering, J.C. (1967c) *J. Sci. Instrum.*, **44**, 430.

Buckmaster, H.A. and Dering, J.C. (1967d) *Can. J. Phys.*, **45**, 107.

Buckmaster, H.A. and Gray, A.L. (1971) *J. Sci. Instrum.*, **42**, 1041.

Buckmaster, H.A. and Kloza, M.J. (1969) *IEEE Trans. Instrum. Meas.* IM-18, 273.

Buckmaster, H.A. and Rathie, R.S. (1971) *Can. J. Phys.*, **49**, 853.

Buckmaster, H.A. and Scovil, H.E.D. (1956) *Can. J. Phys.*, **34**, 711.

Buckmaster, H.A. and Skirrow, J.D. (1969) EPR dispersion spectrometers I. Instrumentation for the Detection of the Fourier Coefficients of the Dispersion Component of Paramagnetic Susceptibility. Unpublished EPR Group Internal Report, Department of Physics, University of Calgary, Calgary, AB, Canada, p. 23. Copies are available from the author.

Buckmaster, H.A. and Skirrow, J.D. (1971) *J. Magn. Reson.*, **5**, 285.

Buckmaster, H.A. and Skirrow, J.D. (1972) *Can. J. Phys.*, **50**, 1092.

Buckmaster, H.A., Fry, D.J.I., and Skirrow, J.D. (1968) *Rev. Sci. Instrum.*, **39**, 930.

Chang, T.-T. (1984) *Magn. Reson. Rev.*, **9**, 65.

Davis, J.L. and Mims, W.B. (1978) *Rev. Sci. Instrum.*, **49**, 1095.

Dering, J.C. (1964) Sensitivity Considerations In Electron Paramagnetic

Resonance Spectrometers. Unpublished M.Sc. Thesis, Department of Physics, University of Alberta, Calgary, AB, Canada.

Dering, J.C. (1967) Theoretical And Instrumental Studies In Electron Paramagnetic Resonance I – An Investigation of the Design of High-Sensitivity and –Stability EPR Spectrometers and a Survey of the Theory and Application of Generalized Spin-Hamiltonians. Unpublished PhD Thesis, Department of Physics, University of Calgary, Calgary, AB, Canada.

Dicke, R.H. (1946) *Rev. Sci. Instrum.*, **17**, 268.

Eaton, G.R. and Eaton, S.S. (2004) EPR Spectrometers at Frequencies Below X-band, in *EPR: Instrumental Methods*, vol. 21 *Biological Magnetic Resonance* (eds C.J. Bender and L.J. Berliner), Kluwer Academic/Plenum Publishing Co., New York, Chapter 2, pp. 59–114.

Erickson, L.E. (1966) *Phys. Rev.*, **143**, 295.

Evans, T.E. and Brey, W.S., Jr (1968) *J. Chem. Phys.*, **49**, 3541.

Feher, G. (1957) *Bell Syst. Tech. J.*, **36**, 449.

Froncisz, W. and Hyde, J.S. (1982) *J. Magn. Reson.*, **47**, 515.

Ginzton, E.L. (1957) *Microwave Measurements*, McGraw-Hill, New York, p. xvii + 515.

Gordon, J.P. (1961) *Rev. Sci. Instrum.*, **32**, 658.

Hardy, W.N. and Whitehead, L.A. (1981) *Rev. Sci. Instrum.*, **52**, 213.

Howarth, D.F., Weil, J.A., and Zemprel, Z.J. (2003) *J. Magn. Reson.*, **161**, 215.

Ishii, T.K. (1966) *Microwave Engineering*, The Ronald Press, New York, p. x + 339.

Lee, T.H. (2000) *IEEE J. Solid State Circuits*, **35**, x + 326.

Llewellyn, P.M. (1957) *J. Sci. Instrum.*, **34**, 236.

Misra, H. (1958) *Rev. Sci. Instrum.*, **29**, 590.

Moreno, T. (1958) *Microwave Transmission Design Data*, Dover Publishing, New York.

Mumford, W.W. and Schelbe, E.H. (1968) *Noise – Performance Factors in Communicating Systems*, Horizon House – Microwave, Inc., Dedeham, MA, USA, p. viii + 89.

Nagatsuma, T. (2009) *IEEE Microwave Mag.*, **10** (4), 64.

Ondria, J.G. (1968) *IEEE Trans. Microw. Theory Tech.*, **MTT-16**, 767.

Peter, M. and Strandberg, M.W.P. (1955) *Proc. Inst. Radio Engrs*, **43**, 869.

Pilbrow, J.R. (1990) *Transiton Ion Electron Paramagnetic Resonance*, Oxford University Press, New York, p. xx + 717.

Poole, C.S., Jr (1967) *Electron Spin Resonance – A Comprehensive Treatise on Experimental Techniques*, Wiley-Interscience, New York, p. xxv + xxx.

Poole, C.S., Jr (1983) *Electron Spin Resonance – A Comprehensive Treatise on Experimental Techniques*, 2nd edn, Wiley-Interscience, New York, p. xxvii + 780.

Pound, R.V. (1946) *Rev. Sci. Instr.*, **17**, 49.

Rathie, R.S. (1970) Noise Measurements Of Microwave Diodes From 102-To-105 Hz Using a Narrow-Band Tunable Switching-Type Radiometer. Unpublished MSc Thesis, Department of Physics, University of Calgary, Calgary, AB, Canada.

Robbins, W.P. (1982) *Phase Noise in Signal Sources (Theory and Applications)*, Peter Peregrinus Ltd., London, p. x + 321.

Schofield, J.H. (1994) *Am. J. Phys.*, **62**, 129.

Shing, Y.H. (1972) Theoretical And Experimental Studies In Electron Paramagnetic Resonance II –An Investigation of the Ground State Splitting of the Gd^{3+} $[4f^7, {}^8S_{7/2}]$ Ion in Lanthanum Ethyl Sulphates and the Design of 1–8 GHz Coaxial Electron Paramagnetic Resonance Spectrometers. Unpublished PhD Thesis, Department of Physics, University of Calgary, Calgary, AB, Canada.

Shing, Y.H. and Buckmaster, H.A. (1976) *J. Magn. Reson.*, **21**, 295.

Smythe, W.R. (1950) *Static and Dynamic Electricity*, 2nd edn, McGraw-Hill, New York, p. xviii + 550xx.

Stesmans, A. and De Vos, G. (1986) *Phys. Rev.*, **B34**, 6499.

Strandberg, M.W.P. (1972) *Rev. Sci. Instrum.*, **43**, 307.

Strandberg, M.W.P., Tinkham, M., Soltz, I.H., Jr, and Davis, C.F., Jr (1956) *Rev. Sci. Instrum.*, **27**, 596.

Talpe, J. (1971) *Theory of Experiments in Paramagnetic Resonance*, Pergamon Press, New York.

Teaney, D.T., Klein, M.P., and Portis, A.M. (1961) *Rev. Sci. Instrum.*, **32**, 721.

Torrey, H.C. and Whitmer, C.A. (1948) *Crystal Rectifiers*, vol. 15, Radiation Laboratory Series, McGraw-Hill, New York.

Tsallis, C. (1997) *Phys. World*, **10** (7), 42.

Wahlquist, H. (1961) *J. Chem. Phys.*, **35**, 17808.

Wang, H., Lin, K.-Y., Tsai, Z.-M., Lu, L.-H., Lu, H.-C., Wang, C.-H., Tsai, J.-H., Huang, T.-W., and Lin, Y.-C. (2009) *IEEE Microw. Mag.*, **10** (1), Feb., 99.

Weil, J.A. and Bolton, J.R. (2007) *Electron Paramagnetic Resonance – Elementary Theory and Practical Applications*, 2nd edn, Wiley-Interscience, New York, p. xx + 664.

Weil, J.A., Bolton, J.R., and Wertz, J.E. (1994) *Electron Paramagnetic Resonance – Elementary Theory and Practical Applications*, Wiley-Interscience, New York, p. xxi + 568.

Yariv, A. and Clapp, F.D. (1959) *Rev. Sci. Instrum.*, **30**, 684.

Zaffarano, F.P. and Galloway, W.C. (1947) Technical Report No. 31, Research Laboratory of Electronics, M. I. T.

Zaghoul, H., Van Kalleveen, T.H.T., Hansen, C.H., and Buckmaster, H.A. (1990) *IEEE Trans. Instrum. Meas.*, **IM-39**, 928.

Appendix 4.2.I

The following is a list of instrumentation subsystems that are frequently found in CW EPR spectrometers.

Frequency counter: The operating frequency of an EPR spectrometer can be measured and monitored to an accuracy of at least $1:10^8$, using an electronic system which compares the frequency to be measured to that of a 5 or 10 MHz high-stability crystal oscillator. Temperature-stabilized crystal oscillators are used because they can have spectral purities and stabilities of better than $1:10^{10}$ and very low drift rates in time as the crystal ages. Such measurements can be made with this accuracy and resolution at 9 GHz in good-quality instrumentation.

Microwave oscillator synchronizer: The spectral purity of a microwave power source can be increased from about $1:10^5$ to $1:10^{8-9}$ through the use of a microwave oscillator synchronizer. This transfers the spectral stability of harmonic of the frequency of one of an array of temperature-stabilized Megahertz crystal oscillators. An error signal from the microwave oscillator synchronizer is used to control the frequency of the microwave power source being stabilized. A stability in excess of $1:10^9 \, \text{s}^{-1}$ is typical, and at least $1:10^8$ at 9 GHz can be achieved for a few hours in a stable temperature environment using this technique. A block diagram of this instrumentation system is shown in Figure 4.2.9.

Power meter: The microwave power incident on the sample resonant structure should be measured and monitored as part of the procedure used to adjust the operating frequency and power output of the microwave power, and to take account of saturation effects. A calibrated microwave bolometer is the active element; the change in current in the bolometer is a measure of the incident microwave power.

Lock-in amplifier/synchronous demodulator: The phase-sensitive detector or lock-in amplifier, as invented by Dicke (1946), enables a signal derived from a known frequency perturbation to be extracted from an extremely noisy environment. The correct description is a "synchronous demodulator," because it relies on the orthogonality of sinusoidal signals to eliminate all signals that are not at the perturbation frequency or a harmonic (modulation frequency in EPR spectrometers), as these are averaged to zero when integrated over a time much longer than its period. This amplifier includes an oscillator from which the magnetic field modulation voltage is derived and as the internal reference frequency for the signal processing. The phase of the internal reference is adjusted until the resultant EPR signal is minimized. The output voltage is quasi-DC, depending on the rate at which the magnetic field is swept. Digital synchronous demodulators are now available, and two-channel versions permit two orthogonal signals to be detected. Synchronous demodulation is superheterodyne demodulation in which the local oscillator frequency and the signal frequency are identical (zero IF or synchronous) and are phase-related. The important feature is that the driving and detected signals are coherent (e.g., phase-related), which is the reason why extraneous signals can be averaged to zero. It is an instrumentation extension of the earlier homodyne demodulator used in early radios to increase their sensitivity, where the local oscillator was at the same frequency as the incoming signal but not phase-related, nor coherent with it. A block diagram of a lock-in or synchronous demodulator is shown in Figure 5.2.3.

Magnetometer: The magnitude of the magnetic field used in an EPR spectrometer can be monitored and stabilized using a Hall effect probe or a proton nuclear magnetic resonance magnetometer. Hall effect probes are commonly used to control the magnetic field. They must be carefully stabilized thermally, and their accuracy is limited to $1:10^{3-4}$. Hall effect probes have been used primarily because they can be used to sweep the magnetic field with good linearity. The proton resonance magnetometer can be used for more accurate magnetic field measurements. The resonance frequency can be measured with an electronic frequency meter to an accuracy exceeding $1:10^7$, depending on the linewidth of the proton resonance. However, the accuracy of the magnetic field calculated from this frequency is determined by the accuracy of the proton gyromagnetic ratio for the sample used and, in general, does not exceed $1:10^5$. Differential magnetic field measurements retain the higher accuracy. The best accuracy is obtained if the probe is inserted into the microwave sample cavity and positioned where the sample is located. The magnetic field difference between measurements made with the probe inside and outside the cavity can be used to increase the field measurement accuracy. Very accurate magnetic field measurements can also be made using the conduction EPR of neutron-irradiated LiF:F samples (Stesmans and De Vos, 1986), because the observed linewidth is 0.065 mT and $g = 2.00229(1)$. These LiF:F samples are very stable and the number of spins in samples can be determined with an accuracy of better than

$1:10^2$ when using them. This approach has the additional advantage that the magnetic field is measured at essentially the same location as the sample under study. Chang (1984) has provided a comprehensive survey of the methods to calibrate magnetic fields in EPR.

Appendix 4.2.II

The following list provides the names and concise descriptions of the microwave/coaxial components used in a CW-EPR spectrometer.

Attenuator: Early waveguide uncalibrated vane flap attenuators evolved into physically larger but precision direct-reading rotary vane devices with higher accuracy and resolution over a wider range of attenuation. Coaxial attenuators employ an array of voltage-controlled microwave diodes and must be calibrated, but are high-speed devices.

Balanced demodulator: A balanced demodulator uses a waveguide or coaxial magic-T in which the input signal enters on the *E*-arm and the reference signal on the *H*-arm. A pair of microwave diodes carefully selected for identical DC and noise properties diodes are located in the symmetric arms (see Figure 4.2.7).

Circulator: A microwave circulator is a three-port device that uses a ferrite disk to determine its directional properties. The losses are lower than when a magic-T is used, but the degree of isolation is limited to less than about 40 dB. Power into port-1 leaves by port-2, and power from port-2 leaves by port-3, etc. The circulator has the same electrical properties as a magic-T in which one of the balanced arms is terminated with a matched load. Four-port circulators also exist that have the same properties as a magic-T. The isolation is degraded by the stray magnetic field from the electromagnet, since it is desirable to minimize the distance from the microwave sample cavity in an EPR spectrometer and the isolation and matching varies as the magnetic field is swept.

Directional coupler: Waveguide and coaxial directional couplers enable a fraction of the microwave power in the straight arm to be sampled with high directivity that exceeds 40 dB. Cross-arm directional couplers are physically smaller in size, but the isolation between the arms is about 20 dB rather than 40 dB. These devices are manufactured with 3, 6, 10, and 20 dB coupling factors.

Ground loop block: Waveguide or coaxial ground loop blocks are frequently used to eliminate ground loops, which can be a serious source of undesirable extraneous noise and signals that can limit the spectrometer sensitivity and overload sensitive amplifiers.

Hybrid-T: Physically, a hybrid-T appears identical to a magic-T, except that it does not include matching posts to make the impedance at each of the four ports equal (see Magic-T). Coaxial hybrid-Ts can be designed to operate at all frequencies below about 26 GHz.

Isolator: Waveguide and coaxial isolators are used to minimize the effect of reflections, because some components such as microwave oscillators are sensitive to small variations in the impedance into which they operate resulting in frequency pulling effects. Care should be taken to ensure that all devices including isolators are not exposed to varying magnetic fields, as they can become the source of problems.

Magic-T: Waveguide magic-Ts are four-port devices with two symmetric and two asymmetric arms or ports. Power introduced into either the E-arm or H-arm does not reach the other arm if the two symmetric arms are accurately matched. One-half of the input power reaches each symmetric arm, and one-half of any signal power reflected from a symmetric arm exits through an asymmetric arm. A complete description is given in the text. Coaxial magic-Ts do not exist, since it is simpler to introduce matching elements separately. See Figure 4.2.7.

Phase shifter: Uncalibrated microwave phase shifters use a shaped dielectric vane that can be moved in the E-plane across the waveguide to vary the phase shift. Their calibration accuracy is rather limited. Precision direct-reading rotary vane phase shifters are physically larger devices that are well-shielded from external radiation. Coaxial phase shifters employ an array of voltage-controlled microwave diodes and must be calibrated, but are high-speed devices.

Appendix 4.2.III

This Appendix is intended to provide some useful hints about problems that may be encountered when operating EPR spectrometers. They are particularly appropriate when laboratory-designed and -constructed spectrometers are involved:

- It is not unusual to observe *fluctuations in the demodulated output* when the experimenter is making adjustments to a spectrometer. These are known colloquially as *"hand-waving effects."* This effect is a common source of signal instability and resultant loss of sensitivity, and is caused by the leakage of microwave power out of one or more of the junctions between the microwave components and its re-entry into other junctions. This effect can arise because a junction is incorrectly connected if the waveguide flanges are not parallel and/or the connecting screws are not uniformly tightened. Careful attention to each joint is very important. The introduction of short waveguide sections with choke flanges on each end is essential to achieve minimum leakage and optimum stability. The O-ring groove in the choke gaskets should be filled with wire mesh gaskets to further improve the quality of the seal between components to reduce leakage in and out of each waveguide connection. A further step is to wrap each joint with adhesive aluminum muffler tape. These steps will ensure that the leakage of power from the microwave power source is reduced to below 160 dBm. Coaxial junctions are much less of a problem, but they can also be wrapped with the muffler tape when necessary. Particular attention should be paid to the waveguide junctions near the megaHertz microwave power source, since the power level is higher and it may have a greater negative effect elsewhere in the instrumentation system.

- *Ground loops* can be a cause of a form of leakage between the microwave power source and the microwave signal demodulator. This problem can be eliminated by using thin Teflon sheets as electrical isolators, and plastic screws to both insulate and isolate the power source and the demodulator from the remainder of the EPR spectrometer waveguide configuration. Special coaxial isolators are also available for this purpose. Ground loop problems are usually associated with ripple at the line voltage frequency (50 or 60 Hz) that arises from the DC voltage power supplies in a spectrometer. Although the use of synchronous demodulation should eliminate this type of extraneous noise, it is preferable to ensure that such noise sources are minimized, as overload problems can cause distortion.

- Another source of signal instability is caused by *acoustic vibrations*. If vocal shouting or hard clapping causes the output signal to fluctuate, then the problem is probably arising because the physical structure of the spectrometer is not sufficiently rigid. Floor vibration due to machinery in another part of a laboratory building can also prove to be troublesome. Special anti-vibration pads are available commercially; these are used routinely in industry by millwrights when machinery is installed, to isolate and dampen machine vibration and reduce noise levels. Their installation under the structure where an EPR spectrometer is mounted may reduce this source of instability. It may be necessary also to isolate the electromagnet if the problem is severe. These pads are designed to be load-sensitive, so the correct area must be employed.

- *Vibration problems* may arise when certain modulation frequencies are employed, as the coaxial or waveguide structure connected to the sample cavity may "sing" at certain frequencies because of resonances in the physical structure at these frequencies. This effect can be remedied by changing the modulation frequency or by stiffening and/or loading the structure.

4.3
High-Frequency EPR Spectrometers
Edward Reijerse

4.3.1
Introduction

During the past few decades, an increasing number of EPR spectrometers have been developed that employ microwave frequencies which are significantly higher than the "conventional" X- and Q-band range – that is, 9 and 35 GHz. The microwave bridge in these instruments typically uses technology deviating from the traditional waveguides and coaxial lines, while the magnetic field is generated

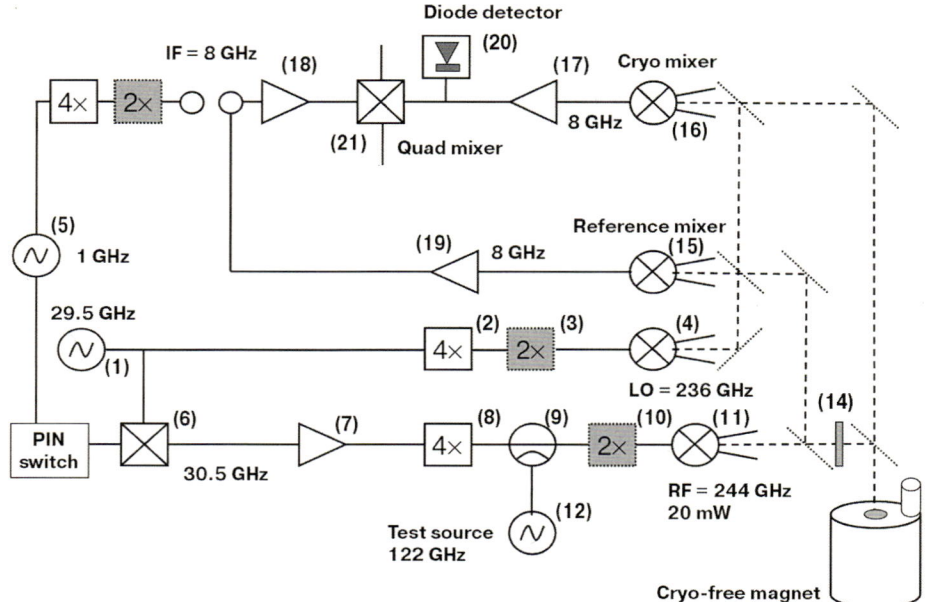

Figure 4.3.5 Basic diagram of a quasi-optical 244 GHz bridge. (1) Dielectric resonator oscillator (DRO) at 29.5 GHz; (2) and (3) Doublers; (4, 11, 15, 16) Corrugated feed horns; (5) 1 GHz voltage-controlled oscillator (VCO); (6) Up-converter; (7) 30.5 GHz power-amplifier; (8, 10) High-power doublers; (9) Waveguide switch; (12) Varactor controlled Gunn-oscillator at 122 GHz; (14) Quasi-optical circulator (Faraday rotator); (15, 16) Single-ended mixers; (17) Cryogenic preamplifier; (18) and (19) Low-noise amplifiers; (20) Diode detector; (21) Quadrature mixer. The bridge can also function with an RF (mm-wave) frequency of 122 GHz (150 mW). In this case, the doublers indicated in gray are omitted and the feed-horns [(4) and (15)] as well as the single-ended mixers [(15) and (16)] are replaced by the corresponding 122 GHz components. The system then functions with an IF frequency of 4 GHz. Figure reproduced with permission from Reijerse et al. (2007).

phase noises of the LO and IF sources are optimally correlated, as the optical path lengths of the signal and reference are equal to each other. In addition, phase drifts due to temperature variations in the probe-head insert are compensated. It is expected that for very high-frequency (pulsed) EPR, these types of quasi-optical system offer the greatest flexibility in configuration as far as sources, detectors, and probe-heads are concerned.

4.3.3
Sensitivity Considerations

The fundamentals of spectrometer sensitivity in relation to the cavity and detector properties have been discussed in great detail by Feher (1956) and by Poole (1983).

microwave power source, since the power level is higher and it may have a greater negative effect elsewhere in the instrumentation system.

- *Ground loops* can be a cause of a form of leakage between the microwave power source and the microwave signal demodulator. This problem can be eliminated by using thin Teflon sheets as electrical isolators, and plastic screws to both insulate and isolate the power source and the demodulator from the remainder of the EPR spectrometer waveguide configuration. Special coaxial isolators are also available for this purpose. Ground loop problems are usually associated with ripple at the line voltage frequency (50 or 60 Hz) that arises from the DC voltage power supplies in a spectrometer. Although the use of synchronous demodulation should eliminate this type of extraneous noise, it is preferable to ensure that such noise sources are minimized, as overload problems can cause distortion.

- Another source of signal instability is caused by *acoustic vibrations*. If vocal shouting or hard clapping causes the output signal to fluctuate, then the problem is probably arising because the physical structure of the spectrometer is not sufficiently rigid. Floor vibration due to machinery in another part of a laboratory building can also prove to be troublesome. Special anti-vibration pads are available commercially; these are used routinely in industry by millwrights when machinery is installed, to isolate and dampen machine vibration and reduce noise levels. Their installation under the structure where an EPR spectrometer is mounted may reduce this source of instability. It may be necessary also to isolate the electromagnet if the problem is severe. These pads are designed to be load-sensitive, so the correct area must be employed.

- *Vibration problems* may arise when certain modulation frequencies are employed, as the coaxial or waveguide structure connected to the sample cavity may "sing" at certain frequencies because of resonances in the physical structure at these frequencies. This effect can be remedied by changing the modulation frequency or by stiffening and/or loading the structure.

4.3
High-Frequency EPR Spectrometers
Edward Reijerse

4.3.1
Introduction

During the past few decades, an increasing number of EPR spectrometers have been developed that employ microwave frequencies which are significantly higher than the "conventional" X- and Q-band range – that is, 9 and 35 GHz. The microwave bridge in these instruments typically uses technology deviating from the traditional waveguides and coaxial lines, while the magnetic field is generated using superconducting magnets. In this chapter, EPR spectrometers operating at frequencies starting from 90 GHz and magnetic fields around 3 T and upwards are defined as high-frequency/high-field (HFHF) EPR spectrometers. The motivation to develop these instruments is driven by the advantages of HFHF EPR, as described in Chapter 2, whereas the history and evolution of the very high-frequency spectrometer is detailed in Chapter 1. The technical aspects and most promising developments in high-frequency EPR instrumentation are discussed in this chapter.

4.3.2
High-Frequency EPR Spectrometer Configurations

The classic CW-EPR spectrometer at X-band is of the homodyne type, as depicted in Figure 4.3.1a. A Gunn oscillator is normally used as a source, this being locked to the resonator by using an automatic frequency control (AFC) circuit. A Schottky diode serves as the detector. A rectangular TE102 cavity or cylindrical TE011 cavity is frequently used as the resonator. This scheme can be scaled to frequencies of up to 150 GHz, as demonstrated Lebedev and coworkers (Galkin et al., 1977; Lebedev, 1994), who used a single-mode cylindrical TE011 resonator. Unfortunately, at these frequencies the Schottky detector suffers from an increased noise figure, and a helium-cooled InSb bolometer or mixer is often used instead. For pulse operation, however, the bandwidth (1 MHz) of this detector is insufficient, and the Schottky detector is preferred, despite its lower sensitivity. An IMPATT (IMPact ionization Avalanche Transit-Time) diode source is used as a high-power source (Bresgunov et al., 1991). The W-band (95 GHz) bridge developed by the Urbana group (Wang et al., 1994; Nilges et al., 1999) is also of the homodyne type, but uses a standard Schottky-diode detector. Especially, in combination with a

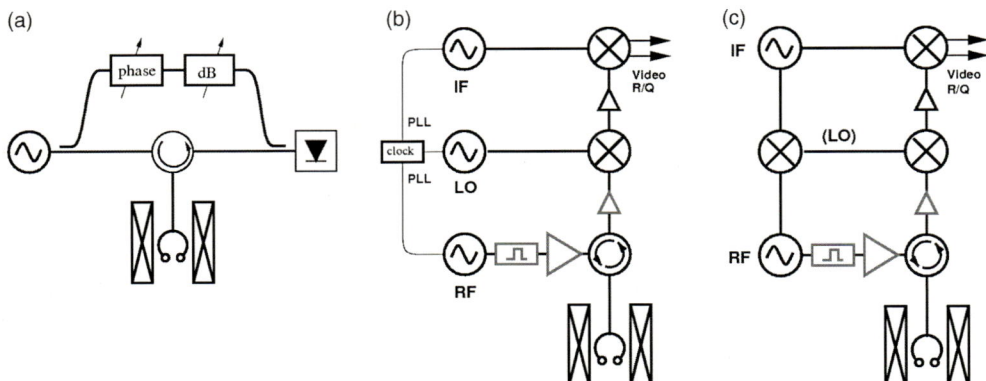

Figure 4.3.1 Basic spectrometer schemes. (a) Homodyne; (b) Heterodyne with PLL locked sources; (c) Heterodyne with mixing scheme. Optional elements (amplifiers or pulse-former unit) are indicated in gray.

low-noise amplifier (Nilges et al., 1999), the sensitivity is compatible with that of the bolometer bridges; that is, the absolute sensitivity approached the theoretical limit of 10^8 spins per Gauss, using a cylindrical TE013 resonator. The availability of a low-noise amplifier also facilitated the development of a homodyne pulsed W-band bridge (Goldfarb et al., 2008).

The homodyne scheme can also be employed at much higher frequencies. Due to the increasing losses in single-mode waveguides, one is, however, forced to switch to oversized propagation systems. When the group of Jack Freed built the first 250 GHz CW-EPR instrument (Lynch, Earle, and Freed, 1988), their "1 mm EPR" spectrometer employed oversized smooth waveguides and polyethylene lenses to propagate the millimeter waves to and from a transmission Fabry–Pérot resonator. Subsequently, the group of Brunel pioneered the development of wide-band, high-frequency EPR instruments up to 600 GHz, in which the millimeter-waves were propagated in oversized smooth guides and transmitted through the sample holder towards a helium-cooled InSb bolometer (Barra, Brunel, and Robert, 1990; Brunel, Barra, and Martinez, 1995; Hassan et al., 2000). It has been shown that smooth circular guides, although very convenient to use, suffer from a poor mode conversion leading to standing waves. These make it difficult to control the local oscillator level at the detector, and render the system very sensitive to vibrations. However, with the systematic introduction of quasi-optics in very high-frequency EPR, Smith's group solved these typical problems for the most part (Smith et al., 1998). Figure 4.3.2 shows the basic layout of their set-up, where the millimeter-waves are propagated in free space and directed by polarizers and off-axis mirrors. Inside the sample cryostat, the millimeter-waves propagate in an oversized circular corrugated guide. The equivalent of a circulator is realized by the application of a polarizer in combination with a Faraday rotator, which rotates the polarization of the millimeter-waves by, for example, 45°. In this manner, a full equivalent of the homodyne scheme of a CW X-band spectrometer (as shown in Figure 4.3.1a) is obtained. In practice, the balanced pair of mixers in Figure 4.3.2 is replaced by a helium-cooled InSb mixer. The spectrometer can operate at 80–100 GHz, as well as 160–200 GHz, by using a doubled source. This set-up can be easily adapted to operate at different frequencies because the quasi-optical elements are usually broadband. In particular, when using the so-called "unity-gain" optics, only the matching layer of the Faraday rotator as well as the feed horns of source and detector will need to be exchanged. Several groups took over this design and constructed their own versions of a homodyne quasi-optical CW-EPR spectrometer (Saylor et al., 2000). The main working principle of the quasi-optical millimeter-wave bridge is the so-called "induction-mode detection." This is illustrated in Figure 4.3.3, which shows the variable temperature insert. The system comprises a corrugated guide, which couples the millimeter-waves to the Fabry–Pérot resonator. The resonance mode between the concave and semi-transparent flat mirror has cylindrical symmetry, which allows both orthogonal polarizations of the millimeter-wave to be present. Due to the circular polarization of the EPR absorption, the signal is transferred to the orthogonal mode, and can be reflected from the polarizer P3. Even without the resonator – for example, by removing the

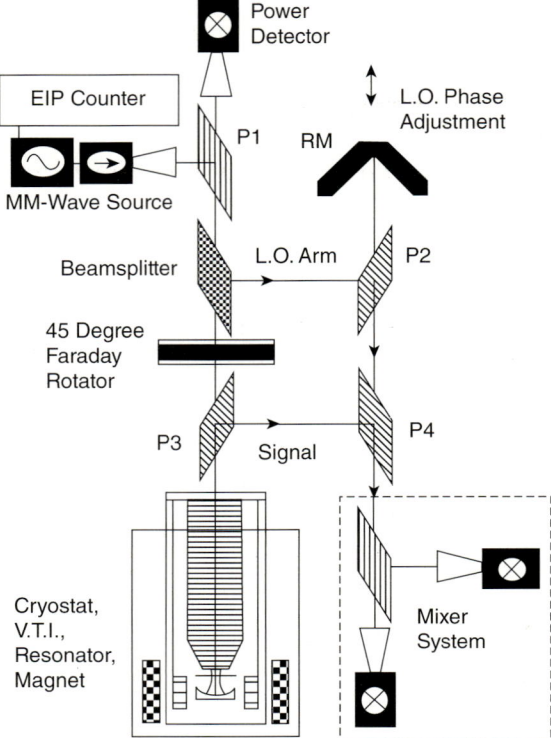

Figure 4.3.2 Quasi-optical induction-mode spectrometer. Power is transmitted to the resonator in the variable temperature insert, by reflecting off polarizer P1 and passing through the beam splitter, Faraday rotator, and polarizer P3. The EPR signal is polarization- and amplitude-encoded onto the RF via the magnetic field modulation, and the induction signal is stripped off using polarizer P3 and combined with the local oscillator signal using polarizer P4. The local oscillator signal is stripped off using the beam splitter, and may have its phase or amplitude changed by rotating or translating the roof mirror (RM) after reflection from polarizer P2. Figure reproduced with permission from Smith et al. (1998).

coupling mesh which acts as semitransparent mirror—a large isolation can be reached between the exciting millimeter-wave signal and the detected EPR signal. This has important implications in suppressing source noise (as discussed in the following section).

For larger detection band-widths, as is necessary in high-frequency pulsed and transient EPR applications, a heterodyne detection scheme is often adopted. This applies particularly when a low-noise amplifier for the frequency of interest does not exist, or specifies a noise figure worse than the conversion loss of the fundamental detection mixer. In the configuration illustrated in Figure 4.3.1b, the high-frequency EPR signal (in heterodyne jargon: the radiofrequency; RF) is first

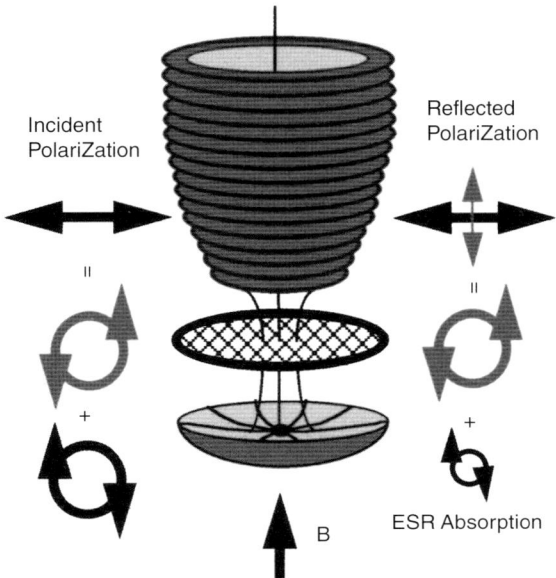

Figure 4.3.3 Schematic diagram to illustrate the principle of induction-mode EPR spectroscopy. The cavity is excited by a linearly polarized beam that can be thought of as two counter-rotating circular polarized beams. In EPR, only one of these circular polarized beams is absorbed, leading to a transfer of power to the orthogonal state. Figure reproduced with permission from Smith et al. (1998).

transformed to an intermediate frequency by mixing with a local oscillator (LO) signal, typically at 2–8 GHz offset to the RF frequency. The EPR signal at the intermediate frequency (IF) – that is, the offset frequency between RF and LO – can be amplified using a low-noise amplifier (LNA). Finally, the signal is mixed back to a video output, using a source at the IF frequency. Some groups lock all sources to the same crystal oscillator, as shown in Figure 4.3.1b, although it emerges that the phase-locked loop (PLL) noise of these locks often produces substantial phase noise at the video output. Consequently, most groups have now adopted mixing schemes similar to those shown in Figure 4.3.1c, which use uncorrelated – but very stable – oscillators. By balancing the optical path lengths of the various branches, all phase noise contributions can be suppressed (at least in theory) within a certain bandwidth. In practice, this way, a very good suppression (>40 dB) can be reached without much problem, as deduced in the following section. The pulsed W-band spectrometers developed by the Leiden group (Weber et al., 1989; Disselhorst et al., 1995), and also by the Berlin group (Prisner, Rohrer, and Möbius, 1994), have adopted this scheme. The Bruker company used a variant of the heterodyne scheme in which the pulses and detection are processed at 9 GHz in their standard X-band pulsed bridge. (The W-band bridge can be obtained as an accessory.) The instrument contains a stable 86 GHz oscillator and an up-converter which transforms the 9 GHz (pulsed) excitation signal to 95 GHz, where it is amplified to 100–400 mW. The resulting EPR signal is led through a LNA and fed into a down-converter. This transforms the 95 GHz EPR signal to 9 GHz, after which the signal is transported to the X-band bridge where it is analyzed in the standard detection system (Schmalbein *et al.*, 1999; Hofer *et al.*, 2004; Carl *et al.*, 2009). This elegant and economic solution affords full use of the advanced electronics already available at X-band.

Currently, at higher frequencies (>100 GHz) no LNAs are currently available, but developments are proceeding at a rapid pace. The quasi-optical bridges can be readily configured to operate in heterodyne mode, in which case the bolometer mixer is replaced by a Schottky mixer. At very high frequencies (>150 GHz), however, fundamental mixers are only available in single-ended configuration, which means that both the LO and RF signals must enter through the same port. This requires that the LO and RF signal are diplexed (i.e., two ports are multiplexed into one) in the quasi-optical bridge. This concept leads to a considerable complication of the bridge layout, since both RF and LO signals require separate quasi-optical paths. This principle is applied in the 275 GHz instrument developed by the Leiden group (Blok *et al.*, 2004; Blok *et al.*, 2005); the same group also constructed a single-mode resonator operating at 275 GHz, which is integrated in the quasi-optical system (Figure 4.3.4), and allows for ENDOR irradiation as well as single-crystal rotation. When the quasi-optics is built according to the unity-gain principle, multifrequency operation is facilitated, as demonstrated by the Tallahassee group (van Tol, Brunel, and Wylde, 2005; Morley, Brunel, and van Tol, 2008), who developed a transient and pulsed EPR instrument which operates at 120 and 240 GHz.

A similar approach was followed in the CW/pulsed EPR bridge operating at 122 and 244 GHz, as constructed by Reijerse *et al.* (Reijerse *et al.*, 2007). The electrical layout and quasi-optics of this system are illustrated in Figure 4.3.5. The local-oscillator channel is based on a fixed-frequency 29.5 GHz dielectric resonator oscillator (DRO) with very low phase noise (−110 dBc @ 100 kHz. A small portion of this signal is coupled out and mixed with the output of a 0.8–1.0 GHz voltage-controlled oscillator (VCO), which has similar phase noise. The resulting signal at 30.5 GHz is the starting point of the excitation (RF) channel. The RF signal is first amplified to 1 W, and then multiplied to 122 GHz (i.e., ×4) or 244 GHz (i.e., ×8), using varactor multipliers. This amounts to signal powers of around 200 mW at 122 GHz, and 20 mW at 244 GHz. The LO signal is also multiplied by a factor of 4 or 8, producing about 1 mW at the fundamental, single-ended, mixer inputs. The IF signal (0.8–1.0 GHz) can be phase- and amplitude-modulated using a pulse-former unit. The IF signal can also be combined with an additional source signal to perform electron–electron double resonance (ELDOR) experiments. The excitation signal is propagated through the quasi-optical bridge, as depicted in Figure 4.3.6, and finally reflected from the polarizer next to the Faraday rotator [as indicated by (14) in Figure 4.3.5] towards the cryogenic insert consisting of a corrugated guide, which can be fitted with either a resonator or a nonresonant probe-head. The mm-wave bridge can function in both induction- and reflection-

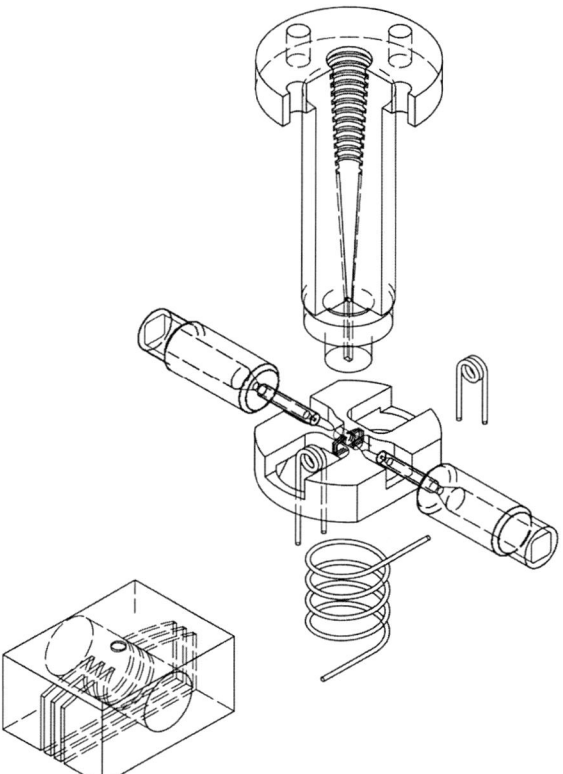

Figure 4.3.4 Single-mode resonator for 275 GHz with ENDOR coils. The cylindrical resonator is at the center of the bronze block, and is shown separately on a larger scale in more detail. The diameter of the resonator is 1.4 mm, and its length can be varied between 1.0 and 1.4 mm with two movable plungers to tune the cavity to the fixed frequency of 275.7 GHz of the oscillator. Three slits with a width of 0.1 mm and at a mutual distance of 0.1 mm are cut in the block, using a circular saw. Two two-turn RF coils with a diameter of 3 mm are positioned outside the cavity. The unit shown at the top is the taper that serves to change the circular shape of the corrugated waveguide to the rectangular waveguide to optimize the coupling of the microwaves into the cavity. The coil below the cavity block serves to modulate the magnetic field in CW-EPR experiments. Figure reproduced with permission from Blok et al. (2005).

mode detection. In CW mode, an induction-mode detection is normally used, in which case the signal is passed through the first polarizer and directed to a cryogenically cooled, single-ended mixer. In order to join the EPR signal with the LO branch, a Martin–Puplett interferometer is used as a diplexer. The signal which is directly reflected from the probe-head (e.g., a nonresonant cuvette) is fed through the Faraday rotator, and reflected towards the reference mixer [(15) in Figure 4.3.5 and (B) in Figure 4.3.6]. In principle, the mixer signal can be used as the reference signal for the IF quad mixer [(21) in Figure 4.3.5]. This has the advantage that the

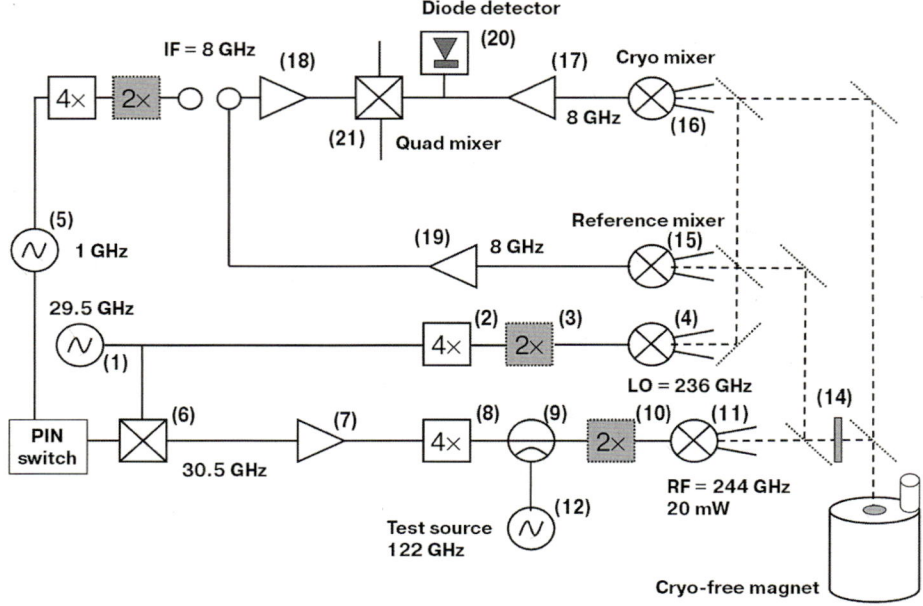

Figure 4.3.5 Basic diagram of a quasi-optical 244 GHz bridge. (1) Dielectric resonator oscillator (DRO) at 29.5 GHz; (2) and (3) Doublers; (4, 11, 15, 16) Corrugated feed horns; (5) 1 GHz voltage-controlled oscillator (VCO); (6) Up-converter; (7) 30.5 GHz power-amplifier; (8, 10) High-power doublers; (9) Waveguide switch; (12) Varactor controlled Gunn-oscillator at 122 GHz; (14) Quasi-optical circulator (Faraday rotator); (15, 16) Single-ended mixers; (17) Cryogenic preamplifier; (18) and (19) Low-noise amplifiers; (20) Diode detector; (21) Quadrature mixer. The bridge can also function with an RF (mm-wave) frequency of 122 GHz (150 mW). In this case, the doublers indicated in gray are omitted and the feed-horns [(4) and (15)] as well as the single-ended mixers [(15) and (16)] are replaced by the corresponding 122 GHz components. The system then functions with an IF frequency of 4 GHz. Figure reproduced with permission from Reijerse et al. (2007).

phase noises of the LO and IF sources are optimally correlated, as the optical path lengths of the signal and reference are equal to each other. In addition, phase drifts due to temperature variations in the probe-head insert are compensated. It is expected that for very high-frequency (pulsed) EPR, these types of quasi-optical system offer the greatest flexibility in configuration as far as sources, detectors, and probe-heads are concerned.

4.3.3
Sensitivity Considerations

The fundamentals of spectrometer sensitivity in relation to the cavity and detector properties have been discussed in great detail by Feher (1956) and by Poole (1983).

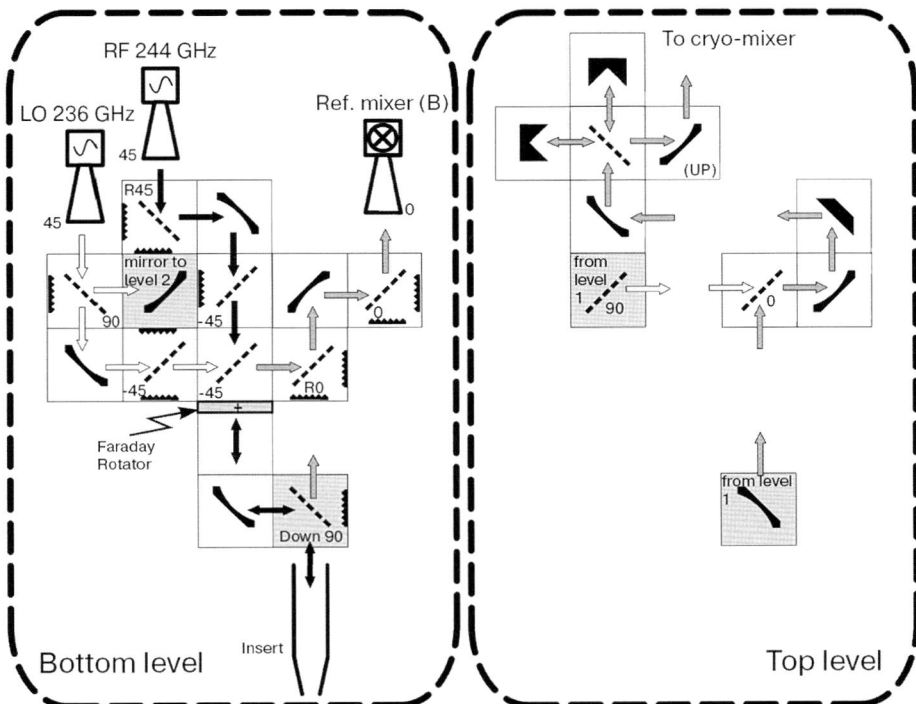

Figure 4.3.6 Quasi-optical millimeter (mm)-wave propagation in the HF-EPR bridge. The black arrows represent the path of the RF (mm-wave) excitation power; the open white arrows indicate the path of the local oscillator (LO), while the gray arrows show the combined LO/RF path containing either the EPR signal (bottom level) or the power directly reflected from the sample insert (top level) directed to the reference mixer. The orientation of the grid wires is indicated in degrees with respect to the vertical. Other symbols indicate the position of mirrors and absorbers. Figure reproduced with permission from Reijerse et al. (2007).

Nevertheless, a more practical analysis is helpful here when comparing the various spectrometer designs. Accordingly, the relevant properties of the different spectrometer components are described in the following subsections.

4.3.3.1 Cavity and Sample Holder

In general, the sample cavity is characterized by its resonance frequency ω_0, the unloaded Q (Q_u), and the filling factor η. For practical purposes, the filling factor can be approximated simply by the ratio of sample to cavity volume, that is,

$$\eta = \frac{\int_S B_1^2 d\tau}{\int_C B_1^2 d\tau} \approx \frac{V_S}{V_C} \tag{4.3.1}$$

At resonance, the imaginary part of the magnetic susceptibility of the sample (χ'') will lead to absorption of microwaves, thus enhancing the losses in the cavity, thereby reducing the Q-value:

$$\frac{1}{Q} = \frac{1}{Q_U} + \frac{1}{Q_\chi} \tag{4.3.2}$$

The contribution due to the sample is given by:

$$Q_\chi = \frac{\int_C B_1^2 d\tau}{\chi'' \int_S B_1^2 d\tau} = \frac{1}{\chi'' \eta} \tag{4.3.3}$$

Assuming that $Q \approx Q_u$, the change in Q is given by:

$$\Delta Q = -\chi'' \eta Q_U^2 \tag{4.3.4}$$

This Q-value change, in turn, results in a change in reflected, or transmitted, power, ΔP, or voltage, ΔE, from the cavity. Poole (1983) describes several situations as follows.

4.3.3.2 Reflection Cavity with Square-Law Detector

In order to optimize the response of the cavity, it should be overcoupled, characterized by the voltage standing wave ratio (VSWR) = 3.73, for which the power response is given by:

$$\frac{\Delta P}{P_W} = 0.193 \frac{\Delta R}{R_C} = 0.193 \frac{\Delta Q}{Q_U} = 0.193 \chi'' \eta Q_U, \tag{4.3.5}$$

where P_w represents the incident power and R_c the cavity resistance.

4.3.3.3 Reflection Cavity with Linear Detector

In this case, the cavity can be tuned critically (VSWR = 1). The change in the reflected voltage (ΔE_r) is now given by:

$$\frac{\Delta E}{E_W} = 0.5 \frac{\Delta R}{R_C} = 0.5 \frac{\Delta Q}{Q_U} = 0.5 \chi'' \eta Q_U. \tag{4.3.6}$$

For transmission cavities, similar relations are obtained. It is clear that the ratio $\Delta P/P_w$, or $\Delta E/E_w$, plays a central role in determining the spectrometer sensitivity. On the one hand, it is compared to the noise figure of the spectrometer, whereas on the other hand, it is related to the minimum number of detectable spins, N_{min}. For the present discussion, the noise figure F of the spectrometer is defined as the (constant) amount of noise, in multiples of kT, introduced by the set-up to the signal over a certain detection bandwidth Δf:

$$\frac{F \cdot kT \Delta f}{P_W} = \left(\frac{\Delta P}{P_W}\right)_{min} = \left(\frac{\Delta E}{E_W}\right)^2_{min}. \tag{4.3.7}$$

According to the Bloch equations (as described in Chapter 10), the resonant susceptibility χ'' of the sample is related to the magnetic susceptibility χ_0 as:

$$\chi'' \approx \frac{\omega_0}{\Delta\omega_0} \chi_0 \qquad (4.3.8)$$

where ω_0 is the resonance frequency and $\Delta\omega_0$ the width of the resonance line. Following Curie's law, one has for the magnetic susceptibility:

$$\chi_0 = \left(\frac{N}{V_S}\right) \frac{g^2 \mu_B^2 S(S+1) \mu_0}{3kT}. \qquad (4.3.9)$$

For power response of a linear detector, as described by Equation 4.3.6, the limiting detectable power change can be obtained by using Equation 4.3.7. Substituting the limiting values for the resonant susceptibility (χ''_{min}), Equations 4.3.8 and 4.3.9 lead to the following expression for the absolute sensitivity of an EPR spectrometer:

$$N_{min} = \frac{6 V_C k T \Delta B}{\mu_0 g^2 \mu_B^2 S(S+1) B_r Q_u} \left(\frac{F \cdot k T \Delta f}{P_W}\right)^{1/2}. \qquad (4.3.10)$$

For the concentration sensitivity, one finds by dividing Equation 4.3.10 by V_S:

$$\frac{N_{min}}{V_S} = \frac{6 k T \Delta B}{\mu_0 g^2 \mu_B^2 S(S+1) B_r Q_u \eta} \left(\frac{F \cdot k T \Delta f}{P_W}\right)^{1/2}. \qquad (4.3.11)$$

The following descriptions and values are used for the various parameters appearing in Equations 4.3.10 and 4.3.11. V_C is the volume (m³) of the resonator, with the typical values: TE102@9.5 GHz = 1.1×10^{-5} m³; TE011@90 GHz = 2×10^{-8} m³; k (Boltzman constant) = 1.38062×10^{-23} J K^{-1}; T(sample temperature) is typically room temperature = 300 K; ΔB is the linewidth of the absorption signal = 0.1 mT; B_r is the resonant field: X-band@340 mT, W-band@3.4 T; Q_u is the unloaded quality factor of the resonator: X-band@5000, W-band@2000; Δf is the detection bandwidth = 1 Hz; P_w is the incident power on the resonator (X-band: 100 mW; W-band: 10 mW); F is the noise factor of the spectrometer (X-band diode: 100; W-band bolometer: 20); μ_0 is the magnetic susceptibility of vacuum = $4\pi \times 10^{-7}$ Vs Am^{-1}; g is the g-factor = 2; μ_B is the Bohr magneton = 9.274×10^{-24} J T^{-1}.

For a typical X-band spectrometer, these values lead to a spin sensitivity of 10^{11} spins per Gauss or 5×10^{11} spins per Gauss cm^{-3}. At W-band, the typical values are $N_{min} = 2 \times 10^7$ spins per Gauss or 10^{11} spins per Gauss cm^{-3}.

4.3.3.4 Spectrometer Bridge and Detector

The spectrometer performance, excluding the cavity, as given by Equation 4.3.7, will now be discussed. From Equation 4.3.7, the first impression is that, in the limit of infinite power, the relative noise contribution is zero. In practice, however, the noise factor, F, is nonzero and power-dependent, and consequently a more detailed description of the system performance parameter $S_m = \Delta P / P_w$ is developed here. It should be noted that an actual spectrometer is also characterized by insertion losses incurred by its components. Thus, the conversion losses in mixers reduce the signal at the detector $\Delta P_d = \Delta P/L$, as well as the available (effective) power $P_e = P_w/L$. The signal at the detector ΔP_d ($= S_m P_w/L = S_m P_e$) should be compared with all the noise contributions experienced by the detector. The following relevant parameters are needed:

- N_D, the noise floor of the detector (a typical value for an InSb bolometer is $-160\,\text{dBm}\,\text{Hz}^{-1}$).
- N_A, the amplitude noise of the millimeter-wave source (a typical value for a Gunn oscillator at 90 GHz is around $-140\,\text{dBc}\,\text{Hz}^{-1}$ at 1–100 kHz offset).
- P_{LO}, the local oscillator power required for the detector (this number varies from $-20\,\text{dBm}$ for a bolometer mixer to $0\,\text{dBm}$ for a Schottky mixer).
- Γ_C, the reflection coefficient of the cavity at resonance (for a critically coupled resonator, values can be reached between $-40\,\text{dB}$ and $-20\,\text{dB}$).
- N_Φ, the phase noise of the millimeter-wave source.
- F_q and F_0, the phase noise conversion with and without cavity (see below).

The noise contributions at the detector now sum up as follows:

$$\Delta P_d = S_m \cdot P_e = N_D + N_A \cdot P_{LO} + P_e \cdot \Gamma_C \cdot N_A \quad (4.3.12)$$

$$+ P_e \cdot \Gamma_C \cdot F_0 \cdot N_\Phi + P_e \cdot F_q \cdot N_\Phi \quad (4.3.13)$$

$$+ P_e \cdot \Gamma_C \cdot N_\Phi \quad (4.3.14)$$

The limiting power ratio is then:

$$\frac{\Delta P}{P_W} = S_m = \{N_D/P_e + N_A \cdot P_{LO}/P_e + \Gamma_C \cdot N_A + \Gamma_C \cdot F_0 \cdot N_\Phi + F_q \cdot N_\Phi + \Gamma_C \cdot N_\Phi\} \quad (4.3.15)$$

The contributions included in Equation 4.3.12 are the most important for a homodyne detection system. It can be easily verified that, for bolometric detection, they are all less than $-160\,\text{dBm}$, assuming the "standard" effective power of $P_e = 0\,\text{dBm}$ and $P_{LO} = -20\,\text{dBm}$. The system-performance parameter S_m, which is also interpretable as "dynamic range", is ideally of the order of 150–160 dB. The noise contributions in Equation 4.3.13 that depend on phase noise N_Φ stem from the millimeter-wave source. Even in homodyne systems, these contributions should be taken into consideration, especially when a reference arm is in operation. The conversion of phase noise to amplitude noise by a cavity and a reference arm is given by (Smith and Lesurf, 1991):

$$F_\Phi = \sin^2(\omega_0 \Delta t) \cdot \left[(\Gamma_C(\omega) - \Gamma_C(\omega_0))^2 + 4\Gamma_C(\omega)\Gamma_C(\omega_0)\sin^2(\Delta\omega\Delta\tau - \Phi_C(\omega))/2 \right], \quad (4.3.16)$$

where $\Delta\tau$ is the time delay between the excitation path and the reference arm; $\Delta\omega = ||\omega_0 - \omega||$ is the offset frequency from the carrier (ω_0); $\Gamma_C(\omega)$ is the reflection coefficient of the cavity at frequency ω; and $\Phi_C(\omega)$ is the phase of the millimeter-wave signal reflected from the cavity at frequency ω.

For absorption measurements, the factor $\sin^2(\omega_0 \Delta t)$, appearing in Equation 4.3.16, is tuned to zero by adjusting the phase, using the time delay Δt in the reference arm, thus eliminating, in theory, all phase-noise contributions. A slight phase mismatch can, however, introduce dispersion components into the signal carrying substantial phase-noise contributions. For an ideally matched cavity for which $\Gamma_C(\omega_0) = 0$, the conversion factor (in dispersion detection) is given by:

$$F_q = \frac{2Q_L \left(\frac{\Delta\omega}{\omega_0}\right)^2}{1 + 2Q_L \left(\frac{\Delta\omega}{\omega_0}\right)^2} \qquad (4.3.17)$$

For $Q_L = 2000$ and $\omega_0 = 95\,\text{GHz}$, $F_q = -84\,\text{dB}$ at 100 kHz offset. A Gunn source locked to a YIG oscillator will produce around $-90\,\text{dBc}$ of phase noise at 100 kHz offset, leading to an effective phase-noise contribution of $-174\,\text{dBm}$ (for $P_e = 0\,\text{dBm}$). For offset frequencies well below the bandwidth of the resonator (0.1%), the FM to AM conversion seems to remain well below the background thermal noise.

The signal directly reflected from the resonator, or from a nonresonant probehead, can also pick up phase-noise contributions from the time delay between the excitation path and the reference arm – that is, that given by the second term between brackets in Equation 4.3.16. By substituting $\Gamma_C(\omega) = \Gamma_C(\omega_0) = 1$ one obtains:

$$F_0 = 4\sin^2(\Delta\omega \Delta t/2) \qquad (4.3.18)$$

For a typical delay of $\Delta\tau = 6\,\text{ns}$, equivalent to 2 m path difference, one obtains $F_0 = -48.5\,\text{dB}$ at 100 kHz offset. For 20 dB cavity isolation, the reduction in phase noise is $-68.5\,\text{dB}$, reaching a safe value of $-158\,\text{dBm}\,\text{Hz}^{-1}$ for the Gunn oscillator ($-90\,\text{dB}\,\text{Hz}^{-1}$) even in dispersion detection. For higher incident power, additional isolation is needed to stay below the detector noise floor ($-160\,\text{dBm}\,\text{Hz}^{-1}$). This can be reached using the induction-mode detection configuration, for example, that achieved by using a Fabry–Pérot resonator. Cross-polar detection can provide at least 20 dB additional isolation.

For heterodyne detection systems, where the phase noise of the two sources is not correlated, the main noise contribution is provided by direct phase noise transfer to the IF signal, as given by Equation 4.3.14. By using the mixing scheme presented in Figure 4.3.1c, the phase noise between the sources can be correlated, leading to transfer functions describe by equations similar to Equation 4.3.18. Considering again the example of a 95 GHz phase-locked Gunn source with $-90\,\text{dBc}\,\text{Hz}^{-1}$ phase noise at 100 kHz and a cavity and/or cross-polar isolation of 30 dB, the $-120\,\text{dB}$ dynamic range is achieved. This requires 30–40 dB to be gained by correlating the phase noise between the LO and RF branches; in most cases, this is easily achieved.

4.3.4
Conclusions and Future Perspectives

A brief, practical overview of high-frequency EPR instrumentation has been provided in this chapter. It has been shown that, up to 150 GHz, conventional waveguide technology can still be used, albeit with some limitations. It is to be expected, however, that quasi-optical technology will rapidly take over as the leading strategy in developing millimeter-wave EPR bridges. The spectacular developments in THz technology and radioastronomy continue to provide an increasing number of components that can be used in high-frequency EPR instruments. In addition, millimeter-wave excitation has attracted the attention of the NMR community such that, during the past few years, many projects involving high-field dynamic nuclear polarization (DNP) have been initiated (Denysenkov et al., 2008). Clearly, future technical developments in high-frequency EPR will occur at an even more rapid pace.

Pertinent Literature

In the recent past, the subject of high-frequency EPR instrumentation has been well documented, with Möbius and Savitsky (2009) describing high-field EPR applications on proteins, together with details of model systems and an excellent review of the instrumental aspects and applications of high-frequency EPR. The earlier reviews of Smith and Riedi (Smith and Riedi, 2000; Riedi and Smith, 2002; Riedi and Smith, 2004) remain valuable, as they detail many specific high-field EPR applications. Likewise, Volume 22 of the series on *Biological Magnetic Resonance* (Grinberg and Berliner, 2004), entitled "Very high Frequency (VHF) ESR/EPR," provides a broad overview on both instrumentation and applications. More dated reviews by Prisner (1997) and by Earle, Budil, and Freed (1996) have focused on the technical aspects of high-frequency EPR. Finally, the classic report of quasi-optic homodyne spectrometers was provided by Smith et al. (1998).

References

Barra, A.L., Brunel, L.C., and Robert, J.B. (1990) *Chem. Phys. Lett.*, **165**, 107–109.

Blok, H., Disselhorst, J.A.J.M., Orlinskii, S.B., and Schmidt, J. (2004) *J. Magn. Reson.*, **166**, 92–99.

Blok, H., Disselhorst, J.A.J.M., van der Meer, H., Orlinskii, S.B., and Schmidt, J. (2005) *J. Magn. Reson.*, **173**, 49–53.

Bresgunov, A.Yu., Dubinskii, A.A., Krymov, V., Petrov, Yu.G., Poluektov, O.G., and Lebedev, Y.S. (1991) *Appl. Magn. Reson.*, **2**, 715–728.

Brunel, L.C., Barra, A.L., and Martinez, G. (1995) *Physica B*, **204**, 298–302.

Carl, P., Heilig, R., Maier, D.C., Hofer, P., and Schmalbein, D. (2009) *Bruker Spin Rep.*, **154/155**, 35.

Denysenkov, V.P., Prandolini, M.J., Krahn, A., Gafurov, M., Endeward, B., and Prisner, T.F. (2008) *Appl. Magn. Reson.*, **34**, 289–299.

Disselhorst, J.A.J.M., Vandermeer, H., Poluektov, O.G., and Schmidt, J. (1995) *J. Magn. Reson. Ser. A*, **115**, 183–188.

Earle, K.A., Budil, D.E., and Freed, J.H. (1996) Millimeter wave electron spin resonance using quasioptical techniques, in *Advances in Magnetic and Optical Resonance* (ed. S.W. Warren), Academic Press, New York, pp. 253–323.

Feher, G. (1956) *Sensitivity in Microwave Paramagnetic Resonance Absorption Techniques*, Bell Telephone System, Technical Publications.

Galkin, A.A., Grinberg, O.Y., Dubinskii, A.A., Kabdin, N.N., Krymov, V.N., Kurochkin, V.I., Lebedev, Y.S., Oranskii, L.F., and Shuvalov, V.F. (1977) *Instrum. Exp. Tech.*, **20**, 1229.

Goldfarb, D., Lipkin, Y., Potapov, A., Gorodetsky, Y., Epel, B., Raitsimring, A.M., Radoul, M., and Kaminker, I. (2008) *J. Magn. Reson.*, **194**, 8–15.

Grinberg, O.Y. and Berliner, L.J. (2004) *Very High Frequency (VHF) ESR/EPR*, Kluwer/Plenum Publishers, New York.

Hassan, A.K., Pardi, L.A., Krzystek, J., Sienkiewicz, A., Goy, P., Rohrer, M., and Brunel, L.C. (2000) *J. Magn. Reson.*, **142**, 300–312.

Hofer, P., Kamlowski, A., Maresch, G.G., Schmalbein, D., and Weber, R.T. (2004) The Bruker ELEXSYS E600/680 W-band spectrometer series, in *Very High Frequency (VHF) ESR/EPR* (eds O.Y. Grinberg and L.J. Berliner), Kluwer/Plenum Publishers, New York, pp. 401–429.

Lebedev, Y.S. (1994) *Appl. Magn. Reson.*, **7**, 339–362.

Lynch, W.B., Earle, K.A., and Freed, J.H. (1988) *Rev. Sci. Instrum.*, **59**, 1345–1351.

Möbius, K. and Savitsky, A. (2009) *High Field EPR Spectroscopy on Proteins and Their Model Systems*, Royal Society of Chemistry, Cambridge.

Morley, G.W., Brunel, L.C., and van Tol, J. (2008) *Rev. Sci. Instrum.*, **79**, 046703-1–046703-5.

Nilges, M.J., Smirnov, A.I., Clarkson, R.B., and Belford, R.L. (1999) *Appl. Magn. Reson.*, **16**, 167–183.

Poole, C.P. (1983) *Electron Spin Resonance*, John Wiley & Sons, Inc., New York.

Prisner, T.F. (1997) Pulsed high-frequency/high field EPR, in *Advances in Magnetic and Optical Resonance* (ed. S.W. Warren), Academic Press, New York, pp. 245–299.

Prisner, T.F., Rohrer, M., and Möbius, K. (1994) *Appl. Magn. Reson.*, **7**, 167–183.

Reijerse, E., Schmidt, P.P., Klihm, G., and Lubitz, W. (2007) *Appl. Magn. Reson.*, **31**, 611–626.

Riedi, P.C. and Smith, G. (2002) Progress in high field EPR, in *Electron Paramagnetic Resonance* (eds B.C. Gilbert, M.J. Davies, and D.M. Murphy), Royal Society of Chemistry, Cambridge, p. 254.

Riedi, P.C. and Smith, G. (2004) Progress in high field EPR: inorganic materials, in *Electron Paramagnetic Resonance* (eds B.C. Gilbert, M.J. Davies, and D.M. Murphy), Royal Society of Chemistry, Cambridge, pp. 338–373.

Saylor, C.A., van Tol, H., Krzystek, J., Smith, G., Wylde, R., and Brunel, L.C. (2000) Multifrequency high-field quasi-optic homodyne EMR spectrometer. American Physical Society, Annual March Meeting, 20–24 March, 2000, Minneapolis, MN, abstract #S36.191.

Schmalbein, D., Maresch, G.G., Kamlowski, A., and Hofer, P. (1999) *Appl. Magn. Reson.*, **16**, 185–205.

Smith, G.M. and Lesurf, J.C.G. (1991) *IEEE Trans. Microw. Theory*, **39**, 2229–2236.

Smith, G.M. and Riedi, P.C. (2000) Progress in high field EPR, in *Electron Paramagnetic Resonance* (eds N.M. Atherton, M.J. Davies, and D.C. Gilberts), Royal Society of Chemistry, Cambridge, p. 164.

Smith, G.M., Lesurf, J.C.G., Mitchell, R.H., and Riedi, P.C. (1998) *Rev. Sci. Instrum.*, **69**, 3924–3937.

van Tol, J., Brunel, L.C., and Wylde, R.J. (2005) *Rev. Sci. Instrum.*, **76**, 074101-1–074101-8.

Wang, W., Belford, R.L., Clarkson, R.B., Davis, P.H., Forrer, J., Nilges, M.J., Timken, M.D., Walczak, T., Thurnauer, M.C., Norris, J.R., Morris, A.L., and Zhang, Y. (1994) *Appl. Magn. Reson.*, **6**, 195–215.

Weber, R.T., Disselhorst, J.A.J.M., Prevo, L.J., Schmidt, J., and Wenckebach, W.T. (1989) *J. Magn. Reson.*, **81**, 129–144.

4.4
Pulsed Techniques in EPR
Sankaran Subramanian and Murali C. Krishna

4.4.1
Introduction

In the so-called continuous-wave (CW) mode EPR spectroscopy, the microwave (MW) frequency is kept constant and the field is swept to obtain a spectrum. Alternatively, it is possible to identify the electron/nuclear energy levels by employing a radiofrequency (RF) pulse applied for a short duration. Such a pulse generates at once a band of frequencies centered about the frequency, with the bandwidth and amplitude governed by the Fourier transform (FT) of the pulse (Farrar and Becker, 1971; Shaw, 1984). For example, the FT of a rectangular time-domain pulse of duration t_p leads to a "sinc" [$\sin(x)/x$] profile of frequencies as shown in Figure 4.4.1, with the width between the first two nodes on either side of the center frequency being $1/t_p$.

In other words, the application of a short-duration pulse generates a band of frequencies, simultaneously subjecting the system to *multifrequency* excitation. This allows the entire spectrum to be excited in 'one-shot' rather than a sequential time-consuming sweep, as in the CW mode. It will then be possible to obtain the entire spectrum by a FT of the time-domain response; this is known as the free induction decay (FID) of the system. For example, an excitation pulse of duration 20 ns at a transmitter frequency, say, of 300 MHz, corresponds to subjecting the system simultaneously to a frequency band of 50 MHz (±25 MHz) centered on the transmitter (carrier) frequency, so that spectral components from 275 to 325 MHz

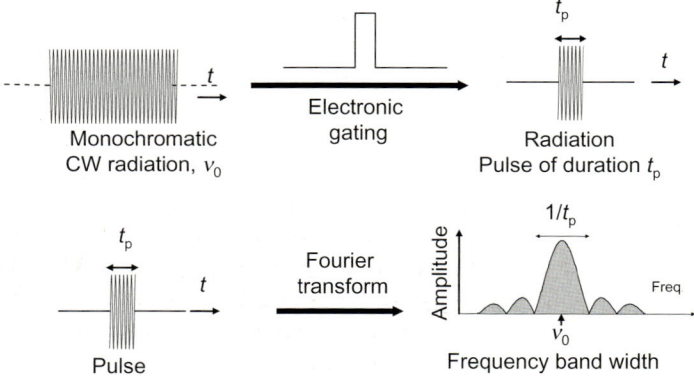

Figure 4.4.1 Formation of a pulse using electronic gating and the power spectrum from a pulse of duration t_p with its characteristic sinc [$\sin(x)/x$] profile with width of the central region approximately equal to $1/t_p$. The magnitude (absolute value) spectrum is shown as a function of frequency.

will be more or less uniformly excited. It was demonstrated by Ernst and Anderson (1966), that the FT of the impulse response obtained from a spin system in a magnetic field and the corresponding spectrum obtained from a slow sweep of the frequency at constant field are absolutely identical. So, what additional advantage is achieved by time-domain methods? The important aspect that makes pulsed spectroscopy efficient and unique is that the free induction decay lasts for only on the order of the spin–spin relaxation time T_2 of the system; the system can then be excited, the FID collected, and several responses added coherently during the time it takes for the CW mode to sweep just once through the spectrum. When n responses are summed coherently, the signal-to-noise ratio (SNR) is improved by a factor of $\sqrt{n}$, since the thermal noise of the spectrometer is a signal with random phase whereas the resonance signal is always phase-coherent. Such an improvement of SNR in going from CW to pulsed mode of detection is even more efficient in EPR, because the transverse relaxation times of most common spin probes are a few hundred nanoseconds. Moreover, it is possible to excite and coherently sum almost a million responses in one second, accomplishing – in principle – nearly three orders of magnitude in SNR improvement. Apart from the main advantage of SNR improvement, pulse methods allow the manipulation of the average Hamiltonian under which the spin system evolves, and hence aids in simplifying the spectra for better analysis (Schweiger and Jeschke, 2001). Specially tailored multiple pulse excitations can provide simplifications in the spectra, and allow the unraveling of specific interactions and quantify them. Pulsed methods also allow direct measurement of the relaxation times of the spin system, and the carrying out of coherence transfer among the electrons, as well as associated nuclear spins (Freed, 2000).

In NMR or magnetic resonance imaging (MRI), where one deals with the spectra whether in uniform field or in the presence of field gradients for spatial encoding, the overall frequency bandwidth that must be dealt with seldom exceeds a few hundred kHz, except in the case of solid-state NMR and in systems with nonzero nuclear quadrupole moment. As shown in Figure 4.4.1, pulses of microsecond duration will eminently cover such a spectral bandwidth in NMR and MRI. However, in EPR the spectral linewidths themselves are on the order of MHz, and even the simplest of organic radicals with hyperfine splitting may have spectra spread over 100 MHz. Spectra from transition metals and inorganic radicals will cover such a large bandwidth (over 1000 MHz) that it is often impossible to cover the full spectral range using a single pulsed excitation. A combination of pulses of nanoseconds duration, and field sweeps/offsets are necessary to address such wide spectral extents. Apart from the bandwidth, the relaxation times of unpaired electron systems are in the micro- to submicrosecond regime, and the FIDs last hardly for a few microseconds (Figure 4.4.2).

In many cases, the so-called *dead time* – the time required for the intense power of an electromagnetic pulse imparted on to the resonator to dissipate down to the thermal noise level of the spectrometer electronics circuit and the spectrometer receiver amplifier to recover – can completely mask the FID, preventing practically any time-domain measurements. Therefore, resonators or resonant cavities must

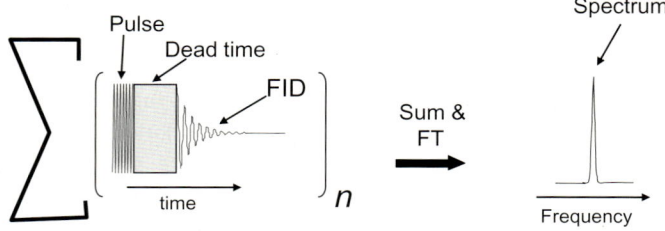

	Pulse width	Dead time	FID length	Repeat time	Spectral Range	Line width
NMR	10-100 μs	1-50 μs	ms - s	ms - s	<MHz	Hz - kHz
EPR	20 – 100 ns	100 – 500 ns	μs	10 - 50μs	>>50 MHz	MHz

Figure 4.4.2 Comparison between various magnetic resonance spectroscopic parameters of protons and unpaired electrons of relevance in time-domain NMR and EPR. It can be seen that the dynamics of unpaired electrons are several orders faster than those of the protons, requiring very specialized electronic components.

be employed which have low ringing characteristics (fast recovery), as well as specially designed amplifiers that recover rapidly following an onslaught from an intense MW or RF pulse. In the following, the standard features of a pulsed EPR spectrometer will be described, with a brief description of the components and their characteristics. The radiation sources, resonators and detection systems vary depending on the frequency range. Traditionally, the most common frequencies for EPR spectroscopy are the X-band (9 GHz and 0.32 T) and Q-band (35 GHz and 1.2 T). Lower frequencies are useful for measuring lossy biological systems, while very high frequencies are useful for measuring very small samples at high sensitivity. The most important thing to remember here is that the sensitivity of magnetic-resonance detection scales approximately as frequency^x, where x varies from 1 to 4, depending on a variety of factors (Eaton, Eaton, and Rinard, 1998) and the penetration of radiation in tissues and biological systems exponentially decreases with frequency (Polk and Postow, 1996). Samples with a very low dielectric constant and deep-frozen aqueous systems are measurable, however, at high frequencies.

EPR spectrometers have been built at a range of frequencies (see Chapter 2 for a list of the various frequency ranges for EPR spectroscopy and the corresponding letter designations). Because the sensitivity and resolution of magnetic anisotropic features improve tremendously with frequency for studying tiny samples at high efficiency, high-frequency spectrometers up to 300 GHz have been developed (Bennati et al., 1999; Burghaus et al., 1992; Davoust, Doan, and Hoffman, 1996; Disselhorst et al., 1995; Earle, Tipikin, and Freed, 1996; Gromov et al., 1999, 2001;

Prisner, Rohrer, and Möbius, 1994; Rohrer *et al.*, 2001; Schmalbein *et al.*, 1999; Lynch, Earle, and Freed, 1988; Astashkin, Enemark, and Raitsimring, 2006). Simultaneously, spectrometers at lower frequencies at S-band (2–4 GHz) and L-band (1–2 GHz) are needed to address large biological or aqueous samples (Quine *et al.*, 1996; Hirata *et al.*, 2002; Willer *et al.*, 2000; Ono *et al.*, 1986; Walczak *et al.*, 2005). For *in vivo* imaging with reasonable penetration, it is necessary to go even below L-band to a RF regime, based on the experience of MRI on biological objects, live animals, and humans (Halpern and Bowman, 1991; Kuppusamy, Chzhan, and Zweier, 1995; Kevan and Bowman, 1990; Subramanian, Mitchell, and Krishna, 2003).

4.4.2
Components of a Pulsed EPR Spectrometer

A grossly simplified schematic of a pulsed EPR spectrometer is shown in Figure 4.4.3. The EPR spectrometer consists of a transmitter arm, a sample cavity or resonator, a receiver arm, and a homogeneous magnetic field that can be subjected to linear field scan around the free-electron resonance field. In the case of a pulsed (time-domain) spectrometer, the transmitter should provide RF/microwave pulses of appropriate amplitude and duration.

The resonator and the detector chain should be isolated better than 40–50 dB, either by the use of a circulator or a diplexer that safeguards the detector amplifier during application of the intense RF pulse. Usually, the pulses are generated using electronic gates that "open" and "close" with the time resolution of 1 ns, and provide high attenuation during the "close" period. A timing device known as a pulse programmer, which can be programmed using an on-line computer, gener-

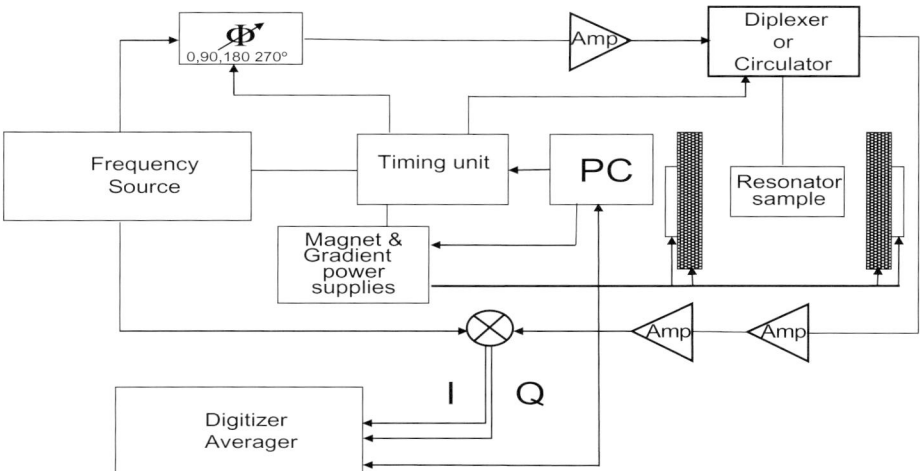

Figure 4.4.3 A grossly simplified schematic of a pulsed EPR spectrometer/imager.

ates timing pulses in a predetermined way to operate the transmitter, isolate the receive arm, open the acquisition gate, and allow the acquisition of data for the desired time periods, and so on. The pulse programmer may allow up to 16 (or more) different pulses and several different time delays to be derived. The phases of the transmitter as well as the receiver may be shifted in 90° intervals (or, in special cases, to any value), allowing for the necessary phase cycling that is helpful in signal averaging, to select a desired pathway of the transverse coherence, and to eliminate background noise (Stoll and Kasumaj, 2008; Jerschow and Muller, 1998; Schweiger, 1995). The signals on the receiver – after suitable amplification, depending on the specific application – can be acquired using devices such as a box-car integrator, data-acquisition system with an analog to digital converter and a summer, or simply a digital oscilloscope with signal-averaging capability. Depending on the frequency of the spectrometer, specialized components are needed for attenuators, preamplifiers and amplifiers, phase shifters, gates, circulators, DC (direct current) blocks, and high/low- and band-pass filters. At frequencies below L-band, the radiations are transmitted using shielded coaxial cables, and the circuits consist of resistors, capacitors, and inductors, as in an FM transmitter/receiver. At microwave frequencies, however, distributed circuit elements and transmission-line theory are more useful methods for design and analysis. At these frequencies, waveguides or transmission lines and cavity resonators are used.

In this chapter, it would be impossible to describe comprehensively all of the available material on pulsed EPR spectrometers. Hence, as representative examples two pulsed EPR spectrometers will be described: one at K_a band (as reported by Astashkin, Enemark, and Raitsimring, 2006); and one at 300 MHz, which was designed and built at the National Cancer Institute (Subramanian *et al.*, 1999). References to several pulsed EPR spectrometers designed and built by leading research groups are provided at the end of the chapter.

4.4.2.1 K_a-Band (26.5–40 GHz) Pulsed EPR Spectrometer

It is possible to have either dedicated transmitter and receiver operating at the K_a band or, alternatively – as in the Bruker E580Q spectrometer (Bruker, 2009) – the X-band pulses can be up-converted to K_a band in the transmit side, whilst at the receiver end the signal can be down-converted back to X-band such that, thereafter, the X-band receiver system can be put to use. The second approach can be a less expensive upgrade of an X-band spectrometer to K_a band. For example, Astashkin, Enemark, and Raitsimring (2006) used an X-band or C-band (6.5–10 GHz) MW source and quadrupled it to the K_a band (26–40 GHz) by using a frequency quadrupler. The K_a-band spectrometer is shown schematically in Figure 4.4.4, and the transmitter and receiver arms are detailed in Figure 4.4.5. Only the major components are shown in the schematics, and described below. In Figures 4.4.4 and 4.4.5 the numbers in bold font correspond to the labeled components.

Transmitter: As seen in Figure 4.4.5a, the low-power MW pulse-forming unit consists of two channels of either an S/C (2–8 GHz) or X/K_u (8–18 GHz) band sources with a single pole-single throw (SPST) switch, a phase shifter, and a bi (0° and 180°) or quadra (0°, 90°, 180° and 270°) phase modulator. Both the channels have a MW quadrupler that would generate respectively either 8–32 or 32–72 GHz MW frequencies. Because of the frequency quadrupler, the phases also become quadrupled; hence, for phase control and quadrature phase cycling a K_a band phase modulator was installed at the transmit end prior to amplification by a traveling-wave tube amplifier (TWTA). Figure 4.4.4 shows the low-frequency source channels directed to an active MW quadrupler (labeled as 1/1a in the figure). The K_a band pulses pass through the bi-phase modulator (2) and a driver amplifier (3), and are amplified by the TWTA (4). The amplified pulses are routed through an isolator, a precision attenuator (5), a phase-shifter (6) and a circulator (7). In the middle of the K_a band, the power output was around 270 W, while the recovery time of the pulse gating circuitry was 95 ns and the gate shut-off time was 10 ns.

Receiver: The signals reflected or emanating from the cavity/resonator pass through the circulator (labeled as 7 in Figure 4.4.4), precision attenuator (8) to the homodyne receiver based on a high-quality mixer (12). The reference frequency, after passing through the quadrupler (1a), is directed to one of two low-noise amplifiers (11a or 11b) operating in the ranges, 26–33 GHz and 33–40 GHz with appropriate band limiters (9) and a single pole double throw (SPDT) switch (9a), with additional protection switches to safeguard the amplifiers (11a and 11b). The two switches or a switch plus limiter gave an attenuation of at least 100 dB, providing good protection of the amplifiers from overload. The intermediate frequency (IF) output from the mixer (12) was then amplified by a video amplifier (13), with a net amplification factor of 80 dB inclusive of conversion and insertion losses.

Timing control: In the Bruker E580 spectrometers, the timing control is achieved through the arbitrary waveform generator (AWG; PatternJet®). Gromov et al. (2001) have described an EPR pulse programmer based on an application-

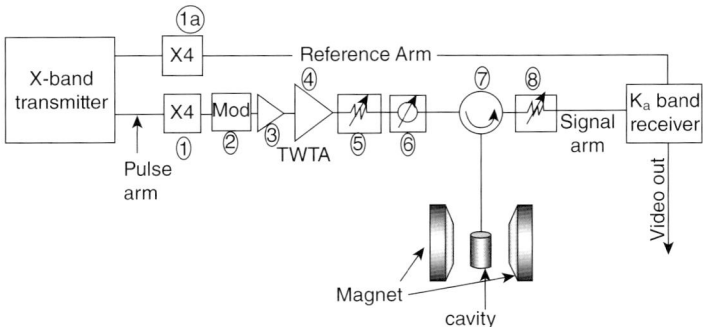

Figure 4.4.4 Schematics of K_a band EPR spectrometer. See text for details. Reprinted with permission from Astashkin, Enemark, and Raitsimring (2006); © 2006, Wiley Interscience.

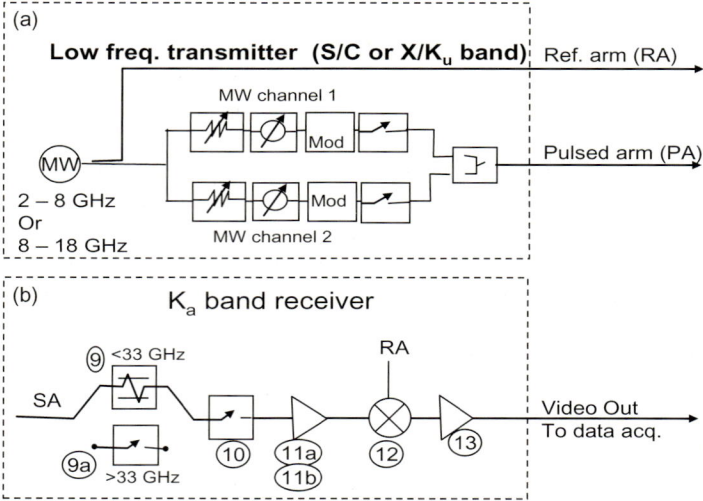

Figure 4.4.5 (a) The low-frequency transmit section and (b) the receiver section of a K_a band spectrometer. Reprinted with permission from Astashkin, Enemark, and Raitsimring (2006); © 2006, Wiley Interscience.

specific integrated circuit and a digital signal processor. For the 300 MHz time-domain EPR spectrometer developed by Subramanian *et al.* (1999), a pulse programmer was used with nanosecond resolution, employing a commercially available timing and input–output port modules and control software based on LabView® (Devasahayam, Subramanian, and Krishna, 2008). This was operated through a personal computer front panel graphic user interface (GUI) that controlled the pulse widths, delays, and the associated acquisition trigger timings. In the K_a-band spectrometer, Ashtaskin *et al.* used an arbitrary waveform generator (AWG) from Chase Scientific Co. (AWG-100) implemented as a peripheral component interconnect, PCI, computer board with 12-bit vertical resolution and 1 ns time resolution. The 12 data bits are available as transistor-transistor logic (TTL) outputs, and are used to control all the drivers for MW switches, phase modulators, TWTA gates, and so on.

Data acquisition: The signal output from the video amplifier is acquired using: (i) a box-car integrator; (ii) direct acquisition on an oscilloscope; or (iii) a fast analog-to-digital converter with an associated summer. In the case of the box-car integrator, the output signal was digitized using a suitable data acquisition (DAQ) board. In this case, the acquisition rate follows the pulse repetition rate of a couple of kHz, allowing the use of moderately slow digitizers with high (16-bit) vertical resolution. This would be useful to achieve an adequate dynamic range when evaluating small variations of large signals with a poor SNR. Alternatively, a digitizer or an oscilloscope can provide the full time sequence of events (FIDs, echoes, etc.), but with a low vertical resolution and reduced dynamic range.

4.4.2.2 Radiofrequency Pulsed EPR Spectrometers Operating at 300, 500, and 750 MHz

Although the sensitivity of EPR is quite low at low frequencies (and low fields), during the past 15 years there has been renewed interest in the design of EPR spectrometers at frequencies at L-band (1–2 MHz) and RF (below 1000 MHz). The main reason for this has been that EPR finds unique applications in the study of free radicals in biological systems (Berliner, 2003). Naturally occurring free radicals *in vivo* are present in low concentrations, and are usually short-lived. Nevertheless, stable redox-sensitive free radicals can be introduced into living system to probe the tissue redox status by using EPR (Kuppusamy et al., 2002; Mitchell et al., 2000; Yamada et al., 2002). Further, many free-radical systems undergo linewidth broadening by dissolved oxygen (*in vivo* partial pressure of oxygen, pO_2) which can be monitored by EPR or, better still, in a spatially resolved manner by EPR imaging, to generate pO_2 maps (Kuppusamy et al., 1994,1998; Subramanian et al., 2002; Velan et al., 2000; Zweier, Thompsongorman, and Kuppusamy, 1991; Zweier et al., 1998; Gallez et al., 1996a, 1996b; Swartz and Walczak, 1998). Whether spectroscopy or imaging is being conducted, frequencies above L-band do not have sufficient penetration when dealing with small animals or aqueous solutions, and consequently EPR spectroscopy *in vivo* must be carried out at low frequencies. As a result, several CW and also time-domain EPR spectrometers have been developed at L-band and radiofrequencies (Walczak et al., 2005; Ono et al., 1986; Colacicchi et al., 1993; Bourg et al., 1993; Alecci et al., 1998; Brivati, Stevens, and Symons, 1991). An example of this, a pulsed RF spectrometer operating in the frequencies 300, 600, and 750 MHz, is described below.

A schematic of the multifrequency pulsed spectrometer designed and built at the National Cancer Institute (NIH, Bethesda, MD) (Devasahayam et al., 2007), operating at 300, 500, and 750 MHz, is shown in Figure 4.4.6. The transmit frequencies are derived from a stable phase-locked master oscillator, leading to several sources of frequencies, including 20, 50, 300, 500, and 750 MHz, all of which maintain a constant, definite phase relationship. It is possible to select any one of the three frequencies – namely 300, 500, or 750 MHz – for operation of the spectrometer. The selected frequency is first amplified to a moderate level, and then passed through quad-phase modulator and a power amplifier (500 W). The output of the amplifier is connected through a diplexer (transmit/receive switch) to a home-built parallel coil resonator tuned to one of the above frequencies. The diplexer used in this study was of a standard $\lambda/4$ configuration, with a series pin-diode switch assembly in the transmitter path and a shunt pin-diode switch assembly in the receiver path (Murugesan et al., 1998). Such an arrangement provided the high-speed switching (ca. 5 ns) that was necessary for time-domain RF FT-EPR experiments. The receiver isolation during transmit mode was 25 dB, with a transmit insertion loss of 2 dB; the insertion loss during receive mode was 0.5 dB.

198 | 4 Spectrometers

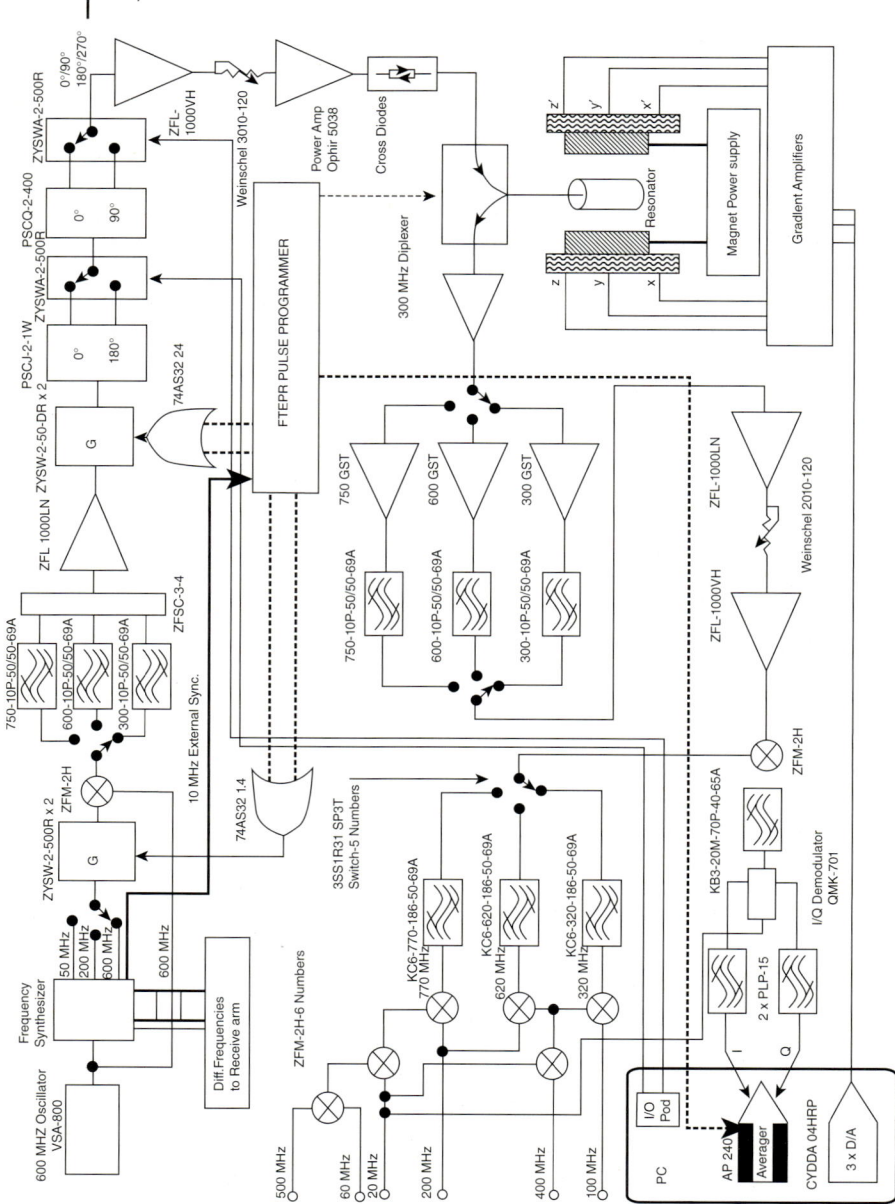

Figure 4.4.6 Schematics of the 300 MHz EPR spectrometer/imager at the National Cancer Institute, NIH, USA.

The signal output of the diplexer was amplified by a low-noise (gallium arsenide field-effect transistor; GaAs FET) amplifier, and down-converted to an IF centered around 20 MHz, by mixing the receiver output to a local oscillator frequency of 20 MHz above the transmitter frequency derived from the same master oscillator source, using frequency-doublers, mixers, and band-pass filters. The spectrometer operates in a quadrature detection mode by using a quad demodulator. Direct conversion of the 300 MHz EPR signals to the base band was avoided to eliminate system feed-through, and to reduce the noise bandwidth before final amplification. However, any imperfection in the two channels is likely to lead to unwanted "ghost" image peaks, especially in samples with a large dynamic range. Similarly, any offset DC voltage difference between the amplifiers is likely to translate into a zero-frequency peak (the "quadrature glitch" at the center of the spectrum). Phase-cycling routines are commonly used in NMR to eliminate artifacts associated with quadrature detection, as well as systematic noise from the spectrometer. A cyclically ordered phase sequence (CYCLOPS; Ernst, Bodenhausen, and Wokaun, 1987) is used to eliminate these artifacts. The spectrometer has a timing unit with pulse width resolution of 1 ns. Signal acquisition is carried out using a two-channel PCI-based data acquisition system (Acqiris AP 240 DAQ; Acqiris USA, Monroe, NY, USA), with maximum sampling frequency of $1\,\mathrm{Gs\,s^{-1}}$ in dual-channel mode, with real time summing of the signals.

4.4.3
Resonators

A resonator is a device that allows the input radiation to be contained in an enclosure for locating the sample, and to obtain maximum sensitivity. An optimal resonator in time-domain EPR should have the following characteristics:

- An adequate bandwidth; one in the range of 10 to 100 MHz might be necessary to produce the satisfactory frequency resolution to provide the spectral coverage in EPR spectroscopy and imaging experiments.
- An efficient conversion of the RF power to the radiofrequency $\mathbf{B}_1$ field.
- A recovery time which is shorter than the response time of the signal.

The overall recovery time of the spectrometer depends on the resonator ring-down time and the receiver recovery time associated with preamplifier overload. The characteristic time constant of the resonator ringing depends on the operating frequency and the so-called quality factor, Q, of the resonator.

Several resonator designs have been described (Hyde, Froncisz, and Oles, 1989; Froncisz and Hyde, 1982; Ono et al., 1986; Sakamoto, Hirata, and Ono, 1995; Pfenninger et al., 1988; Koptioug, Reijerse, and Klaassen, 1997; Alecci et al., 1998; Rinard et al., 1993). Well-known resonators used in MRI, such as solenoid, saddle, surface coils, and birdcage, can also be used for pulsed RF EPR studies, following suitable modification of the Q-value. The very short spin–spin relaxation times of paramagnetic systems preclude the use of inherently high-Q devices. A parallel resonant circuit with inductance L, capacitance C, and resistance R (all in parallel) has a ringing transient of angular frequency ω_r and decay time constant τ_r given by,

$$\omega_r = \omega_0[1 - L/(4Q^2)]^{0.5}$$
$$\tau_r = 2Q/W \quad (4.4.1)$$

where $\omega_0 = (1/2\pi)(LC)^{-0.5}$ and $Q = R/(\omega_r L)$. It should be noted however, that for a conventional resonance circuit with C in parallel with a series combination of L and R, the quality factor is given by $Q = (\omega_0 L)/R$. In all *in-vivo* EPR imaging experiments at 300 MHz at the NCI, a parallel coil resonator (Devasahayam *et al.*, 2000) – where the resonator L, C, and R are *all* in parallel (as shown in Figure 4.4.7a) – was used. A rule-of-thumb for 1 kW incident power is that $30\tau_r$ is required for the resonator energy to decay to a sufficient degree so as to allow the weak induction signals to be detected (Pfenninger *et al.*, 1995). This value will, of course, depend on the specific incident pulse power, the resonator conversion factor, the signal intensity, and the circuit noise power. For the fast-relaxing electron spins, the receiver dead time due to the resonator ringing following the transmitter pulse can severely limit the electron induction signal available for data collection and, when Q is high, may not even survive beyond the dead-time. Equation 4.4.1 further indicates that the characteristic ringing time constant τ_r of a resonator is also related (inversely) to the carrier frequency ω_0. For a frequency bandwidth of 12 MHz, centered on 300 MHz, the optimal Q that would adequately cover this frequency range is 25, and τ_r is 26.5 ns. Hence, immediately following the trailing edge of the RF pulse the system would be "dead" for about 500 ns. This is a much larger value than those encountered at pulsed X- or Q-band operations, and the electron induction signals originating from most biologically useful EPR probes (such as nitroxides) would have decayed by this time. Therefore, the mandatory requirement of a short recovery time and the need to cover a reasonable bandwidth

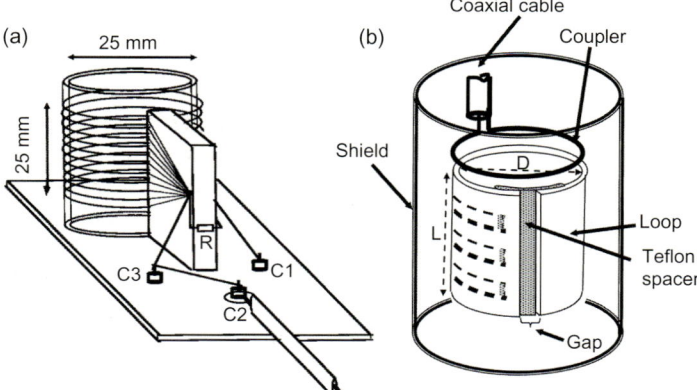

Figure 4.4.7 Schematics of (a) parallel coil resonator and (b) loop–gap resonator.

of excitation (10–12 MHz) dictate the design of low-Q resonators for time-domain RF EPR measurements. However, Q cannot be arbitrarily reduced. The time-domain SNR for a resonant coil with the *sample fully enclosed* is given by Abragam (Abragam, 1961) as:

$$\text{SNR} = (\omega_0^2 N \hbar^2 / 4 k_B T_{\text{sample}}) \cdot (\mu_0 Q / 4 k_B T_{\text{eff}} B V_{\text{coil}})^{1/2} \qquad (4.4.2)$$

In Equation 4.4.2, V_{coil} is the coil volume, N is the total number of spins, Q is the quality factor of the resonant circuit, ω_0 is the resonance frequency, k_B is Boltzmann's constant, and B is the receiver bandwidth. T_{sample} is the temperature of the sample, and T_{eff} is a temperature characterizing the noise of the system. The SNR in Equation 4.4.2 is defined as the time-domain signal amplitude of the free induction decay at its peak, divided by the root mean square Johnson noise. Therefore, with all other things being equal, the SNR will depend on $\sqrt{(Q/V_{\text{coil}})}$.

In the microwave region, rectangular and cylindrical cavity resonators are commonly used. Indeed, many cavity-type reflection and transmission resonators have been used in microwave spectroscopy that can be adapted for pulsed EPR. Cavity resonators are of either cylindrical or rectangular box (parallelepiped) type enclosures, usually machined from brass and coated with gold or silver. The dimensions of the cavity will be on the order of the wavelength of the microwave. A resonant cavity is the equivalent of a tuned RLC circuit used at RF frequencies, and depending on the size and shape, the electric- and magnetic-field vectors inside can persist in a number of standing wave patterns (known as modes), which can be derived from Maxwell's equations. Ideally, the RF magnetic field vector $\mathbf{B}_1$ should be perpendicular to the Zeeman field, and the sample location should be at the site of maximum $\mathbf{B}_1$ field. The cavity will have stubs inserted into it to make it tunable, with the stubs in turn being coupled to the MW transmission wave-guide with an iris coupler consisting of a Teflon insertion screw with a metal tip. For very high frequencies in the MW regime, Fabry–Pérot resonators are used (Tcach *et al.*, 2004).

Other resonators used include dielectric resonators and loop–gap resonators (Roesenbaum, 1964; Walsh and Rupp, 1986; Dykstra and Mrarkham, 1986; Golovina, Geifman, and Belous, 2008; Froncisz and Hyde, 1982; Froncisz, Oles, and Hyde, 1989; Ono *et al.*, 1986; Hyde, Froncisz, and Oles, 1989). A dielectric cavity is made of such a material – for example, fused quartz or sapphire – that tends to concentrate the $\mathbf{B}_1$ field over the volume of the sample and effectively increase the filling factor ($\int_{\text{sample}} \mathbf{B}_1^2 dV / \int_{\text{cavity}} \mathbf{B}_1^2 dV$). Loop–gap resonators (Figure 4.4.7b) consist of one or several cylindrical loops with one or several gaps (cuts). The loops provide the inductance and the gaps the capacitance; moreover, with a suitable choice of both, it is possible to design resonators covering a wide range of frequencies in the RF and MW regions. These are generally low Q-devices with short recovery times, and hence have the high time-resolution required for pulsed EPR. They are ideally suited for small animal imaging, which requires larger bandwidths. A coupling loop acts like an antenna, and can be used to deliver MW/RF power into the loop–gap resonator; then, by adjusting the distance between the loop and the resonator edge, the impedance matching can be accomplished.

Further details on resonator design and construction are provided in the bibliography at the end of the chapter.

4.4.4
Pulsed Excitation and Relaxation

Standard spectroscopic analysis using EPR can be carried out either in the CW mode or in pulsed mode, just as in NMR. The early NMR spectroscopy studies were conducted totally in the CW mode, until it was established that it was possible to generate an impulse response of the nuclear spin system by a RF pulse corresponding to a broadband excitation, and that the resulting FID (upon FT) could provide a spectrum identical to that obtained at constant field with a relatively slow sweep of the frequency (Lowe and Norberg, 1957; Ernst and Anderson, 1966). There are subtle differences between the nature of EPR and NMR spectra, especially with respect to the bandwidth covered by each. The NMR spectra of common nuclei, such as ^{1}H, ^{13}C, ^{31}P, and ^{19}F, cover a range of a maximum of about 200 ppm of the spectrometer frequency, covering at the most a few hundred kHz in the RF region. Therefore, it would be quite easy to excite the complete spectral region by a narrow pulse of a few microseconds duration. On the other hand, EPR spectra of even simple organic radicals may cover a bandwidth (spectral range) of several tens of MHz, and the spectra of transition metal complexes, and of metals with more than one unpaired electron, may have spectra spread over several thousand MHz. As such, a complete spectral excitation of the entire EPR spectrum may be rather difficult, and only portions of the spectrum can be excited. The relaxation times (see Chapter 10) of paramagnetic spin systems can be estimated only indirectly by measuring the spectral linewidth or power saturation effects using CW spectroscopy. On the other hand, the transverse and longitudinal relaxation times of unpaired spin systems can be directly measured in a pulsed EPR spectrometer. It is easy to understand the nature of pulsed excitations and responses in the so-called "rotating frame" (for details, see Chapter 10).

4.4.5
Fourier Transform in Magnetic Resonance

At this point, it is pertinent to examine the so-called "Fourier integral" and Fourier transform that relates the time response of the spin system to its frequency spectrum. Any continuous or piece-wise continuous function can be represented by a sum of infinite series of sines and cosines of a fundamental frequency and its higher harmonics. This series–the "Fourier series"–converts a time function to its frequency counterpart, and likewise the inverse operation will convert a frequency spectrum to its temporal format. The corresponding mathematical operations are the FT given by:

$$F(\omega) = \int f(t)e^{-i\omega t} dt \qquad (4.4.3)$$

and the inverse FT given by

$$f(t) = 1/(2\pi)\int F(\omega)e^{i\omega t}d\omega \qquad (4.4.4)$$

Equations 4.4.3 and 4.4.4 imply that the FT will, in general, produce complex results. The relationship between the time-domain response and its FT also defines the spectral line shapes. An exponential decay of the transverse magnetization upon FT produces a spectrum which has a Lorentzian line shape. The decay of magnetization from noninteracting spins, which is known as the "Bloch decay" (Bloch, Hansen, and Packard, 1946), does follow a single exponential with the characteristic time constant T_2, ($\propto \exp(-t/T_2)$, and the corresponding Lorentzian absorption spectrum has a full-width at half-maximum (FWHM) height given by $(\pi T_2)^{-1}$. From the nature of the FT pairs, it is easy to make the following observations:

- Even functions in time-domain, that is, $f(-t) = f(t)$ will have only a real FT, and the transformation can be accomplished fully by using the cosine series. Likewise, an odd function of time, $f(-t) = -f(t)$, will lead to antisymmetric frequency functions and can be fully synthesized using the sine series. The result of the cosine transform is the absorption spectrum and that of the sine transform is the dispersion spectrum; these are respectively the Bloch's out-of-phase and in-phase AC susceptibility of the spin system with respect to the direction of the application of the oscillatory magnetic field. These are Hilbert transforms of each other.

- As mentioned above, an exponential decay in time domain leads to a Lorentzian function in the frequency domain, while a Gaussian decay in the time domain leads to a Gaussian function in frequency domain, and *vice versa*.

- A very quickly decaying signal will have a broad range of frequencies (which interfere to shorten the lifetime of the FID), and a slowly decaying signal will have a narrow range of frequencies. In other words, systems with large linewidths will have a short FID, while those with a narrow linewidth will persist longer in the time domain.

4.4.6
Simple Pulsed EPR Experiments

Because the FT of the impulse response gives the spectrum that is identical to that obtained by a slow sweep of the field or frequency, pulsed methods provide a unique way of obtaining the spectrum rapidly. Because the decay time of the transverse magnetization and the time required to reaching equilibrium populations are governed by the relaxation times T_1 and T_2 (which are on the order of microseconds), it is possible to acquire the impulse response by using a fast analog-to-digital converter and performing an on-line FT so as to generate a spectrum almost instantly. This has enormous potential in addressing very dilute samples (low spin count), as hundreds of pulses can be applied and the responses coherently summed. This is because the magnetic-resonance-induction signal increases in proportion to the number of responses added coherently, whereas the random noise increases in proportion to the square root of the number. In addition, there is an additional multiplex advantage, namely that all of the frequencies required to cover the spectral range are provided by a narrow pulse, corresponding to a broadband excitation and the response acquired "on-the-fly" consists of all the spectral information and is subsequently unraveled by a fast FT. This leads to the so-called "Felgett advantage" (Fellgett, 1949a, 1949b), which corresponds to multichannel excitation and detection, leading to a very rapid spectral data acquisition analogous to the use of interferometry in FT infrared (IR) spectroscopy.

EPR spectra can be acquired, in principle, by using pulses, followed by the FT of the impulse response (FID), very similar to the practice of NMR. There are, however, certain limitations in the case of EPR. Notably, only organic radicals that have a total spectral spread of up to 20 G (<50 MHz) can be fully excited by pulses of about 20 ns width. Larger pulses do not cover the spectrum adequately, and the power spectrum tends to drop on either extreme of the spectrum, leading to a lower sensitivity. Further, as mentioned previously, the resonator bandwidth is also severely limited by the quality factor, which determines the bandwidth to be covered effectively. Both, inorganic radicals and transition-metal complexes have such a large spectral spread (more than several hundred Gauss) that the spectra can be excited only in parts, and in conjunction with magnetic-field shifts and sweeps. It is for this reason that simple pulsed excitation, followed by FT, is applicable in EPR, mainly to obtain the spectra of free radicals containing a single narrow line or a few hyperfine lines with the overall spectral spread not exceeding a few tens of MHz (1–2 mT). On the other hand, multiple pulse experiments can provide a wealth of data when examining the various weak interactions between the unpaired electron and the surrounding nuclei, or electron–electron interactions in systems with more than one electron (biradicals). These interactions manifest themselves as additional modulations on the electron resonance, and can be of the utmost importance in quantitatively unraveling important structural and dynamic aspects of large paramagnetic molecules. Multifrequency and multiple-pulse EPR spectroscopy have recently been used to unravel the structure of membrane proteins labeled with spin labels, and those which contain (or are reconstituted with) paramagnetic metal entities (Calle *et al.*, 2006; Freed, 2000, Möbius; 1995; Schweiger and Jeschke, 2001). Before describing some of these experiments, however, it is important first to examine two simple pulse experiments that have allowed the quantitation of longitudinal and transverse relaxation times, T_1 and T_2.

4.4.6.1 Inversion Recovery and Hahn Echo Pulse Sequences, T_1 and T_2

In the rotating frame, and *on resonance*, with the magnetization vector and frame being at identical frequency, the effective Zeeman field is zero and therefore the spins feel only the MW/RF magnetic field ($\mathbf{B}_1$) and tend to precess about the $\mathbf{B}_1$ axis. In other words, when the applied field is along the rotating x-axis, the net magnetization will be rotated from the z-axis towards the rotating x–y plane with angular frequency $\gamma_e B_1$ radians per second. If the duration of the MW/RF pulse is t_p, then the magnetization tilt angle θ can be defined as:

$$\theta = \gamma_e B_1 t_p \quad \text{radians} \tag{4.4.5}$$

When $\theta = \pi/2$, the pulse is termed a 90° pulse, while when it is equal to π, the pulse is termed a 180° pulse. The latter is also called an "inversion pulse" since, at the end of the 180° pulse, the magnetization is inverted from its original direction (say, from z to –z). A combination of the 90° and 180° pulses can be used to measure T_1 and T_2 directly using pulsed EPR; however, they must be determined via time-consuming saturation experiments, and with linewidth measurements under nonsaturation conditions in a highly homogeneous Zeeman field, when CW-EPR is used.

The application of a 180° pulse on a system in equilibrium rotates the entire magnetization ($\mathbf{M}_o$) to the –z direction of the rotating frame. Since there is no transverse magnetization, the system tends to reach equilibrium from $-\mathbf{M}_o$ to $+\mathbf{M}_o$ by dissipating the input energy into the degrees of freedom of the surrounding "lattice" (as described in Chapter 10 for relaxation mechanisms). The magnetization ($\mathbf{M}_z$) grows exponentially from $-\mathbf{M}_o$ to $+\mathbf{M}_o$ crossing "zero-magnetization" governed by the characteristic time constant T_1 given by

$$d\mathbf{M}_z(t)/dt = (\mathbf{M}_z(t) - \mathbf{M}_o)/T_1 = \gamma(\mathbf{M} \times \mathbf{B}_1)_z \tag{4.4.6}$$

It should be noted that the magnetization at the end of an inversion pulse is the same as equilibrium magnetization, with a negative sign ($\mathbf{M}_z(0) = -\mathbf{M}_{z,eq}$), and it can be shown that the z-component of magnetization recovers as (Figure 4.4.8):

$$\mathbf{M}_z(t) = \mathbf{M}_{z,eq}(1 - 2 e^{-t/T_1}) \tag{4.4.7}$$

In order to monitor the recovery of magnetization following the inversion pulse, a 90° pulse is applied after a delay τ, which leads to tilting the magnetization as it recovers to the transverse plane to generate a FID, the initial amplitude of which will vary as a function of t, given by the above equation. The time-course of the z-component of magnetization can be used to derive T_1. Instead of inverting the magnetization by using a 180° pulse, it is also possible to apply a burst of five to six 90° pulses at very short intervals, leading to a saturation of the spin system (leading to net zero magnetization due to equalization of the populations in the Zeeman levels). This can be followed by monitoring the magnetization recovery by using a 90° pulse progressively delayed from the saturation "burst" and plotting the initial amplitude of the FID, where the growth of the z-magnetization from zero is governed by (Figure 4.4.8):

$$\mathbf{M}_z(t) = \mathbf{M}_{z,eq}(1 - e^{-t/T_1}) \tag{4.4.8}$$

The measurement of transverse relaxation, T_2 time is carried out using the so-called Hahn spin-echo sequence (Hahn, 1950), which consists of a 90° pulse followed by a delay (τ) and then a 180° pulse. The application of the first pulse tilts the net magnetization to the transverse plane (xy), with no magnetization along

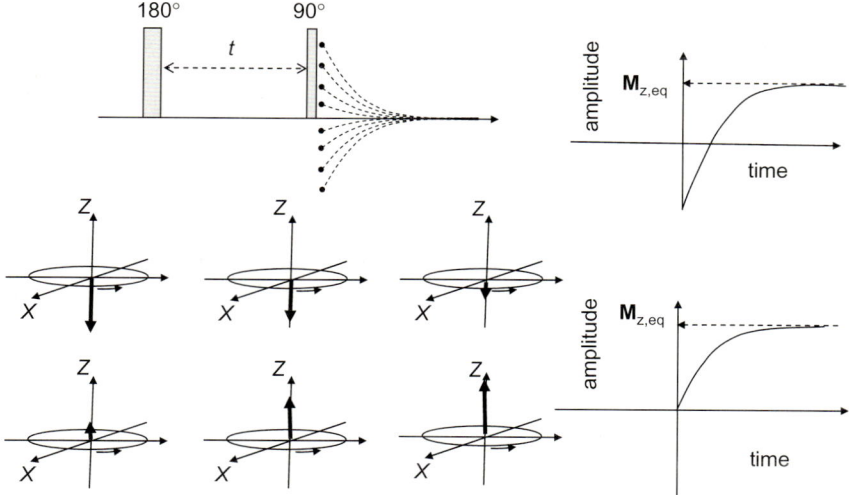

Figure 4.4.8 The inversion recovery (180°-t-90°) sequence creates FIDs with the first point showing a progressive increase in intensity from −M_o to +M_o governed by the spin-lattice relaxation time T_1. T_1 can also be followed by a series of 90° pulses (saturation), followed by a 90° pulse. Here, the signal increases from 0 to M_o, governed by the relaxation time T_1.

the z-axis. The transverse magnetization tends to lose phase-coherence due to the inherent spin–spin relaxation, which is an energy-conserving process. This is further accentuated by the Zeeman-field inhomogeneities, caused by the inherent quality of the Zeeman field, as well as the magnetic susceptibility anisotropy of the spin system. The latter can further disperse the coherence of transverse magnetization by increasing the bandwidth of the processional frequency. The gross decay constant is denoted by T_2^*. The application of the 180° refocusing pulse after the delay τ, as discovered by Erwin Hahn (Hahn, 1950), leads to a reversal of the effect of all other time-dependent Hamiltonians. The intrinsic transverse relaxation, however, cannot be reversed, and this results in the formation of an "echo." In other words, transverse magnetization which will decay to near-zero governed by T_2^*, will be resurrected partially leading to an echo, peaking in amplitude at time 2τ. A series of (90° – τ – 180°) pulse pairs will lead to an estimation of the phase-memory time, T_M, by following the echo peak heights as a function of 2τ. The time course of the echo peak heights follows the relation:

$$\mathbf{M}_{xy}(t)[\propto \text{Echo peak}] = \mathbf{M}_{xy}(0)e^{-t/T_M} \qquad (4.4.9)$$

where $t = 2\tau$. The phase of the echo will be opposite to that of the FID if the two pulses were applied along the same axis (the Carr–Purcell echo sequence; Carr and Purcell, 1954; see Figure 4.4.9) whereas, if one applies the second pulse 90° out-of-phase with respect to the first pulse (Meiboom–Gill echo sequence; Meiboom and Gill, 1958; see Figure 4.4.9), then the echo will be of the same phase as the

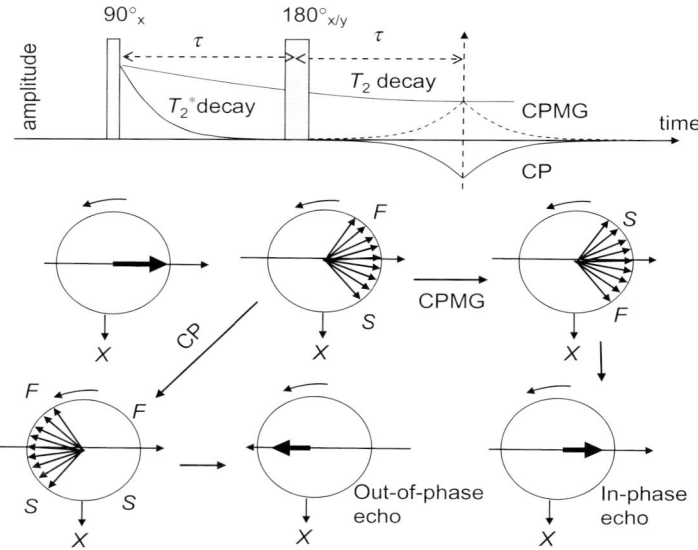

Figure 4.4.9 Schematics of the Hahn (Carr–Purcell) echo and the modified Meiboom–Gill echo sequences. See text for details.

FID. Strictly speaking, the measured T_M depends on the magnitude of the delay between the 90° and 180° pulses, since refocusing of the spin isochromats (groups of transverse magnetization vectors, generally called "spin-packets") is often disturbed by mutual interference and spin diffusion, in addition to molecular diffusion. The measurement of T_M as a function of the inter-pulse delay and extrapolating the value to zero inter-pulse delay can provide the best estimate of transverse relaxation time T_2, without any influence from the other processes mentioned above. In other words, when the transverse magnetization is not defocused due to spin diffusion, the phase memory time T_M and the transverse relaxation time T_2 approach each other. Two-pulse Hahn-echoes are not the only echo phenomenon studied in NMR and EPR. In fact, a series of three pulses, $90 - \tau - 90 - t_1 - 90°$ produces several echoes – five in all, including three Hahn echoes, a stimulated echo, and a refocused echo (see Figure 4.4.10) (Mims, 1972a).

In simple terms, in the three-pulse echo sequence, the second 90° pulse flips part of the magnetization which is along the rotating y' axis into the Zeeman axis (z), where it can decay only by the relatively longer spin-lattice relaxation. When this magnetization is brought back to the transverse plane by the third 90° pulse, the so-called "stimulated echo" results; this will have a peak height which is larger than the two-pulse Hahn echo for the same total delay. The echo height in systems where the electron spin is coupled to magnetic nuclei, via hyperfine or dipolar coupling, exhibits periodic modulation due to these couplings. In fact, a careful analysis of the echo modulation frequencies can lead to quantification of these couplings that may help in identifying the magnetic nuclei in the vicinity of the

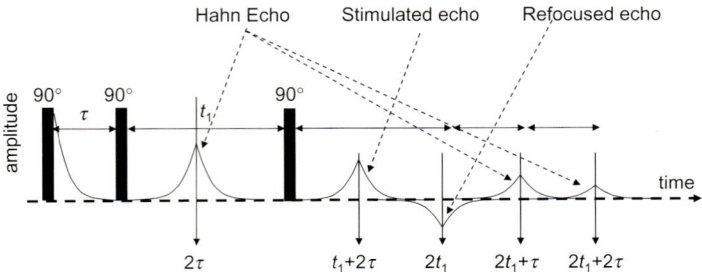

Figure 4.4.10 Three-pulse ($\pi/2 - t - \pi/2 - t_1 - \pi/2$) sequence generating the Hahn echo, the stimulated, and refocused echoes.

unpaired electron, as well as providing an accurate estimate of distances in the case of dipolar modulation. This phenomenon, referred to as electron spin echo envelope modulation (ESEEM), is a very important pulsed EPR tool used for unraveling complex molecular structures (Blumberg, Mims, and Zuckerma, 1973; Blumberg, Zuckerma, and Mims, 1972; Rowan, Hahn, and Mims, 1965; Kang, Eaton, and Eaton, 1994; Davoust, Doan, and Hoffman, 1996; Prisner, Rohrer, and Möbius, 1994; Ionita et al., 2009; Carmieli et al., 2006).

Two-pulse Hahn-echoes are not the only echo phenomenon studied in NMR and EPR. The Hahn echoes decay with a time constant T_M, whereas the stimulated echo (at $t_1 + 2\tau$) decays with a time constant T_2 which is shortened slightly by spin diffusion, but is longer than T_M. Since all echoes exhibit ESEEM, it would be convenient to use a three-pulse experiment and the stimulated echoes in ESEEM studies, because of higher sensitivity and resolution.

4.4.7
Pulsed ENDOR, ESEEM, and HYSCORE

The structural elucidation of large paramagnetic molecules such as metalloenzymes by EPR often involves reconstitution; that is, the substitution of a diamagnetic metal site by a paramagnetic ion of similar size and charge, for example Co(II) by Cu(II), and Mg(II) by Cu(II). In order to arrive at a precise structure, it is necessary to quantitatively evaluate many magnetic interactions, some of which are very small in magnitude and may be unresolved in a conventional CW-EPR spectrum. The g-tensor of the paramagnetic species often helps to establish the symmetry of the unpaired spin distribution and the electronic energy levels of the open shell system. However, weaker interactions such as the hyperfine and superhyperfine interaction of the unpaired electron, through bond and through space, with the surrounding magnetic nuclei, the interaction of the electric quadrupole moment of interacting nuclei which have nonspherical nuclear spin distribution for nuclei with spin $I \geq 1$, and so on, may cause a multitude of interactions, including fine structure, the presence of 'forbidden" transitions, and unequal hyperfine spacing. In addition, in systems with more than one unpaired electron, the dipolar and exchange interactions between the electrons which are present even in the absence of any external field, zero-field interaction, cause additional splitting. All of these fine interactions occur as either resolved or unresolved spectral splitting in CW-EPR, and a complete analysis – which would be very difficult in the absence of good resolution of these features – may provide a wealth of information on the structure and three-dimensional (3-D) topology of large molecules. In pulsed EPR, these interactions manifest themselves as additional modulations on the main Zeeman interaction. Just as time-domain NMR employs multiple pulse sequences that subject the spin system to various selective manipulations, and monitors the evolution of various time-dependent effective Hamiltonians that help to generate highly resolved multidimensional spectra, it is also possible to apply pulsed techniques in EPR to unravel unresolved and complex spectra that can provide valuable quantitative structural information. The various interactions in EPR are represented in the spin Hamiltonian (as described by the various terms discussed in Chapter 3). It was mentioned above, that the spectral width corresponding to pulses of a few tens of nanoseconds would hardly cover a bandwidth of a few tens of MHz and, consequently, that only a part of the entire spectrum would be excited by the pulse. Yet, this situation can be used to advantage in pulsed EPR where, in a randomly distributed spin system with an anisotropic g-tensor, a narrow bandwidth excitation will select a spectral region that corresponds to a very narrow distribution of orientations, thereby providing information from this limited orientation, as in an *oriented* single crystal.

This will help to derive the principal components of many of the interactions which are second-rank tensors fairly accurately, without the need to perform elaborate time-consuming measurements at a large number of orientations of the paramagnetic species in crystals. In addition, it allows such measurements even in systems which do not crystallize at all (Schweiger and Jeschke, 2001) (see Figure 4.4.11).

4.4.7.1 Nuclear Modulation Effects Leading to ENDOR and ESEEM

The energy levels of a paramagnetic system with an interacting electron spin ($S = \frac{1}{2}$) and a magnetic nucleus of spin $I = \frac{1}{2}$ are shown in Figure 4.4.12 (Calle et al., 2006). Two of the energy levels in which the z-component of the electron spin angular momentum component with $m_S = |+\frac{1}{2}\rangle$ is the α-manifold, whereas that with $m_S = |-\frac{1}{2}\rangle$ is the β-manifold, and there are two sublevels in this manifold with the nuclear spin angular momentum components $\pm|\frac{1}{2}\rangle$. Likewise, there are two sublevels in the β-manifold, leading to four levels labeled as $|\alpha\alpha\rangle$, $|\alpha\beta\rangle$, $|\beta\alpha\rangle$ and $|\beta\beta\rangle$, where the first letter denotes the electron spin and the second letter that of the nucleus.

Allowed EPR transitions are two in number, as indicated by vertical arrows (*a*) with $\Delta m_S = \pm 1$, and $\Delta m_I = \pm 0$, $1 \leftrightarrow 3$ and $2 \leftrightarrow 4$. There are two forbidden EPR transitions (*f*) with $\Delta m_S = \pm 1$, and $\Delta m_I = \pm 1$, $1 \leftrightarrow 4$ and $2 \leftrightarrow 3$, the so-called "zero-quantum" and "double-quantum" transitions involving the simultaneous flip of the electron and nuclear spins. There are also two transitions involving exclusively nuclear spin flips ($\Delta m_S = \pm 0$, and $\Delta m_I = \pm 1$) denoted as $\omega_\alpha(1 \leftrightarrow 2)$ and $\omega_\beta(3 \leftrightarrow 4)$.

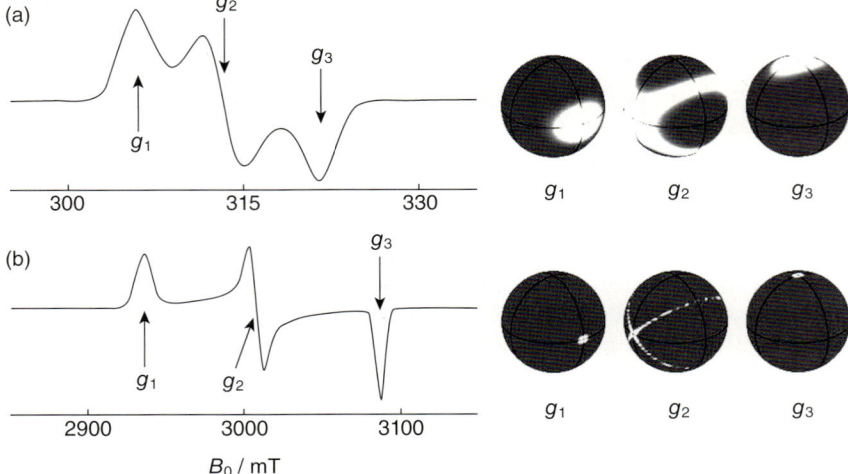

Figure 4.4.11 The principle of orientation selection by exciting narrow region of "powder" spectra. As the frequency of measurement increases, a pulse with limited bandwidth allows the selection of principal components of the EPR tensors, even from a randomly oriented powder sample. The selected regions are indicated as bright regions in a black ellipsoid. Powder spectra from a typical rhombic g-tensor system, $g_1 \neq g_2 \neq g_3$. (a) X-band, 9.8 GHz; (b) W-band, 94.1 GHz. Reprinted with permission from Calle et al. (2006); © 2006, Verlag Helvetica Chimica Acta, Zurich.

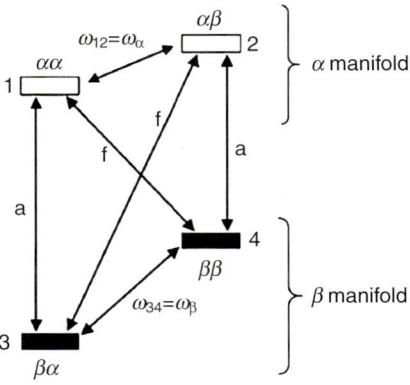
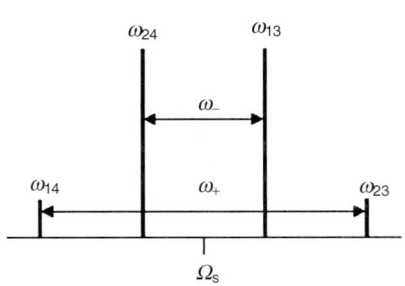

Figure 4.4.12 Energy level diagram (left) and the corresponding schematic EPR spectrum (right) for $S = \frac{1}{2}$, $I = \frac{1}{2}$ model system with $|A_s| < |2\omega_I|$ (weak coupling case). Route a: allowed EPR transitions (1 ↔ 3) and (2 ↔ 4). Route f: forbidden EPR transitions (1 ↔ 4) and (2 ↔ 3). Nuclear transitions are (1 ↔ 2) and (3 ↔ 4). A_s is the hyperfine coupling at an arbitrary orientation. Reprinted with permission from Calle et al. (2006); © 2006, Verlag Helvetica Chimica Acta, Zurich.

If we denote the isotropic part of the hyperfine coupling by a_{iso},

$$a_{iso} = 1/3[A_1 + A_2 + A_3] \qquad (4.4.10)$$

where A_i are the principal components of the observed anisotropic hyperfine tensor, such that the principal components of the dipolar hyperfine tensor is given by:

$$\mathbf{T} = (A_1 - a_{iso}, A_2 - a_{iso}, A_3 - a_{iso}); \text{ and } Tr(\mathbf{T}) = 0 \qquad (4.4.11)$$

The allowed (ω_{24} and ω_{13}) and the forbidden (ω_{14} and ω_{23}) EPR transitions are noted in Figure 4.4.12, and the extreme transitions will be seldom seen except in weak-coupling cases where the magnitude of hyperfine coupling at any orientation $|A_\theta| \ll 2\omega_I$, where ω_I is the NMR transition energy of the nucleus. By setting up the 4×4 matrix representation of the spin Hamiltonians H_{Zee}, H_{HF} and H_{NZee} within the product spin space $|\alpha\alpha\rangle$, $|\alpha\beta\rangle$, $|\beta\alpha\rangle$ and $|\beta\beta\rangle$, and diagonalizing the matrices, one can arrive at the energies of the nuclear (ENDOR) and the electronic (EPR) transitions and their transition probabilities. These are given below:

$$\omega_\alpha = |\omega_{12}| = \left[\left(\omega_I + \frac{A}{2}\right)^2 + \left(\frac{B}{2}\right)^2\right]^{1/2}; \omega_\beta = |\omega_{34}| = \left[\left(\omega_I - \frac{A}{2}\right)^2 + \left(\frac{B}{2}\right)^2\right]^{1/2}$$
$$(4.4.12)$$

And, for an axially symmetric hyperfine tensor ($A_1 = A_2 \neq A_3$), A and B above are given by

$$A = a_{iso} + T(3\cos^2\theta - 1); \quad B = 3T\sin\theta\cos\theta \qquad (4.4.13)$$

where θ is the angle between the electron–nuclear vector and the Zeeman field axis. The transition probabilities of the allowed (a) and forbidden (f) EPR transitions are given by:

$$I_a = \cos^2\eta = \frac{\omega_I^2 - \frac{1}{4}\omega_-^2}{\omega_\alpha\omega_\beta}; \quad I_\beta = \sin^2\eta = \frac{\omega_I^2 - \frac{1}{4}\omega_+^2}{\omega_\alpha\omega_\beta} \qquad (4.4.14)$$

Here, 2η is the angle between the nuclear quantization axes in the two electron manifolds (α and β), ω_I is the nuclear Larmor frequency, and $\omega_+ = \omega_\alpha + \omega_\beta$ and $\omega_- = \omega_\alpha - \omega_\beta$. It is clear from the schematics in Figure 4.4.12 that the nuclear transitions gain intensity when the forbidden transitions (f) have finite probability. In ESEEM, where the MW pulses excite both the allowed and the forbidden EPR transition, one can see (indirectly) the nuclear transition frequencies manifest themselves as modulation of the electron spin echo. In ENDOR, on the other hand, the nuclear transition frequencies are excited directly with RF irradiation at the appropriate frequencies; since nuclear transitions are excited between levels whose polarizations are governed by electron spin polarization, an enhancement in transition on the order of $\gamma_{electron}/\gamma_{nucleus}$ is obtained.

4.4.7.2 Mims and Davis Pulsed ENDOR Sequences

ENDOR can be performed both in CW and pulsed modes. The first experiments of ENDOR, as conducted by Feher, were in fact carried out in the CW mode (Feher,

1956; Feher, Fuller, and Gere, 1957). The experiment consisted of saturating an EPR transition of system with coupled nuclear spins. As the EPR transitions lose intensity due to a reduction in the population difference between the levels, a RF sweep is performed to excite the nuclear transitions. This releases the bottleneck in the population distribution, and allows the EPR transition to regain the intensity; this is referred to as "ENDOR-induced EPR." The nuclear transitions are enhanced and occur at frequencies above and below the nuclear Larmor frequency, depending on the hyperfine coupling and nuclear quadrupole coupling. Since nuclear resonances are quite narrow, hyperfine splitting which are on the order of EPR linewidth, and hence unresolved in the EPR spectrum can be evaluated with high precision in ENDOR. This is particularly true for low gamma nuclei such as ^{13}C, ^{15}N, ^{14}N, and ^{2}H, when measured at high spectrometer frequencies. CW-ENDOR consists of carefully locking the Zeeman field on a designated EPR transition, and then sweeping a range of NMR frequencies, keeping the resonator tuned to the changing RF all the time by servo-regulated automatic tuning and matching. In terms of the instrumentation needed, CW-ENDOR is quite challenging and requires a critical balance of the RF power and relaxation times to achieve optimum sensitivity. As RF and MW are present simultaneously, much effort is required to reduce any interference from low-frequency and the associated artifacts. Pulsed ENDOR, on the other hand, is less susceptible to artifacts as the RF and MW power are applied only for a short duration, and are seldom applied simultaneously. Pulsed ENDOR allows the study of the manifestations of electron T_1, T_2 and nuclear T_1 and T_2, as well as electron nuclear cross-relaxation with good selectivity. Basically, the nuclear resonance frequencies which contain high-resolution hyperfine information are monitored by creating an enormous polarization between the nuclear sublevels by saturating the connected EPR transition. Upon exciting the nuclear transitions by selective RF pulses, the two- or three-pulse echo becomes enhanced when the nuclear resonance frequencies match with the RF frequency.

The two sequences that are used in pulsed ENDOR are those of Davis and Mims (Davies, 1974; Mims, 1965) (see Figure 4.4.13). In Davis ENDOR (Figure 4.4.13a), the first selective MW π pulse inverts the populations in two connected levels of a particular EPR transition (levels 1 ↔ 3; Figure 4.4.14, schemes (a) and (b)); following this, a selective RF π pulse is applied. When the frequency and the bandwidth of this RF pulse corresponds to transitions between nuclear sublevels connected to the inverted EPR transition, the nuclear polarization also becomes inverted (see Figure 4.4.14, schemes (b) and (c)) which, in turn, will alter the population of the "observer" electron spin transition 1 ↔ 3. If the RF does not correspond to any of the nuclear hyperfine sublevels, then no change is seen in the population of the levels 1 and 3 (Figure 4.4.14, scheme (d)). The change in electron polarization brought about by a transition of the connected nuclear sublevels can be monitored by appending the conventional Hahn echo sequence $\pi/2$-τ-π-τ (Figure 4.4.13a) following the selective RF pulse. The echo intensity is monitored as a function of the RF, which is incremented stepwise.

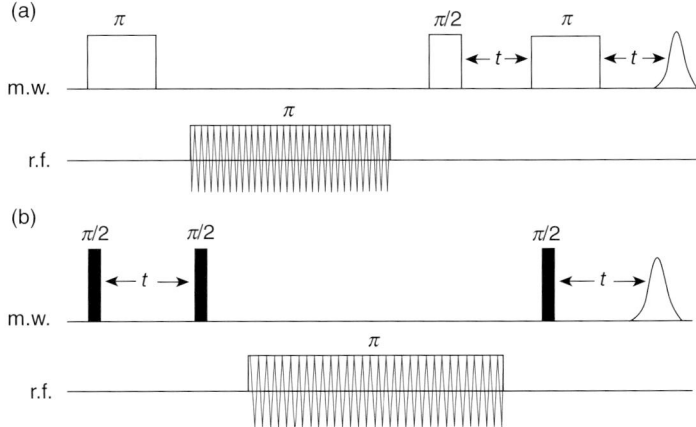

Figure 4.4.13 Pulse sequence for (a) Davies ENDOR and (b) Mims ENDOR. The inter-pulse delays are kept constant, while the RF is incremented over the desired frequency range. Reprinted with permission from Calle et al. (2006); © 2006, Verlag Helvetica Chimica Acta, Zurich.

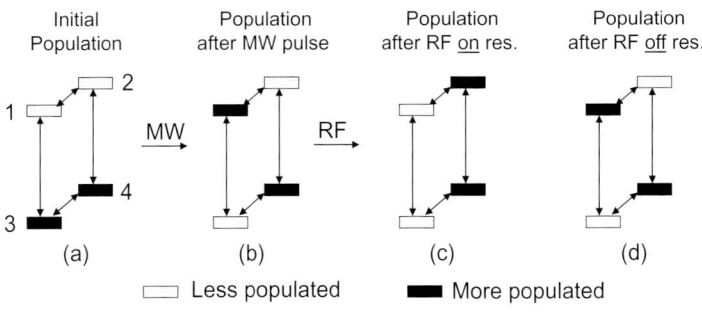

Figure 4.4.14 Energy levels and population in a two-spin $S = \frac{1}{2}$ and $I = \frac{1}{2}$ system during the Davies ENDOR experiment. Scheme (a): Equilibrium populations; Scheme (b): Selective MW π pulse inverts the polarization of EPR transition (1,3); Scheme (c): Populations after the RF π pulse on resonance with nuclear transitions; Scheme (d): Off-resonance with the nuclear transitions (no effect). Reprinted with permission from Calle et al. (2006); © 2006, Verlag Helvetica Chimica Acta, Zurich.

In Mims ENDOR (Mims, 1965), essentially the same concept is applied, except that the nuclear transitions are monitored by the change in the intensity of the *stimulated echo* as against the Hahn echo. Therefore, the Mims ENDOR sequence (Figure 4.4.13b) consists of three nonselective $\pi/2$ MW pulses. The preparation pulse pair $\pi/2$-τ-$\pi/2$ generates electron polarization, depending on the delay τ. If the selective RF pulse that follows this corresponds to a connected nuclear transition, then it will change the electron polarization, and hence the amplitude of the stimulated echo peak generated by the third $\pi/2$ MW pulse, followed by the delay time τ. In Mims ENDOR, the use of a stimulated echo, the intensity of which depends on the transverse relaxation time T_2, makes the system more efficient than Davis ENDOR. (This is because a stimulated echo of which the intensity is governed by the relatively longer T_2 is more intense than the Hahn echo, where the intensity is governed by the shorter phase memory time, T_M.) The efficiency of ENDOR also depends on the magnitude of the hyperfine coupling, $A(\theta)$, and the echo time τ. It can be shown that, in Mims ENDOR, the maximum efficiency occurs when $\tau = (2n + 1)\pi/A(\theta)$, and is zero for $\tau = 2n\pi/A(\theta)$ ($n = 0, 1, 2$). Thus, "blind spots" will appear, just as in ESEEM and HYSCORE (see below).

4.4.8
Electron Spin Echo Envelope Modulation (ESEEM) and Hyperfine Sublevel Correlation Spectroscopy (HYSCORE)

Following the first observations of ESEEM by Rowan et al. (1965), the theory was subsequently developed by Mims (Mims, 1972a, 1972b). Whilst the sequences used closely resemble those of pulsed ENDOR, there are no RF pulses to directly excite nuclear transitions. The basic ESEEM, one-dimensional (1-D) sequence and a two-dimensional (2-D) four-pulse sequence that generates a correlation spectrum known as HYSCORE are shown in Figure 4.4.15.

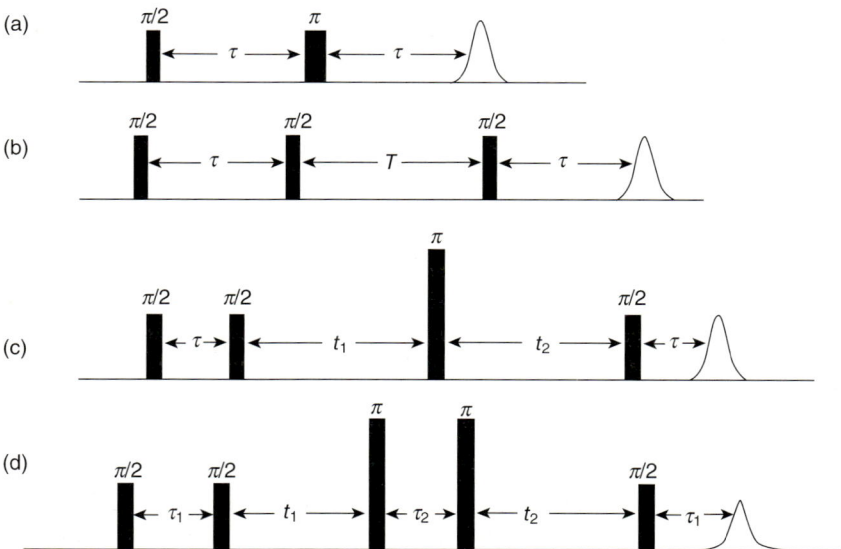

Figure 4.4.15 Pulse sequences making use of the ESEEM effect. (a) Two-pulse Hahn echo sequence with the primary echo; (b) The three-pulse sequence producing stimulated echo; (c) Four-pulse sequence for HYSCORE experiment; (d) Five-pulse sequence for DONUT-HYSCORE. See the text for details. Reprinted with permission from Calle et al. (2006); © 2006, Verlag Helvetica Chimica Acta, Zurich.

In the two-pulse ESEEM experiment (Figure 4.4.15a), the intensity of the two-pulse (primary) echo is monitored as a function of the interval τ of the $\pi/2-\tau-\pi-\tau$ sequence. It can be shown that the amplitude of the echo will be modulated; for an $S = \frac{1}{2}$ and $I = \frac{1}{2}$ system, it is given by

$$V^{2-pulse}(\tau) = 1 - \frac{k(\theta)}{4}[2 - 2\cos(\omega_\alpha \tau) - 2\cos(\omega_\beta \tau) + \cos(\omega_- t) + \cos(\omega_+ t)] \quad (4.4.15)$$

with $\quad k(\theta) = \left(\frac{B\omega_I}{\omega_\alpha \omega_\beta}\right)^2 \quad (4.4.16)$

where $k(\theta)$ is the orientation-dependent modulation depth given in Equation 4.4.16. For isotropic hyperfine coupling and along the principal axis of the hyperfine tensor, the echo modulation is zero, and maximum when $|A(\theta)| = |2\omega_I|$. The two-pulse Hahn echo amplitude is governed by the gross transverse relaxation time T_M, which is much shorter than the true transverse relaxation T_2 and, therefore—as in the case of Mims ENDOR—a three-pulse sequence can be employed (Figure 4.4.15b). In this case, the echo peak amplitude is modulated in terms of the nuclear transition frequencies in the alpha and beta manifolds (ω_α and ω_β; see Equations 4.4.12 and 4.4.13), and are given as a function of τ and T by,

$$V^{3-pulse}(\tau, T) = 1/2[V^\beta(\tau, T) + V^\alpha(\tau, T)] \quad (4.4.17)$$

$$V^{\alpha(\beta)}(\tau, T) = 1 - \frac{k}{2}[1 - \cos(\omega_{\beta(\alpha)} \tau)][1 - \cos(\omega_{\alpha(\beta)}(\tau + T))] \quad (4.4.18)$$

When T is varied, the echo amplitude is modulated as a function of the two frequencies ω_α and ω_β and the sum and difference frequencies ω_+ and ω_- (cf. Equation 4.4.14) do not appear, as is the case for the two-pulse ESEEM (cf. Equation 4.4.16). The three-pulse ESEEM (as in Mims ENDOR) exhibits blind spots at $\omega_{\alpha(\beta)}$ when $\tau = 2\pi n/\omega_{\alpha(\beta)}(n = 1, 2, 3, ...)$. By performing several experiments at different τ-values and coherently summing the results, the blind spots can be removed. In order to generate the nuclear transition frequencies, the value of τ in Equation 4.4.17, and that of T in Equation 4.4.18, are to be systematically incremented, and the echo-envelope modulation is traced point by point. This is, therefore, a very time-consuming experiment.

An alternative approach to the two- and three-pulse ESEEM, as suggested by Hubrich, Hoffmann, and Schweiger (1995), allows the entire envelope modulation to be obtained in a single experiment. In this approach, the last refocusing $\pi/2$ pulse is replaced by a weak, long irradiation for a period of time to cause an extended time excitation (ETE). Each infinitesimal time segment of this irradiation corresponds to a weak pulse that causes a stimulated echo at the corresponding time after the refocusing pulse. The series of these echoes then provides signals which are exactly similar to the two-pulse ESEEM sequence, but in a single experiment.

A variation of the ESEEM sequence involving additional pulses and variable delays leads to 2-D hyperfine selective correlated spectra (HYSCORE), as referred to in the well-known 2-D correlation spectra (e.g., COSY and NOESY) in NMR

(Ernst, Bodenhausen, and Wokaun, 1987). This method was first conducted by Merks and de Beer (1978) and later elaborated by Höfer et al. (Höfer, 1994; Shane et al., 1992). All 2-D correlation experiments in magnetic resonance involve four time periods, namely Preparation; Evolution; Mixing; and Second evolution (Detection):

[Preparation][Evolution......t_1.........][Mixing][2nd Evoluation......t_2....]
[Detection]

(4.4.19)

The spin system is "prepared" by using either a single pulse or a combination of pulses to generate a particular transverse coherence that will "evolve" during the period t_1 under the effective Hamiltonians that are active. As a simplified example, a simple paramagnetic system with $S = ½$ and $I = ½$, under the influence of two $\pi/2$ pulses separated by an interval τ, generates all the coherences corresponding to the electron α-spin manifold and the β-spin manifolds. This corresponds to the highly resolved ENDOR spectrum, which consists of all the echo envelope modulation frequencies and evolves during the period t_1. A π pulse at the end of the t_1 period will interchange the electron spin $\alpha \leftrightarrow \beta$, which corresponds to an instantaneous transformation of a part of all coherences associated with the α spin to that of β spin, and *vice versa*. This so-called "mixing period" is followed by the second evolution period t_2, after which a $\pi/2$ pulse is applied which generates the well-known stimulated echo after a period τ. The periods t_1 and t_2 are systematically incremented, and the echo amplitude is monitored. Because the entire sequence is carried out during a time interval which is much shorter than the electron phase memory time, T_M, all of the frequencies that were present in the two evolution periods t_1 and t_2 continue to persist until generation of the echo. The echo envelope thus becomes doubly modulated as a function of the frequencies during t_1 and those during t_2, similar to the acquired signals in a general 2-D NMR spectroscopic experiment. Upon sequential 2-D FT of the detected signal, a 2-D correlation map is obtained in which the diagonal consists of a highly resolved ENDOR spectrum, while the cross-peaks show coherence transfer (magnetization transfer) between α and β sublevels belonging to the *same* electron. In other words, those coherences which evolve undisturbed right through the *preparation, evolution, mixing*, and *detection* periods have the same frequency fingerprints in both the frequency dimensions (f_1 and f_2), leading to peaks along the diagonal of the 2-D map. In contrast, coherences that are *exchanged* or transferred between spins during the *mixing* period produce cross-peaks (off-diagonal peaks) that are symmetrically disposed on either side of the diagonal at coordinates (f_i, f_j) and (f_j, f_i), corresponding to magnetic coherence transfer between spin i and spin j. This will help clearly to isolate and distinguish the hyperfine subspectra belonging to different electrons in systems containing more than one electron, and also to isolate hyperfine sublevels belonging to different spatially distinct sites in a crystal, which would otherwise produce a highly overlapping conventional 1-D ENDOR spectrum. The data collected will be a function of the systematic incrementation of the time periods t_1 and t_2, the two time-increments deciding the Nyquist frequency bandwidths of the axes of the 2-D spectrum. The detected signal, which is a function of t_1 and t_2, can be represented as (Höfer, Grupp, and Mehring, 1986a; Höfer et al., 1986b):

$$S(t_1, t_2) = \sum_j \left(\sum_k \exp(i\omega_{jk}t_2) \right) \times \left(\sum_l \exp(i\omega_{jl}t_1) \right) \quad (4.4.20)$$

Upon taking the complex double FT, one obtains a 2-D spectrum

$$F(\omega_1, \omega_2) = \sum_j \left(\sum_k f'(\omega_2 - \omega_{jk}) - if''(\omega_2 - \omega_{jk}) \right) \times \left(\sum_k f'(\omega_1 - \omega_{jl}) - if''(\omega_1 - \omega_{jl}) \right) \quad (4.4.21)$$

where

$$f'(\omega - \omega_j) = \frac{\delta_j}{(\omega - \omega_j)^2 + \delta_j^2} \quad \& \quad f''(\omega - \omega_j) = \frac{(\omega - \omega_j)}{(\omega - \omega_j)^2 + \delta_j^2} \quad (4.4.22)$$

The two parts are the real and imaginary part of the complex spectrum correspond to the absorption and dispersion line shapes in a phase-corrected 1-D spectrum. It should be noted that both the real and imaginary part of the 2-D spectrum of Equation 4.4.21 contains absorption and dispersion components, leading to the so-called "twisted" line shape which is well-known in 2-D spectra. Subsequently, one can resort to the usual absolute value mode of presentation:

$$|f| = \sqrt{[(f')^2 + (f'')^2]} \quad (4.4.23)$$

A HYSCORE spectrum arises only when the off-diagonal matrix elements (nonsecular part) of the hyperfine Hamiltonian H_{HF} ($A_{xx}S_xI_x - + - A_{yy}S_yI_y$, etc.) are finite. It helps greatly to identify the ENDOR lines that belong to a particular M_S manifold and as such in unraveling the nuclear modulations, which are determined by hyperfine-interaction constants in a complicated ENDOR spectrum of many spatially distinct sites in a crystal with low (less than a few µT) hyperfine splitting. When the nucleus involved has an electric quadrupole moment, HYSCORE can also help to analyze and quantify the nuclear quadrupole coupling constants. It has also been shown that advanced HYSCORE schemes with "matched pulses" – the so-called "SMART HYSCORE" – lead to the unambiguous analysis of hyperfine coupling from many weakly coupled spins, and can help to unravel correlation information between the different hyperfine tensors and their relative orientations. This leads to HYSCORE being an invaluable tool in the structural determination of large biomolecules with paramagnetic centers, such as metalloenzymes and spin-labeled proteins. More recently, Goldfarb et al. (1998) proposed a 2-D experiment known as "DONUT-HYSCORE" (*double nuclear coherence transfer hyperfine sublevel correlation*) which was designed to obtain correlations between nuclear frequencies belonging to the same electron spin manifold. The sequence employed was $\pi/2 - \tau_1 - \pi/2 - t_1 - \pi - \tau_2 - \pi - t_2 - \pi/2 - \tau_1 -$ echo, and the echo was measured as a function of t_1 and t_2, whereas τ_1 and τ_2 were kept constant. DONUT-HYSCORE is complementary to the standard HYSCORE experiment, which generates correlations between nuclear frequencies belonging to different M_S manifolds of the same electron, and is particularly useful for ^{14}N nuclei. The HYSCORE and DONUT-HYSCORE sequences are also illustrated in Figure 4.4.15.

4.4.9
Electron–Electron Double Resonance (ELDOR), Double Electron–Electron Resonance (DEER), or Pulsed ELDOR (PELDOR)

ELDOR represents an ingenious means of examining magnetization-transfer mechanisms such as Heisenberg spin-exchange, nuclear relaxation, and molecular dynamics such as rotational diffusion, in paramagnetic systems with more than one unpaired electron (Smigel et al., 1974; Kay et al., 2007; Milov et al., 1999; Schosseler, Wacker, and Schweiger, 1994). In this case, a strongly saturating CW MW source (known as the "pump") and a weak, nonsaturating MW source (the "probe") are employed. Alternatively, either two different MW sources, or one source along with an RF modulation of the Zeeman field, can be used. The 2-D manifestation of ELDOR, which is known as either double electron–electron resonance (DEER; Milov, Salikhov, and Shirov, 1981) or pulsed ELDOR (PELDOR; Denysenkov et al., 2006; Hertel et al., 2005), uses two different MW frequencies or a single MW frequency in the so-called "2 + 1" technique (Kurshev, Raitsimring, and Ichikawa, 1991; Kurshev, Raitsimring, and Tsvetkov, 1989) which involves spin-echoes (namely spin-echo ELDOR). DEER allows the quantitative probing of inter-electron dipole–dipole coupling information in the range of 1 to 8 nm, and far surpasses the sensitivity of 2-D nuclear Overhauser enhancement (NOE) spectroscopy (Ernst, Bodenhausen, and Wokaun, 1987), which is limited to distances under 2 nm in high-resolution NMR. Spin-echo ELDOR allows the study of slow motional dynamics, while 2-D FT techniques enable a wide range of combination of pump and probe frequencies with nanosecond timing resolution that allows the study of the microscopic orientation of large biomolecules. Magnetization transfers can also be investigated using multiquantum ELDOR.

The DEER pulse sequence involving two different MW frequencies ω_{mw1} and ω_{mw2} is shown in Figure 4.4.16a (Fedin et al., 2004). The ω_{mw1} pulse train $(\pi/2 - \tau_1 - \pi - \tau_1 + \tau_2 - \pi - \tau_2)$ creates a doubly refocused echo of the spin system, S_1. This coherence experiences a dipolar field from the second electron, S_2, precessing at frequency ω_{mw2} given by $[(3\cos^2\theta - 1) \times g_1 g_2 \mu_B^2 S(+1)/r_{12}^3]$, where θ is the angle between the external field and the vector joining the two electrons and r_{12} is the inter-electron distance. If we can now selectively invert the dipolar field by a weak π pulse at the frequency ω_{mw2} and shift it gradually, the refocused echo peak amplitude will become modulated at a frequency $\mu_B B_{dd}/\hbar$, where B_{dd} is the dipolar field from the electrons precessing at ω_{mw2}. One can then determine the space-dependent dipole–dipole coupling, and hence the distance between the two electrons.

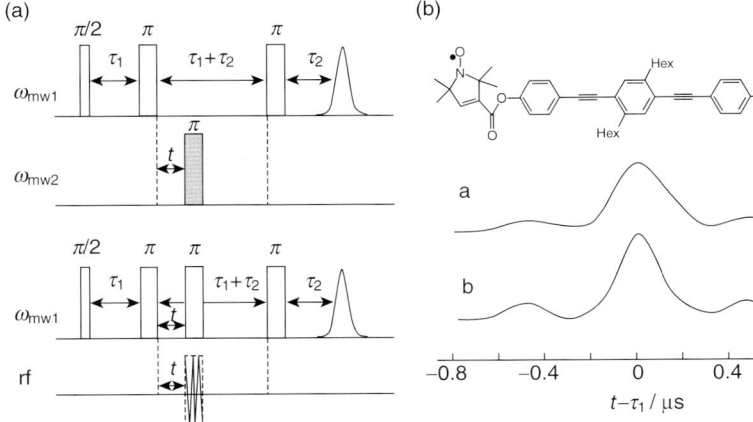

Figure 4.4.16 (a) Standard DEER sequence with two MW frequencies ω_{mw1} and ω_{mw2}; (b) DEER using a "bichromatic" pulse that excites a number of multiphoton transitions at frequencies $\omega_{mw1} \pm n\, \omega_{rf}$; (c) Comparison of four-pulse DEER time-traces obtained with two MW frequencies, and those obtained with bichromatic pulse scheme on the nitroxide radical. Curve a: Standard four-pulse DEER with two MW frequencies with $\Delta\omega/2\pi = 24$ MHz; curve b: Four-pulse DEER with a bichromatic pulse $\omega_{rf} = 24$ MHz. Reprinted with permission from Calle et al. (2006); © 2006, Verlag Helvetica Chimica Acta, Zurich.

It is also possible to replace the pulse at ω_{mw2} by an RF pulse that modulates the Zeeman field with $\omega_{RF} = \omega_{mw1} - \omega_{mw2}$ and shifting the RF pulse gradually, as shown in Figure 4.4.16b. Under such "bichromatic" excitation, a number of multiphoton transitions occur at frequencies $\omega_{mw1} \pm n\omega_{RF}$ ($n = 0, 1, 2, 3, ...$). The effect of the bichromatic pulse can be analyzed by referencing the system to a doubly rotating frame ("toggling frame"), suffice it to say that the echo envelopes obtained by both approaches (with two MW frequencies ω_{mw1} and ω_{mw2}, and one MW frequency ω_{mw1} and $\mathbf{B}_0$ modulated with $\omega_{RF} = (\omega_{mw1} - \omega_{mw2})$ are identical (Figure 4.4.16c). These PELDOR and DEER methods, which collectively are known as pulsed dipolar spectroscopy (PDS), have been developed to isolate the weak space-dependent dipolar effects from electron spin echo (ESE; both Hahn and stimulated) decays, which generally are dominated by relaxation and nuclear hyperfine modulations. Indeed, these techniques have played a crucial role in structural biology through the precise determination of long-range distances between multiple spin-labeled biomolecules, a technique referred to as site-directed spin labeling (SDSL) (McHaourab et al., 1999; Oh et al., 1999; Hubbell and Altenbach, 1994). More recently, EPR-aided distance measurements have proven to be as important as X-ray crystallography and multidimensional NMR, and perhaps even more so in the structural investigation of membrane proteins, which may have molecular weights in excess of 200 kDa (too high for an NMR approach) and are often difficult to crystallize and thus not amenable to X-ray crystallography.

In this context, it is important to mention another novel and very useful approach for investigating intramolecular distances to establish tertiary structures using pulsed EPR. This is known as double quantum coherence (DQC) in site-specific, spin-labeled biomolecules. In fact, PDS and DQC have recently revived the practice of pulsed EPR as an indispensable, reliable and robust structural biology tool (Jeschke, 2002; Berliner, Eaton, and Eaton, 2000; Borbat et al., 2001; Jeschke and Polyhach, 2007; Freed, 2000).

4.4.10
Double-Quantum EPR

Whilst the theory of DQC is discussed in Chapter 12, the following material represents a highly simplified version, and the original articles cited at the end of the chapter should be consulted if greater detail is required. The schematic of the energy levels for a two-electron system relevant in this case, and the various quantum coherences, are shown in Figure 4.4.17. Both, MW and RF pulses, when applied transverse to the Zeeman axis (represented by the Hamiltonian $g\mu_B B_1 I_x$ at the resonance frequency) will tilt the magnetization from the z-axis towards the xy plane, generating transverse components. These correspond to a coherent superposition of levels 1 and 2, 1 and 3, 2 and 4, and 3 and 4 (Figure 4.4.17). These represent the normal "allowed" transitions between Zeeman levels differing in the total z-component of angular momentum ($\Delta M = \Sigma S_{zi}$) by ±1. These are single

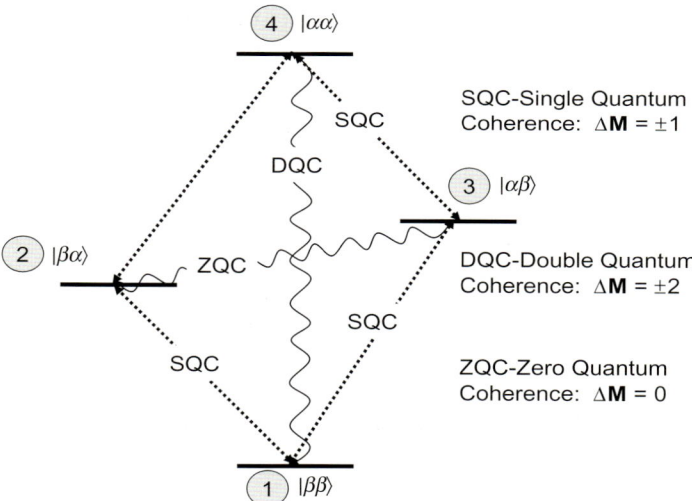

Figure 4.4.17 Schematics of the energy levels for a two-electron system ($S = \frac{1}{2}$) generating four levels. The various transitions involving changes in the total spin quantum number are indicated as single quantum coherence (SQC), double quantum coherence (DQC) and zero-quantum coherence (ZQC) transitions, corresponding to the total change in the total quantum number 1, 2, and 0, respectively.

quantum coherences (SQCs), and are observable. In the density operator formalism, these transverse coherences correspond to the observed signal, and are given by the nonzero traces of $Tr(I_x\sigma)$ or $Tr(I_y\sigma)$. The application of more than one pulse, or transverse magnetization evolving under specific interactions (Hamiltonians), can generate coherences, for example, by connecting levels differing in total $M = \pm 2$ or 0.

These are respectively known as DQCs and zero-quantum coherences (ZQCs). They do not induce a current in the detection coil, although they do have frequencies given by the sum or the differences between the SQCs that involve the connected levels (see Figure 4.4.17). The DCQs undergo a phase shift which is twice that experienced by SQCs, whereas ZQCs are not affected by phase shifts. The DQCs and ZQCs can be measured indirectly by converting them into observable SQCs in a 2-D spectroscopic strategy, where the observed SQCs will exhibit modulations, the FT of which will generate the spectrum corresponding to the directly unobservable coherences. As the various coherences experience phase shifts proportional to their coherence order, it is possible separate them by the so-called "phase-cycling modality" to isolate a particular coherence order, and to eliminate others (Gemperle et al., 1990).

One very important and novel approach to the measurement of distances between various regions of a macromolecule (e.g., proteins, nucleic acids, synthetic polymers) labeled with paramagnetic spin labels is that of EPR DQCs, which involves studying the pure dipolar spectrum obtained by FT of the modulation of stimulated/refocused echoes. The echoes are created exclusively out of DQCs, and involve pairs of electrons in site-directed and selectively doubly spin-labeled biomolecules. These DQCs evolve under the dipolar coupling between pairs of electrons. The dipolar coupling itself is distributed for each pair of electrons, according to the statistically randomly distributed magnitude and orientation (θ) of the inter electron vector (r), and is given by $(3\cos^2\theta - 1) \times [\mu_B^2 S(S+1)/r_{12}^3]$.

For a pair of electrons, the pure DQCs when observed indirectly as modulation on the stimulated echo after conversion to SQC appear as the famous Pake diagram (Pake, 1950). This is characteristic of a pure dipolar spectrum giving two pairs of singularities separated by the "parallel" and "perpendicular" orientation of the inter-electron vector with respect to the magnetic field. Figure 4.4.18 shows four-, five-, and six-pulse sequences of combinations of π and $\pi/2$ pulses with fixed (t_p, $t_m - t_p$) and variable delays (t_1 and t_2) that have been used to generate 2-D spectral data, and the generation of a pure dipolar spectrum The inter-electron distance is derived from the dipolar spectra by using specialized simulation programs involving Tikhonov regularization related to the Levenberg–Marquardt algorithm for nonlinear least-squares problems (Tikhonov and Arsenin, 1997; Jeschke et al., 2002; Chiang, Borbat, and Freed, 2005).

The overall DQC procedure for determining electron–electron distance distributions in double-labeled large molecules is shown schematically in Figure 4.4.19 (Chiang, Borbat, and Freed, 2005). Detailed information on the application of DQC EPR to biomolecular structures is provided in "Pertinent Literature" at the end of the chapter.

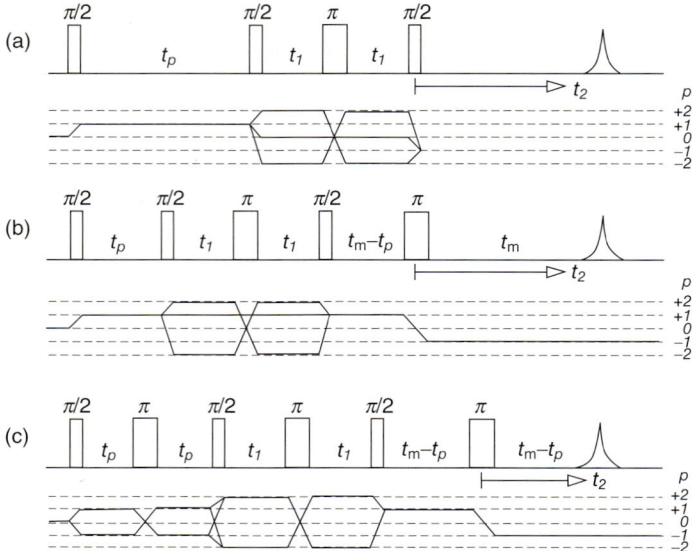

Figure 4.4.18 The four-, five-, and six-pulse DQC sequences used in unraveling long- and short-range distances in large membrane proteins from the pure dipole–dipole modulations of "DQC-filtered" echoes. In each sequence, the pathways of the coherence creation and selection are also indicated. "p" is the coherence order. Reprinted with permission from Borbat, P.P. and Freed, J.H. (2000) Double Quantum ESR and distance measurements, in *Biological Magnetic Resonance*, Vol. 19 (eds L.J. Berliner, S.S. Eaton, and G.R. Eaton), Chapter 9, p. 394; © 2000, Kluwer Academic/Plenum Publishers.

4.4.11
Concluding Remarks

This chapter has provided an overview of pulsed EPR techniques, with emphasis placed mainly on the basic instrumentation and methods involved. An exhaustive coverage of pulsed EPR, with details of mathematical aspects and applications, is clearly beyond the scope of this chapter. However, the literature cited within the text, together with the pertinent reading, will provide more than enough detail. The aim of this chapter was to introduce and highlight the importance of modern pulsed EPR techniques, the relevant instrumentation of which is expensive and currently accessible to very few research groups. Nonetheless, the potential of these methods to elucidate the complex 3-D structures of biomolecules has roused enormous interest. Recently, very high-frequency and time-domain EPR methods have been used to characterize the electron-transfer mechanisms involved in photosynthesis, and indeed much valuable information has been acquired in relation to these phenomena (Lubitz, 2002; Möbius, 2000, 1995; Feher, 1992; Corker,

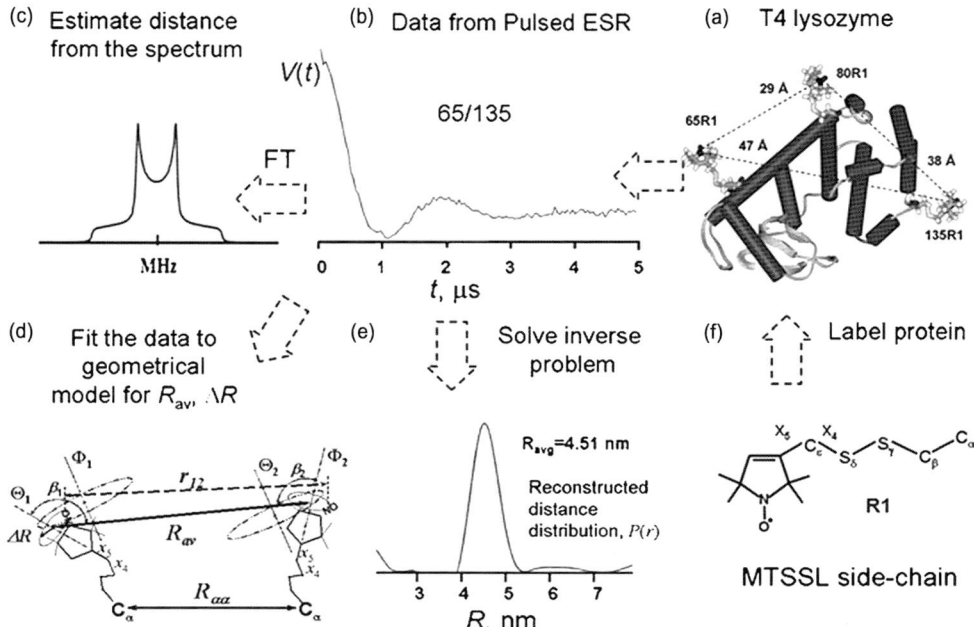

Figure 4.4.19 Distance measurements by pulsed ESR. (a) A cartoon structure of T4 lysozyme with MTSSL; (f) Side-chains shown; (b) Time-domain data $V(t)$ from 17.3 GHz DQC for 65/135 labeling; (C) The dipolar spectrum of panel (b); (d) The geometric model used to fit the data for R_{av} and ΔR; (e) The pair distance distribution reconstructed by solving the corresponding inverse problem using the Tikhonov regularization method. Reprinted with permission from Chiang, Y.-W. et al. (2005) *Journal of Magnetic Resonance*, **172**, 281; © 2005, Elsevier.

1976). The use of pulsed EPR techniques such as ESEEM, ENDOR, ELDOR, and DCQ, in combination with 1-D and 2-D EPR methods has begun to unravel the precise electronic structural and distance information up to a range of 8 nm, and this has proved invaluable to structural molecular biologists. The determination of intra-molecular and inter-segmental distances using these methods has also allowed the elucidation of tertiary and quaternary structures, using distance-constrained energy minimization molecular modeling procedures. Likewise, high-resolution multidimensional NMR spectroscopy continues to develop at a rapid pace, generating vast information on through-bond and through-space connectivity in soluble proteins with masses of up to 200 kDa. With NOE, the maximum through-space distance amenable to quantification is, at most, 2 nm; yet, in selectively labeled macromolecules, unpaired electrons – with an almost three orders of magnitude higher magnetic moment compared to protons – should allow the quantitation of distances of up to 8 nm, notably in water-insoluble (typically membrane) proteins that tend to aggregate (Meiler et al., 2007; Jeschke et al., 2002; Feher, 1992). Clearly, based on these recent applications of high-frequency pulsed techniques, the importance and impact of EPR cannot be overemphasized.

Pertinent Literature

The material presented in this chapter is based on many references, the majority of which are listed below.

T.C. Farrar and E.D. Becker (1971) *Pulse and Fourier transform NMR: Introduction to theory and methods*, Academic Press, New York. This is an early monograph of pulsed EPR theory and applications, and is targeted at "beginners" of the subject.

L. Kevan and M.K. Bowman (eds) (1990) *Modern Pulsed and Continuous-Wave Electron Spin Resonance*. John Wiley & Sons, New York. A source book on both FT and CW-EPR, containing articles on the instrumentation and applications of pulsed EPR, including ESEEM, double quantum 2-D FT-EPR spectroscopy, and pulsed saturation recovery.

A. Schweiger and G. Jeschke (2001) *Principle of Pulse Electron Paramagnetic Resonance*. Oxford University Press, Oxford. A comprehensive treatise dealing with the foundations of pulsed EPR theory and instrumentation. It also provides a thorough exposition of all modern pulse EPR methodologies, such as ENDOR, ELDOR, and ESEEM. This book provides a systematic overview of the entire subject that not only explains the basics and advanced aspects, but also provides sufficient insight to the reader to be able to conduct novel experiments and analyze the results. An excellent treatise indeed.

Other papers mentioned in the reference section include: (i) A.V. Astashkin, J.H. Enemark, and A. Raitsimring (2006) *Conc. Magn. Reson.*, **29B**, 125; (ii) C.E. Davoust, P.E. Doan, and B.M. Hoffman (1996) *J. Magn. Reson. A*, **119**, 38; (iii) J.A.J.M. Disselhorst, H. Vandermeer, O.G. Poluektov, and J. Schmidt (1995) *J. Magn. Reson. Ser. A*, **115**, 183; (iv) I. Gromov, V. Krymov, P. Manikandan, D. Arieli, and D. Goldfarb (1999) *J. Magn. Reson.*, **139**, 8; (v) I. Gromov, J. Shane, J. Forrer, R. Rakhmatoullin, Y. Rozentzwaig, and A. Schweiger (2001) *J. Magn. Reson.*, **149**, 196; (vi) C. Calle, A. Sreekanth, M.V. Fedin, J. Forrer, I. Garcia-Rubio, I.A. Gromov, D. Hinderberger, B. Kasumaj, P. Leger, B. Mancosu, G. Mitrikas, M.G. Santangelo, S. Stoll, A. Schweiger, R. Tschaggelar, and J. Harmer (2006) *Helv. Chim. Acta*, **89**, 2495. The last paper provides a review of most pulsed EPR approaches, the pulse sequences, and their application to molecular structure.

References

Abragam, A. (1961) *The Principles of Nuclear Magnetism*, Clarendon Press, Oxford.

Alecci, M., Brivati, J.A., Placidi, G., and Sotgiu, A. (1998) *J. Magn. Reson.*, **130**, 272.

Astashkin, A.V., Enemark, J.H., and Raitsimring, A. (2006) *Conc. Magn. Reson.*, **29B**, 125.

Bennati, M., Farrar, C.T., Bryant, J.A., Inati, S.J., Weis, V., Gerfen, G.J., Riggs-Gelasco, P., Stubbe, J., and Griffin, R.G. (1999) *J. Magn. Reson.*, **138**, 232.

Berliner, L.J. (ed.) (2003) *In Vivo EPR (ESR)*, Kluwer Academic, New York.

Berliner, L.J., Eaton, G.R., and Eaton, S.S. (2000) *Distance Measurements in Biological Systems by EPR*, Kluwer Academic, New York.

Bloch, F., Hansen, W.W., and Packard, H.E. (1946) *Phys. Rev.*, **69**, 127.

Blumberg, W.E., Zuckerman, D.M., and Mims, W.B. (1972) *Bull. Am. Phys. Soc.*, **17**, 149.

Blumberg, W.E., Mims, W.B., and Zuckerman, D.M. (1973) *Rev. Sci. Instrum.*, **44**, 546.

Borbat, P.P., Costa-Filho, A.J., Earle, K.A., Moscicki, J.K., and Freed, J.H. (2001) *Science*, **291**, 266.

Bourg, J., Krishna, M.C., Mitchell, J.B., Tschudin, R.G., Pohida, T.J., Friauf, W.S., Smith, P.D., Metcalfe, J., Harrington, F., and Subramanian, S. (1993) *J. Magn. Reson. B*, **102**, 112.

Brivati, J.A., Stevens, A.D., and Symons, M.C.R. (1991) *J. Magn. Reson.*, **92**, 480.

Bruker (2009) E580 FT/CW System, http://www.bruker-biospin.com/elexsys_e580.html.

Burghaus, O., Rohrer, M., Gotzinger, T., Plato, M., and Möbius, K. (1992) *Meas. Sci. Technol.*, **3**, 765.

Calle, C., Sreekanth, A., Fedin, M.V., Forrer, J., Garcia-Rubio, I., Gromov, I.A., Hinderberger, D., Kasumaj, B., Leger, P., Mancosu, B., Mitrikas, G., Santangelo, M.G., Stoll, S., Schweiger, A., Tschaggelar, R., and Harmer, J. (2006) *Helv. Chim. Acta*, **89**, 2495.

Carmieli, R., Papo, N., Zimmermann, H., Potapov, A., Shai, Y., and Goldfarb, D. (2006) *Biophys. J.*, **90**, 492.

Carr, H.Y. and Purcell, E.M. (1954) *Phys. Rev.*, **94**, 630.

Chiang, Y.-W., Borbat, P.P., and Freed, J.H. (2005) *J. Magn. Reson.*, **172**, 279.

Colacicchi, S., Alecci, M., Gualtieri, G., Quaresima, V., Ursini, C.L., Ferrari, M., and Sotgiu, A. (1993) *J. Chem. Soc., Perkins Trans. 2*, (11), 2077.

Corker, G.A. (1976) *Photochem. Photobiol.*, **24**, 617.

Davies, E.R. (1974) *Phys. Lett. A*, **A47**, 1.

Davoust, C.E., Doan, P.E., and Hoffman, B.M. (1996) *J. Magn. Reson. A*, **119**, 38.

Denysenkov, V.P., Prisner, T.F., Stubbe, J., and Bennati, M. (2006) *Proc. Natl Acad. Sci. USA*, **103**, 13386.

Devasahayam, N., Subramanian, S., Murugesan, R., Cook, J.A., Afeworki, M., Tschudin, R.G., Mitchell, J.B., and Krishna, M.C. (2000) *J. Magn. Reson.*, **142**, 168.

Devasahayam, N., Subramanian, S., Murugesan, R., Hyodo, F., Matsumoto, K.I., Mitchell, J.B., and Krishna, M.C. (2007) *Magn. Reson. Med.*, **57**, 776.

Devasahayam, N., Subramanian, S., and Krishna, M.C. (2008) *Rev. Sci. Instrum.*, **79**, 026106.

Disselhorst, J.A.J.M., Vandermeer, H., Poluektov, O.G., and Schmidt, J. (1995) *J. Magn. Reson. Ser. A*, **115**, 183.

Dykstra, R.W. and Mrarkham, G.D. (1986) *J. Magn. Reson.*, **69**, 350.

Earle, K.A., Tipikin, D.S., and Freed, J.H. (1996) *Rev. Sci. Instrum.*, **67**, 2502.

Eaton, G.R., Eaton, S.S., and Rinard, G.A. (1998) Frequency dependence of EPR sensitivity, in *Spatially Resolved Magnetic Resonance* (eds P. Blümler, B. Blümich, R. Botto, and E. Fukushima), Wiley-VCH Verlag GmbH, Weinheim, pp. 65–74.

Ernst, R.R. and Anderson, J.R. (1966) *Rev. Sci. Instrum.*, **37**, 93.

Ernst, R.R., Bodenhausen, G., and Wokaun, A. (1987) *Principles of Nuclear Magnetic Resonance in One and Two Dimensions*, Oxford University Press, New York.

Farrar, T.C. and Becker, E.D. (1971) *Pulse and Fourier Transform NMR: Introduction to Theory and Methods*, Academic Press, New York.

Fedin, M., Kalin, M., Gromov, I., and Schweiger, A. (2004) *J. Chem. Phys.*, **120**, 1361.

Feher, G. (1956) *Phys. Rev.*, **103**, 834.

Feher, G. (1992) *J. Chem. Soc., Perkins Trans. 2*, (11), 1861.

Feher, G., Fuller, C.S., and Gere, E.A. (1957) *Phys. Rev.*, **107**, 1462.

Fellgett, P.B. (1949a) *Proc. Phys. Soc. London A*, **62**, 399.

Fellgett, P.B. (1949b) *J. Opt. Soc. Am.*, **39**, 970.

Freed, J.H. (2000) *Annu. Rev. Phys. Chem.*, **51**, 655.

Froncisz, W. and Hyde, J.S. (1982) *J. Magn. Reson.*, **47**, 515.

Froncisz, W., Oles, T., and Hyde, J.S. (1989) *J. Magn. Reson.*, **82**, 109.

Gallez, B., Bacic, G., Goda, F., Jiang, J.J., O'Hara, J.A., Dunn, J.F., and Swartz, H.M. (1996a) *Magn. Reson. Med.*, **35**, 97.

Gallez, B., Debuyst, R., Liu, K.J., Demeure, R., Dejehet, F., and Swartz, H.M. (1996b) *Magn. Reson. Mater. Phys.*, **4**, 71.

Gemperle, C., Aebli, G., Schweiger, A., and Ernst, R.R. (1990) *J. Magn. Reson.*, **88**, 241.

Goldfarb, D., Kofman, V., Libman, J., Shanzer, A., Rahmatouline, R., Doorslaer, S.V., and Schweiger, A. (1998) *J. Am. Chem. Soc.*, **120**, 7020.

Golovina, I., Geifman, I., and Belous, A. (2008) *J. Magn. Reson.*, **195**, 52.

Gromov, I., Krymov, V., Manikandan, P., Arieli, D., and Goldfarb, D. (1999) *J. Magn. Reson.*, **139**, 8.

Gromov, I., Shane, J., Forrer, J., Rakhmatoullin, R., Rozentzwaig, Y., and Schweiger, A. (2001) *J. Magn. Reson.*, **149**, 196.

Hahn, E.L. (1950) *Phys. Rev.*, **80**, 580.

Halpern, H.J. and Bowman, M.K. (eds) (1991) *Low Frequency EPR Spectrometers: MHZ Range*, CRC Press, Boca Raton.

Hertel, M.M., Denysenkov, V.P., Bennati, M., and Prisner, T.F. (2005) *Magn. Reson. Chem.*, **43**, S248.

Hirata, H., Ueda, M., Ono, N., and Shimoyama, Y. (2002) *J. Magn. Reson.*, **155**, 140.

Höfer, P. (1994) *J. Magn. Reson. A*, **111**, 77.

Höfer, P., Grupp, A., and Mehring, M. (1986a) *Phys. Rev. A*, **33**, 3519.

Höfer, P., Grupp, A., Nebenfuhr, H., and Mehring, M. (1986b) *Chem. Phys. Lett.*, **132**, 279.

Hubbell, W.L. and Altenbach, C. (1994) *Curr. Opin. Struct. Biol.*, **4**, 566.

Hubrich, M., Hoffmann, E.C., and Schweiger, A. (1995) *J. Magn. Reson. A*, **114**, 271.

Hyde, J.S., Froncisz, W., and Oles, T. (1989) *J. Magn. Reson.*, **82**, 223.

Ionita, G., Florent, M., Goldfarb, D., and Chechik, V. (2009) *J. Phys. Chem. B*, **113**, 5781.

Jerschow, A. and Muller, N. (1998) *J. Magn. Reson.*, **134**, 17.

Jeschke, G. (2002) *Chem. Phys. Chem.*, **3**, 927.

Jeschke, G., Koch, A., Jonas, U., and Godt, A. (2002) *J. Magn. Reson.*, **155**, 72.

Jeschke, G. and Polyhach, Y. (2007) *Phys. Chem. Chem. Phys.*, **9**, 1895.

Kang, P.C., Eaton, G.R., and Eaton, S.S. (1994) *Inorg. Chem.*, **33**, 3660.

Kay, C.W.M., Mkami, H.E., Cammack, R., and Evans, R.W. (2007) *J. Am. Chem. Soc.*, **129**, 4868.

Kevan, L. and Bowman, M.K. (eds) (1990) *Modern Pulsed and Continuous-Wave Electron Spin Resonance*, John Wiley & Sons, Inc., New York.

Koptioug, A.V., Reijerse, E.J., and Klaassen, A.K. (1997) *J. Magn. Reson.*, **125**, 369.

Kuppusamy, P., Afeworki, M., Shankar, R.A., Coffin, D., Krishna, M.C., Hahn, S.M., Mitchell, J.B., and Zweier, J.L. (1998) *Cancer Res.*, **58**, 1562.

Kuppusamy, P., Chzhan, M., Vij, K., Shteynbuk, M., Lefer, D.J., Giannella, E., and Zweier, J.L. (1994) *Proc. Natl Acad. Sci. USA*, **91**, 3388.

Kuppusamy, P., Chzhan, M., and Zweier, J.L. (1995) *J. Magn. Reson. Ser. B*, **106**, 122.

Kuppusamy, P., Li, H.Q., Ilangovan, G., Cardounel, A.J., Zweier, J.L., Yamada, K., Krishna, M.C., and Mitchell, J.B. (2002) *Cancer Res.*, **62**, 307.

Kurshev, V.V., Raitsimring, A.M., and Tsvetkov, Y.D. (1989) *J. Magn. Reson.*, **81**, 441.

Kurshev, V.V., Raitsimring, A.M., and Ichikawa, T. (1991) *J. Phys. Chem.*, **95**, 3564.

Lowe, I.J. and Norberg, R.E. (1957) *Phys. Rev.*, **107**, 46.

Lubitz, W. (2002) *Phys. Chem. Chem. Phys.*, **4**, 5539.

Lynch, W.B., Earle, K.A., and Freed, J.H. (1988) *Rev. Sci. Instrum.*, **59**, 1345.

McHaourab, H.S., Kalai, T., Hideg, K., and Hubbell, W.L. (1999) *Biochemistry*, **38**, 2947.

Meiboom, S. and Gill, D. (1958) *Rev. Sci. Instrum.*, **29**, 688.

Meiler, J., Bortolus, M., Al-Mestarihi, A., and McHaourab, H. (2007) *Biophys. J.*, Suppl. S, 648A.

Merks, R.P.J. and de Beer, R. (1978) *J. Phys. C*, **11**, L673.

Milov, A.D., Salikhov, K.M., and Shirov, M.D. (1981) *Fiz. Tverd. Tela.*, **23**, 975.

Milov, A.D., Maryasov, A.G., Tsvetkov, Y.D., and Raap, J. (1999) *Chem. Phys. Lett.*, **303**, 135.

Mims, W.B. (1965) *Proc. R. Soc. London A*, **283**, 452.

Mims, W.B. (1972a) *Phys. Rev. B*, **6**, 3543.

Mims, W.B. (1972b) *Phys. Rev. B*, **5**, 2409.

Mitchell, J.B., Russo, A., Kuppusamy, P., and Krishna, M.C. (2000) *Ann. N. Y. Acad. Sci.*, **899**, 28.

Möbius, K. (1995) *Appl. Magn. Reson.*, **9**, 389.

Möbius, K. (2000) *Chem. Soc. Rev.*, **29**, 129.

Murugesan, R., Afeworki, M., Cook, J.A., Devasahayam, N., Tschudin, R., Mitchell, J.B., Subramanian, S., and Krishna, M.C. (1998) *Rev. Sci. Instrum.*, **69**, 1869.

Oh, K.J., Zhan, H.J., Cui, C., Altenbach, C., Hubbell, W.L., and Collier, R.J. (1999) *Biochemistry*, **38**, 10336.

Ono, M., Ogata, T., Hsieh, K.C., Suzuki, M., Yoshida, E., and Kamada, H. (1986) *Chem. Lett.*, (11), 491.

Pake, G.E. (1950) *Am. J. Phys.*, **18**, 473.

Pfenninger, S., Forrer, J., Schweiger, A., and Weiland, T. (1988) *Rev. Sci. Instrum.*, **59**, 752.

Pfenninger, S., Froncisz, W., Forrer, J., Luglio, J., and Hyde, J.S. (1995) *Rev. Sci. Instrum.*, **66**, 4857.

Polk, C. and Postow, E. (1996) *Hand Book of Biological Effects Of Electromagnetic Fields*, 2nd edn, CRC Press, Boca Raton.

Prisner, T.F., Rohrer, M., and Möbius, K. (1994) *Appl. Magn. Reson.*, **7**, 167.

Quine, R.W., Rinard, G.A., Ghim, B.T., Eaton, S.S., and Eaton, G.R. (1996) *Rev. Sci. Instrum.*, **67**, 2514.

Rinard, G.A., Quine, R.W., Eaton, S.S., and Eaton, G.R. (1993) *J. Magn. Reson. Ser. A*, **105**, 137.

Roesenbaum, F.J. (1964) *Rev. Sci. Instrum.*, **35**, 1550.

Rohrer, M., Brugmann, O., Kinzer, B., and Prisner, T.F. (2001) *Appl. Magn. Reson.*, **21**, 257.

Rowan, L.G., Hahn, E.L., and Mims, W.B. (1965) *Phys. Rev.*, **137**, A61.

Sakamoto, Y., Hirata, H., and Ono, M. (1995) *IEEE Trans. Microw. Theory Tech.*, **43**, 1840.

Schmalbein, D., Maresch, G.G., Kamlowski, A., and Höfer, P. (1999) *Appl. Magn. Reson.*, **16**, 185.

Schosseler, P., Wacker, T., and Schweiger, A. (1994) *Chem. Phys. Lett.*, **224**, 319.

Schweiger, A. (1995) *J. Chem. Soc. Faraday Trans.*, **91**, 177.

Schweiger, A. and Jeschke, G. (2001) *Principle of Pulse Electron Paramagnetic Resonance*, Oxford University Press, Oxford.

Shane, J.J., Höfer, P., Reijerse, E.J., and Deboer, E. (1992) *J. Magn. Reson.*, **99**, 596.

Shaw, D. (1984) *Fourier Transform NMR Spectroscopy*, Elsevier, New York.

Smigel, M.D., Dalton, L.R., Hyde, J.S., and Dalton, L.A. (1974) *Proc. Natl Acad. Sci. USA*, **71**, 1925.

Stoll, S. and Kasumaj, B. (2008) *Appl. Magn. Reson.*, **35**, 15.

Subramanian, S., Murugesan, R., Devasahayam, N., Cook, J.A., Afeworki, M., Pohida, T., Tschudin, R.G., Mitchell, J.B., and Krishna, M.C. (1999) *J. Magn. Reson.*, **137**, 379.

Subramanian, S., Mitchell, J.B., and Krishna, M.C. (2003) Time-domain radio frequency EPR imaging, in *Biological Magnetic Resonance* (ed. L.J. Berliner), Kluwer Academic, New York, p. 153.

Subramanian, S., Yamada, K., Irie, A., Murugesan, R., Cook, J.A., Devasahayam, N., Van Dam, G.M., Mitchell, J.B., and Krishna, M.C. (2002) *Magn. Reson. Med.*, **47**, 1001.

Swartz, H.M. and Walczak, T. (1998) *Adv. Exp. Med. Biol.*, **454**, 243.

Tcach, I., Rogulis, U., Greulich-Weber, S., and Spaeth, J.-M. (2004) *Rev. Sci. Instrum.*, **75**, 4781.

Tikhonov, A.N. and Arsenin, V.Y. (1997) *Solutions of Ill-Posed Problems*, Halsted Press (John Wiley & Sons, Inc.), New York.

Velan, S.S., Spencer, R.G.S., Zweier, J.L., and Kuppusamy, P. (2000) *Magn. Reson. Med.*, **43**, 804.

Walczak, T., Lesniewski, P., Salikhov, I., Sucheta, A., Szybinski, K., and Swartz, H.M. (2005) *Rev. Sci. Instrum.*, **76**, 13107.

Walsh, W.M., Jr and Rupp, L.M., Jr (1986) *Rev. Sci. Instrum.*, **57**, 2278.

Willer, M., Forrer, J., Keller, J., Van Doorslaer, S., Schweiger, A., Schuhmann, R., and Weiland, T. (2000) *Rev. Sci. Instrum.*, **71**, 2807.

Yamada, K.I., Kuppusamy, P., English, S., Yoo, J., Irie, A., Subramanian, S., Mitchell, J.B., and Krishna, M.C. (2002) *Acta Radiol.*, **43**, 433.

Zweier, J.L., Chzhan, M., Samouilov, A., and Kuppusamy, P. (1998) *Phys. Med. Biol.*, **43**, 1823.

Zweier, J.L., Thompsongorman, S., and Kuppusamy, P. (1991) *J. Bioenerg. Biomembr.*, **23**, 855.

5
Multifrequency EPR: Experimental Considerations

5.1
Multiarm EPR Spectroscopy at Multiple Microwave Frequencies: Multiquantum (MQ) EPR, MQ-ELDOR, Saturation Recovery (SR) EPR, and SR-ELDOR
James S. Hyde, Robert A. Strangeway, and Theodore G. Camenisch

5.1.1
Introduction

This chapter 5.1 outlines the technology involved in the irradiation of an electron paramagnetic resonance (EPR) sample with more than one microwave frequency, where the frequencies are derived from a single microwave oscillator and a phase-coherent synthesizer array. Also considered are the microwave resonator geometries that will support more than one frequency, and which fall into two classes: bimodal and broadband. If multiple frequencies are incident on the sample, then EPR signals from each frequency can be anticipated, and intermodulation sidebands produced by the sample may also be present.

If the different incident frequencies are derived from a single source, they are said to be "coherent." The key report in the development of this technology was that of Strangeway *et al.* (1995), in a context of multiquantum (MQ) EPR. These studies have subsequently been extended and developed in various ways, and are discussed in this chapter. Consideration is given not only to MQ-EPR but also to three other experiments: (i) MQ electron–electron double resonance (ELDOR) (Mchaourab *et al.*, 1991); (ii) saturation recovery (SR) EPR (Huisjen and Hyde, 1974; Percival and Hyde, 1975); and (iii) SR-ELDOR (Hyde, Froncisz, and Mottley, 1984). The apparatus and results are described at 35 GHz (Q-band) (Camenisch *et al.*, 2004; Klug *et al.*, 2005; Hyde *et al.*, 2004) and 94 GHz (W-band) (Froncisz *et al.*, 2008; Hyde *et al.*, 2007); moreover, an approach is under development to extend the technology to 140 GHz (D-band, from 110 to 170 GHz). The details of these studies are presented in the general context of nitroxide-radical spin-label aqueous samples in the room-temperature range.

Each frequency is created in a separate microwave arm, as shown in the generic multiarm bridge in Figure 5.1.1. Each arm frequency is produced by mixing the microwave oscillator output with an output from the synthesizer array. For

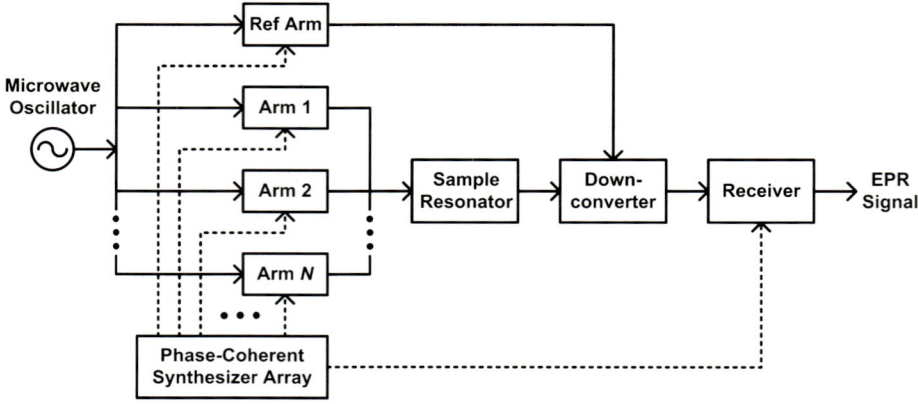

Figure 5.1.1 Generic block diagram of a multiarm, multiple-frequency EPR bridge.

higher-frequency, millimeter-wave EPR bridges, the microwave frequency in each arm is translated again, either by frequency-mixing or multiplication. Throughout the frequency-creation process each arm remains independent to avoid the formation of intermodulation products during power amplification. The microwave frequencies created in arms 1 to N are combined and directed to the sample resonator; the combining process occurs at the transmission line that leads to the sample resonator to preserve signal purity. The EPR signal from the sample resonator is then translated in the downconverter to a frequency range that is suitable for the receiver which, in turn, performs the EPR signal detection process. Direct digital detection methods can be employed in the receiver (Hyde *et al.*, 1998; Froncisz *et al.*, 2001; Jesmanowicz and Hyde, 2006). The same receiver is used regardless of the sample-irradiation frequency band, because the downconverter in any particular bridge is designed to translate the EPR signal to the frequency range of the receiver.

A key aspect in the utilization of synthesizer technology is that of *phase coherency*, which means that the phase relationship is maintained between the different frequencies. This definition of phase coherency will be maintained throughout this chapter 5.1. If the instantaneous phase of one frequency is known, then the instantaneous phases of all the other arm frequencies are predictable. Traditionally, a common time base (clock) has been utilized with multiple synthesizers to achieve quasi-phase-coherent signals. Although the phase relationship is not always absolute because phase drifts are known to occur between synthesizer units, other synthesizer technologies are now available to achieve an improved phase coherency (see Section 5.1.3).

An enabling technology for the studies discussed here is the development of the loop–gap resonator (LGR) at Q- and W-band (Hyde *et al.*, 2004; Froncisz, Oles, and Hyde, 1986; Sidabras *et al.*, 2007), with extension to D-band feasible (this subject is discussed in Chapter 5.2). Briefly, these structures have a sensitivity that is similar to cavity resonators, but with lower Q-values. This high bandwidth permits irradiation of the sample with multiple microwave frequencies, with minimal reflection because of off-resonance irradiation. The potential loss of sensitivity due to the low Q-value is compensated by a high resonator efficiency parameter, Λ, which is a further benefit in that less irradiating microwave power is required (see Chapter 5.2). One ramification of LGR use is the requirement that the microwave bridge be tunable over an adequate frequency range to accommodate the LGR. The LGR frequency is generally not tunable, and the resonant frequency changes with different sample tubes and sample types, especially at millimeter-wave frequencies.

Many of the research groups in this field would consider the methods described in this chapter to be correctly classified as variations of continuous-wave (CW) EPR spectroscopy. It has been suggested, however, that this methodology is extremely general, and encompasses all possible EPR experiments. Rather, the irradiation pattern can be considered in either the frequency domain or the time domain through the Fourier transform, with any frequency arm being either pulsed or modulated in an endless variety of ways. The explicit availability of several frequency arms addresses a significant problem in EPR spectroscopy that does not arise in nuclear magnetic resonance (NMR); namely, the huge microwave power required to deliver a hard pulse to the entire spectrum.

5.1.2
Review of Frequency-Translation Techniques

In order to generate multiple, phase-coherent frequencies, frequency *translation* can be utilized. When the input frequency is translated to a higher frequency, such a translation is termed "upconversion"; when translated to a lower frequency, the term "downconversion" is used. The three most common methods to have been used for frequency translation in EPR bridges are frequency mixing, single-sideband (SSB) modulation, and frequency multiplication (Figure 5.1.2).

In frequency mixing, a mixer is used to translate frequencies, either in upconversion or downconversion. Two frequencies are applied to the mixer as inputs – designated here as the input frequency f_{in} and the local oscillator (LO) frequency f_{LO}. The mixer primarily generates the sum and the difference frequencies. If $f_{in} < f_{LO}$, the sum frequency ($f_{LO} + f_{in}$) is called the upper sideband (USB),

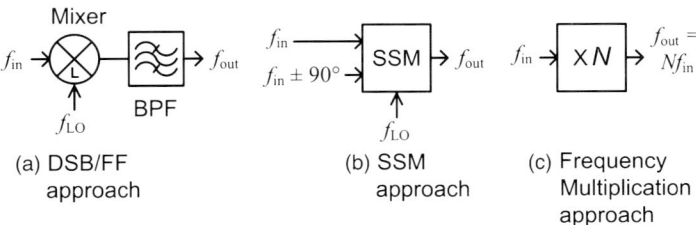

Figure 5.1.2 Frequency-translation approaches.

and the difference frequency ($f_{LO} - f_{in}$) the lower sideband (LSB). Thus, the output is double-sideband. Remnants (leakage) of the LO and input signal will appear in the mixer output. The nonlinearity of the mixer also generates other frequencies ($nf_{LO} \pm mf_{in}$, where n and m are integers) called "spurious products" from the input frequencies. A bandpass filter (BPF) is normally used to complete the frequency-translation process by passing the desired sideband and suppressing the undesired sideband (the "image"), residual leakage of the input and the LO signals, and spurious products. This technique has been termed the double-sideband/fixed-filter (DSB/FF) approach (Strangeway et al., 1995). The filter is "fixed" because it is not tuned; in other words, the range of frequencies (bandwidth) that it passes is constant and, as a consequence, the output spectrum is as clean as the selectivity of the filtering. The undesired frequencies were suppressed 65 dB or more in a Q-band multiarm bridge (Camenisch et al., 2004).

The single-sideband modulator (SSM) contains two mixers configured to generate a single output frequency (sideband) from three input sinusoids, without the need for subsequent bandpass filtering. The SSM requires an LO input and two lower frequency inputs that are 90° apart in phase (quadrature-phase); the resultant sideband frequency at the output is either ($f_{LO} + f_{in}$) or ($f_{LO} - f_{in}$), depending on the phasing of the quadrature inputs. The SSM is particularly useful for generating frequencies close to f_{LO}, where bandpass filtering is not practical. The SSM typically offers a 20–30 dB suppression of the undesired frequencies relative to the desired sideband output, although the suppression can be tuned up to 60 dB. The resultant suppression is, however, highly dependent on f_{LO}, the match between the mixer characteristics inside the SSM, temperature, and also the amplitudes and phases of the lower frequency inputs. Thus, the use of the SSM is problematic when frequencies need to be changed.

Frequency multipliers literally multiply the input frequency by an integer value (N) of two, or more. Whilst this approach is common for the translation of microwave frequencies to millimeter-wave frequencies, multipliers are not suitable for achieving noninteger frequency-translation ratios (details of frequency multiplication are provided in Section 5.1.4, in the context of millimeter-wave frequencies).

Various methods can be utilized to generate multiple closely spaced, phase-coherent microwave frequencies, and some of these have been surveyed by Strangeway et al. (1995). The practical implementations of most methods include limited frequency tuning ranges and/or limited long-term stability. The use of synthesizers to generate arm frequencies directly results in degraded phase-noise performance at higher microwave frequencies, and especially at millimeter-wave frequencies. The DSB/FF translation approach has been successfully implemented in multiarm EPR bridges in the configuration shown in Figure 5.1.1 (Camenisch et al., 2004; Hyde et al., 2007).

In the DSB/FF EPR multiarm bridge configuration, each arm consists of a mixer, BPF, and a power amplifier (PA), as depicted in Figure 5.1.3. The microwave oscillator feeds the LO port of the translation mixer; one output from the array of phase-coherent synthesizers is used as the mixer input, and the subsequent filter selects the desired sideband. The microwave oscillator is adjustable to

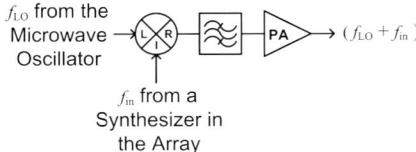

Figure 5.1.3 One arm in the DSB/FF frequency-translation approach.

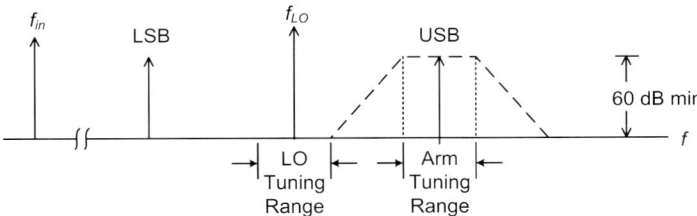

Figure 5.1.4 DSB/FF signals and filtering.

align the arm frequency to the sample resonator frequency. The nominal f_{in} value to the translation mixer must be large enough for bandpass filtering, and also must be greater than the final output frequency tuning range of the arm, plus an additional frequency range to accommodate the finite frequency selectivity characteristics of the filter (the filter "skirt"), as shown by the dashed line in Figure 5.1.4. The PA then increases the output power of the arm to the required level. This single-frequency-per-arm approach minimizes the generation of spurious signals in the power amplification process. The arm topology is replicated for as many frequencies as are required to irradiate the sample, plus one more for the reference arm (see Figure 5.1.1).

5.1.3
Review of Multiarm Bridges

A functional schematic of a multiarm DSB/FF EPR bridge that has been utilized at X- and Q-band (Strangeway et al., 1995; Camenisch et al., 2004; Klug et al., 2005; Hyde et al., 2004) is shown in Figure 5.1.5. In this case, three arms are configured for sample irradiation, while the fourth arm functions as the reference arm. The power from the microwave source, a YIG (yttrium-iron-garnet) oscillator (X-band) or Gunn diode oscillator (Q-band), is split to feed the LO ports of the translation mixers in each arm.

The *I* port of each mixer is driven by a synthesizer, which is used to generate a desired signal from a given reference signal. The latter is usually supplied by a stable oscillator (called a "reference" or a "clock") at 10 MHz, with various methods being used for this purpose. Indirect frequency synthesizers utilize multiple phase-lock-loops (PLLs) to generate the output signal. Whilst these offer superior phase-noise performance and fine frequency resolution, they generally exhibit

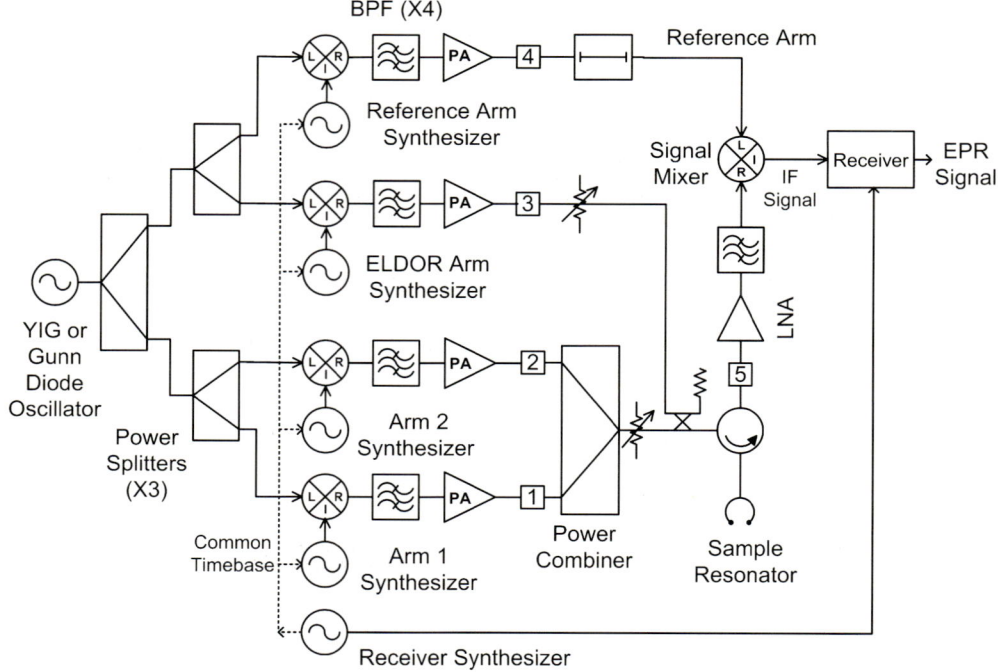

Figure 5.1.5 Multiarm DSB/FF EPR bridge configuration.

some degree of phase drift from the ideal nominal phase (quasi-phase coherency), and also exhibit slow frequency switching speeds. Direct digital synthesizers (DDSs) play back a digital version of the desired signal (usually sinusoidal) through a digital-to-analog converter (DAC), such that the phase of the signal is precisely controlled. Although the DDS frequency switching speeds can be fast, the DDSs have an output frequency which is limited by the internal digital circuits and the DAC. Moreover, they may also generate elevated spurious signals. Arbitrary waveform generators (AWGs) also play back a digital version of a desired signal through a DAC, but the signal is generally not sinusoidal and hence is frequency-limited relative to the sinusoidal DDS. AWGs are often utilized to modulate other synthesizers, employing various combinations of synthesizer technology.

The synthesizer array is formed when individual synthesizers share a common time base, which usually is one of the synthesizer clocks. The common clock establishes (quasi) phase coherency between the arm frequencies. Each synthesizer frequency is nominally the same, but is offset from the others; the synthesizer frequencies are set to obtain the desired frequency differences between the arms in a convenient, agile, and accurate manner.

In these bridges, the BPF selects the upper sideband so that a lower-frequency microwave source is required. The phase noise is generally improved for lower-frequency microwave sources, and the PAs elevate the arm powers to sufficient levels. (The numbered boxes within the arms represent microwave power connections to the W- and D-band assemblies described later in this chapter; see Figure 5.1.5.) All amplified arm outputs, except for the reference arm, are combined and directed to the sample resonator.

The microwave EPR signal from the sample resonator is directed to the low-noise amplifier (LNA), which establishes the receiver noise figure. The EPR signal must be downconverted for signal detection, and this signal is one of the inputs to the signal mixer. The reference arm output drives the LO port of the signal mixer and is phase-coherent with the EPR signal because the same frequency-generation process and common microwave oscillator are utilized in all arms. The lower sideband of the signal mixer output is selected in the downconversion of the microwave EPR signal to a convenient intermediate frequency (IF). The downconverted EPR signal is then detected in the receiver, such as an analog-to-digital (A/D) converter that directly samples the IF (Hyde et al., 1998; Froncisz et al., 2001; Jesmanowicz and Hyde, 2006). Another output from the synthesizer array (receiver synthesizer) is used in the receiver to maintain phase coherency in the signal detection process.

The X- and Q-band bridge configuration in Figure 5.1.5 utilizes only one frequency translation to generate the microwave frequencies that irradiate the sample. In the Bruker ELEXSYS W-band EPR bridge, only one frequency-translation stage is applied from X- to W-band (Hofer et al., 2004), although the use of an SSB upconverter for frequency translation in a W-band electron spin-echo bridge has been described by Weber et al. (1989) and Allgeier et al. (1990). Frequency multiplication to W-band has also been utilized (Gromov et al., 1999; Goldfarb et al., 2008). Multiple frequency translations can also be used to reach higher frequencies (Hofbauer et al., 2004). Frequency translation from Q- to W-band has been implemented in both CW- and pulsed EPR experiments in the authors' laboratory (Froncisz et al., 2008; Hyde et al., 2007). Other approaches to multiarm EPR bridges at higher millimeter-wave frequencies are possible (Grinberg and Berliner, 2004).

A functional block diagram of a circuit that translates the Q-band arms (at the numbered connection points in Figure 5.1.5) to W-band is shown in Figure 5.1.6 (note that one of the Q-band arms is not utilized here). Q-band arms 1 and 3 (in Figure 5.1.5) are translated to W-band by mixing with a nominal 59 GHz LO and selecting the upper sideband. Arm 1 has been utilized as the pump in SR and SR ELDOR experiments by gating either the Q-band power or the synthesizer in arm 1 (see Figure 5.1.5). The receiver protect switch is utilized in pulsed W-band experiments; otherwise, it is replaced by a waveguide of the same length. The W-band LNA primarily establishes the receiver noise figure. The EPR signal is downconverted to Q-band by mixing with the same 59 GHz LO (to ensure phase coherency) and filtering, while the Q-band signal is downconverted to an IF carrier frequency of 1 GHz for EPR detection. This two-stage frequency-translation process is an effective approach for the W-band multiarm EPR bridge.

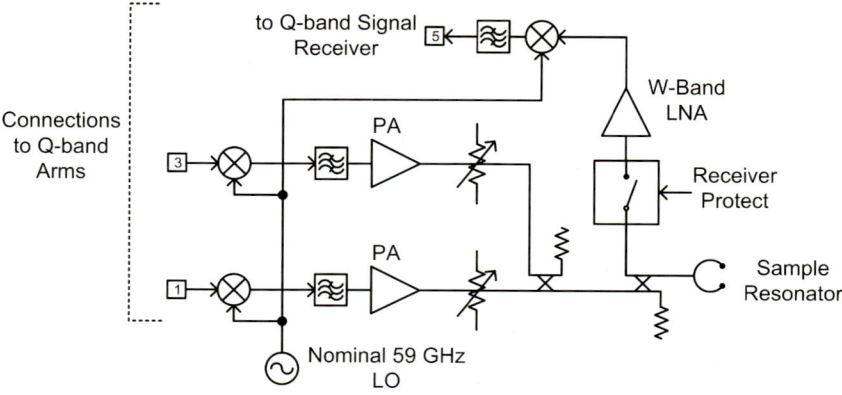

Figure 5.1.6 Second frequency-translation stage from Q- to W-band.

5.1.4
Multiarm Bridges at Higher Millimeter-Wave Frequencies

Whilst the use of a DSB/FF approach is effective at 94 GHz, the extension of this technique to higher millimeter-wave frequencies with adequate frequency tuning ranges becomes more difficult for several reasons. The fundamental microwave sources used for the LO have an elevated phase noise relative to their lower-frequency counterparts. Although, at present, solid-state power amplifiers are commercially unavailable, this situation is changing (Samoska, Peralta, and Hu, 2004). IMPATT diode injection-locked amplifiers are relatively narrow-band, while vacuum-tube amplifier technology is expensive, more difficult to implement, and often noisy. High-selectivity waveguide filters are more difficult to realize because of higher losses, smaller dimensions, and tighter tolerances as the frequency increases. Increasing the nominal synthesizer frequency to accommodate these filter limitations results in elevated phase noise from the synthesizer. Hence, the DSB/FF approach alone is difficult to implement at higher millimeter-wave frequencies for EPR use.

Millimeter-waves above 100 GHz are often generated in EPR bridges with fundamental sources, or by using frequency multiplication of microwave sources. The use of multiple phase-coherent fundamental millimeter-wave sources is problematic at best because of the difficulty of phase-locking multiple independent sources, as well as concerns of elevated noise. The generation of a higher frequency from a lower input frequency can be achieved by frequency multiplication. The phase noise in frequency multiplication fundamentally equals the phase noise-to-carrier of the input signal multiplied by N^2. The phase noise in the DSB/FF approach is fundamentally the sum of the phase noise-to-carrier ratios of the LO and input signal. The DSB/FF approach is generally superior in this regard, because the phase-noise sum is significantly lower than the multiplicative phase noise for higher-order (N) frequency multiplication. However, the current lack of available solid-state power amplifiers at higher millimeter-wave frequencies limits the DSB/FF approach.

A potential approach for higher millimeter-wave frequencies is to combine the DSB/FF approach with *low-order* (n) frequency multiplication; a proposed D-band bridge illustrating this approach is shown in Figure 5.1.7. The low-order frequency multiplication of each Q-band frequency in separate arms is viable because the phase noise does not increase dramatically when N is low. For example, if a fundamental source (Gunn diode oscillator) at 105 GHz is mixed with Q-band bridge frequencies (nominally, 35 GHz), then 140 GHz will be generated. The 105 GHz source degrades the LO phase noise by an estimated 14 dB relative to the Q-band source. The degradation in phase noise due to low-order (×4) multiplication of the Q-band frequency to D-band is 12 dB, slightly better than for the DSB/FF approach. Thus, DSB/FF combined with low-order frequency multiplication appears to be a viable approach to multiple-frequency/multiarm EPR bridges at higher millimeter-wave frequencies.

The D-band circuit in Figure 5.1.7 is driven from the Q-band bridge in Figure 5.1.5 from the indicated connection points. A high-power Q-band amplifier drives each multiplier to produce significant power output, nominally 50 mW each at D-band. Furthermore, these multipliers are quite beneficial in pulsed experiments because they can be turned on and off quickly (<1 ns) when the Q-band input power drops below a certain threshold. Although a D-band solid-state LNA is currently not commercially available, the existence of prototype units has been reported (Samoska, 2006). The reference arm frequency is offset from the other arm frequencies; this mixes with the EPR signal to produce the desired intermediate frequency that is sent to the signal receiver. It should be noted that the phase coherency of the Q-band platform has been preserved in this configuration, because the multiplication process is phase-coherent.

The Q-band bridge has been utilized as the common microwave platform for the first frequency-translation stage in both the W- and D-band millimeter-wave bridge designs. Thus, the multiple-frequency translation strategy is an effective approach for EPR bridges at higher millimeter-wave frequencies.

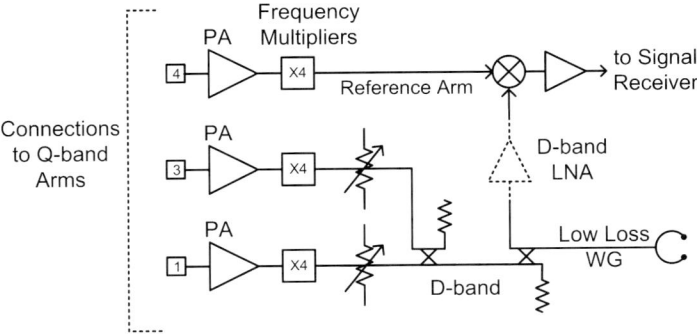

Figure 5.1.7 Low-order multiplication approach to D-band.

5.1.5
Resonator Considerations for Multiarm Experiments

When Hyde, Chien, and Freed (1968) conducted X-band ELDOR experiments on nitroxide radicals in solution, two mechanisms were identified that connected nitrogen nuclear hyperfine lines; namely, a Heisenberg exchange and a fast nitrogen nuclear relaxation. A bimodal cavity was used to support the pumping and observation of microwaves, the frequencies of which were set about (2.8 MHz/G) × (the hyperfine splitting in gauss) apart, and the DC (direct current) magnetic field was swept. Here, 2.8 MHz/G is the gyromagnetic ratio of the free electron. The two cavity modes were rectangular TE_{102} and TE_{103}, where two half-wavelengths of the TE_{103} mode shared the same space as the TE_{102} mode, with fields orthogonal. A tuning screw in the third half-wavelength space permitted adjustment of the resonant frequencies.

The methodology was extended to frequency-swept ELDOR, where the tuning screw was varied by a stepping motor, while an automatic frequency control (AFC) system permitted the locking of a klystron to the swept microwave resonance frequency (Hyde, Sneed, and Rist, 1969). The experimental methodology was applied to DPPH (2,2-diphenyl-1-picrylhydrazyl) in solution. These pioneering multiarm experiments have been reviewed by Kevan and Kispert (1976).

A bimodal cavity was also employed in early SR experiments. Although the pump and the observing microwave frequencies were the same, multiarm technology was employed. This instrument was first described by Huisjen and Hyde (1974), with further developments made later by Percival and Hyde (1975). An improved integrity of the SR signal was obtained by 180° phase modulation of the pump, which canceled out the free induction decay (FID). This method of FID suppression in SR experiments requires two arms: the observe, and the phase-modulated pump.

With the advent of the LGR (Froncisz and Hyde, 1982), the use of bimodal cavities was abandoned both for ELDOR and SR experiments. Because of the very high efficiency parameter, Λ, of the LGR (see Chapter 5.2), the amount of power required for saturation was greatly reduced. Furthermore, because of the low Q-values, the LGR could support frequencies that were fairly far apart. For example, Hyde *et al.* (1985) used an LGR for ELDOR on spin labels, while Hyde, Froncisz, and Mottley (1984) described a pulse ELDOR experiment on spin labels using an LGR. In another series of experiments, J.J. Yin used the LGR in SR (Yin, Pasenkiewicz-Gierula, and Hyde, 1987; Yin and Hyde, 1987; Yin, Feix, and Hyde, 1988; Yin, Feix, and Hyde, 1990).

In the authors' laboratory, MQ-EPR was developed in a series of experiments (for a review, see Hyde, 1998), in which two incident microwave frequencies were required that were phase coherent and about 10 kHz apart. The technology of Strangeway *et al.* (1995) was developed in response to the need for exceptionally pure coherent sample irradiation frequencies. Mchaourab *et al.* (1991) have described an X-band MQ ELDOR experiment in which the first requirements for three frequencies for sample irradiation were noted, but ultimately an LGR was employed.

All of the experiments considered thus far in this chapter were conducted at X-band. Whilst it was natural to consider multifrequency variants of multiarm experiments, an immediate problem was encountered at S-band – namely that the bandwidth of an LGR, when expressed in MHz, would vary linearly with frequency at constant Q-value. Although, typically, the LGR was too narrow-banded to permit S-band ELDOR experiments, Piasecki, Froncisz, and Hyde (1996) overcame this problem by developing a novel bimodal LGR in which two distinct S-band bridges were used together to create the multiarm capability required for that experiment. The Piasecki resonator is a very different design from others described elsewhere, and may represent an enabling technology for future experimentation.

LGR technology has been extended to higher frequencies, notably to 18 GHz by Oles, Hyde, and Froncisz (1989), to 35 GHz by Froncisz, Oles, and Hyde (1986), and to 94 GHz by Sidabras *et al.* (2007). This extension was greatly facilitated by parallel developments in the finite-element modeling of electromagnetic fields (see Chapter 5.2), and by the process of electric discharge machining (EDM). The increase in bandwidth as the frequency is progressively stepped to a higher frequency is of fundamental importance in the development of multiarm EPR technology.

In summary, two resonator strategies have been used to support the outputs of multiarm bridges, bimodal resonators, and broadband resonators such as the LGR. The latter approach appears to function best at X-band and higher frequencies. Bimodal resonators are not in general use because of their complexity, although several have been described previously. As the microwave frequency decreases below X-band, the bandwidth of LGRs decreases and multiarm experiments become problematic. To the authors' knowledge, the experiment of Piasecki, Froncisz, and Hyde (1996) is the only investigation to have used a bimodal LGR.

5.1.6
Reference Arm and Receiver Design Considerations for Multiarm Experiments

In the early X-band CW-ELDOR experiments using bimodal cavities, much concern was expressed that the ordinary EPR signal arising at the pump frequency might induce a signal in the observe mode, and thereby contaminate the ELDOR signal. In addition, drifts of isolation sometimes occurred. An additional degree of isolation was provided in the signal arm by the introduction of a "pump arm trap" (see Figure 5.1.8), which consisted of a tunable high-Q TE_{011} cavity that could be carefully tuned and matched at the pump frequency. This was found to reduce the spurious pump power in the signal arm by about 30 dB.

The use of a trap remained important when the LGR came into common usage. The calculated real and imaginary parts of the resonator reflection coefficient at

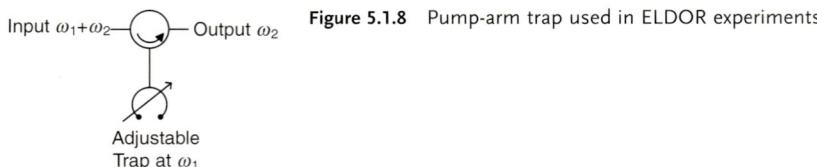

Figure 5.1.8 Pump-arm trap used in ELDOR experiments.

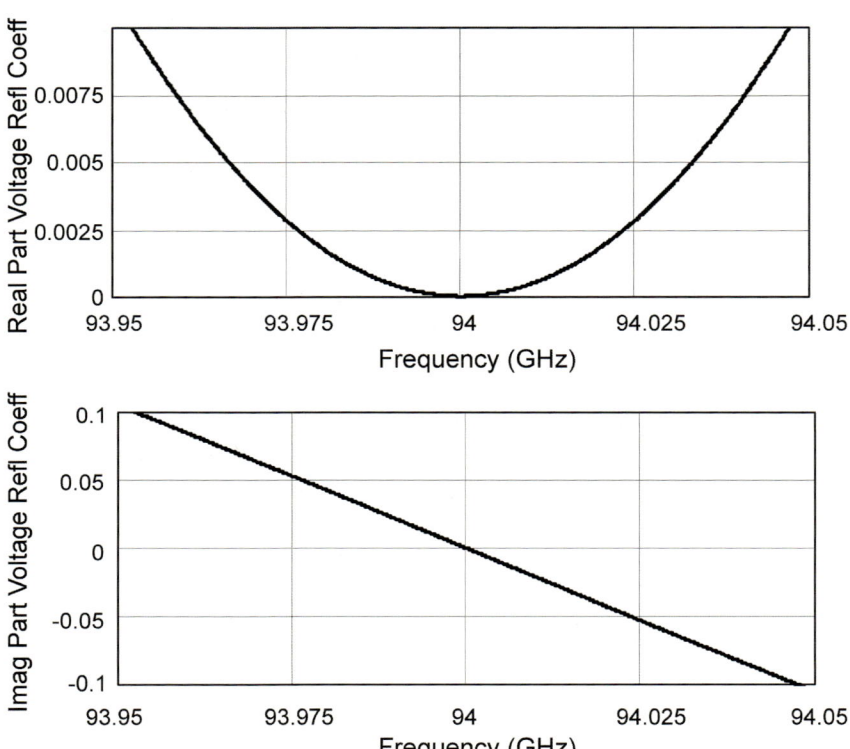

Figure 5.1.9 W-band LGR reflection coefficients. Reprinted with permission from W.J. Ellison, *Journal of Physical and Chemical Reference Data*, Vol. 36, Issue 1, pp. 10; © 2007, American Institute of Physics.

W-band are illustrated in Figure 5.1.9. At X-band, the figure remained the same, except for a scaling factor of 10^{-1} on the horizontal axis. The investigator must decide whether it is more important to provide a match at the pump frequency or at the observe frequency, but the use of a pump frequency trap will ameliorate this problem.

The traps do not function well at higher frequencies (e.g., 35 or 94 GHz), because of the increasingly high bandwidth of the trap cavity. For good filtering of the pump frequency at higher frequencies, the separation of the pump and observe frequencies must be increased in a progressive manner.

It is expected that direct digital detection, followed by computer-based detection to baseband (i.e., near-zero frequency, at field modulation frequency, or at the frequency of an intermodulation sideband of interest) will be useful for separating spurious microwave frequencies from true EPR signals. Clearly, analog detection using a microwave diode will become an obsolete technology.

During the course of research investigations which led to the Q-band SR experiment of Hyde *et al.* (2004), a new means of suppressing FID was identified, which is at the pump frequency, from the observing signal. The latter was shifted about 1000 Hz away, which was still well within the homogeneous linewidth. In the rotating frame at the observe frequency, the FID is at 1000 Hz, where it averages out as SR signals are accumulated. This method was also used in the SR study at W-band conducted by Froncisz *et al.* (2008), and also at 94 GHz (W-band) (Hyde *et al.*, 2007). Additional methods also exist to encode the signals from the various arms (in addition to the actual frequencies involved), which facilitate separation after digital detection and involve explicit modulations of the individual frequencies. For example, it was established that frequency modulation is a viable alternative to field modulation for experiments at W-band (Hyde *et al.*, 2007). It is to be expected that frequency modulation of the observe frequency (but not the pump) in a CW W-band ELDOR experiment would permit an easy separation of the true ELDOR signal from spurious pump signals.

Thus, at least in principle, there appear to be no problems with regards to the detection or separation of signals reflected from the sample resonator at every irradiation frequency. However, an important problem remains which has not yet been fully solved, namely protection of the LNA from damage by intense microwave pulses that occur during the ring-up or ring-down of the resonator. Indeed, the insertion-loss of currently available protection devices must be regarded as unacceptable.

5.1.7
Discussion

The authors' interest in the development of multiarm technology employing coherent microwave frequencies in each arm began with a series of studies that explored EPR analogies to the NMR rotary resonance experiment of Redfield (1955), followed by the closely related NMR experiment of Anderson (1956). In total, three studies were conducted (Froncisz, Sczaniecki, and Hyde, 1989; Sczaniecki, Hyde, and Froncisz, 1990; Hyde, Sczaniecki, and Froncisz, 1989) in which two coherent microwave frequencies of different amplitudes were introduced to the sample, and where the effect of one frequency on the signal observed at the other frequency was investigated. During the course of these studies, a practical EPR method – now termed modulation of saturation (MOS) – was developed (Hyde, Sczaniecki, and Froncisz, 1989), in which the two incident frequencies are sufficiently close together that they can irradiate a single homogeneous line. One frequency – the pump – is at a high amplitude and is amplitude-modulated, while the other frequency – which is observed – is at a low amplitude. When the pump power is high, the EPR signal is low, and when the pump power is off, the EPR signal is high; in this case the lineshape is pure absorption. During the course of these experiments, intermodulation sidebands were discovered which led to MQ-EPR, with the research groups' attention being redirected to this new magnetic resonance phenomena. Thus, it would be appropriate, in the opinion of the authors, to return to the study of rotary resonance and MOS spectroscopy.

Multiarm technology is commonly used for MQ experiments, whereby not only are two closely spaced, spectrally pure microwave frequencies required, but a phase-coherent reference arm is also necessary. Most often, this reference-arm frequency is set midway between irradiating frequencies for so-called "centric" detection. In this case, intermodulation sidebands that are symmetric about the reference frequency are superimposed. The use of a noncentric reference is also possible.

The use of coherent pump and observe frequencies for SR experiments in order to suppress FID signals seems well-established through reports on Q- and W-band (Hyde et al., 2004; Froncisz et al., 2008). This time-domain experiment utilizes two coherent irradiating sources.

The use of coherent pump and observe frequencies in ELDOR has never been reported, although the MQ-ELDOR apparatus described by Mchaourab et al. (1991) employed coherent radiation by three incident frequencies. Not only was the true ELDOR signal arising from transfer of saturation between pumped and observed hyperfine lines found, but additional weak peaks were also found that could be tentatively attributed to rotary resonance effects.

Froncisz and coworkers extended the MQ methodology at X-band by introducing the concept of tetrachromatic radiation, which involved two closely spaced pairs of irradiating frequencies (Jelen and Froncisz, 1998; Dutka et al., 2004). When two of the possible three frequency intervals were set to the same value of about 10 kHz, it was possible to interpret the spectra; this represented a first step towards increasing the number of arms (N in Figure 5.1.1).

Currently, the application of multiarm technology in the multifrequency range of Q- and W-band is at an early stage. Although the measurement of spin-lattice relaxation times of nitroxides in solution and in membranes has been extended to Q- and W-band (Klug et al., 2005; Froncisz et al., 2008), the relaxation processes still seem to be not fully understood. Klug et al. (2005) used MQ at Q-band to distinguish the presence of two different motional components of the spin-labeled visual protein, arrestin. Likewise, Sarewicz et al. (2008) presented a theoretical analysis of the spectra of a protein–protein complex in equilibrium, where one of the proteins had been spin-labeled. These authors made the comment, based on the results of W-band experiments conducted in their laboratory, that W-band appears to be particularly useful when studying protein–protein equilibrium, because of the well-resolved spectra and the increased sensitivity to motion. More recently, various collaborations with other visitors to the National Biomedical EPR Center have been undertaken that are expected to provide information based on the unique multiarm instrumentation described in this chapter.

Pertinent Literature

The treatise by C.P. Poole (1983) *Electron Spin Resonance: A Comprehensive Treatise on Experimental Techniques*, 2nd edition, John Wiley, New York, provides a general introduction to many types of microwave EPR bridges.

Acknowledgments

These studies were supported by grants EB001980 (National Biomedical EPR Center), EB001417 (HFSS Modeling in Aqueous Biological Samples for EPR), and EB002052 (Development of Biomedical EPR Instrumentation), James S. Hyde, PI, from the National Institutes of Health.

References

Allgeier, J., Disselhorst, J.A.J.M., Weber, R.T., Wenckebach, W.Th., and Schmidt, J. (1990) High frequency pulsed electron spin resonance, in *Modern Pulsed and Continuous-Wave Electron Spin Resonance* (eds L. Kevan and M.K. Bollman), John Wiley & Sons, Inc., New York, pp. 267–283.

Anderson, W.A. (1956) *Phys. Rev.*, **102**, 151.

Camenisch, T.G., Ratke, J.J., Strangeway, R.A., and Hyde, J.S. (2004) A versatile Q-band electron paramagnetic resonance spectrometer [abstract]. IEEE Region 4 Electro/Information Technology Conference (EIT2004), 25–27 August 2004, Milwaukee, Wisconsin.

Dutka, M., Gurbiel, R.J., Kozioł, J., and Froncisz, W. (2004) *J. Magn. Reson.*, **170**, 220.

Froncisz, W. and Hyde, J.S. (1982) *J. Magn. Reson.*, **47**, 515.

Froncisz, W., Oles, T., and Hyde, J.S. (1986) *Rev. Sci. Instrum.*, **57**, 1095.

Froncisz, W., Sczaniecki, P.B., and Hyde, J.S. (1989) *Phys. Med.*, **2–4**, 163.

Froncisz, W., Camenisch, T.G., Ratke, J.J., and Hyde, J.S. (2001) *Rev. Sci. Instrum.*, **72**, 1837.

Froncisz, W., Camenisch, T.G., Ratke, J.J., Anderson, J.R., Subczynski, W.K., Strangeway, R.A., Sidabras, J.W., and Hyde, J.S. (2008) *J. Magn. Reson.*, **193**, 297.

Goldfarb, D., Lipkin, Y., Potapov, A., Gorodetsky, Y., Epel, B., Raitsimring, A.M., Radoul, M., and Kaminker, I. (2008) *J. Magn. Reson.*, **194**, 8.

Grinberg, O. and Berliner, L.J. (eds) (2004) *Very High Frequency (VHF) ESR/EPR*, vol. 22, Biological Magnetic Resonance, Kluwer/Plenum, New York.

Gromov, I., Krymov, V., Manikandan, P., Arieli, D., and Goldfarb, D. (1999) *J. Magn. Reson.*, **139**, 8.

Hofbauer, W., Earle, K.A., Dunnam, C.R., Moscicki, J.K., and Freed, J.H. (2004) *Rev. Sci. Instrum.*, **75**, 1194.

Höfer, P., Kamlowski, A., Maresch, G.G., Schmalbein, D., and Weber, R.T. (2004) The Bruker ELEXSYS E600/680 94 GHz spectrometer series, in *Very High Frequency (VHF) ESR/EPR*, vol. 22, Biological Magnetic Resonance (eds O. Grinberg and L.J. Berliner), Kluwer/Plenum, New York, pp. 401–429.

Huisjen, M. and Hyde, J.S. (1974) *Rev. Sci. Instrum.*, **45**, 669.

Hyde, J.S. (1998) Multiquantum EPR, in *Foundations of Modern EPR* (eds S.S. Eaton, G.R. Eaton, and K.M. Salikhov), World Scientific Publishing, New York, pp. 741–757.

Hyde, J.S., Chien, J.C.W., and Freed, J.H. (1968) *J. Chem. Phys.*, **48**, 4211.

Hyde, J.S., Sneed, R.C., Jr, and Rist, G.H. (1969) *J. Chem. Phys.*, **51**, 1404.

Hyde, J.S., Froncisz, W., and Mottley, C. (1984) *Chem. Phys. Lett.*, **110**, 621.

Hyde, J.S., Yin, J.-J., Froncisz, W., and Feix, J.B. (1985) *J. Magn. Reson.*, **63**, 142.

Hyde, J.S., Sczaniecki, P.B., and Froncisz, W. (1989) *J. Chem. Soc. Faraday Trans. I*, **85**, 3901.

Hyde, J.S., Mchaourab, H.S., Camenisch, T.G., Ratke, J.J., Cox, R.W., and Froncisz, W. (1998) *Rev. Sci. Instrum.*, **69**, 2622.

Hyde, J.S., Yin, J.-J., Subczynski, W.K., Camenisch, T.G., Ratke, J.J., and Froncisz, W. (2004) *J. Phys. Chem. B*, **108**, 9524.

Hyde, J.S., Froncisz, W., Sidabras, J.W., Camenisch, T.G., Anderson, J.R., and Strangeway, R.A. (2007) *J. Magn. Reson.*, **185**, 259.

Jelen, M. and Froncisz, W. (1998) *J. Chem. Phys.*, **109**, 9272.

Jesmanowicz, A. and Hyde, J.S. (2006) *Proc. Int. Soc. Mag. Reson. Med.*, **14**, 2027 [abstract].

Kevan, L. and Kispert, L.D. (1976) *Electron Spin Double Resonance Spectroscopy*, John Wiley & Sons, Inc., New York.

Klug, C.S., Camenisch, T.G., Hubbell, W.L., and Hyde, J.S. (2005) *Biophys. J.*, **88**, 3641.

Mchaourab, H.S., Christidis, T.C., Froncisz, W., Sczaniecki, P.B., and Hyde, J.S. (1991) *J. Magn. Reson.*, **92**, 429.

Oles, T., Hyde, J.S., and Froncisz, W. (1989) *Rev. Sci. Instrum.*, **60**, 389.

Percival, P.W. and Hyde, J.S. (1975) *Rev. Sci. Instrum.*, **46**, 1522.

Piasecki, W., Froncisz, W., and Hyde, J.S. (1996) *Rev. Sci. Instrum.*, **67**, 1896.

Redfield, A.G. (1955) *Phys. Rev.*, **98**, 1787.

Samoska, L. (2006) *Proceedings of the IEEE MTT-S International Microwave Symposium Digest*, pp. 333–336.

Samoska, L., Peralta, A., and Hu, M. (2004) *IEEE Microwave Wireless Compon. Lett.*, **14**, 56.

Sarewicz, M., Szytuła, S., Dutka, M., Osyczka, A., and Froncisz, W. (2008) *Eur. Biophys. J.*, **37**, 483.

Sczaniecki, P.B., Hyde, J.S., and Froncisz, W. (1990) *J. Chem. Phys.*, **93**, 3891.

Sidabras, J.W., Mett, R.R., Froncisz, W., Camenisch, T.G., Anderson, J.R., and Hyde, J.S. (2007) *Rev. Sci. Instrum.*, **78**, 034701.

Strangeway, R.A., Mchaourab, H.S., Luglio, J., Froncisz, W., and Hyde, J.S. (1995) *Rev. Sci. Instrum.*, **66**, 4516.

Weber, R.T., Disselhorst, J.A.J.M., Prevo, L.J., Schmidt, J., and Wenckebach, W.Th. (1989) *J. Magn. Reson.*, **81**, 129.

Yin, J.-J. and Hyde, J.S. (1987) *J. Magn. Reson.*, **74**, 82.

Yin, J.-J., Feix, J.B., and Hyde, J.S. (1988) *Biophys. J.*, **53**, 525.

Yin, J.-J., Feix, J.B., and Hyde, J.S. (1990) *Biophys. J.*, **58**, 713.

Yin, J.-J., Pasenkiewicz-Gierula, M., and Hyde, J.S. (1987) *Proc. Natl Acad. Sci. USA*, **84**, 964.

5.2
Resonators for Multifrequency EPR of Spin Labels
James S. Hyde, Jason W. Sidabras, and Richard R. Mett

5.2.1
Introduction

The loop-gap resonator (LGR) was introduced to the electron paramagnetic resonance (EPR) community by Froncisz and Hyde in 1982, since which time extensive reports have been made on the subject. The rationale for this class of resonators was developed in various reviews (Hyde and Froncisz, 1986; Hyde and Froncisz, 1989; Rinard and Eaton, 2005).

Briefly, an LGR is a lumped-circuit resonator that has clearly discernable inductances, capacitances, and resistances, and which has dimensions that are small compared to the wavelength. In particular, in many practical circumstances the LGR is an ideal resonator when the sample is very small. The LGR stands in contrast to a cavity resonator, which is classified as a distributed circuit resonator. The components of the cavity resonator are not readily discernable, while the dimensions are comparable to half of a free-space wavelength, $\lambda_0/2$. It is an appropriate circuit when relatively large amounts of sample are available.

In this chapter, however, this rationale is no longer used, the main theme being the use of finite-element modeling to design resonators. The concept of lumped versus distributed circuit designs does not explicitly enter the design process. Finite-element modeling permits an improved optimization of the resonator, sample configuration, and coupling structure, with some dimensions perhaps being of the order of $\lambda_0/2$, and others much less than $\lambda_0/2$. This point of view revolves around numeric solutions of Maxwell's equations, with no simplifying assumptions regarding the circuit dimensions relative to the wavelength. These new models show important corrections, even for ranges of applicability where lumped circuit and perturbation approaches have previously been regarded as adequate.

Currently, the largest area of application of EPR in biomedical research is the use of nitroxide radical spin labels in molecular-structure determinations which, invariably, will be in the aqueous phase. Hence, this class of application forms the principal target for the resonator developments discussed in the chapter.

Some years ago, Hyde introduced a system of classification of EPR samples whereby three questions were posed: (i) Is the sample saturable with the available microwave power?; (ii) Is the sample limited in either size or availability?; and (iii) Does the sample exhibit significant dielectric loss? (Hyde, Eaton, and Eaton, 2006). Whilst this classification has proved useful in the development of multipurpose resonators, it proved to be less effective for spin-label applications, where the answers to such questions may lie intermediate between a definite "yes" and a definite "no."

As a specific example, consider the EPR of site-directed spin labels (SDSLs) at X-band. During the early history of SDSLs, a few microliters of a spin-labeled mutant protein could be prepared quite easily at a concentration of about 10^{-4} M, and the LGRs were designed accordingly; however, as the result of technical improvements in sample preparation, sample volumes of $20\,\mu l$ may now be commonly available. Today, accessibility studies using dissolved oxygen or paramagnetic metal ions are commonplace with SDSLs, and the available microwave power for saturation in resonators that accommodate $20\,\mu l$ of sample (as is required for these studies) may be insufficient. Moreover, for oxygen-accessibility studies, the oxygen is customarily introduced through thin-walled plastic cuvettes, and the diffusion of oxygen into large samples may take too long. Consequently, at X-band a "yes" or "no" answer to the question of whether or not the sample is limited and saturable is just on the cusp – with "yes" being the appropriate answer for lower microwave frequencies, and "no" for higher frequencies. For frequencies below X-band, LGR technology would generally be preferred, but for higher frequencies the choice between loop–gap and cavity resonators would be more subtle.

Aqueous samples present special problems because of the high dielectric constant and the high loss tangent. Both, Hyde (1972) and Eaton and Eaton (1977) have shown that at X-band, a good sensitivity can be obtained by using flat cells in a rectangular TE_{102} cavity in two different orientations, with the finite-thickness cell lying: (i) in a null of the incident tangential radiofrequency (RF) electric field; and (ii) in an orientation such that the RF electric field is perpendicular to the sample-cell surface. Both principles – which today are well understood – have been incorporated seamlessly into the design of sample tubes for aqueous samples (Mett and Hyde, 2003; Sidabras, Mett, and Hyde, 2005). In addition, unusual sample-tube cross-sections based on these principles can be extruded in polytetrafluoroethylene (PTFE).

At Q- (35 GHz) and W- (94 GHz) band, resonators operating in the cylindrical TE_{011} mode have almost always been used. Recently, LGRs have been developed at these frequencies (Hyde *et al.*, 2004; Sidabras *et al.*, 2007) which have demonstrated some interesting properties. Notably, the sensitivities are about the same, when comparing TE_{011} cavities with LGRs, but the LGR Q-values are lower, and the resonator efficiency parameters, Λ, higher, where Λ is given by:

$$\Lambda \equiv \frac{H_{1r}}{\sqrt{P_0}}. \tag{5.2.1}$$

Here, P_0 is the incident microwave power (in Watts) and H_{1r} is the peak rotating frame RF magnetic field in the sample (Froncisz and Hyde, 1982; Hyde and Froncisz, 1989; Freed, Leniart, and Hyde, 1967; Mett *et al.*, 2008a). The lower Q-values of the LGRs result in greater bandwidths than for cavities at these high frequencies. High Λ- and low Q-values are advantageous for pulse EPR experiments; for example, what could previously be achieved with 1 kW of power at 94 GHz with a Fabry–Pérot resonator can now be achieved at 1 W with an LGR. Thus, ideas regarding resonator design at high frequencies are beginning to change.

In summary, many old conceptual models for resonator design – which, in retrospect, were approximations necessary because the solution of Maxwell's equations in closed form was either impractical or impossible – have become less valued with the advent of finite-element modeling and high-quality computer workstations. As noted above, these old ideas include: (i) the concepts of lumped and distributed circuits; (ii) the division of samples into eight limiting classes; (iii) the preconception that LGRs are limited to use at low frequencies; and (iv) very restricted ways in which aqueous samples can be handled.

In addition, recent studies using finite-element design have led to a number of insights that seem to have no clear analogy in the earlier literature. These include:

- The modeling of eddy currents from field modulation coils (Mett, Sidabras, and Hyde, 2004; Mett *et al.*, 2005).
- The modeling of microwave leakage from modulation slots that allow the penetration of field modulation (Mett and Hyde, 2005).

- The discovery of a capacitive slotted-iris design, which is complementary to conventional inductive slots or holes (Mett, Sidabras, and Hyde, 2008b).
- The use and limitations on over-moding of resonant cavities (or LGR length) by propagation at cut-off.
- The pushing of LGR and cavity designs into the quasi-optical regime.
- Exploration of the interface between LGRs and cavities, noting that re-entrant cavities morph into LGRs.

5.2.2 Methods

5.2.2.1 Computer-Based Simulations

The current approach to EPR resonator design relies on two commercially available programs to calculate the electromagnetic fields for various geometries: (i) Ansoft High Frequency Structure Simulator (HFSS); and (ii) Ansoft Maxwell 3D (Pittsburgh, PA, USA). Ansoft HFSS is a full-wave, three-dimensional (3-D), finite-element modeling program which is used to solve Maxwell's equations for fields in microwave resonators ranging from 0.25 to 250 GHz. HFSS contains two independent approaches to solving Maxwell's equations: the eigenmode method, which finds all of the natural resonant modes of a structure; and the driven-mode method, which mimics the output of a network analyzer. Ansoft Maxwell 3D is a quasi-static, 3-D, finite-element modeling program that is useful for field modulation and eddy current analysis. During field modulation analysis, a Helmholtz pair is used to simulate an ideal field modulation coil, after which the resonator is placed between the pair and the field modulation penetration determined.

By using these programs, an EPR resonator can be fully simulated with all electromagnetic characteristics known. This saves fabrication time and reduces the number of physical iterations required to complete a microwave design; normally, about 50 simulations are carried out before a final design is realized. Both programs are currently run on two identical Dell Precision 690 workstations with dual dual-core 3 GHz Intel 5100 series Xeon processors with 4 MB of shared cache per chip and 16 GB of Double Data Rate generation 2 (DDR2) RAM. A typical simulation now takes about 30 min to complete. For analytical and numerical support, Wolfram's Mathematica (Version 7.0, Champaign, IL, USA) is used. All three programs have been developed using 64-bit technology, which allows this software to fully access all available RAM and have full multiprocessor support.

Two methods to obtain EPR signal intensities from Ansoft HFSS have been devised:

- The first method uses parameters that are easily calculated in the HFSS field calculator to determine a normalized EPR signal intensity under either saturating or nonsaturating conditions. This calculation can be made using eigenmode or driven-mode methods.

- The second method uses the driven-mode solution to calculate the change of reflection coefficient when a change of magnetic susceptibility of the sample is made.

Ansoft HFSS allows a complete characterization of materials through a complex dielectric constant and complex magnetic susceptibility. Simulations are performed to identify the reflection coefficient magnitude difference between the two material "states." This method is used for the characterization of more complex resonant systems.

For a matched reflection cavity and a voltage-sensitive detector, Feher (1957) expressed the microwave signal that characterizes CW-EPR by Equation 5.2.2:

$$S = \chi'' \eta Q_0 \sqrt{P_0}, \qquad (5.2.2)$$

where χ'' is the RF magnetic susceptibility, η is the filling factor, Q_0 is the unloaded (eigenmode) Q-value with sample, and P_0 is the incident microwave power. These terms are further developed in Equations 5.2.3 and 5.2.4:

$$\eta = \frac{\mu_0 \int H_{1r}^2 dV_s}{U} \qquad (5.2.3)$$

and

$$Q_0 = \omega \frac{U}{P_0}, \qquad (5.2.4)$$

where U is the energy stored in the resonator. In Equation 5.2.2, χ'' represents the part of the RF susceptibility that contributes to the EPR absorption signal. It is the ratio of the RF magnetization due to the sample spins to the component of the applied magnetic field, H_{1r}, in the rotating frame. Feher's analysis was performed for a linearly polarized RF magnetic field perpendicular to the static magnetic field. In the present analysis, the signal is expressed explicitly in terms of the component of RF magnetic field that induces the EPR signal, H_{1r}. The RF magnetic field in the resonator can be set to any value, as given by Equation 5.2.1. EPR signals from different resonators can be compared using the output from HFSS by separating the resonator fields from the sample response, χ''. The numerator in Equation 5.2.3 is the stored energy of the circularly polarized component of the RF magnetic field in the sample.

Inserting Equations 5.2.3 and 5.2.4 into Equation 5.2.2 leads to

$$S = \chi'' \omega P_0^{-1/2} \mu_0 \int H_{1r}^2 dV_s. \qquad (5.2.5)$$

If the resonator efficiency parameter given by Equation 5.2.1 is substituted into this equation, we have

$$S = \chi'' \omega \sqrt{P_0} \mu_0 \int \Lambda^2 dV_s \qquad (5.2.6a)$$

and

$$S = M \frac{\omega \sqrt{P_0}}{H_{1r}} \mu_0 \int \Lambda^2 dV_s, \qquad (5.2.6b)$$

where $\chi'' = M/H_{1r}$ and M is the RF magnetization. Implicit in these equations is that Λ is evaluated locally at each point in the sample rather than using the peak value, as in Equation 5.2.1.

If Λ is assumed to be uniform over the sample and the magnetization is assumed to increase linearly with incident H_1,

$$S \sim \Lambda^2 V_s. \tag{5.2.7}$$

This equation has been used by the authors to evaluate the relative sensitivities of samples in uniform field cavities that do not exhibit microwave power saturation with the available microwave power. Under this assumption, Equations 5.2.2 and 5.2.7 are equivalent. Equation 5.2.7 is convenient because the quantity Λ is readily calculated in Ansoft HFSS. Alternatively, η and Q can also be calculated and Equation 5.2.2 used directly, setting the incident power equal to 1 W. Equation 5.2.6a can also be used, since it is equivalent to Equation 5.2.2.

An important feature of Ansoft HFSS is that it incorporates the influence of sample losses on the phase of the RF fields throughout the sample. Sample losses can cause the RF magnetic field in the sample to deviate from linear polarization in the resonator. The RF magnetic field solution output by Ansoft HFSS is in the form of a phasor that is related to the time-harmonic RF magnetic field according to:

$$\mathbf{H}_1(\vec{r}, t) = \mathrm{Re}\left[\tilde{\mathbf{H}}_1(\vec{r}) e^{-i\omega t}\right], \tag{5.2.8}$$

where $\tilde{\mathbf{H}}_1$ is the (complex) phasor. In terms of the phasor, the RF energy in the resonator is given by

$$U = \frac{1}{2}\mu_0 \int \tilde{\mathbf{H}}_1 \cdot \tilde{\mathbf{H}}_1^* dV, \tag{5.2.9}$$

where the integral is over the entire resonator volume, including the sample. The relationship between $\tilde{\mathbf{H}}_1$ and the rotating frame magnetic field component can be found by introducing the complex orthogonal unit vectors (Jackson, 1975),

$$\hat{\varepsilon}_{\pm} = \frac{\hat{x} \pm i\hat{y}}{\sqrt{2}}, \tag{5.2.10}$$

where (x, y, z) are Cartesian unit vectors and the static magnetic field is in the $\hat{z}$ direction. A phasor RF magnetic field of the form $\tilde{H}_{1+}\hat{\varepsilon}_+$ represents a circularly polarized field of phasor amplitude $\tilde{H}_{1+}$ rotating clockwise (right-handed) when facing in the direction of the static magnetic field. Given an arbitrary $\tilde{\mathbf{H}}_1$, the right- or left-handed circularly polarized components can be found from

$$\tilde{H}_{1\pm} = \tilde{\mathbf{H}}_1 \cdot \hat{\varepsilon}_{\pm}^*, \tag{5.2.11}$$

respectively. The magnitude of the RF magnetic field in the (right-handed) rotating frame is the amplitude of $\mathrm{Re}(\tilde{H}_{1+}\varepsilon_+ e^{-i\omega t})$, and, therefore,

$$H_{1r} = \frac{|\tilde{H}_{1+}|}{\sqrt{2}}. \tag{5.2.12}$$

It can be shown that the stored energy in the right-hand circularly polarized component of the magnetic field is

$$U_{H+} = \frac{1}{2}\mu_0 \int \tilde{H}_{1+}\tilde{H}_{1+}^* \hat{\varepsilon}_+ \cdot \hat{\varepsilon}_+^* dV = \mu_0 \int H_{1r}^2 dV, \tag{5.2.13}$$

consistent with the numerator in Equation 5.2.3. For a linearly polarized $\tilde{\mathbf{H}}_1$ perpendicular to $\mathbf{H}_0$, it follows that $H_{1r} = |\tilde{\mathbf{H}}_1|/2$ and $U_{H+} = U/2$.

In treating the polarization explicitly in the present analysis, Equation 5.2.2 accounts for the orientation of $\mathbf{H}_1$ with respect to the applied static magnetic field and its decomposition into circularly polarized components, one of which is EPR-active. However, Equation 5.2.2 has deficiencies in the way in which the susceptibility is handled. Notably, it does not explicitly include the effect of the applied static magnetic field on the signal, nor the possibility of microwave power saturation. By introducing the term M into Equation 5.2.6b (i.e., the magnetization as used in Bloch's original report (Bloch, 1946)), these magnetic resonance phenomena can be included. For example, if comparisons across different applied magnetic fields are desired, the Boltzmann factor is included in an *ad hoc* manner.

The EPR signal from spin labels in aqueous solution is readily saturated, and Equations 5.2.2 and 5.2.6 are not useful in determining the relative sensitivities of the two resonator geometries. Instead, it is appropriate to re-adjust the incident power P_0, such that the incident RF magnetic field on the sample is always the same. If it is further assumed that the available volume of aqueous sample is unlimited, then the comparison of alternative geometries results in the ratio of "concentration sensitivities," which is a useful parameter for the EPR spectroscopist. This is the most commonly encountered circumstance in the studies reported here, which has a focus on nitroxides in water. For this circumstance, it can be imagined that the incident power could be re-adjusted as required to keep H_1 constant at the sample, even though the Q-value might be quite low.

It is convenient in a first stage of the analysis for concentration sensitivity to assume that the RF magnetic field is uniform over the sample. Then, Equations 5.2.6a and b become

$$S = \chi'' \omega \mu_0 H_{1r} \Lambda V_s \tag{5.2.14a}$$

and

$$S = M \omega \mu_0 \Lambda V_s. \tag{5.2.14b}$$

The analysis leading to Equations 5.2.14a and b is essentially the same as given by Hyde and Froncisz (1989) for saturable and nonsaturable samples. Equation 5.2.14a has been used in many of the analyzes discussed here. The term H_{1r} is assumed to be defined only by the experiment, and is a fixed arbitrary constant ($\mu_0 H_{1r} = 1\,\text{G}$) that does not affect the comparison between resonators or alternative sample geometries.

One can allow the value of the RF field to vary over the sample. In this case, Equations 5.2.14a and b become

$$S = \omega\mu_0 \int \chi'' H_{1m} \Lambda dV_s \qquad (5.2.15a)$$

and

$$S = \omega\mu_0 \int M \Lambda dV_s, \qquad (5.2.15b)$$

where H_{1m} is the RF magnetic field normalized to a peak value (e.g., $\mu_0 H_{1m\,max} = 1\,G$) in the sample. With this normalization, the integral in Equation 5.2.15a is independent of the incident power (apart from χ''). Although no analysis in the circumstances where Equation 5.2.15b is appropriate has yet been reported, it is routine to integrate Λ over the sample volume in order to make comparisons of concentration sensitivity. Recently, several reports have been made implementing a form of Equation 5.2.15a (Mett and Hyde, 2003; Sidabras, Mett, and Hyde, 2005; Sidabras et al., 2007; Hyde and Mett, 2002). It is also possible to use Equations 5.2.2 or 5.2.5, provided that the incident power is adjusted for $\mu_0 H_{1r\,max} = 1\,G$. This is equivalent to normalizing the peak field in the sample, as in Equation 5.2.15a.

5.2.2.2 Fabrication and Testing

Ultra-precision electrical discharge machining (EDM) is used to fabricate LGRs and other precise resonator parts. For this, fine sizes of tungsten wire are used down to 0.03 mm diameter (a wire of this size will produce a typical cut of 0.05 mm). Positional tolerances down to ±0.001 mm and true dimensional shapes and corners allow for the ultra-precise resonator fabrication of gaps, irises, and modulation slots. In addition, by taking multiple passes EDM technology can reduce the surface roughness to 4 µm RMS.

AutoDesk Inventor Professional 2009, a 3-D CAD program with full standards compliance, permits the visualization of increasingly complex resonator designs. Sample-support considerations for a practical resonator design are taken into account at the time that the assembly is drawn in AutoDesk Inventor. Within the 3-D environment, it is possible to arrange and align multiple drawings to create a complete assembly. In this case, the sample alignment and handling is incorporated into the assembly, thus providing a complete picture of the finished resonator, which can be viewed from any angle. Mechanical drawings are imported to computer-controlled fabrication machines. The EDM facility uses a 2-D CAD drawing for tolerancing specifications, and inputs a 3-D CAD drawing to the computer control.

When a resonator or new sample tube has been fabricated, it is bench-tested on an Agilent E8363B performance network analyzer (PNA), whereby microwave measurements are made of frequency, Q-values, and coupling range. The E8363B PNA has a frequency range of 40 MHz to 40 GHz, with translation heads available for higher frequencies.

Once a resonator has been characterized and bench-tested, it is evaluated by a spectroscopist using a standard sample. Typically, the sample is TEMPO

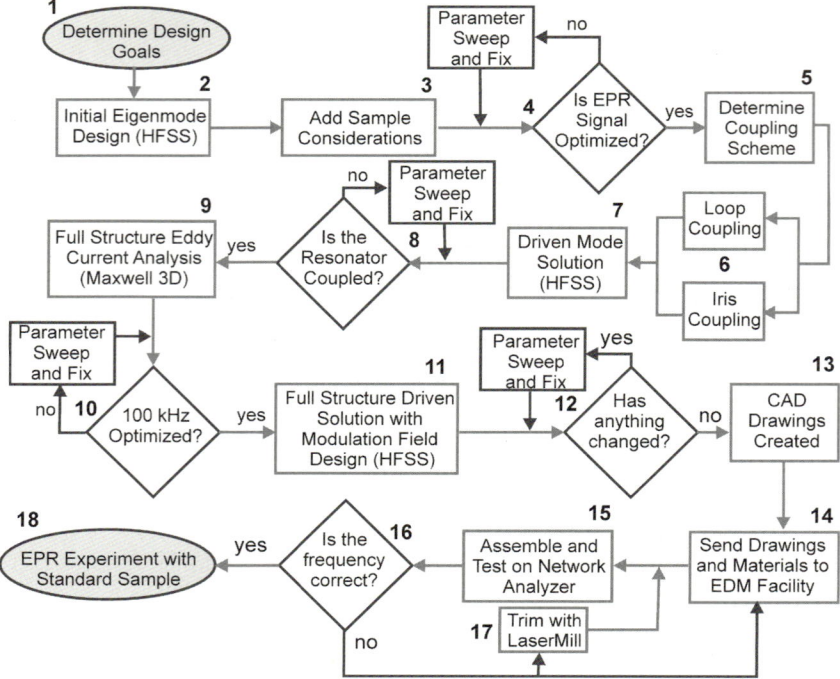

Figure 5.2.1 Flowchart for resonator development. The words "parameter sweep and fix" are shorthand for analysis of the problem and finding a solution.

(2,2,6,6–tetramethylpiperidine-1-oxyl) in water, at a variety of sample concentrations, with the relative EPR signal intensities being recorded and compared with Ansoft calculations. For saturable samples, Equation 5.2.15a is used for the comparison, as each resonator is characterized at constant peak microwave field. If the sample is considered to be unsaturable, Equation 5.2.6a is used for comparison, holding P_0 constant.

This process, which is shown schematically in Figure 5.2.1, has been used for all resonators. The five diamond-shaped areas are critical evaluation-and-decision points, each with feedback if further improvement is necessary.

5.2.3
Aqueous Samples

5.2.3.1 The Complex Dielectric Constant as a Function of Frequency and Temperature

Previously, Ellison (2007) developed an interpolation function to fit aqueous experimental complex permittivity data at atmospheric pressure over the frequency range of 0 to 25 THz, and a temperature range of 0 to 100 °C. Experimental permittivity data are compared to the function in Figure 5.2.2. At frequencies below

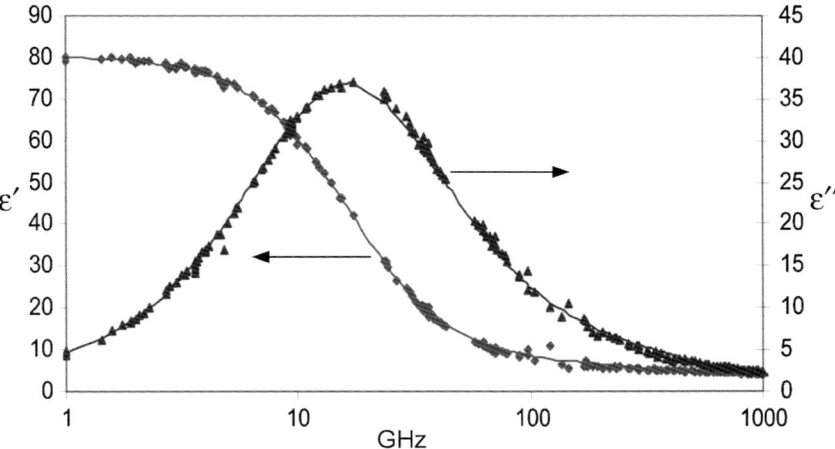

Figure 5.2.2 The 20 °C liquid water relative dielectric permittivity real and imaginary parts as a function of frequency in GHz. Diamonds represent the real part and correspond to the left scale; triangles represent the imaginary part and correspond to the right scale. Reprinted with permission from W.J. Ellison, *Journal of Physical and Chemical Reference Data*, 36 (1), 10; © 2007, American Institute of Physics

X-band (9.5 GHz), the permittivity magnitude is high and the loss tangent relatively low, whilst near X-band the loss tangent increases rapidly with frequency and becomes greater than unity. Above X-band, the loss tangent remains high, up to about 250 GHz, although the permittivity magnitude decreases. Ellison's interpolation function has been entered by the authors into Mathematica for use in Ansoft HFSS.

The interpolation function also permits an analysis of the temperature dependence, as shown in Figure 5.2.3. At 260 GHz, the loss tangent increases rapidly with temperature, doubling from 0 to 50 °C. The magnitude of the permittivity also nearly doubles, due mainly to the increase in the imaginary part. It should be noted that 260 GHz is the appropriate EPR microwave frequency for proton dynamic nuclear polarization (DNP) experiments at 400 MHz. The magnitude of the permittivity determines the sample geometry required to minimize sample losses and maximize the EPR signal, as discussed in the following sections.

5.2.3.2 Dielectric Loss Types and Parallel and Perpendicular E-Field Geometries

The EPR signal strength is maximized by maximizing the sample volume and minimizing the sample loss. The latter is determined by a combination of the complex permittivity and the sample placement and orientation relative to the electric field magnitude. As noted in Section 5.2.1, a familiar strategy for minimizing sample loss is to place the sample in an electric field null; this is referred to as a "parallel" orientation, because the electric field is parallel to the sample surface. If the magnitude of the permittivity is sufficiently high, a perpendicular sample orientation also will minimize sample loss.

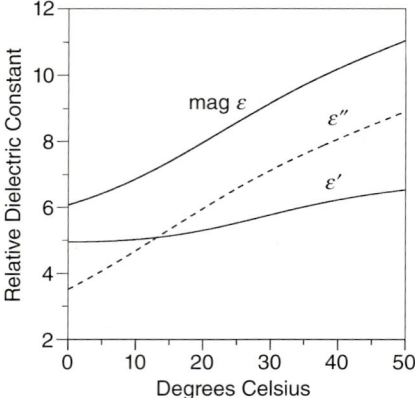

Figure 5.2.3 Dielectric properties of water at 260 GHz versus temperature. Variations of ε'' affect the losses of the sample. At room temperature and higher, placement becomes more critical.

Both, Mett and Hyde (2003) and Sidabras, Mett, and Hyde (2005) introduced a classification scheme for aqueous sample losses, referred to as Types I, II, and III:

- A Type I loss occurs due to an electric field being parallel to the sample surface, noting the continuity of the tangential electric field across the sample surface. It is the only type of loss for a sample placed in an electric field null.

- A Type II loss occurs due to an electric field normal to the sample surface. This type of loss is directly proportional to the sample thickness and, because of the discontinuity in electric field across the sample surface, it is reduced by the magnitude of the relative dielectric constant.

- A Type III loss occurs due to the nonuniform electric field near a sample edge immersed in an electric field.

For both types of sample orientation, the maximum amount of sample is limited by the so-called "sucking-in" effect, as discussed by Stoodley (1963). In this effect, as more sample is added to the resonator, the position of the electric field maximum in the resonator shifts toward the sample, and the electric field gradient and magnitude in the sample increases. A complicating factor here is that the RF magnetic field also tends to increase in the sample, which can in turn increase the filling factor. The effect is caused by a decrease in the electromagnetic wavelength in the sample due to the magnitude of the relative dielectric constant (Sidabras, Mett, and Hyde, 2005; Mett, Sidabras, and Hyde, 2004):

$$\lambda_0 / \mathrm{Re}\left[\sqrt{\varepsilon_r}\right]. \tag{5.2.16}$$

The sample holder can also contribute to this effect. The impact of this on the overall EPR signal is difficult to characterize except with finite-element modeling. The effects discussed here are heavily dependent on the microwave frequency.

5.2.3.3 Results in Commercial Resonators at X-Band Using Extruded Sample Tubes

A typical EPR geometry consists of a quartz capillary placed in the center of a cylindrical TE_{011} cavity, in the region of maximum RF magnetic field. This geometry results in a purely tangential electric field on the sample boundary; the continuity in the tangential electric field causes sample losses and EPR signal degradation.

Unlike the rectangular TE_{102} cavity, the cylindrical TE_{011} cavity is azimuthally symmetric; this permits a flat cell to be oriented at any angle in the center of the cavity, without causing any change in the EPR signal strength. By placing a flat cell in the center of a cylindrical TE_{011}, a 4.2-fold increase in saturable EPR signal strength is realized over the single capillary, though not all of this signal gain can be attributed to the increased sample volume. An increase in the curl of the electric field, resulting in a higher concentrated magnetic field, will also improve the filling factor.

Another approach to increase EPR signal strength is used in Bruker's commercial AquaX sample holder. This structure consists of 19 capillaries, each 0.4 mm in diameter, bundled in a closed-packed geometry. The AquaX yields a 1.1-fold increase in the EPR saturable signal strength over the flat cell (4.6-fold over the standard capillary). The components of the electric field within this structure are shown in Figure 5.2.4, where the black regions are of low electric field magnitude

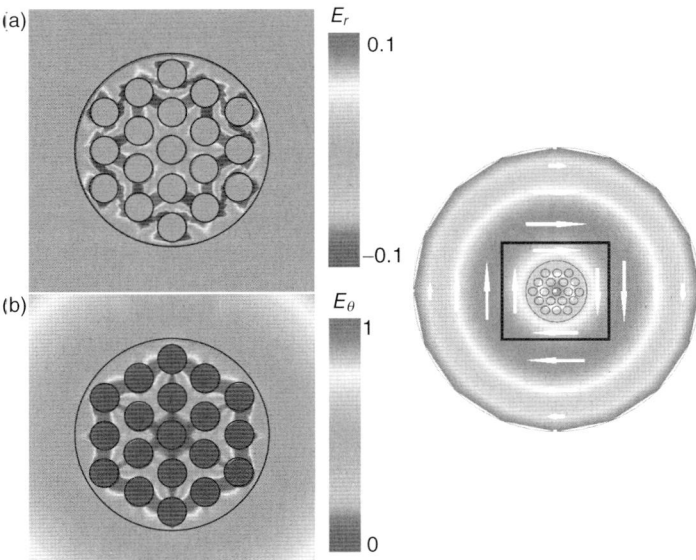

Figure 5.2.4 Electric fields in the Bruker AquaX cell in a cylindrical TE_{011} cavity. (a) E_θ; (b) E_r. Within each capillary, the electric fields rotate with a null at the center. This circulation arises from polarization charges on the surface of the capillary. Fields seen between capillaries also arise from these polarization charges.

Table 5.2.1 Saturable EPR signal strength for multiple capillaries.

Volume (μl)	No. of capillaries	Size of capillary inner diameter (mm)	EPR signal ($W^{1/2}$)	Q_L	Frequency (GHz)
198.5	7	0.96	14.05	473	9.47
	13	0.71	18.69	837	9.47
	19	0.60	21.11	964	9.45
	31	0.46	23.45	1488	9.47
	37	0.42	24.91	1526	9.46

and illustrate the benefits of the finite-element analysis. The high electric field regions between the outer ring of capillaries are caused by the presence of polarization charges on the surfaces of the capillaries.

On studying this close-packed geometry it was found that, by increasing the number of capillaries and keeping the capillary diameter constant, the Q-values became too low for most EPR experiments, even though a theoretical signal gain was shown. By keeping the volume constant, a signal enhancement of 1.18 over Bruker's commercial AquaX was shown when using 37 capillaries. The EPR signal results for a series of close-packed geometries, ranging from seven to 37 capillaries, are listed in Table 5.2.1. The optimization of other packing geometries did not result in any EPR signal improvement with respect to the Bruker AquaX.

5.2.3.4 Multichannel Design

The placement of multiple flat cells radially and perpendicular to the electric field in the cylindrical TE_{011} (as shown in Figure 5.2.5) results in a twofold improvement in saturable EPR signal strength over the Bruker AquaX. Figure 5.2.5 shows a simulation in Ansoft HFSS using 18 flat cells with dimensions of $0.2 \times 2.7 \times 38.8$ mm. Although it would appear advantageous to do so, simulations showed that crossing the flat cells at the center provided no additional benefit, and this can be attributed to the sucking-in effect. With the geometry as described, an electric field null is centered in each of the flat cells, which minimizes loss.

It is possible that multilumen extrusion (Zeus Inc., www.zeusinc.com) could be used to create the multichannel sample structure of Figure 5.2.5. The Zeus Inc. process can be used to extrude multilumen PTFE structures with wall thicknesses as small as 0.1 mm, with high tolerance and in many geometries. With a channel thickness of 0.2 mm, however, this structure would appear to be at the edge of practicality.

A single capillary, a flat cell, Bruker's AquaX, and the Figure 5.2.5 structure are compared in Table 5.2.2. For the Figure 5.2.5 design, the Q-values and frequency shift can be adjusted by limiting how far the sample extends radially; the optimum dimensions are shown, where all sample holders were constructed from PTFE.

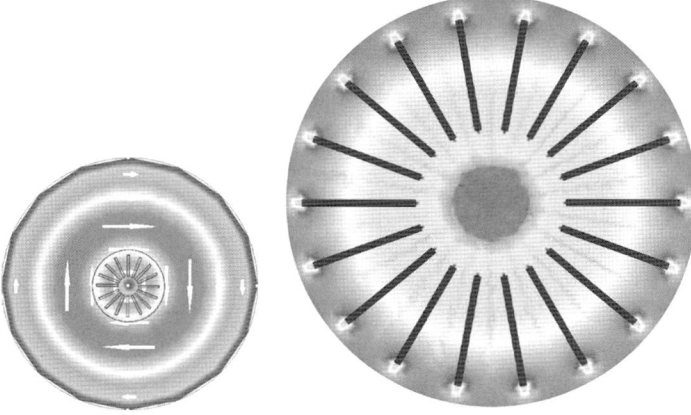

Figure 5.2.5 Electric field with AquaSpoke in a cylindrical TE_{011} cavity, $E_\theta = E_\perp$. This geometry is analogous to a flat cell in perpendicular orientation in a rectangular TE_{102} geometry.

Table 5.2.2 Saturable EPR signal strength for multiple cell configurations in a cylindrical TE_{011} cavity.

Sample type	Dimensions (mm)	Volume (μl)	Normalized signal	Q_L	Frequency (GHz)
Capillary	1.1 diameter	36.87	1.0	2190	9.48
Flat cell	0.4 × 8	124.16	4.2	1820	9.47
AquaX (19 cells)	0.4 diameter each	92.64	4.6	3886	9.48
Close-packed (37 cells)	0.42 diameter each	198.5	5.3	1533	9.47
Multichannel (18 flat cells)	0.2 × 2.7 each	377.14	12.0	928	9.25

The sample holder material is important when designing multilumen structures that are to take advantage of perpendicular electric fields. Simulations have shown that a sample holder with a dielectric constant as close as possible to free space is desirable.

An inspection of Figure 5.2.3 indicates that the results of Sections 5.2.3.3 and 5.2.3.4, which are a consequence of the dielectric properties of water, will depend heavily on the microwave frequency. Moreover, they are expected to be relevant at Q-band, but perhaps not at W-band.

5.2.4
Uniform Field Cavities and Loop–Gap Resonators

A uniform RF field over the sample will provide the best quality EPR signal, because all portions of the sample are uniformly saturated. For a uniform RF field, the EPR signal is proportional to the product of the resonator efficiency parameter Λ and the sample volume for saturable samples (see Equation 5.2.14b). Over the past decade, several practical resonators with a uniform RF field over the sample have been introduced, with the designs that have included many different resonator types, such as cavity resonators of rectangular, cylindrical, and coaxial geometries, of TE and TM polarizations, and also LGRs. Each resonator type has distinct advantages from low to high frequencies, as discussed below.

5.2.4.1   Intrinsic Uniformity

The fields of any cavity mode with zero index are uniform in that dimension. For example, the RF electric and magnetic fields of the cylindrical TE_{011} mode are uniform in azimuth, and in the rectangular TE_{102} mode, uniform in the small rectangular dimension. Because of metal–wall boundary conditions, the only cavity modes that have uniform fields in two dimensions are the TM_{0_0} modes, which are nonuniform only in the radial dimension. These modes are not commonly used for EPR because the RF electric field is peaked along the resonator axis (but see Section 5.2.4.3).

5.2.4.2   Uniform Field Cavities

The circular TE_{011} mode has the magnetic field peaked along the cylinder axis, and is therefore the natural choice for a capillary sample tube. However, with shortened ends the lowest order mode has the index 1 in the axial direction. Three different means of shaping the ends of a TE_{011} cavity to produce an RF open instead of an RF short have been introduced by Anderson, Mett, and Hyde (2002), Hyde, Mett, and Anderson (2002), and Mett, Froncisz, and Hyde (2001). In each design, a central section of the cavity has a uniform RF magnetic field over the sample for any sample length. The mode index is the uniform dimension, given the symbol "U," as in TE_{01U}. A practical cavity design at X-band based on the studies of Rinard and Eaton (2005) is shown in Figures 5.2.6–5.2.8.

5.2.4.3   Uniformity in Two Dimensions

An interesting result of these studies is that it is always possible to produce uniform RF magnetic fields in two dimensions (three dimensions are not possible, because the fields must have at least half a wavelength). Although uniformity in two dimensions can take the form of a plane, a more practical geometry appears to be an azimuthal ring uniformity, as shown in Figure 5.2.9. The cylindrical TM_{020} mode also has a uniform azimuthal ring. Resonators and sample cuvettes for this class of resonator have yet to be developed.

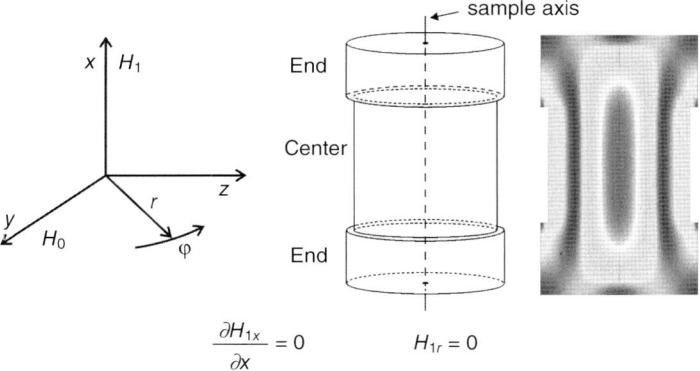

Figure 5.2.6 Radiofrequency magnetic field magnitude in a cylindrical TE_{01U} geometry. The resonator is uniform over two dimensions, x and φ, in the region of interest.

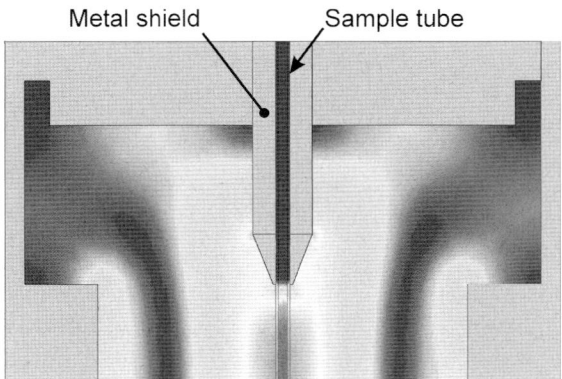

Figure 5.2.7 Radiofrequency magnetic field magnitude in a uniform field X-band cavity with microwave shields in the end regions. A shield hides the sample from the return flux region in each end section, which results in about 95% uniform field over the sample volume.

5.2.4.4 Loop–Gap Resonators

The LGR, as introduced into EPR, was based on a lumped circuit model (Froncisz and Hyde, 1982). The structure was viewed as a solenoid, despite having just a single turn, and therefore the RF magnetic field was expected to be strictly uniform except for minor effects within about one radius of the ends of the loop. In all of the earlier studies, the length of the loop had been much less than a wavelength, but more recently the extension of the LGR to a greater length had been investigated and nonuniformities encountered. The reason for this was the same as for the uniform field cavities; namely, that without special attention being paid to the design of the LGR ends, an infinite wave impedance would not be presented to

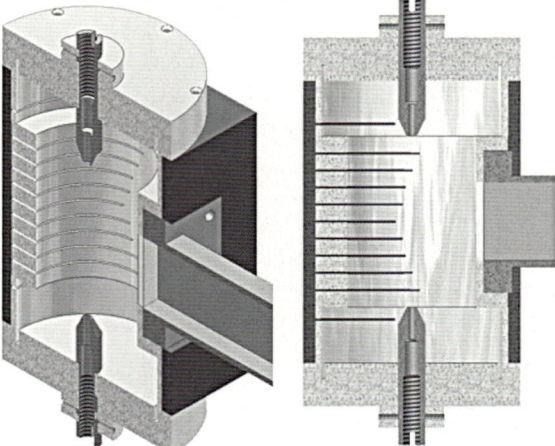

Figure 5.2.8 Uniform field X-band TE$_{01U}$ cavity mechanical design. Movable shield sections allow for fine frequency adjustment to satisfy the cut-off condition.

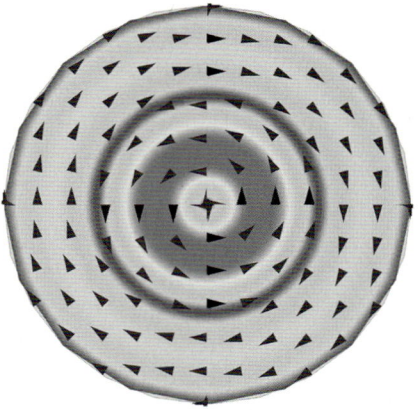

Figure 5.2.9 Radiofrequency electric field in a cylindrical TE$_{02U}$ cavity. A plane of uniform RF magnetic field is present in the azimuthal ring of the null electric field.

the LGR body (Mett, Sidabras, and Hyde, 2007). This would cause a standing wave along the LGR axis, and a cosine shape to the RF magnetic field. With an appropriate end design, however, which could be accomplished by small metallic trimming elements or dielectric end sections, the LGR axial wavelength could be made infinite, producing a uniform axial RF field. This is shown for a Q-band three-loop–two-gap design in Figure 5.2.10. A two-loop–one-gap LGR can be viewed as a section of ridged waveguide at cut-off; by using this analogy, the family of uniform field cavities can be extended to include LGRs.

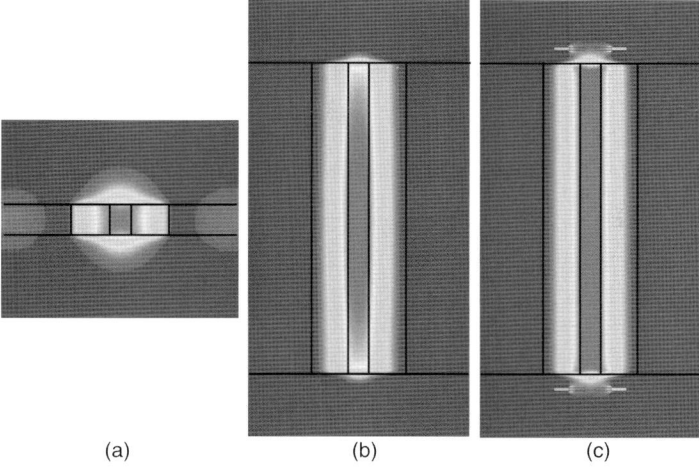

Figure 5.2.10 Cross-sectional profiles of axial RF magnetic field energy H_z^2 in a plane bisecting the three-loop–two-gap resonator through the gaps. (a) 1 mm-long LGR; (b) 10 mm untrimmed; (c) 10 mm trimmed.

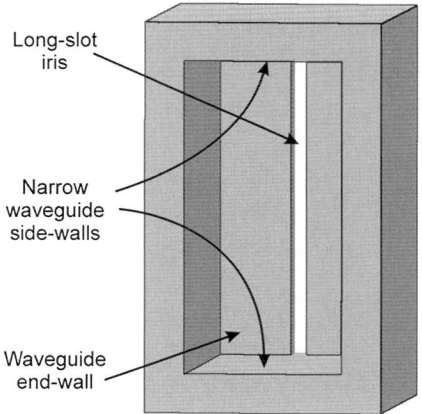

Figure 5.2.11 Long-slot capacitive iris geometry.

5.2.5
Coupling

When the size of the sample increases, the Q-value decreases and the iris size, necessarily, must be increased. A larger iris produces a larger eigenmode perturbation which, in turn, pushes the RF electric field null out of the sample location. This decreases the Q-value further, and coupling becomes more difficult. Coupling can be achieved by an appropriate iris design, which includes both capacitive (Figure 5.2.11) and inductive coupling. In addition, multiple irises instead of a single iris can be used. The analysis of these effects requires finite-element modeling.

Inductive coupling has the general property that the surface currents of the coupler have an almost 90° phase shift relative to the resonator currents near the iris (Mett, Sidabras, and Hyde, 2008b). If the resonator Q-value is relatively low, the coupler currents strongly perturb the resonator currents in the vicinity of the coupler; depending on the resonator design, this can cause RF magnetic field nonuniformities at the sample, and lead to degradation of the EPR signal quality. For couplers of a given coupling strength, the RF magnetic fields of capacitive couplers are lower than inductive couplers. The surface currents of capacitive couplers have minimal phase shifts relative to the cavity surface currents in the vicinity of the coupler, and are generally less perturbing than inductive couplers.

5.2.5.1 Coupling at Low Frequencies

At low frequencies, inductive coupling is weaker due to the frequency term in Faraday's law and, typically, capacitive coupling is used for MRI and low-frequency EPR. Inductive coupling does have the benefit, however, that it is self-balanced, with no balun network being needed and the design being simplified. Inductive coupling is conveniently obtained by using a coaxial transmission line terminated by a sample loop; moreover, it can be enhanced at low frequencies by reducing the self-inductance of the coupling loop while maintaining the mutual inductance of the coupler (Sidabras, Hyde, and Mett, 2006) and the LGR. For a coupling loop, this can be achieved by increasing the axial length of the loop while keeping the cross-sectional area fixed. In the design of Sidabras, Hyde, and Mett (2006), the coupling loop was positioned in the return flux region of a one-loop–two-gap LGR with shield. The coupling loop was inductive and resonant, producing large RF magnetic fields. However, the RF magnetic field uniformity at the sample was not perturbed by the coupler, because the gap structure shielded the interior loop.

5.2.5.2 Coupling at High Frequencies

At high frequencies using a waveguide transmission line, capacitive coupling can be achieved by a thin slotted iris with a large dimension greater than one-quarter of a free-space wavelength (such an iris is shown in Figure 5.2.11). Here, the length of the slot can be the same as the large dimension of the waveguide. The advantages of this type of coupler over a standard inductive iris include: (i) the RF magnetic field mode uniformity is higher; (ii) the frequency deviation of the coupled system from the natural resonance frequency of the resonator is lower; and (iii) the frequency of the coupled system is higher than the natural resonance frequency of the resonator, which tends to cancel the frequency shift that occurs due to the sample presence (Mett, Sidabras, and Hyde, 2008b).

Figure 5.2.12 shows the z-component of the surface currents in a Q-band TE_{011} cylindrical cavity. In a pure modal solution, no z-components exist, but with a round, purely inductive iris, significant z-components are found around the hole, resulting in coupling of power to the nearby TE_{311} mode. The TE_{011} cavity can be

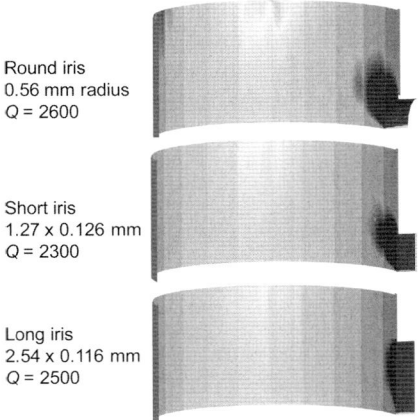

Figure 5.2.12 Comparison of (z) currents in cylindrical TE$_{011}$ at W-band for three iris designs (RGB +, 0, -). Significant coupling to the adjacent TE$_{311}$ mode is found when the iris perturbs the cavity wall. Reduction of this perturbation with the long iris results in a more pure TE$_{011}$.

Round iris
0.56 mm radius
Q = 2600

Short iris
1.27 x 0.126 mm
Q = 2300

Long iris
2.54 x 0.116 mm
Q = 2500

coupled with an iris that is both capacitive and inductive; such an iris would couple less to the problematic TE$_{311}$ mode, which would couple into field modulation slots (Mett and Hyde, 2005), causing microwave leakage. The thin capacitive iris minimizes z-component perturbation.

5.2.6
Field Modulation Penetration

Magnetic field modulation can be introduced into EPR cavities by making the cavity walls electrically thin for modulation, but electrically thick for microwaves (Poole, 1983). Intuition suggests the use of the ratio of conductor thickness to skin depth as a design parameter for introducing modulation, and also achieving a high Q-value. However, a study of eddy currents revealed that a different parameter should be used to predict the modulation penetration (Mett et al., 2005), as discussed below. This parameter quantifies the transition between inductive-limited and resistive-limited eddy currents (Tegopoulos and Kriezes, 1985).

"Skin depth" refers to the attenuation of an electromagnetic plane wave inside a conductor (Ramo, Whinnery, and Van Duzer, 1984),

$$\delta = (\pi f \mu_0 \sigma)^{-1/2}, \tag{5.2.17}$$

where f is the frequency of oscillation, μ_0 is the magnetic permeability of free space, and σ is the conductivity. For the plane wave, the skin depth is the distance for which the amplitude of the electric or magnetic field or the current density drops by a factor of e.

Eddy currents are induced in conductors by time-changing magnetic fields which, when perpendicular to the surface of a conductor, causes the same eddy currents to reduce the magnetic fields from their potential scale in the absence of the conductor. When the time-changing magnetic fields are parallel to the surface, the eddy currents can either increase or decrease the field strength, depending on the location of the currents that produce the applied magnetic field.

A coupled-circuit eddy-current model (Mett et al., 2005) predicts how eddy currents flowing on the surface of a conductor influence the surrounding magnetic field strength. For an eddy path of thickness t and width w in the shape of a flat loop, the resistance and self-inductance of the eddy loop can be quantified and related to each other. The electromagnetic field developed around the loop drives an eddy current that is limited by the resistance and self-inductance. When the eddy magnetic field caused by this current is expressed in terms of a ratio with the applied magnetic field, a dimensionless modulation penetration parameter is obtained.

$$\kappa \equiv \frac{\delta^2}{tw}. \tag{5.2.18}$$

It was found that the eddy magnetic field could be limited by decreasing the thickness or width of the current path until κ was greater than about 4. The resistance of the eddy path then became sufficient to limit the size of the eddy current, and a significant modulation field could penetrate the eddy loop.

Resonator size is found to influence modulation penetration. For resonators of 2 cm and smaller, a significant (75%) 100 kHz modulation penetration is achieved with moderately thin plating thickness (3 μm). This renders the technique particularly suitable for EPR resonators in the range of Q-band (35 GHz) to W-band (95 GHz). For larger resonators, to achieve a significant modulation penetration at reasonable plating thicknesses becomes more difficult, because the increase in the size of the eddy current loop causes, in turn, an increase in the eddy magnetic field and reduces the modulation penetration. In addition, a lower frequency RF requires a thicker plating to maintain the RF Q factor. In this case, slots cut perpendicular to the modulation field can be used to break the eddy current path into smaller regions, and thus increase the modulation penetration. The techniques of Mett et al. (2005) can be used to predict the required slot spacing. Essentially, decreasing the eddy path width w until κ is greater than about 4 produces penetration.

Modulation slots can cause the leakage of microwaves from the resonator (Mett and Hyde, 2005). A Q-band cylindrical TE_{011} cavity with slots cut parallel to wall currents and with mode suppression gaps in the end regions was studied using Ansoft HFSS. The RF leakage was found to be caused by RF field distortions from the waveguide coupling iris centered on the cylindrical sidewall. The distortion was of two distinct types, each of which resulted in comparable leakage levels. One distortion was from the iris fields that had components perpendicular to the fields of the mode of interest. The iris near-field couples to a nonresonant circumferential radial mode in the modulation slots, like that of a sectorial waveguide horn. The other type of RF leakage was caused by coupling of the aperture to nearby cavity modes that leaked from the modulation slots if they had currents across the slots. End gaps have been used to "suppress" the TM modes that, without the gaps, are degenerate with the TE_{011}. It was found that the TM modes were not suppressed, but rather altered, as were most modes. The modes were combinations of coaxial modes that were excited in the end-gap regions coupled to normal cavity modes. The most desirable method of reducing leakage is to use a combination of a thin-slotted capacitive iris, thin modulation slots, and sufficiently thick cavity walls; EDM techniques can be used to produce slots of the required dimensions. A less-desirable approach to reducing the RF leakage is to place a floating conducting shield just outside the modulation slots. If resistive material is placed in the end gaps, then RF leakage from the modulation slots would decrease because the amount of energy coupled to an adjacent cavity mode would be reduced, and this would increases the Q-value of the TE_{011} mode.

5.2.7
Sample Access Stacks

Finite-element modeling has permitted the investigation of microwave leakage from sample access stacks of different size, shape, and asymmetry (Mett, Hyde, and Anderson, 2002). The main conclusion of these studies was that it is possible to have much larger sample access stacks than have been used in the past. For example, in cylindrical geometries, the diameter of the sample access stack can be up to 85% of the diameter of the cavity itself. This is because significant coupling to a TE_{11} mode propagating out of the stack requires far more asymmetry than is generally possible. In addition, it was found that a larger stack could produce *less* TE_{11} mode coupling than a smaller one. The RF leakage from the stack is kept low by making the sample access stack longer than usual, and the use of thin-walled quartz sample tubes also allows larger sample access. These findings have significant consequences for Q- and W-band EPR cavity resonator design.

A primary goal of the studies conducted by Mett, Hyde, and Anderson (2002) was to design a Q-band cylindrical TE_{011} cavity that could accommodate an X-band-size sample tube. This would be useful for freeze–quench samples that require the mechanical packing of snow from an isopentane liquid. Sample-access stacks in traditional cylindrical resonator designs have a diameter at which all cylindrical propagation modes are well below cut-off; that is, no propagating modes can exist in the stack. This makes sample tube diameters at Q- and W-band traditionally quite small.

The fields of a particular propagation mode in a waveguide are sinusoidal if the frequency of the mode is above the cut-off frequency of the particular mode in the particular waveguide. The cut-off frequency, f_c, of any cylindrical propagation mode is given by Equation 5.2.19 from Ramo, Whinnery, and Van Duzer (1984),

$$f_c = \frac{\alpha c}{a\sqrt{\varepsilon_r}}, \qquad (5.2.19)$$

where α represents a numerical factor that depends on the mode, c is the speed of light in vacuum, a is the radius of the cylinder, and ε_r is the relative dielectric constant of the material in the cylinder. Values of α for the lowest three propagation modes in a cylinder are given by: $\alpha_{TE11} = 0.293$, $\alpha_{TM01} = 0.383$, and $\alpha_{TE01} = 0.609$. If the frequency of the mode f is below the cut-off frequency, the fields are evanescent (decay exponentially) with axial distance z up the stack with a decay constant

$$\gamma = \frac{2\pi}{c}\sqrt{\varepsilon_r(f_c^2 - f^2)}. \tag{5.2.20}$$

The Ansoft HFSS eigenmode solution method was used to examine the influence of large stacks on the cylindrical TE_{011} cavity mode. For this, the top of the stack was modeled with a resistive boundary equivalent to the wave impedance of free space. The results are shown in Figure 5.2.13. Due to symmetry, the lowest modes, TE_{11} and TM_{01}, are not present in the stack, but if they were the size of the stack would permit propagation. This would cause significant RF leakage and a reduction in Q, especially for the 8 cm-long stack. However, only the TE_{01} mode is excited in Figure 5.2.13. Since it is evanescent, a longer stack effectively prevents leakage of this mode from the cavity. The axial field profiles in the stack region for several different diameter stacks were examined, and the decay profile compared to the predictions of the field decay equations above.

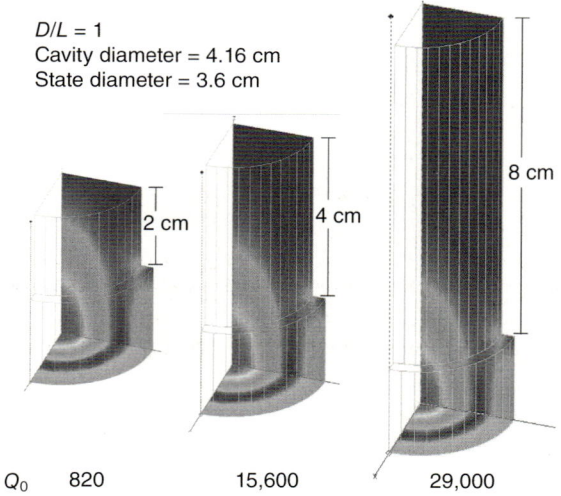

Figure 5.2.13 Influence of stack length on the cavity Q. The TE_{011} cavity mode couples to the TE_{01} propagation mode at the cavity/stack boundary. Since the TE_{01} propagation mode is evanescent in the stack, the field decay is greater for an 8 cm stack than for a shorter one. Leakage is negligible for the longer stack. The top of the stack is modeled with a resistive boundary. Due to symmetry, there is no coupling to the TE_{11} propagation mode, which is not evanescent in the stack. $f = 9.2$ GHz.

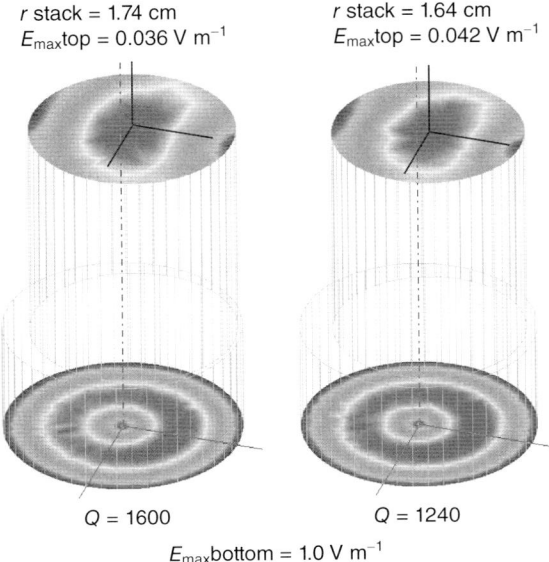

Figure 5.2.14 Influence of asymmetries on Q degradation. Nonconcentricities cause coupling to the TE_{11} propagation mode at the cavity/stack boundary. Since the TE_{11} mode is not evanescent in the stack, large microwave leakage and Q degradation results. More leakage can occur for a smaller stack than a larger stack, due to coupling at the cavity/stack boundary. Significant leakage occurs only with large asymmetries.

Ansoft HFSS was then used to determine how much nonconcentricity was required to produce enough coupling to the TE_{11} mode to cause noticeable Q degradation. A 5 mm quartz rod placed off-center produced very little Q degradation until it was more than 5 mm off axis – which is not practically achievable. Placing the stack off-center also excites the TE_{11} mode (details are shown in Figure 5.2.14). These results may be understood in terms of the theory of reflection and transmission of modes by plane diaphragms in waveguides. Jackson (1975) showed that the transmission coefficient (which quantifies the fraction of the wave energy that couples at the aperture) for a propagating mode excited by a field over a planar aperture of any size is given by

$$T = \int_{aperture} \mathbf{E} \cdot \mathbf{E}_c^* dS, \qquad (5.2.21)$$

where $\mathbf{E}$ represents the normalized electric field that is present over the aperture, and $\mathbf{E}_c$ represents the normalized electric field of the coupled mode over the aperture. One way to view this equation is to imagine $\mathbf{E}$ as the field of the cavity mode and $\mathbf{E}_c$ the field of a propagating TE_{11} mode in the stack. Each mode has its own field symmetry. It can be seen that T will be zero over any aperture if $\mathbf{E}$ and $\mathbf{E}_c$

have orthogonal symmetries. In Figure 5.2.14, it can be seen that the smaller aperture causes a larger T because the cavity fields over the center of the smaller (nonconcentric) aperture are more complementary to TE_{11} than they are for the larger aperture.

The relative dielectric constant of the material filling the cavity has a large influence on the resonant frequency of the cavity. In order to maintain an operating frequency of 35 GHz, a cavity-length adjustability feature was designed whereby a small radial air gap between the cavity wall and end wall was used to prevent the fields from causing RF current losses at metal–metal sliding contacts.

Since f_c is inversely proportional to $\sqrt{\varepsilon_r}$, the material filling the stack has a large influence on the cut-off frequency of propagation modes in the stack. By using quartz, $\varepsilon_r = 3.78$, the maximum stack diameter at Q-band to ensure that the TE_{11} mode cannot propagate was found to be 5.4 mm. The decision was taken to use a common X-band sample tube of 4 mm outer diameter, the stack diameter for which was 0.4 cm. The TE_{11} cut-off frequency for this stack was approximately 43 GHz (empty), 27 GHz when filled with hydrocarbons, and 22 GHz when filled with ice. The cut-off frequency for the TE_{01} mode was 2.1-fold higher than for the TE_{11} mode.

5.2.8
Conclusions

The structure of this chapter, and indeed of the authors' own research, has been to divide resonator design into parts, namely free space mode; sample support geometries; coupling; field modulation strategies; and sample entry stacks. Subsequently, each part has been analyzed independently, with particular attention being paid to the microwave frequency, including the dependence of the complex dielectric constant of water on the microwave frequency. Notably, such an approach will involve the use of finite-element modeling of electromagnetic fields. Whilst many of the results are essentially impossible to verify in the laboratory, the authors have every confidence in the validity of the finite-element methodology in principle – although of course errors may arise in its use! For instance, although Ansoft's default meshing algorithm focuses on regions of maximum electric field, it does allow the user to further mesh other important regions. In addition, special attention is required in the gap regions of LGRs, where loop-to-gap ratios can be very high. The challenge here is to develop a level of understanding that will prevent these errors from occurring, with the final proof lying in the overall performance of the complete structure.

Ultra-precise EDM fabrication is an important advance that couples naturally with theoretical modeling. Yet, an extension to laser milling is foreseen, permitting fabrication at still higher levels of precision. These fabrication tools, when combined with finite-element modeling, should enable the creation of high-frequency resonators that once were out of reach.

Pertinent Literature

All resonator design projects conducted in the authors' laboratory began with a review of the relevant sections of C.P. Poole's treatise published in 1983, *Electron Spin Resonance: A Comprehensive Treatise on Experimental Techniques*, 2nd edition, John Wiley & Sons, New York. Many resonator-design discussions between the authors led to Feher's report in 1957 on EPR sensitivity (*Bell System Technical Journal*, **36**, 449). Other reviews cited in the present chapter include:

- J.S. Hyde and W. Froncisz (1986), in *Electron Spin Resonance* (ed. M.C.R. Symons), The Royal Society of Chemistry, London, pp. 175–184.
- J.S. Hyde and W. Froncisz (1989), in *Advanced EPR: Applications in Biology and Biochemistry* (ed. A.J. Hoff), Elsevier, Amsterdam, pp. 277–306.
- G.A. Rinard and G.R. Eaton (2005), in *Biological Magnetic Resonance* (eds. S.S. Eaton, G.R. Eaton, and L.J. Berliner), Academic Press/Plenum Publishing, New York, pp. 19–52.
- J.S. Hyde (2005) *Handbook of Microwave Technology*, **2**, p 365, which provides a review of the resonator technology developed at Varian.
- O. Grinberg and L.J. Berliner (eds) (2004) *Very High Frequency (VHF) ESR/EPR*, volume 22 of *Biological Magnetic Resonance*. Resonators for use at high frequency are described in several chapters of this book.

Acknowledgments

These studies were supported by grants EB001980, EB001417, and EB002052 from the National Institutes of Health.

References

Anderson, J.R., Mett, R.R., and Hyde, J.S. (2002) *Rev. Sci. Instrum.*, **73**, 3027.

Bloch, F. (1946) *Phys. Rev.*, **70**, 460.

Eaton, S.S. and Eaton, G.R. (1977) *Anal. Chem.*, **49**, 1277.

Ellison, W.J. (2007) *J. Phys. Chem. Ref. Data*, **36**, 1271.

Feher, G. (1957) *Bell Systems Tech. J.*, **36**, 449.

Freed, J.H., Leniart, D.S., and Hyde, J.S. (1967) *J. Chem. Phys.*, **47**, 2762.

Froncisz, W. and Hyde, J.S. (1982) *J. Magn. Reson.*, **47**, 515.

Hyde, J.S. (1972) *Rev. Sci. Instrum.*, **43**, 629.

Hyde, J.S. and Froncisz, W. (1986) Loop-gap resonators, in *Electron Spin Resonance* (ed. M.C.R. Symons), The Royal Society of Chemistry, London, pp. 175–184.

Hyde, J.S. and Froncisz, W. (1989) Loop-gap resonators, in *Advanced EPR: Applications in Biology and Biochemistry* (ed. A.J. Hoff), Elsevier, Amsterdam, pp. 277–306.

Hyde, J.S. and Mett, R.R. (2002) *Curr. Top. Biophys.*, **26**, 7.

Hyde, J.S., Mett, R.R., and Anderson, J.R. (2002) *Rev. Sci. Instrum.*, **73**, 4003.

Hyde, J.S., Yin, J.-J., Subczynski, W.K., Camenisch, T.G., Ratke, J.J., and Froncisz, W. (2004) *J. Phys. Chem. B*, **108**, 9524.

Hyde, J.S., Eaton, G.R., and Eaton, S.S. (2006) *Concepts in Magnetic Resonance*,

EPR at Work [special issue], vol. 28A, issue 1, John Wiley & Sons, Inc., New York, pp. 1–100.

Jackson, J.D. (1975) *Classical Electrodynamics*, 2nd edn, John Wiley & Sons, Inc., New York.

Mett, R.R. and Hyde, J.S. (2003) *J. Magn. Reson.*, **165**, 137.

Mett, R.R. and Hyde, J.S. (2005) *Rev. Sci. Instrum.*, **76**, 014702.

Mett, R.R., Froncisz, W., and Hyde, J.S. (2001) *Rev. Sci. Instrum.*, **72**, 4188.

Mett, R.R., Hyde, J.S., and Anderson, J.R. (2002) Abstract presented at the *44th Rocky Mountain Conference on Analytical Chemistry, Denver, CO.*

Mett, R.R., Sidabras, J.W., and Hyde, J.S. (2004) *Curr. Top. Biophys.*, **28**, 117.

Mett, R.R., Anderson, J.R., Sidabras, J.W., and Hyde, J.S. (2005) *Rev. Sci. Instrum.*, **76**, 094702.

Mett, R.R., Sidabras, J.W., and Hyde, J.S. (2007) *Appl. Magn. Reson.*, **31**, 571.

Mett, R.R., Sidabras, J.W., Golovina, I.S., and Hyde, J.S. (2008a) *Rev. Sci. Instrum.*, **79**, 094702.

Mett, R.R., Sidabras, J.W., and Hyde, J.S. (2008b) *Appl. Magn. Reson.*, **35**, 285.

Poole, C.P., Jr (1983) *Electron Spin Resonance: A Comprehensive Treatise on Experimental Techniques*, 2nd edn, John Wiley & Sons, Inc., New York.

Ramo, S., Whinnery, J.R., and Van Duzer, T. (1984) *Fields and Waves in Communication Electrics*, 2nd edn, John Wiley & Sons, Inc., New York.

Rinard, G.A. and Eaton, G.R. (2005) Loop-gap resonators, *Biological Magnetic Resonance* (eds S.S. Eaton, G.R. Eaton, and L.J. Berliner), Academic/Plenum, New York, 19–52.

Sidabras, J.W., Mett, R.R., and Hyde, J.S. (2005) *J. Magn. Reson.*, **172**, 333.

Sidabras, J.W., Hyde, J.S., and Mett, R.R. (2006) Abstract presented at the *48th Rocky Mountain Conference on Analytical Chemistry, Breckenridge, CO.*

Sidabras, J.W., Mett, R.R., Froncisz, W., Camenisch, T.G., Anderson, J.R., and Hyde, J.S. (2007) *Rev. Sci. Instrum.*, **78**, 034701.

Stoodley, L.G. (1963) *J. Electron. Control*, **14**, 531.

Tegopoulos, J.A. and Kriezes, E.E. (1985) *Eddy Currents in Linear Conducting Media*, Elsevier, New York.

5.3
Multifrequency EPR Sensitivity

George A. Rinard, Richard W. Quine, Sandra S. Eaton, and Gareth R. Eaton

5.3.1
Introduction

This chapter considers the minimum number of spins that can be observed by EPR at various microwave frequencies. It is convenient to discuss the EPR signal in terms of signal voltage. In the presence of a magnetic field, the precession of the magnetic moment of the electron spins in a paramagnetic sample induces a voltage in the materials of the resonator. This is a general principle. To simplify the discussion in this introduction, the resonator is assumed to be made of copper and to be a loop–gap resonator (LGR) (Hyde and Froncisz, 1989). A convenient geometry of LGR, which can be assumed without any loss of generality, is a right-circular cylinder with a capacitive gap parallel to the cylinder axis. This is the geometry of the first LGR described by Froncisz and Hyde (1982).

The sensitivity of EPR spectrometers as a function of microwave frequency is multifaceted. For transitions that can be seen only at high microwave frequencies,

Table 5.3.1 Classes of sample.

Does the sample saturate?	Is the sample size limited?	Does the sample exhibit dielectric loss?
Yes	Yes	Yes
Yes	Yes	No
Yes	No	Yes
Yes	No	No
No	Yes	Yes
No	Yes	No
No	No	Yes
No	No	No

the dependence of sensitivity on frequency is binary – either the microwave frequency is high enough for observation of the transition, or it is not. More commonly, the question of EPR sensitivity is phrased in terms of signal-to-noise ratio (SNR) for an $S = ½$ organic radical as a function of microwave frequency for an ideal spectrometer. The predicted frequency dependence of sensitivity is a function of the characteristics of the sample (Table 5.3.1). There are eight cases to consider, which arise from all possible combinations of three properties (Hyde, 2006): is the sample size limited; is the sample lossy; and does the signal saturate at available microwave powers?

5.3.1.1 Nomenclature

The literature of this field uses three terms: SNR, sensitivity, and minimum detectable number of spins, with much the same intent to describe either the performance of a particular EPR spectrometer or the theoretical limit. In the following discussion the terms SNR, sensitivity, and minimum detectable number of spins will be used largely interchangeably. Usually, SNR is used to describe an experimentally observed performance. For commercial X-band CW-EPR spectrometers a common performance check is the SNR achieved on a "weak pitch" standard sample provided by the manufacturer and run under carefully specified conditions.

Sensitivity and minimum detectable number of spins have identical meanings: they can be either experimental descriptions of a particular spectrometer system, or theoretical predictions. For example, if the construction of the spectrometer were well-characterized, such that the noise added by components was known, it might be calculated that the EPR signal from 10^x spins could be detected with SNR = 1. If the measured SNR for a sample extrapolates to the number of spins expressed as 10^x with SNR = 1, then the sensitivity of this spectrometer (which is equal to the minimum detectable number of spins) would be 10^x. Alternatively, the number of spins might be calculated that, in principle, should be measurable if the only noise was the thermal noise of the resonator. This value will likely be less than the experimental 10^x. By full analysis, as outlined later in this chapter, it would be possible to identify the limiting components and thus establish engineering development goals for incremental improvement in the performance of the spectrometer toward the goal of detecting the theoretical minimum number of spins.

Sensitivity is used in this chapter to designate either the theoretically achievable value (Table 5.3.2) or the extrapolated experimental value (see Table 5.3.5). The frequency dependence of the signal intensity, by contrast, is used only in making theoretical predictions. To characterize a spectrometer, it is necessary to consider both signal intensity and noise, as is in the concepts of SNR, sensitivity, and minimum number of detectable spins. As the spectral linewidth increases, the signal amplitude decreases. If the noise is "white" (random), the noise that passes through the detection system increases in proportion to the square root of the bandwidth. Therefore, the units (G√Hz) take into account the width of the spectral line in gauss and the bandwidth of the detection system in Hz.

5.3.2
Frequency Dependence of Sensitivity for an Ideal Spectrometer, at the Thermal Noise Limit

5.3.2.1 General Expression for SNR

The general expression for CW-EPR signal intensity has been presented in many reports. Its derivation was described in Rinard et al. (1999a, 2004). The signal voltage at the end of the transmission line connected to the resonator, V_s, is given by:

$$V_s = \chi'' \eta Q \sqrt{PZ_0}$$

where η (dimensionless) is the filling factor (Poole, 1967), Q is the loaded quality factor of the resonator, Z_0 is the characteristic impedance of the transmission line, and P is the microwave power to the resonator produced by the external microwave source (Rinard et al., 1999a, 2004). The magnetic susceptibility of the sample, χ'', is the imaginary component of the effective radiofrequency (RF) susceptibility. Optimizing the EPR measurement involves optimizing each of these crucial parameters. The key variables in this discussion of EPR sensitivity – χ'', Q, and η – must be carefully defined.

The experimental SNR is the ratio of the maximum signal amplitude to the root mean square (rms) noise. In the following calculations it is important to distinguish carefully between rms and peak values of noise and of B_1. The estimate of the ultimate achievable sensitivity includes a calculation of the thermal noise power. The resultant noise voltage (square root of power) is an rms value, which is appropriate for the denominator of the sensitivity calculation. The V_s depends on the square root of microwave power, which is proportional to the rms B_1. The filling factor calculation uses the linearly polarized B_1, but only the circularly polarized component creates the EPR signal.[1] The factor of ½ can be included explicitly in the calculation of V_s, or in the calculation of η.

In calculating spin susceptibility, one would ideally use the full line shape function, and the fraction that is detected in the CW measurement. In the derivations for Table 5.3.2, it is assumed that the entire sample magnetization is measured, and that there is no power saturation of the spin system. Except in special cases, these assumptions about line shape are unlikely to be valid over the entire range of multifrequency EPR currently accessible experimentally. Based on these definitions, the spin magnetization, $M_0 \left(= H_0 \chi_0 = \dfrac{B_0}{\mu_0} \chi_0 \right)$, is

$$M_0 = N_0 \frac{\gamma^2 \hbar^2 B_0 S(S+1)}{3 k_B T} \; \text{J T}^{-1} \text{ m}^{-3} \; (= \text{A m}^{-1}),$$

For $S = \frac{1}{2}$,

$$M_0 = N_0 \frac{\gamma^2 \hbar^2 B_0}{4 k_B T} = N_0 \frac{g^2 \beta^2 B_0}{4 k_B T_{sample}}$$

In the above expressions, $g\mu_B = \gamma\hbar$; $\gamma = 1.7608 \times 10^7 \text{ rad s}^{-1} \text{G}^{-1} = 1.7608 \times 10^{11} \text{ s}^{-1} \text{T}^{-1}$; $\hbar = 1.0546 \times 10^{-27}$ erg s rad^{-1}; N_0 is the number of spins per unit volume. (In some publications, the number of spins per unit volume is split into two terms, N, the number of spins in volume V, and V, the volume of sample in m^3.) The static magnetic field $B_0 = \omega_0/\gamma$; S is the electron spin (this will be assumed to be ½ in all of the calculations); $k_B = 1.3806 \times 10^{-16}$ erg K^{-1} is Boltzmann's constant; and T is the temperature of the sample in K. The permeability of a vacuum, $\mu_0 = 4\pi \times 10^{-7} \text{T}^2 \text{J}^{-1} \text{m}^3$. The magnetic susceptibility of the sample, χ'' (dimensionless), is the imaginary component of the effective RF susceptibility. For a Lorentzian line, with width at half height = $\Delta\omega$ at resonance frequency, ω,

$\chi'' = \chi_0 \dfrac{\omega}{\Delta\omega}$ (Although most EPR spectra are not well-resolved Lorentzian lines, there is no loss of generality by making this assumption in these calculations.)

To calculate the SNR, it is necessary to know the distribution of the microwave fields in the resonator, the resonator Q, the filling factor of the sample in the resonator, and the coupling of the resonator to the transmission line, as well as the magnetic properties of the sample.

---

1) Here, more detail is provided concerning the multiplicative factor of ½ that is needed to take account of the fact that only one of the circularly polarized components of B_1 affects the spin magnetization, and therefore only half of the B_1 calculated from the incident power is used. One might think that the B_1 for the sample in the equation for filling factor should be multiplied by ½; however, this would be equivalent to reducing the efficiency of the resonator by ½ and this is not the case. The effect of circular polarization is similar to the effect of reducing the source power so that B_1 has half its previous value. It can be seen that this would reduce the EPR signal by a factor of ½, not (½)^2. Therefore, to reduce confusion, the ½ factor is in the equation for V_s instead of in the equation for η.

To calculate the frequency dependence of SNR, a loop–gap-type resonator with a constant height/radius ratio was used at all frequencies. It was assumed that the resistance of the materials of construction and the geometric size of the resonator determine the resonator Q. Loop-gap resonators have been used at frequencies up to W-band (Oles, Hyde, and Froncisz (1989), Hyde et al. (2007), Sidabras et al. (2007)). At higher frequencies it is common to use a TE_{011} or Fabry-Perot resonator. The properties of a TE_{011} cylindrical resonator are similar enough to a cylindrical LGR that the approximations made in Table 5.3.2 for predictions of frequency dependence of sensitivity can also be reasonably applied to TE_{011} resonators.

The details of derivations of the frequency dependence of each term in the expression for V_S are provided by Rinard et al. (1999a, 2002c, 2004). Care must be taken when interpreting expressions for the frequency dependence of η and Q independently, because the fundamental resonator parameters affect Q, the filling factor, and microwave B_1 distribution in interdependent ways. The results obtained for the three cases determined by Rinard et al. (2004) are listed in Table 5.3.2.

Table 5.3.2 Predicted frequency dependence of EPR sensitivity when resonator resistance dominates[a)b)c)].

		Case 1	Case 2	Case 3
		Constant sample size	Sample size $\propto 1/\omega$	Constant sample size
		Constant LGR size	LGR size $\propto 1/\omega$	LGR size $\propto 1/\omega$
1	L (resonator inductance)	1	ω^{-1}	ω^{-1}
2	R (resonator resistance)	$\omega^{1/2}$	$\omega^{1/2}$	$\omega^{1/2}$
3	Resonator Q	$\omega^{1/2}$	$\omega^{-1/2}$	$\omega^{-1/2}$
4	η (filling factor)	1	1	ω^3
5	EPR SNR at constant P	$\omega^{3/2}$	$\omega^{1/2}$	$\omega^{7/2}$
6	$B_1/\sqrt{P}$	$\omega^{-1/4}$	$\omega^{3/4}$	$\omega^{3/4}$
7	P to maintain constant B_1	$\omega^{1/2}$	$\omega^{-3/2}$	$\omega^{-3/2}$
8	EPR SNR at constant B_1	$\omega^{7/4}$	$\omega^{-1/4}$	$\omega^{11/4a}$

a) Note that in Rinard et al. (1999a), Table 1, Case 3, row 8, there is a typographical error: the EPR SNR is proportional to $\omega^{11/4}$, not $\omega^{5/4}$ as stated in that reference.
b) The table lists overall frequency dependence and does not include proportionality constants. In order to predict frequency dependence without consideration of the details of the EPR spectrum, Table 5.3.2 was constructed assuming that all of the spins are observed at each frequency. Experimentally, all of the spins in standard samples (such as weak pitch) may not be observed, so the custom is to report experimental sensitivity in terms of spins per Gauss, and also to correct for detection system bandwidth, so the units are spins per Gauss per square root Hz.
c) P is the power (in Watts) from the source that is incident on the resonator. B_1 is the microwave magnetic field in the resonator.

5.3.2.2 Explanation of Table 5.3.2

The entries in Table 5.3.2 are based on the frequency dependence of the parameters for LGRs when the diameter and length are either kept constant (Case 1), or when the dimensions of the resonators are changed in proportion to the inverse of the frequency (Cases 2 and 3). The key parameters are described here.

- The inductance, L, of the LGR is the ratio of the cross-sectional area, A, to the length, z, of the cylindrical resonator. The area equals $\pi d^2/4$, where d is the diameter. Including fundamental constants, $L = \dfrac{\eta_0 \pi d^2}{4z}$. If the resonator size is constant, L will remain constant, so the entry for Case 1 is 1. If the resonator is reduced in size inversely proportional to the frequency, L also scales as ω^{-1}, as in the entries for Cases 2 and 3.

- The resistance, R, of the materials of the resonator increases as the square root of the microwave frequency. Hence the entries in row 2 of Table 5.3.2 are all $\omega^{1/2}$.

- There are many ways to express the quality factor, Q, of a resonator. Using the time constant, t, for the cavity ringdown following a pulse, $Q = 2\pi v t$, where v is the resonant frequency. Q can also be expressed as the ratio of the resonant frequency to the bandwidth of the resonator, $Q = \omega/\Delta\omega$. For the purposes of calculating sensitivity, it is convenient to write the loaded Q in the form, $Q = \dfrac{\omega L}{2R}$, which explains why it was necessary to find the frequency dependence of L and R. Row 3 for the three cases in Table 5.3.2 gives the frequency dependence of Q as the ratio of the frequency dependence of L and R in rows 1 and 2.

- The filling factor, η, in row 4 of Table 5.3.2, is treated as a geometric factor. It is assumed that the microwave B_1 is constant over the sample, so the filling factor is proportional to the fraction of the volume of the resonator occupied by the sample. In Cases 1 and 2 this fraction is a constant, but in Case 3 the resonator dimensions are reduced inversely proportional to the frequency, but the sample size remains constant, so the filling factor increases proportional to ω^3.

- The spin magnetization increases linearly with magnetic field (hence, microwave frequency). The EPR signal (V_s) is proportional to $\chi''\eta Q$, so when the microwave power is constant, and the resonator is critically coupled, the frequency dependence of the EPR signal is ω times the frequency dependence of η and Q. Hence, row 5 in Table 5.3.2 is ω times rows 3 and 4. Here, and throughout Table 5.3.2, the noise is assumed to be white thermal noise in the resonator, which is independent of microwave frequency. Hence, the trend in ultimate SNR (sensitivity) is $\chi''\eta Q = (\omega)(1)(\omega^{1/2}) = \omega^{3/2}$ for Case 1 and analogously for Cases 2 and 3 in row 5.

- Since one would strive to compare EPR spectra obtained with the same microwave B_1 at the sample, and not just the same microwave power incident on the resonator, the next three rows in Table 5.3.2 make this conversion. The derivation of $B_1/\sqrt{P}$ (row 6) is beyond the scope of this section (Rinard et al., 2004). To maintain constant B_1 (row 7), the power must be increased by the inverse square of the entries in row 6. The final goal of these calculations, the EPR sensitivity when B_1 is constant (row 8), is the product of rows 5, 6, and 7.

The predictions in Table 5.3.2 assume that the resonator has the same loop–gap structure at all frequencies, and is scaled in size inversely proportional to the frequency. The samples and LGRs used in the studies by Rinard et al. (2002c; 2002b) were designed to satisfy this assumption and to permit the comparison of SNR at 250 MHz, L-band, and X-band. In most experiments, this assumption would not be satisfied. For example it should be noted that, in Table 5.3.2, if the sample is unlimited in size (Case 2), so that the amount of sample can be scaled with the size of the resonator, the EPR SNR is predicted to increase at a lower microwave frequency. On the other hand, if the sample is small enough to fit a high-frequency resonator and the same sample is used at all frequencies (Case 3), the SNR is predicted to increase with the 11/4 power of the microwave frequency.

5.3.2.3 On Beyond the Predictions of Table 5.3.2

The experimental data available for multifrequency spectrometers is for a variety of resonator types and sizes. Based on the equation for V_s, the direction of improved sensitivity can be predicted, even when the construction of the resonators does not fit the assumptions of Table 5.3.2. The sensitivity will be proportional to the filling factor of the sample in the resonator and to the resonator Q:

$$S_i \propto \eta_i Q_i$$

If the resonators and spectrometer systems were designed such that the microwave B_1 at the sample was the same (and uniform over the sample), and ηQ was the same for both resonators, then the sensitivity will increase with the volume, V_i, of sample in the resonator (assuming that the concentration of spins was the same in the samples compared).

Thus, comparing two spectrometers, $i = a$ and b, the sensitivity ratio will be:

$$\frac{S_a}{S_b} = \frac{\eta_a Q_a V_a}{\eta_b Q_b V_b}$$

Several reports on high-frequency EPR refer back to the chapter by Lebedev (1990) for sensitivity, the analysis of which was revised by Rinard et al. (1999a). As pointed out by Rinard et al. (2004), there are alternative sets of assumptions that lead to different dependences of SNR on frequency. One particularly interesting case, considered by Rinard et al. (2004), was that in which a lossy sample reduces the resonator Q (this can occur because the sample loss increases the resistance that determines the resonator Q). When the resonator Q is reduced by the factor R,

Table 5.3.3 Predicted frequency dependence of EPR sensitivity when sample loss dominates[a)b)].

		Constant sample size	Sample size $\propto 1/\omega$	Constant sample size
		Constant LGR size	LGR size $\propto 1/\omega$	LGR size $\propto 1/\omega$
1	SNR at constant P	ρ	$\omega^2 \rho$	ρ
2	$B_1/\sqrt{P}$	$\omega^{-1}\rho^{1/2}$	$\omega^{3/2}\rho^{1/2}$	$\omega^{-1}\rho^{1/2}$
3	P to maintain constant B_1	$\omega^2\rho^{-1}$	$\omega^{-3}\rho^{-1}$	$\omega^2\rho^{-1}$
4	SNR at constant B_1	$\omega\rho^{1/2}$	$\omega^{1/2}\rho^{1/2}$	$\omega\rho^{1/2}$

a) The table lists overall frequency dependence and does not include proportionality constants.
b) ρ is sample resistivity. The sample resistivity usually is frequency dependent, so the full frequency dependence is not only the ω terms.

the SNR is reduced by the same factor. In NMR, and especially in MRI, this effect is often called "sample noise." Formal derivation, including the sample resistivity, ρ, results in the predictions in Table 5.3.3.

5.3.2.4 Dependence of SNR on g-Anisotropy

In the derivations summarized in Tables 5.3.2 and 5.3.3, it was assumed that the linewidth $\Delta\omega$ does not change with frequency. This is a good assumption for the single Lorentzian line used in this model calculation; however, if a real line shape of a magnetically dilute species were dominated by g-anisotropy, the line would become narrower at lower frequency,[2] and the SNR would improve somewhat at the lower frequency as a result. Conversely, for broad lines of magnetically dilute species, the width of which is determined by g-anisotropy, the SNR will decrease at a higher frequency by a factor of ω relative to the terms in Tables 5.3.2 and 5.3.3, due to the proportional decrease in spins per Gauss as the frequency increases (Davoust, Doan, and Hoffman, 1996). Thus, if the spectrometer components do not result in an improvement in SNR that is at least linear in frequency, the linear decrease in spins per Gauss due to g-factor dispersion can reduce the sensitivity at any frequency for which $\Delta g/\Delta B > 1$, where ΔB in this expression represents the linewidth.

5.3.2.5 Source Noise

It was also assumed that source noise does not contribute to the SNR. The source noise could be attenuated by using a bimodal or crossed-loop resonator (see Rinard

2) The typical trend of g-strain-dominated linewidths increasing at higher frequency is not fully general. There are cases in which linewidth does not increase monotonically with frequency. See, for example, the oscillations of linewidth in the frequency range 5 to 10 GHz reported by Popov and Katrich (1993).

et al., 1996, 2000, 2002a). If any other part(s) of the spectrometer contribute noise, the denominator of the SNR expression will contain this noise contribution in addition to the thermal noise discussed here. Various gains, losses, and noise figures in the system are combined in the standard Friis equation (Lee and Dalman, 1994), which relates the overall noise in the spectrometer detection system to the gains (or losses) and noise added by each stage. There are many ways to write the Friis equation. A common method, favored by engineers, is in terms of the noise temperature, which is *not* the physical temperature, of each stage.

$$T_{e1...n} = T_{e1} + \frac{T_{e2}}{g_1} + ... + \frac{T_{en}}{g_1 g_2 \cdots g_{n-1}}$$

where g_i is the power gain of the ith stage and T_{ei} is the noise temperature of the ith stage, and $T_{e1i...n}$ is the overall noise temperature for the system. For each stage, one can write:

$$T_{ei} = T_0(F_i - 1), \text{ where } T_0 \text{ is the standard temperature.}$$

The noise factor (F_i) is the ratio of the output noise power to the portion of the output noise power that is produced by the input thermal noise at $T_0 = 290\,K$. For a noiseless network $F = 1$ and $T_e = 0$, $T_0 = 290\,K$. The noise figure is the log of the noise factor, and equals 0 dB for a noiseless circuit. An alternate–and equivalent–expression for the Friis equation in terms of gains and noise powers is provided in Chapter 4.2.

To understand the importance of the various losses and noise sources in a spectrometer, it would require values of all of the terms in the Friis equation. However, it can be seen from this equation that the overall noise factor (F) of a network is heavily dependent on the gain and noise factor of the first stage amplifier. If the early stages have a low enough noise factor (F_i) and a high enough gain, the later stages are of little importance in the noise calculation, unless they introduce an enormous noise or interference signal (such as 50–60 Hz mains line frequency!). It is especially important to appreciate that all losses prior to the first amplifier increase the effective noise figure by the amount of the losses. For practical calculations of a spectrometer, a mixer, when used as a phase-sensitive detector, has a loss of 1.44 dB (Rinard et al., 1999b).

A noisy microwave source and/or poor noise figures in the signal detection pathway may be major factors confounding comparisons of resonators constructed at different microwave frequencies. The comparison also does not account for the effects of magnetic field modulation, which may introduce microphonics, nor of the problems of environmental noise (such as building vibrations, power line noise, and communications clutter). Furthermore, it is rare that the resonator Q, coupling, or filling factor is well characterized for all of the spectrometers at the frequencies that one would want to compare.

5.3.3
Experimental Validation of Predicted Dependence of Sensitivity on Frequency

5.3.3.1 CW Spectrometers at Frequencies <10 GHz

Although it seems rather forbidding, it is possible to make the necessary calculations and measurements taking into account all spectrometer characteristics. Starting with the number of spins in the sample, one calculates the signal and compares it with the noise expected for the known (measured) gains, losses, and noise figures of each stage in the signal detection path of the spectrometer.

As a practical matter, it is difficult to characterize a spectrometer by measuring every internal loss and gain, and then applying the Friis equation. A more direct approach is to measure the end-to-end (input to output) gain and noise figure, and by so doing treat the spectrometer construction details as a "black box." In the case of spectrometers configured for direct detection (generally pulse or rapid scan spectrometers), this measurement can be made directly by modulating an RF source of known amplitude and measuring the output amplitude. Broadband noise can be measured directly at the output after attaching a 50 Ω load to the input. From this measurement, and the gain obtained by the modulation technique, the noise figure can be calculated. For traditional CW spectrometers that use phase-sensitive detection, a calibrated noise source can be used to make the end-to-end measurement. However, in order to make an accurate calculation of the gain and noise figure of the spectrometer by this method, the noise equivalent bandwidth of the spectrometer must be known (Rinard *et al.*, 1999c).

Comparison of CW-EPR signal intensities at 250 MHz, 1.5 GHz, and 9.1 GHz (Rinard *et al.*, 2002c): The predictions summarized in Table 5.3.2 were tested in a comparison of 250 MHz,[3] L-band (1.5 GHz), and X-band (9.1 GHz) EPR signals. Since these frequencies differ successively by factors of 6, the resonators are characterized by geometries scaled by a factor of 6. One pair of resonators was scaled from 250 MHz to 1.5 GHz, and another pair from 1.5 GHz to 9 GHz. All terms in the comparison were measured directly, and their uncertainties estimated. When both the resonator size and the sample size are scaled with the inverse of RF/microwave frequency, ω, the EPR signal at constant B_1 scales as $\omega^{-1/4}$ (Table 5.3.2, Case 2). The theory predicts that the signal at the lower frequency will be larger than the signal at the higher frequency by the ratio 1.57. For 250 MHz to 1.5 GHz, the experimental ratio was 1.52, while for the 1.5 GHz to 9 GHz comparison the ratio was 1.14 (Rinard *et al.*, 2002c).

5.3.3.2 Pulsed EPR Spectrometers in the Frequency Range 250 MHz to 9.5 GHz

Absolute SNR at S-band (2.7 GHz): The detailed analysis of a spectrometer and absolute SNR were compared at S-band (2.7 GHz) (Rinard *et al.*, 1999b). For a

---
3) Although, in the engineering literature, the band that includes 250 MHz is called VHF, in this book VHF is used to designate frequencies greater than 140 GHz.

particular case, an S-band spin echo was calculated to be 3.0 V at the digital oscilloscope, and the measured echo was 2.9 V. The measured noise agreed with that predicted for thermal noise and the known gains and noise figures of the detection path, using the Friis equation (Rinard et al., 1999c).

Comparison of S-band and X-band spin echo: Measurements of electron spin echo signal and noise in well-characterized S-band and X-band spectrometers agree with predictions of frequency dependence based on first principles (Rinard et al., 1999b).

FID intensities at 250 MHz and L-band: The EPR pulsed free induction decay (FID) of a degassed solution of a triaryl methyl radical, methyl tris(8-carboxy-2,2,6,6-tetramethyl(-$d3$)-benzo[1,2-d:4,5-$d0$]bis(1,3)dithiol-4-yl) tripotassium salt, 0.2 mM in H_2O, was measured at 247.5 MHz and L-band (1.40 GHz). The calculated and observed FID signal amplitudes (in mV) agreed within 1% and 6%, and the ratio of the normalized FID signals at the two frequencies agreed within 5% (Rinard et al., 2002b). In these pulsed EPR spectrometers the noise was dominated by the noise of the high-power pulsed amplifiers, so absolute SNR measurements were not useful for frequency dependence validation.

Rapid scan SNR at 250 MHz: Analogous measurements were made for a rapid scan CW 250 MHz spectrometer at the University of Denver, using 600 kHz bandwidth and direct detection; the results obtained are listed in Table 5.3.4.

5.3.3.3 Summary of Experimental Validation of SNR of CW and Pulsed Spectrometers at Frequencies <10 GHz

The results summarized in this section provide confidence that this approach to the prediction of SNR as a function of frequency is a valid way to identify the major targets for improvement of EPR spectrometer performance. Immediately evident from the Friis equation is that loss prior to the first-stage amplifier, and the noise figure of the first-stage amplifier, can dominate the overall noise figure of the

Table 5.3.4 SNR for rapid scan EPR at 250 MHz.

Parameter	Calculated theoretical value at the resonator	Measured value at the output of the spectrometer reflected to the resonator
Signal amplitude	1.29×10^{-6} V peak	1.24×10^{-6} V peak
Thermal noise	3.5×10^{-7} V rms or -116 dBm	-115.96 dBm
Source phase noise	Use measured value of 0.94 dB above thermal	0.94 dB above thermal
Total noise	3.9×10^{-7} V rms or -115 dBm	-115.02 dBm
Signal/Noise	3.25 signal volts/noise V rms	3.126 signal volts/noise V rms

system. In older CW-EPR spectrometers, the detector and first-stage amplifier had much poorer noise figures than current instruments, which utilize extensively improved technologies. When all else is improved, microwave source noise becomes an important contributor; consequently, for the highest-Q resonators the microwave source must be specially selected to have the lowest noise among those currently available.

Sometimes, it is necessary to trade-off the SNR for other experimental goals. For example, there is almost no loss in short lengths of X-band waveguide, but there is insufficient space in a normal cryostat that fits in a typical magnet gap to use waveguide to transmit microwaves to, and signal from, the cavity. Hence, in cryostats a coaxial cable is used at X-band, despite the fact that a 1 m length of coaxial cable results in about 16% loss of the signal voltage (ca. 1.5 dB). Waveguide can be used in the cryostat at Q-band, however, because the dimensions are smaller.

5.3.4
Reference Samples for SNR: Weak Pitch

The SNRs of standard X-band CW spectrometers have for many years been monitored with a "weak pitch" (0.000 33% pitch in KCl) sample, as defined by Varian Instruments (Eaton and Eaton, 1980; Maier, 1997). Conversion between the experimental weak pitch SNR and sensitivity, in terms of the number of spins, can be performed using the conversion factor derived by Bruker: "A signal/noise of 360:1 corresponds to a sensitivity of 0.8×10^{10} spins/10^{-4}T" (Bruker ESP300 series specifications, 1990). Improvements in CW SNR values over the years have resulted from decreasing the phase noise of the microwave source, increasing the resonator Q and filling factor, increasing the uniformity of B_1 and of the magnetic field modulation over the sample, decreasing the noise figure of the detector, and improving the lock-in amplifier that is used to detect the field-modulated EPR signal (Maier, 1997). The X-band sensitivity performance achievements listed in Table 5.3.5, labeled "SHQ," were obtained with a Bruker high-Q cavity, which has been designed to produce a better performance than the TE_{102} cavity used traditionally for SNR measurements. With high-Q resonators and low-noise sources, modern Bruker spectrometers specify a weak pitch SNR of better than 3000:1 at X-band, and 2000:1 at Q-band. The X-band performance specification of 3000:1 SNR for weak pitch corresponds to 10^9 spins/G√Hz.

There is no universal weak pitch sample for comparing the performance of EPR spectrometers as a function of microwave frequency. Even where standards have been developed for X-band EPR, additional effort is required to develop standards for high-frequency EPR (HFEPR) for g-value, magnetic field linearity, signal-to-noise, line shape, resolution (CW, time-domain, ENDOR), spin count (CW, FID, ESE), power and B_1 at the sample, magnetic field modulation, and temperature. The problems with DPPH as a standard at high field are discussed by Kolaczkowski, Cardin, and Budil (1999).

5.3.5
Performance of High-Frequency (≥94 GHz)/High-Field EPR Spectrometers

5.3.5.1 CW Spectrometers

Möbius and Savitsky (2009) provide an historical development of high-field EPR, and the contributions to sensitivity of many technological improvements, with citations to many early reports not cited in this chapter. Nilges et al. (1999) compared 13 different W-band detector configurations, illustrating a factor of 63 range in SNR. A low-noise preamplifier improved the noise figure of the detector by 10 dB. Rohrer et al. (2001) discussed losses in the various components of a 180 GHz spectrometer, and Earle, Budil, and Freed (1996) similarly discussed the performance of a 250 GHz spectrometer. As an example of the problems involved in the construction of high-frequency EPR spectrometers, Rohrer et al. (2001) estimated that the theoretical sensitivity of their 180 GHz CW spectrometer was about two orders of magnitude better than the experimental sensitivity. Among the causes cited were source noise, microphonics due to magnetic field modulation, and the detection pathway 13 dB noise figure. Wang et al. (1994) compared the phase noise of three 94 GHz sources. Lebedev (1990) compared sensitivities for four types of resonator, and showed that a TE_{011} resonator gave a better sensitivity than a Fabry–Pérot. Möbius and Savitsky (2009) pointed out that cavity resonators yield a better concentration sensitivity for low dielectric loss samples, but that Fabry–Pérot resonators were better for high-dielectric loss samples. Many technical features of available components that limit the sensitivity of high-field spectrometers have been discussed in detail by Möbius and Savitsky (2009). Each of these features must be borne in mind when comparing the absolute and relative sensitivities of various spectrometers.

In addition to the frequency dependence of the characteristics of spectrometer components, the dielectric properties of the sample are also frequency-dependent. Water becomes less lossy above ca. 35 GHz, making aqueous samples feasible at high frequency if an appropriate holder is developed (Nandi, Bhattacharyya, and Bagchi, 2000; Möbius and Savitsky, 2009). Some relevant dielectric properties were presented and discussed by Earle, Zeng, and Budil (2001). Lebedev (1990) plotted the effect of dielectric losses at 2 mm wavelength for various resonators.

5.3.5.2 Pulsed EPR Spectrometers

The frequency dependence of sensitivity of a pulsed EPR spectrometer for small samples that can be fully excited, depends strongly on the resonator conversion efficiency – the conversion of incident pulse power to B_1 at the sample, and the reciprocal conversion of time-domain EPR signal (FID, echo) magnetization into signal strength at the detector. For example, Rohrer et al. (2001) found that the conversion efficiency did not improve as much as predicted when the frequency was above about 100 GHz. They attributed this to larger distortions of the microwave magnetic field due to sample and resonator dimensions at higher microwave

Table 5.3.5 Reported sensitivities of commercial and noncommercial high-field CW-EPR spectrometers.

Frequency (GHz)	Reference	Minimum detectable number of spins per G per $\sqrt{Hz}$[a]	Predicted sensitivity scaled by $\omega^{11/4}$ relative to X-band
9.5	Maier (1997)	6×10^9 TE$_{102}$ cavity 9×10^8 SHQ cavity 2×10^{-12} M SHQ	9×10^8 assumed in comparisons
9.5	P. Höfer (personal communication, 2008)	10^9 for SNR = 3000:1	
9	Blank et al. (2004)	1.2×10^9	
34	P. Höfer (personal communication, 2008)	8×10^7 weak pitch in 1 mm i.d. tube.	2.7×10^7
94	Wang et al. (1994)	$<10^9$ non-lossy; 2.4×10^{10} aqueous	
95	Schmalbein et al. (1999)	2×10^7	1.6×10^6
95	Möbius and Savitsky (2009)	10^7 with TE$_{011}$ cavity; 4×10^8 with Fabry–Pérot	
95	Prisner, Rohrer, and Möbius (1994)	4×10^7	
110	Hassan et al. (2000); J. Krzystek (personal communication, 2008)	10^{12}	1.1×10^6
120	Van Tol (2004); Van Tol, Brunel, and Wylde (2005)	10^{11}	8.4×10^5
140	Lebedev (1990)	4×10^7	5.5×10^5
150 (2 mm, D-band)	Grinberg and Berliner (2004)	4×10^7	4.6×10^5
180	Rohrer et al. (2001)	10^9	2.8×10^5
220	Cardin et al. (1999)	2×10^{10} aqueous sample	1.6×10^5
220	Hassan et al. (2000); J. Krzystek (personal communication, 2008)	10^{12}	

Table 5.3.5 (Continued)

Frequency (GHz)	Reference	Minimum detectable number of spins per G per $\sqrt{Hz}$[a]	Predicted sensitivity scaled by $\omega^{11/4}$ relative to X-band
240	Van Tol (2004); Van Tol, Brunel, and Wylde (2005); Morley, Brunel, and van Tol (2008)	5×10^{10} (transmission) 3×10^8 (Fabry–Pérot) for aqueous samples; 2×10^8 non-lossy	1.3×10^5
250	Earle, Budil, and Freed (1996)	10^{11}	1.1×10^5
250	Barnes and Freed (1997)	2×10^9 aqueous sample	
250	Earle and Freed (1999); Earle and Smirnov (2004); Freed (2004)	10^8 10^7 converted to standard conditions	
330	Hassan et al. (2000); J. Krzystek (personal communication, 2008)	10^{12}	5.2×10^4
360	Möbius and Savitsky (2009)	1.5×10^9 with Fabry–Pérot	4×10^4
Up to 475 (17 T magnet)	Hassan et al. (1999)	10^{11}; 10^9 at 4 K	1.9×10^4
Up to 700 (25 T magnet)	Hassan et al. (1999)	10^{12}	7×10^3

[a] The minimum detectable number of spins is at room temperature unless a temperature is specified in the table. Table 5.3.2 was constructed assuming that all of the spins are observed at each frequency. Experimentally, all of the spins in standard samples (such as weak pitch) may not be observed, so the custom is to report experimental sensitivity in terms of spins per Gauss, and also to correct for detection system bandwidth, so the units are spins per Gauss per square root Hz.

frequencies. Their experimental enhancement of pulse sensitivity at 180 GHz relative to 9 GHz was a factor of 7 lower than the theoretical expected enhancement. Since source noise often contributes importantly in high-frequency CW-EPR SNR, but is not important for pulsed FID or spin echo measurements, pulsed EPR has often been found to exhibit better improvements in SNR as a function of frequency than has CW-EPR (Prisner, Rohrer, and Möbius, 1994). Comparing actual SNR on CW and pulsed spectrometers within a given acquisition time also requires consideration of the detection bandwidths, of the limits due to spin relaxation times, and also of the hardware-limited pulse repetition times. Prisner, Rohrer, and Möbius (1994) and Hofbauer et al. (2004) predicted that the SNR of a pulsed EPR experiment would increase as $\omega^{7/2}$ for a small, narrow-line, sample, but that the SNR would be proportional to $\omega^{-1/2}$ if the sample was not restricted in size, and is broadened by g-anisotropy or g-strain. The shorter dead time after the microwave or laser pulse in a high-frequency spectrometer, for the same Q as a low-frequency spectrometer, makes the high-frequency spectrometer more sensitive for short-time signals in pulsed and in time-resolved (transient response) EPR (Möbius and Savitsky, 2009).

5.3.6
Reported Sensitivities of CW and Pulsed Spectrometers at Various Frequencies

While it is fairly easy to make predictions of sensitivity as a function of frequency, it is difficult to make measurements to compare with the predictions. Rinard et al. (1999b, 2002c) showed that the predicted dependence of SNR on frequency (as discussed in Section 5.3.2.1) agrees with experimental data for 250 MHz to 9.4 GHz. The present authors are unaware of any comparable frequency-dependent sensitivity measurements at higher microwave frequencies. However, Table 5.3.5 lists the minimum detectable number of spins reported by Bruker BioSpin for X-, Q-, and W-band CW spectrometers and by other laboratories at other frequencies. In Table 5.3.5, the last column is the predicted sensitivity based on Case 3 of Table 5.3.2, calculated relative to the Bruker X-band sensitivity reported for a CW spectrometer with a super-high-Q cavity resonator. It is noted that even this carefully engineered commercial X-band EPR spectrometer does not achieve the ultimate possible SNR, due in part to its being designed for flexible use. Commercial spectrometers are designed for the best SNR possible for a non-lossy sample, while also incorporating features that allow a wide range of applications and protect the spectrometer against damage. Sensitivity data for pulse spectrometers are summarized in Table 5.3.6.

Many of the reports cited in Tables 5.3.5 and 5.3.6 were explicit that the sensitivity was stated in terms of the minimum detectable number of spins extrapolated from the experimental results to SNR = 1, and it is assumed that the other reports made analogous extrapolations. Some of the pulse experimental results in Table 5.3.6 were expressed as single-pulse, and some as SNR in a defined time of averaging (the original reports should be consulted for these details).

Direct comparison of the sensitivity numbers in Tables 5.3.5 and 5.3.6 would require a detailed analysis of the detection method in each spectrometer, including such points as digitizer integration time constant, and the amount of averaging per point, so that the correction to the equivalent of a "1 second time constant," as specified for the standard weak pitch measurement, is properly performed for equal data collection times. Such corrections were made in some of the reports cited, so that the sensitivity is presented as a value extrapolated to SNR = 1 and a detection time constant of 1 s.

Table 5.3.6 Reported sensitivities of pulsed EPR spectrometers.

Frequency (GHz)	Reference	Minimum detectable number of spins per G per $\sqrt{Hz}$[a]
9	Blank et al. (2004)	2×10^7 in 60 min
9	Blank and Freed (2006)	10^7, with 60 min of averaging
17	Borbat, Crepeau, and Freed (1997)	6×10^{13}
95	Allgeier et al. (1990)	3.1×10^{11} (30 MHz bandwidth)
95	Hofbauer et al. (2004)	1.1×10^{10} 1.3×10^{11} to compare with Allgeier et al. (1990)
95	Goldfarb et al. (2008)	10^7 (single pulse, 40 K)[b]
240	Van Tol (2004); Van Tol, Brunel, and Wylde (2005); Morley, Brunel, and van Tol (2008)	10^8 at 5 K averaged for 1 s

a) The minimum detectable number of spins is for pulse operation at room temperature unless a temperature is specified in the table. Table 5.3.2 was constructed assuming that all of the spins are observed at each frequency. Experimentally, all of the spins in standard samples (such as weak pitch) may not be observed, so the custom is to report experimental sensitivity in terms of spins per Gauss, and also to correct for detection system bandwidth, so the units are spins per Gauss per square root Hz.
b) The sensitivity reported by Goldfarb et al. (2008), was measured on nitroxyl radicals, using a 25 ns pulse that covered 14 G, and detected with 300 MHz bandwidth, which is converted to 1 Hz bandwidth in the table.

5.3.6.1 Further Details on CW EPR Sensitivity

The experimental CW sensitivity results summarized in Table 5.3.5 are plotted in Figure 5.3.1, where the solid line is calculated for Case 3 in Table 5.3.2. The minimum number of detectable spins at X-band is set at 9×10^8 (the value for the Bruker EMX system) and predicted values at higher frequency are calculated assuming $\omega^{-11/4}$ dependence (minimum detectable = $9 \times 10^8 \, (\omega/9.5 \times 10^9)^{-11/4}$. The lack of agreement between the experimental points and the predicted dependence is an indication of the extent to which spectrometers have not achieved maximum sensitivity.

Blank and Freed (2006) predicted that at 60 GHz and 77 K, a sensitivity of 1000 spins should be achievable – a prediction that surely will stimulate vigorous attempts to attain this goal. As shown in Table 5.3.5, fewer than 10 000 spins should be detectable at room temperature at the highest frequencies at which spectrometers have already been built, if all signal detection components could be optimized.

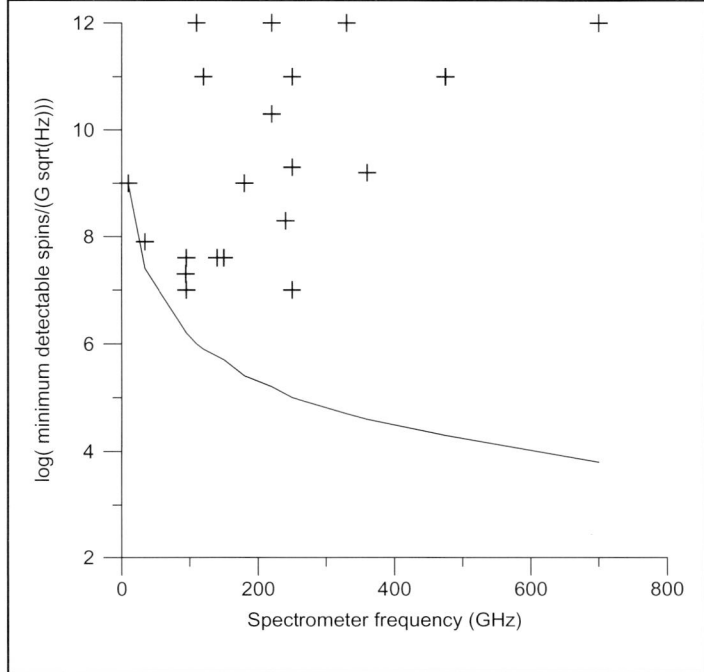

Figure 5.3.1 Plot of experimental sensitivity, defined as minimum number of spins detectable/ (G√Hz), as listed in Table 5.3.5, as a function of spectrometer frequency, relative to X-band. The solid line is the $\omega^{-11/4}$ dependence predicted for Case 3 in Table 5.3.2.

The "Keck" magnet spectrometer at the National High Magnetic Field Laboratory, which could obtain spectra at frequencies up to 700 GHz had a sensitivity in 1999 of 10^{12} spins at room temperature, before a Fabry–Pérot resonator was developed (Hassan et al., 1999). The 120 and 240 GHz spectrometers reported by Van Tol (2004) have a much better sensitivity when operated with a Fabry–Pérot cavity than when they operate in transmission without a cavity. It has been estimated (J. Krzystek, personal communication, 2008) that the sensitivity of a quasioptical homodyne reflection spectrometer is better by a factor of about 20–40 relative to the same spectrometer in transmission mode.

Further discussions of sensitivity and SNR as a function of microwave frequency have been published (Earle, Budil, and Freed, 1996; Borbat, Crepeau, and Freed, 1997; Budil and Earle, 2004; Earle and Smirnov, 2004; Freed, 2004; Hofbauer et al., 2004). The differences between concentration sensitivity and absolute sensitivity for transmission spectrometers operating at 190 and 285 GHz were discussed by Krzystek and Telser (2003). Photo-excited transient EPR signal intensities are less dependent on frequency than ground-state systems (Van Tol, 2004).

5.3.7
Sensitivity Aspects Beyond the Minimum Detectable Number of Spins: Frequency Dependence of Pulse and CW Measurements Related to Distances Between Spins

5.3.7.1 Electron–Electron Coupling

The frequency dependence of EPR sensitivity is the same for pulsed EPR as for CW-EPR. However, there are special experiments in which the frequency dependence of some aspects beyond the minimum detectable number of spins affect the information provided by the experiment.

Dipolar coupling between unpaired electrons results in half-field transitions. The relative intensity of the half-field transitions is predicted to be greater at lower frequency, which was confirmed with S-band and X-band CW spectra (Eaton et al., 1983).

The T_1 of a slowly-relaxing radical is shortened by interaction with a more rapidly relaxing spin. Calculations using the methods of Seiter et al. (1998) predict that measurements at a lower microwave frequency would permit a greater precision in distances at short distances, and would extend the range to longer distances.

Borbat and Freed (2007) considered in detail pulsed dipolar spectroscopy (PDS), which includes double electron–electron resonance (DEER, also termed PELDOR) and double quantum coherence (DQC) spectroscopy.

The DQC experiment (Borbat and Freed, 2000; Borbat and Freed 2007) involves exciting the electron spins such that both spins flip: $\beta\beta \rightarrow \alpha\alpha$. Ideally, the B_1 is large enough to excite all spins in the sample. If the spectral width increases in proportion to the microwave frequency due to g-anisotropy, which occurs at frequencies substantially higher than 35 GHz, the spreading out of the spectrum will cause the sensitivity improvement to be proportional to $\omega^{5/2}$ instead of $\omega^{7/2}$. For a sample with negligible g-anisotropy, the concentration sensitivity is proportional to $\omega^{1/2}$, and for a sample at a sufficiently high frequency that the spectral width is proportional to ω, the concentration sensitivity is proportional to $\omega^{-1/2}$ (Borbat and Freed, 2007). In this case, the concentration sensitivity will not improve at a higher frequency. Borbat and Freed (2007) analyzed the characteristics of the sample and spectrometer that contribute to the SNR in the DQC experiment, and reported a relationship that could be applied if sufficient experimental data were available.

The four-pulse DEER measurement of distances includes a 90°, 180°, 180°-spin–echo pulse sequence at one microwave frequency, and a single 180° pulse at a second microwave frequency. The pulse at the second frequency is stepped in time between the second and third pulses at the first frequency (Jeschke, Pannier, and Spiess, 2000). The two frequencies are separated such that they excite different parts of the spectrum. There is nothing in the DEER method *per se* that is frequency-dependent. Only portions of the spins are excited by each pulse, but the frequencies can be separated to match the g-dispersion, provided this is still within the bandwidth defined by the resonator Q. At a higher frequency, the DEER signal is not attenuated as much relative to that at the center of the resonator "Q-dip" as at X-band.

5.3.7.2 Electron–Nuclear Coupling

In a two-pulse or three-pulse electron spin–echo (ESE) measurement, the microwave frequency is chosen to be resonant with the electron spin Larmor splitting. If the microwave B_1 is large enough to encompass the nuclear hyperfine splitting of the electron transition, then both allowed and forbidden transitions will be excited, and there will be modulation of the amplitude of the ESE signal with a period that is determined by the nuclear Zeeman frequency (Kevan, 1979; Dikanov and Tsvetkov, 1992). The depth of modulation increases at lower frequency, proportional to the inverse square of the magnetic field. When the nuclear Zeeman and hyperfine interactions cancel in the $m_s = +½$ spin state, the nuclear spin Hamiltonian reduces to the pure quadrupolar Hamiltonian (Flanagan and Singel, 1987).

In CW-EPR at low microwave frequency, the anisotropies in g-values and hyperfine couplings make the Cu(II) linewidths in the $g_\parallel$ region of powder (frozen solution) spectra narrower than at higher frequencies (Antholine, 2005).

Both CW and pulsed ENDOR provide much-enhanced information at higher microwave frequencies, because the nuclear Zeeman frequencies become separated by more than the hyperfine couplings, and transitions due to ^1H are more easily separated from those due to heavier nuclei (Hoffman et al., 1993; Möbius, 1993; Goldfarb and Krymov, 2004; Groenen and Schmidt, 2004).

5.3.7.3 Summary

For any of these types of EPR spectroscopy, if the sample is limited in size and the filling factor is high, then the sensitivity will be improved at higher frequency. Orientation selection is improved by using higher microwave frequencies.

5.3.8
Limitations of Sensitivity Considerations

In addition to the hardware aspects of HFEPR technology, there are also methodological limits which, at present, include lack of standards for the assurance of high-quality results, difficulty of performing power-saturation studies, distorted lines due to materials properties, loss of intensity due to spectral dispersion (spins spread out more), difficulty in predicting the optimum frequency due to the trade-offs between spectral dispersion and g-strain and hyperfine anisotropy, and difficulty in defining and maintaining the sample environment.

5.3.8.1 CW Spectrometers

A fundamental problem that plagues most HF-EPR is how to phase the spectrum after acquisition, and how to obtain pure absorption or pure dispersion spectra [e.g., see Nilges et al., (1999); Earle, Zeng, and Budil (2001)]. This is partly a problem with standing waves of unknown phase in the system. Furthermore, there is need for a new resonator for nearly every application, and filling factors of resonators have not been optimized in many cases.

5.3.8.2 Resonators

The need for, and often the advantage of, small or very differently configured resonators at high frequency make it difficult to compare the SNR when using the same sample at X-band and at a high frequency. Schmalbein et al. (1999) showed X-band and W-band spectra of the same 600 µg Mn^{2+}:CaO sample. The X-band spectrum was acquired using a TE_{102} rectangular cavity, while the W-band spectrum was acquired using a smaller resonator. The W-band spectrum had a better SNR, but if a larger sample had been used at X-band it would have had a better SNR. Consideration of the spectrometer conditions for this Mn^{2+} sample were extrapolated to absolute sensitivities of 8×10^9 spins/G√Hz at X-band and 2×10^7 spins/G√Hz at W-band.

5.3.8.3 Samples

The sample interacts strongly with the microwaves, and at high frequencies reflections can be as important as absorptions. Earle and Freed (1999), in a study using a 10 nl sample of spin-labeled muscle fibers, achieved a higher sensitivity than did Barnes and Freed (1997) when using a 140 nl sample of aqueous spin label, even though the same resonator was used in both cases. This improvement was attributed largely to the smaller sample interfering less with the quasioptical properties of the Fabry–Pérot resonator. Filling factors and the overall figure of merit for Fabry–Pérot resonators were compared with other resonators by Budil and Earle (2004).

5.3.8.4 Pulse Spectrometers

There is, as yet, no standard for reporting the SNR or sensitivity of pulsed EPR spectrometers. It is not sensible to continue to try to convert to "CW-like" conditions of 1 Hz bandwidth, etc., especially when the time for the measurement is not also made equivalent. One approach to inter-spectrometer and inter-laboratory comparisons would be to report the sensitivity of a single two-pulse echo with a stated detection bandwidth. Since the T_m of samples used will vary, the sensitivity could be corrected for the interpulse time, τ, based on the actual decay rate if it is known, to a value of τ equal to the deadtime of the instrument. An irradiated fused quartz (SiO_2) sample has been proposed as a spin echo standard Eaton et al. (2010).

5.3.9
Conclusions

The predicted frequency dependence of EPR signal amplitudes, and of the SNR if the noise is limited by thermal noise sources (as summarized in Table 5.3.2), has been validated by measurements on well-characterized spectrometers at X-band and below. Even after six decades of technology development, the sensitivity of EPR spectrometers is far from the theoretical sensitivity limit, except for a few low-frequency instruments. Even X-band spectrometers are still undergoing a period of rapid improvement. At frequencies higher than X-band there remain opportunities for major—indeed, orders of magnitude—improvements in sensitivity.

The clear message from Tables 5.3.5 and 5.3.6, and from Figure 5.3.1, is that the $\omega^{11/4}$ promise is not fulfilled in practice for high-field/high-frequency EPR spectrometers. However, it should be noted that many such instruments are designed for a particular type of sample, and many of them use transmission mode rather than high-Q resonators in reflection mode, which results in a lower SNR. Low-noise components are less readily available at higher microwave frequencies, and transmission line losses are higher at higher frequencies. This results in a higher overall spectrometer noise figure, and hence a lower SNR, than at lower frequencies. As some of the reports summarized in Table 5.3.5 are more than a decade old, it is likely that the laboratories involved have since improved the sensitivity of these or subsequent spectrometers. It is hoped that the brief summary provided in this chapter will stimulate efforts toward improving the spectrometers, and also lead to updated reports.

The focus of this chapter has been that of absolute spin sensitivity. Many types of EPR measurement have frequency-dependent sensitivity. For example, as discussed in detail by Kevan and Schwartz (1979) and Dikanov and Tsvetkov (1992), the depth of ESEEM modulation depends on frequency, and DEER and DQC measurements of inter-spin distances have been shown to be superior at frequencies above X-band (Borbat and Freed, 2000; Bruker BioSpin, 2002).

Acknowledgments

This chapter is based on research supported by NIH NIBIB grants EB000557 and EB02807 (G.R.E. and S.S.E.), EB002034 (Howard Halpern, PI), and RR12183 (Rinard, PI). Communications with Brian Hoffman (Northwestern University), David Budil (Northeastern University), Daniela Goldfarb (Weizmann), Ralph Weber and Peter Höfer (Bruker) helped to identify some of the information, and also stimulated portions of the chapter. Klaus Möbius generously provided a preprint of the instrumentation chapter of his book on high-field EPR. Jurek Krzystek and Hans van Tol (National High Magnetic Field Lab) provided details of the sensitivity of the NHMFL high-field spectrometers (additional information on these spectrometers is available at: http://www.magnet.fsu.edu/usershub/scientificdivisions/emr/facilities/index.html).

Pertinent Literature

The sensitivity achievable in EPR has been a central interest for much of the history of the field. The observation about EPR by Feher that "...its usefulness is limited by the sensitivity of the experimental setup," continues to be true, and continues to drive new developments in EPR instrumentation. Feher's 1957 analysis (Feher, 1957), Ingram's 1958 monograph (Ingram, 1958), and the extensive discussion in Chapter 14 of Poole's 1967 monograph (Poole, 1967) provide the foundation for most subsequent work. For a while, there seemed to be some discrepancy with analogous assessments of sensitivity in NMR, so the relationships were re-derived from first principles by Rinard and coworkers (1999a, 1999c, 2002b, 2002c, 2004). The chapters by Rinard *et al.* (1999a, 2004) should be consulted for detailed engineering derivations and explanation of the equations used to generate Table 5.3.2 in this chapter.

References

Allgeier, J., Disselhorst, A.J.M., Weber, R.T., Wenckebach, W.T., and Schmidt, J. (1990) High-frequency pulsed electron spin resonance, in *Modern Pulsed and Continuous-Wave Electron Spin Resonance* (eds L. Kevan and M.K. Bowman), John Wiley & Sons, Inc., New York, pp. 267–283.

Antholine, W.E. (2005) *Biol. Magn. Reson.*, **23**, 417–454.

Barnes, J.P. and Freed, J.H. (1997) *Rev. Sci. Instrum.*, **68**, 2838–2846.

Blank, A. and Freed, J.H. (2006) *Israel J. Chem.*, **46**, 423–438.

Blank, A., Dunnam, C.R., Borbat, P.P., and Freed, J.H. (2004) *Appl. Phys. Lett.*, **85**, 5430–5432.

Borbat, P.P. and Freed, J.H. (2000) *Biol. Magn. Reson.*, **19**, 383–459.

Borbat, P.P. and Freed, J.H. (2007) *EPR Newslett.*, **17**, 21–33.

Borbat, P.P., Crepeau, R.H., and Freed, J.H. (1997) *J. Magn. Reson.*, **127**, 155–167.

Bruker BioSpin (2002) *ESEEM, ENDOR, and ELDOR at 34 GHz*, vol. 16, Bruker BioSpin, Rheinstetten, Germany.

Budil, D.E. and Earle, K.A. (2004) *Biol. Magn. Reson.*, **22**, 353–399.

Cardin, J.T., Kolaczkowski, S.V., Anderson, J.R., and Budil, D.E. (1999) *Appl. Magn. Reson.*, **16**, 273–292.

Davoust, C.E., Doan, P.E., and Hoffman, B.M. (1996) *J. Magn. Reson. A*, **119**, 38–44.

Dikanov, S.A. and Tsvetkov, Y.D. (1992) *Electron Spin Echo Envelope Modulation (ESEEM) Spectroscopy*, CRC Press, Boca Raton, FL.

Earle, K.A., Budil, D.E., and Freed, J.H. (1996) *Adv. Magn. Reson. Opt. Reson.*, **19**, 253–323.

Earle, K.A. and Freed, J.H. (1999) *Appl. Magn. Reson.*, **16**, 247–272.

Earle, K.A. and Smirnov, A.I. (2004) *Biol. Magn. Reson.*, **22**, 95–143.

Earle, K.A., Zeng, R., and Budil, D.E. (2001) *Appl. Magn. Reson.*, **21**, 275–286.

Eaton, S.S. and Eaton, G.R. (1980) *Bull. Magn. Reson.*, **1**, 130–138.

Eaton, S.S., More, K.M., Sawant, B.M., and Eaton, G.R. (1983) *J. Am. Chem. Soc.*, **105**, 6560–6567.

Eaton, S.S., Eaton, G.R., Quine, R.W., Mitchell, D., Kathirvelu, V., and Weber, R.T. (2010) *J. Magn. Reson.*, **205**, 109–113.

Feher, G. (1957) *Bell Syst. Tech. J.*, **36**, 449–484.

Flanagan, H.L. and Singel, D.J. (1987) *J. Chem. Phys.*, **87**, 5606–5616.

Freed, J.H. (2004) *Biol. Magn. Reson.*, **22**, 19–43.

Froncisz, W. and Hyde, J.S. (1982) *J. Magn. Reson.*, **47**, 515–521.

Goldfarb, D. and Krymov, V. (2004) *Biol. Magn. Reson.*, **22**, 306–351.

Goldfarb, D., Lipkin, Y., Potapov, A., Gorodetsky, Y., Epel, B., Raitsimring, A.M., Radoul, M., and Kaminker, I. (2008) *J. Magn. Reson.*, **195**, 8–15.

Grinberg, O. and Berliner, L.J. (eds) (2004) *Very High Frequency (VHF) ESR/EPR*, vol. 22, Biological Magnetic Resonance, Kluwer Academic/Plenum Publishers, New York.

Groenen, E.J.J. and Schmidt, J. (2004) *Biol. Magn. Reson.*, **22**, 278–304.

Hassan, A.K., Maniero, A.L., Tol, J.V., Saylor, C., and Brunel, L.C. (1999) *Appl. Magn. Reson.*, **16**, 299–308.

Hassan, A.K., Pardi, L.A., Krzystek, J., Sienkiewicz, A., Goy, P., Rohrer, M., and Brunel, L.-C. (2000) *J. Magn. Reson.*, **142**, 300–312.

Hofbauer, W., Earle, K.A., Dunnam, C.R., Moscicki, J.K., and Freed, J.H. (2004) *Rev. Sci. Instrum.*, **75**, 1194–1208.

Hoffman, B.M., DeRose, V.J., Doan, P.E., Gurbiel, R.J., Houseman, A.L.P., and Telser, J. (1993) *Biol. Magn. Reson.*, **13**, 151–218.

Hyde, J.S. (2006) *Concepts Magn. Reson.*, **28A**, 82–83.

Hyde, J.S. and Froncisz, W. (1989) Loop gap resonators, in *Advanced EPR: Applications in Biology and Biochemistry* (ed. A.J. Hoff), Elsevier, Amsterdam, pp. 277–306.

Hyde, J.S., Froncisz, W., Sidabras, J.W., Camenisch, T.G., Anderson, J.E., and Strangeway, R.A. (2007) *J. Magn. Reson.*, **185**, 259–263.

Ingram, D.J.E. (1958) *Free Radicals As Studied by Electron Spin Resonance*, Butterworth Scientific Publications.

Jeschke, G., Pannier, M., and Spiess, H.W. (2000) *Biol. Magn. Reson.*, **19**, 493–512.

Kevan, L. (1979) Modulation of electron spin-echo decay in solids, in *Time Domain Electron Spin Resonance* (eds L. Kevan and R.N. Schwartz), Wiley-Interscience, New York, pp. 279–341.

Kevan, L. and Schwartz, R.N. (eds) (1979) *Time Domain Electron Spin Resonance*, John Wiley & Sons, Inc..

Kolaczkowski, S.V., Cardin, J.T., and Budil, D.E. (1999) *Appl. Magn. Reson.*, **16**, 293–298.

Krzystek, J. and Telser, J. (2003) *J. Magn. Reson.*, **162**, 454–465.

Lebedev, Y.S. (1990) High-Frequency Continuous-Wave Electron Spin Resonance, in *Modern Pulsed and Continuous-Wave Electron Spin Resonance* (eds L. Kevan and M.K. Bowman), John Wiley & Sons, Inc., New York, pp. 365–404.

Lee, C.A. and Dalman, G.C. (1994) *Microwave Devices, Circuits and Their Interaction*, John Wiley & Sons, Inc., New York.

Maier, D. (1997) *Bruker Spin Rep.*, **144**, 13–15.

Möbius, K. (1993) *Biol. Magn. Reson.*, **13**, 253–274.

Möbius, K. and Savitsky, A.N. (2009) *High-Field EPR Spectroscopy on Proteins and Their Model Systems*, Royal Society of Chemistry, Cambridge.

Morley, G.W., Brunel, L.-C., and van Tol, J. (2008) *Rev. Sci. Instrum.*, **79**, 064703.

Nandi, N., Bhattacharyya, K., and Bagchi, B. (2000) *Chem. Rev.*, **100**, 2013–2045.

Nilges, M.J., Smirnov, A.I., Clarkson, R.B., and Belford, R.L. (1999) *Appl. Magn. Reson.*, **16**, 167–183.

Oles, T., Hyde, J.S., and Froncisz, W. (1989) *Rev. Sci. Instrum.*, **60**, 389–391.

Poole, C.P. (1967) *Electron Spin Resonance: A Comprehensive Treatise on Experimental Techniques*, Interscience Publishers, New York.

Popov, V.A. and Katrich, S.A. (1993) *Physica B*, **183**, 211–216.

Prisner, T.F., Rohrer, M., and Möbius, K. (1994) *Appl. Magn. Reson.*, **7**, 167–183.

Rinard, G.A., Quine, R.W., Ghim, B.T., Eaton, S.S., and Eaton, G.R. (1996) *J. Magn. Reson. A*, **122**, 50–57.

Rinard, G.A., Eaton, S.S., Eaton, G.R., Poole, C.P., Jr, and Farach, H.A. (1999a) *Handbook of Electron Spin Resonance*, Springer-Verlag, New York, vol. 2, pp. 1–23.

Rinard, G.A., Quine, R.W., Harbridge, J.R., Song, R., Eaton, G.R., and Eaton, S.S. (1999b) *J. Magn. Reson.*, **140**, 218–227.

Rinard, G.A., Quine, R.W., Song, R., Eaton, G.R., and Eaton, S.S. (1999c) *J. Magn. Reson.*, **140**, 69–83.

Rinard, G.A., Quine, R.W., and Eaton, G.R. (2000) *J. Magn. Reson.*, **144**, 85–88.

Rinard, G.A., Quine, R.W., Eaton, G.R., and Eaton, S.S. (2002a) *Magn. Reson. Engineer.*, **15**, 37–46.

Rinard, G.A., Quine, R.W., Eaton, S.S., and Eaton, G.R. (2002b) *J. Magn. Reson.*, **154**, 80–84.

Rinard, G.A., Quine, R.W., Eaton, S.S., and Eaton, G.R. (2002c) *J. Magn. Reson.*, **156**, 113–121.

Rinard, G.A., Quine, R.W., Eaton, S.S., and Eaton, G.R. (2004) *Biol. Magn. Reson.*, **21**, 115–154.

Rohrer, M., Brügmann, O., Kinzer, B., and Prisner, T.F. (2001) *Appl. Magn. Reson.*, **21**, 257–274.

Schmalbein, D., Maresch, G.G., Kamlowski, A., and Höfer, P. (1999) *Appl. Magn. Reson.*, **16**, 185–205.

Seiter, M., Budker, V., Du, J.-L., Eaton, G.R., and Eaton, S.S. (1998) *Inorg. Chim. Acta*, **273**, 354–366.

Sidabras, J.W., Mett, R.R., Froncisz, W., Anderson, J.R., and Hyde, J.S. (2007) *Rev. Sci. Instrum.*, **78**, 034701-1 to 034701-6.

Van Tol, J. (2004) *Biol. Magn. Reson.*, **22**, 524–538.

Van Tol, J., Brunel, L.-C., and Wylde, R.J. (2005) *Rev. Sci. Instrum.*, **76**, 074101.

Wang, W., Belford, R.L., Clarkson, R.B., Davis, P.H., Forrer, J., Nilges, M.J., Timkin, M.D., Walczak, T., Thurnauer, M.C., Norris, J.R., Morris, A.L., and Zhang, Y. (1994) *Appl. Magn. Reson.*, **6**, 195–215.

**Part Two
Theoretical**

6
First Principles Approach to Spin-Hamiltonian Parameters

Frank Neese

6.1
Introduction

The previous chapters in this book have familiarized the reader with use of the spin Hamiltonian in EPR spectroscopy. In particular, it has become clear which parameters enter into the spin Hamiltonian, and how they are determined by fitting the spectrum. But, the questions that now arise are: What is the physical origin of the spin-Hamiltonian parameters, and how can they be predicted from first principles? Obviously, the spin Hamiltonian in itself is not of a fundamental nature, but rather belongs to the class of "effective Hamiltonians." These are Hamiltonians that describe only a limited part of the spectrum of eigenstates available to a given electronic system (molecule) in a phenomenological way. If the concept of the spin Hamiltonian is a good one, then it is necessary that there is a clean connection between the parameters that enter the spin Hamiltonian and the fundamental physics that govern the electronic structure of matter. Fortunately, it is true that the spin Hamiltonian *is* a good concept, and that such a clean connection can be found. To establish this important link is the subject of this chapter.

Once established, there are many possible uses for a theory of spin-Hamiltonian parameters. Perhaps foremost is the desire to interpret complicated experimental spectra. By calculating the spectroscopic properties of candidate molecules, it is hoped that unknown species can be identified for which only a spectroscopic signature exists. Another important use of quantum chemistry in modern EPR spectroscopy is to deal with the problem of over-parameterization. Owing to the high resolution that can nowadays be obtained when using high-field and pulse techniques, many resonances are observed and many magnetic interaction tensors must be determined. Yet, even if a molecular structure is available, this may not be an easy task because, for example, it may not be clear which nucleus gives rise to which part of a given hyperfine pattern. Simulations may not be possible without knowing the orientation of a given magnetic tensor in the molecular frame, or at least in the reference frame of the spin Hamiltonian. In all of these cases, quantum chemical calculations can be a great help, by providing a set of very reasonable starting parameters for advanced simulation work. Even if the predictions of quantum chemistry may not be accurate enough to calculate spectra that match the experimental spectra next to perfectly (although there are examples for such cases in studies of free radicals), the values are often good enough such that, by allowing for 10–20% changes relative to the theoretical predictions, a satisfactory fit of the experiment is achieved. Finally, there may be interest in magneto-structural correlations that establish the underlying reasons for the sensitivity of spin-Hamiltonian parameters to the details of the molecular geometric and electronic structure. Typical usages here may be a prediction of the sign of the zero-field splitting (ZFS) tensor as a function of d^N electron configuration and coordination number. Many possible correlations can be investigated theoretically that may not be easily accessible experimentally. In this way, it is often possible to bring order and meaning to a large body of experimental observations.

Clearly, the task at hand is very large, and this chapter alone cannot serve as a substitute for a textbook of theoretical chemistry; hence, some familiarity with the concepts of molecular quantum mechanics is assumed. After formulating the general theory in terms of abstract exact eigenstates of the so-called Born–Oppenheimer Hamiltonian, attention will be focused on density functional theory (DFT), which presently dominates the field of applied quantum chemistry. A detailed introduction to the method, philosophy, and implementation of DFT itself has recently been provided (Neese, 2008), while *ab initio* methods are described elsewhere (Neese *et al.*, 2007). In addition, an excellent textbook (Koch and Holthausen, 2000) describes the use of DFT in chemistry, while the underlying theory is carefully discussed elsewhere (Parr and Yang, 1989). It is not the aim of the present chapter to provide a comprehensive review of the original literature, as this already exists in several forms; however, the details of relevant reports are provided for the reader in Section 6.4.

6.2
The Spin Hamiltonian

In order to more precisely state the nature of the problem, the leading spin-Hamiltonian parameters can be briefly introduced at this point. The spin Hamiltonian is an effective Hamiltonian, and contains only spin-variables of a "fictitious" electron spin S and the nuclear spins I_A, I_B,... All reference to the spatial part of the many-electron wavefunction – and therefore to detailed molecular electronic and geometric structure – is implicitly contained in the well-known spin-Hamiltonian parameters **D** (zero-field splitting), **g** (g-tensor), **A** (hyperfine coupling), **Q** (quadrupole coupling), σ (chemical shift) and **J** (spin-spin coupling), all of which are considered as adjustable parameters in the analysis of experiments, and are explained in detail below. Thus, the spin Hamiltonian is "a convenient place to rest" in the analysis of experimental data by theoretical means. The spin Hamiltonian that includes the interactions covered above is:

6.2 The Spin Hamiltonian

$$\hat{H}_{SPIN} = \hat{\mathbf{S}}\mathbf{D}\hat{\mathbf{S}} + \beta \mathbf{B}\mathbf{g}\hat{\mathbf{S}} + \sum_{A}\left[\hat{\mathbf{S}}\mathbf{A}^{(A)}\hat{\mathbf{I}}^{(A)} + \beta_{N}\mathbf{B}\mathbf{g}_{N}^{(A)}\hat{\mathbf{I}}^{(A)} + \hat{\mathbf{I}}^{(A)}\mathbf{Q}^{(A)}\hat{\mathbf{I}}^{(A)}\right] + \sum_{A<B}\left[\hat{\mathbf{I}}^{(A)}\mathbf{J}^{(AB)}\hat{\mathbf{I}}^{(B)}\right]$$

(6.1)

where the sum over "A" refers to the magnetic nuclei, **B** is the magnetic flux density, and μ and μ_N are the electronic and nuclear Bohr magneton, respectively. A schematic representation of the interactions occurring in the spin Hamiltonian is shown in Figure 6.1.

The spin Hamiltonian acts on a basis of product functions $|SM_S\rangle \otimes |I^{(A)}M_I^{(A)}\rangle \otimes \ldots \otimes |I^{(N)}M_I^{(N)}\rangle$. For not too many spins, this basis is often small enough to allow an exact diagonalization of the spin Hamiltonian and, therefore, exact quantum mechanical treatments of the spin-physics in the spin Hamiltonian framework. For high-dimensional spin Hamiltonian problems, both, brute-force and a variety of perturbation theoretical methods can be employed in order to arrive at exact or good approximate solutions.

The role of theory is then to derive the connection of first-principle electronic structure approaches to the spin-Hamiltonian parameters, and to devise practical algorithms for predicting the values of these parameters. In this way, calculations can help to develop the full information content of spectra and thus allow conclusions about the geometric and electronic structure of the system under investigation to be drawn.

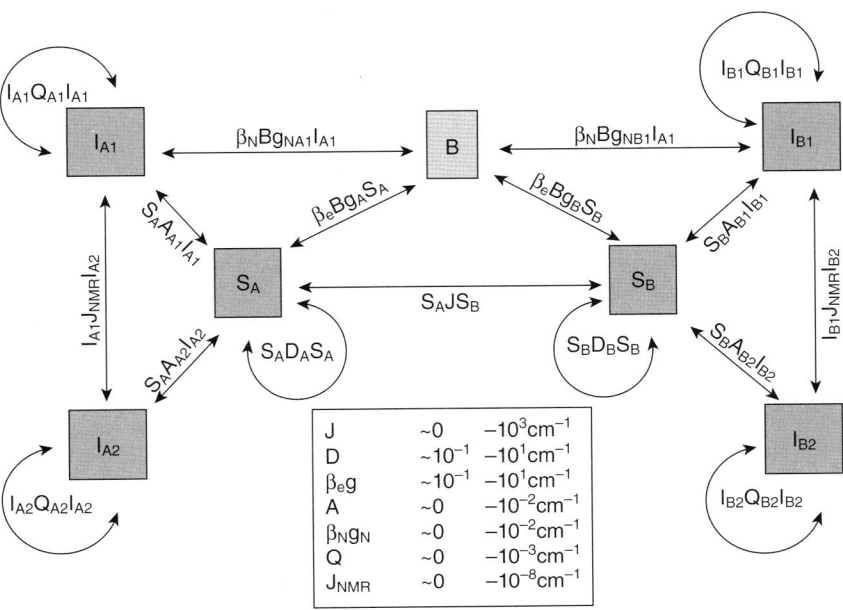

Figure 6.1 Magnetic interactions occurring in the phenomenological spin-Hamiltonian, and their typical order of magnitude.

6.3
Electronic Structure Theory of Spin-Hamiltonian Parameters

In this section, the basic theory of EPR spin-Hamiltonian parameters is presented in the way it is implemented in contemporary electronic structure program packages, such as the ORCA (Neese et al., 2009), Gaussian03 (Frisch et al., 2004), Dalton (Ågren et al., 2005), ADF (Baerends et al., 2005), and MagRespect (Malkin et al., 2003) programs. All of these have significant but differing capabilities in the prediction of EPR properties. Since it is not assumed that the readers are experts in quantum chemistry, it is necessary to briefly introduce the underlying theory of N-electron systems. Some practical advice for performing the calculations will be provided in Section 6.3.6; this should be reasonably self-contained and not require the detailed study of the more formal material contained in Sections 6.3.1–6.3.5.

6.3.1
Electronic Structure Methods

In order to introduce the subject of electronic structure methods, we start from the Born–Oppenheimer (BO) Hamiltonian operator, which considers only the leading electrostatic interactions (in atomic units):

$$\hat{H}_{BO} = -\frac{1}{2}\sum_i \vec{\nabla}^2 - \sum_{i,A} \frac{Z_A}{|\mathbf{r}_i - \mathbf{R}_A|} + \frac{1}{2}\sum_{i \neq j} \frac{1}{|\mathbf{r}_i - \mathbf{r}_j|} + \frac{1}{2}\sum_{A \neq B} \frac{Z_A Z_B}{|\mathbf{R}_A - \mathbf{R}_B|}$$

$$= \hat{T} + \hat{V}_{eN} + \hat{V}_{ee} + \hat{V}_{NN}$$

$$= \hat{h} + \hat{V}_{ee} + \hat{V}_{NN} \tag{6.2}$$

The successive terms in Equation 6.2 describe the kinetic energy of the electrons, the electron–nuclear attraction, the electron–electron repulsion and the nuclear–nuclear repulsion, respectively. In Equation 6.2, i, j sum over electrons at positions $\mathbf{r}_i$, A, B over nuclei with charge Z_A at positions $\mathbf{R}_A$. The nuclear positions are assumed to be fixed and the electrons are supposed to readjust immediately to the positions of these (classical) nuclei. The BO-operator contains just the leading electrostatic interactions between the negatively charged electrons and the positively charged nuclei. This accounts for the vast amount of the molecular total energy, and most problems of chemical structure and energetics can be satisfactorily discussed in terms of these comparatively simple interactions. Yet, the BO-operator contains the coupled motion of N-electrons, and to find the exact eigenfunctions and eigenvalues of the (time-independent) BO-Schrödinger equation:

$$\hat{H}_{BO}\Psi(\mathbf{x}_1, \ldots, \mathbf{x}_N | \mathbf{R}) = E(\mathbf{R})\Psi(\mathbf{x}_1, \ldots, \mathbf{x}_N | \mathbf{R}) \tag{6.3}$$

is a hopelessly complicated task for any but the simplest systems. In Equation 6.3, the many electron wavefunction $\Psi(\mathbf{x}_1, \ldots, \mathbf{x}_N | \mathbf{R})$ has been introduced which depends on the space ($\mathbf{r}$) and spin (σ) variables of the N-electrons ($\mathbf{x}_i \equiv (\mathbf{r}_i, \sigma_i)$. $\mathbf{R}$

collectively denotes the positions of the nuclei on which the many-electron wavefunction and the eigenvalues $E(\mathbf{R})$ depend parametrically. According to the basic quantum theory, everything that can be known about the molecular system in the time-independent case is contained in $\Psi(\mathbf{x}_1, \ldots, \mathbf{x}_N | \mathbf{R})$. It is also important to note that *all* measurements always probe the N-electron system. Molecular orbitals (MOs) to be introduced below are *never* observable and, in fact, the entire theory of molecules can be exactly formulated without any recourse to orbitals. Yet, MOs are very convenient building blocks in the majority of approximate methods that have been developed to date.

MOs first appear in the framework of the Hartree–Fock method, which is a mean-field treatment. The basic idea here is to start from a N-particle wavefunction which is appropriate for a system of non-interacting electrons. Having fixed the *Ansatz* for the N-particle wavefunction in this way, the variational principle is used in order to obtain the best possible approximation for the fully interacting system. Such "independent particle" wavefunctions are Slater-determinants which consist of antisymmetrized products of single-particle wavefunctions $\{\psi(\mathbf{x})\}$ (the antisymmetry brought about by the determinantal form is essential in order to satisfy the Pauli principle). Thus, the Slater-determinant is written:

$$\Psi(\mathbf{x}_1, \ldots, \mathbf{x}_N) = 1/\sqrt{N!} \det|\psi_1 \ldots \psi_N| \tag{6.4}$$

and is abbreviated as $|\psi_1 \ldots \psi_N|$. The Hartree–Fock approach consists of varying the shapes of these orbitals in order to minimize the Rayleigh-functional $\langle \Psi | \hat{H}_{BO} | \Psi \rangle / \langle \Psi | \Psi \rangle$. The result is an upper bound to the total molecular energy. The minimization leads to the following pseudo-single-particle equations (Hartree–Fock equations) which must be fulfilled by the orbitals $\psi_i(\mathbf{x})$:

$$\hat{F}\psi_i = \left(\hat{h} + \sum_j \hat{J}_j - \hat{K}_j \right) \psi_i = \varepsilon_i \psi_i \tag{6.5}$$

Here, ε_i is the orbital energy for the ith MO. The Coulomb and exchange operators $\hat{J}_j$ and $\hat{K}_j$ are defined by their actions on an orbital $\psi_i(\mathbf{x})$:

$$\hat{J}_j \psi_i(\mathbf{x}) = \psi_i(\mathbf{x}) \int \frac{|\psi_j(\mathbf{x}')|^2}{|\mathbf{r} - \mathbf{r}'|} d\mathbf{x}' \tag{6.6}$$

$$\hat{K}_j \psi_i(\mathbf{x}) = \psi_j(\mathbf{x}) \int \frac{\psi_j^*(\mathbf{x}') \psi_i(\mathbf{x}')}{|\mathbf{r} - \mathbf{r}'|} d\mathbf{x}' \tag{6.7}$$

The Hartree–Fock equations describe the motion of electrons in the field of the nuclei and the average field of the other electrons (hence the name "mean-field" treatment). The Hartree–Fock equations are a major simplification of the N-electron problem. However, this simplification comes at the price of introducing a coupled set of nonlinear equations – the Coulomb- and exchange-operators depend on the eigenfunctions of the Fock operator which are unknown before the Hartree–Fock equations are solved; thus, an iterative approach is necessary. In practice, it is not even possible to solve the Hartree–Fock equations exactly, since the orbitals assume fairly complicated shapes. Therefore, one commonly introduces a set of auxiliary one-electron functions $\{\varphi(\mathbf{x})\}$ (basis functions) which are used to expand the orbitals:

$$\psi_i(\mathbf{x}) = \sum_\mu c_{\mu i} \varphi_\mu(\mathbf{x}) \tag{6.8}$$

Now, the minimization is performed with respect to the coefficients $c_{\mu i}$ while the basis functions are held fixed. The expansion is only exact in the limit of a mathematically complete basis set $\{\varphi(\mathbf{x})\}$ which is impossible to obtain in practice. Thus, the results depend on the size and nature of the employed basis functions, but there is a well-defined *basis set limit*. Since the BO-operator is spin-free, it is customary to let the orbitals be eigenfunctions of the single-electron spin-operator by choosing them to be either spin-up or spin-down orbitals. This leads to the spin-unrestricted (UHF) method ($\sigma = \alpha, \beta$):

$$\hat{F}^\sigma \psi_i^\sigma = \left(\hat{h} + \sum_{j\sigma'} \hat{J}_j^{\sigma'} - \delta_{\sigma\sigma'} \hat{K}_j^\sigma \right) \psi_i^\sigma = \varepsilon_i^\sigma \psi_i^\sigma \tag{6.9}$$

For a closed-shell system, the spin-up and spin-down Fock operators are equal, and the spin-orbitals are obtained in pairs of equal shape and energy. Instead of dividing the set of orbitals into spin-up and spin-down orbitals, it is also possible to pursue a division into closed-shell and open-shell orbitals; this leads to the restricted open-shell HF (ROHF) method. Although the formalism for this method is slightly more involved than the UHF formalism, the general ideas are identical. The ROHF wavefunction is an eigenfunction of the total spin squared ($\hat{S}^2$) operator, while the UHF wavefunction does not have this feature. The energy of the UHF wavefunction, on the other hand, is lower than that of the ROHF wavefunction due to the increased variational freedom in the UHF case. The matrix elements of the Fock-operator become in the UHF case:

$$F_{\mu\nu}^\sigma = h_{\mu\nu} + \sum_{\kappa\tau} P_{\kappa\tau}(\mu\nu|\kappa\tau) - P_{\kappa\tau}^\sigma(\mu\kappa|\nu\tau) \tag{6.10}$$

with the one- and two-electron integrals over basis functions being defined as:

$$h_{\mu\nu} = \int \varphi_\mu(\mathbf{r}) \hat{h} \varphi_\nu(\mathbf{r}) d\mathbf{r} \tag{6.11}$$

$$(\mu\nu|\kappa\tau) = \iint \varphi_\mu(\mathbf{r}_1) \varphi_\nu(\mathbf{r}_1) \frac{1}{|\mathbf{r}_1 - \mathbf{r}_2|} \varphi_\kappa(\mathbf{r}_2) \varphi_\tau(\mathbf{r}_2) d\mathbf{r}_1 d\mathbf{r}_2 \tag{6.12}$$

Here, the "density matrix" was introduced:

$$P_{\mu\nu}^\sigma = \sum_i c_{\mu i}^\sigma c_{\nu i}^\sigma \tag{6.13}$$

The sum over i is restricted to the occupied MOs with spin σ, and consists of N_α and N_β terms for spin-up and spin-down, respectively. The total density matrix is $\mathbf{P} = \mathbf{P}^\alpha + \mathbf{P}^\beta$ in terms of which the electron density at point $\mathbf{r}$ can be written:

$$\rho(\mathbf{r}) = \sum_{\mu\nu} P_{\mu\nu}\varphi_\mu(\mathbf{r})\varphi_\nu(\mathbf{r}) \tag{6.14}$$

In the context of EPR spectroscopy, it is also important to define the spin-density matrix $\mathbf{P}^{\alpha-\beta} = \mathbf{P}^\alpha - \mathbf{P}^\beta$ in terms of which $\rho^{\alpha-\beta}(\mathbf{r}) = \sum_{\mu\nu} P_{\mu\nu}^{\alpha-\beta}\varphi_\mu(\mathbf{r})\varphi_\nu(\mathbf{r})$. In the finite basis set, the UHF equations assume the form of a pseudo-eigenvalue problem:

$$\mathbf{F}^\sigma(\mathbf{c})\mathbf{c}^\sigma = \varepsilon\mathbf{S}\mathbf{c} \tag{6.15}$$

where $\mathbf{S}$ is the overlap matrix $S_{\mu\nu} = (\mu|\nu)$ and ε is a diagonal matrix of orbital energies. $\mathbf{F}$ has been written as $\mathbf{F}^\sigma(\mathbf{c})$ to emphasize that the fact that $\mathbf{F}$ depends on its own eigenvectors which emphasizes the iterative nature of the problem. The total energy of the UHF-approximation is:

$$\begin{aligned} E_{UHF} &= \frac{1}{2}\sum_{\mu\nu} P_{\mu\nu}^\alpha(h_{\mu\nu} + F_{\mu\nu}^\alpha) + P_{\mu\nu}^\beta(h_{\mu\nu} + F_{\mu\nu}^\beta) \\ &= \sum_{\mu\nu} P_{\mu\nu}h_{\mu\nu} + \frac{1}{2}\sum_{\kappa\tau}\left[P_{\mu\nu}P_{\kappa\tau}(\mu\nu|\kappa\tau) - (\mu\kappa|\nu\tau)(P_{\mu\nu}^\alpha P_{\kappa\tau}^\alpha + P_{\mu\nu}^\beta P_{\kappa\tau}^\beta)\right] + V_{NN} \end{aligned} \tag{6.16}$$

This equation shows clearly, that the nonlocal exchange "interaction" is only present for electrons with the same spin.

Today, Hartree–Fock calculations can be performed with reasonable basis sets for quite large molecules. On standard personal computers, molecules with up to about 500 atoms are within reach, and by using state-of-the-art linear scaling technology even systems with more than 1000 atoms can be studied. In the basis-set limit (the limit reached when the basis set is approaching mathematical completeness), the Hartree–Fock methods recover typically more than 99% of the total molecular energy. Unfortunately, however, even the remaining error is still very large on the chemical scale and amounts to hundreds of kcal mol^{-1}. Thus, many properties cannot be predicted sufficiently accurately by the Hartree–Fock theory. Since the Hartree–Fock method describes a mean-field approach to the N-electron problem, the residual error is called the "correlation error"; in fact, a great deal of effort has been expended into calculating the so-called "correlation energy," which is the difference between the exact eigenvalue of the BO-Hamiltonian and the HF-energy. If one follows an *ab initio* philosophy, one starts from the HF solution and tries to calculate the correlation energy using a variety of approaches such as configuration interaction (CI), many-body perturbation theory (MBPT). This is also known as the Møller–Plesset perturbation theory; the simplest approximation is the second-order estimate, widely known as MP2, or the powerful coupled-cluster (CC) theory. Such "correlated *ab initio* approaches," if taken far enough, can systematically and reliably approach the exact BO-results, although they are known to be notoriously expensive in terms of computational resource requirements. There is much hope, however, that very good approximations to the rigorous *ab initio* methods will become available in the foreseeable future, even for large molecules, owing to the development of so-called "linear scaling" approaches.

A radically different approach to the N-electron problem is provided by DFT. Owing to the celebrated Hohenberg–Kohn theorems, it is known that – at least in principle – the knowledge of $\rho(\mathbf{r})$ is already sufficient in order to deduce the exact ground-state energy. However, this comes at a price of introducing an unknown exchange-correlation functional denoted as $E_{xc}[\rho]$. Since a systematic procedure to approach the exact $E_{xc}[\rho]$ appears to be unknown, physically motivated guesses must be introduced and, indeed, over the years many such approximations have been suggested, with new functionals appearing the literature on an almost weekly basis. Unfortunately, each functional has its own strengths and weaknesses which must be assessed through extensive series of test calculations.

Without going into great detail, it should be noted that the so-called Kohn–Sham (KS) procedure (Koch and Holthausen, 2000) allows the solution of a set of pseudo-single particle equations which would provide the exact ground-state energy if the exact $E_{xc}[\rho]$ were known. This procedure introduces the so-called "noninteracting reference system," which is described by a single Slater determinant and shares with the physical system the electron density calculated through Equation 6.14. The spin-unrestricted KS equations appear similar to the UHF equations; in a finite basis set, they read (upon dividing the electron density into its spin components $\rho^\alpha(\mathbf{r})$ and $\rho^\beta(\mathbf{r})$:

$$F^\sigma_{\mu\nu} = h_{\mu\nu} + \sum_{\kappa\tau} P_{\kappa\tau}(\mu\nu|\kappa\tau) + \int \varphi_\mu(\mathbf{r})\varphi_\nu(\mathbf{r})V^\sigma_{XC}[\rho^\alpha,\rho^\beta](\mathbf{r})d\mathbf{r} \tag{6.17}$$

Thus, in place of the HF exchange term, there now appears a *local* exchange-correlation potential, which is defined as the functional derivative of $E_{xc}[\rho^\alpha,\rho^\beta]$ (ρ^α and ρ^β are the spin-up and spin-down components of the total electron density $\rho = \rho^\alpha + \rho^\beta$) with respect to the spin-components of $\rho(\mathbf{r})$:

$$V^\sigma_{XC}[\rho^\alpha,\rho^\beta](\mathbf{r}) = \frac{\delta E_{xc}[\rho]}{\delta \rho^\sigma(\mathbf{r})} \quad (\sigma = \alpha,\beta) \tag{6.18}$$

The total KS energy is:

$$E_{UKS} = \sum_{\mu\nu} P_{\mu\nu} h_{\mu\nu} + \frac{1}{2}\sum_{\kappa\tau} P_{\mu\nu} P_{\kappa\tau}(\mu\nu|\kappa\tau) + E_{XC}[\rho^\alpha,\rho^\beta] + V_{NN} \tag{6.19}$$

The second term consists of the coulombic self-interaction of the electron cloud, and this can be written in a perhaps somewhat more illuminating way as:

$$E_J = \frac{1}{2}\iint \rho(\mathbf{r}_1) \frac{1}{|\mathbf{r}_1 - \mathbf{r}_2|} \rho(\mathbf{r}_2) d\mathbf{r}_1 d\mathbf{r}_2 \tag{6.20}$$

Likewise, the Coulomb contribution to the KS matrix is:

$$\begin{aligned} J_{\mu\nu} &= \iint \varphi_\mu(\mathbf{r}_1)\varphi_\nu(\mathbf{r}_1) \frac{1}{|\mathbf{r}_1 - \mathbf{r}_2|} \rho(\mathbf{r}_2) d\mathbf{r}_1 d\mathbf{r}_2 \\ &= \iint \varphi_\mu(\mathbf{r})\varphi_\nu(\mathbf{r}) V_C(\mathbf{r}) d\mathbf{r}, \end{aligned} \tag{6.21}$$

which emphasizes the local nature of the Coulomb potential $V_C(\mathbf{r})$. Since this potential is of long range, its calculation usually dominates the computational effort of a Hartree–Fock or KS calculation. The precise functional forms of the various approximations to $E_{xc}[\rho^\alpha, \rho^\beta]$ are complicated, and involve odd powers of $\rho(\mathbf{r})$ such as $\rho(\mathbf{r})^{4/3}$. If the functional also depends on the gradient of $\rho(\vec{\nabla}\rho(\mathbf{r}))$, one obtains functionals from the "generalized gradient approximation" (GGA) family. Modern functionals may also depend on the Laplacian of the density $(\vec{\nabla}^2 \rho(\mathbf{r}))$ and the kinetic energy density ($\tau(\mathbf{r})$), which leads to the family of "meta-GGA" functionals. In recent years, so-called "hybrid functionals" have become very popular; these involve a fraction of the nonlocal Hartree–Fock exchange, and have been found to improve the results for total energies, as well as many molecular properties.

As there are many important conceptual and practical subtleties in DFT that cannot be discussed within the framework of this section, the interested reader is referred to a recent review that also provides pointers to the specialist literature (Neese, 2008).

6.3.2
Additional Terms in the Hamiltonian

Given an approximation to the ground-state energy of the BO-Hamiltonian by one method, there is a need to introduce the smaller field- and spin-dependent terms in the Hamiltonian which give rise to the interactions that are actually probed with EPR spectroscopy. These terms can be derived through relativistic quantum chemistry (which is beyond the scope of this chapter). Among the many terms that arise, the following interactions will mainly be needed:

- **The spin–orbit coupling (SOC):** Unlike as found in many textbooks, this term in the Hamiltonian is of a *two*-electron nature and reads within the Breit–Pauli approximation:

$$\hat{H}_{SOC} = \hat{H}_{SOC}^{(1)} + \hat{H}_{SOC}^{(2)} \tag{6.22}$$

$$\hat{H}_{SOC}^{(1)} = \sum_i \hat{h}_i^{1el-SOC} = \sum_i \hat{\mathbf{h}}_i^{1el-SOC} \hat{\mathbf{s}}_i = \frac{\alpha^2}{2} \sum_i \sum_A Z_A r_{iA}^{-3} \hat{\mathbf{l}}_{iA} \hat{\mathbf{s}}_i \tag{6.23}$$

$$\hat{H}_{SOC}^{(2)} = \hat{H}_{SSO}^{(2)} + \hat{H}_{SOO}^{(2)} = \sum_i \sum_{j \neq i} \hat{\mathbf{g}}_{i,j}^{2el-SOC} = -\frac{\alpha^2}{2} \sum_i \sum_{j \neq i} r_{ij}^{-3} \hat{\mathbf{l}}_{ij} (\hat{\mathbf{s}}_i + 2\hat{\mathbf{s}}_j) \tag{6.24}$$

Here, $\alpha = c^{-1}$ in atomic units is the fine structure constant (~1/137), $\hat{\mathbf{r}}_i, \hat{\mathbf{p}}_i, \hat{\mathbf{s}}_i$ are the position, momentum and spin operators of the ith electron, and $\hat{\mathbf{l}}_{iA} = (\hat{\mathbf{r}}_i - \mathbf{R}_A) \times \hat{\mathbf{p}}_i$ is the angular momentum of the ith electron relative to nucleus A. The vector $\hat{\mathbf{r}}_{iA} = \hat{\mathbf{r}}_i - \mathbf{R}_A$ of magnitude r_{iA} is the position of the ith electron relative to atom A. Likewise, the vector $\hat{\mathbf{r}}_{ij} = \hat{\mathbf{r}}_i - \hat{\mathbf{r}}_j$ of magnitude r_{ij} is the position of the ith electron relative to electron j, and $\hat{\mathbf{l}}_{ij} = (\hat{\mathbf{r}}_i - \hat{\mathbf{r}}_j) \times \hat{\mathbf{p}}_i$ is its angular momentum relative to this electron. The one-electron term is familiar from many phenomenological treatments; for example, in atomic spectroscopy and ligand field theory. The two-electron term has contributions from the spin-same-orbit (SSO) and spin-other-orbit (SOO) terms, both of which are important for a quantitatively correct treatment of SOC. They essentially provide a screening of the one-electron term, in much the same way as the nuclear–electron attraction and electron–electron repulsion contributions counteract each other in the BO-Hamiltonian. Since the full SOC operator is difficult to handle in large-scale molecular applications, it is desirable to approximate it as accurately as possible. This is possible through the accurate spin-orbit mean-field (SOMF) approximation, as developed by Hess et al. (1996). Without going into the details of the derivation, the form of this operator can be stated as:

$$\hat{h}^{SOMF} = \sum_i \hat{z}^{SOMF}(i)\hat{s}(i) \quad (6.25)$$

with the matrix elements of the kth component of the SOMF operator given by:

$$\langle \varphi_\mu | \hat{z}_k^{SOMF} | \varphi_\nu \rangle = \langle \varphi_\mu | \hat{h}_k^{1el-SO} | \varphi_\nu \rangle$$
$$+ \sum_{\kappa\tau} P_{\kappa\tau} \left[(\varphi_\mu \varphi_\nu | \hat{g}_k^{SO} | \varphi_\kappa \varphi_\tau) - \frac{3}{2}(\varphi_\mu \varphi_\kappa | \hat{g}_k^{SO} | \varphi_\tau \varphi_\nu) - \frac{3}{2}(\varphi_\tau \varphi_\nu | \hat{g}_k^{SO} | \varphi_\mu \varphi_\kappa) \right] \quad (6.26)$$

and:

$$\hat{h}_k^{1el-SO}(\mathbf{r}_i) = \frac{\alpha^2}{2} \sum_i \sum_A Z_A r_{iA}^{-3} \hat{l}_{iA;k} \quad (6.27)$$

$$\hat{g}_k^{SO}(\mathbf{r}_i, \mathbf{r}_j) = -\frac{\alpha^2}{2} \hat{l}_{ij;k} r_{ij}^{-3} \quad (6.28)$$

Here, **P** is the total charge density matrix calculated by some theoretical method. Essentially, in the same way that the Hartree–Fock approximation gives 99% of the total molecular energy, the SOMF operator covers around 99% of the two-electron SOC operator. This will be used exclusively below to approximate the SOC terms which will arise in the equations for the spin-Hamiltonian parameters.

- **The direct magnetic dipolar spin–spin interaction:** This interaction is described by a genuine two-electron operator of the form:

$$\hat{H}_{SS} = \frac{g_e^2 \alpha^2}{8} \sum_{i \neq j} \left(\frac{\hat{s}_i \hat{s}_j}{r_{ij}^3} - 3\frac{(\hat{s}_i \mathbf{r}_{ij})(\hat{s}_j \mathbf{r}_{ij})}{r_{ij}^5} \right) \quad (6.29)$$

where the free-electron g-value $g_e = 2.002319...$ is used.
- **The hyperfine coupling:** This term describes the well-known dipolar-interaction between the electron spin and the nuclear spins.

$$\hat{H}_{SI} = \frac{\alpha}{2} g_e \beta_N \sum_A g_N^{(A)} \sum_i \left(\frac{\hat{\mathbf{s}}_i \hat{\mathbf{I}}^{(A)}}{r_{iA}^3} - 3\frac{(\hat{\mathbf{s}}_i \mathbf{r}_{iA})(\hat{\mathbf{I}}^{(A)} \mathbf{r}_{iA})}{r_{iA}^5} \right) \quad (6.30)$$

Here, $g_N^{(A)}$ is the g-value of the A''th nucleus and $\hat{\mathbf{I}}^{(A)}$ is the spin-operator for the nuclear spin of the A'' th nucleus. While the isotropic Fermi contact term is frequently introduced as a separate operator, it arises naturally as a boundary term in the partial integration of the singular operator in Equation 6.30.

- **The nuclear–orbit interaction:** The interaction of the nuclear spin with the orbital angular momentum of the electrons leads to the following term in the Hamiltonian.

$$\hat{H}_{LI} = \frac{\alpha}{2}\beta_N \sum_A g_N^{(A)} \sum_i \frac{\mathbf{l}_i^A \hat{\mathbf{I}}^{(A)}}{r_{iA}^3} \tag{6.31}$$

- **The quadrupole coupling:** The quadrupole coupling describes the interaction of the electric field gradient (EFG) at a given nucleus with the quadrupole moment of that nucleus (only present for nuclei with spin $I > 1/2$). The electronic quantity of interest is the field gradient operator. The quadrupole interaction may be written as an operator of the following form:

$$\hat{H}_Q = e^2 \sum_{A,i} Q^{(A)} \hat{\mathbf{I}}^{(A)} \hat{\mathbf{F}}^{(A)}(i) \hat{\mathbf{I}}^{(A)} \tag{6.32}$$

where $Q^{(A)}$ is the quadrupole moment of the A'th nucleus, e is the elementary charge, and the field gradient operator is given by:

$$\hat{F}_{\mu\nu}^{(A)}(i) = \frac{r_{iA}^2 \delta_{\mu\nu} - 3 r_{iA;\mu} r_{iA;\nu}}{r_{iA}^5} \tag{6.33}$$

- **The electronic Zeeman-interaction:** The interaction of the electrons with a static external magnetic field is described by:

$$\hat{H}_{LB} = \frac{\alpha}{2} \sum_i \mathbf{B}(\mathbf{l}_i + g_e \hat{\mathbf{s}}_i) \tag{6.34}$$

From the fully relativistic treatment there arises a "kinetic energy correction" (relativistic mass correction) to the spin-Zeeman energy that is given by:

$$\hat{H}_{SB}^{RMC} = \frac{\alpha^3 g_e}{2} \sum_i \nabla_i^2 \mathbf{B} \mathbf{s}_i \tag{6.35}$$

6.3.3
Sum-Over States Theory of Spin Hamiltonian Parameters

In general, a ground state with total spin S gives rise to $2S + 1$ "magnetic sublevels," with $M_S = S, S - 1,...,-S$. At the level of the BO-Hamiltonian, these $2S + 1$ sublevels are all degenerate; however, upon introduction of the additional terms from Section 6.3.2 (collectively denoted as $\hat{H}_1$), this degeneracy is lifted and the task at hand is to describe the splittings of the magnetic sublevels through an effective Hamiltonian of the same form as the spin Hamiltonian. An illuminating approach has been outlined by McWeeny (1992), and is followed here in order to provide formally exact expressions of the various terms that arise in the spin Hamiltonian. Suppose, that the entire spectrum of exact eigenfunctions of the BO-Hamiltonian is available, and that the effect of the much smaller additional terms is to be introduced through perturbation theory. The ground-state, many-electron wavefunction is denoted as $|0SM\rangle$ and the excited states are written as $|bS'M'\rangle$, where b is a compound index that summarizes everything that is necessary in order to unambiguously identify the given excited state of total spin S' and magnetic quantum number M'. In order for the perturbation theory to be valid, it is assumed that the energy separation $\Delta_b = E_b - E_0$, with:

$$E_b = \langle bS'M'|\hat{H}_{BO}|bS'M''\rangle = \delta_{M'M''}\langle bS'S'|\hat{H}_{BO}|bS'S'\rangle \tag{6.36}$$

being much larger than the matrix elements of the additional terms in the Hamiltonian. Since, for a reasonably well isolated ground state, Δ_b is on the order of thousands of wavenumbers and the additional terms are on the order of at most a few hundred wavenumbers, this is in most cases a reasonable assumption. Using the partitioning theory of McWeeny, the effective Hamiltonian in the space of the magnetic sublevels becomes to second order:

$$\langle 0SM|\hat{H}_{\text{eff}}|0SM'\rangle$$
$$= E_0\delta_{MM'} + \langle 0SM|\hat{H}_1|0SM'\rangle - \sum_{bS'M''}\Delta_b^{-1}\langle 0SM|\hat{H}_1|bS'M''\rangle\langle bS'M''|\hat{H}_1|0SM'\rangle \tag{6.37}$$

The first term on the right-hand side is the total energy of the reference state, and might be dropped as it does not lead to splittings between the magnetic sublevels. The second term is of first order in perturbation theory, while the third term is of second order and involves an infinite sum over excited states. The task to deduce the correct expressions of the spin-Hamiltonian parameters from this equations involves a considerable amount of algebra (which is not carried out here). Basically, the idea is to seek operators or pairs of operators which carry the same spin- or field terms as the corresponding terms in the spin Hamiltonian. The next stage is to study whether these terms have the same M, M' and M'' dependence as the corresponding spin Hamiltonian terms; this is best achieved through application of the Wigner–Eckart theorem. The results are the following formally exact second-order expressions for the various spin-Hamiltonian parameters (note that in all the expressions below only the "standard components" of each multiplet with $M = S$ appear):

- **Zero-field splitting:**
  The ZFS consists of a first-order term arising from the direct spin–spin interaction:

$$D_{\mu\nu}^{(SS)} = \frac{1}{2}\frac{\alpha^2}{S(2S-1)}\left\langle 0SS\left|\sum_i\sum_{j\neq i}\frac{r_{ij}^2\delta_{\mu\nu} - 3(r_{ij})_\mu(r_{ij})_\nu}{r_{ij}^5}\{2\hat{s}_{zi}\hat{s}_{zj} - \hat{s}_{xi}\hat{s}_{xj} - \hat{s}_{yi}\hat{s}_{yj}\}\right|0SS\right\rangle \tag{6.38}$$

and a second-order term that arises from SOC. Using the effective one-electron SOC operator described above, the components of the **D**-tensor can be shown to be:

$$D_{kl}^{SOC-(0)} = -\frac{1}{S^2} \sum_{b(S_b=S)} \Delta_b^{-1} \left\langle 0SS \left| \sum_i z_{k;i}^{SOMF} \hat{s}_{i;z} \right| bSS \right\rangle \left\langle bSS \left| \sum_i z_{l;i}^{SOMF} \hat{s}_{i,z} \right| 0SS \right\rangle \tag{6.39}$$

$$D_{kl}^{SOC-(-1)} = -\frac{1}{S(2S-1)} \sum_{b(S_b=S-1)} \Delta_b^{-1} \left\langle 0SS \left| \sum_i z_{k;i}^{SOMF} \hat{s}_{i,+1} \right| bS-1S-1 \right\rangle$$

$$\left\langle bS-1S-1 \left| \sum_i z_{l;i}^{SOMF} \hat{s}_{i,-1} \right| 0SS \right\rangle \tag{6.40}$$

$$D_{kl}^{SOC-(+1)} = -\frac{1}{(S+1)(2S+1)} \sum_{b(S_b=S+1)} \Delta_b^{-1} \left\langle 0SS \left| \sum_i z_{k;i}^{SOMF} \hat{s}_{i,-1} \right| bS+1S+1 \right\rangle$$

$$\left\langle bS+1S+1 \left| \sum_i z_{l;i}^{SOMF} \hat{s}_{i,+1} \right| 0SS \right\rangle \tag{6.41}$$

Thus, the SOC contribution has three terms which arise from excited states with the same total spin as the ground state, and from terms which differ by unit of total spin-angular momentum.

- **g-Tensor:** The g-tensor can be written as a sum of four parts. The first part is the free-electron g-value, and is usually dominant; however, as it is a natural constant it adds nothing to the information content of the g-tensor in analyzing the molecular geometric and electronic structure. The second and third terms are of first order, and usually fairly small. The dominant contribution to $\Delta g = g - 1g_e$ arises from the cross term between the SOC and the orbital Zeeman-interactions (fourth term).

$$g_{zz}^{(SB)} = g_e \tag{6.42}$$

$$g_{kl}^{(RMC)} = \delta_{kl} \frac{\alpha^2}{2} \frac{1}{S} \frac{g_e}{2} \left\langle 0SS \left| \sum_i \nabla_i^2 \hat{s}_{z;i} \right| 0SS \right\rangle \tag{6.43}$$

$$g_{kl}^{(GC)} = \frac{1}{S} \left\langle 0SS \left| \sum_{i,A} \xi(r_{iA}) \{ \mathbf{r}_{iA} \mathbf{r}_i - r_{iA;k} r_{i;l} \} s_{z;i} \right| 0SS \right\rangle \tag{6.44}$$

$$g_{kl}^{(OZ/SOC)} = -\frac{1}{S} \sum_{b(S_b=S)} \Delta_b^{-1} \left\{ \left\langle 0SS \left| \sum_i l_{i;k} \right| bSS \right\rangle \left\langle bSS \left| \sum_i z_{l;i}^{SOMF} s_{z;i} \right| 0SS \right\rangle \right.$$

$$\left. + \left\langle 0SS \left| \sum_i z_{k;i}^{SOMF} s_{z;i} \right| bSS \right\rangle \left\langle bSS \left| \sum_i l_{i;l} \right| 0SS \right\rangle \right\} \tag{6.45}$$

The first term is simply the spin-Zeeman term, the second term is the reduced-mass correction, and the third term is a gauge correction to the SOC which has been written in a somewhat simplified form here that is explained in detail elsewhere (Lushington and Grein, 1996; Neese, 2001)

- **Hyperfine coupling:** The hyperfine coupling consists of three contributions where the Fermi-contact term has been separated from the traceless dipolar contribution:

$$A_{kl}^{(A;c)} = \delta_{kl} \frac{8\pi}{3} \frac{\alpha}{2} \frac{1}{S} g_e \beta_N g_N^{(A)} \left\langle 0SS \left| \sum_i s_{z;i} \delta(r_{iA}) \right| 0SS \right\rangle \quad (6.46)$$

$$A_{kl}^{(A;d)} = \frac{\alpha}{2} \frac{1}{S} g_e \beta_N g_N^{(A)} \left\langle 0SS \left| \sum_i s_{z;i} r_{iA}^{-5} \{ \delta_{kl} r_{iA}^2 - 3 r_{iA;k} r_{iA;l} \} \right| 0SS \right\rangle \quad (6.47)$$

$$A_{kl}^{(A;SO)} = -\frac{\alpha}{2S} g_e \beta_N g_N^{(A)} \sum_{b(S_b=S)} \Delta_b^{-1} \Big\{ \left\langle 0SS \left| \sum_i l_{i;k}^A r_{iA}^{-3} \right| bSS \right\rangle \left\langle bSS \left| \sum_i z_{l;i}^{SOMF} s_{z;i} \right| 0SS \right\rangle$$

$$+ \left\langle 0SS \left| \sum_i z_{k;i}^{SOMF} s_{z;i} \right| bSS \right\rangle \left\langle bSS \left| \sum_i l_{i;l}^A r_{iA}^{-3} \right| 0SS \right\rangle \Big\} \quad (6.48)$$

The first two terms are widely known, while the third term is a contribution from the SOC.

- **Field gradients:** Finally, the computation of field gradients is given by:

$$V_{kl}^{(A)} = \left\langle 0SS \left| \sum_i r_{iA}^{-5} \{ \delta_{kl} r_{iA}^2 - 3 r_{iA;k} r_{iA;l} \} \right| 0SS \right\rangle \quad (6.49)$$

Note that this term is a simple spin-independent first-order property of the system.

6.3.4
Linear Response Theory

The equations listed in Section 6.3.3 are important from a conceptual point of view, as they demonstrate the basic physics that is involved in spin-Hamiltonian parameters most clearly. However, as a basis of actual calculations, they are unfortunately much less useful, owing to the presence of the second-order terms. The evaluation of these terms would require an infinite sum over excited many-electron states, but in practice only a few dozen many-electron states can be calculated, at most. Although quite useful results have been obtained with this approach, the convergence of the perturbation sum is uncertain and can, in the general case, hardly be guaranteed. In the case of DFT, the excited states cannot be obtained explicitly, as the Hohenberg–Kohn theorems apply only to the electronic ground state. Hence, it is important to seek an alternative definition of the various spin-Hamiltonian parameters. An approach of substantial generality and elegance is provided by the so-called linear response theory (LRT) which, in the present author's view, is just one realization of a family of methods that are all formulated in a similar spirit. If time does not explicitly occur in the equations – which is not necessary for the formulation of EPR and NMR parameters – then these methods can also be termed "analytic derivative approaches." In the framework of HF and DFT methods, they are known as coupled-perturbed self-consistent field (CP-SCF) methods or double-perturbation theory (DPT), respectively. Each of these acronyms represents computational methods that provide identical results, and it is simply a matter of taste as to which framework is preferred. These methods, which have been developed to a highly sophisticated level in quantum chemistry, have proven extremely useful in many contexts, including geometry optimization and frequency calculations (geometric derivatives), and the precise prediction of many molecular properties. In fact, the LRT approaches can be shown to implicitly involve an untruncated sum over excited states of the system that are (implicitly) described at the same level of sophistication as the ground state. The key quantities of interest in LRT are the derivatives of the (approximate) total ground state energy with respect to external perturbation parameters $\lambda, \kappa, \ldots$ where λ and κ may denote the components of an external field or a nuclear magnetic moment of an electronic magnetic moment. Formally, the derivative of the perturbation sum in Equation 6.36 could be taken, and the connection then made to the response formalism by with the appropriate derivative of the approximate total energy, calculated using the theoretical method of choice. Since all spin-Hamiltonian parameters are bilinear in external perturbations, the desired quantity is the second partial derivative of the total energy.

In order to appreciate the general concepts that are involved, the linear response equations for a self-consistent-field (SCF) ground state will be sketched below. This description is appropriate if the state of interest is well described by a Hartree–Fock or DFT single determinant (Section 6.3.1). The ground state energy is written here as:

$$E = V_{NN} + \sum_i h_{ii} + \frac{1}{2}\sum_{i,j}(ii|jj) - c_{HF}(ij|ij) + c_{DF}E_{XC}[\rho] \quad (6.50)$$

where the parameters c_{HF} and c_{DF} are scaling parameters for the HF exchange energy and the XC-energy, respectively. Thus, HF theory corresponds to $c_{HF}=1; c_{DF}=0$, "pure" DFT corresponds to $c_{HF}=0; c_{DF}=1$, while hybrid DFT methods choose $0 < c_{HF} < 1$. The energy has been written here in terms of the occupied orbitals $\psi_i(\mathbf{x})$. They are determined self-consistently from the SCF (Hartree–Fock or KS) equations:

$$\left\{\hat{h} + \int \frac{\rho(\mathbf{x}')}{|\mathbf{x}-\mathbf{x}'|}d\mathbf{x}' - c_{HF}\sum_i \hat{K}^{ii} + c_{DF}\frac{\delta E_{XC}[\rho]}{\delta\rho(\mathbf{x})}\right\}\psi_i(\mathbf{x}) = \varepsilon_i\psi_i(\mathbf{x}) \quad (6.51)$$

For illustrating the concepts, it is sufficient to consider the case where the basis functions are chosen to be independent on the external perturbations. To include such a dependence (as is necessary for example for geometric or magnetic field perturbations) is straightforward, but would lead to more lengthy equations which are not of interest for the purpose of this section (see Neese, 2008). Since the MO coefficients c are determined in a variational procedure, one has:

$$\frac{\partial E}{\partial c_{\mu i}} \frac{\partial c_{\mu i}}{\partial \lambda} = 0 \tag{6.52}$$

and therefore the first derivative of the energy with respect to a perturbation λ is:

$$\left.\frac{\partial E}{\partial \lambda}\right|_{\lambda=0} = \sum_{\mu\nu} P_{\mu\nu} \langle \varphi_\mu | \hat{h} | \varphi_\nu \rangle_\lambda \tag{6.53}$$

where:

$$\langle \varphi_\mu | \hat{h} | \varphi_\nu \rangle_\lambda = \left\langle \varphi_\mu \left| \frac{\partial \hat{h}}{\partial \lambda} \right| \varphi_\nu \right\rangle \tag{6.54}$$

if λ is a one-electron perturbation, and if the basis functions are independent of λ. Through an additional differentiation the second partial derivative becomes:

$$\left.\frac{\partial^2 E}{\partial \lambda \delta \kappa}\right|_{\lambda=0;\kappa=0} = \sum_{\mu\nu} P_{\mu\nu} \langle \varphi_\mu | \hat{h} | \varphi_\nu \rangle_{\lambda\kappa} + \sum_{\mu\nu} \frac{\partial P_{\mu\nu}}{\partial \kappa} \langle \varphi_\mu | \hat{h} | \varphi_\nu \rangle_\lambda \tag{6.55}$$

This important equation contains two contributions. The first term is referred to as a first-order contribution as it only depends on the ground-state density. The second term is a second-order contribution, as it requires knowledge of the first derivative of the density matrix with respect to an external perturbation and the first derivative of the Hamiltonian. These two terms substitute the two first- and second-order terms in the perturbation sum of Equation 6.36. It remains to be shown how the perturbed density matrix can be calculated.

A simple approach will be followed here for calculation of the perturbed density matrix. To this end, the perturbed orbitals $\psi_i^{(\lambda)} \equiv |i^{(\lambda)}\rangle$ are calculated in terms of the zeroth order orbitals $\psi_i^{(0)} \equiv |i^{(0)}\rangle$. Differentiation of the SCF equations yields:

$$\{\hat{F}^{(0)} - \varepsilon_i^{(0)}\}|i^{(\lambda)}\rangle + \{\hat{F}^{(\lambda)} - \varepsilon_i^{(\lambda)}\}|i^{(0)}\rangle = 0 \tag{6.56}$$

The perturbed orbitals $|i^{(\lambda)}\rangle$ are expanded as:

$$|i^{(\lambda)}\rangle = \sum_a U_{ai}^{(\lambda)} |a^{(0)}\rangle \tag{6.57}$$

(Note that here and below, the labels i, j, k, l refer to occupied orbitals, and a, b, c, d to unoccupied orbitals.) The unitary matrix $\mathbf{U}$ has only occupied/virtual blocks in the case that the basis functions do not depend on the perturbation. In order to determine the unique elements of $\mathbf{U}$, one uses the perturbed SCF equations:

$$\sum_b U_{bi}^{(\lambda)} \langle a^{(0)} | \hat{F}^{(0)} - \varepsilon_i^{(0)} | b^{(0)} \rangle + \langle a^{(0)} | \hat{F}^{(\lambda)} - \varepsilon_i^{(\lambda)} | i^{(0)} \rangle = 0 \tag{6.58}$$

$$= U_{ai}^{(\lambda)} (\varepsilon_a^{(0)} - \varepsilon_i^{(0)}) + \langle a^{(0)} | \hat{F}^{(\lambda)} | i^{(0)} \rangle = 0 \tag{6.59}$$

However, $\hat{F}^{(\lambda)}$ depends on the perturbed orbitals. Therefore, after taking the derivative of the SCF operator carefully, one obtains

6.3 Electronic Structure Theory of Spin-Hamiltonian Parameters

$$\hat{F}^{(\lambda)} = \hat{h}^{(\lambda)} + \int \frac{\rho^{(\lambda)}(\mathbf{x}')}{|\mathbf{x}-\mathbf{x}'|} - c_{HF} \sum_{jb} U_{bj}^{(\lambda)*} \hat{K}^{bj} + U_{bj}^{(\lambda)} \hat{K}^{jb} + c_{DF} \iint \frac{\delta^2 E_{XC}[\rho]}{\partial \rho(\mathbf{x}) \partial \rho(\mathbf{x}')} \rho^{(\lambda)}(\mathbf{x}') d\mathbf{x} d\mathbf{x}'$$

$$= \hat{h}^{(\lambda)} + \sum_{jb} U_{bj}^{(\lambda)*} \hat{J}^{bj} + U_{bj}^{(\lambda)} \hat{J}^{jb} - c_{HF} U_{bj}^{(\lambda)*} \hat{K}^{bj} - c_{HF} U_{bj}^{(\lambda)} \hat{K}^{jb} + c_{DF} \int f_{xc}[\rho] \rho^{(\lambda)}(\mathbf{x}) d\mathbf{x}$$

$$= \hat{h}^{(\lambda)} + \sum_{jb} U_{bj}^{(\lambda)*} \left[\hat{J}^{bj} - c_{HF} \hat{K}^{bj} + c_{DF} \int f_{xc}[\rho] \psi_b^{(0)*}(\mathbf{x}) \psi_j^{(0)}(\mathbf{x}) d\mathbf{x} \right]$$

$$+ U_{bj}^{(\lambda)} \left[\hat{J}^{jb} - c_{HF} \hat{K}^{jb} + c_{DF} \int f_{xc}[\rho] \psi_j^{(0)*}(\mathbf{x}) \psi_b^{(0)}(\mathbf{x}) d\mathbf{x} \right] \quad (6.60)$$

Here, the "XC-kernel" $f_{xc}[\rho]$ has been defined as the second functional derivative with respect to ρ, and it has been tacitly assumed that for all functionals in use this yields a factor $\delta(\mathbf{x}-\mathbf{x}')$ which reduces the double integral to a single integral. Taken together, this results in the first-order equations:

$$U_{ai}^{(\lambda)}\left(\varepsilon_a^{(0)} - \varepsilon_i^{(0)}\right) + \left\langle a^{(0)} \left| \hat{h}^{(\lambda)} \right| i^{(0)} \right\rangle + \sum_{jb} U_{bj}^{(\lambda)*} \left[(bj|ai) - c_{HF}(ba|ji) + (ai|f_{xc}|jb) \right]$$

$$+ U_{bj}^{(\lambda)}\left[(jb|ai) - c_{HF}(ja|bi) + (ai|f_{xc}|jb)\right] = 0 \quad (6.61)$$

At this point, it is useful to distinguish two different types of perturbation. First, "electric field-like perturbations" yield purely real $\left\langle a^{(0)} \left| \hat{h}^{(\lambda)} \right| i^{(0)} \right\rangle$ and consequently, also purely real and symmetric **U** matrices. In this case one has:

$$\mathbf{A}^{(E)} \mathbf{U}^{(\lambda)} = -\mathbf{V}^{(\lambda)} \quad (6.62)$$

with:

$$A_{ia,jb}^{(E)} = \delta_{ij} \delta_{ab} \left(\varepsilon_a^{(0)} - \varepsilon_i^{(0)}\right) + 2(jb|ia) + 2c_{DF}(ai|f_{xc}|jb) - c_{HF}\{(ba|ji) + (ja|bi)\} \quad (6.63)$$

$$V_{ai}^{(\lambda)} = \left\langle a^{(0)} \left| \hat{h}^{(\lambda)} \right| i^{(0)} \right\rangle \quad (6.64)$$

Note that the A-matrix (the "electric Hessian") is independent of the nature of the perturbation, and that the **U** and **V** matrices have been written as vectors with the compound index, ai.

Second, "magnetic-field-like perturbation" yields purely Hermitian imaginary matrix elements $\left\langle a^{(0)} \left| \hat{h}^{(\lambda)} \right| i^{(0)} \right\rangle$, and consequently also purely imaginary and Hermitian **U** matrices. This leads to:

$$\mathbf{A}^{(M)} \mathbf{U} = -\mathbf{V}^{(\lambda)} \quad (6.65)$$

with

$$A_{ia,jb}^{(M)} = \delta_{ij} \delta_{ab} \left(\varepsilon_a^{(0)} - \varepsilon_i^{(0)}\right) + c_{HF}\{(ib|ja) + (ba|ij)\} \quad (6.66)$$

Thus, magnetic field like perturbation yield much easier response (or "coupled perturbed") equations in which the contributions from any local potential vanish. In fact, in the absence of HF exchange the A-matrix becomes diagonal and the linear equation system is solved in a trivial manner. This then leads to a "sum-over-orbital"-like equation for the second derivative, which resembles in some way a "sum-over-states" equation. One should, however, carefully distinguish the sum-over-states picture from linear response or analytic derivative techniques, as they have a very different origin. For electric-field-like or magnetic-field-like perturbations in the presence of HF exchange, one thus has to solve a linear equation system of the size N(occupied) × N(virtual), which may amount to dimensions of several hundred thousand coefficients in large-scale applications. However, there are efficient iterative techniques to solve such large-equation systems without ever explicitly constructing the full A-matrix. Once the perturbed orbitals have been determined, the perturbed density is found from:

$$P_{\mu\nu}^{(\lambda)} = \sum_i U_{ai}^{(\lambda)*} c_{\mu a}^{(0)} c_{\nu i}^{(0)} + U_{ai}^{(\lambda)} c_{\mu i}^{(0)} c_{\nu a}^{(0)} \tag{6.67}$$

6.3.5
Expression for Spin-Hamiltonian Parameters for Self-Consistent Field Methods

Using the results of the preceding sections, it is now possible to provide explicit expressions for all spin-Hamiltonian parameters.

- **Zero-field splitting:** The ZFS is the least well-developed spin-Hamiltonian parameter in EPR spectroscopy. It is also the most complicated, since the spin–spin (SS) contribution is a genuine two-electron property. For this contribution, McWeeny and Mizuno (1961) have shown:

$$D_{kl}^{(SS)} = -\frac{g_e^2}{16} \frac{\alpha^2}{S(2S-1)} \sum_{\mu\nu} \sum_{\kappa\tau} \{P_{\mu\nu}^{\alpha-\beta} P_{\kappa\tau}^{\alpha-\beta} - P_{\mu\kappa}^{\alpha-\beta} P_{\nu\tau}^{\alpha-\beta}\} \langle \mu\nu | r_{12}^{-5} \{3 r_{12,k} r_{12,l} - \delta_{kl} r_{12}^2\} | \kappa\tau \rangle \tag{6.68}$$

The integrals appearing in Equation 6.68 appear complicated at first glance, but are readily calculated and, owing to factorization of the two-particle spin-density matrix, Equation 6.68 can be implemented for large-scale application without creating any storage of computation time bottlenecks.

It is very interesting to examine the physical content of Equation 6.68 in a little more detail. From the form of the operator and the appearance of the spin-density matrix, it is clear that it describes the (traceless) direct electron–electron magnetic dipole–dipole interaction between unpaired electrons. Such a term is widely used in modeling the EPR spectra of interacting electron spins within the "point-dipole" approximation. Equation 6.68 consists of two parts. In analogy with the Hartree–Fock theory, the first part should be recognized as a "Coulomb" contribution, while the second part is an "exchange" contribution. Thus, even the direct dipolar spin–spin interaction contains an exchange contribution, which is of fundamentally different origin than the "genuine" exchange interaction used in the modeling of interacting spins. This point does not appear to be widely recognized; nevertheless, assuming an exponential decay of the basis functions, it becomes evident that the exchange term would be expected to fall off much more quickly with interspin distance than the Coulomb contribution. The "distributed-point-dipole"-like equations can be recovered from Equation 6.68 by: (i) neglecting the exchange contribution; (ii) assuming that the spin density matrix is diagonal in the chosen basis; and (iii) "compressing" the basis functions to δ-functions centered at the atomic positions. One then obtains:

$$D_{kl} \approx -\frac{g_e^2}{16} \frac{\alpha^2}{S(2S-1)} \sum_{AB} P_A^{\alpha-\beta} P_B^{\alpha-\beta} R_{AB}^{-5} \left[3\mathbf{R}_{AB;k} \mathbf{R}_{AB;l} - \delta_{kl} R_{AB}^2 \right] \tag{6.69}$$

where A and B sum over nuclei, and $P_A^{\alpha-\beta} = \sum_{\mu \in A} P_{\mu\mu}^{\alpha-\beta}$ is the "gross" spin population on atom A. Equation 6.69 describes the interaction of point dipoles centered at atomic positions, where each atom pair is weighted by the product of the spin populations that reside on this atom. Note this gross atomic spin population differs from the usual numbers that are predicted by Mulliken or Löwdin analysis (Jensen, 1999) and that are part of the output of many electronic-structure programs, as the latter contain terms that depend on the basis function overlap, while $P_A^{\alpha-\beta}$ does not. However, since the approximation leading to the point-dipole formula, Equation 6.69, have been rather crude, large additional errors may not be expected if Mulliken or Löwdin spin-populations are inserted into Equation 6.69. If the distance between two spin-carrying fragments is large enough, it may even be possible to reduce Equation 6.69 to a single term, where R_{AB} must then refer to an "effective" distance. It is proposed to calculate this as follows. First, the center of gravity of the spin-density of fragments "F1" and "F2" are defined as:

$$\mathbf{R}^{(F1)} = \sum_{A \in F1} \overline{P}_A^{\alpha-\beta} \mathbf{R}_A \tag{6.70}$$

$$\mathbf{R}^{(F2)} = \sum_{B \in F2} \overline{P}_B^{\alpha-\beta} \mathbf{R}_B, \tag{6.71}$$

after which the vector $\mathbf{R}_{AB}$ in Equation 6.69 is replaced by $\mathbf{R}_{12} = \mathbf{R}^{(F1)} - \mathbf{R}^{(F2)}$. The barred quantity $\overline{P}_A^{\alpha-\beta}$ refers to a normalized spin-population such that the sum $\sum_{A \in F1} \overline{P}_A^{\alpha-\beta} = 1$. This appears to be a slightly more rigorous approach than the commonly used approach, in which the intercenter distance is fixed by subjective plausible choices that may, however, differ between different investigators. Note also that the full g-tensor does *not* enter either in Equation 6.68 or 6.69. We have already criticized the use of the g-tensor in the point-dipole approximations for the hyperfine couplings (Neese, 2003a), and a similar situation also applies to the case of the dipolar ZFS tensor.

The SOC contribution to the ZFS involves the response of the orbitals to the SOC. The formalism to achieve an analytic derivative formulation of this part of the spin Hamiltonian has been elucidated recently by the present author, and is somewhat more involved than the analogous methodology for the g-tensor (Neese, 2007a). Without going into too much detail, the three contributions to the D-tensor in Equations 6.39–6.41 may be rewritten as follows:

$$D_{kl}^{(0)} = -\frac{1}{4S^2}\sum_{\mu\nu}\langle\mu|h_k^{SOC}|\nu\rangle\frac{\partial P_{\mu\nu}^{(0)}}{\partial S_l^{(0)}} \quad (6.72)$$

$$D_{kl}^{(+1)} = \frac{1}{2(S+1)(2S+1)}\sum_{\mu\nu}\langle\mu|h_k^{SOC}|\nu\rangle\frac{\partial P_{\mu\nu}^{(-1)}}{\partial S_l^{(+1)}} \quad (6.73)$$

$$D_{kl}^{(-1)} = \frac{1}{2S(2S-1)}\sum_{\mu\nu}\langle\mu|h_k^{SOC}|\nu\rangle\frac{\partial P_{\mu\nu}^{(+1)}}{\partial S_l^{(-1)}} \quad (6.74)$$

In Equations 6.72–6.79, the components $S_l^{(m)}$ ($m = 0, \pm 1$) are the vector operator components of the total spin. The spin-densities $\mathbf{P}^{(m)}$ are the response densities with respect to a SOC perturbation. They are calculated from a non-standard set of coupled-perturbed equations analogous to the ones described above for the g-tensor as:

$$\frac{\partial P_{\mu\nu}^{(0)}}{\partial S_l^{(0)}} = \sum_{i\alpha a\alpha} U_{a\alpha i\alpha}^{(0);l} c_{\mu i}^\alpha c_{\nu a}^\alpha + \sum_{i\beta a\beta} U_{a\beta i\beta}^{(0);l} c_{\mu i}^\beta c_{\nu a}^\beta \quad (6.75)$$

$$\frac{\partial P_{\mu\nu}^{(+1)}}{\partial S_l^{(-1)}} = \sum_{i\alpha a\beta} U_{a\beta i\alpha}^{(-1);l} c_{\mu i}^\alpha c_{\nu a}^\beta - \sum_{i\beta a\alpha} U_{a\alpha i\beta}^{(-1);l} c_{\mu a}^\alpha c_{\nu i}^\beta \quad (6.76)$$

$$\frac{\partial P_{\mu\nu}^{(-1)}}{\partial S_l^{(+1)}} = -\sum_{i\alpha a\beta} U_{a\beta i\alpha}^{(+1);l} c_{\mu a}^\beta c_{\nu i}^\alpha + \sum_{i\beta a\alpha} U_{a\alpha i\beta}^{(+1);l} c_{\mu i}^\beta c_{\nu a}^\alpha \quad (6.77)$$

with the U-coefficients in Equations 6.75–6.77 being calculated from:

$m = 0$:

$$\left(\varepsilon_{aa}^{(0)} - \varepsilon_{ia}^{(0)}\right)U_{a\alpha i\alpha}^{k(0)} + c_{HF}\sum_{j\alpha b\alpha}U_{b\alpha j\alpha}^{k(0)}\{(b_\alpha i_\alpha | a_\alpha j_\alpha) - (j_\alpha i_\alpha | a_\alpha b_\alpha)\} = -(a_\alpha|h_k^{SOC}|i_\alpha)$$
$$(6.78)$$

$$\left(\varepsilon_{a\beta}^{(0)} - \varepsilon_{i\beta}^{(0)}\right)U_{a\beta i\beta}^{k(0)} + c_{HF}\sum_{j\beta b\beta}U_{b\beta j\beta}^{k(0)}\{(b_\beta i_\beta | a_\beta j_\beta) - (j_\beta i_\beta | a_\beta b_\beta)\} = -(a_\beta|h_k^{SOC}|i_\beta)$$
$$(6.79)$$

$m = +1$:

$$\left(\varepsilon_{aa}^{(0)} - \varepsilon_{i\beta}^{(0)}\right)U_{a\alpha i\beta}^{k(+1)} + c_{HF}\sum_{j\alpha b\alpha}U_{b\beta j\alpha}^{k(+1)}(b_\beta i_\beta | a_\alpha j_\alpha) - c_{HF}\sum_{b\alpha j\beta}U_{b\beta j\alpha}^{k(+1)}(j_\beta i_\beta | a_\alpha b_\alpha)$$
$$= -(a_\alpha|h_k^{SOC}|i_\beta)$$
$$(6.80)$$

$$\left(\varepsilon_{a\beta}^{(0)} - \varepsilon_{i\alpha}^{(0)}\right)U_{a\beta i\alpha}^{k(+1)} + c_{HF}\sum_{j\beta b\alpha}U_{b\alpha j\beta}^{k(+1)}(b_\alpha i_\alpha | a_b j_\beta) - c_{HF}\sum_{b\beta j\alpha}U_{b\beta j\alpha}^{k(+1)}(j_\alpha i_\alpha | a_\beta b_\beta) = 0$$
$$(6.81)$$

6.3 Electronic Structure Theory of Spin-Hamiltonian Parameters

$m = -1$:

$$\left(\varepsilon_{a\beta}^{(0)} - \varepsilon_{i\alpha}^{(0)}\right)U_{a\beta i\alpha}^{k(-1)} + c_{HF}\sum_{j\beta b\alpha}U_{b\alpha j\beta}^{k(-1)}(b_\alpha i_\alpha | a_b j_\beta) - c_{HF}\sum_{b\beta j\alpha}U_{b\beta j\alpha}^{k(-1)}(j_\alpha i_\alpha | a_\beta b_\beta) = -\left(a_\beta|h_k^{SOC}|i_\alpha\right)$$

(6.82)

$$\left(\varepsilon_{a\alpha}^{(0)} - \varepsilon_{i\beta}^{(0)}\right)U_{a\alpha i\beta}^{k(-1)} + c_{HF}\sum_{j\alpha b\alpha}U_{b\beta j\alpha}^{k(-1)}(b_\beta i_\beta | a_\alpha j_\alpha) - c_{HF}\sum_{b\alpha j\beta}U_{b\beta j\alpha}^{k(-1)}(j_\beta i_\beta | a_\alpha b_\alpha) = 0$$

(6.83)

This formalism is the exact analog of those equations used to compute the g-tensor and the SOC contribution to the HFC tensor, and follow directly from the general Equations 6.39–6.41. This is available in the ORCA package, and has been shown to correct some of the deficiencies of earlier formulations of the ZFS tensor in the DFT framework. As the latter procedures are more commonly met in the literature, they are described only briefly here. Pederson and Khanna (1999) have suggested the equation:

$$D_{kl}^{(SOC)} = -\frac{1}{4S^2}\sum_{i\beta,a\beta}\frac{\langle\psi_i^\beta|h_k^{SOC}|\psi_a^\beta\rangle\langle\psi_a^\beta|h_l^{SOC}|\psi_i^\beta\rangle}{\varepsilon_a^\beta - \varepsilon_i^\beta}$$
$$-\frac{1}{4S^2}\sum_{i\alpha,a\alpha}\frac{\langle\psi_i^\alpha|h_k^{SOC}|\psi_a^\alpha\rangle\langle\psi_a^\alpha|h_l^{SOC}|\psi_i^\alpha\rangle}{\varepsilon_a^\alpha - \varepsilon_i^\alpha}$$
$$+\frac{1}{4S^2}\sum_{i\alpha,a\beta}\frac{\langle\psi_i^\alpha|h_k^{SOC}|\psi_a^\beta\rangle\langle\psi_a^\beta|h_l^{SOC}|\psi_i^\alpha\rangle}{\varepsilon_a^\beta - \varepsilon_i^\alpha}$$
$$+\frac{1}{4S^2}\sum_{i\beta,a\alpha}\frac{\langle\psi_i^\alpha|h_k^{SOC}|\psi_a^\alpha\rangle\langle\psi_a^\alpha|h_l^{SOC}|\psi_i^\alpha\rangle}{\varepsilon_a^\alpha - \varepsilon_i^\beta},$$

(6.84)

which is valid in the case of DFT functionals that do not contain the HF exchange ($c_{HF} = 0$). It can be shown to be a special case of the more general treatment outlined above if the prefactors in front of the individual terms in Equation 6.84 are all set to $1/4S^2$ rather than to the more rigorous values in Equations 6.39–6.41. This equation has been previously implemented and compared to the following equation, which had also been motivated from the general Equations 6.39–6.41 (Ray et al., 2005; Schöneboom, Neese, and Thiel, 2005):

$$D_{kl}^{(SOC)} = -\frac{1}{4S^2}\sum_{i,p}\frac{\langle\psi_i|h_k^{SOC}|\psi_p\rangle\langle\psi_p|h_l^{SOC}|\psi_i\rangle}{\varepsilon_p^\beta - \varepsilon_i^\beta}$$
$$-\frac{1}{4S^2}\sum_{p,a}\frac{\langle\psi_p|h_k^{SOC}|\psi_a\rangle\langle\psi_a|h_l^{SOC}|\psi_p\rangle}{\varepsilon_a^\alpha - \varepsilon_p^\alpha}$$
$$+\frac{1}{2}\frac{1}{S(2S-1)}\sum_{p\neq q}\frac{\langle\psi_p|h_k^{SOC}|\psi_q\rangle\langle\psi_q|h_l^{SOC}|\psi_p\rangle}{\varepsilon_q^\beta - \varepsilon_p^\alpha}$$
$$+\frac{1}{2}\frac{1}{(S+1)(2S+1)}\sum_{i,a}\frac{\langle\psi_i|h_k^{SOC}|\psi_a\rangle\langle\psi_a|h_l^{SOC}|\psi_i\rangle}{\varepsilon_a^\alpha - \varepsilon_i^\beta}$$

(6.85)

In this equation, there enters a set of "quasi-restricted" orbitals (QROs) which are explained in detail (Schöneboom, Neese, and Thiel, 2005). It may be appreciated that both formulations involve the terms that are already apparent in the general treatment given by Equations 6.39–6.41. Specifically, the first two terms correspond to the contributions from the spin-conserving excitations, while the third and fourth terms correspond to the contributions from the excited states of lower and higher multiplicity than the ground state, respectively. However, this QRO formalism has now been superseded by the more general development in Equations 6.72–6.74. More information on the relative importance of the individual terms can be found elsewhere (Neese, 2006b).

- **g-Tensor:** The g-tensor is now well studied, with many available implementations and applications. One obtains the following expressions for the four contributions:

$$g_{kl} = g_e \delta_{kl} + \Delta g^{RMC} \delta_{kl} + \Delta g^{GC}_{kl} + \Delta g^{OZ/SOC}_{kl} \tag{6.86}$$

$$\Delta g^{RMC} = -\frac{\alpha^2}{S} \sum_{\mu,\nu} P^{\alpha-\beta}_{\mu\nu} \langle \varphi_\mu | \hat{T} | \varphi_\nu \rangle \tag{6.87}$$

$$\Delta g^{GC}_{kl} = \frac{1}{2S} \sum_{\mu,\nu} P^{\alpha-\beta}_{\mu\nu} \left\langle \varphi_\mu \left| \sum_A \xi(r_A)(\mathbf{r}_A \mathbf{r} - \mathbf{r}_{A,k} \mathbf{r}_l) \right| \varphi_\nu \right\rangle \tag{6.88}$$

$$\Delta g^{(OZ/SOC)}_{kl} = \frac{1}{2S} \sum_{\mu\nu} \frac{\partial P^{(\alpha-\beta)}_{\mu\nu}}{\partial B_k} \langle \varphi_\mu | \hat{z}^{SOMF}_l | \varphi_\nu \rangle \tag{6.89}$$

where $\hat{T}$ is the kinetic-energy operator. It is noted that the g-tensor expressions make, through the operators **r** in Equation 6.88 and **l** (implicit in Equation 6.89; see Section 6.3.4), reference to the global origin of the coordinate system. This would seem to imply the unphysical and unfortunate situation that the results of the computations depend on the choice of origin. This is indeed so in g-tensor calculations, and would only disappear in the basis-set limit which is, in practice, never reached. The way to overcome this artifact would be to employ magnetic field-dependent basis functions ("gauge including atomic orbitals", GIAOs). The GIAOs represent an elegant means of solving the gauge problem, but require some additional computational effort. They have been very successful in the prediction of NMR chemical shifts, where it is essential that the gauge dependence is removed entirely. The alternative independent gauge for localized orbitals (IGLOs), which is also popular in chemical shift calculations, is not successful in EPR spectroscopy as the separate localization of spin-up and spin-down orbitals may introduce artifacts into the results. Fortunately, the gauge problem is not great in EPR spectroscopy, and meaningful results can be obtained even if a slight origin dependence persists. In order to make results comparable, a reasonable choice of origin is still required, and this is conveniently provided by the center of electronic charge. The error made by this approximation is much smaller than other remaining errors due to the functional, the basis set, the molecular model, or the treatment of environmental effects.

- **Hyperfine coupling (HFC):** One finds the following expressions for the three parts of the HFC:

$$A_{kl}^{(A;c)} = \delta_{kl} \frac{8\pi}{3} \frac{P_A}{2S} \rho^{\alpha-\beta}(\mathbf{R}_A) \tag{6.90}$$

$$A_{kl}^{(A;d)} = \frac{P_A}{2S} \sum_{\mu\nu} P_{\mu\nu}^{\alpha-\beta} \langle \varphi_\kappa | r_A^{-5}(r_A^2 \delta_{\mu\nu} - 3r_{A;\mu} r_{A;\nu}) | \varphi_\tau \rangle \tag{6.91}$$

$$A_{kl}^{(A;SO)} = -\frac{P_A}{S} \sum_{\mu\nu} \frac{\partial P_{\mu\nu}^{\alpha-\beta}}{\partial \hat{I}_k^{(A)}} \langle \varphi_\mu | z_l^{SOMF} | \varphi_\nu \rangle \tag{6.92}$$

where $P_A = g_e g_N \mu_e \mu_N$. Thus, the first two terms are straightforward expectation values, while the SOC contribution is a response property. In this case, it is necessary to solve a set of coupled-perturbed equations, with the nucleus–orbit interaction taken as the perturbing operator. Since, the solution of the coupled-perturbed equations becomes time consuming for larger molecules, this should only be carried out for a few selected heavier nuclei. For light nuclei, the SOC correction is usually negligible.

- **Electric field gradient (EFG):** The EFG tensor is straightforwardly calculated from:

$$V_{\mu\nu}^{(A)} = \sum_{\kappa,\tau} P_{\kappa\tau} \langle \phi_\kappa | r_A^{-5}(r_A^2 \delta_{\mu\nu} - 3r_{A;\mu} r_{A;\nu}) | \varphi_\tau \rangle \tag{6.93}$$

Once available, the EFG tensor can be diagonalized. The numerically largest element V_{max} (in atomic units) defines the value of q, which is in turn used to calculate the quadrupole splitting parameter as $e^2 qQ = 235.28 V_{max} Q$, where Q is the quadrupole moment of the nucleus in barn. Transformed to its eigensystem, the quadrupole splitting enters the spin Hamiltonian in the following form:

$$\hat{H}_Q + \hat{I}\mathbf{Q}\hat{I} = \frac{e^2 qQ}{4I(2I-1)} \hat{I} \begin{pmatrix} -(1-\eta) & 0 & 0 \\ 0 & -(1+\eta) & 0 \\ 0 & 0 & 2 \end{pmatrix} \hat{I} \tag{6.94}$$

The asymmetry parameter η is defined as:

$$\eta = \frac{|V_{mid} - V_{min}|}{V_{max}} \tag{6.95}$$

It is to be noted that this is the only term which involves the total electron density rather than the spin density. The field gradient tensor is consequently of a quite different nature than the hyperfine coupling, which depends on the same dipolar interaction integrals. However, in the case of the HFC they are contracted with the spin density instead of the electron density.

It is important to realize that the equations of this section are not only valid in the case of a SCF ground-state description, but are also of a much wider applicability. In the case of correlated *ab initio* methods, the equations to be solved in order to determine the effective density and its response merely become much more complicated than the relatively simple CP-SCF equations sketched above. The general line of thought is identical, however, and it is merely the "mechanics" of the calculation that become more involved.

6.3.6
Practical Aspects

The following practical advice is based on several years of experience with the calculation of EPR properties. However, it is also of a subjective nature, and other investigators may prefer different approaches that may not be less fruitful, or successful, than those outlined below.

6.3.6.1 Choice of Molecular Model

The most important step in a theoretical study of spin-Hamiltonian parameters is the choice of molecular model. Whenever possible, the largest and most realistic model that is compatible with the available computational resources should be chosen, if the aim of the study is to predict the spectroscopic parameters as accurately as possible. However, large models tend to be of low symmetry, and their potentially complicated structure makes it sometimes difficult to understand the physical and chemical origin of the computational results. Thus, in this case, the computed numbers are as incomprehensible to the investigator as the spin-Hamiltonian parameter values obtained from fitting the experimental spectra. If the only aim of the investigation is to obtain theoretical numbers – perhaps as a substitute for experimental measurements that cannot be performed – this may not be considered a drawback. Yet, if the goal of the study is to obtain insight into the origin of the spin-Hamiltonian parameters, one is well advised to choose an as-small-as-possible model and to study on a theoretical basis the structure/spectral relationships in this model system. Such investigations can provide invaluable insight that may even lead to better calculations on the large target system.

6.3.6.2 Choice of Geometry

The second critical step in a theoretical study is to choose a geometry for the calculation. Unless one has a very good reason to the contrary, it is highly desirable to employ quantum chemically optimized geometries – that is to minimize the total energy with respect to all nuclear coordinates **R**. Automatic procedures for performing such calculations are part of virtually all quantum-chemistry programs. Theoretical geometries are well-defined once a suitable "model chemistry" and a starting structure are selected. This is important in order to satisfy the criterion of reproducibility. Second, the theoretical structures do not depend on the resolution of a crystallographic experiment. In particular, the optimized positions of hydrogens are usually more reliable than those obtained from crystal structures. Note that even good protein crystal structures have uncertainties in the bond distances (~0.1 Å) that are not tolerable for quantum-chemical studies. Third, the theoretical structures are in many (if not most) cases of good to excellent quality. However, there are some situations where a fully optimized structure is more harmful than helpful. One such case arises when the theoretical model chemistry fails to predict a certain geometrical feature that is known to be present in the system. A second case is met when the model consists only of a part of the actual system and the remaining part imposes geometric constraints that will be violated in an unconstrained optimization. This situation arises typically if the active sites of proteins are studied. In this case, one may freeze a small amount of geometrical parameters in order to satisfy the constraints provided by the protein pocket. A much more rigorous approach is to model the entire protein using a combined quantum mechanical and molecular mechanical (QM/MM) approach. Here, one treats a small part (the active site) with a QM model, while the rest of the structure is treated with molecular mechanics. Such calculations are expected to accurately reflect the geometric constraints of the protein and furthermore include the leading long-range electrostatic effects of the protein environment together with its first solvation shell. In some situations, it will not be sufficient to look at individual optimized structures, and the dynamics of the system must be taken into account. These situations are best handled by molecular dynamics (MD) calculations. One then selects a sufficiently large number of "snapshots" from the MD trajectory (Asher, Doltsinis, and Kaupp, 2005). Calculations performed at these snapshot geometries provide the average values of the desired spectroscopic parameters and also give, through proper statistics, an estimate of the inhomogeneous linewidth of the spectra. However, the treatment of QM/MM and dynamic effects has not yet reached a "black-box" level, and presently experts are still required in order to carry out such calculations.

6.3.6.3 Choice of Theoretical Method

Within the framework of the present chapter, the choice of a theoretical method is identical to the choice of an appropriate density functional. Owing to the large number of different functionals that have been developed over the years, it is important to select a functional with a well-documented performance in the area of interest. If no calibration studies have been performed for the given functional, an important requisite for a successful study is to perform such a calibration on a series of related molecules with known structure and spectroscopic properties. Perhaps the two best tested functionals in chemistry and EPR spectroscopy are the hybrid B3LYP functional and the "pure" GGA functional BP86. The performance of both functionals in EPR spectroscopy has been extensively studied, and the error bars to be expected from the calculations are known for many types of systems. Overall, the B3LYP functional seems to be the best choice for EPR spectral predictions. This is fortunate since B3LYP is also the *de facto* standard for many chemical applications concerning structure and energetics. Thus, one appears to be on safe ground if this functional is used. This is not to say that other functionals may not perform with very similar quality; among the hybrid functionals, the PBE0 model may be specifically mentioned. The advantage of nonhybrid functionals is that, owing to the absence of HF exchange, the KS equations can be solved much more efficiently and performance gains by a factor of ~10 are possible using modern computational techniques. This is particularly important in geometry optimizations which require typically many individual KS calculations. In the present author's experience, the geometries of organic systems predicted by BP86 are not much worse than those predicted by B3LYP, while the geometries of transition-metal complexes are in many cases even better than the B3LYP structures. Thus, one can invest the savings offered by the GGA functionals in a better basis set or a more realistic chemical model in studying the system of interest.

Some investigators advocate varying the amount of Hartree-Fock exchange in hybrid functionals in an attempt to obtain a better agreement with experimental values. In the opinion of the present author, however, there is little reason to invest computer time in such a study. The "optimal" amount of HF exchange varies from system to system, and from property to property. Thus, it would be difficult to claim that an overall more realistic electronic structure has been obtained by the fitting procedure compared to the results obtained with standard functionals for which extensive calibration calculations are often available. One may also ask the question: What additional physical insight is obtained from the variation of the fractional HF exchange compared to the more traditional approach of correcting for systematic deficiencies of the theoretical procedures by scaling procedures based on linear regression?

6.3.6.4 Choice of Basis Set

In general, one must balance the cost of the calculation against the accuracy of the results. Ideally, the results obtained only reflect the chemical model, and the theoretical method chosen and basis set effects do not show up at all. This is, unfortunately, only possible if very large basis sets are used, and in practice one always has a dependence of the results on the employed basis set. Experience has shown that the smallest reasonable basis sets for the prediction of structures and energies are of the so-called double-ζ plus polarization quality (Jensen, 1999). Such basis sets consist of a "split" representation of each valence orbital, a single basis function for each core orbital, and one set of higher angular momentum "polarization" functions. Typical members of such basis sets are the 6-31G* (or 6-31G**) basis set, the SV(P) (or SVP) basis set, or the DGAUSS-DZVP basis set. Among these, the SVP basis set has the smallest number of primitives, and therefore leads to the fastest calculations. Fairly good results in comparison to the basis set limit are typically already obtained if polarized triple-ζ basis sets are used. Typical members are 6-311G**, cc-pVTZ or TZVP where the latter, again, is the most efficient in terms of computational requirements. If a second and third set of polarization functions is employed (as in cc-pVTZ or in 6-311G(2df,2pd) or in TZVPP), the results become very accurate but the computations may take as much as an order of magnitude longer than those with simple split-valence bases. Thus, a reasonable strategy is to first perform the geometry optimization with a small basis set and to re-optimize with a larger basis set once convergence has been obtained. For the EPR property predictions at the optimized geometries, somewhat different basis sets should be chosen. However, the only parameter that seems to be critical in this respect is the isotropic hyperfine coupling. All other properties appear to be adequately predicted by the standard bases mentioned above. The isotropic hyperfine coupling requires basis sets with additional flexibility in the core region. This flexibility is provided for example by the EPR-II and EPR-III basis sets which are otherwise equivalent to split-valence and polarized triple-ζ basis sets, respectively. These are available for the first-row elements (H-F). Alternatively, the IGLO-II and IGLO-III basis sets have been developed for NMR calculations and are available for most first- and second-row elements (up to Cl). First-row transition metals are usually well treated by the TZVP basis set; however, for hyperfine calculations a "brute force" set termed CP(PPP) has been defined based on the developments of Ahlrichs' group (Schäfer, Horn, and Ahlrichs, 1992; Schäfer, Huber, and Ahlrichs, 1994). For transition metals beyond the first-transition row, the effects of relativity become important and should be treated on an all-electron level if hyperfine or quadrupole couplings are to be predicted. For other properties, the use of effective core potentials may be adequate, but small cores should be employed.

6.3.6.5 Summary and Recommendations

Calculations of EPR parameters can now be performed on a routine basis using highly efficient and user-friendly program packages, together with cheap mass-market personal computers running under the Windows or Linux operating systems. For example, the present author's group has developed the ORCA program (Neese et al., 2009) which is particularly well suited for EPR property predictions, and is available free of charge. Alternative programs have been mentioned before. For most applications it is recommended first to optimize the geometry of the system using the BP86 functional (together with the efficient so-called RI approximation) and the SVP basis set, followed by reoptimization with the TZVP basis set. The B3LYP functional may also be used for the structure optimization, but this will require longer computation times (by a factor of 10–20). For the EPR property predictions, the B3LYP functional is recommended together with the EPR-II basis set for first-row atoms, IGLO-II or TZVP for the second row, and CP(PPP) for the first-row transition metals. For greater accuracy, the EPR-III and IGLO-III basis sets should be employed. The size of the systems that can be treated with this methodology in reasonable computation times is around 100–200 atoms. If parallel computational facilities (such as Linux clusters) are available, even larger systems may already be targeted.

6.4
Concluding Remarks

In this chapter, the theory of spin-Hamiltonian parameters, as implemented today in sophisticated quantum-chemical program packages, has been presented and illustrated through a series of representative applications. Although this chapter already contains a significant number of equations, it has not been possible to describe all important recent developments in detail. It is, nevertheless, hoped that the general lines of reasoning have become transparent to the interested reader in this way. Hopefully, it has also become evident that the theoretical approaches based on DFT are already extremely useful for complementing experimental investigations. The methods have been developed to a stage, where the programs can be readily used by nonexperts and provide results that frequently compare well with experimental measurements. Whilst this is true to a large extent for organic radicals and triplets, the area of transition-metal EPR spectroscopy remains an enormous challenge to theory. Nevertheless, if viewed with sufficient care, the calculations can be of tremendous help when analyzing complicated spectra. For example, the calculations tend to provide tensor orientations that are quite good, and this information is often difficult to determine from the experimental spectra without any additional source of information, such as laborious single-crystal measurements. Thus, a reasonable strategy would be to start extended simulation procedures from a set of theoretically calculated parameters, and then to refine these through nonlinear least-squares fitting. In many cases it will be found that the theoretical values will need to be adjusted. Even if the calculations produce errors on the order of 10–30%, they usually provide correct trends and can, therefore, provide an invaluable aid in the deconvoluting highly complex spectral patterns. It is hoped that this section will contribute to this line of thought in the following ways: (i) to explain the general strategies that lead to the currently used methods; (ii) to document precisely what is being done in the available program packages; (iii) to hint on how the calculations can be done in practice; and (iv) to show what can be achieved on the basis of the available methods. There is no doubt that the development of theoretical methods for the prediction of EPR parameters must be continued in order for "theory" to become an even stronger partner of "experiment." However, the rapid progress that has been made in this field during the past ten years provides an optimistic view into the future of this exciting partnership.

Acknowledgments

The author wishes to express his deep gratitude to the numerous colleagues and coworkers listed in the original literature, who have contributed their incredible expertise, insight and enthusiasm to the respective studies. All of the studies described were performed in the spirit of open collaboration, and it has been a real pleasure to be involved in them. The author is also very grateful for having had the opportunity to work with many highly motivated Ph.D. and postdoctoral students, without which the data presented here would have been impossible to acquire. Financial support from the Deutsche Forschungsgemeinschaft (priority program 1137 "molecular magnetism"), the SFB 663 ("Molecular response to electronic excitation", University of Düsseldorf), SFB 624 ("template effects in chemistry"), SFB 813 ("Chemistry at Spin Centers") and the German-Israeli Foundation (grant I * 746-137.9/2002) is also gratefully acknowledged.

Pertinent Literature

A comprehensive review of the subject can be found in: Neese (2003b, 2004a, 2004b, 2006a, 2007b, 2008); Neese and Munzarova (2004); Neese et al. (2007); Neese and Solomon (2003); and Sinnecker and Neese (2007). Note should also be taken of a recent volume concerning the theory of EPR and NMR parameters that details some of the aspects developed here (Kaupp, Malkin, and Bühl, 2004). High-level theoretical discussions are to be found in the classic texts of McWeeny (1992) and Harriman (1978).

References

Ågren, H., Helgaker, T., Jørgensen, P., Klopper, W., Olsen, J., Ruud, K., Vahtras, O., et al. (2005) DALTON, A Molecular Electronic Structure Program; see http://www.kjemi.uio.no/software/dalton/dalton.html.

Asher, J.R., Doltsinis, N.L., and Kaupp, M. (2005) *Mag. Res. Chem.*, **43**, S237–SS47.

Baerends, E.J., Ziegler, T., van Lenthe, E. et al. (2005) ADF2004.01, SCM, Theoretical Chemistry, Vrije Universiteit, Amsterdam, The Netherlands. Available at: www.scm.com.

Frisch, M.J., Trucks, G.W., Schlegel, H.B., Scuseria, G.E., Robb, M.A., Cheeseman, J.R., Montgomery, J.A., Jr, Vreven, T., Kudin, K.N., Burant, J.C., Millam, J.M., Iyengar, S.S., Tomasi, J., Barone, V., Mennucci, B., Cossi, M., Scalmani, G., Rega, N., Petersson, G.A., Nakatsuji, H., Hada, M., Ehara, M., Toyota, K., Fukuda, R., Hasegawa, J., Ishida, M., Nakajima, T., Honda, Y., Kitao, O., Nakai, H., Klene, M., Li, X., Knox, J.E., Hratchian, H.P., Cross, J.B., Bakken, V., Adamo, C., Jaramillo, J., Gomperts, R., Stratmann, R.E., Yazyev, O., Austin, A.J., Cammi, R., Pomelli, C., Ochterski, J.W., Ayala, P.Y., Morokuma, K., Voth, G.A., Salvador, P., Dannenberg, J.J., Zakrzewski, V.G., Dapprich, S., Daniels, A.D., Strain, M.C., Farkas, O., Malick, D.K., Rabuck, A.D., Raghavachari, K., Foresman, J.B., Ortiz, J.V., Cui, Q., Baboul, A.G., Clifford, S., Cioslowski, J., Stefanov, B.B., Liu, G., Liashenko, A., Piskorz, P., Komaromi, I., Martin, R.L., Fox, D.J., Keith, T., Al-Laham, M.A., Peng, C.Y., Nanayakkara, A., Challacombe, M., Gill, P.M.W., Johnson, B., Chen, W., Wong, M.W., Gonzalez, C., and Pople, J.A. (2004) *Gaussian 03*, Gaussian, Inc., Wallingford CT. Available at: www.gaussian.com.

Harriman, J.E. (1978) *Theoretical Foundations of Electron Spin Resonance*, Academic Press, New York.

Hess, B.A., Marian, C.M., Wahlgren, U., and Gropen, O. (1996) *Chem. Phys. Lett.*, **251**, 365.

Jensen, F. (1999) *Introduction to Computational Chemistry*, John Wiley & Sons, Inc., New York.

Kaupp, M., Malkin, V., and Bühl, M. (2004) *The Quantum Chemical Calculation of NMR and EPR Properties*, Wiley-VCH Verlag GmbH, Heidelberg.

Koch, W. and Holthausen, M.C. (2000) *A Chemist's Guide to Density Functional Theory*, Wiley-VCH Verlag GmbH, Weinheim.

Lushington, G.H. and Grein, F. (1996) *Theor. Chim. Acta*, **93**, 259–267.

McWeeny, R. (1992) *Methods of Molecular Quantum Mechanics*, Academic Press, London.

McWeeny, R. and Mizuno, Y. (1961) *Proc. Roy. Soc. (London)*, **A259**, 554.

Malkin, V.G., Malkina, O.L., Reviakine, R., Arbuznikov, A.V., Kaupp, M., Schimmelpfennig, B., Malkin, I., Helgaker, T., and Ruud, K. (2003) MAG-ReSpect. Würzburg and Bratislava.

Neese, F. (2001) *J. Chem. Phys.*, **115**, 11080–11096.

Neese, F. (2003a) *J. Chem. Phys.*, **118**, 3939.

Neese, F. (2003b) *Curr. Opin. Chem. Biol.*, **7**, 125–135.

Neese, F. (2004a) Biological applications of EPR parameter calculations, in *The Quantum Chemical Calculation of NMR and EPR Properties* (eds M. Kaupp, M. Bühl, and V. Malkin), Wiley-VCH Verlag GmbH, Heidelberg, pp. 581–591.

Neese, F. (2004b) Zero-field splitting, in *The Quantum Chemical Calculation of NMR and EPR Properties* (eds M. Kaupp, M. Bühl, and V. Malkin), Wiley-VCH Verlag GmbH, Heidelberg, pp. 541–564.

Neese, F. (2006a) *J. Biol. Inorg. Chem.*, **11**, 702–711.

Neese, F. (2006b) *J. Am. Chem. Soc.*, **128**, 10213–10222.

Neese, F. (2007a) *J. Chem. Phys.*, **127**, 164112.

Neese, F. (2007b) First Principle Calculation of EPR Parameters, in *Specialist Reports on EPR Spectroscopy* (ed. B. Gilbert), Royal Society of Chemistry, London, pp. 73–95.

Neese, F. (2008) *Biol. Mag. Res.*, **28**, 175–232.

Neese, F. and Munzarova, M.L. (2004) History of EPR parameter calculations, in *The Quantum Chemical Calculation of NMR and EPR Properties* (eds M. Kaupp, M. Bühl, and V. Malkin), Wiley-VCH Verlag GmbH, Heidelberg, pp. 21–32.

Neese, F. and Solomon, E.I. (2003) Calculation and interpretation of spin-Hamiltonian parameters in transition metal complexes, in *Magnetoscience – From Molecules to Materials*, vol. IV (eds J.S. Miller and M. Drillon), Wiley-VCH Verlag GmbH, Weinheim, pp. 345–466.

Neese, F., Petrenko, T., Ganyushin, D., and Olbrich, G. (2007) *Coord. Chem. Rev.*, **251**, 288–327.

Neese, F., Becker, U., Ganyushin, D., Kossmann, S., Hansen, A., Liakos, D., Petrenko, T., Riplinger, C., and Wennmohs, F. (2009) ORCA – An *Ab initio*, Density Functional and Semiempirical Program Package. University of Bonn. Available at: http://www.thch.uni-bonn.de/tc/orca.

Parr, R.G. and Yang, W. (1989) *Density Functional Theory of Atoms and Molecules*, Oxford University Press, Oxford.

Pederson, M.R. and Khanna, S.N. (1999) *Phys. Rev. B*, **60**, 9566–9572.

Ray, K., Begum, A., Weyhermüller, T., Piligkos, S., van Slageren, J., Neese, F., and Wieghardt, K. (2005) *J. Am. Chem. Soc.*, **127**, 4403–4415.

Schäfer, A., Horn, H., and Ahlrichs, R. (1992) *J. Chem. Phys.*, **97**, 2571–2577.

Schäfer, A., Huber, C., and Ahlrichs, R. (1994) *J. Chem. Phys.*, **100**, 5829–5835.

Schöneboom, J.C., Neese, F., and Thiel, W. (2005) *J. Am. Chem. Soc.*, **127**, 5840–5853.

Sinnecker, S. and Neese, F. (2007) Theoretical bioinorganic spectroscopy, in *Atomistic Approaches in Modern Biology: from Quantum Chemistry to Molecular Simulations* (ed. M. Reiher), Topics in Current Chemistry, vol. 268, pp. 47–83.

7
Spin Hamiltonians and Site Symmetries for Transition Ions

Sushil K. Misra

7.1
Introduction

A number of publications have dealt with the forms of the spin Hamiltonian, as well as with the various spin operators that describe the spin Hamiltonian (see, e.g., Painter and Poole, 1968; Koster and Statz, 1959; Hauser, 1963; Leushin, 1964; Grant and Strandberg, 1964; Huang, Lin, and Zhu, 1964; Ray, 1964; Bieri and Kneubühl, 1965; Geru, 1965; Roitsin, 1962, 1963; Deryugina and Roitsin, 1966; Grekhov and Roitsin, 1976; Rose, 1957; Rudowicz, 1987; Misra, 1999). The form of spin Hamiltonian for a transition ion depends on the point symmetry at its crystallographic site in the host crystal, rather than on the space group of the crystal.

The most commonly used spin operators describing spin Hamiltonians are known as Stevens' operator equivalents (see, e.g., Abragam and Bleaney, 1970; Al'tshuler and Kozyrev, 1974; Baker, Bleaney, and Hayes, 1958; Buckmaster, 1962; Elliott and Stevens, 1953; Jones, Baker, and Pope, 1959; Judd, 1955; Stevens, 1952), denoted here as $\hat{Y}_n^m$ ($|m| \leq n$); the caret (^) on Y denotes that it is an operator. These are listed in Appendix 7.I, Section A (Misra, 1999; with corrections noted by Rudowicz and Chung, 2004,). The matrix elements of these operators, which are listed in Appendix 7.I, are used to form the Hamiltonian matrix (Abragam and Bleaney, 1970; Al'tshuler and Kozyrev, 1974; Misra, 1999; with corrections noted by Ryabov, 2009) (see Appendix 7.I, Section B). A systematic listing of the SHs of all 32 crystallographic site symmetries is included in Table 7.1. The descent in symmetry interrelationships of the point groups, depicted in Figure 7.1, provides important and useful information for phase-transition studies.

The general details of the spin Hamiltonian for all point group site symmetries are provided in Section 7.2, whilst procedures to evaluate spin-Hamiltonian parameters are described in Chapter 8. In Section 7.3 are reviewed the terms of the spin Hamiltonians appropriate to all the 32 point groups, including the forms of the required Stevens' operators, as well as their matrix elements applicable to all spin values $S \leq 7/2$, covering almost all cases encountered in practice for transition-metal ions, except for high-spin molecules.

Multifrequency Electron Paramagnetic Resonance, First Edition. Edited by Sushil K. Misra.
© 2011 Wiley-VCH Verlag GmbH & Co. KGaA. Published 2011 by Wiley-VCH Verlag GmbH & Co. KGaA.

Table 7.1 Relationships of 10 derived point groups and 11 inversion point groups with their corresponding 11 proper point group for the seven crystal systems (Misra, Poole, and Farach, 1996).

Crystal system	Spin Hamiltonian	Proper point group	Order (n)	Derived point group (order n)	Inversion point group (order 2n)
Triclinic	H_C^{tric}	C_1 [1]	1		$C_1 xi = C_i$ [$\bar{1}$]
Monoclinic	H_C^{mono}	C_2 [2]	2	$C_1 \times \sigma_h = C_s$ [m]	$C_2 xi = C_{2h}$ [2/m]
Orthorhombic	H_D^{orth}	D_2 [222]	4	$C_2 \times \sigma_v = C_{2v}$ [2mm]	$D_2 xi = D_{2h}$ [2/mmm]
Trigonal	H_C^{trig}	C_3 [3]	3		$C_3 xi = C_{3i}$ [$\bar{3}$]
	H_D^{trig}	D_3 [32]	6	$C_3(+\sigma_v) = C_{3v}$ [3m]	$D_3 xi = D_{3d}$ [$\bar{3}$m]
Tetragonal	H_C^{tetr}	C_4 [4]	4	$C_2(+\sigma_d) = S_4$ [$\bar{4}$]	$C_4 xi = C_{4h}$ [4/m]
	H_D^{tetr}	D_4 [422]	8	$C_4(+\sigma_v) = C_{4v}$ [4mm] $D_2(+\sigma_d) = D_{2d}$ [$\bar{4}$2m]	$D_4 xi = D_{4h}$ [4/mmm]
Hexagonal	H_C^{hexa}	C_6 [6]	6	$C_3(+\sigma_v) = C_{3h}$ [$\bar{6}$]	$C_6 xi = C_{6h}$ [6/m]
	H_D^{hexa}	D_6 [622]	12	$C_6(+\sigma_v) = C_{6v}$ [6mm] $D_3 \times C_s = D_{3h}$ [$\bar{6}$m2]	$D_6 xi = D_{6h}$ [6/mmm]
Cubic	H_C^{cube}	T [23]	12		$T xi = T_h$ [m$\bar{3}$]
	H_D^{cube}	O [432]	24	$T(+\sigma_d) = T_d$ [$\bar{4}$3m]	$O xi = O_h$ [$\bar{3}$m]

Some derived groups are obtained by the direct product operation denoted by x, and some are obtained by adding the symmetry element given in parentheses to the indicated subgroup. Schönflies symbols are used for the point groups with international symbols given in square brackets. Column 2 gives the spin Hamiltonian for the aggregate associated with each proper point group. (Some alternate notations are: $C_s = C_{1h}$, $C_i = S_2$, $C_{3i} = S_6$, $D_2 = V$, $D_{2d} = D_{2v} = V_d$, $D_{2h} = V_h$, $D_{3d} = D_{3v}$, cubic = isometric.)

The specific spin Hamiltonians and features of observed EPR spectra for the various transition-metal ions are described in Section 7.4.

7.2
Spin Hamiltonians

Much of the detail provided in this section, except for a listing of the matrix elements, have been described by Misra, Poole, and Farach (1996). The spin Hamiltonian may be written as a summation over products of terms of the form $B_n^m \hat{O}_n^m$ and $C_n^m \hat{\Omega}_n^m$ in zero magnetic field, where for a total spin $S \leq 7/2$ the factor n is an even integer between 0 and 6 ($n \leq 2S$), and m lies in the range $0 \leq |m| \leq n$ for each n. The $\hat{O}_n^m$ and $\hat{\Omega}_n^m$ are equivalent spin operators; these are listed in Appendix 7.I, Section B, along with their matrix elements for various spin values ($\leq 7/2$), and the coefficients B_n^m and C_n^m depend on the crystal field at the transition metal ion site.

7.2 Spin Hamiltonians

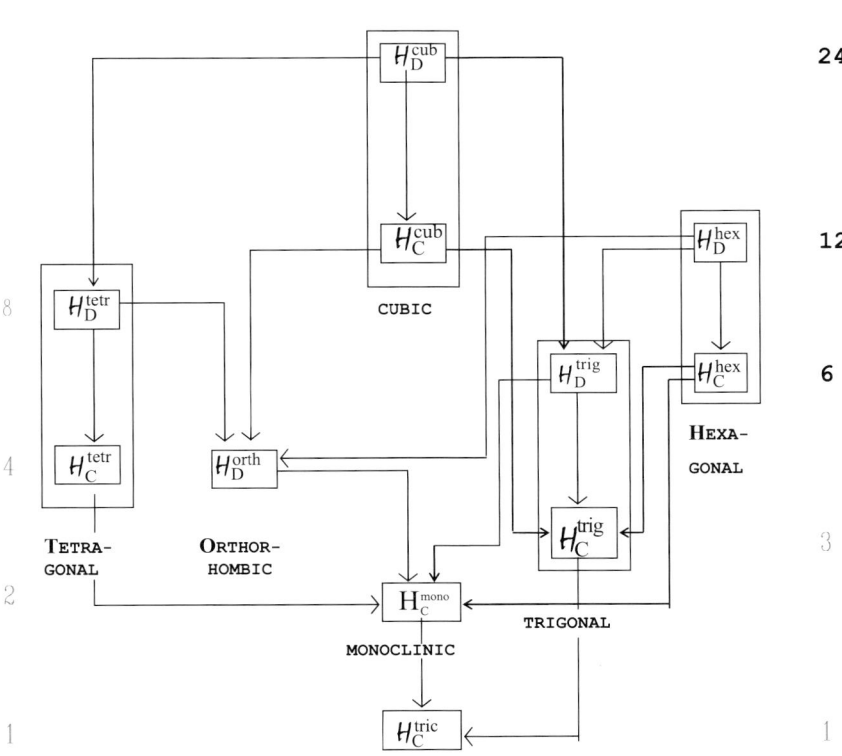

Figure 7.1 The descent of symmetry to lower symmetries, starting from a higher symmetry.

In the presence of an externally applied Zeeman field (B), one also has field-dependent terms of the type $B_\alpha B_{n\alpha}^m \hat{O}_n^m$ and $B_y C_{ny}^m \hat{\Omega}_n^m$ ($\alpha = x,z$).

Finally,

$$H = \sum_{n=1,3,5}^{n}\left(\sum_{m=0}\sum_{\alpha=x,z}^{n} B_\alpha B_{n\alpha}^m \hat{O}_n^m + \sum_{m=1}^{n} B_y C_{ny}^m \hat{\Omega}_n^m\right)$$
$$+ \sum_{n=0,2,4,6}\sum_{m=0}^{n} B_n^m \hat{O}_n^m + \sum_{n=2,4,6}\sum_{m=1}^{n} C_n^m \hat{\Omega}_n^m. \tag{7.1}$$

In the above spin Hamiltonian, the first two terms, containing $B_{n\alpha}^m$ and C_{ny}^m, are the Zeeman field (B)-dependent terms, while the last two terms are the zero-field terms.

There are 11 forms of the spin Hamiltonian, one for each of the 11 proper point groups. The relationships of the 10 derived point groups and the 11 inversion point groups with their corresponding 11 proper point groups for the seven crystal systems are listed in Table 7.1 (Misra, Poole, and Farach, 1996; Altmann and Herzig, 1994). Table 7.2 lists the nonzero coefficients B_n^m and C_n^m in these 11 spin

Table 7.2 Nonzero coefficients of zero-field terms in the spin Hamiltonian.

Spin Hamiltonian	Nonzero coefficients (B_l^m) of real operators O_l^m	Nonzero coefficients (C_l^m) of imaginary operators Ω_l^m
$C^{(1)}$: H_C^{tric}	$B_2^0, B_2^1, B_2^2, B_4^0, B_4^1, B_4^2, B_4^3, B_4^4,$ $B_6^0, B_6^1, B_6^2, B_6^3, B_6^4, B_6^5, B_6^6$	$C_2^1, C_2^2, C_4^1, C_4^2, C_4^3, C_4^4,$ $C_6^1, C_6^2, C_6^3, C_6^4, C_6^5, C_6^6$
$C^{(2)}$: H_C^{mono}; $C_2 \| Z^{a)}$	$B_2^0, B_2^2, B_4^0, B_4^2, B_4^4, B_6^0, B_6^2, B_6^4, B_6^6$	$C_2^2, C_4^2, C_4^4, C_6^2, C_6^4, C_6^6$
$C^{(2)}$: H_C^{mono}; $C_2 \| Y^{a)}$	$B_2^0, B_2^2, B_4^0, B_4^2, B_4^4, B_6^0, B_6^2, B_6^4, B_6^6,$ $B_2^1, B_4^1, B_4^3, B_6^1, B_6^3, B_6^5$	
$C^{(2)}$: H_C^{mono}; $C_2 \| X^{a)}$	$B_2^0, B_2^2, B_4^0, B_4^2, B_4^4, B_6^0, B_6^2, B_6^4, B_6^6$	$C_2^1, C_4^1, C_4^3, C_6^1, C_6^3, C_6^5$
$C^{(3)}$: H_D^{ortho}	$B_2^0, B_2^2, B_4^0, B_4^2, B_4^4, B_6^0, B_6^2, B_6^4, B_6^6$	
$C^{(4)}$: H_C^{tetr}	$B_2^0, B_4^0, B_4^4, B_6^0, B_6^4$	C_4^4, C_6^4
$C^{(5)}$: H_D^{tetr}	$B_2^0, B_4^0, B_4^4, B_6^0, B_6^4$	
$C^{(6)}$: H_C^{trig}	$B_2^0, B_4^0, B_4^1, B_4^3, B_6^0, B_6^1, B_6^3, B_6^6$	C_4^3, C_6^3, C_6^6
$C^{(7)}$: H_D^{trig}	$B_2^0, B_4^0, B_4^1, B_4^3, B_6^0, B_6^1, B_6^3, B_6^6$	
$C^{(8)}$: H_C^{hexa}	$B_2^0, B_4^0, B_6^0, B_6^6$	C_6^6
$C^{(9)}$: H_D^{hexa}	$B_2^0, B_4^0, B_6^0, B_6^6$	
$C^{(10)}$: H_C^{cube}	4-fold symmetry axis: $B_4^0, B_4^4(=5B_4^0), B_6^0, B_6^2,$ $B_6^4(=-21B_6^0), B_6^6(=-B_6^2)$	
$C^{(11)}$: H_D^{cube}	4-fold symmetry axis: $B_4^0, B_4^4(=5B_4^0), B_6^0, B_6^4(=-21B_6^0)$ 3-fold symmetry axis[b]: $B_4^0, B_4^3(=\pm 20\sqrt{2}B_4^0), B_6^0,$ $B_6^3(=\mp(35\sqrt{2}B_6^0/4))$	

(Note: Some common notations for spin-Hamiltonian parameters are as follows:
$D = b_2^0 = 3B_2^0$, $E = \frac{1}{3}b_2^2 = B_2^2$; $b_2^m = 3B_2^m$; $b_4^m = 60B_4^m$, $b_6^m = 1260B_6^m$.

For cubic symmetry: $a = \frac{2}{5}b_4^4$, where a describes the fourth-order term in the spin Hamiltonian:
$\frac{a}{6}\left[S_x^4 + S_y^4 + S_z^4 - \frac{1}{5}S(S+1)(3S^2+3S-1)\right]$, while for axial-symmetry: $F = 2b_4^0$, where F describes the fourth-order term in the spin Hamiltonian: $\frac{F}{180}[35S_z^4 - 30S(S+1)S_z^2 + 25S_z^2 + 6S(+1) + 3S^2(S+1)^2]$.)

a) C_2 axis could be parallel to any one of the magnetic X, Y, Z axes for monoclinic symmetry.
b) The upper and lower signs refer to the "Watanabe" and "Hutchings" systems as discussed in detail by Rudowicz (1987), who also lists other transformations to symmetry axes.

Table 7.3 Nonzero coefficients of field-dependent terms in the spin Hamiltonian as listed by Al'tshuler and Kozyrev (1974) and by Grekhov and Roitsin (1976).

Spin Hamiltonian	X-component	Y-component	Z-component
$C^{(1)}$: H_C^{tric} ($g_{xx} \neq g_{yy} \neq g_{zz}$; $A_{xx} \neq A_{yy} \neq A_{zz}$)	$B_{1x}^1, B_{3x}^1, B_{3x}^2, B_{3x}^3, B_{5x}^1,$ $B_{5x}^2, B_{5x}^3, B_{5x}^4, B_{5x}^5$	$C_{1y}^1, C_{3y}^1, C_{3y}^2, C_{3y}^3, C_{5y}^1,$ $C_{5y}^2, C_{5y}^3, C_{5y}^4, C_{5y}^5$	$B_{1z}^0, B_{3z}^0, B_{3z}^1, B_{3z}^2, B_{3z}^3, B_{5z}^0, B_{5z}^1,$ $B_{5z}^2, B_{5z}^3, B_{5z}^4, B_{5z}^5$
$C^{(2)}$: H_C^{mono} $C^{(3)}$: H_D^{ortho} ($g_{xx} \neq g_{yy} \neq g_{zz}$; $A_{xx} \neq A_{yy} \neq A_{zz}$)	$B_{1x}^1, B_{3x}^1, B_{3x}^3, B_{5x}^1,$ B_{5x}^3, B_{5x}^5	$C_{1y}^1(=B_{1x}^1), C_{3y}^1(=B_{3x}^1),$ $C_{3y}^3(=-B_{3x}^3), C_{5y}^1(=B_{5x}^1),$ $C_{5y}^3(=-B_{5x}^3), C_{5y}^5(=B_{5x}^5)$	$B_{1z}^0, B_{3z}^0, B_{3z}^2, B_{5z}^0,$ B_{5z}^2, B_{5z}^4
$C^{(4)}$: H_C^{tetr} $C^{(5)}$: H_D^{tetr} ($g_{xx} = g_{yy} \neq g_{zz}$; $A_{xx} = A_{yy} \neq A_{zz}$)	$B_{1x}^1, B_{3x}^1, B_{3x}^3, B_{5x}^1,$ B_{5x}^3, B_{5x}^5	$C_{1y}^1(=B_{1x}^1), C_{3y}^1(=B_{3x}^1),$ $C_{3y}^3(=-B_{3x}^3), C_{5y}^1(=B_{5x}^1),$ $C_{5y}^3(=-B_{5x}^3), C_{5y}^5(=B_{5x}^5)$	$B_{1z}^0, B_{3z}^0, B_{5z}^0, B_{5z}^4$
$C^{(6)}$: H_C^{trig} $C^{(7)}$: H_D^{trig} ($g_{xx} = g_{yy} \neq g_{zz}$; $A_{xx} = A_{yy} \neq A_{zz}$)	$B_{1x}^1, B_{3x}^1, B_{3x}^2, B_{5x}^1,$ $B_{5x}^2, B_{5x}^4, B_{5x}^5$	$C_{1y}^1(=B_{1x}^1), C_{3y}^1(=B_{3x}^1),$ $C_{3y}^2(=-B_{3x}^2), C_{5y}^1(=B_{5x}^1),$ $C_{5y}^2(=-B_{5x}^2), C_{5y}^4(=B_{5x}^4),$ $C_{5y}^5(=-B_{5x}^5)$	$B_{1z}^0, B_{3z}^0, B_{3z}^3, B_{5z}^0, B_{5z}^3$
$C^{(8)}$: H_C^{hexa} $C^{(9)}$: H_D^{hexa} ($g_{xx} = g_{yy} \neq g_{zz}$; $A_{xx} = A_{yy} \neq A_{zz}$)	$B_{1x}^1, B_{3x}^1, B_{5x}^1, B_{5x}^5$	$C_{1y}^1(=B_{1x}^1), C_{3y}^1(=B_{3x}^1),$ $C_{5y}^1(=B_{5x}^1), C_{5y}^5(=-B_{5x}^5)$	$B_{1z}^0, B_{3z}^0, B_{5z}^0$
$C^{(10)}$: H_C^{cube} $C^{(11)}$: H_D^{cube} ($g_{xx} = g_{yy} = g_{zz}$; $A_{xx} = A_{yy} = A_{zz}$)	$B_{1x}^1, B_{3x}^1, B_{3x}^3, B_{5x}^1,$ B_{5x}^3, B_{5x}^5 With the following relations: $B_{1x}^1 = C_{1y}^1 = B_{1z}^0$ $B_{3x}^3 = B_{5x}^1 - (1/4)B_{5z}^2;$ $B_{3x}^3 = -(5/3)B_{3x}^1$ $-(2/3)B_{3z}^2;$ $B_{5x}^5 = (7/5)B_{5x}^1$ $-(3/5)B_{5x}^3$ $-(2/5)B_{5z}^2.$	$C_{1y}^1, C_{3y}^1, C_{3y}^3, C_{5y}^1, C_{5y}^3, C_{5y}^5$ With the following relations: $C_{3y}^1 = B_{3x}^1 + (1/2)B_{3z}^2;$ $C_{3y}^3 = (5/3)B_{3x}^1 + (1/6)B_{3z}^2$ $C_{5y}^1 = -B_{5x}^3 - (1/8)B_{5x}^5;$ $C_{5y}^5 = (7/5)B_{5x}^1 - (3/5)B_{5x}^3$ $-(1/40)B_{5z}^2.$	$B_{1z}^0, B_{3z}^0, B_{3z}^2, B_{5z}^0, B_{5z}^2, B_{5z}^4$ With the following relations: $B_{3z}^2 = -(4/3)B_{3x}^1 - (1/3)B_{3z}^2;$ $B_{5z}^0 = -(\sqrt{7}/5)[B_{5x}^1 - (2/3)B_{5x}^3$ $-(1/6)B_{5z}^2];$ $B_{5z}^4 = (7/2)B_{5x}^1 + 3B_{5x}^3$ $-(1/4)B_{5z}^2.$

(Note: $\mu_B g_{xx} = B_{1x}^1, \mu_B g_{yy} = C_{1y}^1, \mu_B g_{zz} = B_{1z}^0.$)

Hamiltonians, while Table 7.3 lists the nonzero coefficients B_{nx}^m, B_{nz}^m, and C_{ny}^m. The first row of Table 7.3 is for the triclinic spin Hamiltonian H_C^{tric} which has all possible terms; the next six rows are for spin Hamiltonians with dominant twofold and fourfold axes, while the following four rows are for spin Hamiltonians with dominant threefold or sixfold axes. Results are given for the two cubic spin Hamiltonians for the quantization direction z-axis chosen in the fourfold direction C_4, and also for z selected along a threefold direction, C_3. The first column of Table 7.3 indicates how the g-factor and hyperfine tensors are isotropic ($g_{xx} = g_{yy} = g_{zz}$; likewise for A_{ij}) for the cubic point groups, how they are axially symmetric ($g_{xx} = g_{yy} \neq g_{zz}$; likewise for A_{ij}) for hexagonal, tetragonal and trigonal site symmetries, and how they are completely anisotropic ($g_{xx} \neq g_{yy} \neq g_{zz}$; likewise for A_{ij}) for the orthorhombic, monoclinic, and triclinic cases. Descent of symmetry expressions for the distortions of the cubic Hamiltonians H_D^{cube} and H_C^{cube} along fourfold (C_4) and threefold (C_3) axes, together with distortions of the hexagonal Hamiltonians H_D^{hex} and H_C^{hex} along threefold (C_3) and twofold (C_2) axes, are given in Appendix 7.II. These descent-of-symmetry expressions can be deduced with the help of Figure 7.1, which describes how one arrives at lower symmetries, starting from a higher symmetry.

7.3
Spin-Hamiltonian Terms for Various Site Symmetries

The aim of this section is to review the terms of the spin Hamiltonians appropriate to all the 32 point groups, and to provide the forms of the required Stevens' operators as well as their matrix elements applicable to all spin values $S \leq 7/2$, covering almost all cases encountered in practice. The terms that can exist in the spin Hamiltonian for a transition-metal ion or other paramagnetic species at a crystallographic site, are determined by the point-group symmetry of that site: the higher the symmetry, the fewer the number of terms. Table 7.1 lists the point group symmetries associated with each of the 11 spin Hamiltonian types. When the symmetry at a particular site is lowered – as by the passage through a structural phase transition – additional terms are added to the spin Hamiltonian. Of particular interest is what is termed the "descent of symmetry" – namely, the lowering of symmetry by an infinitesimal movement of atoms away from the positions that they occupy in the more symmetrical arrangement. Appendix 7.II describes the descent of symmetry for the distortion of: (i) a cubic Hamiltonian along a fourfold (C_4 or S_4) axis; (ii) a cubic Hamiltonian along a threefold (C_3) axis; (iii) a hexagonal Hamiltonian along a sixfold (C_6) axis; and (iv) a hexagonal Hamiltonian along a transverse, twofold (C_2) axis.

Tetrahedral and octahedral coordinations are quite common because closely packed spheres, such as oxygen ions, have one octahedral and two tetrahedral interstitial sites per oxygen sphere, and many commonly studied crystals have their overall structure largely determined by the exact or approximate close-packing arrangement of large anions. Relatively small transition-metal cations ordinarily occupy the interstitial sites between the larger anions. The site symmetries for transition ions in a number of commonly studied host crystals are included in Appendix 7.III (Crystal Hosts). The different approaches of various authors have led to some confusion in the notation; consequently, the results are given here in a common notation along with comparisons with other widely used in the literature. In addition, site symmetries at crystallographic sites are not always readily available; for example, they are not included in the reports by Wyckoff 1964, 1965a,

1965b, but they are available in the International Tables (1944, 1967) and in the book by Jansen and Boon (1967).

Reviews and evaluations of spin Hamiltonian parameters (SHPs), as provided by Misra (1976, 1983, 1984, 1986, 1988a, 1988b), can be used to evaluate such parameters for the various point groups in crystals. For this, the least-squares technique is used, whereas by simultaneously fitting a large number of EPR line positions recorded for several orientations of the external magnetic field, either the eigenvalues and eigenvectors of the spin Hamiltonian matrix (as evaluated on a computer using numerical techniques) can be employed, or eigenvalues can be used that have been calculated to second order in perturbation. Further details of this approach are provided in Chapter 8 of this book, which describes the evaluation of spin-Hamiltonian parameters.

7.4
Transition Ions

7.4.1
Introduction to Transition-Metal Ions

The transition-metal ions are those having unfilled $3d$-electron shells (iron group; or first-transition series), unfilled $4d$-, $5d$-electron shells (palladium and platinum groups, respectively), unfilled $4f$-electron shells (rare-earth group), and unfilled $5f$-electron shells (actinide group).

Table 7.4 (Low, 1960) provides the electron configurations of the neutral ions in the various transition groups. The s- and p-electrons outside the d- and f-shells are the valence electrons of the neutral atoms, which are not present in the ions that are studied in solids. The S-state ions are those for which the unfilled shell is half-filled – that is, with d^5 and f^7 configurations. For an appropriate interpretation of EPR spectra it is important to know the spin Hamiltonians of the various transition ions (Low, 1960; Orton, 1968; Ursu and Lupei, 1986; König and König, 1984; Misra and Upreti, 1986; Kohin, 1979; Chand, Jain, and Upreti, 1988; Misra and Wang, 1990; Biryukov, Voronkov, and Safin, 1969), and these are described as follows.

7.4.2
First-Transition Series Ions ($3d^n$, Iron-Group Ions)

A list of the valences of the ions, the ground state of the free ions, the spin–orbit coupling constant of the free ions, and the spectroscopic properties is provided in Table 7.5 (Low, 1960), whereas the nuclear properties of the ions belonging to the first transition series are listed in Table 7.6 (Abragam and Bleaney, 1970). Lists of the orbital degeneracy in the lowest state in an octahedral symmetry, the spin degeneracy, and the total degeneracy of the lowest level are also included. The scheme of the lowest energy levels due to different crystal-field terms, spin–orbit coupling and Zeeman effect for the various ions d^1 to d^9 are depicted in Figure 7.2

Table 7.4 Ground-state configuration of the electron shells of the neutral atoms belonging to various transition groups (Low, 1960).

Iron group (First transition group)			Palladium group (Second transition group)			Platinum group (Third transition group)		
Atomic no.	Atom	Configuration	Atomic no.	Atom	Configuration	Atomic no.	Atom	Configuration
21	Se	$(Ar)3d4s^2$	39	Y	$(Kr)4d5s^2$	71	Lu	$(La)5d6s^2$
22	Ti	$3d^24s^2$	40	Zr	$4d^25s^2$	72	Hf	$5d^26s^2$
23	V	$3d^34s^2$	41	Nb	$4d^45s$	73	Ta	$5d^36s^2$
24	Cr	$3d^54s$	42	Mo	$4d^55s$	74	W	$5d^46s^2$
25	Mn	$3d^54s^2$	43	Tc	$4d^55s^2$	75	Re	$5d^56s^2$
26	Fe	$3d^64s^2$	44	Ru	$4d^75s$	76	Os	$5d^66s^2$
27	Co	$3d^74s^2$	45	Rh	$4d^85s$	77	Ir	$5d^76s^2$
28	Ni	$3d^84s^2$	46	Pd	$4d^{10}$	78	Pt	$5d^96s$
29	Cu	$3d^{10}4s$	47	Ag	$4d^{10}5s$	79	Au	$5d^{10}6s$
30	Zn	$3d^{10}4s^2$	48	Cd	$4d^{10}5s^2$	80	Hg	$5d^{10}6s^2$

Rare-earth group (Lanthanides)			Actinide group		
Atomic no.	Atom	Configuration	Atomic no.	Atom	Configuration
57	La	$(Xe)5d6s^2$	89	Ac	$(Rn)6d7s^2$
58	Ce	$4f^26s^2$	90	Th	$6d^27s^2$
59	Pr	$4f^36s^2$	91	Pa	$5f^26d7s^2$
60	Nd	$4f^46s^2$	92	U	$5f^36d7s^2$
61	Pm	$4f^56s^2$	93	Np	$5f^46d7s^2$
62	Sm	$4f^66s^2$	94	Pu	$5f^67s^2$
63	Eu	$4f^76s^2$	95	Am	$5f^77s^2$
64	Gd	$4f^75d6s^2$	96	Cm	$5f^76d7s^2$
65	Tb	$4f^85d6s^2$	97	Bk	$5f^86d7s^2$
66	Dy	$4f^{10}6s^2$	98	Cf	$5f^{10}7s^2$
67	Ho	$4f^{11}6s^2$	99	Es	$5f^{11}7s^2$
68	Er	$4f^{12}6s^2$	100	Fm	$5f^{12}7s^2$
69	Tm	$4f^{13}6s^2$			
70	Yb	$4f^{14}6s^2$			

(König and König, 1984; Biryukov, Voronkov, and Safin, 1969). The various ions are discussed below in order of increasing number of unpaired 3d electrons.

$3d^1$ (Ti^{3+}, V^{4+}, Cr^{5+}, Mn^{6+}): In an octahedral crystal field the ground state, $^2T_2$, is split by spin–orbit coupling into a lower quartet, Γ_8, and a higher doublet, Γ_7. The g-factor for the Γ_8 level is zero, since there is no first-order Zeeman effect. The energy levels of the three Kramers' doublets, into which the ground state $^2T_2$ splits due to an axial field and spin–orbit coupling in a field of predominantly octahedral symmetry, superimposed on a field of tetragonal symmetry, are three doublets $E_\pm$ and E_0 with energies:

7.4 Transition Ions

Table 7.5 Spectroscopic electronic properties of the iron group ($3d^n$) ions (Low, 1960).

	$3d^1$	$3d^2$	$3d^3$	$3d^4$	$3d^5$	$3d^6$	$3d^7$	$3d^8$	$3d^9$
1. Configuration									
2. Valence state of the ion	Ti^{3+}	V^{3+}	Cr^{3+}	Mn^{3+}	Cr^+	Fe^{2+}	Fe^+	Co^+	
	V^{4+}	Cr^{4+}	V^{2+}	Cr^{2+}	Fe^{3+}		Ni^{3+}	Ni^{2+}	Cu^{2+}
	Cr^{5+}	Ti^{2+}	Mn^{4+}		Co^{4+}		Co^{2+}	Cu^{3+}	
	Mn^{6+}	Mn^{5+}			Mn^{2+}				
3. Ground state of free ion	$^2D_{3/2}$	$^3F_2$	$^4F_{3/2}$	$^5D_0$	$^6S_{5/2}$	$^5D_4$	$^4F_{9/2}$	$^3F_4$	$^2D_{5/2}$
4. Free ion spin-orbit coupling constants (cm^{-1})	154	104	87 55	85 57		−100	−180	−335	−852
5. Orbital ground state in octahedral symmetry	G_5 Triplet	G_4 Triplet	G_2 Singlet	G_3 Doublet		G_5 Triplet	G_4 Triplet	G_2	G_3
6. Total electronic spin	1/2	1	3/2	2	5/2	2	3/2	1	1/2
7. Degeneracy of the ground state	4	3	4	1	either 4 or 2	3	2	3	2
8. Nature of spin-lattice relaxation rate	Fast	Fast	Slow	Slow	Slow	Fast	Fast	Slow	Slow
9. Number of electrons in the ε and γ orbitals (in accordance with Pauli's principle and Hund's rule)	ε	$\left(\frac{9}{5}\right)\varepsilon\left(\frac{1}{5}\right)\gamma$	$(\varepsilon)^3$	$(\varepsilon)^3\gamma$	$(\varepsilon)^3\gamma^2$	$(\varepsilon)^4\gamma^2$	$\left(\frac{24}{5}\right)\varepsilon\left(\frac{11}{5}\right)\gamma$	$(\varepsilon)^6\gamma^2$	$(\varepsilon)^6\gamma^3$

(The ε and γ orbitals represent the energy levels corresponding to the threefold degenerate d-wavefunctions xy/r^2, yz/r^2, zx/r^2 and twofold degenerate wavefunctions $(x^2-y^2)/r^2$ and $(3z^2-r^2)/r^2$, respectively, in a cubic field.)

$$E_{\pm} = -\frac{1}{6}\delta - 28\mu_B \pm \frac{1}{2}W, \text{ and } E_0 = \frac{1}{3}\delta - 28\mu_B - \frac{1}{2}\lambda. \tag{7.2}$$

In Equation 7.2, δ represents the splitting of the orbital levels in the absence of spin–orbit coupling, μ_B is the Bohr magneton, and λ is the spin–orbit coupling constant. In particular, W and δ are given by

$$W = \left[\left(\delta + \frac{1}{2}\lambda\right)^2 + 2\lambda^2\right]^{1/2}; \delta = -(60B + 9A), \tag{7.3}$$

Table 7.6 3d-group atoms, and nuclear properties of the stable isotopes with nonzero nuclear spins.

Z	Atom	Mass number	Abundance (%)	Nuclear spin (I)	Nuclear magnetic moment (n.m.)	Nuclear electric quadrupole moment Q (barns)
21	Sc	45	100	7/2	+4.7564	−0.22
22	Ti	47	7.28	5/2	−0.7884	
		49	5.51	7/2	−1.1040	
23	V	50	0.24	6	+3.3470	
		51	99.76	7/2	+5.148	−0.052
24	Cr	53	9.55	3/2	−0.4744	−0.03
25	Mn	55	100	5/2	+3.443	+0.4
26	Fe	57	2.19	1/2	+0.0903	
27	Co	59	100	7/2	+4.469	0.36
28	Ni	61	1.19	3/2	±0.75	(+)0.16
29	Cu	63	60.09	3/2	+2.226	−0.18
		65	30.91	3/2	+2.385	−0.19
					(a)	(a)

The nuclear moments are expressed to four or five significant figures, as determined by NMR, and include the diamagnetic correction. (Abragam and Bleaney, 1970, p. 416).
a) From Fuller and Cohen (1965).

where A and B are the coefficients of the operators $O_2^0(L_z) = 3L_z^2 - L(L+1)$ and $O_4^0(L_z) = 35L_z^4 - 30L(L+1)L_z^2 + 25L_z^2 - 6L(L+1) + 3L^2(L+1)^2$, respectively (Biryukov, Voronkov, and Safin, 1969) in the spin Hamiltonian describing the tetragonal field. For $\delta > 0$, the doublet E_- lies the lowest, while for $\delta < 0$ E_0 lies the lowest. When E_0 lies the lowest, $g_\parallel = g_\perp = 0$, whereas when E_- lies the lowest

$$g_\parallel = \left(\frac{\delta + \frac{1}{2}\lambda}{S}\right) - 1, \text{ and } g_\perp = \left(\frac{\delta - \frac{3}{2}\lambda}{S}\right) + 1. \tag{7.4}$$

For a weak trigonal distortion of the octahedral field, the results are the same as those given by Equation 7.4, provided that δ is appropriately defined. When the trigonal field is not much weaker than the cubic field, one no longer has the g-factors corresponding to the E_0 doublet equal to zero; in fact, g may be appreciably different from zero.

The separation between the doublets is ~100 cm^{-1} for slightly distorted octahedral complexes. Therefore, due to broadening of the EPR line by rapid spin-lattice relaxation, EPR has only been observed at very low temperatures. On the other hand, in complexes of low molecular symmetry, as in solutions containing Ti^{3+} ions, in the presence of complexing agents or in vanadyl and chromyl compounds, EPR may be observed even at room temperature due to transitions between

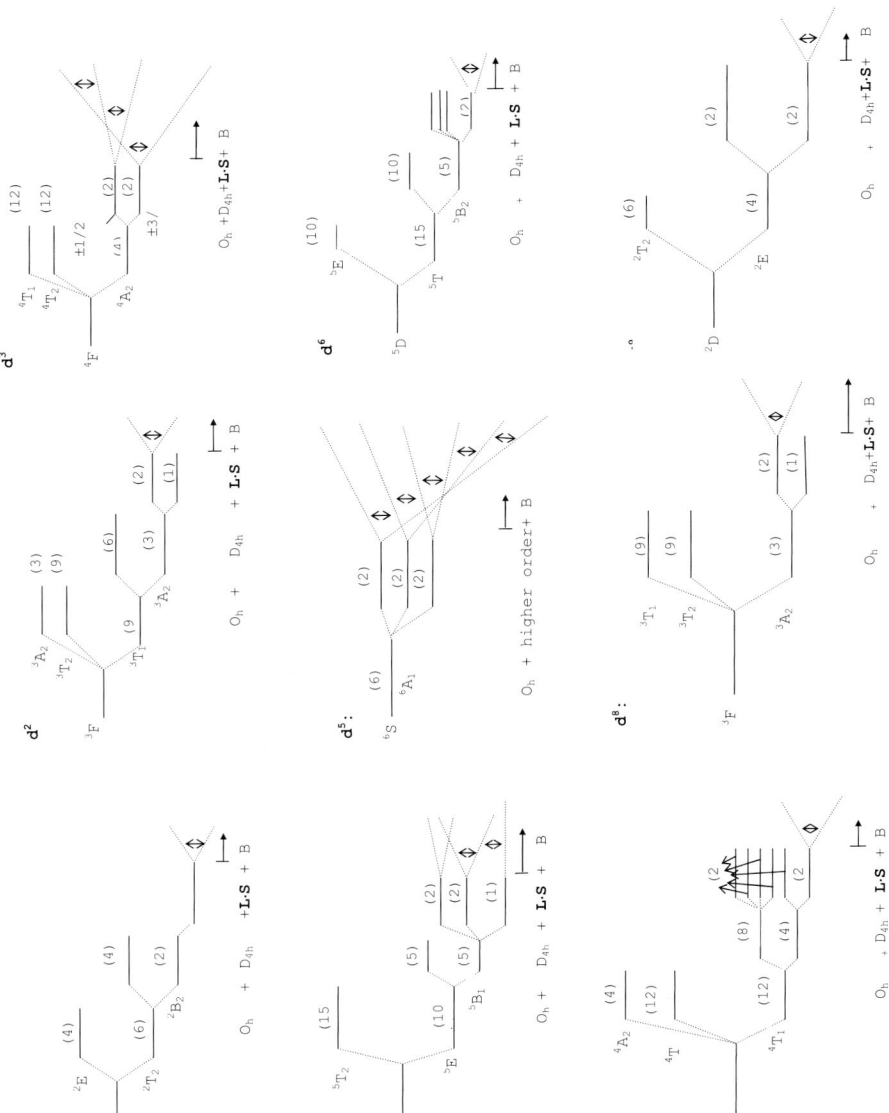

Figure 7.2 The ground state splittings for the various $3d^n$ ions (arbitrary scale) under the action of octahedral (O_h), tetragonal (D_{4h}), spin-orbit (**L·S**), and an external magnetic field (**B**).

components of the lowest doublets, described by $S = ½$ and the spin Hamiltonian

$$H = g_\| \mu_B B_z S_z + g_\perp \mu_B (B_x S_x + B_y S_y) + A S_z I_z + B(S_x I_x + S_y I_y). \quad (7.5)$$

In some cases, Ti^{3+} and VO^{2+} complexes show a pair exchange interaction between Ti^{3+} and VO^{2+} ions, respectively, resulting in a lower singlet and a higher triplet state. For these cases $S = 1$, and the following spin Hamiltonian applies:

$$H = g_\| \mu_B B_z S_z + g_\perp \mu_B (B_x S_x + B_y S_y) + b_2^0 (S_z^2 - S(S+1)/3)$$
$$+ A S_z I_z + B(S_x I_x + S_y I_y) \quad (7.6)$$

An alternative approach to this exchange interaction effect, particularly applicable to VO^{2+} ions, is discussed below in context with the Cu^{2+} ion ($3d^9$).

For the case of the Mn^{6+} ion $S = ½$ and $I = 5/2$, the ground state is a doublet in a tetrahedral field. The applicable spin Hamiltonian is that described by Equation 7.6.

EPR of the VO^{2+} ion has been reviewed by Kohin (1979) and by Chand, Jain, and Upreti (1988). The ground-state configuration of both V^{4+} and the molecular ion VO^{2+}, and its complexes, is $3d^1$; however, a multiple bond exists between V^{4+} and O^{2-}. The ground state, in a tetragonally distorted octahedral field, is described by the $^2B_{2g}$ (d_{xy}) atomic orbital. The effect of the ligand electrons is taken into account by considering, in addition, to crystal-field theory, the molecular-orbital (MO) approach, since the crystal-field theory alone is not sufficient.

The MO approach, in which the individual electron orbitals are constructed to be linear combinations of the atomic orbitals, is well described by Kohin (1979).

The EPR spectra of the VO^{2+} ion, $S = ½$, $I = 7/2$, are described by the following spin Hamiltonian:

$$H = \mu_B S \cdot \tilde{g} \cdot B + S \cdot \tilde{A} \cdot I + I \cdot \tilde{Q} \cdot I - \mu_n B \cdot \tilde{g}_n \cdot I \quad (7.7)$$

Here, the successive terms represent the electronic Zeeman, the hyperfine, the quadrupole, and the nuclear–Zeeman interactions, respectively. When the principal axes (x, y, z) of the $\tilde{g}, \tilde{A}, \tilde{Q}$ and $\tilde{g}_n$ matrices coincide, the spin Hamiltonian in Equation 7.7 simplifies to:

$$H = \sum_{i=x,y,z} (\mu_B g_i B_i S_i + A_i S_i I_i - \mu_n g_n B_i I_i)$$
$$+ Q'[I_z^2 - I(I+1)/3] + Q''(I_x^2 - I_y^2) \quad (7.8)$$

The solution of Equation 7.8 to second-order perturbation can be found in Abragam and Bleaney (1970). The fitting of EPR line positions, using the least-squares technique and second-order perturbed eigenvalues, to evaluate the spin-Hamiltonian parameter, is described by Misra (1988a), for the general case where the principal axes of the $\tilde{g}$ and $\tilde{A}$ matrices are not coincident (additional details are provided in Chapters 8 and 14.)

EPR provides information on the symmetry and orientation of the vanadyl complex in the host lattice. Usually, the vanadyl complex takes either the form

[VOX$_4$]$^{n-}$, or [VOX$_5$]$^{n-}$, where the fivefold coordination about the metal ion is square-pyramidal and the axial vanadyl oxygen atom remains significantly closer to the metal ion. The five-coordinated species can weakly add a sixth axial ligand *trans* to the vanadyl oxygen atom to form the [VOX$_5$]$^{n-}$ species; wherein, the effective symmetry appears to be C$_{4v}$, with both $\tilde{g}$ and $\tilde{A}$ matrices being axially symmetric. It is expected that the Z axes of the $\tilde{g}$ and $\tilde{A}$ matrices coincide with the V=O bond of the vanadyl ion. When the symmetry is lower than C$_{4v}$, both $\tilde{g}$ and $\tilde{A}$ matrices become nonaxial.

The electronic energy and bonding parameters can be related to $\tilde{g}$ and $\tilde{A}$ matrices in transition-metal complexes. The following simplified expressions, appropriate to C$_{4v}$ symmetry, neglecting small overlap terms, can be used for the estimation of bonding parameters (Assour, Goldmacher, and Harrison, 1965; Boucher, Tynan, and Yen, 1969):

$$\Delta g_\| = \frac{8\lambda \beta_1^{*2}\beta_2^{*2}}{\Delta E(b_2^* \to b_1^*)};$$

$$\Delta g_\perp = \frac{2\lambda e_\pi^{*2}\beta_2^{*2}}{\Delta E(b_2^* \to e_\pi^*)};$$

$$\frac{-A_\|}{P} = \beta_2^{*2}\left(K + \frac{4}{7}\right) + \Delta g_\| + \frac{3}{7}\Delta g_\perp;$$

$$\frac{-A_\perp}{P} = \beta_2^{*2}\left(K - \frac{2}{7}\right) + \frac{11}{14}\Delta g_\perp. \tag{7.9}$$

In Equation 7.9, $\Delta g_{\|,\perp} = 2.0023 - g_{\|,\perp}$; β_1^{*2}, β_2^{*2} and e_π^{*2} are the MO bonding parameters, ΔE are the energy splittings of the corresponding orbitals, P ($= 2.0023 g_n \mu_B \mu_n <r^{-3}>$) is the dipolar coupling parameter, λ is the spin–orbit coupling constant for the vanadyl ion, and K is the Fermi-contact interaction constant. Because of the uncertainties in the various quantities, EPR data do not provide an unambiguous determination of the electronic energy levels. The parameters β_1^{*2} and β_2^{*2} indicate the in-plane σ- and in-plane π-bonding, respectively; while e_π^{*2} denotes the out-of-plane π-bonding. β_1^{*2}, depending upon the σ-donor strength of the ligand, decreases as the covalent bonding increases. The parameter β_2^{*2}, being affected by the delocalization of the electron on to the ligands, decreases as the covalent bonding increases. Finally, the π-bond strength decreases as a result of the increased charge on the metal, and, in general, the stronger the in-plane donor atom the less covalent is the vanadyl V=O bond. The parameter K is sensitive to the 4s character of the orbitals. Additional information about bonding can be obtained by an analysis of the ligand hyperfine structure.

3d^2 (V^{3+}, Cr^{4+}, Mn^{5+}, Fe^{6+}): This configuration corresponds to non-Kramers' ions. The ^{3}F free-ion ground state is split by an octahedral field with an orbital triplet (^{3}T$_1$) lying lowest, which is further split by a trigonal field, resulting in an orbital-singlet ground state, as shown in Figure 7.2. The threefold spin degeneracy of this singlet is lifted by the combined action of the trigonal field and the spin–orbit coupling. EPR is observed at a field B_0 for the transition within the $M = \pm 1$

doublet with $g_\perp = 0$, and $g_\parallel = \dfrac{h\nu}{2B_0\mu_B\cos\theta}$, where θ is the angle between the crystal-field axis and the Zeeman field, B_0.

The spin–orbit interaction splits the $^3T_1$ octahedral ground state into $\Gamma_1 + \Gamma_3 + \Gamma_4 + \Gamma_5$, of which the lowest triplet Γ_4 splits into a lowest-lying singlet ($m = 0$) and a doublet ($m = \pm 1$) lying several cm^{-1} higher in energy. For the triplet $S = 1$, the applicable spin Hamiltonian is the same as that given by Equation 7.6. The resonance involves the levels $m = \pm 1$, and is thus weak. The best-known examples are those of V^{3+} ions in the lattices of Al_2O_3 and CdS single crystals.

In the case of a tetrahedral field, the $^3A_2$ ground state is also split into a singlet and a doublet, resulting in $S = 1$, characterized by the spin Hamiltonian (Equation 7.6). EPR is easily observed even at room temperature, since the lowest level is an orbital singlet, for example Cr^{4+} in tetralkyls, such as $Cr(CH_3)_4\cdot[MnO_4]^{3-}$, is an inorganic ion of tetrahedral symmetry in this configuration.

$3d^3$ (V^{2+}, Cr^{3+}, Mn^{4+}): The octahedral field produces a $^4A_2$ term lowest, which is split by axial fields into two Kramers' doublets, with the total spin $S = 3/2$, which are described by the spin Hamiltonian

$$H = g_\parallel \mu_B B_z S_z + g_\perp \mu_B (B_x S_x + B_y S_y) + b_2^0 [S_z^2 - S(S+1)/3]$$
$$+ b_2^2 (S_x^2 - S_y^2) + A(S_z I_z + S_y I_y + S_z I_z). \tag{7.10}$$

The g-value is almost isotropic. The hyperfine term $H\cdot\tilde{A}\cdot I$ applies to all the ions with nonzero nuclear spin – that is, V^{2+}, Cr^{3+}, and Mn^{4+}.

$3d^4$ (Cr^{2+}, Mn^{3+}): The ground state for these ions in the free state is ^5D. The strength of the Ligand field determines the multiplicity of the ground state for ions in coordination compounds. In a weak ligand field $S = 2$, whereas in strong ligand fields $S = 1$.

- Weak ligand fields ($S = 2$): A cubic (octahedral) field causes an orbital doublet ^5E to lie the lowest for this non-Kramers' configuration, as shown in Figure 7.2. A tetragonal distortion splits this doublet further, only the fivefold spin degeneracy remains. EPR may be observed between the five singlets, which are produced by the splitting of these levels, anywhere from 1 to 10 cm^{-1}, by the joint effect of a rhombic crystal field and spin–orbit coupling. Thus, $S = 2$, with the spin Hamiltonian

$$H = g\mu_B(B_x S_x + B_y S_y + B_z S_z) + b_2^0 [S_z^2 - S(S+1)/3] + b_2^2(S_x^2 - S_y^2). \tag{7.11}$$

An example is the Cr^{2+} ion in $CrSO_4\cdot 5H_2O$.

For a slightly distorted tetragonal field (D_{2d} symmetry), the total spin $S = 2$, characterized by the spin Hamiltonian

$$H = g\mu_B(B_x S_x + B_y S_y + B_z S_z) + b_2^0 (S_z^2 - S(S+1)/3) + b_2^2(S_x^2 - S_y^2)$$
$$+ b_4^0 \left[35S_z^4 - 30S(S+1)S_z^2 + 25S_z^2 + \frac{1}{6}a(S_1^4 + S_2^4 + S_3^4)\right]. \tag{7.12}$$

The cubic axes 1, 2, and 3, and the principal axes x, y, and z, coincide for cubic crystals, as well as $b_2^2 = 0$ in Equation 7.12, for example, Cr^{2+}-doped ZnS and CdS.

In a distorted tetrahedral field, the orbital triplet lies the lowest; the combined effect of the lower symmetry fields and the spin–orbital coupling results in a slightly split non-Kramers' doublet below all other levels. The EPR for this doublet ($S = \frac{1}{2}$) is described by following spin Hamiltonian:

$$H = g'_\parallel \mu_B B_z S_z + \Delta S_x, \tag{7.13}$$

where Δ is the separation within the lowest doublet.

- Strong ligand fields ($S = 1$): There is hardly any ion that is found to belong to this case; hence, it will not be discussed.

$3d^5$ (Cr^+, Mn^{2+}, Fe^{3+}, Co^{4+}): These are S-state ions, described in detail in Section 7.4.6.

$3d^6$ (Fe^{2+}): The ground state of this free ion is $^5D_4$. In coordination compounds, the strength of the ligand fields determines the multiplicity of the ground state. $S = 2$ in weak ligand fields, $S = 0$ in strong ligand fields, and $S = 1$ in intermediate spin situations.

- Weak ligand fields ($S = 2$): An octahedral field produces the ground state $^5T_2$. Spin–orbit coupling along with an axial field are responsible for a singlet ($M = 0$) and a doublet ($M = \pm 1$) lying lowest. The doublet is further split by a rhombic field. Mostly, EPR signals are observed at low temperatures as lines are broadened by spin-lattice relaxation at high temperatures. The lowest doublet is described by the effective spin $S = \frac{1}{2}$, with the spin Hamiltonian given by Equation 7.13; an example of this is FeF_2.

- Strong ligand fields ($S = 0$): No EPR is observed, since this case corresponds to diamagnetism as $S = 0$.

- Intermediate spin cases ($S = 1$): EPR is not observed, since the singlet ($M = 0$) lies lowest, separated by a large zero-field splitting from the doublet ($M = \pm 1$).

$3d^7$ (Fe^+, Co^{2+}, Ni^{3+}): The ground state for free $3d^7$ ions is ^4F. The multiplicity in coordination compounds is determined by the strength of the ligand field: $S = 3/2$ in weak fields, and $S = \frac{1}{2}$ in strong fields.

- Strong ligand fields: ($S = \frac{1}{2}$; low spin). The EPR spectra are analyzed using the spin Hamiltonian

$$H = \mu_B(g_x B_x S_x + g_y B_y S_y + g_z B_z S_z) + A_x S_x I_x + A_y S_y I_y + A_z S_z I_z. \tag{7.14}$$

In general, $g_\parallel \sim g_e = 2.0023$ and $g_\perp > g_\parallel$ since $\lambda < 0$; for example, planar complexes of Co^{2+} and Ni^{3+}, such as dithiolates. Quadrupole nuclear interactions are important in Co^{2+} ($I = 7/2$), and the spin Hamiltonian for axial symmetry becomes

$$H = g_\parallel \mu_B B_z S_z + g_\perp \mu_B (B_x S_x + B_y S_y) + A S_z I_z$$
$$+ B(S_x I_x + S_y I_y) + Q'[I_z^2 - I(I+1)/3] - g_n \mu_n \mathbf{B} \cdot \mathbf{I} \tag{7.15}$$

The unpaired electron is essentially in the $d_{3z^2-r^2}$ orbital, which is perpendicular to the plane of the complex. For the strong-field configuration ($t_2^6 e$), the ground state is an orbital doublet, in an octahedral field split by the tetragonal field into two Kramers' doublets, described by the g-factors

$$g_\parallel = 2 - \frac{8\lambda}{\Delta}, \quad g_\perp = 2 - \frac{2\lambda}{\Delta} \quad (d_{x^2-y^2} \text{ configuration}), \tag{7.16}$$

and

$$g_\parallel = 2, \quad g_\perp = 2 - \frac{6\lambda}{\Delta} \quad (d_{3z^2-r^2} \text{ configuration}). \tag{7.17}$$

In the above equations, Δ is the cubic field splitting. The sign of the tetragonal field determines which one of the above doublets lies lowest.

In a trigonal field, or in a cubic field, the orbital doublet is not expected to split. However, the Jahn–Teller effect causes a spontaneous distortion to reduce the energy by splitting of the orbital levels. At temperatures above a certain value (critical temperature), the dynamic Jahn–Teller effect occurs; this is interpreted as being due to the ionic complex oscillating between several distorted configurations of equal energy. This results in an isotropic g-value, which is the mean of the distorted values given by Equations 7.16 and 7.17, that is, $g = 2 - \frac{4\lambda}{\Delta}$. Below the critical temperature, the distortions are frozen-in, giving anisotropic spectra, consistent with the $d_{x^2-y^2}$ doublet being the lowest.

- Weak ligand fields ($S = 3/2$; high spin): In an octahedral field, a $^4T_1$ state lies lowest, split by the spin–orbit interaction. The lowest level is a Kramers' doublet, and the effective spin is 1/2, associated with an isotropic EPR spectrum. In a field of lower symmetry, there are six Kramers' doublets that contribute to the spectrum, which becomes very anisotropic, characterized by $S = ½$. The spectra are described by the spin Hamiltonian (Equation 7.14). The g-factor, to first order, is predicted to be 4.33 in an octahedral field since an effective spin is used. When distortions to the octahedral field (for example, in MgO and CaO) are considered, very different g-values are expected. For example, Co^{2+} in TiO_2 has $g_{xx} = 2.090$, $g_{yy} = 3.725$, and $g_{zz} = 5.860$.

In an octahedral symmetry, Fe^+ and Co^{2+} ($t_2^5 e^2$) experience the weak crystal-field interaction. The lowest doublet of the $t_2^5 e^2$ configuration in an axially distorted octahedral field is described by following g-values ($S = ½$), in terms of a parameter x ($0 \leq x \leq \infty$):

$$g_\parallel = 2 + 4(\alpha+2)\left[\left(\frac{3}{x^2} - \frac{4}{(x+2)^2}\right) \bigg/ \left(1 + \frac{6}{x^2} + \frac{8}{(x+2)^2}\right)\right], \tag{7.18}$$

$$g_\perp = 4\left[\left(1 + \frac{2\alpha}{x+2} + \frac{4}{x(x+2)}\right) \bigg/ \left(1 + \frac{6}{x^2} + \frac{8}{(x+2)^2}\right)\right]. \tag{7.19}$$

where the trigonal, or tetragonal, splitting parameter δ is given by

$$\delta = \alpha\lambda\left[\frac{x+3}{2} - \frac{3}{x} - \frac{4}{x+2}\right]. \tag{7.20}$$

Here, α is expected to lie between 1.0 and 1.5. Co^{2+} also exhibits a hyperfine structure described by $S \cdot \tilde{A} \cdot I$; the matrix $\tilde{A}$ is expected to be quite anisotropic.

In a ligand field of tetrahedral symmetry, the $^4A_2$ ground state is split into two Kramers' doublets in fields of lower symmetry, for example, in Cs_3CoCl_5 and Cs_3CoBr_5. The applicable spin Hamiltonian is ($S = 3/2$):

$$H = g_\parallel\mu_B B_z S_z + g_\perp\mu_B(B_x S_x + B_y S_y) + b_2^0[S_z^2 - S(S+1)/3]. \tag{7.21}$$

The Ni^{3+} ion experiences the strong field case due to its extra charge ($t_2^6 e$).

$3d^8$ (Co^+, Ni^{2+}, Cu^{3+}): An octahedral field splits the ^3F ground state of the free $3d^8$ ion into two triply degenerate states and the orbitally nondegenerate $^3A_2$ singlet state which lies the lowest. The spin degeneracy (Γ_5) is not removed by the spin–orbit coupling. The lowest level is split by axial fields into a singlet and a doublet, and a rhombic field splits it into three singlets. The EPR spectra are described the same way as those for the $3d^3$ configuration using the spin Hamiltonian (Equation 7.11), except that here $S = 1$, rather than 2. The Co^+ and Ni^{2+} spectra sometimes exhibit "double quantum" transitions, for which $\Delta M = \pm 2$, where M is the electronic magnetic quantum number. EPR of $3d^8$ ions can be observed at 77 K, or even at room temperature, since there are no excited states close to the ground state. The broad EPR lines of the Ni^{2+} ions may sometimes be difficult to detect.

$3d^9$ (Ni^+, Cu^{2+}): The details of the spin Hamiltonian, ground-state wave functions and the Cu^{2+} Jahn–Teller effect have been well-described in a review on Cu^{2+} by Misra and Wang (1990). The ground state of the configuration is $t_2^6 e^3$ in an octahedral field, which may be regarded as a single hole in an e orbital. The theory is the same as that for a $3d^7$ ion in a strong octahedral field for which the g-factors have been given by Equations 7.18 and 7.19. The Jahn–Teller distortion also takes place in the same way as that for a $3d^7$ ion in a trigonal, or cubic, symmetry. Thus, the orbital degeneracy is lifted by the Jahn–Teller effect in the case of trigonal symmetry although the ^2E state is not split. The Jahn–Teller effect may distort the molecular complex at higher temperatures, but at low temperature it leads to a static distortion, resulting in a spectrum due to three magnetic complexes, each characterized by tetragonal symmetry. The applicable spin Hamiltonian is ($S = \frac{1}{2}$):

$$H = \mu_B g_x B_x S_x + g_y B_y S_y + g_z B_z S_z + Q'[I_z^2 - I(I+1)/3] + Q''(I_x^2 - I_y^2). \tag{7.22}$$

In a rhombic symmetry with sixfold coordination, the ground state wavefunction is a superposition of the $|x^2-y^2>$ and $|3z^2-r^2>$ states. In an octahedral sixfold coordination, the ^2D orbital splits into an E_g (Γ_3) doublet and a T_{2g} (Γ_5) triplet, the ground state being the T_g state, which can be further split by the Jahn–Teller effect. In a tetragonally distorted octahedral field, there are found admixtures of the excited states $|xy>$, and $\{|yz>, |zx>\}$ in the ground state wavefunction, which is either $|x^2-y^2>$, or $|3z^2-r^2>$. The principal g-values satisfy the relations $g_\| > g_\perp > g_e$ for an elongated tetragonal distortion and $g_\perp > g_\| \approx g_e$ for a compressed-tetragonal distortion, with g_e being the free-electron value. In a square-planar symmetry, which is the limiting case of an elongated tetragonally distorted octahedron, the ground-state wavefunction is predominantly $|x^2-y^2>$. In a tetrahedral symmetry, with fourfold coordination, the orbital state of the Cu^{2+} ion splits into a E_g (Γ_3) doublet and a T_{2g} (Γ_5) triplet, with the ground state being T_{2g}. In a trigonally distorted tetrahedral field the T_{2g} orbital of Cu^{2+} splits further into a ^2E doublet and $^2B_2$ singlet, the ground state being the ^2E doublet, for which a Jahn–Teller splitting can take place.

In an octahedral symmetry, the $\tilde{g}$ and $\tilde{A}$ matrices have the following principal values:

$$g_\| = g_e - \frac{8\lambda}{\Delta}; \quad A_\| = p\left\{-K - \frac{4}{7} - \frac{6\lambda}{7\Delta_2} - \frac{8\lambda}{\Delta_1}k\right\};$$

$$g_\perp = g_e - \frac{2\lambda}{\Delta}; \quad A_\perp = p\left\{-K + \frac{2}{7} - \frac{11\lambda}{7\Delta_2}\right\}; \quad (7.23)$$

whereas for an octahedral symmetry with a compressed tetragonal distortion

$$g_\| = g_e; \quad A_\| = p\left\{-K + \frac{4}{7} + \frac{6\Delta}{7\Delta_3}\right\};$$

$$g_\perp = g_e - \frac{6\lambda}{\Delta_3}; \quad A_\perp = p\left\{-K - \frac{2}{7} - \frac{45\lambda}{7\Delta_3}\right\}; \quad (7.24)$$

In Equations 7.23 and 7.24, Δ, Δ_1, Δ_2, and Δ_3 are the energy splittings of the Cu^{2+} ion in a tetragonally distorted octahedral crystal field, $p = 2g_n\mu_B\mu_n <r^{-3}>$; where $<r^{-3}>$ is the average of the inverse-cube radius of the Cu^{2+} ion, k is the orbital-reduction factor, and K is the Fermi-contact interaction constant.

The Jahn–Teller effect of the Cu^{2+} ion can be described in terms of a CuL_6 complex, where L is a surrounding ligand ion. Bates and Chandler (1971, 1973) have reviewed the Jahn–Teller experimental data on the Cu^{2+} ion at tetrahedral sites.

EPR spectra of the Cu^{2+} ion sometimes exhibit an exchange interaction between isolated Cu^{2+} ion pairs, resulting in a lower singlet and a higher triplet state. The spin Hamiltonian is described as follows ($S = 1$):

$$H = \mu_B(g_xB_xS_x + g_yB_yS_y + g_zB_zS_z) + A_xS_xI_x + A_yS_yI_y$$
$$+ A_zS_zI_z + Q'[I_z^2 - I(I+1)/3] + Q''(I_x^2 - I_y^2). \quad (7.25)$$

Alternatively, the interaction between the pair of ions 1 and 2 can be expressed by the spin Hamiltonian

$$H = H_1 + H_2 + H_{int}, \quad (7.26)$$

where H_1 and H_2 are the spin Hamiltonians of the isolated ions, while H_{int} represents the interaction between them. For the case of the two ions in identical sites of axial symmetry with their z axes parallel to the internuclear vector, **r**, one has

$$H = g_\parallel \mu_B B_z(S_{1z} + S_{2z}) + g_\perp \mu_B [B_x(S_{1x} + S_{2x}) + B_y(S_{1y} + S_{2y})]$$
$$+ A(S_{1z}I_{1z} + S_{2z}I_{2z}) + B(S_{1x}I_{1x} + S_{2x}I_{2x} + S_{1y}I_{1y} + S_{2y}I_{2y}) + H_D, \quad (7.27)$$

where,

$$H_D = (\mu_B^2/r^3)[g_x^2 S_{1x}S_{2x} + g_y^2 S_{1y}S_{2y} - 2g_z^2 S_{1z}S_{2z}]. \quad (7.28)$$

In Equation 7.28 only the magnetic dipole–dipole interaction is operative. Diagonalization of the corresponding Hamiltonian matrix yields – in addition to the conventional EPR parameters $g_\parallel$, $g_\perp$, A, B – a value for the internuclear distance, r, between the metal ions.

7.4.3
Second and Third Transition Series (The 4d, Palladium and 5d, Platinum Groups)

Much fewer EPR data have been reported for the 4d and 5d groups, as these are much less abundant in Nature, and because low temperatures are often required for the observation of EPR. It is less easy, as compared to the iron-group ions, to find suitable crystal lattices into which they can be introduced. These ions tend to form strong covalent bonds with neighboring ions, giving rise to a strong crystal field. Thus, for ions containing up to three unpaired electrons, the resonance results are similar to those for the corresponding 3d ions, but for the d^n ions with $n \geq 4$, there always exists the strong-field configurations, similar to those for the Fe^{3+} and Ni^{2+} ions. The majority of the these ions possess isotopes with nonzero nuclear spins. Table 7.7 (Orton, 1968) contains the details of nuclear data for those ions, for which EPR have been observed. An earlier review of the 4d, 5d elements has been published by Griffiths, Owen and Ward (1953).

$4d^1$ (Nb^{4+}, Mo^{5+}), $5d^1$ (Ta^{4+}, W^{5+}, Re^{6+}): The details of EPR for these configurations are the same as those for the $3d^1$ ion. An isolated Kramers' doublet lies lowest in a distorted octahedral symmetry; the g-factor is expected to be in the range 1.0–2.0. Low temperatures are usually, but not always, required to observe EPR. In general, the complexes show axial symmetry in the various samples studied. The applicable spin Hamiltonian is, thus, described by Equation 7.5.

$4d^2$ (Mo^{4+}), $5d^2$ (Re^{5+}, Ru^{6+}): The features of EPR resonance for these ions are similar to those of $3d^2$ ions. For example, the $(ReO_4)^{3-}$ complex has a similar configuration to the $3d^2$ ion complex $(MnO_4)^{3-}$.

$4d^3$ (Mo^{3+}, Tc^{4+}), $5d^3$ (Re^{4+}): In an octahedral field, an orbital-singlet lies the lowest, the threefold spin degeneracy being raised by the spin–orbit and lower symmetry terms. In the case of the Mo^{3+} ion and the majority of d^3 ions in an external

Table 7.7 Nuclear properties of the palladium ($4d^n$) and platinum ($5d^n$) groups (Orton, 1968; Abragam and Bleaney, 1970, p. 473).

Ion	Z	Configuration	Isotope	Abundance (%)	Nuclear spin (I)	Nuclear magnetic moment (n.m.)	Nuclear electric quadrupole moment Q (barns)
Nb^{4+}	41	$4d^1$	Nb^{93}	100	9/2	+6.167	−0.2
Mo^{5+}	42		Mo^{95}	15.6	5/2	−0.9135	0.12
	42		Mo^{97}	9.6	5/2	−0.9327	1.1
Mo^{3+}	42	$4d^3$	Mo^{95}	15.7	5/2	−0.9135	0.12
	42		Mo^{97}	9.6	5/2	−0.9327	1.1
Tc^{4+}	43		Tc^{99}	~100	9/2	+5.680	+0.3
Ru^{3+}	44	$4d^5$	Ru^{99}	12.8	5/2	−0.63	
	44		Ru^{101}	17.0	5/2	−0.69	
Rh^{2+}	45	$4d^7$	Rh^{103}	100	1/2	−0.0883	
Rh^0	45	$4d^9$	Rh^{103}	100	1/2	−0.0883	
Ag^{2+}	47	$4d^9$	Ag^{107}	51.3	1/2	−0.1136	
	47		Ag^{109}	48.6	1/2	−0.1305	
Ta^{4+}	181	$5d^1$	Ta^{181}	100	9/2		
W^{5+}	74		W^{183}	14.4	1/2	+0.117	
Re^{4+}	75	$5d^3$	Re^{185}	37.1	5/2	+3.172	+2.6
	75		Re^{187}	62.9	5/2	+3.204	+2.6
Ir^{4+}	77	$5d^5$	Ir^{191}	38.5	3/2	+0.1440	+1.5
	77		Ir^{193}	61.5	3/2	+0.1568	+1.5
Pt^{3+}	195	$5d^7$	Pt^{195}	33.8	1/2	+0.6060	

field, the spin Hamiltonian (Equation 7.10) applies. In order to analyze EPR data in a cubic (octahedral) symmetry site – for example, for the Re^{4+} ion – extra terms are needed in the spin Hamiltonian ($S = 3/2$, $I = 3/2$):

$$H = g\mu_B B \cdot S + \mu_B u \left[B_x S_x^3 + B_y S_y^3 + B_z S_z^3 - \frac{1}{5}(B \cdot S)(3S^2 - 1) \right]$$
$$+ AS \cdot I + U \left[I_x S_x^3 + I_y S_y^3 + I_z S_z^3 - \frac{1}{5}(S \cdot I)(3S^2 - 1) \right]. \quad (7.29)$$

$4d^5$ (Mo^+, Ru^{3+}), $5d^5$ (W^+, Re^{2+}, Os^{3+}, Ir^{4+}): These are S-state ions; the details are given in Section 7.4.6.

$4d^6$ (Pd^{4+}, Rh^{3+}), $5d^6$ (Pt^{4+}): Pd^{4+} and Pt^{4+} are diamagnetic, hence no EPR can be observed. EPR for Rh^{3+} in NaCl (γ-irradiated) and $Na_3RhCl_6 \cdot 12H_2O$ have been reported at 77 K, with the values of g and A (Misra, 1999, p. 240).

$4d^7$ (Rh^{2+}, Pd^{3+}), $5d^7$ (Pt^{3+}): In a trigonal symmetry a Jahn–Teller splitting is experienced by this configuration, similar to that for a $3d^7$ ion. Because of the occurrence of the dynamical Jahn–Teller effect, at higher temperatures an isotropic spectrum ($g_\parallel = g_\perp$) is observed, while at low temperatures, the Jahn–Teller distortions are frozen-in, resulting in an anisotropic spectrum ($g_\| \neq g_\perp$). In monoclinic symmetry, the orbital doublet for this configuration is split, thus no Jahn–Teller effect is observed. Most of the compounds in which EPR has been observed are dithiolates and related complexes. The applicable spin Hamiltonian is described by Equation 7.14 with $S = \frac{1}{2}$.

$4d^8$ (Pd^{2+}), $5d^8$ (Pt^{2+}): These are diamagnetic, hence no EPR can be observed.

$4d^9$ (Rh^0, Ag^{2+}), $5d^9$ (Au^{2+}): Dynamic and static Jahn–Teller effects are expected for this configuration at high and low temperatures, respectively. The theory to interpret the EPR spectra is the same as that described for $3d^9$ ions.

7.4.4
Rare-Earth Ions

For these ions, the $4f$ shell contains the magnetic electrons. The configuration consists of the terms characterized by well-defined L and S values, which are coupled by the spin–orbit interaction to a well-defined value of the total angular momentum, J. In some rare-earth ions a partial breakdown of Russell–Saunders coupling causes an admixture of levels having the same angular momentum, J. The values of spin–orbit coupling constants for the rare-earth group ions, as listed in Table 7.8, are much larger than those for the iron-group ions. However, the crystal fields seen by the rare-earth ions, which are smaller than the spin–orbit couplings, are, in turn, very much smaller than those in the iron-group ions. This is due to:

- The average values, $<r^n>$, for the f^n, rare-earth, wavefunctions, are smaller than those for the d^n, iron-group, wavefunctions since, unlike the $3d^n$ functions, the $4f^n$ functions do not extend very far.

- The rare-earth $4f$ electrons are screened by the outer electrons which, in turn, attenuate the effect of the crystal field potential.

- The crystals of most rare-earth ions possess low symmetries, so that the axial and cubic contributions to the electrostatic crystal field potentials are partially, or even fully, suppressed.

For a $4f^n$ ($n = 1$ to 13) rare-earth ion in a crystal, the ground state is that determined by Hund's rule with the total angular momentum J. This is a good quantum number, since the separation between adjacent J levels (100–1000 cm^{-1}) due to the spin–orbit interaction generally exceeds the overall splitting of the ground-state manifold due to the crystal field by at least an order of magnitude. The crystalline electric field (CEF) Hamiltonian within the ground state (J) multiplet of a rare-earth ion, in terms of Stevens' operators, is (Abragam and Bleaney, 1970)

$$H_{CEF} = \sum_{\ell,m} B_\ell^m O_\ell^m, \qquad (7.30)$$

where $\ell = 2, 4, 6$ ($|m| \leq \ell$).

Table 7.8 Spectroscopic and nuclear properties of rare-earth ions (Low, 1960).

Trivalent ion	Ground state of free ion	Spin-orbit coupling in cm^{-1}	$\langle r^{-3} \rangle$ in (A)$^{-3}$	Isotope	Nuclear spin	Nuclear magnetic moment	Larmor frequency MHz at 10^4 gauss	Quadrupole moment Q ×10^{-24} cm^{-2}
^{58}Ce	$^2F_{5/2}$	640.0	32.5					
^{59}Pr	$^3H_4$	800.0	37.0	141	5/2	3.92	11.95	-5.4×10^{-2}
^{60}Nd	$^4I_{9/2}$	900.0	42.0	143	7/2	−1.03	2.24	<1.2
				145	7/2	0.64	1.4	<1.2
^{61}Pm	$^5I_4$	(1070.0)	47.0					
^{62}Sm	$^6H_{5/2}$	1200.0	51.0	147	7/2	−0.83	1.8	0.72
				149	7/2	−0.68	1.5	0.72
^{63}Eu	$^7F_0$	1410.0	57.0	151	5/2	3.4	10.0	~1.2
				153	5/2	1.5	4.6	~2.5
^{64}Gd	$^8S_{7/2}$	1540.0	62.0	155	3/2	−0.24	1.2	1.1
	$^2F_{5/2}$			157	3/2	−0.32	1.6	1.0
^{65}Tb	$^7F_4$	1770.0	68.0	159	3/2	1.52	7.72	
^{66}Dy	$^6H_{15/2}$	1860.0	74.0	161	5/2	0.38	1.2	
				163	5/2	0.53	1.6	
^{67}Ho	$^5I_8$	2000.0	80.0	165	7/2	3.29	7.17	2.0
^{68}Er	$^4I_{15/2}$	2350.0	86.0	167	7/2	0.48	1.04	~10.0
^{69}Tm	$^3H_6$	2660.0	92.0	169	1/2	−0.20	3.05	
^{70}Yb	$^2F_{7/2}$	2940.0	98.0	171	1/2	0.43	6.6	
				173	5/2	−0.6	1.8	3.9

The CEF parameters, B_ℓ^m, can be, alternatively, expressed as:

$$B_\ell^m = A_\ell^m \langle r^\ell \rangle \alpha_\ell, \tag{7.31}$$

where the α_ℓ are Stevens' operator-equivalent factors (Abragam and Bleaney, 1970) and $\langle r^\ell \rangle$ are the radial integrals which have been calculated for the most rare-earth ions using the Hartree–Fock procedure (Abragam and Bleaney, 1970). The A_ℓ^m are often assumed to be the same for all the rare-earth ions in an isostructural series of compounds (Assour, Goldmacher, and Harrison, 1965). The required α_ℓ and $\langle r^\ell \rangle$ values for the various trivalent 4f^n rare-earth ions (Abragam and Bleaney, 1970) are listed in Table 7.9 (Misra, Chang, and Felsteiner, 1997).

In general, for a rare-earth ion in a crystal, an EPR transition may be observed at microwave frequencies at low temperatures, between those two lowest levels of the ground state J multiplet, which are significantly populated and possess a nonzero magnetic moment corresponding to an effective spin $S = \frac{1}{2}$. The effective g-values corresponding to these levels ($|+\rangle$ and $|-\rangle$) can be calculated from the matrix elements of the components of the angular momentum operators J_x, J_y, J_z within these two levels.

7.4 Transition Ions

Table 7.9 List of the Hund's rule ground state $^{2S+1}L_J$ (J, L, S denote the total, orbital and spin angular momentum, respectively), g_J (as defined by Equation 7.31), α_j and $\langle r^i \rangle$ values for the various $4f^n$ rare-earth ions.

Ion	$4f^n$	$^{2S+1}L_J$	g_J	α_2	α_4	α_6	$\langle r^2 \rangle$	$\langle r^4 \rangle$	$\langle r^6 \rangle$	Ref.
Ce^{3+}	$4f^1$	$^2F_{5/2}$	6/7	$\frac{-2}{5\cdot 7}$	$\frac{2}{3^2\cdot 5\cdot 7}$	0	1.200	3.455	21.226	a,b
Pr^{3+}	$4f^2$	$^3H_4$	4/5	$\frac{-2^2\cdot 13}{3^2\cdot 5^2\cdot 11}$	$\frac{-2^2}{3^2\cdot 5\cdot 11^2}$	$\frac{2^4\cdot 17}{3^4\cdot 5\cdot 7\cdot 11^2\cdot 13}$	1.086	2.822	15.726	a,b
Nd^{3+}	$4f^3$	$^4I_{9/2}$	8/11	$\frac{-7}{3^2\cdot 11^2}$	$\frac{-2^3\cdot 17}{3^3\cdot 11^3\cdot 13}$	$\frac{-5\cdot 17\cdot 19}{3^3\cdot 7\cdot 11^3\cdot 13^2}$	1.001	2.401	12.39	a,b
Pm^{3+}	$4f^4$	$^5I_4$	3/5	$\frac{2\cdot 7}{3\cdot 5\cdot 11^2}$	$\frac{2^3\cdot 7\cdot 17}{3^3\cdot 5\cdot 11^3\cdot 13}$	$\frac{2^3\cdot 17\cdot 19}{3^3\cdot 7\cdot 11^3\cdot 13^2}$	0.937*	2.124*	10.480	a,c
Sm^{3+}	$4f^5$	^6H	2/7	$\frac{13}{3^2\cdot 5\cdot 7}$	$\frac{2\cdot 13}{3^3\cdot 5\cdot 7\cdot 11}$	0	0.883	1.897	8.775	a,b
Eu^{3+}	$4f^6$	$^7F_0$		0	0	0	0.816*	1.695*	7.429*	a,b,c
Gd^{3+}	$4f^7$	$^8S_{7/2}$		0	0	0	0.785	1.515	6.281	a,b
Tb^{3+}	$4f^8$	$^7F_6$	3/2	$\frac{-1}{3^2\cdot 11}$	$\frac{2}{3^3\cdot 5\cdot 11^2}$	$\frac{-1}{3^4\cdot 7\cdot 11^2\cdot 13}$	0.739*	1.150*	5.238*	a,b,c
Dy^{3+}	$4f^9$	$^6H_{15/2}$	4/3	$\frac{-2}{3^2\cdot 5\cdot 7}$	$\frac{-2^3}{3^3\cdot 5\cdot 7\cdot 11\cdot 13}$	$\frac{2^2}{3^3\cdot 7\cdot 11^2\cdot 13^2}$	0.726	1.322	5.102	a,b
Ho^{3+}	$4f^{10}$	$^5I_8$	5/4	$\frac{-1}{2\cdot 3^2\cdot 5^2}$	$\frac{-1}{2\cdot 3\cdot 5\cdot 7\cdot 11\cdot 13}$	$\frac{-5}{3^3\cdot 7\cdot 11^2\cdot 13^2}$	0.692	1.221	4.489	d
Er^{3+}	$4f^{11}$	$^4I_{15/2}$	6/5	$\frac{2^2}{3^2\cdot 5^2\cdot 7}$	$\frac{2}{3^2\cdot 5\cdot 7\cdot 11\cdot 13}$	$\frac{2^3}{3^3\cdot 7\cdot 11^2\cdot 13^2}$	0.666	1.126	3.978	a,b
Tm^{3+}	$4f^{12}$	$^3H_6$	7/6	$\frac{1}{3^2\cdot 11}$	$\frac{2^3}{3^4\cdot 5\cdot 11^2}$	$\frac{-5}{3^4\cdot 7\cdot 11^2\cdot 13}$	0.646	1.119	3.646	d
Yb^{3+}	$4f^{13}$	$^2F_{7/2}$	8/7	$\frac{2}{3^2\cdot 7}$	$\frac{-2}{3\cdot 5\cdot 7\cdot 11}$	$\frac{2^2}{3^3\cdot 7\cdot 11\cdot 13}$	0.613	0.960	3.104	a,b

The last column gives the references from where the various $\langle r^i \rangle$ value have been obtained. More details are given in Abragam and Bleaney, (1970, p. 276) and Misra, Chang, and Felsteiner (1997; Misra, 1999).
* Estimated from the plots of $\langle r^i \rangle$ against atomic number of these values are given by Freeman and Watson (1962).
a) Abragam and Bleaney (1970).
b) Freeman and Watson (1962).
c) Schumacher and Hollingsworth (1996).

$$g_\alpha \sigma_\alpha = g_J \begin{pmatrix} \langle +\|J_\alpha\|+ \rangle & \langle +\|J_\alpha\|- \rangle \\ \langle -\|J_\alpha\|+ \rangle & \langle -\|J_\alpha\|- \rangle \end{pmatrix}; \alpha = x, y, z; \quad (7.32)$$

where,

$$g_J = \langle J\|J_\alpha\|J \rangle = 1 + \frac{J(J+1) + S(S+1) - L(L+1)}{2J(J+1)} \quad (7.33)$$

is the Landé factor, $|\pm\rangle$ denote the orthonormal eigenvectors corresponding to the two lowest levels considered for the calculation of the magnetic moment, and σ_α

are Pauli spin matrices. Finally, the effective g-values are given by Misra, Chang, and Felsteiner (1997):

$$g_x = g_J\varepsilon_x; \quad g_y = g_J\varepsilon_y; \quad g_z = g_J\varepsilon_z. \tag{7.34}$$

In Equation 7.34, ε_x, ε_y, and ε_z denote, respectively, the differences of the eigenvalues of J_x, J_y, and J_z for these two levels. The value of g_J, as calculated using Equation 7.33 for different rare-earth ions, are also listed in Table 7.9. As for the $|\pm\rangle$ levels, one chooses the set of two levels which lie the lowest and for which all ε_α ($\alpha = x, y, z$) are not zero. For Kramers' ions, these belong to the lowest-lying doublet. On the other hand, for non-Kramers' ions, care must be taken when choosing these two levels, as follows. When these levels belong to different eigenvalues, the $|+\rangle$ and $|-\rangle$ eigenvectors should be weighted by appropriate Boltzmann factors, which are, respectively, $1/N\exp(-E_+/k_BT)$ and $1/N\exp(-E_-/k_BT)$; $N = \exp(-E_+/k_BT) + \exp(-E_-/k_BT)$. The g-values are then temperature-dependent.

The crystal field splittings for the rare-earth ions are rather small ($<k_BT$), as indicated by the almost nearly equal values of the magnetic susceptibility of the rare-earth ions in crystals and those of the free ions at room temperature.

7.4.4.1 Odd Number of 4f Electrons

The rare-earth ions with an odd number of electrons, except for the S-state ions Eu^{2+} and Gd^{3+}, have a single Kramers' doublet lowest in an axial, or lower, symmetry field. Their EPR spectra are described by the spin Hamiltonian

$$H = \mu_B g B \cdot S + A I \cdot S \tag{7.35}$$

7.4.4.2 Even Number of 4f Electrons

The Jahn–Teller effect is expected to remove the degeneracy of the splitting of each J level into singlets and doublets in C_{3h} or C_{3v} symmetry. However, this effect is rather small since the 4f electrons, being inner electrons, are screened. It is the characteristic of accidental (non-Kramers') doublets that a first-order Zeeman effect is observed if the magnetic field is applied parallel to the axis of symmetry. On the other hand, only a second-order effect is expected if it is applied normal to the symmetry axis. The non-Kramers' rare-earth ions, with an even number of electrons, may show EPR in an axial symmetry, which is due to the unsplit doublet. In that case, a Δ term is included in the spin Hamiltonian to take into account small departures from axial symmetry due to crystalline imperfections:

$$H = \mu_B g B \cdot S + AS \cdot I + \Delta_x S_x + \Delta_y S_y + Q'[I_z^2 - I(+1)/3] - \mu_n g_n B \cdot I, \tag{7.36}$$

with $S = \frac{1}{2}$. Usually, a parameter Δ is used: $\Delta = (\Delta_x^2 + \Delta_y^2)^{1/2}$, and the condition of resonance is

$$h\nu = \sqrt{(g_z\mu_B B_z + Am)^2 + \Delta^2}.$$

There is expected a distribution in the values of Δ_x, Δ_y, which in turn determines the line shape. This is due to random variations of the crystal-field parameters of lower symmetry. These transitions are sometimes quite intense.

7.4 Transition Ions

Excepting the S-state ions, the rare-earth ions require a temperature of 20 K (or lower) for the observation of their EPR spectra. The nuclear data of the rare-earth ions are listed in Table 7.8.

The details of EPR spectra for the various $4f^n$ ions are described as follows.

$4f^1$ ($^2F_{5/2}$, Ce^{3+}): For a tetragonal symmetry, the lowest doublet is described as $(a|\pm 5/2> + b|\mp 3/2>)$ and the g-values are mostly close to $g_\parallel \approx 3.0$ and $g_\perp \approx 1.5$. Sometimes, the states $|\pm 1/2>$ are also admixed into the lowest doublet by the crystal field.

In threefold symmetry, the EPR is either observed from the doublet $|\pm 5/2>$ for which $g_\parallel \approx 4.0$ and $g_\perp \approx 0$, or from the doublet $|\pm 1/2>$, for which $g_\parallel \approx 0.95$ and $g_\perp \approx 2.185$. For trigonal symmetry, $g_\parallel \approx 2.4$ and $g_\perp \approx 0.1$.

In cubic symmetry, the ground state is fourfold degenerate (Γ_8), of which the levels are unequally spaced in a magnetic field. For **B** parallel to a cubic axis, the energies are $\pm (6/2) g_L \mu_B B$ and $\pm (11/6) g_L \mu_B B$, where g_L is the Landé g-factor (= 6/7 for Ce^{3+}), and B is the intensity of the magnetic field.

In an axial field, there is a splitting of the $J = 5/2$ state into three Kramers' doublets. The corresponding g factors are $g_\parallel = 6/7$, $g_\perp = 18/7$; $g_\parallel = 18/7$, $g_\perp = 0$; and $g_\parallel = 30/7$, $g_\perp = 0$ for the doublets $J_z = \pm 1/2, \pm 3/2$, and $\pm 5/2$, respectively.

In the case of ethyl sulfates, there are two levels – the lower level characterized by $g_\parallel = 0.955$, $g_\perp = 2.185$ corresponding to the doublet $J_z = \pm \frac{1}{2}$, while the upper level is characterized by $g_\parallel = .72$, $g_\perp = 0.2$ corresponding to the doublet $J_z = \pm 5/2$. For a better description, the effect of the crystal field term A_6^6 should be taken into account. For double nitrates, the crystal field potential terms A_4^3 and A_6^3 produce considerable admixtures of $\Delta J_z = \pm 3$. Thus, $J_z = \pm 5/2$ level becomes $N[\cos\theta|5/2, 7/2> + \sin\theta|5/2, -5/2>] + a|7/2, 7/2> + b|7/2, -5/2> + c|7/2, 1/2>$. For certain trichlorides, the lowest doublet is characterized by $J_z = \pm 5/2$, for which $g_\parallel = 30/7$ and $g_\perp = 0$. The best fits are obtained for the ground-state doublet $a|\pm 2, \pm > - b|\pm 3, \mp > + c|\mp 3, \mp >$, with $a = 0.373$, $b = 0.925$, and $c = 0.071$ (Low, 1960; Hutchinson and Wong, 1958).

$4f^2$ ($^3H_4$, Pr^{3+}): The experimental data suggest that the lowest levels are described by $\cos\theta|4, \pm 2> -\sin\theta|4, \mp 4>$ for C_{3h} site symmetry (ethyl sulfates) with the g-values given by $\Lambda(4\cos^2\theta - 8\sin^2\theta)$; and by $a|4, \pm 2> + b|4, \mp 4> + c|4, \mp 1>$ for C_{3v} site symmetry (for double nitrates and trichlorides). These expressions do not take into account the admixture from the $J = 5$ manifold.

$4f^3$ ($^4I_{9/2}$, Nd^{3+}): In a tetragonal symmetry the ground-state doublet of this Kramers' ion is described by $(a|\pm 9/2> + b|\pm 1/2> + c|\mp 7/2>)$ (the $J = 9/2$ manifold lies the lowest), while in a threefold symmetry the ground state is $(a|\pm 7/2> + b|\mp 5/2>)$. When the symmetry is cubic, the lowest level is fourfold degenerate (Γ_8), the same as that for a $4f^1$ ion.

$4f^4$ ($^5I_4$, Pm^{3+}): In a threefold symmetry the ground-state doublet is described by $(a|J = 5/2, J_z = \pm 1/2> + b|J = 7/2, J_z = \pm 1/2>)$, the g-values being nearly isotropic. The third-order crystal-field terms mix the state $|\pm 1/2>$ with $|\pm 5/2>$, rendering the $\tilde{g}$ tensor anisotropic.

$4f^5$ ($^6H_{5/2}$, Sm^{3+}): The experimental results are described by the doublet $\cos\theta \,|\, 5/2, \pm 1/2 > \pm \sin\theta \,|\, 7/2, \pm 1/2 >$ for ethyl sulfates, characterized by predominantly axial symmetry, with $\theta \sim 4°$, since the ratio $Bg_\parallel/Ag_\perp$ is about 4, revealing a considerable admixture of the various J levels. On the other hand, in the case of double nitrates, characterized by a predominant threefold symmetry, the doublet is described by:

$$N[\cos\theta\,|\,5/2,\pm1/2> \pm \sin\theta\,|\,5/2,\pm1/2>] + a\,|\,7/2,\pm1/2> + b\,|\,7/2,\mp5/2> + c\,|\,7/2,\pm7/2>,$$

where N is the normalization constant.

$4f^7$ ($^8S_{7/2}$; Eu^{2+}, Gd^{3+}): These S-state ions and are described in detail in Section 7.4.6.3.

$4f^8$ ($^7F_6$, Tb^{3+}): In axial symmetry, a non-Kramers' doublet lies lowest, described predominantly by the states $|\,J_z = \pm 6>$ with an admixture of the $|\,0>$ state: $a\,|\pm 6> + b\,|\,0> + c\,|\mp 6>$, for which the g-values are $g_\parallel = 18$, $g_\perp \approx 0$. The doublet is split somewhat due to the A_6^6 crystal field term, which mixes the $J_z = \pm 6$ and $J_z = 0$ states.

$4f^9$ ($^6H_{15/2}$, Dy^{3+}): For this Kramers' ion, in a cubic symmetry, a Γ_8 quartet lies lowest, responsible for EPR. In addition, the first excited doublet also gives an isotropic EPR line. In an octahedral coordination (cubic symmetry) a doublet level lies lowest, for which the g-value is expected to be 6.60. However, this value is modified due to covalency effects, which are, however, small for rare-earth ions due to the screening of the $4f$ electrons by the outer electrons. In the double nitrates $g_\parallel = 4.281$, $g_\perp = 8.923$, described by the doublet $a\,|\pm 13/2> + b\,|\pm 7/2>) + c\,|\pm 1/2> + d\,|\mp 5/2> + e\,|\mp 11/2)$. The admixture from the $J = 13/2$ manifold is expected to be small, since $Ag_\perp/Bg_\parallel \sim 1$.

$4f^{10}$ ($^5I_8$, Ho^{3+}): For this ion in a tetragonal (fourfold), or a threefold symmetry, $g_\parallel$ lies in the range 13.5–16.0 and $g_\perp \approx 0$. The hyperfine splitting is very large, of the order of an X-band microwave quantum (~9.5 GHz).

The complexity of the Ho^{3+} EPR spectrum originates from two factors:

- The ground state is the doublet $(a\,|\pm 7> + b\,|\pm 1>) + c\,|\mp 5>)$ situated close to a singlet $(a\,|+6> + b\,|-6> + c\,|\,0>)$. Thus, transitions between a doublet level to the singlet level can be observed, in addition to the transitions between the doublet levels. This results, to some extent, in a spectrum explained by the spin Hamiltonian characteristic of axial symmetry that splits the singlet away from the doublet for an effective spin $S = 1$:

$$H = \mu_B[g_\parallel B_z S_z + g_\perp(B_x S_x + B_y S_y)] + b_2^0[S_z^2 - S(S+1)/3]$$
$$+ AS_z I_z + B(S_x I_x + S_y I_y) + Q'[I_z^2 - I(I+1)/3]$$
$$- R_\parallel I_z B_z - R_\perp(I_x B_x + I_y B_y) - g_N \mu_N(I_x B_x + I_y B_y + I_z B_z)$$

- Due to large hyperfine structure, additional complexity arises, due to appearance of the hyperfine transitions $\Delta m = \pm 1$, where m is the nuclear magnetic quantum number.

The salient features of Ho^{3+} EPR spectra are as follows (Note: Below the letter B has been used for both the magnetic field and the hyperfine parameter, depending on the context):

i) When the microwave magnetic field, B_{rf}, is parallel to the z axis, $\Delta M = 2$ transitions, where M is the electronic magnetic quantum number, are observed, associated with eight hyperfine lines each. The line positions are described by:

$$h\nu = 2g_{\parallel}\mu_B B + (2A - B^2/D)m - (B^2 h\nu/2D^2)[I(I+1) - m^2].$$

ii) When $\mathbf{B} \parallel$ z axis and $\mathbf{B}_{rf} \perp$ z axis, seven hyperfine lines, each placed midway between pairs described in (i) above, are observed. This corresponds to forbidden hyperfine transitions $\Delta m = \pm 1$. The line positions are described by:

$$h\nu = 2g_{\parallel}\mu_B B + 2(A - Q')(m + 1/2)$$
$$- [B^2 h\nu/2D^2][I(I+1) + 1/4 - (m + 1/2)^2].$$

iii) When $\mathbf{B} \perp$ and $\mathbf{B}_{rf} \parallel$ to z-axis, six unequally spaced lines are observed, corresponding to transitions between $|\pm\rangle$ levels. These are further split as a second-order effect when $\mathbf{B} \perp$ z-axis. The hyperfine structure is rather large, of the same order as the energy difference between levels.

iv) When $\mathbf{B} \parallel$ z-axis, a number of lines are observed at high frequencies and high magnetic fields. The hyperfine line separation becomes lesser at higher magnetic fields. There exists a second group of lines whose separation decreases towards low field. These transitions correspond to those between $|1\rangle$ and $|0\rangle$ levels. Of the two sets, one appears before these levels cross. The transitions are described by

$$h\nu = \pm b_2^0 - g_{\parallel}\mu_B B - Am[1 + B^2/2(h\nu)^2 \pm mB^2/(\pm 4D - 2 h\nu)$$
$$+ B^2[I(I+1) - m^2][1/h\nu + (\pm 4D - 2h\nu)^{-1}].$$

The + and − signs are used depending on the sign of b_2^0 being negative or positive, respectively.

$4f^{11}$ ($^4I_{15/2}$; Er^{3+}, Ho^{2+}): In a tetragonal site symmetry, the ground-state doublet is described by $|J_z = \pm \frac{1}{2}\rangle$, while in a trigonal (threefold) symmetry the ground-state doublet is of the form $(a|\pm 7/2\rangle + b|\mp 5/2\rangle)$. In a cubic symmetry, either a Γ_7 doublet lies lowest, described by $(a|\pm 13/2\rangle + b|\pm 5/2\rangle + c|\mp 3/2\rangle + d|\mp 11/2\rangle)$, corresponding to an isotropic g-value of 6.772, or a Γ_6 doublet lies lowest described by $(a|\pm 15/2\rangle + b|\pm 7/2\rangle + c|\pm 1/2\rangle d|\pm 9/2\rangle)$, with the g-value ≈ 5.9. In an octahedral coordination (cubic field) a Γ_8 quartet lies lowest.

$4f^{13}$ ($^2F_{7/2}$; Yb^{3+}, Tm^{2+}): This configuration corresponds to a single hole in the 4f shell. In an eightfold cubic coordination a Γ_7 doublet lies the lowest, described by ($\sqrt{3/2}\,|\pm 5/2\rangle - 1/2\,|\pm 3/2\rangle$), for which the expected g-value is 3.429. In a tetragonal field, a Γ_7 doublet is expected to lie lowest. In an octahedral symmetry a Γ_6 doublet lies lowest, with the theoretical g-value of 2.667. The covalency effects may modify this value. In a tetragonal site symmetry, the ground-state doublet is probably ($a\,|\pm 5/2\rangle + b\,|\pm 3/2\rangle$).

7.4.5
Actinide Ions (5f^n)

For these ions, the unpaired electrons are in the 5f-shell, behaving like a second rare-earth group. The electronic properties of these ions are influenced by actinide contraction and by relativistic effects. The contraction causes the energy separation of the 5f, 6d and 7s electrons to be small; thus, a strong competition of these electrons takes place in determining the electronic properties of electrons. Because of relativistic effects, the Coulomb interelectronic repulsion decreases and the spin–orbit coupling is increased, and this leads to a severe breakdown of the Russell–Saunders coupling. The crystal-field effects are stronger for the actinide ions as compared with those of the lanthanide ions, because of larger ionic radii of the actinide ions than those of the lanthanide ions. EPR results indicate that the 5f wave function, unlike the 4f wave function, extends on the surrounding diamagnetic neighbors.

The EPR data on actinide ions are meager. This is because it is difficult to grow single crystals with elements that are strongly radioactive, as the intense radioactivity causes destruction of the crystal in the vicinity of the paramagnetic ion, which gives rise to EPR of free radicals, obscuring the EPR spectrum of the paramagnetic ion.

The valences of the actinide group are listed in Table 7.10.

A review article including the EPR results on some uranium compounds was published by Ursu and Lupei (1986).

7.4.5.1 5f^1 Configuration
This configuration corresponds to the ions Np^{3+} and U^{5+}.

Neptunyl ion: This ion has one magnetic electron. The states $m = M = \pm 3, \pm 2, \pm 1, 0$, and $M = m = 1/2$ are possible (here, m refers to the individual electrons). The lowest energy, in the presence of an axial field, is for $M = \pm 3/2$, followed next by $M = \pm 2$, which lies higher by about 10^4 cm^{-1}. The spin–orbit coupling splits the fourfold degeneracy into two Kramers' doublets with $M_j = M + m = \pm 7/2$ and $\pm 5/2$. The resulting lowest state is a doublet, for which one can employ a spin Hamiltonian of the form:

$$H = \mu_B[g_\parallel B_z S_z + g_\perp(B_x S_x + B_y S_y)] + b_2^0[S_z^2 - S(S+1)/3]$$
$$+ AS_z I_z + B(S_x I_x + S_y I_y) + Q'[I_z^2 - I(I+1)/3]$$
$$- R_\parallel I_z B_z - R_\perp(I_x B_x + I_y B_y) - g_N \mu_N(I_x B_x + I_y B_y + I_z B_z)$$

Table 7.10 Valencies in the actinide group (Low, 1960).

No. of electrons	Ac	Th	Pa	U	Np	Pu	Am	Cm	Bk	Cf
0	Ac^{3+}	Th^{4+}	Pa^{5+}	U^{6+} UO_2^{2+}						
1		$Th^{3+}?$	Pa^{4+}	U^{5+} UO_2^+	Np^{6+} NpO_2^{2+}					
2		Th^{2+}	Pa^{3+}	$\underline{U}^{4+}$	Np^{5+} NpO_2^+	Pu^{6+} PuO_2^{2+}				
3				U^{3+}	$\underline{Np}^{4+}$	Pu^{5+} PuO_2^+	Am^{6+} AmO_2^{2+}			
4				U^{2+}	Np^{3+}	Pu^{4+}	Am^{5+} AmO_2^+			
5						Pu^{3+}	Am^{4+}			
6							Am^{3+}			
7								$\underline{Cm}^{3+}$	Bk^{4+}	Cf^{5+}
8									Bk^{3+}	Cf^{4+}
9										Cf^{3+}

Ions with question marks denote doubtful valencies. Ions which are underlined are those for which paramagnetic resonance has been reported.

with $S = 1/2$. The g-values of the lowest-lying doublet are given by the matrix elements

$$g_\parallel = 2\langle a|kL_z + 2S_z|a\rangle,$$
$$g_\perp = 2\langle a|kL_z + 2S_x|b\rangle, \quad (7.37)$$

where $|a\rangle (=|3,-1/2\rangle)$ and $|b\rangle (=|-3,1/2\rangle)$ are the basic states describing the doublet. Thus, $g = 6k - 2$. In the presence of a threefold symmetry, the g-factors given by Equation 7.37 are modified:

$$g_\parallel = \frac{k(6 - 6p^2 + 4q^2) - (2 + 2p^2 - 2q^2)}{1 + p^2 + q^2},$$
$$g_\perp = \frac{4p - 2\sqrt{6}kpq}{1 + p^2 + q^2}. \quad (7.38)$$

In a first approximation, assuming $k = 1$ (no covalency), it is found that $p \approx 0.1$ and $q \geq 0.1$.

U^{5+} ion: The first excited state of this ion ($6d^1$) lies at about $10\,000\,\text{cm}^{-1}$ above the ground state ($5f^1$). The spin–orbit coupling splits the ground configuration into two levels, $^2F_{5/2}$ and $^2F_{7/2}$, the former being the lowest. The interaction of the $5f$ electron with the crystal field depends strongly on the coordination of the ions in the crystal. Two major cases have been found for cubic symmetry:

i) For an eightfold, or tetrahedral, coordination, the crystal field interaction is weaker than the spin–orbit coupling. In a cubic crystal field, the following splitting of levels takes place:

$$^2F_{7/2} \rightarrow \Gamma_6(\text{doublet}) + \Gamma_7(\text{doublet}) + \Gamma_8(\text{quartet})$$
$$^2F_{5/2} \rightarrow \Gamma_6(\text{doublet}) + \Gamma_8(\text{quartet}); \qquad (7.39)$$

the crystal field components belonging to each electronic level are closely bunched together, which enhances relaxation. Thus, EPR can be detected only at very low temperatures. The ground crystal-field level can be Γ_6, or Γ_8, depending on the strength of interaction, resembling the situation for Ce^{3+}.

ii) In an octahedral coordination, the crystal-field interaction is stronger than the spin-orbit coupling, it affects only the ^2F term of the $5f^1$ configuration:

$$^2F(5f^1) \rightarrow \Gamma_2(\text{singlet}) + \Gamma_4(\text{triplet}) + \Gamma_5(\text{triplet}) \qquad (7.40)$$

Additional splitting is caused by the spin–orbit coupling; a lower symmetry also causes further splitting. The ground level of the crystal field is usually Γ_7, originating from the Γ_2 singlet, containing admixtures from all the Γ_7 states. Since the ground level is sufficiently far separated from the excited levels, the relaxation processes are weak and EPR can only be observed at liquid nitrogen, or slightly higher, temperatures.

UO_2^+ constitutes another class of U^{5+} complexes; here, the uranium ion is strongly bonded to two oxygens. The electronic structure of this complex is similar to that of the neptunyl complex NpO_2^+. Here, EPR can, in principle, be only observed at low temperatures.

7.4.5.2 $5f^2$ Configuration

This configuration corresponds to plutonyl $(PuO_2)^{2+}$ and U^{4+} ions.

Plutonyl ions: The lowest states of $2f$-electron configuration, consisting of two electrons with parallel spins ($M_s = \pm 1$) and in the orbits of $M_L = \pm 3$, $M_L = \pm 2$, respectively, are given by $M_L = \pm 5$, $M_s = \pm 1$. Here, M_L and M_s are the magnetic quantum numbers corresponding to the total orbital and spin angular momenta, while M_L, M_s correspond to the individual orbital and spin angular momenta. Spin–orbit coupling splits these two states, and the ground state is $^3H_4$. The principal g-factors are

$$g_\| = 2(kM_L + 2M_s) = 10k - 4;$$
$$g_\perp = 0. \qquad (7.41)$$

The off-diagonal matrix elements of the nonaxial crystal-field terms and the orbital reduction factor k cause a reduction in the value of the g factor.

The non-Kramers' ground-state doublet does not exhibit the usual microwave absorption, because $g_\perp = 0$. However, if the microwave and Zeeman fields are

Tetravalent uranium (U^{4+}) ion: For this ion, the ground level is $^3H_4$. The spin–orbit coupling is stronger than the crystal field, regardless of the symmetry and coordination. In a cubic crystal field the ground level splits as follows:

$$^3H_4 \to \Gamma_1(\text{singlet}) + \Gamma_3(\text{doublet}) + \Gamma_4(\text{triplet}) + \Gamma_5(\text{triplet}) \tag{7.42}$$

In an octahedral symmetry the ground state is always a Γ_1 singlet. The Γ_4 level lies more than $1000\,\text{cm}^{-1}$ above this; thus, the Γ_1 level will become populated only at low temperatures. No EPR is expected for this case.

In cases of eightfold and tetrahedral (cubic) symmetry, the ground level can be a Γ_1 singlet, or a Γ_5 triplet, depending on the ratio of the fourth to sixth order terms in the crystal field potential; in lower symmetries the ground level can be either a singlet, or a non-Kramers' doublet. The crystal-field levels are closely bunched, the relaxation processes are strong, and EPR can be observed only at very low temperatures.

7.4.5.3 $5f^3$ Configuration ($^4I_{9/2}$; U^{3+}, Np^{4+})

U^{3+} **ion:** The ground level for this ion is $^4I_{9/2}$. The spin–orbit coupling is stronger than the crystal-field interaction, regardless of the symmetry and coordination. A cubic field causes the splitting

$$^4I_{9/2} \to \Gamma_6(\text{doublet}) + 2\Gamma_8(\text{quartet}) \tag{7.43}$$

A distortion of an axial, or a lower, symmetry splits the quartets in doublets. EPR can always be observed, except for those situations where the ground state is a nonmagnetic doublet with $g_\perp = 0$, for example, a weak trigonal distortion.

The hyperfine-structure of U^{235} and U^{233} shows the $\Delta m = \pm 1$ and $\Delta m = \pm 2$ transitions. The hyperfine structure of U^{3+} has been used mainly to determine the nuclear moments of odd uranium isotopes.

The EPR spectra of U^{3+} and Np^{3+} are very similar. (Np^{3+} has the configurational $4f^3$, with the same ground state, $^4I_{9/2}$.) Under a C_{3h} symmetry, the ground state doublet is:

$$\cos\theta |J_z = \pm 7/2 \rangle + \sin\theta |J_z = \mp 5/2 \rangle. \tag{7.44}$$

When there is axial symmetry around one of the cubic axes, the ground-state doublet is represented by

$$a|J_z = \pm 9/2\rangle + b|J_z = \pm 1/2\rangle + c|J_z = \mp 7/2\rangle, \tag{7.45}$$

where c is expected to be small.

$5f^4$ **divalent uranium ion (U^{2+}):** The ground-state configuration of divalent uranium has not been uniquely determined. The three possible configurations that have been proposed are $5f^4$, $5f^66d$, or $5f^37s$; these configurations have either the $^5I_4$, or $^5L_6$, as the lowest level. The ground level of $^5I_4$ is a singlet, and that of $^5L_6$ is either a Γ_1 singlet, or a Γ_3 nonmagnetic doublet, in an eightfold cubic crystal field. EPR of these configurations cannot, therefore, be recorded.

$5f^5$ ($^6H_{5/2}$; Pu^{3+}): In a field of threefold symmetry, EPR is observed from a single Kramers' doublet $|J_z = \pm 1/2\rangle$, similar to the spectrum of Sm^{3+} ($4f^5$ configuration). A doublet hyperfine structure from Pu^{239} can also be observed.

$5f^7$ ($^8S_{7/2}$): Cm^{3+} corresponds to the S-state configuration, which is discussed in the following section.

7.4.6
S-State Ions

7.4.6.1 Introduction

The S-state transition ions have the configurations $3d^5$ (Ti^{-1}, V^0, Cr^+, Mn^{2+}, Fe^{3+}, Co^{4+}), $4d^5$, $5d^5$, $4f^7$ (Gd^{3+}, Eu^{2+}) and $5f^7$ (Cm^{3+}). For d^5 ions the ground state is $^6S_{5/2}$, while for f^7 ions it is $^8S_{7/2}$. When these ions are placed in a medium, or weak, crystal field, the ground state will remain an S state. Neither the crystal field, nor the spin–orbit coupling, can remove the six-, or eightfold degeneracies of the d^5 and f^7 ions, respectively. Even the cubic field cannot remove these degeneracies according to the group theory. Higher-order perturbations, including simultaneously the crystal-field and spin–orbit coupling, are necessary to split the S-state. Being of a higher order, these perturbations produce a small splitting of the ground state. The spectra of S-state ions can be observed with ease even at room temperature; for this reason, considerable EPR data have been reported on these ions. Nevertheless, the mechanisms responsible for the splitting of S-state ions are not yet completely understood.

7.4.6.2 Spin Hamiltonian

The spin Hamiltonian for the S-state ions, $S = 5/2$ (d^5) and $7/2$ (f^7), can, for an arbitrary symmetry, be expressed as

$$H = \mu_B S \cdot \tilde{g} \cdot B + \sum_m B_2^m O_2^m + \sum_m B_4^m O_4^m + \sum_m B_6^m O_6^m; \qquad (7.46)$$

where the last term, involving sixth-order spin operators is to be included only for f^7 ions. In Equation 7.46, $O_n^m(|m| = n)$ are the spin operators as given by Abragam and Bleaney (1970). Not all B_n^m are nonzero for different symmetries. A list of the nonzero B_n^m-values for the various symmetries is provided in Table 7.11 (Misra, 1999).

Today, sophisticated numerical techniques are available for the determination of $\tilde{g}$ and B_l^m from a knowledge of EPR line positions, and these have been extensively discussed by Misra (1986). Here, the EPR line positions, for several orientations of the Zeeman field, are simultaneously fitted in a least-squares manner,

Table 7.11 Nonvanishing zero-field operator coefficients B_k^q the spin Hamiltonian for the various symmetries (Misra, 1986, 1999).

Cubic:	$B_4^0, B_4^4 = 5B_4^0, B_6^4 = -21B_6^0$; $A_z = A_x = A_y$.
Axial:	B_2^0, B_4^0, B_6^0; $A_z, A_x = A_y$.
Hexagonal:	$B_2^0, B_4^0, B_6^0, B_6^6$; $A_z, A_x = A_y$.
Tetragonal:	$B_2^0, B_4^0, B_4^4, B_6^0, B_6^4$; $A_z, A_x = A_y$.
Trigonal:	$B_2^0, B_4^0, B_4^3, B_6^0, B_6^3, B_6^6$; $A_z, A_x = A_y$.
Orthohombic:	B_k^m; $k = 2, 4, 6$; $0 \leq m$ (even) $\leq k$; A_z, A_y, A_x.
Monoclinic:	Orthorhombic + B_1, where $B_1(C_2 \| x) = B_k^{-(2n-1)}$, $B_1(C_2 \| y) = B_k^{2n-1}$, $B_1(C_2 \| z) = B_k^{-2n}$ and $k = 2, 4, 6$; $1 \leq n \leq k/2$; $A_z, A_x = A_y$.
Triclinic:	B_k^q; $k = 2, 4, 6$; $-k \leq q \leq k$; A_z, A_y, A_x.

Also included is the relationship between the principle values of the HF tensor, describing the interaction $\mathbf{S} \cdot \tilde{A} \cdot \mathbf{I} = A_x S_x I_x + A_y S_y I_y + A_z S_z I_z$. For monoclinic symmetry C_2 represents the orientation of the twofold axis, while x, y, z represent the magnetic axes.

utilizing numerical diagonalization of the spin-Hamiltonian matrix on a digital computer (for more detail, see Chapter 8).

7.4.6.3 Theoretical Considerations

The experimental results on the S-state ions reveal the following:

- There are large variations of the ground-state splittings for any ion in different host lattices.

- The magnitude of the axial zero-field-splitting parameter (b_2^0) generally varies considerably with temperature, increasing with temperature.

- The g-factors deviate only slightly from the free-electron value ($g_e = 2.0023$). This deviation is always negative for the rare-earth ions (Gd^{3+}, Eu^{2+}), but can be positive for the iron-group ions (Mn^{2+}, Fe^{3+}).

- A large anomalous hyperfine structure is found for Mn^{2+}, and smaller structure for Eu^{2+} and Gd^{3+}, the hyperfine-structure constant showing a regular variation with covalency.

The nature of the splitting of the S-state ions, which can explain the above observations, is not yet completely understood. A few possible mechanisms have been proposed which involve higher-order perturbations due to, for example, spin–orbit, crystal-field interactions, either within the d^5 or f^7 configuration, or involving excited states of higher-lying configurations.

An overlap, or superposition, model has been applied for the *ab initio* calculation of the spin-Hamiltonian parameters (SHPs), originally proposed by Newman (1971; Newman and Urban, 1975; see also Chapter 3). The model has been applied extensively to the estimation of SHPs of the Gd^{3+} ion (Misra and Upreti, 1986). The point-charge and induced-dipole models have also been successfully applied to the estimation of SHP of the Gd^{3+} ion in some host lattices (Misra and Upreti, 1986). However, a clear-cut microscopic model is not yet available for the estimation of SHP.

In the case of the anomalous hyperfine structure, as exhibited by the S-state ions, extensive hyperfine structures have been observed for all such cases, although a half-filled shell – that is, that of an S-state ion – should not show any hyperfine structure. The origin of the hyperfine structure has been proposed to lie in the configurational interaction of the form $3s3d^54s$ with $3s^23d^5$ for iron-group ions, and similarly that of $4s4f^55s$ with $4s^24f^7$ for the rare-earth ions. Only a small amount of admixture is necessary to explain the splitting, since unpaired s electrons contribute considerably to the hyperfine-structure splitting due to the Fermi-contact interaction. The contribution of the unpaired s-electron states can be expressed as $K\mathbf{S}\cdot\mathbf{I}$, where K is positive for iron-group ions. The sign of K has not been determined for the rare-earth group. The hyperfine-structure splitting varies from crystal to crystal, being larger the more ionic the crystal. The calculation of the hyperfine splitting from first principles is not easy.

$3d^5$ (Cr^+, Fe^{3+}, Mn^{2+}, Co^{4+}): For this half-filled 3d shell, the free ion ground state is $^6S_{5/2}$. There are two main cases to be considered – the weak field ($S = 5/2$) and the strong field ($S = \frac{1}{2}$) – although there is also an intermediate spin case $S = 3/2$.

i) **Weak field case:** The sixfold spin degeneracy ($S = 5/2$) is lifted by the spin–orbit coupling and higher-order perturbations, resulting in three Kramers' doublets separated by energies of ~ 0.1–1.0 cm^{-1} in zero magnetic field. In general, there are $2S = 5$ allowed fine-structure ($\Delta M = \pm 1$) lines, and often forbidden lines $\Delta M = \pm 2, \pm 3$ are also observed. For Mn^{2+}, which shows a characteristic hyperfine spectrum of six components, there are 30 allowed hyperfine lines ($\Delta M = \pm 1$, $\Delta m = 0$) observed by EPR. The g-values for both Fe^{3+} and Mn^{2+} are also close to 2.0. Hyperfine splittings for Mn^{55} are usually in the range 50–90×10^{-4} cm^{-1}, while those for Fe^{57} (2% abundant) are $\sim 10^{-3}$ cm^{-1}. The cubic-field splitting for the Cr^+ ion is $\sim 3 \times 10^{-4}$ cm^{-1}. A hyperfine structure from Cr^{53} can also be observed (A ~ 14×10^{-4} cm^{-1}). As the ground state is only very loosely coupled to the crystal field, the relaxation times are long. Thus, EPR is readily observed at room temperature.

7.4 Transition Ions

For trigonal distortions, as in alums, the spin Hamiltonian has the form ($S = 5/2$):

$$H = g\mu_B(B_xS_x + B_yS_y + B_zS_z) + (b_4^4/15)(S_\xi^4 + S_\eta^4 + S_\zeta^4 - 707/16)$$
$$+ b_2^0(S_z^2 - 35/12) + b_4^0[S_z^4 - (95/14)S_z^2 + 81/16]$$
$$+ A(S_xI_x + S_yI_y + S_zI_z) \quad (7.47)$$

Here, the z-axis is considered to be along the [111] direction with respect to (ξ, η, ζ), which is a set of mutually perpendicular axes. The last term in Equation 7.47 applies to the Mn^{2+} ion for which the nuclear spin $I = 5/2$. For the cases of the Fe^{3+} ion, the hyperfine interaction due to the Fe^{57} nucleus ($I = \frac{1}{2}$) is not observed, except in compounds enriched with Fe^{57}. Further, the fourth order terms in electron spin operators in Equation 7.47 are usually small. Thus, usually the spin Hamiltonian, Equation 7.46, without the fourth and sixth order terms, applies.

For orthorhombic symmetry, as in Tutton salts, the spin Hamiltonian in Equation 7.46 applies, but without the sixth-order terms $b_6^m O_6^m$. If b_2^0, b_2^2 are finite but small (0.001 to 0.1 cm^{-1}), five EPR lines are observed. When $b_2^0, b_2^2 \gg g\mu_B B$, as in the large majority of well-accounted for EPR spectra of high-spin d^5 compounds, the eigenvalues of the spin Hamiltonian in zero magnetic field corresponds to three Kramers' doublets. There are two simplifying cases. If $b_2^0 \neq 0$, $b_2^2 = 0$, the lowest doublet can be characterized by the effective g values $g_\| = 2$, $g_\perp = 6$. When $b_2^0 = 0$, $b_2^2 \neq 0$, the middle doublet is characterized by an isotropic value $g = 4.29$.

For a general solution, the characteristic secular determinant of the matrix of the spin Hamiltonian

$$H = \mu_B(g_xB_xS_x + g_yB_yS_y + g_zB_zS_z) + b_2^0(S_z^2 - 35/12) + b_2^2(S_x^2 - S_y^2),$$

within the manifold of $S = 5/2$, may be solved numerically. The g-tensor may be taken to be equal to 2.0, and one can calculate the eigenvalues by using b_2^0 and $\lambda = b_2^2/b_2^0$. (There is axial symmetry for $\lambda = 0$, while for $\lambda = 1/3$ one has the maximum rhombic symmetry. Values of $\lambda > 1/3$ represent a convergence towards axial symmetry.) Dowsing and Gibson (1969) prepared a graphical display of computed EPR transitions as functions of b_2^0 and λ (in the range $0 \leq \lambda \leq 1/3$) exhibiting the magnetic fields at X-band frequencies.

ii) **Strong-field case:** When the crystal field is strong enough, for example, Mn^{2+} in $K_4Fe(CN)_6$, Fe^{3+} in $K_3Co(CN)_6$, Co^{4+} in Al_2O_3, the coupling between the d electrons is broken down, resulting in a low-spin ground state. All five electrons occupy t orbitals; the system may be considered as a single hole in an orbital-triplet level. There is an isolated Kramers' doublet which lies lowest in a field of low symmetry, and corresponds to an effective spin $S' = 1/2$ with a strongly anisotropic g-factor (~0.7–2.5). This situation is prevalent in the ions Ti^-, Cr^+, Mn^{2+} and Fe^{3+} in the low-spin state. In a strong ligand field characterized by octahedral symmetry, the ground state is $^2T_2$. The following spin Hamiltonian applies ($S = \frac{1}{2}$):

$$H = \mu_B(g_xB_xS_x + g_yB_yS_y + g_zB_zS_z) + A_xS_xI_x + A_yS_yI_y + A_zS_zI_z.$$

The rapid decrease of spin-lattice relaxation times of these ions with increasing temperature results in line broadening at temperatures >20 K. Thus, low temperatures are required to observe EPR.

iii) **Intermediate spin case (S = 3/2):** The spin Hamiltonian applicable in this case is:

$$H = g\mu_B B \cdot S + b_2^0(S_z^2 - 5/4) + b_2^2(S_x^2 - S_y^2).$$

The typical example is that of bis(diisopropyldithiocarbamato) Fe^{3+} chloride (König and König, 1984) for which the value $g = 2.00$ was assumed.

$4d^5$ (Mo^+, Ru^{3+}), $5d^5$ (W^+, Re^{2+}, Os^{3+}, Ir^{4+}): In a strong octahedral field this configuration may be considered as a single hole in a t_2 orbital. The theory is then similar to that for a d^1 ion, but with a negative spin–orbit coupling. The g-values are

$$g_\parallel = -1 + \frac{3(\delta - \lambda/2)}{W}$$

$$g_\perp = 1 + \frac{(\delta + 3\lambda/2)}{W}, \qquad (7.48)$$

where $W = [(\delta - \lambda/2)^2 + 2\lambda^2]^{1/2}$. For an accurate interpretation, covalent effects should be taken into account. Ru^{3+} shows hyperfine structure from Ru^{99} and Ru^{101} with $A \sim 50 \times 10^{-4}\,cm^{-1}$, whereas Ir^{4+} shows a four-line structure from Ir^{191} and Ir^{193} with $A \sim 25 \times 10^{-4}\,cm^{-1}$. Temperatures of 4 K, or less, are required for EPR to be observed.

$4f^7$ (Eu^{2+}, Gd^{3+}, Tb^{4+}): The ground state of this half-filled shell is $^8S_{7/2}$, which is retained in crystals, because the crystal field is weak for rare-earth ions. There are four Kramers' doublets which are split in zero magnetic field (even in a cubic field) by energies of the order of $0.1\,cm^{-1}$. As the effective spin $S = 7/2$, there will be observed seven allowed fine-structure transitions ($\Delta M = \pm 1$). The g-values are typically ~ 1.99. Both, Gd^{3+} and Eu^{3+} show hyperfine structure, but the splitting parameters for Gd^{155} and Gd^{157} are usually rather small ($\sim 5 \times 10^{-4}\,cm^{-1}$), and the structure is not always resolved. Typical values for Eu^{2+} are $A^{151} \sim 30 \times 10^{-4}\,cm^{-1}$, and $A^{153} \sim 15 \times 10^{-4}\,cm^{-1}$. A clear-cut linear dependence of the values of the spin Hamiltonian parameters of Gd^{3+} on the rare-earth host-ion radius has been found, in a series of iso-structural hosts (Misra and Upreti, 1986).

The results of EPR studies on Gd^{3+} have been summarized in a review by Misra and Upreti (1986), covering the literature from 1977 to 1983.

The spin-lattice relaxation times are usually long for $4f^7$ ions, similar to that for $3d^5$ ions, and EPR may be observed at room temperature. Tb^{4+} is an exception, however, with temperatures less than 20 K being necessary for the observation of EPR.

$5f^7$ (Cm^{3+}): Earlier EPR measurements suggest that the splitting of the four spin doublets in zero magnetic field is small, similar to that for Gd^{3+} and Eu^{2+} ($4f^7$ configuration). However, later measurements were in disagreement, indicating large differences in $g_\parallel$ and $g_\perp$; this has been interpreted to be arising from the $\pm 1/2$ doublet, which lies lower than the next doublet by an energy much greater than the measuring quantum. If the host crystal fluoresces strongly, for example, $LaCl_3$, the crystal symmetry is destroyed within a few hours after crystal growth by the intense α radiation from the Cm nucleus.

7.5
Concluding Remarks

This chapter does not include the tabulation of SH parameters for transition ions. A detailed listing of SHP has been reported by Misra (1999), which covers the literature up to the mid-1990s.

Acknowledgments

The author is grateful to Professor C.P. Poole, Jr, for helpful comments to improve the presentation of this chapter, and to the Natural Sciences and Engineering Research Council (NSERC) of Canada for partial financial support.

Pertinent Literature

The sources of information of the type detailed in this chapter are the *ESR Handbook*, volume 2 (Misra, 1999), and the books on EPR by Abragam and Bleaney (1970), Low (1960), and Pilbrow (1990).

References

Abragam, A. and Bleaney, B. (1970) *Electron Paramagnetic Resonance of Transition Ions*, Clarendon, Oxford.

Al'tshuler, S.A. and Kozyrev, B.M. (1974) *Electron Paramagnetic Resonance in Compounds of Transition Elements*, Academic Press, New York.

Altmann, S.I. and Herzig, P. (1994) *Point Group Theory Tables*, Clarendon Press, Oxford.

Assour, J.M., Goldmacher, J., and Harrison, S.E. (1965) *J. Chem. Phys.*, **43**, 159.

Baker, J.M., Bleaney, B., and Hayes, W. (1958) *Proc. R. Soc.*, **A247**, 141.

Bates, C.A. and Chandler, D.E. (1971) *J. Phys.*, **C4**, 2713.

Bates, C.A. and Chandler, D.E. (1973) *J. Phys.*, **C6**, 1975.

Bieri, A. and Kneubühl, K.F. (1965) *Phys. Kond. Mater.*, **4**, 230.

Biryukov, I.P., Voronkov, M.G., and Safin, I.A. (1969) *Tables of Nuclear Quadrupole Resonance Frequencies*, IPST Press, Jerusalem.

Boucher, L.J., Tynan, E.C., and Yen, T.F. (1969) *Electron Spin Resonance of Metal Complexes* (ed. T.F. Yen), Plenum Press.

Buckmaster, H.A. (1962) *Can. J. Phys.*, **40**, 1670.

Chand, P., Jain, V.K., and Upreti, G.C. (1988) *Magn. Reson. Rev.*, **14**, 49.

Deryugina, N.I. and Roitsin, A.B. (1966) *Ukr. fiz. Zh.*, **11**, 594.

Dowsing, R.D. and Gibson, J.F. (1969) *J. Chem. Phys.*, **50**, 294.

Elliott, R.J. and Stevens, K.W.H. (1953) *Proc. R. Soc.*, **A219**, 387.

Freeman, A.J. and Watson, R.E. (1962) *Phys. Rev.*, **127**, 2058.

Fuller, G.H. and Cohen, V.W. (1965) Nuclear Moments: Appendix 1 to Nuclear Data Sheets.

Geru, I.I. (1965) *Ukr. fiz. Zh.*, **10**, 726.

Grant, W.J.C. and Strandberg, M.W.P. (1964) *J. Phys. Chem. Sol.*, **25**, 635.

Grekhov, A.M. and Roitsin, A.B. (1976) *Phys. Status Solidi B*, **74**, 323.

Griffiths, J.H.E., Owen, J., and Ward, I.M. (1953) *Proc. R. Soc. (London)*, **A219**, 526.

Hauser, W. (1963) *Paramagnetic Resonance*, Academic Press, New York, p. 297.

Huang, W.H., Lin, F.C., and Zhu, J.K. (1964) *Proc. Phys. Soc.*, **84**, 661.

Hutchinson, C.A. and Wong, E. (1958) *J. Chem. Phys.*, **29**, 754.

International Tables (1944) *Internationale Tabellen zur Bestimmung von Kristalstrukturen*, vol. 1, Gebrüder Borntrager, Berlin, 1935, Photo-lithoprint reproduction by Edwards Bros., Ann Arbor, MI.

International Tables (1967) *International Tables for X-Ray Crystallography*, vol. 1, Kynoch Press, Birmingham, England, 1959–1965.

Jansen, L. and Boon, M. (1967) *Theory of Finite Groups; Applications to Physics*, North Holland, Amsterdam.

Jones, D.A., Baker, J.M., and Pope, F.D. (1959) *Proc. Phys. Soc.*, **74**, 249.

Judd, B.R. (1955) *Proc. R. Soc.*, **A227**, 552.

Kohin, R.P. (1979) *Magn. Reson. Rev.*, **5**, 75.

König, E. and König, G. (1984) *Electron Paramagnetic Resonance of Coordination and Organometallic Transition Compounds*, Landolt Börnstein, New Series, vol. 12, (eds K.-H. Hellewege and A.M. Hellewege), Springer-Verlag, Berlin.

Koster, G.F. and Statz, H. (1959) *Phys. Rev.*, **113**, 445.

Leushin, A.M. (1964) *Paramagnitnyi Rezonans*, Kazan University, p. 42.

Low, W. (1960) *Paramagnetic Resonance in Solids*, Academic Press, New York.

Misra, S.K. (1976) *J. Magn. Reson.*, **23**, 403.

Misra, S.K. (1983) *Physica*, **121B**, 193.

Misra, S.K. (1984) *Physica*, **124B**, 53.

Misra, S.K. (1986) *Magn. Reson. Rev.*, **10**, 285.

Misra, S.K. (1988a) *Physica*, **B151**, 433.

Misra, S.K. (1988b) *Arab. J. Sci. Eng.*, **13**, 255.

Misra, S.K. (1999) *Handbook of Electron Spin Resonance*, vol. 2 (eds C.P. Poole, Jr and H.A. Farach), A. I. P. Press, New York.

Misra, S.K. and Upreti, G.C. (1986) *Magn. Reson. Rev.*, **10**, 333.

Misra, S.K. and Wang, C. (1990) *Magn. Reson. Rev.*, **14**, 157.

Misra, S.K., Poole, C.P., Jr, and Farach, H.A. (1996) *Appl. Magn. Reson.*, **11**, 29.

Misra, S.K., Chang, Y., and Felsteiner, J. (1997) *J. Phys. Chem. Solids*, **58**, 1–11.

Newman, D.J. (1971) *Adv. Phys.*, **24**, 197.

Newman, D.J. and Urban, W. (1975) *Adv. Phys.*, **24**, 793.

Orton, J.W. (1968) *Electron Paramagnetic Resonance*, Iliffe Books, London.

Painter, G.S. and Poole, C.P., Jr (1968) *J. Phys. Chem. Solids*, **29**, 1754.

Pilbrow, J.R. (1990) *Transition Ion Electron Paramagnetic Resonance*, Clarendon Press, Oxford.

Ray, T. (1964) *Proc. R. Soc.*, **A227**, 76.

Roitsin, A.B. (1962) *Fiz. Tverd. Tela*, **4**, 2948.

Roitsin, A.B. (1963) *Fiz. Tverd. Tela*, **5**, 151.

Rose, M.E. (1957) *Elementary Theory of Angular Momentum*, John Wiley & Sons, Inc., New York.

Rudowicz, C. (1987) *Magn. Reson. Rev.*, **13**, 1.

Rudowicz, C. and Chung, C.Y. (2004) *J. Phys. Condens. Matter*, **16**, 5825.

Ryabov, I.D. (2009) *Appl. Magn. Reson.*, **35**, 481.

Schumacher, P.P. and Hollingsworth, C.A. (1996) *J. Phys. Chem. Solids*, **27**, 749.

Stevens, K.W.H. (1952) *Proc. Phys. Soc.*, **A65**, 209.

Ursu, I. and Lupei, V. (1986) *Mag. Reson. Rev.*, **10**, 253.

Wyckoff, R.W.G. (1964) *Crystal Structures*, John Wiley & Sons, Inc., New York, 2nd edn, vol. 1.

Wyckoff, R.W.G. (1965a) *Crystal Structures*, John Wiley & Sons, Inc., New York, 2nd edn, vol. 2.

Wyckoff, R.W.G. (1965b) *Crystal Structures*, John Wiley & Sons, Inc., New York, 2nd edn, vol. 3.

Appendix 7.1 Spin Operators and Their Matrix Elements

(A) Spin Operators

The operators $\hat{Y}_n^m$ ($|m| \le n$) depending on the components $S_\pm (\equiv S_x \pm iS_y)$ and S_z of the spin angular momentum S are listed below (Misra, 1999; with corrections by Rudowicz and Chung, 2004).

$$\hat{Y}_0^0 = 1$$

$$\hat{Y}_1^0 = S_z$$

$$\hat{Y}_1^{\pm 1} = S_\pm$$

$$\hat{Y}_2^0 = 3S_z^2 - S(S+1)$$

$$\hat{Y}_2^{\pm 1} = \frac{1}{2}(S_z S_\pm + S_\pm S_z)$$

$$\hat{Y}_2^{\pm 2} = S_\pm^2$$

$$\hat{Y}_3^0 = 5S_z^3 - \{3S(S+1) - 1\}S_z$$

$$\hat{Y}_3^{\pm 1} = \frac{1}{2}\left[\left\{5S_z^2 - S(S+1) - \frac{1}{2}\right\}S_\pm + S_\pm\left\{5S_z^2 - S(S+1) - \frac{1}{2}\right\}\right]$$

$$\hat{Y}_3^{\pm 2} = \frac{1}{2}(S_z S_\pm^2 + S_\pm^2 S_z)$$

$$\hat{Y}_3^{\pm 3} = S_\pm^3$$

$$\hat{Y}_4^0 = 35S_z^4 - \{30S(S+1) - 25\}S_z^2 + 3S^2(S+1)^2 - 6S(S+1)$$

$$\hat{Y}_4^{\pm 1} = \frac{1}{2}[\{7S_z^3 - \{3S(S+1) + 1\}S_z\}S_\pm + S_\pm\{7S_z^3 - \{3S(S+1) + 1\}S_z\}]$$

$$\hat{Y}_4^{\pm 2} = \frac{1}{2}[\{7S_z^2 - S(S+1) - 5\}S_\pm^2 + S_\pm^2\{7S_z^2 - S(S+1) - 5\}]$$

$$\hat{Y}_4^{\pm 3} = \frac{1}{2}(S_z S_\pm^3 + S_\pm^3 S_z)$$

$$\hat{Y}_4^{\pm 4} = S_\pm^4$$

$$\hat{Y}_5^0 = 63S_z^5 - \{70S(S+1) - 105\}S_z^3 + \{15S^2(S+1)^2 - 50S(S+1) + 12\}S_z$$

$$\hat{Y}_5^{\pm 1} = \frac{1}{2}\left[\left\{21S_z^4 - 14S(S+1)S_z^2 + S^2(S+1)^2 - S(S+1) + \frac{3}{2}\right\}S_\pm \right.$$
$$\left. + S_\pm\left\{21S_z^4 - 14S(S+1)S_z^2 + S^2(S+1)^2 - S(S+1) + \frac{3}{2}\right\}\right]$$

$$\hat{Y}_5^{\pm 2} = \frac{1}{2}[\{3S_z^3 - \{S(S+1) + 6\}S_z\}S_\pm^2 + S_\pm^2\{3S_z^3 - \{S(S+1) + 6\}S_z\}]$$

$$\hat{Y}_5^{\pm 3} = \frac{1}{2}\left[\left\{9S_z^2 - S(S+1) - \frac{33}{2}\right\}S_\pm^3 + S_\pm^3\left\{9S_z^2 - S(S+1) - \frac{33}{2}\right\}\right]$$

$$\hat{Y}_5^{\pm 4} = \frac{1}{2}(S_z S_\pm^4 + S_\pm^4 S_z)$$

$$\hat{Y}_5^{\pm 5} = S_\pm^5$$

$$\hat{Y}_6^0 = 231S_z^6 - \{315S(S+1) - 735\}S_z^4 + \{105S^2(S+1)^2 - 525S(S+1) + 294\}S_z^2$$
$$- 5S^3(S+1)^3 + 40S^2(S+1)^2 - 60S(S+1)$$

7 Spin Hamiltonians and Site Symmetries for Transition Ions

$$\hat{Y}_6^{\pm 1} = [\{33S_z^5 - \{30S(S+1) - 15\}S_z^3 + \{5S^2(S+1)^2 - 10S(S+1) + 120\}S_z\}S_\pm$$
$$+ S_\pm\{33S_z^5 - \{30S(S+1) - 15\}S_z^3 + \{5S^2(S+1)^2 - 10S(S+1) + 120\}S_z\}]$$

$$\hat{Y}_6^{\pm 2} = \frac{1}{2}[\{33S_z^4 - \{18S(S+1) + 123\}S_z^2 + S^2(S+1)^2 + 10S(S+1) + 102\}S_\pm^2$$
$$+ S_\pm^2\{33S_z^4 - \{18S(S+1) + 123\}S_z^2 + S^2(S+1)^2 + 10S(S+1) + 102\}]$$

$$\hat{Y}_6^{\pm 3} = \frac{1}{2}[\{11S_z^3 - \{3S(S+1) + 59\}S_z\}S_\pm^3 + S_\pm^3\{11S_z^3 - \{3S(S+1) + 59\}S_z\}]$$

$$\hat{Y}_6^{\pm 4} = \frac{1}{2}[\{11S_z^2 - S(S+1) - 38\}S_\pm^4 + S_\pm^4\{11S_z^2 - S(S+1) - 38\}]$$

$$\hat{Y}_6^{\pm 5} = \frac{1}{2}(S_z S_\pm^5 + S_\pm^5 S_z)$$

$$\hat{Y}_6^{\pm 6} = S_\pm^6$$

(B) Matrix Elements

The matrix elements of the operators $\hat{O}_n^m$ ($m \geq 0$), defined as $\hat{O}_n^m = \frac{1}{2}(\hat{Y}_n^{+m} + \hat{Y}_n^{-m})$, appearing in the spin Hamiltonian, are listed below (Abragam and Bleaney, 1970; Al'tshuler and Kozyrev, 1974; Misra, 1999; with corrections noted by Ryabov, 2009).

In order to obtain the unlisted nonzero matrix elements of $\hat{O}_n^m$, the following relations can be used:

$$\langle M|\hat{O}_n^m|M-m\rangle = (-1)^{m+n}\langle -M|\hat{O}_n^m|-M+m\rangle = \langle M-m|\hat{O}_n^m|M\rangle.$$

In order to obtain the matrix elements of $\hat{\Omega}_n^m$, defined as $\hat{\Omega}_n^m = \frac{1}{2i}(\hat{Y}_n^{+m} - \hat{Y}_n^{-m})$, the matrix elements of $\hat{O}_n^m$ can be exploited as follows:

$$\langle M|\hat{\Omega}_n^m|M-m\rangle = (-1)^{m+n+1}\langle -M|\hat{\Omega}_n^m|-M+m\rangle = -\langle M-m|\hat{\Omega}_n^m|M\rangle,$$

In general, $\langle M|\hat{\Omega}_n^m|M\pm m\rangle = \pm i\langle M|\hat{O}_n^m|M\pm m\rangle$.

In all the tables below, the numbers in column F are factors multiplying the matrix elements appearing in the corresponding rows.

O_2^0

S	F	$\langle 0\|0\rangle$	$\langle 1\|1\rangle$	$\langle 2\|2\rangle$	$\langle 3\|3\rangle$	$\langle 4\|4\rangle$	$\langle 5\|5\rangle$	$\langle 6\|6\rangle$	$\langle 7\|7\rangle$	$\langle 8\|8\rangle$
1	1	−2	1							
2	3	−2	−1	2						
3	3	−4	−3	0	5					
4	1	−20	−17	−8	7	28				
5	3	−10	−9	−6	−1	6	15			
6	3	−14	−13	−10	−5	2	11	22		
7	1	−56	−53	−44	−29	−8	19	52	91	
8	3	−24	−23	−20	−15	−8	1	12	25	40

Appendix 7.1 Spin Operators and Their Matrix Elements

S	F	⟨1/2‖1/2⟩	⟨3/2‖3/2⟩	⟨5/2‖5/2⟩	⟨7/2‖7/2⟩	⟨9/2‖9/2⟩	⟨11/2‖11/2⟩	⟨13/2‖13/2⟩	⟨15/2‖15/2⟩
3/2	3	−1	1						
5/2	2	−4	−1	5					
7/2	3	−5	−3	1	7				
9/2	6	−4	−3	−1	2	6			
11/2	1	−35	−29	−17	1	25	55		
13/2	6	−8	−7	−5	−2	2	7	13	
15/2	3	−21	−19	−15	−9	−1	9	21	35

O_2^1

S	F	⟨1‖0⟩	⟨2‖1⟩	⟨3‖2⟩	⟨4‖3⟩	⟨5‖4⟩	⟨6‖5⟩	⟨7‖6⟩	⟨8‖7⟩
1	1/4	$\sqrt{2}$							
2	1/4	$\sqrt{6}$	6						
3	1/4	$2\sqrt{3}$	$3\sqrt{10}$	$5\sqrt{6}$					
4	1/4	$2\sqrt{5}$	$9\sqrt{2}$	$5\sqrt{14}$	$14\sqrt{2}$				
5	1/4	$\sqrt{30}$	$6\sqrt{7}$	$10\sqrt{6}$	$21\sqrt{2}$	$9\sqrt{10}$			
6	1/4	$\sqrt{42}$	$6\sqrt{10}$	30	$7\sqrt{30}$	$9\sqrt{22}$	$22\sqrt{3}$		
7	1/4	$2\sqrt{14}$	$9\sqrt{6}$	$25\sqrt{2}$	$14\sqrt{11}$	54	$11\sqrt{26}$	$13\sqrt{14}$	
8	1/4	$6\sqrt{2}$	$3\sqrt{70}$	$5\sqrt{66}$	$14\sqrt{15}$	$18\sqrt{13}$	$11\sqrt{42}$	$13\sqrt{30}$	60

S	F	⟨1/2‖−1/2⟩	⟨3/2‖1/2⟩	⟨5/2‖3/2⟩	⟨7/2‖5/2⟩	⟨9/2‖7/2⟩	⟨11/2‖9/2⟩	⟨13/2‖11/2⟩	⟨15/2‖13/2⟩
3/2	1/2	0	$\sqrt{3}$						
5/2	1	0	$\sqrt{2}$	$\sqrt{5}$					
7/2	1/2	0	$\sqrt{15}$	$4\sqrt{3}$	$3\sqrt{7}$				
9/2	1	0	$\sqrt{6}$	$\sqrt{21}$	6	6			
11/2	1/2	0	$\sqrt{35}$	$8\sqrt{2}$	$9\sqrt{3}$	$8\sqrt{5}$	$5\sqrt{11}$		
13/2	1	0	$2\sqrt{3}$	$3\sqrt{5}$	$3\sqrt{10}$	$2\sqrt{33}$	$5\sqrt{6}$	$3\sqrt{13}$	
15/2	1/2	0	$3\sqrt{7}$	$4\sqrt{15}$	$3\sqrt{55}$	$16\sqrt{3}$	$5\sqrt{39}$	$12\sqrt{7}$	$7\sqrt{15}$

O_2^2

S	F	⟨1‖−1⟩	⟨2‖0⟩	⟨3‖1⟩	⟨4‖2⟩	⟨5‖3⟩	⟨6‖4⟩	⟨7‖5⟩	⟨8‖6⟩
1	1	1							
2	1	3	$\sqrt{6}$						
3	1	6	$\sqrt{30}$	$\sqrt{15}$					
4	1	10	$3\sqrt{10}$	$3\sqrt{7}$	$2\sqrt{7}$				
5	1	15	$\sqrt{210}$	$2\sqrt{42}$	$6\sqrt{3}$	$3\sqrt{5}$			
6	1	21	$2\sqrt{105}$	$6\sqrt{10}$	$3\sqrt{30}$	$\sqrt{165}$	$\sqrt{66}$		
7	1	28	$6\sqrt{21}$	$15\sqrt{3}$	$5\sqrt{22}$	$6\sqrt{11}$	$3\sqrt{26}$	$\sqrt{91}$	
8	1	36	$6\sqrt{35}$	$\sqrt{1155}$	$3\sqrt{110}$	$2\sqrt{195}$	$\sqrt{546}$	$3\sqrt{35}$	$2\sqrt{30}$

S	F	⟨3/2‖−1/2⟩	⟨5/2‖1/2⟩	⟨7/2‖3/2⟩	⟨9/2‖5/2⟩	⟨11/2‖7/2⟩	⟨13/2‖9/2⟩	⟨15/2‖11/2⟩
3/2	1	$\sqrt{3}$						
5/2	1	$3\sqrt{2}$	$\sqrt{10}$					
7/2	1	$2\sqrt{15}$	$3\sqrt{5}$	$\sqrt{21}$				
9/2	1	$5\sqrt{6}$	$3\sqrt{14}$	$2\sqrt{21}$	6			
11/2	1	$3\sqrt{35}$	$2\sqrt{70}$	$6\sqrt{6}$	$3\sqrt{15}$	$\sqrt{55}$		
13/2	1	$14\sqrt{3}$	$6\sqrt{15}$	$15\sqrt{2}$	$\sqrt{330}$	$3\sqrt{22}$	$\sqrt{78}$	
15/2	1	$12\sqrt{7}$	$3\sqrt{105}$	$5\sqrt{33}$	$2\sqrt{165}$	$6\sqrt{13}$	$\sqrt{273}$	$\sqrt{105}$

O_3^0

S	F	⟨0‖0⟩	⟨1‖1⟩	⟨2‖2⟩	⟨3‖3⟩	⟨4‖4⟩	⟨5‖5⟩	⟨6‖6⟩	⟨7‖7⟩	⟨8‖8⟩
2	6	0	−2	1						
3	30	0	−1	−1	1					
4	6	0	−9	−13	−7	14				
5	6	0	−14	−23	−22	−6	30			
6	30	0	−4	−7	−8	−6	0	11		
7	6	0	−27	−49	−61	−58	−35	13	91	
8	30	0	−7	−13	−17	−18	−15	−7	7	28

Appendix 7.1 Spin Operators and Their Matrix Elements | 369

S	F	⟨1/2‖1/2⟩	⟨3/2‖3/2⟩	⟨5/2‖5/2⟩	⟨7/2‖7/2⟩	⟨9/2‖9/2⟩	⟨11/2‖11/2⟩	⟨13/2‖13/2⟩	⟨15/2‖15/2⟩
3/2	1/2	−9	3						
5/2	3	−4	−7	5					
7/2	15/2	−3	−7	−5	7				
9/2	3	−12	−31	−35	−14	42			
11/2	15/2	−7	−19	−25	−21	−3	33		
13/2	3	−24	−67	−95	−98	−66	11	143	
15/2	3/2	−63	−179	−265	−301	−267	−143	91	455

O_3^1

S	F	⟨1‖0⟩	⟨2‖1⟩	⟨3‖2⟩	⟨4‖3⟩	⟨5‖4⟩	⟨6‖5⟩	⟨7‖6⟩	⟨8‖7⟩
2	2	$\sqrt{6}$	3						
3	10	$-\sqrt{3}$	0	$\sqrt{6}$					
4	6	$-3\sqrt{5}$	$-2\sqrt{2}$	$\sqrt{14}$	$7\sqrt{2}$				
5	2	$-7\sqrt{30}$	$-9\sqrt{7}$	$\sqrt{6}$	$24\sqrt{2}$	$18\sqrt{10}$			
6	10	$-2\sqrt{42}$	$-3\sqrt{10}$	−3	$\sqrt{30}$	$3\sqrt{22}$	$11\sqrt{3}$		
7	6	$-9\sqrt{14}$	$-11\sqrt{6}$	$-10\sqrt{2}$	$\sqrt{11}$	23	$8\sqrt{26}$	$13\sqrt{14}$	
8	10	$-21\sqrt{2}$	$-3\sqrt{70}$	$-2\sqrt{66}$	$\sqrt{15}$	$3\sqrt{13}$	$4\sqrt{42}$	$7\sqrt{30}$	42

S	F	⟨1/2‖−1/2⟩	⟨3/2‖1/2⟩	⟨5/2‖3/2⟩	⟨7/2‖5/2⟩	⟨9/2‖7/2⟩	⟨11/2‖9/2⟩	⟨13/2‖11/2⟩	⟨15/2‖13/2⟩
3/2	1	−3	$\sqrt{3}$						
5/2	3	−4	$-\sqrt{2}$	$2\sqrt{5}$					
7/2	5	−6	$-\sqrt{15}$	$\sqrt{3}$	$3\sqrt{7}$				
9/2	1	−60	$-19\sqrt{6}$	$-2\sqrt{21}$	42	84			
11/2	15	−7	$-\sqrt{35}$	$-2\sqrt{2}$	$\sqrt{3}$	$3\sqrt{5}$	$3\sqrt{11}$		
13/2	1	−168	$-86\sqrt{3}$	$-42\sqrt{5}$	$-3\sqrt{10}$	$16\sqrt{33}$	$77\sqrt{6}$	$66\sqrt{13}$	
15/2	1	−252	$-87\sqrt{7}$	$-43\sqrt{15}$	$-9\sqrt{55}$	$34\sqrt{3}$	$31\sqrt{39}$	$117\sqrt{7}$	$91\sqrt{15}$

O_3^2

S	F	⟨1∥−1⟩	⟨2∥0⟩	⟨3∥1⟩	⟨4∥2⟩	⟨5∥3⟩	⟨6∥4⟩	⟨7∥5⟩	⟨8∥6⟩
2	1	0	$\sqrt{6}$						
3	1	0	$\sqrt{30}$	$2\sqrt{15}$					
4	3	0	$\sqrt{10}$	$2\sqrt{7}$	$2\sqrt{7}$				
5	1	0	$\sqrt{210}$	$4\sqrt{42}$	$18\sqrt{3}$	$12\sqrt{5}$			
6	1	0	$2\sqrt{105}$	$12\sqrt{10}$	$9\sqrt{30}$	$4\sqrt{165}$	$5\sqrt{66}$		
7	3	0	$2\sqrt{21}$	$10\sqrt{3}$	$5\sqrt{22}$	$8\sqrt{11}$	$5\sqrt{26}$	$2\sqrt{91}$	
8	1	0	$6\sqrt{35}$	$2\sqrt{1155}$	$9\sqrt{110}$	$8\sqrt{195}$	$5\sqrt{546}$	$18\sqrt{35}$	$14\sqrt{30}$

S	F	⟨3/2∥−1/2⟩	⟨5/2∥1/2⟩	⟨7/2∥3/2⟩	⟨9/2∥5/2⟩	⟨11/2∥7/2⟩	⟨13/2∥9/2⟩	⟨15/2∥11/2⟩
3/2	1/2	$\sqrt{3}$						
5/2	3/2	$\sqrt{2}$	$\sqrt{10}$					
7/2	1/2	$2\sqrt{15}$	$9\sqrt{5}$	$5\sqrt{21}$				
9/2	1/2	$5\sqrt{6}$	$9\sqrt{14}$	$10\sqrt{21}$	42			
11/2	1/2	$3\sqrt{35}$	$6\sqrt{70}$	$30\sqrt{6}$	$21\sqrt{15}$	$9\sqrt{55}$		
13/2	1/2	$14\sqrt{3}$	$18\sqrt{15}$	$75\sqrt{2}$	$7\sqrt{330}$	$27\sqrt{22}$	$11\sqrt{78}$	
15/2	1/2	$12\sqrt{7}$	$9\sqrt{105}$	$25\sqrt{33}$	$14\sqrt{165}$	$54\sqrt{13}$	$11\sqrt{273}$	$13\sqrt{105}$

O_3^3

S	F	⟨2∥−1⟩	⟨3∥0⟩	⟨4∥1⟩	⟨5∥2⟩	⟨6∥3⟩	⟨7∥4⟩	⟨8∥5⟩
2	6	1						
3	6	$\sqrt{10}$	$\sqrt{5}$					
4	6	$5\sqrt{2}$	$\sqrt{35}$	$\sqrt{14}$				
5	6	$5\sqrt{7}$	$2\sqrt{35}$	$2\sqrt{21}$	$\sqrt{30}$			
6	6	$7\sqrt{10}$	$2\sqrt{105}$	$10\sqrt{3}$	$\sqrt{165}$	$\sqrt{55}$		
7	6	$14\sqrt{6}$	$5\sqrt{42}$	$5\sqrt{33}$	$5\sqrt{22}$	$\sqrt{286}$	$\sqrt{91}$	
8	6	$6\sqrt{70}$	$\sqrt{2310}$	$5\sqrt{77}$	$\sqrt{1430}$	$\sqrt{910}$	$\sqrt{455}$	$2\sqrt{35}$

Appendix 7.1 Spin Operators and Their Matrix Elements

S	F	⟨3/2‖−3/2⟩	⟨5/2‖−1/2⟩	⟨7/2‖1/2⟩	⟨9/2‖3/2⟩	⟨11/2‖5/2⟩	⟨13/2‖7/2⟩	⟨15/2‖9/2⟩
3/2	3	1						
5/2	3	4	$\sqrt{10}$					
7/2	3	10	$4\sqrt{5}$	$\sqrt{35}$				
9/2	3	20	$5\sqrt{14}$	$4\sqrt{14}$	$2\sqrt{21}$			
11/2	3	35	$4\sqrt{70}$	$2\sqrt{210}$	$4\sqrt{30}$	$\sqrt{165}$		
13/2	3	56	$14\sqrt{15}$	$20\sqrt{6}$	$5\sqrt{66}$	$4\sqrt{55}$	$\sqrt{286}$	
15/2	3	84	$8\sqrt{105}$	$5\sqrt{231}$	$20\sqrt{11}$	$2\sqrt{715}$	$4\sqrt{91}$	$\sqrt{455}$

O_4^0

S	F	⟨0‖0⟩	⟨1‖1⟩	⟨2‖2⟩	⟨3‖3⟩	⟨4‖4⟩	⟨5‖5⟩	⟨6‖6⟩	⟨7‖7⟩	⟨8‖8⟩
1	0	0	0							
2	12	6	−4	1						
3	60	6	1	−7	3					
4	60	18	9	−11	−21	14				
5	420	6	4	−1	−6	−6	6			
6	60	84	64	11	−54	−96	−66	99		
7	12	756	621	251	−249	−704	−869	−429	1001	
8	420	36	31	17	−3	−24	−39	−39	−13	52

S	F	⟨1/2‖1/2⟩	⟨3/2‖3/2⟩	⟨5/2‖5/2⟩	⟨7/2‖7/2⟩	⟨9/2‖9/2⟩	⟨11/2‖11/2⟩	⟨13/2‖13/2⟩	⟨15/2‖15/2⟩
1/2	0	0							
3/2	0	0	0						
5/2	60	2	−3	1					
7/2	60	9	−3	−13	7				
9/2	84	18	3	−17	−22	18			
11/2	120	28	12	−13	−33	−27	33		
13/2	60	108	63	−13	−92	−132	−77	143	
15/2	60	189	129	23	−101	−201	−221	−91	273

O_4^1

S	F	⟨1‖0⟩	⟨2‖1⟩	⟨3‖2⟩	⟨4‖3⟩	⟨5‖4⟩	⟨6‖5⟩	⟨7‖6⟩	⟨8‖7⟩
2	3	$-\sqrt{6}$	1						
3	3	$-5\sqrt{3}$	$-4\sqrt{10}$	$5\sqrt{6}$					
4	3	$-9\sqrt{5}$	$-30\sqrt{2}$	$-5\sqrt{14}$	$35\sqrt{2}$				
5	21	$-\sqrt{30}$	$-5\sqrt{7}$	$-5\sqrt{6}$	0	$6\sqrt{10}$			
6	3	$-10\sqrt{42}$	$-53\sqrt{10}$	-195	$-21\sqrt{30}$	$15\sqrt{22}$	$165\sqrt{3}$		
7	3	$-27\sqrt{14}$	$-111\sqrt{6}$	$-250\sqrt{2}$	$-91\sqrt{11}$	-99	$44\sqrt{26}$	$143\sqrt{14}$	
8	21	$-15\sqrt{2}$	$-7\sqrt{70}$	$-10\sqrt{66}$	$-21\sqrt{15}$	$-15\sqrt{13}$	0	$13\sqrt{30}$	130

S	F	⟨1/2‖−1/2⟩	⟨3/2‖1/2⟩	⟨5/2‖3/2⟩	⟨7/2‖5/2⟩	⟨9/2‖7/2⟩	⟨11/2‖9/2⟩	⟨13/2‖11/2⟩	⟨15/2‖13/2⟩
3/2	1/2	0	0						
5/2	3	0	$-5\sqrt{2}$	$2\sqrt{5}$					
7/2	6	0	$-3\sqrt{15}$	$-5\sqrt{3}$	$5\sqrt{7}$				
9/2	21	0	$-3\sqrt{6}$	$-2\sqrt{21}$	-2	12			
11/2	12	0	$-4\sqrt{35}$	$-25\sqrt{2}$	$-15\sqrt{3}$	$3\sqrt{5}$	$15\sqrt{11}$		
13/2	3	0	$-90\sqrt{3}$	$-114\sqrt{5}$	$-79\sqrt{10}$	$-20\sqrt{33}$	$55\sqrt{6}$	$110\sqrt{13}$	
15/2	6	0	$-45\sqrt{7}$	$-53\sqrt{15}$	$-31\sqrt{55}$	$-100\sqrt{3}$	$-5\sqrt{39}$	$65\sqrt{7}$	$91\sqrt{15}$

O_4^2

S	F	⟨1‖−1⟩	⟨2‖0⟩	⟨3‖1⟩	⟨4‖2⟩	⟨5‖3⟩	⟨6‖4⟩	⟨7‖5⟩	⟨8‖6⟩
1	1	0							
2	3	-4	$\sqrt{6}$						
3	3	-20	$-\sqrt{30}$	$6\sqrt{15}$					
4	3	-60	$-11\sqrt{10}$	$10\sqrt{7}$	$30\sqrt{7}$				
5	21	-20	$-\sqrt{210}$	0	$10\sqrt{3}$	$12\sqrt{5}$			
6	3	-280	$-22\sqrt{105}$	$-24\sqrt{10}$	$23\sqrt{30}$	$24\sqrt{165}$	$45\sqrt{66}$		
7	3	-504	$-94\sqrt{21}$	$-130\sqrt{3}$	$15\sqrt{22}$	$116\sqrt{11}$	$121\sqrt{26}$	$66\sqrt{91}$	
8	7	-360	$-54\sqrt{35}$	$-6\sqrt{1155}$	$-3\sqrt{110}$	$12\sqrt{195}$	$15\sqrt{546}$	$78\sqrt{35}$	$78\sqrt{30}$

Appendix 7.1 Spin Operators and Their Matrix Elements

S	F	⟨3/2‖–1/2⟩	⟨5/2‖1/2⟩	⟨7/2‖3/2⟩	⟨9/2‖5/2⟩	⟨11/2‖7/2⟩	⟨13/2‖9/2⟩	⟨15/2‖11/2⟩
5/2	3	$-5\sqrt{2}$	$3\sqrt{10}$					
7/2	6	$-4\sqrt{15}$	$\sqrt{5}$	$5\sqrt{21}$				
9/2	21	$-5\sqrt{6}$	$-\sqrt{14}$	$2\sqrt{21}$	18			
11/2	12	$-8\sqrt{35}$	$-3\sqrt{70}$	$5\sqrt{6}$	$13\sqrt{15}$	$9\sqrt{55}$		
13/2	3	$-210\sqrt{3}$	$-62\sqrt{15}$	$-15\sqrt{2}$	$13\sqrt{330}$	$95\sqrt{22}$	$55\sqrt{78}$	
15/2	6	$-120\sqrt{7}$	$-23\sqrt{105}$	$-15\sqrt{33}$	$8\sqrt{165}$	$80\sqrt{13}$	$25\sqrt{273}$	$39\sqrt{105}$

O_4^3

S	F	⟨2‖–1⟩	⟨3‖0⟩	⟨4‖1⟩	⟨5‖2⟩	⟨6‖3⟩	⟨7‖4⟩	⟨8‖5⟩
2	3	1						
3	3	$\sqrt{10}$	$3\sqrt{5}$					
4	3	$5\sqrt{2}$	$3\sqrt{35}$	$5\sqrt{14}$				
5	3	$5\sqrt{7}$	$6\sqrt{35}$	$10\sqrt{21}$	$7\sqrt{30}$			
6	3	$7\sqrt{10}$	$6\sqrt{105}$	$50\sqrt{3}$	$7\sqrt{165}$	$9\sqrt{55}$		
7	3	$14\sqrt{6}$	$15\sqrt{42}$	$25\sqrt{33}$	$35\sqrt{22}$	$9\sqrt{286}$	$11\sqrt{91}$	
8	3	$6\sqrt{70}$	$3\sqrt{2310}$	$25\sqrt{77}$	$7\sqrt{1430}$	$9\sqrt{910}$	$11\sqrt{455}$	$26\sqrt{35}$

S	F	⟨5/2‖–1/2⟩	⟨7/2‖1/2⟩	⟨9/2‖3/2⟩	⟨11/2‖5/2⟩	⟨13/2‖7/2⟩	⟨15/2‖9/2⟩
5/2	3	$\sqrt{10}$					
7/2	3	$4\sqrt{5}$	$2\sqrt{35}$				
9/2	3	$5\sqrt{14}$	$8\sqrt{14}$	$6\sqrt{21}$			
11/2	$12\sqrt{5}$	$\sqrt{14}$	$\sqrt{42}$	$3\sqrt{6}$	$\sqrt{33}$		
13/2	3	$14\sqrt{15}$	$40\sqrt{6}$	$15\sqrt{66}$	$16\sqrt{55}$	$5\sqrt{286}$	
15/2	6	$4\sqrt{105}$	$5\sqrt{231}$	$30\sqrt{11}$	$4\sqrt{715}$	$10\sqrt{91}$	$3\sqrt{455}$

O_4^4

S	F	⟨2‖−2⟩	⟨3‖−1⟩	⟨4‖0⟩	⟨5‖1⟩	⟨6‖2⟩	⟨7‖3⟩	⟨8‖4⟩
2	12	1						
3	12	5	$\sqrt{15}$					
4	12	15	$5\sqrt{7}$	$\sqrt{70}$				
5	12	35	$5\sqrt{42}$	$3\sqrt{70}$	$\sqrt{210}$			
6	12	70	$21\sqrt{10}$	$15\sqrt{14}$	$5\sqrt{66}$	$3\sqrt{55}$		
7	12	126	$70\sqrt{3}$	$5\sqrt{462}$	$15\sqrt{33}$	$5\sqrt{143}$	$\sqrt{1001}$	
8	12	210	$6\sqrt{1155}$	$15\sqrt{154}$	$5\sqrt{1001}$	$\sqrt{15015}$	$5\sqrt{273}$	$2\sqrt{455}$

S	F	⟨5/2‖−3/2⟩	⟨7/2‖−1/2⟩	⟨9/2‖1/2⟩	⟨11/2‖3/2⟩	⟨13/2‖5/2⟩	⟨15/2‖7/2⟩
3/2	12						
5/2	12	$\sqrt{5}$					
7/2	12	$5\sqrt{3}$	$\sqrt{35}$				
9/2	$12\sqrt{7}$	$5\sqrt{3}$	$5\sqrt{2}$	$3\sqrt{2}$			
11/2	$12\sqrt{2}$	35	$3\sqrt{105}$	$5\sqrt{21}$	$\sqrt{165}$		
13/2	12	$42\sqrt{5}$	$35\sqrt{6}$	$15\sqrt{22}$	$15\sqrt{11}$	$\sqrt{5\cdot 143}$	
15/2	12	$42\sqrt{15}$	$10\sqrt{231}$	$15\sqrt{77}$	$5\sqrt{429}$	$\sqrt{5005}$	$\sqrt{1365}$

O_5^0

S	F	⟨0‖0⟩	⟨1‖1⟩	⟨2‖2⟩	⟨3‖3⟩	⟨4‖4⟩	⟨5‖5⟩	⟨6‖6⟩	⟨7‖7⟩	⟨8‖8⟩
3	180	0	5	−4	1					
4	420	0	9	4	−11	4				
5	2520	0	4	4	−1	−6	3			
6	1080	0	20	26	11	−18	−33	22		
7	60	0	675	1000	751	−44	−979	−1144	1001	
8	1260	0	55	88	83	36	−39	−104	−91	104

S	F		⟨1/2‖1/2⟩	⟨3/2‖3/2⟩	⟨5/2‖5/2⟩	⟨7/2‖7/2⟩	⟨9/2‖9/2⟩	⟨11/2‖11/2⟩	⟨13/2‖13/2⟩	⟨15/2‖15/2⟩
5/2	30	10	−5	1						
7/2	90	15	17	−23	7					
9/2	630	6	11	1	−14	6				
11/2	420	20	44	29	−21	−57	33			
13/2	270	60	145	139	28	−132	−187	143		
15/2	630	45	115	131	77	−33	−143	−143	143	

Appendix 7.1 Spin Operators and Their Matrix Elements

O_5^1

S	F	⟨1‖0⟩	⟨2‖1⟩	⟨3‖2⟩	⟨4‖3⟩	⟨5‖4⟩	⟨6‖5⟩	⟨7‖6⟩	⟨8‖7⟩
3	6	$10\sqrt{3}$	$-9\sqrt{10}$	$5\sqrt{6}$					
4	42	$6\sqrt{5}$	$-5\sqrt{2}$	$-5\sqrt{14}$	$10\sqrt{2}$				
5	84	$4\sqrt{30}$	0	$-10\sqrt{6}$	$-15\sqrt{2}$	$9\sqrt{10}$			
6	36	$20\sqrt{42}$	$12\sqrt{10}$	-90	$-29\sqrt{30}$	$-15\sqrt{22}$	$110\sqrt{3}$		
7	30	$90\sqrt{14}$	$65\sqrt{6}$	$-83\sqrt{2}$	$-106\sqrt{11}$	-374	$-11\sqrt{26}$	$143\sqrt{14}$	
8	42	$330\sqrt{2}$	$33\sqrt{70}$	$-5\sqrt{66}$	$-94\sqrt{15}$	$-150\sqrt{13}$	$-65\sqrt{42}$	$13\sqrt{30}$	780

S	F	⟨1/2‖−1/2⟩	⟨3/2‖1/2⟩	⟨5/2‖3/2⟩	⟨7/2‖5/2⟩	⟨9/2‖7/2⟩	⟨11/2‖9/2⟩	⟨13/2‖11/2⟩	⟨15/2‖13/2⟩
5/2	6	10	$-5\sqrt{2}$	$\sqrt{5}$					
7/2	6	60	$\sqrt{15}$	$-40\sqrt{3}$	$15\sqrt{7}$				
9/2	210	6	$\sqrt{6}$	$-\sqrt{21}$	-6	6			
11/2	84	40	$4\sqrt{35}$	$-10\sqrt{2}$	$-25\sqrt{3}$	$-12\sqrt{5}$	$15\sqrt{11}$		
13/2	18	420	$170\sqrt{3}$	$-9\sqrt{5}$	$-111\sqrt{10}$	$-80\sqrt{33}$	$-55\sqrt{6}$	$165\sqrt{13}$	
15/2	42	360	$105\sqrt{7}$	$16\sqrt{15}$	$-27\sqrt{55}$	$-220\sqrt{3}$	$-55\sqrt{39}$	0	$143\sqrt{15}$

O_5^2

S	F	⟨1‖−1⟩	⟨2‖0⟩	⟨3‖1⟩	⟨4‖2⟩	⟨5‖3⟩	⟨6‖4⟩	⟨7‖5⟩	⟨8‖6⟩
3	6	0	$-\sqrt{30}$	$\sqrt{15}$					
4	6	0	$-7\sqrt{10}$	$-5\sqrt{7}$	$10\sqrt{7}$				
5	6	0	$-4\sqrt{210}$	$-10\sqrt{42}$	0	$42\sqrt{5}$			
6	12	0	$-6\sqrt{105}$	$-27\sqrt{10}$	$-9\sqrt{30}$	$3\sqrt{165}$	$15\sqrt{66}$		
7	30	0	$-10\sqrt{21}$	$-41\sqrt{3}$	$-13\sqrt{22}$	$-4\sqrt{11}$	$11\sqrt{26}$	$11\sqrt{91}$	
8	6	0	$-66\sqrt{35}$	$-19\sqrt{1155}$	$-63\sqrt{110}$	$-28\sqrt{195}$	$5\sqrt{546}$	$117\sqrt{35}$	$182\sqrt{30}$

S	F	⟨3/2‖−1/2⟩	⟨5/2‖1/2⟩	⟨7/2‖3/2⟩	⟨9/2‖5/2⟩	⟨11/2‖7/2⟩	⟨13/2‖9/2⟩	⟨15/2‖11/2⟩
5/2	1/2	$-15\sqrt{2}$	$3\sqrt{10}$					
7/2	3	$-4\sqrt{15}$	$-9\sqrt{5}$	$5\sqrt{21}$				
9/2	15/2	$-7\sqrt{6}$	$-9\sqrt{14}$	$-2\sqrt{21}$	42			
11/2	6	$-8\sqrt{35}$	$-13\sqrt{70}$	$-35\sqrt{6}$	$7\sqrt{15}$	$21\sqrt{55}$		
13/2	9/2	$-70\sqrt{3}$	$-78\sqrt{15}$	$-225\sqrt{2}$	$-7\sqrt{330}$	$45\sqrt{22}$	$55\sqrt{78}$	
15/2	3	$-120\sqrt{7}$	$-81\sqrt{105}$	$-175\sqrt{33}$	$-56\sqrt{165}$	0	$55\sqrt{273}$	$143\sqrt{105}$

O_5^3

S	F	⟨2‖−1⟩	⟨3‖0⟩	⟨4‖1⟩	⟨5‖2⟩	⟨6‖3⟩	⟨7‖4⟩	⟨8‖5⟩
3	36	$-\sqrt{10}$	$2\sqrt{5}$					
4	12	$-35\sqrt{2}$	$2\sqrt{35}$	$20\sqrt{14}$				
5	72	$-10\sqrt{7}$	$-\sqrt{35}$	$5\sqrt{21}$	$7\sqrt{30}$			
6	216	$-7\sqrt{10}$	$-\sqrt{105}$	$5\sqrt{3}$	$2\sqrt{165}$	$4\sqrt{55}$		
7	60	$-70\sqrt{6}$	$-16\sqrt{42}$	$2\sqrt{33}$	$29\sqrt{22}$	$13\sqrt{286}$	$22\sqrt{91}$	
8	36	$-66\sqrt{70}$	$-8\sqrt{2310}$	$-10\sqrt{77}$	$7\sqrt{1430}$	$19\sqrt{910}$	$34\sqrt{455}$	$104\sqrt{35}$

S	F	⟨3/2‖−3/2⟩	⟨5/2‖−1/2⟩	⟨7/2‖1/2⟩	⟨9/2‖3/2⟩	⟨11/2‖5/2⟩	⟨13/2‖7/2⟩	⟨15/2‖9/2⟩
5/2	12	−5	$\sqrt{10}$					
7/2	36	−10	$-\sqrt{5}$	$2\sqrt{35}$				
9/2	180	−7	$-\sqrt{14}$	$\sqrt{14}$	$2\sqrt{21}$			
11/2	12	−280	$-23\sqrt{70}$	$2\sqrt{210}$	$49\sqrt{30}$	$28\sqrt{165}$		
13/2	108	−70	$-14\sqrt{15}$	$-5\sqrt{6}$	$5\sqrt{66}$	$11\sqrt{55}$	$5\sqrt{286}$	
15/2	36	−420	$-34\sqrt{105}$	$-10\sqrt{231}$	$35\sqrt{11}$	$14\sqrt{715}$	$55\sqrt{91}$	$22\sqrt{455}$

O_5^4

S	F	⟨2‖−2⟩	⟨3‖−1⟩	⟨4‖0⟩	⟨5‖1⟩	⟨6‖2⟩	⟨7‖3⟩	⟨8‖4⟩
3	12	0	$\sqrt{15}$					
4	12	0	$5\sqrt{7}$	$2\sqrt{70}$				
5	12	0	$5\sqrt{42}$	$6\sqrt{70}$	$3\sqrt{210}$			
6	36	0	$7\sqrt{10}$	$10\sqrt{14}$	$5\sqrt{66}$	$4\sqrt{55}$		
7	60	0	$14\sqrt{3}$	$2\sqrt{462}$	$9\sqrt{33}$	$4\sqrt{143}$	$\sqrt{1001}$	
8	12	0	$6\sqrt{1155}$	$30\sqrt{154}$	$15\sqrt{1001}$	$4\sqrt{15015}$	$25\sqrt{273}$	$12\sqrt{455}$

S	F	⟨5/2‖−3/2⟩	⟨7/2‖−1/2⟩	⟨9/2‖1/2⟩	⟨11/2‖3/2⟩	⟨13/2‖5/2⟩	⟨15/2‖7/2⟩
5/2	6	$\sqrt{5}$					
7/2	6	$5\sqrt{3}$	$3\sqrt{35}$				
9/2	30	$\sqrt{21}$	$3\sqrt{14}$	$3\sqrt{14}$			
11/2	6	$35\sqrt{2}$	$9\sqrt{210}$	$25\sqrt{42}$	$7\sqrt{330}$		
13/2	18	$14\sqrt{5}$	$35\sqrt{6}$	$25\sqrt{22}$	$35\sqrt{11}$	$3\sqrt{715}$	
15/2	6	$42\sqrt{15}$	$30\sqrt{231}$	$75\sqrt{77}$	$35\sqrt{429}$	$9\sqrt{5005}$	$11\sqrt{1365}$

O_5^5

S	F	⟨3‖−2⟩	⟨4‖−1⟩	⟨5‖0⟩	⟨6‖1⟩	⟨7‖2⟩	⟨8‖3⟩
3	60	$\sqrt{6}$					
4	60	$3\sqrt{14}$	$2\sqrt{14}$				
5	120	$7\sqrt{6}$	$3\sqrt{21}$	$3\sqrt{7}$			
6	120	42	$21\sqrt{3}$	$3\sqrt{77}$	$3\sqrt{22}$		
7	60	$126\sqrt{2}$	$28\sqrt{33}$	$6\sqrt{462}$	$3\sqrt{858}$	$\sqrt{2002}$	
8	60	$42\sqrt{66}$	$36\sqrt{77}$	$6\sqrt{2002}$	$7\sqrt{858}$	$3\sqrt{2002}$	$4\sqrt{273}$

S	F	⟨5/2‖−5/2⟩	⟨7/2‖−3/2⟩	⟨9/2‖−1/2⟩	⟨11/2‖1/2⟩	⟨13/2‖3/2⟩	⟨15/2‖5/2⟩
5/2	60	1					
7/2	60	6	$\sqrt{21}$				
9/2	60	21	$4\sqrt{21}$	$3\sqrt{14}$			
11/2	60	56	$21\sqrt{6}$	$6\sqrt{42}$	$\sqrt{462}$		
13/2	180	42	$28\sqrt{2}$	$7\sqrt{22}$	$4\sqrt{33}$	$\sqrt{143}$	
15/2	60	252	$42\sqrt{33}$	$24\sqrt{77}$	$3\sqrt{3003}$	$2\sqrt{3003}$	$\sqrt{3003}$

O_6^0

S	F	⟨0‖0⟩	⟨1‖1⟩	⟨2‖2⟩	⟨3‖3⟩	⟨4‖4⟩	⟨5‖5⟩	⟨6‖6⟩	⟨7‖7⟩	⟨8‖8⟩
1	0	0	0							
2	0	0	0	0						
3	180	−20	15	−6	1					
4	1260	−20	1	22	−17	4				
5	2520	−40	−12	36	29	−48	15			
6	7560	−40	−20	22	43	8	−55	22		
7	3780	−200	−125	50	197	176	−55	−286	143	
8	13860	−120	−85	2	93	128	65	−78	−169	104

S	F	⟨1/2‖1/2⟩	⟨3/2‖3/2⟩	⟨5/2‖5/2⟩	⟨7/2‖7/2⟩	⟨9/2‖9/2⟩	⟨11/2‖11/2⟩	⟨13/2‖13/2⟩	⟨15/2‖15/2⟩
1/2	0	0							
3/2	0	0	0						
5/2	0	0	0	0					
7/2	1260	−5	9	−5	1				
9/2	5040	−8	6	10	−11	3			
11/2	7560	−20	4	25	11	−31	11		
13/2	2160	−200	−25	185	227	−11	−319	143	
15/2	13860	−75	−25	45	87	59	−39	−117	65

O_6^1

S	F	⟨1‖0⟩	⟨2‖1⟩	⟨3‖2⟩	⟨4‖3⟩	⟨5‖4⟩	⟨6‖5⟩	⟨7‖6⟩	⟨8‖7⟩
3	15	$10\sqrt{3}$	$-3\sqrt{10}$	$\sqrt{6}$					
4	45	$14\sqrt{5}$	$21\sqrt{2}$	$-13\sqrt{14}$	$14\sqrt{2}$				
5	30	$28\sqrt{30}$	$96\sqrt{7}$	$-14\sqrt{6}$	$-231\sqrt{2}$	$63\sqrt{10}$			
6	90	$20\sqrt{42}$	$84\sqrt{10}$	126	$-35\sqrt{30}$	$-63\sqrt{22}$	$154\sqrt{3}$		
7	45	$150\sqrt{14}$	$525\sqrt{6}$	$735\sqrt{2}$	$-42\sqrt{11}$	−1386	$-231\sqrt{26}$	$429\sqrt{14}$	
8	165	$210\sqrt{2}$	$87\sqrt{70}$	$91\sqrt{66}$	$70\sqrt{15}$	$-126\sqrt{13}$	$-143\sqrt{42}$	$-91\sqrt{30}$	1092

Appendix 7.1 Spin Operators and Their Matrix Elements

S	F	⟨1/2‖−1/2⟩	⟨3/2‖1/2⟩	⟨5/2‖3/2⟩	⟨7/2‖5/2⟩	⟨9/2‖7/2⟩	⟨11/2‖9/2⟩	⟨13/2‖11/2⟩	⟨15/2‖13/2⟩
7/2	30	0	$7\sqrt{15}$	$-14\sqrt{3}$	$3\sqrt{7}$				
9/2	120	0	$14\sqrt{6}$	$2\sqrt{21}$	-42	21			
11/2	540	0	$4\sqrt{35}$	$14\sqrt{2}$	$-7\sqrt{3}$	$-14\sqrt{5}$	$7\sqrt{11}$		
13/2	360	0	$50\sqrt{3}$	$45\sqrt{5}$	$6\sqrt{10}$	$-17\sqrt{33}$	$-44\sqrt{6}$	$33\sqrt{13}$	
15/2	330	0	$75\sqrt{7}$	$70\sqrt{15}$	$21\sqrt{55}$	$-56\sqrt{3}$	$-49\sqrt{39}$	$-78\sqrt{7}$	$91\sqrt{15}$

O_6^2

S	F	⟨1‖−1⟩	⟨2‖0⟩	⟨3‖1⟩	⟨4‖2⟩	⟨5‖3⟩	⟨6‖4⟩	⟨7‖5⟩	⟨8‖6⟩
3	24	15	$-2\sqrt{30}$	$\sqrt{15}$					
4	30	84	0	$-36\sqrt{7}$	$24\sqrt{7}$				
5	120	84	$2\sqrt{210}$	$-11\sqrt{42}$	$-42\sqrt{3}$	$42\sqrt{5}$			
6	72	420	$22\sqrt{105}$	$-63\sqrt{10}$	$-84\sqrt{30}$	$-14\sqrt{165}$	$70\sqrt{66}$		
7	1080	70	$10\sqrt{21}$	$-7\sqrt{3}$	$-14\sqrt{22}$	$-21\sqrt{11}$	0	$11\sqrt{91}$	
8	264	630	$78\sqrt{35}$	$\sqrt{1155}$	$-42\sqrt{110}$	$-49\sqrt{195}$	$-20\sqrt{546}$	$39\sqrt{35}$	$182\sqrt{30}$

S	F	⟨3/2‖−1/2⟩	⟨5/2‖1/2⟩	⟨7/2‖3/2⟩	⟨9/2‖5/2⟩	⟨11/2‖7/2⟩	⟨13/2‖9/2⟩	⟨15/2‖11/2⟩
7/2	24	$7\sqrt{15}$	$-21\sqrt{5}$	$5\sqrt{21}$				
9/2	240	$7\sqrt{6}$	$-3\sqrt{14}$	$-5\sqrt{21}$	21			
11/2	216	$12\sqrt{35}$	$-\sqrt{70}$	$-35\sqrt{6}$	$-14\sqrt{15}$	$14\sqrt{55}$		
13/2	720	$35\sqrt{3}$	$3\sqrt{15}$	$-36\sqrt{2}$	$-4\sqrt{330}$	$-3\sqrt{22}$	$11\sqrt{78}$	
15/2	1320	$30\sqrt{7}$	$3\sqrt{105}$	$-7\sqrt{33}$	$-7\sqrt{165}$	$-21\sqrt{13}$	$\sqrt{273}$	$13\sqrt{105}$

O_6^3

S	F	⟨2‖−1⟩	⟨3‖0⟩	⟨4‖1⟩	⟨5‖2⟩	⟨6‖3⟩	⟨7‖4⟩	⟨8‖5⟩
2	0	0						
3	18	$-3\sqrt{10}$	$2\sqrt{5}$					
4	90	$-7\sqrt{2}$	$-2\sqrt{35}$	$4\sqrt{14}$				
5	180	$-6\sqrt{7}$	$-5\sqrt{35}$	$-\sqrt{21}$	$7\sqrt{30}$			
6	36	$-63\sqrt{10}$	$-43\sqrt{105}$	$-175\sqrt{3}$	$14\sqrt{165}$	$84\sqrt{55}$		
7	90	$-70\sqrt{6}$	$-64\sqrt{42}$	$-70\sqrt{33}$	$-21\sqrt{22}$	$21\sqrt{286}$	$66\sqrt{91}$	
8	198	$-18\sqrt{70}$	$-8\sqrt{2310}$	$-50\sqrt{77}$	$-7\sqrt{1430}$	$3\sqrt{910}$	$22\sqrt{455}$	$104\sqrt{35}$

S	F	⟨5/2‖−1/2⟩	⟨7/2‖1/2⟩	⟨9/2‖3/2⟩	⟨11/2‖5/2⟩	⟨13/2‖7/2⟩	⟨15/2‖9/2⟩
5/2	0	0					
7/2	36	$-7\sqrt{5}$	$2\sqrt{35}$				
9/2	360	$-2\sqrt{14}$	$-\sqrt{14}$	$2\sqrt{21}$			
11/2	$36\sqrt{5}$	$-27\sqrt{14}$	$-16\sqrt{42}$	$7\sqrt{6}$	$28\sqrt{33}$		
13/2	360	$-14\sqrt{15}$	$-29\sqrt{6}$	$-4\sqrt{66}$	$6\sqrt{55}$	$6\sqrt{286}$	
15/2	1980	$-2\sqrt{105}$	$-2\sqrt{231}$	$-7\sqrt{11}$	0	$3\sqrt{91}$	$2\sqrt{455}$

O_6^4

S	F	⟨2‖−2⟩	⟨3‖−1⟩	⟨4‖0⟩	⟨5‖1⟩	⟨6‖2⟩	⟨7‖3⟩	⟨8‖4⟩
3	60	-6	$\sqrt{15}$					
4	180	-14	$-\sqrt{7}$	$2\sqrt{70}$				
5	60	-168	$-13\sqrt{42}$	$12\sqrt{70}$	$15\sqrt{210}$			
6	180	-168	$-35\sqrt{10}$	$8\sqrt{14}$	$21\sqrt{66}$	$28\sqrt{55}$		
7	60	-1260	$-546\sqrt{3}$	$-6\sqrt{462}$	$147\sqrt{33}$	$126\sqrt{143}$	$45\sqrt{1001}$	
8	660	-252	$-6\sqrt{1155}$	$-6\sqrt{154}$	$3\sqrt{1001}$	$2\sqrt{15015}$	$19\sqrt{273}$	$12\sqrt{455}$

S	F	⟨5/2‖−3/2⟩	⟨7/2‖−1/2⟩	⟨9/2‖1/2⟩	⟨11/2‖3/2⟩	⟨13/2‖5/2⟩	⟨15/2‖7/2⟩
5/2	60						
7/2	60	$-7\sqrt{3}$	$3\sqrt{35}$				
9/2	$60\sqrt{7}$	$-16\sqrt{3}$	$6\sqrt{2}$	$30\sqrt{2}$			
11/2	$60\sqrt{18}$	-63	$-\sqrt{105}$	$13\sqrt{21}$	$7\sqrt{165}$		
13/2	360	$-56\sqrt{5}$	$-21\sqrt{6}$	$13\sqrt{22}$	$46\sqrt{11}$	$6\sqrt{715}$	
15/2	660	$-42\sqrt{15}$	$-6\sqrt{231}$	$3\sqrt{77}$	$7\sqrt{429}$	$3\sqrt{5005}$	$5\sqrt{1365}$

O_6^5

S	F	⟨3‖−2⟩	⟨4‖−1⟩	⟨5‖0⟩	⟨6‖1⟩	⟨7‖2⟩	⟨8‖3⟩
3	30	$\sqrt{6}$					
4	90	$\sqrt{14}$	$2\sqrt{14}$				
5	60	$7\sqrt{6}$	$9\sqrt{21}$	$15\sqrt{7}$			
6	180	14	$21\sqrt{3}$	$5\sqrt{77}$	$7\sqrt{22}$		
7	90	$42\sqrt{2}$	$28\sqrt{33}$	$10\sqrt{462}$	$7\sqrt{858}$	$3\sqrt{2002}$	
8	30	$42\sqrt{66}$	$108\sqrt{77}$	$30\sqrt{2002}$	$49\sqrt{858}$	$27\sqrt{2002}$	$44\sqrt{273}$

S	F	⟨5/2‖−5/2⟩	⟨7/2‖−3/2⟩	⟨9/2‖−1/2⟩	⟨11/2‖1/2⟩	⟨13/2‖3/2⟩	⟨15/2‖5/2⟩
5/2	1	0					
7/2	60	0	$\sqrt{21}$				
9/2	120	0	$2\sqrt{21}$	$3\sqrt{14}$			
11/2	180	0	$7\sqrt{6}$	$4\sqrt{42}$	$\sqrt{462}$		
13/2	360	0	$14\sqrt{2}$	$7\sqrt{22}$	$6\sqrt{33}$	$2\sqrt{143}$	
15/2	60	0	$42\sqrt{33}$	$48\sqrt{77}$	$9\sqrt{3003}$	$8\sqrt{3003}$	$5\sqrt{3003}$

O_6^6

S	F	⟨3‖−3⟩	⟨4‖−2⟩	⟨5‖−1⟩	⟨6‖0⟩	⟨7‖1⟩	⟨8‖2⟩
3	360	1					
4	360	7	$2\sqrt{7}$				
5	360	28	$14\sqrt{3}$	$\sqrt{210}$			
6	360	84	$14\sqrt{30}$	$7\sqrt{66}$	$2\sqrt{231}$		
7	360	210	$42\sqrt{22}$	$28\sqrt{33}$	$2\sqrt{3003}$	$\sqrt{3003}$	
8	360	462	$42\sqrt{110}$	$12\sqrt{1001}$	$14\sqrt{429}$	$7\sqrt{715}$	$2\sqrt{2002}$

S	F	⟨7/2‖−5/2⟩	⟨9/2‖−3/2⟩	⟨11/2‖−1/2⟩	⟨13/2‖1/2⟩	⟨15/2‖3/2⟩
7/2	360	$\sqrt{7}$				
9/2	360	14	$2\sqrt{21}$			
11/2	360	$28\sqrt{3}$	$7\sqrt{30}$	$\sqrt{462}$		
13/2	720	$21\sqrt{10}$	$7\sqrt{66}$	$7\sqrt{33}$	$\sqrt{429}$	
15/2	$360\sqrt{11}$	$42\sqrt{5}$	84	$4\sqrt{273}$	$7\sqrt{39}$	$\sqrt{455}$

Appendix 7.II Descent of Symmetry

a) Descent of symmetry of cubic Hamiltonian along fourfold axis

$$H_D^{cub} \begin{array}{c} \Rightarrow H_C^{cub} \\ \Rightarrow H_D^{tetr} \end{array} \Rightarrow H_D^{orth} \Rightarrow H_C^{mono} \Rightarrow H_C^{tric}$$

$$H_D^{cub} \Rightarrow H_D^{tetr} \begin{array}{c} \Rightarrow H_C^{tetr} \\ \Rightarrow H_D^{orth} \end{array} \Rightarrow H_C^{mono} \Rightarrow H_C^{tric}$$

b) Descent of symmetry of cubic Hamiltonian along threefold axis

$$H_D^{cub} \genfrac{}{}{0pt}{}{\Rightarrow H_C^{cub}}{\Rightarrow H_D^{trig}} \Rightarrow H_C^{trig} \Rightarrow H_C^{tric}$$

c) Descent of symmetry of hexagonal Hamiltonian along threefold axis

$$H_D^{hex} \Rightarrow H_D^{trig} \genfrac{}{}{0pt}{}{\Rightarrow H_C^{trig}}{\Rightarrow H_C^{mono}} \Rightarrow H_C^{tric}$$

$$H_D^{hex} \Rightarrow H_C^{hex} \genfrac{}{}{0pt}{}{\Rightarrow H_C^{trig}}{\Rightarrow H_D^{mono}} \Rightarrow H_C^{tric}$$

d) Descent of symmetry of hexagonal Hamiltonian along twofold axis

$$H_D^{hex} \Rightarrow H_D^{orth} \Rightarrow H_C^{mono} \Rightarrow H_C^{tric}$$

Appendix 7.III Site Symmetries of Host Crystals

The following table gives the characteristics, site symmetries and spin-Hamiltonian types of transition-metal ions in commonly studied host lattices. The table contains representatives from a wide range of space group types, site symmetries, and spin Hamiltonians. Of particular interest is the nearest-neighbor coordination complex given in column 5, and in most cases this is tetrahedral, MX_4, or octahedral, MX_6. Ordinarily, the symmetry of this complex is lower than cubic, but the deviation from cubic can be small, and in many cases the dominant terms in the spin Hamiltonian are cubic. The transition ion site is noted in column 6, and column 7 gives the site symmetry. Column 8 gives the Hamiltonian, and column 9 gives the Wyckoff reference (volume and page) to the structure description. (R.W.G. Wyckoff, *Crystal Structures*, Wiley, New York, Second Edition, Vol.1, 1964; Vol. 2, 1965; Vol. 3, 1965.) Both International and Schönflies group notations are listed.

Appendix 7.III Site Symmetries of Host Crystals

Host lattice	Formula	Crystal system	Space group	Complex	Crystal site	Point symmetry	Hamiltonian	Wyckoff vol-page	Comment
α-Alum	$KAl(SO_4)_2 \cdot 12H_2O$	Cubic	T_h^6 (Pa3)	$Al(H_2O)_6$ $K(H_2O)_6$	4a 4b	C_{3i} ($\bar{3}$) C_{3i} ($\bar{3}$)	$\mathcal{H}_C^{trig}$ $\mathcal{H}_C^{trig}$	3-872	
β-Alum	$CsCr(SO_4)_2 \cdot 12H_2O$	Cubic	T_h^6 (Pa3)				$\mathcal{H}_C^{trig}$	3-873	Cs coordination is 12
Calcite	$CaCO_3$	Trigonal	D_{3d}^6 (R3c)	Ca	2b	C_{3i} ($\bar{3}$)	$\mathcal{H}_C^{trig}$	2-359	
Cesium chloride	CsCl	Cubic	O_h^1 (Pm3m)	$CsCl_8$	1a	O_h (m3m)	$\mathcal{H}_D^{cube}$	1-103	Each atom at center of other atom group
Corundum	Al_2O_3	Trigonal (≈ hcp)	D_{3d}^6 (R3c)	AlO_6	4c	C_3 (3)	$\mathcal{H}_C^{trig}$	2-6	Slightly-distorted hcp oxygens
Fluorite	CaF_2	Cubic	O_h^5 (Fm3m)	CaF_8	4a	O_h (m3m)	$\mathcal{H}_D^{cube}$	1-239	
Lanthanum chloride	$LaCl_3$	Hexagonal	C_{6h}^2 (P6$_3$/m)	$LaCl_9$	2d	C_{3h} ($\bar{6}$)	$\mathcal{H}_C^{hexa}$	2-77	Two different La-Cl distances
Nickel arsenide	NiAs	Hexagonal	C_{6v}^4 (C6mc)	$NiAs_6Ni_2$ $AsNi_6$	2a 2b	C_{3v} (3m) C_{3v} (3m)	$\mathcal{H}_D^{trig}$ $\mathcal{H}_D^{trig}$	1-122	
Periclase	MgO	Cubic (fcc)	O_h^5 (Fm3m)	MgO_4	4a	O_h (m3m)	$\mathcal{H}_D^{cube}$	1-85	
Perovskite (cubic)	$BaTiO_3$	Cubic	O_h^1 (Pm3m)	BaO_{12} TiO_6	1a 1b	O_h (m3m) O_h (m3m)	$\mathcal{H}_D^{cube}$ $\mathcal{H}_D^{cube}$	1-391	O + Ba together form fcc lattice
Perovskite (orthorhombic)	$BaTiO_3$	Orthorhombic	C_{2v}^{14} (Amm2)	BaO_{12} TiO_6	2a 2b	C_{2v} (mm) C_{2v} (mm)	$\mathcal{H}_D^{orth}$ $\mathcal{H}_D^{orth}$	2-405	Distortion is of tetragonal form
Rutile	TiO_2	Tetragonal	D_{2h}^{14} (P4/mmm)	TiO_6	2a	D_{2h} (mmm)	$\mathcal{H}_D^{orth}$	1-250	Two different Ti–O distances

(Countined)

7 Spin Hamiltonians and Site Symmetries for Transition Ions

Host lattice	Formula	Crystal system	Space group	Complex	Crystal site	Point symmetry	Hamiltonian	Wyckoff vol-page	Comment
Scheelite	$CaWO_4$	Tetragonal	C_{4h}^6 $(4_1/a)$	CaO_8	4b	S_4 $(\bar{4})$	$\mathcal{H}_C^{tetr}$	3-19	
Sphalerite or Zinc blende	ZnS	Cubic	T_d^2 (F43m)	ZnS_4	4a	T_d (43m)	$\mathcal{H}_D^{cube}$	1-108	ZnS_4 and SZn_4 tetrahedra
Spinel	$MgAl_2O_4$	Cubic ($\approx$ fcc)	O_h^7 (Fd3m)	MgO_4 AlO_6	8a 16b	(43m) D_{3d} $(\bar{3}m)$	$\mathcal{H}_D^{cube}$ $\mathcal{H}_D^{trig}$	3-75	Almost perfect fcc oxygen packing
Spinel, inverse	AB_2O_4	Cubic ($\approx$fcc)	O_h^7 (Fd3m)	BO_4 $(A, B)O_6$	8a 16b			3-76	All A and half B atom in 16b site
Tutton salt	$(NH_4)_2SO_4 \cdot MgSO_4 \cdot 6H_2O$	monoclinic	C_{2h}^5 $(P2_1/a)$	$Mg(H_2O)_6$	2a	C_i $(\bar{1})$	$\mathcal{H}_C^{tric}$	3-821	All Mg-OH$_2$ distances not the same
Zincite (Wurtzite ZnS)	ZnO	trigonal	C_{6v}^4 (C6mc)	ZnO_4	2b	C_{3v} (3m)	$\mathcal{H}_D^{trig}$	1-111	Also OZn_4 tetrahedra
Zircon	$ZrSiO_4$	tetragonal	D_{4h}^{19} (I4/amd)	ZrO_4 SiO_4	2a 2b	D_{2d} $(\bar{4}2m)$ D_{2d} $(\bar{4}2m)$	$\mathcal{H}_D^{tetr}$ $\mathcal{H}_D^{tetr}$	3-15	

8
Evaluation of Spin-Hamiltonian Parameters from Multifrequency EPR Data
Sushil K. Misra

8.1
Introduction

The evaluation of spin-Hamiltonian parameters (SHPs) from EPR data of single crystals is a topic of central importance in EPR investigations. Conventionally, perturbation expressions giving the theoretical dependence of resonant EPR line positions on SHPs have been used for this purpose. This has the disadvantage that, since the EPR line positions are obtained only for the special orientations of the external magnetic field, for example, parallel to the X, Y, or Z magnetic axes, it is not possible to estimate the individual values of all SHPs. This is because the line positions for a chosen orientation of the external magnetic field depend only upon certain combinations of some SHPs. In addition, only the absolute values of some of the parameters can be determined from perturbation expressions, since that is how they appear in the perturbation expressions. Furthermore, one is limited here to approximate results, since the perturbation technique yields only approximate values. A historical review of the various reports published using perturbation technique is provided in chronological order in Appendix 8.I, which also includes reports on least-squares fitting (LSF) and brute-force methods, where one uses perturbation expressions.

It is certainly possible to estimate accurately the individual values of all SHPs by computer evaluation of SHPs, by using an exact numerical diagonalization of the spin-Hamiltonian (SH) matrix, provided that the data (line positions and their intensities) corresponding to several orientations of the external magnetic field are simultaneously fitted in a rigorous LSF manner. Since the number of simultaneously fitted data points is large, errors of estimated SHPs are reduced considerably. This can be accomplished by brute-force and LSF techniques; the former approach, which was proposed initially is now obsolete, but is briefly reviewed here for the sake of completeness.

The aim of this chapter is to provide a unified description of the various methods used for the evaluation of SHPs from single-crystal EPR data.

8.2
Perturbation Approach

Many reports have dealt with the perturbation calculation of eigenvalues of SH for electron–nuclear spin coupled systems, as listed in Appendix 8.I in chronological order, describing their features. The eigenvalues calculated by perturbation can, in principle, be used to evaluate SHPs from EPR or ENDOR data. The resonance condition requires that for an observed transition, the difference in the pair of eigenvalues participating in resonance be equal to $h\nu$, where ν is the frequency of resonance and h is Planck's constant. This includes "allowed" ($\Delta M = \pm 1, \Delta m = 0$), and sometimes "forbidden" ($\Delta M \neq \pm 1, \Delta m = 0; \Delta M = \pm 1, \Delta m \neq 0$, or $\Delta M \neq \pm 1, \Delta m \neq 0$) transitions. (Here, M and m denote, respectively, the electronic and nuclear magnetic quantum numbers.) The resonant magnetic field values corresponding to a transition can be obtained by equating the energy eigenvalue difference $|E_{M,m} - E_{M',m'}|$ to $h\nu$, the quantum of energy absorbed in the transition $M, m \leftrightarrow M', m'$, responsible for the resonance. Many reports contain the expressions for the eigenvalues or the resonant-field values. Chambers, Datars, and Calvo, (1964) gave expressions for the resonant-field values appropriate for a spin Hamiltonian corresponding to rhombic or lower symmetry, and containing the fine-structure terms up to fourth order, including an isotropic hyperfine interaction term. Tennant (1976) and McGavin et al. (1980) gave eigenvalues for any orientation of the external magnetic field for a fine-structure interaction, including terms up to sixth order in the crystal field; particular attention was paid to inclusion of cross-terms (the simultaneous contribution of two crystal-field potentials to eigenvalues, that is, terms depending on products of two SHPs corresponding to different crystal field potentials). The authors claimed excellent agreement with data, even for cases where the off-diagonal terms in the spin Hamiltonian were large. Meirovitch and Poupko (1978) gave expressions for fine-structure and hyperfine-structure eigenvalues for an arbitrary orientation of the external magnetic field corresponding to orthorhombic symmetry. They considered terms only up to second order in fine-structure crystal-field potential, but took into account hyperfine, nuclear Zeeman and nuclear quadrupole interactions in detail. Their eigenvalue expressions, from which the resonant-field values for an EPR transition can be obtained, for an arbitrary orientation of the external magnetic field with respect to the crystal-field axes, are described below.

The following discussion will be concerned specifically with EPR data. However, the details apply equally well to ENDOR data where, instead of measuring transition magnetic fields (resonant line positions) at a fixed microwave frequency, one measures resonant frequencies between the pair of energy levels participating in resonance, keeping the external magnetic field constant. (The evaluation of SHP from EPR/ENDOR data for the low-symmetry situation, characterized by noncoincident $\tilde{g}^2, \tilde{A}^2$ tensors in the SH, using perturbation expressions is provided in Chapter 14.)

8.2.1
Spin Hamiltonian

The following spin Hamiltonian, including terms up to sixth order in spin operators good for $S \leq 7/2$, covers most of the cases encountered in practice

$$H = H_f + H_{hf} \tag{8.1}$$

where,

$$\begin{aligned}H_f &= \mu_B S \cdot \tilde{g} \cdot B + B_2^0 O_2^0 + B_2^2 O_2^2 + B_4^0 O_4^0 + B_4^2 O_4^2 + B_4^3 O_4^3 + C_4^3 \Omega_4^3 \\ &\quad + B_4^4 O_4^4 + B_6^0 O_6^0 + B_6^2 O_6^2 + B_6^4 O_6^4 + B_6^6 O_6^6 + B_6^3 O_6^3 + C_6^3 \Omega_6^3\end{aligned} \tag{8.1a}$$

$$H_{hf} = S \cdot \tilde{A} \cdot I - g_N \mu_N B \cdot I + Q'\left\{I_Z^2 - \frac{1}{3}I(I+1)\right\} + Q''(I_x^2 - I_y^2). \tag{8.1b}$$

In Equation 8.1, H_f represents the fine-structure part, whereas H_{hf} represents the hyperfine structure part of the spin Hamiltonian. S and I are the electronic and nuclear spins, respectively; μ_B and μ_N are the Bohr and nuclear magnetons, respectively; g_N is the nuclear g-factor; $\tilde{g}$ is the g-matrix; $\tilde{A}$ is the hyperfine-interaction matrix; and Q', Q'' represent the nuclear–quadrupole-interaction constants. In the expression for H_f, the O_l^m are the operator equivalents as defined by Abragam and Bleaney (1970) and listed in Chapter 7, and Ω_4^3 and Ω_6^3 are imaginary spin operators (Tennant, 1976; see also Chapter 7), often neglected due to their small contribution.

After transforming the crystal field and hyperfine terms of Equation 8.1 to the Zeeman diagonal coordinates, the perturbation technique can be employed to arrive at the eigenvalues $E_{M,m}$. In particular, to second order in perturbation,

$$E_{M,m} = B'M + E_f(M) + E_{hf}(M, m), \tag{8.2}$$

$$\begin{aligned}E_f(M) &= \frac{1}{2}\{(3a_{33}^2 - 1)B_2^0 + C_2^{2'}B_2^2\}f_1(S, M) + \frac{1}{8}\{(35a_{33}^4 - 30a_{33}^2 + 3)B_4^0 \\ &\quad + (7a_{33}^2 - 1)C_2^{2'}B_4^2 + a_{13}a_{33}C_3^{3'}B_4^3 + a_{23}a_{33}S_3^{3'}C_4^3 + C_4^{4'}B_4^4\}f_2(S, M) \\ &\quad + \frac{\sqrt{231}}{4}\{(231a_{33}^6 - 315a_{33}^4 + 105a_{33}^2 - 5)B_6^0 \\ &\quad + (33a_{33}^4 - 18a_{33}^2 + 1)C_2^{2'}B_6^2 + (11a_{33}^2 - 1)C_4^{4'}B_6^4 + C_6^{6'}B_6^6\}f_{60}(S, M) \\ &\quad + \frac{XX^*}{4B'}f_3(S, M) + \frac{YY^*}{8B'}f_4(S, M) + \frac{AA^*}{56B'}f_5(S, M) + \frac{BB^*}{56B'}f_6(S, M) \\ &\quad + \frac{CC^*}{24B'}f_7(S, M) + \frac{DD^*}{64B'}f_8(S, M) \\ &\quad + \frac{(AX^* + A^*X)}{4\sqrt{14}B'}f_9(S, M) + \frac{(BY^* + B^*Y)}{8\sqrt{7}B'}f_{10}(S, M) \\ &\quad + FF^*(g_{6-2}^2/e_4 + g_{62}^2/e_3) + (FY^* + F^*Y)(g_{2-2}g_{6-2}/e_4 + g_{22}g_{62}/e_3) \\ &\quad + (DH^* + D^*H)(g_{4-4}g_{6-4}/e_8 + g_{44}g_{64}/e_7) + KK^*(g_{6-6}^2/e_{10} + g_{66}^2/e_9)\end{aligned} \tag{8.3}$$

Here, the magnetic flux density, **B**, is at the angle $\cos^{-1}(l, m, n)$ to the directions of the principal-direction values g_x, g_y, g_z, of the diagonal g-matrix, $B' = g\mu_B B$, $a_{13} = lg_x/g$, $a_{23} = mg_y/g$, $a_{33} = ng_z/g$; $C_n^{n'} + iS_n^{n'} = (a_{13} - ia_{23})^n$; and $g = (l^2 g_x^2 + m^2 g_y^2 + n^2 g_z^2 + 2lmg_x g_y + 2lng_x g_z + 2mng_y g_z)^{1/2}$. However, if one is dealing with an arbitrary coordinate frame then the direction cosines $a_{i3}(i = 1-3)$ become:

$$a_{13} = (\ell g_{xx} + mg_{xy} + ng_{xz})/g$$
$$a_{23} = (\ell g_{yx} + mg_{yy} + ng_{yz})/g$$
$$a_{33} = (\ell g_{zx} + mg_{zy} + ng_{zz})/g$$

or, $a_{i3} = e_i^T \cdot \tilde{g} \cdot \eta/g$ where $\eta = \begin{pmatrix} \ell \\ m \\ n \end{pmatrix}$ and $e_i (i = 1, 3)$ are the unit vectors in the original x, y, and z directions; η is the unit vector along **B** (Golding and Tennant, 1973; McGavin et al., 1980). For symmetric $\tilde{g}$:

$$g^2 = \eta^T \cdot g^T \cdot g \cdot \eta$$

The $f_n(S, M)$; $n = 1-10$, $f_{60}(S, M)$ as well as the $g_{l\pm m}$, appearing in Equation 8.3, are given in Table 8.1. In a plane containing the g-matrix principal directions the coef-

Table 8.1 Values of $f_n(S, M)$ and $g_{l\pm m}$ required in Equation 8.3.

Function	Value
$f_1(S, M)$	$3M^2 - S(S+1)$
$f_2(S, M)$	$35M^4 + 25M^2 - 30M^2 S(S+1) - 6S(S+1) + 3S^2(S+1)^2$
$f_3(S, M)$	$S_{0-1}^2 (2M-1)^2 - S_{01}^2 (2M+1)^2$
$f_4(S, M)$	$S_{0-1}^2 S_{-1-2}^2 - S_{01}^2 S_{12}^2$
$f_5(S, M)$	$S_{0-1}^2 \{f_5^-(S, M)\}^2 - S_{01}^2 \{f_5^+(S, M)\}^2$
$f_5^{\mp}(S, M)$	$14M^3 \mp 21M^2 + 19M - 6MS(S+1) \pm 3S(S+1) \mp 6$
$f_6(S, M)$	$S_{0-1}^2 S_{-1-2}^2 \{f_6^-(S, M)\}^2 - S_{01}^2 S_{12}^2 \{f_6^+(S, M)\}^2$
$f_6^{\mp}(S, M)$	$7M^2 \mp 14M - S(S+1) + 9$
$f_7(S, M)$	$S_{0-1}^2 S_{-1-2}^2 S_{-2-3}^2 (2M-3)^2 - S_{01}^2 S_{12}^2 S_{23}^2 (2M+3)^2$
$f_8(S, M)$	$S_{0-1}^2 S_{-1-2}^2 S_{-2-3}^2 S_{-3-4}^2 - S_{01}^2 S_{12}^2 S_{23}^2 S_{34}^2$
$f_9(S, M)$	$S_{0-1}^2 (2M-1) f_5^-(S, M) - S_{01}^2 (2M+1) f_5^+(S, M)$
$f_{10}(S, M)$	$S_{0-1}^2 S_{-1-2}^2 f_6^-(S, M) - S_{01}^2 S_{12}^2 f_6^+(S, M)$

$f_{60}(S, M) = 231M^6 + 105[7 - 3S(S+1)]M^4 + 21[14 - 25S(S+1) + 5S^2(S+1)^2]M^2$
$\qquad + 5[-12S(S+1) + 8S^2(S+1)^2 - S^3(S+1)^3]$

$g_{2\pm2} = \frac{1}{2} S_{0\pm1} S_{\pm1\pm2}$,

$g_{4\pm2} = \frac{1}{2} \sqrt{\frac{1}{7}} S_{0\pm1} S_{\pm1\pm2} \{7M^2 \pm 14M + [9 - S(S+1)]\}$,

$g_{4\pm4} = \frac{1}{4} S_{0\pm1} S_{\pm1\pm2} S_{\pm2\pm3} S_{\pm3\pm4}$,

$g_{6\pm2} = \frac{1}{8} \sqrt{\frac{5}{11}} S_{0\pm1} S_{1\pm2} \{33M^4 \pm 132M^3 + 3[91 - 6S(S+1)]M^2$
$\qquad \pm 6[47 - 6S(S+1)]M + [120 - 26S(S+1) + S^2(S+1)^2]\}$,

$g_{6\pm4} = \frac{1}{4} \sqrt{\frac{3}{22}} S_{0\pm1} S_{\pm1\pm2} S_{\pm2\pm3} S_{\pm3\pm4} \{11M^2 \pm 44M + [50 - S(S+1)]\}$,

$g_{6\pm6} = \frac{1}{8} S_{0\pm1} S_{\pm1\pm2} S_{\pm2\pm3} S_{\pm3\pm4} S_{\pm4\pm5} S_{\pm5\pm6}$

$S_{ab} \qquad \{S(S+1) - (M+a)(M+b)\}^{1/2}$.

ficients X, Y, A, ... may be written explicitly in terms of a_{13}, a_{23}, a_{33}, or (ℓ, m, n) and the crystal-field coefficients $B_2^0, B_2^2, B_4^0, \ldots$. In a general direction, this can only be done for the modulus square coefficients $XX^*, YY^*, AA^*, \ldots$. Unfortunately, this cannot be done generally (McGavin et al., 1980), but in practice one can readily obtain a suitable (non-unique) transformation that leads to the correct eigenvalues (Keijzers, Paulussen, and de Boer, 1975; Lin, 1973). The expressions for X, Y, A, B, C, and D in the (g_x, g_y) and (g_x, g_z) planes are given in Table 8.2, and should cover most cases of practical interest. The explicit forms for the coefficients Y, B, D, F, H, and K, that are required for inclusion of the parameters B_6^m are given in Table 8.3 for the magnetic field orientations in the directions g_x, g_y, and g_z. Finally, the e-denominators in Equation 8.3 are as follows: $e_3 = -e_4 = 2B'$, $e_7 = -e_8 = 4B'$, and $e_9 = -e_{10} = 6B'$.

Table 8.2 Values of the coefficients X, Y, A, B, C, D, in the (a) $g_x g_y$ and (b) $g_x g_z$ planes. $a_{13}, a_{23}, a_{33}, C_n^{n'}$, and $S_n^{n'}$ are defined in Equation 8.3.

Coefficient	Value
(a)	
X	$-iS_2^2 B_2^2$
Y	$\frac{1}{2} B_2^0 + \frac{1}{2} C_2^{2'} B_2^2$
A	$-\frac{i\sqrt{7}}{2\sqrt{2}} \{ S_2^{2'} B_4^2 - 2 S_4^{4'} B_4^4 \} - \frac{\sqrt{7}}{4\sqrt{2}} \{ C_3^{3'} B_4^3 + S_3^{3'} C_4^3 \}$
B	$\frac{\sqrt{7}}{2} \{ -5 B_4^0 + C_2^{2'} B_4^2 + C_4^{4'} B_4^4 \} - \frac{i\sqrt{7}}{4} \{ S_3^{3'} B_4^3 - C_3^{3'} C_4^3 \}$
C	$\frac{\sqrt{i}}{2\sqrt{2}} \{ 7 S_2^{2'} B_4^2 + 2 S_4^{4'} B_4^4 \} - \frac{3}{4\sqrt{2}} \{ C_3^{3'} B_4^3 - S_3^{3'} C_4^3 \}$
D	$\frac{1}{4} \{ 35 B_4^0 + 7 C_2^{2'} B_4^2 + C_4^{4'} B_4^4 \} - \frac{i}{4} \{ S_3^{3'} B_4^3 - C_3^{3'} C_4^3 \}$
(b)	
X	$a_{13} a_{33} (-3 B_2^0 + B_2^2)$
Y	$\frac{3}{2} a_{13}^2 B_2^0 + \frac{1}{2} (1 + a_{33}^2) B_2^2$
A	$\frac{\sqrt{7}}{4\sqrt{2}} a_{13} \{ -20 a_{33} (7 a_{33}^2 - 3) B_4^0 + 4 a_{33} (7 a_{33}^2 - 4) B_4^2 + a_{13} (4 a_{33}^2 - 1) B_4^4 \}$
	$+ 4 a_{13}^2 a_{33} B_4^4 \} - \frac{i 3\sqrt{7}}{4\sqrt{2}} a_{13}^2 a_{33} C_4^3,$
B	$\frac{\sqrt{7}}{2} \{ 5 a_{13}^2 (7 a_{33}^2 - 1) B_4^0 + (7 a_{33}^4 - 6 a_{33}^2 + 1) B_4^2 + a_{13} a_{33}^3 B_4^3$
	$+ a_{13}^2 (1 + a_{33}^2) B_4^4 \} - \frac{i\sqrt{7}}{4} a_{13} (3 a_{33}^2 - 1) C_4^3,$
C	$\frac{1}{4\sqrt{2}} \{ -140 a_{13}^3 a_{33} B_4^0 - 28 a_{13} a_{33}^3 B_4^2 + (4 a_{33}^4 + 3 a_{33}^2 - 3) B_4^3$
	$+ 4 a_{13} a_{33} (a_{33}^2 + 3) B_4^4 \} - \frac{i}{4\sqrt{2}} a_{33} (9 a_{33}^2 - 5) C_4^3,$
D	$\frac{1}{4} \{ 35 a_{13}^4 B_4^0 + 7 a_{13}^2 (1 + a_{33}^2) B_4^2 - a_{13} a_{33} (a_{33}^2 + 3) B_4^3$
	$+ (a_{33}^4 + 6 a_{33}^2 + 1) B_4^4 \} + \frac{i}{4} a_{13} (3 a_{33}^2 + 1) C_4^3.$

Table 8.3 Values of the coefficients of Equation 8.3 in principal directions of the g-tensor.

Coefficient	g_x direction (upper sign) / g_y direction (lower sign)	g_z direction
diag(6)	$\frac{\sqrt{231}}{4}(-5B_6^0 \pm B_6^2 - B_6^4 \pm B_6^6)$	$4\sqrt{231}B_6^0$
Y	$\frac{3}{2}B_2^0 \pm \frac{1}{2}B_2^2$	$\frac{1}{2}B_2^2$
B	$\frac{\sqrt{7}}{2}(-5B_4^0 \pm B_4^2 + B_4^4)$	$\sqrt{7}B_4^2$
D	$\frac{1}{4}(35B_4^0 \pm 7B_4^2 + B_4^4)$	$2B_4^4$
F	$\frac{\sqrt{11}}{8\sqrt{5}}(105B_6^0 \mp 17B_6^2 + 5B_6^4 \pm 15B_6^6)$	$\frac{4\sqrt{11}}{\sqrt{5}}B_6^2$
H	$\frac{\sqrt{22}}{8\sqrt{3}}(-63B_6^0 \pm 3B_6^2 + 13B_6^4 \pm 3B_6^6)$	$\frac{2\sqrt{22}}{\sqrt{3}}B_6^4$
K	$\frac{1}{8}(231B_6^0 \pm 33B_6^2 + 11B_6^4 \pm B_6^6)$	$4B_6^6$

8.2.1.1 S = 7/2

The expressions for the resonant fields for the "allowed" transitions for $S = 7/2$ (e.g., Gd^{3+}, Eu^{2+}) can be expressed to fourth order in perturbation for the magnetic-field orientation along the z-axis as follows (Singh and Venkateswarlu, 1967)

$$g\mu_B B = g\mu_B B^{(2)} + g\mu_B B^{(4)}, \tag{8.4}$$

where, to second-order in perturbation, the $g\mu_B B^{(2)}$ in Equation 8.4 are:

$$\pm\frac{7}{2} \leftrightarrow \pm\frac{5}{2}: g\mu_B B^{(2)} = h\nu \mp (6b_2^0 + 20b_4^0 + 6b_6^0) + E[45/(1\pm 3y) - 21/(1\pm 5y)],$$

$$\pm\frac{5}{2} \leftrightarrow \pm\frac{3}{2}: g\mu_B B^{(2)} = h\nu \mp (4b_2^0 - 10b_4^0 - 14b_6^0) + E[60/(1\pm y)$$
$$- 21/(1\pm 5y) - 45/(1\pm 3y)],$$

$$\pm\frac{3}{2} \leftrightarrow \pm\frac{1}{2}: g\mu_B B^{(2)} = h\nu \mp (2b_2^0 - 12b_4^0 + 14b_6^0) + E[21/(1\pm 5y)$$
$$- 45/(1\pm 3y) - 120y/(1-y^2)],$$

$$+\frac{1}{2} \leftrightarrow -\frac{1}{2}: g\mu_B B^{(2)} = h\nu + E[90/(1-9y^2) - 120y/(1-y^2)]. \tag{8.5}$$

In Equation 8.3 $y = b_2^0/g\mu_B B^{(2)}$, $E = (b_2^2)^2/18g\mu_B B^{(2)}$. The parameters b_{2r}^s are related to the parameters B_{2r}^s of Equation 8.1 in the following way: $b_2^s = 3B_2^s (s = 0, 2)$; $b_4^s = 60B_4^s (s = 0, 2, 4)$, $b_6^s = 1260B_6^s (s = 0, 2, 4, 6)$. To fourth order in perturbation, $g\mu_B B^{(4)}$ of Equation 8.4 are:

$$\pm\frac{7}{2} \leftrightarrow \pm\frac{5}{2}: g\mu_B B^{(4)} = g\mu_B B^{(2)} + x\left[150/(1\pm 3y)^2(1\pm y) - 225/(1\pm 3y)^3\right.$$
$$\left. - 70/(1\pm 5y)^2(1\pm 3y) + 49/(1\pm 5y)^3\right],$$

$\pm\frac{5}{2} \leftrightarrow \pm\frac{3}{2} : g\mu_B B^{(4)} = g\mu_B B^{(2)} + x\left[150/(1\pm y)^2 (1\mp y) - (7/(1\pm 5y)^2\right.$
$\qquad\qquad + 20/(1\pm y)^2)(-7/(1\pm 5y) + 20/(1\pm y))$
$\qquad\qquad - 150/(1\pm 3y)^2 (1\pm y) + 225/(1\pm 3y)^3 \Big];$

$\pm\frac{3}{2} \leftrightarrow \pm\frac{1}{2} : g\mu_B B^{(4)} = g\mu_B B^{(2)} + x\left[70/(1\mp y)^2 (1\mp 3y)\right.$
$\qquad\qquad - 150/(1\pm y)^2 (1\mp y) - (15/(1\pm 3y)^2$
$\qquad\qquad + 20/(1\mp y)^2)(-15/(1\pm 3y) + 20/(1\mp y))$
$\qquad\qquad + (7/(1\pm 5y)^2 + 20/(1\pm y)^2)(-7/(1\pm 5y) + 20/(1\pm y))\Big];$

$+\frac{1}{2} \leftrightarrow -\frac{1}{2} : g\mu_B B^{(4)} = g\mu_B B^{(2)} - x\left[70/(1+y)^2 (1+3y)\right.$
$\qquad\qquad + 70/(1-y)^2 (1-3y) + (20/(1+y^2)$
$\qquad\qquad + 15/(1-3y)^2)(-20/(1+y) + 15/(1-3y))$
$\qquad\qquad + (20/(1-y)^2 + 15/(1+3y)^2)(-20/(1-y) + 15/(1+2y))\Big];$
(8.6)

In Equation 8.6 $x = E^2/18g\mu_B B^{(4)}$, $y = b_2^0/g\mu_B B^{(4)}$, where $E = (b_2^2)^2/18g\mu_B B^{(4)}$. (The x and y used in Equation 8.5 for $g\mu_B B^{(2)}$, appearing on the right-hand sides in Equation 8.6, should here be replaced by these expressions for x and y.)

In order to express the resonant fields for the external magnetic field orientation along the x-axis, one needs to transform the SHP in Equations 8.5 and 8.6 as follows (Jones, Baker, and Pope, 1959):

$b_2^0 \to (-b_2^0 + b_2^2)/2;\ b_2^2 \to -(3b_2^0 + b_2^2)/2;\ b_4^0 \to (3b_4^0 - b_4^2 + b_4^4)/8;$
$b_4^2 \to (5b_4^0 - b_4^2 - b_4^4)/2;\ b_4^4 \to (35b_4^0 + 7b_4^2 + b_4^4)/8;$
$b_6^0 \to (-5b_6^0 + b_6^2 - b_6^4 + b_6^6)/16;\ b_6^2 \to (-105b_6^0 + 17b_6^2 - 5b_6^4 - 15b_6^6)/32;$
$b_6^4 \to (-63b_6^0 + 3b_6^2 + 13b_6^4 + 3b_6^6)/16;\ b_6^6 \to (-231b_6^0 - 33b_6^2 - 11b_6^4 - b_6^6)/32.$

8.2.1.2   S = 5/2 (Fe^{3+})

The expressions for the energy differences $W(= h\nu)$ to fourth order in perturbation containing only the second order SHP are available for the various allowed transitions (Chambers, Datars, and Calvo, 1964). They are

$W\left(M: \pm\frac{5}{2} \leftrightarrow \pm\frac{3}{2}\right) = B' \mp 12 B_2^0 + 4(B_2^2)^2/B' \pm 18 B_2^0 (B_2^2)^2/B'^2$
$\qquad\qquad + 325(B_2^0)^2 (B_2^2)^2/B'^3 + 28(B_2^2)^4/B'^3,$

$W\left(M: \pm\frac{3}{2} \leftrightarrow \pm\frac{1}{2}\right) = B' \mp 6 B_2^0 - 5(B_2^2)^2/B' \pm 99 B_2^0 (B_2^2)^2/B'^2$
$\qquad\qquad + 405(B_2^0)^2 (B_2^2)^2/B'^3 - \frac{5}{4}(B_2^2)^4/B'^3,$

$W\left(M: \frac{1}{2} \leftrightarrow -\frac{1}{2}\right) = B' - 8(B_2^2)^2/B' + 648(B_2^0)^2 (B_2^2)^2/B'^3 + 56(B_2^2)^4/B'^3.$ (8.7)

In Equation 8.7, $B' = \mu_B g B$. One should substitute for B_2^0 and B_2^2, when $\mathbf{B} \parallel x$ or $\mathbf{B} \parallel y$, respectively, as follows: $\frac{3}{2}(B_2^2 - B_2^0)$ and $-\frac{1}{2}(3B_2^0 + B_2^2)$ for $\mathbf{B} \parallel x$; and $\frac{3}{2}(B_2^0 + B_2^2)$ and $\frac{1}{2}(3B_2^0 - B_2^2)$ for $\mathbf{B} \parallel y$

$S = 5/2$ (Mn^{2+}): This case further requires consideration of the hyperfine interaction for the Mn^{2+} ion, whose nuclear spin is $I = 5/2$. The expressions for the energies of the Mn^{2+} ion in a single crystal, calculated up to third order in perturbation, along with allowed hyperfine line positions are listed here (Meirovitch and Poupko, 1978; Markham, Rao, and Reed, 1979). [Upreti (1974) has only given expressions to second-order in perturbation, although he did consider the hyperfine interaction.] The spin Hamiltonian for the Mn^{2+} ion in a quadratic crystal field for orthorhombic distortion, neglecting terms of fourth order in electronic spin operator, is given by

$$H = g\mu_B \mathbf{B} \cdot \mathbf{S} - g_N \mu_N \mathbf{B} \cdot \mathbf{I} + A\mathbf{I} \cdot \mathbf{S} + H_{ZFS}. \tag{8.8}$$

In Equation 8.8, the first three terms represent the electronic-Zeeman, nuclear-Zeeman and electron-nuclear hyperfine interactions, respectively. H_{ZFS} in Equation 8.8 is the zero-field splitting (ZFS) term, which can be written for a nonaxial situation as:

$$H_{ZFS} = D\left\{S_z^2 - \frac{1}{3}S(S+1)\right\} + \frac{E}{2}\left\{S_+^2 + S_-^2\right\} \tag{8.9}$$

In Equation 8.9, z denotes the principal axis of the second order ZFS tensor; D and E are the axial and orthorhombic ZFS parameters; and $S_\pm = S_x \pm iS_y$. The $\tilde{g}$ and $\tilde{A}$ matrices for the S-state Mn^{2+} ion, have here been assumed isotropic. The spin Hamiltonian can now be written as the sum of zero-order and perturbation parts:

$$H = H^{(0)} + H', \tag{8.10}$$

with

$$H = \mu_B g B S_z - \mu_N g_N \mathbf{B} \cdot \mathbf{I}; \; H' = H_{ZFS} + A\mathbf{I} \cdot \mathbf{S}, \tag{8.11}$$

where the z-axis, assumed parallel to the direction of the external (Zeeman) magnetic field, B, has been chosen as the axis of quantization. Expressions for the perturbation energies, up to third-order in perturbation, are (Meirovitch and Poupko, 1978; Markham, Rao, and Reed, 1979):

$$E_{Mm} = E_{Mm}^{(0)} + E_{Mm}^{(1)} + E_{Mm}^{(2)} + E_{Mm}^{(3)},$$

with

$$E_{Mm}^{(0)} = g\mu_B BM; \tag{8.12}$$

$$E_{Mm}^{(1)} = AMm - g_N \mu_N Bm + a\left\{M^2 - \frac{1}{3}S(S+1)\right\}; \tag{8.13}$$

$$E^{(2)}_{Mm} = \frac{A^2}{2G}\{MI(I+1) - mS(S+1) + M^2m - Mm^2\}$$
$$+ \frac{2b_+b_-}{G}\{8M^3 + M - 4MS(S+1)\} - \frac{2c_+c_-}{G}\{2M^3 + M - 2MS(S+1)\};$$

(8.14)

$$E^{(3)}_{Mm} = \frac{Ab_+b_-}{G^2}\{S^+_M(2M+1)^3 - S^-_M(2M-1)^3\}$$
$$+ \frac{Ac_+c_-}{G^2}\{(M+1)S^+_M S^+_{M+1} - (M-1)S^-_M S^-_{M-1}\}$$
$$+ \frac{Re(b^2_+c_-)}{G^2}\{(2M+1)(2M+3)S^+_M S^+_{M+1}$$
$$+ (2M-1)(2M-3)S^-_M S^-_{M-1} - 2(2M-1)(2M+1)S^-_M S^+_M\}$$
$$+ \frac{A^3}{4G^2}\{(-M+m-1)S^+_M I^-_m + (M-m-1)S^-_M I^+_m\}$$
$$+ \frac{Ab_+b_-}{G^2}\frac{2m}{M}\{[S(S+1) - M^2]^2 - M^2\} + \frac{Ac_+c_-}{2G^2}m\{S^+_M S^+_{M+1} - S^-_M S^-_{M-1}\}$$
$$+ \frac{A^2 a}{4G^2}\{(2M+1)S^+_M I^-_m - (2M-1)S^-_M I^+_m\}$$

(8.15)

where

$$a = D\left\{\left(\frac{3\cos^2\theta - 1}{2}\right) + \frac{3}{2}\eta\sin^2\theta\cos 2\phi\right\};$$
$$b_\pm = \frac{D}{4}\{-\sin 2\theta + \eta\sin 2\theta\cos 2\phi \pm i2\eta\sin\theta\sin 2\phi\};$$
$$c_\pm = \frac{D}{4}\{\sin^2\theta + \eta(\cos^2\theta + 1)\cos 2\phi \pm i2\eta\cos\theta\sin 2\phi\}; \eta = \frac{E}{D};$$
$$G = g\mu_B B; S^\pm_M = S(S+1) - M(M\pm 1); I^\pm_m = I(I+1) - m(m\pm 1) \quad (8.16)$$

In the above equations, the order of perturbation is indicated by the number n (= 0, 1, 2, 3) within brackets in the superscripts on E_{Mm}; θ and ϕ are, respectively, the polar and azimuthal angles of the principal axis of the ZFS tensor with respect to the external magnetic field; and Re denotes the real part. Because of its small magnitude, only the first-order term of the nuclear Zeeman energy is taken into account. The range for the asymmetry parameter η is $0 \leq \eta \leq 1/3$.

The positions of the allowed, (1/2, $m \leftrightarrow$ −1/2, m), and forbidden, $M, m \leftrightarrow M, m-1$ ($\Delta M = 0, \Delta m = -1$), or $M, m-1 \leftrightarrow M, m$ ($\Delta M = 0, \Delta m = 1$), hyperfine transitions belonging to the Mn^{2+} central sextet can be expressed from Equations 8.12–8.15, by the use of the resonance condition $h\nu = E_{1/2,m} - E_{-1/2,m} = G_o = g\mu_B B_o$, where B_o is the magnetic field corresponding to the middle of the central sextet, as follows:

$$B_c(m, \theta, \phi) = \left[1 - \frac{A}{G_o} m - \frac{4A^2 m}{G_o^3} Df_1(\theta, \phi) \right.$$
$$+ \frac{D^2}{G_o^3} \{(4G_o - 36Am) f_3(\theta, \phi) - 2(G_o - Am) f_2(\theta, \phi)\} \quad (8.17)$$
$$\left. - \frac{A^2}{8G_o^3} \{(35 - 4m^2)G_o + m(4m^2 - 65)A\} \right] B_o$$

In Equation 8.17, the expressions for $f_1(\theta, \varphi)$, $f_2(\theta, \varphi)$, and $f_3(\theta, \varphi)$ are given as follows:

$$f_1(\theta, \varphi) = \{(3\cos^2\theta - 1) + 3\eta \sin^2\theta \cos 2\varphi\} \quad (8.18a)$$
$$f_2(\theta, \varphi) = \{[\sin^2\theta + \eta \cos 2\varphi(1 + \cos^2\theta)]^2 + 4\eta^2 \cos^2\theta \sin^2 2\varphi\} \quad (8.18b)$$
$$f_3(\theta, \varphi) = \{(1 - \eta \cos 2\varphi)^2 \sin^2 2\theta + 4\eta^2 \sin^2\theta \sin 2\varphi\} \quad (8.18c)$$

8.3
Brute-Force Methods to Evaluate SHP

In these methods, although exact numerical diagonalization of the spin Hamiltonian matrix is performed on a diagonal computer, no well-defined criteria are used to arrive at the final set of SHP corresponding to the absolute minimum chi-square, starting from an initial set of SHP. (Hence, the title "brute-force" is used.) Thus, these methods require long computer times, although exact eigenvalues are computed at each stage. The methods described in Sections 8.3.1 and 8.3.2 below were applied specifically to the cases of fine-structure splittings only. However, there is no reason why they cannot be applied to the cases exhibiting both fine and hyperfine structures.

8.3.1
Variation of One Parameter at a Time

In this procedure (Buckmaster et al., 1971), the value of one parameter is varied at a time over a selected range in several steps, starting with a chosen value, while the values of other parameters are kept fixed. The resonant fields measured for only one orientation of the external magnetic field are then considered, such that all of the parameters are covered in descending order of their magnitudes. However, this method has serious deficiencies, as pointed out by Misra and Sharp (1976). Most notable among these is the lack of orthogonal-axis consistency, implying that the SHPs determined for one orientation of **B** are not consistent with those determined for an orthogonal orientation of **B**. This point is not discussed further here.

8.3.2
Variation of Parameters in Subgroups

This brute-force method (Misra and Sharp, 1976) does not suffer from the disadvantages of the method described in Section 8.3.1 above, where only one parameter is varied at a time. Rather, it is capable of handling cases in which there are large off-diagonal elements in the SH matrix, as well as providing the orthogonal-axis consistency. In addition, economy in computer time is achieved here, as in practice one does not have to make as many diagonalizations of the SH matrix as are required when only one parameter is varied at a time. Briefly, in this method, those parameters whose values are of the same order of magnitude are varied simultaneously, and resonant line positions for several orientation of **B** are fitted simultaneously. Thus, the SHPs are divided into subgroups according to orders of magnitude of their values; the various subgroups are treated in descending order of magnitude of their values. Such a grouping is obtained by including in the same subgroup all B_l^m with the same l value. The details are described by Misra and Sharp (1976), who applied it successfully to the evaluation of SHPs for Gd^{3+}-doped $SmCl_3 \cdot 6H_2O$, characterized by large off-diagonal elements in the SH matrix. The orthogonal-axis consistency was, in fact, achieved. (This case could not be satisfactorily treated by the technique described in Section 8.3.1 by varying one parameter at a time.) The LSF procedure described below, which is much quicker, yielded the same values of the parameters as those found by Misra and Sharp (1976) for this case.

8.4
Least-Squares Fitting (LSF) Method

8.4.1
Introduction

In the LSF method, a mathematical criterion is used to arrive at the final set of parameters corresponding to the minimum value of the chi-square (χ^2)-function in an iterative manner, starting from an initial set of parameters (Scarborough, 1966). However, here one requires the values of the first and second derivatives of the χ^2-function, which in turn depend on the derivatives of the eigenvalues with respect to the SHPs. The latter are difficult to evaluate rigorously, as analytic expressions for the eigenvalues are not available, especially for arbitrary orientations of the external magnetic field, and for cases characterized by the presence of large off-diagonal elements in the SH matrix. However, this difficulty can be circumvented by the use of Feynman's theorem (Feynman, 1939), using numerically evaluated eigenvalues and eigenvectors of the SH matrix on a digital computer as shown below. This economizes the computer time substantially, as the vector, whose components are the SHPs, is transformed simultaneously in one step in each iteration to decrease the χ^2 value in the space of SHPs in each iteration.

The various details of the LSF method, outlined by Misra (1976), are provided in Section 8.4.2, as applied specifically to the fitting of EPR line positions. A brief review of the general LSF procedure is presented in this section, essentially to illustrate the underlying technique and to introduce the notation used.

In the LSF technique, the vector a^m, whose components are the values of the SH parameters and the linewidth corresponding to the absolute minimum of the χ^2-function, can be obtained from a^i, the vector whose components are the initially chosen values of the SH parameters and the linewidth:

$$a^m = a^i - (D''(a^m))^{-1} D', \tag{8.19}$$

In Equation 8.19, D' is the column vector whose elements are the first derivatives of the χ^2-function with respect to the parameters evaluated at a^i, and D'' is the matrix whose elements are the second derivatives with respect to the parameters evaluated at a^m:

$$D'_m = \left(\frac{\partial \chi^2}{\partial a_m}\right)_{a^i} \tag{8.20}$$

$$D''_{nm} = \left(\frac{\partial^2 \chi^2}{\partial a_n \partial a_m}\right)_{a^m} \tag{8.21}$$

For application of the LSF technique proposed here, the χ^2-function is defined as the sum of the weighted squares of the differences between the calculated and measured first-derivative absorption resonances at the magnetic field values B_k within the magnetic-field interval considered:

$$\chi^2 = \sum_k [F_c(x_k, a) - F_m(x_k, a)]^2 / \sigma_k^2, \tag{8.22}$$

where $F_c(x_k, a)$ and $F_m(x_k, a)$ are, respectively, the calculated values using the SHP and experimental values of the data points for EPR resonance signal, and σ_k is the corresponding weight factor (related to standard deviation of the datum k, and a is the vector whose components are SHPs).

Since a^m is not known initially, the elements of the matrix D'' are, in practice, evaluated with respect to the chosen initial values of the parameters, a^i, referred to as $D''(a^i)$. A new set of parameters, denoted by the vector a^f is then calculated:

$$a^f = a^i - (D''(a^i))^{-1} D', \tag{8.23}$$

which replaces the vector a^m, given by Equation 8.19. The vector a^f is calculated iteratively using Equation 8.23, and then calculating the χ^2-function using Equation 8.22 until a sufficiently small value of the χ^2-function, consistent with experimental errors is obtained. The difference vector $\Delta a = a^i - a^f$, which will be used later, can here be expressed as

$$\Delta a = D''(a^i)^{-1} D', \tag{8.24}$$

which gives, upon multiplying both sides by $D''(a^i)$

$$D''(a^i)\Delta a = D'. \tag{8.25}$$

8.4.2
Details of the LSF Method as Applied to EPR

The single-ion spin Hamiltonian H can be expressed, in the notation of Abragam and Bleaney (1970) as

$$H = \mu_B B \cdot \tilde{g} \cdot S + \sum_{\ell,m} B_\ell^m O_\ell^m \quad (l \text{ even}, |m| \leq \ell). \tag{8.26}$$

In the analysis of EPR data, one is interested in evaluating the values of the parameters a_j [g_{ij} and B_ℓ^m of Equation 8.26] from the resonant magnetic field values obtained for one, or more, orientations of the external magnetic field.

The χ^2- function for EPR data is defined as

$$\chi^2 = \sum_i \left[(|\Delta E_i| - h v_i)^2 / \sigma_i^2 \right], \tag{8.27}$$

where h is Planck's constant, $\Delta E_i = E_{i'} - E_{i''}$ ($E_{i'}$ and $E_{i''}$ are the energies of the levels between which the microwave energy $h v_i$ is absorbed at resonance, and v_i is the corresponding klystron frequency which is usually the same for all i. The $E_{i'}$ and $E_{i''}$ are obtained by substituting the (numerical) values of the parameters and the resonant field values B_i into the spin Hamiltonian, Equation 8.26, and diagonalizing it on a computer. In Equation 8.27, σ_i is the weight factor for the ith line position.

Note: In the form given by Equation 8.27, one can fit multifrequency data, obtained for several frequencies, all together to obtain the best values obtained from multifrequency EPR.

When using the LSF method, one needs the values of the first (D') and second (D'') derivatives of χ^2 with respect to the parameters. This enables one to construct D' and D'', as discussed in Section 8.4.1. From Equation 8.27 these are given by

$$\frac{\partial \chi^2}{\partial a_j} = 2 \sum_i \frac{\Delta E_i}{|\Delta E_i|} \frac{(|\Delta E_i| - h v_i)}{\sigma_i^2} \left(\frac{\partial E_i'}{\partial a_j} - \frac{\partial E_i''}{\partial a_j} \right), \tag{8.28}$$

$$\frac{\partial^2 \chi^2}{\partial a_j \partial a_l} = 2 \sum_i \frac{1}{\sigma_i^2} \left[\left(\frac{\partial E_i'}{\partial a_j} - \frac{\partial E_i''}{\partial a_j} \right) \left(\frac{\partial E_i'}{\partial a_l} - \frac{\partial E_i''}{\partial a_l} \right) \right.$$
$$\left. + \frac{\Delta E_i}{|\Delta E_i|} (|\Delta E_i| - h v_i) \left(\frac{\partial^2 E_i'}{\partial a_j \partial a_l} - \frac{\partial^2 E_i''}{\partial a_j \partial a_l} \right) \right]. \tag{8.29}$$

From Equations 8.28 and 8.29 it is seen that the problem reduces to the evaluation of the first and second derivatives of $E_{i'}$ and $E_{i''}$ with respect to the parameters. These can be evaluated, numerically, from Feynman's theorem (Feynman, 1939), according to which

$$\frac{\partial E_i}{\partial a_j} = \left\langle \phi_i \left| \frac{\partial H}{\partial a_j} \right| \phi_i \right\rangle, \tag{8.30}$$

where the $|\phi_i\rangle$ is the eigenvector of the SH matrix H having the eigenvalue E_i:

$$H|\phi_i\rangle = E_i |\phi_i\rangle. \tag{8.31}$$

Exploitation of Equation 8.30 leads to (Misra, 1976):

$$\frac{\partial^2 E_i}{\partial a_j \partial a_l} = \sum_n{}' \left\langle \phi_i \left| \frac{\partial H}{\partial a_m} \right| \phi_n \right\rangle \left\langle \phi_n \left| \frac{\partial H}{\partial a_n} \right| \phi_i \right\rangle \Big/ (E_i - E_n) + c.c., \tag{8.32}$$

In Equation 8.32 the sum over n goes over all the eigenvalues of H, the prime over the summation sign indicates that $n \neq i$, and c.c. stands for complex conjugate.

Finally, the derivatives can be expressed in a form which is amenable to computer evaluation:

$$\frac{\partial E_i}{\partial a_j} = Tr\left[\frac{\partial H}{\partial a_j}(|\phi_i\rangle \otimes \langle\phi_i|)\right], \tag{8.33}$$

and

$$\frac{\partial^2 E_i}{\partial a_j \partial a_l} = \sum_n{}' Tr\left[\frac{\partial H}{\partial a_l}(|\phi_n\rangle \otimes \langle\phi_n|) \frac{\partial H}{\partial a_n}(|\phi_i\rangle \otimes \langle\phi_i|)\right] \Big/ (E_i - E_n) + c.c., \tag{8.34}$$

In Equations 8.33 and 8.34, c.c. denotes complex conjugate, Tr stands for the trace of a matrix, $|\phi_i\rangle \otimes \langle\phi_i|$ is the $(2S+1)\times(2S+1)$ matrix obtained by taking the outer product of the column (eigen) vector $|\phi_i\rangle$ with itself, whose elements are

$$(|\phi_i\rangle \otimes \langle\phi_i|)_{pq} = |\phi_i\rangle_p |\phi_i\rangle_q^*, \tag{8.35}$$

where * denotes complex conjugate and $\partial H/\partial a_j$ are the spin operators either proportional to the components of the spin operator S or O_l^m as found from Equation 8.26.

The following steps may, then, be used for the evaluation of SHPs from EPR data by the LSF technique:

i) Make an initial estimate of the SHP (components of a^i) by using the second-order perturbation expressions for eigenvalues. Choose the values of SHPs that cannot be so determined to be zero.

ii) For orthogonal-axis consistency, fit simultaneously the resonant field values corresponding to two, or more, orientations of the external magnetic field. If an insufficient number of "allowed" ($|\Delta M|=1$) lines is available, consider "forbidden" ($|\Delta M|>1$) lines also. The greater the number of data points used for fitting the smaller is the parameter error. For each resonant-field value B_i, calculate the elements of the SH matrix using the a^i, determined in step (i) above. Compute the eigenvalues E_i and the eigenvectors $|\phi_i\rangle$ using an appropriate computer subroutine.

iii) Using the E_i and $|\phi_i\rangle$, evaluate χ^2, $\partial \chi^2/\partial a_j$, and $\partial^2 \chi^2/\partial a_j \partial a_l$ from Equations 8.27–8.29 and 8.33–8.35.

iv) Calculate the matrices D' and D'', defined in Equations 8.20 and 8.21 above.

v) Compute the eigenvalues and eigenvectors of D''. To avoid nonconvergence, if any eigenvalue (say m^{th}) is negative or zero, then replace it as follows:

$$\begin{aligned}\left(D_d'^{-1}\right)_{mm} &= 1/|(D_d)_{mm}| \quad \text{for } (D_d)_{mm} < 0, \\ &= 0 \quad \text{for } (D_d)_{mm} = 0,\end{aligned} \quad (8.36)$$

where D_d is the diagonal matrix obtained by the diagonalization of D'', that is, $D_d = UD''(a^i)U^\dagger$, where U is an appropriate unitary matrix, so $D''(a^i)^{-1} = U^\dagger D_d^{-1} U$ to be used in Equation 8.21. Normally, if a^i is sufficiently close to a^m all the eigenvalues of $D''(a^i)$ are positive definite and the LSF procedure quickly leads to convergence.

Calculate a new set of parameters constituting the vector a^f from Equation 8.23, using the initial set of values of a^i of step (i), and with the resulting a^f calculate a new χ^2. If this χ^2 is less than the old χ^2, go to the next step (vi); otherwise, use the strategies described in Section 8.4.4 below.

vi) If the last $\chi^2 \leq \chi^2_{min}$ where χ^2_{min} is a small value consistent with experimental uncertainties, terminate the computation; a^f as obtained in step (v) above then describes the values of parameters giving the best overall fit to the data. Otherwise, go back to step (ii) using the a^f obtained in step (v) above in the place of a^i.

The LSF method was successfully applied (Misra, 1976) to the fitting of EPR data for Gd^{3+}-doped $SmCl_3 \cdot 6H_2O$, characterized by large off-diagonal elements in the SH matrix. This yielded the same set of parameters as was obtained by using the brute-force method and varying parameters in the subgroups (Section 8.4.2). The benefit was that using the LSF method required only $\frac{1}{68}$ th of the time on a CYBER/CDC computer of that taken for the brute-force method, varying parameters in the subgroups as described in Section 8.3.2 (Misra and Sharp, 1976).

8.4.3
Determination of Parameter Errors

The procedure for determining parameter errors in the fitting of EPR data, based on standard statistical analysis (Bevington, 1969; Wolberg, 1967), has been well described by Misra and Subramanian (1982). The details are as follows. According to statistical analysis the parameter errors can be expressed as:

$$\delta a_j = \sqrt{\varepsilon_{jj}}, \quad (8.37)$$

where, ε_{jj} is the jth diagonal element of the error matrix ε which is defined as follows:

$$\varepsilon = A^{-1}, \; A_{ij} = \frac{1}{2}\left(\partial^2 \chi^2 / \partial a_i \partial a_j\right)_a, \quad (8.38)$$

where χ^2 is as defined by Equation 8.22 above; the required second derivatives can be computed for EPR by using Equations 8.29, 8.33, and 8.34. When all σ_i are equal, the parameter errors are estimated from the following expression (Bevington, 1969):

$$(\delta a_j)^2 = [\chi^2(1)/(N'-N)]\varepsilon_{jj}(1), \quad (8.39)$$

where $\varepsilon(1)$s the error matrix obtained when all $\sigma_i = 1$, N' is the number of lines fitted simultaneously, and N is the number of parameters used. This equation also expresses the fact that when N' is large, the parameter errors are inversely proportional to the square root of N'. Hence, the greater the number of lines fitted, the smaller is the parameter error.

In general, the σ_i-values must be appropriately estimated for EPR data for use in Equation 8.38. Here σ_i represents the uncertainty in the determination of the quantity $(|\Delta E_i| - h v_i)$ appearing in Equation 8.27. This may arise due to uncertainty in the determination of the center of the first-derivative lineshape, as well as that in the determination of the orientation θ_i of the magnetic field with respect to an axis in a chosen plane. Finally,

$$\sigma_i = \left[\left(\frac{\partial \Delta E_i}{\partial B}\right)^2 \delta B_i^2 + \left(\frac{\partial \Delta E_i}{\partial \theta}\right)^2 \delta \theta_i^2 \right]^{1/2} \quad (8.40)$$

where

$$\frac{\partial \Delta E_i}{\partial B} = \mathrm{Tr}\left[\left(\frac{\partial H}{\partial B}\right)(|\phi_{i'}\rangle\langle\phi_{i'}| - |\phi_{i''}\rangle\langle\phi_{i''}|)\right] \quad (8.40a)$$

and

$$\frac{\partial \Delta E_i}{\partial \theta} = \mathrm{Tr}\left[\left(\frac{\partial H}{\partial \theta}\right)(|\phi_{i'}\rangle\langle\phi_{i'}| - |\phi_{i''}\rangle\langle\phi_{i''}|)\right] \quad (8.40b)$$

In Equations 8.40a and b, i' and i'' designate the levels participating in resonance and $H|\phi_i\rangle = E_i|\phi_i\rangle$.

8.4.4
General Strategies for Achieving Convergence

If the initial set of parameters a^i is close enough to that corresponding to the absolute minimum of χ^2, a^m, the LSF iteration procedure quickly leads to a convergence to a^m. If this is not the case, the following strategies may be used to achieve convergence. However, if these do not work one should carefully re-estimate the initial values of the parameters a^i.

8.4.4.1 Use of Interpolated Fields: Calculation of Resonant Field Values
This technique has been proposed for the fitting of LSF data. The basis of this method is the computation of theoretical resonant fields corresponding to the actual resonant fields to be fitted by using the set of initial parameters. These may be calculated either by using perturbation expressions, or by a brute-force technique. In the latter case, numerical diagonalization of the SH matrix is

performed for successively increasing magnetic field values until the calculated energy difference for a chosen field value ΔE_i becomes equal to $h\nu_i$; this field value is the calculated field value corresponding to the given transition. If this does not happen, then differences between the closest calculated fields and the experimental field values are then divided into parts by a judiciously chosen integer n_I (>1). The fitting to the parameters is then made successively with respect to these interpolated fields, starting with the interpolated field values closest to the calculated fields until such time one finally converges to the observed field values. The parameters a^i are thus modified in a step-wise manner toward a^m, using the sets of interpolating fields. The disadvantage of this technique is the excessive computer time taken in calculating the theoretical fields by the brute-force method.

8.4.4.2 Use of Interpolated Frequencies

In recognizing the difficulty in computing theoretical fields, it appears easier to compute theoretical frequencies corresponding to the calculated energy levels participating in resonance to be compared with the observed klystron frequencies, as these theoretical frequencies are proportional to the differences between the energies corresponding to the levels participating in resonance. One can then approach toward a^m starting with a^i in a step-wise fashion using the interpolating frequencies, in the same way as one proceeded with the use of interpolated fields described in Section 8.4.4.1 above. It should be noted that this technique does not always bring convergence.

8.4.4.3 Use of Binary Chop (Misra, 1976)

When the initial values of the set of parameters chosen for an iteration, to be used in Equation 8.23, are far from the absolute minimum of the χ^2-value, the resulting value of parameters may yield a χ^2-value which is greater than that obtained in the previous iteration being corrected to a local minimum. One may then invoke the binary-chop technique, in which the LSF correction $-\Delta a^i$ to a^i to calculate a^f is reduced by a factor of 2 in Equation 8.24; that is, it is now $-(D''(a^i))^{-1}D'/2$. If the resulting χ^2-value is still greater than that obtained in the previous iteration, this difference is again chopped by 2, and so on, until the resulting χ^2-value is less than that in the previous iteration. Finally, it is noted that the binary-chop technique may not always work if the initial values chosen are far apart from those corresponding to the absolute minimum of the χ^2-function. In that case, one must choose different initial values.

8.5
Other Applications of the LSF Method

The LSF method described in Section 8.4 was developed specifically for a spin Hamiltonian with fine-structure terms only, with the assumption that the principal axes of the $\tilde{g}^2$-tensor are coincident with those of the crystal-field b_2^m

tensor (i.e., one had to consider only the diagonal elements g_z, g_x, and g_y). The LSF technique can, however, be extended to cover the cases when there are hyperfine terms present in the spin Hamiltonian, as well as to the case of noncoincident anisotropic $\tilde{g}^2$- and $\tilde{A}^2$-tensors. On the other hand, one could compute rigorously the intensities of EPR lines for an arbitrary orientation of the external magnetic field; since the intensities depend on SHPs, the line intensities, along with the line positions, can then be fitted in an LSF manner to evaluate SHPs. The line positions can also be estimated by using the *homotopy* technique, wherein one exploits the known value of the line position at a close orientation of the external magnetic field by the LSF technique to compute the line position at a desired orientation of **B** (Misra and Vasilopoulos, 1980). (For further details on this topic, see Chapter 9, in which the calculation of angular variation of EPR spectra is described.)

8.5.1
Electron–Nuclear Spin-Coupled Systems (Misra, 1983)

When the SH contains both the fine structure (FS) and hyperfine structure (HFS) terms, certain difficulties will be encountered in the application of the LSF method, as follows

i) The HFS terms cause further splitting of the FS energy levels. Thus, the various energy levels for these electron–nuclear spin-coupled (ENSC) systems are not at all sufficiently spaced, unlike those for systems with only electron spins. Thus, there may be many pairs of energy levels whose differences are reasonably close to $h\nu$. Then, it may be extremely difficult to identify the energy levels participating in resonances for a given EPR line on the basis of energy difference, especially when the initial set of SHPs is far from that corresponding to the absolute minimum of the χ^2-function. Thus, in order to achieve convergence of the LSF procedure, it is essential that the energy levels participating in resonance are correctly identified.

ii) Since here one is dealing with a rather large SH matrix (of dimension $(2S+1)(2I+1)\times(2S+1)(2I+1)$; for example, 36×36 for Mn^{2+}, $S=I=\frac{5}{2}$) it is important to choose good initial values of SHPs to save computer time in arriving at convergence quickly, as discussed below.

8.5.1.1 Estimation of Initial Values of FS SHPs
These can be obtained by using the line positions that correspond to the presence of FS terms only. These lines are fitted by the LSF technique to evaluate the FS SHPs. This is done by taking the line positions $B(M,m)$ for multiplets corresponding to HFS terms (same M, various m) and averaging them to obtain the resulting FS line position, had the HFS interaction been turned off to zero. To this end, one can take advantage of the perturbation expressions as given in Section 8.2 above. Specifically, the contribution B_{hf} to the resonant-field value for

8.5 Other Applications of the LSF Method

the allowed transition $M, m \rightarrow M-1, m$ by the HFS terms $AS_z I_z + B(S_x I_x + S_y I_y)$ for $\mathbf{B} \parallel z$ is given by (Misra, 1984; Bleaney and Ingram, 1951):

$$B_{hf}(M, m) = -Am - \frac{B^2}{2B_0}[I(I+1) - m^2 + m(2M-1)], \qquad (8.41)$$

where $B_0 = (h\nu/g\mu_B)$ [when $\mathbf{B} \parallel x$ replaces A by B and B^2 by $(A^2 + B^2)/2$ in Equation 8.41]. The resulting FS line position $B_f(M)$ as deduced from the HFS line position $[(B(M, m) = B_f(M) + B_{hf}(M, m))]$ from the HFS multiplet by turning the HFS interaction off can now be obtained from Equation 8.41. For example, for Mn^{2+}:

$$B_f(M) = \frac{3}{4}\left[B_{hf}\left(M, \frac{5}{2}\right) + B_{hf}\left(M, -\frac{5}{2}\right)\right] - \frac{1}{8}\left\{\begin{array}{l}\left[B_{hf}\left(M, \frac{3}{2}\right) + B_{hf}\left(M, -\frac{3}{2}\right)\right]\\+\left[B_{hf}\left(M, \frac{1}{2}\right) + B_{hf}\left(M, -\frac{1}{2}\right)\right]\end{array}\right\}.$$

8.5.1.2 Estimation of HFS Parameters

The initial values of the large HFS parameters A and B should be carefully chosen, with the initial values of the remaining parameters set equal to zero. As seen from Equation 8.41, the value of A is very closely the average of all the separations of the HFS transitions for $\mathbf{B} \parallel z$ in the $2S$ resonance lines for the FS group, as the average of m for $m = -I, -(I-1), \ldots, (I-1), I$ is zero. The HFS parameter B can be similarly estimated for $\mathbf{B} \parallel x$.

8.5.1.3 Identification of Energy Levels Participating in Resonance

After fitting $B_f(M)$ to evaluate the FS parameters the electronic quantum numbers for the transitions corresponding to each $B_f(M)$ are identified. It is now easy to decide for an observed resonant EPR line position which particular hyperfine energy levels participate in resonance when the HFS interaction is turned on. Depending upon the sign of A and B the $(2I+1)$ hyperfine energy levels corresponding to a chosen value of M for $\mathbf{B} \parallel z$ (or nearly $\parallel z$) occur in order of increasing energy as follows (Bleaney and Ingram, 1951):

$$m = -I, -(I-1), \ldots, (I-1), I \quad \text{if} \quad MA > 0,$$

whereas they will be

$$m = I, (I-1), \ldots, -(I-1), -I \quad \text{if} \quad MA < 0,$$

for $\mathbf{B} \parallel x$, A is replaced by B.

8.5.1.4 Construction of the SH Matrix for ENSC Systems

For ENSC systems the SH $H(S, I)$ consists of terms of the type $H'(S, I) = [H'(S)H'(I)], H''(S)$ and $H''(I)$. The SH matrix elements corresponding to these terms can then be expressed in the direct-product space as

$$\langle M'm' | H'(S, I) | M'', m'' \rangle = \langle M' | H'(S) | M'' \rangle \langle m' | H'(I) | m'' \rangle,$$

$$\langle M'm' | H''(S) | M'', m'' \rangle = \langle M' | H''(S) | M'' \rangle \langle m' | 1_I | m'' \rangle,$$

$$\langle M'm'|H''(I)|M'',m''\rangle = \langle M'|1_S|M''\rangle\langle m'|H''(I)|m''\rangle,$$

where 1_S and 1_I represent the unit operators in the electronic and nuclear spin spaces, respectively; so that $\langle m'|1_I|m''\rangle = \delta_{m'm''}$ and $\langle M'|1_S|M''\rangle = \delta_{M'M''}$, where $\delta_{m'm''}$ and $\delta_{M'M''}$ are Kronecker delta functions.

8.5.1.5 Absolute Signs of SH Parameters

The LSF technique provides the correct relative signs of all the FS, as well as all of the HFS, parameters. Thus, knowing the absolute sign of one FS parameter (i.e., the largest ZFS parameter B_2^0) and one HFS parameter (i.e., A), the absolute signs of all parameters can be deduced. The absolute sign of B_2^0 can be determined from the relative intensities of lines at liquid helium temperature (LHT) (Abragam and Bleaney, 1970, p. 161). Explicitly, the intensity of the high-field lines for $\boldsymbol{B}\|z$ increases (decreases) relative to the low-field lines; as the temperature is reduced, the absolute sign of B_2^0 is positive (negative). The absolute sign of A can be determined either from hyperfine-interaction data, or from the overall splitting of the hyperfine sets (Bleaney and Ingram, 1951). If the overall splitting of the hyperfine sets for $\boldsymbol{B}\|z$ decreases with the increasing field, then B_2^0 and A have the same sign; otherwise, they have opposite signs. For $\boldsymbol{B}\|x$, similar considerations apply for the sign of B.

The LSF technique as described in this section has been successfully applied, for example, to Mn^{2+}-doped $ZnSiF_6 \cdot 6H_2O$ (Misra and Jalochowski, 1983).

8.5.2
Fitting of ENDOR Data

The considerations of Section 8.5.1 above for the determination of SHPs from EPR data for ENSC systems may be applied to the determination of SHPs from electron–nuclear double resonance (ENDOR) data. In ENDOR, one saturates a FS transition $M' \leftrightarrow M''$, and observes the ENDOR frequencies corresponding to absorption between the HF levels m, corresponding to the FS levels M' and M''. Thus, the data to be fitted are the ENDOR frequencies, which depend on the differences between appropriate energy levels for the ENSC system for a given orientation of the external magnetic field with respect to the crystal-field axes. Hence, in the expression for χ^2-function, ΔE_i are the energy differences between the levels participating in ENDOR, while v_i are the observed ENDOR frequencies. With this cognizance, the LSF technique as developed above for the evaluation of SHPs from EPR data for ENSC systems can be applied similarly to the analysis of ENDOR data. For example, it was successfully applied to the case of Fe^{3+}-doped guanidinium aluminum sulfate hexahydrate (Misra and Van Ormondt, 1984).

8.5.3
Calculation and Fitting of Line Intensities to SHP

By computing EPR line intensities and comparing them with those observed, one obtains an extra check on the computed SHPs. On the other hand, line intensities themselves can be used in the LSF technique to evaluate SHPs, especially in those situations where an insufficient number of lines is available for the evaluation of SHPs. Then, by using both the intensities and line positions, one doubles the number of data points used for the evaluation of parameters, thus decreasing the parameter errors. Such a situation occurs, for example, at liquid-helium temperatures, where one obtains data only for one or two orientations of the external magnetic field. Such a procedure has been exploited (Misra, 1979), the details of which are provided below.

8.5.3.1 The Intensity Operator

By definition, the intensity operator Γ_{mn} is such that its trace with the density matrix ρ gives the intensity I_{mn} of the line corresponding to the transition $m \leftrightarrow n$ (Misra, 1979).

$$I_{mn} = \text{Tr}(\rho \Gamma_{mn}). \tag{8.42}$$

In Equation 8.42, $\rho = \sum_j p_j P_j$ is the density matrix, where $P_j = |\phi_i\rangle \otimes \langle\phi_i|$ is the projection operator, and $p_j = \exp(-E_j/k_B T)/\sum \exp(-E_i/k_B T)$ is the normalized population of the jth energy level with the energy E_j; k_B and T are the Boltzmann constant and absolute temperature, respectively. At high temperatures, $\rho = \sum_j P_j/(2S+1)$, as the populations of all the levels are equal for the presence of FS terms only. When both FS and HFS terms are present, $\rho = \sum_j P_j/[(2S+1)(2I+1)]$.

Finally, the intensity operator is (Misra, 1979):

$$\Gamma_{mn} = K(H_{rf} P_n H_{rf} P_m - H_{rf} P_m H_{rf} P_n), \tag{8.43}$$

where the constant K is the same for all the lines and depends only upon the amplitude of the radiofrequency (RF) magnetic field and the characteristics of the electronic detection system of the spectrometer, and H_{rf} denotes the interaction Hamiltonian of the time-independent RF field inducing transitions with the magnetic moment of the spins. In general,

$$H_{rf} = [g_\| \mu_B (B_{rf})_z S_z + g_\perp \mu_B (B_{rf})_x S_x] \cos \omega t, \tag{8.44}$$

where $g_\|$ and $g_\perp$ are the g-factors parallel and perpendicular to the z-axis, respectively; B_{rf} and ω are the amplitude and angular frequency of the RF field, respectively.

8.5.3.2 Fitting of Line Intensities and Line Positions to SHP

The LSF technique as described in Section 8.4.2 can be applied to the inclusion of both the line intensities and the line positions, provided that χ^2 is appropriately described, and one knows how to compute its first and second derivative for the evaluation of $\Delta \mathbf{a}$, as given by Equation 8.24 (Misra, 1979). As evaluated explicitly

$$\chi^2 = \chi^2_{\text{pos}} + \chi^2_{\text{int}},$$

where χ^2_{pos} is the chi-square corresponding to the EPR line positions; the expressions for its derivatives as required in the LSF method have already been described in Section 8.4.2. As for χ^2_{int}, the contribution of the line intensities, the following are noted (Misra, 1979):

$$\chi^2_{\text{int}} = \sum_{mn} (I_i^{cn} - I_i^{mn})^2 / \sigma_i''^2. \tag{8.45}$$

In Equation 8.45, I_i^{cn}, I_i^{mn} and σ_i'' are, respectively, the normalized calculated intensity (normalized for each external-field orientation separately), normalized measured intensity (normalized the same way as I_i^{cn}), and the weight factor of the ith line intensity. The required derivatives are

$$\frac{\partial \chi^2_{\text{int}}}{\partial a_k} = \sum_i \frac{2(I_i^{cn} - I_i^{mn})}{\sigma_i''^2} \frac{\partial I_i^{cn}}{\partial a_k} \tag{8.46}$$

and

$$\frac{\partial^2 \chi^2_{\text{int}}}{\partial a_k \partial a_l} = \sum_i \left(\frac{2}{\sigma_i''^2} \right) \left[(I_i^{cn} - I_i^{mn}) \frac{\partial^2 I_i^{cn}}{\partial a_k \partial a_l} + \frac{\partial I_i^{cn}}{\partial a_k} \frac{\partial I_i^{cn}}{\partial a_l} \right] \tag{8.47}$$

For use in Equations 8.46 and 8.47, the expressions, given below are required.

8.5.3.3 Normalized Intensity and its Derivatives

The calculated normalized intensity I_i^{cn} and the measured normalized intensity I_i^{mn} for a given orientation of the external magnetic field are

$$I_i^{cn} = I_i^c / \sum_j I_j^c, \text{ and } I_i^{mn} = I_i^m / \sum_j I_j^m \tag{8.48}$$

In Equation 8.48, j covers all the lines considered for a given orientation of the external magnetic field. One obtains from Equation 8.48, using Equations 8.42 and 8.43:

$$\frac{\partial I_i^{cn}}{\partial a_k} = \frac{\left(\sum_j I_j^c \right) \frac{\partial I_i^c}{\partial a_k} - I_i^c \sum_j \frac{\partial I_j^c}{\partial a_k}}{\left(\sum_j I_j^c \right)^2}$$

and

$$\frac{\partial^2 I_i^{cn}}{\partial a_k \partial a_l} = \left[\left(\sum_j I_j^c \right) \left(\sum_j I_j^c \right) \left(\frac{\partial^2 I_i^c}{\partial a_k \partial a_l} \right) + \left(\sum_j \frac{\partial I_j^c}{\partial a_l} \right) \frac{\partial I_i^c}{\partial a_k} - \left(\sum_j \frac{\partial I_j^c}{\partial a_k} \right) \frac{\partial I_i^c}{\partial a_k} - I_i^c \left(\sum_j \frac{\partial^2 I_j^c}{\partial a_k \partial a_l} \right) \right]$$
$$- 2 \left\{ \left(\sum_j I_j^c \right) \frac{\partial I_i^c}{\partial a_k} - I_i^c \left(\sum_j \frac{\partial I_j^c}{\partial a_k} \right) \right\}$$
$$- I_i^c \left(\sum_j \frac{\partial^2 I_j^c}{\partial a_k \partial a_l} \right) - 2 \left\{ \left(\sum_j I_j^c \right) \frac{\partial I_i^c}{\partial a_k} - I_i^c \left(\sum_j \frac{\partial I_j^c}{\partial a_k} \right) \right\} \left(\sum_j \frac{\partial I_j^c}{\partial a_l} \right) / \left(\sum_j I_j^c \right)^3$$

$$\tag{8.49}$$

8.5 Other Applications of the LSF Method

For use in Equation 8.49, as obtained from Equation 8.43, we have

$$\frac{\partial I_i^c}{\partial a_k} = \left[\left(\frac{\partial p_{i'}}{\partial a_k} - \frac{\partial p_{i''}}{\partial a_k}\right) Tr\left(H_{rf} P_{i''} H_{rf} P_{i'} + (p_{i'} - p_{i''})\right) \frac{\partial f_{ij}}{\partial a_k}\right], \qquad (8.50)$$

where

$$\frac{\partial f_{ij}}{\partial a_k} = Tr\left(\frac{\partial H_{rf}}{\partial a_k} P_{i''} H_{rf} P_{i'} + H_{rf} \frac{\partial P_{i''}}{\partial a_k} H_{rf} P_{i'} + H_{rf} P_{i''} \frac{\partial H_{rf}}{\partial a_k} P_{i'} + H_{rf} P_{i''} H_{rf} \frac{\partial P_{i'}}{\partial a_k}\right). \qquad (8.51)$$

For use in the above equations the derivatives of the projection operators P_j and the normalized thermal populations p_j are required; these are given later in Equations 8.53–8.56.

From Equation 8.50 one obtains for the second derivatives

$$\frac{\partial^2 I_i^c}{\partial a_k \partial a_l} = \left(\frac{\partial^2 p_{i'}}{\partial a_k \partial a_l} - \frac{\partial^2 p_{i''}}{\partial a_k \partial a_l}\right) Tr\left(H_{rf} P_{i''} H_{rf} P_{i'}\right) + \left(\frac{\partial p_{i'}}{\partial a_k} - \frac{\partial p_{i''}}{\partial a_k}\right)$$

$$\times Tr\left(\frac{\partial H_{rf}}{\partial a_l} P_{i''} H_{rf} P_{i'} + H_{rf} \frac{\partial P_{i''}}{\partial a_l} H_{rf} P_{i'} + H_{rf} P_{i''} \frac{\partial H_{rf}}{\partial a_l} P_{i'}\right. \qquad (8.52)$$

$$\left. + H_{rf} P_{i''} H_{rf} \frac{\partial P_{i'}}{\partial a_l}\right)\left(\frac{\partial p_{i'}}{\partial a_l} - \frac{\partial p_{i''}}{\partial a_l}\right) \frac{\partial f_{ij}}{\partial a_k} + (p_{i'} - p_{i''}) \frac{\partial^2 f_{ij}}{\partial a_k \partial a_l},$$

where

$$\frac{\partial^2 f_{ij}}{\partial a_k \partial a_l} = Tr\left(\frac{\partial H_{rf}}{\partial a_l} \frac{\partial P_{i''}}{\partial a_k} H_{rf} P_{i'} + H_{rf} \frac{\partial^2 P_{i''}}{\partial a_k \partial a_l} H_{rf} P_{i'}\right.$$

$$+ H_{rf} \frac{\partial P_{i''}}{\partial a_k} \frac{\partial H_{rf}}{\partial a_l} P_{i'} + H_{rf} \frac{\partial P_{i''}}{\partial a_k} H_{rf} \frac{\partial P_{i'}}{\partial a_l} + \frac{\partial H_{rf}}{\partial a_l} P_{i''} H_{rf} \frac{\partial P_{i'}}{\partial a_k}$$

$$+ H_{rf} \frac{\partial P_{i''}}{\partial a_l} H_{rf} \frac{\partial P_{i'}}{\partial a_k} + H_{rf} P_{i''} \frac{\partial H_{rf}}{\partial a_l} \frac{\partial P_{i'}}{\partial a_k} + H_{rf} P_{i''} H_{rf} \frac{\partial^2 P_{i'}}{\partial a_k \partial a_l}$$

$$+ \frac{\partial H_{rf}}{\partial a_k} \frac{\partial P_{i''}}{\partial a_l} H_{rf} P_{i'} + \frac{\partial H_{rf}}{\partial a_k} P_{i''} \frac{\partial H_{rf}}{\partial a_l} P_{i'} + \frac{\partial H_{rf}}{\partial a_k} P_{i''} H_{rf} \frac{\partial P_{i'}}{\partial a_l}$$

$$\left. + \frac{\partial H_{rf}}{\partial a_l} P_{i''} \frac{\partial H_{rf}}{\partial a_k} P_{i'} + H_{rf} \frac{\partial P_{i''}}{\partial a_l} \frac{\partial H_{rf}}{\partial a_k} P_{i'} + H_{rf} P_{i''} \frac{\partial H_{rf}}{\partial a_k} \frac{\partial P_{i'}}{\partial a_l}\right).$$

For use in Equations 8.50–8.52, the derivatives of the projection operators are

$$\frac{\partial P_i}{\partial a_k} = -(H - E_i 1)^{-1} \left(\frac{\partial H}{\partial a_k} - \frac{\partial E_i}{\partial a_k}\right) P_i \qquad (8.53)$$

and

$$\frac{\partial^2 P_i}{\partial a_k \partial a_l} = -(H - E_i 1)^{-1} \left[\frac{\partial^2 E_i}{\partial a_k \partial a_l} P_i - \left(\frac{\partial H}{\partial a_l} - \frac{\partial E_i}{\partial a_l} 1\right) \frac{\partial P_i}{\partial a_k}\right] \qquad (8.54)$$

In Equations 8.53 and 8.54 1 is the unit matrix. The derivatives of eigenvalues as required in Equations 8.53 and 8.54 can be computed using Equations 8.33 and

8.34. Finally, the derivatives of the normalized populations as required in Equations 8.51 and 8.52 are:

$$\frac{\partial p_i}{\partial a_k} = \frac{p_i}{k_B T}\left\{\left(\sum_j e^{-E_j/k_B T}\frac{\partial E_j}{\partial a_k}\right)\Big/z - \frac{\partial E_i}{\partial a_k}\right\}, \text{ where } z = \sum_j e^{-E_j/k_B T}, \quad (8.55)$$

and

$$\frac{\partial^2 p_i}{\partial a_k \partial a_l} = \frac{\partial p_i/\partial a_l}{k_B T}\left\{\left(\sum_j e^{-E_j/k_B T}\frac{\partial E_j}{\partial a_k}\right)\Big/z - \frac{\partial E_i}{\partial a_k}\right\}$$

$$+ \frac{p_i}{k_B T}\left\{-\sum_j \frac{e^{-E_j/k_B T}}{k_B T}\frac{\partial E_j}{\partial a_l}\frac{\partial E_j}{\partial a_k} + e^{-E_j/k_B T}\frac{\partial^2 E_j}{\partial a_k \partial a_l}\Big/z\right\}$$

$$+ \left(\sum_j e^{-E_j/k_B T}\frac{\partial E_j}{\partial a_k}\right)\left(\sum_j \frac{e^{-E_j/k_B T}}{k_B T}\frac{\partial E_j}{\partial a_l}\right)\Big/z^2 - \frac{\partial^2 E_i}{\partial a_k \partial a_l}. \quad (8.56)$$

8.5.3.4 Limits of Applicability of the Method

The line-intensity expression and its application to LSF as described is only applicable to the S-state ions and EPR induced by the RF magnetic field. It is not applicable to the cases of Jahn–Teller ions, as in these cases the line is broadened by strain; the peak position comes from nonzero strain sites and the transition probabilities in different regions of the line are also sometimes significantly strain dependent.

The intensity operator described above has been successfully applied to the computation of the line intensities for Gd^{3+}-doped $SmCl_3 \cdot 6H_2O$ (Misra, 1979).

8.6
Concluding Remarks

Among several aspects of evaluation of SHPs from EPR data using the LSF techniques in conjunction with numerical diagonalization of the SH matrix, the following are noteworthy:

i) When the eigenvalues and eigenvectors of the SH matrix are computed, no problem arises if all of the elements of this matrix are real. For this case, well-known subroutines, such as JACOBI, which deal with real symmetric matrices can be used. When the elements of the SH matrix are complex due to the presence of spin operators like S_y, or the presence of $O_l^{-|m|}$ spin operators in the SH, then computer subroutines such as EISPACK can be used to determine the real eigenvalues and complex eigenvectors of the SH matrix. If these subroutines are not available, it is still possible to use a symmetric matrix with real elements, A', the dimension of which is twice that of the SH matrix

$$\begin{pmatrix} A & -B \\ B & A \end{pmatrix},$$

where **A** and **B** are the matrices composed of the real and imaginary parts of the SH matrix, respectively, constructed in a one-to-one correspondence with the SH matrix. The eigenvalues of this $2n \times 2n$ matrix are degenerate in n pairs; each pair has the eigenvalues of the matrix $\mathbf{A} + i\mathbf{B}$, and the eigenvectors (in columns with $2n$ elements) corresponding to any pair are such that the sets of upper and lower n elements can be chosen to be, respectively, the real and imaginary parts of the corresponding eigenvalue of the matrix $A+iB$, such that $(\mathbf{A} + i\mathbf{B})(\mathbf{u} + i\mathbf{v}) = \lambda(\mathbf{u} + i\mathbf{v})$ leads to

$$\begin{pmatrix} A & -B \\ B & A \end{pmatrix}\begin{pmatrix} u \\ v \end{pmatrix} = \lambda \begin{pmatrix} u \\ v \end{pmatrix}, \quad \text{and} \quad \begin{pmatrix} A & -B \\ B & A \end{pmatrix}\begin{pmatrix} -v \\ u \end{pmatrix} = \lambda \begin{pmatrix} -v \\ u \end{pmatrix}.$$

ii) The SH matrix can be expressed in either the "crystal" or "magnetic" coordinate system. It is recommended that the "crystal" representation be used, because the number of terms in the SH is then minimum. Many more spin-operator terms are present in the SH in "magnetic" coordinates than for "crystal" coordinates, and it is then necessary to use complicated tables of trigonometric functions based on "rotational" matrices required to transform the spin operators from "crystal" coordinates to the magnetic coordinates (Buckmaster, 1962). However, one of the advantages of using "magnetic" coordinates is for perturbation calculations, because the Zeeman term, which is the largest, is taken to be the zero-order term. Another advantage of using "magnetic" coordinates is that the coefficient of the Zeeman term determines the intensity of HF "forbidden" transition relative to the "allowed" HF transition. A good discussion of operator equivalents has been provided by Rudowicz (1985a, 1985b), as well as for the relations between arbitrary SHP in various axes systems.

iii) The use of perturbation expressions has almost become obsolete, because exact matrix diagonalization is now easy to carry out on PCs. However, perturbation expressions are still useful to investigate the role and relative importance of individual energy terms, which depend on SHPs. They also provide initial values to be used in LSF using eigenvalues and eigenvectors calculated on a computer.

iv) The angular variation of resonant fields can be calculated using known field values at a given orientation of the magnetic field (initial resonant fields) by the method of homotopy (Misra and Vasilopoulos, 1980) to calculate a polycrystalline spectrum (Misra, 1999a). The initial resonant fields are easy to calculate by perturbation when the orientation of the magnetic field is parallel to one of the symmetry axes. As for an arbitrary orientation of the magnetic field, the *eigenfield* method (Belford, Belford, and Burkhalter, 1973; Belford et al., 1974; Drew, 2002), as described in Chapter 9, can be used to calculate the resonant field for an arbitrary orientation of the magnetic field, to be used as initial values, or to simulate a glassy spectrum (Chapter 9). Although the eigenfield method is time consuming, being a brute-force method, it is the only alternative for this purpose. Once the polycrystalline spectrum has been calculated, one can exploit the least-squares fitting procedure to evaluate the spin-Hamiltonian parameters from a polycrystalline (powder) spectrum, as described by Misra (1999b) and outlined in Chapter 9.

v) Recently, McGavin and Tennant (2009) have published expressions for higher-order Zeeman and spin terms in the SH for EPR, and their description in irreducible form using Cartesian, tesseral spherical tensor, and Stevens' operator expressions.

Acknowledgments

The author is grateful to Professor C.P. Poole, Jr, for helpful comments to improve the presentation of this chapter, and to the Natural Sciences and Engineering Research Council (NSERC) of Canada for partial financial support.

Pertinent Literature

The first article summarizing the various techniques described above on computer evaluation of SHP was written by Misra (1986). The inclusion of intensity of EPR lines to evaluate SHP was proposed by Misra (1979). The estimation of errors in evaluating SHP from EPR data was outlined by Misra and Subramanian (1982). The *homotopy* technique was first proposed in context with EPR by Misra and Vasilopoulos (1980). The list of references provided below includes the most important reports on the subject of estimating SHP from EPR data. More related details are provided in Chapters 7, 9, and 14. The various perturbation techniques as applied to EPR, in chronological order of their development, are listed in Appendix 8.I.

References

Abragam, A. and Bleaney, B. (1970) *Electron Paramagnetic Resonance of Transition Ions*, Clarendon Press, Oxford.

Belford, G.G., Belford, R.L., and Burkhalter, J.F. (1973) *Magn. Reson.*, **11**, 251.

Belford, R.L., Davis, P.H., Belford, G.G., and Lenhardt, T.M. (1974) Extended interactions between metal ions. *Am. Chem. Soc. Symp. Ser.*, **5**, 40.

Bevington, P.B. (1969) *Data Reduction and Error Analysis for the Physical Sciences*, McGraw-Hill, New York.

Bleaney, B. and Ingram, D.J.E. (1951) *Proc. R. Soc. A*, **205**, 336.

Buckmaster, H.A. (1962) *Can. J. Phys.*, **40**, 1670.

Buckmaster, H.A., Chatterjee, R., Dering, J.C., Fry, D.J.I., Shing, Y.H., Skirrow, J.D., and Venkatesan, B. (1971) *J. Magn. Reson.*, **4**, 113.

Chambers, J.G., Datars, W.R., and Calvo, C. (1964) *J. Chem. Phys.*, **41**, 806.

Drew, S.C. (2002) Selected topics in modern continuous-wave and pulsed electron paramagnetic resonance. PhD Thesis. Monash University, Melbourne, Australia.

Feynman, R.P. (1939) *Phys. Rev.*, **56**, 340.

Golding, R.M. and Tennant, W.C. (1973) *Mol. Phys.*, **25**, 1163.

Jones, D.A., Baker, J.M., and Pope, D.F.D. (1959) *Proc. Phys. Soc.*, **74**, 249.

Keijzers, C.P., Paulussen, G.F.M., and de Boer, E. (1975) *Mol. Phys.*, **29**, 1973.

Lin, W.C. (1973) *Mol. Phys.*, **25**, 247.

McGavin, D.G., Palmer, R.A., Singers, W.A., and Tennant, W.C. (1980) *J. Magn. Reson.*, **40**, 69.

McGavin, D.G. and Tennant, W.C. (2009) *J. Phys. Condens. Matter*, **21**, 245501.

Markham, G.D., Rao, B.D.N., and Reed, G.H. (1979) *J. Magn. Reson.*, **33**, 595.

Meirovitch, E. and Poupko, R. (1978) *J. Phys. Chem.*, **82**, 1920.

Misra, S.K. (1976) *J. Magn. Reson.*, **23**, 403.

Misra, S.K. (1979) *J. Phys. C*, **12**, 5221.

Misra, S.K. (1983) *Physica*, **121B**, 193.

Misra, S.K. (1984) *Physica*, **124B**, 53.

Misra, S.K. (1986) *Magn. Reson. Rev.*, **10**, 285.

Misra, S.K. (1988) *Arab. J. Sci. Eng.*, **13**, 255.

Misra, S.K. (1999a) *J. Magn. Reson.*, **137**, 83.

Misra, S.K. (1999b) *J. Magn. Reson.*, **140**, 179.

Misra, S.K. and Jalochowski, M. (1983) *Physics B*, **119**, 295.

Misra, S.K. and Sharp, G.R. (1976) *J. Magn. Res.*, **23**, 191.

Misra, S.K. and Subramanian, S. (1982) *J. Phys. C*, **15**, 7199.

Misra, S.K. and Van Ormondt, D. (1984) *Phys. Rev. B*, **30**, 6327.

Misra, S.K. and Vasilopoulos, P. (1980) *J. Phys. C*, **13**, 1083.

Rudowicz, C. (1985a) *J. Magn. Reson.*, **63**, 95.

Rudowicz, C. (1985b) *J. Phys. C*, **18**, 1415.

Scarborough, J.B. (1966) *Numerical Mathematical Analysis*, Johns Hopkins Press, Baltimore.

Singh, G.B. and Venkateswarlu, P. (1967) *Proc. Ind. Acad. Sci.*, **A65**, 211.

Tennant, W.C. (1976) Report No. CD2209, Chemistry Division, Department of Scientific and Industrial Research, New Zealand.

Upreti, G.C. (1974) *J. Magn. Reson.*, **13**, 336.

Wolberg, J.R. (1967) *Prediction Analysis*, Van Nostrand, Princeton, NJ.

Appendix 8.1 Historical Review

This appendix lists, in ascending chronological order, the particular features of the various reports on perturbation calculations of eigenvalues of spin Hamiltonian for electron–nuclear spin-coupled systems (Misra, 1988); the relevant reference is included in square brackets in the comments column. The terms "tensors" and "matrices" have been used interchangeably in context with g and A factors. To be correct, $\tilde{g}, \tilde{A}$ are matrices, whereas $\tilde{g}^2, \tilde{A}^2$ are tensors.

Features of perturbation calculation

1) The energy levels and transition probabilities for a spin-only electron coupled to neighboring nuclei are given.

   G.T. Trammell, H. Zeldes, and R. Livingston, *Phys. Rev.*, **110**, 630 (1958).

2) This article shows that the $\tilde{A}$ matrix may not be symmetric.

   H.M. McConnell, *Proc. Natl Acad. Sci. USA*, **44**, 766 (1958).

3) First-order perturbation energy of electron-nuclear spin coupled systems is calculated.

   H. Zeldes and R. Livingston, *J. Chem. Phys.*, **35**, 1410 (1961).

4)	Hyperfine energy is calculated to first-order in perturbation corresponding to one electron spin interacting with a number of neighboring nuclear spins.	J.A. Weil and J.H. Anderson, *J. Chem. Phys.*, **35**, 1410 (1961).
5)	The point is made that when there is no crystal field splitting, measurement of resonant magnetic fields, or resonant frequencies, gives no indication of the asymmetric nature of the g matrix unless circularly polarized radiation is used.	F.K. Kneubühl, *Physik der Konden. Matt.*, **1**, 410 (1963).
6)	Second-order perturbation energies are calculated for the case of coincident **g**, **A** matrices, using axes of quantization along $g \cdot B$ and $A \cdot S$ for electron and nuclear spins, respectively.	R.E.D. McClung, *Can. J. Phys.*, **46**, 2271 (1968).
7)	Details of analysis of ligand ENDOR data are given to evaluate SHP.	J.M. Baker, E.R. Davies, and J.P. Hurrell, *Proc. Roy. Soc. (London) A* **308**, 403 (1968).
8)	This is a well-known (albeit old) book on EPR	A. Abragam and B. Bleaney, *Electron Paramagnetic Resonance of Transition Ions*, Oxford: Clarendon Press, (1970). It discusses the properties of $\tilde{g}$, $\tilde{A}$ matrices; it also gives eigenvalue expressions, based on perturbation theory; see, in particular, Tables 16 and 17, pp. 863, 864.
9)	EPR transition fields for a general spin Hamiltonian are calculated, using seventh-order degenerate perturbation theory, and used for the evaluation of $\tilde{g}$, $\tilde{A}$ matrices.	C.R. Byfleet, D.P. Chong, J.A. Hebden, and C.A. McDowell, *J. Magn. Reson.*, **2**, 69 (1970).
10)	Third-order perturbation expressions are given for the eigenvalues for the case of coincident tensor axes.	R.M. Golding, R.H. Newman, A.D. Rae, and W.C. Tennant, *J. Chem. Phys.*, **57**, 1912 (1971).

11) Second-order perturbed eigenvalues are given for noncoincident tensors, ignoring the nuclear Zeeman term. Application to the study of powder spectra is included. R.M. Golding and W.C. Tennant, *Mol. Phys.*, **25**, 1163 (1972).

12) Second-order perturbation expressions are given for the case of coincident (diagonal) tensors appropriate to orthorhombic symmetry. U. Sakaguchi, Y. Arata, and S. Fujiwara, *J. Magn. Reson.*, **9**, 118 (1973).

13) Second-order expressions for the case of non-diagonal $\tilde{g}, \tilde{D}, \tilde{A}$ matrices are given. A convenient choice of quantization axes for nuclear and electron spins for eigenvalue calculations is provided. W.C. Lin, *Mol. Phys.*, **25**, 247 (1973).

14) Second-order eigenvalues are given for the SH of low symmetry, that is, for non-coincident tensors. Angular dependence of transition probability up to first-order is given. A. Rockenbauer and P. Simon, *J. Magn. Reson.*, **11**, 217 (1973).

15) Second-order perturbation expressions are used to evaluate SHP corresponding to non-coincident tensors. A method has been developed which removes the necessity to measure the spectra on the same crystal and in three mutually orthogonal planes. Specific application is made to the case of Cu^{2+}-doped nickel di-*n*-butyl dithiocarbamate. C.P. Keijzers, G.F.M. Paulussen, and E. de Boer, *Mol. Phys.*, **29**, 1973 (1975).

16) General third-order solutions are given, neglecting the nuclear Zeeman term. Particular application to the study of 'half-field' spectra ($\Delta M = \Delta 2$; $\Delta m = 0, \pm 1$) is outlined. R.M. Golding and W.C. Tennant, *Mol. Phys.*, **28**, 167 (1974).

17)	Second-order perturbation solutions are given for the SH for the case of non-coincident and asymmetric $\tilde{g}, \tilde{A}$ matrices; the nuclear Zeeman term is not considered. Angular dependence of intensities for allowed and first-order forbidden transition is described.	A. Rockenbauer and P. Simon, *Mol. Phys.*, **28**, 1113 (1974).
18)	Second-order perturbation treatment is given for the case of non-coincident tensors, where the $\tilde{g}, \tilde{A}$ matrices may be asymmetric. The transition probability is given to the first-order and complementarily to the second-order for some important cases.	M. Iwasaki, *J. Magn. Reson.*, **16**, 417 (1974).
19)	Second-order perturbation energy for noncoincident tensors, and for more than one nucleus, is given. The nuclear Zeeman terms, as well certain cross-terms, missing in the previous publications, have also been included.	J.A. Weil, *J. Magn. Reson.*, **18**, 113 (1975).
20)	SHP are evaluated assuming noncoincident $\tilde{g}, \tilde{A}$ matrices. However, the term depending on $g^T.A^T.g$ has been ignored.	S.D. Pandey and P. Venkateswarlu, *J. Magn. Reson.*, **17**, 137 (1975).
21)	SH is transformed to coordinates in which electron Zeeman interaction is diagonal. The eigenvalues are calculated to second-order in perturbation; the effect of previously ignored second-order and cross-terms is illustrated.	W.C. Tennant, *Mol. Phys.*, **31**, 1505 (1976).
22)	SH for low symmetry is given. Expressions for EPR resonant magnetic fields are provided.	A.M. Grekhov and A.B. Roitsin, *Phys. Status Solidi (B)*, **74**, 323 (1976).
23)	It is pointed out that the various tensors in the spin Hamiltonian may be non-coincident for monoclinic or triclinic point symmetry. The cited references are also relevant.	J.R. Pilbrow, *J. Magn. Reson.*, **31**, 5041 (1978).

24)	Effective g values for purely electronic systems with $S = 3/2$ and $S = 5/2$ are calculated using third-order perturbation.	J.R. Pilbrow, *J. Magn. Reson.*, **31**, 479 (1978).
25)	Evaluation of SHP from ligand ENDOR data is described.	C.A. Hutchison Jr and T.E. Orlowski, *J. Chem. Phys.*, **73**, 1 (1980).
26)	A method has been described to compute SHP for noncoincident tensors. Both perturbation and matrix diagonalizations have been employed. The final values of parameters are evaluated by a brute-force step-wise variation of parameters in succession to yield the lowest value of χ^2. The discussion has been restricted to symmetric tensors only.	D.C. Mcgavin, R.A. Palmer, W.A. Singers, and W.C. Tennant, *J. Magn. Reson.*, **40**, 69 (1980).
27)	Both, the historical development of low-symmetry ideas and their theoretical bases have been reviewed. Major experimental results are tabulated and the discussion includes some of the attempts to interpret low-symmetry EPR data using crystal field and molecular orbital models. It is concluded that the role of excited-state orbitals, under the influence of a low-symmetry field, is largely responsible for the different orientations of the principal directions of the various interactions experienced by a paramagnetic ion.	J.R. Pilbrow and M.R. Lowrey, *Rep. Prog. Phys.*, **43**, 433 (1980).
28)	A review article on a generalized SH and low-symmetry effects in EPR.	A.B. Roitsin, *Phys. Status Solidi (B)*, **104**, 11 (1981).
29)	This is a general review article on the evaluation of SHP from EPR data.	S.K. Misra, *Magn. Reson. Rev.*, **10**, 285 (1986).

30)	Specific application of LSF technique to evaluate non-coincident g^2, A^2 tensors using first-order perturbation is illustrated.	S.K. Misra, *Physica*, **124B**, 53 (1984).
31)	Application of LSF procedure for $S = 1/2$, $I = 3/2$ using a first-order perturbation approximation, has been illustrated.	S.K. Misra, J. Bandet, G. Bacquet, and T.E. McEnally, *Phys. Status Solidi (A)*, **80**, 581 (1983).
32)	Determination of SHP including high-spin Zeeman terms of form BS^3 and allowing (g, D) matrices to be noncoincident, as well as considering nonuniaxial anisotropy of the g matrix.	M. J. Mombourquette, W. C. Tennant and J. A. Weil, *J. Chem. Phys.*, **85**, 68 (1986).
33–36)	Energy expressions for the Mn^{2+} ion up to third-order in perturbation, including both fine and hyperfine terms in SH, are given.	E. Meirovitch and R. Poupko, *J. Phys. Chem.*, **82**, 1920 (1978); G.D. Markham, B.D.N. Rao and G.H. Reed, *J. Mag. Reson.*, **33**, 595 (602); S.K. Misra, *Physica B*, **203**, 193 (1994); S.K. Misra, *Physica B*, **240**, 183 (1997).

9
Simulation of EPR Spectra

Sushil K. Misra

9.1
Introduction

The simulation of polycrystalline, or powder (these two terms, will hereafter be used interchangeably) spectra of a transition-metal ion has been of great interest, especially in samples for which it is not possible to grow single crystals easily (Reed and Markham, 1984; Chiswell, McKenzie, and Lindboy, 1987) and there is no choice but to analyze a powder-sample spectrum to estimate spin-Hamiltonian parameters. The straightforward way to simulate a powder spectrum is to use perturbation expressions if the zero-field splitting (ZFS) parameters are not large. The alternative is to use the brute-force technique, wherein diagonalization of the spin-Hamiltonian (SH) matrix is performed to compute the required energy levels for the values of the external magnetic field varied in small steps over a chosen range. This is done for selected orientations of the external magnetic-field (**B**) over the unit sphere, as described in Chapter 8. Evidently, if the ZFS parameters are large and the spin (S) of the transition-metal ion is also large – for example, $S = 5/2$ for the Mn^{2+} ion and $S = 7/2$ for the Gd^{3+} ion – the brute-force technique requires exorbitant computer times. On the other hand, for large ZFS the perturbation approximation is not valid. To this end, the use of an homotopy technique, in conjunction with matrix diagonalization (as described in Chapter 8, Section 8.5) will lead to considerable savings in the required computer time.

This chapter is devoted to simulation of EPR spectra; coverage begins with single-crystal spectral simulation, and this is developed further to simulate polycrystalline spectrum, followed by spectra in glassy materials. Finally, details of spectral simulation in random networks are included.

9.2
Simulation of Single-Crystal Spectrum

Experimentally, one records a set of allowed and forbidden EPR lines for a chosen orientation of the external magnetic field (**B**) with respect to the symmetry axes of the crystal. Simulation of EPR spectrum requires knowledge of both positions and intensities of EPR lines, as well as the line-shape function, $F(B_{ri}, B)$, where B_{ri} denotes the resonance line positions for the various possible transitions $i' \leftrightarrow i''$, which can be Gaussian, Lorentzian, or a complicated function appropriate to the sample. It is easier to calculate the line positions for orientations of **B** parallel to the symmetry axes X, Y, and Z by perturbation expressions, as described in Chapter 8. For an arbitrary orientation of **B**, the task becomes more difficult, however. One can then either use the time-consuming brute-force method by varying the magnetic field in small steps, diagonalizing the spin-Hamiltonian matrix at each step, and comparing the difference in the eigenvalues of the levels participating in resonance with the energy of the quantum of radiation inducing resonance to calculate the resonance field; alternatively, the "eigenfield" method can be used (Belford, Belford, and Burkhalter, 1973; Belford et al., 1974). For this, each line position is weighted in proportion to its transition probability, $P(i, \theta, \varphi, v_c)$ for the ith transition, between the levels i' and i'', participating in resonance at the microwave frequency v_c at the orientation (θ, φ) of **B** over the unit sphere. Thus, the single-crystal simulated spectrum can be expressed as:

$$S(B, v_c) = \sum_i P(i, \theta, \phi, v_c) F(B_{ri}, B), \tag{9.1}$$

In Equation 9.1, the summation is over all the lines observed for the particular orientation of **B**.

9.2.1
Transition Probability

The probability $P(i, \theta, \varphi, v_c)$ of a transition, i, between the energy levels i' and i'' is expressed as follows (Misra, 1999a):

$$P(i, \theta, \varphi, v_c) \propto |\langle \Phi_{i'} | (B_{1x}S_x + B_{1y}S_y + B_{1z}S_z) | \Phi_{i''} \rangle|^2. \tag{9.2}$$

In Equation 9.2, S_α and $B_{1\alpha}$; $(\alpha = x, y, z)$ represent the components of the electron spin operator, **S**, and the modulation radiofrequency field B_1. $|\Phi_{i'}\rangle$ and $|\Phi_{i''}\rangle$ are the eigenvectors of the SH (spin Hamiltonian) matrix, **H**, corresponding to the energy levels $E_{i'}$ and $E_{i''}$ participating in resonance $[H|\Phi_k\rangle = E_k |\Phi_k\rangle]$. Each eigenvector in Equation 9.2 can be expressed in terms of its real (Re) and imaginary (Im) parts as follows:

$$|\Phi_{i'}\rangle = Re|\Phi_{i'}\rangle + Im|\Phi_{i'}\rangle. \tag{9.3}$$

For example, a term on the right-hand side of Equation 9.2 can be expressed, as,

$$|\langle \Phi_{i'} | B_1 S_x | \Phi_{i''} \rangle|^2 = B_1^2 \Big[\{Re\langle \Phi_{i'} | S_x | \Phi_{i''} \rangle\}^2 + \{Im\langle \Phi_{i'} | S_x | \Phi_{i''} \rangle\}^2 \Big]$$
$$= B_1^2 \Big[\{ReTr[S_x(|\Phi_{i''}\rangle \otimes \langle \Phi_{i'}|)]\}^2$$
$$+ \{ImTr[S_x(|\Phi_{i''}\rangle \otimes \langle \Phi_{i'}|)]\}^2 \Big]. \tag{9.4}$$

In Equation 9.4, Tr is the trace of a matrix, and $|\Phi_{i'}\rangle \otimes \langle \Phi_{i'}|$ represent the outer product of the two eigenvectors $|\Phi_{i'}\rangle$ and $|\Phi_{i''}\rangle$ with the matrix elements:

$$(|\Phi_{i'}\rangle \otimes \langle\Phi_{i'}|)_{jk} = |\Phi_{i'}\rangle_j |\Phi_{i'}\rangle_k^*, \qquad (9.5)$$

where the * denotes a complex conjugate. Further, taking into account the fact that S_x is a real symmetric matrix such that the only nonzero elements in the representation in which the matrix for S_z is diagonal (so that possible values for j are over the electronic magnetic quantum number ($M = S, S-1, \ldots -(S-1), -S$) are:

$$Re(S_x)_{j,j+1} = Re(S_x)_{j+1,j}, \text{ and } Im(S_x)_{j,k} = 0, \qquad (9.6)$$

and that

$$\begin{aligned} Re(|\Phi_{i'}\rangle \otimes \langle\Phi_{i'}|)_{jk} &= Re|\Phi_{i'}\rangle_j Re|\Phi_{i'}\rangle_k + Im|\Phi_{i'}\rangle_j Im|\Phi_{i'}\rangle_k \\ Im(|\Phi_{i'}\rangle \otimes \langle\Phi_{i'}|)_{jk} &= -Re|\Phi_{i'}\rangle_j Im|\Phi_{i'}\rangle_k + Im|\Phi_{i'}\rangle_j Re|\Phi_{i'}\rangle_k \end{aligned} \qquad (9.7)$$

One obtains,

$$\begin{aligned} &Re Tr[S_x(|\Phi_{i''}\rangle \otimes \langle\Phi_{i'}|)] \\ &= \sum_j (S_x)_{j,j+1} \left\{ Re(|\Phi_{i''}\rangle \otimes \langle\Phi_{i'}|)_{j+1,j} + Re(|\Phi_{i''}\rangle \otimes \langle\Phi_{i'}|)_{j,j+1} \right\} \end{aligned} \qquad (9.8)$$

$$\begin{aligned} &Im Tr[S_x(|\Phi_{i''}\rangle \otimes \langle\Phi_{i'}|)] \\ &= \sum_j (S_x)_{j,j+1} \left\{ Im(|\Phi_{i''}\rangle \otimes \langle\Phi_{i'}|)_{j+1,j} + Im(|\Phi_{i''}\rangle \otimes \langle\Phi_{i'}|)_{j,j+1} \right\} \end{aligned} \qquad (9.9)$$

In Equations 9.8 and 9.9 the sum over j covers $M = -S, -(S-1), \ldots, (S-1)$. Thus, the right-hand side of Equation 9.2 can be calculated by the use of Equations 9.8 and 9.9. Similarly, for the $B_1 S_y$ and $B_1 S_z$ terms on the right-hand side of Equation 9.2, one has an equation similar to Equation 9.4, with S_x replaced by S_y and S_z, respectively.

To evaluate the corresponding expression in S_y, one notes that S_y contains only imaginary elements with the nonzero elements:

$$\begin{aligned} Im(S_y)_{j,j+1} &= -Im(S_y)_{j+1,j} \\ Re(S_y)_{j,k} &= 0 \end{aligned} \qquad (9.10)$$

One then obtains

$$\begin{aligned} &Re Tr[S_y(|\Phi_{i''}\rangle \otimes \langle\Phi_{i'}|)] \\ &= -\sum_j Im(S_y)_{j,j+1} \left\{ Im(|\Phi_{i''}\rangle \otimes \langle\Phi_{i'}|)_{j+1,j} - Im(|\Phi_{i''}\rangle \otimes \langle\Phi_{i'}|)_{j,j+1} \right\} \end{aligned} \qquad (9.11)$$

$$\begin{aligned} &Im Tr[S_y(|\Phi_{i''}\rangle \otimes \langle\Phi_{i'}|)] \\ &= \sum_j Im(S_y)_{j,j+1} \left\{ Re(|\Phi_{i''}\rangle \otimes \langle\Phi_{i'}|)_{j+1,j} - Re(|\Phi_{i''}\rangle \otimes \langle\Phi_{i'}|)_{j,j+1} \right\} \end{aligned} \qquad (9.12)$$

In Equations 9.11 and 9.12, the required imaginary and real parts of the outer products on the right-hand sides are given by Equation 9.7. As for the corresponding expression in S_z, one notes that S_z has only diagonal elements nonzero, which are real. Thus,

$$\begin{aligned} Im(S_z)_{jk} &= 0 \\ Re(S_z)_{jk} &= (S_z)_{jk} \delta_{jk} \end{aligned} \qquad (9.13)$$

where δ_{jk} is the Kronecker-delta symbol, such that $\delta_{ij} = 0$ for $i \neq j$, $\delta_{ij} = 1$ for $i = j$. Finally,

$$Re Tr[S_z(|\Phi_{i''}\rangle \otimes \langle\Phi_{i'}|)] = \sum_j (S_z)_{j,j} \{Re|\Phi_{i''}\rangle_j Re|\Phi_{i'}\rangle_j + Im|\Phi_{i''}\rangle_j Im|\Phi_{i'}\rangle_j\} \quad (9.14)$$

$$Im Tr[S_z(|\Phi_{i''}\rangle \otimes \langle\Phi_{i'}|)] = \sum_j (S_z)_{j,j} \{-Re|\Phi_{i''}\rangle_j Im|\Phi_{i'}\rangle_j + Im|\Phi_{i''}\rangle_j Re|\Phi_{i'}\rangle_j\} \quad (9.15)$$

When the orientation of the $\mathbf{B}_1$ field responsible for inducing transitions is parallel to any one of the x, y, z directions, one can use the appropriate expression similar to Equation 9.4 to calculate the relative intensity of the $i' \leftrightarrow i''$ transition. For an arbitrary orientation of $\mathbf{B}_1$ field, one can sum the real and imaginary parts of the three terms together. The required square of the absolute value in Equation 9.2 is then obtained as follows:

$$|\langle\Phi_{i'}|B_{1x}S_x + B_{1y}S_y + B_{1z}S_z|\Phi_{i''}\rangle|^2 = B_1^2 \left[\left| \sum_{\alpha=x,y,z} a_\alpha Re\{S_\alpha(|\Phi_{i''}\rangle \otimes \langle\Phi_{i'}|)\} \right|^2 + \left| \sum_{\alpha=x,y,z} a_\alpha Im\{S_\alpha(|\Phi_{i''}\rangle \otimes \langle\Phi_{i'}|)\} \right|^2 \right] \quad (9.16)$$

In Equation 9.16, $a_x = \sin\theta\cos\varphi$, $a_y = \sin\theta\sin\varphi$, $a_z = \cos\theta$.

9.2.2
Single-Crystal Lineshape Function $F(B_{ri}, B_k)$

The spectrum is calculated by performing the sum in Equation 9.1 with $P(i, \theta_j, \varphi_j, v_c)$ centered at $B_r(i, \theta_j, \varphi_j, v_c)$ with the lineshape function $F(B_{ri}, B_k)$, extended over a magnetic-field interval $\pm\Delta B$, related to $\Delta B_{1/2}$ (full-width at half-maximum, FWHM) characteristic of the lineshape. (For example $\Delta B = \pm 10 \Delta B_{1/2}$ to obtain good precision for Lorentzian lineshape.) The two most common line shapes are:

- **Gaussian lineshape, F_G:**

$$F_G(B_{ri}, B_k) = K_G \exp[-(B_k - B_{ri})^2/\sigma^2], \quad (9.17)$$

where B_{ri} is the resonant field value for the ith transition, σ is the linewidth, and $K_G \{= (1/B_\Delta)(\ln 2/\pi)^{1/2}\}$ is the normalization constant for the lineshape (Poole, 1967; Buckmaster and Dering, 1968). Here, $B_\Delta = (1/2) B_{1/2}$ is the half-width at half-maximum (HWHM).

- **Lorentzian lineshape, F_L:**

$$F_L(B_{ri}, B_k) = K_L \Gamma [\Gamma^2 + (B_k - B_{ri})^2]^{-1}, \quad (9.18)$$

where Γ is the Lorentzian linewidth (HWHM = $(3)^{1/2}\Delta B_{pp}/2$, with ΔB_{pp} being the peak-to-peak first-derivative linewidth (Poole, 1967)).

More complicated lineshapes appropriate to glassy samples are discussed in Section 9.5.

9.3
Simulation of a Polycrystalline Spectrum

Reed and Markham (1984), Zhang and Buckmaster (1992, 1993, 1994), and Coffino and Peisach (1996) have discussed the computational strategies involved in the simulation of polycrystalline spectra; especially using matrix diagonalization. The EPR spectrum in a polycrystalline material can be obtained by overlapping simulated single-crystal spectra computed for a large number of orientations (θ, φ) of the external magnetic (Zeeman) field, **B**, over the unit sphere weighted in proportion of $\sin\theta d\theta d\varphi$ to take into consideration the distribution of various constituting crystallites whose principal axes are oriented in the interval $d\theta$, $d\varphi$ about (θ, φ). Thus, the simulated polycrystalline spectrum can be expressed as:

$$S(B, v_c) = \int_{\theta=0}^{\pi/2} \int_{\phi=0}^{2\pi} \sum_i P(i, \theta, \phi, v_c) F(B_{ri}, B) d(\cos\theta) d\phi \tag{9.19}$$

It is clear from Equation 9.19 that, in particular, one needs to know the resonant field values for the various transitions, as well as their transition probabilities. The most direct way to calculate resonant fields for arbitrary orientations of the magnetic field is to use the time-consuming brute-force technique, as explained in Section 9.2. However, a considerable saving of computer time can be accomplished by using the *homotopy* technique, as described next.

9.3.1
Angular Variation of EPR Spectra: Homotopy Technique

Here, for a given set of SHPs, one computes the resonant fields for different orientations of the external magnetic field in a given plane. For $\boldsymbol{B} \| z$, one can calculate these fields from perturbation expressions, provided that the off-diagonal elements in the SH matrix are not large. However, for arbitrary orientations of the external magnetic field it is difficult to calculate perturbation expressions. Another method is the application of the brute-force technique, as described in Chapter 8 (see Section 8.4.4.1, ENSC systems). However, this requires exorbitantly long computer times, especially when S and I are large. Another time-consuming technique is the eigenfield method (Belford, Belford, and Burkhalter, 1973; Belford et al., 1974).

The least-squares fitting (LSF) technique not only provides accurate values of the calculated fields, but it is also computationally efficient (Misra and Vasilopoulos, 1980). The procedure, called the *homotopy* technique, for calculating the resonant line position B_r at the orientation, $(\theta + \delta\theta, \phi + \delta\phi)$, from the knowledge of the line position at the orientation (θ, ϕ), using the LSF technique and Taylor-series expansion, is as follows:

$$B_r(\theta + \delta\theta, \phi + \delta\phi) = \text{Iterative limit of} \left[B_r(\theta, \phi) - \left(\frac{\partial^2 \chi^2}{\partial B^2}\right)_{B'_r}^{-1} \left(\frac{\partial \chi^2}{\partial B}\right)_{B'_r} \right] \quad (9.20)$$

In Equation 9.20, one starts with $B'_r = B_r(\theta, \varphi)$, with χ^2 defined as follows:

$$\chi^2 = \sum_i \left[(|\Delta E_i| - h\nu_i)^2 / \sigma_i^2 \right], \quad (9.21)$$

where h is Planck's constant, $\Delta E_i = E_{i'} - E_{i''}$, where $E_{i'}$ and $E_{i''}$ are the energies of the levels between which the microwave energy $h\nu_i$ is absorbed at resonance, and ν_i is the corresponding klystron frequency, which is usually the same for all i. The $E_{i'}$ and $E_{i''}$ are obtained by substituting the (numerical) values of the parameters and the resonant field values B_i into the SH, and diagonalizing it on the computer. In Equation 9.21, σ_i is the weight factor for the ith line position. Thus, by starting with $B_r(\theta, \varphi)$, one can determine $B_r(\theta + \delta\theta, \phi + \delta\phi)$ which, in turn, can be used to compute $B_r(\theta + 2\delta\theta, \phi + 2\delta\phi)$. In this way, starting with an initial set of $B_r(\theta, \phi)$ the LSF technique can be used to calculate resonance fields for various magnetic field orientations in a given plane, by choosing a sufficiently small steps of $\delta\theta$ or $\delta\varphi$.

9.3.1.1 Computation of the Initial Resonant Fields $B_r(\theta, \phi)$

These can be either computed by the use of perturbation expressions, as described in Section 8.2 for $\mathbf{B} \parallel \mathbf{z}$, or $\mathbf{B} \parallel \mathbf{x}$, provided that the off-diagonal elements in the SH matrix are not large, or by the brute-force technique mentioned above, especially when the off-diagonal elements in the SH are large. Another time-consuming technique is the eigenfield method (Belford, Belford, and Burkhalter, 1973; Belford et al., 1974).

9.3.1.2 Computation of the First and Second Derivatives of χ^2 with Respect to B

For use in Equation 9.20, applying the procedure outlined in Section 8.4.2, these derivatives can be expressed as follows:

$$\frac{\partial \chi^2}{\partial B} = 2(|\Delta E_i| - h\nu_i) \left(\frac{\partial E_{i'}}{\partial B} - \frac{\partial E_{i''}}{\partial B} \right) \frac{\Delta E_i}{|\Delta E_i|}, \quad (9.22)$$

$$\frac{\partial^2 \chi^2}{\partial B^2} = 2\left(\frac{\partial E_{i'}}{\partial B} - \frac{\partial E_{i''}}{\partial B} \right)^2 + 2(|\Delta E_i| - h\nu_i) \frac{\Delta E_i}{|\Delta E_i|} \left(\frac{\partial^2 E_{i'}}{\partial B^2} - \frac{\partial^2 E_{i''}}{\partial B^2} \right) \quad (9.23)$$

and $\Delta E_i = E_{i'} - E_{i''}$. The derivatives of $E_{i'}$, $E_{i''}$ required in Equations 9.22 and 9.23 can be expressed, using the details given in Chapter 8 (Section 8.4.2), as follows:

$$\frac{\partial E_i}{\partial B} = Tr\left[\left(\frac{\partial H}{\partial B}\right) (|\phi_i\rangle \otimes \langle\phi_i|) \right], \quad (9.24)$$

$$\frac{\partial^2 E_i}{\partial B^2} = \sum_n Tr\left[\left(\frac{\partial H}{\partial B}\right) (|\phi_n\rangle \otimes \langle\phi_n|) \left(\frac{\partial H}{\partial B}\right) (|\phi_i\rangle \otimes \langle\phi_i|) \right] / (E_i - E_n). \quad (9.25)$$

In Equations 9.24 and 9.25, $|\varphi_i\rangle$ is the eigenvector of H with the eigenvalue E_i: $H|\phi_i\rangle = E_i|\phi_i\rangle$; and the index n runs over all levels except for i (indicated by a prime over the summation sign).

Further, the derivative of an eigenvalue $E_{i'}$, or $E_{i''}$, can be evaluated by the use of Feynman's theorem (Feynman, 1939):

$$\frac{\partial E_{i'}}{\partial B} = \langle \Phi_{i'} | \frac{\partial H}{\partial B} | \Phi_{i'} \rangle \qquad (9.26)$$

$$= \mu_B \langle \Phi_{i'} | g_{zz} \cos\theta S_z + g_{xx} \sin\theta \cos\phi S_x + g_{yy} \sin\theta \sin\phi S_y | \Phi_{i'} \rangle$$

where $g_{\alpha\beta}$ ($\alpha, \beta = x, y, z$) are the diagonal components of the $\tilde{g}$-matrix. (Here, it has been assumed that the $\tilde{g}^2$-*tensor* is diagonal in the coordinate axes chosen.) In writing the above equation, the fact that only the Zeeman term:

$$H_z = \mu_B S \cdot \tilde{g} \cdot B \qquad (9.27)$$

depends on the external field has been taken into account. The components of B are $B_z = B\cos\theta$, $B_x = B\sin\theta\cos\phi$, $B_y = B\sin\theta\sin\phi$. Here, θ, ϕ are the angles describing the orientation of B with respect to z.

The above procedure was successfully applied (Misra and Vasilopoulos, 1980) to the calculation of the angular variation of EPR line positions for the purely electronic systems of Gd^{3+}-doped $NdCl_3 \cdot 6H_2O$ at X- and Q-band frequencies $(S = \frac{7}{2})$, X-band EPR spectrum of Fe^{3+}-doped guanidinium aluminum sulfate hexahydrate $(S = \frac{5}{2})$, and the ENSC system of Mn^{2+}-doped $Ni(CH_3COO)_2 \cdot 4H_2O$ $(S = I = \frac{5}{2})$.

9.3.1.3 Problems Encountered in the Application of Homotopy Method, and their Solutions

The following problems are worthy of discussion.

i) When two or more lines converge to the same resonant-field value at a particular orientation θ or ϕ (case of "degeneracy"), then for a subsequent orientation $\theta + \delta\theta$, or $\phi + \delta\phi$, of the magnetic field, only one of the lines can be computed by the use of Equation 9.20, unless one checks the various eigenvalues of the SH matrix extremely carefully. This arises because there is now only one initial value available to be used in Equation 9.20. In order to overcome this problem, one judiciously changes the interval $\delta\theta$ to a different (smaller) value at the preceding orientation $\theta - \delta\theta$ so the situation of degeneracy is circumvented. Another way is to use the brute-force method at the subsequent orientation $\theta + \delta\theta$ to find two or more corresponding resonant fields. (In the following, the angle ϕ is not mentioned, for which the same considerations as those for θ apply.)

ii) There are some lines which do not continue up to $\theta = 0°$ (e.g., Z-axis) or $\theta = 90°$ (e.g., X-axis), or both, for example, the "loop" or "lobe" lines. The existence of these lines can be checked be the application of the brute-force method for a judiciously chosen θ (say θ_0), at which these lines exist. The field values thus obtained may be used for studying their angular variation by the use of Equation 9.20 by decreasing and increasing the value of θ in steps of $\delta\theta$, starting from θ_0 at which the resonant field values are known.

iii) When a resonant-field value, computed by the use of Equation 9.42, becomes negative even for a sufficiently small $\delta\theta$, this indicates that this line terminates at that orientation θ, for example, the "lobe" lines do not continue beyond a certain value of θ.

iv) If it is found that a computed resonant-field value changes by a relatively large magnitude for a particular choice of $\delta\theta$ greater accuracy may be obtained by decreasing $\delta\theta$ so that the change in the field value is not "too" large.

v) At low magnetic fields, sometimes the eigenvalues, arranged in increasing or decreasing values, corresponding to a given transition do not occur in the same order for $\theta + \delta\theta$ as for θ. This is not surprising, as the computer output gives the eigenvalues in increasing, or decreasing values. The correct pairs of eigenvalues (i' and i'') can be checked by examining the expectation value $\langle \phi_{i'} | H' | \phi_{i''} \rangle$, where $H' = \mu_B B_{rf} \cdot \tilde{g} \cdot S$ represents the interaction of the microwave magnetic field, B_{rf}, with the electronic magnetic moment. The value of this matrix element for incorrect pairs of eigenvalues will be negligible.

9.3.2
Lineshapes

Specific details are provided here for the Gaussian and Lorentzian resonance line shapes. The simulated spectrum is computed by the use of Equation 9.19, wherein the integrals are converted to discrete sums. The various additional parameters/techniques required in the computation are described below.

9.3.3
Transition Probabilities

To simulate a polycrystalline spectrum, one needs to calculate the transition probabilities $P(i, \theta, \varphi, v_c)$ required in Equation 9.19 at various orientations of the external magnetic field which, in turn, depend on the eigenvectors $|\Phi_i(\theta, \varphi)\rangle$; $i = i', i''$ corresponding to the energy levels, $E_{i'}$ and $E_{i''}$ participating in resonance. The transition probability at the infinitesimal orientation $\theta + \delta\theta$ and $\varphi + \delta\varphi$ of **B** can be obtained from the eigenvectors $|\Phi_i(\theta + \delta\theta, \varphi + \delta\varphi)\rangle$. The latter can be calculated using $B_r(i, \theta + \delta\theta, \phi + \delta\phi)$ as obtained using homotopy, and then diagonalizing the spin Hamiltonian expressed at $(\theta + \delta\theta, \varphi + \delta\varphi)$ for this value of the magnetic field. The procedure for a numerical calculation of transition probability for the general case when the elements of the SH matrix are complex has been described above in Section 9.2.1.

9.3.4
Resonance Eigenpairs

The same resonance eigenpair, consisting of energy levels characterized by the electronic magnetic quantum numbers M and $M-1$, which describe the eigenvectors of H_{Ze}, the Zeeman part of the SH, should be used to calculate the resonance fields corresponding to the allowed fine-structure transitions as the orientation

(θ, φ) of **B** is changed incrementally to ($\theta + \delta\theta$, $\varphi + \delta\varphi$) until the unit sphere is covered. This is accomplished by first diagonalizing the matrix for the Zeeman interaction, $H_{ze} = \mu_B \mathbf{B} \cdot \tilde{g} \cdot \mathbf{S}$, with the eigenvalues arranged either in decreasing, or increasing, order of their values (M-order), then transforming the matrix of H_{ZFS}, the ZFS part of the spin Hamiltonian by the matrix **V**, formed by the eigenvectors of H_{ze} as columns, corresponding to the eigenvalues of H_{ze}, either in decreasing, or increasing, M-order. H_{ze} is diagonal in this representation, since the diagonal elements are the eigenvalues of H_{ze}. The transformed SH matrix $H^T = V(H_{Ze} + H_{ZFS})V^{\dagger}$, so obtained in the basis of the eigenvectors of H_{ze}, is then diagonalized to find its eigenvalues and eigenvectors, this time without ordering the eigenvalues, using an appropriate subroutine. This ensures that that the resonance eigenpair M, M−1 retains the same label during the homotopy procedure. Otherwise, it is frequently impossible to follow the same eigenpair for a given transition as the **B** orientation is changed, particularly when this orientation is more than $\pi/6$ radians away from the crystal principal axis. Alternatively, since the M, M−1 eigenpair is unequivocally identified when the perturbation expressions are used, and H_{ze} is chosen as the zero-order term, one can make a correspondence between the eigenvalues of the full Hamiltonian ($H_{Ze} + H_{ZFS}$) as obtained by matrix diagonalization and those calculated by perturbation theory, in which case the label M is well defined. This is done by finding the closest perturbation eigenvalues to the eigenvalues calculated by matrix diagonalization, the two sets not being identical.

9.3.5
Integrals

The integral for the polycrystalline spectrum $S(B, v_c)$, as given by Equation 9.19, can be expressed as a sum over different orientations (θ_j, φ_j) of **B** distributed over the unit sphere divided into grids whose intersections for successive grids are infinitesimally close to each other, and over the values of **B** divided into channels, B_k, distributed over the range of the magnetic field considered. Thus, Equation 9.19 can be expressed, using a constant C, as the following sum:

$$S(B, v_c) = C \sum_{i,q_j,j_j,k} P(i, \theta_j, \phi_j, v_c) F(B_r(i, \theta_j, \phi_j, v_c), B_k) \sin\theta_j \tag{9.28}$$

In Equation 9.28, the values of θ_j are distributed over the range 0 to $\pi/2$, while those of φ_j are over 0 to π, taking into account the fact that the EPR spectrum remains unchanged when the magnetic field orientation is reversed in direction due to time-reversal invariance. Also, $\sin\theta_j$ takes into account the uniform distribution of the crystallites constituting the powder such that the number of crystallites with their axes along θ_j is proportional to $\sin\theta_j$. The summation over k takes into account the probability of the amplitude of absorption at the magnetic field value B_k due to the lineshape distribution $F(\omega_i, B_r(i, \theta_j, \varphi_j, v_c), B_k)$ for the ith transition for the orientation of B along the (θ_j, φ_j) direction.

9.3.6
(θ_j, φ_j) Grid

One can conveniently choose a (θ_j, φ_j) grid where the θ–value is changed from 0 to $\pi/2$ in n_θ steps, where n_θ is a sufficiently large number, say 30 (i.e., every 3°, or even every degree, if sufficient computer time is available), depending on the convergence of $B_r(i, \theta, \varphi)$ values computed using Equation 9.19. Similar considerations apply to changes in φ-values in n_φ steps, say 120 (i.e., every 3°, or even every degree). Thus, one can start at $\theta = 0°$, $\varphi = 0°$, and then increase θ by $\pi/2n_\theta$ and compute the resonant fields $B_r(i, \theta_j, \varphi_j, v_c)$, and the corresponding transition probabilities $P(i, \theta_j, \varphi_j, v_c)$ for φ-values, incrementing from $\varphi = 0°$ in $2\pi/n_\varphi$ steps. The interval between successive (θ_j, φ_j) values should be chosen sufficiently small such that the resonant field values at these two orientations of **B** do not differ from each other by relatively large magnitudes. The various (θ_j, φ_j) values required in the sum (Equation 9.28) are, thus, taken into account by the grid structure. When there appear "crossing" or "looping" transitions – for example, in the case of the Fe^{3+} ion (Misra and Vasilopoulos, 1980), as discussed in Section 9.3.1.3 – problems arise when two transitions cross each other between two successive (θ_j, φ_j) values (considered a "crossing" transition), or a transition does not occur at the adjacent (θ_j, φ_j) values ("looping" transition). In order to overcome these, certain strategies may be employed, as discussed in Section 9.3.1.3. In addition, an improved partitioning scheme of the grid may be used. To this end, Wang and Hanson (1995, 1996) developed a novel scheme – named the SOPHE (Sydney Opera House) partitioning scheme – which involved a combination of cubic spline and linear interpolations; in this case, the unit sphere is partitioned into triangularly shaped convexes subtending nearly the same solid angles. These reports also provided reference to earlier proposed partitioning schemes, such as the "Igloo method" for partitioning an octant of the unit sphere (Nilges, 1979; Belford and Nilges, 1979; Maurice, 1980).

The spectrum is then calculated by performing the sum in Equation 9.20 with $P(i, \theta_j, \varphi_j, v_c)$ centered at $B_r(i, \theta_j, \varphi_j, v_c)$ with the lineshape function $F(B_r(i, \theta_j, \varphi_j, v_c), B_k)$ described in Section 9.2. More complicated line shapes appropriate to polycrystalline samples are discussed by Misra (1996), and references therein of other relevant reports.

9.3.7
Steps Required in Simulation of Powder Spectrum

The following steps are recommended to simulate a powder spectrum using the proposed homotopy method.

i) The allowed resonance fields ($|\Delta M| = 1$) to be used as initial values are calculated at $\theta = \varphi = 0°$ using the third-order perturbation expressions or matrix diagonalization by the brute-force method for the eigenvalues $E(M)$ (Lynch, Boorse, and Freed, 1993; Misra, 1997), and the resonance condition

$$hv_c = |E(M) - E(M-1)| \qquad (9.29)$$

ii) Using the resonance fields obtained for $\theta = \varphi = 0°$ in step (i) as initial values, the resonance fields at $\theta = 2°$, $\varphi = 0°$ are then calculated by the application of Equation 3.1, abandoning the iteration procedure when the difference of the resonance field estimated in an iteration becomes less than, a small value, say 0.01 G, from that estimated in the previous iteration. (Only $\varphi = 0°$ is needed for $\theta = 0°$, since for this orientation the x and the y components of **B** are identically zero.) The required eigenvalues and eigenvectors of the spin-Hamiltonian matrix are computed by the use of the subroutine JACOBI, which diagonalizes real symmetric matrices, and is particularly efficient when the off-diagonal elements in the SH matrix are infinitesimally small, as is naturally the case in homotopy. [Briefly, the diagonalization in the JACOBI algorithm is accomplished by successive rotation to annihilate the off-diagonal elements of the 2 × 2 submatrix constituted by the largest off-diagonal element and corresponding diagonal elements of the SH matrix at any stage of successive rotations; for details, see *Numerical Recipes* (Press et al., 1992).] These values obtained at $\theta = 2°$, $\varphi = 0°$ are then used as initial values for calculating resonance fields at $\theta = 2°$, but with $\varphi \neq 0$: for example, at $\varphi = 360/15 = 24°$. Thereafter, the resonance fields are calculated using homotopy as described above in Section 9.3.1, for $\theta = 2°$, $\varphi = n_\varphi \times 24°$ for n_φ taking successively the values 2, 3, ... , 14, so as to span the φ interval over 360° in equal steps of 24° from $\theta = 2°$, using the set of resonant values obtained in the preceding calculation as initial values. The resonance fields are thereafter calculated for $\theta = 4°$, $\varphi = 0°$ using those calculated for $\theta = 2°$, $\varphi = 0°$ as initial values. Thereafter, for $\theta = 4°$ and for $\varphi \neq 0$: $\varphi = n_\varphi \times 24°$; for n_φ assuming successively the values 1, 2, 3, ... , 23, the resonance fields are calculated in the same way as that for $\theta = 2°$ and $\varphi \neq 0°$, etc. These calculations are repeated by increasing θ-values in steps of 2°; and for each new θ, with $\varphi = 0°$ the resonance fields for the previous θ-value with $\varphi = 0°$ are used. The φ step of 24° is changed to 2° when $|\sin \theta| \times 180 \geq 15$, for the simple reason that the number of crystallites at an orientation θ is proportional to $\sin \theta$ (Misra, 1997), thus not requiring too many values of φ when θ is small. In sum, the resonance fields are calculated for θ-values in steps of 2° from 0° to $\theta = 90°$, and for each θ-value appropriate φ-values are chosen uniformly distributed over 360°. For the resonance fields so calculated at each orientation (θ_j, φ_j) of **B**, the corresponding relative intensity of the $i' \leftrightarrow i''$ transition, $P(i, \theta, \varphi, v_c)$, is also calculated by the using the procedure described in Section 9.2.1. The resonance fields for the various transitions and the corresponding intensities for various (θ_j, φ_j) combinations are then stored, to be used later for simulation of powder spectrum.

iii) The magnetic-field range over which the spectrum is to be simulated is divided into equally-spaced intervals, B_k referred to as channels (k); for example, into 8000 channels. Each resonance field is then changed, if necessary, to assume the value of the closest channel, and the intensity of that resonance field, as given by Equation 9.2, is assigned to be the y-value at that channel, with the channel representing the x-value. All resonance fields are, thus, taken into account. The lineshape distribution is, thereafter, incorporated by distributing the intensity about a channel corresponding to a resonance line according to the lineshape function centered at that channel over a reasonably large interval ΔB, for example, 5000 G, on both the higher and lower field sides of the resonance field. Channel intensities corresponding to all the resonance fields calculated for the various (θ_j, φ_j) values are, thus, taken into account by extending the minimum field of the range, B_{min}, to $B_{min} - \Delta B$; and the maximum field of the range, B_{max}, to $B_{max} + \Delta B$. The simulated absorption spectrum, $S(B, v_c)$ is calculated in this manner.

9.3.7.1 Calculation of First-Derivative EPR Spectrum

Most experimental EPR data are obtained as the first derivative of the absorbed microwave power as functions of the external magnetic field intensity. This is true when $B_{mod} \ll (1/2)\delta B_{1/2}$, where B_{mod} is the amplitude of the modulation magnetic field, and $\delta B_{1/2}$ is the HWHM of the EPR absorption line. The simulated first-derivative spectrum is calculated by taking the derivative of $S(B, v_c)$, given by Equation 9.28, with respect to B, along with that of the lineshape. Specifically, for the Gaussian and Lorentzian lineshapes, given by Equations 9.17 and 9.18, one has, respectively, for the first-derivative:

$$\partial F_G(B_{ri}, B_k)/\partial B_k = -2K_G(B_k - B_{ri})\exp(-(B_k - B_{ri})^2/\sigma^2)/\sigma^2, \tag{9.30}$$

$$\partial F_L(B_{ri}, B_k)/\partial B_k = -2K_L\Gamma[\Gamma^2 + (B_k - B_{ri})^2]^{-2}(B_k - B_{ri}). \tag{9.31}$$

The simulated first-derivative absorption spectrum is expressed from Equation 9.28, as

$$F_c(B_k, V_c) = \partial S(B_k, V_c)/\partial B_k =$$
$$= C \sum_{i, q_j, j_j} P(i, \theta_j, \varphi_j) \partial F(w_i, B_r(i, j, \varphi_j, n_c), B_k)/\partial B_k \sin\theta_j \tag{9.32}$$

From Equation 9.32, one has for the two lineshapes, using Equations 9.30 and 9.31:

- **Gaussian lineshape:**

$$F_c(B_k, V_c)$$
$$= \sum_i N |\langle \Phi_{i'} | B_{1x} S_x + B_{1y} S_y + B_{1z} S_z | \Phi_{i''} \rangle|^2 \exp[-(B_k - B_{ri})^2/\sigma^2](B_k - B_{ri})/\sigma^2$$

$$\tag{9.33}$$

- **Lorentzian lineshape:**

$$F_c(B_k, V_c) = \sum_i N |\langle \Phi_{i'} | B_{1x} S_x + B_{1y} S_y + B_{1z} S_z | \Phi_{i''} \rangle|^2 (B_k - B_{ri})\Gamma[\Gamma^2 + (B_k - B_{ri})^2]^{-2}$$

$$\tag{9.34}$$

In Equations 9.33 and 9.34, the normalization constant, N, may be appropriately chosen, for example, $|F_c(B_k, V_c)|_{max} = 1$, where $|F_c(B_k, V_c)|_{max}$ is the largest magnitude of all the calculated values.

9.3.8
Illustrative Example

The proposed homotopy technique was illustrated successfully (Misra, 1999a) by application to the case of spectrum of the Mn^{2+} ion ($S = 5/2$) in $Mn(\gamma\text{-picoline})_4I_2$ powder sample, characterized by a rather large value of the ZFS parameter, D, as measured by Lynch, Boorse, and Freed (1993) at 249.9 GHz (far-infrared band). It was shown there that simulation with matrix diagonalization is significantly different from that by perturbation approximation, and provides a much better agreement with the experimental spectrum.

9.3.9
Additional remarks

It should also be noted that:

- The procedure of simulating a powder spectrum as outlined in this section, using the technique of homotopy and numerical diagonalization of the spin-Hamiltonian matrix, is rigorous, since the eigenvalues and eigenvectors of the SH matrix are used rather than perturbation expressions. This is particularly useful when the ZFS parameter, D, is relatively large, in which case for smaller klystron frequencies, for example, 95 GHz (W-band) perturbation approximation is not justified. In the illustrative example, $D/h\nu \sim 0.1$ at 249.9 GHz, while $D/h\nu \sim 0.25$ at 95 GHz. The proposed matrix diagonalization procedure, using JACOBI diagonalization algorithm, is particularly efficient in computer time required when the off-diagonal terms of SH matrix are infinitesimally small, as in the present case of homotopy.

- This procedure to simulate a powder EPR spectrum can be easily extended to simulate powder spectra for electron–nuclear double resonance (ENDOR), solid-state nuclear magnetic resonance (NMR), nuclear quadrupole resonance (NQR), electron spin-echo (ESE) envelope modulation, as well as to include hyperfine structure, if there is any.

- Finally, it is noted that the (θ_j, φ_j) grid as chosen here is quite appropriate for the Mn^{2+} ion, since there do not exist any "crossing," "looping," or "lobe" transitions in this case (Misra and Vasilopoulos, 1980). However, when these do exist – for example, in the case of the Fe^{3+} ion – certain strategies are required. These are discussed in Section 9.3.1.3.

9.4
Evaluation of Spin-Hamiltonian (SH) Parameters and the Linewidth from a Polycrystalline EPR Spectrum

There exist many molecules, such as metalloproteins (Reed and Markham, 1984; Chiswell, McKenzie, and Lindboy, 1987), for which it is not easy to grow single crystals of sufficient size to enable them to be mounted with their axes oriented in chosen directions for angular variation of EPR spectrum for varying orientations of **B** with respect to the crystal symmetry axis in order to evaluate SH parameters accurately. Moreover, it is not always possible to rotate either the crystal, or the magnetic field – for experimental reasons – or because the orientations of the axes change when structural phase-transitions occur, and it is not possible to reorient the crystal inside the cryostat at low temperatures. It is then preferable to record a polycrystalline (powder) spectrum. The possibility of fitting simultaneously a large number of EPR resonance line positions is then excluded, since only one spectrum, which is insensitive to the orientation of the external magnetic field, is available for a polycrystalline sample.

To date, SH parameters have been estimated for ions of larger spins by the use of brute-force techniques, wherein a polycrystalline spectrum is simulated by arbitrarily adjusting the parameter values. The set of values that provides the most reasonable fit is then considered to be the set representing the best values of the SH parameters. The intensities were also calculated using the zero-order wave-functions that are insensitive to the orientation of **B**, while perturbation expressions were used for the eigenvalues. However, this may not be quite precise enough when the orientation of **B** and of the symmetry axis of the single crystal deviate significantly from each other and the ZFS parameter, D, is relatively large. In this section, techniques of perturbation and matrix-diagonalization, in conjunction with homotopy (Misra, 1999b), are described for the direct evaluation of SH parameters from a polycrystalline EPR spectrum.

9.4.1
Estimation of Spin-Hamiltonian Parameters from a Polycrystalline Spectrum

This can be accomplished in two ways. The first approach is based on using the shape of the central hyperfine sextet, in particular the separation of hyperfine forbidden doublets, noting that the outer fine-structure sextets are most often broadened out in a polycrystalline spectrum. A second, more rigorous, approach is based on matrix diagonalization and least-squares fitting. The two procedures are described below for the Mn^{2+} ion; similar applications can be made to other transition-metal ions with spins ≥ 1.

9.4.1.1 Estimation of D and E Parameters for the Mn^{2+} Ion from Forbidden Hyperfine Doublet Separations in Polycrystalline Samples in the Central Sextet

Simulated spectra at X and K bands for the central hyperfine sextet, by using third-order perturbation expressions (Meirovitch and Poupko, 1978), have been illustrated by Misra (1997) for various values of D. In addition, it is possible to use the relative intensities of the various hyperfine lines in the EPR spectra to estimate D, as illustrated by Allen (1965) and shown in Figure 9.1.

In a polycrystalline material, the constituting crystallites are randomly oriented, so that all values of θ which the axis of symmetry of a crystallite makes with respect to **B**, from 0° to 90°, are possible. The number of crystallites with their axes ori-

9.4 Evaluation of SH Parameters and the Linewidth from a Polycrystalline EPR Spectrum

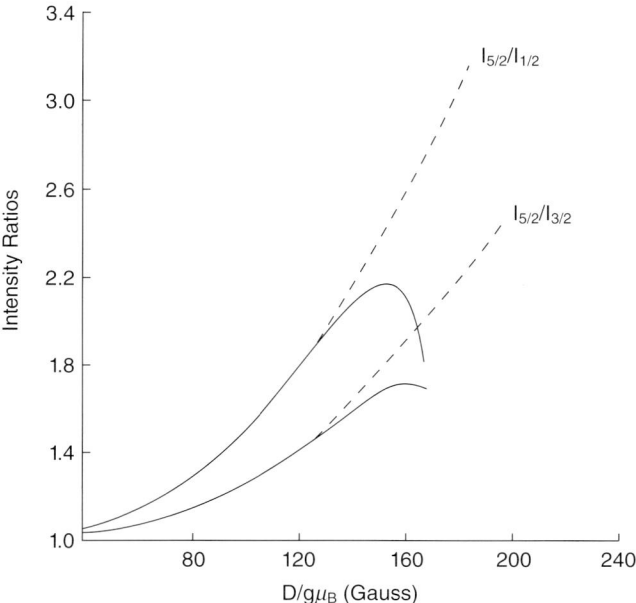

Figure 9.1 Ratios of line intensities versus $D/g\mu_B$ (Gauss). The solid lines show the second-order ratios calculated for D-values up to $D/g\mu_B = 170\,G$; the dashed lines exhibit the extrapolated curve from the calculated curve up to $D/g\mu_B = 120\,G$. Adapted from Allen (1965).

ented between θ and $\theta + d\theta$ is proportional to $\sin\theta d\theta$. Similarly, all values of ϕ, the azimuthal angle of the axis of symmetry of a crystallite, between 0 and 2π, are possible; and the number of crystallites with their axes oriented between ϕ and $\phi + d\phi$ is proportional to $d\phi$. The peaks (maxima) of forbidden hyperfine lines for a polycrystalline sample corresponding to the angle $\theta = \theta_o$, at which the maximum value of the intensity of a forbidden hyperfine line $M, m \leftrightarrow M, m-1$ ($\Delta M = 0$, $\Delta m = -1$), or $M, m-1 \leftrightarrow M, m$ ($\Delta M = 0$, $\Delta m = 1$), relative to the allowed hyperfine line $M, m \leftrightarrow M-1, m$ ($\Delta M = 1$, $\Delta m = 0$) in the central sextet ($M = 1/2$) occur. It is given by the following expression (Meirovitch and Poupko, 1978):

$$R(\theta, \varphi) = \left(\frac{3D}{4g\mu_B B_0(M)}\right)^2 f_o(\theta, \varphi)\left\{1+\left[\frac{S(S+1)}{3M(M-1)}\right]\right\}^2 [I(I+1)-m(m-1)], \tag{9.35}$$

where $f_o(\theta, \varphi) = \sin^2 2\theta\,(1-\eta\cos\varphi)^2$.

A simple calculation then yields the values of the angular-dependent factors in the forbidden line positions:

$$B_+(M, m) = B(M, m \leftrightarrow M-1, m-1);\ B_-(M, m) = B(M, m-1 \leftrightarrow M-1, m),$$

corresponding to the maximum of intensity, which occurs for $\theta = \theta_o$, that is, when

$$\cos^2 \theta_o = \frac{4(1-\eta^2)}{5(2+\eta^2)}. \qquad (9.36)$$

Finally, the separation of the peaks (maxima of intensity) corresponding to doublets of hyperfine forbidden transitions in the central hyperfine sextet ($M = \frac{1}{2}$, $m \leftrightarrow M = -\frac{1}{2}, m$) in a polycrystalline sample has been calculated by Misra (1994, 1997), expressed as follows:

$$\Delta B^{(PC)}\left(\frac{1}{2}, m\right) = \left[\frac{17A^2}{2G_0^2} + \frac{9D^2 AF(\eta)}{25G_0^3} + \frac{4A^2 D}{5G_0^3}(2m-1) - \frac{67A^3}{4G_0^3}(2m-1)\right] B_o, \qquad (9.37)$$

$$\text{with} \quad F(\eta) = \left\{\left(\frac{2+3\eta^2}{2+\eta^2}\right)^2 + \eta^2\left[\frac{25}{18} + \frac{10(2-\eta^2)}{3(2+\eta^2)} + \frac{4}{9}\left(\frac{2-\eta^2}{2+\eta^2}\right)^2\right]\right\}, \qquad (9.38)$$

where $\eta = \frac{E}{D}$; $G_0 = g\mu_B B_0$, B_0 being the magnetic field at the middle of the central hyperfine sextet.

Equation 9.37 is the key equation to be used to estimate the values of D and E from a Mn^{2+} EPR spectrum in a polycrystalline material. Figure 9.2 shows plots of values of $|D|$ as functions of average hyperfine forbidden-doublet separation in the Mn^{2+} central sextet for: (a) $\eta = 0.0$; (b) $\eta = 0.12$; (c) $\eta = 0.23$; and (d) $\eta = 0.33$. Further, taking into account the first term, which has the largest value, it is concluded that the hyperfine doublet separation $\Delta B^{(PC)}\left(\frac{1}{2}, m\right)$, as given by Equation 9.37, is positive unless the value of D exceeds D_o for which $\Delta B^{(PC)}\left(\frac{1}{2}, m\right) = 0$ (Misra, 1997). It is also noted that, because of the presence of G_o^2 [$\approx (h\nu)^2$, where h and ν are Planck's constant and klystron frequency, respectively] in the denominator of the first term in Equation 9.37, the hyperfine doublet separation decreases with increasing microwave frequency; for example, it will be more than 12-fold smaller at Q-band (~35 GHz) than that at X-band (~10 GHz) for the central sextet. Thus, X-band data are preferable over Q-band data to estimate D from the forbidden hyperfine doublet separation. To this end, it should be noted that the lower the microwave frequency, the greater is the resolution of the hyperfine doublet separation.

9.4.1.2 Rigorous Evaluation of SH Parameters from a Polycrystalline Spectrum by Using Matrix Diagonalization and Least-Squares Fitting (LSF)

For polycrystalline samples, it is possible to follow the procedure described in Chapter 8 (see Section 8.4), using the LSF technique for single-crystal line positions. The specific details of a LSF procedure, which can be used to estimate SH parameters from a polycrystalline spectrum to provide SH parameters and the linewidth by an exploitation of the homotopy and matrix diagonalization technique (HTMD) are now described. In addition, extension to the case of two magnetically inequivalent paramagnetic species is discussed.

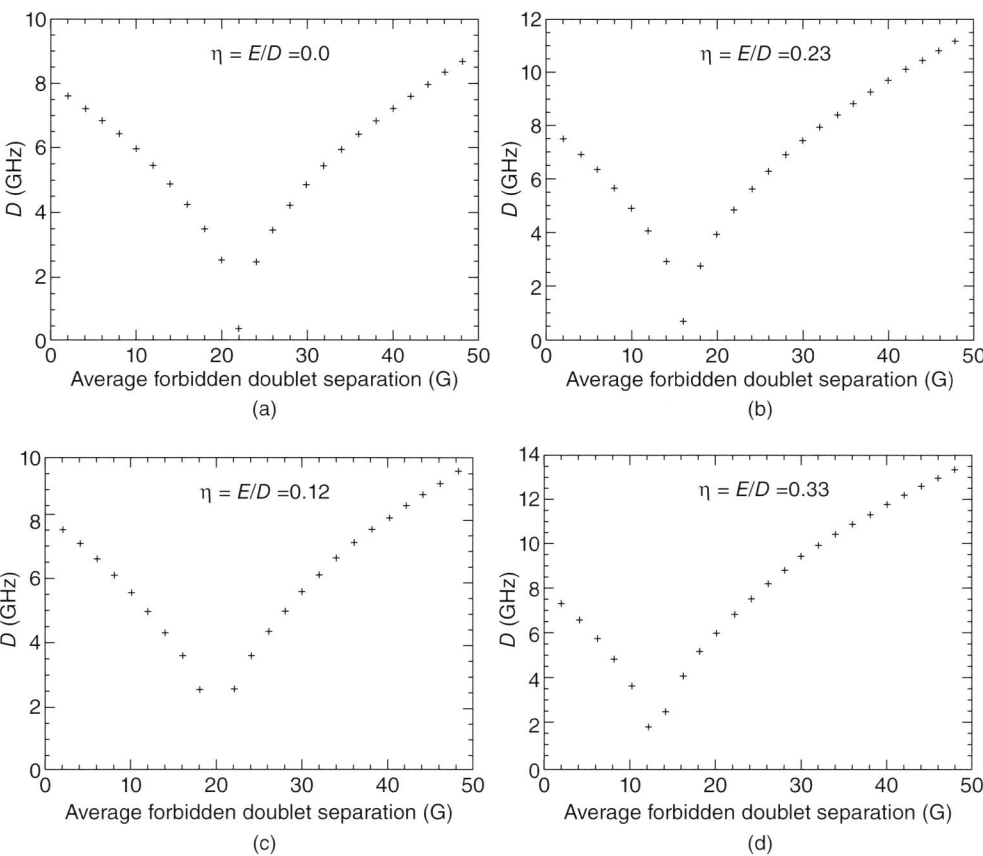

Figure 9.2 Plots of values of |D| as functions of the average hyperfine forbidden doublet separation in the central sextet for the Mn^{2+} ion for (a) $\eta = 0.0$; (b) $\eta = 0.12$; (c) $\eta = 0.23$; and (d) $\eta = 0.33$.

The χ^2-Function for a Polycrystalline EPR Spectrum and its First and Second Derivatives For application of the LSF technique in this context, the χ^2-function is defined as the sum of the weighted squares of the differences between the calculated and measured first-derivative absorption resonances at the magnetic field values B_k within the magnetic-field interval considered:

$$\chi^2 = \sum_k [F_c(B_k, v_c) - F_m(B_k, v_c)]^2 / \sigma_k^2, \tag{9.39}$$

where $F_c(B_k, v_c)$ and $F_m(B_k, v_c)$ are, respectively, the normalized calculated, as given by Equation 9.19 and the measured values of the first-derivative EPR resonance signal, and σ_k is the weight factor, related to standard deviation of the datum k. The measured/calculated values may be normalized in such a way that the maximum of each is equal to 1.

In the LSF technique, the vector $\mathbf{a}^f$ as given by (Misra, 1976):

$$\mathbf{a}^f = \mathbf{a}^i - (\mathbf{D}''(\mathbf{a}^i))^{-1}\mathbf{D}', \tag{9.40}$$

is then calculated iteratively until a sufficiently small value of the χ^2-function, consistent with experimental errors is obtained. The errors in the values of the parameters can be calculated statistically from the matrix elements of $\mathbf{D}''$.

Calculation of the Matrix Elements of the First-Derivative Matrix D′ and the Second-Derivative Matrix D″ with Respect to the Initial Parameters $\mathbf{a}^i$, The Vector Whose Components are the Initially Chosen Values of the SH Parameters and the Linewidth From Equation 9.35 one obtains:

$$D'_m = \left(\frac{\partial \chi^2}{\partial a_m}\right)_{\mathbf{a}^i} = 2\sum_k [F_c(B_k, v_c) - F_m(B_k, v_c)] \left(\frac{\partial F_c(B_k, v_c)}{\partial a_m}\right) \Big/ \sigma_k^2 \tag{9.41}$$

$$D''_{nm} = \left(\frac{\partial^2 \chi^2}{\partial a_n \partial a_m}\right)_{\mathbf{a}^i} = 2\sum_k \left\{ [F_c(B_k, v_c) - F_m(B_k, n_c)] \left(\frac{\partial^2 F_c(B_k, v_c)}{\partial a_n \partial a_m}\right) \right.$$
$$\left. + \left(\frac{\partial F_c(B_k, v_c)}{\partial a_n}\right)\left(\frac{\partial F_c(B_k, v_c)}{\partial a_m}\right) \right\} \Big/ \sigma_k^2 \tag{9.42}$$

The first- and second-derivatives of $F_c(B_k, v_c)$, given by Equations 9.17 and 9.18, with respect to the parameters, appearing in Equations 9.41 and 9.42, for the Gaussian and Lorentzian lineshapes are given as follows.

- **Gaussian lineshape**

$$\frac{\partial F_c(B_k, v_c)}{\partial a_m} = \sum_i N |\langle \Phi_{i'}|B_{1x}S_x + B_{1y}S_y + B_{1z}S_z|\Phi_{i''}\rangle|^2 \exp\left[-(B_k - B_{ri})^2/\sigma^2\right]$$
$$\times 2(B_k - B_{ri})\left\{\frac{\partial B_{ri}}{\partial a_m} + (B_k - B_{ri})\frac{\partial \sigma}{\partial a_m}\Big/\sigma\right\}\Big/\sigma^2 \tag{9.43}$$

In Equation 9.43, $\partial \sigma/\partial a_m = 1/w_G$, for $a_m = w_G\sigma$, $= 0$ otherwise.
From Equation 9.41 the second derivative of $F_c(B_k, v_c)$ is

$$\frac{\partial^2 F_c(B_k, v_c)}{\partial a_n \partial a_m} = \sum_i N |\langle \Phi_{i'}|B_{1x}S_x + B_{1y}S_y + B_{1z}S_z|\Phi_{i''}\rangle|^2 \exp\left[-(B_k - B_r)^2/\sigma^2\right]$$
$$\times \left\{[2(B_k - B_{ri})/\sigma^2]\left\{\frac{\partial^2 B_{ri}}{\partial a_n \partial a_m} - \frac{\partial B_{ri}}{\partial a_n}\frac{\partial \sigma}{\partial a_m}\Big/\sigma - (B_k - B_{ri})\frac{\partial \sigma}{\partial a_n}\frac{\partial \sigma}{\partial a_m}\Big/\sigma^2\right\}\right.$$
$$-\left\{2\left(\frac{\partial B_{ri}}{\partial a_n} + 2(B_k - B_{ri})\frac{\partial \sigma}{\partial a_n}\Big/\sigma\right) - [2(B_k - B_{ri})/\sigma]^2\right.$$
$$\left.\left[\frac{\partial B_{ri}}{\partial a_n} + (B_k - B_{ri})\frac{\partial \sigma}{\partial a_n}\Big/\sigma\right]\right\} \times \left[\frac{\partial B_{ri}}{\partial a_m} + (B_k - B_{ri})\frac{\partial \sigma}{\partial a_m}\Big/\sigma\right]\Big/\sigma^2\right\}, \tag{9.44}$$

9.4 Evaluation of SH Parameters and the Linewidth from a Polycrystalline EPR Spectrum

where account has been taken of the fact that the second derivative of σ with respect to the parameters, a_n, is zero.

- **Lorentzian lineshape**

$$\frac{\partial F_c(B_k, v_c)}{\partial a_m} = \sum_i N |\langle \Phi_{i'} | B_{1x} S_x + B_{1y} S_y + B_{1z} S_z | \Phi_{i''} \rangle|^2$$

$$\times \frac{1}{\left[\Gamma^2 + (B_k - B_{ri})^2\right]^2} \left\{ \frac{\partial \Gamma}{\partial a_m} (B_k - B_{ri}) - \Gamma \frac{\partial B_{ri}}{\partial a_m} \right\} \quad (9.45)$$

$$+ \Gamma (B_k - B_{ri}) \left\{ \frac{\partial}{\partial a_m} \left(1 / \left[\Gamma^2 + (B_k - B_{ri})^2\right]^2 \right) \right\}$$

where,

$$\left\{ \frac{\partial}{\partial a_m} \left(1 / \left[\Gamma^2 + (B_k - B_{ri})^2\right]^2 \right) \right\} = \left(-4 \left[\Gamma \frac{\partial \Gamma}{\partial a_m} - (B_k - B_{ri}) \frac{\partial B_{ri}}{\partial a_m} \right] \bigg/ \left[\Gamma^2 + (B_k - B_{ri})^2\right]^3 \right)$$

(9.46)

In Equations 9.45 and 9.46, $\partial \Gamma / \partial a_m = 1/w_L$, for $a_m = w_L \Gamma$, = 0 otherwise.

From Equation 9.45, one has

$$\frac{\partial^2 F_c(B_k, v_c)}{\partial a_n \partial a_m} = \sum_i N |\langle \Phi_{i'} | B_{1x} S_x + B_{1y} S_y + B_{1z} S_z | \Phi_{i''} \rangle|^2$$

$$\times \frac{\partial}{\partial a_n} \left(1 / \left[\Gamma^2 + (B_k - B_{ri})^2\right]^2 \right) \left\{ \frac{\partial \Gamma}{\partial a_m} (B_k - B_{ri}) - \Gamma \frac{\partial B_{ri}}{\partial a_m} \right\}$$

$$- \left(1 / \left[\Gamma^2 + (B_k - B_{ri})^2\right]^2 \right) \left\{ \frac{\partial \Gamma}{\partial a_m} \frac{\partial B_{ri}}{\partial a_n} + \frac{\partial \Gamma}{\partial a_n} \frac{\partial B_{ri}}{\partial a_m} + \Gamma \frac{\partial^2 B_{ri}}{\partial a_n \partial a_m} \right\}$$

$$+ \left\{ \frac{\partial \Gamma}{\partial a_n} (B_k - B_{ri}) - \Gamma \frac{\partial B_{ri}}{\partial a_n} \right\} \left\{ \frac{\partial}{\partial a_m} \left(1 / \left[\Gamma^2 + (B_k - B_{ri})^2\right] \right) \right\}$$

$$+ \Gamma (B_k - B_{ri}) \left\{ \frac{\partial^2}{\partial a_n \partial a_m} \left(1 / \left[\Gamma^2 + (B_k - B_{ri})^2\right] \right) \right\},$$

(9.47)

where, from Equation 9.46,

$$\frac{\partial^2}{\partial a_n \partial a_m} \left(1 / \left[\Gamma^2 + (B_k - B_{ri})^2\right] \right) = -4 \left[\frac{\partial \Gamma}{\partial a_n} \frac{\partial \Gamma}{\partial a_m} + \frac{\partial B_{ri}}{\partial a_n} \frac{\partial B_{ri}}{\partial a_m} - (B_k - B_{ri}) \frac{\partial^2 B_{ri}}{\partial a_n \partial a_m} \right]$$

$$- 12 \left[\Gamma \frac{\partial \Gamma}{\partial a_m} - (B_k - B_{ri}) \frac{\partial B_{ri}}{\partial a_m} \right] \frac{\partial}{\partial a_n} \left[\Gamma^2 + (B_k - B_{ri})^2 \right]^2 \bigg/ \left[\Gamma^2 + (B_k - B_{ri})^2\right]$$

(9.48)

In expressing Equation 9.48, account has been taken of the fact that the second derivative of Γ with respect to the parameters, a_n, is zero.

The optimum SH parameters and linewidth can now be estimated from a polycrystalline spectrum using the LSF technique by following the steps outlined in Chapter 8, Section 8.4.2.

9.4.1.3 Evaluation of SH Parameters and Linewidths for the Case of Two Magnetically Inequivalent Species

The spectrum to be considered in this case is a superposition of the spectra corresponding to two species, with the weight factors proportional to the populations of the two species in the sample when two magnetically inequivalent species are present in a polycrystalline sample. Consequently, there are a total of $(n_1 + n_2 + 1)$ parameters to be fitted, where n_1 and n_2 are the number of parameters characterizing the two species, since each species is characterized by its own set of parameters, and the additional parameter gives the fraction (f) of one of the species. The resultant spectrum is expressed as:

$$F_c(a_{1i}, a_{2j}, f) = f F_{c1}(a_{1i}) + (1-f) F_{c2}(a_{2j}) \tag{9.49}$$

where $i = 1, 2, \ldots, n_1$, and $j = 1, 2, \ldots, n_2$.

The experimental spectrum can be fitted to the parameters, using the derivatives of F_c with respect to the parameters a_{1i}, a_{2j}, f. It follows from Equation 9.49 that

$$\partial F_c/\partial a_{1i} = f \partial F_{c1}/\partial a_{1i} \tag{9.50}$$

$$\partial F_c/\partial a_{2j} = (1-f) \partial F_{c2}/\partial a_{2j}; \tag{9.51}$$

$$\partial F_c/\partial f = F_{c1} - F_{c2}. \tag{9.52}$$

The derivatives, given by Equations 9.50 or 9.51, can be calculated in the same way as for the case of one species, except for the introduction of the weighting factors of f and $(1-f)$, respectively, for species 1 and 2. Using Equations 9.50–9.52 for the required derivatives, the values of the parameters a_{1i}, a_{2j}, and the fraction f can be evaluated by the LSF technique in conjunction with the HTMD method, using the approach outlined in Section 9.4.1.2 for only one species.

The procedure for two species outlined here can be similarly extended to the case of presence of more than two magnetically inequivalent species.

9.4.1.4 Illustrative Example

The LSF technique was successfully applied to estimate SH parameters and linewidth from the 249.9 GHz EPR spectrum for Mn(γ-picoline)$_4$I$_2$ polycrystalline sample by Misra (1999b). In addition, successful application was made of the LSF method described here to evaluate the Mn^{2+} parameters from 249.9- and 95-GHz data on tetrahedrally distorted compounds: dichloro-, dibromo- diiodo-bis triphenyl phosphine dioxide manganese complexes by Wood et al. (1999). It is noted that the number of iterations required is reduced if the initially chosen values of the parameters are close to those corresponding to the absolute minimum of the χ^2-value. For example, one can deduce the values of the parameters D and E from the separation of the peaks of the hyperfine doublets using eigenvalues calculated to third order in perturbation in the central sextet, as described in Section 9.4.1.1. The positions of the peaks for the two extreme cases $\eta = |E/D| = 0$ (axial symmetry) and $\eta = 1$ (maximum rhombic distortion), as listed by Misra (1997), from which an educated estimate can be made by extrapolation for intermediate values of η.

9.4.1.5 General Remarks
The following points are noteworthy:

Although the procedure has been described for the "field-swept" EPR spectrum, it is equally applicable to "frequency-swept" spectrum, since the resonance condition, $|\Delta E_i| = h\nu_c$, still remains the same.

The procedure outlined here can be easily extended to evaluate the parameters describing ENDOR, solid-state NMR, NQR, and ESE envelope modulation spectra from polycrystalline samples, as well as to include hyperfine structure (Misra, 1983; Misra, 1986).

9.5 Simulation of EPR Spectra in Disordered Materials: Application to Glassy Materials

9.5.1 Introduction

Glasses possess the characteristics of disordered materials. The simulation of EPR spectra in these materials will be illustrated here for the Mn^{2+} ion ($S = 5/2, I = 5/2$). Glasses are characterized by distributions of spin-Hamiltonian parameters around mean values. The Mn^{2+} ion possesses certain advantages over other ions. Being an S-state ion, its $\tilde{g}$ and $\tilde{A}$ matrices can be considered to be isotropic in glasses, as their small anisotropies are smeared out in glassy spectra. On the other hand, the hyperfine structure of its spectrum adds some complexity to the spectra. By possessing the electronic spin larger than ½, its spectrum reflects interaction with the environment due to spin–orbit coupling. In glassy materials, one takes into account the distribution of the ZFS parameters D and E, as well as random orientations of Mn^{2+} principal magnetic axes with respect to the external magnetic field.

There are two approaches to simulate and interpret Mn^{2+} EPR spectra in disordered materials. A simple approach, when applicable, takes into account perturbation expressions for eigenvalues and relative intensities of EPR lines for cases of different site symmetries, for example, axial (Misra, 1994) and orthorhombic (Misra, 1997). A more rigorous approach is to simulate the spectrum on a computer, using appropriate line shapes and distribution of parameters, as discussed by Kliava and Purans (1980), and by Misra (1996). This section provides details of how to simulate spectra in glassy materials, taking into account the distribution of D, E values. In addition, it includes details of computer simulation of spectra and comparison with experimental spectra.

9.5.2 Computer Simulation of EPR Spectra in Glasses

The simulation of a powder spectrum on a computer, using matrix diagonalization, was described in Section 9.3.7. Glasses are characterized by structural

9 Simulation of EPR Spectra

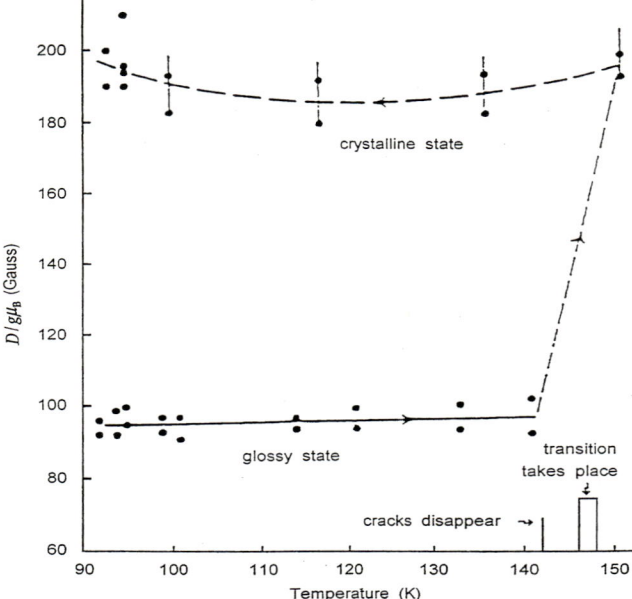

Figure 9.3 Variation of the value of $D/g\mu_B$ of the Mn^{2+} ion in 12 N HCl, with temperature. The arrows show the direction of the temperature changes in the experiment. Adapted from Allen (1965).

disorders, subjecting paramagnetic ions to experience random surroundings. The spin-Hamiltonian parameters thus vary rather widely. There are three types of glass in which Mn^{2+} spectra have been mainly studied: borate (Misra, 1996); phosphate; and silicate (Kliava and Purans, 1980; further examples are provided by Misra, 1996 and de Wijn and Van Balderin, 1967). Some materials pass from the glassy to the disordered state with variation of temperature, as illustrated in Figure 9.3 (Allen, 1965).

The simulation of EPR spectra in glasses requires the use of rather precise Mn^{2+} EPR line shapes, taking into account the distribution of spin-Hamiltonian parameters (Misra, 1996). Only the parameters values restricted to $|D| \ll g\mu_B B$, $|E| \ll g\mu_B B$, $|A| \ll g\mu_B B$ will be considered here. The resonance magnetic fields for transitions between the states M, m and $M-1$, $m+i$, denoted by $B_0 = B_{M,m;M-1,m+i}(g, A, D, E, \theta, \varphi)$, have been calculated by de Wijn and Van Balderin (1967), Bleaney and Rubins (1961), Bir (1964), and Upreti (1974). In the expression for B_0, $i = 0$ indicates allowed, while $i = \pm 1, \pm 2, \cdots$ correspond to forbidden transitions; θ, φ are the polar and azimuthal angles formed by the magnetic field with the principal axes of the second-order crystal-field tensor in the spin Hamiltonian. Finally, a total of nine parameters is to be determined: g_0, A_0, D_0, E_0, Δg, ΔA, ΔD, ΔE, and ΔB_{PP}, where the subscript 0 denotes the mean values g_0, A_0, D_0, and E_0, characterized by the widths Δg, ΔA, ΔD, and ΔE,

9.5 Simulation of EPR Spectra in Disordered Materials: Application to Glassy Materials

respectively, along with the peak-to-peak linewidth, ΔB_{PP}. The parameters g and A for Mn^{2+} are much less sensitive to a change in environment in comparison to D and E. Thus, Δg and ΔA are relatively much smaller, and their variation can be taken into account by an appropriate value of ΔB_{PP}, which also takes into account spin–spin interaction and other terms not included in the spin Hamiltonian. The value of $g_0 = 2.0$ can be safely assumed for the S-state ion, Mn^{2+}.

Considering only the variations ΔD and ΔE to be significant, describing Gaussian spreads with variances $\Delta D^2/2$ and $\Delta E^2/2$, respectively, and a correlation coefficient r $(-1 \leq r \leq 1)$, the joint statistical probability then becomes (de Wijn and Van Balderin, 1967):

$$P(D, E) = \left(\pi \Delta D \Delta E \sqrt{(1-r^2)}\right)^{-1} \tag{9.53}$$
$$\times \exp\left\{-\frac{1}{(1-r^2)}\left[\left(\frac{D-D_0}{\Delta D}\right)^2 - 2r\frac{(D-D_0)(E-E_0)}{\Delta D \Delta E} + \left(\frac{E-E_0}{\Delta E}\right)^2\right]\right\}$$

When $r = 0$, D and E are totally uncorrelated; $P(D, E)$ is expressed as the product of two Gaussian distribution which are independent of each other:

$$P(D) = \frac{1}{\sqrt{\pi}\Delta D}\exp\left[-\left(\frac{D-D_0}{\Delta D}\right)^2\right]; P(E) = \frac{1}{\sqrt{\pi}\Delta E}\exp\left[-\left(\frac{E-E_0}{\Delta E}\right)^2\right] \tag{9.54}$$

The values $r = \pm 1$ represent a total correlation between D and E. $P(D, E)$, as given by Equation 9.53 is thus nonzero only when $(E - E_0)/(D - D_0) = \pm \Delta E/\Delta D$, in which case, only one variation $P(D)$ or $P(E)$ is to be used in Equation 9.54.

Assuming that ensembles of randomly-oriented identical sites and randomly-distorted sites are mutually independent, the EPR spectra in a glassy material can be expressed taking into account all fine-structure transitions and all orientations of Mn^{2+} ions, following Taylor and Bray (1972), as

$$P(H) = \sum_{m=-3/2}^{5/2} \sum_{\substack{i \\ |m-i|\leq 5/2}} \sum_{m=-5/2}^{5/2} \int_{-\infty}^{\infty} dD \int_{-\infty}^{\infty} dE \int_{0}^{\pi} d\theta \int_{0}^{2\pi} d\phi \cdot \tag{9.55}$$

$$P(D, E) \cdot \sin\theta \cdot W_{M,m,M-1,m+i}(D, E, \theta, \varphi) \cdot F\left(\frac{B-B_0}{\Delta B_{PP}}\right)$$

In Equation 9.55, $F[(B - B_0)/\Delta B_{PP}]$ is a lineshape function which can be considered to be the first derivative of either a Gaussian or a Lorentzian; $W_{M,m;M-1,m+i}(D, E, \theta, \varphi)$ is the probability of a transition between the states M, m and $M-1, m + i$ averaged over different orientations of the microwave magnetic field. W can be calculated using the method of Bir (1964), which is superior to the traditional perturbation approach (Bleaney and Rubins, 1961; Kliava and Purans, 1978), unless exact eigenvectors are used as obtained by computer diagonalization of the spin-Hamiltonian matrix (as described in Section 9.2.1). The limits of integration over D and E can be taken to be $D_0 \pm 2\Delta D$ and $E_0 \pm 2\Delta E$. Finally, the factor $\partial B_0/\partial h\nu$ that should be included in the integrand of Equation 9.55 has here been approximated to be unity as its value deviates, in fact, from unity only by a few percent (Aasa and Vanngard, 1975).

In practice, the calculation is confined to the transitions belonging to the central hyperfine sextet $\frac{1}{2}, m \leftrightarrow -\frac{1}{2}, m + i$ ($m = \frac{5}{2}, \frac{3}{2}, \frac{1}{2}, -\frac{1}{2}, -\frac{3}{2}, -\frac{5}{2}$), since the noncentral lines are much less intense as compared to the central lines because they have a stronger angular dependence. In fact, in glasses parameter distributions totally smear out all noncentral transition lines as their resonant fields contain terms linear in D and E, which is not the case for the resonant fields for the central transition lines.

9.5.3
Computer-Simulated Spectra and Comparison with Experiment

Figures 9.4–9.11 display various simulated and experimental spectra, as described in the figure captions. The salient features of these are as follows:

i) Figure 9.5 indicates a drastic discrepancy between the two simulated spectra – one which does and the other which does not – take into account the different transition probabilities of the various transitions in the simulated spectrum. This leads to the conclusion that transition probabilities are extremely important in arriving at a reasonable fit.

ii) Changing the signs of D_0 or E_0 does not lead to appreciable variation of simulated spectra, since the third-order terms in resonant field linear in D_0 and E_0 are small.

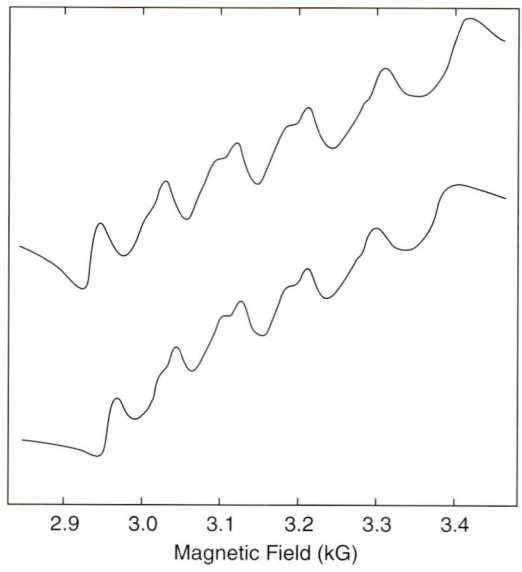

Figure 9.4 EPR spectra of Mn^{2+} observed at $\nu = 8.9\,GHz$ in phosphate ($ZnO \cdot P_2O_5$, upper trace) and silicate ($K_2O \cdot 4Si_2O$, lower trace) glasses. Adapted from Kliava and Purans (1980).

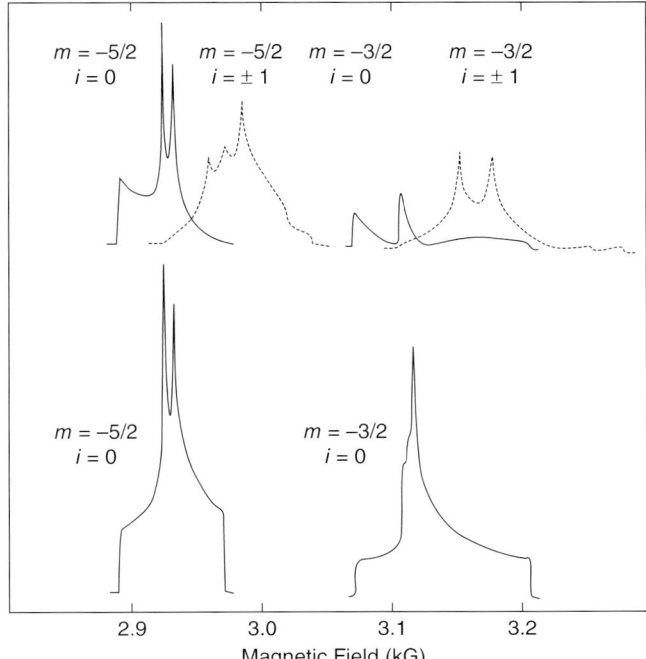

Figure 9.5 Mn^{2+} powder spectra simulated to third order in perturbation for some allowed (solid lines) and forbidden (dashed lines) hyperfine transitions belonging to the central fine-structure transition: $g = 2.0$, $A/g\mu_B = -95\,G$, $D/g\mu_B = 210\,G$, $E/g\mu_B = 70\,G$, $\nu = 8.9\,GHz$. In the upper diagram the transition probability is calculated using Bir's method (Bir, 1964), whilst in the lower diagram the transition probability equals 1. The difference in the two diagrams shows the importance of taking into account the transition probabilities in spectral simulation. Adapted from Kliava and Purans (1980).

iii) The best-fit spin-Hamiltonian parameters are found to be the same for the two glasses from Figures 9.9 and 9.10 for phosphate and silicate glasses with the exception of A: $g = 2.0$, $|D_0|/g\mu_B = 220 \pm 20\,G$, $|E_0|/g\mu_B = 70 \pm 15\,G$, $\Delta D/g\mu_B = 80 \pm 20\,G$, $\Delta E/g\mu_B = 30 \pm 10\,G$, $r = 0.0 \pm 0.2$, while $A/g\mu_B = -93 \pm 1\,G$ for the phosphate glass and $A/g\mu_B = -87 \pm 1\,G$ for the silicate glass. Outside the limits of error for D_0, E_0, ΔD, ΔE and A, the variation of one parameter degrades the other parameters. The limits of error are determined for the correlation coefficient r for fixed values of all other parameters. Finally, it is noted that only the absolute values of D_0 and E_0 can be determined in glasses when D_0 is small. On the other hand, when D_0 is large ($D_0/g\mu_B \geq 10\,000\,G$), then variable temperature measurements will allow determination of the signs for D, E. Zero-field EPR may also be used to determine the signs of D, E.

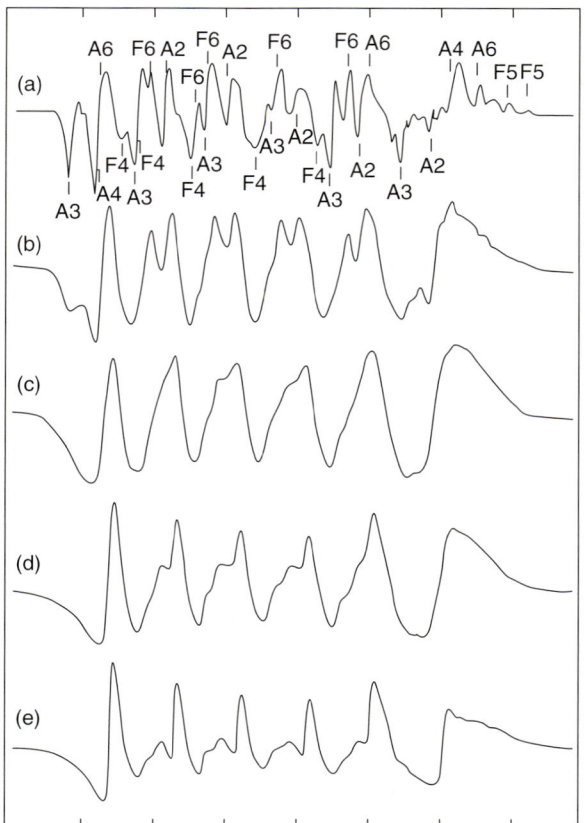

Figure 9.6 A series of spectra computed for different distributions of the fine structure parameter (for the central fine-structure transition): $g = 2.0$, $A/g\mu_B = -93$ G, $D_0/g\mu_B = 220$ G, $E_0/g\mu_B = 73$ G, $\nu = 8.9$ GHz, $\Delta H_{pp} = 7$ G (Lorentzian lineshape function). The following values (in gauss) of $\Delta D/g\mu_B$, $\Delta E/g\mu_B$ have been used for different traces: (a) 0, 0; (b) 40, 20; (c) 80, 40; (d) 120, 60; (e) 160, 80. In trace (a), the attribution of some of the spectral features to definite critical points is shown. The numbers 1 to 6 indicate the six types of critical points (see Equation 9.61) for the allowed (A) and forbidden ($i = \pm 1$) (F) hyperfine-structure transitions. Adapted from Kliava and Purans (1980).

9.5.4
Shape of EPR Spectra in Glasses: Effect of SH Parameters

9.5.4.1   Distribution of the Fine-Structure Parameters D and E

The effect of the distributions of the fine-structure parameter on the shape of EPR spectra is illustrated in Figure 9.6. The "lines" and "peaks" that appear in the absence of distribution of D and E (curve **a**) broaden at first as ΔD and ΔE are increased (curves **b–e**); some of them broaden out completely. As ΔD and ΔE are

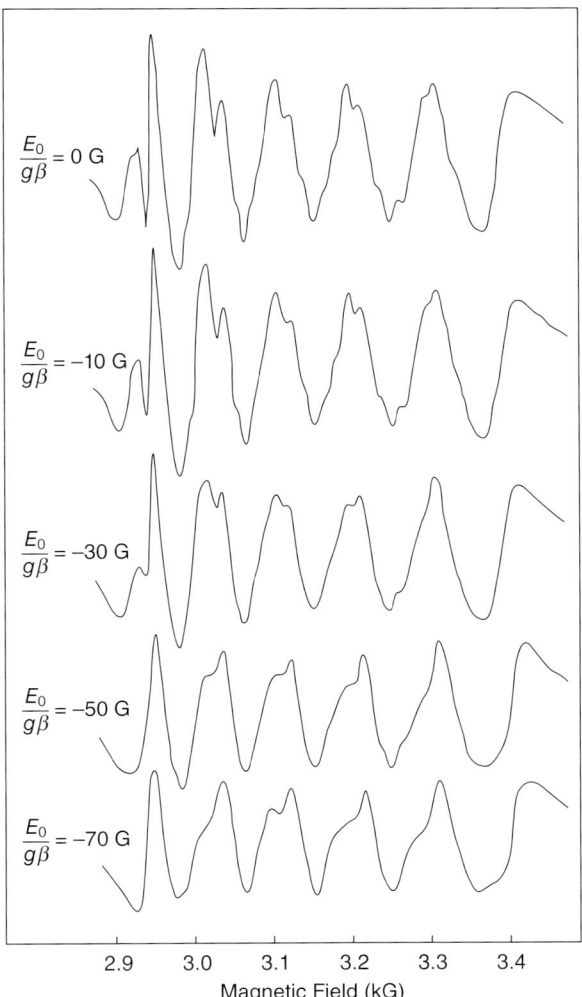

Figure 9.7 A series of spectra computed for different values of the central fine-structure transition: $g = 2.0$, $A/g\mu_B = -93\,G$, $D_0/g\mu_B = 220\,G$, $\Delta D/g\mu_B = 80\,G$, $\Delta E/g\mu_B = 30\,G$, $\nu = 8.9\,GHz$, $\Delta B_{PP} = 6\,G$ (Lorentzian lineshape). Adapted from Kliava and Purans (1980).

increased further, the remaining "lines" and "peaks" no longer broaden, contrary to expectation, and even become narrower (curves **d–e**). This can be explained by expressing Equation 9.55 as follows (Kneubuhl, 1960):

$$P(B) = \sum_{M,i,m} \int dB_0 \int_s d\sigma \cdot p(d,e) \cdot W_{M,m;M-1,m+i}(d,e,\theta,\phi) \cdot F\left(\frac{B-B_0}{\Delta B_{PP}}\right)/|\text{grad } B_0|$$

(9.56)

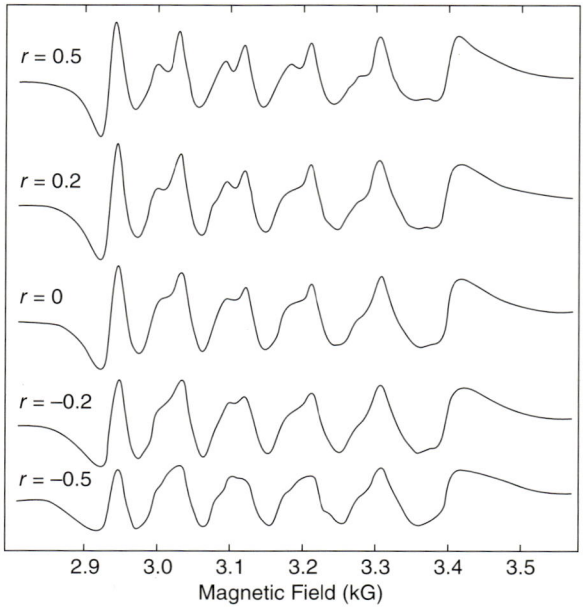

Figure 9.8 A series of spectra computed for different values of the correlation coefficient r for the central fine-structure transition: $g = 2.0$, $A/g\mu_B = -93\,G$, $D_0/g\mu_B = 220\,G$, $E_0/g\mu_B = 73\,G$, $\Delta D/g\mu_B = 80\,G$, $\Delta E/g\mu_B = 30\,G$, $\nu = 8.9\,GHz$, $\Delta B_{PP} = 6\,G$ (Lorentzian lineshape). Adapted from Kliava and Purans (1980).

In Equation 9.56, the integration is over B_0, the three-dimensional surface S is defined by $B_0(d, e, \theta, \varphi) = $ constant, $d = D/\Delta D$ and $e = E/\Delta E$,

$$P(d,e) = \frac{1}{\pi(1-r^2)^{1/2}} \exp\left\{-\frac{1}{(1-r^2)}\left[(d-d_0)^2 - 2r(d-d_0)(e-e_0) + (e-e_0)^2\right]\right\} \tag{9.57}$$

In Equation 9.57, $d_0 = D_0/\Delta D$, $e_0 = E_0/\Delta E$. Since B_0 depends on d, e, θ, and φ, one has

$$|\text{grad } B_0| = \left[(\text{grad}_{\theta,\varphi} B_0)^2 + \left(\frac{\partial B_0}{\partial d}\right)^2 + \left(\frac{\partial B_0}{\partial e}\right)^2\right]^{1/2}, \tag{9.58}$$

where

$$|\text{grad}_{\theta,\varphi} B_0| = \left[\left(\frac{\partial B_0}{\partial \theta}\right)^2 + \frac{1}{\sin^2\theta}\left(\frac{\partial B_0}{\partial \varphi}\right)^2\right]^{1/2}. \tag{9.59}$$

When $\Delta D = \Delta E = 0$, it is seen from Equation 9.56 that the "lines" and "peaks" in the powder EPR spectra occur at resonance-field values for which $|\text{grad}_{\theta,\varphi} B_0| = 0$, that is,

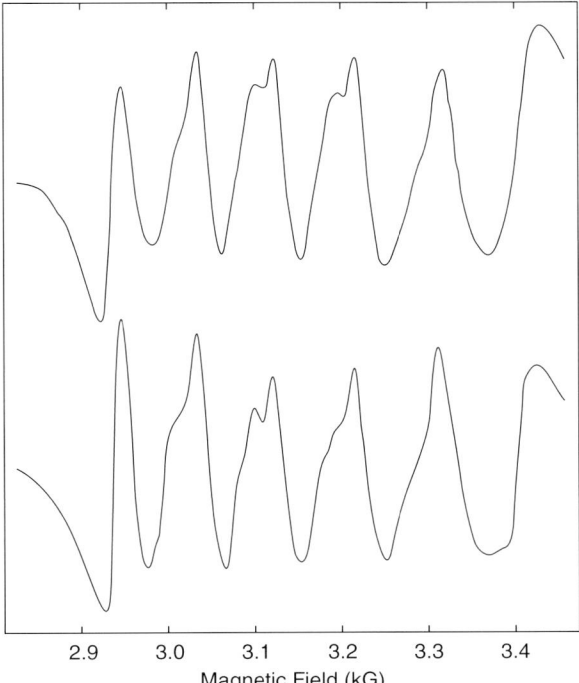

Figure 9.9 Comparison of the experimental and computer-simulated spectra for the zinc-phosphate glass (ZnO·P$_2$O$_5$). The upper trace represents the experimental spectrum after subtracting the broad underlying resonance. The lower trace is the computed best-fit spectrum (for the central fine-structure transition) with $g = 2.0$, $A/g\mu_B = -93\,G$, $D_0/g\mu_B = 220\,G$, $E_0/g\mu_B = 70\,G$, $\Delta D/g\mu_B = 80\,G$, $\Delta E/g\mu_B = 30\,G$, $\nu = 8.9\,GHz$, $\Delta B_{PP} = 6\,G$ (Lorentzian lineshape). Adapted from Kliava and Purans (1980).

$$\frac{\partial B_0}{\partial \theta} = 0, \quad \frac{1}{\sin\theta}\frac{\partial B_0}{\partial \varphi} = 0. \tag{9.60}$$

It can be shown that for each hyperfine transition six different types of critical points occur at definite values of θ_0, φ_0, which are as follows:

1. $\theta_0 = 0$;
2, 3. $\theta_0 = \pi/2$, $\varphi = \pi/4 \mp \pi/4$ (or $5\pi/4 \mp \pi/4$);
4. $\theta_0 = \pi/2$, $\cos 2\varphi_0 = \alpha D/E$, where

$$\alpha = \frac{2 - (2m+i)A/B_0 + 3(2m+i)(1+i)A^2/DB_0}{18 - 73(2m+i)A/B_0}$$

5, 6. $\cos^2\theta_0 = x$,
$\varphi_0 = \pi/4 \mp \pi/4$ (or $5\pi/4 \mp \pi/4$), where

$$x = \frac{10D \pm 6E - (2m+i)(37D \pm 35E)A/B_0 - 3(2m+i)(1+i)A^2/B_0}{[18 - 73(2m+i)A/B_0](D \pm E)}. \tag{9.61}$$

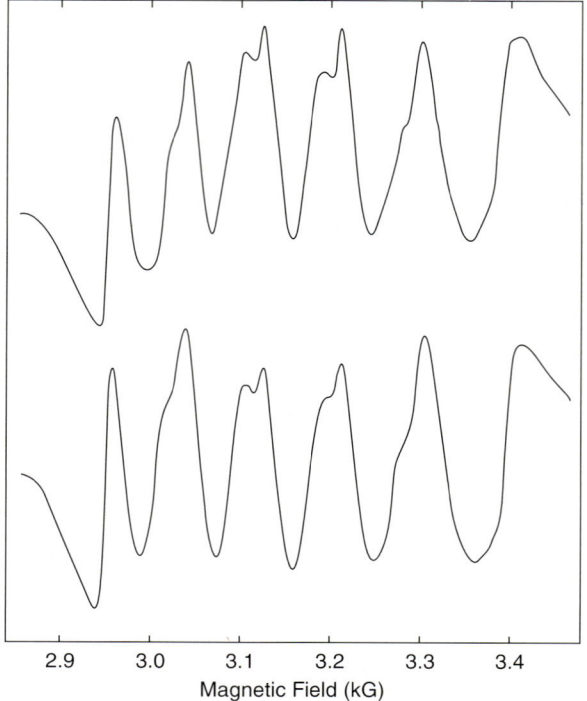

Figure 9.10 Comparison of the experimental and computer-simulated spectra for the potassium silicate glass (K$_2$O·4SiO$_2$). The upper trace represents the experimental spectrum after subtracting the broad underlying resonance. The lower trace is the computed best-fit spectrum (for the central fine-structure transition) with $g = 2.0$, $A/g\mu_B = -87\,G$, $D_0/g\mu_B = 220\,G$, $E_0/g\mu_B = 70\,G$, $\Delta D/g\mu_B = 80\,G$, $\Delta E/g\mu_B = 30\,G$, $\nu = 8.9\,GHz$, $\Delta B_{pp} = 7\,G$ (Lorentzian lineshape). Adapted from Kliava and Purans (1980).

The assignment of some "lines" and "peaks" to define critical points is illustrated in Figure 9.6 (curve a).

Supposing now that $\Delta D \neq 0$, $\Delta E \neq 0$. If the parameter distributions are small enough to satisfy the conditions

$$\left|\frac{\partial B_0}{\partial D}\right|_c \cdot \Delta D, \left|\frac{\partial B_0}{\partial E}\right|_c \cdot \Delta E \ll \frac{\partial}{\partial \theta}|grad_{\theta,\varphi} B_0|_c, \frac{\partial}{\partial \varphi}|grad_{\theta,\varphi} B_0|_c, \qquad (9.62)$$

where the index c means that all the distributions are taken at the critical point (θ_0, φ_0, D_0, E_0), the value of $\partial B_0/\partial \theta$ and $\partial B_0/(\sin\theta \cdot \partial\varphi)$ increase rapidly as one moves away from the critical point. Finally, Equation 9.50 becomes

$$|grad\, B_0|_c = \left[\left(\frac{\partial B_0}{\partial D}\right)_c^2 (\Delta D)^2 + \left(\frac{\partial B_0}{\partial E}\right)_c^2 (\Delta E)^2\right]^{1/2} \qquad (9.63)$$

Equation 9.63 reveals that the amplitudes of the "lines" and "peaks" decrease, while their widths increase, in proportion to $|grad\, B_0|_c$. This broadening becomes

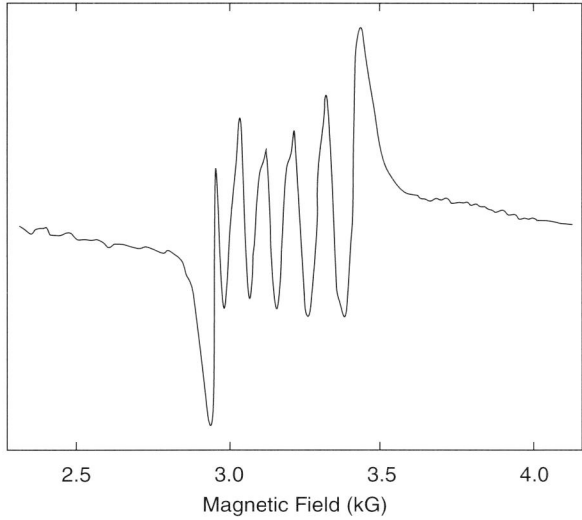

Figure 9.11 A spectrum computed taking into account all the fine-structure transitions: $g = 2.0$, $A/g\mu_B = -93$ G, $D_0/g\mu_B = 220$ G, $E_0/g\mu_B = 70$ G, $\Delta D/g\mu_B = 80$ G, $\Delta E/g\mu_B = 30$ G, $\nu = 8.9$ GHz, $\Delta B_{pp} = 11$ G. Adapted from Kliava and Purans (1980).

significant when it approaches or exceeds the value of ΔB_{pp} which includes all other sources of line broadening.

9.5.4.2 Sharp Features in Spectra

The salient features of computer-simulation, as far as sharp features in the spectra in the region $g \approx 2.0$ are concerned, are as follows:

i) The values D, $E \approx 0$ account for sharp features in computer-simulated spectra at broad parameter distributions (Figure 9.6, curves **d** and **e**).

ii) For the case of intermediate parameter distributions, no definite statement can be made on parameter distribution without a complicated analysis.

iii) As for the dependence of the spectra on the ratio E_0/D_0, even at rather broad parameter distributions ($\Delta D/g\mu_B = 80$ G, $\Delta E/g\mu_B = 30$ G) an increase of E_0/D_0 produces a strong variation in spectral shape. Computer fitting to the experimental spectra enables the determination of the absolute value of the ratio E_0/D_0 along with other spin-Hamiltonian parameters uniquely for Mn^{2+} in glasses with sufficient accuracy. Great care must be taken to "qualitatively" interpret EPR spectra of Mn^{2+} in disordered systems. For example, a spectrum with broad parameter distributions can be easily mistaken for one with small values of D_0 and E_0 (Taylor and Bray, 1972), since there are present some sharp features in both cases. This mistake is not possible with computer simulation of EPR spectra as the overall features of computed spectra are quite different in the two cases.

9.5.4.3 Broad Resonances in Spectra

Computer simulations support the claim of Griscom and Griscom (1967) that the broad resonance does not arise from noncentral fine-structure transitions, since the contribution of these transitions to the total EPR spectrum is, in fact, insufficient to account for the observed background resonance (see Figure 9.11). As seen from Figure 9.4, the peak-to-peak intensities of the broad resonance relative to sharp central features are about 3.3 and 4.0, respectively, for the phosphate and silicate glasses.

Griscom and Griscom (1967) concluded from the fact that broad underlying resonances observed at X-band collapse at Ka band, and that a broad distribution of sites with values $|D|/h$ ranging from near zero to as high as 7 GHz (~2.5 kG) and $|E_0/D_0| \approx 1/3$ is responsible for this behavior; sites with $|D|/h \geq 2$ GHz (~0.7 kG) give rise to background resonances. Computer simulations (Kliava and Purans, 1980), however, yield sharp $g = 2.0$ features with $|D_0|/g\mu_B \approx 220$ G (~0.6 GHz) with the same ratio $|E_0/D_0| \approx 1/3$, but without continuous broad distribution of sites. Thus, it is concluded that in addition to the large site-to-site distortion range, sites for which short-range order is much better preserved do exist in glasses (Tucker, 1962).

9.6
Simulation of EPR Spectra in Disordered Random Network Materials

The modeling of EPR spectra in glasses in Section 9.5 was based on a distribution of spin Hamiltonian parameters to take into account the disorder. However, this is not the most general case, in that it does not take into account the randomness inherent in glasses with short-range order and site-to-site variations, for example, that described by Thorpe, Djordevic, and Jacobs (1996) and Vacher and Courtens (1996). This means that in disordered network materials, such as glasses, different spin-probe sites are characterized by different spin-Hamiltonian parameters, possessing the lowest, triclinic symmetry. The following subsections provide the details of such an approach, as outlined by Pilbrow and Drew (1999).

9.6.1
Introduction

It is noted that only the sharp features of continuous-wave (CW) EPR spectra in glasses of transition-metal and rare-earth ions have been accounted for by considering spectra as powder-like (as discussed in Section 9.5). This was accomplished by considering a "high-symmetry" spin Hamiltonian, with a random distribution of parameters, which leads to large linewidths, rather than producing the whole spectrum which extends over many kilogauss for glasses, as described by Griscom (1980). For example, the commonly occurring $g = 4.3$ peak due to the Fe^{3+} ion ($S = 5/2$) in silicate glasses has been explained by invoking two different models:

- The middle Kramers' doublet, when the ratio $|E/D| = 1/3$, in appropriate orientations of the magnetic axes, has an isotropic $g \sim 4.0$ (as discussed by Pilbrow, 1990, p. 137).

- The $g = 4.3$ peak appears as a consequence of the lowest Kramers' doublet, assuming axial symmetry (Peterson, Kurkjian, and Carnevale, 1974; Carnevale, Peterson, and Kurkjian, 1976), due to the use of a joint distribution function for correlated effective $g_\parallel$ and $g_\perp$-values for the lowest $S = \frac{1}{2}$ doublet, where the principal g-factors are 2.0 and 6.0, respectively (Pilbrow, 1990, p. 137).

These two models were shown to be incompatible with each other by Pilbrow (1990, p. 137). The first model requires special constraints that do not represent all Fe^{3+} ions in glasses, whereas the second model is not consistent with the low symmetry of the Fe^{3+} sites. Another example is the occurrence of $g = 6.0$ peak for Gd^{3+} and Eu^{2+} ions ($S = 7/2$) for a range of $|E/D|$ values, explained by a broad distribution of spin Hamiltonian parameters, in addition to an arbitrary broad linewidth of ~80 G.

9.6.2
CW-EPR Spectrum for Random Distribution of SH Parameters at Various Sites in Glasses

The procedure described here is that outlined by Pilbrow and Drew (1999 and Drew and Pilbrow (2002), whose objective was to simulate the broad absorption spectrum due to transition-metal ions in glasses, rather than reproduce the "sharp" features observed in the usual first-derivative CW-EPR absorption spectrum. The underlying philosophy was that a glass should be treated differently from a polycrystalline powder, because the microcrystallite model assumes that the relative orientation of the various low-symmetry tensor interactions at each spin site is identical. The use of a distribution of spin Hamiltonians was therefore advocated, rather than using a statistical distribution of SH parameters in conjunction with a "powder average." Approaching the problem by beginning with a broad distribution of bond angles and bond lengths of transition-metal or rare-earth ions in tetrahedral coordination with oxygen ions was deemed unfeasible at the time. However, the state of the art of density functional theory (DFT) is such that, in the near future, it may prove possible to study the connection between the random bond angles and lengths at each spin site and the low-symmetry SH parameters (see Chapter 6 for further details on this subject). For example, in a recent theoretical study conducted by Drew and Hanson, the effects were investigated of varying the metal–dithiolate fold angle on the low-symmetry SH parameters of Mo(V) centers, using the ORCA program of Neese (2009). A similar approach may provide the needed insight into the nature of the distributions of principal values and directions appropriate for modeling the random geometric structure of glasses. Such insight was lacking in earlier treatments (Pilbrow and Drew, 1999; Drew and Pilbrow, 2002; Drew, 2002), where uncorrelated Gaussian distributions of the electron Zeeman and fine-structure parameters were generated using the

Monte Carlo technique. The spectrum was built up using the following expression, assuming no lineshape, but rather that each line would be a delta-function with its height being proportional to its intensity (transition probability):

$$S(B, v_c) \propto \sum_i P(i, v_c)\delta(B - B_i(v_c)), \quad \text{where} \tag{9.64}$$

$$P(i, v_c) \propto |\langle \Phi_{i'}|(B_{1x}S_x + B_{1y}S_y + B_{1z}S_z)|\Phi_{i''}\rangle|^2 \tag{9.65}$$

is the transition probability of the ith transition between the eigenpairs i' and i'', participating in resonance.

9.6.2.1 Calculation of Eigenvectors $|i'\rangle$ and $|i''\rangle$ Required in Equation 9.57

The *eigenfields method*, as described by Belford, Belford, and Burkhalter (1973) and by Belford *et al.* (1974), can be used to compute the resonant fields and transition probabilities, using the eigenvalues and eigenvectors of a generalized matrix. Briefly, in this technique (which is described in Appendix 9.I) the pairs of eigenvalue equations for the two energy levels participating in resonance, whose energy difference are equal to the quantum of microwave energy, are coupled. This results in a generalized eigenvalue equation with the eigenvalues being the resonant magnetic fields. Thus, the two coupled equations of dimension $n \times n$ become a single $n^2 \times n^2$ eigenvalue equation. The resonance fields are obtained from the generalized eigenvalues, while the eigenvectors required to calculate the transition probability given by Equation 9.65 can be obtained from the generalized $n^2 \times 1$ eigenvectors. This eigenfields method does not require any iteration to locate the resonance fields, but it becomes computationally intensive for high electron and/or nuclear spin. Some time saving can be achieved if only the generalized eigenvalues (the resonant field positions) are computed using the eigenfields method, and these magnetic fields are substituted into the ordinary $n \times n$ eigenvalue equation to solve for the eigenvectors. It should be mentioned that the eigenfields method is faster if one chooses only to solve for the generalized eigenvalues, and not both the eigenvalues and eigenvectors. Once the eigenvalues (resonance fields) have been computed, it is less computationally intensive to obtain the resonance intensities by substituting the resonance fields into the ordinary $n \times n$ spin Hamiltonian. Alternatively, one can use the quick LSF technique iteratively (as described in Chapter 8, Section 8.4), choosing initial values for which one can make an educated guess by trial, until such time that convergence is achieved.

9.6.3
Limitations of the Original Implementation and its Assumptions

The method of simulation used here has, in principle, the ability to distinguish between:

- A micro-crystallite, which is simpler to visualize and define, represented by a model in which there exist random strains defined by distributed spin Hamiltonian parameters of a polycrystalline powder characterized by a single spin Hamiltonian; and

- (ii) a random network of a glass, which is far more difficult to define, because of the possibility of having a distribution of C_1-symmetry Hamiltonians in addition to distributed spin Hamiltonian parameters.

Acknowledgments

The author is grateful to Professor C.P. Poole, Jr, for helpful comments to improve the presentation of this chapter, to Dr Simon Drew on numerous constructive suggestions to improve Section 9.6 (random networks), and to the Natural Sciences and Engineering Research Council (NSERC) of Canada for partial financial support.

Pertinent Literature

The present work provides essential insights into the simulation of Mn^{2+} EPR spectra in polycrystalline samples (Misra, 1999a), glassy materials (Misra, 1996; Kliava and Purans, 1978, 1980), and random networks (Pilbrow and Drew, 1999, Drew and Pilbrow, 2002). A description is also provided of how to estimate spin Hamiltonian parameters from the forbidden hyperfine doublet separations (Misra, 1994, 1997) using perturbation energies, and by the application of LSF and matrix diagonalization (Misra, 1999b). Whilst no attempt has been made to provide full details of the many publications on this subject, the references included, and those listed therein, would provide a rather exhaustive supply of information. An interesting review of both NMR and EPR spectra in polycrystalline materials, including several values of electronic and nuclear spins, has been produced by Taylor, Bangher, and Kriz (1975).

References

Aasa, R. and Vanngard, T. (1975) *J. Magn. Reson.*, **19**, 308.

Allen, B.T. (1965) *J. Chem. Phys.*, **43**, 3820.

Belford, R.L. and Nilges, M.J. (1979) Computer simulation of Powder Spectra. EPR Symposium, 21st Rocky Mountain conference, Denver, Colorado.

Belford, G.G., Belford, R.L., and Burkhalter, J.F. (1973) *Magn. Reson.*, **11**, 251.

Belford, R.L., Davis, P.H., Belford, G.G., and Lenhardt, T.M. (1974) Extended interactions between metal ions. *Am. Chem. Soc. Symp. Ser.*, **5**, 40.

Bir, G.L. (1964) *Sov. Phys. Solid State*, **5**, 1628.

Bleaney, B. and Rubins, R.S. (1961) *Proc. Phys. Soc. (London)*, **77**, 33; corrigendum, **78**, 78.

Buckmaster, H.A. and Dering, J.C. (1968) *J. Appl. Phys.*, **39**, 4486.

Carnevale, A., Peterson, G.E., and Kurkjian, C.R. (1976) *J. Non-Cryst. Solids*, **22**, 269.

Chiswell, B., McKenzie, E.D., and Lindboy, L.F. (1987) Manganese, in *Comprehensive Coordination Chemistry*, vol. 4 (eds G. Wilkinson, R.D. Gillard, and J.A.

McCleverty), Pergamon Press, Oxford, pp. 1–122.

Coffino, A.R. and Peisach, J. (1996) *J. Magn. Reson. B*, **111**, 127.

de Wijn, H.W. and Van Balderin, R.F. (1967) *J. Chem. Phys.*, **46**, 1381.

Drew, S.C. (2002) Selected topics in modern continuous-wave and pulsed electron paramagnetic resonance. PhD Thesis, Monash University, Melbourne, Australia.

Drew, S.C. and Pilbrow, J.R. (2002) Continuous wave and pulsed EPR spectroscopy of paramagnetic ions in some fluoride, silicate and meta-phosphate glasses, in *EPR in the 21st Century* (eds A. Kawamori, J. Yamauchi, and H. Ohta), Elsevier, New York, pp. 39–47.

Feynman, R.P. (1939) *Phys. Rev.*, **56**, 340.

Griscom, D.L. (1980) *J. Non-Cryst. Solids*, **40**, 211.

Griscom, D.L. and Griscom, R.E. (1967) *J. Chem. Phys.*, **47**, 2711.

Kliava, J.G. and Purans, J. (1978) *Phys. Status Solidi A*, **49**, K43.

Kliava, J.G. and Purans, J. (1980) *J. Magn. Reson.*, **40**, 33.

Kneubuhl, F.K. (1960) *J. Chem. Phys.*, **33**, 1074.

Lynch, W.B., Boorse, R.S., and Freed, J.H. (1993) *J. Am. Chem. Soc.*, **115**, 10909.

Maurice, A.M. (1980) PhD Thesis, University of Illinois, Urbana, Illinois.

Meirovitch, E. and Poupko, R. (1978) *J. Phys. Chem.*, **82**, 1920.

Misra, S.K. (1976) *J. Magn. Reson.*, **23**, 403.

Misra, S.K. (1983) *Physica*, **121B**, 193.

Misra, S.K. (1986) *Magn. Reson. Rev.*, **10**, 285.

Misra, S.K. (1994) *Physica B*, **203**, 193.

Misra, S.K. (1996) *Appl. Magn. Reson.*, **10**, 193.

Misra, S.K. (1997) *Physica B*, **240**, 183.

Misra, S.K. (1999a) *J. Magn. Reson.*, **137**, 83.

Misra, S.K. (1999b) *J. Magn. Reson.*, **140**, 179.

Misra, S.K. and Vasilopoulos, P. (1980) *J. Phys. C Condens. Matter*, **13**, 1083.

Moler, C.B. and Stewart, G.W. (1973) *SIAM J. Numer. Anal.*, **10**, 241.

Neese, F. (2009) *Inorg. Chem.*, **48**, 2224–2232.

Nilges, M.J. (1979) PhD thesis, University of Illinois, Urbana, Illinois.

Peterson, G.E., Kurkjian, C.R., and Carnevale, A. (1974) *J. Phys. Chem. Glasses*, **15**, 52.

Pilbrow, J.R. (1990) *Transition Ion Electron Paramagnetic Resonance*, Clarendon Press, Oxford.

Pilbrow, J.R. and Drew, S.C. (1999) An idea for modeling EPR due to spin probes in disordered systems. *Mol. Phys. Rep.*, **26**, 109–116.

Poole, C.P., Jr (1967) *Electron Spin Resonance: A Comprehensive Treatise on Experimental Techniques*, 1st edn, John Wiley & Sons, Inc., New York, p. 776.

Press, W.H., Teukolsky, S.A., Vetterling, W.T., and Flannery, B.P. (1992) *Numerical Recipes in Fortran*, 2nd edn, Cambridge University Press, Cambridge, pp. 456–462. The correct JACOBI subroutine is available on the website: http://netlib2.cs.utk.edu/.

Reed, G.H. and Markham, G.D. (1984) EPR of Mn(II) complexes with enzymes and other proteins, in *Biological Magnetic Resonance*, vol. 6 (eds L.J. Berliner and J. Reuben), Plenum, New York, pp. 73–142.

Taylor, P.C., Bangher, J.F., and Kriz, H.M. (1975) *Chem. Rev.*, **75**, 203.

Taylor, P.C. and Bray, P.J. (1972) *J. Phys. Chem. Solids*, **33**, 43.

Thorpe, M.F., Djordevic, B.R., and Jacobs, D.J. (1996) The structure and mechanical properties of networks. Proceedings of the NATO Advanced Institute on Amorphous Insulators and Semiconductors, Sozopol, Bulgaria, 26 May–8 June 1996, (eds M.F. Thorpe and M.I. Mitkova), NATO, ASI Series 3, *High Technology*, vol. 23, pp. 289–328.

Tucker, R.F. (1962) *Advances in Glass Technology*, Plenum, New York, pp. 103–114.

Upreti, G.C. (1974) *J. Magn. Reson.*, **13**, 336.

Vacher, R. and Courtens, E. (1996) Porous silica. Model fractal materials and their vibrations. Proceedings of the NATO Advanced Study Institute on Amorphous Insulators and Semiconductors, Sozopol, Bulgaria, 26 May–8 June 1996, (eds M.F. Thorpe and M.I. Mitkova), NATO, ASI Series 3 *High Technology*, vol. 23, pp. 255–288.

Wang, D. and Hanson, G.R. (1995) *J. Magn. Reson. Ser. A*, **117**, 1.

Wang, D. and Hanson, G.R. (1996) *Appl. Magn. Reson.*, **11**, 401.

Wood, R.M., Stucker, D.M., Bryan Lynch, W., Misra, S.K., and Freed, J.H. (1999) *Inorg. Chem.*, **38**, 5384–5388.

Zhang, Y.P. and Buckmaster, H.A. (1992) *J. Magn. Reson.*, **99**, 533.

Zhang, Y.P. and Buckmaster, H.A. (1993) *J. Magn. Reson. Ser. A*, **102**, 151.

Zhang, Y.P. and Buckmaster, H.A. (1994) *J. Magn. Reson. Ser. A*, **109**, 241.

Appendix 9.I The Eigenfield Equation

The objective is to find all the energy levels of the spin Hamiltonian H with a field-dependent separation, which is equal in energy to the microwave quantum ω_{mw}. This requires the existence of an eigenstate $|i\rangle$ with energy λ and another state $|f\rangle$ with energy $\lambda + \omega_{mw}$:

$$H|i\rangle = \lambda |i\rangle \tag{9.I.1}$$

$$H|f\rangle = (\lambda + \omega_{mw})|i\rangle \tag{9.I.2}$$

or, in matrix representation,

$$H \cdot c_1 = \lambda c_1 \tag{9.I.3}$$

$$H \cdot c_2 = (\lambda + \omega_{mw}) c_2 \tag{9.I.4}$$

where c_1 and c_2 are n-dimensional column vectors whose components form the linear coefficients of the basis states

$$|i\rangle = \sum_{k=1}^{n} c_{1k} |k\rangle \tag{9.I.5}$$

$$|f\rangle = \sum_{k=1}^{n} c_{2k} |k\rangle \tag{9.I.6}$$

To rewrite Equation 9.I.3 in the form of the generalized eigenvalue equation: $A \cdot Z = xB \cdot Z$, one begins by taking the direct product of Equation 9.I.3 with c_2^*:

$$(H \cdot c_1) \otimes (I_n \cdot c_2^*) = \lambda (I_n \cdot c_1) \otimes (I_n \cdot c_2^*), \tag{9.I.7}$$

where I_n is the n-dimensional identity matrix. Using the relation: $(AB) \otimes (CD) = (A \otimes B)(C \otimes D)$, one obtains

$$(H \otimes I_n)(c_1 \otimes c_2^*) = \lambda (I_n \otimes I_n)(c_1 \otimes c_2^*) \tag{9.I.8}$$

Forming the direct product of c_1 with the complex conjugate of Equation 9.I.4 similarly leads to

$$(I_n \otimes H^*)(c_1 \otimes c_2^*) = (\lambda + \omega_{mw})(I_n \otimes I_n)(c_1 \otimes c_2^*) \tag{9.I.9}$$

Subtracting Equation 9.I.8 from Equation 9.I.9, and denoting $Z = c_1 \otimes c_2^*$, one obtains

$$(I_n \otimes H^* - H \otimes I_n - \omega_{mw} I_n \otimes I_n) Z = 0 \tag{9.I.10}$$

One now needs to specify the form of the spin Hamiltonian. If one writes it as

$$H = F + xG, \tag{9.I.11}$$

where x is used to represent the magnetic field, then the first term of Equation 9.I.11 corresponds to field-independent terms, whereas the second represents terms linear in the magnetic field. Using this form of the Hamiltonian in Equation 9.I.10 gives

$$A \cdot Z = xB \cdot Z \tag{9.I.12}$$

with

$$A = (I_n \otimes F^* - F \otimes I_n - \omega_{mw} I_n \otimes I_n) \tag{9.I.13}$$

$$B = G \otimes I_n - I_n \otimes G^* \tag{9.I.14}$$

It is noted that F and G are the $n \times n$ matrix representatives of F and G in the basis $|k\rangle$. A and B are $n^2 \times n^2$ Hermitian matrices, because H is Hermitian.

It can be seen that in this procedure an $n \times n$ ordinary eigenvalue equation has been converted into an $n^2 \times n^2$ generalized eigenvalue equation. Thus, for high-spin systems with hyperfine structure, the Eigenfields approach becomes unwieldy (Drew, 2002). The solution of Equation 9.I.13 may be accomplished by the (complex) QZ algorithm (Moler and Stewart, 1973; Drew, 2002).

10
Relaxation of Paramagnetic Spins

Sushil K. Misra

10.1
Introduction

This chapter deals with the relaxation of paramagnetic spins. The earlier history of relaxation is described in Appendix 10.I. In the discussion that follows, the information that can be deduced from relaxation data is first described, along with the terminology used in the context of relaxation. The characteristic spin-lattice (T_1) and spin–spin (T_2) relaxation times are defined with the help of Bloch's equations. Thereafter, the various experimental techniques to measure relaxation times are briefly described. This is followed by a discussion of the various relaxation processes effective in crystalline and amorphous solids, both dilute and concentrated in spins, the effect of intramolecular dynamics of molecular species, and relaxation in liquids. Finally, the pertinent literature is reviewed.

The relaxation of a paramagnetic center is determined by its structure and its interaction with the environment. (Hereafter, the terms paramagnetic ion, paramagnetic spin, and paramagnetic center will be used interchangeably.) This can be exploited to obtain, among others, information about the following properties: excited states of the paramagnetic center; molecular motions; hyperfine and Zeeman anisotropies; potential for use as a magnetic resonance imaging (MRI) contrast reagent or a spin-probe broadening reagent; distinguishing local concentration from bulk concentration based on diffusion of local concentration that affects relaxation time, local concentration of nuclear spins from spin–echo decays, distance between a slowly relaxing spin and a more rapidly relaxing spin; and intramolecular and intermolecular interactions with methyl protons from echo–decay curves.

The following technical terms are used in context with paramagnetic relaxation:

Spin-lattice relaxation (SLR) time (T_1): This is the time constant for arriving at the equilibrium population at a given temperature once the equilibrium populations are disturbed; further details are given in Section 10.3.1 in terms of Bloch's equations.

Multifrequency Electron Paramagnetic Resonance, First Edition. Edited by Sushil K. Misra.
© 2011 Wiley-VCH Verlag GmbH & Co. KGaA. Published 2011 by Wiley-VCH Verlag GmbH & Co. KGaA.

Spectral diffusion time (T_D): Spectral diffusion is also referred to as *spin diffusion*. This takes into account the effect of all nonresonance processes that transfer magnetization from one paramagnetic ion to another paramagnetic ion. This has the semblance of relaxation. The various processes that can affect this are: motion of an anisotropic paramagnetic center, exchange interaction between two paramagnetic ions, cross-relaxation between an electron spin and a nuclear spin, and nuclear spin flip-flops. (The term flip-flop means that two coupled spins flip in opposite directions.) These processes can affect continuous-wave (CW) saturation, saturation transfer, spin echo and inversion–recovery measurements. They become especially significant when EPR lines due to different paramagnetic ions overlap. Sometimes, the terms spectral diffusion and spin diffusion are used to describe different physical processes. For example, spectral diffusion has been defined as the shape of EPR frequency of a single radical (Salikhov and Tsvetkov, 1979; Bowman and Kevan, 1979), whereas spin diffusion has been defined as the transfer of spin from one paramagnetic center to another paramagnetic center located at a different spatial position. This results in changing the dipolar field at the location of a particular spin (Salikhov and Tsvetkov, 1979; Bowman and Kevan, 1979). In addition, the term "cross-relaxation" is used to describe the transfer of energy between spins with different Zeeman frequencies (Salikhov and Tsvetkov, 1979; Bowman and Kevan, 1979).

Nuclear spin diffusion rate: This is mutual nuclear spin flip-flop rate due to the $I_{1+}I_{2-}$ interaction between two nuclear spins I_1 and I_2; the subscripts + and − represent the raising and lowering operators (Anderson, 1958; Wolf, 1966; Tse and Hartman, 1968). This becomes significant when magnetically dilute electron spins are surrounded by magnetically concentrated nuclear spins. Exchange of magnetization between electronic and nuclear spins takes place due to the dipolar coupling between them. Nuclear spin diffusion is also referred to as *spatial diffusion* because mutual nuclear spin flips move the magnetization through the sample.

Spin–spin relaxation time (T_2): This describes the time constant for dephasing of the magnetization in the transverse (x, y) plane. It is, in fact, the rate for mutual electron spin flips effected by the spin flip-flop interaction operator $S_{1+}S_{2-}$. T_2 is described more quantitatively by Bloch's equations. In liquids, T_2 is attributed to the linewidth of a Lorentzian line, and is caused by a rapid fluctuation in a local field (Atherton, 1993). Another inequivalent definition to this is the effect of rotational motion perturbing the hyperfine and Zeeman anisotropies (Goldman et al., 1972). In order to take into account inhomogeneous broadening one introduces the term $\frac{1}{T_2'}$, so that the inverse T_2 time is expressed as (Abragam, 1961).

$$\frac{1}{T_2} = \frac{1}{T_2'} + \frac{1}{2T_1}$$

Spin-echo dephasing time (T_m): This encompasses all processes that disturb electron spin phase coherence. The various processes are: spin–spin relaxation time, T_2, nuclear spin diffusion, and librational motion of paramagnetic species. It has been pointed out (Norris, Thurnauer, and Bowman, 1980) that, in the case when phase-memory decay function is a simple exponential, one can use T_m in place of T_2 in Bloch's equations.

Instantaneous diffusion: This occurs when the concentration of electron spins is sufficiently high so that in a spin–echo experiment the second pulse flips both the observed spin and a neighboring spin coupled to it by the dipolar interaction. This results in *dephasing*, and causes the phase to change the resonant energy of the observed spin. This process can be used to determine local spin concentrations (Eaton and Eaton, 1993). It is sometimes also referred to as *spin diffusion*.

Cross-relaxation: This describes the mutual spin-flip of two unlike spins whose spectral lines overlap (Bloembergen, 1949). It is effected by the dipolar interaction between an electron spin and a nuclear spin. It is called an "allowed" cross-relaxation when two electron spins flip simultaneously in two different transitions (Wenckebach and Poulis, 1973). The resulting energy difference is passed on to the surrounding nuclei. Cross-relaxation contributes to spectral diffusion.

Saturation factor: This is derived from Bloch's equations. It is defined as

$$s = \frac{1}{1+\gamma^2 B_1^2 T_1 T_2},$$

where B_1 is the amplitude of the irradiating sinusoidal microwave field. For an undistorted EPR line shape, it is required that $s \approx 1$. The saturation factor is most commonly estimated by CW power saturation measurements. For a single EPR line spectrum with no nuclear hyperfine interaction, the values of T_1 and T_2 can be obtained from s, knowing the value of B_1. In systems where spectral diffusion is faster than spin-lattice relaxation, one should replace T_1 and T_2 by T_D and T_m, respectively.

10.2
Equilibrium Magnetization of a Paramagnetic Spin System

For a gram molecular weight of a spin system, the magnetization M_0 along the external magnetic field, B, at temperature T is expressed as (Kittel, 1996):

$$M_0 = \frac{N_A g_{eff}^2 \mu_B^2 S(S+1) B}{3 k_B T}. \tag{10.1}$$

In Equation 10.1, N_A is Avogadro's number (6.02×10^{23}), g_{eff} is the g-factor of the paramagnetic spin, μ_B ($=9.274\,0901 \times 10^{-24}$ J/T) is the Bohr magnetron, k_B

($= 1.380\,622 \times 10^{-23}$ J/K) is the Bolztmann's constant, and T is the temperature in Kelvin (K).

10.3
Relaxation Phenomena: Spin–Lattice and Spin–Spin Relaxation Times

The relaxation of a spin system—that is, its approach to equilibrium—is best described phenomenologically by Bloch's equations, which detail the rate of change of the magnetization, $\vec{M}$, of the spin system. Under the action of an external magnetic field, $\vec{B}$, the magnetization will precess about $\vec{B}$. In addition, if the magnetization, M, does not possess its equilibrium value $M_0 = \chi_0 H_0$, it will approach this value characterized by the relaxation times. The z-component of magnetization, M_z, along $\vec{B}$ will approach its equilibrium value in an exponential manner. This is described as (Abragam, 1961; Kittel, 1996; Pake and Estle, 1973):

$$\frac{dM_z}{dt} = \frac{M_0 - M_z}{T_1} \tag{10.2}$$

In Equation 10.2, T_1 is called the spin-lattice, or longitudinal, relaxation time. Indeed, M_z is related to T_1, since the energy density, U, of the spin system due to interaction with the lattice is governed by the longitudinal component, M_z, of the magnetization density:

$$U = -N < \vec{\mu} \cdot \vec{B} >_{av} = -N < \vec{\mu} >_{av} \cdot \vec{B} = -M_z B \tag{10.3}$$

In Equation 10.3, N is the number of spins, each possessing the magnetic moment, μ, so that $\vec{M} = N\vec{\mu}$.

The transverse components of the magnetization M_x and M_y do not determine the energy of the spin. Therefore, they vary with time without interaction with the lattice, and are governed by a different relaxation time, T_2, due to interaction with the immediate environment. This is called spin–spin, or transverse, relaxation time. The mechanism for relaxation of the transverse components M_x and M_y is due to the varying local fields experienced by a spin due to the spin dipoles of its neighbors, leading to different precessional frequencies. These cause change in the transverse components of the magnetization, which can be assumed to be exponential, and are described by

$$\frac{dM_x}{dt} = -\frac{M_x}{T_2} \tag{10.4}$$

$$\frac{dM_y}{dt} = -\frac{M_y}{T_2} \tag{10.5}$$

10.3.1
Bloch's Equations

In the presence of an external magnetic field, Equations 10.2, 10.4, and 10.5 should be modified to take into account the precessional motion of the spin about the external magnetic field $\gamma(\vec{B}\times\vec{M})$, where γ is the electron's gyromagnetic ratio [$=g_e\mu_B/\hbar$, with g_e ($=2.0023$) being the free electronic g factor and $\hbar$ ($=1.0546\times10^{-34}$ J.s) being Planck's constant, divided by 2π]. The total rate of change of the various components of the magnetization, is expressed by the following equations, known as Bloch's equations (Abragam, 1961; Kittel, 1996; Pake and Estle, 1973):

$$\frac{dM_x}{dt} = \gamma(\vec{B}\times\vec{M})_x - \frac{M_x}{T_2}$$
$$\frac{dM_y}{dt} = \gamma(\vec{B}\times\vec{M})_y - \frac{M_y}{T_2} \qquad (10.6)$$
$$\frac{dM_z}{dt} = \gamma(\vec{B}\times\vec{M})_z + \frac{M_0 - M_z}{T_1}$$

These equations provide a simple description of the rate of change of magnetization of spin systems. It should, however, be noted that more complicated expressions will be required when there is an interaction among dipoles with spin $S > 1/2$ at lower temperatures in the presence of a weak external magnetic field, or when there exists saturation.

Finally, it is noted that Bloch's equations involve the spin–lattice (T_1) and spin–spin (T_2) relaxation times to describe the time variation of the magnetization. In turn, spin–lattice and spin–spin couplings affect both T_1 and T_2, although spin–lattice coupling determines T_1 predominantly, whereas spin–spin couplings determine T_2 predominantly.

10.4
Rotating Frame

The T_1, T_2 times can be best described in the rotating frame where the coordinate system is rotating about the z-axis at the Larmor frequency $\omega = \gamma B$. In the *free-induction decay*, a 90°- or $\pi/2$-pulse rotates the initial magnetization M_0 by 90° from the z-axis, so that it is along the x' axis in the rotating frame at $t=0$. (The rotation angle is determined by the value of $\gamma B_1 t_p$, where B_1 is the amplitude of the microwave field and t_p is the duration of the microwave pulse; for a $\pi/2$-pulse: $\gamma B_1 t_p = \pi/2$) The Bloch equations in this frame with the axes denoted by $x', y', z'(=z)$ are (Pake and Estle, 1973):

$$\frac{d'M'_x}{dt} = -\frac{M'_x}{T_2}$$
$$\frac{d'M'_y}{dt} = -\frac{M'_y}{T_2} \qquad (10.7)$$
$$\frac{dM_z}{dt} = \gamma(\vec{B}\times\vec{M})_z + \frac{M_0 - M_z}{T_1}$$

For $t > 0$, the solutions of these equations are:

$$M_{x'} = M_0 \exp\left(\frac{-t}{T_2}\right)$$
$$M_{y'} = 0.$$
$$M_z = M_0\left[1 - \exp\left(\frac{-t}{T_1}\right)\right]$$
(10.8)

When transformed to the laboratory frame, they become

$$M_x = M_0 \exp\left(\frac{-t}{T_2}\right)\cos(\omega t)$$
$$M_y = M_0 \exp\left(\frac{-t}{T_2}\right)\sin(\omega t).$$
$$M_z = M_0\left[1 - \exp\left(\frac{-t}{T_1}\right)\right]$$
(10.9)

It is seen from Figure 10.1, which depicts the time change of M_z, that the approach of M_z to the equilibrium magnetization is with the characteristic time T_1. Similarly, Figure 10.2 shows that M_x approaches zero with the characteristic time T_2, oscillating at the Larmor frequency (M_y behaves similarly.) The damped oscillation of M_x, represents its free-induction decay.

10.5
Experimental Techniques to Measure Relaxation Times

Two types of technique are used for EPR measurements, based on CW and pulse approaches. Alternatively, non-EPR techniques are also available; these utilize

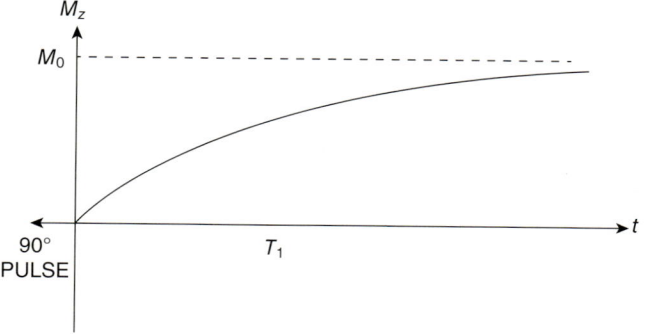

Figure 10.1 The growth of the component of the magnetization M along B_0 following a 90° pulse. At $t = T_1$, the spin lattice relaxation time, M_z has reached a value of $(1 - 1/e)M_0$, where M_0 is the thermal equilibrium value (Pake and Estle, 1973).

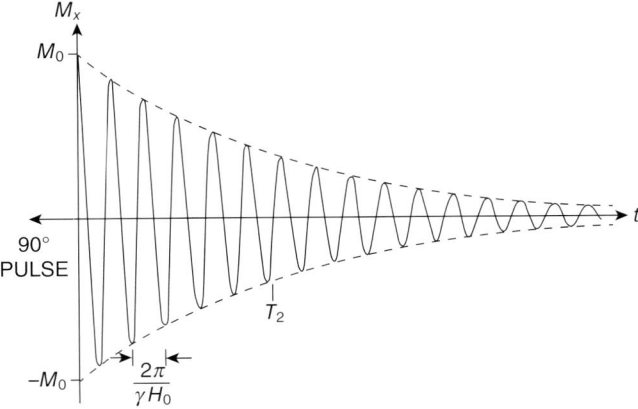

Figure 10.2 A component of M transverse to B_0 following a 90° pulse. This free induction decay has a characteristic time, T_2, the spin–spin relaxation time (Pake and Estle, 1973).

NMR, which can provide an indirect but rather precise means of estimating T_1, T_2 times. Other nonresonance techniques, such as Mössbauer, differential susceptibility and optical spectroscopies, can also be used.

10.5.1
CW-EPR Techniques (Bertini, Martini, and Luchinat, 1994)

Saturation methods can be exploited to measure both T_1 and T_2 times. Unfortunately, very short relaxation times cannot be measured with these techniques because, with the microwave powers available, it is not possible to saturate the signal in this case. On the other hand, line-shape analysis enables the estimation of T_2 times, provided that one takes into account inhomogeneous broadening. The analysis in this case is quite complicated, however, and the signal detection is limited. In fact, the signal width may be as large as the total spectral width; for example, for a lanthanide ion with a T_2 of the order of 10^{-13} s, one expects a linewidth of about 3 GHz, comparable to the X-band frequency (~9.5 GHz).

10.5.1.1 CW Saturation
When the spins are irradiated at resonance frequency (ω_0) with sufficient power for a long time, the intensity of the signal, saturated to various degrees, characterized by a Lorentzian lineshape can be expressed as

$$I = \frac{I_0}{1 + \gamma^2 B_1^2 T_1 g(\omega)}, \tag{10.10}$$

where

$$g(\omega) = \frac{T_2}{1 + (\omega_0 - \omega)^2 T_2^2}. \tag{10.11}$$

When the line is irradiated by a frequency within the natural linewidth such that $(\omega - \omega_0)T_2 \ll 1$, then Equation 10.10 becomes (by using Equation 10.11):

$$I = \frac{I_0}{1 + \gamma^2 B_1^2 T_1 T_2}, \qquad (10.12)$$

where I_0 is the observed intensity when the saturation factor, $\gamma^2 B_1^2 T_1 T_2$, is negligible, that is when the B_1 field is small. With saturation, as B_1 is increased, the intensity, I, attains the steady-state value with the time constant

$$T_{sat} = \frac{T_1}{1 + \gamma^2 B_1^2 T_1 g(\omega)}, \qquad (10.13)$$

which becomes at resonance:

$$T_{sat} = \frac{T_1}{1 + \gamma^2 B_1^2 T_1 T_2}. \qquad (10.14)$$

When the saturation factor becomes substantial ($\gamma^2 B_1^2 T_1 T_2 \gg 1$), Equation 10.14 for the time constant becomes

$$T_{sat} = \frac{1}{\gamma^2 B_1^2 T_2}. \qquad (10.15)$$

One can, therefore, measure T_1 values using CW-EPR, by setting the field at resonance on the selected line and recording the intensity of the signal as a function of the microwave power, P, proportional to B_1^2 (Standley and Vaughan, 1969; Makinen and Wells, 1987):

$$P = \frac{h\gamma^2 B_1^2}{4\pi^2} \qquad (10.16)$$

Thus, one obtains from Equation 10.12:

$$T_1 T_2 = \frac{h}{4\pi^2 P}\left(\frac{I_0}{I} - 1\right) \qquad (10.17)$$

Using Equation 10.17, one can determine T_1 provided that T_2 is known from elsewhere, which is not always straightforward. This technique can be used in the absence of having sufficient power to saturate the line entirely. Rather, small values of P can thus be used to determine T_1.

When saturation is large ($\gamma^2 B_1^2 T_1 T_2 \gg 1$), Equations 10.15–10.17 provide a means to determine T_1, since then

$$T_1 = T_{sat}\left(\frac{I_0}{I} - 1\right). \qquad (10.18)$$

10.5.2
Longitudinally Detected Paramagnetic Resonance (LODEPR) to Measure Short Relaxation Times (10^{-8} s) (Giordano et al., 1981)

If a CW spectrometer is modified such that it can be subjected to two microwave fields perpendicular to the directions of the external magnetic field (z) with slightly larger (ω_l) and slightly small (ω_s) microwave frequencies as compared to the Larmor frequency, then the magnetization along z has an amplitude which is proportional to the product $T_1 T_2$. It oscillates at the frequency $\omega_l - \omega_s$. Since, under nonsaturating conditions, the corresponding CW-EPR intensity depends only on T_2, the relaxation time T_1 can be directly estimated by comparing the LODEPR and EPR signals. This technique can be exploited to investigate inhomogeneously broadened lines, which consists of a family of overlapping spin packets, provided that all spin packets have the same T_1. Finally, it is possible to measure T_1 values down to 10^{-8} s with this technique. The pulsed version of this technique, LODPEPR, is described in Section 10.5.4.

10.5.3
Amplitude Modulation Technique to Measure Very Short Relaxation Times ($10^{-6} - 10^{-9}$ s) (Misra, 2005)

In this technique, as developed by Hervé and Pescia (1960, 1963), the amplitude of the microwave field is modulated at varying frequencies, Ω. The resulting signal proportional to M_z, the z-component of magnetization where the z-axis is along the direction of the external magnetic field, is measured as a function of Ω. The signal versus Ω plot can be analyzed to estimate both T_1 and T_2. It has been proposed recently to exploit the CW-EPR signal in the transverse direction (M_y), or to use a dielectric resonator (Schweiger and Jeschke, 2001) which does not suffer from the drawbacks of using a pick-up coil in the original Hervé–Pescia arrangement.

10.5.4
Pulsed EPR Techniques to Measure Relaxation Times

Microwave pulsed experiments can be used to measure relaxation times. The main techniques are *saturation recovery* and *inversion recovery*, as well as electron spin echo (ESE). With the lower limit of duration of pulses of about 10 ns, the measurement of relaxation times longer than 10 ns should be feasible. This limit of relaxation time is larger than the relaxation times encountered at temperatures above liquid-helium temperatures, for example, for some transition-metal ions. However, it can be used up to relatively higher temperatures depending on paramagnetic species. For example, one can measure T_1 for many Cu^{2+} complexes at 100 K by either inversion or recovery techniques, and therefore this technique is limited to rather low temperatures. Another drawback of the pulse technique is the ring-down time, which overloads the detecting system to about 200–300 ns, and is heavily dependent on the Q and microwave frequency. The ring-down time can be reduced to about 100 ns by using anti-phase-delayed pulses or bimodal cavities. [The ring-down time is defined as being equal to $-Q_L \ln(P_{noise}/P_0)/\omega_s$, with Q_L, P_{noise}, P_0, and ω_s being the quality factor of the resonator, noise power, incident power, and excitation frequency, respectively (Schweiger and Jeschke, 2001)]. Finally, it should be noted that relaxation times much shorter than 100 ns cannot be measured by using pulsed techniques.

10.5.5
Long Pulse Saturation Recovery Using CW Detection (Huisjen and Hyde, 1974; Percival and Hyde, 1975; Eaton and Eaton, 2000)

In this method, which is referred to as *saturation recovery*, a microwave pulse is applied for a sufficiently long time to saturate the EPR transition at resonance. One then waits for the end of switching transients and the ring-down time of the resonator; thereafter, the recovery of the EPR signal is detected with low-power CW microwaves. The recovery-time constant depends on spectral diffusion if it is a function of the length of saturating pulse. Consequently, the limiting value of the apparent relaxation time, as a function of the pulse duration, is considered to be the best approximation to T_1. This method suffers from the disadvantage that CW detection is less sensitive than ESE (electron spin echo) detection. On the other hand, CW detection has the advantage that it can be used for very short T_2 values, and for samples in which the echo-envelope modulation becomes so deep that it is difficult to observe an echo.

10.5.6
Inversion Recovery (Eaton and Eaton, 2000)

In this technique, an initial π-pulse ($B_1 \gamma t_p = \pi$) is first applied that inverts the direction of the spins; one then waits for an interval of time T, during which the spins relax. Thereafter, a second $\pi/2$ pulse is applied, and one again waits for a time τ and then applies a π pulse. An echo is then produced due to these two pulses after the time interval τ. In this $\pi - T - \pi/2 - \tau - \pi - \tau$ pulse sequence, due to the initial π pulse, the echo starts out inverted and recovers as a function of T through zero approaching the equilibrium value equal to that produced by the standard two-pulse ($\pi/2 - \tau - \pi - \tau$) echo at time τ. The time T is stepped in the experiment to monitor relaxation as a function of T. Spectral diffusion effects are significant here due to shortness of the inverting pulse. On the other hand, an intense short pulse is capable of exciting the maximum number of spins. In estimating T_1, one must also take into account the effect of the diffusion rate.

10.5.7
Electron Spin Echo (ESE) Technique (Schweiger and Jeschke, 2001)

Here, one records an echo after the sequence ($\pi/2 - \tau - \pi - \tau$), with τ stepped, providing a decay curve for the echo signal as a function of τ with the time constant T_2. In practice, this is limited to be not much less than 100 ns.

10.5.8
Long-Pulse Saturation with Spin-Echo Detection

The ESE spectrometer can be used to operate with a long saturation pulse in front of the spin-echo sequence as follows: $sat - \tau - \pi/2 - \tau_{fixed} - \pi - \tau_{fixed}$. As a consequence, the z-component of the magnetization recovers during the time τ after saturation, after the spin-echo sequence, with τ_{fixed} kept as small as possible. Stepping the value of τ provides the recovery of the echo signal as a function of τ, characterized by the time constant, which is the relaxation time T_1. When the value of $\gamma B_1 \gg (T_1 T_2)^{-1/2}$, one achieves full saturation. For the case when the signal is not fully saturated, a corresponding decrease in sensitivity occurs. A limiting factor here is the observability of an undistorted echo. A more commonly used variation of this technique employs decomposition of the π-pulse into two $\pi/2$-pulses separated by the time τ, which is stepped: $\pi/2 - \tau_{fixed} - \pi/2 - \tau - \pi/2 - \tau_{fixed}$. During τ the spins are held in a "waiting" state along the z-axis, but thereafter they are refocused into the xy-plane by the second $\pi/2$-pulse. Here, the echo intensity is proportional to T_1. This method avoids the problems caused by insufficient saturation, and allows the determination of T_1 values close to 100 ns (Bertini, Martini, and Luchinat, 1994).

10.5.9
Picket-Fence Excitation (Eaton and Eaton, 2000)

The effects of spectral diffusion can be suppressed by using "picket-fence" excitation, wherein a series of $\pi/2$ pulses with spacing greater than a few times the phase-memory time, T_m, is used to saturate the spin system instead of the initial π pulse. The picket-fence sequence suppresses the spectral diffusion by repeatedly turning the spins to the x-y plane after intervals during which spectral diffusion occurs. An alternative to the picket-fence sequence is the application of an initial saturating pulse with a duration that is long compared to T_D (the time constant for spectral diffusion), followed by two-phase echo detection. The dependence of the recovery time on spectral diffusion is eliminated when the initial saturating pulse burns a hole wider than the inverse diffusion time. Another way to eliminate this dependence is to use a range of frequencies in the echo-detection pulse that is wider in comparison to the spectral diffusion.

10.5.10
Echo Repetition Rate (Eaton and Eaton, 2000)

Here, one repeats the ensemble of the two-pulse system used for echo, and measures the echo amplitude as a function of the time interval (T), between successive ensembles of two pulses. The echo amplitude decreases with decreasing T due to the time required for recovery of the z-magnetization that was turned into the x-y plane by the $\pi/2$ pulse. This decrease then yields T_1, provided that there is no spectral diffusion.

10.5.11
Three-Pulse-Stimulated Echo (Eaton and Eaton, 2000)

Here, one employs a three-pulse-stimulated echo sequence, which is used for electron spin–echo envelope modulation (ESEEM), $\pi/2 - \tau - \pi/2 - T - \pi/2 - \tau - echo$, wherein the time T is stepped. The intensity of the τ-echo should decay with the time constant T_1, provided that all pulses are perfect and that no spectral diffusion occurs. In reality, the decay is faster than T_1, and consequently this technique does not provide a good measure of T_1.

10.5.12
Longitudinally Detected Pulsed EPR (LODPEPR) (Schweiger, 1991; Schweiger and Ernst, 1988)

Here, T_1 is measured by applying the pulse sequences $\pi - \tau - \pi$. The first pulse turns the magnetization along the z-axis, whereas the second pulse brings the residual magnetization back along the z-axis. As compared to the spin–echo technique, the main advantage here is that the detection is performed orthogonal to the exciting pulses, which eliminates ring-down in the detection system. The LODPEPR technique extends the range of T_1 values that be measured to as low as ~10 ns.

10.5.13
Other Pulse Techniques

There are other less commonly used techniques employing pulses to measure T_1 which will not be described here (further details are available in Eaton and Eaton, 2000). These include: (i) use of the Ernst angle (Fukushima and Roeder, 1981; Canet, 1996; Van de Ven, 1995); (ii) pulse techniques, such as electron-electron double resonance (ELDOR), two-dimensional (2-D) Fourier transform (FT) EPR, correlation spectroscopy (COSY), and spin–echo-correlated spectroscopy (SECSY) (Lee, Patyal, and Freed, 1993; Sastry et al., 1996a, 1996b; Saxena and Freed, 1997); and (iii) multi-quantum EPR (MQEPR) (Hyde, Sczanicki, and Froncisz, 1989; Sczianecki, Hyde, and Froncisz, 1990; Mchaourab et al., 1991; Mchaourab and Hyde, 1993a, 1993b, 1993c; Mchaourab et al. 1993; Eaton, 1993; Strangeway et al., 1995).

10.5.14
Measurements of Relaxation Time by Line-Shape Analysis: Linewidth and Spin–Spin Relaxation Time

The EPR linewidth is a measure of the indeterminacy in the energy difference between the two levels participating in resonance. By Heisenberg's uncertainty relation, this indeterminacy is inversely proportional to the lifetime of the participating energy levels in resonance. The EPR line has Lorentzian shape when the transverse relaxation is described by a single exponential with time constant T_2, which describes a "spin-packet" in the absence of unresolved hyperfine splitting, *g*- and *A*-strains. In that case, the peak-to-peak first derivative, which is also the line width of the absorption line shape at the height of half-peak has the value in frequency unit $\Delta v = 1/(\pi T_2)$. It is first noted that the line-shape function contains information on the natural linewidth. With correct analysis, T_2 can be estimated as one of the parameters to fit the spectral lineshape with the theory (Bertini, Martini, and Luchinat, 1994; Schneider and Freed, 1989), though this may be difficult if there exists any extensive unresolved hyperfine structure. It should, however, be noted that T_1, by definition, does not determine any lineshape. However, in the extreme narrowing limit $T_1 = T_2$, which provides an indirect measure of T_1 from the lineshape, the reason for which can be explained as follows. If the T_1 relaxation returns the spin polarization from the *x-y* plane to the *z*-axis rapidly, the net *x-y* polarization will be negligible. As T_1 decreases with increasing temperature, the lifetime broadening caused by the finite lifetime of the transition to the ground state by phonon emission due to a short T_1 will exceed the contributions to CW linewidth, rendering $T_1 = T_2$. It should also be noted that even if $T_1 \neq T_2$, one can estimate T_2 for a spin packet from the linewidth.

10.5.15
Temperature-Dependent Contribution to EPR Linewidth (Poole and Farach, 1971)

For a single Lorentzian line, the peak-to-peak derivative linewidth can be expressed as a function of T_1, T_2, and B_1:

$$\Delta B_{PP}^2 = \frac{4}{3}\left(\frac{1}{\gamma^2 T_2^2} + B_1^2 \frac{T_1}{T_2}\right), \quad (10.19)$$

which becomes when the saturation factor $\gamma^2 B_1^2 T_1 T_2 \ll 1$:

$$\Delta B_{PP}^2 = \frac{4}{3}\frac{1}{\gamma^2 T_2^2}. \quad (10.20)$$

Thus, the increase in linewidth with temperature can be used to estimate T_2, which in turn is equal to T_1, as given by Equation 10.20. Finally, it should be noted that, at low temperatures, the EPR linewidth is determined by inhomogeneous broadening, whereas at intermediate temperatures the linewidth can be expressed as a convolution of the relaxation-broadened spin-packet linewidths with inhomogeneous broadening.

10.5.16
Non-EPR Techniques to Measure Relaxation Times

These techniques, most of which are beyond the scope of this book, include (Eaton and Eaton, 2000): (i) line-shape changes in the Mossbauer spectrum (Schulz, Brandon, and Debrunner, 1990); and (ii) NMR relaxation techniques (LaMar, Horrocks, and Hahn, 1973; Banci, Bertini, and Luchinat, 1991).

10.6
Relaxation Mechanisms

There is no single relaxation mechanism that applies to all systems; rather the effective mechanisms of relaxation differ from material to material. The various mechanisms that are effective in dilute ionic solids in the crystalline state are discussed in Section 10.6.1; these include general background, the *direct process*, the *Orbach process*, and the *two-phonon Raman process*. Thereafter, their generalizations to amorphous (disordered) materials will be described in Section 10.6.2, the relaxation behavior of which emerges as being significantly different from that of single crystals. Relaxation in dilute liquid solutions is discussed in Section 10.6.3, followed by a brief review of the effect of intramolecular dynamics of molecular species on the relaxation of organic radicals and transition-metal ions in partially or totally frozen states (Section 10.6.4). Relaxation due to interaction among different paramagnetic spins in concentrated solutions is discussed in Section 10.6.5, while spin-fracton relaxation in fractal materials is outlined in Section 10.6.6. The frequency – and therefore, the magnetic field-dependence of relaxation – is reviewed in Section 10.6.7.

10.6.1
Spin-Lattice Relaxation in Diluted Ionic Solids in the Crystalline State

The various mechanisms responsible for spin-lattice relaxation (SLR) in solids are depicted in Figure 10.3, and are described below.

10.6.1.1 General Background

In diluted systems, spin–spin interactions are not significant. The principal mechanism of spin-lattice relaxation in this case is phonon modulation of the ion-ligand potential (Abragam and Bleaney, 1970). The crystal field splitting is modulated by fluctuation of the spin–ligand distance, and this perturbation, being electrical in origin, interacts indirectly with spin through spin–orbit coupling. In the theoretical treatment of this mechanism, the distance between the paramagnetic ion and each atom closely interacting with it (hereafter termed the ligand distance) is expanded about the equilibrium distance in powers of the strain tensor. When dealing with a statistical distribution of lattice vibrations, it is convenient to express the Hamiltonian in terms of phonon creation and annihilation operators. In addition, one uses strain as the basic variable in the spin Hamiltonian. The displacement q_{Ri} of the ith atom in the unit cell located at the lattice point $\vec{R}$ can be expressed in terms of the normal modes with the wave vector $\vec{k}$ and polarization indices μ by (Abragam and Bleaney, 1970):

$$q_{Ri} = \frac{1}{\sqrt{NM_i}} \sum_{\vec{k},\mu} e^{i\vec{k}\cdot\vec{R}_i} Q_{\vec{k}\mu} \varepsilon_{\vec{k}\mu i}; \qquad (10.21)$$

In Equation 10.10, N is the number of unit cells in the crystal, M_i is the mass of the ith atom in the unit cell, and $Q_{\vec{k}\mu} = (\hbar/2\omega_{\vec{k}\mu})^{1/2}[a^*_{-\vec{k}\mu} + a_{\vec{k}\mu}]$ with $a^*_{\vec{k}\mu}$ and $a_{\vec{k}\mu}$ being

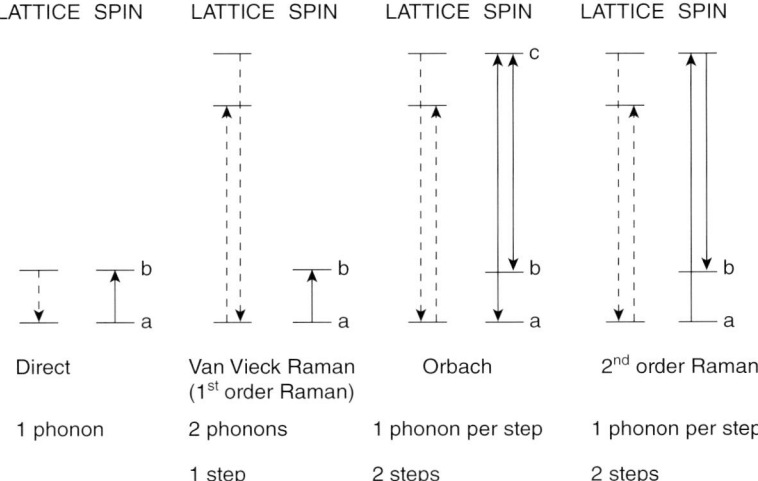

Figure 10.3 The various mechanisms for electron relaxation in the solid state. Solid lines represent spin transitions, dotted lines represent lattice transitions, a and b indicate the electronic spin ground state levels between which relaxation occurs; c is the electronic excited state (Bertini, Martini, and Luchinat, 1994).

the creation and annihilation operators respectively, whereas $\omega_{\vec{k},\mu}$ is the characteristic frequency for phonons with the wave vector $\vec{k}$ and the polarization index μ. The displacement $q_{\vec{R}_i}$ can be used to calculate the strain tensor: $e_{\alpha\beta} = (\partial q_\alpha/\partial R_\beta + \partial q_\beta/\partial R_\alpha)/2$; $\alpha,\beta = x,y,z$. The basic interaction for electron relaxation is the phonon modulation of the ion-ligand potential. The orders of phonon emissions depend on different powers of the strain tensor. In the initial treatment, only the nearest ligand neighbors are considered together to contribute to the crystalline potential acting on the paramagnetic ion. The normal modes of vibration of the resulting cluster are expanded in terms of the phonons of the host crystal. If this is not sufficient, increasing numbers of ligands are included in the cluster which, in turn, makes the calculations more complicated, losing analogy with the static crystal-field potential. The total Hamiltonian (van Vleck, 1939, 1940) leads to the eigenvalues: $E_\ell = \sum_{\vec{k},\mu} \hbar\omega_{\vec{k},\mu}(n_{\vec{k},\mu} + 1/2)$, where $n_{\vec{k},\mu} = 1/[\exp(\hbar\omega_{\vec{k},\mu}/k_B T - 1)]$ is the Bose factor, which represents the occupation number of the state $\vec{k}, \mu$. The orbit-lattice coupling is introduced by including cluster vibrations. The vibrational normal modes are expressed in terms of running phonon waves (Orbach and Tachiki, 1967). For simplification, the crystal is approximated by an isotropic solid described by a linear phonon dispersion curve with the Debye cut-off wave vector, k_D (Orbach and Stapleton, 1972).

10.6.1.2 The Direct Process

The first term in the expansion shown in Equation 10.21, linear in the strain, is responsible for single phonon emission or absorptions. In the direct process, the spin system changes its energy by the Zeeman splitting $\hbar\omega_0 (= g_e\mu_B B)$ at resonance, while the lattice energy reduces by the same amount for conservation of energy. The associated relaxation process, termed the *direct process*, is characterized by the relaxation rate (Orbach, 1961a, 1961b):

$$\frac{1}{T_1} = \frac{3}{2\pi\hbar\rho v^5} \left(V^{(1)}\right)^2 \omega_0^3 \coth\left(\frac{\hbar\omega_0}{2k_B T}\right) \tag{10.22}$$

In Equation 10.22, ρ is the density of the crystal, v is the velocity of sound, $V^{(1)}$ is the potential due to the lattice stress to first order, and $\left(V^{(1)}\right)^2$ is the absolute value squared of the matrix element for the transition. The relaxation rate given by Equation 10.22 is very sensitive to the velocity of sound, being proportional to v^{-5}; for example, a mere 2.5-fold change in v leads to a change in the relaxation rate by two orders of magnitude. However, at these elevated temperatures, only a very small part of the phonon spectrum with an even smaller density of states is effective, as compared to that at room temperature. At high temperatures, $T \gg \frac{\hbar\omega_0}{k_b}$, the relaxation rate (T_1^{-1}) depends linearly on temperature, being proportional to $\omega_0^2 T$, whereas at lower temperatures, $T \ll \frac{\hbar\omega_0}{k_b}$, the relaxation rate is independent of temperature, being proportional to ω_0^3. It should be noted that, at lower temperatures, the concept of temperature of the lattice is not well defined. This is because the dissipation of spin heat of a concentration of spins in a magnetic field produced by microwave energy is blocked and becomes confined to only a relatively small fraction of effective phonons. Thus, extra time will be required to redistribute this heat to the whole phonon spectrum representing the entire lattice. This process, which limits the entire relaxation rate, is referred to as the *bottleneck process* (van Vleck, 1941a, 1941b).

For *Kramers' doublets*, the matrix element of the electronic potential vanishes within the states of the doublet in zero magnetic field. For the lowest-lying Kramers' doublet, when a magnetic field is present, the Zeeman term causes admixing of the excited states with the ground state on the order of $\frac{\hbar\omega_0}{\Delta}$, where Δ is the energy difference between the excited and ground states due to the crystal field. Equation 10.22 then leads to (Kronig, 1939; van Vleck, 1940):

$$\frac{1}{T_1} = \frac{3\hbar}{2\pi\rho v^5} \frac{1}{\Delta^2} \left(V^{(1)}\right)^2 \omega_0^5 \coth\left(\frac{\hbar\omega_0}{2k_B T}\right) \tag{10.23}$$

The relaxation rate, given by Equation 10.23, is found to depend on $\omega_0^4 T$ at higher temperatures, whereas it is proportional to ω_0^5 at low temperatures. This effect is significant only at low temperatures, due to small phonon density at the energy $\hbar\omega_0 (= g_e\mu_B B_0)$. Assuming that $V^{(1)} \approx 10 \text{ cm}^{-1}$, $\hbar\omega_0 = 1 \text{ cm}^{-1}$, $T = 1 \text{ K}$, the experimental estimates of T_1 from the data are 10^{-3} s for non-Kramers' ions, and 1 s for Kramers' ions, using $\Delta = 3V^{(1)}$. These estimates are found to agree well with the data on lanthanides. (As for 3d ions, the values of $V^{(1)}$ are higher, leading to shorter T_1 times.)

When the *bottleneck process* is effective, one finds for both Kramers' and non-Kramers' ions the following expression for the relaxation rate (van Vleck, 1941a, 1941b; Geschwind, 1972):

$$\frac{1}{T_1} = \coth^2\left(\frac{\hbar\omega_0}{2k_BT}\right). \tag{10.24}$$

10.6.1.3 The Orbach Process (Orbach and Stapleton, 1972; Orbach, 1961a, 1961c)

In this process, an ion from the higher energy state |b⟩ makes a transition to a higher intermediate state |c⟩ by absorbing a phonon of appropriate frequency by a direct process. It then arrives at the ground state |a⟩ after emitting a phonon. When the state |c⟩ is real, the process is called the Orbach (or resonant Raman) process, and when the state |c⟩ is virtual, it is called Raman process. For the Orbach process, the relaxation rate can be expressed as (Finn, Orbach, and Wolf, 1961):

$$\frac{1}{T_1} = \frac{3|V^{(1)}|^2 \Delta^3}{2\pi\hbar^4 \rho v^5} \frac{1}{\left[\exp\left(\frac{\Delta}{k_BT}\right) - 1\right]}. \tag{10.25}$$

In Equation 10.25, Δ is the energy separation between the states |c⟩ and the average of the states |a⟩ and |b⟩, and the factor $1/[\exp(\Delta/k_BT) - 1]$ describes the phonon occupation number. When $k_BT \ll \Delta \ll k_B\theta_D$, where θ_D is the Debye temperature, one obtains

$$\frac{1}{T_1} \propto \exp\left(-\frac{\Delta}{k_BT}\right). \tag{10.26}$$

When $\Delta \approx k_B\theta_D$, the relaxation time becomes

$$\frac{1}{T_1} \propto \frac{\Delta}{\left[\exp\left(\frac{\Delta}{k_BT}\right) - 1\right]} \tag{10.27}$$

One must be very careful not to be get caught in the possible pitfalls that exist in the use of Equation 10.25, as pointed out by Orbach and Stapleton (1972), and by Abragam and Bleaney (1970). In any case, for the rare-earth group, one can simplify the relaxation rate to describe rather well the experimental values (Abragam and Bleaney, 1970)

$$\frac{1}{T_1} \approx 10^4 \Delta^3 \exp\left(-\frac{\Delta}{k_BT}\right) \tag{10.28}$$

In most cases, the Orbach mechanism is most efficient for $\frac{\Delta}{k_BT} < 6$ (Bertini, Martini, and Luchinat, 1994).

10.6.1.4 Two-Phonon Raman Process

When the second-order terms in the strain tensor in the expansion of the ligand distance (Equation 10.21) are considered, there result inelastic two-phonon emission and absorption processes. These are referred to as Raman-like processes (Waller, 1932). In this process, which involves Raman scattering of phonons by the ion, all abundant ions at room temperature participate, with each scattered phonon differing in energy by $\hbar\omega_0$ from that of the incident ion. Here, two lattice distortions oscillating at frequencies ω_1 and ω_2 combine together to generate a frequency of vibration $|\omega_2 - \omega_1|$. The intermediate excited state $|c\rangle$ lies well above the continuum of allowed phonon frequencies (as shown in Figure 10.3), separated by the crystal-field energy, Δ. The transition probability is expressed in terms of the matrix elements of the potentials $V_1^{(1)}$ and $V_2^{(1)}$ between the excited state $|c\rangle$ and the states $|a\rangle$, $|b\rangle$, respectively. This mode of relaxation is termed the "second-order Raman process," because the required fluctuation occurring at the frequency $|\omega_2 - \omega_1|$ appears due to the second-order terms in the strain tensor.

For non-Kramers' ions, the relaxation rate can be expressed as (van Vleck, 1939, 1940; Orbach, 1961c):

$$\frac{1}{T_1} = \frac{9}{4\pi^3 \rho^2 v^{10}} \frac{\left(V^{(1)}\right)^4}{\Delta^2} \left(\frac{k_B T}{\hbar}\right) I_6, \tag{10.29}$$

where

$$I_n = \int_0^{\theta_D/T} \frac{x^n e^x dx}{(e^x - 1)^2}. \tag{10.30}$$

Now, $I_6 = 6!$, since the integral in Equation 10.30 can be simplified to I'_n when $x \,(= \Delta/k_B T) \gg 1$ and $\theta_D/T \gg 1$:

$$I'_n = \int_0^\infty x^n e^x dx = n!. \tag{10.31}$$

A first-order perturbation term results when the second-order orbit-lattice interaction term, $V^{(2)}$, has nonzero matrix elements directly between the states $|a\rangle$ and $|b\rangle$ (Figure 10.3). The second-order term has the form $\varepsilon_1 \varepsilon_2 V^{(2)}$, where ε_1 and ε_2 are the strains caused by the loss of an ε_1 phonon and at the same time creation of an ε_2 phonon caused by the fluctuation in the difference of frequencies $|\omega_2 - \omega_1|$. For non-Kramers' ions, the relaxation rate due to the second term is similar to that given by Equation 10.29 (van Vleck, 1939, 1940; Orbach, 1961c):

$$\frac{1}{T_1} = \frac{9(6!)}{4\pi^3 \rho^2 v^{10}} \left(V^{(2)}\right)^2 \left(\frac{k_B T}{\hbar}\right)^7. \tag{10.32}$$

It is practically impossible to distinguish between the two relaxation rates for non-Kramers' ions given by Equations 10.29 and 10.32. If $V_1^{(1)} \approx V_2^{(1)} \approx \Delta$, the two rates are equal to each other. These processes contribute to the relaxation rate which is $10^{-6} T^6$ times that determined by the direct process.

For Kramers' ions, the first-order Raman process is nonzero only when the Zeeman term is taken into account, like the direct process. Inclusion of the Zeeman term introduces a factor to $\dfrac{\hbar\omega_0}{\Delta'}$, where Δ' (which may be different from

Δ), is the energy separation of the excited level from the energy of the levels with which the Zeeman operator has nonzero matrix elements. Finally, the relaxation rate can be expressed as (van Vleck, 1939, 1940; Orbach, 1961a):

$$\frac{1}{T_1} = \frac{9(6!)}{4\pi^3 \rho^2 v^{10}} \left(\frac{\hbar \omega_0}{\Delta'}\right)^2 (V^{(2)})^2 \left(\frac{k_B T}{\hbar}\right)^7 \tag{10.33}$$

In a similar manner, the second-order Raman process for Kramers' ions yields (van Vleck, 1939, 1940; Kronig, 1939; Pilbrow, 1990):

$$\frac{1}{T_1} = \frac{9\hbar^2}{\pi^3 \rho^2 v^{10}} \left(\frac{V^{(1)4}}{\Delta^4}\right)\left(\frac{k_B T}{\hbar}\right)^9 I_8, \tag{10.34}$$

which becomes at low temperatures:

$$\frac{1}{T_1} = \frac{(9!)\hbar^2}{\pi^3 \rho^2 v^{10}} \left(\frac{V^{(1)4}}{\Delta^4}\right)\left(\frac{k_B T}{\hbar}\right)^9. \tag{10.35}$$

In Equations 10.34 and 10.35, Δ is energy separation between the energies of the ground states and the excited state. This process emerges as being $10^{-5} T^9$ times more effective than the direct process, and is thus predominant at 5 K. The first-order Raman process is never the dominant process. In fact, although at low temperatures it dominates the second-order process, it is less effective than the direct process.

Finally, for $T > 10$ K, the Orbach term prevails when there exists a low-lying excited level; when this is not the situation, a Raman process prevails. Further, at temperatures approaching ambient (295 K), the phonon description of an idealized crystal is no longer valid. At these temperatures, optical phonons corresponding to short-range vibrations become the dominant electron relaxation mechanism.

It should be noted that, in the two-phonon process when there exists a low-lying excited state, this causes a resonance that dominates the usual nonresonant, two-phonon relaxation process (Finn, Orbach, and Wolf, 1961). This requires phonons with energy equal to the difference of the energies of the excited and ground-state crystal field. Thus, it has the temperature dependence of a thermally activated process, and is independent of the magnetic field.

10.6.1.5 SLR due to Exchange Interaction

The details of how the SLR of members of exchange-coupled pairs is affected by lattice vibrations in single crystals was discussed by Gill (1962), adapted from the theories developed, among others, by Waller (1932) and Al'tshuler (1956). The spin Hamiltonian for a pair of spins S_1 and S_2 is expressed as:

$$\begin{aligned}H = {} & g\mu_B(\mathbf{S}_1 + \mathbf{S}_2)\cdot \mathbf{B} + D[S_{1z}^2 + S_{2z}^2 - (2/3)S(S+1)] + J\mathbf{S}_1\cdot \mathbf{S}_2 \\ & + (g^2\mu_B^2/r_{12}^3)[\mathbf{S}_1\cdot \mathbf{S}_2 - (3/r_{12}^2)(\mathbf{S}_1\cdot \mathbf{r}_{12})(\mathbf{S}_2\cdot \mathbf{r}_{12})] \\ & + \text{anisotropic exchange terms} \end{aligned} \tag{10.36}$$

In Equation 10.36, $S (= S_1 = S_2)$ is the magnitude of spins S_1 and S_2, g is the Landé factor, r_{12} is the vectorial distance from spin 1 to spin 2, D is the zero-field splitting parameter, and J is the exchange-interaction constant. The anisotropic exchange terms are due to exchange splitting in excited orbital levels arising via the spin–orbit coupling; these terms may be ignored since the largest one of them is of the order $J\lambda/\Delta$, where Δ is the energy of these levels. If $J \gg D$, the energy levels form multiples characterized by effective spins S'. Microwave transitions may be observed between levels belonging to the same multiplet S'.

Relaxation of Members of Exchange-Coupled Pairs at Low Temperatures as Effected by Modulation of the Exchange Parameter J by Lattice Vibrations (Phonons) in Single Crystals This process is similar to that proposed by van Vleck (1940) to account for the relaxation of single ions. It is expected to be faster than that of the single ion by a factor $\sim (E_p/E_s)^2$, where E_p, E_s are relevant phonon energies. In this process, lattice vibrations are instrumental in inducing relaxation of an ion coupled to another by exchange interaction at lower temperatures. The probability of relaxation from level i to level j is (Gill, 1962; Al'tshuler, 1956; van Vleck, 1940; Stevens, 1967):

$$P_{ij}^D = \frac{8\pi^3}{3h^4} \frac{r^2}{\rho v^5} \frac{E_{ij}^3}{[1-\exp(-E_{ij}/kT)]} \sum_{q=x,y,z} \left| \left\langle i \left| \frac{dU}{dq} \right| j \right\rangle \right|^2, \quad (10.37)$$

where E_{ij} is the energy difference $(E_i - E_j)$, ρ is the density of the crystal, v is the average velocity of sound, r is the distance between the two ions of the pair, and U is the interaction whose modulation by displacements in direction q due to lattice vibrations brings about relaxation. If J is sufficiently small for anisotropic exchange to be negligible, the interaction dU/dq may be identified with (dJ/dr) $(\mathbf{S}_1 \cdot \mathbf{S}_2)$. When $J \gg D$, the largest off-diagonal elements of $\mathbf{S}_1 \cdot \mathbf{S}_2$ are of the order D/J, and couple certain states whose effective spins S' differ by 2. It is also noted that at those temperatures for which $E_{ij} \gg kT$, Equation 10.37 yields a linear dependence of the relaxation rate upon temperature.

10.6.2
Relaxation in Amorphous Systems

Since about 1970, rather unique behaviors of SLR times have been reported in amorphous systems; indeed, in contrast to ionic crystals, the strength and temperature dependence of SLR rates have been found to be very unusual. The temperature dependence of the relaxation rate is different in amorphous materials from that in single crystals. Amorphous materials differ from single crystals, in that in these materials there exists a distribution of the various parameters characterizing interactions among magnetic ions.

The reported data on SLR studies are rather scarce for amorphous materials, although a large volume of experimental data has been reported on electron SLR in ionic crystals. Some interesting relaxation-time studies in amorphous materials are listed in Table 10.1. These include the observation of a T^2 followed by T-dependence of T_1^{-1} as the temperature was increased (Kurtz and Stapleton, 1980) in irradiated β-alumina at very low temperatures. In addition, rather strong dependence of T_1^{-1} on temperature in glasses doped by Yb^{3+}, as reported by Antipin, Kochalev, and Shleklin (1984), and by Stevens and Stapleton (1990) and by Vergnoux et al. (1996), seems to be very unusual in such systems.

The relaxation rates are much faster when the exchange interaction is the dominant mechanism for SLR, as compared to the case when tunneling-level state (TLS) centers, involving electron–nuclear dipolar interaction and Fermi-contact hyperfine interaction, effect the SLR process (Misra, 1998a, 1998b). In fact, the relaxation rates due to the exchange interaction are on the order of two to three magnitudes higher than those expected on the basis of TLS centers, as evidenced by the measured values of SLR rates listed in Table 10.1 (Misra, 1998a, 1998b). It was deduced that the SLR rates due to the exchange interaction, and those effected by TLS centers, exhibit a quadratic temperature dependence at very low temperatures, but a linear temperature dependence at intermediate temperatures (Misra, 1998a, 1998b). Relaxation due to phonon modulation of the crystal field is insignificant in the case of amorphous systems doped with paramagnetic centers characterized by partially, or totally, quenched orbital angular momentum ($L \sim 0$) (e.g., Fe^{3+}, Mn^{2+}, Cr^{3+}), since there exists no direct interaction between the paramagnetic center and the crystal field. For amorphous materials, the three principal mechanisms for SLR are due to: (i) interaction between exchange-coupled pairs, directly modulated by lattice vibrations; (ii) cross-relaxation between an isolated ion and an exchange-coupled electron pair, the latter being perturbed by lattice vibrations; and (iii) electron–nuclear dipolar (END) (Bowman and Kevan, 1977a), or Fermi-contact (FC) hyperfine (HF) interaction (Bowman and Kevan, 1977a; Kurtz and Stapleton, 1980) between the electron magnetic moment and TLS centers, the latter being modulated by lattice vibrations or phonons. According to mechanisms (i) and (iii), T_1^{-1} varies as T^2 at very low temperatures, and then as T at intermediate temperatures. The latter is in contrast to the direct process, which takes place only at very low temperatures in ionic crystals, much below 4 K. On the other hand, mechanism (ii), in which cross-relaxations play an important role, leads to a plateau in the relaxation rate at higher temperatures. The distribution over a wide range of energy (E) of TLS centers and that of strength (J) of the exchange interaction responsible for the formation of exchange pairs, are characteristics unique to amorphous materials. In ionic crystals, on the other hand, E and J have rather well-defined, sharp values. In amorphous materials, T_1 has been observed to be several orders of magnitude shorter than those reported for single crystals at low temperatures; examples include molecular glasses (Berman, 1949; Choy, Salinger, and Chiang, 1970; McTaggart and Slack, 1969) and irradiated β-alumina (Kurtz and Stapleton, 1980).

10.6.2.1 Relaxation via TLS Centers

Applicable specifically to amorphous materials (Phillips, 1972; Anderson, Halperin, and Varma, 1972), in the localized-tunneling-level states model, a group of atoms in amorphous materials resides in asymmetric double-well potentials, characterized by two closely spaced energy levels, as shown in Figure 10.4. Tunneling occurs through the barrier that separate the two wells. Two mechanisms

Table 10.1 Experimentally measured spin-lattice relaxation times in various amorphous hosts at different temperatures, and the mechanisms responsible for them.

Sample	1.5–10	10 K	20 K	50 K	100 K	200 K	Effected via (ion)
Irradiated ethanol glasses			25	60	500	3000	TLS (END) (trapped electron) (Bowman and Kevan, 1977a)
β-alumina irradiated			7	40			TLS (FC) (Color center) (Kurtz and Stapleton, 1980)
Amorphous silicon			10^5	5×10^5	2×10^6	10^7	10^8 — Exchange (electron) (Gourdon, Fretier, and Pescia, 1981)
Amorphous silicon			5×10^4				Exchange (electron) (Askew et al., 1984)
Borate glass (0.1% Fe_2O_3)				$\sim 10^6$	7×10^6	2×10^7	4×10^7 — Exchange (Fe^{3+}) (Zinsou et al., 1996)
Borate glass (0.5% Fe_2O_3)				$\sim 10^7$	3×10^7	4.5×10^7	5×10^7 — Exchange (Fe^{3+}) (Misra, 1998a)
Borate glass (0.1% and 1.0% Fe_2O_3)	$\sim 10^4$						Exchange (Fe^{3+}) (Misra and Pilbrow, 2007)
Silicate glass (0.1% Fe_2O_3)	$\sim 10^4$						Exchange (Fe^{3+}) (Misra and Pilbrow, 2004)
$MgO \cdot P_2O_5$ (0.2% Mn)							2.5×10^6 — Exchange (Mn^{2+}) (Misra, 1998a)
Irradiated molecular crystal			10	50	10^3	5×10^3	TLS (trapped electron) (Dalton, Kwiram, and Owen, 1972)
$Cu_{2x}Cr_{2x}Sn_{2-2x}S_4$ (x = 0.8)			5×10^8	8×10^8	10^9	5×10^9	4×10^9 (a plateau) — Exchange (Cr^{3+}) (Sarda et al., 1989)

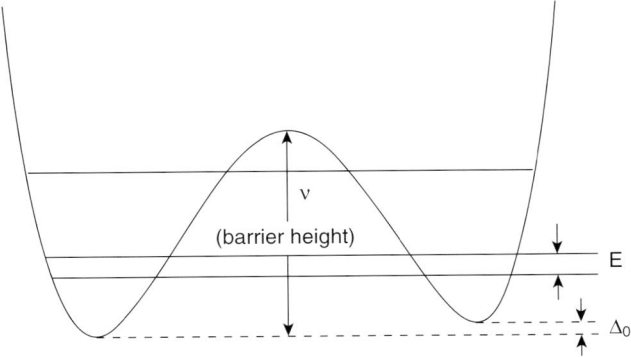

Figure 10.4 Two-level system in an atomic double-well potential (asymmetric).

of SLR involving TLS centers have been identified, namely END and the FC HF interaction of the relaxing electron with the neighboring nuclei which serve as TLS centers. It follows that the FC and END interactions lead to a T^2 followed by a T-dependence of the SLR rate on temperature at low temperatures. The SLR mechanisms via TLS centers explain satisfactorily the magnitude and temperature dependence observed in a number of experiments. The main factors that determine the relaxation behavior are: (i) the distribution function of the disorder modes; (ii) tunneling and activated-disorder-mode relaxation; and (iii) coupling between phonons and disorder modes. The nature of the TLS-spin interaction remains an open question, however, since as yet no information on the detailed mechanism of this interaction has been deduced from the available experimental data.

10.6.2.2 SLR Effected by Electron–Nuclear Dipolar Coupling to a TLS Center

Bowman and Kevan (1977a) measured the SLR rates of trapped electrons in three frozen ethanol glass samples: C_2H_5OH, C_2D_5OH, and C_2H_5OD between 5 and 150 K. The results obtained showed that: (i) $T_1^{-1} \propto T$ for trapped electrons in ethanol glasses between 7 and 100 K; (b) T_1 is sensitive to the deuteration of glasses; and (iii) the relaxation rate of trapped electrons is about 10^5-fold larger for the $T_1^{-1} \propto T$ process in ethanol glasses than in KCl crystals. These authors were the first to invoke the TLS model to explain the relaxation mechanism. Basically, in this process, an electron spin interacts by the END interaction with a neighboring nucleus characterized by TLS. The coupling of this nucleus to lattice vibrations (phonons) effects the SLR. This mechanism leads to a T^2-dependence at very low temperatures, which changes to a linear (T)-dependence as the temperature is increased.

There are several simple TLS-state distribution functions affecting electron SLR (Bowman and Kevan, 1977a): (i) Delta-function distribution: $n(E) = \delta(E - E')$ (Gamble, Miyagawa, and Hartman, 1968; Murphy, 1966); (ii) Inverse-E distribution: $n(E) = 0$ for $E < \Delta_0$, $n(E) = E^{-1}$ for $E \geq \Delta_0$; (iii) Step-function distribution:

$n(E) = 0$ for $E < \Delta_0$; $n(E) =$ constant for $\Delta_0 \leq E \leq E_m$; $n(E) = 0$ for $E > E_m$; (iv) Low-energy distribution: $n(E) \neq 0$, for $E \ll kT$; $= 0$ otherwise. All of these functions have one feature in common which is that, above a certain temperature, the SLR rate is linearly proportional to the temperature of the sample. The first three distributions yield much stronger temperature dependences at very low temperatures. This is in strong contrast to the relaxation mechanisms due to lattice phonons, which have been well studied in single crystals doped with paramagnetic transition-metal ions.

10.6.2.3 SLR due to Fermi-Contact Hyperfine Interaction with a TLS Center

Spin-lattice relaxation measurements on color centers in K, Li, and Na type β-alumina (Kurtz and Stapleton, 1980) revealed an exceptionally fast relaxation, with anomalous temperature and microwave frequency dependences. This behavior was quantitatively described by a mechanism involving the coupling via the FC interaction of a color center to the phonon-induced relaxation of a nearby localized TLS (as discussed above in context with the END interaction), except that here the interaction is FC instead of END. This leads to a quadratic dependence on temperature at very low temperatures, and a linear dependence on temperature of the SLR rate at intermediate temperatures.

10.6.2.4 Temperature Dependence of Relaxation Rate in Amorphous Materials due to Exchange Interaction

The effect of exchange interaction in amorphous materials is different from that in single crystals, because the strength of the exchange interaction depends on the separation between the ions constituting exchange-coupled pairs, which varies for different pairs. A quadratic temperature dependence can be obtained when a three-dimensional uniform distribution of exchange-coupled pair separation is taken into account. This occurs for exchange energies less than thermal energies at very low temperatures (Stapleton et al., 1980). On the other hand, at relatively higher temperatures for which $J \gg kT$, a linear dependence of SLR on temperature is obtained. Experimental data on relaxation rates of localized electrons in amorphous silicon (Gourdon, Fretier, and Pescia, 1981; Askew et al., 1984) and those of Cr^{3+} ions in $Cu_{2x}Cr_{2x}Sn_{2-2x}S_4$ (Sarda et al., 1989) are in accordance with those predicted quadratic and linear temperature dependences, as analyzed by Misra (1998a, 1998b). This is in contrast to a linear dependence of relaxation rate on temperature which is found for single crystals, and for which no distribution of exchange-interaction values exists.

10.6.2.5 Relaxation for the Case of Strong Cross-Relaxation and Weak Spin-Lattice Relaxation of Single Ions in Amorphous Materials (Al'tshuler, 1956)

Low Temperatures In this case, a single ion is coupled to an exchange-coupled pair. When a cross-relaxation occurs at low temperatures, in which a single ion flips from one energy level to the other, accompanied by a simultaneous spin flip of an exchange-coupled pair at an energy Δ' above that of a single ion, a suitably averaged value of the temperature-dependent part of Equation 10.37 yields an approximate temperature dependence as T^2 for a distribution of values of Δ' as shown by Gill, (1962), when there are present a variety of clusters, as in amorphous samples.

High Temperatures (Schulz and Jeffries, 1966) For the case of strong relaxation of the excited pair, at an energy Δ' above that of a single ion, which becomes increasingly populated as the temperature is increased, the temperature dependence of SLR emerges as $T_1^{-1} \propto c_1^2/[1+\exp(\Delta'/kT)]$, where c_1 is the concentration of paramagnetic ions. This leads to a plateau in the relaxation rate at higher temperatures; an example is that observed in a borate glass sample doped with Fe_2O_3, characterized by a high concentration of paramagnetic ions and a larger antiferromagnetic exchange interaction ($J = -0.98$ K) (Zinsou et al., 1996).

10.6.3
Relaxation in Diluted Liquid Solutions

In order to describe relaxation in diluted solutions, one needs to consider the orientation-dependent part of the spin Hamiltonian in conjunction with a correlation function characterized by a time constant, referred to as the reorientation correlation time. The theoretical treatment is different for $S = ½$ and $S > ½$ systems.

$S = ½$: The theory for this case has been described by Kivelson (1960). The applicable spin Hamiltonian is:

$$H = \hbar[-S \cdot G \cdot B_0 + \sum_k J_k \cdot F_k \cdot S + \sum_{j,k} I_{jk} \cdot J_{jk} \cdot I_{jk} + \sum_i J_i S \cdot S_i] \quad (10.38)$$

In Equation 10.38, in the first term on the right-hand side, G is the magnetogyric ratio tensor, while in the second term in on the right-hand side, F_k and J_k describe the hyperfine-interaction tensor and nuclear spin operator, respectively, wherein the subscript k refers to the kth group of equivalent nuclei, with

$$J_k = \sum_j I_{jk}. \quad (10.39)$$

Further, in Equation 10.38, in the third term on the right-hand side, J_{jk} is the quadrupolar-interaction tensor for the jth nucleus belonging to the kth group of equivalent nuclei, and in the fourth term J_i is the integral for the exchange interaction connecting the spin with the ith molecule.

The exchange and quadrupolar terms become significant when the system is dilute in spins; on the other hand, information on translational diffusion can be obtained in solutions concentrated in spins. The spin–orbit effects are included in the G matrix, which is predominantly isotropic, for example, in organic radicals. When the reorientation time is short with respect to the various anisotropies – that is, when the motional narrowing is in effect, T_2^{-1} is expressed as (Kivelson, 1960; McConnell, 1956; McGarvey, 1957; Atkins and Kivelson, 1966):

$$T_2^{-1}(M_J) = \alpha + \alpha' + \beta M_J + \gamma M_J^2 + \delta M_J^3 \tag{10.40}$$

In Equation 10.40, M_J is the magnetic quantum number corresponding to the angular momentum J in the hyperfine-interaction space; α, β, γ, and δ are parameters that depend on the magnetogyric and hyperfine tensors, as well as on the spectral density for rotational motion. All other interactions, such as Heisenberg exchange and spin-rotational interaction are included in α'. Explicitly, the values of the parameters in Equation 10.40 were calculated by Bruno, Harrington, and Eastman (1977), as listed below:

$$\alpha = \frac{2\tau}{\sqrt{3}|\gamma_e|} \left\{ \begin{array}{l} \frac{F^2}{5}\left(1 + \frac{3}{4}u + \frac{3F}{14\omega_x}\right) + \frac{I(I+1)}{5} \times \\ \left[\Delta^2 p + \frac{7}{3}\Delta^2 u - \frac{F\Delta^2}{\omega_x}\left(\frac{20}{21} - \frac{5}{7}p\right) + \right] \\ \sqrt{\frac{2}{3}}F\Delta a(\omega^{-1} + \omega_x^{-1})\left(1 - \frac{3}{4}p\right) \end{array} \right\}; \tag{10.40a}$$

$$\beta = \frac{2\tau}{\sqrt{3}|\gamma_e|} \left[\begin{array}{l} -\frac{4}{5}\sqrt{2/3}F\Delta\left(1 + \frac{3}{4}\mu + \frac{9}{28}\frac{F}{\omega_x}\right) + \\ \frac{I(I+1)\Delta^2 a}{5}(\omega^{-1} + \omega_x^{-1})\left(p - \frac{4}{3}\right) + \\ \frac{8}{21}\sqrt{2/3}I(I+1)\frac{\Delta^3}{\omega_x}\left(1 - \frac{3}{4}p\right) + \\ \frac{1}{7}\sqrt{2/3}\frac{\Delta^3}{\omega_x}p - \frac{\Delta^2 a}{5\omega}p \end{array} \right]; \tag{10.40b}$$

$$\gamma = \frac{2\tau}{\sqrt{3}|\gamma_e|} \left\{ \begin{array}{l} \frac{8\Delta^2}{15}\left[1 - \frac{3}{8}p - \frac{u}{8} + \frac{F}{\omega_x}\left(1 - \frac{15}{56}p\right)\right] - \\ \frac{1}{5}\sqrt{2/3}F\Delta a(\omega^{-1} + \omega_x^{-1})\left(1 - \frac{3}{4}p\right) \end{array} \right\}; \tag{10.40c}$$

$$\delta = \frac{2\tau}{\sqrt{3}|\gamma_e|}\left[\frac{4}{15}\Delta^2 a(\omega^{-1} + \omega_x^{-1})\left(1 - \frac{3}{4}p\right) - \frac{64}{105}\sqrt{2/3}\frac{\Delta^3}{\omega_x}\left(1 - \frac{15}{32}p\right)\right]; \tag{10.40d}$$

where

$$F = \frac{2}{3}(g_{//} - g_\perp)\frac{\mu_B B_0}{\hbar}; \tag{10.40e}$$

$$u = 1/(1 + \omega^2 \tau_c^2); \tag{10.40f}$$

$$\omega_x^{-1} = \frac{\omega \tau_c^2}{(\omega \tau_c)^2 + 1}; \tag{10.40g}$$

$$\Delta = (|\gamma_e|\sqrt{6})(A_\perp - A_{//}); \tag{10.40h}$$

and

$$p = \left[1 + \frac{a^2 \tau_c^2}{4}\right]^{-1}. \qquad (10.40\mathrm{i})$$

In the above, τ_c is the correlation time for modulation of the interaction between the electron and the lattice. The a term, which is proportional to M_j^4, manifests itself when the quadrupolar terms are taken into account. The correlation is then τ_r, called the "rotational correlation time." The Stokes–Einstein relationship yields this to be $4\pi r^3 \eta/(3k_B T)$, where r is the molecular radius, and η is the viscosity. τ_r emerges as being proportional to $\exp(E_a/k_B T)$ as the viscosity exhibits Arrhenius-type dependence on temperature, with E_a being the activation energy. This equation is applicable provided that the hyperfine coupling occurs only with a single set of equivalent nuclei. When this is not the situation, hyperfine coupling with different sets of equivalent nuclei should be taken into account. (Further details may be found in Banci, Bertini, and Luchinat, 1986; Muus and Atkins, 1972; Campbell and Freed, 1980; Hyde and Froncisz, 1982.)

In the motional-narrowing limit, equations simpler than Equations 10.40 will result (Kivelson, 1960), listed as follows:

$$T_1^{-1} = \frac{2}{5\hbar^2} \left\{ \begin{array}{l} (\Delta g^2/3 + \delta g^2)\mu_B^2 B_0^2 + 1/8 \left(\begin{array}{c} \Delta A^2 \\ + \delta A^2 \end{array} \right) [7I(I+1) - m_I^2] \\ -(\Delta g \Delta A + 3\delta g \delta A)\mu_B B_0 m_I \end{array} \right\} \frac{\tau_c}{1 + \omega_s^2 \tau_c^2}$$

(10.41)

$$T_2^{-1} = \frac{1}{5\hbar^2} \left(\begin{array}{l} \left\{ \begin{array}{l} 4/3(\Delta g^2/3 + \delta g^2)\mu_B^2 B_0^2 + 1/8(\Delta A^2 + \delta A^2)[3I(I+1) + 5m_I^2] - \\ 4/3(\Delta g \Delta A + 3\delta g \delta A)\mu_B B_0 m_I \end{array} \right\} \tau_c \\ + \left\{ \begin{array}{l} (\Delta g^2/3 + \delta g^2)\mu_B^2 B_0^2 + 1/8(\Delta A^2 + \delta A^2)[7I(I+1) - m_I^2] \\ -(\Delta g \Delta A + 3\delta g \delta A)\mu_B B_0 m_I \end{array} \right\} \frac{\tau_c}{1 + \omega_s^2 \tau_c^2} \end{array} \right)$$

(10.42)

where,

$\Delta A = A_{zz} - 1/2(A_{xx} + A_{yy});$

$\delta A = 1/2(A_{xx} - A_{yy})$ [For axial symmetry $\delta A = 0$.]

The correlation function cannot be defined for electron relaxation in cases when the perturbation approximation is not applicable. This occurs when τ_c^{-1} is not sufficiently larger than the hyperfine splitting, or Zeeman anisotropy expressed in frequency units. Although the electron relaxation cannot be expressed as an analytic function, it can be computed numerically in terms of EPR linewidths, using the Liouville superoperators as described by Moro and Freed (1981) and Vasavada, Schneider, and Freed (1987).

In addition, the rotation of a molecule can bring about relaxation. In this case, the electrons do not rigidly follow the movement of the nuclear framework. Rather, the resultant dynamic change in rotating charges produces an overall magnetic moment which interacts with both the nuclear and electronic spins in the molecule and creates a nuclear pathway; this is termed "spin rotation," the magnitude of which is described in terms of the spin-rotation coupling tensor C, as investigated by Atkins and Kivelson (1966) and by Hubbard (1963). The resulting inverse relaxation times are:

$$T_1^{-1} = T_2^{-1} = (2Ik_BT\tau_\omega/3\hbar^2)(2C_\perp^2 + C_\parallel^2) \tag{10.43}$$

In Equation 10.43, I is the molecular moment of inertia and τ_ω is the rotational correlation time characteristic of the relaxation of rotational angular momentum, expressed as

$$\tau_\omega = 3ID/4r^2k_BT = I/8\pi r^3\eta, \tag{10.44}$$

where D is the translational diffusion constant, and $C_\perp$ and $C_\parallel$ are the diagonal components of the spin-rotational interaction tensor. Such values are reasonably proportional to components of the tensor Δg. Finally, Equation 10.43 can be written according to Atkins and Kivelson (1966) as:

$$T_1^{-1} = T_2^{-1} = (12\pi r^3)^{-1}(\Delta g_\parallel^2 + 2\Delta g_\perp^2)k_BT/\eta. \tag{10.45}$$

This mechanism is expected to be significant for dilute $S = \frac{1}{2}$ systems in liquids. Over and above the effects considered so far, Kivelson (1966) has analyzed the effect of oscillations of solvent molecules surrounding the metal complexes. The correlation time τ_c is defined by

$$\langle q_i(\tau)q_i(0)\rangle = q_0^2\exp(-\tau/\tau_c) \tag{10.46}$$

In Equation 10.46, q_0^2 is a typical r.m.s. value of the amplitudes of the intermolecular interactions, q_i. The Orbach mechanism may be important when there are accessible excited levels. If $\hbar\delta_{0n}/k_BT$ is not too large (<6), an Orbach process will contribute efficiently (Kivelson, 1966):

$$T_1^{-1} = 16\left(\frac{\lambda}{\Delta}\right)^2\left(\frac{\Phi'q_0}{\Delta r_0}\right)^2\left(\frac{\Delta}{\delta_{0n}}\right)^2\frac{\tau_c^{-1}}{\{\exp(\hbar\delta_{0n}/k_BT-1\}}. \tag{10.47}$$

In Equation 10.47, $\Phi' = \Sigma_i|\delta V_{cry}\delta q_ir_0|$ is a potential that includes the approximate magnitude of the time-dependent interaction; q_i are the lattice or ligand modes; r_0 is a characteristic intermolecular distance; and Δ is the energy separation between the levels participating in the transition.

The second-order interaction in the zero-field term in electron spin, $\hat{S}\cdot\tilde{D}\cdot\hat{S}$, in the spin Hamiltonian should be taken into account when $S > \frac{1}{2}$. Because of the constraint that $\tilde{D}$ be a traceless tensor, spin relaxation is effected due to the molecular reorientation, or distortion produced by collision with solvent molecules. For $S = 1$, the relevant expressions are provided by Bloembergen and Morgan (1961), and by Rubinstein, Baram, and Luz (1971):

$$T_1^{-1} = \frac{\Delta^2}{5}\left(\frac{\tau_v}{1+\omega_s^2\tau_v^2} + \frac{4\tau_v}{1+4\omega_s^2\tau_v^2}\right), \tag{10.48}$$

where $\Delta^2 = (D_{xx}^2 + D_{yy}^2 + D_{zz}^2)/\hbar^2$, τ_v is the correlation time for the electron–lattice interaction, which can be a rotational correlation time for any other fluctuation of the ZFS tensor. Similarly, as derived by Rubinstein, Baram, and Luz (1971):

$$T_2^{-1} = \frac{\Delta^2}{10}\left(3\tau_v + \frac{5\tau_v}{1+\omega_s^2\tau_v^2} + \frac{2\tau_v}{1+4\omega_s^2\tau_v^2}\right), \tag{10.49}$$

For $S = 3/2$, two electronic relaxation times should be taken into account. These are: (i) transition I: $+1/2 \leftrightarrow -1/2$ and (ii) transition II: the other $\Delta m_s = 1$ transitions. The longitudinal relaxation rates are (Rubinstein, Baram, and Luz, 1971):

$$T_{1(I)}^{-1} = \frac{\Delta^2}{5}\left(\frac{12\tau_v}{1+\omega_s^2\tau_v^2}\right), \tag{10.50}$$

$$T_{1(II)}^{-1} = \frac{\Delta^2}{5}\left(\frac{12\tau_v}{1+4\omega_s^2\tau_v^2}\right), \tag{10.51}$$

and the transverse relaxation rates are (Rubinstein Baram, and Luz, 1971):

$$T_{2(I)}^{-1} = \frac{\Delta^2}{5}\left(\frac{6\tau_v}{1+\omega_s^2\tau_v^2} + \frac{6\tau_v}{1+4\omega_s^2\tau_v^2}\right), \tag{10.52}$$

$$T_{2(II)}^{-1} = \frac{\Delta^2}{5}\left(6\tau_v + \frac{\tau_v}{1+\omega_s^2\tau_v^2}\right), \tag{10.53}$$

As for $S = 5/2$, the analytical expressions for the three transitions cannot be obtained (Rubinstein, Baram, and Luz, 1971). An averaged longitudinal relaxation rate is given by Bloembergen and Morgan (1961):

$$T_1^{-1} = \frac{2\Delta^2}{50}[4S(S+1)-3]\left(\frac{\tau_v}{1+\omega_s^2\tau_v^2} + \frac{4\tau_v}{1+4\omega_s^2\tau_v^2}\right). \tag{10.54}$$

10.6.4
Effect of Intramolecular Dynamics of Molecular Species on Relaxation

Intramolecular interactions and motion affect relaxation times. The various processes are described as follows (Eaton and Eaton, 2000).

10.6.4.1 Dephasing by Methyl Groups in Solvent or Surroundings

Any process that takes a spin outside of the observation window to monitor relaxation will eliminate it from contributing to a T_1 recovery. If the resonance frequency is sufficiently shifted, it will prevent the spin from refocusing to form the echo, and will therefore constitute a dephasing mechanism. The rotation of molecular species surrounding a paramagnetic center in solutions modifies relaxation rates due to their rotation. For example, in methyl-containing solvents, the rotation of methyl molecules is an additional effective process for relaxation, over and above spin diffusion, in typical proton-containing solvents. This is evidenced, for example, by the special role played by the solvent methyl groups in electron-spin echo dephasing for nitroxyl radicals (Salikhov and Tsvetkov, 1979), and $S = 1/2$ transitional metal complexes (Bowman and Kevan, 1979). The rotation of methyl groups is observed at low temperatures down to a few degrees Kelvin. As many of the organic solvents that are used in common mixtures and form good glasses when frozen at low temperatures contain methyl groups (e.g., toluene, methyl-THF, ethanol, methanol), and many protein side chains also contain methyl groups, the effect of methyl groups on T_m is significant, and must be taken into account. Both, spin labels and metals in proteins may be in the vicinity of rotating $-CH_3$ and $-NH_3$ groups and other side-chain groups with correlation times in the range 10^{-7} to 10^{-12} s (Torchia, 1984). The packing in a crystal has a strong influence on the methyl rotational correlation times in crystalline amino acids and peptides (Torchia, 1984). It has been observed that an increase in the length of the amino-acid chain decreases the rotational correlation time (Keniry et al., 1984). In proteins, large-amplitude motions of the side chains occur above temperatures of 283 K (Keniry et al., 1983).

10.6.4.2 Shape of the Echo-Decay Curve

In methyl-containing solvents, the shape of the echo decay is different from that in nonmethylated solvents. This is indicated by the particular values of the exponent (x), which are significantly less than 2, when the shape of echo decay is fitted to the expression (Zhidomirov and Salikhov, 1969): $E(2\tau) = \exp[-(2\tau/T_m)^x]$. The values of the exponents $x < 2$ characterize those cases for which the correlation time of the process dominates over dephasing is comparable to or shorter than τ, the time between pulses.

10.6.4.3 Averaging of Electron-Nuclear Couplings due to Rotation of Methyl Groups

With increasing rotation rate of methyl groups within a radical or transition-metal complex, the electron-nuclear hyperfine coupling begins to be averaged out. This causes dephasing, and results first in a characteristic decrease, followed by a rise in T_m as the temperature is increased.

10.6.4.4 Effect of a Rapidly Relaxing Partner on Electron–Electron Spin–Spin Coupling

The relaxation of a more rapidly relaxing partner spin-coupled to a slower relaxing spin (for example, a nitroxyl radical) leads to a dephasing mechanism. This process results in a distinctive decrease and subsequent increase in T_m as the relaxation rate of the more rapidly relaxing partner increases.

10.6.4.5 Librational Motion

Small-amplitude molecular motions are known as *librations* for an anisotropic paramagnetic center. They can cause significant changes in the resonance energy, and are responsible for echo dephasing when the changes occur on the time scale of an echo experiment. These affect the value of T_m, with the effect being greater in glassy solvents than in crystalline lattices (relevant experimental data are available in Eaton and Eaton, 2000).

10.6.4.6 Molecular Tumbling

For an anisotropic paramagnetic center, tumbling moves spins off resonance, causing dephasing. This occurs when a glass softens, which enhances tumbling, affecting T_m.

10.6.4.7 Biomolecules

A listing of relaxation times for organic radicals and metals for species of interest in biomedical EPR is provided by Eaton and Eaton (2000).

10.6.4.8 Macromolecules

Macromolecules in fluid solutions tumble so slowly that the paramagnetic centers embedded in them act as if they are in the solid state. As a result, it may appear that the SLR rate for a paramagnetic center in a protein does not depend on the correlation time for the protein. On the other hand, it may be considered that electron SLR occurs via the vibration of a protein, thus providing a microcrystalline environment for the center, as in solids and glasses, even when the protein is in solution at room temperature (Eaton and Eaton, 2000).

10.6.5
Relaxation among Different Paramagnetic Centers in Concentrated Solution

In concentrated systems, the electron–electron interactions among different paramagnetic centers affect will have significant effects on the SLR rates. When the spins are alike, relaxation occurs due to a modulation of their interactions; an example is that of exchange, superexchange, dipolar, which occurs through phonons or fluctuations in crystal vibrations. In the case of solutions, the molecular collisions or lifetime of aggregates provide efficient pathways, with the SLR rates of radicals exchanging by orders of magnitudes in concentrated solutions when compared to those in dilute solutions (Kreilick, 1968; Giordano et al., 1981). An additional feature of relaxation in concentrated systems is the simultaneous electronic transition of two spins (flip-flop transitions), without a net energy change. If the two electron spins are coupled to nuclear spins, there is a dramatic effect on T_2 – and thus on the linewidth – and this must be taken into account in the lineshape analysis. When the spins are not alike, one should consider the effect of their dissimilar SLR rates, over and above those applicable to like spins. This results in a decrease in the relaxation rate of the slow-relaxing spin coupled to a fast-relaxing spin, when the exchange interaction between them is such that $J/\hbar > \sqrt{1/(T_1^{fast} T_1^{slow})}$ (Bertini, Martini, and Luchinat, 1994).

10.6.6
Spin-Fracton Relaxation

The dependence of the relaxation rate on temperature as $1/T_1 \propto T^{6.3}$ in fractal material can be explained by the theories developed by Orbach and Alexander and coworkers (Alexander, Entin-Wohlman, and Orbach, 1985a, 1985b, 1986) on SLR in fractals. Fractal material is more porous than the ordinary Euclidean matter encountered in everyday life, for which: (i) the density of number of occupied sites in a sphere of radius r is proportional to r^d, where the dimensionality $d = 3$; and (ii) the mass M is proportional to L^3. On the other hand, for fractal matter $d < 3$, and in terms of this d the mass M is proportional to L^d (Pietronero and Tosatti,

1986). An important concept in context with fractals is the self-similarity of matter, which dictates following self-avoiding paths; for example, Cayley's tree and Sierpenski's gasket.

Some relevant dimensions are now defined for fractals: $\bar{d}$ is defined to as the "effective dimensionality" (anomalous dimension); $\bar{\delta}$ is the exponent that provides the dependence of the diffusion constant on distance; and $\bar{\bar{d}} = 2\bar{d}/(2+\bar{\delta})$, referred to as the "new intrinsic fracton dimensionality," which is required because proper mode counting requires a reciprocal space with a new intrinsic fracton dimensionality.

The background theory when calculating SLR is reviewed here, and is similar to that used for consideration of SLR in Euclidean matter. The time-dependent perturbation theory (Fermi's golden rule), when applied to an electron-vibrational coupling relaxation process due to transitions between the levels a and b, leads to

$$\frac{1}{T_1} = P_{a \to b} + P_{b \to a} = \frac{2\pi}{\hbar}|H_{int}|^2 \frac{dn}{dE}, \qquad (10.55)$$

where $\frac{dn}{dE}$ is the vibrational density of states of fractons, analogous to phonons in Euclidean matter. In the context of fractal matter, one considers exchange of one phonon and two phonons for both Kramers' doublets for ions with half-odd integral spin, characterized by two intermediate time-reversed states, and non-Kramers' doublets for ions with integer spin.

This relaxation rate has been calculated for the emission or absorption of localized vibrational quanta by a localized electronic state. Here, it is characterized by a probability density calculated for two cases: (i) Vibrational localization, as on a fractal network, which is geometric in origin with the fracton being the quantized vibrational states; and (ii) a consequence of scattering with localized phonons being the quantized vibrational states. The relaxation rate is characterized by a probability density which has been calculated for both types of localization, (i) and (ii), and for two extreme limits: (a) the sum of the electronic and vibrational energy widths is independent of the spatial distance between the electronic and vibrational states; and (b) the sum of the two energy widths equals $\hbar$ times the relaxation rate.

10.6.6.1 One-Fracton Emission

There are two cases to be considered in this context, based on the particular time profile (Alexander, Entin-Wohlman, Orbach, 1985a).

Case (i): Here, the time-profile for interaction with fractons is calculated to be

$$F(t) \propto t^{-c_1 (\ln t)^{(D/d_\phi)-1}}, \qquad (10.56)$$

where c_1 is a constant, D is The fractional dimensionality, and d_ϕ is defined by the range dependence of the fracton wavefunction:

$$\phi \propto \exp(-r^{d_\phi}). \qquad (10.57)$$

This decay is faster than any power law, but slower than exponential. This profile, which is known as a "stretched exponential," expresses the decay of a localized excitation by the emission of a single localized vibrational quantum (referred to as a "fraction"). The main effect of localization is on the relaxation–time profile, which is a strongly nonexponential decay. Different spatial sites can have very different relaxation rates because of their different distances from the closest relaxing vibrational mode.

Case (ii): Here, the time profile for the interaction with fractons is given by:

$$F(t) \propto constant + c_2(1/t)(\ln t)^{(D/d_\phi)-1}, \tag{10.58}$$

where c_2 is a constant.

10.6.6.2 Two-Fracton Inelastic Scattering (Localized Electronic State) (Alexander, Entin-Wohlman, and Orbach, 1985a)

This leads to the calculation of a two localized-vibrational quanta (Raman) relaxation process for a localized electronic state, which is relevant to relaxation at high temperatures with high-energy vibrational excitations, characterized by short-length scale. Vibrational localization (Alexander et al., 1986) can be geometric in origin, similar to that on a fractal network with quantized vibrational states or fractons, a consequence of scattering analogous to Anderson localization with the quantized vibrational states localized phonons. The long-term behavior begins as:

$$t^{[1/(2a-1)]} \exp[-c_3(t)^{\frac{1}{a-1}}], \tag{10.59}$$

where c_3 is a constant; $a = 4q + 2\bar{\bar{d}}$ for Kramers' transitions and $a = 4q + 2\bar{\bar{d}} - 2$ for non-Kramers transitions; $\bar{\bar{d}}$ is the new intrinsic fracton dimensionality; and $q = d = d_\phi/D$, with D being the fracton dimensionality. After a cross-over time, the long-term behavior varies as:

$$(\ell n\ t)^{\eta - 1/2} t^{-c_1 (\ell n\ t)^{2\eta}}, \tag{10.60}$$

where c_1 is a constant and $\eta = D/d_\phi$. This particular time decay is faster than any power law, but slower than exponential or stretched exponential.

In the presence of rapid electronic processes of cross-relaxation, the time profile is exponential with the low-temperature relaxation time behavior:

for Kramers' ions: $$\left(\frac{1}{T_1}\right)_{ave} \propto T^{2\bar{\bar{d}}[2+2d_\phi/D]-1}; \tag{10.61}$$

and

for non-Kramers' ions: $$\left(\frac{1}{T_1}\right)_{ave} \propto T^{2\bar{\bar{d}}[1+2d_\phi/D]-3}. \tag{10.62}$$

These results may be exploited to explain the fractional temperature exponents found for electronic SLR in macromolecules and nuclear spin-lattice relaxation in glasses (Alexander, Entin-Wohlman, Orbach, 1986).

Stapleton et al. (1980) assume that $q = 1$, whence $a = 4 + 2\bar{\bar{d}}$ (Kramers') and $a = 2 + 2\bar{\bar{d}}$ (non-Kramers'), which are different from the ansatz of Alexander,

Entin-Wohlman, and Orbach (1985b), according to which $q = \bar{\bar{d}} \, d_{min}/\bar{d}$, which leads to the following expressions for inverse SLR times:

for Kramers' ions: $$\frac{1}{T_1^{ave}} \propto T^{2\bar{\bar{d}}[1+2d_{min}/\bar{d}]-1}, \qquad (10.63)$$

and

for non-Kramers' ions: $$\frac{1}{T_1^{ave}} \propto T^{2\bar{\bar{d}}[1+2d_{min}/\bar{d}]-3}. \qquad (10.64)$$

One may quote Alexander, Entin-Wohlman, and Orbach (1985b) on the difference between the interpretation of Stapleton et al. (1980) and their own: "*Our ansatz for q seems more natural than that of Stapleton et al. (1980) in the context of a scaling description.*" One can, however, also have physical conditions where $q = 1$ would be correct, as in Stapleton et al. (1980). Since $\bar{\bar{d}}$, $\bar{d}$ and d_{min} are not known independently, there is, at present, no way of determining q for systems investigated by Stapleton et al. (1980).

The analytic forms for temperature dependence of Raman spin-lattice relaxation rate for localized fractons, fractons with localization neglected and localized (or extended) phonons for $T \ll \Theta_D$, are listed in Table 10.2. It has been noted (Alexander, Entin-Wohlman, and Orbach, 1985b) that, for Kramers' ions there are always two time-reversed intermediate states; this results in the twofold difference in T dependence in the exponent for Kramers' and non-Kramers' ions.

The experimental data of Stapleton et al. (1980) on heme-protein in frozen solutions of myoglobin azide revealed the relaxation rate to be governed by $1/T_1 \propto T^{6.3}$, which these authors explained as being due to spin-fracton relaxation. However,

Table 10.2 Analytic forms for temperature dependences of relaxation times for fractal material.

Transition ion	Character of thermal excitation	Temperature dependence	Temperature dependence using percolation exponents for $d = 3$: $\bar{d} = 4/3$, $\bar{\bar{d}} = 2.5$, $\bar{d}/d_{min} = 1.8$
Kramers'	Fractons	$T^{2\bar{\bar{d}}[1+2\frac{d_{min}}{\bar{d}}]-1}$	$T^{4.63}$
Non-Kramers'	Fractons	$T^{2\bar{\bar{d}}[1+2\frac{d_{min}}{\bar{d}}]-3}$	$T^{2.63}$
Kramers'	Fractons (without localization)	$T^{2\bar{\bar{d}}+3}$	$T^{5.66}$
Non-Kramers'	Fractons (without localization)	$T^{2\bar{\bar{d}}+1}$	$T^{3.66}$
Kramers'	Localized (or extended phonons)	T^{2d+3}	T^9
Non-Kramers'	Localized (or extended phonons)	T^{2d+1}	T^7

their subsequent data on metalloproteins could not be explained by fracton relaxation (Eaton and Eaton, 2000). Further, the temperature dependence of their data was very similar to that in nonprotein hosts, for example, low-spin Fe^{3+} in small-molecule porphyrin complexes, and in methemoglobin and myoglobin, which do not have the same fractal dimensions as that of heme-protein. The non-integer exponent to T describing spin-fracton relaxation can sometimes arise due to overlapping contributions from different competing relaxation processes characterized by different integer exponents. One has, therefore, to be careful in concluding unequivocally the occurrence of spin-fracton relaxation. Evidence of spin-fracton relaxation in resins doped by rare-earth ions was recently reported by Pescia, Misra, and Zaripov (1999).

10.6.7
Frequency/Field Dependence of Paramagnetic Relaxation

The frequency of the microwave field used in EPR today ranges anywhere from about ~1 GHz to ~1000 GHz, and is sensitive to molecular motion, vibration, and oscillatory motion. Likewise, with increasing microwave frequency the energy-level separations increase as the Zeeman term increases, due to an increasing magnetic field. Since at resonance one observes transitions between various energy levels, their intensities depend upon the population difference between the levels participating in resonance, and thus on the magnetic field. As a consequence, the degree of sensitivity of relaxations is affected by the magnetic field, and hence the microwave frequency (Eaton and Eaton, 2000).

With regards to the field dependence, it has been noted that the SLR rate due to the direct process depends on the field as B^2 for non-Kramers' ions and as B^4 for Kramers' ions when the frequency, v, is such that $hv \ll k_B T$ (Standley and Vaughan, 1969; Davids and Wagner, 1964). On the other hand, the Raman process is independent of the magnetic field, whereas the Orbach process is independent of the magnetic field, except for its dependence on the magnetic field, which determines the excited-state energy levels. Further, relaxation via the modulation of tunneling modes in glassy matrices depends on B^{-2} (Bowman and Kevan, 1977b, 1977c).

In general, the frequency dependence of the SLR rate is determined by the spectral density function of the form $\tau_c/(1+\omega^2\tau_c^2)$, where $\omega = 2\pi v$. Thus, with increasing frequency, the correlation time τ_c due to random motions may become comparable to $1/\omega$, resulting in an enhanced relaxation rate. On the other hand, with increasing frequency the relaxation rate decreases. This is important for the process that effects relaxation in the motion that occurs near the softening point of a glass depending on the value of $\omega^2\tau_c^2$, which would imply an increase or decrease in relaxation rate depending on whether it is greater or less than 1, respectively, with the maximum effect occurring when $\omega^2\tau_c^2 \sim 1$. Some data on the frequency dependence of relaxation times have been reported by Prisner (1997).

In particular, relaxation is more sensitive to faster motions at higher microwave frequencies, implying higher magnetic fields. This results in enhanced spectral diffusion at higher frequencies, and is supported by experimental data on EPR linewidth studies at 250 GHz as compared to that at 9.5 GHz (Barnes *et al.*, 1999; Budil, Earle, and Freed, 1993; Earle, Budil, and Freed, 1993; Earle, Moscicki, and Freed, 1997).

Pertinent literature

The material on paramagnetic relaxation in this chapter is based on the following references:

- Abragam, A. and Bleaney, B. (1970) *Electron Paramagnetic Resonance of Transition Ions*, Oxford University Press, Oxford. This is an old classic on EPR theory, including Bloch's equations, and background theory on spin-lattice relaxation.

- Bertini, G. Martini, and Luchinat, C. (1994) in *Handbook on Electron Spin Resonance*, vol. 1 (eds C.P. Poole, Jr and H.A. Farach), American Institute of Physics, New York. Chapter III, entitled "Relaxation, Background, and Theory," reviews the general theory of spin-lattice relaxation and provides an outline of pulse techniques. Chapter IV, entitled "Relaxation Data Tabulation," lists relaxation data on EPR of transition-metal ions, free electrons, inorganic radicals, free organic radicals, and spin labels in the solid state, solution, frozen state, and adsorbed onto solids. Here, the exchange effects, temperature dependence, and motional aspects are discussed at length.

- Eaton, S.S. and Eaton, G.R. (2000) in *Biological Magnetic Resonance*, vol. 19, *Distance Measurements in Biological Systems by EPR* (eds L.J. Beriner, S.S. Eaton, and G.R. Eaton), Kluwer Academic/Plenum Publishers, New York. Chapter 2, entitled "Relaxation Times of Organic Radicals; and Transitional Metal Ions," presents a detailed review, including data tabulation on relaxation times in organic radicals and metals. Also included is an exhaustive list of publications, citing the relaxation literature comprehensively from a modern view of relaxation in biological EPR.

- Misra, S.K. (1998) Spin-lattice relaxation times in amorphous materials as effected by exchange interactions, tunneling level states (TLS) centres, and fractons, *Spectrochimica Acta A*, **54**, 2257. This is a review of the particular SLR mechanisms (as suggested in the title), including some data tabulation.

- Pake, G.E. and Estle, T.I. (1973) *The Physical Principles of Electron Paramagnetic Resonance*. Benjamin Publications, Reading, Massachusetts. This presents the general theory of SLR, Bloch's equations, and rotating frame.

- Manenkov, A.A. and Orbach, R. (eds) (1966) *Spin-lattice Relaxation in Ionic Solids*. Harper and Row, Publishers, New York. This book provides a good collection of classic reports published on SLR during the period 1932–1965. As well, it provides an early history of paramagnetic SLR.

- Poole, C.P., Jr and Farach, H.A. (1971) *Relaxation in magnetic resonance: dielectric and Mössbauer applications.* Academic Press, New York. This book deals with the experimental and theoretical aspects of relaxation processes in magnetic resonance.

Acknowledgments

The author is grateful to Professors C.P. Poole, Jr, Gareth Eaton, and Sandra Eaton for helpful comments in improving the presentation of this chapter, and to the Natural Sciences and Engineering Research Council (NSERC) of Canada for partial financial support.

References

Abragam, A. (1961) *The Principles of Nuclear Magnetism,* Oxford University Press, Oxford, pp. 443–445.

Abragam, A., and Bleaney, B. (1970) *Electron Paramagnetic Resonance of Transition Ions,* Oxford University Press, Oxford.

Alexander, S., Entin-Wohlman, O., and Orbach, R. (1985a) *Phys. Rev. B,* **32,** 6447.

Alexander, S., Entin-Wohlman, O., and Orbach, R. (1985b) *J. Phys. Lett.,* **46,** L555–L560.

Alexander, S., Entin-Wohlman, O., and Orbach, R. (1986) *Phys. Rev. B,* **33,** 3935.

Al'tshuler, S.A. (1956) *Bull. Acad. Sci. USSR,* **20,** 1207.

Anderson, P.W. (1958) *Phys. Rev.,* **109,** 1492.

Anderson, P.W., Halperin, B.I., and Varma, C.M. (1972) *Phil. Mag.,* **25,** 1.

Antipin, A.A., Kochalev, B.I., and Shleklin, B.I. (1984) *JETP Lett.,* **39,** 182.

Askew, T.R., Muench, P.J., Stapleton, H.J., and Brower, K.L. (1984) *Solid State Commun.,* **49,** 667.

Atherton, N.W. (1993) *Principles of Electron Spin Resonance,* Prentice Hall, New York, Chapter 1.

Atkins, P.W. and Kivelson, D. (1966) *J. Chem. Phys.,* **44,** 169.

Banci, L., Bertini, I., and Luchinat, C. (1986) *Magn. Reson. Rev.,* **11,** 1.

Banci, L., Bertini, I., and Luchinat, C. (1991) *Nuclear and Electron Relaxation,* Wiley-VCH Verlag GmbH, Weinheim.

Barnes, J.P., Liang, Z., Mchaourab, H.S., Freed, J.H., and Hubbell, W.L. (1999) *Biophys. J.,* **76,** 3298.

Berman, R. (1949) *Phys. Rev.,* **76,** 315.

Bertini, I., Martini, G., and Luchinat, C. (1994) Relaxation, background, and theory, in *Handbook on Electron Spin Resonance,* vol. 1 (eds C.P. Poole and H.A. Farach), American Institute of Physics, New York, Chapter III, pp. 51–77.

Bleaney, B., Boyle, G.S., Cooke, A.H., Duffus, R.J., O'Brien, M.C.M., and Stevens, K.W.H. (1955) *Proc. Phys. Soc. (London) A,* **68,** 57.

Bloembergen, N. (1949) *Physica,* **15,** 386.

Bloembergen, N. and Morgan, L.O. (1961) *J. Chem. Phys.,* **34,** 842.

Bloembergen, N., Shapiro, S., Pershan, P.S., and Artman, J.O. (1959) *Phys. Rev.,* **114,** 445.

Bowman, M.K. and Kevan, L. (1977a) *J. Phys. Chem.,* **81,** 456.

Bowman, M.K. and Kevan, L. (1977b) *Discuss. Faraday Soc.,* **63,** 7.

Bowman, M.K. and Kevan, L. (1977c) *J. Phys. Chem.,* **81,** 456.

Bowman, M.K. and Kevan, L. (1979) Electron spin-lattice relaxation in nonionic solids, in *Time Domain Electron Spin Resonance* (eds L. Kevan and R.N. Schwartz), John Wiley & Sons, Inc., New York, p. 68.

Bruno, V., Harrington, J.K., and Eastman, M.P. (1977) *J. Phys. Chem.,* **81,** 1111.

Brya, W. and Wagner, P.E. (1965) *Phys. Rev. Lett.*, **14**, 431.

Budil, D.E., Earle, K.A., and Freed, J.H. (1993) *J. Chem. Phys.*, **97**, 1294.

Campbell, R.F. and Freed, J.H. (1980) *J. Chem. Phys.*, **84**, 2668.

Canet, D. (1996) *Nuclear Magnetic Resonance: Concepts and Methods*, John Wiley & Sons, Inc., Chichester, p. 106.

Choy, C.L., Salinger, G.L., and Chiang, Y.C. (1970) *J. Appl. Phys.*, **41**, 597.

Dalton, L.R., Kwiram, A.L., and Owen, J.A. (1972) *Chem. Phys. Lett.*, **17**, 495.

Davids, D.A. and Wagner, P.E. (1964) *Phys. Rev. Lett.*, **12**, 141.

Earle, K.A., Budil, D.E., and Freed, J.H. (1993) *J. Phys. Chem.*, **97**, 13289.

Earle, K.A., Moscicki, J.K., and Freed, J.H. (1997) *J. Phys. Chem.*, **106**, 9996.

Eaton, G.R. (1993) *Biophys. J.*, **64**, 1373.

Eaton, S.S. and Eaton, G.R. (1993) *J. Magn. Reson. A*, **102**, 354. (N.B. The field range should be 3263 to 3273 G in Fig. 1, and not 2869 to 2879 G, as published in this paper).

Eaton, S.S. and Eaton, G.R. (2000) Relaxation Times of Organic Radicals and Transition Metal Ions, in *Biological Magnetic Resonance*, Vol. 19, (eds L.J. Berliner, S.S. Eaton, and G.R. Eaton), Kluwer Academic/Plenum Publishers, New York.

Estalji, S., Kanert, O., Steiner, J., Jian, H., and Ngai, K.L. (1991) *Phys. Rev. B*, **43**, 7481.

Fierz, M. (1938) *Physica*, **5**, 433.

Finn, C.B.P., Orbach, R., and Wolf, W.P. (1961) *Proc. Phys. Soc.*, **296**, 261.

Fukushima, E. and Roeder, S.B.W. (1981) *Experimental Pulse NMR: A Nuts and Bolts Approach*, Addison-Wesley Publishing Co., Reading, MA.

Gamble, W.I., Miyagawa, I., and Hartman, R.L. (1968) *Phys. Rev. Lett.*, **20**, 415.

Geschwind, S. (1972) Optical Techniques in EPR in Solids, in *Electron Paramagnetic Resonance* (ed. S. Geschwind), Plenum, New York, pp. 121–216, 353–425.

Gill, J.C. (1962) *Proc. Phys. Soc. (London)*, **79**, 58.

Giordano, M., Martinelli, M., Pardi, L., and Santucci, S. (1981) *Mol. Phys.*, **42**, 523.

Goldman, S.A., Bruno, G.V., Polanszek, C.F., and Freed, J.H. (1972) *J. Chem. Phys.*, **56**, 716.

Gorter, C.J. (1947) *Paramagnetic Relaxation*, Elsevier, New York.

Gourdon, J.C., Fretier, P., and Pescia, J. (1981) *J. Phys. Lett.*, **42**, 21, based on J. C. Gourdon, PhD thesis (Thèse d'Etat, no. 69), Université Paul Sabatier, Toulouse, France (1975).

Hasagawa, H. (1960) *Phys. Rev.*, **118**, 1523.

Heitler, W. and Teller, E. (1936) *Proc. R. Soc. London*, **A155**, 629.

Hervé, J. and Pescia, J. (1960) *Compt. Rend.*, **251**, 665.

Hervé, J. and Pescia, J. (1963) *Compt. Rend.*, **255**, 2926.

Hubbard, P.S. (1963) *Phys. Rev.*, **131**, 1155.

Huisje, M. and Hyde, J.S. (1974) *J. Chem. Phys.*, **60**, 1682.

Hunklinger, S. (1987) *Phil. Mag. B*, **56**, 199.

Hyde, J.S. and Froncisz, W. (1982) *Annu. Rev. Biophys. Bioeng.*, **11**, 391.

Hyde, J.S., Sczanicki, P.B., and Froncisz, W. (1989) *J. Chem. Soc. Faraday Trans. I*, **85**, 3901.

Jollenbeck, G.B., Kanert, O., Strinert, J., and Jain, H. (1988) *Sol. St. Commun.*, **65**, 30.

Keniry, M.A., Rothgeb, T.M., Smith, R.L., Gutowsky, H.S., and Oldfield, E. (1983) *Biochemistry*, **22**, 1917.

Keniry, M.A., Kintanar, A., Smith, R.L., Gutowsky, H.S., and Oldfield, E. (1984) *Biochemistry*, **22**, 288.

Kittel, C. (1996) *Introduction to Solid State Physics*, 7th edn, John Wiley & Sons, New York.

Kivelson, D. (1960) *J. Chem. Phys. Soc.*, **33**, 1094.

Kivelson, D. (1966) *J. Chem. Phys.*, **45**, 1324.

Kohler, W. and Friedrich, J. (1987) *Phys. Rev. Lett.*, **59**, 2199.

Kreilick, R.W. (1968) *J. Am. Chem. Soc.*, **90**, 2711.

Kronig, R.L. (1939) *Physica*, **6**, 33.

Kurtz, S.R. and Stapleton, H.J. (1980) *Phys. Rev. B*, **22**, 4223.

LaMar, G.N., Horrocks, W.D., Jr, and Hahn, R.H. (eds) (1973) *NMR of Paramagnetic Molecules*, Academic Press, New York, p. 288, 568, 611, 638.

Lee, S., Patyal, B.R., and Freed, J.H. (1993) *J. Chem. Phys.*, **98**, 3665.

Lyo, S.K. and Orbach, R. (1980) *Phys. Rev. B*, **22**, 4223.

McConnell, H.M. (1956) *J. Chem. Phys.*, **25**, 709.

McGarvey, B.R. (1957) *J. Phys. Chem.*, **61**, 1232.
Mclaughlin, A.C. and Leigh, J.S. Jr. (1973) *J. Magn. Reson.*, **9**, 206.
McTaggart, J.H. and Slack, G.A. (1969) *Cryogenics*, **9**, 384.
Makinen, M.W. and Wells, G.B. (1987) Application of ESR Saturation Methods to Paramagnetic Metal Ions in Proteins, in *Metal Ions in Biological Systems* (ed. H. Sigel), Marcel Dekker, New York, pp. 129–206.
Manenkov, A.A. and Prokhorov, A.M. (1962) *Zh. Eksp. i Teor. Fiz.*, **42**, 1371.
Mchaourab, H.S. and Hyde, J.S. (1993a) *J. Chem. Phys.*, **98**, 1786.
Mchaourab, H.S. and Hyde, J.S. (1993b) *J. Magn. Reson. B*, **101**, 178.
Mchaourab, H.S. and Hyde, J.S. (1993c) *J. Chem. Phys.*, **99**, 4975.
Mchaourab, H.S., Christidis, T.C., Froncisz, W., Sczianecki, P.B., and Hyde, J.S. (1991) *J. Magn. Reson.*, **92**, 429.
Mchaourab, H.S., Pfenninger, H.S., Antholine, W.E., Felix, C.C., Hyde, J.S., and Kronek, P. (1993) *Biophys. J.*, **64**, 1576.
Misra, S.K. (1998a) *Spectrochim. Acta A*, **54**, 2257.
Misra, S.K. (1998b) *Phys. Rev. B*, **58**, 14971–14977.
Misra, S.K. (2005) *Appl. Magn. Reson.*, **28**, 55.
Misra, S.K. and Pilbrow, J. (2004) *Phys. Rev. B*, **69**, 212411.
Misra, S.K. and Pilbrow, J.R. (2007) *J. Mag. Reson.*, **185**, 38–41.
Moro, G. and Freed, J.H. (1981) *J. Chem. Phys.*, **74**, 3757.
Murphy, J. (1966) *Phys. Rev. B*, **145**, 241.
Muus, L.T. and Atkins, P.W. (1972) *Electron Spin Relaxation in Liquids*, Plenum, New York.
Norris, J.R., Thurnauer, M.C., and Bowman, M.K. (1980) *Adv. Biol. Med. Phys.*, **17**, 365.
Orbach, R. and Tachiki, M. (1967) *Phys. Rev.*, **158**, 524.
Orbach, R. (1961a) *Proc. R. Soc. London Ser. A*, **264**, 458.
Orbach, R. (1961b) *Proc. R. Soc. London Ser. A*, **264**, 485.
Orbach, R. (1961c) *Proc. Phys. Soc. A*, **77**, 821.
Orbach, R. and Stapleton, H.J. (1972) Electron Spin Relaxation, in *Electron Paramagnetic Resonance* (ed. S. Geschwind), Plenum, New York, pp. 121–126.
Overhauser, A.W. (1953) *Phys. Rev.*, **89**, 689.
Pake, G.E. and Estle, T.L. (1973) *Electron Paramagnetic Resonance*, Benjamin Publications Inc., Reading, MA.
Percival, P.W. and Hyde, J.S. (1975) *Rev. Sci. Instrum.*, **46**, 1522.
Pescia, J., Misra, S.K., and Zaripov, M. (1999) *Phys. Rev. Lett.*, **83**, 1866.
Phillips, W.A. (1970) *Proc. Roy. Soc. (London)*, **A319**, 565.
Phillips, W.A. (1972) *J. Low-Temp. Phys.*, **7**, 351.
Phillips, W.A. (1981) *Amorphous Solids, Low-Temperature Properties*, Vol. 24 of *Topics in Current Physics*, Springer, Berlin.
Phillips, A. (1990) Phonons 89. Proceedings of the Third International Conference on Phonon Physics, Vol. 1, (eds S. Hunklinger, W. Ludwig, and G. Weiss), World Scientific, Singapore, p. 367.
Pietronero, L. and Tosatti, E. (eds) (1986) Fractals in Physics. Proc. 6th International Symposium on Fractals in Physics, International Centre for Theoretical Physics, Trieste, Italy, 9–12 July 1985, North Holland, Amsterdam.
Pilbrow, J.R. (1990) *Transition Ion Electron Paramagnetic Resonance*, Clarendon Press, Oxford, p. 387.
Poole, C.P., Jr and Farach, H.A. (1971) *Relaxation in Magnetic Resonance*, Academic Press, New York, p. 267.
Prisner, T.F. (1997) *Adv. Magn. Optic. Reson.*, **20**, 245.
Roth, L.M. (1960) *Phys. Rev.*, **118**, 1534.
Rubinstein, M., Baram, A., and Luz, Z. (1971) *Mol. Phys.*, **20**, 67.
Salikhov, K. and Tsvetkov, Yu.D. (1979) Electron spin-echo studies of spin-spin interactions in solids, in *Time Domain Electron Spin Resonance* (eds L. Kevan and R.N. Schwartz), John Wiley & Sons, Inc., New York, Ch. 7, pp. 231–277.
Sarda, I., Colombet, P., Ablart, G., and Pescia, J. (1989) *C. R. Acad. Sci. Paris*, **308**, Series II, 159.
Sastry, V.S.S., Polimeno, A., Crepeau, R.H., and Freed, J.H. (1996a) *J. Chem. Phys.*, **105**, 5753.
Sastry, V.S.S., Polimeno, A., Crepeau, R.H., and Freed, J.H. (1996b) *J. Chem. Phys.*, **105**, 5773.

Saxena, S. and Freed, J.H. (1997) *J. Phys. Chem.*, **101**, 7998.

Schneider, D.J. and Freed, J.H. (1989) Spin Relaxation and Material Dynamics, in *Lasers, Molecules, and Methods*, vol. 73 (eds J.O. Hirschfelder, R.E. Wyatt, and R.D. Coalson), John Wiley & Sons, Inc., New York, pp. 387–527.

Schulz, C.E., Brandon, S., and Debrunner, P.G. (1990) *Hyperfine Interactions*, **58**, 2399.

Schulz, M.B. and Jeffries, C.D. (1966) *Phys. Rev.*, **149**, 270.

Schweiger, A. (1991) *Advances in Pulsed and C. W. EPR*, Eds. L. Kevan and X. Beckman.

Schweiger, A. and Ernst, R.R. (1988) *J. Magn. Reson.*, **77**, 512.

Schweiger, A. and Jeschke, G. (2001) *Principles of Pulse Electron Paramagnetic Resonance*, Oxford University Press, Oxford.

Scott, P.L. and Jeffries, C.D. (1962) *Phys. Rev.*, **127**, 32.

Sczianecki, P.B., Hyde, J.S., and Froncisz, W. (1990) *J. Chem. Phys.*, **93**, 3891.

Standley, K.J. and Vaughan, R.A. (1969) *Electron Spin Relaxation Phenomena in Solids*, Adam Hilger, London.

Stapleton, H.J., Allen, J.P., Flynn, C.P., Stinson, D.G., and Kurtz, S. (1980) *Phys. Rev. Lett.*, **45**, 1456.

Stevens, K.W.H. (1967) *Rep. Prog. Phys.*, **30**, 189.

Stevens, S.B. and Stapleton, H.J. (1990) *Phys. Rev. B*, **42**, 9794.

Strangeway, R.A., Mchaourab, H.S., Luglio, H.S., Froncisz, W., and Hyde, J.S. (1995) *Rev. Sci. Instrum.*, **66**, 4516.

Swanenburg, T.J.B., Poulis, N.J., and Drewes, W.J. (1963) *Physica*, **24**, 713.

Torchia, D.A. (1984) *Annu. Rev. Biophys. Bioeng.*, **13**, 125.

Tse, D. and Hartman, S.R. (1968) *Phys. Rev. Lett.*, **21**, 511.

van Vleck, J.H. (1939) *J. Chem. Phys.*, **7**, 72.

van Vleck, J.H. (1940) *Phys. Rev.*, **57**, 426.

van Vleck, J.H. (1941a) *Phys. Rev.*, **59**, 724.

van Vleck, J.H. (1941b) *Phys. Rev.*, **59**, 730.

Van de Ven, F.J.M. (1995) *Multidimensional NMR in Liquids: Basic Principles and Experimental Methods*, Wiley-VCH Verlag GmbH, New York, p. 109.

Vasavada, K.V., Schneider, D.J., and Freed, J.H. (1987) *J. Chem. Phys.*, **86**, 647.

Velter-Stefanescu, M., Grosecu, R., Ursu, I., and Nistor, S. (1986) Proc. 23rd Congress Ampère, Rome, p. 234.

Vergnoux, D., Zinsou, P.K., Zaripov, M., Ablart, G., Pescia, J., Misra, S.K., and Orlinskii, S. (1996) *Appl. Magn. Reson.*, **11**, 493.

Waller, I. (1932) *Z. Phys.*, **79**, 370.

Wenckebach, W.Th., and Poulis, N.J. (1973) Spin dynamics in paramagnetic crystals. 1 Proceedings 17th Congress Ampère (ed. V. Hovi), North-Holland Publishing, p. 120.

Wolf, E.L. (1966) *Phys. Rev.*, **142**, 555.

Yafet, Y. (1957) *Phys. Rev.*, **106**, 679.

Yafet, Y. (1963) g-Factors and Spin-Lattice Relaxation of Conduction Electrons, in *Solid State Physics*, vol. 14 (eds F. Seitz and D. Turnbull), Academic Press, New York, p. 1.–98.

Zhidomirov, G.M. and Salikhov, K.M. (1969) *Sov. Phys. JETP*, **29**, 1037.

Zinsou, P., Vergnoux, D., Ablart, G., Pescia, J., Misra, S.K., and Berger, R. (1996) *App. Mag. Reson.*, **11**, 487.

Appendix 10.I Early History of Paramagnetic Spin-Lattice Relaxation

Paramagnetic relaxation originated with the classic paper of Waller (1932), based on simple modulation of the dipolar interaction by lattice vibrations (phonons) to effect relaxation transitions, well before resonant or nonresonant methods of detection were developed. The relaxation times predicted by this mechanism for the direct process ($T_1 \propto B^{-2}T^{-1}$) and Raman process ($T_1 \propto T^{-7}$) were later found to be much greater than those observed experimentally. This was soon followed by studies of radiofrequency (RF) electronic susceptibilities, in particular those of C.J.

Gorter's group in Leiden, Holland, which were initially thought to be in rough agreement with the general form of Waller's predictions. However, it soon became clear that the actual relaxation mechanism was much stronger and more complex than the simple phonon modulation of the dipole–dipole interaction. Heitler and Teller (1936) pointed out that modulation of the crystalline electric field was the more important process for electron paramagnets. The theory for this mechanism was developed in sufficient quantitative detail by Fierz (1938), Kronig (1939), and van Vleck (1940) to warrant a direct comparison with the data obtained by Gorter's group, and the conclusion according to van Vleck was that the agreement with experimental data was "miserable." Notwithstanding this, the theory of van Vleck had great success in describing the behavior of a wide variety of paramagnetic salts. In fact, the failure in explaining the Leiden data on relaxation times measured for the direct process in titanium alum was due to the chemistry of the alum itself (Bleaney et al., 1955). The difficulties in explaining the alum data led van Vleck to propose the mechanism of the "phonon bottleneck", the heating of the phonons tuned to the motion of the spin system, which was settled subsequently by the data of Scott and Jeffries (1962), and by Brya and Wagner (1965). With the advent of resonance-detection techniques and the direct observation of transient behavior, it became possible to investigate microscopic relaxation processes. Cross-relaxation, proposed by Bloembergen et al. (1959) proved to be a very effective energy-transfer mechanism in multilevel systems that could produce significant departure from the behavior of an isolated paramagnetic ion. Many of the early anomalies in relaxation behavior encountered by the nonresonant research groups in concentrated paramagnetic salts were accounted for by the cross-relaxation mechanism. The discovery of the resonance relaxation process via close-lying intermediate states (Orbach, 1961a; Manenkov and Prokhorov, 1962) was brought about by the data on the temperature-dependence of relaxation times in multilevel systems. There was found to be a close connection between the usual static crystalline-field theory and the dynamic orbit-lattice interaction (Scott and Jeffries, 1962; Orbach, 1961a). Remarkably accurate and internally consistent values for the magnitude, temperature, and field dependence of relaxation times in a variety of iron-group and rare-earth salts had been obtained experimentally to warrant a detailed comparison of the orbit-lattice interaction with that calculated using a specific model. Details of microscopic relaxation mechanisms have been obtained by such comparisons. An excellent summary of nonresonant relaxation processes was produced by C.J. Gorter (1947). The relaxation of conduction electrons was reported by Overhauser (1953) and Yafet (1957, 1963), whereas the relaxation of impurities in semiconductors was reported by Roth (1960) and Hasagawa (1960).

11
Molecular Motions

Sushil K. Misra and Jack H. Freed

11.1
Introduction

In this chapter we will discuss the study of molecular motion by EPR, using both continuous-wave (CW) and pulsed EPR, in particular two-dimensional electron–electron double resonance (2-D-ELDOR). In recent years, the study of protein dynamics using site-directed spin labels (SDSL) by EPR has become an important subject.

The EPR spectra of a paramagnetic probe provide information on the motion of its environment. As compared to NMR, the information obtained from EPR of a spin label offers the following advantages:

- EPR is much more sensitive per spin than NMR.

- The time scale in time-domain experiments of EPR is nanoseconds, whereas it is milliseconds for NMR.

- EPR is capable of focusing on a limited number of spins, as the spin-label spectrum is simple.

- EPR spectra change dramatically with the tumbling motion of the probe, being extremely sensitive to the local fluidity; this is not the usual case in NMR, where nearly complete averaging of the spectra occurs, allowing only residual rotational effects dictated by the T_1 and T_2 relaxation times.

- Fast EPR "snapshots" can be taken with high-frequency EPR, whereas slow EPR "snapshots" are taken with low-frequency EPR. Thus, multifrequency EPR helps one to unravel the complex dynamics of biosystems occurring on different time scales.

- Pulsed EPR provides a tool to distinguish homogeneous broadening, which characterizes the dynamics of the entire environment, from inhomogeneous broadening, which displays the effect of the local structure.

The organization of the chapter is as follows. A historical introduction is provided in Section 11.2, after which some relevant experimental data are described in

Multifrequency Electron Paramagnetic Resonance, First Edition. Edited by Sushil K. Misra.
© 2011 Wiley-VCH Verlag GmbH & Co. KGaA. Published 2011 by Wiley-VCH Verlag GmbH & Co. KGaA.

Section 11.3 to illustrate the importance of multifrequency EPR in studying motion. The theory of slow motion as studied by EPR is outlined in Section 11.4; this will include discussion of: (i) application of the stochastic Liouville equation (SLE); and (ii) the slowly relaxing local structure (SRLS) model together with the related microscopic ordered-macroscopic disordered (MOMD) model to analyze multifrequency EPR data. In Section 11.5 will be discussed the application of molecular dynamics to the prediction of multifrequency EPR spectra. Some concluding remarks are made in Section 11.5, while Section 11.6 includes the details of literature pertinent to molecular motion, as studied using EPR.

11.2
Historical Background

An account of the historical development of lineshape theory is given by Freed (2005). Historically, the basic theories for magnetic resonance lineshapes mainly for the motional narrowing limit were developed, among others, by Kubo and Tomita (1954), Wangness and Bloch (1953) and Redfield (1957, 1965).

The "anomalous alternating linewidth" effect, wherein the electrochemically generated EPR spectrum of the *p*-dinitrotetramethylbenzene anion showed that well-resolved proton superhyperfine splittings (shfs) appeared on the first, third, and fifth lines of the hf splitting from the two equivalent ^{14}N nuclei, while the second and fourth lines were so broad that the proton shfs were completely masked, was resolved by the application of a more generally inclusive theory (Freed, 2005). This is the theory of EPR linewidths for organic radicals (Freed and Fraenkel, 1963), and this is still valid today for spectra in the motional-narrowing regime. This theory has been improved since then, with the incorporation of more precise and detailed models of the molecular dynamics into the formulation. An example is the incorporation of anisotropic rotational diffusion into the linewidth theory, and its illustration by reinterpreting a linewidth study on *p*-dinitrobenzene (Freed, 1964).

When Hyde and Maki (1964) first observed the electron nuclear double resonance (ENDOR) of organic radicals in liquids, there was no theory to explain why ENDOR could occur in liquids, and the reason for their successful observation was a mystery. However, a reformulation and generalized theory of EPR saturation, by analogy with the Freed and Fraenkel (1963) theory, led to a general theory of EPR saturation and double resonance (Freed, 1965). To date, this formulation – and its later extensions – serve as the basis for interpreting EPR saturation and ENDOR experiments for motionally narrowed spectra (Leniart, Conner, and Freed, 1975; Dorio and Freed, 1979; Kurreck, Kirste, and Lubitz, 1988; Möbius, Lubitz, and Freed, 1989; Möbius, Lubitz, and Plato, 1989). Using the saturation and double resonance theory, an appropriate theory for electron–electron double resonance (ELDOR) in liquids was developed by Freed (Hyde, Chien, and Freed, 1968) to explain the data obtained by Hyde and Chien. ELDOR serves, among other things, as a powerful means of studying spin relaxation for exploring rotational and translational motions in liquids.

Freed (1968) provided a formal, general formulation of the Redfield theory valid to all orders, using Kubo's method of generalized cumulant expansions in statistical mechanics (Kubo, 1962, 1963). However, in order to study spin-labeled macromolecules, which yield slow-motional EPR spectra, the most powerful method for simulating them is based on the SLE, which only requires the assumption that, for a complete analysis, the motional dynamics can be described by a (general) Markov process. The matrix representation of the SLE, and its solution, then requires computational methods. Such a theory was elucidated by Freed, Bruno, and Polnaszek (1971a) for the relevant cases of g-tensor and hyperfine anisotropy, and included saturation phenomena. The slow-motional spectra of peroxylamine-disulfonate (PADS) in ice, which did not show the substantial inhomogeneous broadening of typical spin labels, could be fitted with jump-type reorientations rather than with simple Brownian motion (Goldman et al., 1972). This was an indication that the slow-motional spectra were more sensitive to the microscopic molecular dynamics than were the fast-motional spectra. The SLE approach was also extended to a complete solution of slow-tumbling triplets (Freed, Bruno, and Polnaszek, 1971b), generalizing the findings of Norris and Weissman (1969).

Slow-motional EPR spectra were exploited during the 1970s to acquire new insights into molecular rotational motions in isotropic fluids, in liquid crystals, and in model membranes (Hwang et al., 1975; Polnaszek and Freed, 1975; Lin and Freed, 1979; Smectics, 1979). In addition, the SLE approach was exploited to provide a quantitative theory for the then new phenomena of chemically induced dynamic electron polarization (CIDEP) and chemically induced dynamic nuclear polarization (CIDNP) (Freed and Pederson, 1976). The SLE was also employed to provide a theory for Hyde's saturation-transfer technique, which was useful for studying very slow motions (Hyde and Dalton, 1979; Beth and Robinson, 1989).

These rather tedious slow-motional calculations were challenging, however, and to address this a very efficient method of computing solutions to the SLE was developed during the 1980s, namely the Lanczos algorithm (LA). This drastically reduced the computation time, by at least an order of magnitude, greatly reduced storage requirements, (Moro and Freed, 1981) and also ultimately led to the versions that could be made generally available (see below). The strength of the LA technique is that it takes full advantage of the sparsity of the SLE matrix; also after just a few Lanczos projections it produces a greatly reduced matrix sub-space that includes very effectively what is important to describe the EPR lineshapes. This is partly because the initial or "starting" vector represents essentially the EPR transition moments that include the physics of the EPR experiment. Further, the slower-decaying eigenvalues of the SLE, which dominate the EPR experiment, are obtained accurately from the small Lanczos sub-space. Unnecessary eigenvalues which are poorly represented in this small sub-space are automatically projected out of the solution when the specific EPR observable – that is, the lineshape – is calculated. Finally, a plaguing problem of the LA approach, namely the accumulation of round-off error, is overcome because the calculation is terminated before it becomes serious, as one needs only a small sub-space generated by a relatively small number of Lanczos projections. To further enhance the usage of SLE algorithms, very powerful methods have been developed for selecting the minimum basis set to represent the SLE (Schneider and Freed, 1989a, 1989b), and for determining reliably that sufficient Lanczos projections have been utilized (Schneider and Freed, 1989a, 1989b). These references provide the most effective computational algorithm to date.

11.3
High-Field Multifrequency CW-EPR Experiments to Unravel Molecular Motion

Figure 11.1 shows the experimental EPR spectra of PDT/toluene at 250 GHz in various motional regimes: motional narrowing, slow motion, and the rigid lattice limit. Figure 11.2 shows a series of simulated multifrequency spectra covering the range of 15 to 2000 GHz for a spin-bearing molecule with a rotational correlation time of 1.7 ns. This shows that a motional process that appears fast at lower

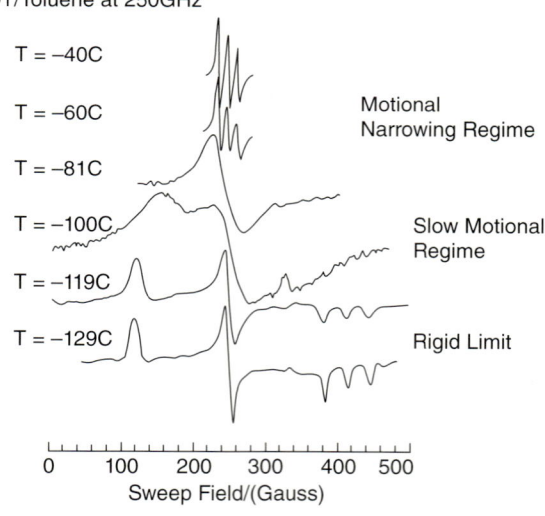

Figure 11.1 EPR spectra of PDT/toluene at 250 GHz in various motional regimes. Motional narrowing (−40 °C, −60 °C), slow motion (−81 °C, −100 °C), and rigid limit (−119 °C, −129 °C).

Figure 11.2 (a) Simulated first-derivative EPR spectra at 9.1 and 250 GHz for a dilute powder containing a cholesterol-like nitroxide (CSL; short vertical lines. The magnetic-field values where CSL absorbs when its x'-, y'-, and z'-axes are parallel to B_0 (Barnes and Freed, 1998); (b) Simulated first-derivative multifrequency EPR spectra for a nitroxide, reorienting with a rotational diffusion constant $R = 10^8 \, s^{-1}$ (corresponding to rotational correlation time $\tau_R = 1.67$ ns) in the range 15 to 2000 GHz. From this, it is clear that a motional process that appears fast at lower frequencies will appear slow at higher frequencies.

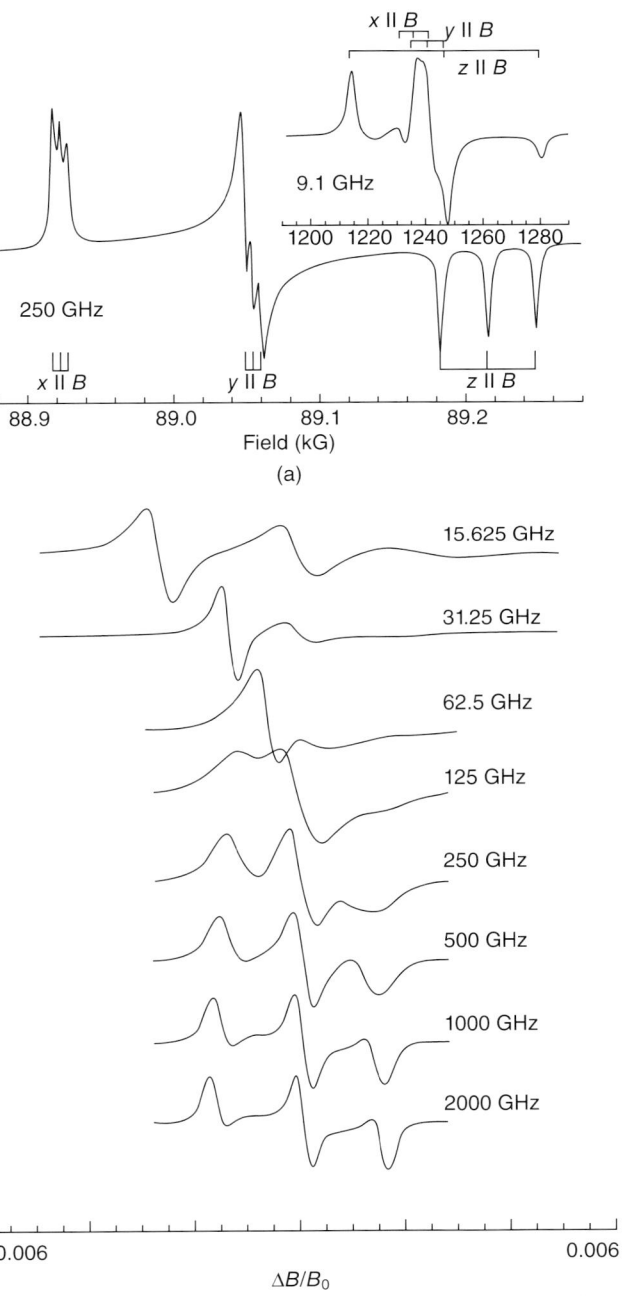

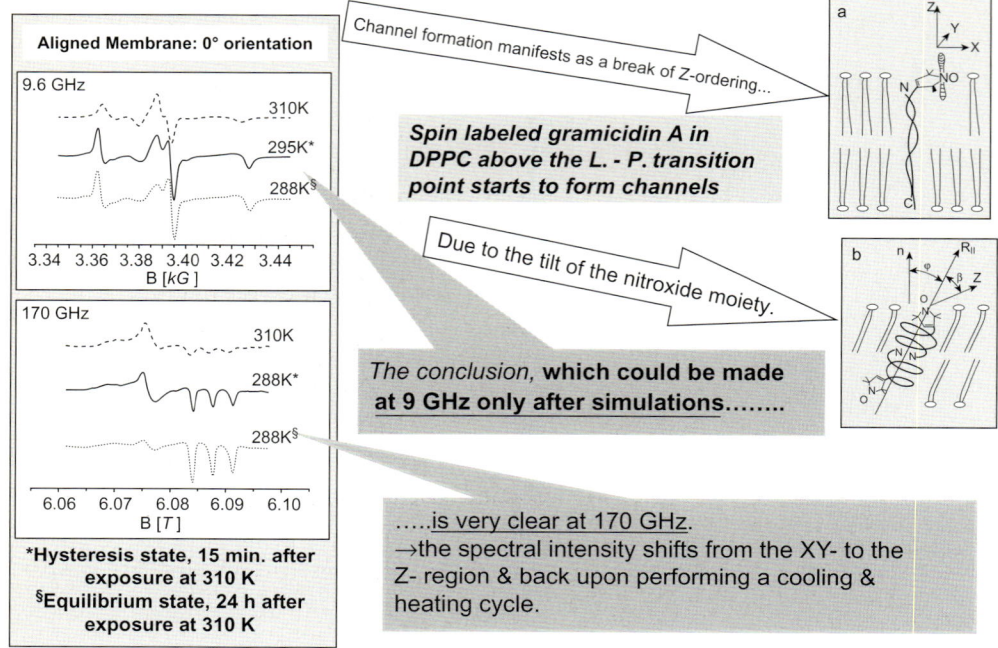

Figure 11.3 EPR spectra at 9 and 170 GHz, showing detection of the formation and dissociation of head-to-head dimers by GAsl in aligned samples of DPPC. Channels begin to form above the $L_\beta - P_\beta$ transition point in spin-labeled gramicidin A in DPPC.

frequencies will seem slow at higher frequencies. Thus, for complex systems, such as proteins or membranes, the slow overall and collective motions will be displayed better at lower frequencies, whereas the fast motions will be more sensitively demonstrated at higher frequencies. An example of monitoring the anisotropy of the motion can be seen in Figure 15.5, which shows how high-frequency (170 GHz) EPR spectra can demonstrate convincingly the anisotropy of motion for molecular rotations about the X-, Y-, and Z-molecular axes (cf. Dzikovski, et al., 2009). An example of the simple detection of a biological process in a model membrane is shown in Figure 11.3, which includes spectra at 9.6 and 170 GHz. Here, the formation and dissociation of head-to-head dimers by spin-labeled gramicidin A (GAsl) in aligned samples of DPPC is observed. Dimer channels of GAsl form above the main transition point in a lipid membrane. Channel formation in the sample manifests as a breaking of Z-ordering of monomers, due to the tilt of the nitroxide moiety upon dimer formation. The conclusion which could be made at 9.6 GHz, after detailed simulation, is very clearly evident from a visual inspection of the spectrum at 170 GHz, namely that the spectral intensity shifts from the Z- to the XY-spectral region and back upon performing a heating and cooling cycle. Another example of how multifrequency EPR distinguishes motion at different

11.3 High-Field Multifrequency CW-EPR Experiments to Unravel Molecular Motion | 503

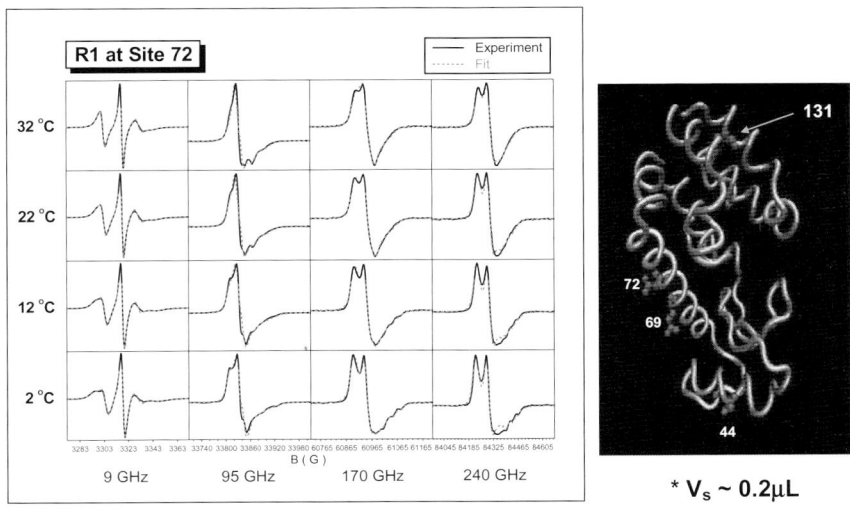

Figure 11.4 An example of how multifrequency EPR distinguishes motion at different temperatures, as exhibited by the EPR spectra of T4 lysozyme spin-labeled at mutant site 131 at 9, 95, 170, and 240 GHz at 2, 12, 22, and 32 °C.

temperatures is exhibited by the spectra in Figure 11.4 of T4 lysozyme spin-labeled at mutant site 131 at 9, 95, 170, and 240 GHz at 2, 12, 22, and 32 °C.

11.3.1
Determination of the Axes of Motion from High-Field, High-Frequency (HFHF) EPR Spectra: Orientational Resolution

Since the EPR spectra at millimeter-wave frequencies exhibit very high sensitivity to the g-tensor, the spectra provide excellent orientational resolution as compared to that achieved at conventional microwave frequencies (Budil et al., 1989; Earle, Budil, and Freed, 1993; Earle et al., 1997, 1998, and the previous paragraph). This is discussed more in detail in Chapter 15 (see Figure 15.3a and b). Therefore, at 250 GHz for example (Earle, Budil, and Freed, 1993), one can discern about which axis, or axes, the motion occurs in the CW-EPR spectrum. An excellent demonstration of the orientational resolution at 250 GHz in studies utilizing nitroxide spin labels was provided by the data on aligned membranes containing a mixture of headgroups: zwitterionic phosphatidylcholine (PC) and negatively charged phosphatidylserine (PS), using the cholesterol-like spin label (CSL) (Barnes and Freed, 1998). The macroscopic alignment of the membranes further enhanced the orientational resolution at 250 GHz, allowing for an orientation-dependent study (this topic is discussed in greater detail in Chapter 15).

11.3.2
Observation of Motion as a Function of Frequency

It is found that, the higher the EPR frequency, the slower the motion appears for a given diffusion rate, as shown in Figure 11.2. It is found that at the low-frequency end one observes simply the motionally narrowed spectra, which become slow-motional, consistent with the rigid limit (powder-like), at the high-frequency end. This is tantamount to the higher-frequency spectra acting as a faster "snapshot" of the motion (Earle, Budil, and Freed, 1993; Earle et al., 1997), due to the enhanced role of the g-tensor (Zeeman) term, linear in the magnetic field, B_0, in the spin Hamiltonian. This is explained as due to the fact that the condition for motional narrowing, $|H_1(\Omega)|^2 \tau_R^2 \ll 1$ is no longer valid, rendering the spectra slow motional; here, τ_R is the rotational relaxation time and $H_1(\Omega)$ is the orientation-dependent part of the spin Hamiltonian, with Ω referring to the Euler angles describing the molecular orientation, which increases in magnitude with increasing frequency, ω_0 and magnetic field, B_0.

11.3.3
Virtues of Multifrequency EPR in Studying Molecular Motion

By exploiting multifrequency EPR spectra, one can decompose complex modes of motions of proteins and DNA according to their different timescales (Liang and Freed, 1999). This also applies to the study of the dynamics of complex fluids, such as glass-forming fluids (Earle et al., 1997) and liquid crystals (Lou et al., 2001). This would result, for example, in "freezing-out" the slow overall tumbling motions in protein spectra at higher frequencies, leaving only the faster internal modes of motion. On the other hand, at lower frequencies one can observe clearly the motions at a slower timescale with the faster motions averaged out. In glass-forming fluids, the faster motions consist of reorientations of the probe molecules, whereas the slower motions relate to the dynamics of the solvent cage (Earle et al., 1997).

This was convincingly demonstrated in the case of proteins by Barnes et al. (1999), using the 9 and 250 GHz CW-EPR spectra of spin-labeled mutants of the stable protein T4 lysozyme in aqueous solution. At 250 GHz, the overall rotation was too slow to significantly affect the spectrum, so that it was satisfactorily described by the simpler MOMD model (Meirovitch, Nayeem, and Freed, 1984; as discussed in Section 11.3.4.2.1), because the overall motion was perceived to be so slow at 250 GHz that it corresponded to the rigid limit, and a good resolution of the internal dynamics was achieved. Using the internal motion parameters so obtained at 250 GHz, the 9 GHz lineshape data were fitted to the SRLS model (as described in Section 11.3.4.2.1) to successfully obtain the rates for the global dynamics. In this manner, with multifrequency data, the two types of motion were separated and the spectral resolution of these motions was significantly enhanced. The details of the SRLS model as applied to protein dynamics are illustrated in Figure 11.5. Very recently, an extensive multifrequency study on spin-labeled T4L covering four frequencies showed how simultaneous quantitative fits could be obtained with the SRLS model (Zhang et al., 2010).

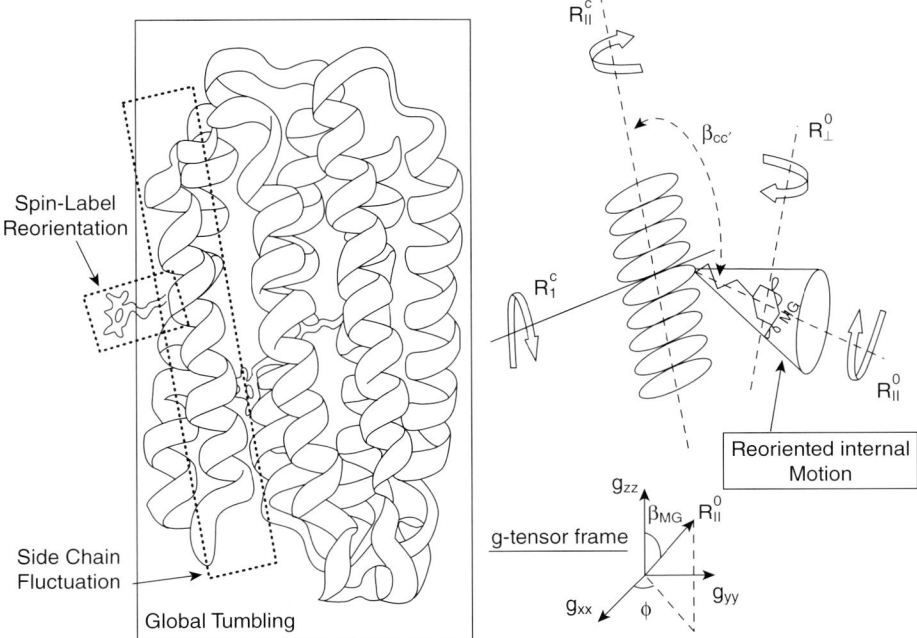

Figure 11.5 Protein dynamics of spin-labeled protein (left), showing three types of motion: spin-label reorientation; side-chain fluctuations; and global tumbling. The SRLS model is illustrated, including relevant motional parameters (Liang and Freed, 1999).

The same multifrequency approach was applied to a study of the dynamic structure of model membranes using an end-chain lipid (Lou, Ge, and Freed, 2001), where the 250 GHz data were exploited in terms of the MOMD model relating to just the internal dynamics and ordering of the ends of the acyl-chains, whereas the 9 GHz spectra are affected by both the internal and overall motions, and analyzed in terms of the SRLS model. It should be pointed out here that if the 250 GHz spectra are not taken into account, then the 9 GHz spectra, which provide only limited resolution to the dynamics, could be fitted to the simpler MOMD model. However, the dynamic and ordering parameters obtained must be interpreted as a composite of both the internal and overall motions, with no obvious way of separating them (Lou, Ge, and Freed, 2001).

11.3.4
Stochastic Liouville Equation (SLE) to Describe Slow-Motional EPR Spectra

A quantitative treatment of slow-motional EPR is accomplished by solving the SLE, using the combined spin and orientational distribution function, $\rho(\Omega,t)$, which is composed of both the spin density matrix, $\rho(t)$, and the orientational distribution function, $P(\Omega,t)$, governed by the differential equation:

$$\frac{\partial \rho(\Omega, T)}{\partial t} = -i[\hat{H}, \rho] - \hat{\Gamma}_\Omega \rho(\Omega, t), \tag{11.1}$$

where $\hat{\Gamma}$ is the operator for the rotational diffusion, and Ω again represents the Euler angles between the fluctuating molecular frame and the laboratory frame.

It should be noted that: (i) the normal density matrix is obtained by averaging $\rho(\Omega,t)$ over all Ω: $\rho(t) = \langle \rho(\Omega, t) \rangle_\Omega$; and (ii) $\rho(\Omega,t)$ reduces to $P(\Omega,t)$ when there are no electron or nuclear spins: $S = I = 0$. Schneider and Freed (1989a) describe the details of calculating slow-motional EPR lineshapes for a nitroxide radical in solution, and Misra (2007) describes similar details for an electron spin ($S = \frac{1}{2}$) coupled to two nuclei with arbitrary spins, whereas Zerbetto et al. (2007) describe details for a doubly spin-labeled system.

11.3.4.1 Calculation of Slow-Motion Spectrum

The EPR lineshape function: This is expressed as follows:

$$I(\omega) = \left\langle\!\!\left\langle v \left[i(\omega - \omega_0)\hat{I} + (\Gamma_\Omega - i\hat{H}^\times) \right]^{-1} \middle| v \right\rangle\!\!\right\rangle, \qquad (11.2)$$

where $|v\rangle\!\rangle$ is the "starting vector", which contains the information that: (i) the spectrum is an isotropic average over all orientations (except, of course for macroscopically aligned fluids, e.g. liquid crystals), which implies that only the components of $|v\rangle$ which contain $D^0_{0,0} = 1$ are nonzero; and (ii) only the components of $|v\rangle$ which correspond to $p_s = 1$; $m'_I = m_I$, (where p^s refers to the coherence order of the electron spins cf. below) excited by the radiation, are nonzero:

$$|v\rangle\!\rangle = \phi^0_{0,0}(\Omega) \times |p_s = 1, p_I = 0, m_I \rangle\!\rangle, \qquad (11.3)$$

with $m_I = +1, 0, -1$ for ^{14}N.

Finally, the spectrum is calculated to be;

$$I(\omega) = \frac{1}{\pi} \text{Re} \left\{ \sum_{j=1}^{N} \frac{C_j^2}{\lambda_j + i(\omega - \omega_0)} \right\}, \qquad (11.4)$$

where λ_j is the jth complex eigenvalue of the SL operator, and C_j is the component of the jth eigenvector along the direction of the starting vector $|v\rangle\!\rangle$. The real part of λ_j; $\text{Re}\lambda_j$ corresponds to a T_2^{-1}-like decay and $\text{Im}\lambda_j$; to the resonant frequency of the jth "mode." Equation 11.2 yields the Redfield motional narrowing result (Redfield, 1965) for fast motions, whereas it yields the rigid-limit (solid-like) powder pattern in the very slow motion limit.

The Stochastic Liouville Equation Using the superoperator notation for the spin Hamiltonian, $H^\times$, the SLE (Equation 11.1 above) can be expressed as

can be expressed as

$$\frac{\partial \rho(\Omega, T)}{\partial t} = -i[\hat{H}, \rho] - \hat{\Gamma}_\Omega \rho(\Omega, t) = [i\hat{H}^\times - \hat{\Gamma}_\Omega]\rho(\Omega, t) \qquad (11.5)$$

The last equality of Equation 11.5 defines the superoperator form, $H^\times$, by relating it to the second term of Equation 11.5. [For more details, see Schneider and Freed

(1989a,b) and Misra (2007).] The spin Hamiltonian in Equation 11.5, which consists of hyperfine and Zeeman terms for $S = 1/2$, can be expressed as:

$$\hat{H} = \sum_{\substack{sum \\ over \\ indices}} \hat{A}^{(\ell,m)}_{\mu,\ell} D^{\ell}_{mm'}(\Omega_{LG}) F^{(l,m')*}_{\mu,G}, \quad (11.6)$$

In Equation 11.6, $\hat{A}^{(\ell,m)}_{\mu,\ell}$ is the irreducible spin tensor with spin operators quantized in the laboratory (L) frame in which the z-axis is along the external magnetic field, B_0; $\hat{F}^{(\ell,m^*)*}_{\mu,G}$ are molecular functions quantized in the molecular (G) frame, which is fixed in the molecular frame; and $D^{\ell}_{m,m'}(\Omega_{LG})$ are the Wigner rotation coefficients, which effect transformations of the matrix elements between the L and G frames (for more detail, see Section 11.3.4.1.8).

Simple Anisotropic Tumbling with Axial Symmetry In the simplest model of rotational diffusion such as the motions of rods or discs, one can express the rate of change of the orientational distribution function as follows:

$$\frac{\partial P(\Omega, t)}{\partial t} = -\nabla_\Omega \cdot R \cdot \nabla_\Omega P(\Omega, t) = -\Gamma_\Omega P(\Omega, t), \quad (11.7)$$

where R is the rotational diffusion tensor, diagonal in the appropriate molecular frame. The solution of this diffusion equation has well-defined eigenfunctions, which are for the case of axial symmetry:

$$\Phi^L_{M,K}(\Omega) = \sqrt{\frac{2L+1}{8\pi^2}} D^L_{M,K}(\Omega), \quad (11.8)$$

These are normalized Wigner rotation coefficients, whose eigenvalues are given by the damping rates:

$$E^L_{M,K} = R_\perp L(L+1) + (R_\parallel - R_\perp)K^2, \quad (11.9)$$

where $R_{xx} = R_{yy} = R_\perp$ and $R_{zz} = R_\parallel$. These can be generalized to nonaxial diffusion (Freed, 1994).

Matrix Representation of the SLE Operator $(i\hat{H}^\times - \hat{\Gamma})$ First, the matrix elements of the SLE operator must be expressed in a convenient orthonormal basis set, which is in the direct-product space of the orientational and spin functions, as follows:

$$|\sigma\rangle\rangle = \Phi^L_{M,K}(\Omega) \otimes (|S, m_s\rangle\langle S, m'_s|) \otimes (|I, m_I\rangle\langle I, m'_I|), \quad (11.10)$$

For the nitroxides being considered here, $S = \frac{1}{2}$, and $I = 1$ (^{14}N) or $I = \frac{1}{2}$ (^{15}N) (with this understanding, the spin indices S and I will be dropped hereafter). For ease of considering coherence order, modified spin magnetic quantum numbers will now be used, as follows:

$$p^s = m_s - m_{s'}; \quad q^s = m_s + m_{s'} \quad (11.11)$$

Here, p^s defines the coherence order: $p^s = 0$ corresponds to the diagonal elements of the density matrix, whereas $p^s = \pm 1$ corresponds to off-diagonal matrix elements between which the radiation field induces transitions.

Basis Sets of the SL Operator: Lanczos Algorithm The basis set required to represent the SL operator is rather large, of dimension N, which requires rather exorbitant times to diagonalize the SL matrix. One can achieve order-of-magnitude and even greater reduction in computation time by employing the LA, given that the SL matrix is sparse. This is achieved by providing an objective criterion to determine when a sufficient sub-space, with dimension $n_s \ll N$, has been generated by exploiting the starting vector $|v \gg$ to select out a small sub-set of vectors, known as Lanczos vectors, which span the sub-space required to calculate the EPR lineshape. This subspace is then projected out, and the SL matrix is converted to tri-diagonal form, which is easily diagonalized, or else solved by the method of continued fractions (Schneider and Freed, 1989a,b). In this manner, a greatly reduced number of multiplications is required. In a modified form, the LA can be used to provide an objective method to prune the original basis set to go from N to a minimum N_{min} needed to represent the relevant eigenvectors.

Diffusion in Anisotropic Media In an anisotropic medium, such as liquid crystals or membranes (or in the presence of side-chain motion in proteins), the orientational distribution of the spin probe is not isotropic. In that case, its equilibrium distribution, $P_{eq}(\Omega)$, can be derived from an orientational potential energy, $U(\Omega)$, which is the potential of mean torque experienced by it:

$$P_{eq}(\Omega) = \frac{\exp[-U(\Omega)/k_B T]}{\int d\Omega \exp[-U(\Omega)/k_B T]}, \tag{11.12}$$

where k_B is Boltzmann's constant and T is the temperature.

The diffusion operator then becomes:

$$\hat{\Gamma}_\Omega = \nabla_\Omega \cdot R \cdot \left[\nabla_\Omega + \frac{1}{k_B T}(\nabla_\Omega \cdot U) \right] P(\Omega, t). \tag{11.13}$$

Equation 11.13 is known as a Smoluchowski equation. It has the property that for any initial $P(\Omega, 0)$, $\lim_{t \to \infty} P(\Omega, t) = P_{eq}(\Omega)$; in other words, $P_{eq}(\Omega)$ is an eigenfunction of $\hat{\Gamma}_\Omega$ with the zero eigenvalue. $\hat{\Gamma}_\Omega$, as given by Equation 11.12, is nonsymmetric but can be converted into the symmetric form by the following transformation:

$$\tilde{\Gamma}_\Omega = P_{eq}^{-1/2}(\Omega) \hat{\Gamma}_\Omega(\Omega) P_{eq}^{1/2}(\Omega), \tag{11.14}$$

which yields:

$$\tilde{\Gamma}_\Omega = [\nabla_\Omega - (\nabla_\Omega U)/2k_B T] \cdot R \cdot [\nabla_\Omega + (\nabla_\Omega U)/2k_B T]. \tag{11.15}$$

The diffusion Equation 11.7 may be solved for $\tilde{P}(\Omega, t) = P_{eq}^{-1/2}(\Omega) P(\Omega, t)$. The symmetric matrix $\tilde{\Gamma}_\Omega$ can be diagonalized after calculating its matrix elements explicitly in the basis formed by the functions $\Phi_{K,M}^L(\Omega)$. The new SL operator becomes:

$$i\hat{H}^{\times} - \tilde{\Gamma}_{\Omega}, \tag{11.16}$$

for which the new starting vector is:

$$|\tilde{v}\rangle\rangle = P_{eq}^{1/2}(\Omega)|v\rangle\rangle, \tag{11.17}$$

Finally, the slow-motional EPR lineshape is given by

$$I(\omega) = \pi^{-1}\langle\langle\tilde{v}|[i(\omega - \omega_0)\hat{I} + (\tilde{\Gamma}_{\Omega} - i\hat{H}^{\times})]^{-1}|\tilde{v}\rangle\rangle, \tag{11.18}$$

where $H^{\times}$ is defined in Equation 11.2. Equation 11.18 is solved using the same procedure as before, by diagonalization of this SL operator and using the original basis set. For more details see Schneider and Freed (1989a,b) and Misra (2007).

The Potential Function, $U(\Omega)$, and the Ordering Tensor S The potential energy operator, $U(\Omega)$, can be expanded in terms of the Wigner rotation matrix elements $D_{M,K}^L(\Omega)$ as follows:

$$-U(\Omega)/k_B T = \sum_{L,M,K} c_{M,K}^L D_{M,K}^L(\Omega). \tag{11.19}$$

The resulting ordering tensor elements can be obtained by using $P_{eq}(\Omega)$ as follows:

$$\begin{aligned} S_0 &= \langle D_{0,0}^2(\Omega)\rangle = \int d\Omega\, P_{eq}(\Omega) D_{0,0}^2(\Omega); \\ S_2 &= \langle D_{0,2}^2(\Omega) + D_{0,-2}^2(\Omega)\rangle. \end{aligned} \tag{11.20}$$

It should be noted that only S_0 and S_2 are utilized in most cases of interest here.

The Spin Hamiltonian Operator This can be expressed as:

$$\hat{H} = \sum_{\mu=g,A}\sum_{l=0,2}\sum_{m=-\ell}^{\ell}\sum_{m'=-\ell}^{\ell}\sum_{m''=-\ell}^{\ell} A_{\mu,\ell}^{(\ell,m)} D_{mm'}^{\ell}(\Omega_{LM}) D_{m'm''}^{\ell}(\Omega_{MG}) F_{\mu,G}^{(l,m'')*}. \tag{11.21}$$

In Equation 11.21, the tensor elements $\hat{A}_{\mu,\ell}^{(\ell,m)}$ are constituted by the external magnetic field B_0, and the spin operators **S**, **I**, **S·I**, expressed in the laboratory frame, whereas the tensor elements $F_{\mu,G}^{(\ell,m'')}$ are constituted by the diagonal matrix elements $g_{xx}, g_{yy}, g_{zz}, A_{xx}, A_{yy}, A_{zz}$ of the g- and A-matrices expressed in the g-matrix frame. The reference frames: L, M, and G in Equation 11.20 are defined next.

Reference Frames Used in the MOMD Model to Define the Orientation of a Spin Probe to Study its Structural and Dynamic Properties The various frames used for this purpose are defined as follows, and illustrated in Figure 11.6:

- **Lab frame (L):** This is defined with respect to the external magnetic field, whose direction is used as its z-axis.

- **Director frame (D):** The director, $\hat{n}$, parallel to the membrane normal, defines this frame, which is tilted relative to the magnetic field by the angle ψ, and is obtained by transformation by the set of Euler angles $\Psi_{L\rightarrow D}$ from LF to DF.

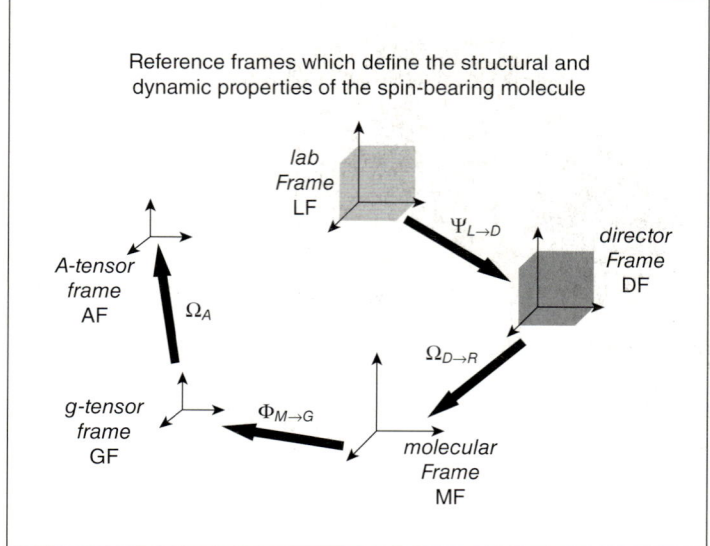

Figure 11.6 Reference frames used to define the orientation of a sample to study its structural and dynamic properties. (i) Lab frame (LF), defined with respect to the external magnetic field, whose direction is used as its z-axis; (ii) Director frame (DF), defined by the director, $\hat{n}$, of the membrane, tilted relative to the magnetic field by the angle ψ, and obtained by the transformation by the set of Euler angles $\Psi_{L \to D}$ from LF to DF; (iii) Molecular frame (MF), fixed within the molecule; (iv) g-tensor frame (GF), the principal-axes frame of the g-tensor of the paramagnetic ion, and is obtained by the transformation by the set of Euler angles $\Phi_{M \to G}$ from MF to GF; (v) A-tensor frame, defined by the principal-axes of the A-tensor, obtained by the transformation by the set of Euler angles Ω_A from GF to AF.

- **Molecular frame (M):** This is defined by the principal axes of the molecular diffusion tensor (or molecular ordering frame), and is fixed within the molecule.
- **g-Tensor frame (G):** This is the principal-axis frame of the g-matrix, and is obtained by transformation by the set of Euler angles $\Phi_{M \to G}$ from MF to GF.
- **A-tensor frame:** This is defined by the principal-axes of the A-tensor, and is obtained by the transformation by the set of Euler angles Ω_A from GF to AF.

In order to define the orientation of the spin-label, the typical molecular magnetic tensor in irreducible tensor notation is transformed from the MF to LF frame as follows:

$$F^{(2,m)*}_{\mu,\ell} = \sum_{m',m'',m'''} F^{(2,m''')*}_{\mu,m} D^2_{m,m'}(\Psi_{L \to D}) D^2_{m',m''}(\Omega_{D \to M}) D^2_{m'',m'''}(\Phi_{M \to G}), \qquad (11.22)$$

where $\Psi_{L \to D} = (0, \psi, 0)$ is a sufficient set of transformation Euler angles, from LF to DF.

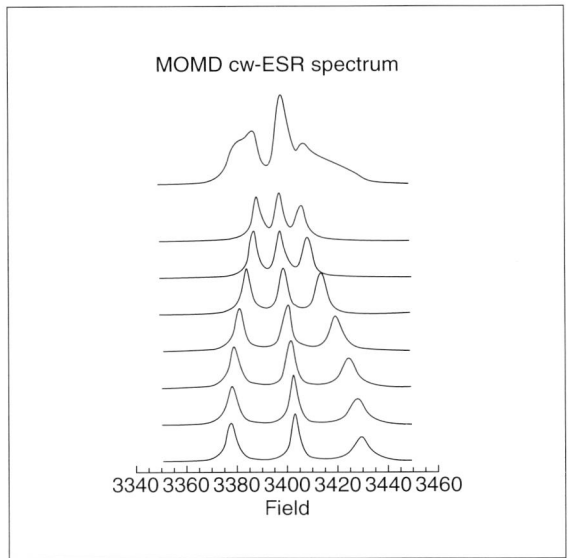

Figure 11.7 X-band CW-EPR spectra for the NO radical, shown as a function of the director tilt (ψ) in the second and subsequent plots from the top in descending order, for $\psi = 90°$, 75°, 60°, 45°, 30°, 15°, 0°, respectively, along with its MOMD spectrum displayed at the top.

Figure 11.7 shows a simulated X-band CW-EPR spectrum for the nitroxide radical as a function of the director tilt (ψ), along with its MOMD spectrum (as discussed in Section 11.3.4.2.2).

11.3.4.2 MOMD and SRLS Models

Slow-Motional EPR Lineshape This is calculated using Equation 11.1.

MOMD This is applicable, for example, to membrane vesicles or a very slowly tumbling protein with internal (side-chain) motion (Meirovitch, Nayeem, and Freed, 1984). In order to take into account random macroscopic disorder, for the case when there exists microscopic order, one should take an average of the spectra from all orientations, ψ, which define the transformation angles $\Psi_{L \to D}$ that appear in Equation 11.23, to obtain the composite MOMD spectrum, as follows:

$$I(\omega) = \int I(\omega, \psi) \sin \psi \, d\psi \tag{11.23}$$

By definition, this spectrum is inhomogeneously broadened, but it happens in a characteristic manner, which depends on the ordering potential, or equivalently upon the ordering tensor S – for example, that given by Equation 11.21.

Diffusion Operator This operator used in the SLE, as described by Equations 11.7 and 11.13, based on the assumption of over-damped motions inherent in

Smoluchowski equations as well as axially symmetric diffusion, is expressed as follows (see Schneider and Freed, 1989a; Misra, 2007 for more details):

$$\hat{\Gamma}_\Omega = \hat{\Gamma}(\Omega_{LM}, \Omega_{LC}) = \hat{\Gamma}^o(\Omega_{LM}) + \hat{F}^c(\Omega_{LC}) + \hat{F}^o(-\Omega_{C'M}) + \hat{F}^c(-\Omega_{C'M}), \quad (11.24)$$

where,

$$\hat{\Gamma}^o(\Omega_{LM}) = R_\perp^o \hat{J}^{o2} + (R_\parallel^o - R_\perp^o)\hat{J}_z^{o2}; \quad (11.24a)$$

$$\hat{\Gamma}^c(\Omega_{LC}) = R_\perp^c \hat{J}^{c2} + (R_\parallel^c - R_\perp^c)\hat{J}_z^{c2}; \quad (11.24b)$$

$$F^o(-\Omega_{C'M}) = \frac{1}{2}\left[R_\perp^o \hat{J}^{o2} u(\Omega_{C'M}) + (R_\parallel^o - R_\perp^o)\hat{J}_z^{o2} u(\Omega_{C'M}))\right]$$
$$- \frac{1}{4}\left[R_\perp^o \hat{J}_+^o u(\Omega_{C'M})\hat{J}_-^o u(\Omega_{C'M}) + R_\parallel^o (\hat{J}_z^o{}^2 u(\Omega_{C'M}))^2\right]; \quad (11.24c)$$

$$F^c(-\Omega_{C'M}) = \frac{1}{2}\left[R_\perp^c \hat{J}^{c2} u(\Omega_{C'M}) + (R_\parallel^c - R_\perp^c)\hat{J}_z^{c2} u(\Omega_{C'M}))\right]$$
$$- \frac{1}{4}\left[R_\perp^c \hat{J}_+^c u(\Omega_{C'M})\hat{J}_-^c u(\Omega_{C'M}) + R_\parallel^c (\hat{J}_z^{c2} u(\Omega_{C'M}))^2\right]. \quad (11.24d)$$

Equations 11.24 implies $\hat{\Gamma}_\Omega$ depends on several quantities, as follows:

$$\hat{\Gamma}_\Omega = \hat{\Gamma}(R_\perp^o, R_\parallel^o, R_\perp^c, R_\parallel^c, c_0^2, c_2^2).$$

Internal-Ordering Potential The explicit form of this potential, to be used in Equations 11.24c and d, is

$$u(\Omega_{C'M}) \equiv -U(\Omega_{C'M})/k_B T$$
$$= c_0^2 D_{00}^2(\Omega_{C'M}) + c_2^2 \left[D_{02}^2(\Omega_{C'M}) + D_{0-2}^2(\Omega_{C'M})\right].$$

<u>**Slowly Relaxing Local Structure (SRLS) Model**</u> With the enhanced resolution offered by the HF-HF EPR, more sophisticated models of molecular reorientation have been proposed to fit these EPR spectra. Now, the many-body problem of dealing with the microscopic details of fluids is approximated by a set of collective degrees of freedom which represent the main effects of the solvent on a rotating solute. These collective variables are modeled as a loose solvent "cage," which is considered to be relaxing slowly and within which the solute is assumed to be reorienting more rapidly. This so-called slowly relaxing local structure (SRLS) is obtained by generalizing the MOMD model by allowing the Euler angles $\Psi_{L \to D}$ to fluctuate in time due to some slow overall process; this may be a slow tumbling of the vesicle or an overall rotation of the protein. The enhanced sensitivity of very high frequency (VHF) EPR to rotational dynamics was exploited successfully to explore the details of the dynamic solvent cage in a 250 GHz EPR study of the dynamics of several nitroxide spin probes dissolved in the glass-forming solvent *ortho*-terphenyl (OTP) (Earle et al., 1997). As shown in Figure 11.8, the SRLS model adequately fits the model-sensitive regions of the 250 GHz spectra, leading to a coherent picture of the dynamics. The rotational diffusion tensors of the various probes appropriately conform to the simple expectation that the diffusion constant becomes larger as the probe becomes smaller. The relaxation rate of the cage is

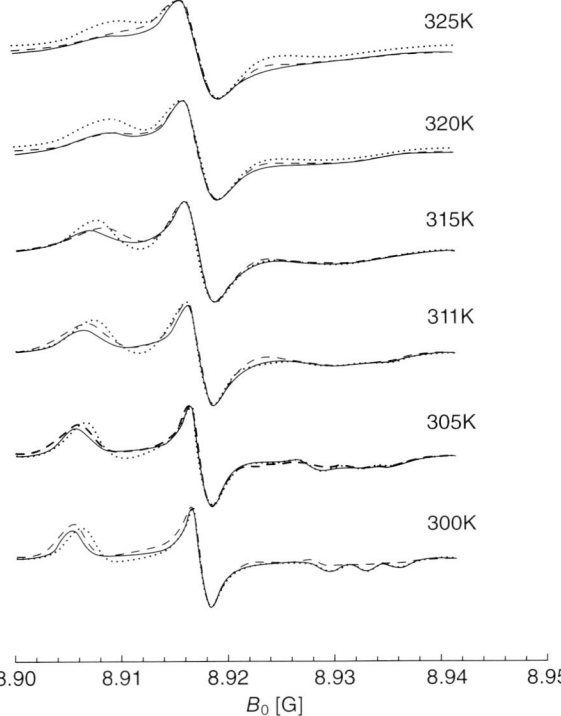

Figure 11.8 Comparison of two models for fitting effects of rotational diffusion on 250 GHz EPR spectra of spin probe of a cholesterol-like nitroxide (CSL) in ortho- terphenyl solvent (solid line) experiment. The dashed line indicates the SRLS model, and the dashed-dotted line simple Brownian diffusion (Earle et al., 1997).

found to be the slowest, and independent of the particular probe, consistent with the fact that the cage relaxation involves primarily the movement of the OTP solvent molecules. Further, it was possible to estimate the magnitude and directionality of the cage-orienting potential. In addition, the dynamics affected the slow-motional EPR spectra in a nonlinear manner. This enables one to discern between two limiting cases: (i) a homogeneous liquid characterized by a complex motional dynamics, such as described by the SRLS model; and (ii) an inhomogeneous liquid characterized by a distribution of simple relaxation times, for example, Brownian tumbling, with (ii) shown to be incompatible with the 250 GHz spectra.

The SLE remains valid in this augmented SRLS model, since the combined system of solute plus cage is represented by collective Markovian equations. The Lanczos projections then effectively determine the extent to which the cage variables are needed to analyze the EPR spectrum (Polimeno and Freed, 1995). The various details applicable to the consideration of a membrane vesicle are shown in Figure 11.5, together with details of the various rotational motions, and restricted internal motion, as well as the orientation of the g-tensor frame. It should be noted that in the limit when $R_\parallel^c$ and $R_\perp^c \to 0$, the SRLS model becomes the MOMD model.

11 Molecular Motions

The coordinate systems used in considering the SRLS model are shown in Figure 11.6. These are: (i) Laboratory (LF); (ii) Global diffusion (CF), obtained via transformation by the set of Euler angles $\Omega_{LC}(t)$ from LF; (iii) Internal director (C'F), obtained by transformation by the set of Euler angles $\Omega_{CC'}$ from CF; (iv) Internal diffusion (MF), obtained by transformation by the set of Euler angles $\Omega_{C'M}(t)$ from C'F; and (v) Magnetic-tensor frame (GF), obtained by transformation by the set of Euler angles Ω_{MG} from MF.

According to an earlier version of the SRLS model (Freed, 1977; Meirovitch et al., 2010 review), the spectral density is derived to be:

$$J_{K,M}(\omega) = \frac{\kappa(K,M)\tau_R}{1+\omega^2\tau_R^2} + \frac{1}{5}[5\kappa(0,M)]^2 \delta_{K,0}\langle|S_t|^2\rangle \times \left[\frac{\tau_x}{1+\omega^2\tau_x^2} - \frac{\tau'_R}{1+\omega^2\tau'^2_R}\right],$$

where $\tau'^{-2}_R = \tau_R^{-1} + \tau_x^{-1} \cong \tau_R^{-1} \gg \tau_x^{-1}$. For an isotropic medium $\kappa(K,M) = 1/5$, which is equivalent to what is known as the model-free form in NMR (Meirovitch et al., 2010 review).

SRLS lineshape function: The dependencies of this are expressed as follows:

$$I(\omega)_{SRLS} = I(R^o_\perp, R^o_\parallel, R^c_\perp, R^c_\parallel, c^2_0, c^2_2, \beta_{MG}, \omega).$$

11.4
Pulsed EPR Study of Molecular Motion

A major drawback of CW-EPR for relaxation studies is its inability to extract homogeneous line broadening reliably from inhomogeneously broadened EPR spectra, such as those obtained with nitroxide spin labels. This homogeneous line broadening is due to the motional modulation of the hyperfine and g-tensors, as well as that from the other spin-relaxation processes. The inhomogeneous broadening, which is typically due to the undesirable effects of unresolved proton superhyperfine splitting and local ordering in the MOMD model, obscures the homogeneous line broadening. However, by using pulsed EPR – specifically electron spin echo (ESE) – the inhomogeneous broadening can be canceled and the homogeneous linewidths, i.e. the inverse of T_2, determined. ESE spectrometers of this type have been constructed by Stillman et al. (1980) and in much refined form by Borbat, Crepeau, and Freed (1997) and reviewed by Freed (2000).

It emerges that there occurs a homogeneous T_2 minimum as a function of temperature, an example being that observed by Brown (1974), Stillman, Schwartz, and Freed (1980) and Millhauser and Freed (1984). This occurs because T_2 depends differently on the rotational correlation time ($T_R = 1/6R$), for fast and slow motions. Specifically, for fast motion, T_2 exhibits the well-known inverse dependence on correlation time, whereas for slow motion the homogeneous T_2 depends on the correlation time to a positive (usually fractional) power. A detailed explanation of this in terms of the SLE was provided by Schwartz, Stillman, and Freed (1982). In the slow-motional regime, the T_2 is affected in two limiting cases as follows. First, in the strong jump reorientation limit, each jump causes a large change in the resonant frequency, which leads to an uncertainty in lifetime broadening, and T_2 becomes equal to the correlation time (Mason and Freed, 1974). On the other hand, in the limit of simple Brownian motion (the infinitesimal jump limit), T_2 is roughly proportional to the square-root of the correlation time, as interpreted by Kivelson and Lee (1982) in an heuristic manner.

11.4.1
T_2-Type Field-Swept 2D ESE

It is obviously not possible to extract much information on motional dynamics from a single value of T_2. In the fast-motional regime one observes different T_2's for the different hf lines. In the slow-motional regime, one can study the variation of T_2 across the spectrum to obtain information on motional models. This would be superior to studying the CW lineshape of a slow-motional EPR experiment, in that the T_2 relates solely to the dynamic processes. This advantage is dispelled in the regime of very slow motions, however, where "solid-state" relaxation processes, such as spin diffusion, become dominant in T_2. In one approach, the homogeneous T_2 can be measured by using pulsed EPR, wherein the magnetic field is swept and the spin echo is collected from weak, i.e. highly selective, microwave pulses (Millhauser and Freed, 1984). The Fourier transform of these signals in the echo-decay time, τ, provides a 2-D spectrum in which the homogeneous lineshape appears along the frequency axis, while the EPR lineshape essentially appears along the field axis. This is shown for tempone in 85% glycerol/H_2O at $-75\,°C$ in Figure 11.9. Thus, the homogeneous T_2 variation across the spectrum can be studied, and explained quite successfully by a Brownian reorientational model. Subsequently, it was found that the patterns of T_2 variation across the spectrum, when plotted in a normalized contour fashion, could be used to distinguish the model of rotation and also the degree of rotational anisotropy (see Figure 11.9). This technique found further application to spin labels in oriented model membranes and to labeled proteins, as well as to slow motions on surfaces (Millhauser et al., 1987; Freed, 1987).

11.4.2
Magnetization Transfer by Field-Swept 2-D-ESE

This technique, which can be used to determine the magnetization transfer rates across the EPR spectrum, is performed in the same manner as a T_2-type 2-D-ESE experiment, but where a stimulated echo sequence $\pi/2-\pi/2-\pi/2$ replaces the echo sequence $\pi/2-\pi$, and the time T is stepped out between the second and third pulses (Schwartz, Millhauser, and Freed, 1986). The variation with T is governed by two exponential decays according to the approximate theory; one is in T_1, and the other is in T_A, an effective magnetization transfer time (for $T_A \ll T_1$). The spin-bearing molecules irradiated by the first two $\pi/2$ pulses are shifted by slow-rotational reorientations to frequencies outside the irradiated region, and are therefore not detected by the third $\pi/2$ pulse. Thus, this magnetization transfer process leads to

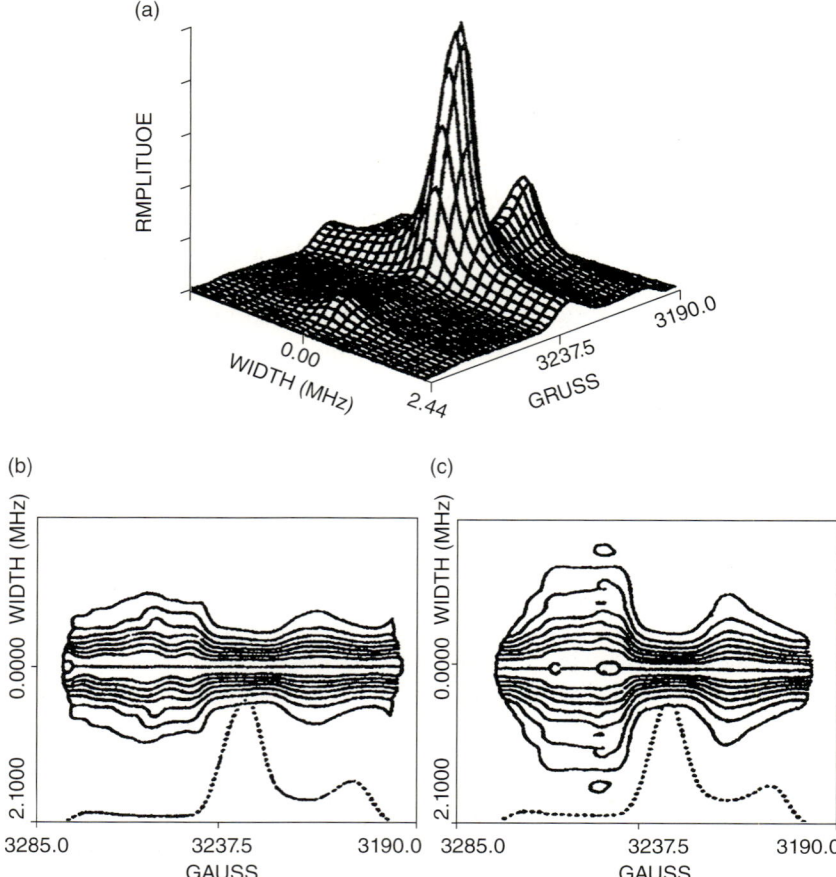

Figure 11.9 (a) The 2-D-ESE spectrum of tempone in 85% glycerol/H$_2$O at −75 °C. The slices along the width axis provide the homogeneous lineshape for the different magnetic field positions of the EPR spectrum; (b) Normalized contours for panel (a) as well as the spectral slice from panel (a) taken along the width = 0 axis; (c) The analogous contours for cholesterol in n-butylbenzene at −135 °C. These show the different contour patterns from the near-spherical tempone compared to that from the cigar-shaped cholestane (Millhauser and Freed, 1984).

a more rapid decay of the stimulated echo as a function of T. A Brownian rotation model also predicts a T_A variation across the spectrum. Dramatic variations of T_A across the spectrum for NO$_2$ adsorbed onto crushed vycor, attributed to very anisotropic rotational motion on the surface (Freed, 1987). There was found an enhanced T_A for the spectral regimes corresponding to x- and z-molecular axes being parallel to the magnetic field, implying more rapid rotation about the y-axis, which is parallel to the line connecting to the two oxygen atoms. This motional anisotropy was clearly visible from the 2-D contours, without requiring further detailed analysis.

11.4.3
Stepped-Field Spin-Echo ELDOR

This is a more informative method of studying magnetization transfer. In this technique, an alternative to using two microwave frequencies, the magnetic field is stepped out during the time between the first inverting π pulse and the detecting $\pi/2$–π spin-echo sequence (Hornak and Freed, 1983; Dzuba et al., 1984). The comprehensive theory of spin relaxation in ESE for fast and slow motions includes both longitudinal and cross-relaxation in liquids (Schwartz, 1984; Schwartz, Millhauser, and Freed, 1986). In ELDOR, one observes the transitions out of a certain spectral region, and the spectral region to which the transition is made.

11.4.4
2-D Fourier Transform EPR

Two-dimensional (2-D) NMR was first developed by Richard Ernst and coworkers in 1976 (Aue and Ernst, 1976), and this led in 1979 to the study of magnetization transfer. In 2D NMR one uses nonselective radiofrequency (rf) pulses to successfully irradiate the entire spectrum and to collect the data shortly after pulse application. This process introduced coherences simultaneously to all spectral components, and enabled the observation of coherence transfer between these components. Ernst and Jeener subsequently showed how magnetization transfer could also be studied in this manner (Jeener et al., 1979). Nonetheless, it took another ten years for 2-D-EPR to incorporate these ideas (Gorcester and Freed, 1986), for the simple reason that the EPR experiment is much more difficult to carry out. In the case of EPR, microwaves are used rather than rf used in NMR. As the EPR relaxation times are orders of magnitudes faster, pulse widths are required that are orders of magnitude shorter, and the spectral bandwidths that must be orders of magnitude wider. Consequently, it proved necessary to develop modern FT techniques in EPR as a requirement for developing 2-D-EPR. Modern FT-ESR appeared at Bowman's laboratory in Argonne (Angerhofer, Massoth, and Bowman, 1988), at Dinse's laboratory in Dortmund, Germany (Dobbert, Prisner, and Dinse, 1986), at Lebdev's laboratory in Moscow (Panferov et al., 1984), and in Freed's laboratory at Cornell (Eliav et al., 1984; Gorcester and Freed, 1986; Freed, 1986). The 2-D FT-EPR experiments conducted at Cornell consisted of a 2-D-ESE experiment, appropriately called spin-echo-correlated spectroscopy (SECSY) which utilizes two $\pi/2$ pulses, and a free induction decay (FID)-based 2-D-exchange experiment which utilizes three $\pi/2$ pulses, now referred to as 2-D-ELDOR (Gorcester and Freed, 1986). With SECSY, it was possible to obtain homogeneous T_2^{-1} values from the whole spectrum simultaneously from an (inhomogeneously broadened) EPR signal. In contrast, the (first) FT-based 2-D-ELDOR experiment exhibited crosspeak development that had resulted from a Heisenberg spin-exchange. To make the technique of 2-D-FT-EPR generally applicable, sophisticated phase-cycling was introduced on the technical side, whilst on the theoretical side a full analysis was developed for the fast-motional 2-D spectra, taking into account the generation of cross-peaks by the Heisenberg exchange (HE) and electron-nuclear dipolar (END) terms. Additional studies involved how to distinguish between the respective contributions to enable quantitative measurements could be made of HE in an anisotropic fluid (Gorcester and Freed, 1988), and of END terms in a liquid crystal (Gorcester, Ranavare, and Freed, 1989). The measurement of END terms led to sophisticated insights being acquired into molecular dynamics in ordered fluids, that could not be obtained with CW-EPR. The measurement of rates of chemical exchange in a semi-quinone system was also demonstrated, by using 2-D-ELDOR (Angerhofer, Massoth, and Bowman, 1988).

Subsequently, 2-D-FT-EPR was further developed to address the slow-motion regime by introducing further improvements. This was accomplished by increasing the spectral coverage to 250 MHz, enhancing the data-acquisition rates, significantly reducing the spectrometer dead times (Patyal et al., 1990a in the case of SECSY:ESR, and Patyal et al., 1990b in the case of 2-D-ELDOR), and developing the general theory for the quantitative analysis of 2-D spectra (Lee, Budil, and Freed, 1994b). Complex fluids could then be studied in detail, including phospholipid membrane vesicles (Lee et al., 1994a; Crepeau et al., 1994), liquid crystalline solutions (Sastry et al., 1996a, 1996b), and liquid-crystalline polymers (Xu et al., 1996). The dead times here were reduced to approximately 50–60 ns. Simultaneous fits of 2-D-ELDOR data at several mixing times, T_m, provided a third dimension in that one could monitor how the cross-peaks would grow in relation to the auto peaks with increasing mixing time, as shown in Figure 11.10 (Costa-Filho, Shimoyama, and Freed, 2003a) for a liquid-crystalline phase of lipid vesicles, and in Figure 11.11 (Costa-Filho, Shimoyama, and Freed, 2003b) for the effects of the peptide gramicidin A of lipid vesicles. This information provides quantitative information on the nuclear spin-flip-inducing processes of both HE, which are related to translational diffusion, and the intramolecular electron–nuclear dipolar interaction, which is related to tumbling motions. In those studies further technical improvements had brought the dead-times down to 25–30 ns. The more recent work of Chiang et al. (2007) on lipid-cholesterol mixtures of varying compositions and temperatures is a tour de force that illustrates the great power of 2D-ELDOR in extracting *all* the available dynamical information for complex systems. It was aided by an improved method of 2D-ELDOR data analysis called the "full S_{c-}" method (Chiang and Freed, 2006).

11.4.4.1 Lineshapes of the Auto and Cross-Peaks: Homogeneous (HB) and Inhomogeneous Broadening (IB)

Two types of line shape can be obtained from the correlation spectroscopy (COSY) and 2-D-ELDOR data. Depending on the coherence pathway, one can obtain either the FID-like signal (S_{c+}), which is sometimes referred to as the "anti-echo," or the echo-like (S_{c-}) signal, wherein there is refocusing of the inhomogeneous broadening terms in the spin Hamiltonian, which leads to their cancellation in the echo that is formed. The echo-like S_{c-} 2-D signal can be Fourier transformed and rearranged to obtain the homogeneous broadening along one frequency dimension, and essentially the CW spectrum along the other frequency dimen-

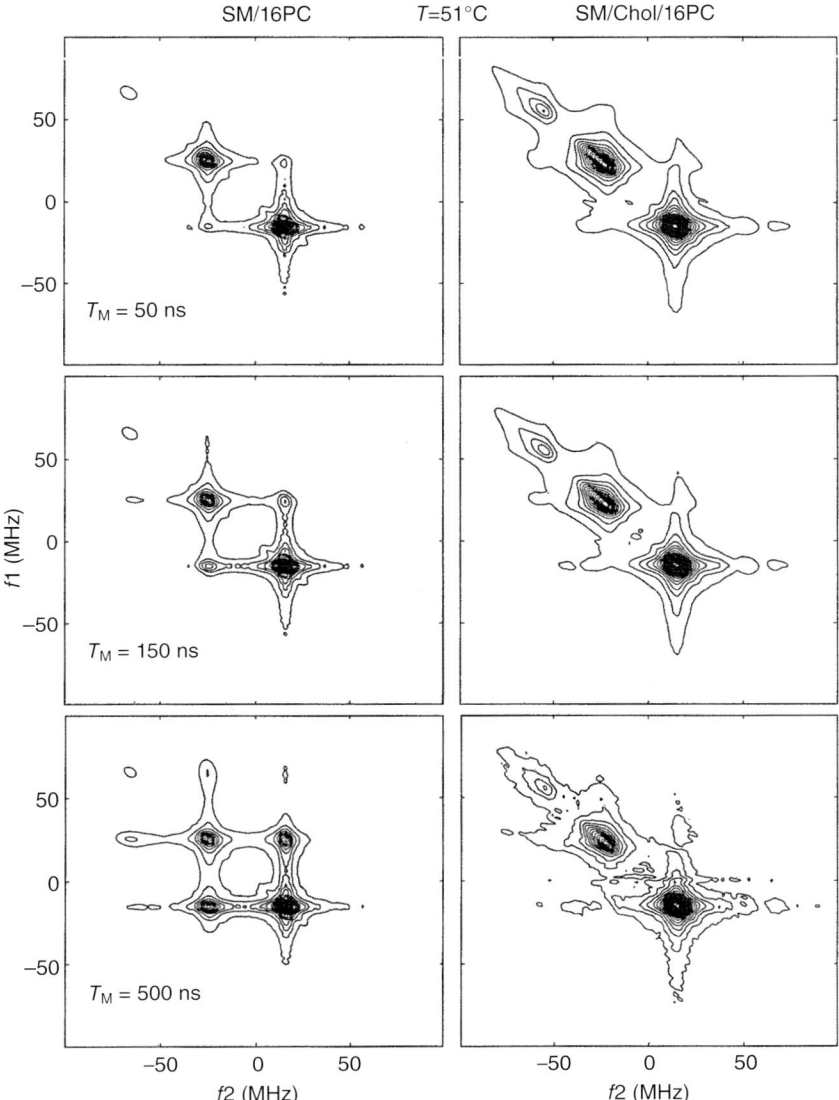

Figure 11.10 2-D-ELDOR signals at 17.3 GHz versus mixing time, T_m, of 16-PC in liquid-crystalline phase from pure lipid vesicles (left column) compared with 16 PC in liquid-ordered phase from 1:1 ratio lipid to cholesterol (right column) at 50 °C (Costa-Filho, Shimoyama, and Freed, 2003a).

sion. In this way, one obtains the 2-D-SECSY format from the COSY format by this transformation. In the case of 2-D-ELDOR, the same transformation provides the HB for the auto peaks in the 2-D-ELDOR S_{c-} spectrum, whereas the cross-peaks are affected by any differences in the IB that exist between the two spectral lines

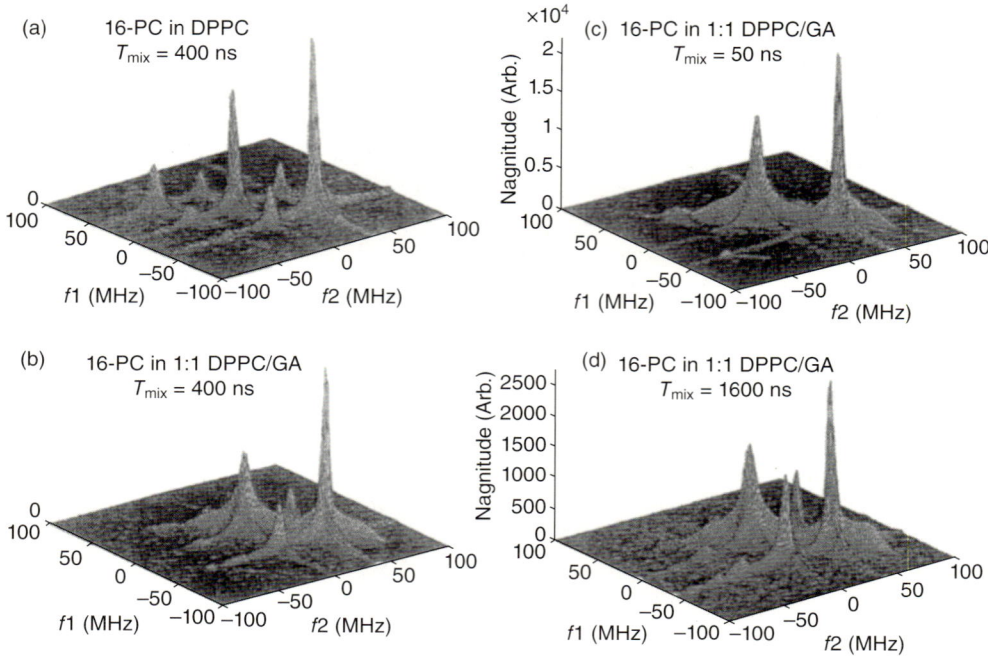

Figure 11.11 2-D-ELDOR signals at 17.3 G, showing the effect of peptide gramicidin A (GA) on the dynamic structure of a lipid membrane containing (end chain) nitroxide-labeled lipid (16-PC) at 75 °C. (a) Pure lipid, mixing time, $T_m = 400$ ns; (b–d) 1:1 lipid to GA with $T_m = 400$ ns, 50 ns, and 1.6 μs, respectively (Costa-Filho et al., 2003b).

connected by the particular cross-peak. The information so obtained from the auto and cross-peaks can be further exploited to study spin relaxation in detail. The S_{c-} format is particularly useful for studying ultra-slow motions, e.g. macromolecules in viscous media (Saxena and Freed, 1997). On the other hand, the FID-like S_{c+} 2-D spectra contain the full effects of inhomogeneous broadening.

11.4.5
MOMD and SRLS Models and 2-D-ELDOR

From the above discussion, it is clear that the dynamics and structure of complex fluids can be studied in great detail by exploiting the S_{c-} 2-D-ELDOR spectra. This enables one to study the microscopic alignment in lipid membranes, which gives rise to a superposition of the "single crystal-like" spectra obtained for each orientation of the membrane normal with respect to the static magnetic field. This orientational alignment is provided by the microscopic structure that typically characterizes complex fluids, about which the molecular tumbling occurs. Membrane vesicles exhibit "powder-like" spectra, as they possess membrane components at all angles with respect to the magnetic field, which is referred to as MOMD. The local ordering determines the IB and the information on dynamics can be obtained from the S_{c-} spectra, which yields homogeneous T_2-values, as well as the development of the cross peaks with mixing time. [See Crepeau et al. (1994); Patyal, Crepeau, and Freed (1997) and Chiang et al. (2007), who studied several different nitroxide spin labels in phospholipid membrane vesicles to obtain accurate dynamics and ordering parameters in the context of MOMD.]

2-D-ELDOR is extremely sensitive to the properties of membrane vesicles, with the data acquired showing dramatic changes in the membranes' properties. Moreover, such changes can even be detected visually from the spectral patterns by a simple inspection; an example is seen in Figure 11.10, which shows the 2-D-ELDOR contour plots as a function of the mixing time, T_m, for the spin-labeled lipid, 1-palmitoyl-2-(16-doxyl stearoyl) phosphatidylcholine (16-PC) in pure lipid vesicles, in a standard liquid-crystalline phase and also for a lipid-cholesterol mixture in 1:1 ratio, in a "liquid-ordered" (LO) phase (Ge et al., 1999). The qualitative difference in the spectra indicate that the LO phase exhibits a significantly greater ordering than the liquid crystalline phase, due to its increased microscopic ordering. In addition, the LO phase exhibits a much slower development of cross-peaks as a function of T_m, due to a restricted range of orientational motion as a result of the presence of microscopic ordering (Costa-Filho, Shimoyama, and Freed, 2003a).

In addition to the complex inhomogeneous lineshapes that are caused by MOMD (the theory of which was provided by Meirovitch, Nayeem, and Freed, 1984), there exists another often-encountered source of IB. This, specifically, is the slow-motional regime that does not average out the rigid limit line shapes completely, as the motions are too slow. This problem is dealt with effectively in the MOMD theory. As the slow-motional spectra have a comparable time scale to that for molecular dynamics, they provide a greater insight into the microscopic details of the molecular dynamics. As with complex fluids, it was found that a more sophisticated model than the MOMD model—specifically, the SRLS model as described above—was needed to analyze the 2-D-ELDOR spectra in order to achieve a reasonably good agreement with experiment. The SRLS model was tested in studies on a macroscopically aligned liquid crystal solvent, called 4O,8 (Sastry et al., 1996a, 1996b). This solvent exhibits many phases as a function of temperature, including isotropic, nematic, liquid-like smectic A, solid-like smectic B, and crystalline phases. This model, in addition to using the macroscopic liquid crystalline-orienting potential, has provided consistently better fits than were obtained with the simpler MOMD model, and does not include any local structure. Hence, one could, using the macroscopically aligned samples, obtain very extensive relaxation, dynamic, and structural information which includes virtually all of the parameters obtainable from any EPR experiments on spin relaxation in a complex fluid! These ten parameters are as follows: the two-term (asymmetric) macroscopic ordering potential in the liquid crystalline phases; the axially symmetric diffusion tensor for the probe; its two-term orienting potential in the local structure or cage; the relaxation rate for the cage; the residual homogeneous T_2^{-1} due to processes other than the reorientational modulation of the ^{14}N dipolar and g-tensors; the residual (Gaussian) inhomogeneous broadening not due to the specific slow-motional contributions from the ^{14}N hf- and g-tensors; and the overall T_1 for the electron spins.

When investigating the effects of the peptide gramicidin A (GA) on the dynamic structure of model membranes, the changes in 2-D-ELDOR spectra – when as compared to those in the CW-EPR spectra – were found to be more dramatic, thus demonstrating the much greater sensitivity of 2-D-FT-EPR to molecular dynamics (Patyal, Crepeau, and Freed, 1997). It emerges that in these studies, performed at 9.3 GHz and with dead times of 50–60 ns, one could only be used to study the bulk lipids and not the boundary lipids that coated the peptide, evidence from which was provided in the CW-EPR spectra, albeit at very limited resolution. This problem was overcome, however, by invoking the higher-frequency, 17.3 GHz, 2-D-ELDOR, with an increased signal-to-noise ratio (SNR) and reduced dead times (~25–30 ns) to demonstrate the presence of two components (Costa-Filho et al., 2003b). These were: (i) the bulk component, as reported by Patyal, Crepeau, and Freed (1997), which exhibited relatively fast dynamics; and (ii) the boundary lipid, which grows in as the GA is added, and whose 2-D-ELDOR spectrum is undoubtedly that of a more slowly reorienting lipid, as expected. Moreover, these spectra could be simulated with a model of bending the end-chain of the lipid as it coated the GA. Such details of the dynamic structure of complex membrane systems can only be obtained using 2-D-ELDOR. The recent studies of membrane systems (Chiang et al., 2007) carry even further the capabilities of 2-D-ELDOR.

11.4.6
Extension of 2-D-ELDOR to Higher Frequencies

Just as in CW-EPR described earlier in this chapter, one can hope to perform multifrequency 2-D-ELDOR studies. That this is feasible was demonstrated by Hofbauer et al. (2004) and Earle et al. (2005) at 95 GHz. We show in Fig. 11.12 just such an example. The challenges here are the much greater spectral bandwidths to irradiate and the much shorter T_2 decays. In addition, these higher frequency spectra with greater orientational resolution pose a much greater challenge to their theoretical simulation in the slow-motional regime. However, a new computational algorithm has very recently been developed (Chiang and Freed, 2011) which promises to overcome these difficulties.

11.5
Simulation of Multifrequency EPR Spectra Using More Atomistic Detail Including Molecular Dynamics and Stochastic Trajectories

11.5.1
Augmented SLE

Improved modeling has been used for the stochastic modeling of the side-chain dynamics of spin-labeled proteins, an example being MTSSL (1-oxyl-2,2,5,5-tetramethyl-Δ^3-pyrroline-3-(methyl)methanethiosulfonate spin label) linked to poly α-helix domain (Tombolato, Ferrarini, and Freed, 2006a). The features of this model are described briefly as follows. Here, one considers stable conformers as

11.5 Simulation of Multifrequency EPR Spectrum Using Molecular Dynamics

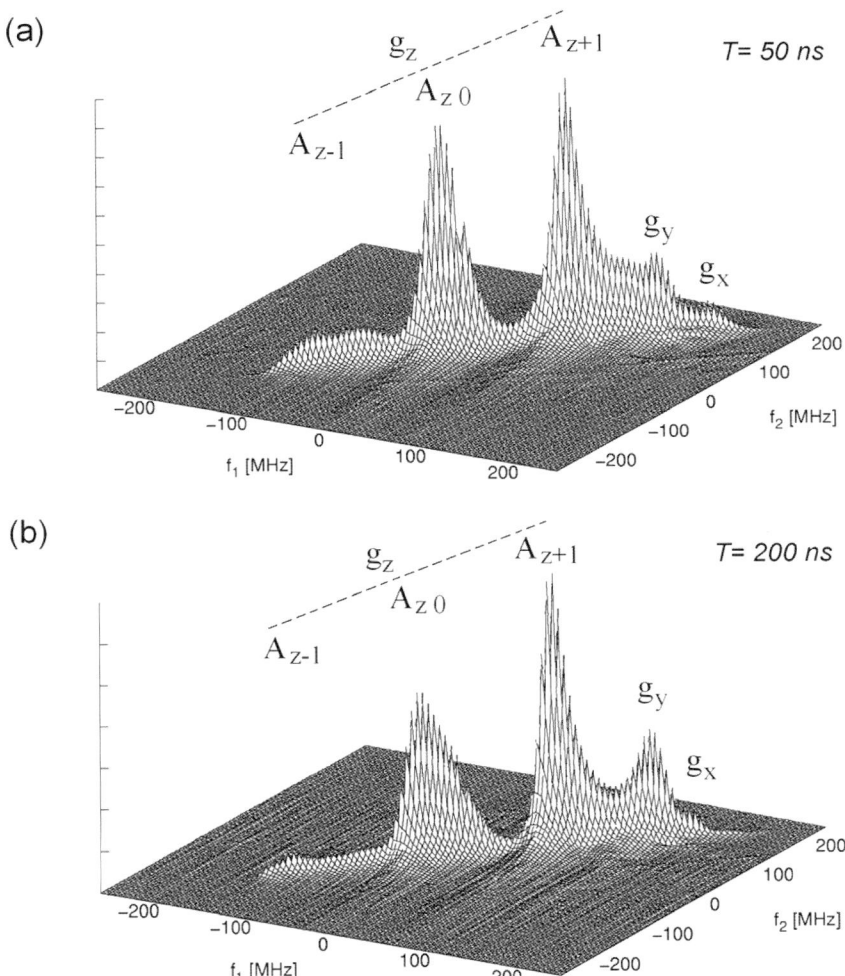

Figure 11.12 2D-ELDOR spectra of Gramicidin A spin label (GASl) in aligned DPPC membranes at 7°C: with the director parallel to the applied field ($\psi = 0°$) with mixing times (a) 50 and (b) 200ns, respectively. The spectral extent is approximately 350 MHz, corresponding to ±6 mT (±60 G). The B_1 in this case is about 1.7 mT (17 G).

determined from quantum mechanical calculations and utilizes estimates of the chain dynamics. The simplifying assumption is made that conformers with low barriers exhibit a fast exchange, while those with high barriers exhibit no exchange in the EPR time scale. There are no free parameters used in these calculations. The modified SLE used here is:

$$\frac{\partial \rho(\Omega_D, t)}{\partial t} = -i\overline{L(\Omega_D)}\rho(\Omega_D, t) - \Gamma_\Omega \rho(\Omega_D, t) - \left[T_2^{-1}(\Omega_D) + \Gamma(\Omega_D)\right]\rho(\Omega_D, t),$$

(11.25)

where $\overline{L(\Omega_D)}$ is the Liouville superoperator with the magnetic tensors partially averaged out by chain dynamics, $\Gamma(\Omega_D)$ is the diffusion operator for overall protein tumbling, and $T_2^{-1}(\Omega_D)$ is the linewidth contribution from chain dynamics, as calculated using Redfield theory (Redfield, 1965), i.e., motional narrowing theory. The torsional energy profiles were obtained using quantum mechanics, taking into account the constraints imposed by the local environment. Torsional motions about each of the five dihedral angles were taken as independent. The details of the potential energy provide the description of the system in terms of the significant rotamers undergoing conformational jumps and the librations which occur about the minima of the side-chain torsional potentials. The approximate diffusive treatment of the dynamics used provides a reasonable account of energetic and frictional features of the tether. This analysis enables one to estimate the amplitude and time-scale of the chain motions. Then, by Equation 11.25, it is possible to derive some general considerations on the effect of the tether dynamics on EPR spectra. This approach is thus based on a simple model, which leads to an easier interpretation of the determining factors of conformational dynamics of the side chain. The theory contains many realistic features, predicting some general results on the geometry and kinetics of this dynamics. These results have been exploited to directly introduce the dynamics of the nitroxide on lineshape analysis, for example, to interpret the EPR spectra of mutants of T4 lysozyme (Tombolato, Ferrarini, and Freed, 2006b). Another approach for obtaining detailed simulations is to employ either Monte-Carlo simulations (Sale et al., 2002, 2005), which are not covered here, or MD simulations using dynamic trajectories. General procedures for simulating the EPR spectra of nitroxide spin labels from MD and stochastic trajectories are described below.

11.5.2
MD Simulations Using Trajectories

The EPR spectrum may be calculated from a time evolution of the transverse magnetization, which is determined from the time dependence of the spin Hamiltonian. In order to take into account the dynamics of the spin label, one needs to consider appropriate models, such as the SLE technique described by Schneider and Freed (1989a), as outlined above. The dynamics of the electron–nuclear spin-coupled system of the spin label is treated quantum mechanically, whereas the rotational dynamics of the spin label can be treated classically. When one deals directly with the probability density to take into account the dynamic stochastic processes, the coupled classical-quantum evolution is described by the SLE, as discussed above. However, another approach was developed more recently by Sezer, Freed, and Roux (2008a, 2008b; Sezer, Freed, and Roux, 2009), and used successfully to simulate EPR spectra. This involves the use of dynamic trajectories with explicit realizations of the process in the time domain.

11.5.3
Use of Dynamic Trajectories to Simulate Multifrequency EPR Spectra

The simulation of EPR spectra using trajectories has been carried out in the past, for example by Saunders and Johnson (1968), Pederson (1972), and Robinson, Slutsky, and Auteri (1992). It has been claimed that this approach has the following advantages:

- It is possible to generate trajectories for more complicated stochastic models that can be treated by the SLE formalism, as suggested by Hakansson, Persson, and Westlund (2002) and by Persson et al. (2002). But, the SLE approach does allow for a wide range of sophisticated models (Meirovitch et al., 2010), and is orders of magnitude faster (Sezer et al., 2008a).

- It is possible to simulate EPR spectra directly from atomistic MD trajectories, without invoking any stochastic model (Eviatar, van der Heide, and Levine, 1995; Steinhoff and Hubbell, 1996; Hakansson et al., 2001; Stoica, 2004; and Beier and Steinhoff, 2006), although in much of this work the quantum spin dynamics has been overly simplified (Sezer et al., 2008a). A major disadvantage is the much greater computation time by orders of magnitude for computing EPR spectra by trajectories vs. use of the SLE (Sezer et al., 2008a).

High-frequency EPR provides an increased sensitivity to dynamics on the sub-nanosecond time scale, and therefore provides experimental spectra which can be exploited to establish a tighter connection with MD simulations.

The treatment of Sezer, Freed, and Roux (2008a) will be detailed here (albeit only in outline due to restricted space) to provide a description of the procedure of using MD trajectories to simulate multifrequency EPR spectra. In general, many, long trajectories are required for the convergence of spectra (Robinson, Slutsky, and Auteri, 1992; Hakansson et al., 2001; Stoica, 2004; Eviatar, van der Heide, and Levine, 1995). Alternatively, it is possible to use MD trajectories to estimate the parameters of a preselected stochastic dynamic model, and then to use these parameters either by solving the SLE (Budil et al., 2006) or by generating trajectories (Steinhoff and Hubbell, 1996; Beier and Steinhoff, 2006). Previously, spectra from more sophisticated rotational dynamical models, such as MOMD and SRLS (as discussed above), were not simulated by applying the trajectory-based approach. Only simple isotropic diffusion (Pederson, 1972; Robinson, Slutsky, and Auteri, 1992) or isotropic diffusion in a cone (Fedchenia, Westlund, and Cegrell, 1993) were taken into account, by employing rotational diffusion trajectories to simulate the EPR spectra. The trajectory-based approach was not exploited to its full potential, due to a lack of any rigorous formalism to simulate the trajectories for anisotropic diffusion.

An efficient numerical integrator to generate trajectories for sophisticated anisotropic rotational diffusion models, such as MOMD and SRLS, was developed to achieve this. In the practical algorithm, the gap between the small time steps at which the snapshots along the MD trajectories are available, and the longer time steps required for numerical propagation of the stochastic or quantal dynamics is accounted for by using time-averaging procedures. One describes the quantal spin dynamics in relation to the numerical propagation of the relevant part of the density matrix in Hilbert space, and the classical anisotropic Brownian diffusion in a potential, so as to develop an accurate and efficient numerical integrator for general rotational diffusion. Finally, one can compare the spectra for free and restricted rotational diffusion models simulated by using the developed time-domain integrators with the SLE. To this end, one utilizes time-averaging arguments to bridge the gap between the various integration time steps. Multifrequency spectra may then be simulated by using rotational diffusion or MD trajectories.

11.5.4
Numerical Integrators

11.5.4.1 Integration of the Quantal Spin Dynamics

Since the CW and FID spectra are formally equivalent (Abragam, 1961), the discussion here will be based on terms of the latter, which is numerically more appropriate.

The Spin Hamiltonian and the Interaction Picture The case of the nitroxide spin probe, which consists of an unpaired electron spin $S = \tfrac{1}{2}$ and a ^{14}N nucleus ($I = 1$), will be considered. (The case of 15N follows in a similar manner.) In units of angular frequency, using the standard notation, the spin Hamiltonian of a nitroxide is

$$H(t) = \gamma_e (B \cdot G(t) \cdot S + I \cdot A(t) \cdot S), \qquad (11.26)$$

where γ_e is the electronic gyromagnetic ratio, A is the hyperfine tensor (in units of magnetic field) and $G(t) \equiv g(t)/g_e$, with g_e being the free-electron g-factor. (Note this is the same Hamiltonian as in Equations 11.4 and 11.5.) Typically, the G and A tensors are diagonal in the same coordinate frame, referred to as N. They are also time-dependent due to the motion of the frame N with respect to the laboratory frame, L, (this explicitly time dependence is only implicit in Equations 11.4 and 11.5.) defined with respect to the external magnetic field $\mathbf{B}$: $(0,0,B_0)$, in which the electronic spin is quantized, so that all the vector and tensor components in Equation 11.26 are defined with respect to L. The nuclear Zeeman and quadrupolar interactions are neglected here, but they can be easily included if required; the coupling with other spins is also ignored. The electronic and nuclear spins localized on a single spin label are described by the state vector $|\psi(t)\rangle$, the dynamics of which is governed by the spin Hamiltonian via the Schrödinger equation. The spin Hamiltonian given by Equation 11.26 can be broken into two parts: (i) a large and time-independent part H; and (ii) the remaining time-dependent part, which is denoted as $V(t)$. One can express

$$H \equiv \gamma_e G_0 B_0 S_z = \omega_0 S_z; \quad G_0 \equiv \frac{1}{3} Tr\{G\}$$

The state vector oscillates with the Larmor frequency ω_0 in the absence of $V(t)$, whereas in its presence the instantaneous frequency of precession fluctuates around ω_0 by a time-dependent modulation, much smaller than ω_0.

It is more convenient to transform to the coordinate frame rotating at the Larmor frequency, i.e. what is referred to as the "interaction picture," so that one does not have to use extremely small integration steps required to resolve the fast oscillations at ω_0. To go into the interaction picture, one transforms the state vector and the operators as follows:

$$|\psi'(t)\rangle \equiv e^{iHt}|\psi(t)\rangle, \quad V'(t) \equiv e^{iHt}V(t)e^{-iHt},$$

which transforms the Schrödinger equation to

$$|\dot{\psi}'(t)\rangle = -iV'(t)\psi'(t), \tag{11.27}$$

where the dot indicates a derivative with respect to time. The spin operators are transformed as

$$S'_z = S_z, \quad S'_+ = S_+ e^{+i\omega_0 t}, \quad S'_- = S_+ e^{-i\omega_0 t} \tag{11.28}$$

Finally, the time-dependent part of the Hamiltonian becomes, in the interaction picture,

$$V'(t) = V_z(t) + \sum_{\kappa=\pm} V_\kappa(t) e^{i\kappa\omega_0 t}, \tag{11.29}$$

where the operators $V_\nu(t) \equiv (b_\nu(t) + a_\nu(t))S_\nu, \nu = z, \pm,$ are defined as follows;

$$a_z(t) \equiv \gamma_e \sum_{i=x,y,z} A_{iz}(t) I_i,$$

$$a_\pm(t) \equiv \gamma_e \sum_i \frac{1}{2}(A_{ix}(t) \mp iA_{iy}(t)) I_i,$$

which act only on the nuclear spin state, and the scalars

$$b_z(t) \equiv \gamma_e B_0 G'_{zz}(t),$$

$$b_\pm(t) \equiv \gamma_e B_0 \frac{1}{2}(G'_{zx}(t) \mp iG'_{zy}(t)),$$

are expressed in terms of the traceless tensor:

$$G'(t) = G(t) - G_0 E,$$

where E denotes the identity matrix in the electronic space.

The High-Field (HF) Approximation In the interaction picture–or the rotating frame in the present case–the fast-varying term responsible for Larmor precession has been removed in the effective Hamiltonian. This means that $|\psi'(t)\rangle$ varies on a time scale which is now much longer than the Larmor precession time scale. However, there are parts of the Hamiltonian which oscillate at the Larmor frequency; these are the terms containing $e^{i\kappa\omega_0 t}$ in Equation 11.29, which average out the effect of the slowly varying coefficients $V_\pm(t)$, that depend in turn on the magnetic tensors $G(t)$ and $A(t)$. Then, in order to calculate the slowly varying quantity, the transverse magnetization in this frame, one can consider only the slowly varying part $V_z(t)$ of Equation 11.29. This leads to the high-field approximation, derived as the zeroth-order term in the expansion of the Schrödinger Equation 11.27 in powers of $\varepsilon = 1/\omega_0$, by seeking a solution of Equation 11.27 in the form:

$$|\psi'(t)\rangle = |\psi^0(t)\rangle + \varepsilon \sum_{\kappa=\pm} |\psi^\kappa(t)\rangle e^{i\kappa t/\varepsilon},$$

where both $|\psi^0(t)\rangle$ and $|\psi^\kappa(t)\rangle$ are slowly varying. Proceeding in this fashion, one tries to derive an equation of motion for $|\psi^0(t)\rangle$. Finally, one obtains an equation for the slowly varying part of the state vector, correct to the first order in ε:

$$|\dot\psi^0(t)\rangle = -iH_S(t)|\psi^0(t)\rangle, \qquad (11.30)$$

where the effective slow Hamiltonian is as follows:

$$H_S(t) \equiv V_z(t) + \varepsilon [V_+(t), V_-(t)], \qquad (11.31)$$

In carrying out further analysis of $H_S(t)$, neglecting the terms which depend on $1/\omega_0$ as justified in the high-field approximation, and retaining only the V_z part of the Hamiltonian, one is led to the effective Hamiltonian:

$$H_{HF}(t) = \gamma_e(B_0 G_{zz'}(t) + I \cdot a(t) \cdot S_z); \; a_i(t) \equiv A_{iz}(t). \qquad (11.32)$$

It should be noted that an equivalent form of Equation 11.32 is the starting point for the SLE analysis of slow motion for the unsaturated line shapes (Freed, Bruno, and Polnaszek, 1971a; Polnaszek and Freed, 1975; Meirovitch, Nayeem, and Freed, 1984; Schneider and Freed, 1989a, b; Polimeno and Freed, 1993; Polimeno and Freed, 1995; Budil et al., 1996; Liang and Freed, 1999). In the HF approximation, the contribution of spin flips to the decay of transverse magnetization has been neglected; this is due to ignoring the terms in $S_\pm$ in the spin Hamiltonian. It emerges that the slow Hamiltonian, given by Equation 11.31 and its lowest-order approximation, expressed by Equation 11.32, are diagonal in the electronic Hilbert space, and do not allow spin flips. As a consequence, these Hamiltonians are not suitable to describe the arrival to equilibrium of the longitudinal magnetization, which is effected by these spin flips. Thus, in order to treat the phenomena which cause T_1 relaxation, it would be necessary to consider fast dynamics at the time scale of the Larmor precession.

The HF approximation decouples the spin dynamics of the $m_s = \frac{1}{2}(+)$ and $m_s = -\frac{1}{2}(-)$ sectors of the Hilbert space, seen by introducing the state vector $|\psi'(t)\rangle = \begin{pmatrix} |\psi'^+(t)\rangle \\ |\psi'^-(t)\rangle \end{pmatrix}$, and the Hamiltonian, given by Equation 11.32 in Equation 11.30, obtaining

$$\begin{pmatrix} |\dot\psi'^+(t)\rangle \\ |\dot\psi'^-(t)\rangle \end{pmatrix} = -i \begin{pmatrix} H_{HF}^{++}(t) & 0 \\ 0 & H_{HF}^{--}(t) \end{pmatrix} \begin{pmatrix} |\psi'^+(t)\rangle \\ |\psi'^-(t)\rangle \end{pmatrix}, \qquad (11.33)$$

where the slow state vector $|\psi^0\rangle$ was replaced by the state vector $|\psi'\rangle$ in the interaction picture. In view of Equation 11.33, numerical integration of the quantum dynamics is achieved by keeping track of the temporal development of the two parts $|\psi'^{\pm}\rangle$ separately, following the short-time propagation scheme, as follows:

$$|\psi'^{\pm}(t+\Delta t)\rangle = e^{\mp i\Delta t H_{HF}^{++}(t)}|\psi'^{\pm}(t)\rangle, \quad (11.34)$$

where the equivalence $H_{HF}^{--} = -H_{HF}^{++}$, valid in the HF approximation, was used. It is noted that the quantum integrator summarized by Equation 11.34 was also used by Eviatar, van der Heide, and Levine (1995), with their vectors P and Q corresponding to $|\psi'^{\pm}\rangle$.

Calculation of the Spectrum by the Use of Reduced Density Operator The CW spectrum, the object of the present simulation, is the Fourier–Laplace transform of the transverse magnetization $M_+ = M_x + iM_y$:

$$\tilde{M}_+(\omega) = \int e^{-i\omega t} M_+(t) dt,$$

where $M_+(t) = \langle \hat{M}_+(t) \rangle$ is the quantum-mechanical expectation value of the operator $\hat{M}_+(t) \propto \hat{S}_+$. It is noted, using Equation 11.28 that

$$M_+(t) = \langle \psi'(t)|\hat{M}'_+|\psi'(t)\rangle = e^{i\omega_0 t}\langle \psi'(t)|\hat{M}_+|\psi'(t)\rangle,$$

which shows that equivalently one can use the Schrödinger picture by sandwiching the operator $\hat{M}_+$ with the state vector $\langle\psi'(t)|$ in the interaction picture and simply shifting the spectrum by the Larmor frequency by multiplying with $e^{i\omega_0 t}$. Accordingly, one gets

$$\tilde{M}_+(\omega+\omega_0) = \int_0^\infty e^{i\omega t}\langle\psi'(t)|\hat{M}_+|\psi'(t)\rangle dt = \int_0^\infty e^{i\omega t}\langle\psi'^+(t)|\hat{M}_+|\psi'^-(t)\rangle dt, \quad (11.35)$$

where the last equality has been written in view of S_+, proportional to $\hat{M}_+$, being a raising operator. Defining the reduced density matrix:

$$\rho'^{-+}(t) \equiv |\psi'^-(t)\rangle\langle\psi'^+(t)|, \quad (11.36)$$

one obtains

$$\langle\psi'(t)|M_+|\psi'(t)\rangle = Tr\{M_+\rho'^{-+}(t)\}.$$

It should be noted here that, in the HF approximation, the time evolution of ρ'^{-+} is independent of the time evolution of the other sectors of the reduced spin density matrix (ρ'^{++}, ρ'^{+-}, and ρ'^{--}), defined analogously to ρ'^{-+} as in Equation 11.36. It follows from the short-time propagator described by Equation 11.34 and the definition of ρ'^{-+}, Equation 11.36, that the short-term dynamics of ρ'^{-+} is:

$$\rho'^{-+}(t+\Delta t) = e^{i\Delta t H_{HF}^{++}(t)}\rho'^{-+}(t)e^{i\Delta t H_{HF}^{++}(t)} \quad (11.37)$$

It is also noted that the same matrix acts on both sides of ρ'^{-+} in the above equation; this is different from the evolution of the full density matrix in the Hilbert space.

Equation 11.37 is the key expression for the integrator for the relevant sector of the quantum spin dynamics being developed here. In order to evaluate this

11 Molecular Motions

efficiently, one needs to quickly calculate the matrix in the exponential part of Equation 11.37:

$$e^{i\Delta t \hat{H}_{HF}^{++}(t)} = e^{i\Delta t(1/2)\gamma_e(B_0 G'_{zz}(t) + a(t) \cdot I)}, \tag{11.38}$$

at each step. It is easy to take into account the first term in the second parenthesis in Equation 11.38, as it amounts to a simple, time-dependent phase factor. However, the second term in this parenthesis is more complicated to evaluate. The straightforward method to evaluate is by first diagonalizing the matrix $a(t) \cdot I$ in the nuclear space by a similarity transformation, exponentiating its eigenvalues, and then applying the inverse similarity transformation. Sezer, Freed, Roux (2008a) discuss a more efficient alternative by invoking the relation between the nuclear spin matrices and the three-dimensional (3-D) representation of the rotation group, so that

$$\hat{N} = \sum_i n_i \hat{I}_i,$$

where $\hat{N} = (n_x, n_y, n_z)$ is a unit vector that satisfies the exponential expansion:

$$e^{-i\theta \hat{N}} = E_I - i(\sin\theta)\hat{N} - (1-\cos\theta)\hat{N}^2, \tag{11.39}$$

where E_I denotes the identity operator in the 3-D Hilbert space of the nuclear spin. As a consequence, one can avoid solving the eigenvalue problem of $a(t) \cdot I$ at each time step, and instead calculate the magnitude a and the direction $\mathbf{n}$ of the vector $\mathbf{a}(t)$. One needs to use the angle $\theta = \gamma_e \Delta t \frac{1}{2} a$ and the unit vector $\mathbf{n}$ to construct the short-term propagator (Equation 11.39), as shown explicitly by Sezer, Freed, Roux (2008a), who used the following equations:

$$\mathrm{Re}(e^{-i\theta\hat{N}}) = I_3 + \begin{pmatrix} c_\theta\left[n_z^2 + \left(\frac{1}{2}\right)(n_x^2+n_y^2)\right] & \left(\frac{1}{\sqrt{2}}\right)[s_\theta n_y + c_\theta n_z n_x] & c_\theta\left(\frac{1}{2}\right)(n_x^2-n_y^2) \\ \left(\frac{1}{\sqrt{2}}\right)[-s_\theta n_y + c_\theta n_z n_x] & c_\theta\cdot(n_x^2+n_y^2) & \left(\frac{1}{\sqrt{2}}\right)[s_\theta n_y - c_\theta n_z n_x] \\ c_\theta\left(\frac{1}{2}\right)(n_x^2-n_y^2) & \left(\frac{1}{\sqrt{2}}\right)[-s_\theta n_y - c_\theta n_z n_x] & c_\theta\left[n_z^2 + \left(\frac{1}{2}\right)(n_x^2+n_y^2)\right] \end{pmatrix}$$

and

$$\mathrm{Im}(e^{-i\theta\hat{N}}) = \begin{pmatrix} s_\theta n_z & \left(\frac{1}{\sqrt{2}}\right)[s_\theta n_x - c_\theta n_z n_y] & -c_\theta n_x n_y \\ \left(\frac{1}{\sqrt{2}}\right)[s_\theta n_x + c_\theta n_z n_y] & 0 & \left(\frac{1}{\sqrt{2}}\right)[s_\theta n_x + c_\theta n_z n_y] \\ c_\theta n_x n_y & \left(\frac{1}{\sqrt{2}}\right)[-s_\theta n_y - c_\theta n_z n_x] & -s_\theta n_z \end{pmatrix}$$

Sezer, Freed, and Roux (2008a) make the argument that it is preferable to work with the density matrix, ρ'^{-+}, rather than using the state vector $|\psi'^{\pm}\rangle$, since the former represents the ensemble average of all the state vectors consistent with the macroscopic initial condition, when calculating the FID after a $\pi/2$ (90°) pulse

11.5 Simulation of Multifrequency EPR Spectrum Using Molecular Dynamics

applied at time $t = 0$, which renders $M_+(t = 0^+) = 1$, which does not uniquely determine the state vector $|\psi(0^+)\rangle$. Thus, if one used the state vector, an additional averaging over all the possible starting state vectors that give the correct initial magnetization is required. One, therefore, eliminates the sampling noise associated with averaging over a finite number of initial state vectors by propagating the density matrix rather than using the state vector. This justifies the extra computational cost of propagating a 3×3 matrix as compared with a 3×1 vector.

Equilibrium and Time-Dependent Density Matrix Starting with the decoupled initial conditions, one can express the density operator in terms of the average Hamiltonian:

$$\rho^{eq} = \rho(0) \propto \exp(-\hbar \overline{\hat{H}}/k_B T) \equiv a(\hat{E} - bS_z), \tag{11.40}$$

where $\hat{E}$ is the identity operator in Hilbert space and a and b are scaler coefficient. In writing the last term, the fact that the sample is equilibrated under the influence of a constant magnetic field, and the average Hamiltonian is less than 1% of $k_B T$ so that the exponential can be expanded to the first order only, have been taken into account. At a later time, given that $\hat{E}$ commutes with the Hamiltonian, one can write the density matrix in the form:

$$\rho(t) \equiv a(\hat{E} + \sigma(t)).$$

It should also be noted that $\hat{E}$ does not affect the expectation value of the magnetization, since $Tr\{\hat{M}\hat{E}\} = 0$, as $\hat{M}$ is proportional to $\hat{S}$. Thus, one need only keep track of $\sigma(t)$, which is, in fact, the only relevant part of the density matrix. It is further noted from Equation 11.40 that $\sigma(0) = \sigma^{eq} \propto S_z$. After applying the 90° pulse, $\sigma(0^+) \propto S_y$, which means that $\sigma^{-+}(0^+) \propto E_I$.

11.5.4.2 Generation of Stochastic Trajectories for Rotational Diffusion

In this section, an explanation is provided of how to develop an efficient numerical integrator for the rotational Brownian diffusion of a body-fixed frame (B) with respect to a space fixed frame (S). If there exists an ordering potential $U(\Omega)$, it can be parameterized by using the Euler angles, Ω; here, $\Omega = (\alpha, \beta, \gamma)$ describes the instantaneous orientation of B with respect to S. The basic model considered here forms the basis for more sophisticated motional models such as MOMD and SRLS.

Use of Quaternions to Treat Rotational Dynamics The kinematics of rotations required here can be conveniently treated by using quaternions (Lynden-Bell and Stone, 1989), rather than Euler angles. The components of the quaternion for the orientation of B with respect to S being given in terms of the Euler angles $\Omega = \{\alpha, \beta, \gamma\}$ are calculated as follows (Lynden-Bell and Stone, 1989):

$$\begin{aligned} q_0 &= \cos(\beta/2)\cos((\gamma + \alpha)/2), \\ q_1 &= \sin(\beta/2)\sin((\gamma - \alpha)/2), \\ q_2 &= \sin(\beta/2)\cos((\gamma - \alpha)/2), \\ q_3 &= \cos(\beta/2)\sin((\gamma + \alpha)/2), \end{aligned} \tag{11.41}$$

In terms of the components of the quaternion, the 3×3 rotation matrix emerges as (Biedenharn and Louck, 1981):

$$R = \begin{pmatrix} q_0^2 + q_1^2 - q_2^2 - q_3^2 & 2q_1q_2 - 2q_0q_3 & 2q_1q_3 + 2q_0q_2 \\ 2q_1q_2 + 2q_0q_3 & q_0^2 - q_1^2 + q_2^2 - q_3^2 & 2q_2q_3 - 2q_0q_1 \\ 2q_1q_3 - 2q_0q_2 & 2q_2q_3 + 2q_0q_1 & q_0^2 - q_1^2 - q_2^2 + q_3^2 \end{pmatrix}. \quad (11.42)$$

The bottom row of Equation 11.42 contains just the components of the vector z of the stationary coordinate system with respect to the axes of B, which are denoted, for later use, as:

$$X \equiv R_{zx} = (z)_{x'}; \quad Y \equiv R_{zy} = (z)_{y'}; \quad Z \equiv R_{zz} = (z)_{z'}. \quad (11.43)$$

The orientation of B with respect to S is described by the 2×2 unitary matrix, which can be expanded in terms of the Pauli spin matrices $\sigma_1, \sigma_2, \sigma_3$ and the 2 × 2 identity matrix σ_0 (not to be confused with σ, the density matrix used above), as follows:

$$Q = \begin{pmatrix} q_0 - iq_3 & -q_2 - iq_1 \\ q_2 - iq_1 & q_0 + iq_3 \end{pmatrix} = q_0\sigma_0 - i\sum_{i=1,2,3} q_i\sigma_i, \quad (11.44)$$

characterized by unit determinant:

$$q_0^2 + q_1^2 + q_2^2 + q_3^2 = 1. \quad (11.45)$$

The components of the quaternion corresponding to the transformation relating B to S are the real numbers q_i. Q becomes time-dependent when there is motion of B with respect to S, described by the equation of motion:

$$\frac{d}{dt}Q(t) = W(t)Q(t), \quad \text{where } W(t) = -i\frac{1}{2}\sum_i w_i(t)\sigma_i, \quad (11.46)$$

where $\omega(t)$ is the instantaneous angular velocity of B. $Q(t)$, as given by Equation 11.46, can be numerically integrated to generate the time series of Q, in the same way as achieved by Fedchenia, Westlund, and Cegrell (1993) in their rigorous treatment of isotropic rotational diffusion restricted to a conical region.

When considering anisotropic diffusion, one needs to work with the components of ω with respect to B, denoted as $\tilde{\omega}_{i'}$, rather than with respect to S, as in the treatment of isotropic rotational diffusion described above. The equation of motion of Q becomes:

$$\frac{d}{dt}Q(t) = Q(t)\tilde{W}(t), \quad \tilde{W}(t) = -i\frac{1}{2}\sum_i \tilde{w}_i(t)\sigma_{i'}. \quad (11.47)$$

It should be noted that, in Equation 11.47, the components of the angular velocity of the rotating frame are with respect to the stationary frame, whereas in Equation 11.46 they are with respect to the body-fixed frame. Integration of Equation 11.47 yields:

$$Q(t + \Delta t) = Q(t)e^{\Delta t \tilde{W}(t)}, \quad (11.48)$$

which preserves the determinant of Q, and in turn the normalization of the quaternion, as given by Equation 11.45. Now, in close analogy to the evaluation of

Equation 11.38 using Equation 11.39, one can exponentiate the matrix in Equation 11.48 calculating only the trigonometric functions:

$$\exp\left(-i\sum_i \frac{\tilde{\omega}_i \Delta t}{2}\sigma_i\right) = \cos\theta\sigma_0 - i\sin\theta\sum_i u_i\sigma_i$$

$$= \begin{pmatrix} \cos\theta - iu_z\sin\theta & -(u_y + iu_x)\sin\theta \\ (u_y - iu_x)\sin\theta & \cos\theta + iu_z\sin\theta \end{pmatrix}. \quad (11.49)$$

In Equation 11.49, θ and $u = (u_x, u_y, u_z)$ denote, respectively, the magnitude and the direction of the vector $\tilde{\omega}(t)\Delta t/2$. The propagation of the quaternion Q_{SB} describing the orientation of the coordinate system B with respect to the system S, is described by Equations 11.48 and 11.49, provided that one knows the physics of the orientational dynamics to determine how $\tilde{\omega}(t)$ changes with time. This is described in the following subsection.

Consideration of Anisotropic Brownian Diffusion in an External Potential Here, one takes into account the rotational diffusion in the presence of a potential $U(\Omega)$ (cf. Equation 11.19). Hereafter, the tilde over ω will be dropped, and it will be assumed that all the vector and tensor components are taken with respect to the coordinate system B. The components of the instantaneous angular velocity $\omega(t)$ in B follow the equation of motion (Kalmykov, 2001 and Coffey, Kalmykov, and Waldron, 2004):

$$\omega(t) = -\mathbf{D}\nabla u(\Omega(t)) + \xi(t), \quad (11.50)$$

in the limit of high friction, so that inertial terms are neglected. In Equation 11.50, the first term on the right-hand side corresponds to the systematic torque due to the potential $u(\Omega) \equiv U(\Omega)/k_B T$, whereas the second term, $\xi(t)$, is the random torque that leads to the orientational diffusion. The other symbols in Equation 11.50 are **D**, the rotational diffusion tensor, which is diagonal in B, and $\nabla = \left(\frac{\partial}{\partial\phi_x}, \frac{\partial}{\partial\phi_y}, \frac{\partial}{\partial\phi_z}\right)$, with ϕ_i being the angle of rotation around the ith axis of B. The components of the random torque satisfy the conditions (Kalmykov, 2001; Coffey, Kalmykov, and Waldron, 2004):

$$\mathbf{E}\{\xi_i(t)\} = 0, \quad \mathbf{E}\{\xi_i(t_1), \xi_j(t_2)\} = 2D_{ii}\delta_{ij}\delta(t_1 - t_2), \quad (11.51)$$

where **E** denotes the expectation value over the Gaussian probability density of ξ. Here, D_{ii} are the components of **D** with respect to B. The conditions in Equation 11.51 are valid only when the components of ξ are expressed in the coordinate frame in which the diffusion tensor is diagonal, which is the only frame in which the components of the diffusion tensor, and therefore, the intensities of the random torque decouple. When the diffusion tensor is isotropic this is naturally true in any coordinate system, including the space-fixed frame, so that it is possible, in this case, to exclusively express all the vector components with respect to S. However, for the general anisotropic case, one has to work with the components of the diffusion tensor with respect to B so that, as noted above, it is imperative to use

Equation 11.47 instead of Equation 11.46 for the equation of motion for the quaternion.

One can use the angular momentum operator J to describe the torque $-\nabla u(\Omega)$ (Polimeno and Freed, 1993; Kalmykov, 2001 and Coffey, Kalmykov, and Waldron, 2004):

$$-\nabla u(\Omega) = -iJu(\Omega),$$

which can be written in the component form as

$$\omega_i(t) = -iD_{ii}J_i u(\Omega(t)) + \xi_i(t), \qquad (11.52)$$

where the partial differential operators corresponding to the components J_i expressed in B are (Hakansson, Persson, and Westlund, 2002):

$$J_z = -i\frac{\partial}{\partial \gamma}; \quad J_\pm = e^{\mp i\gamma}\left[-i\cot\beta\frac{\partial}{\partial \gamma} \pm \frac{\partial}{\partial \beta} + \frac{i}{\sin\beta}\frac{\partial}{\partial \alpha}\right]; \quad J_\pm = J_x \pm iJ_y.$$

In order to facilitate operating with J_i on the potential, it is convenient to express the latter as an expansion over the eigenfunctions of J_i (Polnaszek et al., 1973; Meirovitch, Nayeem, and Freed, 1984; Polimeno and Freed, 1993, 1995):

$$u(\Omega) = -\sum_{j,m} c_j^m D_{0m}^j(\Omega).$$

The Wigner functions:

$$D_{nm}^j(\Omega) = e^{-in\alpha} d_{nm}^j e^{-im\gamma}$$

are the eigenfunctions of J_z:

$$J_z D_{nm}^j(\Omega) = -m D_{nm}^j(\Omega). \qquad (11.53)$$

They also have the property that

$$J_\pm D_{nm}^j(\Omega) = -\sqrt{j(j+1) - m(m \pm 1)} D_{nm\pm 1}^j(\Omega). \qquad (11.54)$$

With the use of Equations 11.53 and 11.54, the problem of differentiation of the potential transforms to straightforward algebraic manipulation of the components of the quaternion.

Finally, three specific forms of the potential used by Sezer, Freed, and Roux (2008a), are listed below:

i) for the potential $u(\Omega) = -c_0^2 D_{00}^2(\Omega) = -c_0^2 \frac{1}{2}(3Z^2 - 1)$, which favors those orientations for which z- and z'-axes are either parallel or antiparallel ($Z = \pm 1$):

$$-iJ_x u = -i\sqrt{\frac{3}{2}}c_0^2[D_{01}^2 + D_{0-1}^2] = -3c_0^2 YZ;$$

$$-iJ_y u = -\sqrt{\frac{3}{2}}c_0^2[D_{01}^2 - D_{0-1}^2] = 3c_0^2 XZ; \quad -iJ_z u = 0 \qquad (11.55)$$

(It is noted that the primed axes refer to those in the body-fixed frame B.)

ii) for the potential $u(\Omega) = -c_2^2[D_{02}^2(\Omega) + D_{0-2}^2(\Omega)] = -c_2^2 \frac{\sqrt{6}}{2}(X^2 - Y^2)$, which favors orientations in which the z-axis is parallel or antiparallel to x' ($X = \pm 1$) and disfavors orientations in which the z-axis is parallel or antiparallel to y' ($Y = \pm 1$):

$$-iJ_x u = -ic_2^2[D_{0-1}^2 + D_{01}^2] = -\sqrt{6}c_2^2 YZ; \quad -iJ_y u = -c_2^2[D_{0-1}^2 - D_{01}^2] = -\sqrt{6}c_2^2 XZ;$$

$$-iJ_z u = -2ic_2^2[D_{02}^2 + D_{0-2}^2] = 2\sqrt{6}c_2^2 XY$$

iii) from (i) and (ii), one has for the general potential:

$$u(\Omega) = -c_0^2 D_{00}^2 - c_2^2[D_{02}^2(\Omega) - D_{0-2}^2(\Omega)], \tag{11.56}$$

$$-iJ_x u = \left(-3c_0^2 - \sqrt{6}c_2^2\right)YZ; \quad -iJ_y u = \left(3c_0^2 - \sqrt{6}c_2^2\right)XZ; \quad -iJ_z u = 2\sqrt{6}c_2^2 XY, \tag{11.57}$$

where X, Y, and Z are defined in Equation 11.43.

Equation 11.52 can be numerically integrated by generating three random numbers $N_i(t)$ with Gaussian distribution of zero mean and unit standard deviation, taking into account the statistical properties of the random term in Equation 11.52, as described by Equation 11.51, so that

$$\frac{\omega_i(t)\Delta t}{2} = -iJ_i u(\Omega(t))\frac{D_{ii}(t)\Delta t}{2} + \sqrt{\frac{D_{ii}(t)\Delta t}{2}}N_i(t), \tag{11.58}$$

which is the necessary input to calculate propagation of the quaternion Q_{SB} using Equations 11.48 and 11.49, which describe the orientation of the coordinate system B with respect to the system S.

Spherical Grid to Incorporate the Initial Conditions for Rotational Diffusion These can be generated as random orientations of B with respect to S, weighted by the Boltzmann factor $\exp(-u(\Omega)/k_B T)$. Ponti (1999) has shown that systematically covering the surface of a sphere with a homogeneously distributed grid is much more efficient than a random choice. To this end, it has been demonstrated convincingly that distributing the points along a spiral that twists from the north pole to the south pole provides the most efficient grid with a high convergence rate (Ponti, 1999). Accordingly, the spherical polar coordinates of the points along the spiral are: $\theta_i = \arccos(s_i)$; $\phi_i = \sqrt{\pi N} \arcsin(s_i)$, where $s_i \in (-1, 1)$; $i = 1, \ldots, N$, parameterizes the spiral and N is the number of points on the spiral. It is further noted that since the potentials $u(\Omega)$ used here are proportional to the Wigner functions $D_{0m}^j(\Omega)$, as given by Equation 11.57, which are independent of the Euler angle α, the initial conditions for the Euler angles are chosen as $\alpha = 0, \beta = \theta_i, \gamma = \phi_i$; the corresponding quaternion is calculated by using Equation 11.41.

11.5.4.3 Testing the Integrators: Generation of Trajectories for Typical Stochastic Models of Spin-Label Dynamics

Using the framework of rotational dynamics described above, one can generate trajectories for typical stochastic models of the spin-label dynamics, for example, Brownian rotational diffusion (BD), MOMD (Meirovitch, Nayeem, and Freed,

1984), and SRLS (Polimeno and Freed, 1993, 1995). These are schematically represented as follows:

BD: $L \to$ free (an)isotropic diffusion $\to M$ (fixed) $\to N$;
MOMD: $L \to$ powder $\to D \to$ restricted (an)isotropic diffusion $\to M$ (fixed) $\to N$;
SRLS: $L \to$ free isotropic diffusion $\to D \to$ restricted (an)isotropic diffusion $\to M$ (fixed) $\to N$

Here, M and D refer to the body-fixed frame (B) and the stationary frame (S), respectively, as defined above. In the MOMD and SRLS models, the molecular frame M can diffuse with respect to the director frame D, which can itself be either randomly oriented in the MOMD model, or it can undergo free isotropic diffusion with respect to L in the SRLS model. In considering the diffusive motion of D with respect to L, D plays the role of the body-fixed frame B, whereas L plays the role of the stationary frame S. The intermediate director frame is skipped in the BD model, as there is no external ordering potential. L is identified with S, and M with B when simulating this model by using the formalism described in Section 11.5.3.2. In a given model, the initial conditions for each of the diffusion parts are chosen from the points distributed on a spherical grid.

Broadening of Spectral Lines due to Additional Relaxation Mechanisms To date, these have not been accounted for in the simulation, although they can be included phenomenologically in the form of Lorentzian and Gaussian relaxation times. To take into account Lorentzian broadening with relaxation time constant T_L, one multiplies the magnetization $M_+(t)$ by e^{-t/T_L}. On the other hand, one convolutes the spectral line with a Gaussian to take into account Gaussian broadening, that is multiply $M_+(t)$ by $e^{-t^2/8T_G^2}$, where T_G is the derivative peak-to-peak linewidth of the Gaussian, since convolution in the frequency domain is equivalent to multiplication in the time domain. Finally, since the trajectories are of some finite duration T, the resultant appearance of high frequencies in the Fourier transform can be suppressed by multiplying by the Hamming window (Ernst, Bodenhausen, and Wokaun, 1987):

$$h_T(t) = 0.54 + 0.46\cos(\pi t/T).$$

Thus, taking into account all these considerations, a derivative-mode absorption spectrum is calculated as

$$\frac{d\tilde{M}_+(\omega)}{d\omega} = \text{Im} \int_0^T dt\, t e^{-i\omega t} h_T(t) e^{-t/T_L} e^{-t^2/8T_G^2} M_+(t). \tag{11.59}$$

Illustrative Examples (BD and MOMD Models) Using $B_0 = 0.34\,\text{T}$, and the following values for the nitroxide magnetic tensors: $g^N = \text{diag}(2.008\,09, 2.005\,85, 2.002\,02)$ and $A^N = \text{diag}(6.2, 4.3, 36.9)\,\text{G}$, the time domain spectra, as simulated by the trajectory-based approach described above for the BD model with the isotropic diffusion are shown in Figure 11.13, which also shows, for comparison, the spectra simulated by using the SLE-based approach developed by Freed and coworkers

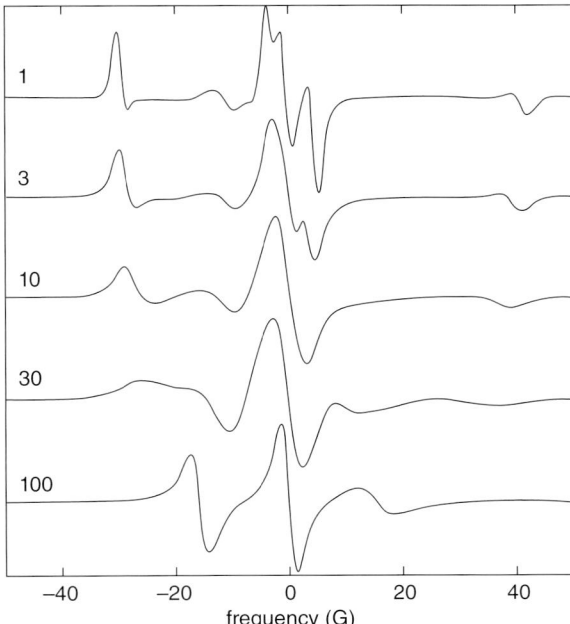

Figure 11.13 Spectra of isotropic free diffusion for various diffusion rates in units of $10^6 \, s^{-1}$ (indicated next to each spectrum), simulated by using the trajectory-based approach (dashed lines) and the SLE (continuous lines). The magnetic-tensor components are $g^N = \text{diag}(2.008\,09, 2.005\,85, 2.002\,02)$ and $A^N = \text{diag}(6.2, 4.3, 36.9)$; $B_0 = 0.34\,\text{T}$ (Sezer, 2008a).

over motional regimes from slow ($D = 1 \times 10^6 \, s^{-1}$; correlation time $\tau = 167$ ns) to fast ($D = 100 \times 10^6 \, s^{-1}$; correlation time $\tau = 1.67$ ns). [For relating D to τ, it is noted that the correlation time τ is inversely proportional to D ($\tau = 1/(6D)$).] There is excellent agreement between the trajectory-based and SLE-based approaches over the whole motional regime. In Figure 11.14 is shown the effect of the anisotropy of the diffusion tensor for both the trajectory-based and SLE-based approaches, using the same values of the nitroxide magnetic tensors as those used for Figure 11.13 and, again, the agreement between the two is excellent. It should be noted from Figure 11.14, that a fast rotational diffusion about the nitroxide z-axis ($D_{zz} > D_{yy} > D_{xx}$), as shown in the top spectrum in Figure 11.14, does not mix the larger A_{zz} component with the smaller components A_{xx} and A_{yy}, unlike that in the fast rotation about the x- and y-axes as seen in the bottom two spectra of Figure 11.14. In other words, the resulting spectrum is more slow-like in the former case, as compared to that for the latter two, for which the averaging of A_{zz} is more efficient. Figure 11.15 displays the effect of an ordering potential on the spectra, using the same values of the nitroxide magnetic tensors as those used for Figure 11.13, for two cases:

- The upper spectrum in Figure 11.15 is simulated for the potential $u(\Omega) = -c_0^2 D_{00}^2(\Omega) = -c_0^2 \dfrac{1}{2}(3Z^2 - 1)$, which favors those orientations for which

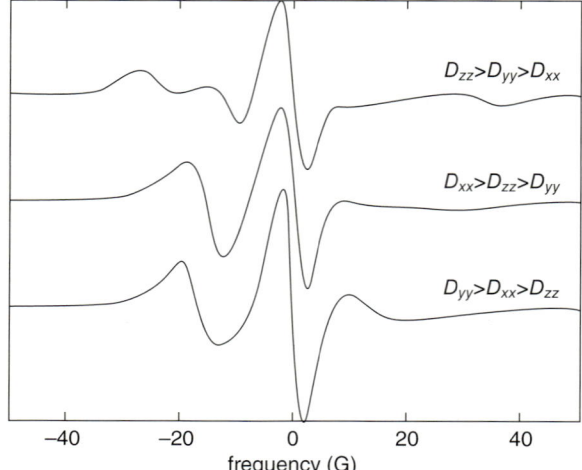

Figure 11.14 Simulated time-domain (dashed) and frequency-domain (continuous) spectra of anisotropic-free diffusion. The components of the diffusion tensor, $10 \times 10^6\,\text{s}^{-1}$, $30 \times 10^6\,\text{s}^{-1}$, and $100 \times 10^6\,\text{s}^{-1}$, were assigned in the order indicated in the plot. The magnetic tensors are given in the caption to Figure 11.12; $B_0 = 0.34\,\text{T}$ (Sezer, 2008a).

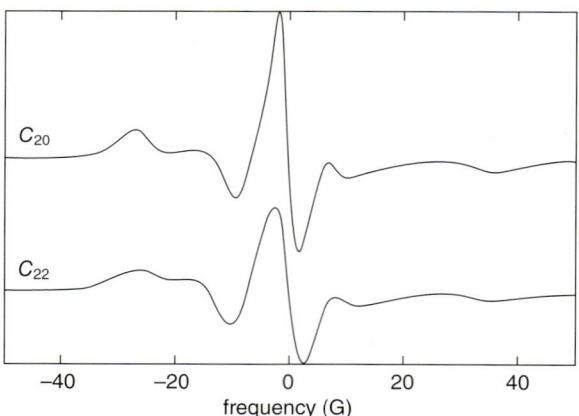

Figure 11.15 Comparison of time-domain (dashed) and SLE (continuous) spectra for two MOMD models with $(c_0^2, c_2^2) = (2.0, 0)$ and $(0, 2.0)$, respectively. The nonzero coefficient is indicated next to the spectrum. $D = 30 \times 10^{-6}\,\text{s}^{-1}$. The magnetic tensors are given in the caption to Figure 11.12; $B_0 = 0.34\,\text{T}$ (Sezer, 2008a).

the z- and z'-axes are either parallel or antiparallel ($Z = \pm 1$), given by Equation 11.55 above with $c_0^2 = 2.0$.

- The lower spectrum in Figure 11.15 is simulated for the potential $u(\Omega) = -c_2^2 [D_{02}^2(\Omega) + D_{0-2}^2(\Omega)] = -c_2^2 \dfrac{\sqrt{6}}{2}(X^2 - Y^2)$, which favors orientations in which the z-axis is parallel or antiparallel to x' ($X = \pm 1$) and disfavors orienta-

11.5 Simulation of Multifrequency EPR Spectrum Using Molecular Dynamics

Table 11.1 Parameters used in the simulation of spectra shown in Figures 11.13–11.15.

Model	B_0 (T)	stpN	Δt (ns)	freN	rstN	T_G^{-1} (G)
BD	0.34	800	1.0	1600	800	1.0
MOMD	0.34	2000	0.4	3200	1600	1.0

tions in which the z-axis is parallel or antiparallel to y' ($Y = \pm 1$), given by Equation 11.56 above with $c_2^2 = 2.0$.

The value of $D = 30 \times 10^6 \, \text{s}^{-1}$, describing isotropic diffusion, was used for both the simulations. Again, excellent agreement was found between the trajectory-based and SLE-based approaches (see also Beth et al., 2008).

Discussion of Simulation Parameters (Figures 11.13–11.15) The details of these parameters, which are listed in Table 11.1, are as follows. The duration of each trajectory is the product of "stpN" and Δt, which are, respectively, the number of simulation steps over which each stochastic trajectory lasted and the integration time step. "freN" is the number of spherical grid points used for free diffusion of M (BD) and the random distribution of D (MOMD) with respect to L, whereas "rstN" is the number of spherical grid points used for restricted diffusion of M with respect to D (MOMD). In the case of BD, since this restricted diffusion is not present, "rstN" indicates the number of independent trajectories initiated from each of the "freN" spherical grid points. The final column of Table 11.1 lists the value of the inhomogeneous Gaussian broadening introduced in the spectra by hand. It should be noted that the integration time step Δt (see Table 11.1) used to simulate the spectra in Figures 11.13 and 11.14 is much smaller than all but one of the correlation times mentioned above (1.67 ns) for the correlation time scales (1.67 ns to 167 ns) of the rotational diffusion. It should, then, be sufficient to follow the dynamics, except for 1.67 ns for $D = 100 \times 10^6 \, \text{s}^{-1}$. However, the excellent agreement of these results in all cases with those calculated using the SLE approach shows that, even in this case, the integration time step is sufficient. In order to ensure adequate resolution of the gradient of the potential energy a smaller integration step was chosen for the two MOMD models, for simulations using MD and SLE procedures (see also DeSensi et al., 2008).

With regards to the times of computation, at least a 1000-fold longer computer time was required when using stochastic trajectories than when using SLE. Consequently, the use of trajectories is worthwhile only when the dynamics cannot be treated with the SLE approach, at which time the MD simulations should be used. Otherwise, when simulations of spectra based on BD, MOMD, and SRLS models are required, the SLE method should be the method of choice, on the basis of its greater efficiency.

Combination of MD and Stochastic Trajectories In order that the experimental spectra are realistically reproduced by simulations, it is necessary to introduce the effect of the rotational diffusive dynamics, in addition to the dynamics of the spin labels present in the MD trajectories, to sample the slower global macromolecular dynamics, for example, the tumbling of a protein in solution. This is accomplished by allowing the coordinate system M, which is attached to the macromolecule, to undergo isotropic or anisotropic rotational diffusion with respect to the laboratory-fixed coordinate frame L, as shown below:

$$L \rightarrow \text{rotational diffusion} \rightarrow M \rightarrow \text{MD trajectories} \rightarrow N \qquad (11.60)$$

In this procedure, MD trajectories provide the dynamics of the coordinate frame N with respect to M, while the use of time-domain formalism developed above generates the dynamics of M with respect to L. Sezer, Freed, and Roux (2008b) illustrate the methodology to combine MD with stochastic trajectories to a spin labeled, polyanaline α-helix in explicit solvent by specifically resolving some formal issues related to the application of such a stochastic/MD trajectory-based approach, in which stochastic trajectories were used to take into account the tumbling dynamics which are slow and poorly sampled in atomistic MD simulations. Three methodological prerequisites were resolved:

- An accurate and efficient numerical scheme for propagating the quantum dynamics of the spins, achieved by working with the reduced density matrix in Hilbert space.

- An accurate and efficient numerical scheme for the treatment of rotational Brownian diffusion.

- The general case of restricted anisotropic diffusion, treated by using quaternions instead of Euler angles to parameterize the relative orientation of two coordinate systems, to which fits naturally the familiar restricting potential, written as a sum of a few spherical harmonics.

The time averaging of the magnetic tensors was also considered to bridge the gap between the fast time scale of the MD trajectories and the slow time scale of the quantum propagation. To this end, averaging time windows appropriate for the simulations of spectra at different magnetic fields were estimated.

It should be noted that, although MD and stochastic trajectories are used together, as proposed by Sezer, Freed, and Roux (2008a), the demands on the number and duration of the MD trajectories are largely unrealistic for routine MD simulations of solvated spin-labeled proteins. An alternative is to build stochastic, discrete-state Markov-chain models from the MD trajectories (Sezer et al., 2008c) and then use them to simulate the EPR spectra. This follows the scheme:

$$L \rightarrow \text{rotational diffusion} \rightarrow M \rightarrow \text{Markov chain} \rightarrow N.$$

For this model, the time-domain integrators and the time averaging arguments proposed here remain equally valid.

A comparison with, and a relevant review of, other reports on MD simulations is provided by Sezer, Freed, and Roux (2008a). For the developments in MD simulations to spin-labeled proteins, which are beyond the scope of this chapter, the reader is referred (in addition to the above citations), to the publications by Sezer, Freed, and Roux (2008a, 2008b, 2008c) and Sezer et al. (2009).

11.6
Concluding Remarks

As compared with CW-EPR, pulsed EPR has proven itself to be a much more powerful technique for the study molecular dynamics in a large variety of chemical, physical, and biological systems. However, whereas NMR enables the study of residual effects of motion-dependent terms – as reflected in the values of T_1 and T_2 in the spin Hamiltonian – that are completely averaged out by the molecular motion, dramatic lineshape variations are often found in CW-EPR spectra, which are particularly sensitive to the molecular motions. Moreover, these features are significantly enhanced when a multifrequency approach is adopted. Further, by using 2-D-ELDOR, molecular dynamics can be studied in much greater detail. Likewise, it is possible uniquely to resolve homogeneous from inhomogeneous broadening, and to clearly distinguish among cross-relaxation processes, in addition to determining T_1 values. As is clear from the above discussions, it is desirable to extend 2-D-ELDOR to higher frequencies in order to perform multifrequency studies that will provide even more detailed information on molecular motion than has hitherto been attained. To this end, Hofbauer et al. (2004) have developed a coherent pulsed high-power spectrometer operating at 95 GHz. It is also important to develop spin labels with a more limited flexibility and well-defined conformations, particularly with regards to the study of protein dynamics (Columbus and Hubbell, 2002). This is necessary in order to reduce the effects of the internal motions of the spin label's tether, which interferes when identifying the more relevant features of molecular dynamics from the data acquired. In recent developments involving the study of molecular motion by EPR, molecular-dynamics simulations using stochastic trajectories have been successfully applied to simulate the EPR spectra of nitroxide spin labels, an example being the analysis of the side-chain dynamics of spin-labeled proteins.

Acknowledgments

S.M. is grateful to Professor C. P. Poole, Jr for providing constructive comments on this chapter. This work was supported in part by NIH/NCRR grant #5P41RR016292 (J.H.F.).

Pertinent Literature

Many reviews have been produced on molecular motion studied with EPR, the most notable being that of Freed (2005), which has provided the bulk of the material included in this chapter. Other related reviews include those by Freed (1998, 2000, 2002) and by Borbat et al. (2001). The details of the solution of slow-motion EPR spectra for a nitroxide radical ($S = ½$, $I = 1$) are described by Schneider and Freed (1989a,b), while those for an electron spin ($S = ½$) coupled to two nuclear spins with arbitrary spins are provided by Misra (2007) and the two nitroxide case is given by Zerbetto *et al.* (2007). With regards to molecular dynamics simulations, the article of Sezer, Freed, and Roux (2008a) incorporates a host of information, much of which is included in the theory presented in this chapter. The additional papers by Sezer *et al.* (2008b, 2008c, 2009) describe further how this procedure is applied to calculate EPR spectra of spin-labeled proteins from MD simulations.

References

Abragam, A. (1961) *Principles of Nuclear Magnetism*, Oxford University Press, Oxford.

Angerhofer, A., Massoth, R.J., and Bowman, M.K. (1988) *Israel J. Chem.*, **28**, 227.

Aue, W.P. and Ernst, R.R. (1976) *J. Chem. Phys.*, **64**, 2229.

Barnes, J., Liang, Z., Mchaourab, H., Freed, J.H., and Hubbell, W.L. (1999) *Biophys. J.*, **76**, 3298.

Barnes, J.P. and Freed, J.H. (1998) *Biophys. J.*, **75**, 2532.

Beier, C. and Steinhoff, H.J. (2006) *Biophys. J.*, **91**, 2647.

Beth, A.H. and Robinson, B.H. (1989) *Biol. Magn. Reson.*, **8**, 179.

Beth, A.H., Lybrand, T.P., and Hustedt, E.J. (2008) *Biophys. J.*, **94**, 3798.

Biedenharn, L.C. and Louck, J.D. (1981) *Angular Momentum in Quantum Physics: Theory and Applications*, Addison-Wesley, Reading, MA.

Borbat, P.P., Costa-Filho, A.J., Earle, K.A., Moscicki, J.K., and Freed, J.H. (2001) Electron spin resonance in studies of membranes and proteins. *Science*, **291**, 266.

Borbat, P.P., Crepeau, R.H., and Freed, J.H. (1997) *J. Magn. Reson.*, **127**, 155.

Brown, I.M. (1974) *J. Chem. Phys.*, **60**, 4930.

Budil, D.E., Earle, K.A., Lynch, W.B., and Freed, J.H. (1989) Electron paramagnetic resonance at 1 millimeter wavelengths, in *Advanced EPR Applications in Biology and Biochemistry*, vol. 8 (ed. A. Hoff), Elsevier, Amsterdam, p. 307.

Budil, D.E., Lee, S., Saxena, S., and Freed, J.H. (1996) *J. Magn. Reson.*, **A120**, 155.

Budil, D.E., Sale, K.L., Khairy, K.A., and Fajer, P.G. (2006) *J. Phys. Chem. A*, **110**, 3703.

Chiang, Y.-W., Costa-Filho, A.J., and Freed, J.H. (2007) *J. Phys. Chem. B*, **111**, 11260.

Chiang, Y.-W. and Freed, J.H. (2011) *J. Chem. Phys.* (in press).

Coffey, W.T., Kalmykov, Y.P., and Waldron, J.T. (2004) *The Langevin Equation with Applications to Stochastic Problems in Physics, Chemistry, and Electrical Engineering*, 2nd edn, World Scientific, Singapore.

Columbus, L. and Hubbell, W.L. (2002) A new spin on protein dynamics. *Trends Biochem. Sci.*, **27**, 288.

Costa-Filho, A.J., Shimoyama, Y., and Freed, J.H. (2003a) *Biophys. J.*, **84**, 2619.

Costa-Filho, A.J., Crepeau, R.H., Borbat, P.P., Ge, M., and Freed, J.H. (2003b) *Biophys. J.*, **84**, 3364.

Crepeau, R.H., Saxena, S.K., Lee, S., Patyal, B.R., and Freed, J.H. (1994) *Biophys. J.*, **66**, 1489.

DeSensi, S.D., Rangel, D.P., Beth, A.H., Lybrand, T.P., and Hustedt, E.J. (2008) *Biophys. J.*, **94**, 3798–3809.

Dobbert, O., Prisner, T., and Dinse, K.P. (1986) *J. Magn. Reson.*, **70**, 173.

Dorio, M. and Freed, J.H. (eds) (1979) *Multiple Electron Resonance Spectroscopy*, Plenum Press, New York.

Dzikovski, B., Tipikin, D., Livshits, V., Earle, K., and Freed, J.H. (2009) *Phys. Chem. Chem. Phys.*, **11**, 6676.

Dzuba, S.A., Maryasov, A.G., Salikhov, K.M., and Tsvetkov, Yu.D. (1984) *J. Magn. Reson.*, **58**, 95.

Earle, K.A., Budil, D.E., and Freed, J.H. (1993) *J. Phys. Chem.*, **97**, 13289.

Earle, K.A., Moscicki, J., Polimeno, A., and Freed, J.H. (1997) *J. Chem. Phys.*, **106**, 9996.

Earle, K.A., Moscicki, J., Polimeno, A., and Freed, J.H. (1998) *J. Chem. Phys.*, **109**, 10525.

Earle K.A., Hofbauer W., Dzikowski, B., Moscicki, J.K., and Freed, J.H. (2005). *Magn. Res. in Chem.*, **43**, S256.

Eliav, U. and Freed, J.H. (1984) *J. Phys. Chem.*, **88**, 1277.

Ernst, R.R., Bodenhausen, G., and Wokaun, A. (1987) *Principles of Nuclear Magnetic Resonance in One and Two Dimensions*, Oxford University Press, Oxford.

Eviatar, H., van der Heide, U., and Levine, Y.K. (1995) *J. Chem. Phys.*, **102**, 3135.

Fedchenia, I.I., Westlund, P.-O., and Cegrell, U. (1993) *Mol. Simul.*, **11**, 373.

Freed, J.H. (1964) *J. Chem. Phys.*, **41**, 2077.

Freed, J.H. (1965) *J. Chem. Phys.*, **43**, 2312.

Freed, J.H. (1968) *J. Chem. Phys.*, **49**, 376.

Freed, J.H. (1977) *J. Chem. Phys.*, **66**, 4183.

Freed, J.H. (1987) Molecular Rotational Dynamics in Isotropic and Ordered Fluids by ESR, in *Rotational Dynamics of Small and Macromolecules in Liquids, Lecture Notes in Physics* **293**, (ed. T. Dorfmüller and R. Pecora), Springer–Verlag, Berlin, Ger., p. 89.

Freed, J.H. (1998) Linewidths, Lineshapes, and Spin Relaxation in the One and Two Dimensional ESR of Organic Radicals and Spin Labels, in *Foundations of Modern EPR* (eds G. Eaton, S. Eaton, and K. Salikhov), World Scientific, NJ, USA, pp. 658–683.

Freed, J.H. (2000) *Annu. Rev. Phys. Chem.*, **51**, 655.

Freed, J.H. (2002) Modern ESR methods in studies of the dynamic structure of proteins and membranes, in *EPR in the 21st Century* (eds A. Kawamori, J. Yamauchi, and H. Ohta), Elsevier Science, Amsterdam, The Netherland, pp. 719–730.

Freed, J.H. (2005) ESR and molecular dynamics, in *Biomedical EPR – Part B; Methodology, Instrumentation, and Dynamics* (eds G. Eaton, S.S. Eaton, and L.J. Berliner), Kluwer, New York, pp. 239–268.

Freed, J.H. and Frankel, G.K. (1963) *J. Chem. Phys.*, **39**, 326.

Freed, J.H. and Pederson, J.B. (1976) *Adv. Magn. Reson.*, **8**, 1.

Freed, J.H., Bruno, G.V., and Polnaszek, C. (1971a) *J. Phys. Chem.*, **75**, 3385.

Freed, J.H., Bruno, G.V., and Polnaszek, C. (1971b) *J. Chem. Phys.*, **55**, 5270.

Ge, M., Field, K.A., Aneja, R., Holowka, D., Baird, B., and Freed, J.H. (1999) *Biophys. J.*, **77**, 925.

Goldman, S.A., Bruno, G.V., Polnaszek, C., and Freed, J.H. (1972) *J. Chem.*, **56**, 716.

Gorcester, J. and Freed, J.H. (1986) *J. Chem. Phys.*, **85**, 5375.

Gorcester, J. and Freed, J.H. (1988) *J. Chem. Phys.*, **88**, 4678.

Gorcester, J., Ranavare, S.R., and Freed, J.H. (1989) *J. Chem. Phys.*, **90**, 5764.

Hakansson, P., Westlund, P.-O., Lindahl, E., and Edholm, O. (2001) *Phys. Chem. Chem. Phys.*, **3**, 5311.

Hakansson, P., Persson, L., and Westlund, P.-O. (2002) *J. Chem. Phys.*, **117**, 8634.

Hofbauer, W., Earle, K.A., Dunnam, C., Moscicki, J.K., and Freed, J.H. (2004) *Rev. Sci. Instrum.*, **75**, 1194.

Hornak, J.P. and Freed, J.H. (1983) *Chem. Phys. Lett.*, **101**, 115.

Hornak, J.P. and Freed, J.H. (1986) *J. Magn. Reson.*, **67**, 501.

Hwang, J.S., Mason, R.P., Hwang, L.P., and Freed, J.H. (1975) *J. Phys. Chem.*, **79**, 489.

Hyde, J.S., Chien, J.C.W., and Freed, J.H. (1968) *J. Chem. Phys.*, **48**, 4211.

Hyde, J.S. and Dalton, L.R. (1979) Saturation-transfer spectroscopy, in *Spin Labeling II. Theory and Applications* (ed. L.J. Berliner), Academic Press, NY, pp. 3–70.

Hyde, J.S. and Maki, A.H. (1964) *J. Chem. Phys.*, **40**, 3117.

Jeener, J., Meier, B.H., Bachman, P., and Ernst, R.R. (1979) *J. Chem. Phys.*, **71**, 4546.

Kalmykov, Y.P. (2001) *Phys. Rev. E*, **65**, 021101.

Kivelson, D. and Lee, S. (1982) *J. Chem. Phys.*, **76**, 5746.

Kubo, R. (1962) *J. Phys. Soc. Jpn*, **17**, 1100.

Kubo, R. (1963) *J. Math. Phys.*, **4**, 174.

Kubo, R. and Tomita, K. (1954) *J. Phys. Soc. Jpn*, **9**, 888.

Kurreck, H., Kirste, B., and Lubitz, W. (1988) *Electron Nuclear Double Resonance Spectroscopy of Radicals in Solution*, VCH Verlag GmbH, Weinheim, Germany.

Lee, S., Patyal, B.R., Saxena, S., Crepeau, R.H., and Freed, J.H. (1994a) *Chem. Phys. Lett.*, **221**, 397.

Lee, S., Budil, D.E., and Freed, J.H. (1994b) *J. Chem. Phys.*, **101**, 5529.

Leniart, D.S., Connor, H.D., and Freed, J.H. (1975) *J. Chem. Phys.*, **63**, 165.

Liang, Z.C. and Freed, J.H. (1999) *J. Phys. B*, **103**, 6384.

Lou, Y., Ge, M., and Freed, J.H. (2001) *J. Chem. Phys. B*, **105**, 11053.

Lynden-Bell, R.M. and Stone, A.J. (1989) *Mol. Simul.*, **3**, 271.

Mason, R.P. and Freed, J.H. (1974) *J. Phys. Chem.*, **78**, 1321.

Meirovitch, E., Igner, D., Igner, E., Moro, G., and Freed, J.H. (1982) *J. Chem. Phys.*, **77**, 3915.

Meirovitch, E., Nayeem, A., and Freed, J. (1984) *J. Phys. Chem.*, **88**, 3454.

Meirovitch, E., Polimeno, A., and Freed, J. (2010) *J. Chem. Phys.*, **132**, 207101.

Meirovitch, E., Shapiro, Y.E., Polimeno, A., and Freed, J.H. (2010) *Progress in NMR Spectroscopy*, **56**, 360.

Millhauser, G.L. and Freed, J.H. (1984) *J. Chem. Phys.*, **81**, 37.

Millhauser, G.L., Gorcester, J., and Freed, J.H. (1987) New Time-Domain ESR Methods for the Study of Slow Motions on Surfaces, in *Electron Magnetic Resonance of the Solid State* (ed. J.A. Weil), Can. Chem. Soc. Publ., Ottawa, Ont., p. 571.

Misra, S.K. (2007) *J. Magn. Reson.*, **189**, 59–77.

Möbius, K., Lubitz, W., and Freed, J.H. (1989) Liquid-state ENDOR and triple resonance, in *Advanced EPR, Applications in Biology and Biochemistry* (ed. A.J. Hoff), Elsevier, Amsterdam, The Netherlands, p. 441.

Möbius, K., Lubitz, W., and Plato, M. (1989) Liquid-state ENDOR and triple resonance, in *Advanced EPR, Applications in Biology and Biochemistry* (ed. A.J. Hoff), Elsevier, Amsterdam, The Netherlands, p. 441.

Moro, G. and Freed, J.H. (1981) *J. Phys. Chem.*, **84**, 2837.

Norris, J.R. and Weissman, S.I. (1969) *J. Phys. Chem.*, **73**, 3119.

Panferov, P.F., Grinberg, O.Y., Dubinskii, A.A., and Lebdev, Y.S. (1984) *Dokl. Phys. Chem.*, **278**, 888.

Patyal, B.R., Crepeau, R.H., and Freed, J.H. (1997) *Biophys. J.*, **73**, 2201.

Patyal, B.R., Crepeau, R.H., Gamliel, D., and Freed, J.H. (1990a) *Chem. Phys. Lett.*, **175**, 445.

Patyal, B.R., Crepeau, R.H., Gamliel, D., and Freed, J.H. (1990b) *Chem. Phys. Lett.*, **175**, 453.

Pederson, J.B. (1972) *J. Chem. Phys.*, **57**, 2680.

Persson, L., Cegrell, U., Usova, N., and Westlund, P.-O. (2002) *J. Math. Chem.*, **31**, 65.

Polnaszek, C.F. and Freed, J.H. (1975) *J. Phys. Chem.*, **79**, 2283.

Polimeno, A. and Freed, J. (1993) *Adv. Chem. Phys.*, **83**, 89.

Polimeno, A. and Freed, J. (1995) *J. Phys. Chem.*, **99**, 10995.

Ponti, A. (1999) *J. Magn. Reson.*, **138**, 288.

Redfield, A.G. (1957) *IBM J.*, **1**, 19.

Redfield, A.G. (1965) *Adv. Magn. Res.*, **1**, 1.

Robinson, B.H., Slutsky, L.J., and Auteri, F.P. (1992) *J. Chem. Phys.*, **96**, 2609.

Sale, K., Sar, C., Sharp, K.A., Hideg, K.A., and Fajer, P.G. (2002) *J. Magn. Reson.*, **156**, 104.

Sale, K., Song, L., Liu, Y.-S., Perozo, E., and Fajer, P.G. (2005) *J. Am. Chem. Sc.*, **127**, 9334.

Sastry, V.S.S., Polimeno, A., Crepeau, R.H., and Freed, J.H. (1996a) *J. Chem. Phys.*, **105**, 5753.

Sastry, V.S.S., Polimeno, A., Crepeau, R.H., and Freed, J.H. (1996b) *J. Chem. Phys.*, **105**, 5773.

Saunders, M. and Johnson, C.S. Jr (1968) *J. Chem. Phys.*, **48**, 534.

Saxena, S.K. and Freed, J.H. (1997) *J. Phys. Chem. A*, **101**, 7998.

Schneider, D.J. and Freed, J.H. (1989a) Calculating slow-motional magnetic resonance spectra: a user's guide, in *Spin Labeling: Theory and Applications, Vol. III*, vol. 8 (ed. L.J. Berliner), Biological Magnetic Resonance, Plenum Publishing Corp., New York, pp. 1–75.

Schneider, D.J. and Freed, J.H. (1989b) *Adv. Chem. Phys.*, **73**, 387.

Schwartz, L.J. (1984) Molecular rotation and time-domain ESR. PhD Thesis, Cornell University.

Schwartz, L.J., Stillman, A.J., and Freed, J.H. (1982) *J. Chem. Phys.*, **77**, 5410.

Schwartz, L.J., Millhauser, G.L., and Freed, J.H. (1986) *Chem. Phys. Lett.*, **127**, 60.

Sezer, D., Freed, J.H., and Roux, B. (2008a) *J. Chem. Phys.*, **128**, 165106.

Sezer, D., Freed, J.H., and Roux, B. (2008b) *J. Phys. Chem. B*, **112**, 5755.

Sezer, D., Freed, J.H., and Roux, B. (2008c) *J. Phys. Chem. B*, **112**, 11014.

Sezer, D., Freed, J.H., and Roux, B. (2009) *J. Am. Chem. Soc.*, **131**, 2597.

Steinhoff, H.J. and Hubbell, W.J. (1996) *Biophys. J.*, **71**, 2201.

Stillman, A.E., Schwartz, L.J., and Freed, J.H. (1980) *J. Phys. Chem.*, **73**, 3502.

Stoica, I. (2004) *J. Phys. Chem. B*, **108**, 1771.

Tombolato, A., Ferrarini, A., and Freed, J.H. (2006a) *J. Phys. Chem. B*, **110**, 26248.

Tombolato, A., Ferrarini, A., and Freed, J.H. (2006b) *J. Phys. Chem. B*, **110**, 26260.

Wangness, R.K. and Bloch, F. (1953) *Phys. Rev.*, **102**, 728.

Xu, D., Crepeau, R.H., Ober, C.K., and Freed, J.H. (1996) *J. Phys. Chem.*, **100**, 15873.

Zerbetto, M., Carlotto, S., Polimeno, A., Corvaja, C., Franco, L., Toniolo, C., Formaggio, F., Barone, V., and Cimino, P. (2007) *J. Phys. Chem. B*, **111**, 2668.

12
Distance Measurements: Continuous-Wave (CW)- and Pulsed Dipolar EPR

Sushil K. Misra and Jack H. Freed

12.1
Introduction

Information on long-range distances between selected sites is a prerequisite to understanding the detailed structure of complex systems. In NMR utilizing a spin label, it is possible to measure distances in the range between 8 Å to, at most, 25 Å. On the other hand, with EPR, distances in the range between about 10 and 90 Å can and have been measured. EPR spectroscopy may be the best practical technique for complex systems, since methods such as small-angle scattering, X-ray scattering and small-angle neutron scattering have insufficient contrast whereas fluorescence resonant energy transfer (FRET) lacks key virtues of EPR noted below. Distance measurements in EPR are based on exploitation of the dipolar interaction (DI), that depends on the distance between two paramagnetic probes, and which may be identical or may differ from each other, and are located at different spatial positions in the sample. They can be introduced using site-directed mutagenesis (also known as site-directed spin labeling; SDSL), a technique which enables the investigation of, for example, proteins and other biomolecules. In fact, the signature of the DI can be clearly seen in CW-EPR spectra under favorable circumstances for short distances. On the other hand, in pulsed EPR [hereafter referred to as pulsed dipolar spectroscopy; PDS], experiments can be designed specifically to clearly distinguish the dipolar interaction. The two commonly used PDS techniques are: (i) pulsed electron double resonance (PELDOR; this term was originally coined by the Russians, who first developed the technique), which is also referred to as double electron-electron resonance (DEER; this term was introduced later in the USA, and will be used hereafter in this chapter); and (ii) double quantum coherence (DQC)-EPR.

DEER requires an experimental arrangement which is relatively simple to implement and is currently available commercially. On the other hand, DQC requires instrumentation that can provide short, intense pulses and extensive "phase-cycling". The quantitative results so obtained can be exploited to obtain distances, their distributions, and orientational correlations. Neither the DEER (PELDOR) nor DQC acronyms indicate that these techniques deal exclusively with dipolar couplings; hence the acronym PDS will be used collectively hereafter for these techniques.

It is relevant at this point to compare distance measurements by PDS with those by the other commonly used techniques, specifically X-ray, NMR, and fluorescence resonance energy transmission (FRET). It should first be noted that PDS is often helped by – and provides information supplementary to – the structural information available from X-ray crytallography and NMR. One of the greatest virtues of using EPR is that one only requires trace amounts (in the case of proteins or biomolecules, from nanomoles to picomoles; Klug et al., 2005; Bhatnagar et al., 2010; Georgieva, et al., 2010) of the sample due to its EPR spin sensitivity. EPR is also amenable to study in diverse environments, such as dilute solutions, micelles, lipid vesicles, native membranes, supported lipid bilayers. Such measurements are frequently not possible with NMR or X-ray crystallography, where one requires, for example, larger quantities of samples, high-quality crystals (for x-rays), as well as high solubility (for NMR), and smaller molecules (NMR). FRET is also used for distance measurements, as it is much more sensitive per fluorophore than EPR, and can operate at biological temperatures. That is, FRET can be applied to fluid solutions at room temperature, whereas the PDS-EPR experiments are conducted with frozen solutions. The main advantages of PDS over FRET are:

- The molecular size of the probes is smaller, so that the original structure is less distorted. In PDS, there is often used a methanethiosulfonate spin label (MTSSL), which introduces only a small perturbation to the structure and function of the protein. Since the nitroxides, used in PDS, are smaller in size than most fluorescent labels used in FRET, there is less uncertainty in their positions relative to the backbone of the protein.

- Attaching two similar paramagnetic labels synthetically is less demanding than attaching two different donor and acceptor labels.

- The EPR technique can also be applied to opaque materials.

- The distances between the nitroxide spin-labels used in EPR is determined more accurately, as they are directly obtained from frequency measurements. This contrast to the distances between chromophores used in FRET, where there exist uncertainties in the parameter κ^2 used for distance determination.

- Also, the distance distributions are readily obtained in PDS-EPR.

This chapter is organized as follows. The theory for distance measurements in EPR using the dipolar interaction is provided in Section 12.2, the CW-EPR methods used for distance determination are discussed in Section 12.3, and pulsed dipolar techniques in general are detailed in Section 12.4. The details of three- four-pulse DEER techniques, and their merits and disadvantages as compared to CW-EPR and FRET, are described in Section 12.5. The density matrix calculations of echo signals for three- and four-pulse DEER sequences are given in detail in Appendices 12.I and 12.II, respectively. The DQC technique is described in Section 12.6, along with the density-matrix evolution algorithm used to calculate the DQC signal, while sensitivity considerations and their multifrequency aspects are discussed in Section 12.7. The subject of distance distributions, including Tikhonov regularization, is covered in Section 12.8, and the pertinent literature is reviewed in Section 12.9.

12.2
The Dipolar Interaction and Distance Measurements

The general expression for a dipolar interaction between two magnetic moments is given later in this chapter (see also Chapter 10.) For EPR, the relevant part of the dipolar interaction between two electrons, 1 and 2, treated as point dipoles, can be expressed in frequency units as follows (Borbat and Freed, 2007):

$$\frac{H_{dd}}{\hbar} = \frac{\gamma_e^2 \hbar}{r^3}(3\cos^2\theta - 1)\left[S_{1z}S_{2z} - \frac{1}{4}(S_1^+S_2^- + S_1^-S_2^+)\right]. \tag{12.1}$$

In Equation 12.1, γ_e is the electronic gyromagnetic ratio; r is the distance between the two electrons; θ is the angle between the vector joining the two electrons and the external magnetic field; and S_1, S_2 are the two electron spins, whose subscripts may be x, y, or z, indicating the components along the three axes, and the superscripts $+$ and $-$ indicating the raising and lowering operators, respectively: $S_i^\pm = S_{ix} \pm iS_{iy}$; $i = 1, 2$. The first term inside the square brackets is called the *secular* term, whereas the second term is called the *pseudosecular* term. The *non-secular* term is omitted for high fields and frozen samples. It is clear from Equation 12.1 that, from a knowledge of the expectation value of H_{dd}, one can estimate the distance, r, between the spin probes.

As a consequence of Equation 12.1, the CW-EPR line for each spin is split into two due to the dipolar interaction. The splitting of this doublet, $|A(r,\theta)|$, which can be expressed from Equation 12.1 as:

$$A(r,\theta) = \omega_d(1 - 3\cos^2\theta), \tag{12.2}$$

In Equation 12.2 the value of ω_d depends on whether the spins are "unlike" or "like". $A(r,\theta)$ also depends on the angle θ, varying over the range of values from $-2\omega_d$ ($\theta = 0°$) to ω_d ($\theta = 90°$).

12.2.1
Unlike Spins

This is the case when $\omega_d \ll |\omega_1 - \omega_2|$, where ω_1 and ω_2 are the resonant frequencies of the two electron spins in the absence of dipolar coupling. Equation 12.2 is obtained when only the secular term in Equation 12.1 is taken into account, and ignoring the second term in square brackets (pseudosecular term). For this case:

$$\omega_d = \gamma_e^2 \hbar / r^3, \tag{12.3}$$

It should be noted that, in the case of nitroxide spin labels, there usually exists the situation of "unlike" spins due to the different orientations of the magnetic, g, and A (hyperfine) tensors with respect to the external magnetic field, of any two nitroxides. The "unlike" limit is valid for $r \geq 20\,\text{Å}$.

12.2.2
Like Spins

Here, $\omega_d \gg |\omega_1 - \omega_2|$, and both the secular and pseudosecular terms in Equation 12.1 are taken into account, so that

$$\omega_d = 3\gamma_e^2\hbar/2r^3, \tag{12.4}$$

12.2.3
Intermediate Case

Here, $\omega_d \approx |\omega_1 - \omega_2|$, and one has to carry out a careful simulation of the spectrum taking into account both the secular and pseudosecular terms in Equation 12.1, as well as using the full spin Hamiltonian.

Distance measurements in EPR are made by either using a CW method or pulsed techniques. These are described as follows.

12.3
CW EPR Method to Measure Distances

In the past, nitroxide spin probes have been the most popular CW-EPR probes for distance determination with regards to their dipolar interaction, mainly because they are stable and easy to attach to systems (such as cysteine residues in proteins). Their powder spectra are governed by the inhomogeneous broadenings, which differ from site to site, due to the variation of nitrogen hyperfine (hf)- and g-tensors largely from the $\cos\theta$ angular variation in eq. 12.1, as well as to unresolved proton superhyperfine couplings. Current CW methods include techniques based on calculating the ratios of peak intensities (Kokorin et al., 1972; Sun et al., 1999), the relative intensity of half-field transition (Eaton et al., 1983), Fourier deconvolution of dipolar interactions (Rabenstein and Shin, 1995), and the computer simulation of lineshapes (Hustedt et al., 1993). All of these methods depend on the observation of reasonably significant broadening of the lineshape due to electron–electron dipolar interaction, limited by inhomogeneous line broadening due to unresolved hyperfine couplings, g-anisotropy, and relaxation. For distances of up to $20\,\text{Å}$ between spin labels, the dipolar broadening is still measurable for CW-EPR. Since, the dipolar interaction between nitroxide spin labels produces a relatively small broadening effect, as compared to that due to other interactions, it must be extracted from the CW-EPR powder spectra, either by a rigorous multi-parameter fit (Hustedt et al., 1997), or by a spectral deconvolution (Rabenstein and Shin, 1995). For these procedures, one also needs spectra from a singly-labeled sample as a reference for the background broadening, although this will produce complications that are compounded by incomplete spin-labeling (as discussed by Persson et al., 2001). As the magnitude of the dipolar coupling increases – specifically for distances less than 15 Å, when it becomes comparable to other inhomogeneous spectral broadenings – it can be more easily deduced from CW-EPR spectra. As a consequence, CW-EPR is practical for shorter distances up to a maximum of about 15–20 Å, with the more reliable values being obtained for distances less than 15 Å (Persson et al., 2001); in particular, this applies to half-field transitions (5–10 Å), lineshape simulations (up to 15 Å), and Fourier deconvolution (8–20 Å). On the other hand, with PDS (see below) one can measure distances in the range of 10 to 90 Å currently.

12.4
Pulsed Dipolar EPR Spectroscopy (PDS)

For smaller dipolar interactions corresponding to longer distances, pulsed techniques are utilized where the distance scale is limited by the rate constant (i.e., inverse of phase memory time, T_M) for echo dephasing, which is much smaller than the inhomogeneously broadened continuous-wave linewidth. This allows for dipolar oscillations in the time domain corresponding to much lower frequencies, hence longer distances (cf. eq. 12.3) than can be measured by cw means. Currently, two PDS techniques are in use, namely DEER (that is, PELDOR) and DQC. Three-pulse DEER was developed originally as an improvement over spin-echoes to utilize the dipolar interaction to measure distances between spin probes, whereas the more recent, more improved techniques are four-pulse DEER and DQC. The spectroscopic details of these techniques are detailed in Chapter 4, Section 4.4, whereas the theoretical details of their principles of operation with regards to distance determination are described below. It should be noted here that in DEER one purposely excites only a part of the spectrum using pulses that are weaker and of relatively longer duration. This means that different parts of the CW-EPR spectrum, resonating at different frequencies ω_A and ω_B, can be excited separately. In contrast, in the case of DQC, the pulses are intense and of short duration, so that the entire CW-EPR spectrum is excited under the action of the various pulses at the same frequency applied at different times. In PDS, a spin-echo is detected, in which the inhomogeneous spectral broadenings are canceled out due to refocusing. The temporal evolution of this spin-echo is governed by T_M, the phase relaxation time, which is a weaker effect than typical inhomogeneous broadenings. It is important to note here that the dipolar and exchange couplings are discriminated from all other interactions by employing the suitable pulsed sequences. In this manner, the signal caused by the presence of single-labeled molecules is obviated. Although, the direct signal from single-labeled molecules is filtered out in PDS, they can still contribute to the background *intermolecular* dipolar signal; however, the latter can be minimized by working at low concentrations.

Pulsed dipolar spectroscopy is now used routinely to measure distances greater than 15–20 (Banham et al., 2006; Borbat et al., 2006; Borbat et al., 2004; Cai et al.,

2006; Jeschke, 2002; Park et al., 2006), to distances as long as 60 to 90 Å (Georgieva et al., 2008; Georgieva, et al., 2010; Bhatanagar, et al. 2010), and is also quite efficient down to 10 Å (Fafarman et al., 2007; Borbat and Freed, 2007), significantly overlapping the range covered by CW-EPR. However, unlike CW-EPR, PDS is much less susceptible to inefficient labeling, and it can also readily yield distance distributions; its sensitivity is also very high, as the following discussion reveals.

The main goal of PDS is to solve structures of biomolecules by providing distance constraints. It has been exploited to accomplish the following: (i) single-distance measurements; ii) multiple-distance measurements; (iii) triangulation; (iv) oligomeric proteins; (v) protein complexes; (vi) embedding PDS constraints and rigid-body modeling; (vii) investigation of difficult labeling cases; and (viii) structural and conformational heterogeneity, protein folding (Borbat and Freed, 2007).

12.5
Double Electron–Electron Resonance (DEER)

DEER (also known as PELDOR), was developed in Russia during the 1980s by Milov and collaborators (Milov, Salikhov, and Shirov, 1981; Milov, Ponomarev, and Tsvetkov, 1984) with improvements introduced in 1993 (Larsen and Singel, 1993). Years later (Martin et al., 1998; Narr, Godt, and Jeschke, 2002), a deadtime-free variant of the method was proposed–termed four-pulse DEER (Pannier et al., 2000)–which enables commercial instrumentation, and with advances in site-directed spin labeling, resulted in a widespread application to structural biology.

In order to describe the DEER experiment, an isolated coupled pair of spins in a (disordered) solid must first be considered. The DEER signal is a modulation of the echo amplitude of the observing spins, S_1, resonating at the frequency, ω_1, by another set of spins, S_2, called "pumping" spins, resonating at the frequency, ω_2. When excited by respective microwave pulses, the spins S_1 experience the dipolar field generated by the spins S_2; the echo of spins S_1 will then be modulated in a manner that is a function of the time at which spins S_2 are excited. The resulting DEER signal can be expressed as a product of two parts: (i) decay due to the intermolecular interactions of the unpaired spins; and (ii) the periodic oscillations generated by the intramolecular dipolar interactions of the paired spins, expressed as

$$I(t) = I_{intra}(t) \times I_{inter}(t) \tag{12.5}$$

The modulation of the echo arises from the observing spins S_1 experiencing a local magnetic field arising from the dipolar interactions with the nearby spins S_2. Since the direction of this field at S_1 is determined by the spin state of the coupled spin, S_2, changing the latter selectively affects the contribution of the dipolar field to the total magnetic field at S_1.

Distance measurements between two spins involve excitations of both S_1 and S_2 spins. The expected spectrum of the spin pair consists of a pair of doublets, centered at ω_1 and ω_2, each with equal splitting due to the dipolar interaction, as shown in Figure 12.1.

12.5 Double Electron–Electron Resonance (DEER)

We write eqs. 12.2 and 12.3 for dipolar splitting, ω_{12}, between two "unlike" spins S_1 and S_2 at distance r is given by:

$$A(r, \theta) \equiv \omega_{12} = \frac{\mu_B^2 g_1 g_2}{\hbar r^3}(3\cos^2\theta - 1), \tag{12.6}$$

where μ_B is the Bohr magneton, g_1 and g_2 are the g-factors for spins 1 and 2, as shown in Figure 12.1a. For the samples randomly distributed with respect to the field, the integration of ω_{12} over the angle θ on the unit sphere produces a symmetric Pake pattern, in which the singularities ($\theta = 0°$) are separated by $2\frac{\mu_B^2 g_1 g_2}{\hbar r^3}$, as shown in Figure 12.1b.

Distance measurements between two spins involve excitations of both S_1 and S_2 spins. The expected spectrum of the spin pair consists of a pair of doublets, centered at ω_1 and ω_2, each with equal splitting, ω_{ee}, due to the dipolar interaction, as shown in Figure 12.2.

The pulse exciting the S_1 spins is called the "observer" pulse, and that exciting the S_2 spins is called the "pump" pulse. The pulses ω_1 and ω_2 must be sufficiently

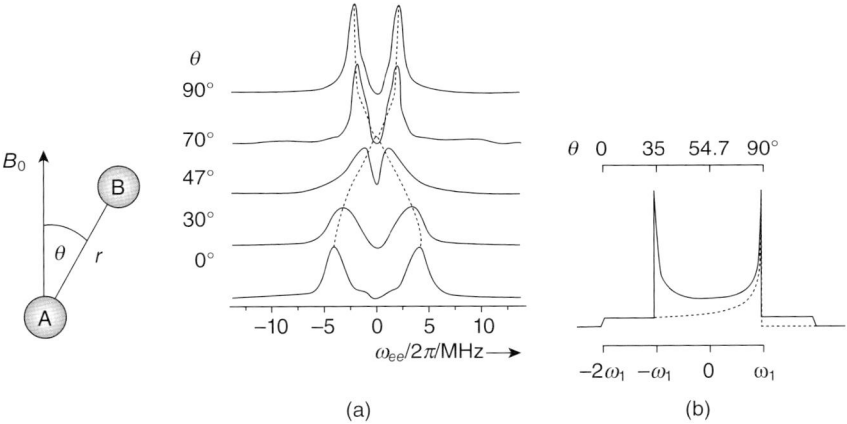

Figure 12.1 (a) The dependence of the dipolar spectrum on the angle θ between the magnetic field direction and the spin–spin vector; (b) Simulated Pake pattern for an isotropic sample. The θ scale refers to one line of the dipolar doublet. Adapted from Bhatnagar (2005).

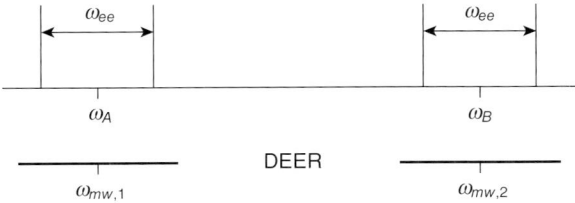

Figure 12.2 EPR line spectrum of a pair of coupled electron spins consisting of two doublets. Adapted from Bhatnagar (2005).

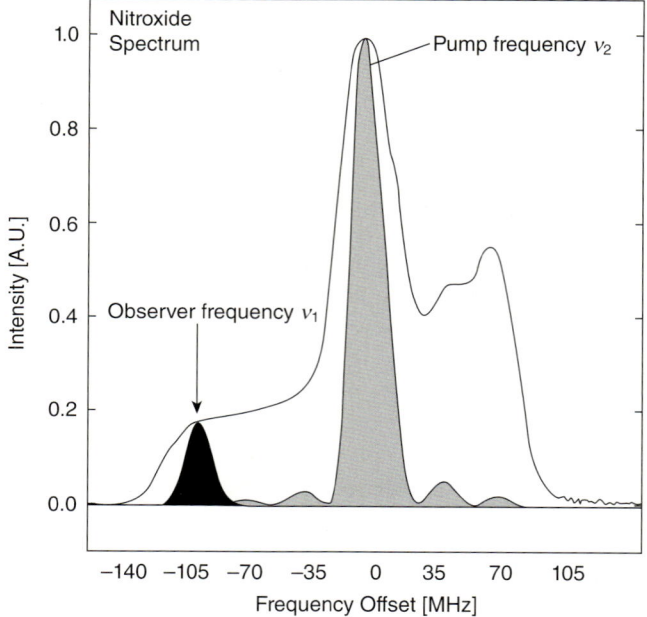

Figure 12.3 Excitation profiles due to pump and observer pulses in a nitroxide spectrum. Adapted from Pannier et al. (2000).

small that the excite the two levels of the doublet of the respective spin, S_1 or S_2. In a DEER sequence, one monitors the observer pulse when a pump pulse is applied at a variable time in between the $\pi/2$ and π observer pulses. For the selective excitation of S_1 and S_1 spins, it is necessary that $|\omega_1 - \omega_2|$ be sufficiently large and the pulses weak enough to avoid overlap. In most cases, two pulses at different frequencies can be applied to avoid overlap if the total width of the EPR spectrum is more than 50 MHz (Jeschke, 2002). In the case of nitroxides, a difference of 60 MHz in the observer and pump frequencies is sufficient to avoid overlap of excitation ranges (Jeschke, Pannier, and Spiess, 2000); an example of this is shown in Figure 12.3.

12.5.1
Orientation-Selection Considerations in DEER

The Pake pattern in the frequency domain for an isotropic sample is shown in Figure 12.3b. In practice, when the pulses are not sufficiently intense and narrow to excite the whole spectrum, this is not obtained, and only parts of the nitroxide spectrum will be excited, leading to "orientation selection" of the probe. The most commonly used probe, the nitroxide radical, exists in many different orientations in a disordered sample. When a pulse is applied, not all the orientations are equally excited, so that, only a part of the spectrum corresponding to the S_1 and S_2 spins is excited. These selected orientations for S_1 and S_2 spins may be correlated with each other by a certain set of angles between the vector connecting the electron pairs and the static field direction. Thus, the angle θ in Equation 12.1 does not necessarily represent an isotropic distribution (Jeschke, Pannier, and Spiess, 2000). The effect of orientation selection on the signal is significant if the excitation bandwidth is much smaller than the anisotropy of the g or hyperfine tensor. In such cases, the orientation selection may suppress the $\theta = 0$ feature from the Pake pattern, as the observer and pump frequencies are clearly distinct from each other.

12.5.2
Three-Pulse DEER

The three-pulse DEER sequence, which was originally developed by Milov, Salikhov, and Shirov (1981), consists of a two-pulse Hahn echo sequence $\pi/2-\tau'-\pi$ at the "observer" frequency, as shown in the upper part of Figure 12.4. The observer frequency is kept fixed, while an additional π pulse at the "pump" frequency is applied after a delay, τ, which is variable, subsequent to the observer frequency pulse at the time $t = 0$ (as shown in the lower part of Figure 12.4). The signal is obtained by recording the Hahn echo amplitude at the observer frequency as a function of τ.

The three-pulse DEER sequence is:

$$X_1(\pi/2) - \tau - X_2(\pi/2) - (\tau' - \tau) - X_1(\pi) - (t - \tau')$$

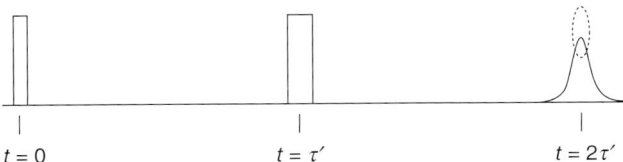

$t = 0$ $t = \tau'$ $t = 2\tau'$

$t = \tau$

Figure 12.4 A schematic drawing of the three-pulse DEER sequence, consisting of a refocused echo sequence at the observer frequency v_1 and of a π pulse at the pump frequency v_2. (Adapted from Larsen and Singel (1993)).

In Appendix 12.I are given the details of the calculation of a three-pulse DEER signal using a density-matrix approach The following three-pulse echo amplitude, given by the expectation value of S_1^+, is calculated in Appendix 12.I:

$$<S_1^+(2\tau')> = -iI_0 \cos(\omega_{12}\tau). \quad (12.7)$$

Equation 12.7 contains the dipolar interaction between two spins, ω_{12}. The cosine Fourier transformation of this modulation will give rise to a Pake pattern when the exchange coupling is ignored. The details of the Pake pattern can be exploited to estimate the distance between two spins by extracting the values of $v_{ee} = \omega_{ee}/2\pi$ for $\theta = 90°$ from the Pake pattern. Here $\omega_{ee} = \omega_d (3\cos^2\theta - 1) + J$, where J is the exchange interaction between the two spins (which will be neglected here), and ω_d is defined by Equation 12.2. The distance is given by the formula (Larsen and Singel, 1993):

$$r = (53.041 \, \text{MHz}/v_{ee})^{1/3} \, nm$$

The derivation of the echo amplitude given by Equation 12.6 is based on the assumption of an isolated pair of spins. In reality, a sample is characterized by a statistical distribution of spins. As a consequence, its observed spectrum is the superposition of the various spectra associated with all the constituents of the sample. The DEER spectrum is affected by both the remote (intermolecular) and near (intra-molecular)-pair dipolar interactions. The contribution due to the near pairs typically exhibits oscillatory behavior with time. On the other hand, the interactions due to remote pairs are governed by broadly distributed dipolar couplings that have the effect of a damping of the time-dependent signal. The damping and oscillatory components of the signal are clearly seen in Figure 12.5, which

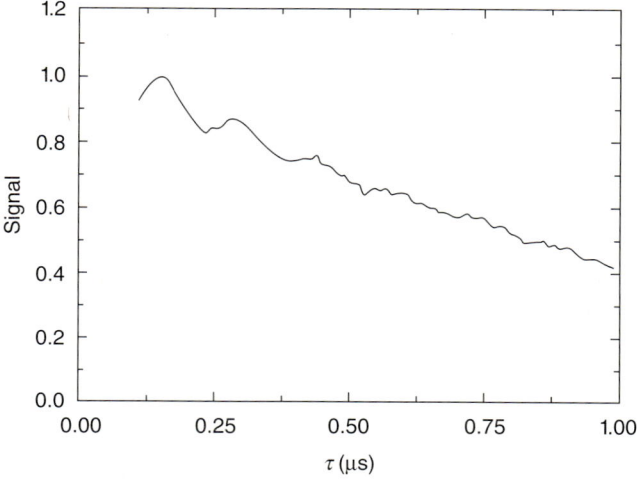

Figure 12.5 Time domain data (spin-echo envelope) of a 2 mM, frozen toluene solution of a biradical at 77 K from a three-pulse DEER sequence. Adapted from Larsen and Singel (1993).

12.5 Double Electron–Electron Resonance (DEER)

shows an observed three-pulse DEER spectrum. Finally, a three-pulse DEER signal is expressed by the following expression (Klauder and Anderson, 1962; Borbat and Freed, 2007):

$$V(t) = V_0 e^{-kCF_B t}[1 - \lambda\{1 - u(\omega_d t)\}];$$
$$u(\omega_d t) = \int_0^{\pi/2} \cos[\omega_d(1 - 3\cos^2\theta)t]d(\cos\theta)\}] \quad (12.8)$$

where, C is the concentration of sample; F_B is the fraction of spins S_2 excited by the pump pulse; λ is the modulation-depth parameter that depends on the fraction of S_1 spins excited by the observer pulse, and $k = 8\pi^2 \mu_B^2 g_1 g_2/(9\sqrt{3}\hbar)$.

According to Equation 12.8 for $V(t)$, it is seen that in order to avoid overdamping of the DEER signal, one should use a small bulk concentration, C.

12.5.3
Four-Pulse DEER

This was developed by Pannier et al. (2000). In the three-pulse case there are dead-time effects, as well as pulse overlap issues. Thus, it is not possible to measure signals from spin pairs, which have such a distance distribution that they decay completely during the microwave pulses of the order of few tens of nanoseconds. Also, in practice, the signal is distorted for times less than the dead time, which is typically in the range from 30 to 100 ns. It is possible to eliminate these dead time issues and achieve "dead-time free" signals by using an additional pulse. This is accomplished in the four-pulse DEER sequence, as shown in Figure 12.6. This four-pulse sequence can be expressed schematically by the following scheme:
$X_1(\pi/2)$ ---- τ_1 ---- $X_1(\pi)$ ---- t ---- $X_2(\pi/2)$ ---- $(\tau_1 + \tau_2 - t)$ ---- $X_1(\pi)$ ---- $(t' - \tau_2)$.
The echo signal detected is the expectation value $< I^+(t) >$. The effect of the pulses is described as follows. When the observing spins S_1, are oriented by the $\pi/2$ pulse along the y-axis following the first microwave pulse of frequency ω_1, which tips

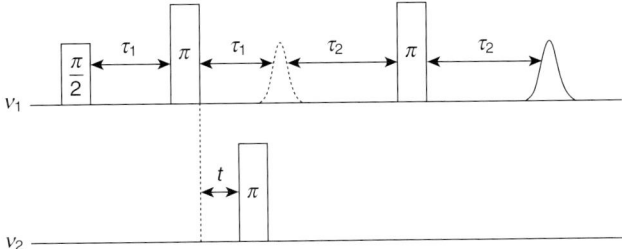

Figure 12.6 Four-pulse DEER sequence, which produces a refocused echo sequence for the observer spins. A π pulse at the pump frequency v_2 is applied at time t after the application of the first π pulse to observer spins. The second π pulse at the observer frequency is applied after the time τ_2 subsequent to the formation of the first echo of observer spins. The time t is varied, whereas τ_1 and τ_2 are fixed. Adapted from Bhatnagar (2005) and Pannier et al. (2000).

these spins into the *xy*-plane, they precess with the angular rate of ω_1. Spins S_1 at different sites experience slightly different angular rates due to the presence of field inhomogeneities and different resonance fields, which dephase the spins S_1 from each other. For example, the dipolar field due to nearby spins S_2 adds or subtracts its angular rate by $\pm\frac{1}{2}\omega_{12}$, depending on the spin state of spins S_2. A spin echo is created when a π-pulse is applied to spins S_1, which reverses the dephasing and refocuses spin S_1 along the *x*-axis. This echo obtained by application of $\pi/2$–π pulse sequence, which is called a "Hahn echo", plays a key role in most pulsed EPR experiments. After the echo, spins S_1 continue to precess in the *xy*-plane, and the subsequent π pulse again refocuses, leading to the "refocussed echo," but with a smaller amplitude as the phase coherence continues to be lost due to the spin relaxation. The dipolar contribution is now reversed in its superimposition on the total field experienced by spins S_1, by applying a π pulse to spins S_2, which is known as a "pumping" or "ELDOR" pulse (as it is the second frequency to excite spins S_2). It reverses the direction of S_2 spins, which, in turn, reverses the dipolar contribution experienced by the S_1 spins from $+\frac{1}{2}\omega_{12}$ to $-\frac{1}{2}\omega_{12}$ and *vice versa*. S_1 spins accrue a phase lag due to the change of the dipolar contribution to the angular rate of precession of spins *A* in between the two π-pulses; these are then no longer refocused along the *x*-axis and this results in a decrease in the intensity of the echo. The accrued phase lag reflects the strength of the dipolar interactions, ω_{12} (as given above in Equation 12.6), as well as the time at which the pumping pulse is applied, which determines how long spins S_1 experience the dipolar field $+\frac{1}{2}\omega_{12}$ versus $-\frac{1}{2}\omega_{12}$. The total accrued phase difference is $\omega_{12}(\tau-t)$, where τ is the interval between the $\pi/2$ and π pulses and *t* is the timing of the pumping pulse with respect to the initial echo, which is 2τ after the first $\pi/2$ pulse. The DEER signal is the modulation of the echo intensity as a function of the time of application of the pumping pulse between the π pulses. Finally, the echo intensity oscillates as

$$I_{intra}(t) = I_0 \cos(\omega_{12}\{\tau_1 - t\}), \tag{12.9}$$

where I_0 is the echo in the absence of the dipolar interactions.

The density-matrix treatment of a four-pulse DEER sequence is given in Appendix 12.II, where the following result is calculated for the echo signal for an isolated pair of spins:

$$<I^+(2\tau_1 + 2\tau_2)> = iI_0 \cos(a[t-\tau_1]) = iI_0 \cos(\omega_d[t-\tau_1]) \tag{12.10}$$

The four-pulse DEER signal consists of a damping component when averaged over a real sample, taking into account statistical distributions of the various pairs of spins, similar to that considered for the three-pulse DEER sequence. To be used in Equation 12.5, for a homogeneous distribution of spins, such as in a glass solution (Milov, Salikhov, and Shirov, 1981; Milov, Ponomarev, and Tsvetkov, 1984), one has:

$$I_{inter}(t) = \exp(-kCF_B|\tau_1 - t|), \tag{12.11}$$

where *C* is the concentration of spins S_1 that interact with each other via intermolecular interactions, F_B is the fraction of spins *B* excited by the pumping microwave with frequency, v_p, and *k* is given by:

$$k = \frac{8\pi^2 \mu_B^2 g_A g_B}{9\sqrt{3}\hbar}.$$

It is also superimposed by an oscillatory part, as shown in Figure 12.7, similar to that found in a three-pulse sequence. However, the refocussed echo amplitude is smaller in the four-pulse sequence as compared to that due to a three-pulse sequence, since here an additional pulse has been introduced, so a longer time evolution is required for the refocussed echo. On the other hand, since the values of $(t-\tau_1)$ are smaller here as compared to that due to a three-pulse sequence, they lead to enhanced modulation depths.

The four-pulse DEER signals, as obtained from six oligonucleotides, are shown in Figures 12.8a and b (Schiemann et al., 2004). These reveal that time dependencies of five doubly labeled oligonucleotides DNA1–DNA5 show oscillations except for DNA 6, which is singly labeled, missing the second spin for dipolar interaction. The echo decay observed for all DNA results from intramolecular spin–spin coupling. The Fourier transform of the DEER signal due to DNA1 in the frequency domain, as obtained after subtraction of echo decay, is shown in Figure 12.9; here, the Pake pattern is clearly visible, allowing the distances to be calculated in the same way as in a three-pulse DEER experiment.

12.5.4
Merits and Limitations of DEER as Compared to CW-EPR and FRET

These are as follows:

- The DEER method has the advantage that the characteristic oscillations in the signal due to the dipolar interaction are obtained only from doubly labeled protein; there is no contribution from singly labeled protein. Further, the rate

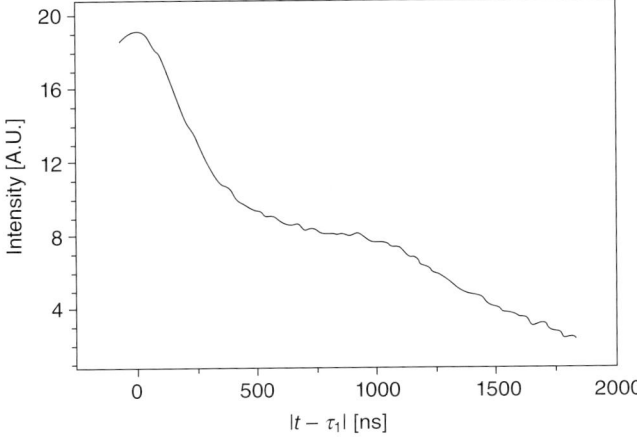

Figure 12.7 Experimental time domain signal obtained from four-pulse DEER experiment. Adapted from Bhatnagar (2005).

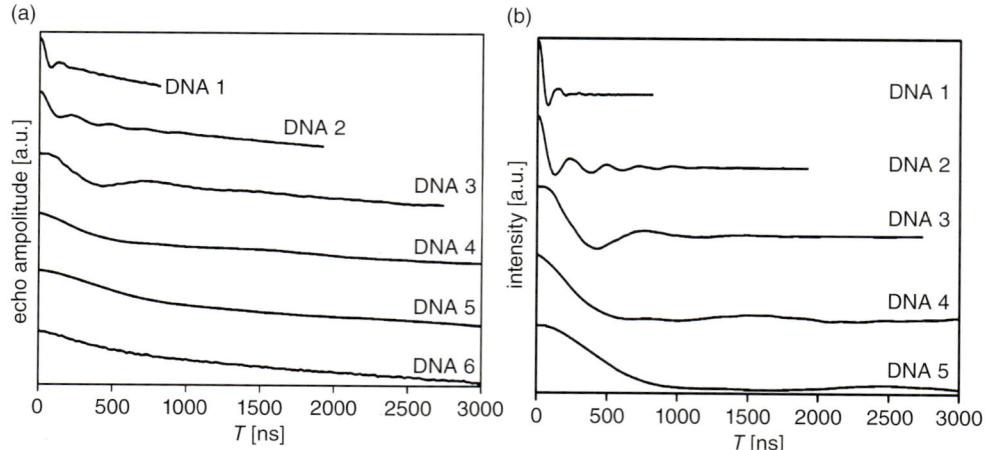

Figure 12.8 (a) DEER time traces of DNAs 1–6; (b) DEER time traces of DNAs 1–5 after subtraction of the echo decay. Adapted from Schiemann et al. (2004).

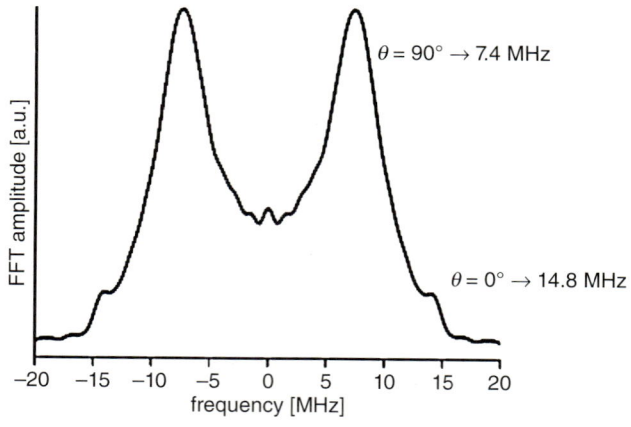

Figure 12.9 Fourier-transformed spectrum for DNA 1. Adapted from Schiemann et al. (2004).

at which the oscillation is damped provides a direct indication of the widths of distance distribution, which can only be obtained from CW lineshapes by either simulation or deconvolution.

- Distance measurements by DEER are limited on the upper side by the T_2 relaxation time, whereas the lower limit of ca. 15–20 Å is determined by the failure of the weaker DEER pulses to irradiate both parts of the Pake doublet, whereas DQC does not have this limitation, see Figure 12.10.

- One can measure larger distances, up to 80 Å (see Figure 12.11, from Georgieva et al., 2010), with DEER, than are possible with CW-EPR.

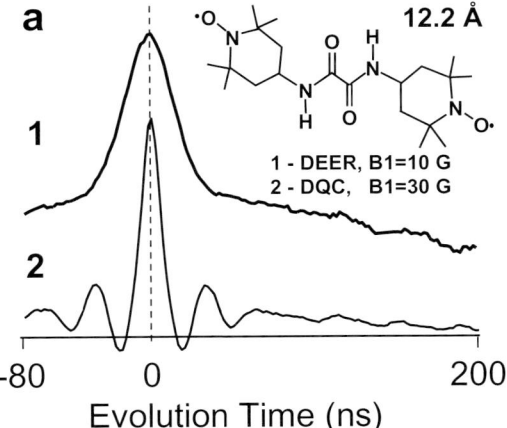

Figure 12.10 The challenges of short distances. DQC and DEER were applied to a rigid 12.2 Å nitroxide biradical. Detection pulses in DEER were 16/32/32ns, the pumping pulse was 18ns (B_1~10G). This is found to be insufficient to properly excite the dipolar spectrum. DQC using 6.2ns π-pulse (B_1~30G) develops the ~30MHz oscillations very cleanly. The longer pulses of DEER lead to a spread in the refocusing point of different spin packets, and the weaker B_1, both smear out the high-frequency dipolar oscillations. (From Borbat and Freed, 2007).

- A four-pulse DEER sequence is especially suitable when there exist broad distributions of small and intermediate distances. Furthermore, no comparison with the spectra due to monoradicals is required in DEER, unlike that in CW-EPR.

- As mentioned above, in DEER, the difference $|\omega_1 - \omega_2|$ needs to be significant in order to accomplish the required selective excitation. This restricts the use of those probes which satisfy the condition for DEER.

- Unlike FRET, orientation-dependence in DEER is well defined. In contrast, with FRET one can measure distances at the single-molecule scale, whereas at least 10 pmol of sample is required for DEER distance measurements.

12.6 Six-Pulse DQC

This technique for distance measurements was developed at Cornell University by Freed and coworkers (Freed, 2000; Borbat et al., 2001; Borbat, Mchaourab, and Freed, 2002, Borbat and Freed, 1999; Borbat and Freed, 2000). It is superior in a number of ways to DEER, but requires more intense microwave pulses. These include greater sensitivity especially for low concentration samples since (nearly) all the spins are excited; absence of orientational effects in its standard 1D version; filtering of single quantum signals and other noise by the double quantum filter; and the ability to obtain good results for distances as short as 10 Å. (See

Figure 12.11, from Borbat and Freed, 2007.) The double quantum coherence phenomenon has been extensively used in NMR (Ernst, Bodenhausen and Wokaun, 1987). We show in Figure 12.12 the six pulse sequence used in DQC-ESR. It transforms the initial density matrix under the successive action of six pulses, and six subsequent free-evolutions. In the ideal limit of perfect hard pulses, one may readily derive a simple expression for DQC-ESR by the product operator method analogous to those given in Appendix II.1 and II.2 for DEER (Borbat and Freed, 1999; 2000). It yields a cosine expression for the signal similar to those of Equations 12.7 and 12.10 for DEER (cf. Figure 12.12 caption). The following discussion is mostly taken from the recent publication on rigorous six-pulse DQC simulations by Misra, Borbat, and Freed (2009).

When a sample containing bilabeled proteins is subjected to a sufficiently strong microwave pulse, the nitroxide EPR spectrum is (almost) uniformly excited, and any orientational selection is (largely) suppressed; that is, it does not modify the

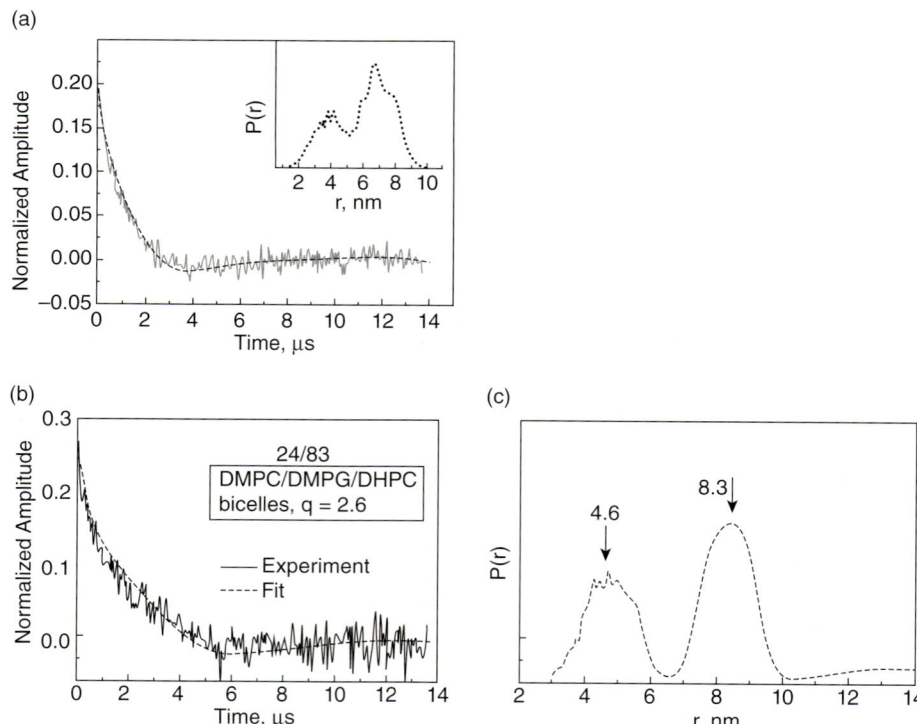

Figure 12.11 (a) The experimental time-domain data and distance distribution for 70% deuterated A30P alpha-synuclein (αS) mutant spin-labeled at positions 24 and 72 and reconstituted in micellar SDS-d25 using deuterated NMR buffer. (b) The experimental time-domain data (green) for 70% deuterated WT αS spin-labeled at positions 24 and 83 and reconstituted into bicelles. The fit (red) is based on distance distribution (c) produced by Maximum Entropy Method (MEM). Protein deuteration allowed recording dipolar signal on the time scale as long as 14 μs. (Distribution centered at 4.6 nm in C due to protein in solution unassociated with bicelles.) (From Georgieva et al., 2010).

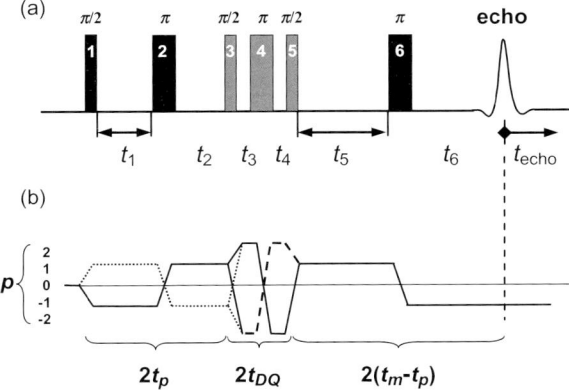

Figure 12.12 (a) The six-pulse DQC sequence; (b) Here, the coherence pathways correspond with the pulses shown in (a), in that a transition from one p state to another p state is generated by a pulse; the horizontal lines show coherence orders during the evolutions in the absence of a pulse. As for the timing between the various pulses the following is noted. The time interval $t_1 = t_2 = t_p$ is increased in equal steps, Δt_p, typically ranging from 1 to 10 ns, over a period of $t_m = t_p + t_5$ (200–4000 ns in this case). The time $t_3 = t_4 = t_{DQ}$ is kept fixed, typically at 20 ns; $t_5 = t_6$ is stepped by $-\Delta t_p$ to maintain a constant t_m; this starts from the initial time t_m. The echo signal is recorded in a window $t_w \sim 80\text{–}160$ ns, centered at a time delay $2t_m + 2t_{DQ}$ after the first pulse—that is, at about $t_6 = t_m$ after the sixth pulse. Note that the width of the echo sampling window limits the minimal values of t_6 and t_p by about $t_w/2$ and their maximum values to $(t_m - t_w/2)$. The dipolar evolution is recorded as a symmetric signal with respect to $t_{dip} \equiv t_m - 2t_p$ over the range of $\pm t_m$ in steps of $2\Delta t_p$. $t_{dip} = 0$ when the pulse separations are $t_1 = t_2 = t_5$. In practice, t_p starts with t_{p0} (~400 ns in this case), so that the last pulse and the echo window do not overlap. A simple analysis shows that the 1D signal goes as: $\cos\omega_i t_m - \cos\omega_i t_{dip}$. The signal in the 2-D DQC experiment is recorded (or computed) over $\pm (t_m - t_{p0})$, with t_{p0} always greater than $t_w/2$. Adapted from Misra, Borbat, and Freed (2009).

echo amplitude (except for the effect of pseudosecular dipolar terms, which is essential for short distances). Also, in high B_1-fields, the effect of dipolar coupling during the pulses becomes relatively weak. Therefore, for not very short distances and in sufficiently strong rf excitation fields (B_1s), the information on orientations of the magnetic tensors of the spin-label moieties, is virtually excluded from the time-domain dipolar evolution of the echo amplitude. However, as shown by Borbat and Freed (2000) and Misra, Borbat, and Freed (2009), it is still retained in the spin-echo evolution, and can be retrieved by recording the 2-D time-domain data as a function of the spin-echo time (t_{echo}) and the dipolar evolution time (t_{dip}). This can then be converted into a two-dimensional Fourier transform (2-D-FT) spectrum which, after making a "shear" transformation (Lee, Budil, and Freed, 1994), separates the dipolar dimension from the spectral dimension. Rigorous computations of 1-D and 2-D signals were carried out by Misra, Borbat, and Freed (2009), some of which are presented here. Efficient but approximate analytical expressions to this end were developed for 1-D signals by Borbat and Freed (2000) (see Appendix 12.V), by omitting the dipolar coupling during the pulses and assuming an ideal double-quantum (DQ) filter (cf. Figure 12.10, pulses 3, 4, and

5). Whereas, such expressions are quite useful for practical purposes and are computationally very efficient, they do not always supplant the rigorous calculations, especially in the case of short distances, for example, less than 15.0 Å. Numerical simulations of 1-D spectra were first carried out rigorously using these new codes in order to test the nature and extent of deviations from the exact results, and to establish the scope of applicability. The pulse propagators are calculated, using accurate numerical diagonalizations of the Hamiltonians involved. Although the computational approach is necessarily time-consuming, it does provide useful insights into the features of DQC spectroscopy.

12.6.1
Theoretical Background and Computation of Six-Pulse DQC Signal

The six-pulse DQC sequence is shown in Figure 12.10, indicating the pulse pattern and the relevant coherence pathways (Borbat and Freed, 1999; Borbat and Freed, 2000). The computational method for 1-D and 2-D-DQC spectra is outlined as follows.

The initial density matrix operator in thermal equilibrium for the two nitroxides is determined by the static spin-Hamiltonian ($\hat{H}$):

$$\hat{\rho}(0) = \frac{\exp(-\hat{H}_0/kT)}{\text{Tr}[\exp(-\hat{H}_0/kT)]} \rightarrow \hat{S}_{1z} + \hat{S}_{2z},$$

where the z-axis is chosen to be aligned along the direction of the external magnetic field, and the subscripts number the two electron spins. The arrow points to the relevant portion of $\hat{S}(0)$ assuming a high-temperature approximation. The time evolution of the spin density matrix, $\rho(t)$, is governed by the Liouville–von Neumann equation:

$$\frac{d\hat{\rho}}{dt} = -\frac{i}{\hbar}\hat{\hat{H}}(t)\hat{\rho}(t) - \hat{\hat{\Gamma}}(\hat{\rho}(t) - \hat{\rho}(0)).$$

where $\hat{\hat{H}}\rho \equiv [\hat{H}, \rho]$, $\hbar$ is Plank's constant divided by 2π, $\hat{\hat{\Gamma}}$ is the relaxation operator, $i^2 = -1$. Neglecting the relaxation, the density matrix evolves under the action of $\hat{H}$, as:

$$\frac{d\rho}{dt} = -\frac{i}{\hbar}[\hat{H}(t), \rho(t)],$$

the solution of which, after a period of time Δt, yields the density matrix, $\rho(t)$:

$$\rho(t + \Delta t) = e^{-\frac{i\hat{H}\Delta t}{\hbar}} \rho(t) e^{\frac{i\hat{H}\Delta t}{\hbar}} \equiv \hat{\hat{U}}(\hat{H}, \Delta t)\rho(t) \tag{12.12}$$

Numerical computation using Equation 12.12 of the DQC EPR signal is described below.

The un-normalized relevant part of the initial density matrix for the two coupled nitroxides, with electron spins $S_k = 1/2$, in thermal equilibrium at high temperatures is:

$$\rho(0) \rightarrow S_{1z} + S_{2z} \tag{12.13}$$

12.6 Six-Pulse DQC

(The normalization will be performed at the end of the calculation). The action of the six-pulse DQC sequence is illustrated as follows:

$$R_1\left(\frac{\pi}{2}\right) \to Q_1(t_p) \to R_2(\pi) \to Q_2(t_p) \to R_3\left(\frac{\pi}{2}\right) \to Q_3(t_{DQ}) \to R_4(\pi) \to Q_4(t_{DQ}) \to$$
$$R_5\left(\frac{\pi}{2}\right) \to Q_5(t_m - t_p) \to R_6(\pi) \to Q_6(t_m - t_p + t_{echo}), \quad (12.14)$$

where R_k (k = 1, 2, ... 6) are the six pulse propagators, and Q_k (k = 1, 2, ..., 6) are free-evolution propagators. Now, the kth pulse applied at the time t and acting during the period of time, τ_k, in the frame rotating with the angular frequency of the circular component of microwave magnetic field resonant with Larmor frequency of the nitroxide electron spin, transforms the density matrix, $\rho(t)$, according to:

$$\rho(t) - (R_k) \to \rho(t + \tau_k) = e^{-i\hat{H}_k \tau_{ki}} \rho(t) e^{i\hat{H}_k \tau_{ki}},$$

with R_k being the kth pulse propagator due to the effective Hamiltonian $\hat{H}_k$ acting during the period of time τ_k. The action of a π-pulse changes the sign of a coherence order, p (defined in Table 12.IV.1 of Appendix 12.IV), and the $\pi/2$ pulse generates other coherence orders so that $p \to p \pm 1$ (Gemperle et al., 1990). In order to follow the coherence pathways of interest, the density matrix is then projected onto the coherence pathways $\mathbf{p}_k$, which are chosen after the pulse according to Figure 12.10, as follows:

$$\rho'(t + \tau_k) = P(\mathbf{p}_k)\rho(t + \tau_k), \quad (12.15)$$

where the idempotent operator $P(\mathbf{p}_k)$ projects the density matrix on the coherence pathways $\mathbf{p}_k$ chosen after the kth pulse. As shown in Figure 12.10, the successive coherence pathways of interest are [(1, −1); (−1, 1); (2, −2); (−2, 2); (1); (−1)], that are chosen after the actions of the six pulses, with two branching points that lead to a total of four distinct pathways. The subsequent free evolution during the time t_k transforms $\rho'(t + \tau_k)$ as $\rho'(t + \tau_k) \xrightarrow{Q(t_k)} \rho(t + \tau_k + t_k)$ according to:

$$\rho(t + \tau_k + t_k) = e^{-i\hat{H}t_k} \rho'(t + \tau_k) e^{i\hat{H}t_k} \quad (12.16)$$

Table 12.IV.1 Coherence pathways and respective matrix elements for two coupled spins.

Coherence order, p	Corresponding matrix elements (i, k) in the electronic subspace of ρ
+2	(1,4)
+1	(1,2), (1,3), (2,4), (3,4)
0	(1,1), (2,2), (2,3), (3,2), (3,3), (4,4)
−1	(2,1), (3,1), (4,2), (4,3)
−2	(4,1)

In Equation 12.16, $\hat{H}$ is the spin-Hamiltonian in the absence of a pulse referred to as $\hat{H}_0$ below. The density matrix $\rho(t + \tau_k + t_k)$ is next used in place of $\rho(t)$ in Equation 12.15 for the calculation of the density matrix under the action of the $(k + 1)$-pulse. The steps defined by Equations 12.15 and 12.16 are then repeated to successively transform the density matrix to calculate the final density matrix, which becomes a function of several arguments, $\rho_f = \rho(\tau, t, p, t_{echo}, \eta, \lambda_1, \lambda_2)$. These arguments are defined as follows: $\tau = (\tau_1, \ldots, \tau_6)$ are the pulse durations; $t = (t_1, \ldots, t_6)$ are the subsequent free evolution periods; $p = (p_1, \ldots, p_6)$ are the relevant coherence orders during the evolution periods; t_k, t_{echo} are time variables used to record the dipolar evolution and to produce the spin echo envelope. The remaining arguments are the Euler angles $\eta = (\chi, \theta, \varphi)$, which define the orientation of the vector **r** connecting the magnetic dipoles associated with the electron spins in the laboratory frame (the angle χ was chosen to be zero as the medium is isotropic). In the dipolar (molecular) frame, whose z-axis is coincident with **r**, the Euler angles $\lambda_k = (\alpha_k, \beta_k, \gamma_k)$ define the principal axis of the nitroxide magnetic tensors (Figure 12.13) with α_1 chosen to be 0. Finally, the complex echo signal is given by:

$$F_+ = -2Tr[S_+ \tilde{\rho}_f].$$

where $\tilde{\rho}_f$ is the normalized density matrix.

The various propagators responsible for the evolution of the density matrix, depend on the exact form determined by $\hat{H}_0$ in the absence of a pulse, or by $\hat{H}_0 + \hat{H}_p$ in the presence of a pulse. Appropriate pulse time intervals τ_k are chosen to achieve nominal flip angles of $\pi/2$ ($k = 1, 3, 5$) and π ($k = 2, 4, 6$), respectively. The various Hamiltonians are:

$$\hat{H}_0 = \hat{H}_{01} + \hat{H}_{02} + \hat{H}_{12} \tag{12.17}$$

with

$$\hat{H}_{0k} = S_{kz} g_k \cdot \mathbf{B}_0 - \gamma_n I_{kz} B_0 + S_{kz} \mathbf{A}_k \cdot \mathbf{I}_k; \tag{12.18}$$

where $k = 1, 2$ denotes nitroxides 1 and 2, and H_{12} describes their coupling

$$\hat{H}_{12} = \hat{H}_D + \hat{H}_J = \frac{D}{2}(3\cos^2\theta - 1)(S_z^2 - \frac{1}{3}S^2) + J(\frac{1}{2} - 2\mathbf{S}_1 \cdot \mathbf{S}_2). \tag{12.19}$$

Here, J is the electron exchange constant and D is the dipolar coupling constant:

$$D = \frac{3\gamma_e^2 \hbar}{2r^3}, \tag{12.20}$$

which is equivalent to ω_d in eq. 12.4. The dipolar constant $d = 2D/3$ will most often be used throughout the text. In Equation 12.18, $I_{1,2}$ are the nuclear spins of the nitrogen (^{14}N or ^{15}N) nuclei on the two nitroxides.

The interaction of a nitroxide with the radiation field due to the applied microwave pulse k in the reference frame rotating with the carrier frequency ω_{rf}, usually set at or near the Larmor frequency, is expressed as:

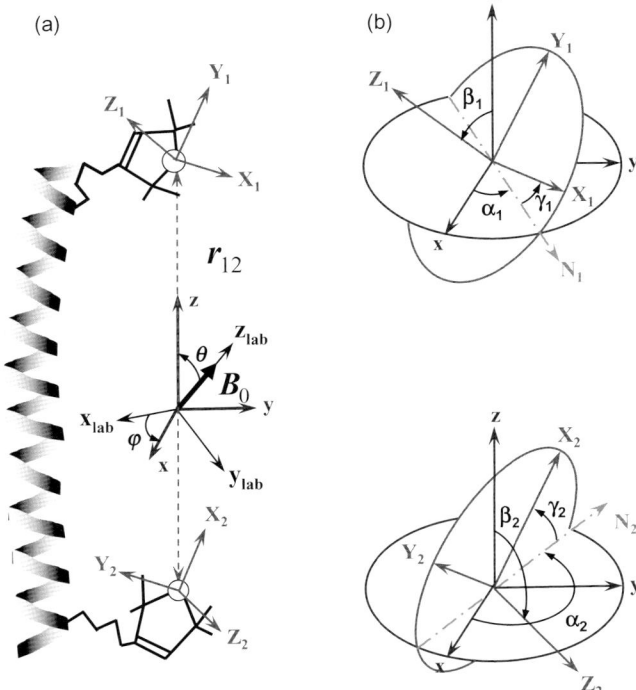

Figure 12.13 The set of Euler angles $\lambda_k = (\alpha_k, \beta_k, \gamma_k)$, $(k = 1, 2)$, which define the orientations of the hf and g-tensor tensor principal axes for nitroxides 1 and 2 in the dipolar (molecular) frame of reference. In this frame, the z-axis is chosen to coincide with the vector **r**, connecting the magnetic dipoles of the nitroxides. The orientation of the dipolar frame in the laboratory frame (with z-axis parallel to the external magnetic field B_0) is defined by the Euler angles $\eta = (0, \theta, \varphi)$. Adapted from Misra, Borbat, and Freed (2009).

$$\hat{H}_{pk} = \frac{\gamma_e B_{1k}}{2}(e^{-i\varphi_k} S_+ + e^{i\varphi_k} S_-); \tag{12.21}$$

where B_{1k} is the amplitude of the circular magnetic component. The phases, φ_k, can be set to zero for all the pulses for purposes of the present calculations and consequently $\hat{H}_{pk} = \gamma_e B_{1k} S_x$. The amplitudes, B_{1k}, are assumed to be equal here, but they do not have to be equal for different k-values. $\hat{H}_{0k}$, as given in Appendix 12.III, describing the spin Hamiltonian for the two nitroxides, becomes in the laboratory frame as

$$\hat{H}_{0k} = C_k S_{kz} + G_k I_{kz} + A_k S_{kz} I_{kz} + B_k S_{kz} I_{k+} + B_k^* S_{kz} I_{k-}$$

The time-domain DQC signal is calculated for a chosen set of λ_k and η, using appropriate variations of the time intervals following the various pulses (as given in the caption of Figure 12.12 for typical values used for the simulation). The calculations were carried out in the product space $S_1 \otimes S_2 \otimes I_1 \otimes I_2$ with the dimension $N = (2S_1 + 1)(2S_2 + 1)(2I_1 + 1)(2I_2 + 1)$. Accordingly, $\hat{H}_0$ and $\hat{H}$ are represented by order $N \times N$ matrices $\mathbf{H}_0$ and $\mathbf{H}$; that is, with the size of 36 × 36 for the two coupled (^{14}N) nitroxides, ($S_{1,2} = 1/2$, $I_{1,2} = 1$). The procedure to calculate $\rho_f(t_{dip}, t_{echo}, \theta, \varphi)$ is outlined in Appendix 12.IV.

Finally, the complex echo signal is given by:

$$F_+(t_{dip}, t_{echo}, \theta, \phi) = -2Tr[S_+\rho_f(t_{dip}, t_{echo}, \theta, \varphi)]/Tr[1_N].$$

where $\mathbf{1}_N$ is the unit matrix in the product space. For a powder sample, the echo signal is the average of the signals over the orientations of the molecule in the laboratory frame:

$$S(t_{dip}, t_{echo}) = \int_0^{2\pi} d\varphi \int_0^{\pi} F_+(t_{dip}, t_{echo}, \theta, \varphi) P_\Omega \sin\theta d\theta. \tag{12.22}$$

In Equation 12.22, P_Ω is the angular distribution of molecular axes in the laboratory frame ($P_\Omega = 1/4\pi$ for an isotropic distribution). In performing powder averaging in isotropic medium, the integration limits to be used are $[0, \pi]$ in axial angles (φ) and $[0, \pi/2]$ in polar angles (θ).

12.6.2
Illustrative Examples

The reader is referred to the publication by Misra, Borbat, and Freed (2009) for a detailed description of the illustrative examples based on the simulation procedure described here. Four of these examples are included in this chapter, as shown in Figures 12.14–12.17, the captions of which provide detailed descriptions.

12.6.3
Conclusions and Future Prospects of Six-Pulse DQC Echo Signal Simulation

The main features and conclusions from the DQC simulations presented here are as follows:

- The simulations for cases of short distances (10–15 Å) are rigorously performed utilizing the full spin Hamiltonian during the pulse.

- The results show that the application of a very strong B_1 field leads to clean Pake doublets in one dimension, that enables one to determine the dipolar (and exchange) couplings with the effects of correlations with the nitroxide magnetic tensors largely suppressed, in most cases. Then, in 2-D format, one may examine the "fingerprint" and make distinctions among different orientations of the principal-axis systems of the magnetic tensors of the nitroxides when correlations are present.

- It is clearly demonstrated from the simulations, that the concept of increased correlation sensitivity in 2-D FT spectra is indeed valid.

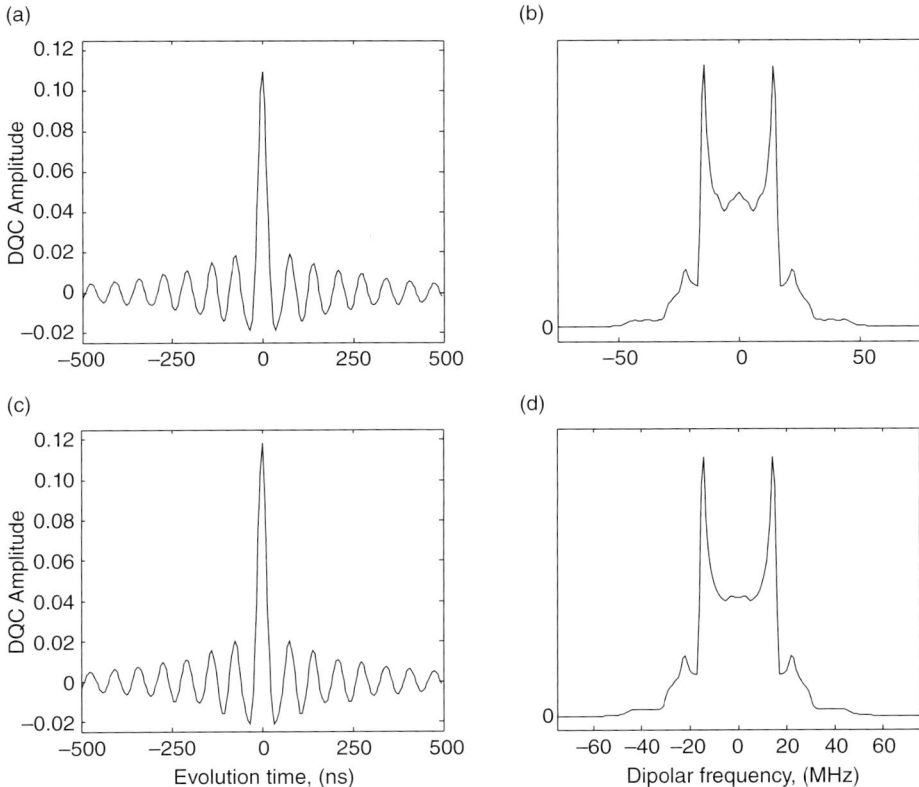

Figure 12.14 (a) Time-domain 1-D DQC signals and their Fourier transforms for ^{14}N nitroxides with their magnetic tensor axes orientations distributed isotropically in the molecular frame (i.e., referred to as uncorrelated case). (Top) – A computation result based on analytical approximation (cf. Equations 12.V.1–12.V.3 in Appendix 12.V) and (bottom) that computed rigorously. $B_0 = 6200\,G$, $B_1 = 30\,G$, and dipolar coupling (d) is 15 MHz (15.1 Å). This figure shows the 2-D time-domain data in dipolar and echo times and its 2-D FT. A small peak at $3d/2$ and a weak shoulder extending up to $3d$ are manifestations of the pseudosecular terms in H_{12}, as given by Equation 12.19. The difference between the two cases is quite small, being mostly caused by using simplified amplitude factors. Adapted from Misra, Borbat, and Freed (2009).

- It is also shown that increased DQC signal strengths are obtained by performing experiments with stronger pulses.
- The criterion for using the approximate analytical approach versus rigorous 1-D simulations at conventional frequencies (up to Q-band) has been established, as discussed in Appendix 12.V.
- For all practical purposes, rigorous DQC simulations should be utilized for the 2-D domain, strong coupling cases, and the millimeter-wave range. Simula-

(a)

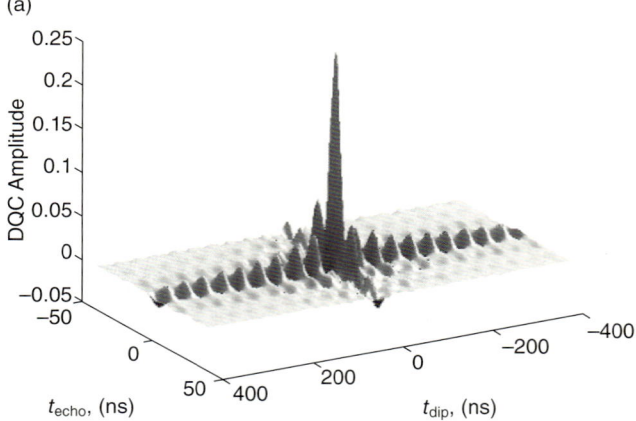

(b)

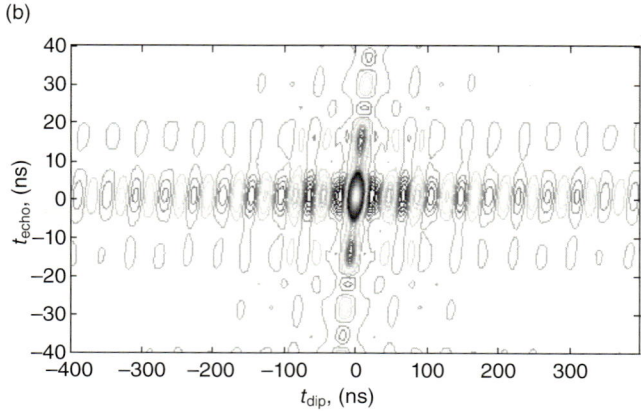

Figure 12.15 Time domain 2-D DQC signal shown as 3-D stack plot and contour plot. The simulations were carried out rigorously for $B_0 = 6200\,G$, $B_1 = 60\,G$, $d = 25\,MHz$ and uncorrelated ^{14}N nitroxides. The tilt of the spin-echo refocusing line is clearly visible; this is due to the fact that the spin-echo envelope is recorded over the time period where only one point corresponds to the dipolar interaction refocusing. A shift by Δt in the spin-echo time corresponds to a shift by $\Delta t/2$ in the position of the dipolar coupling refocusing point. Adapted from Misra, Borbat, and Freed (2009).

tions based on analytic approaches are two to three orders of magnitude computationally more efficient and virtually linearly scalable in multiprocessor-systems. This makes it possible to apply them to more complex cases that include averaging over multiple parameters, data fitting, or to multispin systems.

- The pseudosecular terms exhibit their effects clearly in 1-D and 2-D dipolar data. The pseudosecular part of dipolar coupling is responsible for the spectral

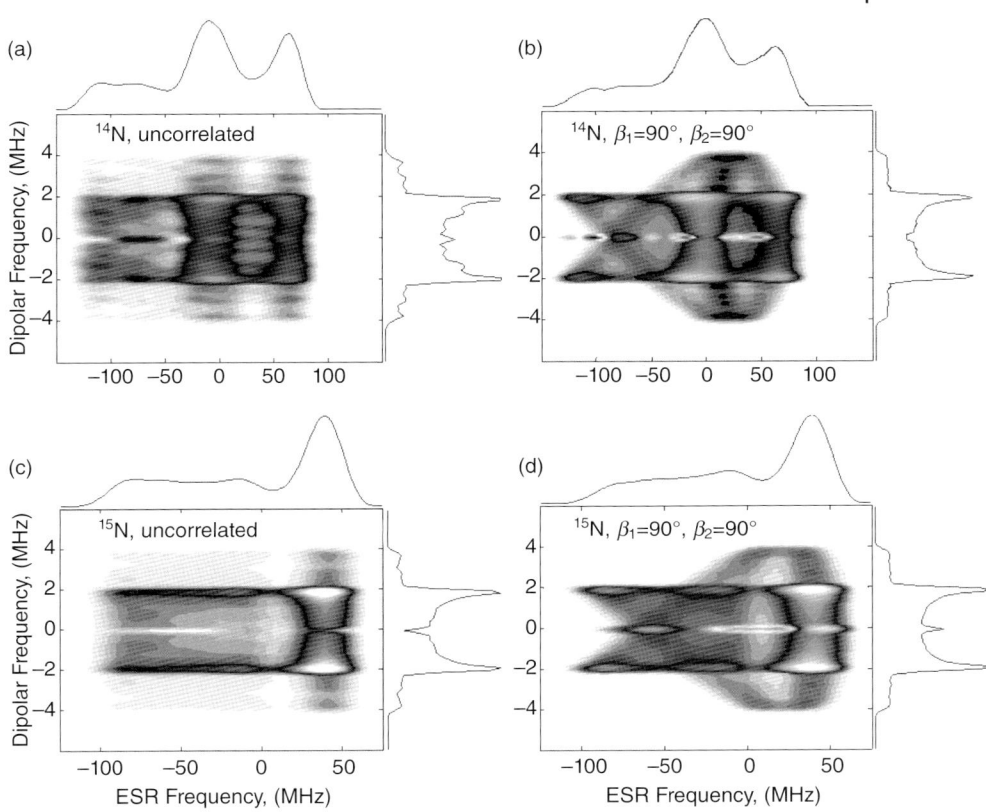

Figure 12.16 2-D DQC (filled) magnitude contour plots obtained by 2-D FT with respect to t_{dip} and t_{echo}. Top row: ^{14}N uncorrelated (a) and correlated (b) case. Bottom row: ^{15}N uncorrelated (c) and correlated (d) cases. $B_0 = 6200$ G, $d = 2$ MHz. B_1 was set to infinity (i.e., perfect pulses), pseudosecular terms were neglected. In (b, d) angles beta were (90°, 90°); the other angles were set to zero. Note the similarity of the 1-D dipolar spectra obtained by integration along the EPR frequency. These all are classic Pake doublets, but in the 2-D representation the differences are striking. For the uncorrelated cases, the dipolar spectrum is uniform for different slices along the EPR frequency axis, whereas for the correlated case they show a distinct "fingerprint" of this type of correlation. Since pseudosecular terms are neglected, the results are just applicable to long distances, such as the present case. Adapted from Misra, Borbat, and Freed (2009).

peaks with $3d/2$ splitting when two spins resonate at sufficiently close frequencies. This depends differently on orientational correlations than that for the secular part, leading to a richer 2-D spectrum.

- The 1-D spectrum does not show orientational correlations in most cases. On the other hand, the 2-D spectrum does exhibit patterns that are distinct from those obtained in the absence of correlation.

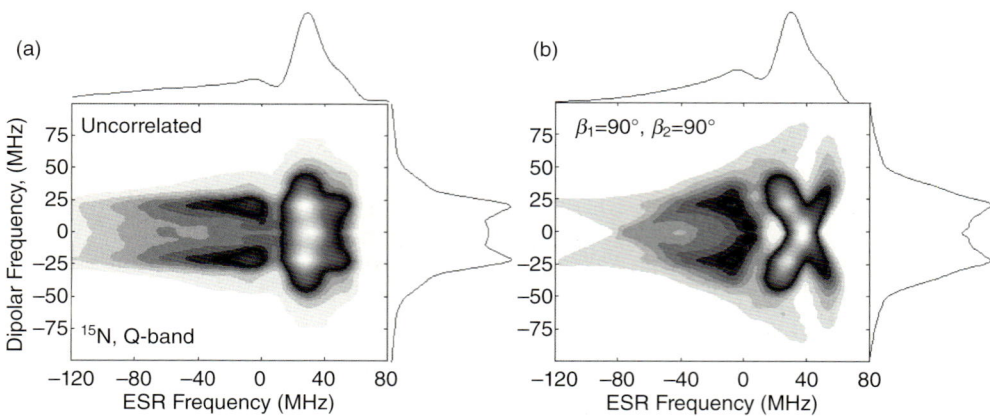

Figure 12.17 2-D DQC (filled) magnitude contour plots computed for ^{15}N nitroxides using $B_0 = 12\,500$, $B_1 = 60\,G$, $d = 25\,MHz$ with a Gaussian distribution in d (FWHM = 5 MHz). Panel (a) represents an uncorrelated case, whereas in (b) the angles beta were (90°, 90°) and alphas and gammas were set to zero. 4×10^4 Monte Carlo trials on a random set in $\{\cos\theta, \varphi, d\}$ were used to generate the data for (a); 180×180 mesh in $\{\cos\theta, \varphi\}$ was used to generate (b). The 1-D dipolar spectra on the right-hand sides of (a, b) are nearly completely smeared, and may be suited only to estimate d and its variance. The 2-D spectra, however, are quite different. The 2-D spectrum in (b) exhibits a distinct fingerprint of orientational correlation, but the 2-D spectrum for the uncorrelated case in (a) is similar to that in Figure 12.5, in that it tends to streak parallel to EPR frequency axis, as would be expected for such a case, where any point in the EPR spectrum corresponds to all possible orientations. Adapted from Misra, Borbat, and Freed (2009).

- The electron spins are treated here in the point dipole approximation ignoring spin-density delocalization. However, for distances less than about 10 , one should take into account spin-density delocalization, which is significant for example, on tyrosyl or flavin radicals, leading to a rhombic dipolar tensor.

- Relaxation effects have not been considered here. Phase relaxation can be introduced phenomenologically, as described by Saxena and Freed (1997) and Borbat and Freed (2000, 2007a), but sufficiently fast spin-lattice relaxation does require treatment with full rigor in Liouville space. However, there exist simplified versions that can be used in Hilbert space (see, for example, Lee, Patyal, and Freed, 1993).

12.7
Sensitivity Considerations: Multifrequency Aspects

The main criterion for sensitivity of PDS (Borbat and Freed, 2000) is based on the ability to measure reliably a distance in a reasonable period of time. An acceptable signal-to-noise ratio (SNR) to this end is nominally taken as a S_{acc} of 10, which

12.7 Sensitivity Considerations: Multifrequency Aspects

should be attained in an acceptable time of experiment nominally taken as 8h of signal averaging. However, a S_{acc} of 10 is a bare minimum, and one usually requires a SNR of at least 50 (Chiang, Borbat, and Freed, 2005a). An experimental calibration in the spirit of Borbat and Freed (2000) was made based on a measurement of the spin-echo amplitude using a two-pulse primary echo (PE), which provides the SNR, S_1 (PE), for a single shot. The ratio of the echo amplitudes relevant for DQC versus DEER was ca. 6.5, and the ratio of the SNRs of the single-shot signals at the condition of optimal signal reception was $S_1 \approx 0.42\,\mu M^{-1}$ (DEER) and $S_1 \approx 1.25\,\mu M^{-1}$ (DQC) (for more details, see Borbat and Freed, 2007). Accordingly, the estimates of the dipolar signals for the two techniques are summarized as follows. The S_1 value for DQC, based on a 3/6/3/6/3/6 ns pulse sequence, is greater by a factor of 3.6 than that for a four-pulse DEER with 16/32/32 ns pulses in the detection mode and a 32 ns pump pulse, based on the experimental observations at Cornell. The SNR of the raw data of the full PDS experiment, using the sensitivity analysis given by Borbat and Freed (2000), was estimated to be

$$\mathrm{SNR} = 2S_1 x^2 C\eta_c K(f, T_1)(ft_{\exp}/n)^{1/2} \exp\left(-\frac{2t_{\max}}{T_m} - 2kxCGt_{\max}\right), \quad (12.23)$$

where, $t_{\max}$ is the duration of the acquisition of experimental data; f is the frequency of the pulse sequence repetition; n is the number of data points in the record (more details are given by Borbat and Freed, 2000), indicating that Equation 12.23 gives a conservative estimate); C is the concentration of doubly labeled protein (μM); and η_c is the ratio of the sample volume to that used in the calibration. As for the exponential in Equation 12.23, the first term accounts for the phase relaxation, and the second for instantaneous diffusion; G is method specific, which is 0.14 for DEER and ca. 0.52 in DQC; x is the spin-labeling efficiency, which modifies the fraction of both spins that need to be flipped in PDS, exhibiting its strong effect on the outcome of an experiment (assumed to be 1 for complete labeling); $K(f, T_1) = 1 - \exp(-1/fT_1)$ represents the effect of incomplete spin-lattice relaxation for a given relaxation time, T_1, and the repetition rate, f. The following regimes, supported by experiment, are worthy of note in this context as discussed by Borbat and Freed (2000, 2007).

Short distances, low concentrations: Using Equation 12.23 and the various parameters listed by Borbat and Freed (2000, 2007) for a short distance of 20 Å ($T_{dip} \equiv v_d^{-1} = 154$ ns), just 4 min of signal averaging of the DQC signal provides a SNR of 10 for a C of 1 μM, whereas DEER will require nearly 60 min to achieve this result. Finally, a high SNR of 100 could be attained for DQC in 6.5 h for the same amount of protein.

Long distances: Using the various parameters listed by Borbat and Freed (2000, 2007) for this case, a SNR of 10 is achieved in 8 h for a C of 2.1 μM for DQC, whereas for DEER one would need 104 h. $R_{\max}$ is found to be 59 Å by using one period of T_{dip}, whereas for half of the period, $R_{\max}$ is 75 Å. For larger distances, one has a smaller SNR.

Distances in the optimal PDS range: 50 Å is considered to be an upper limit for the "optimal PDS" distance range. Then, T_{dip} is 2.4 μs, for which a t_{max} of 2.4 μs is sufficient to provide the distance reasonably accurately for a structure constraint. For the challenging case of $T_m = 1.5$ μs, a good SNR of 50 is achieved in 16 min by DQC, whereas one requires nearly 3.5 h to achieve the same result by DEER. Shorter distances of 20–45 Å are measured faster, or else yield a better SNR, or better resolution. The spin sensitivity is closely related to the concentration sensitivity, and increases rapidly with an increase in the working frequency due to the small volume of the resonator used at higher frequency. DQC is better suited to handle smaller amounts.

12.7.1
Frequency Dependence of Sensitivity of PDS

The single-shot SNR of the dipolar signal, S_1, in Equation 12.23 in the absence of relaxation with a view to estimate its frequency dependence is now considered. This is determined by the SNR of the relevant echo signal, which depends on the fraction of the participating A spins giving rise to the echo, further modified by a factor (of less than 1) which depends on the fraction of the B spins flipped by the pump pulse. (In DQC, the B spins are the same as the A spins.) A maximum SNR is achieved when nearly all spins are excited, the resonator Q matches the bandwidth of the echo and that of the excitation pulses, and the signal reception is optimized, for example, by matched filtering. As shown by Mims (1965), the single-shot SNR of the part of the echo modified by dipolar coupling, S_1, is

$$S_1 = \beta_0 \omega C V_s G H (Q\omega / V_c F_N \Delta f)^{1/2}, \tag{12.24}$$

where β_0 is a constant, $\omega = 2\pi f$, with f being the working frequency; C is the spin concentration in the sample; V_c is the resonator effective volume; $V_s = V_c \eta$ is the sample volume, with η being the filling factor of the resonator; G and H are the spectral excitations of spins A and B, respectively; Q is the loaded Q-value of the resonator; F_N is the system noise figure; and Δf is the receiver bandwidth. A discussion of the typical values of the parameters required in DQC and DEER is given by Borbat and Freed (2000, 2007). Taking all these into account, the SNR that can be achieved for the integrated dipolar signal, turns out to be

$$S_1 \propto \omega^2 C V_c^{1/2} \eta B_s^{-1} K K_2 K_1^{1/2}. \tag{12.25}$$

The frequency dependence of V_c is given by $V_c = \alpha \omega^{-3}$, with α depending on the resonator design, so that the concentration dependence of SNR is

$$S_1(C) \propto C \alpha^{1/2} \omega^{1/2} \eta B_s^{-1} K K_2 K_1^{1/2}. \tag{12.26}$$

On the other hand, the dependence of the absolute sensitivity on the number of spins (N) is

$$S_1(N) \propto N \alpha^{-1/2} \omega^{7/2} B_s^{-1} K K_2 K_1^{1/2}. \tag{12.27}$$

At very high frequencies, for which $B_s^{-1} \propto \omega$, the concentration and spin-number dependencies become

$$S_1(C) \propto C\alpha^{1/2}\omega^{-1/2}\eta KK_2K_1^{1/2}; \qquad (12.28)$$

$$S_1(N) \propto N\alpha^{-1/2}\omega^{5/2}KK_2K_1^{1/2}. \qquad (12.29)$$

Equations 12.28 and 12.29 imply that $S_1(C) \propto \omega^{-1/2}$ and $S_1(N) \propto \omega^{5/2}$, which means that the concentration sensitivity is not significantly benefited by going to higher frequencies, for example, in the millimeter range; however, the absolute sensitivity should be improved. More detailed considerations are required for designing of resonators with a larger value of α.

Recently, Ghimire et al. (2009) have found that substantial increase in DEER sensitivity can be obtained by collecting DEER data at Q-band (34 GHz) on a Bruker spectrometer, in comparison with that obtained at X-band (9 GHz) on a Bruker spectrometer. Specifically, in their experiment a 169-fold decrease in data collection time was associated with a huge boost in sensitivity by a factor of 13. They do not fully address instrumental factors responsible for this, nor do they note that the 17 GHz home-made spectrometer of Freed and co-workers has been delivering comparable high sensitivity for a number of years (Borbat and Freed, 2007).

12.8 Distance Distributions: Tikhonov Regularization

In the past, various methods have been proposed to determine the distance distributions of paramagnetic centers in solids (see, for example, Chiang, Borbat, and Freed, 2005a, 2005b; Bowman et al., 2004; and Jeschke et al., 2004). The Tikhonov regularization method (Tikhonov and Arsenin, 1997) has now become routine for extracting distance distributions from the data from both DEER and DQC. It is described as follows.

The time-domain dipolar signal for uniform spin distributions in the sample can be generally expressed as $V_{intra}A_{inter} + B_{inter}$, wherein B_{inter} originates from singly labeled molecules and free label or pairs where one of the spins does not participate. After removing the A and B terms as much as possible, the remainder is a reasonably accurate representation of V_{intra}, which is then subjected to inverse reconstruction by Tikhonov regularization or related methods. One represents the ideal-case problem by a Fredholm integral equation of the first kind

$$V_{intra}(t) = V_0 \int_0^\infty P(r)K(r,t)dr, \qquad (12.30)$$

where the kernel $K(r, t)$ for an isotropic sample, using Equations 12.2 and 12.3, is given by

$$K(r,t) = \int_0^1 \cos[\omega_d t(1-3x^2)]dx. \qquad (12.31)$$

The distance distribution, $P(r)$, is obtained by inversion of the signal V_{intra}, given by Equation 12.30. This can be achieved by using standard numerical methods, such as singular value decomposition (SVD), which in this case is an ill-posed problem, requiring regularization methods to arrive at a stable solution for $P(r)$. The actual form of the kernel $K(r, t)$ may differ from the ideal form given by Equation 12.31, since the data are discrete and available only over a limited time interval in the practical implementation.

The full distribution in distance, $P(r)$, can be recovered by Tikhonov regularization (Chiang, Borbat, and Freed, 2005a; Chiang, Borbat, and Freed, 2005b; Jeschke et al., 2004). This is accomplished by seeking an optimum $P(r)$, which tries to minimize the residual form of the fit to the data, and at the same time trying to maximize the stability of $P(r)$ by reducing its oscillations. The parameter λ, known as the regularization parameter, determines the relative importance of the two. It is optimized by the L-curve method (Hansen, 1992; Chiang et al., 2005a), which is computationally very efficient and the most reliable available to date. This regularization removes the contributions of the small singular values, σ_i in the SVD that are corrupted by the noise by introducing the filter function:

$$f_i \equiv \frac{\sigma_i^2}{\sigma_i^2 + \lambda^2}, \qquad (12.32)$$

which filters out those contributions for which $\sigma_i^2 \ll \lambda_i^2$. One may then use the maximum-entropy method (MEM) to refine $P(r)$ further, though this computationally more time-consuming (Chiang, Borbat, and Freed, 2005b). One is able to simultaneously fit and remove the effects of A_{inter} and/or B_{inter}, while optimizing the $P(r)$ from raw experimental data (Chiang, Borbat, and Freed, 2005a, 2005b) by using the latest versions of MEM and Tikhonov regularization. In this manner, distance distributions are recovered faithfully from the test data, simulated using the ideal kernel of Equation 12.31, even in the presence of significant noise, for example a SNR of 10 (Chiang, Borbat, and Freed, 2005a, 2005b; Bowman et al., 2004; Jeschke et al., 2004). In practice, real data deviate from this ideal picture, and there appears increased uncertainty, which requires a significantly higher SNR.

Recently, Jeschke (2009) has discussed the possibility of overcoming the ill-posed problem of determining distances from DEER data by implementing constraints, such as Tikhonov regularization (Tikhonov, 1955; Jeschke et al., 2004; Chiang, Borbat, and Freed, 2005b). This hides the influence of noise and of other distortions in noise data, and may result in reasonably looking distributions which are actually devoid of any information. Jeschke discusses criteria when DEER data are reliable, as well as a Monte Carlo approach to the validation of distance distributions. Such an approach is achieved by using the software DeerAnalysis2008 (Jeschke et al., 2006; http://www.epr.ethz.ch/software/index).

12.9
Additional Technical Aspects of DEER and DQC

Some noteworthy points are summarized as follows (for more details, see Borbat and Freed, 2007, 2007b):

- In the modern approach to PDS, one preferably uses a loop–gap resonators (LGR) or dielectronic resonators (DR) to achieve a higher sensitivity, and also a much smaller sample size when required.

- In three-pulse DEER, use of a single amplifier could lead to the problem of not having insufficient power at X-band but was not a problem at Ku-band.

- Since the pulses in four-pulse DEER are not required to be close (unlike those in three-pulse DEER), the significant dead-time effects inherent in three-pulse DEER are avoided in four-pulse DEER, thereby achieving greater sensitivity.

- Although DEER can be used, in principle, without phase cycling or even with incoherent pulses, it is recommended to use spectrometers with high instrument stability, such as are commercially available.

- Suppression of the baseline (background signal) is a key virtue of DQC, and is achieved by extensive phase-cycling, in particular by the use of a double-quantum filter. This diminishes any unwanted modulation of the signal due to low-frequency noise and drifts in phase or gain, as well as nuclear electron spin echo envelope modulation (ESEEM) effects, arising out of modulation of the large background from the single-order coherence signals.

- ESEEM effects: nuclear spin effects are minimized in three-pulse DEER, since the excitation and detections regions are well separated. However, ESEEM effects cannot be neglected in a typical four-pulse DEER experiment with a single power amplifier at X-band. The standard suppression techniques are very successful in both DQC and DEER. In addition, the proton ESEEM is virtually eliminated by increasing the frequency from 9 GHz to 17 GHz, but not the deuterium ESEEM in DQC.

- Orientation selection in DEER and DQC: As discussed by Larsen and Singel (1993) and by Maryasov, Tsvetkov, and Rapp (1998), orientation selection in DEER occurs due to the anisotropy of the nitroxide magnetic tensors, and their orientations relative to the inter-spin vector, which arises out of use of selective pulses. On the other hand, DQC is much less sensitive to orientational selectivity due to use of hard (intense) pulses. If required, however, orientational correlations can be revealed in considerable detail in a 2-D model, as illustrated recently by Misra, Borbat, and Freed (2009). In any case, the flexibility of side-chain labels, such as MTSSL, decreases the correlation effects considerably, whereas at high field it can be exploited to obtain some additional information on orientation of nitroxide side chains, and endogenous radical centers (Denysenkov et al., 2006; Polyhach et al., 2007).

12.10
Concluding Remarks

It has been demonstrated by at Cornell (Borbat et al., 2006; Park et al., 2006; Borbat et al., 2007; Upadhyay et al., 2008; Georgieva et al., 2008; Georgieva et al., 2010; Bhatnagar et al., 2010) as well as in other labs that PDS is clearly capable of being applied to extensive protein mapping. Future developments will enable EPR distance restraints, combined with modeling, nitroxide side-chain geometry simulation, and structure prediction to be applied to identify the detailed 3-D structures of large proteins, and of their complexes. Additional technical improvements are expected in PDS; in particular, DQC has not yet achieved optimum performance, and it is expected that both DQC and DEER will be further developed at a higher frequency. It is hoped that, in the near future, PDS – both as DEER and DQC – will become a standard technique for structural determinations, based on the realization that the technique has many virtues.

Acknowledgments

The authors are very indebted to Peter Borbat for his many seminal contributions to the subjects treated in this chapter and to the many conversations with him about them. S.M. is grateful to Professor C.P. Poole, Jr for constructive comments on this chapter. Partial support for this work is from NIH/NCRR grant # P41-RR016292 (J.H.F.)

Pertinent Literature

The articles by Borbat and Freed (2007a,b) provide an excellent review of PDS, along with a listing of pertinent reports. A recent article by Misra, Borbat, and Freed (2009) treats the simulation of DQC signal on a quantitative basis, the results of which can serve as a standard. The key articles on three-pulse DEER are from Milov, Salikhov, and Shirov (1981) and Larsen and Singel (1993), and the references cited therein. In the case of four-pulse DEER, a key article is that from Pannier et al. (2000), and the references cited therein.

References

Banham, J.E., Timmel, C.R., Abbot, R.J.M., Lea, S.M., and Jeschke, G. (2006) *Angew. Chem. Int. Ed.*, **45**, 1058.

Bhatnagar, J. (2005) Essay on DEER. Unpublished (Private Communication). For details, contact: skmisra@alcor.concordia.ca.

Bhatnagar, J., Borbat, P.P., Pollard, A.M., Freed, J.H., and Crane, B.R. (2010) *Biochemistry*, **49**, 3824.

Borbat, P.P. and Freed, J.H. (1999) *Chem. Phys. Lett.*, **313**, 145.

Borbat, P.P. and Freed, J.H. (2000) Double-Quantum ESR and distance measurements, in *Biological Magnetic Resonance*, vol. 19 (eds L.J. Berliner, S.S. Eaton, and G.R. Eaton), Kluwer Academic/Plenum Publications, New York, pp. 383–459.

Borbat, P.P. and Freed, J.H. (2007a) *EPR Newslett.*, **17** (2–3), 21.

Borbat, P.P. and Freed, J.H. (2007b) *Methods in Enzymology*, **423**, 52.

Borbat, P.P., Costa-Filho, A.J., Earle, K.A., Moscicki, J.K., and Freed, J.H. (2001) *Science*, **291**, 266.

Borbat, P.P., Mchaourab, H.S., and Freed, J.H. (2002) *J. Am. Chem. Soc.*, **124**, 5304.

Borbat, P.P., Davis, J.H., Butcher, S.E., and Freed, J.H. (2004) *J. Am. Chem. Soc.*, **126**, 7746.

Borbat, P.P., Ramlall, T.F., Freed, J.H., and Eliezer, D. (2006) *J. Am. Chem. Soc.*, **128**, 10004.

Borbat, P.P., Surendhran, K., Bortolus, M., Zou, P., Freed, J.H., and Mchaourab, H.S. (2007) *PLoS Biology*, **5**, 2211.

Bowman, M.K., Maryasov, A.G., Kim, N., and DeRose, V.J. (2004) *Appl. Magn. Reson.*, **26**, 223.

Cai, Q., Kuznetzow, A.K., Hubbell, W.L., Haworth, I.S., Gacho, G.P.C., Eps, N.V., Hideg, K., Chambers, E.J., and Qin, P.Z. (2006) *Nucleic Acids Res.*, **34**, 4722.

Chiang, Y.-W., Borbat, P.P., and Freed, J.H. (2005a) *J. Magn. Reson.*, **172**, 184.

Chiang, Y.-W., Borbat, P.P., and Freed, J.H. (2005b) *J. Magn. Reson.*, **172**, 279.

Denysenkov, V.P., Prisner, T.F., Stubbe, J., and Bennati, M. (2006) *Proc. Natl Acad. Sci. USA*, **103**, 13386.

Eaton, S.S., More, K.M., Sawant, B.M., and Eaton, G.R. (1983) *J. Am. Chem. Soc.*, **105**, 6560.

Ernst, R.R., Bodenhausen, G., and Wokaun, A. (1987) *Principles of Nuclear Magnetic Resonance in One and Two Dimensions*, Clarendon Press, Oxford.

Fafarman, A.T., Borbat, P.P., Freed, J.H., and Kirshenbaum, K. (2007) *Chem. Commun.*, (**4**), 377.

Freed, J.H. (1976) The theory of Slow Tumbling ESR Spectra for Nitroxides, in *Spin Labeling: Theory and Application* (ed. L.J. Berliner), Academic Press, New York, Chapter 2, pp. 53–132.

Freed, J.H. (2000) *Annu. Rev. Phys. Chem.*, **51**, 655.

Gemperle, C., Aebli, G., Schweiger, A., and Ernst, R.R. (1990) *J. Magn. Reson.*, **88**, 241.

Gemperle, C., Aebli, G., Schweiger, A., and Hansen, P.C. (1992) *SIAM Rev.*, **34**, 561.

Georgieva, E.R., Ramlall, T.F., Borbat, P.P., Freed, J.H., and Eliezer, D.J. (2008) *Am. Chem. Soc.* **130**, 12856.

Georgieva, E.R., Ramlall, T.F., Borbat, P.P., Freed, J.H., and Eliezer, D.J. (2010) *J. Biol. Chem.* **285**, 28261.

Ghimire, H., McCarrick, R.M., Budil, D.E., and Lorigan, G.A. (2009) *Biochemistry (Rapid Report)*, **48**, 5782.

Hansen, P.C. (1992) *SIAM Rev.*, **34**, 561.

Hustedt, E.J., Cobb, C.E., Beth, A.H., and Beechem, J.M. (1993) *Biophys. J.*, **74**, 1861.

Hustedt, E.J., Smirnov, A.I., Laub, C.F., Cobb, C.E., and Beth, A.H. (1997) *Biophys. J.*, **72**, 1861.

Jeschke, G. (2002) *Chem. Phys. Chem.*, **3**, 927.

Jeschke, G. (2009) *EPR Newslett.*, **18**, 15.

Jeschke, G., Chechlik, V., Ionita, P., Godt, A., Zimmerman, H., Banham, J., Timmel, C.R., Hilger, D., and Jung, H. (2006) *Appl. Magn. Reson.*, **30**, 473.

Jeschke, G., Pannier, M., and Spiess, H.W. (2000) Double Electron Electron Resonance, in *Biological Magnetic Resonance*, vol. 19 (eds L.J. Berliner, G.R. Eaton, and S.S. Eaton), Plenum, New York, p. 493.

Jeschke, G., Panek, G., Godt, A., Bender, A., and Paulsen, H. (2004) *Appl. Magn. Reson.*, **26**, 223.

Klauder, J.R. and Anderson, P.W. (1962) *Phys. Rev.*, **125**, 912.

Klug, C.S., Camenisch, T.G., Hubbell, W.L., and Hyde, J.S. (2005) *Biophys. J.*, **88**, 3641.

Kokorin, A.I., Zamaraev, K.I., Grigoryan, G.L., Ivanov, V.P., and Rozantsev, E.G. (1972) *Biofizika*, **17**, 34.

Larsen, R.G. and Singel, D.J. (1993) *J. Chem. Phys.*, **98**, 5134.

Lee, S., Patyal, B.R., and Freed, J.H. (1993) *J. Chem. Phys.*, **98**, 3665.

Lee, S., Budil, D.E., and Freed, J.H. (1994) *J. Chem. Phys.*, **99**, 7098.

Libertini, L.J., and Griffith, O.H. (1970) *J. Chem. Phys.*, **53**, 1359–1367.

Martin, R.E., Pannier, M., Diederich, F., Gramlich, V., Hubrich, M., and Spiess, H.W. (1998) *Angew. Chem. Int. Ed.*, **37**, 2834.

Maryasov, A.G., Tsvetkov, Y.D., and Rapp, J. (1998) *Appl. Magn. Reson.*, **14**, 101.

Milov, A.D., Salikhov, K.M., and Shirov, M.D. (1981) *Sov. Phys. Solid State*, **23**, 565.

Milov, A.D., Ponomarev, A.B., and Tsvetkov, Y.D. (1984) *Chem. Phys. Lett.*, **110**, 67.

Mims, W.B. (1965) *Rev. Sci. Instrum.*, **36**, 1472.

Misra, S.K. (2007) *J. Magn. Reson.*, **189**, 59.

Misra, S.K., Borbat, P.P., and Freed, J.H. (2009) *Appl. Magn. Reson.*, **36**, 237.

Narr, E., Godt, A., and Jeschke, G. (2002) *Angew. Chem. Int. Ed.*, **41**, 3907.

Pannier, M., Veit, S., Godt, A., Jeschke, G., and Spiess, H.W. (2000) *J. Magn. Reson.*, **142**, 331.

Park, S.-Y., Borbat, P.P., Gonzalez-Bonet, G., Bhatnagar, J., Pollard, A.M., Freed, J.H., Bilwes, A.M., and Crane, B.R. (2006) *Nat. Struct. Mol. Biol.*, **13**, 400.

Persson, M., Harbridge, J.R., Hammerstrom, P., Mitri, R., Martensson, L.-G., Carlsson, U., Eaton, G.R., and Eaton, S.S. (2001) *Biophys. J.*, **80**, 2886.

Polyhach, Y., Godt, A., Bauer, C., and Jeschke, G. (2007) *J. Magn. Reson.*, **185**, 118.

Press, W.H., Teukolsky, S.A., Vetterling, W.T., and Flannery, B.P. (1992) *Numerical Recipes in Fortran*, 2nd edn, Cambridge University Press, pp. 456–462.

Rabenstein, M.D. and Shin, Y.-K. (1995) *Proc. Natl Acad. Sci. USA*, **92**, 8239.

Saxena, S. and Freed, J.H. (1997) *J. Chem. Phys.*, **107**, 1317–1340.

Schiemann, O., Piton, N., Mu, Y., Stock, G., Engels, J.W., and Prisner, T.F. (2004) *J. Am. Chem. Soc.*, **126**, 5722.

Schneider, D.J. and Freed, J.H. (1989a) Calculating Slow Motional Magnetic Resonance Spectra: A User's Guide, in *Spin Labeling: Theory and Applications*, Vol. III, Biological Magnetic Resonance 8, (Plenum, NY), pp. 1–76.

Schneider, D.J. and Freed, J.H. (1989b) Spin Relaxation and Motional Dynamics, in *Advances in Chemical Physics*, vol. 73 (eds J.O. Hirschfelder, R.E. Wyatt, and R.D. Coalson), John Wiley & Sons, Inc., New York, pp. 387–528.

Sun, J., Voss, J., Hubbell, W.L., and Kaback, H.R. (1999) *Biochemistry*, **38**, 3100.

Tikhonov, A.N. (1955) *Numerical Methods for the Solution of Ill-Posed Problems*, Kluwer Academic Publishers, Dordrecht, Boston.

Tikhonov, A.N. and Arsenin, V.Y. (1997) *Solution of Ill-Posed Problems*, Halsted Press, John Wiley & Sons, Inc., New York.

Upadhyay, A.K., Borbat, P.P., Wang, J., Freed, J.H., and Edmondson, D.E. (2008) *Biochemistry*, **47**, 1554.

Appendix 12.I
Density-Matrix Derivation of Echo Signal for Three-Pulse DEER

Here, calculations are described for a coupled pair of isolated spins in a disordered solid. These calculations are carried out in Hilbert space, wherein the spins are assumed to be quantized along the external magnetic field, B, directed along the Z axis. Relaxation effects are neglected in the present consideration (Bhatnagar, 2005).

The spin Hamiltonian for coupled nitroxides is expressed in frequency units as:

$$H/\hbar = \omega_{01}S_{1z} + \omega_{02}S_{2z} + aS_{1z}S_{2z} \qquad (12.I.1)$$

In Equation 12.I.1, $\omega_{01} = \gamma_1 B_0$ is the resonance frequency of the isolated S_1 spins; $\omega_{02} = \gamma_2 B_0$ is the resonance frequency of the isolated S_2 spins; a is the dipolar coupling frequency expressed by Equation 12.2 ignoring the pseudosecular term in Equation 12.1; B_0 is the external magnetic field; and γ_1, γ_2 are the gyromagnetic ratios for S_1 and S_2 spins, respectively. The following considerations are made in the rotating frame, in which the S_1 spins are off-resonance by ΔB_{01} and the S_2 spins by ΔB_{02}. The corresponding chemical shifts are: $\Delta \omega_{01} = \gamma_1 \Delta B_{01} - \omega_{01}$ and $\Delta \omega_{02} = \gamma_2 \Delta B_{02} - \omega_{02}$.

The pulse sequence in three-pulse DEER is: $X_I(\pi/2)$ ---- τ ---- $X_S(\pi/2)$ ---- $(\tau' - \tau)$ ---- $X_I(\pi)$ ---- $(t - \tau')$. In other words, at the time $t = 0^-$, a $\pi/2$ pulse, $X_I(\pi/2)$,

Appendix 12.I Density-Matrix Derivation of Echo Signal for Three-Pulse DEER

is applied on S_1 spins about the X axis. Thereafter, at time τ, a $\pi/2$, $X_2(\pi/2)$, pulse is applied on S_2 spins about the X-axis, followed by the application of a π-pulse on S_1 spins about the X axis, $X_1(\pi)$, after a time delay of $(\tau'-\tau)$. The subsequent time evolution of the density matrix for $t > \tau'$ produces an echo for S_1 spins at $t = 2\tau'$, which is the signal of interest here.

The echo signal for S_1 spins at any time is given by the expectation value of the transverse magnetization, proportional to $<S_1^+(t)> = Tr\{S_1^+\rho(t)\}$, where Tr stands for trace. The initial value of the density matrix, considering the Zeeman terms only, ignoring the dipolar and exchange terms, is:

$$\rho(0^-) = \exp(-H/k_BT)/Z \approx \{1-(\hbar\omega_{01}S_{1z}+\hbar\omega_{02}S_{2z})/k_BT\}/Z, \quad (12.I.2)$$

where $Z = 1/\sum_i \exp(-E_i/k_BT)$, the sum being over the four energy states of the two coupled nitroxides with spin ½ each; T is the temperature; and k_B is the Boltzmann constant.

The $\pi/2$ pulse, applied on S_1 spins about the X-axis at $t = 0^-$, transforms the density matrix to:

$$\rho(0^+) = X_1(\pi/2)\rho(0^-)X_1^{-1}(\pi/2)$$
$$= X_1(\pi/2)(1/Z)\{1-(\hbar\omega_{01}S_{1z}+\hbar\omega_{02}S_{2z})/k_BT\}X_1^{-1}(\pi/2) \quad (12.I.3)$$

where $X_1(\pi/2) = e^{i\frac{\pi}{2}S_{1x}}$. The $X_1(\pi/2)$ pulse rotates the S_1 magnetization from the Z axis to Y axis, and has no effect on S_2 spins: $e^{i\frac{\pi}{2}S_{1x}}S_{1z}e^{-i\frac{\pi}{2}S_{1x}} = S_{1y}$.

Now, since only the echo signal from S_1 spins is detected, the constant term and the term in S_{2z} in the initial value of the density matrix does not contribute to it, so they are hereafter ignored in the density matrix. Thus, considering only the term with S_{1y}, one has

$$\rho(0^+) = -\hbar\omega_{01}S_{1y}/(Zk_BT) \quad (12.I.4)$$

The evolution of the density matrix under the effect of the static Hamiltonian,

$$H_s = \Delta\omega_1 S_{1z} + \Delta\omega_2 S_{2z} - aS_{1z}S_{2z} \quad (12.I.5)$$

over the time interval τ is now considered for $0 < t < \tau$:

$$\rho(t) = e^{i(\Delta\omega_1 S_{1z}+\Delta\omega_2 S_{2z}-aS_{1z}S_{2z})t}\rho(0^+)e^{-i(\Delta\omega_1 S_{1z}+\Delta\omega_2 S_{2z}-aS_{1z}S_{2z})t} \quad (12.I.6)$$

After the application of a π-pulse about the X-axis on S_2 spins at $t = \tau^-$, one obtains from Equations 12.I.4 and 12.I.5 the resulting density matrix:

$$\rho(\tau^+) = X_2(\pi)\rho(\tau^-)X_2^{-1}(\pi)$$
$$= (-\hbar\omega_{01}/Zk_BT)X_2(\pi)e^{i(\Delta\omega_1 S_{1z}+\Delta\omega_2 S_{2z}-aS_{1z}S_{2z})\tau}S_{1y}e^{-i(\Delta\omega_1 S_{1z}+\Delta\omega_2 S_{2z}-aS_{1z}S_{2z})\tau}X_2^{-1}(\pi)$$
$$(12.I.7)$$

Inserting the unit operator $(X_2^{-1}(\pi)X_2(\pi))$ in Equation 12.I.7 before and after S_{1y}, and writing it as a product of three terms, one obtains:

$$\rho(\tau^+) = (-\hbar\omega_{o1}/Zk_BT)A*B*C,$$

where $A = X_2(\pi)e^{i(\Delta\omega_1 S_{1z}+\Delta\omega_2 S_{2z}-aS_{1z}S_{2z})\tau}X_2^{-1}(\pi)$; $B = X_2(\pi)S_{1y}X_2^{-1}(\pi)$; and $C = X_2(\pi)$ $e^{-i(\Delta\omega_1 S_{1z}+\Delta\omega_2 S_{2z}-aS_{1z}S_{2z})\tau}X_2^{-1}(\pi)$.

Now, since the S_1 and S_2 operators commute, the B-term simply reduces to S_{1y}. As for the terms A and C, the $X_2(\pi)$ pulse will simply reverse the signs of the exponentials of the terms $e^{i\Delta\omega_2 S_{2z}}$ and $e^{-iaS_{1z}S_{2z}}$, while leaving $e^{i\Delta\omega_1 S_{1z}}$ unchanged, since the pulse is selective to S_2 spins only, so that $A = e^{i(\Delta\omega_1 S_{1z}-\Delta\omega_2 S_{2z}+aS_{1z}S_{2z})\tau}$ and $C = e^{-i(\Delta\omega_1 S_{1z}-\Delta\omega_2 S_{2z}+aS_{1z}S_{2z})\tau}$. One, therefore, obtains for the density matrix after the application of the $X_2(\pi)$ pulse at $t = \tau^-$

$$\rho(\tau^+) = (-\hbar\omega/Zk_BT)e^{i(\Delta\omega_1 S_{1z}-\Delta\omega_2 S_{2z}+aS_{1z}S_{2z})\tau}S_{1y}e^{-i(\Delta\omega_1 S_{1z}-\Delta\omega_2 S_{2z}+aS_{1z}S_{2z})\tau} \quad (12.1.8)$$

Since the operators S_1 and S_2 commute, the above expression can be simplified by putting the terms containing only the S_{2z} operators in the exponentials on the left and right sides of S_{1y} equal to zero. Then, one obtains

$$\rho(\tau^+) = (-\hbar\omega_{01}/Zk_BT)e^{i(\Delta\omega_1 S_{1z}+aS_{1z}S_{2z})\tau}S_{1y}e^{-i(\Delta\omega_1 S_{1z}+aS_{1z}S_{2z})\tau} \quad (12.1.9)$$

Now, the system evolves during the time $(\tau'-\tau)$ before a π-pulse is applied to S_1 spins, that is for $\tau < t < \tau'$, during which the density matrix evolves under the action of the static Hamiltonian H_s:

$$\rho(t) = e^{i(\Delta\omega_1 S_{1z}+\Delta\omega_2 S_{2z}-aS_{1z}S_{2z})(t-\tau)}\rho(\tau^+)e^{-i(\Delta\omega_1 S_{1z}+\Delta\omega_2 S_{2z}-aS_{1z}S_{2z})(t-\tau)} \quad (12.1.10)$$

Using the expression for $\rho(\tau^+)$ from Equation 12.I.9, one obtains for the density matrix at $t = \tau'^-$,

$$\rho(\tau'^-) = (-\hbar\omega_{01}/Zk_BT)e^{i\Delta\omega_1 S_{1z}\tau'}e^{-iaS_{1z}S_{2z}(\tau'-2\tau)}S_{1y}e^{-i\Delta\omega_1 S_{1z}\tau'}e^{-i\Delta\omega}e^{iaS_{1z}S_{2z}(\tau'-2\tau)} \quad (12.1.11)$$

In simplifying Equation 12.I.11 it is noted that, after substituting the expression for $\rho(\tau^+)$ from Equation 12.I.9, the terms involving $e^{i\Delta\omega_2 S_{2z}(\tau'-2\tau)}$ on the left of S_{1y} and $e^{-i\Delta\omega_2 S_{2z}(\tau'-2\tau)}$ on the right side cancel out, since S_{1y} and S_{2z} commute with each other.

Now, at $t = \tau'^-$, the application of a π-pulse to S_1 spins about the X-axis transforms it according to:

$$\rho(\tau'^+) = X_1(\pi)\rho(\tau'^-)X_1^{-1}(\pi) = X_1(\pi)(-\hbar\omega_{01}/Zk_BT)e^{i\Delta\omega_1 S_{1z}\tau'}e^{-iaS_{1z}S_{2z}(\tau'-2\tau)}$$
$$S_{1y}e^{-i\Delta\omega_1 S_{1z}\tau'}e^{iaS_{1z}S_{2z}(\tau'-2\tau)}X_1^{-1}(\pi) \quad (12.1.12)$$

Now, by inserting the unity operator $[= X_1(\pi)X_1^{-1}(\pi)]$ twice on the right-hand side of the above equation, and using the same procedure as that used to derive Equation 12.I.8 for $\rho(\tau^+)$, one obtains

$$\rho(\tau'^+) = (-\hbar\omega_{01}/Zk_BT)A*B*C, \text{ where}$$

$$A = X_1(\pi)e^{i\Delta\omega_1 S_{1z}\tau'}e^{-iaS_{1z}S_{2z}(\tau'-2\tau)}X_1^{-1}(\pi) = e^{-i\Delta\omega_1 S_{1z}\tau'}e^{iaS_{1z}S_{2z}(\tau'-2\tau)};$$

$$B = X_1(\pi)S_{1y}X_1^{-1}(\pi) = -S_{1y}; \text{ and}$$

$$C = X_1(\pi)e^{-i\Delta\omega_1 S_{1z}\tau'}e^{iaS_{1z}S_{2z}(\tau'-2\tau)}X_1^{-1}(\pi) = e^{i\Delta\omega_1 S_{1z}\tau'}e^{-iaS_{1z}S_{2z}(\tau'-2\tau)}$$

In simplifying B to the second term on the right in the above equation, the fact that a rotation by 180° about the X-axis will orient the magnetization of S_1 spins

Appendix 12.I Density-Matrix Derivation of Echo Signal for Three-Pulse DEER

along the negative Y axis has been taken into account. Further, as for simplifications of the terms A and C, the selective X_1 π-pulse for S_1 spins will reverse the orientations of S_1 spins. This will result in reversals of the signs of the exponents of $e^{i\Delta\omega_1 S_{1z}\tau'}$ and $e^{-iaS_{1z}S_{2z}(\tau'-2\tau)}$ in the A term, and those of $e^{-i\Delta\omega_1 S_{1z}\tau'}$ and $e^{iaS_{1z}S_{2z}(\tau'-2\tau)}$ in the C term. Finally,

$$\rho(\tau'^+) = (\hbar\omega_{01}/Zk_BT)e^{-i\Delta\omega_1 S_{1z}\tau'}e^{iaS_{1z}S_{2z}(\tau'-2\tau)}S_{1y}e^{i\Delta\omega_1 S_{1z}\tau'}e^{-iaS_{1z}S_{2z}(\tau'-2\tau)} \quad (12.I.13)$$

Now, the system evolves under the action of the static Hamiltonian for time $t > \tau'$, during which

$$\rho(t) = e^{i(\Delta\omega_1 S_{1z}+\Delta\omega_2 S_{2z}-aS_{1z}S_{2z})(t-\tau')}\rho(\tau'^+)e^{-i(\Delta\omega_1 S_{1z}+\Delta\omega_2 S_{2z}-aS_{1z}S_{2z})(t-\tau')} \quad (12.I.14)$$

By substituting Equation 12.I.13 for $\rho(\tau'^+)$ in Equation 12.I.14, and simplifying, one obtains

$$\rho(t) = (\hbar\omega_{01}/Zk_BT)e^{i\Delta\omega_1 S_{1z}(t-2\tau')}e^{-iaS_{1z}S_{2z}(t+2\tau-2\tau')}S_{1y}e^{-i\Delta\omega_1 S_{1z}(t-2\tau')}e^{iaS_{1z}S_{2z}(t+2\tau-2\tau')} \quad (12.I.15)$$

At the time $t = 2\tau'$, when the echo is formed, the exponentials involving $\Delta\omega_1$ simplify to unity, leaving only the terms with $S_{1z}S_{2z}$. Then the density matrix becomes

$$\rho(2\tau') = (\hbar\omega_{01}/Zk_BT)e^{iaS_{1z}S_{2z}2\tau}S_{1y}e^{-iaS_{1z}S_{2z}2\tau} \quad (12.I.16)$$

The echo amplitude at the time $t = 2\tau'$ will then be

$$<S_1^+(2\tau')> = Tr\{S_1^+\rho(2\tau')\} = (\hbar\omega_{01}/Zk_BT)Tr\{S_1^+ e^{iaS_{1z}S_{2z}2\tau}S_{1y}e^{-iaS_{1z}S_{2z}2\tau}\} \quad (12.I.17)$$

The term in the curly brackets in Equation 12.I.17 depends on the spin–spin coupling only, since the effects of the chemical shifts and field inhomogeneities have been canceled out due to refocusing, and do not affect the magnetization (echo) at $t = 2\tau'$.

For a system consisting of two spin -½, S_1 and S_2, the operator for the magnetization of spins S_1 is proportional to $S_1^+ 1_2$, where $S_1^+ = S_{1x} + iS_{1y}$ and 1_2 is the unit operator in the space of S_2. In order to evaluate the trace in Equation 12.I.17, using the direct-product space of spins S_1 and S_2, one obtains

$$Tr\{(S_1^+ 1_2)e^{iaS_{1z}S_{2z}2\tau}(S_{1y}1_2)e^{-iaS_{1z}S_{2z}2\tau}\}$$

$$= \sum_{M_1M_2}\langle M_1M_2|(S_1^+ 1_2)e^{iaS_{1z}S_{2z}2\tau}(S_{1y}1_2)e^{-iaS_{1z}S_{2z}2\tau}|M_1M_2\rangle$$

$$= \sum_{M_2}\langle 1/2M_2|S_1^+ 1_2|-1/2M_2\rangle e^{-iaM_2\tau}\langle -1/2M_2|\frac{i}{2}S_1^- 1_2|1/2M_2\rangle e^{-iaM_2\tau}$$

$$= \frac{i}{2}(e^{ia\tau}+e^{-ia\tau}) = i\cos(a\tau) \quad (12.I.18)$$

In Equation 12.I.18 M_1 (= ±1/2) and M_2 (= ±1/2) are the electronic spin magnetic quantum numbers for spins S_1 and S_2, respectively. The third line in Equation 12.I.18 is obtained from the preceding line by: (i) introducing the unity operators $\sum_{M_1'M_2'}|M_1'M_2'\rangle\langle M_1'M_2'| = 1$, $\sum_{M_1''M_2''}|M_1''M_2''\rangle\langle M_1''M_2''| = 1$, and $\sum_{M_1'''M_2'''}|M_1'''M_2'''\rangle\langle M_1'''M_2'''| = 1$ after the operator S_1^+, before and after the operator S_{1y}, respectively; and (ii) taking into account the nonzero matrix elements of the operators in the S_1 space, specifically $\langle M_1 = 1/2|S_1^+|M_1 = -1/2\rangle = \langle M_1 = -1/2|S_1^-|M_1 = 1/2\rangle = 1$ and $\langle M_2 = 1/2|S_{2z}|M_2 = 1/2\rangle = -\langle M_2 = -1/2|S_{2z}|M_2 = -1/2\rangle = 1/2$, and (iii) $S_{1y} = (S_1^+ - S_1^-)/2i$.

Finally, one obtains from Equation 12.I.17

$$<S_1^+(2\tau')> = i[\hbar\omega_{01}\cos(a\tau)/Zk_BT] = iI_0\cos(a\tau), \tag{12.I.19}$$

where $I_0 = \hbar\omega_{01}/Zk_BT$.

Equation 12.I.19 shows that the imaginary part of the echo amplitude of three-pulse DEER at $t = 2\tau'$ is modulated by the factor $\cos(a\tau)$.

Appendix 12.II
Density-Matrix Derivation of the Echo Signal for Four-Pulse DEER

Schematically, the Four-pulse sequence can be written in the following form:

$$X_1(\pi/2) - \tau_1 - X_1(\pi) - t - X_S(\pi) - (\tau_1 + \tau_2 - t) - X_1(\pi) - (t' - 2\tau_1 - \tau_2).$$

In order to find out the form of the detected signal, as mentioned before, one needs to calculate $<S_1^+(2\tau_1 + 2\tau_2)>$. A procedure similar to that used with three-pulse sequence in Appendix 12.I will be followed here (Bhatnagar, 2005). Accordingly, the initial density matrix, in the high-temperature approximation, is

$$\rho(0^-) \approx \{1 - (\hbar\omega_{01}S_{1z} - \hbar\omega_{02}S_{2z})/k_BT\}/Z \tag{12.II.1}$$

and

$$\rho(0^+) \approx -\hbar\omega_{01}S_{1y}/(Zk_BT) \tag{12.II.2}$$

Now, the system evolves under the action of the static Hamiltonian, given by Equation 12.I.5 for a time τ_1 after which a S_1-selective π pulse is applied about the X axis. For $0 < t' < \tau_1$,

$$\rho(t') = e^{i(\Delta\omega_1 S_{1z} + \Delta\omega_2 S_{2z} - aS_{1z}S_{2z})t'}\rho(0^+)e^{-i(\Delta\omega_1 S_{1z} + \Delta\omega_2 S_{2z} - aS_{1z}S_{2z})t'} \tag{12.II.3}$$

After a π pulse is applied on S_1 spins at $t = \tau_1^-$, one obtains

$$\rho(\tau_1^+) = X_1(\pi)\rho(\tau^-)X_1^{-1}(\pi)$$
$$(-\hbar\omega_{01}/Zk_BT)X_1(\pi)e^{i(\Delta\omega_1 S_{1z} + \Delta\omega_2 S_{2z} - aS_{1z}S_{2z})\tau_1}S_{1y}e^{-i(\Delta\omega_1 S_{1z} + \Delta\omega_2 S_{2z} - aS_{1z}S_{2z})\tau_1}X_1^{-1}(\pi) \tag{12.II.4}$$

Inserting the unit operator $(X_1^{-1}(\pi)X_1(\pi))$ in Equation 12.II.4 before and after S_{1y}, and writing it as a product of three terms, following the same procedure as that used in Appendix 12.I after Equation 12.I.7, one obtains:

$$\rho(\tau_1^+) = (-\hbar\omega_{01}/Zk_BT)e^{-i(\Delta\omega_1 S_{1z} - \Delta\omega_2 S_{2z} - aS_{1z}S_{2z})\tau_1}(-S_{1y})e^{i(\Delta\omega_1 S_{1z} - \Delta\omega_2 S_{2z} - aS_{1z}S_{2z})\tau_1} \tag{12.II.5}$$

Now, the system evolves for time t, after which a π pulse is applied on S_2 spins. During this time, $\tau_1 < t' < \tau_1 + t$, one obtains:

Appendix 12.II Density-Matrix Derivation of the Echo Signal for Four-Pulse DEER | 583

$$\rho(t') = e^{i(\Delta\omega_1 S_{1z} + \Delta\omega_2 S_{2z} - aS_{1z}S_{2z})(t'-\tau_1)} \rho(\tau_1^+) e^{-i(\Delta\omega_1 S_{1z} + \Delta\omega_2 S_{2z} - aS_{1z}S_{2z})(t'-\tau_1)} \quad (12.\text{II}.6)$$

Substituting now $t' = \tau_1 + t$, one obtains from Equation 12.II.6

$$\rho(\tau_1 + t^-) = (-\hbar\omega_{01}/Zk_BT) e^{i(\Delta\omega_1 S_{1z} - aS_{1z}S_{2z})(t-\tau_1)}(-S_{1y}) e^{-i(\Delta\omega_1 S_{1z} - aS_{1z}S_{2z})(t-\tau_1)} \quad (12.\text{II}.7)$$

In writing the above expression, the terms with $\Delta\omega_2 S_{2z}$ have been canceled out since S_{2z} commutes with S_{1y}.

Now a π pulse is applied to S_2 spins, which transforms the density matrix to

$$\rho(\tau_1 + t^+) = X_2(\pi)\rho(\tau_1 + t^-)X_2^{-1}(\pi)$$

$$= (-\hbar\omega_{01}/Zk_BT) X_2(\pi) e^{i(\Delta\omega_1 S_{1z} - aS_{1z}S_{2z})(t-\tau_1)}(-S_{1y}) e^{-i(\Delta\omega_1 S_{1z} - aS_{1z}S_{2z})(t-\tau_1)} X_2^{-1}(\pi)$$
$$(12.\text{II}.8)$$

Inserting now the unit operator $(X_2^{-1}(\pi)X_2(\pi))$ in Equation 12.II.8 before and after S_{1y}, and writing it as a product of three terms, following the same procedure as that used in Appendix 12.I after Equation 12.9, one obtains:

$$\rho(\tau_1 + t^+) = (-\hbar\omega_{01}/Zk_BT) e^{i(\Delta\omega_1 S_{1z} + aS_{1z}S_{2z})(t-\tau_1)}(-S_{1y}) e^{-i(\Delta\omega_1 S_{1z} + aS_{1z}S_{2z})(t-\tau_1)} \quad (12.\text{II}.9)$$

Subsequent to this, the system evolves under the action of static Hamiltonian during the period $(\tau_1 + \tau_2 - t)$. Then, one can express for $\tau_1 + t < t' < 2\tau_1 + \tau_2$:

$$\rho(t') = e^{i(\Delta\omega_1 S_{1z} + \Delta\omega_2 S_{2z} - aS_{1z}S_{2z})(t'-\tau_1 - t)} \rho(\tau_1 + t^+) e^{-i(\Delta\omega_1 S_{1z} + \Delta\omega_2 S_{2z} - aS_{1z}S_{2z})(t'-\tau_1 - t)}$$
$$(12.\text{II}.10)$$

From Equation 12.II.10 at $t' = 2\tau_1 + \tau_2$, one obtains

$$\rho(2\tau_1 + \tau_2^-) = e^{i(\Delta\omega_1 S_{1z} + \Delta\omega_2 S_{2z} - aS_{1z}S_{2z})(\tau_2 + \tau_1 - t)} \rho(\tau_1 + t^+) e^{-i(\Delta\omega_1 S_{1z} + \Delta\omega_2 S_{2z} - aS_{1z}S_{2z})(\tau_2 + \tau_1 - t)}$$
$$(12.\text{II}.11)$$

Substituting now $\rho(\tau_1 + t^+)$ from Equation 12.II.9, one obtains

$$\rho(2\tau_1 + \tau_2^-) = (-\hbar\omega_{01}/Zk_BT) e^{i\Delta\omega_1 S_{1z}\tau_2} e^{iaS_{1z}S_{2z}(2t - 2\tau_1 - \tau_2)}(-S_{1y}) e^{-i\Delta\omega_1 S_{1z}\tau_2} e^{-iaS_{1z}S_{2z}(2t - 2\tau_1 - \tau_2)}$$
$$(12.\text{II}.12)$$

An S_1-selective π pulse is now applied at $t' = 2\tau_1 + \tau_2$, which yields:

$$\rho(2\tau_1 + \tau_2^+) = X_1(\pi)\rho(2\tau_1 + \tau_2^-)X_1^1(\pi) \quad (12.\text{II}.13)$$

Inserting now the unit operator $(X_1^{-1}(\pi)X_1(\pi))$ in Equation 12.II.9 before and after S_{1y}, and writing it as a product of three terms, following the same procedure as that used in Appendix 12.I after Equation 12.I.7, one obtains:

$$\rho(2\tau_1 + \tau_2^+) = e^{-i\Delta\omega_1 S_{1z}\tau_2} e^{-iaS_{1z}S_{2z}(2t - 2\tau_1 - \tau_2)} S_{1y} e^{i\Delta\omega_1 S_{1z}\tau_2} e^{iaS_{1z}S_{2z}(2t - 2\tau_1 - \tau_2)} \quad (12.\text{II}.14)$$

The system now evolves under the action of the static Hamiltonian. One obtains for the density matrix for the time $t' > 2\tau_1 + \tau_2$:

$$\rho(t') = e^{i(\Delta\omega_1 S_{1z} + \Delta\omega_2 S_{2z} - iaS_{1z}S_{2z})(t' - 2\tau_1 - \tau_2)} \rho(2\tau_1 + \tau_2^+) e^{-i(\Delta\omega_1 S_{1z} + \Delta\omega_2 S_{2z} - iaS_{1z}S_{2z})(t' - 2\tau_1 - \tau_2)}$$
$$(12.\text{II}.15)$$

After calculating the density matrix at $t' = 2\tau_1 + 2\tau_2$ and substituting the value of $\rho(2\tau_1 + \tau_2^+)$ from Equation 12.II.14, one obtains the following expression:

$$\rho(2\tau_1 + 2\tau_2) = (-\hbar\omega_{01}/Zk_BT)e^{-iaS_{1z}S_{2z}(2t-2\tau_1)}S_{1y}e^{iaS_{1z}S_{2z}(2t-2\tau_1)} \quad (12.II.16)$$

In order to calculate $<S_1^+(2\tau_1 + 2\tau_2)>$, the echo due to the four-pulse sequence, one proceeds in the same way as that for a three-pulse DEER sequence, obtaining

$$<S_1^+(2\tau_1 + 2\tau_2)> = -iI_0 \cos(a[t-\tau_1] = -iI_0 \cos[\omega_{ee}(t-\tau_1)], \quad (12.II.17)$$

where $I_0 = \hbar\omega_{01}/(Zk_BT)$.

Appendix 12.III
Spin Hamiltonian for Coupled Nitroxides Used in Six-Pulse DQC Calculation

One can express the Zeeman and hyperfine part of the nitroxide spin Hamiltonian, H_0, in the irreducible spherical tensor operator (ISTO) representation (Freed, 1976; Schneider and Freed, 1989a,b; Misra, 2007) as follows:

$$H_0 = \sum_{\mu_k, L, M} F^{L,M*}_{\mu_k,\ell} A^{L,M}_{\mu_k,\ell},$$

where L is the tensor rank (0 or 2); $M = -L, \ldots L$; $k = 1, 2$ numbers the nitroxides; and l denotes the reference frame where the tensors are defined, that is, magnetic frame (g_k), molecular (dipolar) frame (d), or laboratory frame (l). $A^{L,M}_{\mu_k,\ell}$ are the spin operators with μ_k referring to the kind of magnetic interaction, electron Zeeman (g), nuclear Zeeman (N), or hyperfine (A), and they are usually defined in the laboratory frame, $F^{L,M}_{\mu_j,\ell}$ is proportional to the ISTO of the magnetic interaction and is most conveniently defined in the g-frame. The transformation of $F^{L,M}_{\mu_k,g_k}$ to the laboratory frame yields the H_0. In the high-field limit, where the nonsecular terms ($S_\pm, S_\pm I_z, S_\pm I_\pm, S_\pm I_\mp$) can be omitted, the ISTO form of the g-tensor reduces to:

$$F^{0,0}_{g_k,g_k} = -\sqrt{\frac{1}{3}}\frac{\mu_B}{\hbar}(g^{(k)}_{xx} + g^{(k)}_{yy} + g^{(k)}_{zz}), \quad A^{0,0}_{g_k,\ell} = -\sqrt{\frac{1}{3}}B_0 S_{kz};$$

$$F^{2,0}_{g_k,g_k} = -\sqrt{\frac{2}{3}}\frac{\mu_B}{\hbar}\left[g^{(k)}_{zz} - \frac{1}{2}(g^{(k)}_{xx} + g^{(k)}_{yy})\right], \quad A^{2,0}_{g_k,\ell} = -\sqrt{\frac{2}{3}}B_0 S_{kz};$$

$$F^{2,\pm 2}_{g_k,g_k} = \frac{1}{2}\frac{\mu_B}{\hbar}(g^{(k)}_{xx} - g^{(k)}_{yy}), \quad A^{2,\pm 2}_{g_k,\ell} = 0.$$

Similarly, the relevant components of the hyperfine tensor are:

$$F^{0,0}_{A_k,g_k} = -\sqrt{\frac{1}{3}}\frac{\mu_B}{\hbar}(A^{(k)}_{xx} + A^{(k)}_{yy} + A^{(k)}_{zz}), \quad A^{0,0}_{A_k,\ell} = -\sqrt{\frac{1}{3}}S_{kz}I_{kz};$$

$$F^{2,0}_{A_k,g_k} = -\sqrt{\frac{2}{3}}\frac{\mu_B}{\hbar}\left[A^{(k)}_{zz} - \frac{1}{2}(A^{(k)}_{xx} + A^{(k)}_{yy})\right], \quad A^{2,0}_{A_k,\ell} = -\sqrt{\frac{2}{3}}S_{kz}I_{kz};$$

$$F^{2,\pm 1}_{A_k,g_k} = 0, \quad A^{2,\pm 1}_{A_k,\ell} = \mp\frac{1}{2}S_{kz}I_{k\pm};$$

Appendix 12.III Spin Hamiltonian for Coupled Nitroxides Used in Six-Pulse DQC Calculation

$$F^{2,\pm 2}_{A_k,g_k} = \frac{1}{2}\frac{\mu_B}{\hbar}(A^{(k)}_{xx} - A^{(k)}_{yy}), \quad A^{2,\pm 2}_{A_k,\ell} = 0.$$

In addition, when considered, the nuclear Zeeman term is given by $\sum_k F^{0,0}_{N_k,I} A^{0,0}_{N_k,I}$ with

$$F^{0,0}_{N_k,I} = \sqrt{3}\frac{g_n \mu_n}{\hbar}; \quad A^{0,0}_{N_k,I} = -\sqrt{\frac{1}{3}}B_0 I_{kz}.$$

The nuclear quadrupole term for ^{14}N nitroxide is neglected here. The second-rank tensors $F^{2,M}_{\mu,g_k}$ are transformed in two steps: first, from the kth nitroxide g-tensor axes to the dipolar frame, with its z-axis coincident with the vector **r** connecting magnetic dipoles, and then to the laboratory frame. The transformations from the dipolar frame to g-frame are defined by the Euler angles $\lambda_k \equiv (\alpha_k, \beta_k, \gamma_k)$ as shown in Figure 12.11 and the transformation from the laboratory frame to the dipolar frame, which is defined by $\eta \equiv (0, \theta, \phi)$. The transformed tensors are thus written as:

$$F^{L,M*}_{\mu_k,I} = \sum_{m',m''} D^L_{m,m'}(\eta) D^L_{m',m''}(\lambda_k) F^{L,M}_{\mu_k,g_k}.$$

The Hamiltonian in the laboratory frame becomes:

$$H_0 = \sum_k [S_{kz}(C_k + A_k I_{kz} + B_k I_{k+} + B^*_k I_{k-}) + G_k I_{kz}].$$

where the coefficients C_k, A_k, G_k, and B_k are expressed as follows:

$$C_k = \sum_{m'} D^2_{0,m'}(\eta_k) K_{g_k,m'}(\lambda_k), \quad A_k = \sum_{m'} D^2_{0,m'}(\eta_k) K_{A_k,m'}(\lambda_k);$$

$$2B_k = \sum_{m'} D^2_{1,m'}(\eta_k) K_{A_k,m'}(\lambda_k), \quad G_k = \gamma_{nk} B_0,$$

with

$$K_{\mu_k,m'}(\lambda_k) = [D^2_{m',2}(\lambda_k) + D^2_{m',-2}(\lambda_k)]F^{2,2}_{\mu_k,g} + D^2_{m',0}(\lambda_k) F^{2,0}_{\mu_k,g}$$

includes the transformations $D^2_{m',m''}(\lambda_k)$ from the dipolar frame to the kth magnetic frame. The explicit expressions for C_k, A_k, and B_k are unnecessary, since all the transformations were carried out numerically. They are listed by Saxena and Freed (1997). It is noted that the terms C_k, A_k, and B_k contain all anisotropies in the g and hf tensors as well as the Euler angles needed for their transformation from the respective principal-axes system to the laboratory frame. It is also noted that C_k, G_k, and A_k are real, whereas B_k is complex. Finally, it is noted that in carrying out the computations for B_0 well up to Q band (35 GHz), the nuclear Zeeman term can be safely omitted and the H_0 in real form can be obtained to a high accuracy based on the useful approximation described by Libertini and Griffith (1970).

Appendix 12.IV
Algorithm to Calculate Six-Pulse DQC Signal

The details of how to calculate the final density matrix, ρ_f, from which the DQC echo signal can be calculated as given by Equations 12.IV.3–12.IV.4 below, are given in this Appendix. Basically, this requires a series of transformations of the density matrix by a propagator, pulse or free evolution, and choosing the matrix elements of the density matrix after the application of a pulse on a coherence pathway. Here, the direct-product representation for $S_1 \otimes S_2 \otimes I_1 \otimes I_2$ will be used to express the matrix elements of the various operators as follows. To this end, the following details are required:

i) **Matrix representation and notation:** For each electron, with spin $S = 1/2$, the matrix dimension is 2, whereas for each nucleus, with spin I, it is $(2I + 1)$, so that for the electron–nuclear spin coupled system of the nitroxide pair the size of the product space is $N \times N$ with $N = 4(2I_1 + 1)(2I_2 + 1)$, which becomes 36×36 for ^{14}N ($I = 1$) nitroxides. The Zeeman basis with the basis vectors $|k\rangle \equiv |m_1, m_2; M_1, M_2\rangle$ is employed here, where the $|k\rangle$-values are the eigenvectors of the z-components of the electron and nuclear spin operators: $S_z|m\rangle = m|m\rangle$ and $I_z|M\rangle = M|M\rangle$. The basis states in the product space are described by lower-case Roman letters. Greek letters are used to describe the eigenvectors of H: $H|\alpha\rangle = \omega_\alpha |\alpha\rangle$. The diagonalization of the Hamiltonian, H, is accomplished by the unitary transformation $V^\dagger H V = E$, where E is a diagonal matrix of eigenvalues of H, and $V^\dagger$ is the Hermitian adjoint of V. The eigenvectors of H are the columns of V: $|\alpha\rangle_k = \langle k|\alpha\rangle \equiv V_{k\alpha}$. In the computations the matrices E and V are the outputs of the matrix diagonalization subroutine, such as JACOBI (Press et al., 1992), used here. (A better version of the JACOBI subroutine than that given by Press et al., 1992 can be found on the Netlib website http://netlib.org or else is available from the authors.)

ii) **Initial density matrix in product space:** Using the expression for $\rho(0)$ as given by Equation 12.13, the initial density matrix is expressed as

$$\rho(0) = (S_{1z} \otimes 1_2 + 1_1 \otimes S_{2z}) \otimes 1_{I_1} \otimes 1_{I_2}, \quad (12.\text{IV}.1)$$

where 1_{I_1} and 1_{I_2} are 3×3 unit matrices in the respective nuclear-spin spaces for ^{14}N nitroxides. A diagonal matrix of order 4×4 on the right-hand side of Equation 12.IV.1 represents $S_z \equiv S_{1z} + S_{2z}$ in the product space $S_1 \otimes S_2$ for the two nitroxide electron spins, that is $(\sigma_z \otimes 1_2 + 1_1 \otimes \sigma_z)/2 = \text{diag}(1, 0, 0, -1)$, where 1_1 and 1_2 are 2×2 unit matrices in the spin spaces for electrons 1 and 2, respectively.

iii) **Transformation of the density matrix by a propagator.** The transformed density matrix, ρ', under the action of a propagator is expressed as:

$$\rho' = e^{-iHt/\hbar} \rho\, e^{iHt/\hbar}. \quad (12.\text{IV}.2)$$

The following propagators are required: (i) pulse propagators, due to a $\pi/2$, or a π pulse; and (ii) free-evolution operators in the absence of a pulse. Figure 12.10

shows the effects of the various propagators, calculated using Equation 12.IV.3 below, employing the appropriate Hamiltonians and their durations. As an illustration, the procedure for a given density matrix, ρ, and the Hamiltonian, H, acting during the time period, t, is described here, which can be specialized for the various propagators by appropriate substitutions, with appropriate times t_k ($t_{1,2} = t_p$; $t_{3,4} = t_{DQ}$; $t_5 = t_m - t_p$; $t_6 = t_5 + t_{echo}$) and Equations 12.17–12.21, which describe the Hamiltonians used in the analysis of the six-pulse DQC sequence.

The matrix elements of ρ' in Equation 12.IV.2 can be expressed as $\rho'_{jk} = e^{-iH_{jm}t} \rho_{mn} e^{iH_{nk}t} = V_{j\alpha} e^{-i\omega_\alpha t} V^*_{m\alpha} \rho_{mn} V_{n\beta} e^{-i\omega_\beta t} V^*_{k\beta}$, since $e^{-iH_{km}t} = V_{k\alpha} e^{-i\omega_\alpha t} V^*_{\alpha m}$. The required summations are carried out over the repeating indexes, or explicitly

$$\rho'_{jk} = \sum_{\alpha,\beta,m,n} \rho_{mn} V^*_{m\alpha} V_{n\beta} V_{j\alpha} V^*_{k\beta} e^{-i\omega_{\alpha\beta}t} \quad (12.\text{IV}.3)$$

with $\omega_{\alpha\beta} \equiv \omega_\alpha - \omega_\beta$. Equation 12.IV.3 can be written using a short-hand notation as $\rho' = L\rho$. In this notation, **L** is an operator, which is **Q** for a free evolution period or **R** for the action of a pulse. Coherence pathway selection implies retaining only those elements of ρ', which belong to the pathways of interest, with the subsequent summation conducted over all pathways that contribute to the echo of interest. In computations this is accomplished by retaining only those matrix elements that correspond to the selected pathway, setting the rest to zero. This may be expressed as the application of a projection operator **P** (which, in reality, does not need to be constructed). The final density matrix after application of N pulses and subsequent evolution periods is then calculated as

$$\rho_f(t) = \sum_{\{p_k\}} (\mathbf{Q}_N \mathbf{P}_{p_N} \mathbf{R}_N, \ldots, \mathbf{Q}_1 \mathbf{P}_{p_1} \mathbf{R}_1) \rho(0) \quad (12.\text{IV}.4)$$

The product is computed for the full set of coherence pathways $\{p_k\}$ that contribute to the echo and the sum is then taken to be finally used in computing of $\text{Tr}[\rho_f S_+]$

iv) **Coherence pathway selection:** Subsequent to the action of a pulse propagator, of the matrix elements as calculated in (iii) above, all but those in the electronic product subspace of the density matrix ρ that correspond to selected coherence order p should be set to zero. The correspondence of ρ_{ik} to p is compiled in Table 12.IV.1 pertinent to the coherence pathways depicted in Figure 12.10 illustrating the coherence pathways of the six-pulse DQC sequence. This selection of coherence pathways is achieved experimentally through phase cycling (Borbat and Freed, 2000), or in computations is based on Table 12.IV.1.

Appendix 12.V
Approximate Analytic Expressions for 1-D DQC Signal

For completeness, the equation given by Borbat and Freed (2000), used for making the comparison with 1-D DQC signals produced in rigorous computations, is included here. The echo amplitude, V, is a function of $t_{dip} = 2t_p - t_m$, and is given by

$$V(t_{dip}) = K(\omega_1)K(\omega_2)F(t_p)F(t_m - t_p). \tag{12.V.1}$$

The time variables are defined in accordance with Figure 12.10, and the notations are defined in the text. $F(t)$ is expressed as:

$$F(t) = (p^2 + q^2 \cos Rt)\cos At - q\sin Rt \sin At. \tag{12.V.2}$$

In Equation 12.V.2, $q = b/R$ and $p^2 = 1 - q^2$; $A = d(1 - 3\cos^2\theta)$ and $b = -A/2$, where A and b represent the secular and pseudosecular parts of the dipolar coupling; $R^2 = \Delta\omega^2 + b^2$, where $\Delta\omega = \omega_1 - \omega_2$ is the difference between the Larmor frequencies ω_1 and ω_2 of the nitroxide's electron spins in the frame of reference rotating with the frequency ω_{rf} of the excitation pulses, set to coincide are expressed in the simplest form as

$$K(\omega_k) = \left(\frac{\omega_{1rf}^2}{\Delta\omega_k^2 + \omega_{1rf}^2}\sin^2\left(\frac{\pi}{2}\sqrt{1 + \Delta\omega_k^2/\omega_{1rf}^2}\right)\right)^3, \tag{12.V.3}$$

where $\omega_{1rf} = \gamma_e B_1$ for π pulses, which are all taken to be equal to each other, although in general they need not be.

The powder averaging is carried out essentially in the same way as in the rigorous computations, with ω_k determined for each set of (M_1, M_2), since $\omega_k = \omega_k(\lambda_k, \eta)$. This was determined using an approximation given by Libertini and Griffith (1970).

Part Three
Applications

13
Determination of Large Zero-Field Splitting
Sushil K. Misra

13.1
Introduction

In the context of multifrequency EPR, the determination of large zero-field splitting (ZFS) is a very relevant topic in view of the availability of very high-frequency (VHF) EPR spectrometers. Today, it is possible to access frequencies up to about 1000 GHz to match energy-level spacing between levels participating in resonance to measure very large ZFS – for example, that characterized by the value of the parameter D as large as ~10 cm^{-1} (~300 GHz) . This was not possible when spectrometers involving only X (~9.5 GHz)- and Q (~35 GHz) -bands and magnetic fields up to 2 T were available some four decades ago, such that EPR transitions for many transition-metal ions characterized by large ZFS (e.g., Fe^{2+}, Mn^{3+}) could not be induced, since the energy of the quantum of microwave energy then available was insufficient to match the energy-level spacing between the pairs of levels participating in resonance. Hence, these ions were termed "EPR silent." In this chapter the ions characterized by large ZFS, analysis of relevant EPR data, and the values of the reported ZFS parameters, will be reviewed. For additional relevant information the reader is also referred to Chapters 2, 8, 3 and Section 4.3 of this book, which deal respectively with the "multifrequency aspects of EPR," the "evaluation of spin-Hamiltonian parameters from multifrequency EPR data," crystal-field splittings for transition-metal ions, and "high-frequency spectrometers." Many examples of high-frequency spectra are shown in Chapter 2, while how to evaluate spin-Hamiltonian parameters rigorously, using the least-squares fittings from multifrequency EPR data, and in particular from polycrystalline spectra, is described in detail in Chapter 8. To date, this approach has not been exploited by research groups. It should be noted, as discussed in Chapter 2, that the evaluation of parameters becomes simpler at the higher frequencies used to measure large ZFS, due to the predominant size of the Zeeman term, which serves as the zero-order Hamiltonian, as compared to the ZFS part of the spin Hamiltonian.

It should also be mentioned that in Tallahassee, Florida, USA, at the National High Magnetic Field Laboratory (NHMFL), a tunable high-frequency EPR (HFEPR)

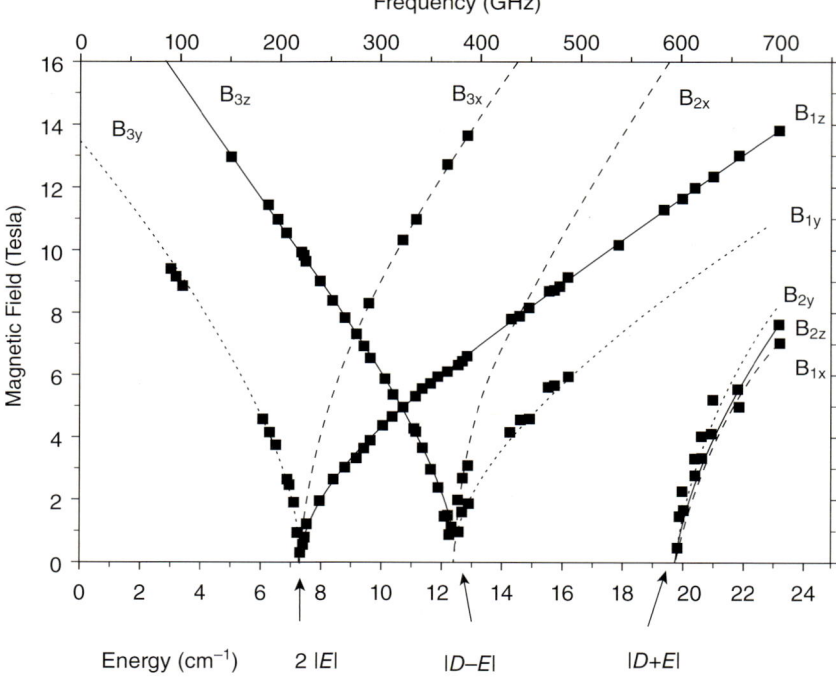

Figure 13.1 Frequency versus magnetic field plot of resonance lines recorded using a tunable-frequency spectrometer for a $S = 1$ system (V^{3+} ion), obtained for $VBr_3(thf)_3$ at 10 K. The lines show simulated variation using the parameters listed in Table 13.1, whereas the squares represent experimental data. The x-, y-, and z-turning points are shown by dashed, dotted, and solid lines, respectively. The arrows indicate the zero-field resonances. Adapted from Krzystek, Ozarowski, and Tesler (2006).

spectrometer has been developed which makes available frequencies in the range 150 to 700 GHz, with a magnetic field that can be varied by up to 25 T. This is particularly suitable for measuring large ZFS. Three examples of resonant fields so obtained are shown in Figures 13.1–13.3 for $S = 1$, 3/2, and 2, respectively (Krzystek, Ozarowski, and Tesler, 2006).

13.2
ZFS of Kramers and Non-Kramers Ions in Different Environments

The spin Hamiltonian (for $S \geq 1$) relevant to the present discussion is:

$$H = \mu_B B \cdot \tilde{g} \cdot S + D\left(S_z^2 - \frac{1}{3}S(S+1)\right) + E(S_x^2 - S_y^2) + \text{higher-order terms},$$

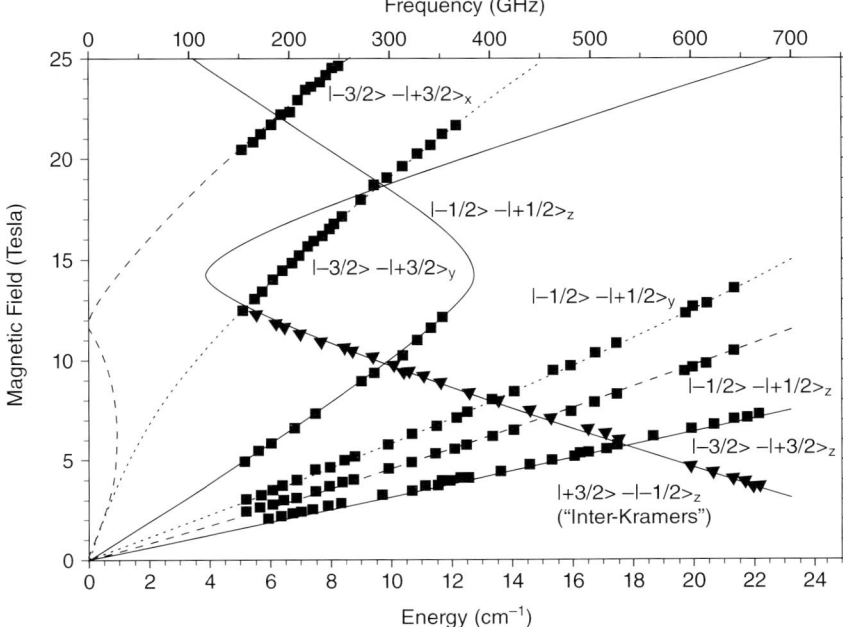

Figure 13.2 Frequency-versus-magnetic field plot of resonance lines recorded using a tunable-frequency spectrometer for an $S = 3/2$ system (Co^{2+} ion), obtained for polycrystalline $CoCl_2(PPh_3)_2$. The squares indicate experimental intra-Kramers' resonances, whereas the triangles show an inter-Kramers' turning point branch. The lines show simulated variation using the parameters listed in Table 13.1. The x-, y-, and z-turning points are shown by dashed, dotted, and solid lines, respectively. Particular transition branches are identified and labeled accordingly. The only zero-field resonance appears at ~900 GHz, which is outside the experimental range. However, the $M_s = -1/2 \leftrightarrow +3/2$ inter-Kramers' turning point leading to it, as marked by triangles, yields a very accurate estimate of the ZFS. Adapted from Krzystek, Ozarowski, and Tesler (2006).

where D and E are usually called ZFS parameters. Only the ions with spin $S \geq 1$ possess a ZFS; these are mostly non-Kramers' ions, possessing integral spins, and some Kramers' ions, possessing half odd-integral spins, in the high-spin state, which are characterized by large ZFS values. In general, ZFS for non-Kramers' ions are significantly larger than those for Kramers' ions. As for the environments where large ZFS occur, some examples include: metalloproteins; polynuclear sites such as iron–sulfur (Fe–S) proteins; coordination complexes of paramagnetic ions; and biological samples. It should be noted that in isostructural halides, the iodides have the largest values of ZFS. The crystal-field splitting under the action of spin–orbit coupling and interactions with crystal fields of lower symmetries for the various transition-metal ions are discussed in detail in Chapter 7. In the main, these are the 3d ions for which large ZFS have been reported; a few examples of

594 | *13 Determination of Large Zero-Field Splitting*

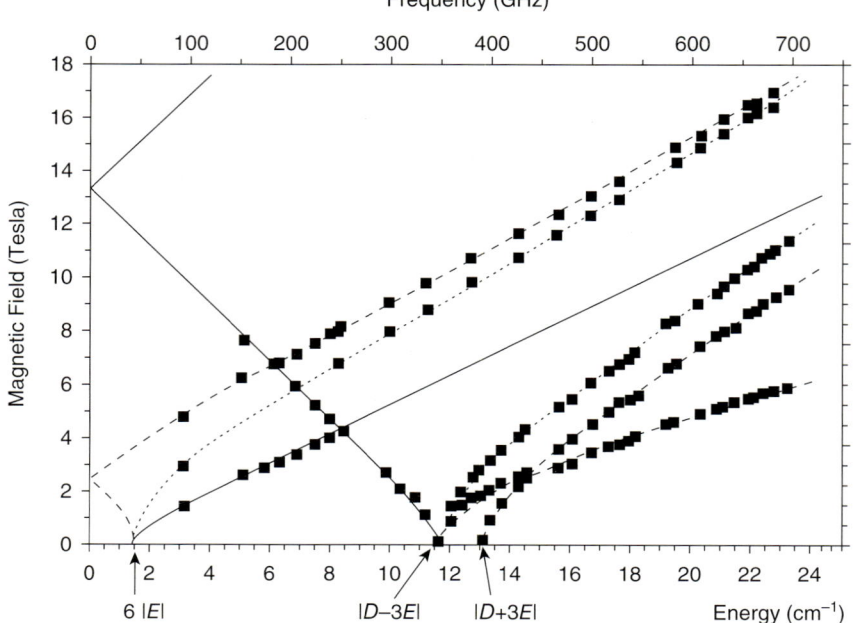

Figure 13.3 Frequency-versus-magnetic field plot of resonance lines recorded using a tunable-frequency spectrometer for a $S = 2$ system (Fe^{2+} ion), obtained for polycrystalline $[Fe(btz)_2(SCN)_2]$. The lines show simulated variation using the parameters listed in Table 13.1, whereas the squares represent experimental data. The x-, y-, and z-turning points are shown by dashed, dotted, and solid lines, respectively. The arrows indicate the three zero-field resonances, of which two are directly detectable in the 95–700 GHz range. Adapted from Krzystek, Ozarowski, and Tesler (2006).

rare-earth (4f) ions, characterized by large ZFS are also available. Because of their peculiarities in the context of large ZFS, as distinct from those of the other ions, the cases of the ions Mn^{3+}, Fe^{2+}, and Fe^{3+} are briefly discussed here, as noted by Krzystek, Ozarowski, and Tesler (2006). Typical ZFS values for these ions are listed in Table 13.1.

Mn^{3+} (d^4, $S = 2$): There is a large number of Mn^{3+} complexes, which are generally air-stable. An important subset of Mn^{3+} compounds consists of porphyrinic (tetrapyrrole) lignads—that is, complexes with the general formula Mn(P)X, where P is the dianionic porphyrin macrocycle and X is an axial, usually anionic ligand, such as Cl. Another important class of Mn^{3+} compounds are those with six-coordinate geometry derived from $[Mn(H_2O)_6]^{3+}$ and other complexes with oxygen donors, such as $Mn(acac)_3$, and related species.

Fe^{2+} (d^6, $S = 2$): High-spin Fe^{2+}, of all the integer-spin metal ions, has the greatest relevance to bioinorganic chemistry. The system of six-coordinate Fe(II) complexes has proven very difficult to study by HF-EPR, despite the fact that there

Table 13.1 A list of selected representative values of the ZFS parameters D and E for various transition-metal ions in a variety of hosts. Based on the detailed list provided by Krzystek, Ozarowski, and Tesler (2006).

Transition-metal ion	Coordination sphere	Host complex(es)	D (cm^{-1})	E (cm^{-1})
V^{3+}, d^2, $S = 1$	O_6	Al_2O_3	7.85–8.29	0
	O_6	$XGa(SO_4)_2 \cdot 12Y_2O$ (X = Rb, Cs, Cu); Y = H,D V(acac)$_3$	3.39–4.91	0
	O_6	VBr$_3$(thf)$_3$	7.47	1.92
	O_6Br_3		−16.16	−3.70
Cr^{3+}, d^3, $S = 3/2$	O_6	Mg_2SiO_4	−0.98	−0.30
Cr^{2+}, d^4, $S = 2$	O_6	$CrSO_4 \cdot 5H_2O$	2.24	0.1
		Mg_2SiO_4 (site M1)	−2.49	0.15
		Mg_2SiO_4 (site M1)	2.23	0.4
	F_8	SrF_2	2.79	~0.7
Mn^{3+}, d^4, $S = 2$	O_6	TiO_2, [Mn(dbm)$_2$(CH$_3$OH)$_2$]Br	−3.4, −3.46	0.12, 0.13
	O_6	Mn(dbm)$_3$, Mn(acac)$_3$	−4.35, −4.52	0.26, 0.25
	O_6	CsMn(SO$_4$)$_2 \cdot 12Y_2O$ (Y = H, D)	−4.43–−4.52	0.25–0.28
	N_6	Mn((terpy)(N$_3$)$_3$	−3.29	0.51
	N_6	Mn((bpea)(N$_3$)$_3$	+3.50	0.82
	N_6	Mn(taa)	−5.90	0.50
Mn^{2+}, d^5, $S = 5/2$	N_4I_2	Mn(o-phen)$_2$I$_2$	0.59	0.15
	P_2I_2	Diiodobis(triphenylphosphine-oxide)Mn(II)	0.91	0.22
	N_3I_2	Mn(terpy)I$_2$	+1.00	0.19
	N_3O_2	MnSOD	0.36	0.01
Fe^{3+}, d^5, $S = 5/2$	O_6	PbTiO$_3$	1.176	0
	N_2O_4	EDTA	0.8	0.27
	N_4I	Deutero porphyrin IX dimethyl ester (X = axial ligand)	16.4	~0
Fe^{2+}, d^6, $S = 2$	O_6	FeSiF$_6 \cdot 6H_2O$	11.78	0.67
	O_6	[Fe(H$_2$O)$_6$](ClO$_4$)$_2$	11.34	0.69
	O_6	(NH$_4$)$_2$[Fe(H$_2$O)$_6$](SO$_4$)$_2$	14.94	3.78
	O_6	[Fe(H$_2$O)$_6$]SO$_4 \cdot H_2O$	10.32	2.23
	N_6	Fe(bithiazoline)$_2$(SCN)$_2$	12.43	0.24
	S_4	[PPh$_4$]$_2$[Fe(SPh)$_4$]	5.84	1.42

Table 13.1 (Continued)

Transition-metal ion	Coordination sphere	Host complex(es)	D (cm^{-1})	E (cm^{-1})
Co^{2+}, d^7, $S = 3/2$	Br$_4$	Cs$_3$CoBr$_5$	−5.3	0
	P$_2$Cl$_2$	Co(PPh$_3$)$_2$Cl$_2$	−14.76	1.14
Ni^{2+}, d^8, $S = 1$	O$_6$	NiSO$_4$·7H$_2$O	−3.5	−1.5
	O$_4$NCl	Ni(hmp)$_4$(dmb)$_4$Cl$_4$	−5.3	1.20
	N$_4$O$_2$	[Ni-(HIM2-py)$_2$NO$_3$]NO$_3$	−10.1	0.02
	N$_6$	[Ni(sarcophagine)](ClO$_4$)$_2$	1.4	0
Gd^{3+}, f^7, $S = 7/2$	O$_6$	Al$_2$O$_3$	0.10	0

exist a large number of such compounds. The next system to be mentioned is that of four-coordinated complexes. This includes (PPH$_4$)$_2$[Fe(SPh)$_4$], which is a model for the reduced form of the active site of the iron–sulfur protein, rubredoxin (Rdred) that comprises an Fe ion coordinated by four cysteine thiolate residues (Knapp et al., 2000).

Fe^{3+} (d^5, $S = 5/2$): There exist a large number of high-spin complexes of Fe^{3+}, including many of biological importance, such as heme proteins with weak axial ligands such as aqua or fluoro, and many non-heme Fe enzymes. Despite this, HF-EPR has not been applied much to this system. Particular attention should be paid to met-hemoglobin (Fe^{3+}), for which a tunable-frequency spectrometer provided data over 45–430 GHz (Alpert et al., 1973).

The values of the ZFS parameters as reported in the literature are reviewed and listed in detail by Krzystek, Ozarowski, and Tesler (2006) for the various transition-metal ions. Table 13.1 represents a summary of representative values of the ZFS parameters D and E listed by them for various transition metal ions. The reader is referred to this article for a very comprehensive listing and related references.

13.3
Concluding Remarks

This chapter discusses large ZFSs, which are now easily measured using high-frequency sources, covering the range of ~1000 GHz, and high magnetic fields, up to 25 T. The so called "EPR-silent" species are no longer unobservable. Some representative values of large ZFS for the various metal ions are listed in this chapter. It is noted that EPR provides the most accurate determination of ZFS parameters, as compared to those determined by the techniques of magnetic-susceptibility measurements, Mossbauer spectroscopy, magnetic circular dichroism (MCD), and inelastic neutron scattering (INS).

Acknowledgments

The author is grateful to Professor C. P. Poole for a critical reading of this chapter and useful comments.

Pertinent Literature

High-frequency high-field EPR (HF-EPR) has been reviewed, for example, by Hagen (1999). Technical aspects of HF-EPR are reviewed by Smith and Reidi (2002). This chapter is largely based on the review article by Krzystek, Ozarowski, and Tesler (2006).

References

Alpert, Y., Couder, J., Tuchendler, J., and Thome, H. (1973) *Biochim. Biophys. Acta*, **332**, 34.

Hagen, W. (1999) *Coord. Chem. Rev.*, **190**, 209.

Knapp, M.J., Krzystek, J., Brunel, L.-C., and Hendrickson, D.N. (2000) *Inorg. Chem.*, **39**, 281.

Krzystek, J., Ozarowski, A., and Tesler, J. (2006) *Coord. Chem. Rev.*, **250**, 2308.

Smith, G.M. and Reidi, P.C. (2002) Progress in High-Field EPR, in *Electron Paramagnetic Resonance* (eds B.C. Gilbert, M.J. Davies, and D.M. Murphy), Royal Society of Chemistry, Cambridge, UK, Ch. 9, p. 254.

14
Determination of Non-Coincident Anisotropic $\tilde{g}^2$, $\tilde{A}^2$, $\tilde{D}$, and $\tilde{P}$ Tensors: Low-Symmetry Considerations

Sushil K. Misra

14.1
Introduction

This chapter discusses in detail the rigorous evaluation of non-coincident anisotropic $\tilde{g}^2$, $\tilde{A}^2$, $\tilde{D}$, $\tilde{P}$, and $\tilde{g}_n^2$ tensors from EPR/ENDOR data for cases of low symmetry, characterized by tensors with non-coincident principal axes in the spin Hamiltonian (SH). Usually, the $\tilde{g}_n^2$ tensor is considered isotropic, but the present consideration will treat the general case when it is anisotropic. Although the following discussion will be concerned specifically with EPR data, the details apply equally well to ENDOR data where, instead of measuring transition magnetic fields (resonant line positions) at a fixed microwave frequency, resonant frequencies are measured between the pair of energy levels participating in resonance, keeping the external magnetic field constant.

14.2
Spin Hamiltonian

For the present considerations the following spin Hamiltonian, representing the interaction of one electron spin with several (N) nuclear spins, will be used (Weil, 1975):

$$H = H_{ze} + H_{zf} + H_{hf} + H_{nq} + H_{zn}. \qquad (14.1)$$

In Equation 14.1, $H_{ze}, H_{zf}, H_{hf}, H_{nq}$, and H_{zn} are, respectively, the electron Zeeman term, the zero-field splitting (ZFS) term, the hyperfine term, the nuclear quadrupole term, and the nuclear Zeeman term. These are expressed individually as follows:

$$H_{ze} = \mu_B S^T \cdot \tilde{g} \cdot B, \qquad (14.2a)$$

$$H_{zf} = S^T \cdot \tilde{D} \cdot S, \qquad (14.2b)$$

Multifrequency Electron Paramagnetic Resonance, First Edition. Edited by Sushil K. Misra.
© 2011 Wiley-VCH Verlag GmbH & Co. KGaA. Published 2011 by Wiley-VCH Verlag GmbH & Co. KGaA.

$$H_{hf} = \sum_{r=1}^{N} I_r^T \cdot \tilde{A}_r \cdot S, \tag{14.2c}$$

$$H_{nq} = \sum_{r=1}^{N} I_r^T \cdot \tilde{P}_r \cdot I_r, \tag{14.2d}$$

$$H_{zn} = -\sum_{r=1}^{N} \mu_N I_r^T \cdot \tilde{g}_{nr} \cdot B. \tag{14.2e}$$

In Equation 14.2a–e, the summation runs over N nuclei, T denotes transpose, and $\tilde{g}$, $\tilde{A}_r$ and $\tilde{g}_{nr}$ are matrices, whereas $\tilde{D}$, and $\tilde{P}_r$ are tensors.

It should be noted that, as discussed by Abragam and Bleaney (1970), $\tilde{g}$, $\tilde{g}_{nr}$, $\tilde{A}_r$, which connect different vectors in the SH, are not really tensors; thus, they should be referred to as *matrices*; on the other hand, $\tilde{g}^2 = \tilde{g}^T \cdot \tilde{g}$, $\tilde{A}^2 = \tilde{A}^T \cdot \tilde{A}$, and $\tilde{g}_{nr}^2 = \tilde{g}_{nr}^T \cdot \tilde{g}_{nr}$, are, indeed, symmetric tensors (Misra, 1984). Furthermore, $\tilde{D}$ and $\tilde{P}_r$, which connect the same vectors in the SH are also symmetric (Kneubühl, 1963) and traceless; that is:

$D_{xx} + D_{yy} + D_{zz} = 0$ and $P_{rxx} + P_{ryy} + P_{rzz} = 0$. Thus, $\tilde{D}$ and $\tilde{P}_r$ are each represented by five independent elements.

In order to compute the eigenvalues of Equation 14.1 using perturbation theory, the electron and nuclear spins are chosen to be quantized along the directions $\tilde{g} \cdot \eta / g$ and $\tilde{A}_r \cdot \tilde{g} \cdot \eta / gK_r$, respectively; here, $g = (\eta^T \cdot \tilde{g}^T \cdot \tilde{g} \cdot \eta)^{1/2}$, $K_r = (\eta^T \cdot \tilde{g}^T \cdot \tilde{A}_r^T \cdot \tilde{A}_r \cdot \tilde{g} \cdot \eta)^{1/2} / g$ and η is the unit vector along B. For these quantization axes, the various terms of the SH, given by Equation 14.1, can be expressed as (Iwasaki, 1974; Weil, 1975):

$$H_{ze} = \mu_B g B S_u, \tag{14.3a}$$

$$H_{zf} = (u^T \cdot \tilde{D} \cdot u) S_u^2 + \tfrac{1}{4} D_0 (S_+ S_- + S_- S_+) + \tfrac{1}{2} D_1^* (S_u S_+ + S_+ S_u)$$
$$+ \tfrac{1}{2} D_1 (S_u S_- + S_- S_u) + \tfrac{1}{4} D_2^* S_+^2 + \tfrac{1}{4} D_2 S_-^2 \tag{14.3b}$$

$$H_{hf} + H_{zn} = \sum_r \{K_r (S_u) I_{kr} + \tfrac{1}{2} A_{1r}^* S_+ I_{kr} + \tfrac{1}{2} A_{1r} S_- I_{kr}$$
$$+ \tfrac{1}{4} A_{+r} S_+ I_{-r} + \tfrac{1}{4} A_{+r}^* S_- I_{+r} + \tfrac{1}{4} A_{-r}^* S_+ I_{+r} + \tfrac{1}{4} A_{-r} S_- I_{-r}\} \tag{14.3c}$$

$$H_{nq} = \sum_r \{(k_r^T \cdot \tilde{P}_r \cdot k_r) I_{kr}^2 + \tfrac{1}{4} P_{0r} (I_{+r} I_{-r} + I_{-r} I_{+r})$$
$$+ \tfrac{1}{2} P_{1r}^* (I_{kr} I_{+r} + I_{+r} I_{kr}) + \tfrac{1}{2} P_{1r} (I_{kr} I_{-r} + I_{-r} I_{kr})$$
$$+ \tfrac{1}{4} P_{2r}^* I_{+r}^2 + \tfrac{1}{4} P_{2r} I_{-r}^2\} \tag{14.3d}$$

In Equations 14.3a–d, $u = \tilde{g} \cdot \eta / g$, i, j are mutually perpendicular unit vectors orthogonal to η, $S_\pm = (S_x \pm i S_y)$, $D_0 = -(u^T \cdot \tilde{D} \cdot u) = (i^T \cdot \tilde{D} \cdot i) + (j^T \cdot \tilde{D} \cdot j)$, $D_1 = (u^T \cdot \tilde{D} \cdot i) + i(u^T \cdot \tilde{D} \cdot j)$, $D_2 = (i^T \cdot \tilde{D} \cdot i) - (j^T \cdot \tilde{D} \cdot j) + 2i(i^T \cdot \tilde{D} \cdot j)$, $K_r(S_u) = (\tilde{A}_r \cdot \tilde{g}/g) S_u - g_{nr} \mu_N B E$ (E is the unit tensor, and $S_u = S \cdot u$ is the component along u); $k_r(S_u) = K_r(S_u) \cdot \eta / (\eta^T \cdot K_r^T(S_u) \cdot K_r(S_u) \cdot \eta)^{1/2}$ (abbreviated as k_r in the above equation), $A_{1r} = (k_r^T \cdot \tilde{A}_r \cdot i) + i(k_r^T \cdot \tilde{A}_r \cdot j)$; $A_{\pm r} = [(\xi_r^T \cdot \tilde{A}_r \cdot i) \pm (\eta_r^T \cdot \tilde{A}_r \cdot j)] + i[(\eta_r^T \cdot \tilde{A}_r \cdot i) \mp (\xi_r^T \cdot \tilde{A}_r \cdot j)]$ (the mutually perpendicular unit vectors ξ_r and η_r are orthogonal to $k_r(S_u)$ and $I_{\pm r} = (I_{\xi r} \pm iI_{\eta r})$; $P_{0r} = -(k_r^T \cdot \tilde{P}_r \cdot k_r)$; $P_{1r} = (k_r^T \cdot \tilde{P}_r \cdot \xi_r) + i(k_r^T \cdot \tilde{P}_r \cdot \eta_r)$; and $P_{2r} = (\xi_r^T \cdot \tilde{P}_r \cdot \xi_r) - (\eta_r^T \cdot \tilde{P}_r \cdot \eta_r) + 2i(\xi_r^T \cdot \tilde{P}_r \cdot \eta_r)$. In the above definitions the multiplying factor $i = \sqrt{-1}$, the subscripts x and y denote components along **i** and **j**, respectively, and * denotes a complex conjugate.

With the expressions given by Equations 14.3a–d, it is easy to calculate the eigenvalues in a perturbation approximation; the terms $H_{ze} + H_{hf} + H_{zn}$ are used as zero-order terms and the direct-product representation in the space of electron-nuclear spin-coupled systems is used (Iwasaki, 1974; Weil, 1975; Weil and Bolton, 2007).

14.3 Eigenvalues

14.3.1 Perturbation Approach

The eigenvalues of the SH, given by Equation 14.1, with the various terms expressed by Equations 14.3a–d, have been calculated to various perturbation approximations, which are valid for low-symmetry, even triclinic, situations. To a second order of perturbation, taking into account nuclear Zeeman and certain cross-terms, these can be expressed as follows:

$$E(M;\{m\}) = E^{(0)}(M;\{m\}) + E^{(1)}(M;\{m\}) + E^{(2)}(M;\{m\}). \tag{14.4}$$

In Equation 14.4 the superscripts 0, 1, and 2 represent, respectively, the zero-, first-, and second-order perturbation contributions, $\{m\}$ represents $m_1, m_2, \ldots, m_N$; M for the electronic magnetic quantum number; and $m_1, m_2, \ldots$ for the nuclear magnetic quantum numbers for nuclei 1, 2, ..., respectively. Explicitly (Weil, 1975; corrected expressions in Weil and Bolton, 2007, p. 199; see also Iwasaki, 1974 for more details):

$$E^{(0)}(M;\{m\}) = \mu_B gBM - \sum_r G_r \mu_N Bm_r, \tag{14.5a}$$

$$E^{(1)}(M;\{m\}) = \tfrac{1}{2} d_1 \left[3M^2 - S(S+1) \right]$$
$$+ \sum_r \left\{ K_r M m_r + \tfrac{1}{2} p_{1r} [3m_r^2 - I_r(I_r+1)] \right\} \tag{14.5b}$$

$$E^{(2)}(M;\{m\}) = (d_2 - d_1^2)[8M^2 + 1 - 4S(S+1)]M/(2\mu_B gB)$$
$$+ \left[Tr(\tilde{D}^2) - 2d_2 + d_1^2 - 2d_{-1}Det(\tilde{D})\right]$$
$$[2S(S+1) - 2M^2 - 1]M/(8\mu_B gB)$$
$$+ \sum_r \{[\tfrac{1}{2}\{Tr(\tilde{A}_r^T \cdot \tilde{A}_r) - k_r^2\}M\{I_r(I_r+1) - m_r^2\}$$
$$- Det(A_r)\{S(S+1) - M^2\}m_r/K_r + (k_r^2 - K_r^2)Mm_r^2$$
$$+ 2(e_r - d_1)K_r\{3M^2 - S(S+1)\}m_r]/(2\mu_B gB)$$
$$+ Tr(P_r)[I_r(I_r+1) - m_r^2]/2\} + \sum_{i>j}(L_{ij} - K_i K_j)Mm_i m_j \Big/ (\mu_B gB)$$
$$+ M^{-1}\sum_r \{\mu_N^2 B^2 (g_r^2 - G_r^2)m_r \Big/ (2K_r)$$
$$+ (p_{2r} - p_{1r}^2)[8m_r^2 - 4I_r(I_r+1) + 1]m_r/(2K_r)$$
$$+ \left[Tr(\tilde{P}_r^2) - 2p_{2r} + p_{1r}^2 - 2p_{-1r}Det(\tilde{P}_r)\right]$$
$$[2I_r(I_r+1) - 2m_r^2 - 1]m_r/(8K_r)$$
$$- \mu_N B(q_r - G_r p_{1r})[3m_r^2 - I_r(I_r+1)]/K_r\} \quad (14.5c)$$

where

$\eta = B/|B|,$

$g^2 = \eta^T \cdot \tilde{g}^T \cdot \tilde{g} \cdot \eta,$

$g^2 d_n = \eta^T \cdot \tilde{g}^T \cdot \tilde{D}^n \cdot \tilde{g} \cdot \eta$ (n = integer),

$g^2 K_r^2 = \eta^T \cdot \tilde{g}^T \cdot \tilde{A}_r^T \cdot \tilde{A}_r \cdot \tilde{g} \cdot \eta,$

$g^2 K_r^2 k_r^2 = \eta^T \cdot \tilde{g}^T \cdot \tilde{A}_r^T \cdot \tilde{A}_r \cdot \tilde{A}_r^T \cdot \tilde{A}_r \cdot \tilde{g} \cdot \eta,$

$g^2 K_r^2 e_r = \eta^T \cdot \tilde{g}^T \cdot (\tilde{D} \cdot \tilde{A}_r^T \cdot \tilde{A}_r + \tilde{A}_r^T \cdot \tilde{A}_r \cdot \tilde{D}) \cdot \tilde{g} \cdot \eta/2,$

$g_r^2 = \eta^T \cdot \tilde{g}_{nr}^T \cdot \tilde{g}_{nr} \cdot \eta,$

$gK_r G_r = \eta^T (\tilde{g}^T \cdot \tilde{A}_r^T \cdot \tilde{g}_{nr} + \tilde{g}_{nr}^T \cdot \tilde{A}_r \cdot \tilde{g}) \cdot \eta/2,$

$g^2 K_r^2 p_{nr} = \eta^T \cdot \tilde{g}^T \cdot \tilde{A}_r^T \cdot \tilde{P}_r^n \cdot \tilde{A}_r \cdot \tilde{g} \cdot \eta$ (n = integer),

$gK_r q_r = \eta^T \cdot (\tilde{g}^T \cdot \tilde{A}_r^T \cdot \tilde{P}_r^T \cdot \tilde{g}_{nr} + \tilde{g}_{nr}^T \cdot \tilde{P}_r \cdot \tilde{A}_r \cdot \tilde{g}) \cdot \eta/2,$

$gK_i K_j L_{ij} = \eta^T \cdot \tilde{g}^T \cdot (\tilde{A}_i^T \cdot \tilde{A}_i \cdot \tilde{A}_j^T \cdot \tilde{A}_j + \tilde{A}_j^T \cdot \tilde{A}_j \cdot \tilde{A}_i^T \cdot \tilde{A}_i) \cdot \tilde{g} \cdot \eta/2.$ \quad (14.6)

In Equations 14.5b and c, Det is the determinant, and Tr the trace of a matrix. The Zeeman term (Equation 14.5a) has been assumed to be dominant. The term in Equation 14.5c, with the factor M^{-1} has been derived under the assumption that the first-order hyperfine energy $K_r M m_r$, for each nucleus is larger than the corresponding nuclear Zeeman term $\mu_N I_r^T \cdot g_{nr} \cdot B$, or the quadrupole interaction $I_r^T \cdot \tilde{P}_r \cdot I_r$.

It is easily seen from Equation 14.5a–c that from EPR data only the tensors $\tilde{g}^T \cdot \tilde{g}$, $\tilde{A}_r^T \cdot \tilde{A}_r$, $\tilde{g}_{nr}^T \cdot \tilde{g}_{nr}$ (rather than $\tilde{g}$, $\tilde{A}_r$ and $\tilde{g}_{nr}$), $\tilde{D}$, and $\tilde{P}_r$ can be evaluated. These determine the single-crystal EPR line positions, calculated from the resonance condition $\Delta E = E_1 - E_2 = h\nu$, where E_1, E_2 are the eigenvalues of the pair of levels participating in resonance at frequency ν.

By using the eigenvalues (Equation 14.5a–c), the resonance condition for the "allowed" transitions $(M, m_i, m_j \leftrightarrow M-1, m_i, m_j)$ can be expressed as

$$\Delta E^{(a)} = h\nu = \mu_B g B + 3d_1(2M-1)/2$$
$$+ (d_2 - d_1^2)\left[1 - 4S(S+1) + 8(3M^2 - 3M + 1)\right]/(2\mu_B g B)$$
$$+ [Tr(\tilde{D}^2) - 2d_2 + d_1^2 - 2d_{-1}Det(\tilde{D})][2S(S+1)$$
$$-1 - 2(3M^2 - 3M + 1)]/(8\mu_B g B)$$
$$+ \sum_r \left\{ K_r m_r + \left[\left[\left\{ Tr(\tilde{A}_r^T \cdot \tilde{A}_r) - k_r^2 \right\} \left\{ I_r(I_r+1) - m_r^2 \right\} \right] \middle/ 2 \right. \right.$$
$$+ Det(A_r)(2M-1)m_r/K_r + (k_r^2 - K_r^2)m_r^2$$
$$+ 2(e_r - d_1)K_r\{3(2M-1)\}m_r]/(2\mu_B g B)\}$$
$$+ \sum_{i>j} (L_{ij} - K_i K_j) m_i m_j \Big/(\mu_B g B)$$
$$- \sum_r \{\mu_N^2 B^2 (g_r^2 - G_r^2) m_r \Big/(2K_r)$$
$$+ (p_{2r} - p_{1r}^2)[8m_r^2 - 4I_r(I_r+1) + 1]m_r/(2K_r)$$
$$+ [Tr(\tilde{P}_r^2) - 2p_{2r} + p_{1r}^2 - 2p_{-1r}Det(\tilde{P}_r)][2I_r(I_r+1) - 2m_r^2 - 1]m_r/(8K_r)$$
$$- \mu_N B(q_r - G_r p_{1r})[3m_r^2 - I_r(I_r+1)]/K_r\}/\{M(M-1)\} \qquad (14.7)$$

The resonance condition for the "forbidden" transitions $(M, m_i, m_j \leftrightarrow (M-1), m_i \pm n_i, m_j \pm n_j; n_i, n_j > 0)$ can likewise be expressed as follows:

$$\Delta E^{(f)} = h\nu = \Delta E^{(a)} +$$

$$\sum_i \left\{ -\frac{1}{2\mu_B g B} \begin{bmatrix} K_i(m_i \mp Mn_i) \\ \begin{bmatrix} (3/2)\{Tr(\tilde{A}_i^T \cdot \tilde{A}_i) - k_i^2\} \\ \{m_i^2 - (M-1)n_i(n_i \pm 2m_i)\} - 3(Det(\tilde{A}_i)/K_i) \\ \{\pm[M(M-2) - S(S+1)/3 + 1]n_i + m_i(1-2M)\} \\ + (K_i^2 - k_i^2)\{m_i^2 - (M-1)n_i(n_i \pm 2m_i)\} \\ -2(e_i - d_1)K_i\{\pm\left[3(M-1)^2 + S(S+1)\right]n_i \\ + m_i(1-2M)\} \end{bmatrix} \\ \pm G_i \mu_N B n_i + (n_i/2)\{Tr(\tilde{P}_i) - 3p_{1i}\}[n_i \pm 2m_i] \end{bmatrix} \right\}$$
$$- (1/(\mu_B g B)) \sum_{i>j} (L_{ij} - K_i K_j)[(M-1)(\pm m_j n_i \pm m_i n_j + n_i n_j) - m_i m_j]$$

$+$ contribution of terms containing the factor M^{-1} in the second order energy terms given by Eq. (5c) \qquad (14.8)

14.3.1.1 Complexities Associated with the Use of Second-Order-Perturbed Eigenvalues in the Application of Least-Squares Fitting (LSF) Procedure

The LSF procedure to determine SH parameters (SHPs) from EPR line positions and intensities has been described in Chapter 8. In this section, the intricacies of the energy differences $\Delta E^{(a)}$ and $\Delta E^{(f)}$ for the "allowed" and "forbidden" transitions, respectively, are discussed with the use of several coordinate systems. "Forbidden" transitions can be treated in terms of a special coordinate system for the presence of one nucleus only. (The specific details of the application of the LSF procedure to determine SHP, using this coordinate system, will be given later.)

"Allowed" transitions: An examination of the eigenvalue difference $\Delta E^{(a)}$ for "allowed" transitions (Equation 14.7) reveals that $\Delta E^{(a)}$ depends on the quantities g^2, d_n, k_i^2, K_i^2 and e_i, as defined by Equation 14.6. It follows, then, that these are the tensors $\tilde{g}^2 = \tilde{g}^T \cdot \tilde{g}$ and $\tilde{A}_i^2 = \tilde{A}_i^T \cdot \tilde{A}_i$ (rather than $\tilde{g}$, $\tilde{A}_i$) which determine the single-crystal "allowed" EPR line positions. Thus, *only these tensors can be evaluated from the "allowed" line positions.* Further, the calculation of K_i^2 and k_i^2 involves the tensors $\tilde{g}^T \cdot \tilde{A}_i^T \cdot \tilde{A}_i \cdot \tilde{g} (= \tilde{g}^T \cdot \tilde{A}_i^2 \cdot \tilde{g})$ and $\tilde{g}^T \cdot \tilde{A}_i^T \cdot \tilde{A}_i \cdot \tilde{A}_i^T \cdot \tilde{A}_i \cdot \tilde{g} (= \tilde{g}^T \cdot \tilde{A}_i^4 \cdot \tilde{g})$, respectively. In order to calculate them, it may appear at first sight that they depend upon the elements of the $\tilde{g}$ matrix. That this is not true can easily be seen by the use of the coordinate system in which the $\tilde{g}^2$ tensor is diagonal (i.e., the principal-axes coordinate system of the $\tilde{g}^2$-tensor); more details can be found in Weil and Anderson (1961). In this coordinate system, the $\tilde{g}$ matrix is also diagonal; its diagonal elements are the square roots of the corresponding elements of the diagonal tensor $\tilde{g}^2$ (Misra, 1984). The calculation of the quantities g^2, d_n, k_i^2, K_i^2 and e_i, as given by Equation 14.6, requires the use of the direction cosines of η (the unit vector along B), the three diagonal elements of the (diagonal) matrix $\tilde{g}^2$, the six independent elements of the symmetric tensor $\tilde{A}^2$, and the five independent elements of the symmetric tensor $\tilde{D}$ (since $Tr(\tilde{D}) = D_{x'x'} + D_{y'y'} + D_{z'z'} = 0$) (Kneubühl, 1963). Then, by using the "allowed" EPR line positions, it is possible to evaluate the three non-zero elements of the diagonal matrix obtained by the diagonalization of $\tilde{g}^2$ and the three Euler angles (θ, φ, ψ) required to rotate the laboratory axes to coincide with the coordinate system, in which the $\tilde{g}^2$-tensor is diagonal [the angles (θ, φ, ψ) also determine the direction cosines of η in the coordinate system in which the $\tilde{g}^2$-tensor is diagonal, as discussed in Section 14.3.2 below], the six independent elements of the $\tilde{A}_i^2$, and the five independent elements of the $\tilde{D}$ tensor.

"Forbidden" transitions: These transitions correspond to the transition energies $\Delta E^{(f)} (\equiv E(M, \{m_i\}) - E(M-1), \{m_i \pm n_i\})$, where $\{\ \}$ covers the indices $i = 1, 2, \ldots, N$), which can easily be calculated from Equations 14.5 and 14.6, and is expressed by Equation 14.8. An examination of $\Delta E^{(f)}$, given by Equation 14.8, reveals that $\Delta E^{(f)}$ is determined by the symmetric tensors $\tilde{g}_{ni}^2$ (each characterized by six independent elements) and $\tilde{P}_i$ (five independent elements since $Tr(\tilde{P}_i) = 0$) (Kneubühl, 1963), along with the $\tilde{g}^2$ and $\tilde{A}_i^2$ tensors. $\tilde{g}_{ni}^2$ and $\tilde{P}_i$ can only be determined from the "forbidden" line positions as the "allowed" line positions do not depend upon them; on the other hand, $\tilde{g}^2$ and $\tilde{A}_i^2$ can, indeed, be determined from the "allowed" EPR line positions as described previously. It is seen that the various quantities G_i, p_{ni} and q_i, describing $\Delta E^{(f)}$, given by Equation 14.8, can easily be calculated by using the coordinate system in which the $\tilde{g}_{ni}^2$ tensor is diagonal, provided that there is only one nuclear spin present ($L_{ij} = 0$ for this case, and the summation over i disappears). Otherwise, it is necessary to use an arbitrary coordinate system, which is common to all spins. In order to calculate the quantities G_i, p_{ni}, g_{ni} and L_{ij}, given by Equation 14.6, the tensor $\tilde{A}_i \cdot \tilde{g}$ (and its transpose $\tilde{g}^T \cdot \tilde{A}_i^T$), $\tilde{P}_i$, $\tilde{A}_i^2$ and $\tilde{g}_{ni}^2$ is required. For the presence of only one nuclear spin (say the ith nuclear spin), in the coordinate system in which $\tilde{g}_{ni}^2$ is diagonal, $\tilde{A}_i \cdot \tilde{g}$ can be expressed as $U_i^{\dagger} (\tilde{A}_i \cdot \tilde{g})_d U_i$, where U_i is the 3×3 matrix of direction cosines relating to the coordinate system in which the tensor $(\tilde{A}_i \cdot \tilde{g})^2 = \{\tilde{g}^T \cdot \tilde{A}_i^T \cdot \tilde{A}_i \cdot \tilde{g} = \tilde{g}^T \cdot \tilde{A}_i^2 \cdot \tilde{g}\}$ is diagonal. In particular, U_i depends on the Euler angles $(\Theta_i', \Phi_i', \Psi_i')$ required to rotate the coordinate system in which $\tilde{g}^T \cdot \tilde{A}_i^2 \cdot \tilde{g}$ is diagonal to coincide with the coordinate system in which $\tilde{g}_{ni}^2$ is diagonal, while $(\tilde{A}_i \cdot \tilde{g})_d$ is the diagonal matrix, the elements of which are the square roots of the matrix obtained by the diagonalization of $\tilde{g}^T \cdot \tilde{A}_i^2 \cdot \tilde{g}$. As for $\tilde{P}_i$, its five independent elements are to be expressed in the coordinate system in which $\tilde{g}_{ni}^2$ is diagonal (further details are provided later in the chapter).

14.3.2
Exact Matrix Diagonalization

The eigenvalues and eigenvectors of the SH matrix, given by Equations 14.1 and 14.2a–e, can be rigorously computed numerically by the use of a suitable diagonalizaing subroutine (e.g., JACOBI or EIGERS) on a digital computer. This is possible provided that, as seen above, the six independent elements of each of the $\tilde{g}^2$, $\tilde{A}_r^2$, and $\tilde{g}_{nr}^2$ tensors, as well as the five independent elements of each of the traceless symmetric matrices $\tilde{D}$ and $\tilde{P}_i$, are known. For, as seen in context with the perturbation calculation of eigenvalues, these particular tensors, indeed determine the eigenvalues. This requires that the matrices $\tilde{g}$, $\tilde{A}_r$ and $\tilde{g}_{nr}$ do not, in fact, determine eigenvalues. This is equivalent to requiring that the matrices $\tilde{g}$, $\tilde{A}_r$ and $\tilde{g}_{nr}$ in Equation 14.2a, c, and e should be replaced by the corresponding symmetric matrices, defined as:

$$O^{(s)} = L^{-1} (O_d^2)^{1/2} L, \quad (14.9)$$

where $(O_d^2)^{1/2}$ is the diagonal matrix, the elements of which are the square roots of the diagonal matrix $O_d^2 (= L O^2 L^{-1})$, obtained by the diagonalization of $O^2 = O^T \cdot O$ ($O = \tilde{g}, \tilde{A}_r, \tilde{g}_{nr}$), and L is the matrix that diagonalizes O^2. Thus, the column vectors of L are the direction cosines of the principal directions of O^2. Then, each $O^{(s)}$, given by Equation 14.9, is a symmetric matrix, characterized by six independent elements. Alternatively, each $O^{(s)}$ can be described in terms of the three diagonal elements of $(O_d^2)^{1/2}$ and three Euler angles (θ, φ, ψ) corresponding to the three successive rotations required to bring the laboratory frame in coincidence with the principal axes of the tensor O^2 (Goldstein, 1951).

Explicitly,

$$L^{-1} = \begin{pmatrix} \cos\psi\cos\varphi - \cos\theta\sin\varphi\sin\psi & -\sin\psi\cos\varphi - \cos\theta\sin\varphi\cos\psi & \sin\theta\sin\varphi \\ \cos\psi\sin\varphi + \cos\theta\cos\varphi\sin\psi & -\sin\psi\sin\varphi + \cos\theta\cos\varphi\cos\psi & -\sin\theta\cos\varphi \\ \sin\theta\sin\psi & \sin\theta\cos\psi & \cos\theta \end{pmatrix}$$
(14.10)

Finally, in order to find the eigenvalues by numerical diagonalization of the SH matrix, the following SH is used:

$$H = \mu_B S^T \cdot \tilde{g}^{(s)} \cdot B + S^T \cdot \tilde{D} \cdot S + \\ + \sum_{i=1}^{N} \left(I_i^T \cdot \tilde{A}_i^{(s)} \cdot S - \mu_N I_i^T \cdot \tilde{g}_{ni}^{(s)} \cdot B + I_i^T \cdot \tilde{P}_i \cdot I_i \right).$$
(14.11)

It should be noted that in Equation 14.11, each of $\tilde{g}^{(s)}$, $\tilde{A}_i^{(s)}$ and $\tilde{g}_{ni}^{(s)}$ is characterized by six parameters in terms of three eigenvalues each of $\tilde{g}^2$, $\tilde{A}^2$ and $\tilde{g}_{ni}^2$ and three corresponding Euler angles, while each of $\tilde{D}$, $\tilde{P}_i$ is characterized by five elements, instead of six, due to the fact that $\text{trace}(\tilde{D}) = \text{trace}(\tilde{P}_i) = 0$. These independent parameters can be evaluated by the application of the LSF technique, as described in the following section.

14.4
Evaluation of SHPs by the LSF Technique

The application of the LSF technique will now be illustrated for the case of electron spin $S = \frac{1}{2}$, which is not characterized by the ZFS tensor D (i.e., the term $S^T \cdot \tilde{D} \cdot S$ does not exist for this case), coupled to one nuclear spin $I \geq \frac{1}{2}$.

14.4.1
First-Order Perturbation

This procedure has been well described by Misra (1984) for the presence of one nuclear spin. Here, the eigenvalues, calculated to first order in perturbation, and neglecting the nuclear quadrupole Zeeman interactions, for $S = \frac{1}{2}$, $I \geq \frac{1}{2}$, are (from Equation 14.5b):

$$E^{(1)}(M,m) = \mu_B g B M + K M m,$$
(14.12)

where the electron magnetic quantum number $M = \pm\frac{1}{2}$ and the nuclear quantum number $m = -I, -(I-1), \ldots, (I-1), I$. For the "allowed" transitions ($\Delta M = \pm 1$; $\Delta m = 0$, i.e., $M = \frac{1}{2} \leftrightarrow -\frac{1}{2}$, $m \leftrightarrow m$) the energy differences are

$$E^{(1)}(m) = \mu_B g B + K m$$
(14.13)

Now, using Equation 14.6, the parameter g can be expressed as

$$g = \left\{ \left(\tilde{g}^2\right)_{xx} \ell^2 + 2\left(\tilde{g}^2\right)_{xy} \ell m + \left(\tilde{g}^2\right)_{yy} m^2 + 2\left(\tilde{g}^2\right)_{yz} mn + \left(\tilde{g}^2\right)_{zz} n^2 + 2\left(\tilde{g}^2\right)_{xz} \ell n \right\}^{1/2}. \tag{14.14}$$

In Equation 14.14, l, m, and n are the direction cosines of $\boldsymbol{\eta} = \mathbf{B}/|\mathbf{B}|$ with respect to the laboratory frame (x, y, z). On the other hand, when using Equation 14.6 for $g^2 K_r^2$ and dropping the subscript r, one can express K in the principal-axes system (x', y', z') of the tensor $\tilde{g}^2$ as

$$K = \left\{ (l'g_{x'})^2 \left(\tilde{A}^2\right)_{x'x'} + 2(l'g_{x'})(m'g_{y'})\left(\tilde{A}^2\right)_{x'y'} \right.$$
$$+ (m'g_{y'})^2 \left(\tilde{A}^2\right)_{y'y'} + 2(l'g_{x'})(n'g_{z'})\left(\tilde{A}^2\right)_{x'z'}$$
$$\left. + (n'g_{z'})^2 \left(\tilde{A}^2\right)_{z'z'} + 2(m'g_{y'})(n'g_{z'})\left(\tilde{A}^2\right)_{y'z'} \right\}^{1/2}. \tag{14.15}$$

In Equation 14.15, $g_{x'}, g_{y'}, g_{z'}$ are the square roots of the principal values of the $\tilde{g}^2$ tensor, whereas l', m', n' are the direction cosines of $\boldsymbol{\eta}$ in the principal axes system of $\tilde{g}^2$.

From Equation 14.7 it is seen that there are $(2I+1)$ "allowed" hyperfine lines for any orientation of $\mathbf{B}$, the center of which lies at the magnetic field value $B_0 = h\nu/\mu_B g$, as found from the resonance condition. Thus, in the first-order perturbation approximation, the centers of hyperfine sets depend only on $\tilde{g}^2$, they do not depend on $\tilde{A}^2$. When examining Equation 14.14 for g, it is seen that at least six data points in three mutually perpendicular planes are required to determine the elements of the symmetric tensor $\tilde{g}^2$. In practice, as many centers of hyperfine sets for the magnetic field orientation are fitted in three mutually perpendicular planes as possible, but this is certainly greater than, or equal to, six in order to evaluate the elements of the $\tilde{g}^2$ tensor. The $\tilde{g}^2$ tensor, so estimated, can then be diagonalized to obtain its principal values and its eigenvectors; the square roots of the principal values are the elements $g_{x'}, g_{y'}, g_{z'}$, while the eigenvectors represent the direction cosines of the principal axes of the $\tilde{g}^2$ tensor with respect to the laboratory frame (x, y, z). By knowing these direction cosines, it is now possible to calculate l', m', n', the direction cosines of $\boldsymbol{\eta}$. The values of the set $(g_{x'}, g_{y'}, g_{z'}; l', m', n')$ so obtained can now be used in Equation 14.15 as initial values to evaluate the elements of $\tilde{A}^2$ tensor in an LSF fitting of hyperfine lines observed for several orientations of the external magnetic field in three mutually perpendicular planes, as described above. The matrix of $\tilde{A}^2$ so obtained can then be diagonalized to obtain the principal values of $\tilde{A}^2$, the resulting eigenvectors are the direction cosines of the principal axes of $\tilde{A}^2$ with respect to the principal axes of the $\tilde{g}^2$ tensor. For further detail, see Misra et al. (1983) and Misra (1984).

14.4.2
Second-Order Perturbation

For the case of $S = \frac{1}{2}$, neglecting the nuclear Zeeman interaction, considering the case of one electron spin coupled to one nuclear spin (the index r is dropped), the energy differences for "allowed" transitions ($\Delta M = \pm 1; \Delta m = 0$), calculated from Equation 14.5a–c, are:

$$\Delta E^{(2)}(m) = \mu_B gB + Km + [\{Tr(\tilde{A}^2) - k^2\} \\ \times \{I(I+1) - m^2\}/2 + (k^2 - K^2)m^2]/(2\mu_B gB) \tag{14.16}$$

The quantities g and K have been explicitly expressed by Equations 14.14 and 14.15. In order to express k^2 explicitly, as given by Equation 14.6, it appears that, in general, it is necessary to know the nine elements of the $\tilde{g}$ matrix, if it is asymmetric. However, in any case, only six parameters are required; these are the three eigenvalues of the symmetric $\tilde{g}^2$ tensor and the three Euler angles defining the direction cosines of the three principal directions of the $\tilde{g}^2$ tensor with respect to the laboratory frame. These direction cosines are the column vectors of the matrix L^{-1}, expressed in terms of the three Euler angles (θ, φ, ψ) given by Equation 14.10.

Finally, k^2 can be expressed as:

$$k^2 = \{(l'g_{x'})^2 (\tilde{A}^4)_{x'x'} + 2(l'g_{x'})(m'g_{y'})(\tilde{A}^4)_{x'y'} \\ + (m'g_{y'})^2 (\tilde{A}^4)_{y'y'} + 2(l'g_{x'})(n'g_{z'})(\tilde{A}^4)_{x'z'} + (n'g_{z'})^2 (\tilde{A}^4)_{z'z'} \\ + 2(m'g_{y'})(n'g_{z'})(\tilde{A}^4)_{y'z'}\}/(g^2 K^2) \tag{14.17}$$

In writing Equation 14.17, the fact that $\tilde{A}^4$ is a symmetric tensor has been taken into account. In Equation 14.17, (l', m', n') are the direction cosines of **B** in the principal axis system (x', y', z') of $\tilde{g}^2$. These can be expressed as

$$\begin{pmatrix} l' \\ m' \\ n' \end{pmatrix} = \begin{pmatrix} l_{x''} & m_{x''} & n_{x''} \\ l_{y''} & m_{y''} & n_{y''} \\ l_{z''} & m_{z''} & n_{z''} \end{pmatrix} \begin{pmatrix} l \\ m \\ n \end{pmatrix}, \tag{14.18}$$

where (l, m, n) are the direction cosines of **B** in the laboratory frame (x, y, z) and $(l_{\alpha''}, m_{\alpha''}, n_{\alpha''}; \alpha = x, y, z)$ are the direction cosines of the principal directions of the $\tilde{g}^2$ tensor with respect to the laboratory frame. The $\tilde{A}^4$ matrix, appearing in Equation 14.17 is defined as $\tilde{A}^4 = \tilde{A}^2 \cdot \tilde{A}^2 = \tilde{A}^T \cdot \tilde{A} \cdot \tilde{A}^T \cdot \tilde{A}$. Thus:

$$(\tilde{A}^4)_{\alpha\beta} = \sum_{\gamma} \tilde{A}^2_{\alpha\gamma} \tilde{A}^2_{\gamma\beta}; \quad \alpha, \beta, \gamma = x, y, z \tag{14.19}$$

(It is seen from Equation 14.19 that $\tilde{A}^4$ is a symmetric tensor, since $\tilde{A}^2$ is a symmetric tensor.)

Equations 14.16–14.18 reveal that the energy differences $\Delta E^{(2)}(m)$ depend on 12 parameters; these are the three eigenvalues of the $\tilde{g}^2$ tensor and the three Euler angles required to rotate the laboratory axes to coincide with the principal axes of the $\tilde{g}^2$ tensor, and the six independent elements of the symmetric tensor $\tilde{A}^2$. The derivatives of the eigenvalues required in the LSF procedure can then be appropriately calculated using the dependence of $\Delta E^{(2)}(m)$ on these parameters. Further, it is seen that, because of the dependence of the third term on the right-hand side of Equation 14.16 on m^2, the centers of the hyperfine sets (hyperfine resonant line positions) for any orientation of **B** depend both on $\tilde{g}^2$ and $\tilde{A}^2$. This is unlike the case of the first-order perturbed eigenvalues, where the dependence of the centers of the hyperfine sets is only on $\tilde{g}^2$. Thus, for the application of the LSF procedure, using second-order perturbed eigenvalues, it is necessary to fit simultaneously the 12 parameters $g_{x'}, g_{y'}, g_{z'}; \theta, \varphi, \psi; \tilde{A}^2_{\alpha\beta}$ ($\alpha, \beta = x, y, z$). The initial values of the first six parameters may be chosen to be those obtained by the LSF method using the expressions for the eigenvalues, calculated by a first-order perturbation approximation, and fitting the centers of the hyperfine sets, as described in Chapter 8, Section 8.5. The initial values of the Euler angles may be obtained from the eigenvectors (direction cosines) of the principal values of the $\tilde{g}$ matrix, by solving Equation 14.9.

In order to use the LSF procedure, the various quantities that determine the energy differences between the pair of levels participating in resonance should first be identified. These are, in turn, functions of the particular SHPs for the "allowed" and "forbidden" transitions (as seen in Sections 14.3.1 and 14.3.2 above).

It is found helpful in some cases to use different special coordinate axes for the fitting of "allowed" and "forbidden" transitions (this is described below). Although an arbitrary coordinate system can also be used, application of the LSF procedure becomes much more complicated in this case (the details are provided later in the chapter).

14.4.3
Use of Special Coordinate Axes

The following discussion is valid for ions with $S > \frac{1}{2}$, so that the ZFS tensor $\tilde{D}$ should be taken into account.

14.4.3.1 "Allowed" Line Positions

One first needs to express explicitly the quantities η, $\tilde{g}^2$, K_i^2, and k_i^2, on which the $\Delta E^{(a)}$s depend, as seen from Equation 14.3, giving the energy differences $\Delta E^{(a)}$ between the levels participating in resonance for the "allowed" transitions. These quantities, in turn, depend on SHPs, and are described (in the principal-axes system of the tensor $\tilde{g}^2$) as follows. Since η is the unit vector along the Zeeman field **B**, its direction cosines (l, m, n) in the laboratory axes are known. In order to express (l', m', n'), its direction cosines in the principal-axes system of $\tilde{g}^2$, the transformation matrix U, which is a function of the Euler angles (θ, φ, ψ) required to rotate the laboratory axes to coincide with the $\tilde{g}^2$ principal axes, are required:

$$\eta = \begin{pmatrix} l' \\ m' \\ n' \end{pmatrix} = U \begin{pmatrix} l \\ m \\ n \end{pmatrix}. \tag{14.20}$$

In Equation 14.20, the transpose of the matrix, U, which is related to the matrix L^{-1}, given by Equation 14.10, is expressed as follows:

$$U = (L^{-1})^{\dagger} \tag{14.21}$$

where † denotes the adjoint. Then, by using Equation 14.6 which defines $\tilde{g}$, K_i^2, k_i^2, d_n and e_i, one can write for these quantities:

$$g^2 = \sum_{\alpha'} a_{\alpha'}^2, \tag{14.22a}$$

$$K_i^2 g^2 = \sum_{\alpha',\beta'} (2 - \delta_{\alpha'\beta'}) A_{\alpha'\beta'}^2 a_{\alpha'} a_{\beta'}, \tag{14.22b}$$

$$k_i^2 K_i^2 g^2 = \sum_{\alpha',\beta'} (2 - \delta_{\alpha'\beta'}) A_{\alpha'\beta'}^4 a_{\alpha'} a_{\beta'} \tag{14.22c}$$

where

$$A_{\alpha'\beta'}^4 = (A^2 \cdot A^2)_{\alpha'\beta'} = \sum_{\gamma'} A_{\alpha'\gamma'}^2 A_{\gamma'\beta'}^2, \tag{14.22d}$$

$$e_i K_i^2 g^2 = \tfrac{1}{2} \sum_{\alpha',\beta'} a_{\alpha'} a_{\beta'} \sum_{\gamma'} (D_{\alpha'\gamma'} A_{i\alpha'\beta'} + A_{i\alpha'\gamma'} D_{\alpha'\beta'}), \tag{14.22e}$$

and

$$d_n g^2 = \sum_{\alpha',\beta'} a_{\alpha'} a_{\beta'} (D^n)_{\alpha'\beta'}, \quad n = -1, 1, 2, \tag{14.22f}$$

In Equations 14.22a–f the summations on $(\alpha', \beta', \gamma')$ are over x', y', z', which label the principal axes of the $\tilde{g}^2$ tensor. The tensors $\tilde{A}^2$ and $\tilde{A}^4 (\equiv \tilde{A}^2 \cdot \tilde{A}^2)$ and, $\tilde{D}$ are symmetric, that is, $\tilde{A}_{\alpha'\beta'}^2 = \tilde{A}_{\beta'\alpha'}^2$, $\tilde{A}_{\alpha'\beta'}^4 = \tilde{A}_{\beta'\alpha'}^4$, $\tilde{D}_{\alpha'\beta'} = \tilde{D}_{\beta'\alpha'}$. In order to calculate $(\tilde{D}^{-1})_{\alpha'\beta'}$, required in the calculation of d_{-1} Equation 14.22f, one first notes that $\tilde{D}$ is a 3×3 matrix. Now, the inverse R of a matrix F can be expressed as (Goldstein, 1951):

$$R = F^{-1} = (Det(F))^{-1} \begin{pmatrix} F_{yy}F_{zz} - F_{yz}F_{zy} & F_{xz}F_{zy} - F_{xy}F_{zz} & F_{xy}F_{yz} - F_{xz}F_{yy} \\ F_{yz}F_{zx} - F_{yx}F_{zz} & F_{xx}F_{zz} - F_{xz}F_{zx} & F_{xz}F_{yx} - F_{xx}F_{yz} \\ F_{yx}F_{zy} - F_{yy}F_{zx} & F_{xy}F_{zx} - F_{xx}F_{zy} & F_{xx}F_{yy} - F_{xy}F_{yx} \end{pmatrix}, \tag{14.23}$$

where the matrix multiplying $(Det(F))^{-1}$ on the right-hand side of Equation 14.23 is defined as $Adj(F)$, the matrix adjoint to the matrix F. Since d_{-1} depends on $\tilde{D}^{-1}$, it would, therefore, be undefined according to Equation 14.23, if the determinant of $\tilde{D}$ were to vanish. If this were to occur, then one can use the alternative expression, $d_{-1}Det(\tilde{D}) = Tr(Adj(\tilde{D})) - d_1 Tr(\tilde{D}) + d_2$ (Weil, 1975; Iwasaki, 1974).

Finally, the $a_{\alpha'}$ terms required in Equations 14.22a–c, e, and f are given as

$$a_{z'} = l' g_{x'} \quad a_{y'} = m' g_{y'} \quad a_{z'} = n' g_{z'} \tag{14.24}$$

In Equation 14.24, $g_{x'}, g_{y'}, g_{z'}$ are the square roots of the principal values of $\tilde{g}^2$, while (l', m', n') are the direction cosines of B (the components of η) in the principal-axes system of $\tilde{g}^2$; these are calculated by the use of Equations 14.20 and 14.21.

The 17 parameters (for the presence of one nuclear spin) which can be evaluated from the LSF fitting of the "allowed" EPR line positions (for $S \geq 1$), observed in three mutually perpendicular planes are thus the three diagonal elements $\tilde{g}^2_\alpha$ ($\alpha' = x', y', z'$), the three Euler angles (θ, φ, ψ), the components of $\tilde{A}^2_{\alpha'\beta'}$ (six independent elements), and the components $D_{\alpha'\beta'}$ (five independent elements). To this end, one needs to use the expression for $\Delta E^{(a)}$, given by Equation 14.7, along with the definitions of the various quantities therein, as given by Equations 14.22a–f, 14.23, and 14.24.

14.4.3.2 "Forbidden" Line Positions

The energy difference $\Delta E^{(f)}$, given by Equation 14.8, between the pairs of levels participating in resonance for "forbidden" transitions requires quantities, over and above those described previously in context with the "allowed" line positions. These are given as follows, for the presence of only one nuclear spin (denoted by i), using the coordinate axes ($x_{i'}, y_{i'}, z_{i'}$) in which the $\tilde{g}^2_{ni}$ tensor is diagonal:

$$g^2_{ni} = \sum_{\alpha'_i} b^2_{\alpha'_i}, \tag{14.25a}$$

$$G_i K_i g = \tfrac{1}{2} \sum_{\alpha'_i, \beta'_i} \eta_{\alpha'_i} \eta_{\beta'_i} \left\{ (\tilde{A}_i \cdot \tilde{g})_{\beta'\alpha'} (\tilde{g}_{ni})_{\beta'} + (\tilde{A}_{i'} \cdot \tilde{g})_{\alpha'\beta'} (\tilde{g}_{ni})_{\alpha'} \right\}, \tag{14.25b}$$

$$p_{ni} K^2_i g^2 = \sum_{\alpha'_i, \beta'_i} \eta_{\alpha'_i} \eta_{\beta'_i} \sum_{\gamma'_i \delta'_i} (\tilde{A}_i \cdot \tilde{g}_{ni})_{\gamma'_i \alpha'_i} (\tilde{A}_i \cdot \tilde{g}_{ni})_{\delta'_i \beta'_i} (\tilde{P}_i)^n_{\gamma'_i \delta'_i}, \tag{14.25c}$$

$$g_i K_i g = \tfrac{1}{2} \sum_{\alpha'_i, \beta'_i} \eta_{\alpha'_i} \eta_{\beta'_i} \left\{ (\tilde{A}_i \cdot \tilde{g})_{\gamma'_i \alpha'_i} (\tilde{P}_i)_{\gamma'_i \beta'_i} (\tilde{g}_{ni})_{\beta'_i} + (\tilde{A}_{i'} \cdot \tilde{g})_{\gamma'_i \beta'_i} (\tilde{P}_i)_{\alpha'_i \gamma'_i} (\tilde{g}_{ni})_{\alpha'_i} \right\}, \tag{14.25d}$$

In Equation 14.25a–d, $\eta_{\alpha'_i}$ are the components of $\boldsymbol{\eta}$, the unit vector along the Zeeman field $\mathbf{B}$; these are, thus, the direction cosines of the external Zeeman field in the principal axes system of the $\tilde{g}^2_{ni}$ tensor. The unit vector $\boldsymbol{\eta}$ can be expressed in terms of the direction cosines (l, m, n) of $\mathbf{B}$ in the laboratory axes as follows:

$$\boldsymbol{\eta}_i = \begin{pmatrix} l'_i \\ m'_i \\ n'_i \end{pmatrix} = U_i \begin{pmatrix} l \\ m \\ n \end{pmatrix}. \tag{14.26}$$

In Equation 14.26, the matrix U_i can be described in the same way as the matrix U, given by Equation 14.21, if one replaces θ, φ, ψ by Θ_i, Φ_i, Ψ_i respectively; here (Θ_i, Φ_i, Ψ_i) are the Euler angles required to rotate the laboratory axes to coincide with the principal axes of the the $\tilde{g}^2_{ni}$ tensor. In Equation 14.25a, the $b_{\alpha'_i}$ are given as $b_{x'_i} = l'_i g_{nx'_i}$, $b_{y'_i} = m'_i g_{ny'_i}$, $b_{z'_i} = n'_i g_{nz'_i}$, where $g_{nx'_i}, g_{ny'_i}, g_{nz'_i}$ are the square roots of the principal values of $\tilde{g}^2_{ni}$. The $\tilde{A}_i \cdot \tilde{g}$ (and thus $\tilde{g}^T \cdot \tilde{A}_i^T$), required in Equations 14.25b–d, to be expressed in the $\tilde{g}^2_{ni}$ principal-axes system, are obtained from

$\tilde{g}^T \cdot \tilde{A}_i^2 \cdot \tilde{g}$ as evaluated from the "allowed" EPR line positions in the $\tilde{g}^2$ principal-axes system, as follows:

$$\tilde{A}_i \cdot \tilde{g} = U_i U U' \left(\tilde{g}^T \cdot \tilde{A}_i^2 \cdot \tilde{g} \right)_d^{1/2} U'^\dagger U^\dagger U_i^\dagger. \tag{14.27}$$

In Equation 14.27, $\left(\tilde{g}^T \cdot \tilde{A}_i^2 \cdot \tilde{g} \right)_d^{1/2}$ is the diagonal matrix, the elements of which are the square roots of the matrix obtained from the diagonalization of the unitary matrix $\tilde{g}^T \cdot \tilde{A}_i^2 \cdot \tilde{g}$, as determined in the $\tilde{g}^2$ principal-axes system, using the "allowed" EPR line positions:

$$\left(\tilde{g}^T \cdot \tilde{A}_i^2 \cdot \tilde{g} \right)_d = U'^\dagger \left(\tilde{g}^T \cdot \tilde{A}_i^2 \cdot \tilde{g} \right) U'. \tag{14.28}$$

The matrices U', U_i, and U used in Equation 14.27 are defined by Equation 14.28, 14.26, and 14.21, respectively.

14.4.4
Use of Arbitrary Coordinate Axes

The use of special coordinate systems, as described above, simplified the expressions of the various quantities required to determine $\Delta E^{(a)}$ and $\Delta E^{(f)}$. Whilst it is possible to use an entirely arbitrary coordinate system, its use is more complicated than those of the special coordinate systems (as described in the preceding section). Further, for the presence of more than one nucleus, no special coordinate system has been found to be particularly useful. For the case of an arbitrary coordinate system, the following observations, as made from an examination of Equations 14.5–14.8 and applicable to the case of the presence of more than one nuclear spin, are found to be helpful. The quantities that can be determined from the "allowed" EPR line positions in an arbitrary coordinate system are the components of the symmetric tensors $\tilde{g}^T \cdot \tilde{g}$, $\tilde{g}^T \cdot \tilde{A}_i^2 \cdot \tilde{g}$ $(= \tilde{g}^T \cdot \tilde{A}_i^T \cdot \tilde{A}_i \cdot \tilde{g})$, and $\tilde{g}^T \cdot \tilde{D} \cdot \tilde{g}$. On the other hand, the "forbidden" line positions enable the determination of the components of $\tilde{g}_{ni}^2$ and $\tilde{P}_i$, provided that one uses

$$\tilde{A}_i \cdot \tilde{g} = U'' \left(\tilde{g}^T \cdot \tilde{A}_i^2 \cdot \tilde{g} \right)_d^{1/2} U''^\dagger, \tag{14.29a}$$

and

$$\tilde{g}_{ni} = U_i \left(\tilde{g}_{ni}^2 \right)_d^{1/2} U_i^\dagger. \tag{14.29b}$$

In Equation 14.29a, $\left(\tilde{g}^T \cdot \tilde{A}_i^2 \cdot \tilde{g} \right)_d^{1/2}$ is the diagonal matrix, the elements of which are the square roots of the elements of the matrix obtained by the diagonalization of $\tilde{g}^T \cdot \tilde{A}_i^2 \cdot \tilde{g}$, already determined from the "allowed" EPR line positions; that is,

$$\left(\tilde{g}^T \cdot \tilde{A}_i^2 \cdot \tilde{g} \right)_d = U''^\dagger \left(\tilde{g}^T \cdot \tilde{A}_i^2 \cdot \tilde{g} \right) U'', \tag{14.30}$$

where U'' is the unitary matrix effecting the diagonalization of $\tilde{g}^T \cdot \tilde{A}_i^2 \cdot \tilde{g}$. Furthermore, in Equation 14.25b, $\left(\tilde{g}_{ni}^2 \right)_d^{1/2}$ is the diagonal matrix, the elements of which are the square roots of the elements of the matrix obtained by the diagonalization of $\tilde{g}_{ni}^2$; that is,

$$\left(\tilde{g}_{ni}^{2}\right)_{d}=U_{i}^{\dagger}\left(\tilde{g}_{ni}^{2}\right)U_{i}, \tag{14.31}$$

where U_i is the unitary matrix that effects the diagonalization of $\tilde{g}_{ni}^2$.

Explicitly, for use in Equations 14.4–14.8, in terms of the above-mentioned quantities $\tilde{g}^2$, $\tilde{g}^T \cdot \tilde{A}^2 \cdot \tilde{g}$, $\tilde{g}^T \cdot \tilde{D} \cdot \tilde{g}$, $\tilde{g}_{ni}^2$, $\tilde{P}_i$ are determinable from EPR data, using $\left(\tilde{g}^2\right)^{-1} = \left(\tilde{g}^T \cdot \tilde{g}\right)^{-1} = \tilde{g}^{-1} \cdot \left(\tilde{g}^T\right)^{-1}$, one can express

$$g^2 = \eta^T \cdot \left(\tilde{g}^2\right) \cdot \eta, \tag{14.32a}$$

$$d_n g^2 = \eta^T \cdot \left(\tilde{g}^T \cdot \tilde{D} \cdot \tilde{g}\right) \cdot \left(\tilde{g}^2\right)^{-1} \cdot \left(\tilde{g}^T \cdot \tilde{D} \cdot \tilde{g}\right) \cdot \left(\tilde{g}^2\right)^{-1} \cdots \left(\tilde{g}^T \cdot \tilde{D} \cdot \tilde{g}\right) \cdot \eta, \tag{14.32b}$$

(NB: In Equation 14.32b, $(\tilde{g}^T \cdot \tilde{D} \cdot \tilde{g})$ appears n times with $(\tilde{g}^2)^{-1}$ in between):

$$K_i^2 g^2 = \eta^T \cdot \left(\tilde{g}^T \cdot \tilde{A}_i^2 \cdot \tilde{g}\right) \cdot \eta, \tag{14.32c}$$

$$k_i^2 K_i^2 g^2 = \eta^T \cdot \left(\tilde{g}^T \cdot \tilde{A}_i^2 \cdot \tilde{g}\right) \cdot \left(\tilde{g}^2\right)^{-1} \cdot \left(\tilde{g}^T \cdot \tilde{A}_i^2 \cdot \tilde{g}\right) \cdot \eta \tag{14.32d}$$

$$e_i K_i^2 g^2 = \tfrac{1}{2} \eta^T \cdot \left[\left(\tilde{g}^T \cdot \tilde{D} \cdot \tilde{g}\right) \cdot \left(\tilde{g}^2\right)^{-1} \cdot \left(\tilde{g}^T \cdot \tilde{A}_i^2 \cdot \tilde{g}\right)\right.$$
$$\left. + \left(\tilde{g}^T \cdot \tilde{A}_i^2 \cdot \tilde{g}\right) \cdot \left(\tilde{g}^2\right)^{-1} \cdot \left(\tilde{g}^T \cdot \tilde{D} \cdot \tilde{g}\right)\right] \cdot \eta, \tag{14.32e}$$

$$g_{ni}^2 = \eta^T \cdot \left(\tilde{g}_{ni}^2\right) \cdot \eta, \tag{14.32f}$$

$$G_i K_i g = \tfrac{1}{2} \eta^T \cdot \left[\left(\tilde{A}_i \cdot \tilde{g}\right)^T \cdot \left(\tilde{g}_{ni}\right) + \left(\tilde{g}_{ni}^T\right) \cdot \left(\tilde{A}_i \cdot \tilde{g}\right)\right] \cdot \eta, \tag{14.32g}$$

$$p_{ni} K_i^2 g^2 = \eta^T \cdot \left(\tilde{A}_i \cdot \tilde{g}\right)^T \cdot \left(\tilde{P}_i\right)^n \cdot \left(\tilde{A}_i \cdot \tilde{g}\right) \cdot \eta, \tag{14.32h}$$

$$q_i K_i g = \tfrac{1}{2} \eta^T \cdot \left[\left(\tilde{A}_i \cdot \tilde{g}\right)^T \cdot \left(\tilde{P}_i\right) \cdot \left(\tilde{g}_{ni}\right) + \left(\tilde{g}_{ni}\right)^T \cdot \left(\tilde{P}_i\right) \cdot \left(\tilde{A}_i \cdot \tilde{g}\right)\right] \cdot \eta, \tag{14.32i}$$

$$L_{ij} K_i K_j g^2 = \tfrac{1}{2} \eta^T \cdot \left[\left(\tilde{g}^T \cdot \tilde{A}_i^2 \cdot \tilde{g}\right) \cdot \left(\tilde{g}^2\right)^{-1} \cdot \left(\tilde{g}^T \cdot \tilde{A}_j^2 \cdot \tilde{g}\right)\right.$$
$$\left. + \left(\tilde{g}^T \cdot \tilde{A}_j^2 \cdot \tilde{g}\right) \cdot \left(\tilde{g}^2\right)^{-1} \cdot \left(\tilde{g}^T \cdot \tilde{A}_i^2 \cdot \tilde{g}\right)\right] \cdot \eta \tag{14.32j}$$

The p_{-1i}, as given by Equation 14.32h requires the inverse matrix $\tilde{P}_i^{-1}$. This can be calculated by using Equation 14.23. However, if $Det(\tilde{P}_i) = 0$, the alternative expression $Tr(Adj(\tilde{P}_i) - P_1 Tr(\tilde{P}_i) + P_2)$ can be used in place of $p_{-1i} Det(\tilde{P}_i)$ (Weil, 1975; Iwasaki, 1974). The quantities that can be determined from EPR data are enclosed within parentheses in Equation 14.32a–i. As far as the evaluation of the required traces and determinants is concerned, using the property that $Tr(EF) = Tr(FE)$ and $Det(EF) = Det(E) \cdot Det(F)$, the following are noted:

$$Tr\left(\tilde{A}^2\right) = Tr\left[\left(\tilde{g}^T \cdot \tilde{A}^2 \cdot \tilde{g}\right)\left(\tilde{g}^T \cdot \tilde{g}\right)^{-1}\right] = Tr\left[\left(\tilde{g}^T \cdot \tilde{A}^2 \cdot \tilde{g}\right) \cdot \left(\tilde{g}^2\right)^{-1}\right] \tag{14.33a}$$

$$Det(\tilde{A}) = \left\{Det\left(\tilde{g}^T \cdot \tilde{A}^2 \cdot \tilde{g}\right) / Det\left(\tilde{g}^T \cdot \tilde{g}\right)\right\}^{1/2}. \tag{14.33b}$$

14.4.5
Simultaneous LSF Fitting of Both the "Allowed" and "Forbidden" Line Positions

It should be noted that both the "allowed" and "forbidden" line positions can be simultaneously fitted, in the LSF procedure, to evaluate the various tensor components. As for the choice of the coordinate system, one could either use any one of the two special coordinate systems described above (principal-axes systems of

$\tilde{g}^2$, or those of $\tilde{g}_{ni}^2$), or an entirely arbitrary coordinate system. Although, the simultaneous fitting of both the "allowed" and "forbidden" line-position groups is considerably more complicated than the LSF fitting of only one of these alone, it is possible to determine the components of all the tensors at once by a simultaneous fitting of both these groups of lines. In addition, the parameter errors for this case are also reduced, as the numbers of lines fitted has been increased (Misra and Subramanian, 1982).

The relevant steps to be followed for the application of the LSF procedure, as well as an illustrative example, are well described by Misra (1988).

14.5
Numerical Evaluation of the Derivatives Required in the LSF Procedure

Application of the LSF technique, using matrix diagonalization, to the case of non-coincident tensors, poses certain problems as the eigenvalues—which determine the resonant line positions—do not depend on the elements of the $\tilde{g}$, $\tilde{A}$ matrices, which, in general, are not symmetric. Rather, they depend on the elements of the symmetric tensors $\tilde{g}^2$, $\tilde{A}^2$, $\tilde{g}_n^2$ which do not appear in the SH given by Equations 14.1 and 14.2a–c. Thus, these are the elements of the symmetric $\tilde{g}^2$, $\tilde{A}^2$ tensors which can be determined from the data, rather than the elements of the matrices $\tilde{g}$, $\tilde{A}$. In fact, one can only determine those quantities from a physical measurement on which the measured property depends. On the other hand, if one were to use the symmetric matrices $O^{(s)}\{=g^{(s)}, A^{(s)}, g_n^{(s)}\}$, as defined by Equation 14.9, the six elements (each of which consists of three principal eigenvalues of the $O^2 = O^T \cdot O$ matrices) and the three corresponding Euler angles, required to rotate the laboratory axes (x, y, z) to coincide with the principal axes of the O^2 tensor, then the SH given by Equation 14.11 could be used for the purpose of matrix diagonalization. In that case, six independent elements are required defining each of the matrices $g^{(s)}$, $A^{(s)}$, and $g_n^{(s)}$ and five independent elements of each of the symmetric matrices $P^{(s)}$ and $Q^{(s)}$, as their traces are zero—that is, $P_{xx} + P_{yy} + P_{zz} = Q_{xx} + Q_{yy} + Q_{zz} = 0$. Consequently, the SH given by Equation 14.11 can be used (in an LSF manner) to evaluate SHPs from resonant line positions obtained for the magnetic field orientation in three mutually perpendicular planes.

For the applicability of the LSF procedure, the required derivatives can be calculated using Feynman's theorem (Feynman, 1939):

$$\partial E_i / \partial a_j = \langle i | \partial H / \partial a_j | i \rangle$$
$$= Tr[(\partial H / \partial a_j)(|i\rangle \otimes \langle i|)]. \quad (14.34)$$

In Equation 14.34, $|i\rangle$ is the eigenvector of H corresponding to the eigenvalue E_i; that is, $H|i\rangle = E_i|i\rangle$; and $|i\rangle \otimes \langle i|$ is the outer product of the eigenvector $|i\rangle$ with itself; its elements are thus given as

$$(|i\rangle \otimes \langle i|)_{kl} = |i\rangle_k |i\rangle_l^*. \quad (14.35)$$

14.5 Numerical Evaluation of the Derivatives Required in the LSF Procedure

In Equation 14.35, $|i\rangle_k$ denotes the kth element of column vector $|i\rangle$. As for the operator $\partial H/\partial a_j$, which is the derivative of H with respect to the parameter a_j, it can be evaluated for different terms of the SH, which depend on different parameters, as follows; H_{zf} can be expressed as:

$$H_{zf} = S^T \cdot \tilde{D} \cdot S = \sum_{\alpha,\beta=x,y,z} D_{\alpha\beta} S_\alpha S_\beta \tag{14.36}$$

Then,

$$\partial(S^T \cdot \tilde{D} \cdot S)/\partial D_{\alpha\beta} = S_\alpha S_\beta + S_\beta S_\alpha \quad (\alpha \neq \beta), \tag{14.37a}$$

and

$$\partial(S^T \cdot \tilde{D} \cdot S)/\partial D_{\alpha\alpha} = S_\alpha^2 - S_z^2 \quad (\alpha = x,y) \tag{14.37b}$$

In writing Equation 14.37a it has been taken into account that D is symmetric; that is, $D_{\alpha\beta} = D_{\beta\alpha}$. In Equation 14.37b, the fact that $D_{zz} = -D_{xx} - D_{yy}$ has been taken into account. In a similar manner, noting that $P_{zz} = -P_{xx} - P_{yy}$, the derivatives of H_{nq} for the presence of a single nucleus are:

$$\partial(I^T \cdot \tilde{P} \cdot I)/\partial P_{\alpha\beta} = I_\alpha I_\beta + I_\beta I_\alpha \quad (\alpha \neq \beta), \tag{14.38a}$$

and

$$\partial(I^T \cdot \tilde{P} \cdot I)/\partial P_{\alpha\alpha} = I_\alpha^2 - I_z^2 \quad (\alpha = x,y). \tag{14.38b}$$

As for the derivatives of the electronic Zeeman term $H_{ze} = \mu_B S^T \cdot g^{(s)} \cdot B$ with respect to the elements of $g^{(s)}$, Equation 14.9 defining $g^{(s)}$ is taken into account. Then, each element of the matrix $g^{(s)}$ is a function of six parameters $(g_1^d, g_2^d, g_3^d; \theta, \varphi, \psi)$ where g_i^d $(i=1,2,3)$ are the square roots of the principal values of the g^2 matrix, and (θ, φ, ψ) are the Euler angles needed to rotate the laboratory axes to coincide with the principal axes of the $\tilde{g}^2$-tensor. Explicitly,

$$H_{ze} = \sum_{\alpha,\beta=x,y,z} \tfrac{1}{2}\mu_B (B_\alpha S_\beta + B_\beta S_\alpha) g_{\alpha\beta}^{(s)}. \tag{14.39}$$

In writing Equation 14.39, the fact that $g^{(s)}$ is a symmetric matrix has been used. The derivatives of H_{ze} can now be expressed as:

$$\frac{\partial H_{ze}}{\partial a_j} = \sum_{\alpha,\beta} \tfrac{1}{2}\mu_B (B_\alpha S_\beta + B_\beta S_\alpha) \partial g_{\alpha\beta}^{(s)}/\partial a_j. \tag{14.40}$$

Finally, using the derivatives $\partial g_{\alpha\beta}^{(s)}/\partial a_j$ from the known dependence of $g_{\alpha\beta}^{(s)}$ on a_j as given previously, the right-hand side of Equation 14.40 can be calculated.

Similarly, for the hyperfine term $H_{hf} = S^T \cdot A^{(s)} \cdot I$, one has

$$\frac{\partial H_{hf}}{\partial a_k} = \sum_{\alpha,\beta} \tfrac{1}{2}(S_\alpha I_\beta + S_\beta I_\alpha) \partial A_{\alpha\beta}^{(s)}/\partial a_k, \tag{14.41}$$

where a_k stands for any of the six parameters $\{a_k\} = (A_1^d, A_2^d, A_3^d; \alpha, \beta, \gamma)$; with A_i^d, $i=1,2,3$, being the square roots of the eigenvalues of the matrix $\tilde{A}^2$; and

(α, β, γ) are the Euler angles required to rotate the laboratory axes to coincide with the principal axes of the $\tilde{A}^2$ tensor.

Finally, the derivatives of the nuclear Zeeman term $H_{zn} = -\mu_N I^T \cdot g_n^{(s)} \cdot B$, can be evaluated with respect to the six elements $(g_{ni}^d, i = 1, 2, 3; \xi, \eta, \zeta)$, where $g_{ni}^d, i = 1, 2, 3$, are the square roots of the eigenvalues of the symmetric tensor $\tilde{g}_n^2$, and (ξ, η, ζ) are the Euler angles required to rotate the laboratory axes to coincide with the principal axes of the $\tilde{g}_n^2$ tensor, in the same manner as that used to evaluate the derivatives of H_{ze}.

14.6
General Remarks

In the evaluation of SHPs from EPR line positions for the case of non-coincident tensors, appropriate to low-symmetry, one can either use the perturbation approach, or an exact numerical diagonalization of the SH matrix. However, in either case, one can only estimate those parameters from the data on which the eigenvalue differences actually depend. For the case of perturbation, for "allowed" transitions, these are the $\tilde{g}^2$ and $\tilde{A}^2$ tensors for $S = \frac{1}{2}$; $I \geq \frac{1}{2}$. The principal values of $\tilde{g}^2$ and $\tilde{A}^2$ tensors and their direction cosines can be obtained by numerical diagonalizations of the corresponding matrices. The eigenvalues of a matrix represent its principal values, while the normalized eigenvectors represent the corresponding direction cosines. The square roots of the eigenvalues of $\tilde{g}^2$ and $\tilde{A}^2$ matrices are the principal values of $\tilde{g}$ and $\tilde{A}$ matrices, respectively.

For the case of $S \geq \frac{1}{2}$, $I \geq \frac{1}{2}$, one can estimate the components of the tensors $\tilde{g}^2$, $\tilde{g}^T \cdot \tilde{A}^2 \cdot \tilde{g}$, and $\tilde{g}^T \cdot \tilde{D} \cdot \tilde{g}$ from the "allowed" line positions if one takes into account the corresponding differences of eigenvalues, as given by Equation 14.5a and c, and the definitions of the quantities $\tilde{g}^2$, d_n, K^2, k^2, as given by Equation 14.6. For the case of only one nucleus, the following relationships are found to be useful:

(i) $\tilde{g}^T \cdot \tilde{A}^2 \cdot \tilde{A}^2 \cdot \tilde{g} = \tilde{g}^T \cdot \tilde{A}^2 \cdot \tilde{g} \cdot (\tilde{g}^T \cdot \tilde{g})^{-1} \cdot \tilde{g}^T \cdot \tilde{A}^2 \cdot \tilde{g}$

$= (\tilde{g}^T \cdot \tilde{A}^2 \cdot \tilde{g}) \cdot (\tilde{g}^2)^{-1} \cdot (\tilde{g}^T \cdot \tilde{A}^2 \cdot \tilde{g})$;

(ii) $\tilde{g}^T \cdot \tilde{D}^2 \cdot \tilde{g} = \tilde{g}^T \cdot \tilde{D} \cdot \tilde{g} \cdot (\tilde{g}^T \cdot \tilde{g})^{-1} \cdot \tilde{g}^T \cdot \tilde{D} \cdot \tilde{g}$

$= (\tilde{g}^T \cdot \tilde{D} \cdot \tilde{g}) \cdot (\tilde{g}^2)^{-1} \cdot \tilde{g}^T \cdot \tilde{D} \cdot \tilde{g}$;

(iii) $Tr(\tilde{A}^2) = Tr\{(\tilde{g}^T \cdot \tilde{A}^2 \cdot \tilde{g}) \cdot (\tilde{g}^2)^{-1}\}$;

(iv) $Det(\tilde{A}) = \{Det(\tilde{A}^2)\}^{1/2} = \{Det(\tilde{g}^T \cdot \tilde{A}^2 \cdot \tilde{g})/\{Det(\tilde{g}^2)\}\}^{1/2}$; and

(v) $\tilde{A}^2 = \tilde{V}^{-1} \cdot (\tilde{g}_d)^{-1} \cdot \tilde{V} \cdot \tilde{V}^{-1} \cdot (\tilde{g}^T \cdot \tilde{A}^2 \cdot \tilde{g}) \cdot \tilde{V} \cdot \tilde{V}^{-1} \cdot (\tilde{g}_d)^{-1} \cdot \tilde{V}$.

In expressing (iii), the facts that $(\tilde{g}^2)^{-1} = (\tilde{g}^T \cdot \tilde{g})^{-1}$ and $Tr(CD) = Tr(DC)$ have been used. For (iv), the fact that $Det(CD) = Det(C) \times Det(D)$ has been used. In (iv) the matrix $\tilde{V}$ diagonalizes the $\tilde{g}^2$-tensor; that is, $\tilde{g}_d^2 = \tilde{V} \cdot \tilde{g}^2 \cdot \tilde{V}^{-1}$. The above relationships are useful when the laboratory axes are employed for the description of the various matrices. From these relationships, it is seen that for the second-order perturbed eigenvalues (Equation 14.5c) the eigenvalue differences for the "allowed" transitions, neglecting the nuclear Zeeman term, depend on the tensors $\tilde{g}^2$ and $\tilde{g}^T \cdot \tilde{A}^2 \cdot \tilde{g}$. The quantities g_n^2, G, p_n, and q, as given by Equation 14.6, cannot be determined from the "allowed" line positions, as the corresponding eigenvalue differences do not depend on them. However, the line positions for the hyperfine "forbidden" transitions do depend on them; they can, then, be evaluated from the "forbidden" line positions, using the values of the $\tilde{g}^2$, $\tilde{A}^2$, and $\tilde{D}$ tensors already determined from the fitting of "allowed" line positions, from the second-order eigenvalues (as given by Equation 14.5c). The following relationship is useful for the presence of more than one nucleus (indicated by subscripts i and j):

$$\tilde{g}^T \cdot (\tilde{A}_i^T \cdot \tilde{A}_i \cdot \tilde{A}_j^T \cdot \tilde{A}_j + \tilde{A}_j^T \cdot \tilde{A}_j \cdot \tilde{A}_i^T \cdot \tilde{A}_i) \cdot \tilde{g}$$
$$= (\tilde{g}^T \cdot \tilde{A}^2 \cdot \tilde{g}) \cdot (\tilde{g}^2)^{-1} \cdot (\tilde{g}^T \cdot \tilde{A}_j^2 \cdot \tilde{g}) + (\tilde{g}^T \cdot \tilde{A}_j^2 \cdot \tilde{g}) \cdot (\tilde{g}^2)^{-1} \cdot (\tilde{g}^T \cdot \tilde{A}_i^2 \cdot \tilde{g})$$

Finally, it should be noted that, as discussed in Sections 14.2 and 14.3, the EPR data cannot be used to determine the individual elements of the $\tilde{g}$ and $\tilde{A}$ matrices; only the eigenvalues and eigenvectors (direction cosines of the principal values) of the $\tilde{g}^2$ and $\tilde{A}^2$ tensors can be determined from EPR data.

Although the present description has been confined to the second-order electron spin operator $S^T \cdot \tilde{D} \cdot S$, higher-order electron spin operators $O_l^m, l = 4, 6, \ldots (-l \leq m \leq l)$ for $S \geq 2$ can be easily incorporated in the LSF procedure (Abragam and Bleaney, 1970). Consideration of higher-order O_l^m in the LSF procedure has been well described in a general way previously (Misra, 1976). Mcgavin et al. (1980) have also considered the presence of several non-coincident tensors of higher than second order (i.e., fourth order) in S and I:

$$H = \mu_B B^T \cdot \tilde{g} \cdot S + S^T \cdot \tilde{D} \cdot S + I^T \cdot \tilde{A} \cdot S + I^T \cdot \tilde{P} \cdot I - \mu_n B \cdot \tilde{g}^{(I)} \cdot I$$
$$+ \sum_m B_4^m O_4^m (S) + \sum_m C_4^m O_4^m (I), \quad (14.42)$$

and were successful in applying the LSF technique for successive refinement of SHPs:

$$\mu_B B \cdot \tilde{g} \cdot S \gg I^T \cdot \tilde{A} \cdot S \sim S^T \cdot \tilde{D} \cdot S \gg I^T \cdot \tilde{P} \cdot I \sim \mu_n B^T \cdot \tilde{g}^{(I)} \cdot I \quad (14.43)$$

Basically, in their approach, perturbation expressions are used for the eigenvalues, and for their dependences on the matrices/tensors $\tilde{g}, \tilde{D}, \tilde{A}, \tilde{P}$, and $\tilde{g}_n$, is exploited. EPR line positions for a minimum of six orientations (two in each of three orthogonal planes) of the external magnetic field are required to determine the six independent elements of each of the tensors $\tilde{g}^2, K^2, d_n, p_{nr}, G_r$, as given by Equation 14.6. First, only the tensor $\tilde{g}^2$ is determined neglecting other terms in the SH, and it is then diagonalized:

$$(\tilde{g}^2)_{diag} = L^{-1} \cdot (\tilde{g}^2) \cdot L \quad (14.44)$$

The components of this tensor are then calculated in the laboratory frame by the inverse transformation:

$$\tilde{g} = L \cdot \left(\tilde{g}^2\right)^{1/2}_{\text{diag}} \cdot L^{-1}. \tag{14.45}$$

This procedure is then repeated, in turn, to determine the elements of the tensors $\tilde{A}^2$, $\tilde{D}$, $\tilde{P}$, and $\tilde{g}_n^2$, in the successive refinement carried out by using Equation 14.43. Finally it is noted that, for the case of the lowest – that is, triclinic symmetry – all possible m values consistent with a given ℓ occur in the SH.

Acknowledgments

The author is grateful to Professor C. P. Poole, Jr for helpful comments to improve the presentation of the chapter, and to the Natural Sciences and Engineering Research Council (NSERC) of Canada for partial financial support.

Pertinent Literature

The energies calculated to second order in perturbation are given by Iwasaki (1974) and Weil (1975). These were further exploited by Misra (1984) to determine SHPs corresponding to non-coincident anisotropic $\tilde{g}^2$, $\tilde{A}^2$ tensors. Higher-order SHPs were considered by Mcgavin et al. (1980).

References

Abragam, A. and Bleaney, B. (1970) Electron Paramagnetic Resonance of Transition Ions, Clarendon Press, Oxford.

Feynman, R.P. (1939) Phys. Rev., **56**, 340.

Goldstein, H. (1951) Classical Mechanics, Addison-Wesley, Cambridge.

Iwasaki, M. (1974) J. Magn. Reson., **16**, 417.

Kneubühl, F.K. (1963) Physik der Konden. Mater., **1**, 410.

Mcgavin, D.C., Palmer, R.A., Singers, W.A., and Tennant, W.C. (1980) J. Magn. Reson., **40**, 69.

Misra, S.K. (1976) J. Magn. Reson., **23**, 403.

Misra, S.K. (1984) Physica, **124B**, 53.

Misra, S.K. (1988) Physica B, **151**, 433.

Misra, S.K., Bandet, J., Bacquet, G., and McEnally, T.E. (1983) Phys. Status Solidi A, **80**, 581.

Misra, S.K. and Subramanian, S. (1982) J. Phys. C, **15**, 7199.

Weil, J.A. (1975) J. Magn. Reson., **18**, 113.

Weil, J.A. and Anderson, J.H. (1961) J. Chem. Phys., **35**, 1410.

Weil, J.A. and Bolton, J.R. (2007) Electronic Paramagnetic Resonance: Elementary Theory and Practical Applications, John Wiley & Sons, Inc., New Jersey.

15
Biological Systems

Boris Dzikovski

15.1
Introduction

In this chapter the applications of multifrequency EPR to study biological systems are discussed. Today, multifrequency EPR has established itself as a valuable approach to study the structure and molecular dynamics of biological molecules. It is shown here how crucial insights into particular biological systems can be obtained by very high-frequency EPR (VHF- EPR), whereas EPR at a lower frequency is less informative. This is accomplished by resolving the g-factors of: (i) nitroxide spin labels attached as reporter groups to biological macromolecules; (ii) biologically generated radicals, such as semiquinones, tyrosine radicals; and (iii) metal cofactors, such as multinuclear copper centers. It is shown here how the improved g-factor resolution achieved by VHF-EPR leads to a dramatic improvement in orientational resolution for spin labels, which in turn provides an enhanced sensitivity to molecular motion over that achieved by the standard X-Band (9.5 GHz) EPR. Since the g-factor resolution is proportional to the EPR frequency, the resolution of the g-tensor components for paramagnetic centers with low anisotropic g-factors is facilitated by an increase in EPR frequency. Occasionally, frequencies above 350 GHz are required to resolve the principal values, as demonstrated by discussions on flavin and biliverdin radicals.

The g-factors of many organic radicals, in general – and of nitroxide spin labels in particular – are extremely sensitive to the polarity of the local environment. This makes VHF-EPR an invaluable tool for probing local polarity, with important applications in investigations of protein folding and membrane research. During recent years, EPR has been used successfully to study both biological and model membranes. Whilst the majority of EPR studies on membranes are still carried out at low frequencies, VHF-EPR offers unique opportunities that are discussed in this chapter in greater detail.

Since VHF-EPR provides a "faster snapshot" of molecular motion, the recording and analysis of EPR spectra of the same system over a range of EPR frequencies can effectively separate different modes of molecular motion, such as the overall tumbling of the entire molecule versus the relative motion of its parts, or the specific mobility in nitroxide tethers. Although an extensive discussion of this topic, using slowly relaxing local structure (SRLS) has already been provided in Chapter 11, some simplified methods for the analysis of complex dynamics are discussed in this chapter.

A promising – though not yet sufficiently explored – application of VHF-EPR to metalloproteins is related to its unique capability for determining large zero-field splitting (ZFS) for the high-spin transition metal ions. For $S \geq 1$, the Zeeman splitting at X-band is often much smaller than the ZFS for non-Kramers ions, which are usually "EPR-silent" at this frequency, or for intra Kramers-level transitions in Kramers ions with $S > 1/2$. In these cases, VHF-EPR is indispensible for the determination of ZFS parameters.

It should be pointed out, however, that VHF-EPR is not always advantageous. For example, in most cases of metalloproteins, EPR at the standard frequency, X-band, or lower is preferable as the use of VHF-EPR there leads to extremely broad lines and, hence, to a reduced sensitivity. In addition, despite substantial progress in VHF-EPR instrumentation, a high-frequency experiment is often more challenging and expensive to set up than an X-band experiment. Furthermore, some biological applications, such as *in-vivo* EPR, require a lower than 9 GHz frequency. Therefore, a specific biological system can be best studied by EPR at a set of particular optimal frequencies suitable to its characteristics. The criteria for the choice of frequencies are discussed in this chapter, and examples of successful VHF and multifrequency studies in biology are also provided.

15.2
VHF EPR as the g-Resolved EPR Spectroscopy

15.2.1
Spectral Resolution of g-Factor Differences

With an increase in EPR frequency, the spectral resolution of paramagnetic centers with different g-factors is greatly improved, as the difference in the field positions of their resonances is proportional to the EPR frequency, ν:

$$\Delta B_0 = \frac{h\nu}{\mu_B}\left(\frac{1}{g_1} - \frac{1}{g_2}\right).$$

Here, h is the Planck constant, μ_B is the Bohr magneton, and g_1 and g_2 are the g-factors of two paramagnetic species present in the sample. For organic radicals – nitroxide spin labels for example – the expression can be written as: $\Delta B_0 \approx \frac{h\nu\Delta g}{4\mu_B}$, as both g_1 and g_2 are approximately equal to 2; here, Δg is the difference between the two g-values.

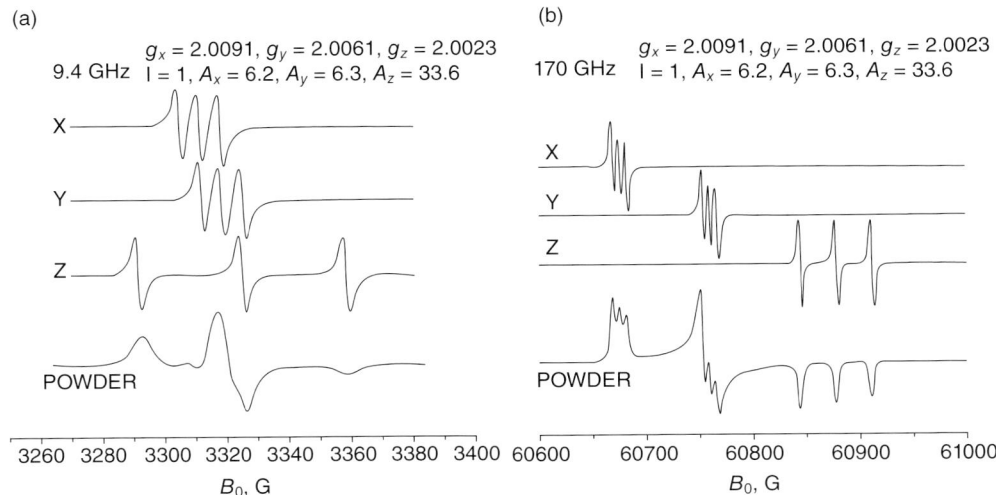

Figure 15.1 Origin of high orientational resolution for VHF-EPR, rigid-limit spectra, simulation. (a) At low frequency (9 GHz), the hyperfine splitting dominates. Lines corresponding to different orientations of the radical relative to the magnetic field overlap in the rigid-limit spectrum. The spectral extent is determined by the largest component of the A-tensor (A_{zz} in the case of nitroxide radicals); (b) At high frequency (170 GHz), the effect of the g-factor anisotropy overwhelms the effect of the A-tensor. Different orientations of the radical in the magnetic field are separated in the spectrum.

15.2.2
Precise Determination of the g-Tensor Principal Values

The use of EPR at higher frequencies has significantly improved the g-factor resolution, in particular for rigid-limit spectra, and allowed for the precise determination of the g-tensor components. Figure 15.1 shows simulations of the rigid-limit powder spectra at both 9.4 and 170 GHz for a typical nitroxide radical that arise from a superposition of all possible orientations of the magnetic g- and A-tensors of the nitroxide radical relative to B_0. The figure illustrates that the full extent of the 9 GHz spectrum is determined by the largest of the principal values of the hyperfine splitting, which is A_{zz}. Hence, at X-band the distance between the outer extrema for a well-resolved spectrum in the rigid limit is exactly $2A_{zz}$, which is usually 65–75 G, depending on the chemical structure of the nitroxide and its local environment. Further, the maximum possible difference in the field position due to the g-tensor differences: $\Delta B_0 = \dfrac{h\nu \Delta g}{4\mu_B} \sim 11\,\text{G}$, where $\Delta g = g_{xx} - g_{zz} \sim 7 \times 10^{-3}$, is relatively small compared to this value. Different orientations of the radical with different g-tensor values may resonate at the same field. The most reliable way to obtain g-tensor values from EPR at low frequencies is to use single crystals. In

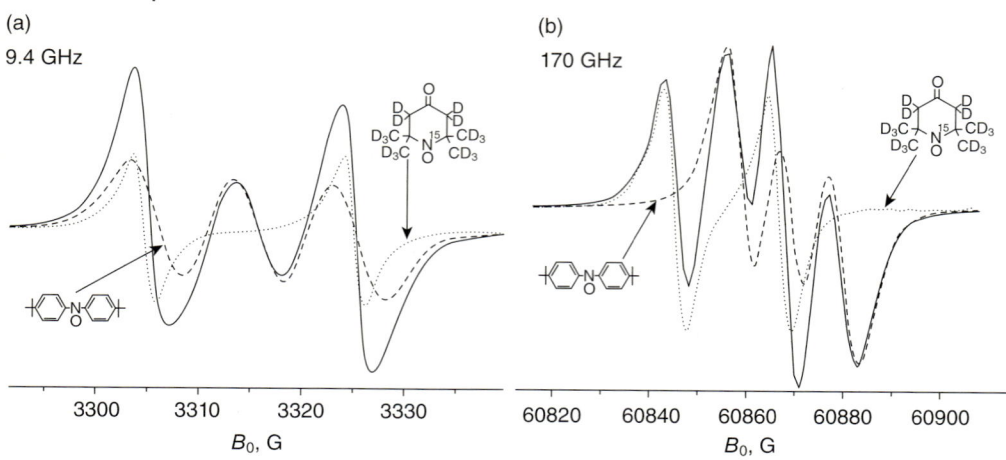

Figure 15.2 (a) 9 GHz EPR spectroscopy exhibits a poor g-factor resolution. For typical nitroxide radicals, the g-values usually differ in the fourth decimal point, which is insufficient for effective separation. Example, an EPR spectrum for a mixture of two nitroxide radicals of different structure in toluene, ^{15}N-PD-Tempone, g_{iso} = 2.005 86 (dots), and ^{14}N-di-*tert*-butyl-phenyl nitroxide, g_{iso} = 2.005 52 (dashes). The resulting superposition spectrum (solid line) is almost symmetrical due to the poor g-resolution at X-band; (b) The same mixture at 170 GHz. The doublet is clearly shifted relative to the triplet, showing sufficient g-factor resolution.

contrast, VHF-EPR can be considered as the g-resolved EPR spectroscopy, wherein different orientations of the radical relative to B_0 resonate at different field values. The principal g-tensor values g_{xx}, g_{yy}, and g_{zz} can be determined by VHF-EPR from the corresponding features of the powder spectrum with high precision, thus eliminating the need to determine these values by using a single crystal. Knowledge of the principal values of g- and A-tensors is a key prerequisite for a successful simulation of EPR spectra at any frequency. The principal values of the g-tensors for a number of organic radicals involved in enzymatic catalysis have been reviewed by Jeschke (2005), while values of the g-tensor components and hyperfine-coupling constants for biologically relevant semiquinone radicals have been summarized by Burghaus et al. (2002).

15.2.3
Resolution of g-Factors of Different Paramagnetic Centers

The EPR spectra recorded at 9.4 and 170 GHz for a toluene solution containing two nitroxide radicals, namely ^{15}N-PDT (4-oxo-2,2,6,6-tetramethylpiperidine-d$_{16}$-1-oxyl) and ^{14}N-di-(*p-tert*-butyl-phenyl)nitroxide, with slightly different isotropic g-factors ($\Delta g \sim 0.0003$), are shown in Figure 15.2. At 9 GHz the difference in g-factors gives a 0.5 G shift between the centers of ^{15}N-doublet and ^{14}N-triplet, which is insufficient for an effective g-factor separation. At this frequency, the superposition of the spectra produces an almost symmetrical pattern, whereas at 170 GHz the magnetic-field offset due to this Δg-value is more than 9 G.

The ability of VHF-EPR to reliably separate paramagnetic centers of different natures has been applied, for example, in the separation of spin-trapping adducts (Smirnova et al., 1997). A pulsed W-band EPR allowed resolution of the radical pair $P_{700}^{+}A_1^{-}$ in highly purified photosystem I (PS I) preparations from *Synechococcus elongates*, and an accurate determination of the g-tensor components for A_1^{-} (van der Ast et al., 1997). VHF-EPR also played an essential role in resolving small g-factor differences in multicopper proteins (c.f. Section 15.5).

15.3
Effect of Polarity of the Environment on the g-Factor

More common in biological applications of EPR, however, are cases that require the resolution of differences in components of the g-factor for the same paramagnetic species in different environments. In the case of nitroxides, for example, the magnetic parameters are known to be sensitive to the polarity of the local environment. In general, changing the local environment of the nitroxide moiety from water (polar) to hydrocarbon (nonpolar) causes a decrease in the values of the components of the hyperfine tensor, and a concomitant increase in the g-tensor components. This effect is most pronounced for the tensor components g_{xx} and A_{zz}.

15.3.1
Examples

15.3.1.1 Derivatives of 2,2,6,6-tetramethylpiperidine-1-oxyl (TEMPO)

For the PDT (4-oxo-2,2,6,6-tetramethylpiperidine-d_{16}-1-oxyl) probe transferred from the polar glycerol/water solvent to toluene, the g_{xx}-value increases by 0.0011 (from 2.00853 to 2.00963; Budil et al., 1989), while A_{zz} decreases from 36.2 to 33.5 G. As seen in Figure 15.3, the fraction of the TEMPO spin label in the water phase has a ~2 G larger value of the isotropic hyperfine constant $a_{iso} = (A_{xx} + A_{yy} + A_{zz})/3$, whereas the value of the isotropic g-factor, $g_{iso} = (g_{xx} + g_{yy} + g_{zz})/3$ decreases by ~0.0004 compared to that in the less-polar environment. At 170 GHz, the g-factors of the two components corresponding to the partition of the spin probe into the areas with different polarity are well resolved, whereas at 9 GHz the combined effect of g and A on the position of the EPR lines is close to the linewidth of the individual hyperfine components. During the 1980s, Lebedev and coworkers conducted extensive studies on the effects of solvent polarity, and demonstrated a high degree of correlation between the values of A_{zz} and g_{xx} (Ondar et al., 1985). Earlier studies performed by Kawamura, Matsunami, and Yonezawa (1967) showed that a di-*tert*-butyl nitroxide spin could have different g versus A plots, depending on whether the local environment was protic or aprotic. In this case, proticity refers to the propensity to donate hydrogen bonds,

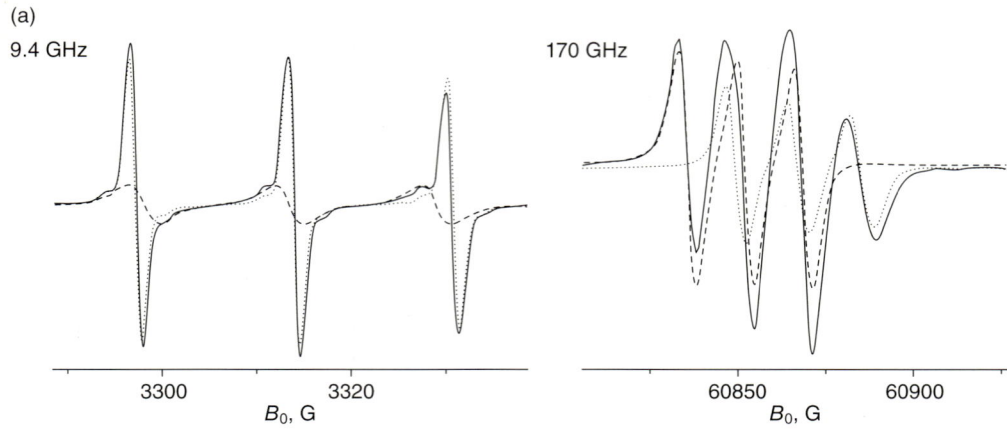

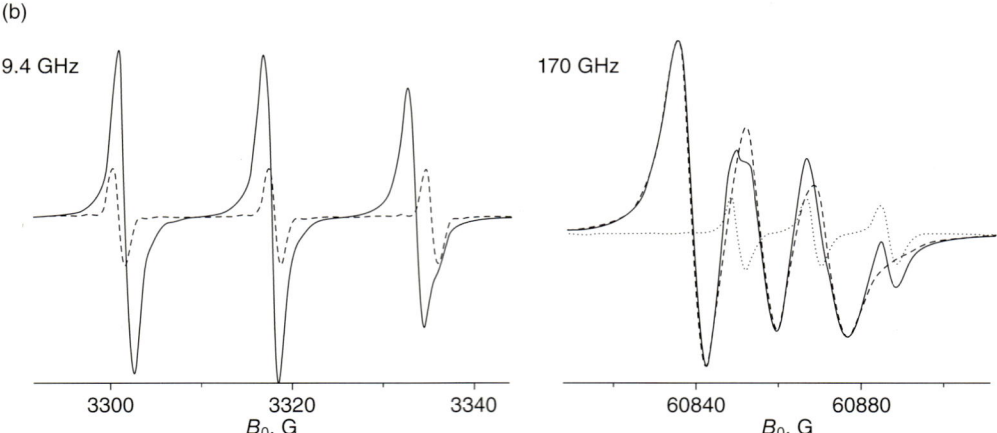

Figure 15.3 Dependence of the g-factor value for the nitroxide radical on the polarity of the local environment. A better g-factor resolution of VHF-EPR allows separation of the EPR signals corresponding to the partition of spin-labeled molecules into different phases and determination of partition coefficients. (a) TEMPO in emulsion toluene/SDS/water (solid line). At 170 GHz the g-factor resolution is sufficient to separate signals from TEMPO molecules in the water phase (dots) and in the emulsion particles (dashes); (b) Partition of TEMPO in the liquid crystal phase of DLPC between the water phase (dots) and lipid phase (dashes). The 170 GHz signal for TEMPO in the lipid bilayer (dashes) is obtained by spectral subtraction.

whereas aprotic refers to solvents which cannot donate a hydrogen bond. The g–A plot for protic solvents is described by a linear relationship with a steeper slope than that for aprotic solvents (Owenius et al., 2001). However, as shown in a recent study using TEMPO nitroxide, if the correlations are indeed different for TEMPO in protic and aprotic solvents, then the difference is rather small (Smirnov and Smirnova, 2001).

15.3.1.2 Spin-Labeled Phospholipid Membranes: 1,2-Dipalmitoyl-*sn*-Glycero-3-Phosphocholine (DPPC) and 1-Palmitoyl-2-Oleoyl-*sn*-Glycero-3-Phosphocholine (POPC)

The enhanced magnetic tensor resolution achieved with 250 GHz EPR has been exploited (Earle et al., 1994) when studying the polarity gradient along the aliphatic chains in frozen DPPC and POPC membranes, using several spin-labeled phosphatidylcholine (n-PC) labels, 1-palmitoyl-2-stearoyl(*n*-DOXYL)-*sn*-glycero-3-phosphocholines as a function of the labeling position. Quantitatively, in the DPPC membrane the g_{xx}-value was increased from 2.008 69 (as reported by the 5-PC spin label) to 2.009 29 at labeling position 16; at the same time, the A_{zz}-value was decreased from 35.0 to 33.5 G. At 250 GHz, the spectral shift due to the difference in g_{xx}-values was 27 G, while the difference in A_{zz}-values was 1.5 G. In contrast, at 9 GHz the expected ~1 G shift due to changes in g_{xx} was buried in the central hyperfine component and could not be reliably estimated. The g-versus-A dependence in POPC is indicative of hydrogen-bonding interactions with water, whereas in the case of DPPC the gentler overall slope of the g_{xx}-versus-A_{zz} dependence is suggestive of an aprotic environment (Earle et al., 1994). Changes in the polarity of the local environment were attributed to different degrees of water penetration along the hydrophobic alkyl chains of the membrane bilayer. A high value of $\partial g_{xx}/\partial A_{zz}$, characteristic of hydrogen-bonded spin labels, was also found in a detailed VHF study of cholesterol-containing 1,2-dimyristoyl-*sn*-glycero-3-phosphocholine (DMPC) membranes that used eleven PC spin labels labeled systematically at positions 4 to 14 (Kurad, Jeschke, and Marsh, 2003). The results of this study, which was carried out simultaneously on frozen and fluid membranes, supported the conclusion that the transmembrane polarity profile registered by spins labels is caused by water penetration into the membrane. More recently, Smirnov and coworkers (Smirnova et al., 2009) used pH-sensitive nitroxides (which were derivatives of imidazole) to profile the heterogeneous dielectric gradient and hydrogen-bonding environment in phospholipid membranes, by using VHF-EPR. In this study a series of transmembrane WALP peptides with a point cysteine mutation were labeled by using a cysteine-specific spin label, methanethiosulfonic acid S-(1-oxyl-2,2,3,5,5-pentamethylimidazolidin-4-ylmethyl) ester (IMTSL; Smirnov et al., 2004). WALP peptides are artificial α-helices formed by alanine–leucine sequences of various length, and flanked by two tryptophan residues at each end. EPR titrations of such peptides, when reconstituted into anionic lipid bilayers, indicate the magnitude of the relative changes in the effective dielectric constant across a bilayer in the vicinity of the peptide α-helix. The results are discussed in terms of proton penetration (Smirnova et al., 2009).

15.3.1.3 Bacteriorhodopsin (BR)

A combination of 94 GHz EPR (W-band) and site-directed spin labeling was used to determine the polarity and proticity profiles of the proton channel in BR (Steinhoff et al., 2000). The g_{xx} tensor elements were determined in a frozen purple membrane for 10 spin labeling positions located in the cytoplasmic loop region and in the protein interior along the BR channel, and the value of g_{xx} was used as a polarity index to follow the hydrophobic barrier of the BR proton channel. The highest observed polarity corresponded to spin-labeled residues that were fully exposed to the aqueous phase, while the minimal polarity close to the middle of the membrane was still considerably higher than that observed for the same nitroxide ring in hydrocarbon solvents. The local pH-values within the proton channel of BR were also investigated using the pH-sensitive IMTSL (Möbius et al., 2005; see above). VHF-EPR is essential when separating the spectra of protonated and unprotonated states in frozen BR samples. By using W-band EPR, both the g_{xx} and A_{zz} parameters could be determined quantitatively, while the g_{xx} tensor components of protonated and unprotonated forms could also almost be resolved. The corresponding field values differed by ~11 G, whereas the linewidth of the g_{xx} resonance line was ~16 G. It has also been shown that, for IMTSL, a further increase in the EPR frequency beyond W-band would probably not yield any improvement in the resolution of g_{xx} due to g-strain effects.

15.3.1.4 Azurin

A similar approach was used for studying the polarity/proticity of four surface sites on azurin (Finiguerra et al., 2006). In this case, a 275 GHz EPR study revealed a small but significant difference in the polarity at these sites, and resolved two spectral components that could not be separated when using W-band EPR. An analysis of the g_{xx}–A_{zz} plot suggested that the difference between the two components corresponded to one additional hydrogen bond.

15.3.1.5 Tyrosyl and Tryptophan Radicals

The use of VHF-EPR was essential in studies of tyrosyl and tryptophan radicals in a variety of biological systems (Bleifuss et al., 2001; Schünemann et al., 2004; Un et al., 1995). The g-tensor anisotropy of tyrosyl radicals was approximately the same as that for nitroxides, and was well resolved at high frequencies. Similar to nitroxides, the g_{xx} principal value for tyrosyl radicals is sensitive to local polarity and the presence of an H-bond in their phenoxyl groups (Figure 15.4). For example, VHF-EPR at 245 and 285 GHz showed no evidence of H-bonding for tyrosyl radicals from ribonucleotide reductase (RNR) R2 proteins purified from prokaryotes, *Escherichia coli* and *Salmonella typhimurium* (Allard et al., 1996; Gerfen et al., 1993). In contrast, tyrosyl radicals from the RNR proteins of herpes simplex virus Type I, *Arabidopsis thaliana*, cysteine and mice (van Dam et al., 1998) were each clearly H-bonded. It was suggested that the hydrogen bond might play a role in stabilizing the tyrosyl radical. For the prokaryote *Mycobacterium tuberculosis*, however, two distinct peaks (2.0080 and 2.0092) of the g_{xx} component were later detected (Liu et al., 2000) and interpreted in terms of a non-H-bonded and a weakly H-bonded fraction of the radical.

15.3.1.6 Flavin

The principal g-factor values were determined by VHF-EPR for another biologically generated radical, *flavin* (Fuchs et al., 2002; Okafuji et al., 2008; Schnegg et al., 2006). The use of an EPR frequency as high as 360 GHz was extremely beneficial in these studies, because of the relatively small anisotropy. It has been shown that

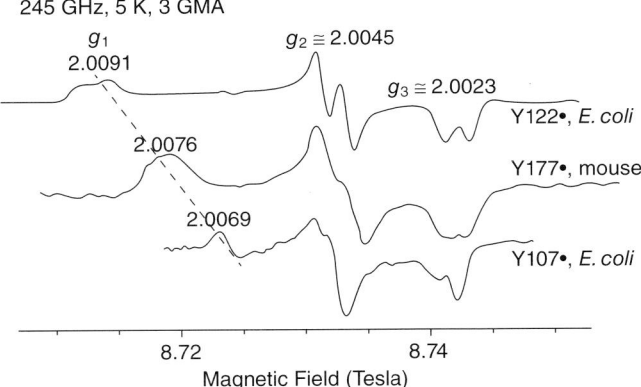

Figure 15.4 Effect of hydrogen bonding on the g_{xx} value of tyrosyl radicals in mutated and wild-type RNRR2. From top to bottom: Y122* in wild-type *E. coli* RNRR2 without H-bonding; Y177* in wild-type mouse RNR-R2 with weak H-bond; Y107* in Y122F W107Y double-mutant *E. coli* RNR-R2 with strong H-bond (Andersson et al., 2003).

in a variety of biological systems for both noncovalently bound (e.g., DNA photolyase of *E. coli*; (6-4) photolyase from *Xenopus laevis*) and covalently protein-bound (e.g., *Aspergillus niger* glucose oxidase) flavin radicals, the g_{yy} component was moderately sensitive to protonation–deprotonation of the nitrogen at position 5 and more subtle differences in the protein environment experienced by the flavin, whereas the g_{xx} and g_{zz} components were much less affected by changes in the environment.

15.3.1.7 Biliverdin Radical

The use of VHF-EPR at a frequency of up to 406 GHz was indispensable for resolving even the less-anisotropic g-factor of the biliverdin radical intermediate in the cyanobacterial enzyme phycocyanobilin:ferredoxin oxidoredictase (Stoll et al., 2009).

15.3.2
Polarity Measurements Outside of Rigid Limit Conditions

If the determination of polarity is required outside of rigid-limit conditions for samples at biologically relevant temperatures, the application of VHF-EPR becomes even more essential. First, as mentioned above, VHF-EPR provides a faster "snapshot" of the molecular dynamics for a given diffusion rate (Freed, 2000). At 250 GHz, for example, the slow-motional spectral regime is reached for motions that are about an order of magnitude faster than those for conventional 9 GHz EPR (Freed, 2004). This means that in many cases, when a 9 GHz spectrum is in the intermediate motional regime and does not allow for direct determination of either A_{zz} or the isotropic hyperfine constant a_{iso}, EPR at a higher frequency can still yield a nearly rigid-limit spectrum with the positions of the peaks determined solely by the local environment polarity. Second, compared to EPR at lower frequencies, VHF-EPR allows the simultaneous determination of both g_{xx} and A_{zz}, and also to separate the effects of molecular motion and polarity. In fact, if molecular motion begins to affect a nearly rigid-limit VHF-EPR spectrum, the motional averaging of the tensor components will decrease the apparent values of both g_{xx} and A_{zz}, while a change in local polarity will have the opposite effects on the values of g and A with respect to each other.

15.4
Improvement in Orientational Resolution for Spin Labels

One of the main virtues of VHF-EPR over EPR at conventional frequencies is the excellent orientational resolution for studies utilizing nitroxide spin labels (Budil et al., 1989; Freed, 2004). As seen in Figure 15.1, at 170 GHz the spectral regions corresponding to molecules with their magnetic X-axis parallel to B_0 (X-region), the Y-axis parallel to B_0 (Y-region), and the Z-axis parallel to B_0 (Z-region) are well separated due to the dominant role played by the g-tensor at this high frequency. As a result, once motional effects are discernible in the spectrum, it is possible to determine about which axis (or axes) the motion occurs. This point is illustrated in Figure 15.5, using an example of spin-labeled molecules included in the solid phase of cyclodextrins (CDs). Cyclodextrins, the cyclic oligomers of D-glucopyranose, have a hydrophobic cavity within their structure. As a consequence, spin-labeled molecules may be readily included into polycrystalline CDs, and show patterns of anisotropic molecular motion about various axes (Birrel et al., 1973; Dzikovski et al., 2009). The pattern of anisotropic rotation about the Z-axis is obvious for 5-doxyl stearic acid 5-sasl from the 170 GHz spectrum, just by a simple visual inspection (Dzikovski et al., 2009). A fast rotation about the magnetic Z-axis of the nitroxide causes partial averaging of the tensor components g_{xx} and g_{yy}. The peaks corresponding to the X- and Y-areas of the rigid-limit spectrum will merge with each other, whilst at the same time the Z-region will remain unaffected by the averaging and retains the triplet appearance seen in the rigid limit. Similarly, for an X-rotating spin probe (e.g., 4-hydroxy-2,2,6,6-tetramethylpiperidine-1-oxyl caprylate), the averaging includes only the Y- and Z-components (see Figure 15.5). Simulations of VHF-EPR spectra recorded at several different temperatures allow the rate of molecular motion and the potential barriers of the rotation about the Z- and X-axes to be obtained. On the other hand, the g-factor resolution at 9 GHz is insufficient for nitroxides (see above), and anisotropic molecular motion in this case can only be observed as an averaging of the components of the hyperfine tensor. However, the A_{xx} and A_{yy} values for nitroxides are virtually equal to each other. In the case of Z-rotation, the A_{zz} component—which determines the outer splitting of the spectrum at 9 GHz—does not participate in the averaging. In the limiting case of an extremely fast anisotropic rotation about the magnetic Z-axis, the resultant spectrum can be obtained as a rigid-limit spectrum, with the effective

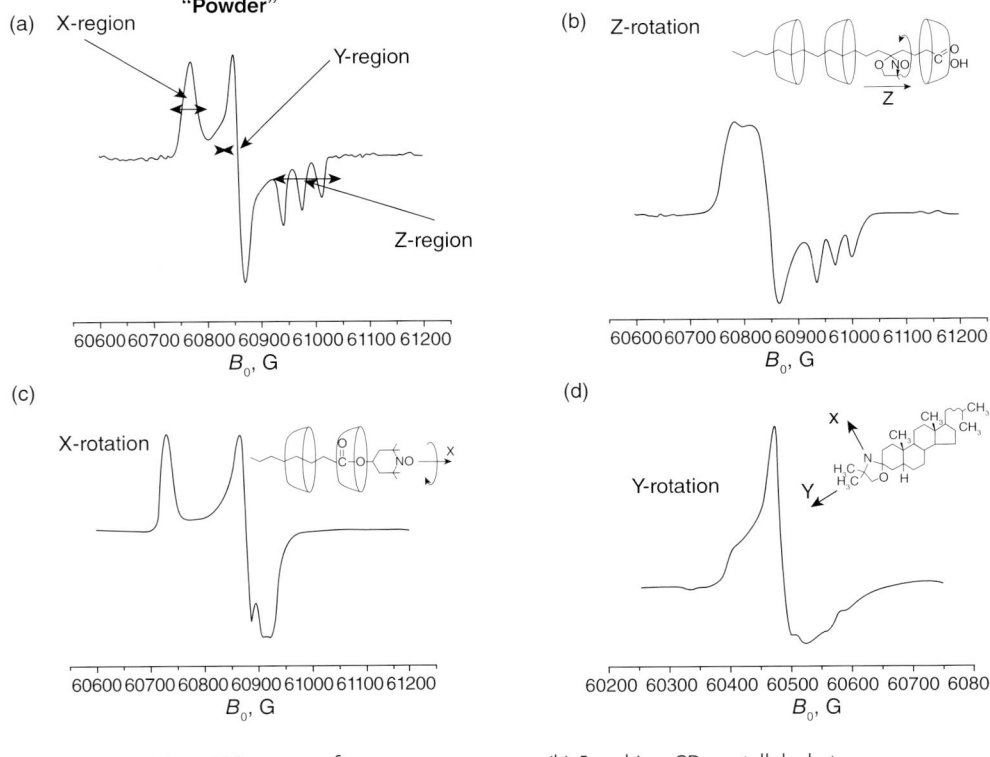

Figure 15.5 170 GHz EPR spectra of nitroxide radicals corresponding to different modes of molecular motion. (a) 5-sasl in γ-CD with crystallohydrate water removed by overnight evacuation at 293 K: rigid-limit spectrum; (b) 5-sasl in γ-CD crystallohydrate at 292 K; (c) TEMPOyl-caprylate in β-CD, 293 K; (d) CSL spin label in the DPPC membrane at 295 K (Dzikovski et al., 2009).

A- and g-tensor components $(g_{xx} + g_{yy})/2$, $(g_{xx} + g_{yy})/2$, g_{zz}, $(A_{xx} + A_{yy})/2$, $(A_{xx} + A_{yy})/2$, and A_{zz} instead of the corresponding rigid limit values of g_{xx}, g_{yy}, g_{zz}, A_{xx}, A_{yy}, and A_{zz}. This makes the 9 GHz spectrum for a Z-rotating nitroxide almost identical to that observed in the rigid limit. Thus, determination of the rate of Z-rotation becomes practically impossible, while VHF-EPR – because of its superior g-factor resolution – can clearly resolve the two cases (Figure 15.6). For the common case in membrane EPR (PC spin labels, etc.), with the magnetic Z-axis directed along the long diffusion axis, the 9 GHz EPR is practically insensitive to axial rotation; hence, in order to determine the rate of this rotation a VHF-EPR experiment is required.

In another limiting case of extremely fast anisotropic rotation about the magnetic X-axis, the spectrum can also be simulated as a rigid-limit spectrum with effective components of the magnetic tensors g_{xx}, $(g_{yy} + g_{zz})/2$, $(g_{yy} + g_{zz})/2$, A_{xx}, $(A_{yy} + A_{zz})/2$, and $(A_{yy} + A_{zz})/2$ instead of g_{xx}, g_{yy}, g_{zz} and A_{xx}, A_{yy} and A_{zz} correspondingly. Patterns of fast X-rotation, although less obvious than those observed at

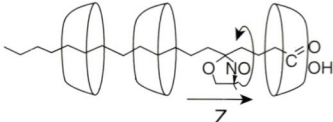

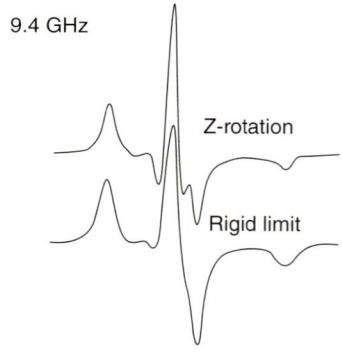 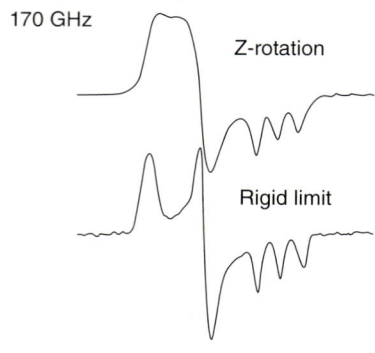

Figure 15.6 The insensitivity of the EPR spectrum to rotation about the axis collinear with the magnetic Z-axis at 9 GHz, due to the almost axially symmetric A-tensor for nitroxides and the insufficient g-factor resolution. Determination of the corresponding rotational diffusion rate requires VHF-EPR.

VHF, are recognizable at 9 GHz. However, as in the case of fast X-rotation, the reliable determination of the parameters of molecular dynamics from 9 GHz spectra (Dzikovski et al., 2009) is often more difficult than from VHF spectra.

15.5
Simulation of EPR Spectra at Various Frequencies: Simple Limiting Cases

In general, the simulation of EPR spectra resulting from a superposition of complex motional modes requires the so-called slowly relaxing local structure (SRLS) model (Freed, 1977; Polimeno and Freed, 1995; Polnaszek and Freed, 1975) and a large-scale computational resource (applications of this rigorous approach for simulations of multifrequency EPR spectra are discussed in Chapter 11). However, there are two simple limiting cases of the SRLS model, which are very common and allow one to analyze EPR spectra using convenient and efficient software (Budil et al., 1996; Schneider and Freed, 1989). The first limiting case is the fast internal motion (FIM) model, wherein the faster internal motion is considered to be so rapid that one observes a partial averaging of magnetic tensors; the second model is the microscopic order macroscopic disorder (MOMD) model

(Hubbell and McConnell, 1971; Liang and Freed, 1999), which neglects the slower mode – for example, the global tumbling of macromolecules – as if it were in the rigid limit. The slower motional mode in the FIM case and the faster internal motion, for MOMD, are parameterized in terms of diffusion constants. In general, the MOMD model has been shown to be a better approximation for EPR spectra obtained at high frequency, whereas the FIM model works better at 9 GHz (Liang and Freed, 1999). As examples of simulation tecqnique, in the multifrequency EPR studies of spin-labeled gramicidin in aligned membranes, the data at X-band were well explained by the FIM model, by using an heuristic adjustment of the magnetic tensor parameters (Dzikovski et al., 2006). The corresponding VHF spectrum, however, was well described by a fixed set of magnetic-tensor parameters determined at the rigid-limit conditions, and a derived set of ordering and diffusion constants from the MOMD model.

15.6
Macroscopically Aligned Phospholipid Membranes

A macroscopically aligned membrane is typically formed by a stack of flat parallel bilayers, where the orientation of the membrane is that of its normal vector relative to a chosen direction in space. For EPR, the most relevant direction is the direction of the external magnetic field. In an aligned membrane the orientation remains the same on the macroscopic scale, usually for the whole sample. An example of macroscopically unaligned membrane is a suspension or a pellet of vesicles, either multilamellar or unilamellar. Although on the microscopic scale comparable to the dimensions of lipid molecules (~30–40 Å) the neighboring bilayer-forming molecules retain the same orientation in space, on the macroscopic (micron) scale the membrane curvature incudes all the possible orientations. In unaligned membrane samples, all orientations of the membrane normal relative to any chosen direction in space are equally probable within any macroscopic volume. These samples can be described in terms of MOMD. A different case of ordered bilayers, called "cylindrical alignment," as defined by Smirnov and Poluektov (2003), is also discussed here.

Traditionally, spin-labeled EPR has been successfully used as a method of choice in membrane studies. The use of well-aligned membrane samples not only improves the EPR spectral resolution dramatically but also provides valuable information on the orientation of the nitroxide moiety relative to the membrane normal. Such information – particularly in the slow-tumbling regime – is difficult to extract from the spectra obtained from vesicles. Because all orientations of the membrane normal relative to the magnetic field are equally probable in vesicles, the orientation of the nitroxide moiety manifests itself only as a result of anisotropic molecular motion about the principal axes of the molecular frame. As one approaches the rigid limit, however, the vesicular spectrum converges to a "powder" spectrum, which is not sensitive to any properties of molecular structure, except for the magnetic tensors of the nitroxide. The vesicle spectrum is a superposition of spectra corresponding to different orientations of the membrane normal relative to the external magnetic field (MOMD model). Ambiguity in model parameters derived from spectral fitting is not unusual in vesicles; on the other hand, the simultaneous fitting of the EPR spectra of an aligned membrane sample in different orientations is in general less susceptible to such ambiguity (Budil et al., 1996; Ge et al., 1994).

A combination of excellent orientational resolution of VHF-EPR and advantages of utilizing macroscopically aligned membranes has a strong potential to provide further insights into membrane structure and function. As seen in Figure 15.1, at 170 GHz the rhombic g-tensor dominates the nearly axial A-tensor splitting; thus, there are three well-resolved spectral regions corresponding to g_{xx}, g_{yy}, and g_{zz}. The effect of changing the orientation of the nitroxide ring relative to the magnetic field manifests itself in a shift of spectral density between these regions. In aligned membranes, this affords a conceptually simple visualization of the orientation of the magnetic axis of the nitroxide moiety relative to the membrane normal, and substantially simplifies the spectral analysis. Using VHF-EPR on oriented samples of manganese and non-heme iron-depleted PS II membranes made it possible to separately determine the orientation of the Tyrosyl D, pheophytin anion, and semiquinone Q_A^- radicals in the PS II assembly (Dorlet et al., 2000). There is a challenge, however, in using macroscopically aligned membrane samples in VHF-EPR studies at biological temperatures above freezing. At 9 GHz, EPR spectra corresponding to different orientations of the membrane normal relative to the magnetic field direction, can be obtained merely by rotating the sample in the resonator. Higher frequencies, however, require very thin (<100 µm), flat samples with B_0 directed strictly perpendicular to the plane of the sample in order to minimize dielectric losses in the resonant structure.

15.6.1
A "Shunt" Fabry–Pérot Resonator. The study of DMPC and DMPS (1,2-dimyristoyl-sn-glycero-3-phospho-L-serine) Membranes with 3-doxyl-5(-cholestane) (CSL) Spin Label

In an elegant study conducted by Barnes and Freed (Barnes and Freed, 1998b), a "shunt" Fabry–Pérot resonator for a 250 GHz EPR spectrometer was designed to accommodate a thin sample that must rest with its flat surface perpendicular to the optical axis of the incident far infra-red (FIR) beam. This geometry made it possible to obtain VHF-EPR spectra of macroscopically ordered samples at membrane director tilts of 0° and 90° relative to the magnetic field. As seen in Figure 15.7, in the macroscopically aligned model membrane with a 80:20 molar ratio of DMPC to DMPS, aligned with its normal directed parallel to the magnetic field, the CSL spin label (a spin-labeled analog of cholesterol) with a 3% molar content in the membrane showed a single sharp peak; this peak demonstrated that the CSL orients with its Y-axis parallel to the membrane normal. However, when the sample was placed in the shunt resonator with the membrane normal perpendicular to the magnetic field, the spectral intensity was equally distributed between the

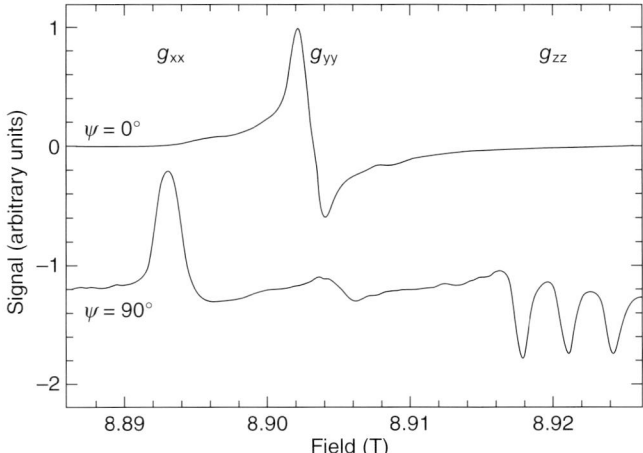

Figure 15.7 VHF-EPR on aligned membranes: 250 GHz EPR spectra of a microscopically aligned model phospholipid membrane at two different values of the director tilt (Barnes and Freed, 1998b).

X- and Z-regions of the principal axes of the g-tensor, with very little in the Y-area of the g-tensor principal axis. This meant that the CSL was then oriented with its X- and Y-axis parallel to the magnetic field with equal probability. A further study, using the "shunt" resonator (Barnes and Freed, 1998a), showed that the behavior of CSL in phosphatidyl serine (PS)-rich membranes was dramatically different from that of phosphatidyl choline (PC)-rich membranes, and could be interpreted only in terms of a strong local biaxial environment. While predicted from molecular dynamics simulations, this appeared to be the first experimental evidence for local biaxiality.

15.6.2
Microtome Technique on Isopotential Spin-Dry Ultracentrifugation (ISDU)-Aligned Membranes

Another method for recording spectra from various membrane orientations in the magnetic field for VHF-EPR is to apply a microtome technique to ISDU-aligned samples (Dzikovski et al., 2006). This simple technique allows for any director tilt value, does not require any special instrumentation, and can be used on EPR spectrometers at any frequency. A schematic graphic representation of the microtome cuts of aligned membrane samples at 90° and 45° is shown in Figure 15.8. The EPR spectra at 9 GHz for the microtome cuts of spin-labeled membranes at 90° and 45° matched well with the corresponding orientations obtained by rotating a thin aligned ("0° angle") sample in the resonator. Spectra at 170 GHz of spin-labeled gramicidin A (GAsl) in the DPPC membrane, a system analyzed previously in detail (Dzikovski et al., 2004), are shown in Figure 15.9. Several inferences can

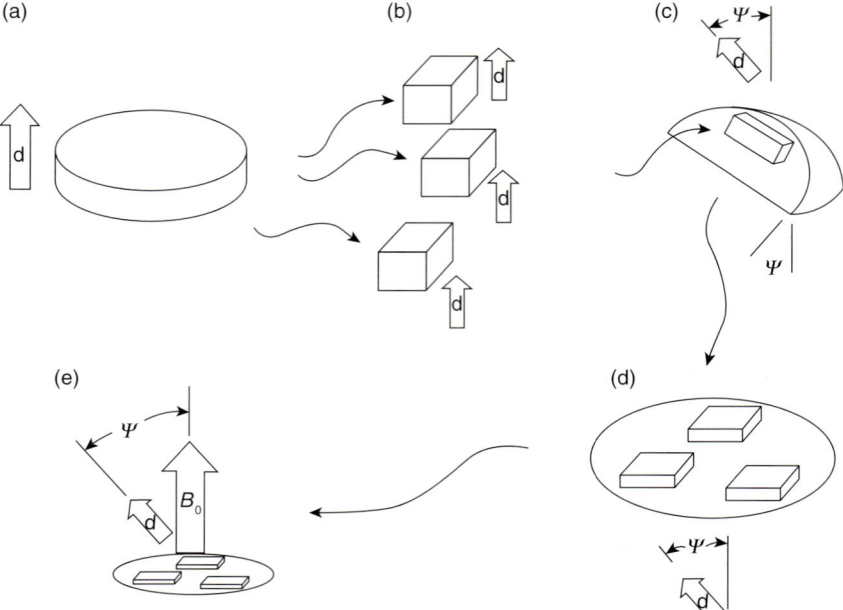

Figure 15.8 Schematic diagram of the microtome process. (a) The first step in the process involves preparing a multilamellar-oriented lipid bilayer membrane with appropriate spin-labeled membrane components using isopotential spin-dry ultracentrifugation (ISDU) or some other technique, such as pressure annealing. The ordering is characterized by the director **d**; (b) Conveniently sized sections of the sample are then removed for mounting on a small block of ice (c), with a face at a predetermined inclination from the normal (ψ) which defines the tilt angle of the director **d** after the microtome slice has been prepared; (d) After cutting, the microtome slices with tilt are mounted on a millimeter wave transparent sample stage; (e) Once a hermetically sealed sample has been prepared, it may be mounted in the resonator, and spectra obtained for this value of tilt angle, ψ (Earle et al., 2005).

be drawn from the comparison of the EPR spectra of the system at 9 and 170 GHz. Since at 170 GHz the effect of the g-tensor overwhelms the A-tensor splitting to a great extent, there are three distinct spectral regions attributed to g_{xx}, g_{yy}, and g_{zz}. In the aligned membranes, this visualizes the orientation of the magnetic axis of the nitroxide moiety relative to the membrane normal, and substantially simplifies the spectral analysis. For example, Z-ordering for GAsl and Y-ordering for CSL in DPPC at 22 °C are very distinct from each other at 170 GHz, and easily recognizable just by inspection (Figure 15.9). At 35 °C in DPPC, at the L_β–P_β phase transition point, an abrupt change in the GAsl spectrum occurs. The observation that the change is very pronounced in aligned membrane samples, but less obvious in vesicles, hints of a change in the nitroxide fragment orientation. The transition at 35–36 °C has a substantial hysteresis; after cooling from 35 °C to 22 °C, it may take hours for the spectrum to regain its initial shape before heating. It has been

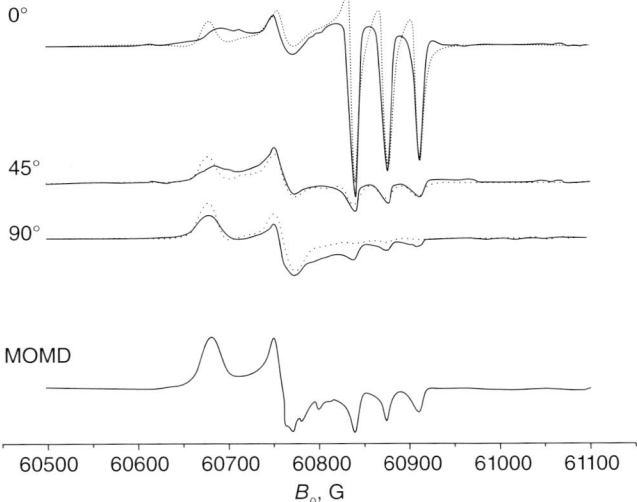

Figure 15.9 170 GHz EPR spectra of spin-labeled gramicidin A (GAsl) in aligned DPPC membranes and vesicles at 16 °C. The angular-dependent spectra are obtained using microtome cuts, simulations are shown in dots. The spectrum in vesicles contains a small admixture of the free spin label (Dzikovski et al., 2006).

shown (Dzikovski et al., 2004) that, in DPPC in the P_β phase, the GAsl starts to form channels. The gramicidin channel is formed by two gramicidin molecules attached to each other by their N-termini; channel formation manifests itself as a disruption of Z-ordering due to tilting of the nitroxide moiety. This conclusion is very clear at 170 GHz, even by a visual inspection without detailed simulation analysis, but it could be inferred from the 9 GHz spectra only after extensive spectral simulations. It is possible to see the spectral intensity shifting from the Z-region to the XY-region of the 170 GHz spectrum and back when performing a cooling/heating cycle (Figure 15.10). The considerable hysteresis time in the cycle could be attributed to a slow dissociation of the GAsl dimer forming the channel in the L_β phase.

15.6.3
Other Membrane-Alignment Techniques

Several other techniques for studying aligned membranes by VHF EPR have also been reported. The previously used NMR bicellar technique has been adapted for EPR (Mangels et al., 2001), where the DMPC/DHPC bicelles align spontaneously in the magnetic field of 3.4 T ($g = 2$ resonance at W-Band) with the bilayer normal perpendicular to the external magnetic field. The lanthanide ions Tm^{3+} or Yb^{3+} that bind to the bilayer in the polar head region change the sign of the magnetic susceptibility anisotropy and flip the bicelles by 90°, directing their normal parallel

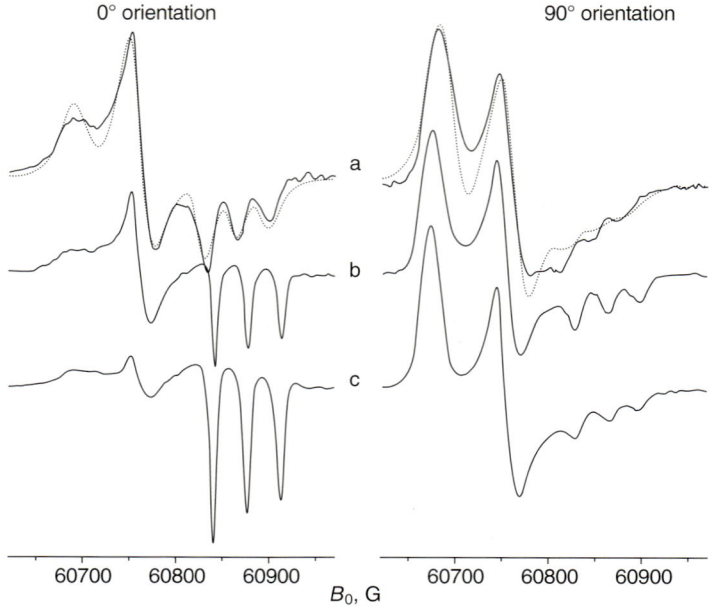

Figure 15.10 Channel formation and dissociation of the gramicidin channel detected by 170 GHz EPR (Dzikovski et al., 2006). Spectrum a: 36 °C; spectrum b, at 16 °C, 20 min after cooling from 36 °C; spectrum c, at 16 °C, 24 h after exposure at 36 °C.

to the magnetic field, while Dy^{3+} helps to better align them in the perpendicular direction. Using the bicellar technique is much more convenient at higher magnetic fields than at X-band, as it eliminates the need to cycle the magnetic field and requires a much lower concentration of Yb^{3+} to obtain a parallel alignment. Recently, it has been shown that lipids assemble into nanotubular bilayers when placed inside nanoporous anodic aluminum oxide (AAO) membranes (Smirnov and Poluektov, 2003). Spin-labeled lipids in arrays of these nanotubes, oriented with their cylindrical axes parallel to the external magnetic field, show VHF-EPR spectra corresponding to those of flat-aligned lipid bilayers, with the bilayer normal perpendicular to the field ("perpendicular orientation"). If the axis of the nanotube is directed perpendicular to B_0, all orientations of the membrane normal relative to B_0 are present in the spectrum. This spectrum is, however, different from that in membrane vesicles because of different weighting factors to take into account the contributions of different orientations.

15.7
Metalloproteins

EPR represents one of the most useful and powerful tools for studying metalloproteins. Indeed, for many metalloproteins that contain iron, copper, cobalt, nickel, manganese, and molybdenum, EPR spectroscopy has provided crucial insights into the structure of the active site and protein function. Most of these experiments were carried out using standard 9 GHz EPR (X-band). Since many metal centers show a large g-anisotropy, X-band EPR is usually sufficient for resolution, whereas the use of VHF-EPR often leads to extremely wide spectral ranges and, hence, to a reduced sensitivity. Furthermore, in some cases frequencies lower than X-band give better resolved spectra. A detailed review of multifrequency EPR studies at frequencies below 9 GHz on copper proteins kas been provided by Antholine (2005). Among other complications in using VHF-EPR for metalloproteins is an additional line broadening due to g-strain and D-strain effects (i.e., a distribution of local g- and D-parameters) (Hagen, 1999).

Ln some cases, however, VHF-EPR can be very useful for studying metalloproteins, and provide information that cannot be acquired at lower frequencies. For a paramagnetic ion, the energy levels can be determined using the spin Hamiltonian:

$$H = \mu_B B \cdot g \cdot S + D\left(S_z^2 - \frac{1}{3}S(S+1)\right) + E(S_x^2 - S_y^2) + \text{higher rank terms (for } S \geq 1)$$

where D and E, with $0 \leq |E/D| \leq 1/3$, are the tetragonal and rhombic ZFS parameters describing axial and rhombic distortion from the cubic symmetry, respectively. The ZFS parameters can be used to obtain valuable information on bonding in transition metal complexes through the use of ligand-field theory. For the high-spin states of transition ions ($S > 1/2$), the Zeeman splitting at X-band is often much smaller than the ZFS. Thus, non-Kramers ions (ions with integer S) are often EPR-silent at X-band, and VHF-EPR becomes essential for the determination of their ZFS parameters. Although a Kramers ion is, in principle, never EPR-silent, the intra-Kramers transitions within its $M_S = \pm 1/2$ doublet are typically uninformative with regard to the ZFS parameters (Krzystek et al., 2006), and it is preferable to detect inter-Kramers resonances. (Here, M_S is the electronic magnetic quantum number.) Thus, the application of VHF-EPR is essential for the determination of ZFS parameters for any transition ion, whether Kramers or non-Kramers, in a high-spin state.

In the case of relatively small ZFS, the values of D and E can be obtained directly for $S > 1/2$ systems from the complete VHF-EPR spectrum in the high-field limit (Lynch et al., 1993). If the Zeeman splitting is much larger than the crystal field, only so-called "allowed" transitions are observed, and the spectra can be analyzed using perturbation theory of spin Hamiltonian (Lynch et al., 1993), although matrix diagonalization yields more accurate values of the parameters (Wood et al., 1999). For larger ZFS, the precise values of D and E can be obtained from multifrequency measurements, using the field dependence for the effective g-values. A simple approximate expression for the effective g-factor value was deduced from third-order perturbation theory for $S = 5/2$ systems (Slappendel et al., 1980):

$$\Delta g = \frac{3}{2}\frac{g^3}{D^2}\left[\left(\frac{h\nu_2}{g_{\nu 2}}\right)^2 - \left(\frac{h\nu_1}{g_{\nu 1}}\right)^2\right],$$

wherein v_1 and v_2 are two microwave frequencies with the corresponding apparent g-values g_{v1} and g_{v2}, and Δg is the deviation of the observed g-value from the real value g. Indeed, VHF-EPR measurements for hemoglobin and myoglobin ($S = 5/2$) showed that the effective g-value moves towards the real g-value with increasing EPR frequency (Albert et al., 1973; Kan et al., 1998; Van Doorslaer and Vinck, 2007).

To date, most VHF-EPR studies on metalloproteins have been carried out on systems containing high-spin iron, manganese, and copper.

15.7.1
Fe^{3+} Systems

Apart from extensive studies on heme proteins, VHF-EPR has been used to resolve the spectra of non-heme iron centers interacting with other paramagnetic species, such as a tyrosyl radical (Assarsson et al., 2001) or another metal center (Seravalli et al., 2004).

15.7.2
Mn^{2+} Systems

From the point of view of EPR, manganese occurs in biology in two mononuclear forms – the large-D form ($D \gg 0.3\,cm^{-1}$) and the small-D form ($D \approx 0.3\,cm^{-1}$) (Hagen, 1999). In the active sites of enzymes, for example, in the Mn superoxide dismutase (SOD), the large-D form is present. VHF-EPR has been used to characterize the magnetic parameters of the Mn center in different SODs (Un et al., 2001). One remarkable property of Mn(II) with $S = 5/2$ is that, at high frequencies, its EPR line becomes narrower, causing a drastic increase in sensitivity compared to that at lower frequencies. The $^{55}Mn^{2+}$ ion with 100% natural abundance has electron spin $S = 5/2$ and nuclear spin $I = 5/2$. Since its g-factor is effectively isotropic and the transitions corresponding to $M_S = \pm 5/2 \leftrightarrow \pm 3/2$ and $M_S = \pm 3/2 \leftrightarrow \pm 1/2$ are extremely broad due to the anisotropy of electron–electron ZFS interactions, in powder (orientationally disordered) samples the EPR of Mn^{2+} shows only six narrow lines. These originate from $M_S = \pm 1/2 \leftrightarrow \pm 1/2$ transition and the hyperfine interaction with the ^{55}Mn nucleus. The shape of these lines is Lorentzian, with the relaxational linewidth determined primarily by second-order effects from the zero field coupling. These second-order contributions decrease in approximately reverse proportion to the external magnetic field B_0, and become negligible when $g\mu_B B_0/h \gg D$, resulting in a sharp Mn sextet at VHF. Some other half-integer spin systems, such as Gd(III) with $S = 7/2$, also show dramatic line narrowing with increasing EPR frequency (Smirnova et al., 1998). This increase in sensitivity for Mn^{2+} often leads in biological preparations to an accidental six-line pattern from manganese impurity, which is distinct at high frequency, but barely detectable at 9 GHz. The paramagnetic Mn^{2+} ion is commonly used as a probe to replace Zn^{2+}, Mg^{2+}, or Ca^{2+} ions when studying other biological systems by EPR. These studies benefit from VHF-EPR because of the increase in sensitivity, a better resolution of g-factor differences, and simplicity of the spectra due to the suppression of forbidden transition in high-field approximation. Studying interactions of the introduced manganese ions with their ligand environment and other paramagnetic cofactors in the biomolecule (Carmieli et al., 2003; Rohrer et al., 2001) can provide insight into the structure and/or provide an independent confirmation of distances calculated from X-ray structure (Kappl et al., 2005; Käss et al., 2000). A more exotic case of Fe/Mn substitution with a following multifrequency study has been reported for sperm whale metmyoglobin (Yashiro et al., 2006), with Fe^{3+} substituted by an integer-spin Mn^{3+} atom ($S = 2$). In a number of studies, VHF-EPR has been applied to study the dinuclear and multinuclear active sites of enzymes of redox-active manganese enzymes, such as the Mn_4Ca metal center of PS II (Teuteloff et al., 2006) and dimanganese catalases (Meier et al., 1996; Teuteloff et al., 2005).

15.7.3
Cu^{2+} Systems

Copper is a key cofactor in a diverse array of biological oxidation–reduction reactions. These involve either outer-sphere electron transfer (as in the blue copper proteins and the Cu_A site of cytochrome oxidase and nitrous oxide reductase), or inner-sphere electron transfer, as in the binding, activation, and reduction of dioxygen, superoxide, nitrite, and nitrous oxide (Solomon et al., 1996). Copper proteins – plant chloroplastic plastocyanins – are responsible for the electron transfer between PS II and PS I in photosynthesis. Historically, copper sites have been divided into three classes by the coordination of the metal center: type 1 or *blue copper*; type 2 or *normal copper*; and type 3 or *coupled binuclear copper* centers. More recently, several other types of copper center (binuclear, trinuclear, and mixed heme–Cu) have been characterized and included into the classification. EPR spectroscopy is invaluable in this classification and assignment of copper centers, because the g- and A-values depend heavily on the ligand environment of the copper atom (see Figure 15.11). The EPR of type 1 copper proteins, or single copper cupredoxins, is very informative for understanding the relationship between the biological activity of the enzymes and their electronic structure, as recent developments in quantum chemistry have placed the calculation of their EPR properties via the distribution of their spin density over the metal site within reach of *ab-initio* methods (Jaszewski and Jezierska, 2001). Typically, for a type 1 copper site the g_{xx} and g_{yy} are not resolved at X-band, and determination of the rhombicity of the g-factor requires VHF-EPR (Figure 15.12). Likewise, 95 GHz pulsed and CW-EPR have been used to study the type 1 small protein azurin from *Pseudomonas aeruginosa*, wild-type and some mutants (Coremans et al., 1994; Coremans et al., 1996; van Gastel et al., 2000), and the type 1 copper site of the enzyme nitrite reductase from *Alcaligenes faecalis* (van Gastel et al., 2001). An analysis of the accurate g-factor values showed the dominant d_{xy} character of the singly occupied molecular orbital for wild-type azurin with appreciable contributions of the d_{z^2} for the M121Q mutant and of d_{yz} (d_{xz}) for M121H (van Gastel

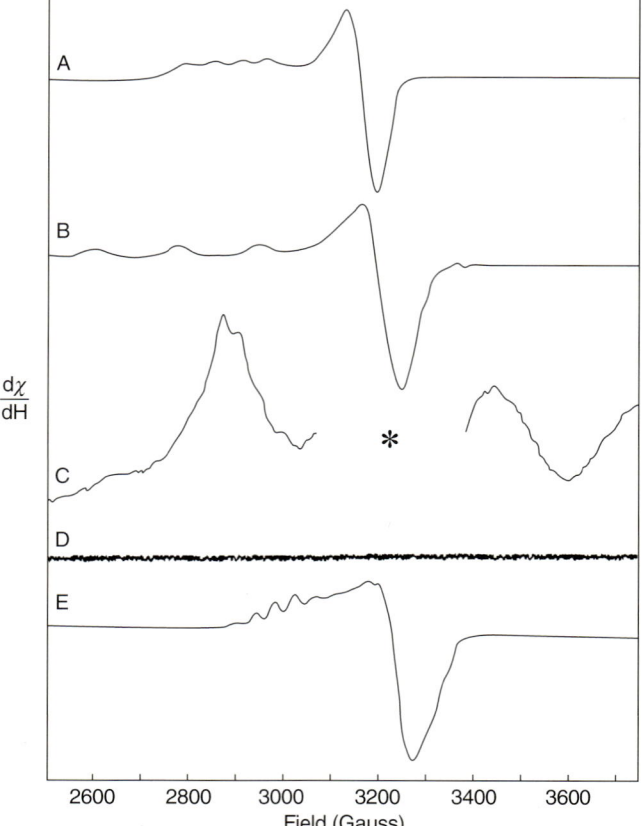

Figure 15.11 EPR spectra (X-band) of different types of copper proteins. Spectrum A: the blue or type 1 copper protein plastocyanin. Spectrum B: the normal copper protein dopamine β-hydroxylase. Spectrum C: the uncoupled T3 coppers with a 4 Å Cu⋯Cu separation present in met-N_3^- T2D laccase. (The 3100–3300 G region also includes overlapping spectral features from the T1 site which have been excluded for clarity.) Spectrum D: the coupled binuclear copper protein oxyhemocyanin. Spectrum E: the Cu_A site in nitrous oxide reductase (Solomon, Sundaram, and Machonkin, 1996).

et al., 2000). In addition to the principal values, this conclusion is supported by experiments on single crystals, which provided the direction of the principal axis of the g-tensors. For the azurin mutants M121Q and M121H, 95 GHz VHF-EPR helped to resolve an interesting conformational bistability of the copper site. For other copper sites, multifrequency pulse EPR has been applied to separate the Cu_A signals and signals from the additional type 2 Cu site (Slutter et al., 1999). The use of VHF-EPR in the case of dicupric lactoferrin showed that splittings in the X-band spectrum, which previously were attributed to superhyperfine interactions of nitrogens with the copper center, were in fact due to a Mn(II) impurity

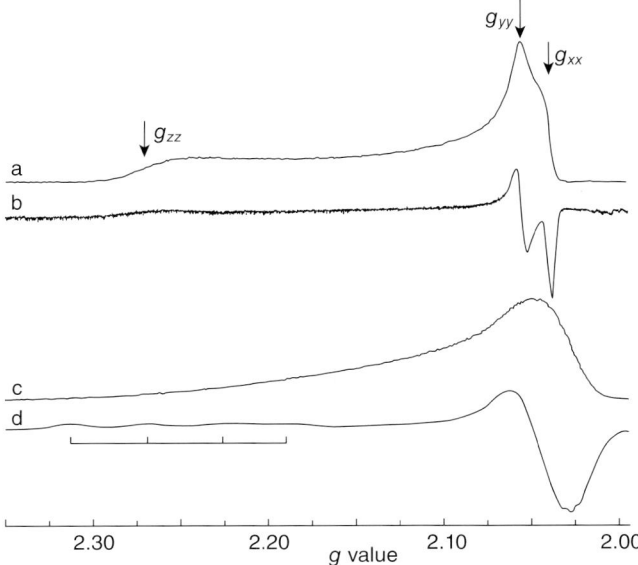

Figure 15.12 Electron spin echo (ESE)-detected and CW-EPR spectra at W-band (spectra a and b) and at X-band (spectra c and d) of *Pseudomonas aeruginosa* azurin. Rhombicity of the g factor (g_{xx} and g_{yy} components) is resolved at W-band, but not at X-band (spectrum b). In contrast, a copper hyperfine interaction is visible only in the low-field part of the X-band spectrum (d) (Ubbink et al., 2002).

(Ubbink et al., 2002). Examples of using multifrequency EPR (9, 115, and 285 GHz) for multicopper oxidases with three different types of copper site are provided by Andersson et al. (2003). The spectra contain overlapping signals which allow several interpretations. VHF-EPR increased the resolution of two signals – a type 1 copper with an axial g tensor $g_{\|} = 2.30$ and $g_{\perp} = 2.06$, and a type 2 copper with a rhombic g tensor of g_{xx}, g_{yy}, g_{zz} = 2.24, 2.05, 2.04, respectively. The use of two high frequencies helped to assign a feature, which could be easily misinterpreted at 115 GHz as a hyperfine split signal, to the g-tensor anisotropy of two different Cu^{2+} $S = 1/2$ centers.

15.8
Concluding Remarks

This chapter has provided an overview of multifrequency studies of biological samples by EPR, with the examples having been chosen specifically to illustrate the value of EPR in such investigations. A combination of the details available in the "Pertinent Literature" section of the chapter provides access to this vast, rapidly emerging, "hot" topic in great detail.

Acknowledgments

This project was supported by Grant Number 5P41RR016292 (at Cornell) from the National Center for Research Resources (NCRR), a component of the National Institutes of Health (NIH). The contents of the chapter are the sole responsibility of the author, and do not necessarily represent the official views of the NCRR or NIH. The author thank Sushil Misra and Alex Smirnov for many substantive suggestions to improve the chapter.

Pertinent Literature

More details relevant to biological applications of VHF and multifrequency EPR can be found in following monographs, reviews, and original papers:

- T.I. Smirnova and A.I. Smirnov (2007) High-Field EPR Spectroscopy in Membrane and Protein biophysics, in *Biological Magnetic Resonance*, vol. **27**, (eds M.A. Hemminga and L.J. Berliner), New York, p. 165. This a detailed review of the applications of VHF-EPR, with the primary focus being on membrane studies and site-directed spin labeling.

- M. Benatti and T. Prisner (2005) New developments in high-field electron paramagnetic resonance with applications in structural biology. *Rep. Prog. Phys.* **68**, 411. This is a thorough review of recent applications of pulse and CW-EPR at high frequencies, and includes more than 160 references.

- K. Möbius *et al.* (2005) This is a detailed account of using VHF-EPR to study polarity/proticity inside the bacteriorhodopsin and Colicin A ion channels, describing a new method of probing site-specific pK_a-values using pH-sensitive nitroxide spin labels. The paper concludes with an outlook of ongoing VHF-EPR experiments on site-specific protein mutants at FU Berlin and Osnabrück.

- B.G. Dzikovski *et al.* (2009) This study of molecular dynamics of spin-labeled molecules, including into solid cyclodextrins, provides a simple and visually clear demonstration for the virtues of VHF-EPR versus EPR at standard frequencies. The paper includes many figures and examples on how to extract parameters of molecular dynamics from VHF-EPR spectra.

- K.K. Andersson *et al.* (2003) This mini-review of high-frequency EPR studies in bioinorganic chemistry provides many good examples of VHF-EPR studies on metalloproteins and tyrosine radicals.

- O. Grinberg and L.W. Berliner (eds) (2004) Very High Frequency (VHF) ESR/EPR, in *Biological Magnetic Resonance*, vol. **22**, Kluwer, New York. This textbook covers the full scope of VHF in a single volume, including chapters on theory, VHF instrumentation, methodology, and a number of biological applications. The contributors of the book are leaders in their fields.

- G. Hanson and L.W. Berliner (eds) (2009) High-resolution EPR applications to metalloproteins and metals in medicine, in *Biological Magnetic Resonance*, vol. **28**, Springer, New York; and G. Hanson and L.W. Berliner (eds) (2009) Metals in biology, in *Biological Magnetic Resonance*, vol. **29**, Springer, New York. The first book of this two-volume set covers high-resolution EPR methods and their application to iron proteins, nickel, copper enzymes and metals in medicine. The second book includes iron–sulfur clusters in "Radical SAM" enzymes, molybdenium-containing hydroxylases, and EPR studies of manganese, zinc and vanadyl proteins and their model complexes.

References

Albert, Y., Gouder, Y., Tuchendler, J., and Thome, H. (1973) *Biochim. Biophys. Acta*, **322**, 34.

Allard, P., Barra, A.-L., Andersson, K.K., Schmidt, P.P., Atta, M., and Gräslund, A. (1996) *J. Am. Chem. Soc.*, **118**, 895.

Andersson, K.K., Schmidt, P.P., Katterle, B., Strand, K.R., Palmer, A.E., Lee, S.-K., Solomon, E.I., Gräslund, A., and Barra, A.-L. (2003) *J. Biol. Inorg. Chem.*, **8**, 235.

Antholine, W.E. (2005) Low-frequency EPR of Cu^{2+} in proteins, in *Biomedical EPR: Free Radicals, Metals, Medicine, and Physiology* (eds S.S. Eaton, G.R. Eaton, and L.J. Berliner), Springer, New York, p. 417.

Assarsson, M., Andersson, M.E., Högbom, M., Perrson, B.O., Sahlin, M., Barra, A.-L., Sjöberg, B.-M., Nordlund, P., and Gräslund, A. (2001) *J. Biol. Chem.*, **276**, 26852.

Barnes, J.P. and Freed, J.H. (1998a) *Biophys. J.*, **75**, 2532.

Barnes, J.P. and Freed, J.H. (1998b) *Rev. Sci. Instrum.*, **69**, 3022.

Birrel, G.B., Van, S.P., and Griffith, O.H. (1973) *J. Am. Chem. Soc.*, **95**, 2451.

Bleifuss, G., Kolberg, M., Pötsch, S., Hofbauer, W., Bittl, R., Lubitz, W., Gräslund, A., Lassman, G., and Lendzian, F. (2001) *Biochemistry*, **40**, 15362.

Budil, D.E., Earle, K.A., Lynch, W.B., and Freed, J.H. (1989) Electron paramagnetic resonance at 1 millimeter wavelengths, in *Advanced EPR: Applications in Biology and Biochemistry* (ed. A. Hoff), Elsevier, Amsterdam, p. 307.

Budil, D.E., Lee, S., Saxena, S., and Freed, J. (1996) *J. Magn. Reson. A*, **120**, 155.

Burghaus, O., Plato, M., Rohrer, M., MacMillan, F., Möbius, K., MacMillan, F., and Lubitz, W. (2002) *J. Phys. Chem.*, **97**, 7639.

Carmieli, R., Manikandan, P., Epel, B., Kalb, J.A., Schnegg, A., Savitsky, A., Möbius, K., and Goldfarb, D. (2003) *Biochemistry*, **42**, 7863.

Coremans, J.W.A., Poluektov, O.G., Groenen, E.J.J., Canters, G.W., Nar, H., and Messerschmidt, A. (1994) *J. Am. Chem. Soc.*, **118**, 4726.

Coremans, J.W.A., Poluektov, O.G., Groenen, E.J.J., Warmerdam, G.C.M., and Canters, G.W. (1996) *J. Phys. Chem.*, **100**, 19706.

Dorlet, P., Rutherford, A.W., and Un, S. (2000) *Biochemistry*, **39**, 7826.

Dzikovski, B.G., Borbat, P.P., and Freed, J.H. (2004) *Biophys. J.*, **87**, 3504.

Dzikovski, B.G., Earle, K.A., Pachtchenko, S., and Freed, J.H. (2006) *J. Magn. Reson.*, **179**, 273.

Dzikovski, B.G., Tipikin, D.S., Livshits, V.A., Earle, K.A., and Freed, J.H. (2009) *Phys. Chem. Chem. Phys.*, **11**, 6676.

Earle, K.A., Dzikovski, B.G., Hofbauer, W., Moscicki, J.K., and Freed, J.H. (2005) *Magn. Res. Chem.*, **43**, S256.

Earle, K.A., Moscicki, J.K., Ge, M., Budil, D.E., and Freed, J.H. (1994) *Biophys. J.*, **66**, 1213.

Finiguerra, M.G., Blok, H., Ubbink, M., and Huber, M. (2006) *J. Magn. Reson.*, **180**, 197.

Freed, J.H. (1977) *J. Chem. Phys.*, **66**, 4183.

Freed, J.H. (2000) *Annu. Rev. Phys. Chem.*, **51**, 655.

Freed, J.H. (2004) The development of high-field/high-frequency ESR: Historical overview, in *Very High Frequency (VHF) ESR/EPR. Biological Magnetic Resonance* (eds O.Y. Grinberg and L.J. Berliner), Kluver, New York, p. 19.

Fuchs, M.R., Schleicher, E., Schnegg, A., Kay, C.W.M., Tirring, J.T., Bittl, R., Bacher, A., Richter, G., Möbius, K., and Weber, S. (2002) *J. Phys. Chem. B*, **106**, 8885.

Ge, M., Budil, D.E., and Freed, J.H. (1994) *Biophys. J.*, **67**, 2326.

Gerfen, G.J., Bellew, B.F., Un, S., Bollinger, J.M.J., Stubbe, J., Griffin, R.G., and Singel, D.J. (1993) *J. Am. Chem. Soc.*, **115**, 6420.

Hagen, W.R. (1999) *Coord. Chem. Rev.*, **190–192**, 209.

Hubbell, W. and McConnell, H. (1971) *J. Am. Chem. Soc.*, **93**, 314.

Jaszewski, A.R. and Jezierska, J. (2001) *Chem. Phys. Lett.*, **343**, 571.

Jeschke, G. (2005) *Biochim. Biophys. Acta*, **1701**, 91.

Kan, P.J.M., van der Horst, E., Reijerse, E.J., van Bentum, P.J.M., and Hagen, W.R. (1998) *Chem. Soc. Faraday Trans.*, **94**, 2975.

Kappl, R., Ranguelova, K., Koch, B., Duboc, C., and Hüttermann, J. (2005) *Magn. Reson. Chem.*, **43**, S65.

Käss, H., MacMillan, F., Ludwig, B., and Prisner, T. (2000) *J. Phys. Chem. B*, **104**, 5362.

Kawamura, T., Matsunami, S., and Yonezawa, T. (1967) *Bull. Chem. Soc. Jpn*, **40**, 1111.

Krzystek, J., Ozarowski, A., and Telser, J. (2006) *Coord. Chem. Rev.*, **250**, 2308.

Kurad, D., Jeschke, G., and Marsh, D. (2003) *Biophys. J.*, **85**, 1025.

Liang, Z. and Freed, J.H. (1999) *J. Phys. Chem. B*, **103**, 6384.

Liu, A., Barra, A.-L., Rubin, H., Lu, G., and Gräslund, A. (2000) *J. Am. Chem. Soc.*, **122**, 1974.

Lynch, B., Boorse, S., and Freed, J. (1993) *J. Am. Chem. Soc.*, **115**, 10909.

Mangels, M.L., Harper, A.C., Smirnov, A.I., Howard, K.P., and Lorigan, G.A. (2001) *J. Magn. Reson.*, **151**, 253.

Meier, A.E., Whittaker, M.M., and Whittaker, J.W. (1996) *Biochemistry*, **35**, 348.

Möbius, K., Savitsky, A., Wegener, C., Plato, M., Fuchs, M., Schnegg, A., Dubinskii, A.A., Grishin, Y.A., Grigor'ev, I.A., Kühn, M., Duché, D., Zimmermann, H., and Steinhoff, H.-J. (2005) *Magn. Res. Chem.*, **43**, 3S4.

Okafuji, A., Schnegg, A., Schleicher, E., Möbius, K., and Weber, S. (2008) *J. Phys. Chem. B*, **112**, 3568.

Ondar, M.A., Grinberg, O.Y., Dubinskii, A.A., and Lebedev, Y.S. (1985) *Sov. J. Chem. Phys.*, **3**, 781.

Owenius, R., Engström, M., Lindgren, M., and Huber, M. (2001) *J. Phys. Chem. A*, **105**, 10967.

Polimeno, A. and Freed, J.H. (1995) *J. Phys. Chem.*, **99**, 10995.

Polnaszek, C.F. and Freed, J.H. (1975) *J. Phys. Chem.*, **79**, 2283.

Rohrer, M., Prisner, T.F., Brügmann, O., Käss, H., Spoerner, M., Wittinghofer, A., and Kalbitzer, H.R. (2001) *Biochemistry*, **40**, 1884.

Schnegg, A., Kay, C.W.M., Schleicher, E., Hitomi, K., Todo, T., Möbius, K., and Weber, S. (2006) *Mol. Phys.*, **104** (10–11), 1627.

Schneider, D.J. and Freed, J.H. (1989) Calculating slow motional magnetic resonance spectra: A user's guide, in *Biological Magnetic Resonance, Spin Labeling, Theory and Application* (eds L.J. Berliner and J. Reuben), Plenum Press, New York, p. 1.

Schünemann, V., Lendzian, F., Jung, C., Contzen, J., Barra, A.-L., Sligar, S.G., and Trautwein, A.X. (2004) *J. Biol. Chem.*, **279**, 10919.

Seravalli, J., Xiao, Y., Gu, W., Cramer, S.P., Antholine, W.E., Gerfen, G.J., and Ragsdale, S.W. (2004) *Biochemistry*, **43**, 3944.

Slappendel, S., Veldnik, G.A., Vliegenthart, J.F.G., Aasa, R., and Malmström, B. (1980) *Biochim. Biophys. Acta*, **642**, 30.

Slutter, C.E., Gromov, I., Epel, B., Pecht, I., Richards, J.H., and Goldfarb, D. (1999) *J. Am. Chem. Soc.*, **123**, 5325.

Smirnov, A.I. and Poluektov, O.G. (2003) *J. Am. Chem. Soc.*, **125**, 8434.

Smirnov, A.I., Ruuge, A., Reznikov, V.A., Voinov, M.A., and Grigor'ev, I.A. (2004) *J. Am. Chem. Soc.*, **126**, 8872.

Smirnov, A.I. and Smirnova, T.I. (2001) *Appl. Magn. Reson.*, **21**, 453.

Smirnova, T.I., Smirnov, A.I., Belford, R.L., and Clarkson, R.B. (1998) *J. Am. Chem. Soc.*, **120**, 5060.

Smirnova, T.I., Smirnov, A.I., Clarkson, R.B., Belford, R.L., Kotake, Y., and Janzen, E. (1997) *J. Phys. Chem. B*, **101**, 3877.

Smirnova, T.I., Voynov, M.A., Poluektov, O.G., and Smirnov, A.I. (2009) Heterogeneous dielectric and hydrogen bonding environment of transmembrane α-helical peptides: CW X-band, D-band, and HYSCORE EPR of spin-labeled WALP. 51st Rocky Mountain Conference on Analytical Chemistry, Snowmass, Colorado, USA. Abstr. #147.

Solomon, E.I., Sundaram, U.M., and Machonkin, T.E. (1996) *Chem. Rev.*, **96**, 2563.

Steinhoff, H.-J., Savitsky, A., Wegener, C., Pfeiffer, M., Plato, M., and Möbius, K. (2000) *Biochim. Biophys. Acta*, **1457**, 253.

Stoll, S., Gunn, A., Brynda, M., Sughrue, W., Kohler, A.C., Ozarowski, A., Fisher, A.J., Lagarias, J.C., and Britt, R.D. (2009) *J. Am. Chem. Soc.*, **131**, 1986.

Teuteloff, C., Keßen, S., Kern, J., Zouni, A., and Bittl, R. (2006) *FEBS Lett.*, **580**, 3605.

Teuteloff, C., Schäfer, K.-O., Sinnecker, S., Barynin, V., Bittl, R., Wieghardt, K., Lendzian, F., and Lubitz, W. (2005) *Magn. Res. Chem.*, **43**, S51.

Ubbink, M., Worrall, J.A.R., Canters, G.W., Groenen, E.J.J., and Huber, M. (2002) *Annu. Rev. Biophys. Biomol. Struct.*, **31**, 393.

Un, S., Atta, M., Fontecave, M., and Rutherford, A. (1995) *J. Am. Chem. Soc.*, **117**, 10713.

Un, S., Dorlet, P., Guillaume, V., Tabares, L.C., and Cortez, N. (2001) *J. Am. Chem. Soc.*, **123**, 10123.

van Dam, P.J., Willems, J.P., Schmidt, P.P., Pötsch, S., Barra, A.-L., Hagen, W.R., Hoffmann, B.M., Andersson, K.K., and Gräslund, A. (1998) *J. Am. Chem. Soc.*, **120**, 5080.

van der Ast, A., Prisner, T.F., Bittl, R., Fromme, P., Lubitz, W., Möbius, K., and Stehlik, D. (1997) *J. Phys. Chem. B*, **101**, 1437.

Van Doorslaer, S. and Vinck, E. (2007) *Phys. Chem. Chem. Phys.*, **9**, 4620.

van Gastel, M., Boulanger, M.J., Messerschmidt, A., Canters, G.W., Huber, M., Murphy, M.E.P., Verbeet, M.P., and Groenen, E.J.J. (2001) *J. Phys. Chem. B*, **105**, 2236.

van Gastel, M., Canters, G.W., Krupka, H., Messerschmidt, A., de Waal, E.C., Warmerdam, G.C.M., and Groenen, E.J.J. (2000) *J. Am. Chem. Soc.*, **122**, 2322.

Wood, R.M., Stucker, D.M., Jones, L.M., Lynch, W.B., Misra, S.K., and Freed, J.H. (1999) *Inorg. Chem.*, **38**, 5384.

Yashiro, H., Kashiwagi, T., Horitani, M., Hobo, F., Hori, H., and Hagiwara, M. (2006) *J. Phys. Conf. Ser.*, **51**, 576.

16
Copper Coordination Environments

William E. Antholine, Brian Bennett, and Graeme R. Hanson

16.1
Introduction

EPR spectra are often complex and arise through a range of interactions involving one or more unpaired electrons, the external magnetic field, and one or more nuclei (Pilbrow, 1990; Hanson and Berliner, 2009; Hanson and Berliner, 2010; Mabbs and Collison, 1992; Weil, Bolton, and Wertz, 2007; Abragam and Bleaney, 1970; Hanson, Noble, and Benson, 2009; Bencini and Gatteschi, 1990; Smith and Pilbrow, 1974). Mathematically, these interactions can be written using the spin Hamiltonian formalism. For an isolated paramagnetic Cu(II) center, A, a general spin Hamiltonian (Pilbrow, 1990; Hanson and Berliner, 2009; Hanson and Berliner, 2010; Mabbs and Collison, 1992; Weil, Bolton, and Wertz, 2007; Abragam and Bleaney, 1970; Hanson, Noble, and Benson, 2009) is:

$$H_A = \beta \, B \cdot g \cdot S + S \cdot A \cdot I + I \cdot P \cdot I - \gamma(1-\sigma) B \cdot I \tag{16.1}$$

where S and I are the electron and nuclear spin operators respectively, g and A are the electron Zeeman and hyperfine coupling matrices respectively, P the quadrupole tensor, γ the nuclear gyromagnetic ratio, σ the chemical shift tensor, β the Bohr magneton, and B the applied magnetic field. Additional hyperfine, quadrupole and nuclear Zeeman interactions will be required when superhyperfine splitting from distant nuclei (which maybe coordinated to the Cu(II) ion) is resolved in the experimental EPR spectrum.

When two or more paramagnetic Cu(II) centers ($A_{i,j}$ i,j = 1, ... , N) interact, the EPR spectrum is described by a total spin Hamiltonian (H_{Total}), which is the sum of the individual spin Hamiltonians (H_{Ai}; Equation 16.1) for the isolated centers (A_i) and the interaction Hamiltonian (H_{Aij}) which accounts for the isotropic exchange, antisymmetric exchange, and the anisotropic spin-spin (dipole–dipole coupling) interactions between a pair of paramagnetic centers (Bencini and Gatteschi, 1990; Smith and Pilbrow, 1974; Hanson, Noble, and Benson, 2009):

$$H_{Total} = \sum_{i=1}^{N} H_{A_i} + \sum_{i,j=1, i\neq j}^{N} H_{A'_{ij}}$$

$$H_{A'_{ij}} = J_{A'_{ij}} S_{A_i} \cdot S_{A_j} + G_{A'_{ij}} S_{A_i} \times S_{A_j} + S_{A_i} \cdot D_{A'_{ij}} \cdot S_{A_j} \tag{16.2}$$

Traditionally, the structural (geometric and electronic) characterization of metal ion binding sites in metalloenzymes and transition metal ion complexes has relied upon an accurate determination of the spin Hamiltonian parameters (Equations 16.1 and 16.2) through computer simulation of the continuous-wave (CW) EPR spectra (Hanson, Noble, and Benson, 2009; Smith and Pilbrow, 1974; Pilbrow, 1990; Gaffney, 2009). Comparison of these parameters with well-characterized structures, or the calculation of these parameters from a well-defined structure using quantum chemistry calculations (crystal and ligand field theories, *ab initio* methods and, more recently, density functional theory; DFT) have been successfully used to determine the structure of metal ion-binding sites (Figure 16.1) (Pilbrow, 1990; Hanson and Berliner, 2009; Hanson and Berliner, 2010; Mabbs and Collison, 1992; Weil, Bolton, and Wertz, 2007; Abragam and Bleaney, 1970; Hanson, Noble, and Benson, 2009; Bencini and Gatteschi, 1990; Smith and Pilbrow, 1974; Gaffney, 2009; Neese, 2009; Frisch *et al.*, 2004; te Velde *et al.*, 2001; Comba and Martin, 2005; Deeth, 2010).

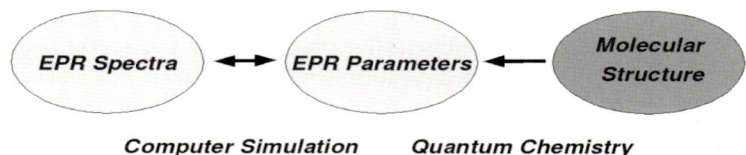

Figure 16.1 Traditional approach to EPR structure determination.

Importantly, the anisotropic components of the superhyperfine (**A**, Equation 16.1) and the anisotropic exchange ($D_{Ai,j}$, Equation 16.2) allow, through the point dipole–dipole approximation, the internuclear distance and relative orientation of the nuclear (or second electron) spin from the electron spin to be determined. Unfortunately, these superhyperfine interactions are often hidden underneath the inhomogeneously broadened resonances. As mentioned earlier in the chapter, one approach to resolving superhyperfine splittings is low-frequency (1–4 GHz) EPR (Basosi, Antholine, and Hyde, 1993; Drew *et al.*, 2007; Drew, Young, and Hanson, 2007; Hanson *et al.*, 1987; Mobius and Savitsky, 2009; Grinberg and Berliner, 2004; Froncisz and Hyde, 1980; Hyde and Froncisz, 1982; Wilson *et al.*, 1988; Wilson *et al.*, 1991; Nielsen *et al.*, 2009). Another involves high-resolution multidimensional pulsed EPR and electron nuclear double (triple) resonance (END(T)OR) spectroscopy, in conjunction with orientation selective experiments. While electron spin echo envelope modulation (ESEEM) and hyperfine sublevel correlation (HYSCORE) experiments are particularly sensitive for extremely weak couplings, from 4–6 Å away from the paramagnetic center, the ENDOR experiment is far more sensitive to strongly coupled nuclei 2–4 Å away from the paramagnetic center. Performing these pulsed orientation-selective experiments in conjunction with a computer simulation allows both the complete hyperfine matrix and also the distance and relative orientation of the nucleus from the paramagnetic center to be determined. Thus, these techniques are capable of determining crystallographic information on noncrystalline samples, and also provide information on the electronic structure of the metal center.

16.2
Multifrequency EPR Toolkit

While X-band frequencies (9–10 GHz) are most commonly used to measure the EPR spectra of a metal ion center, the use of multiple frequencies (lower and higher) offers a number of distinct advantages which enable an accurate determination of the spin Hamiltonian parameters for the metal ion center. These parameters can subsequently be used to define the three-dimensional (3-D) geometric and electronic structure of a metal ion center. Factors governing the choice of microwave frequency for a particular metal ion center include (Pilbrow, 1990; Hanson, Noble, and Benson, 2009; Basosi, Antholine, and Hyde, 1993; Drew et al., 2007; Drew, Young, and Hanson, 2007; Hanson et al., 1987; Mobius and Savitsky, 2009; Grinberg and Berliner, 2004; Froncisz and Hyde, 1980; Hyde and Froncisz, 1982; Wilson et al., 1988; Wilson et al., 1991; Nielsen et al., 2009):

- g-value resolution
- magnitude of the microwave energy
- state mixing of the various spin Hamiltonian interactions
- angular anomalies or off-axis extrema
- orientation selection
- nuclear frequency separation for electron nuclear double (triple) resonance END(T)OR experiments
- distributions of spin Hamiltonian parameters.

A brief discussion of multifrequency CW- and high-resolution pulsed EPR methods, computer simulation and quantum chemistry calculations is presented below, prior to discussing the application of this toolkit to the structural characterization of metal ion-binding sites in Cu(II) containing metalloproteins, cyclic peptides, and transition metal ion complexes.

16.2.1
g-Value Resolution and Orientation Selection

g-Value resolution relates to the separation (in field space) of the anisotropic components of g and/or the spectral resolution of different species with differing **g** matrices. Rearrangement of the resonance condition for EPR spectroscopy allows the resonant field position (B) to be calculated (Equation 16.3), where h and β are Planck's constant and the Bohr magneton, respectively, and v is the microwave frequency. The field separation ($B_i - B_j$, $B_i > B_j$) of two resonances (i, j) can then be calculated and shown to be proportional to the microwave frequency (Equation 16.3).

$$\Delta E = h\nu = g\beta B$$

$$B_i = h\frac{\nu}{g_i \beta B_i}, \quad (i, j = x, y, z; i \neq j)$$

$$B_i - B_j = K\nu \quad \text{where} \quad K = \frac{h(g_j - g_i)}{g_i g_j \beta} \tag{16.3}$$

Provided that $B_i - B_j$ is greater than the sum of the line widths for the individual resonances, then increasing the frequency improves g-value resolution. For example, the multifrequency EPR spectra of a copper(II) cyclic peptide complex reveals a rhombic **g**-matrix at Q-band frequencies (Figure 16.2) (Comba et al., 2008). This is also reflected in the corresponding transition roadmaps (Figure 16.2).

At higher microwave frequencies, the electron Zeeman interaction dominates the hyperfine interaction, yields an increased g-value resolution (Figure 16.2, especially the roadmaps) which reduces overlap of the hyperfine resonances enabling a simpler spectral interpretation, provides greater orientation selection for pulsed EPR and ENDOR experiments, and also allows a determination of the nuclear quadrupole interaction (Pilbrow, 1990; Basosi, Antholine, and Hyde, 1993; Mobius and Savitsky, 2009; Hanson et al., 1987; Drew et al., 2007; Wilson et al., 1988; Wilson et al., 1991).

16.2.2
Magnitude of the Microwave Frequency

Often, the microwave energy at X-band frequencies (~0.3 cm^{-1}) is insufficient to excite transitions within or between Kramers' doublets in high-spin or exchange-coupled spin systems, especially for even electron spin systems. For example, at X-band frequencies, the EPR spectra of di- and tri-nuclear coupled copper(II) complexes may not be observed as the zero-field splitting (ZFS) and/or the isotropic exchange coupling may be greater than the microwave quantum. A similar situation is observed for high-spin Fe(III), where resonances arise from transitions within the three Kramers' doublets and not between them. For even electron spin systems ($S = 2$: Fe(II), Mn(III); $S = 1$ Ni(II), V(III)), the EPR transitions are normally only observed at frequencies greater than 95 GHz, though a distribution of ZFS parameters can sometimes lead to the observation of formally forbidden resonances at lower frequencies. The EPR spectra of Fe(II)-containing metalloproteins have revealed $\Delta M_S = \pm 4$ transitions at $g_{\text{eff}} \sim 16$.

16.2.3
State Mixing

When two or more of the spin Hamiltonian interactions in Equation 16.1 are of a similar order of magnitude, then state mixing can occur which results in

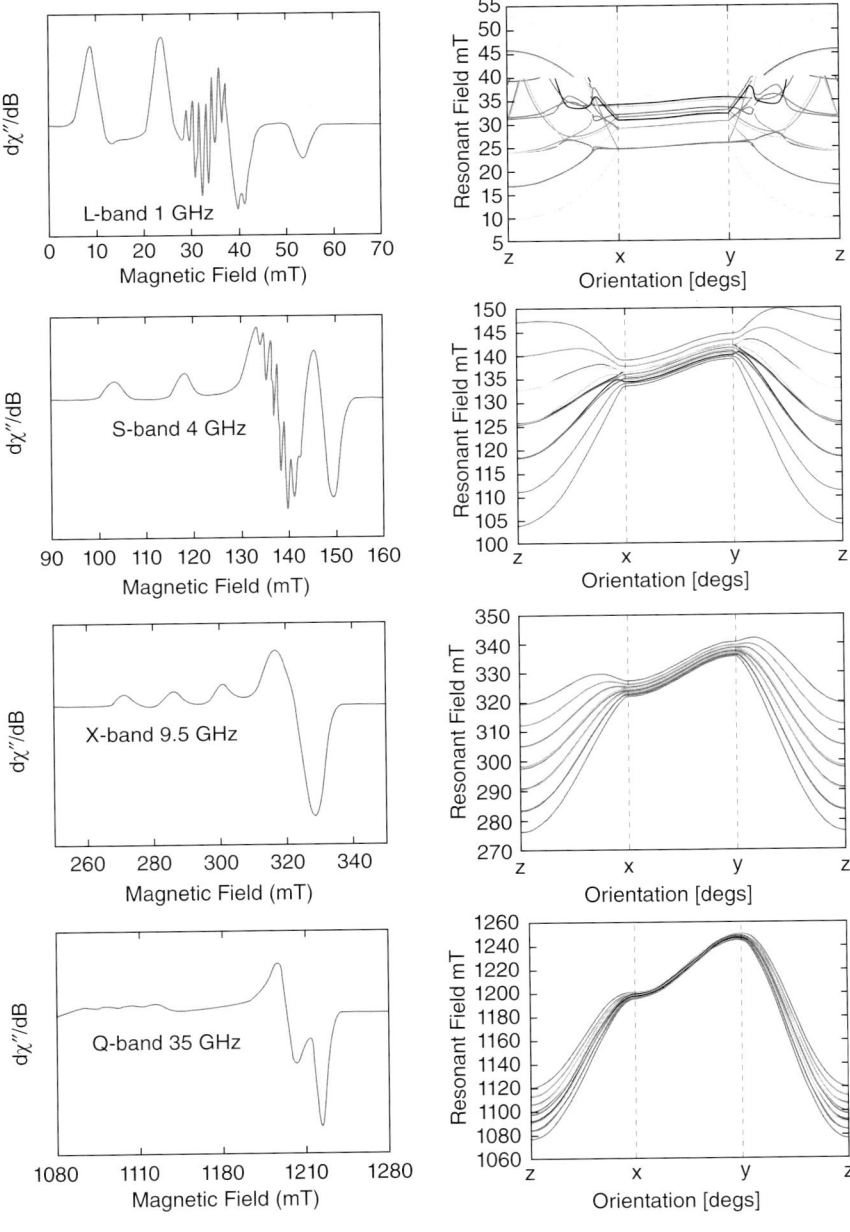

Figure 16.2 Multifrequency EPR spectra (left column) and corresponding transition roadmaps (right column) of a copper(II) complex in which three nitrogen nuclei are equatorially coordinated to the copper(II) ion. $g_x = 2.0884$, $g_y = 2.0512$, $g_z = 2.2780$; A (^{63}Cu): $A_x = 16.95$, $A_y = 15.43$, $A_z = 153.40$; A (^{14}N) × 2: $A_x = 14.54$, $A_y = 7.07$, $A_z = 8.99$; A (^{14}N) × 1: $A_x = 13.22$, $A_y = 15.2$, $A_z = 9.50$ (10^{-4} cm^{-1}). The transition roadmaps show the orientation dependence of the ^{63}Cu(II) hyperfine resonances arising from allowed and forbidden transitions (Comba et al., 2008).

nonlinearity of the energy levels as a function of magnetic field, looping transitions and energy level crossings and anticrossings. At low microwave frequencies (S-band, 4 GHz or lower), the hyperfine and electron Zeeman interactions for an $S = 1/2$ spin system [e.g., Cu(II)] are of a similar order of magnitude, and this leads to increased state mixing that results in complex transition roadmaps, particularly at L-band frequencies (Figure 16.2). There are, however, several advantages, including: (i) spectral redistribution; (ii) larger second-order contributions to the hyperfine interaction (proportional to $1/B$); and (iii) the presence of intense angular anomalies. Spectral redistribution and increased second-order hyperfine terms have enabled a complete determination of the monoclinic (C_s symmetry) hyperfine matrix and its orientation with respect to the principal components of the g matrix for a series of Mo(V) transition metal ion complexes (Drew et al., 2007). In contrast, at higher frequencies these second-order terms are minimized, making it more difficult – if not impossible – to extract the orientation of A with respect to g, especially as the resonances have a larger line width at higher frequencies. The exploitation of intense angular anomalies in determining the metal ions' coordination sphere is discussed further in Sections 16.2.4 and 16.3.3.

State mixing is also readily observed when the fine structure and/or exchange interaction is of a similar order of magnitude as the electron Zeeman interaction (Equations 16.1 and 16.2). While this is inherently important for di- and tri-nuclear Cu(II) complexes, it is perhaps more easily seen in a multifrequency single-crystal EPR study of a high-spin Fe(III) center. At low frequencies, the electron Zeeman and fine structure interactions are of a similar order of magnitude (Figure 16.3a and b), which results in the appearance of looping transitions and energy level crossings and anticrossings. At high microwave frequencies, the fine structure interaction is smaller than the electron Zeeman interaction, and can be treated as a perturbation (similar to the hyperfine interaction at X-band frequencies), resulting in a much simpler spectrum (Figure 16.3c and d) (Hanson, Noble, and Benson, 2009; Gaffney, 2009; Grinberg and Berliner, 2004).

16.2.4
Angular Anomalies

The appearance of an additional resonance (angular anomaly or off-axis extremum; see Figure 16.2, X-band transition roadmap) in randomly oriented anisotropic EPR spectra is especially common for Cu(II) complexes (Pilbrow, 1990) (e.g., Figure 16.2). Indeed, Ovchinnikov and Konstantinov (1978) showed, through a closed-form expression (Equation 16.4), to second order with respect to the hyperfine coupling, that the presence of an off-axis extremum occurred over a finite microwave frequency range for the plane i,j, such that $i \neq j$ and $A_i^2 > A_j^2$:

$$\frac{M_I(2A_i^2 - \Delta_{oij})}{h(A_i)} < \nu < \frac{M_I(2A_j^2 - \Delta_{oij})}{h(A_j)}$$

$$\Delta_{oij} = (g_i^2 A_i^2 - g_j^2 A_j^2) \tag{16.4}$$

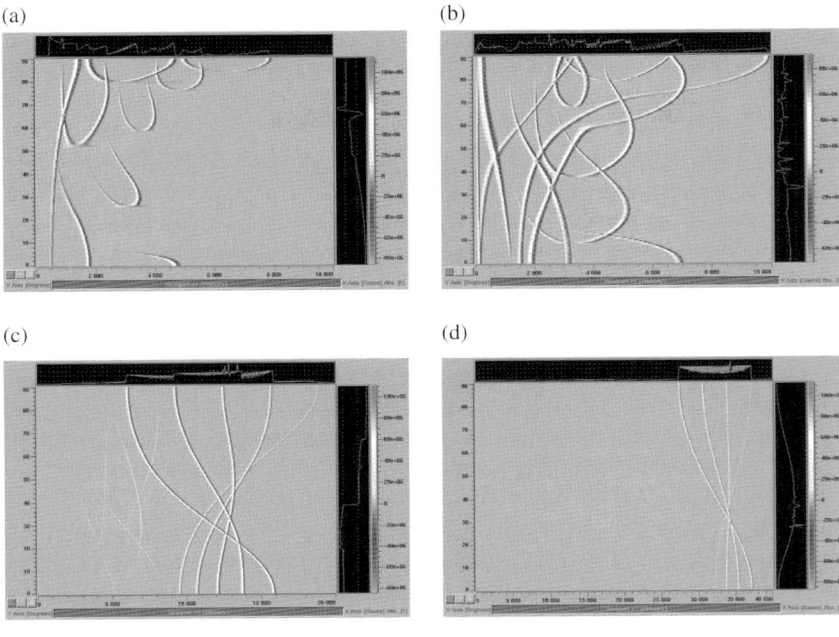

Figure 16.3 Multifrequency single-crystal EPR spectra of high-spin Fe(III) ($S = 5/2$, $D = -4.5$ GHz, $E/D = 0$, $g = 2.0$) showing the orientational dependence of resonances in a plane perpendicular to the "z" axis. (a) S-band, $v = 4.0$ GHz; (b) X-band, $v = 9.75$ GHz; (c) Q-band, $v = 35.0$ GHz; (d) W-band, $v = 95.0$ GHz (Hanson, Noble, and Benson, 2009).

While angular anomalies are readily observed in the EPR spectra of mononuclear copper(II) centers at X-band frequencies (328.4 mT, Figure 16.2), increasing the frequency to Q-band almost removes the angular anomaly (Figure 16.2). At S-band frequencies the angular anomaly may not be resolved, but at lower frequencies (≤ 1 GHz) spectral redistribution allows this to be resolved (see Section 16.3.3).

16.2.5
Distribution of Spin Hamiltonian Parameters

Statistical distributions of spin Hamiltonian parameters are often found in frozen solution samples where, ideally, the paramagnetic center is randomly oriented. However, poor-quality glasses and solvate association with the metal ion center can lead to Gaussian distributions of spin Hamiltonian parameters. Hyde and Froncisz were the first to note that the line width of the parallel copper hyperfine resonances in anisotropic spectra varied not only as a function of frequency, but also as a function of the nuclear spin quantum number (M_I) (Froncisz and Hyde, 1980; Hyde and Froncisz, 1982; Pilbrow, 1990). These authors showed that the line width could be expressed as:

16 Copper Coordination Environments

$$\sigma_i = \sqrt{[\sigma_{R_i}^2 + ((\sigma g_i/g_i)v_0(B))^2 + (\sigma A_i M_I)^2 + 2\varepsilon(\sigma g_i/g_i)v_0(B)\sigma A_i M_I]}$$

$$\sigma_v^2 = \left[\sum_{i=x,y,z} \sigma_i^2 l_i^2 g_i^2 / g^2\right] \quad (16.5)$$

where $\sigma g/g$ and σA correspond to the Gaussian half-widths of the distribution of g and A values respectively, σ_R is residual line width arising from spin lattice and spin–spin relaxation processes, ε is a correlation coefficient between g and A (normally equal to 1, but not necessarily 1), M_I is the nuclear spin quantum number, $v_0(B)$ the resonant field position, and l_i are the direction cosines relating the principal line width axes to the external magnetic field. Clearly, the line width is proportional to the microwave frequency and M_I. While increasing the frequency improves g-value resolution, the line widths of the resonances will also increase (Figure 16.2, loss of ^{14}N superhyperfine coupling at higher frequencies). Conversely, decreasing the microwave frequency reduces not only g-value resolution but also the line width, enabling the resolution of ligand hyperfine coupling. The line width dependence on M_I^2 indicates that the distribution of g and A values will be minimized for each Cu(II) hyperfine resonance at different frequencies (Figure 16.4).

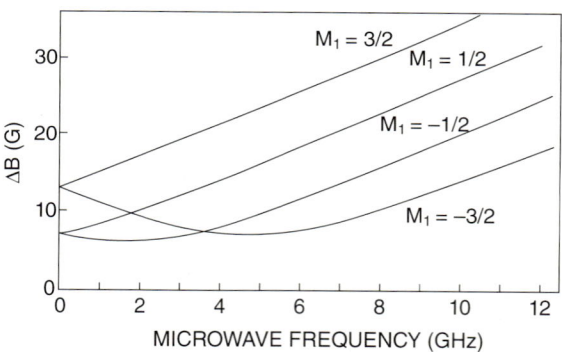

Figure 16.4 Linewidth dependence of parallel Cu(II) hyperfine resonances governed by Equation 16.5 (Froncisz and Hyde, 1980; Hyde and Froncisz, 1982).

S- (2–4 GHz) and L- (1–2 GHz) band EPR spectroscopy has indeed enabled the observation of ^{35}Cl, ^{17}O, ^{77}Se, ^{33}S, ^{2}H and ^{14}N ligand hyperfine coupling in a variety of Mo(V) and Cu(II) complexes and metalloenzymes (e.g., Figure 16.4) (Pilbrow, 1990; Basosi, Antholine, and Hyde, 1993; Doonan et al., 2008; Drew et al., 2009; Hanson et al., 1987; Froncisz and Hyde, 1980; Hyde and Froncisz, 1982; Wilson et al., 1988; Wilson et al., 1991). Often, the decreased g-value resolution produces a redistribution of the hyperfine resonances and, in conjunction with reduced line widths, spectral resolution is enhanced enabling the complete hyperfine matrix to be determined.

16.2.6
Numerical Differentiation and Fourier Filtering

Another useful technique for resolving small hyperfine couplings in CW-EPR spectra, normally measured as first-derivative spectra, is to either measure or numerically calculate second- or higher derivative spectra where spectral resolution is enhanced (Basosi, Antholine, and Hyde, 1993; Comba et al., 2008). Unfortunately, these higher derivative spectra have poor signal-to-noise ratios (SNRs), and attention must be directed either to accumulating a large number of scans or to apply a numerical filter to increase the SNR. One such technique which has proven successful is to: (i) differentiate the CW-EPR spectrum; (ii) baseline-correct the spectrum to avoid artifacts in the Fourier transform; (iii) Fourier-transform the spectrum; (iv) apply a Hamming filter to remove the high-frequency noise without affecting the lower frequencies that give rise to the EPR resonances; and finally (v) to perform an inverse Fourier transform to yield a CW-EPR spectrum with a significantly higher SNR. Examples of this approach can be found in Section 16.5.2, where second-derivative spectra have been calculated to resolve the ^{63}Cu and ^{14}N hyperfine coupling in the perpendicular region (Comba et al., 2008). A careful choice of the hamming window can also be used to directly measure superhyperfine coupling constants. Basosi et al. have taken this approach a step further in showing that a particular region of the Fourier-transformed spectrum is sensitive to the parity of the number of nitrogen nuclei, giving rise to the ligand hyperfine coupling (Lunga, Pogni, and Basosi, 1995).

16.2.7
High-Resolution EPR Techniques

Hyperfine interactions between transition metal ions and ligand nuclei are normally too weak to be fully resolved using CW-EPR, and simply contribute to the inhomogeneous line broadening of resonances in the CW-EPR spectrum (Pilbrow, 1990; Schweiger and Jeschke, 2001; Harmer, Mitrikas, and Schweiger, 2009). Weak metal–ligand hyperfine interactions can, however, be resolved by employing the high-resolution EPR techniques such as END(T)OR, ESEEM, and two-dimensional (2-D) variants such as HYSCORE. Both, ESEEM and HYSCORE spectroscopy enable the observation of small hyperfine couplings from nuclei up to 6–8 Å away from the metal ion, while END(T)OR allows the observation of large hyperfine couplings of nuclei directly coordinated to the metal ion. Examples of some common pulse sequences used in pulsed-EPR are shown in Figure 16.5.

The microwave excitation bandwidth is approximately 40 MHz, and is insufficient to excite the complete anisotropic EPR spectrum, which can be greater than 1 GHz. Consequently, pulsed-EPR/END(T)OR measurements can be considered as burning a "hole" in the CW-EPR spectral line shape. If the pulsed experiment is performed on a randomly oriented frozen solution (or powder sample) as a function of the magnetic field (Figure 16.5a), which is equivalent to orientation ("z" → "x" → "y"), then in conjunction with a computer simulation the complete

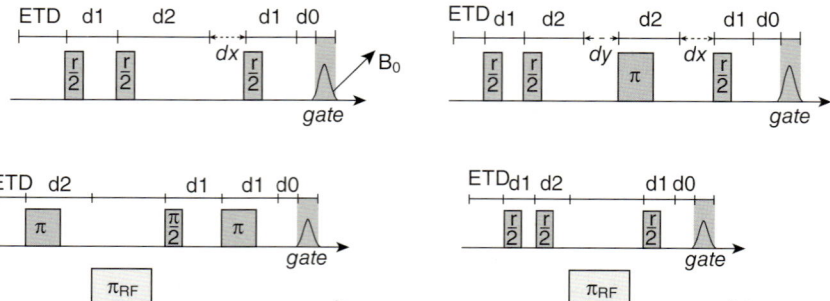

Figure 16.5 Examples of some common pulse sequences. (a) Orientation-selective three-pulse ESSEM; (b) HYSCORE; (c) Davies ENDOR; (d) MIMS ENDOR.

nuclear hyperfine matrix can be determined. By employing the single point dipole–dipole approximation, the distance and orientation of the nucleus with respect to the electron spin can be determined. For further insights into the theoretical and experimental aspects of pulsed-EPR and END(T)OR, the reader is referred especially to the book by Schweiger and Jeschke (2001).

16.2.8
Computer Simulation

An integrated approach "Molecular Sophe" for the computer simulation of CW- and pulsed-EPR and ENDOR spectra, energy level diagrams, transition roadmaps and transition surfaces for a general spin system (including high-spin and exchange-coupled spin systems) has been developed by Hanson, Noble, and Benson (2009). This approach, which is based on molecular structure (Figure 16.6), allows the 3-D molecular (geometric and electronic) characterization of paramagnetic materials using high-resolution multifrequency CW- and pulsed-EPR spectroscopy and quantum chemistry calculations. Until now, the analysis of complex CW- and pulsed EPR spectra has been based on a spin system rather than on molecular structure, and the analysis of pulsed-EPR spectra has mainly relied on analytical expressions involving perturbation theory for an $S = 1/2$ spin system.

A computer simulation of randomly oriented or single-crystal EPR spectra from isolated or coupled paramagnetic centers is required to accurately determine the spin Hamiltonian parameters (Equations 16.1 and 16.2) and the electronic and geometric structure of the paramagnetic center. The simulation of randomly oriented EPR spectra is performed in frequency space through the following integration (Pilbrow, 1990; Hanson, 2003; Hanson et al., 2003; Hanson et al., 2004; Hanson, Noble, and Benson, 2009; Griffin et al., 1999; Heichel et al., 2000):

$$S(B,v_c) = C \int_{\theta=0}^{\tilde{\pi}} \int_{\phi=0}^{\tilde{\pi}} \sum_{i=0}^{N} \sum_{j=i+1}^{N} |\mu_{ij}|^2 f(v_c - v_0(B), \sigma v) d\cos\theta d\phi \qquad (16.6)$$

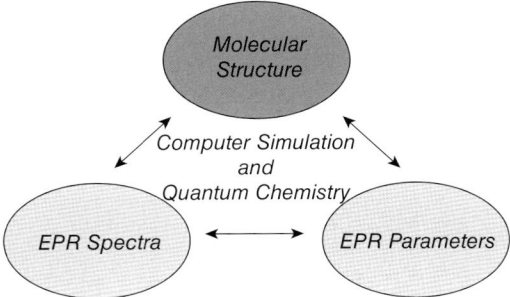

Figure 16.6 Integrated approach to EPR structure determination.

where $S(B, v_c)$ denotes the spectral intensity, $|\mu_{ij}|^2$ is the transition probability, v_c the microwave frequency, $v_o(B)$ the resonant frequency, σ_v the spectral line width, $f(v_c - v_o(B), \sigma_v)$ a spectral lineshape function which normally takes the form of either Gaussian or Lorentzian, and C a constant which incorporates various experimental parameters. The summation is performed over all the transitions (i, j) contributing to the spectrum, and the integrations are performed numerically over half of the unit sphere for species possessing triclinic symmetry, a consequence of time-reversal symmetry (Pilbrow, 1990; Abragam and Bleaney, 1970). For paramagnetic centers exhibiting orthorhombic or monoclinic symmetry, the integrations in Equation 16.6 need only be performed over one or two octants, respectively. Paramagnetic centers with axial symmetry only require integration over θ between 0 and $\pi/2$, while those possessing cubic symmetry require only a single orientation. While the resonant frequencies can be calculated using either perturbation theory or matrix diagonalization, the latter approach is preferred (despite being computationally more expensive) as state mixing results in the perturbation expressions for the resonant field positions no longer being valid (Pilbrow, 1990; Hanson et al., 2004; Hanson, Noble, and Benson, 2009).

Experimentally, the CW-EPR experiment is a field-swept experiment in which the microwave frequency (v_c) is held constant and the magnetic field varied. Computer simulations performed in field space assume a symmetric line shape function [f in Equation 16.6; ($f(B - B_{res}), \sigma_B)$], which must be multiplied by dv/dB and assume a constant transition probability across a given resonance (Pilbrow, 1990; Weil, Bolton, and Wertz, 2007; Hanson, Noble, and Benson, 2009). Sinclair and Pilbrow have described the limitations of this approach in relation to the asymmetric line shapes observed in high-spin Cr(III) spectra and the presence of a distribution of g-values or g-strain broadening (Pilbrow, 1984; Sinclair, 1988; Pilbrow et al., 1983). Performing computer simulations in frequency space produces asymmetric line shapes (without having to artificially introduce an asymmetric line shape function), and in the presence of a large distribution of g-values will also correctly reproduce the downfield shifts of resonant field positions (Sinclair, 1988; Pilbrow et al., 1983). Unfortunately, this approach cannot be used in conjunction with matrix diagonalization, as a very large number of matrix diagonalizations would be required to calculate f, and the transition probability across a particular resonance would result in unacceptably long computational times.

The general-purpose computer simulation software suite, XSophe-Sophe-XeprView® (Hanson et al., 2003, 2004; Griffin et al., 1999; Heichel et al., 2000; Hanson, 2003) enables the analysis of isotropic, randomly oriented and single-crystal CW-EPR spectra, energy level diagrams, transition roadmaps, and transition surfaces. The software suite consists of: XSophe, an X-windows graphical user interface; the Sophe authentication and Common Object Request Broker Architecture (CORBA) daemons; Sophe, a state-of-the-art computational program for simulating CW-EPR spectra and XeprView®, Bruker Biospin's program for visualizing and comparing experimental and simulated CW-EPR spectra. Energy level diagrams and transition roadmaps can be visualized through a web browser, and transition surfaces with an OpenInventor viewer. Further details concerning the functionality of the XSophe-Sophe-XeprView® software suite can be found in the literature (Hanson et al., 2003, 2004; Griffin et al., 1999; Heichel et al., 2000; Hanson, 2003).

The Molecular Sophe (MoSophe) software suite consists of a graphical user interface, the computational program Sophe, and a variety of software tools (XeprView®, gnuplot, ivview and ghostview) for visualizing and comparing simulations and experimental EPR spectra (Hanson, Noble, and Benson, 2009). This provides a powerful research tool for determining the geometric and electronic structures of magnetically isolated and coupled paramagnetic centers within metalloproteins, transition metal ion complexes, and other paramagnetic molecules.

Molecular Sophe's graphical user interface (Figure 16.7) consists of an explorer tree, and several tool bars and windows related to the label selected in the explorer tree. MoSophe is project-oriented, with a project defined in the explorer tree consisting of a number of simulations, each of which contains a sample with one or more molecules consisting of atoms and bonds (interactions). Apart from defining the spin Hamiltonian interactions for each atom, a user can also define the positional coordinates for each atom. Once the sample has been defined, a choice can be made from a range of experiments, including CW-EPR, pulsed-EPR (FID, two-pulse ESEEM, three-pulse ESEEM, SECSY, HYSCORE, two-pulse Echo, MIMS and Davies ENDOR), Reports (Energy Level Diagrams, Transition Roadmaps, Transition Surfaces), and multidimensional experiments (by adding additional temperature, microwave frequency, goniometer angle and magnetic field abscissas) to elucidate the geometric and electronic structures of the molecule(s). Quantum chemistry and molecular modeling calculations will be added in the near future, allowing the molecular (geometric and electronic) structures of metal centers to be determined through a range of EPR experiments and simulations.

16.2.9
Computational Chemistry

As mentioned previously, determination of the geometric and electronic structures has traditionally relied on the calculation of the spin Hamiltonian parameters from

16.2 Multifrequency EPR Toolkit

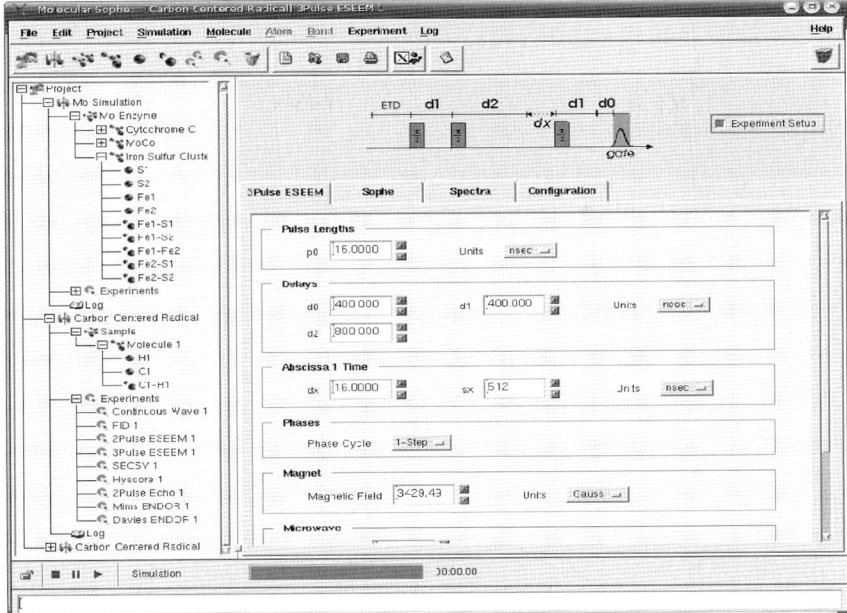

Figure 16.7 The Molecular Sophe graphical user interface, showing the explorer tree, tool bars and the 3-pulse ESEEM experiment.

a given structure, using quantum chemistry calculations that have become increasingly sophisticated over the past decade. Gradually, crystal and ligand-field calculations have been replaced by *ab initio* methods (self-consistent field Xα) and, more recently, by DFT (ORCA, Gaussian and ADF), which allows the calculations of larger molecules (Neese, 2009; Frisch *et al.*, 2004; te Velde *et al.*, 2001). While DFT calculations have been extremely successful for many metal ions and radicals, others have been less successful, for example in the case of copper(II) ions. For these cases, *ab initio* approaches may be significantly better.

Molecular modeling (MM) is another approach for determining geometric structures, and Comba *et al.* have over many years developed a MM-EPR approach that employs MM in conjunction with constraints (internuclear distance and orientation) provided by a computer simulation of the EPR spectra of dinuclear Cu(II) complexes to determine their structures (Comba and Martin, 2005; Deeth, 2010; Bernhardt *et al.*, 1992; Comba and Hilfenhaus, 1995; Comba *et al.*, 1996; Comba *et al.*, 2006; Bernhardt *et al.*, 2002; Comba *et al.*, 1998). Ideally, single-point quantum chemistry calculations should be performed on these optimized structures; however, an accurate calculation of the anisotropic exchange matrix (dipole–dipole coupling) is not possible with DFT calculations at present.

16.3
Multifrequency EPR Simulation of Square–Planar-Based Cu(II)

16.3.1
EPR of Square–Planar-Based Cu(II)

Square–planar-based Cu(II) is important in many biological systems. Many of the properties of the EPR spectra of Cu(II) are well understood (Hathaway and Billing, 1970). Natural-abundance copper is a mixture of ^{63}Cu and ^{65}Cu, both with $I = 3/2$ and with similar, but not identical, gyromagnetic ratios. Cu(II) is a d^9 ion with obligatory $S = \frac{1}{2}$, while square–planar Cu(II) exhibits essentially axial **g** and **A**[$^{63/65}$Cu(II)] tensors with $g > 2$, $g_{\parallel} > g_{\perp}$, and $A_{\parallel} \gg A_{\perp}$. Relationships between spin–orbit coupling and **g** for tetragonal Cu(II) exist (Pilbrow, 1990): $g_{\parallel} = g_e(1 - 4\alpha^2\lambda/\Delta_1)$ and $g_{\perp} = g_e(1 - \alpha^2\lambda/\Delta_2)$, where λ is the spin–orbit coupling ($-830\,\text{cm}^{-1}$), α^2 is the contribution of the $dx^2 - dy^2$ orbital angular momentum, and the Δ terms are the differences between the energies, E, between d orbitals: $[\Delta_1 = E(dx^2 - dy^2) - E(dxy)]$ and $[\Delta_2 = E(dx^2 - dy^2) - E(dxz, dyz)]$. Analogous relationships to those for **g** exist for **A**: $A_{\parallel} = P[-k(4\alpha^2/7) + (g_{\parallel} - g_e) + 3(g_{\perp} - g_e)/7]$ and $A_{\perp} = P[-k(2\alpha^2/7) + 11(g_{\perp} - g_e)/14]$, where $P \sim 0.035\,\text{cm}^{-1}$ and $k \sim 3.5 \times 10^{-4}\,\text{cm}^{-1}$. Both isotopes of copper also exhibit different and non-negligible quadrupolar interactions (**I.P.I**), which can be very difficult to obtain from EPR (White and Belford, 1976; Abrakhmanov and Ivanova, 1978; Belford and Duan, 1978). In their landmark paper, Peisach and Blumberg (1974) demonstrated that Cu(II) EPR spectra can be used to estimate (with low precision) the numbers of N, O, and S ligand donor atoms to tetragonal Cu(II) from $g_{\parallel}$ and $A_{\parallel}$, circumventing the need to know $g_{\perp}$, $A_{\perp}$, or **P**.

16.3.2
Multifrequency EPR of Square–Planar-Based Cu(II): S- and L-Band EPR

Multifrequency EPR, particularly EPR at <9.5 GHz, has long been employed to better characterize the (equatorial) ligand coordination of Cu(II) (Rakhit et al., 1985; Kroneck et al., 1988; Antholine et al., 1992; Neese et al., 1996; Aronoff-Spencer et al., 2000; Burns et al., 2002; Burns et al., 2003; Chattopadhyay et al., 2005). The rationale for the use of EPR at S- and L-band frequencies for the resolution of hyperfine and superhyperfine components in Cu(II) spectra arises from the frequency-dependence of the line width (Equation 16.5; Froncisz et al., 1979; Froncisz, Sarna, and Hyde, 1980; Hyde and Froncisz, 1982). In this expression, σg and σA represent distributions ("strains") in g and A due to microheterogeneity in the electronic structures of the Cu(II) ions in the sample. For two of the four $I = 3/2$ lines, the g and A strains cancel each other in a frequency-dependent manner, and an optimum frequency for each of these two lines will exist.

In Cu(II), $A_{\parallel} \gg A_{\perp}$ and the cancellation effects are noted in the $g_{\parallel}$ region (Froncisz et al., 1979). This phenomenon has proven particularly useful for the estimation of nitrogen ligand donor atoms of Cu(II). A $g_{\parallel}$ line in the powder spectrum of Cu(II) can be considered as identical to an absorption line, due to a single-crystal orientation (Froncisz and Hyde, 1980; Hyde and Froncisz, 1982). Where the line width is sufficiently small, the number and intensities of the superhyperfine features can, in principle, simply be measured to yield the nitrogen coordination number if a sufficient degree of equivalence of $A_\parallel$ (^{14}N) is exhibited; the latter condition is usually met by N ligands of Cu(II) because the contact term is by far the major contributor to the superhyperfine interaction (Froncisz et al., 1979; Hyde and Froncisz, 1982). S-band EPR, ~3.3 GHz, has traditionally been the frequency of choice. The quantum numbers at S-band can still be assumed to be "good" quantum numbers, and perturbation theory is sufficient for interpretation. S-band spectra are only barely contaminated by "forbidden" transitions, and these can be neglected. The availability of a single copper isotope has been essential to the success of the S-band method (Neese et al., 1996); the system of interest must be amenable to synthesis, substitution, or the biological incorporation of ^{63}Cu or ^{65}Cu. Although the substitution of ^{15}N is not necessary, it is desirable, as the spectrum is simplified and the individual lines are more intense and better resolved because $A(^{15}N) \approx 1.4 \times A(^{14}N)$, and because of a lack of ^{14}N quadrupolar coupling (Neese et al., 1996; Burns et al., 2002, 2003). The use of perdeuterated systems is also desirable, but often impracticable with biological systems. S-band EPR can often clearly distinguish two nitrogen ligands from three, but three nitrogens often cannot reliably be distinguished from four, due to the very low intensity of the outermost lines of a four-nitrogen superhyperfine pattern. Analysis of the second-derivative $(\partial^2\chi''/\partial B^2)$ spectrum (Basosi et al., 1986) and superhyperfine pattern intensity analysis via Fourier transform methods (Pasenkiewicz-Gierula et al., 1987) have been applied in an attempt to better interpret the $g_\parallel$ lines of S-band EPR spectra of Cu(II) in terms of the number of coupled magnetic nuclei.

16.3.3
Multifrequency EPR of Square–Planar-Based Cu(II): Very Low-Frequency EPR

The analysis of (usually) the $M_I = -1/2$ $g_\parallel$ line of S-band EPR spectra has been of great use in providing a more precise determination of the number of nitrogen nuclei coupled to Cu(II) than a Peisach–Blumberg analysis. However, cases of 3N and 4N equatorial nitrogen coordination cannot be easily distinguished. As can be seen from Figure 16.8, both the minimum line width attainable and the frequency at which this minimum line width is observed, are functions of A-strain, σA. Lowering σA has the effects of decreasing the lowest attainable line width and decreasing the frequency at which the lowest line width is observed. In this case, the maxim "the lower the frequency, the better" will apply, at least in principle, for the limiting case of vanishing A-strain. The value of $A_\perp(^{63/65}Cu)$ is substantially lower than that of $A_\parallel$ for square–planar-based Cu(II), and $\sigma A_\perp$ is correspondingly smaller than $\sigma A_\parallel$. It is expected, therefore, that the optimum frequency for resolution of a superhyperfine pattern on a $g_\perp$ line will be very low ($\ll 3.3$ GHz).

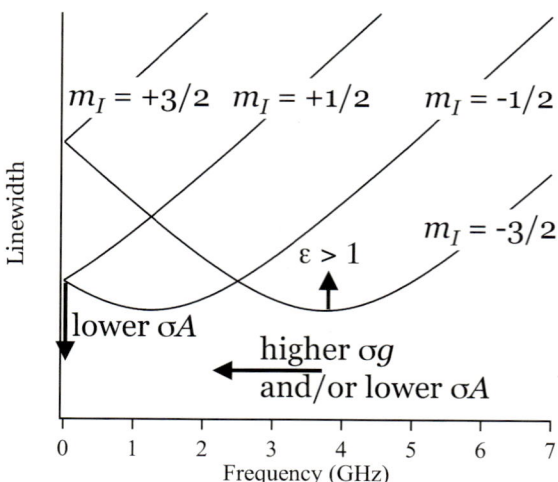

Figure 16.8 Widths of g- and A-strained square–planar Cu(II) $g_\parallel$-lines as a function of frequency. As g-strain increases relative to A-strain, the frequency at which the minimum line width for $M_I = -1/2$ and $-3/2$ is observed decreases. The absolute minimum line width is dependent on A-strain. The frequency at which the lowest line width is observed increases with A-strain, and decreases with g-strain. A g-strain/A-strain correlation parameter, ε, dictates the relative minimum line widths for $M_I = -1/2$ and $-3/2$.

The rationale for the use of EPR at sub-2 GHz frequencies (very low-frequency EPR; VLF-EPR) to study Cu(II) was first presented by Hyde and Froncisz (1982). A subsequent analysis identified the superior sensitivity of EPR at 1 GHz over higher frequencies for a number of spin Hamiltonian parameters for Cu(II) (Hyde, Antholine, and Basosi, 1985). The spectra of inorganic Cu(II) systems have been recorded in the VLF-EPR frequency range (Abrakhmanov and Ivanova, 1978; Basosi et al., 1984; Rothenberger et al., 1986; Pasenkiewicz-Gierula et al., 1987; Basosi, Antholine, and Hyde, 1993), and a few examples of VLF-EPR spectra of Cu(II), and of low-spin Co^{II}, in proteins have been presented (Hori et al., 1983; Antholine, Hanna, and McMillin, 1993). In these cases, the motivation has been either to detect forbidden transitions that have higher transition probabilities at lower frequencies or, simply, to extend the frequency range as part of an overall multifrequency approach. More recently, the use of VLF-EPR has been accompanied by an analysis of the frequency-dependence of the M_I manifold overlap, with an intent to obtain otherwise intractable information on the nitrogen–ligand–donor coordination number (Hyde et al., 2009). This latter study highlighted the ability of VLF-EPR analysis of the $g_\perp$ region of the spectrum to distinguish among 3N and 4N equatorial coordination. An additional application of VLF-EPR is in the direct determination of quadrupolar splittings and "forbidden" transitions; both will contribute to the spectrum in the normal $\mathbf{B}_0 \perp \mathbf{B}_1$ mode, and can be characterized more thoroughly at very low-frequency, particularly when using parallel mode, $\mathbf{B}_0 \parallel \mathbf{B}_1$, EPR (Abrakhmanov and Ivanova, 1978; Rothenberger et al., 1986).

16.3.4
Multifrequency EPR of Square–Planar-Based Cu(II): Experimental Considerations for Low-Frequency EPR

As with EPR at higher frequencies, EPR at low frequencies (0.5–4 GHz) requires a radiofrequency (rf) source, a magnet, a modulation scheme for phase-sensitive detection, and a resonant rf structure. Octave-band microwave bridges covering 0.5 to 4 GHz have been constructed at the National Biomedical EPR Center at the Medical College of Wisconsin (Hyde and Froncisz, 1986). High-performance broadband rf sources (such as the Agilent E4438C synthesizer) are now available, and provide a reduced phase noise compared to mechanically tuned oscillators, as well as providing frequency modulation. Field modulation is more commonly employed (Hyde and Froncisz, 1986; Rothenberger et al., 1986; Antholine, Hanna, and McMillin, 1993; Hyde et al., 2009), but frequency modulation has also been used (Hirata et al., 2003). The development of the loop-gap resonator (LGR) is essential to the practicality of low-frequency EPR for the examination of biological materials. Cavity resonators are unsuitable at very low frequencies because of: (i) a low, $v^{-5/2}$-dependent resonator efficiency (low B_1); (ii) an enhanced demodulation of oscillator phase noise due to the $\sqrt{(1/v)}$ dependence of Q; (iii) their large physical size; and (iv) a poor filling factor when the sample is limited (Froncisz and Hyde, 1982; Hyde and Froncisz, 1986). A typical sample requirement for a cavity resonator at 2.6 GHz is 1 μmol Cu(II) (Froncisz et al., 1979; Froncisz and Hyde, 1980), while at 750 MHz this scales to ≈100 μmol Cu(II), or about 25 ml of 4 mM Cu(II). In contrast, good-quality spectra have been recorded on 70 μl samples of copper-proteins at 1.1 to 2.3 GHz, using an LGR-equipped National Biomedical EPR Center spectrometer (Antholine, Hanna, and McMillin, 1993). A multimode, low-frequency resonant cavity is shown alongside a 1 GHz LGR in Figure 16.9.

A detailed analysis by Eaton and coworkers of LGR performance from 250 MHz to 9 GHz, has shown that scaling factors are much smaller with LGRs than with cavity resonators; a fourfold increase in the number of spins is sufficient to maintain the SNR upon lowering the frequency from 1.5 GHz to 250 MHz, and the required number of spins at 1.5 GHz is only fivefold that at 9 GHz (Rinard et al., 2002). Nevertheless, the use of biological material is often limited, and the long scan times needed at low frequencies require that particular attention be paid to electromagnetic shielding and both mechanical and thermal stabilities. The use of single-isotope ^{63}Cu or ^{65}Cu is essential, and the substitution of ^{14}N with ^{15}N and ^{1}H with ^{2}H may be desirable. Due to the higher $\mathbf{B}_1$ (oscillating magnetic field due to the microwave) of the LGR, care should be taken to identify and employ non-saturating conditions.

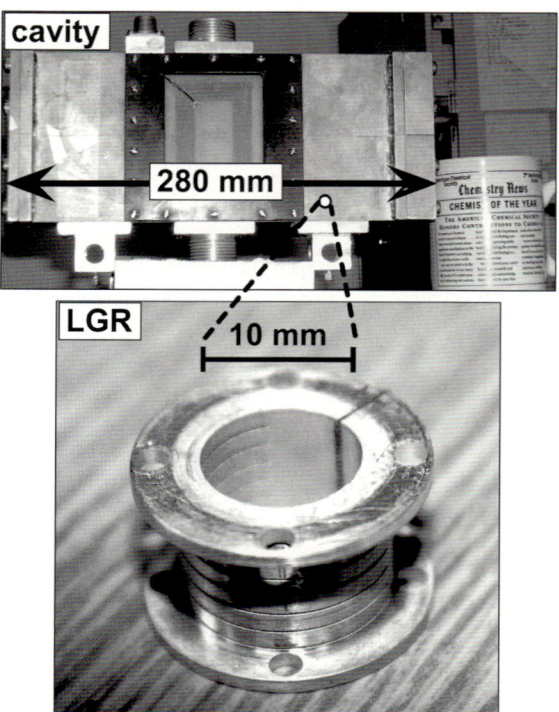

Figure 16.9 Top panel: A multiple mode, low-frequency cavity for L- to S-band EPR. The sample stack is 30 mm in diameter and has an active length of 65 mm (45 ml volume). Lower panel: A 1 GHz LGR with an active volume of 0.6 ml.

16.3.5
Introduction to Multifrequency EPR Simulations of Square–Planar Cu(II)

It seems intuitive that EPR spectra should be collected and simulated at multiple frequencies. The best resolution of the different lines in a spectrum ($g_\perp$, $g_\parallel$, "overshoot," different M_I) will occur at different frequencies because of strain effects, and also because the interplay of field-dependent (Zeeman) and field-independent (hyperfine) shifts in resonance positions will change the extent of overlap of the lines. In addition, spectral features that report on particular spin Hamiltonian parameters may also be measurable with different accuracy and/or precision at different frequencies. Strain effects (Froncisz and Hyde, 1980; Hyde and Froncisz, 1982) may preclude the resolution of superhyperfine splittings at high frequency, and favor a low-frequency approach. On the other hand, the complicated behavior of energy levels at low fields may render determination of the Zeeman interaction imprecise at low frequencies without a full description of the spin Hamiltonian, whereas the Zeeman terms may be obtained at high frequency even with an incomplete spin Hamiltonian.

The selection of multiple experimental EPR frequencies, such that the optimum resolution of all the spectral features is obtained, would appear to provide the benefit of a larger number of observables without any corresponding increase in the number of variables – that is, spin Hamiltonian and lineshape parameters. This is certainly true if the goal is one of comprehensively determining the spin Hamiltonian parameters. However, in some cases, a partial description may be all that is required to address a particular question regarding a chemical or biological Cu(II) system. In such cases, the benefit of extra information provided by multifrequency EPR may be offset by the additional spin Hamiltonian parameters needed to describe the whole multifrequency dataset; some of these parameters may not be directly determinable at any one given frequency.

If the assumption is made that it is a good thing to be in possession of as much experimental information as possible, then the question becomes one of how to deal with the experimental data. Three somewhat separate approaches have been adopted in the literature, and these will be discussed next.

16.3.6
Optimum Frequency Selection

Perhaps the simplest multifrequency approach is to identify the best frequency to use for the particular problem that is faced. The determination of the number of equatorial nitrogens coordinated to Cu(II) can be taken as an example. In the presence of strains in **g** and **A** tensors, it is expected that the feature in which the ^{14}N superhyperfine pattern will be best resolved will be the $g_\perp$ region. The inherently low $A_\perp$-strain, and the collapse of g-strain and rhombicity in $g(x)$ and $g(y)$, at low frequencies would seem to favor the lowest frequency available. However, a better indication of the optimum frequency can be obtained from computer simulations. Based on an approximate and partial spin Hamiltonian that is easily derived from simulation of the experimental spectrum at almost any commonly available frequency, $H = \beta \mathbf{g}.\mathbf{B}.\mathbf{S} + SA_zI(^{63}Cu)$, simulations as a function of frequency can be carried out for each M_I manifold, where M_I is the nuclear magnetic quantum number, using matrix diagonalization software such as XSophe (Wang and Hanson, 1995; Hanson et al., 2004) or EasySpin (Stoll and Schweiger, 2006; Stoll and Schweiger, 2007). An example is shown in Figure 16.10, where model parameters (Hyde et al., 2009) were used to generate the corresponding pure absorption spectra at different frequencies.

Certain particularly intense features can be seen in the $g_\perp$ regions of the $M_I = +1/2$ and the $M_I = +3/2$ manifolds. These features are "overshoot" (angular/hyperfine anomaly; extra absorption [EA]) lines that are a result of a build-up of intensity that arises from the interplay of anisotropies in **g** and **A** (Ovchinnikov and Konstantinov, 1978). An examination of the resonance positions and intensities of the $g_\perp$-lines for the $M_I = +1/2$ and the $M_I = +3/2$ manifolds suggests, in this case, that the *high-field* side of the $M_I = +1/2$ $g_\perp$-line at either 1 or 2 GHz provides a relatively intense feature that does not overlap significantly with any other feature likely to exhibit an intense superhyperfine pattern. (Note that a plateau in the

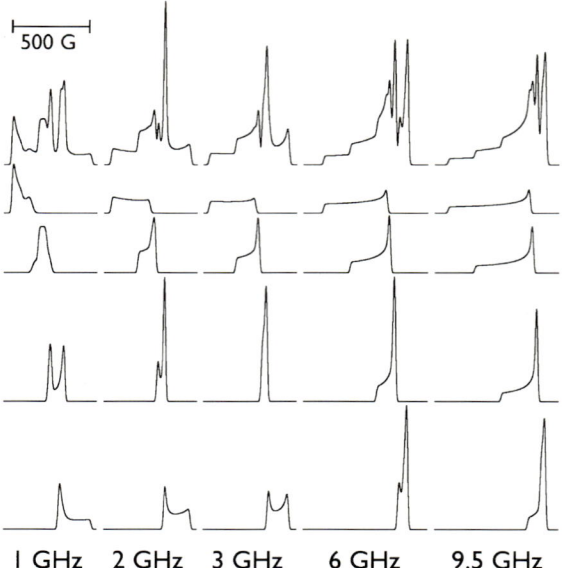

Figure 16.10 Model simulations with $H = \beta \mathbf{B}\cdot\mathbf{g}\cdot\mathbf{S} + \mathbf{S}\cdot\mathbf{A}\cdot\mathbf{I}(^{63}\text{Cu})$, $g_{x,y,z} = 2.079, 2.079, 2.298$, $|A_{x,y,z}(^{63}\text{Cu})| = 2.4, 2.4, 18.5 \times 10^{-3}\,\text{cm}^{-1}$. Frequencies are shown below each column. From top to bottom, rows correspond to: (i) whole absorption spectrum; (ii) $M_I = -3/2$ manifold; (iii) $M_I = -1/2$ manifold; (iv) $M_I = +1/2$ manifold; and (v) $M_I = +3/2$ manifold.

absorption spectrum corresponds to a region of essentially zero intensity in the experimentally obtained derivative spectrum.) In each case, the *low-field* side of that line does overlap somewhat with the $g_\perp$-line of the $M_I = +3/2$ manifold. At 3 GHz, the overlap of the $g_\perp$-lines of the $M_I = +1/2$ and the $M_I = +3/2$ lines is more extensive, and the lineshape of the $M_I = +3/2$ manifold itself is complex. At 6 GHz and above, both the $M_I = +1/2$ and the $M_I = +3/2$ manifolds exhibit intense overshoot lines, which overlap and complicate the $g_\perp$-line of the spectrum. In addition, these lines may not be well resolved at such high frequencies due to g-strain.

Based on this manifold analysis, one would therefore choose to carry out a determination of the number of nitrogens equatorially coordinated to Cu(II) by an analysis of the superhyperfine pattern on the high-field side of the $g_\perp$ line, at frequencies of either 1 or 2 GHz.

Figure 16.11 shows the calculated spectra for Cu(II)-histidine at 1, 2, and 9.5 GHz, assuming either 3N or 4N equatorial coordination. The spectra are shown normalized for the peak-to-trough intensity of the $g_\perp$-line. At 1 and 2 GHz, superhyperfine patterns on the $M_I = -1/2$ $g_\|$-line, which has traditionally been used for superhyperfine analysis, are almost indistinguishable for 3N and 4N. The three center lines for 3N and 4N have relative intensities, normalized for the intensity

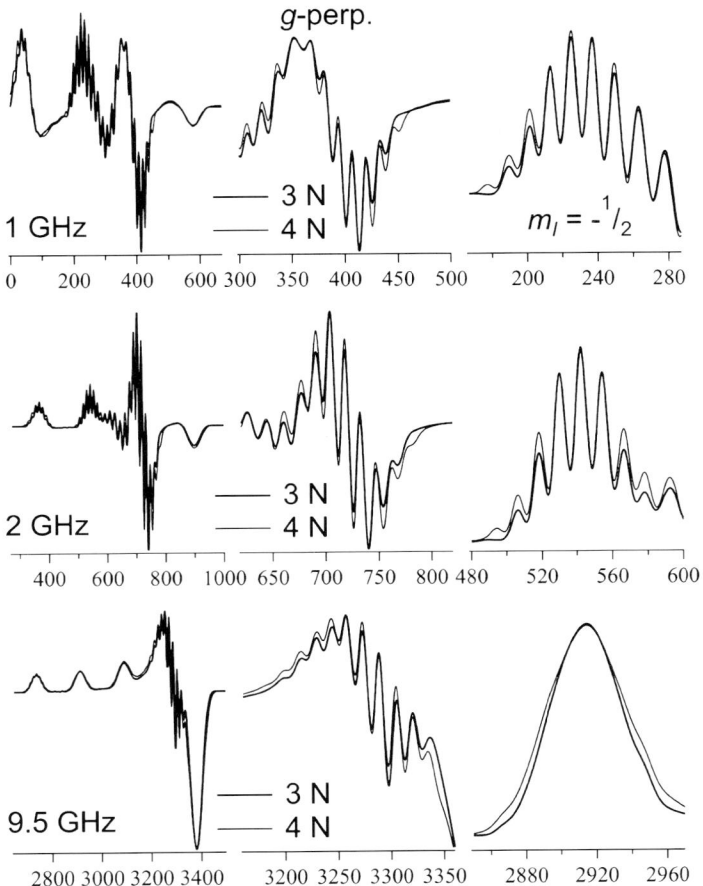

Figure 16.11 Simulations of Cu(II)-histidine assuming either three- or four-equatorial nitrogen coordination. Parameters were $g_{x,y,z} = 2.056, 2.056, 2.261$, $|A_{x,y,z}(^{63}Cu)| = 2.0$, 2.9, 18.5×10^{-3} cm^{-1}, $|A_{x,y,z}(^{14}N)| = 1.43, 1.43$, 1.27×10^{-3} cm^{-1}. g-strain $\sigma g/g = 0.0007$, 0.0007, 0.0005 and A-strain $\sigma A = -0.5, -0.5$, -0.8×10^{-3} cm^{-1} were included, along with intrinsic line widths 0.50, 0.50, 0.25×10^{-3} cm^{-1}. In each panel, spectra corresponding to 3N and 4N coordination are overlaid, normalized to the $g_\perp$ peak-to-trough line intensity.

of the center line of $M_I = -1/2\ g_\parallel$, that differ by only 2%. Greater differences appear on the low-field side of the pattern, but the absolute intensities are small. In practice, distinguishing between 3N and 4N often relies on being able to detect the outermost line, for 4N, or being able to convincingly demonstrate its absence for 3N. On the other hand, there are much clearer differences in intensities of the features on the high-field side of the $g_\perp$-line of the spectra at 1 and 2 GHz, and the fivefold increase in absolute intensity of $g_\perp$ superhyperfine features over those on the $M_I = -1/2$ $g_\parallel$-line at 2 GHz make observation of the outermost line more likely. Finally, as predicted from the manifold analysis, at 9.5 GHz the superhyperfine pattern, although well resolved for Cu(II)-histidine, does not distinguish well between 3N and 4N due to the complexity of the $g_\perp$ region at this frequency.

16.3.7
Sensitivity Analysis

"Sensitivity analysis" is a term defined as "... *the analysis by computer of the sensitivity of a simulated spectra* [sic] *to each of the input parameters in the program* (Hyde, Antholine, and Basosi, 1985)." The approach includes: (i) defining the quantity to fit; (ii) obtaining the best possible fit to that quantity; (iii) systematically varying a given input parameter; and (iv) determining the "sensitivity" to that parameter under the regime employed. Later incarnations included an extension of item (iv), in which the sensitivity of two parameters to each other was determined (Hyde et al., 1989).

Defining the quantity that will be fit is a subjective matter that depends somewhat on the parameters of interest and the system. Quantities used include the standard ($\partial\chi''/\partial B$) EPR spectrum, the absorption spectrum, the square of either the $\partial\chi''/\partial B$ or the $\partial^2\chi''/\partial B^2$ spectrum, and the modulus of either the $\partial\chi''/\partial B$ or the $\partial^2\chi''/\partial B^2$ spectrum (Figure 16.12).

Not surprisingly, Hyde et al. (1989) concluded that differences are found in sensitivity that depend on which of these quantities is used. A further variation is either that the whole spectrum can be analyzed, or that selected regions can be chosen, each of which may be analyzed using a different base spectrum of the forms shown in Figure 16.12. Perhaps in an attempt to stimulate an effort toward a general approach, it has been postulated that a strategy that simultaneously fits the absorption and one or more higher derivative spectra might be developed (Hyde et al., 1989).

Input parameters can include not only spin Hamiltonian parameters but also experimental parameters (e.g., frequency, modulation amplitude), as well as motional parameters in the liquid state, or the addition of a contribution from a second species. The correlation between two parameters can be obtained by simultaneous variation of each over a two-parameter space.

If it is assumed that a good fit can be accomplished to whichever spectral display is chosen, then some goodness of fit (GoF) parameter needs to be specified. GoF parameters have included a least-squares difference (Hyde, Antholine, and Basosi, 1985) and an integral over the difference (Hyde et al., 1989). It then becomes straightforward to determine the sensitivity of a spectrum, or part of a spectrum, to any given parameter. Performing a sensitivity analysis for spin Hamiltonian and strain parameters as a function of microwave frequency, and for specific regions of the spectrum, is a complementary analysis to the manifold analysis described above. Sensitivity analyses at various frequencies can also allow the determination of a more precise set of parameters by indicating at which frequency particular parameters are best obtained by computer simulation.

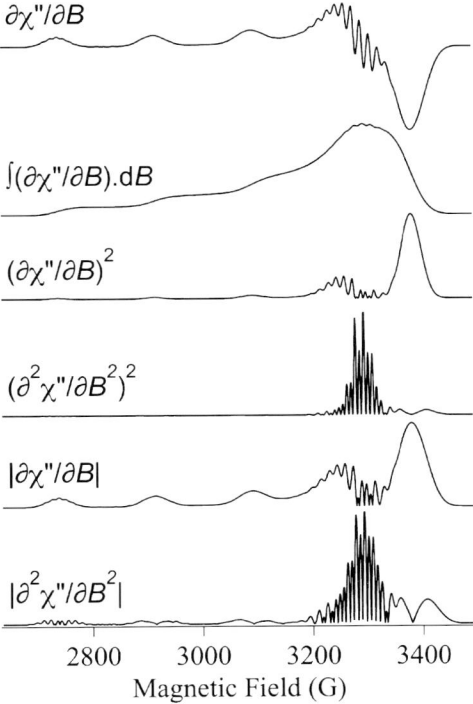

Figure 16.12 Some displays used for fitting in sensitivity analysis. From top to bottom: (i) The normal EPR spectrum; (ii) the absorption spectrum; (iii) the square of the EPR spectrum; (iv) the square of the second-derivative EPR spectrum; (v) the modulus of the EPR spectrum; and (vi) the modulus of the second-derivative EPR spectrum.

16.3.8
Global Fitting

The global fitting of multifrequency spectra is, in principle, desirable and has been successfully employed in certain systems, including the dipolar coupling analysis of nitroxide spin labels (Hustedt et al., 1997). The need for some form of global fitting of Cu(II) systems arises primarily from noncollinearity of **g** and **A**, and fits to L-, S-, and X-band spectra have been needed to characterize the Euler angles of noncoincidence in Cu(II) (Antholine, Hanna, and McMillin, 1993). An additional rationale for global fitting arises from the unfortunate coincidence that $A_\perp$ ($^{63/65}$Cu) can exhibit rhombicity such that A_x($^{63/65}$Cu) ≈ A(av.)(^{14}N) and A_y($^{63/65}$Cu) ≈ 2 × A(av.) (^{14}N) in square–planar-based Cu(II) (Rist and Hyde, 1970; Ammeter, Rist, and Günthard, 1972). This can lead to a frequency-dependent modulation of interference between the $^{63/65}$Cu and ^{14}N patterns, so that at some frequencies the spectra are very sensitive to $A_\perp$ ($^{63/65}$Cu), whereas at others they are not (Figure 16.13).

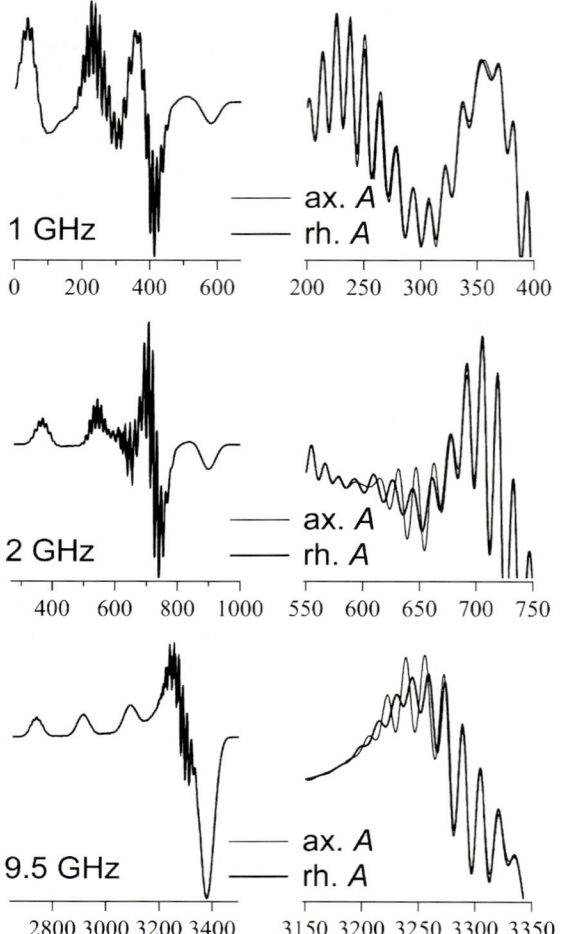

Figure 16.13 Frequency dependence of EPR spectra of Cu(II) to rhombicity in $A_\perp(^{63}Cu)$. In each panel, two computed spectra are overlaid. The parameters for the spectrum labeled "rh. A" were the same as those for Figure 16.11. Those for "ax. A" were identical except that $|A_{x,y}(^{63}Cu)| = 2.45, 2.45 \times 10^{-3}\,cm^{-1}$ instead of $|A_{x,y}(^{63}Cu)| = 2.0, 2.9 \times 10^{-3}\,cm^{-1}$.

It is perhaps initially surprising then that global fitting is not routinely carried out. A careful examination of the sensitivity analysis data for Cu(II), however, indicates that there could be problems with a global approach. For square–planar Cu(II) complexes in the rigid limit analyzed from 1 to 35 GHz, the optimum frequency for determination of the Zeeman parameters was 35 GHz, whereas for the hyperfine interaction, it was 1 GHz. This provides a challenge – namely, how are the fits to spectra at various frequencies to be weighted? The challenge becomes even greater when a superhyperfine structure is considered. In that case, the global fitting must successfully deal simultaneously with: (i) lines such as $g_\perp$ at 35 GHz, which have high relative intensity but a width of some hundreds of Gauss; and (ii) superhyperfine lines at 1 to 9 GHz, some of which have low intensity and widths of only a few Gauss, but are very important for understanding ligand coordination.

Hyde et al. (1989) proposed that such a strategy may involve iterating back and forth between 1 to 2 GHz and 35 GHz, varying the Zeeman parameters to fit the Q-band spectrum, and varying the hyperfine parameters to fit the L-band spectrum. The sensitivity analysis showed that X-band is the most sensitive frequency overall, but that it is not the most sensitive for any one parameter. The Zeeman and hyperfine values obtained by fits to Q- and L-band spectra, respectively, would be used to restrain these parameters in an overall fit at L-, S-, X-, and Q-band. This approach has been employed in the analysis (computer simulation) of multifrequency (S-, X-, and Q-band) EPR spectra of a series of low-symmetry (C_s) thiomolybdenyl and molybdenyl analogues (see Section 16.3.8.1). An extension of this approach, using a combination of Octave and the computational program Sophe to perform a global analysis of a multifrequency (S-, X-, and Q-band) EPR data sets (first- and second-derivative, six spectra in all), has been applied to a low-spin Co(II) complex (see Section 16.3.8.2).

16.3.8.1 Mo(V) Complexes

Multifrequency (S-, X-, Q-band) EPR spectra of a range of thiomolybdenyl and molybdenyl complexes ([Tp*Mo^{V}O(cat)] [Tp* = hydrotris(3,5-dimethylpyrazol-1-yl) borate; X = 2-(ethylthio)phenolate, 2-(n-propyl)phenolate, phenolate; X$_2$ = benzene-1,2-dithiolate, 4-methylbenzene-1,2-dithiolate and benzene-1,2-diolate (cat.)) as models for the active sites in mononuclear molybdenum-containing enzymes were acquired at 130 K and 295 K (Figure 16.14) (Drew et al., 2007). A computer simulation of the spectra yielded a spin Hamiltonian of C_s symmetry or lower, with $g_{zz} < g_{yy} < g_{xx} < g_e$ and $A_{z'z'} > A_{x'x'} \approx A_{y'y'}$, and a non-coincidence angle in the range of β = 24–39° (Table 16.1). The S- and Q-band spectra, were found to be particularly valuable in the unambiguous assignment of spin Hamiltonian parameters for these low-symmetry sites. While Q-band measurements provided greater g-value resolution, larger state mixing and reduced **g**- and **A**-strain at low frequencies (S-band) allowed an accurate determination of the hyperfine matrix, including its orientation with respect to the g matrix. This single set of **g** and **A** parameters (Table 16.1) were then used to simulate the S-, X-, and Q-band spectra. The weaker π-donor terminal sulfido ligand yields a smaller HOMO–LUMO gap and reduced g-values for the thiomolybdenyl complexes compared with molybdenyl analogues (Table 16.1).

Crystal field description of spin Hamiltonian parameters (Drew et al., 2007): Large non-coincidence angles can be explained by a model in which extensive mixing among Mo 4d orbitals takes place. Although ligand-to-metal charge transfer and metal-to-ligand charge transfer states of appropriate symmetry may also contribute, for transition metals the dominant contribution to g_{ij} is usually Δg_{ij}^{d-d}, which arises from transitions within the Mo 4d manifold. In C_s symmetry with a $\sigma^{(xz)}$

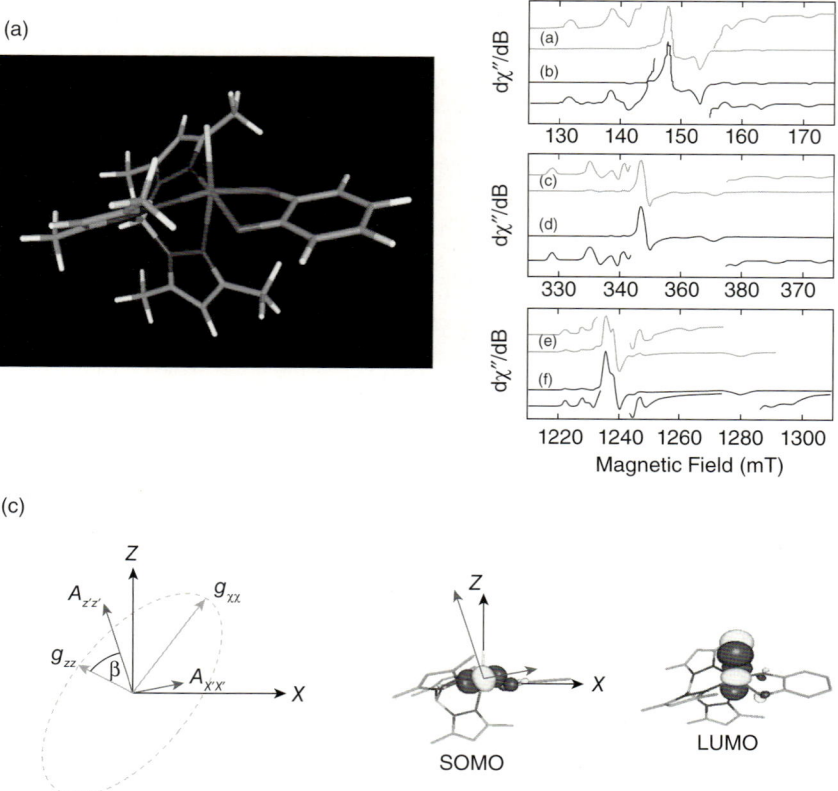

Figure 16.14 Geometric and electronic structural characterization of [Tp*Mo(V)S(cat)]. (a) X-ray crystal structure; (b) Multifrequency EPR studies; (c) DFT calculations showing the orientation of the g and A principal axes, the SOMO and LUMO orbitals (Drew et al., 2007; Drew, Young, and Hanson, 2007).

mirror plane, in which the X axis lies between the metal–ligand bonds, the $d_{X^2-Y^2}$, d_{XZ} and d_{Z^2} orbitals transform as A' and the d_{XY} and d_{YZ} orbitals transform as A''. The metal based anti-bonding wavefunctions are therefore:

$$\psi^{a'*}_{X^2-Y^2} = \alpha[a_1 d_{X^2-Y^2} + b_1 d_{XZ} + c_1 d_{Z^2}]$$
$$\psi^{a'*}_{XZ} = \beta[a_2 d_{XZ} + b_2 d_{X^2-Y^2} + c_2 d_{Z^2}]$$
$$\psi^{a'*}_{Z^2} = \gamma[a_3 d_{Z^2} + b_3 d_{X^2-Y^2} + c_3 d_{XZ}]$$
$$\psi^{a''*}_{XY} = \delta[a_4 d_{XY} + b_4 d_{YZ}]$$
$$\psi^{a''*}_{YZ} = \varepsilon[a_5 d_{YZ} + b_5 d_{XY}] \tag{16.7}$$

where, by definition, $a_i > b_i, c_i$ (i = 1, 2, ...). Here, covalency appears only implicitly through the metal-centered orbital coefficients $\alpha, \ldots, \varepsilon$. Since the molecular X and

Table 16.1 Spin Hamiltonian parameters determined from computer simulation of the multifrequency EPR spectra shown in Figure 16.14b and from DFT calculations (Drew et al., 2007; Drew, Young, and Hanson, 2007).

Method	g_{xx}	g_{yy}	g_{zz}	A_{xx}	A_{yy}	A_{zz}	α	β	γ
Tp* MoV O (cat)									
BP86	1.9867	1.9796	1.9402	9.2	9.5	40.5	10	43	44
BP86+A^{SO}				11.7	11.1	46.8	1	39	−10
B3LYP	1.9790	1.9725	1.9276	16.1	16.4	50.2	10	42	34
B3LYP+A^{SO}				19.7	18.6	57.4	0	37	−2
Experiment	1.9680	1.9660	1.9194	27.0	26.0	64.2	0	36	0
Tp* MoV S (cat)									
BP86	1.9796	1.9757	1.9042	13.3	12.7	42.6	0	36	10
BP86+A^{SO}				14.8	16.6	49.6	2	34	6
B3LYP	1.9698	1.9651	1.8874	20.5	19.6	51.8	1	36	9
B3LYP+A^{SO}				25.0	22.6	59.9	2	34	6
Experiment	1.9646	1.9595	1.8970	30.0	29.0	67.5	0	34.5	0

Y axes are placed between the metal–ligand bonds, the ground state wavefunction is $\Psi_{x^2-y^2}$.

DFT calculations (Drew, Young, and Hanson, 2007): The electronic g matrix and ^{95,97}Mo hyperfine matrix were calculated as second-order response properties from the coupled-perturbed Kohn–Sham equations. The scalar relativistic zero-order regular approximation (ZORA) was used with an all-electron basis and an accurate mean-field spin–orbit operator which included all one- and two-electron terms. A comparison of the principal values and relative orientations of the g and A interaction matrices (Table 16.1), obtained from unrestricted Kohn–Sham calculations at the BP86 and B3LYP level with the values obtained from experimental spectra, shows an excellent agreement at the B3LYP level using ORCA (Neese, 2009). A quasi-restricted approach has been used to analyze the contributions of the various molecular orbitals to g and A. In all complexes, the ground state magnetic orbital is $d_{x^2-y^2}$-based, and the orientation of the A matrix is directly related to the orientation of this orbital through admixture of the d_{xz} orbital (Figure 16.14c). The largest single contribution to the orientation of the g matrix arises from the spin–orbit coupling of the d_{yz}-based LUMO into the ground state (Figure 16.14c). A number of smaller, cumulative charge-transfer contributions augment the d–d contributions.

16.3.8.2 Low-Spin Co(II) Crossover Complexes

A computer simulation of the multifrequency (S-, X-, and Q-band) EPR spectra (Figure 16.15) of the spin crossover complex [Co(terpyRX)$_2$] (BPh$_4$)$_2$·H$_2$O (terpyRX = 4′-alkoxy-2,2′:6′,2″-terpyridine, X = 8) employed the XSophe-Sophe-XeprView computer simulation software suite (Hanson et al., 2004), in conjunc-

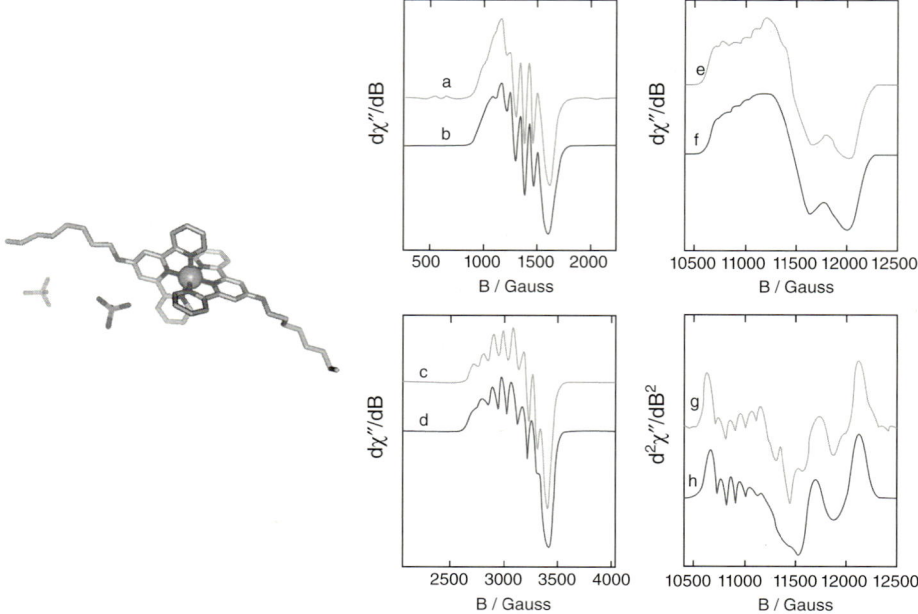

Figure 16.15 Left panel: Multifrequency EPR spectra of [Co(terpyRX)$_2$] (BPh$_4$)$_2$·H$_2$O, measured in chloroform. Spectra identified as: (a) experimental and (b) simulated S-band spectra, (ν = 4.038 GHz), T = 15.0 K; (c) experimental and (d) simulated X-band spectra (ν = 9.375308 GHz), T = 25.0 K; (e) experimental and (f) simulated first-derivative and (g) experimental and (h) simulated second-derivative Q-band spectra (ν = 34.082 GHz, T = 6.07 K). Simulated spectra computed with the spin Hamiltonian and line width parameters given in the text.

tion with Octave, to optimize the spin Hamiltonian and linewidth parameters by fitting the S-, X-, and Q-band (first- and second-derivative) spectra simultaneously (Nielsen et al., 2009). Through selective weighting of the four spectra, it was possible to search the spin Hamiltonian and linewidth parameter space and to show that non-coincident angles between the principal components of g- and A- were not required to adequately simulate the EPR spectra. In obtaining the unique set of orthorhombic spin Hamiltonian and line width parameters (g_i 2.206, 2.141, 2.033; A_i 91.28, 60.30, 36.42 × 10^{-4} cm^{-1}; σ_{Ri} 44.6, 26.7, 33.5; $\sigma g_i/g_i$ 0.0029, 0.1042, 0.0069; σA_i − 14.6, 22.32, 3.27 (i = x, y, z)), it was important to fit not only the first-derivative Q-band spectrum, but also the second-derivative spectrum, as this provided a better resolution of the Co(II) hyperfine structure. The g- and A-strain linewidth model (Equation 16.5) developed by Hyde and Froncisz was employed in the simulations. The magnitudes of the g-values and the largest hyperfine interaction observed here are similar to those found by Kremer, Henke, and Reinen (1982) for single crystals of the parent low-spin [Co(terpy)$_2$](Y)$_2$·nH$_2$O complexes, although the other two hyperfine interactions appear to be quite different. The labeling of the principal components of **g** and **A** matrices from the frozen solution EPR spectra follows the convention for an elongated tetragonal Co(II) system, with the z-axis being that of the smallest g-value and the longest bond length, directed along the axis of elongation.

16.3.8.3 Future Developments

The "holy grail" of analysis of Cu(II) spectra is the automatic extraction of structural information from global simulation of multiple EPR datasets linked to computational analysis (DFT, etc.). One complication with global fitting to Cu(II) EPR spectra is that, as with sensitivity analysis, no one spectral display suggests itself as being the ideal one to fit in all cases. A final complication arises from the variation in SNR over frequency. The realized SNRs at different frequencies will depend on factors that include: (i) whether the sample is limited in quantity or concentration; (ii) whether it is saturable or nonsaturable; (iii) what type of resonator is employed (LGR, resonant cavity, etc.); and (iv) whether the resonator can practicably be scaled to the inverse of the frequency (Rinard et al., 2004). A weighting strategy that took the frequency-dependent SNR into account would need to be evaluated and implemented on a case-by-case basis.

Although these issues have not been – and indeed, never may be – rigorously and generally treated, the above studies and, more recently, the global optimization of EPR data sets involving dipole–dipole coupled EPR spectra of dinuclear Cu(II) cyclic peptide complexes, is providing insights into how to weight particular spectra and parameters to refine a global set of spin Hamiltonian parameters from multiple EPR experiments and computational chemistry calculations. The incorporation of these approaches into both Molecular Sophe (Hanson, Noble, and Benson, 2009) and iResonanz (G. Hanson, C. Noble, and S. Benson, private communication) which have been specifically designed to incorporate a global analysis of spin Hamiltonian parameters (Equations 16.1 and 16.2) from multiple experiments (see Figure 16.7) will enable the geometric and electronic structures of metal centers in metalloproteins and other paramagnetic centers to be elucidated more readily.

16.3.9
Heterogeneity

An additional application of multifrequency EPR and simulations is in detecting and characterizing spectra that arise from mixtures of species. This is a common problem in biological systems where multiple Cu(II)-binding modes may exist, such as in the prion protein (Chattopadhyay et al., 2005). Simulations of Cu(II) species using parameters for 4N (Basosi et al., 1986; Hyde et al., 2009), 2N2O, and 4O (Peisach and Blumberg, 1974) coordination are shown for 1 to 35 GHz in Figure 16.16.

A comparison of the 4N species with a 1:1 mixture of 4N:2N2O (Figure 16.16, right panels) clearly shows how different regions of the spectrum are sensitive to the presence of more than one species at different frequencies. At 1 GHz, the distortion of the $M_I = -3/2$ line at low field is a clear indication of more than one

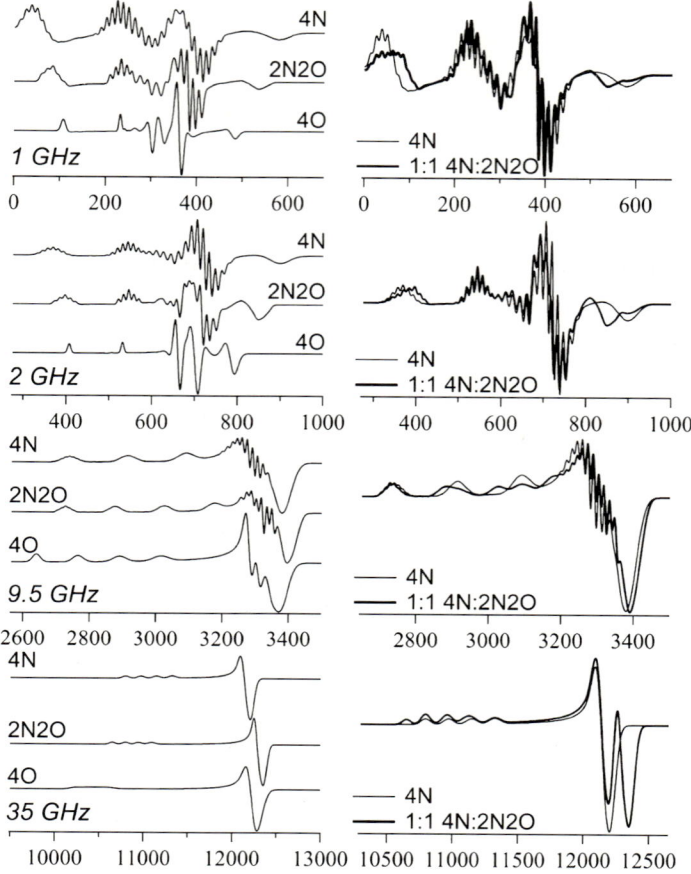

Figure 16.16 Computed spectra assuming single species with either 4N, 2N2O, or 4O equatorial coordination (left), and overlaid spectra of a 4N single species and a 1:1 2N2O mixture (right). The parameters for the 4N spectrum were the same as those for Figure 16.11. The parameters for the 2N2O spectrum were as for 4N, except #^{14}N = 2, $g_{\|}$ = 2.30, $g_\perp$ = 2.03, and $A_{\|}(^{63}\text{Cu})$ = 16.0 × 10^{-3} cm^{-1}. The parameters for the 4O spectrum were as for 4N, except #^{14}N = 0, $g_{\|}$ = 2.40, $g_\perp$ = 2.04, and $A_{\|}(^{63}\text{Cu})$ = 14.0 × 10^{-3} cm^{-1}. 1:1 mixtures were calculated based on the doubly integrated intensities of the individual components.

species. At both 1 and 2 GHz, there is also a noticeable asymmetry in the $M_I = \pm 3/2$ $g_{\|}$-lines. Each of these $M_I = \pm 3/2$ $g_{\|}$-lines lies at an outer edge of the spectral envelope, are uncluttered by overlap with other lines, and therefore are very sensitive markers for multiple species. Interestingly, the intensity of the central lines of the $g_\perp$ feature for the 4N:2N2O mixture are significantly lower than for 4N alone, and the inability to reproduce this phenomenon by simulation assuming a single species may be diagnostic of a mixture (Hyde et al., 2009). The spectrum at 35 GHz is also very sensitive to multiple species, as long as the g-strain is insufficient to render the spectral features unresolved. Multiple species can be readily detected by splitting, or asymmetric broadening, of the $g_\perp$ feature and by the presence of EPR intensity to the low field of the expected $g_\parallel$ pattern. It can be seen from Figure 16.16 that it is relatively straightforward to estimate $A_\parallel(Cu)$, $g_\parallel$, and $g_\perp$ for the two species, 2N2O and 4N, from the 1, 2, and 35 GHz spectra. At X-band, however, the lowest- and highest-field features for 4N and 2N2O are superimposed. While it may be possible to infer a mixture of species from the $A_\parallel(Cu)$ pattern, the extraction of reliable spin Hamiltonian parameters would be very difficult.

16.4
Copper-Coordination Environments: Multifrequency EPR of Three-Coordinate Copper and Mixed-Valence Dinuclear Copper [Cu(1.5$^+$) ... Cu(1.5$^+$)]

16.4.1
Introduction: Spectrum and Structure

16.4.1.1 X-Band EPR Spectrum for Mononuclear, Light Blue Cu^{2+}

This review primarily covers the past seven years of research related to the low-frequency EPR of copper complexes (Hyde and Froncisz, 1982; Basosi, Antholine, and Hyde, 1993; Antholine, 2005). There is some overlap for purposes of introducing copper compounds and placing some of the better spectra in a single review, but the intent is to review low-frequency EPR studies completed subsequent to the material discussed in previous reviews. Spectra for Cu^{2+} in acetylacetonate (Figure 16.17) (Vänngård, 1972) illustrate how typical EPR parameters are determined from a typical EPR spectrum obtained at the conventional 9 GHz frequency for a typical square–planar, Type 2 cupric complex. In Figure 16.17, the EPR parameters $g_\parallel$, $g_\perp$, g_{iso}, $A_\parallel$, $A_\perp$, and A_{iso} are labeled. The dashed lines indicate extra absorption (EA) lines, which are sometimes referred to as "hyperfine anomaly lines" or "overshoot lines." EA lines arise because numerous orientations have such values of EPR parameters that their lines are superimposed. Spectra for Type 2, or light-blue, copper complexes and EA lines have already been discussed in detail in Section 16.3.

16.4.1.2 Peisach–Blumberg-Like Table (EPR Parameters Assembled by the Author)

Table 16.2 lists compounds and EPR parameters $g_\parallel$ and $A_\parallel$ as a function of donor atoms. This table, which has been used in a previous chapter (Antholine, 2005), has been updated to include the parameters from spectra obtained since the last review. Plots of $g_\parallel$ against $A_\parallel$ are known as Peisach–Blumberg plots. For Type 2 copper complexes (see Table 16.2), the $g_\parallel$ value increases from about 2.09 when donor atoms for the cupric complex are all sulfurs (4S) to about 2.14 when the donor atoms are two sulfurs and two nitrogens (2S2N), to about 2.24 for four nitrogens (4N), and to about 2.40 for four oxygen donor atoms (4O). $A_\parallel$ increases when a nitrogen donor atom replaces a sulfur donor atom, and decreases when

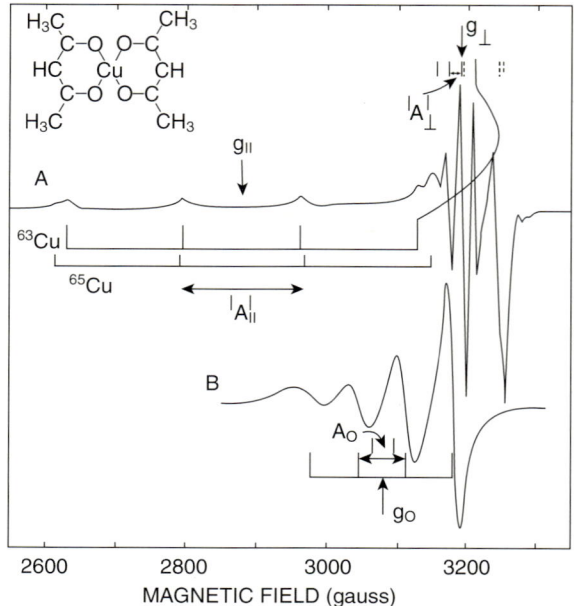

Figure 16.17 EPR spectra of a solution of a Cu^{2+} complex (A) frozen at 77 K and (B) at room temperature. The g factors and hyperfine constants are measured as described in the text, and their values are $g_{\parallel} = 2.285$, $g_{\perp} = 2.060$, $g_0 = 2.135$, $A_{\parallel} = 167\,G$, $A_{\perp} = 17\,G$, and $A_0 = 67\,G$. In (A), the dashed lines indicate the position of the overshoot lines, which occur from the particular angular dependence of the copper hyperfine lines, as shown for one line—the full curved line to the right. The small peaks at the highest field in (A) arise from a second complex existing in low concentration. The sample is 5 mM in Cu^{2+} and 40 mM in acetylacetonate ($CH_3COCH_2COCH_3$) in a 1:1 water:dioxane solvent at alkaline pH. The same tube was used for both spectra with a fivefold higher spectrometer gain in (B). Other settings were microwave power, 2 mW; frequency, 9.2 GHz; field modulation, 7 G; time constant, 0.3 s; and sweep time, 4 min (Vänngård, 1972). Reprinted with permission from T. Vänngård, in *Biological Applications of Electron Spin Resonance* (H.M. Swartz, J.R. Bolton, D.C. Borg, eds), pp. 411–447; © 1972, John Wiley & Sons, Inc.

an oxygen donor atom replaces a nitrogen donor atom. It is sometimes difficult to distinguish between two nitrogen donor atoms, a sulfur donor atom plus an oxygen donor atom (S2NO, $g_{\parallel} = 2.21$, $A_{\parallel} = 185\,G$; see Table 16.2), and four nitrogen donor atoms (4N, $g_{\parallel} = 2.21$, $A_{\parallel} = 175\,G$), but $g_{\parallel}$ and $A_{\parallel}$ values for copper will give an initial reading for the donor atoms.

16.4.1.3 Type 1 (Blue) Copper Centers, Three-Coordinate Cu

Type 1 complexes are three-coordinate, using two nitrogens from the imidazoles of two histidines and one sulfur from cysteine amino acids in the protein to form a trigonal plane (Figure 16.18). An axial ligand that distorts the trigonal geometry is usually from a methione. The color of samples with a Type 1 Cu center is dark

Table 16.2 Parameters for copper complexes.

	$g_\|$	$A_\|(G)$	Donor atoms
Type 2 Complexes (light blue)			
Cu(dtc)$_2$	2.09	130	S-S-S-S
CuKTS	2.14	187	S-S-N-N
CuPTS+GSH	2.14	180	S-S-N-N
CuPTS+ascites cells	2.13	178	S-S-N-N
Cu(PTS)(His)	2.19	169	S-N-N-N
CuPTS+en	2.19	169	S-N-N-N
CuPTS	2.20	186	S-N-N-O
Cu(PTS)(OH$_2$)	2.21	185	S-N-N-O
CuBLM	2.21	175	N-N-N-N
CuTPP	2.18	176	N-N-N-N
Cu-serum albumin	2.18	213	N-N-N-N
Cu-camosine	2.20	190	N-N-N-N
Cu+excess histidine	2.22	183	N-N-N-N
Cu-pMMO	2.24	185	N-N-N-N
Cu^{2+}ascites cells	2.25	180	N-N-N-N (possibly)
Cu-prion (component 3)	2.25	183	N-N-N-N
Cu-prion (component 1)	2.24	157	N-N-N-O
Cu-prion (component 2)	2.27	167	N-N-O-O
CuEDTA(pH 6)	2.29	155	N-N-O-O
Cu^{2+} in excess acetylacetonate	2.28	167	O-O-O-O
Cu(OH)$_4$	2.40	113	O-O-O-O
CuSO$_4$ in DMF	2.40	128	O-O-O-O
Type 1 complexes (dark blue)			
Type 2-depleted fungal laccase	2.19	90	N-N-S
Laccase	2.30	40	N-N-S
Azurin	2.25	57	N-N-S
Cu amicyanin	2.24	53	N-N-S
Cu M98Q amicyanin	2.27	~23	N-N-S

Table 16.2 (Continued)

	$g_\parallel$	$A_\parallel(G)$	Donor atoms
Mixed-valence Cu-Cu (purple)			
Cu_A-N_2OR	2.18	38	N-Cu-S_2-Cu-N
Cu_A-CcO (bovine)	2.18	38	N-Cu-S_2-Cu-N
Cu_A-CcO (R. sphaeroides)	2.19	30	N-Cu-S_2-Cu-N
Cu_A (engineered in P. aeruginosa azurin)	2.18	56	N-Cu-S_2-Cu-N
Cu_2L^{3+}	2.02($g_\parallel < g_\perp$)	~20	N_4-Cu-Cu-N_4
$\{Cu(SNS)\}_2^+$	2.00($g_\parallel < g_\perp$)		S_2-Cu-N_2-Cu-S_2
$\{Cu(PNP)\}_2^+$	2.00($g_\parallel < g_\perp$)		P_2-Cu-N_2-Cu-P_2
$\{Cu(PPP)\}_2^+$	2.00(g_{iso})		P_2-Cu-P_2-Cu-P_2
Cu clusters			
Cu_z: N_2OR:Cu_4(S)	2.16	60 and 24 (two Cu couplings)	
$[Cu_3(\mu S)_2]^{3+}$		$S = 1$ state, g_{iso} 2.05 from room temperature	
Native intermediate laccase $[Cu_3]$		Broad signal with rhombic g-values ~2.15, 1.86, 1.65	

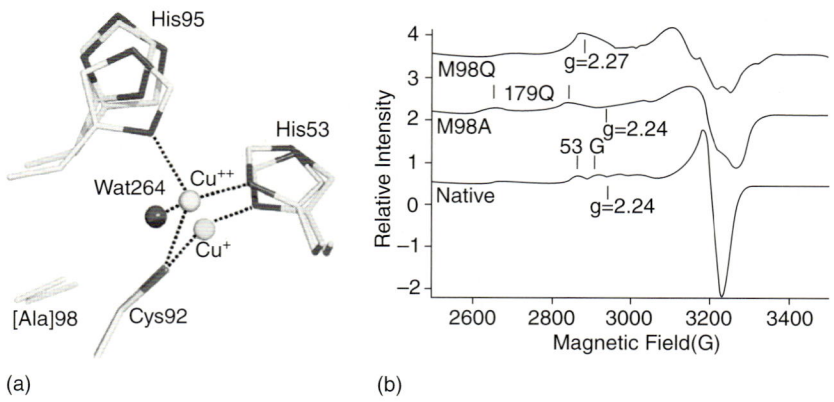

Figure 16.18 (a) Structure of the cupric and cuprous site for mutant M98A; (b) X-band EPR spectra for native, M98A, and M98Q mutants (Carrell et al., 2007). Reprinted with permission from C.J. Carrell, J.K. Ma, W.E. Antholine, J.P. Hosler, F.S. Mathews, V.L. Davidson, *Biochemistry*, **46** (7), 1900–1912; © 2007, American Chemical Society.

blue compared to light blue for square–planar, Type 2 copper complexes. Extensive literature is available for Type 1 Cu centers (Adman et al., 1978; Nar et al., 1991; Werst, Davoust, and Hoffman, 1991; and references therein). The present authors have collaborated on only a few studies on the multifrequency EPR of Type 1 Cu (Table 16.2); the latest study concerns spectra from mutation of the axial ligand of amicyanin (Table 16.2) (Carrell et al., 2007). The crystal structure of oxidized and reduced M98A amicyanin is attributed to a Type 1 center, but with the axial ligand provided by water (Figure 16.18, left). Upon freezing, the blue color is lost, and the EPR spectrum is assigned to a Type 2 copper center (Figure 16.18, right). The structure of reduced M98A amicyanin is broken, and the coordination is almost linear (Figure 16.18, left). The replacement of Met98 in amicyanin with glutamine yields a Type 1 copper site with increased rhombicity, and an increase in the distance from Cu to the trigonal equatorial plane (Carrell et al., 2007). Complicating the EPR method, a signal from a weakly bound Type 2 Cu is superimposed under the Type 1 signal. When the Type 2 signal has been removed by washing with EDTA, the spectrum is similar to the spectrum from stellacyanin, which naturally possesses a glutamine in the axial position.

A single isotope of copper, ^{63}Cu, was used in order to decipher g- and A-values from the Q- (35 GHz) and S-band (3.4 GHz) spectra for ^{63}Cu M98Q (Figure 16.19). Taking the first harmonic emphasized sharp lines (Figure 16.20).

16.4.2
EPR for New Three-Coordinate Copper Complexes

16.4.2.1 Three-Coordinate CuL(SCPh₃) and Copper(II)Phenolate Complexes

The structure and EPR signal of CuL(SCPh$_3$) ($g_\parallel$ = 2.17, $A_\parallel$ = 111 × 10^{-4} cm^{-1}) are similar to Type 1 blue copper signals, with $A_\parallel$ values of ~90 × 10^{-4} cm^{-1} (Figures 16.21 and 16.22) (Holland and Tolman, 1999). This report was the first example of a synthetic Cu(II) Type 1 complex for which the EPR parameters mimic the parameters for fungal laccase and ceruloplasmin Type 1 sites. Continuing this work, Cu^{2+}-phenolate compounds represent a new class of three-coordinate Cu^{2+} compounds (Jazdzewski et al., 2001). The X-band EPR spectrum for Cu^{2+}LCliPrCl is an axial signal with $g_\parallel$ = 2.20, $g_\perp$ = 2.05, and $A_\parallel$ = 129 G. The spectrum for Cu^{2+}Lipr(OC$_6$H$_4$OMe) is more rhombic, with a reduction in $A_\parallel$ (g_z = 2.22, g_y = 2.07, g_x = 2.02, $A_\parallel$ = 90 G).

16.4.2.2 CuPPN, Three-Coordinate Copper Amido and Aminyl Complexes (More Like a Free Radical)

The unpaired electron in a recently reported Cu^{2+}amide complex, through delocalization of the unpaired electron, is described as a Cu^{1+}-aminyl radical (see the structure shown in Figure 16.23) (Mankad et al., 2009a). The complex, [Ph$_2$BP$_2^{tBu}$]Cu(NTol$_2$), is referred to as CuPPN. The EPR parameters $g_\parallel$ and $A_\parallel$ are not obvious as in previous spectra for Type 1 Cu sites. The S- and X-band spectra for CuPPN are dominated by six lines each, with several shoulders (Figure 16.24). To emphasize the sharp features and shoulders on the lines in the Q-, X-, and

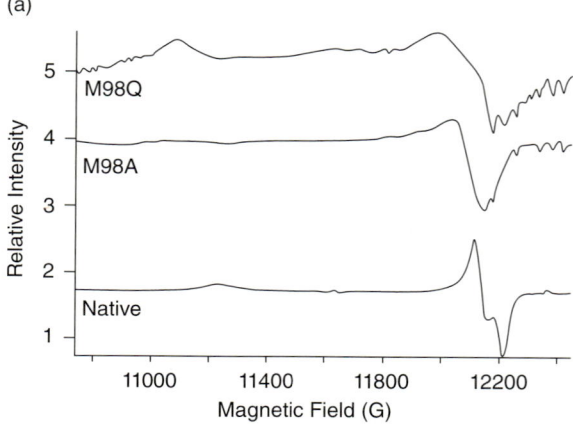

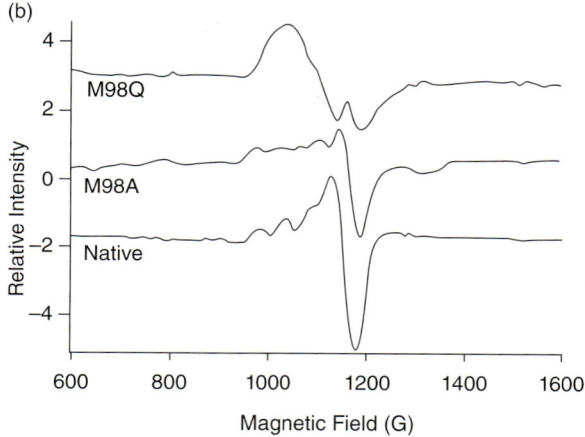

Figure 16.19 (a) Q- (50 K) and (b) S-band (133 K) EPR spectra of oxidized native amicyanin (200 μM), M98A amicyanin (200 μM), and M98Q amicyanin (200 μM) in 10 mM potassium phosphate, pH 7.4, containing 5% glycerol (Carrell et al., 2007). Reprinted with permission from C.J. Carrell, J.K. Ma, W.E. Antholine, J.P. Hosler, F.S. Mathews, V.L. Davidson, *Biochemistry*, **46** (7), 1900–1912; © 2007, American Chemical Society.

S-band spectra, the first harmonic for spectra in Figure 16.24 was taken (Figure 16.25). It is difficult to assign EPR parameters without these spectra at different frequencies, but therein lies the virtue of multifrequency EPR.

16.4.2.3 Simulation of Spectra for CuPPN (Quenched EPR Parameters Expected for a Radical)

Assuming the outer lines are hill-shaped in Figure 16.24, assigning these lines as pure copper lines leads to unsatisfactory simulations (see Figure 16.26). If A^P for

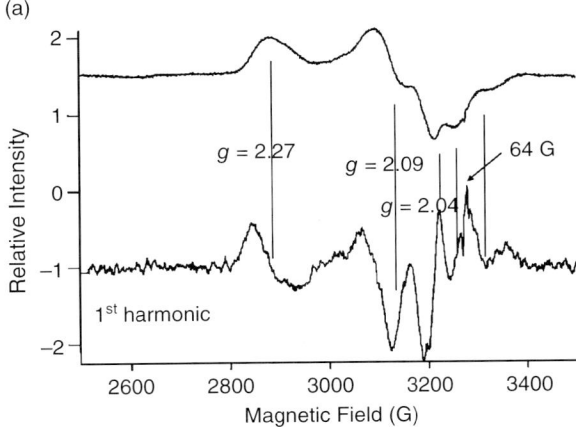

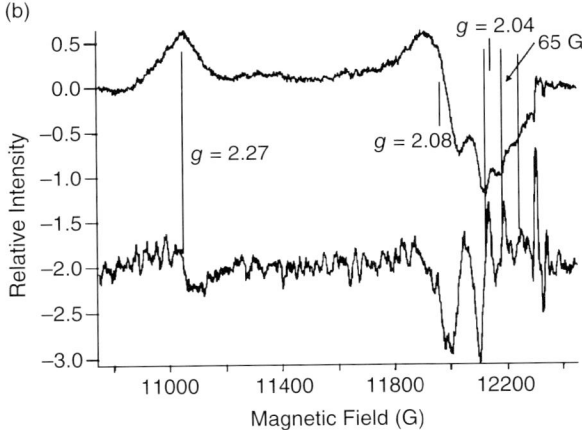

Figure 16.20 (a) X- (77 K) and (b) Q-band (50 K) EPR spectra of ^{63}Cu M98Q amicyanin (270 μM) in 10 mM potassium phosphate, pH 7.4, containing 5% glycerol. The values for g_z, g_y, and g_x and the hyperfine splittings for A_x^{Cu} are marked in both the X- and Q-band spectra to show the precision of the values from the two frequencies. The spike on the high-field side of the Q-band spectra is due to a background signal. The first harmonic of (a) X- and (b) Q-band spectra emphasizes the hyperfine lines (Carrell et al., 2007). Reprinted with permission from C.J. Carrell, J.K. Ma, W.E. Antholine, J.P. Hosler, F.S. Mathews, V.L. Davidson, *Biochemistry*, **46** (7), 1900–1912; © 2007, American Chemical Society.

2Ps is equal to A^{Cu}, then a six-line pattern, 1:3:4:4:3:1, dominates the spectra (Figures 16.24–16.26). To include the possibility that there is substantial electron density on the nitrogen, simulations—in which all other parameters are held constant and A_y^N is varied—put shoulders on the six lines, as seen on the experimental spectra (Figure 16.27).

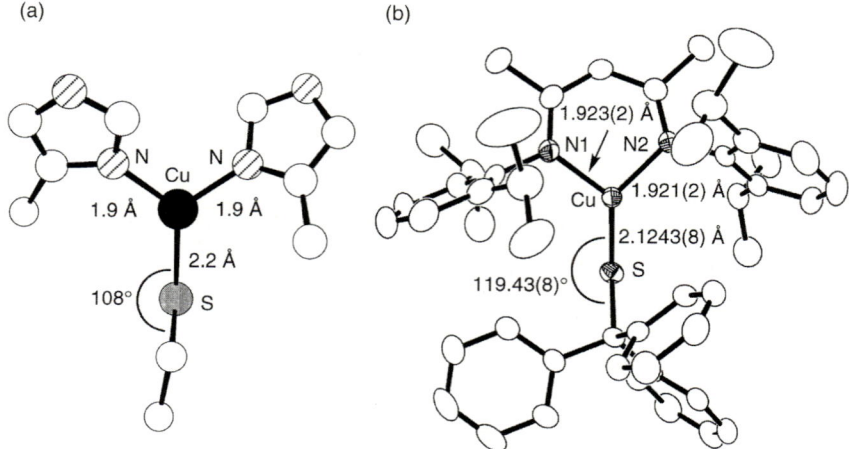

Figure 16.21 (a) Type 1 copper site in fungal laccase as determined by X-ray crystallography; (b) Thermal ellipsoid (50%) diagram of the X-ray structure of CuL(SCPh$_3$) (hydrogen atoms omitted) (Holland and Tolman, 1999). Reprinted with permission from P.L. Holland, W.B. Tolman, *Journal of the American Chemical Society*, **121** (31), 7270–7271; © 1999, American Chemical Society.

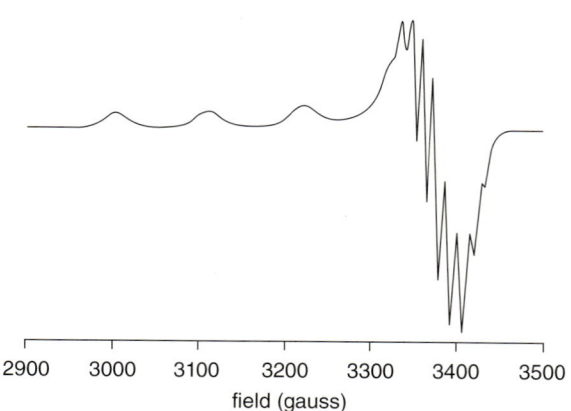

Figure 16.22 X-band EPR spectrum of CuL(SCPh$_3$) in toluene, where L$^-$ is 2,4-bis(2,6-diisopropylphenylimido) pentane (9.61 GHz, 20 K): $g_\parallel = 2.17$, $A_\parallel = 111 \times 10^{-4}$ cm^{-1}, $g_\perp = 2.04$, $A_\perp = 13 \times 10^{-4}$ cm^{-1} (Holland and Tolman, 1999). Reprinted with permission from P.L. Holland, W.B. Tolman, *Journal of the American Chemical Society*, **121** (31), 7270–7271; © 1999, American Chemical Society.

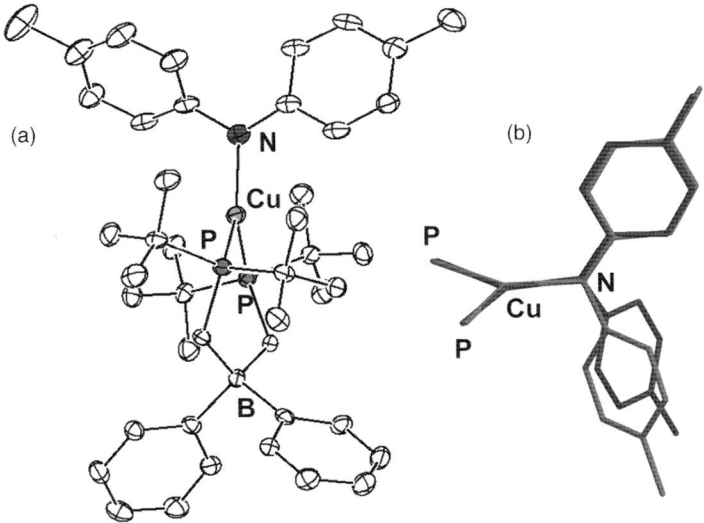

Figure 16.23 (a) Solid-state structure of [Ph$_2$BP$_2^{tBu}$]Cu(NTol$_2$), CuPPN; (b) Structural overlay of anionic Cu^{1+} amido complex {[Ph$_2$BP$_2^{tBu}$]Cu(NTol$_2$)}{Li(12-crown$_2^{-4}$)}, where {[Ph$_2$BP$_2^{tBu}$] = Ph$_2$B(CH$_2$P$_2^{tBu}$)$_2$, Tol = tolyl, and Cu^{2+}PPN (blue), with only phosphorus atoms of [Ph$_2$BP$_2^{tBu}$] shown (Mankad et al., 2009a). Reprinted with permission from N.P. Mankad, W.E. Antholine, R.K. Szilagyi, J.C. Peters, *Journal of the American Chemical Society*, 36 (11), 3878–3880; © 2009, American Chemical Society.

16.4.2.4 EPR Parameters for CuPPN (Unpaired Electron Density Delocalized as Expected for a Radical)

It is difficult to obtain EPR parameters for CuPPN. In this case, multifrequency spectra and simulations give a set of parameters that fit the extremes of the multifrequency spectra and the shoulders on the most intense lines. The parameters are $g_{x,y} = 2.008$, $g_z = 2.030$; $A_{x,y}^{Cu} = 12$ G, $A_z^{Cu} = 61$ G; $A_{x,y}^{P} = 53$ G, $A_z^{P} = 62$ G; $A_{x,z}^{N} = 8.6$ G, $A_y^{N} = 36$ G. Parameters from these simulations are consistent with–but are not proof of–the actual parameters. The intensities in the spectra should also fit better than at present, while the numbers of variables are overdetermined, leading to uncertainties. Nevertheless, these multifrequency spectra and simulations provide "ballpark" EPR parameters. Certainly, these parameters are better than could be determined only from the X-band spectrum. Together with supporting data from non-EPR methods, it is concluded that EPR spectra–which are very unusual for a Cu^{2+} complex–are representative of a highly delocalized electron approaching a free-radical-like character (Mankad et al., 2009a).

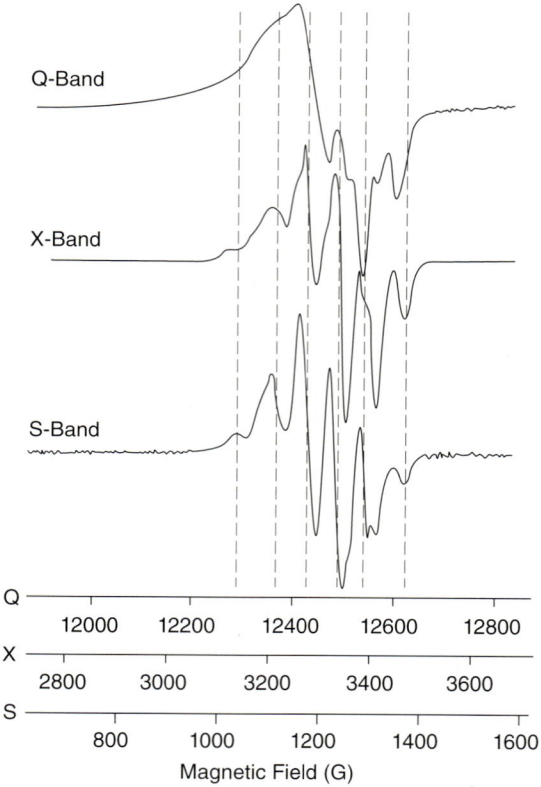

Figure 16.24 Multifrequency EPR spectroscopy of CuPPN. Spectra were recorded in a frozen dichloromethane/toluene glass. Microwave frequency (temperature): 35.100 GHz (123 K), 9.188 GHz (77 K), and 3.392 GHz (123 K) for Q-, X-, and S-bands, respectively (Mankad et al., 2009a). Reprinted with permission from N.P. Mankad, W.E. Antholine, R.K. Szilagyi, J.C. Peters, Journal of the American Chemical Society, **36** (11), 3878–3880; © 2009, American Chemical Society.

16.4.3
Spectra for Mixed-Valence Dinuclear Copper Complexes

16.4.3.1 Nitrous Oxide Reductase, N_2OR (^{15}N Example)

A seven-line pattern in the spectra for N_2OR is indicative of the $[Cu^{1.5+}Cu^{1.5+}]$ mixed-valence complex for which one electron is delocalized over two coppers (Figures 16.28 and 16.29) (Neese et al., 1996; Kroneck, 2001). A reduction of line-width in the spectrum for ^{65}Cu and ^{15}N-histidine double-labeled N_2OR, compared to the spectrum for ^{65}Cu labeled N_2OR, indicates the presence of unresolved hyperfine lines from N_{His} coordinated to the Cu_A site. The lines with the double-labeled N_2OR are not significantly better resolved at lower frequencies, although in earlier

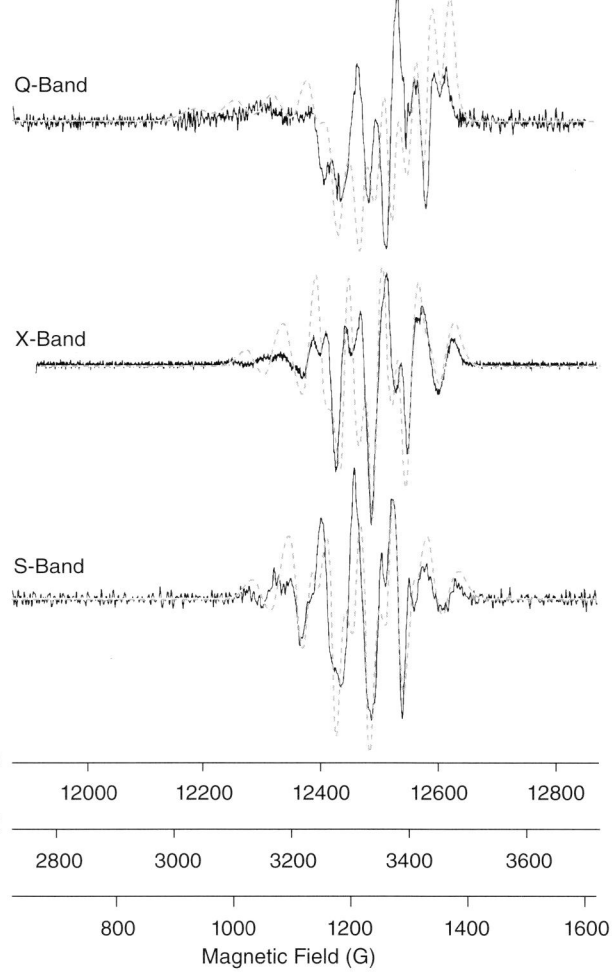

Figure 16.25 The first harmonic for experimental (black) and simulated (red) EPR spectra at Q-, X-, and S-bands for CuPPN (Mankad et al., 2009a). Reprinted with permission from N.P. Mankad, W.E. Antholine, R.K. Szilagyi, J.C. Peters, *Journal of the American Chemical Society*, **36** (11), 3878–3880; © 2009, American Chemical Society.

studies lines with a singly labeled ^{63}Cu N$_2$OR were better resolved at lower microwave frequencies. In this way, the spectrum can be readily assigned to a seven-line pattern from a mixed-valence complex (Antholine et al., 1992, and references therein). The EPR parameters from the simulated spectrum for N$_2$OR (Figure 16.30; Table 16.3) are as follows (Neese et al., 1996).

Table 16.3 Best fit parameters for the simulation of the X-band EPR spectrum of ^{65}Cu- and [^{15}N]histidine-enriched N$_2$OR from *P. stutzeri*[a] (Neese et al., 1996).

	g	A^{Cu} (MHz)	W (G)	c_2 (G MHz^{-1})	ε (G^2 MHz^{-1})
g_{min}	2.003	105	13.0	1.0	−0.002
g_{mid}	2.005	116	12.0	2.0	0.001
g_{max}	2.203	235	13.5	7.9	−0.0078

a) The [Cu$^{1.5+}$ ... Cu$^{1.5+}$] site is described by two identical interacting monomers, Cu$_{A1}$ and Cu$_{A2}$; the mixing parameter $\alpha^2 = 0.51$ refers to the probability of finding the system in the state where the unpaired electron is localized on one-half of the dimer. Tilt angle $\beta = 38°$.

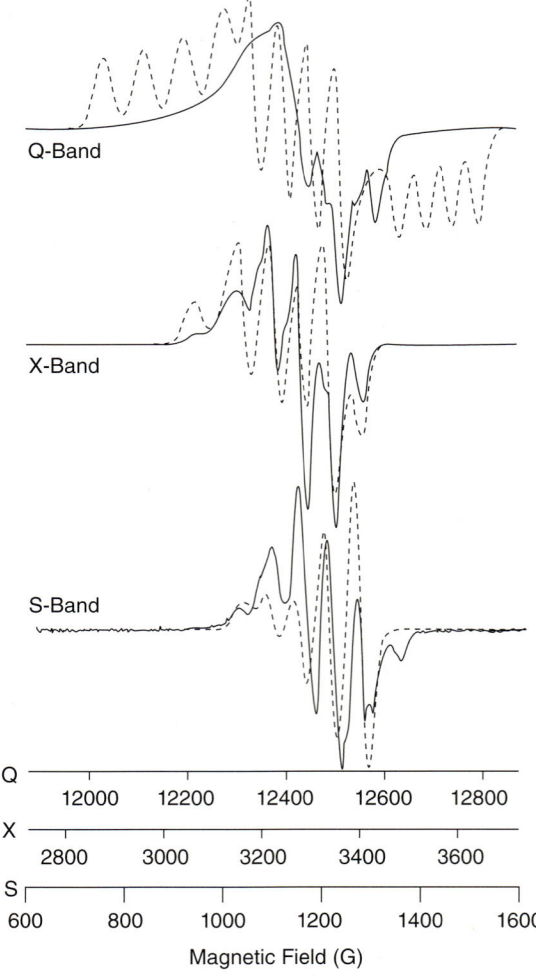

Figure 16.26 Experimental (black) and simulated (red) EPR spectra for CuPPN resulting from assigning outer lines as derived purely from Cu (Mankad et al., 2009a). Reprinted with permission from N.P. Mankad, W.E. Antholine, R.K. Szilagyi, J.C. Peters, *Journal of the American Chemical Society*, **36** (11), 3878–3880; © 2009, American Chemical Society.

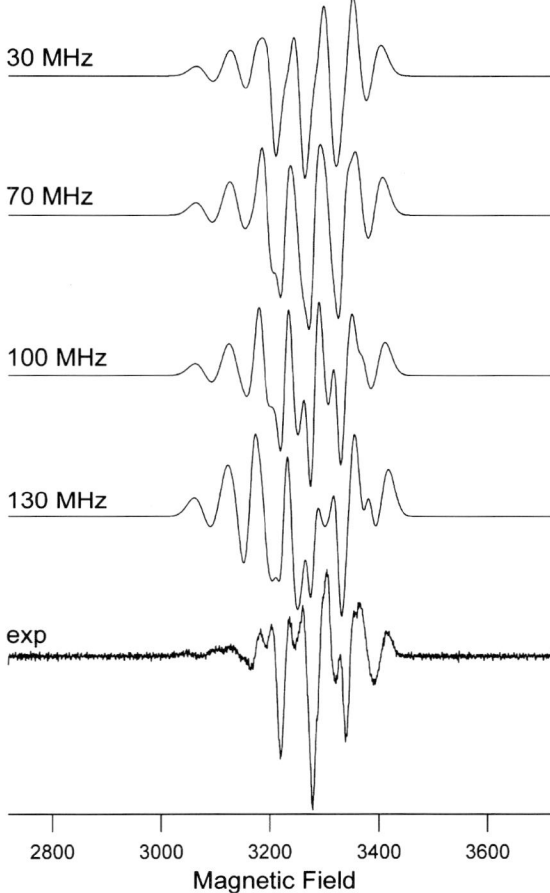

Figure 16.27 Simulated first harmonic X-band spectra for CuPPN resulting from maintaining all other simulation parameters while varying A_y^N as indicated (Mankad et al., 2009a). Reprinted with permission from N.P. Mankad, W.E. Antholine, R.K. Szilagyi, J.C. Peters, *Journal of the American Chemical Society*, **36** (11), 3878–3880; © 2009, American Chemical Society.

16.4.3.2 Perturbation of the EPR Spectrum of Cu$_A$, H120X

In earlier studies, there was a discrepancy as to whether the unpaired electron of the Cu$_A$ site of H120N was localized on a single copper. A change from a seven-line pattern in wild-type (i.e., native protein) to a four-line pattern in the spectrum for H120N was consistent with the unpaired electron localized on a single copper (Figure 16.31) (Lukoyanov et al., 2002). EPR parameters from the multifrequency spectra are given in Table 16.4. However, Q-band electron-nuclear double

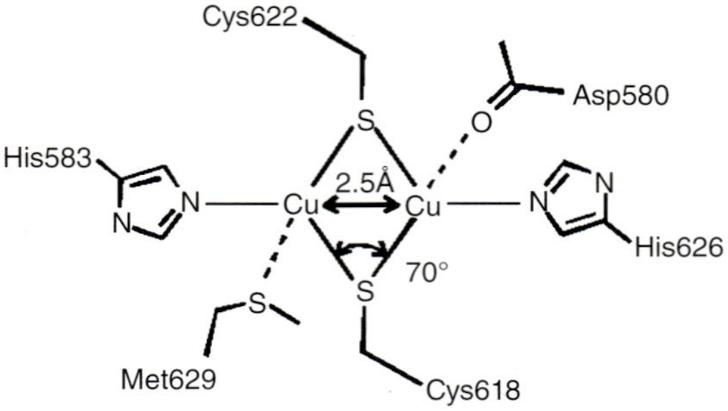

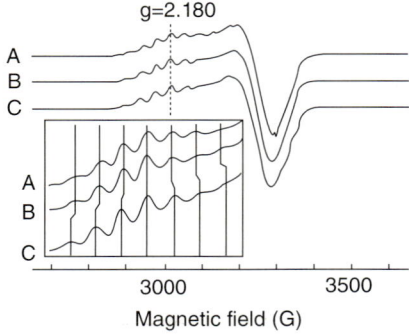

Figure 16.28 Spectroscopic model of the Cu_A center in N_2OR from *Pseudomonas stutzeri*. The ligand assignment was primarily based on the crystallographic data of Cu_A (Kroneck, 2001). Reprinted with permission from P.M.H. Kroneck, Binuclear Copper A, *Handbook of Metalloproteins*, Vol. 2; © 2001, John Wiley & Sons, Inc.

Figure 16.29 X-band continuous wave (CW) EPR spectra of (A) $^{63/65}Cu$ natural abundance and (B) ^{63}Cu- and (C) ^{65}Cu-enriched N_2OR from *P. stutzeri* cells, washed in cold 50 mM Tris–HCl, pH 7.5, and 50 mM $MgCl_2$: temperature, 10 K; power, 200 µW; microwave frequency, 9.241 GHz; modulation amplitude, 4 G. The inset shows an expansion of the $g_{\|}$ region with line shifts due to Cu isotope substitution (Neese et al., 1996). Reprinted with permission from F. Neese, W.G. Zumft, W.E. Antholine, P.M.H. Kroneck, *Journal of the American Chemical Society*, **118** (36), 8692–8699; © 1996, American Chemical Society.

resonance (ENDOR) lines of cysteine C_β protons and lines for the His46 nitrogen ligand provide hyperfine couplings that resemble those of mixed-valence Cu_A (Lukoyanov et al., 2002). Recently, this discrepancy was solved through the characterization of wild-type and mutant Cu_A sites, using a variety of techniques (Xie et al., 2008). Spectra taken from wild-type, wild-type at a low pH, and H120A again show a seven-line pattern for wild-type and a four-line pattern for wild-type at a low pH and mutant (Figure 16.32). The EPR parameters from simulations using

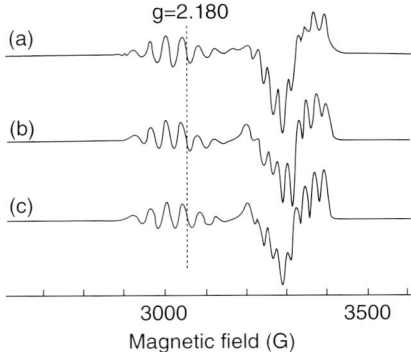

Figure 16.30 (a) Experimental and simulated X-band CW-EPR spectra of ^{65}Cu- and [^{15}N] histidine-enriched N$_2$OR from *P. stutzeri*; (b) Simulation with noncollinear g and A axes, and Cu hyperfine tensors tilted by the Euler angles (1/2)β = ±17, respectively (parameters in Table 16.2); (c) Simulation with collinear g and A axes (Neese et al., 1996). Reprinted with permission from F. Neese, W.G. Zumft, W.E. Antholine, P.M.H. Kroneck, *Journal of the American Chemical Society*, **118** (36), 8692–8699; © 1996, American Chemical Society.

XSophe-Sophe-XeprView (Hanson et al., 2004; a matrix-diagonalization simulation program developed by Graeme Hanson at the Center for Advanced Imaging, University of Queensland, and available from Bruker Biospin) fit a decrease in A_z from $53 \times 10^{-4}\,\mathrm{cm}^{-1}$ to $7 \times 10^{-4}\,\mathrm{cm}^{-1}$ for H120A (Xie et al., 2008) (Table 16.5). For H120A, however, a 1% 4S orbital mixing due to the distorted ligand field of Cu$_A$ adds a large and positive direct Fermi contact contribution to the hyperfine coupling that results in a smaller value for A_z^{Cu} (Xie et al., 2008). This is consistent with ENDOR, resonance Raman, and magnetic circular dichroism (MCD) evidence for delocalization (Xie et al., 2008). In this case, the four-line EPR spectrum in the $g_{\parallel}$ region is misleading with respect to delocalization.

16.4.3.3 Cytochrome c Oxidase (CcO): Best Demonstration of the Use of Low-Frequency for Mixed-Valence Sites

The spectrum in Figure 16.33 for CcO illustrates the value of low frequency to extract EPR parameters (Antholine, 1997). The X-band spectrum gives two g-values whereas, in contrast, the second-derivative spectrum at 4.5 GHz is nicely resolved into multiple hyperfine lines. It is still difficult to determine the pattern for hyperfine lines. Assuming that the two high-field lines are copper hyperfine lines, spectra at multiple frequencies were used to determine whether the pattern at high field is a seven-line pattern for a dinuclear mixed-valence copper site with intensities of 1:2:3:4:3:2:1, or a four-line pattern for a single copper with intensities of 1:1:1:1. The calculated g-values remain constant for a mixed-valence [Cu$^{1.5}$Cu$^{1.5}$]

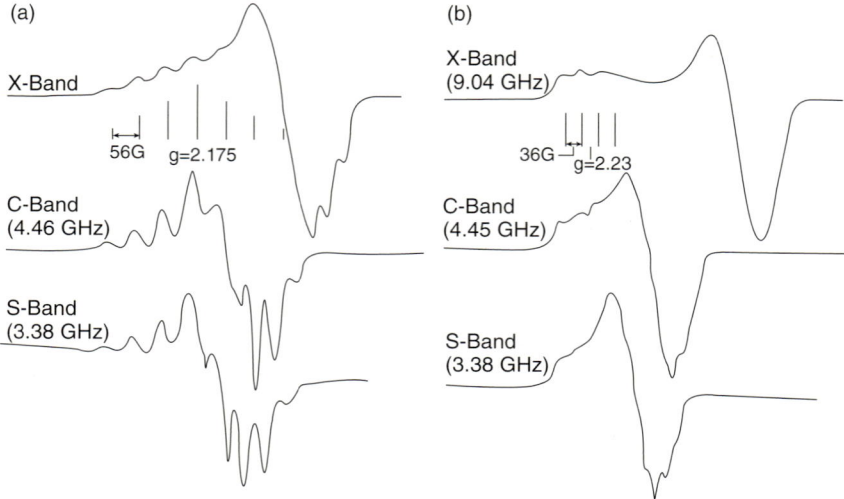

Figure 16.31 Comparison of S-, C-, and X-band spectra of (a) Cu$_A$-azurin and (b) H120N in 50% glycerol, 0.1 M phosphate buffer, pH 5.2. Experimental conditions: temperature, 15 K; microwave power, 23 dB down from an incident power of 7 mW; time constant, 0.1 s; scan time, 4 min; modulation amplitude, 5 G; gain, 3.2 × 10²; cavity, loop-gap resonator at S- and C- bands. A standard TE$_{102}$ resonator was used at X-band with 5 mW power (Lukoyanov et al., 2002). Reprinted with permission from D. Lukoyanov, S.M. Berry, Y. Lu, W.E. Antholine, C.P. Scholes, *Biophysical Journal*, **82** (5), 2758–2766; © 2002, Elsevier.

Table 16.4 Multifrequency EPR findings (Lukoyanov et al., 2002).

Sample	$g_z^{Cu\text{-}Cu\,a)}$	$g_z^{Cu\,b)}$	$A_z^{c),d)}$ (G)	v_e (GHz)
Wild-type				
S-Cu$_A$	2.18	2.36	58	3.38
C-Cu$_A$	2.17	2.31	55	4.46
X-Cu$_A$	2.18	2.24	56	9.06
Mutants				
S-H120G	2.13	2.23	34	3.38
S-H120N	2.12	2.23	36	3.38
C-H120G	2.15	2.24	36	4.45
C-H120N	2.15	2.23	36	4.45
X-H120G	2.18	2.24	40	9.08
X-H120N	2.19	2.24	40	9.05

a) $g_z^{Cu\text{-}Cu}$ was calculated assuming a seven-line pattern beginning with the first resolved line in the $g_\parallel$ region.
b) g_z^{Cu} was calculated assuming a four-line pattern, also beginning with the first resolved line on the low field side in the $g_\parallel$ region.
c) The error in A_z is ±4 G, determined from the average of different spectra taken on different spectrometers.
d) An unusual feature of the lines in the $g_\parallel$ region is broadening of the high-field line in the $g_\parallel$ region. This is attributed to g- and A-strain and is a reason for difficulty in distinguishing between a seven-line pattern and a four-line pattern at a single frequency.

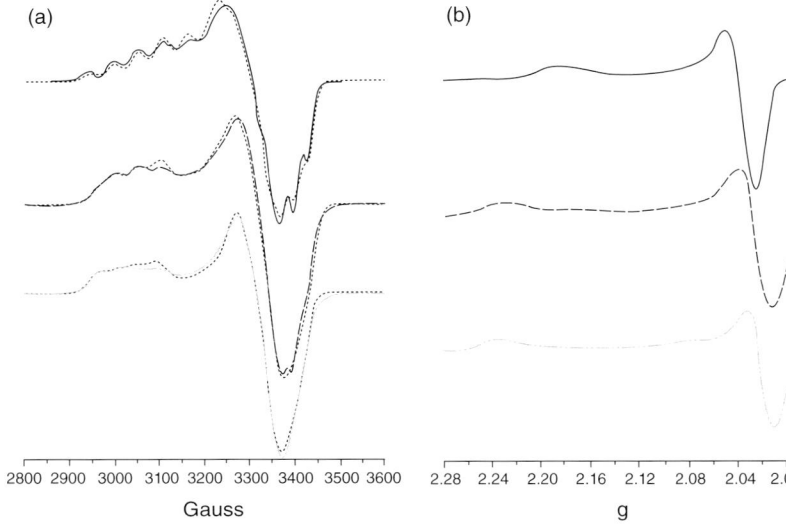

Figure 16.32 (a) X-band EPR spectra: microwave frequency, 9.46 GHz; temperature 77 K; (b) Q-band EPR spectra: microwave frequency, 33.86 GHz; temperature 77 K. Spectra: high-pH form (black), low-pH form (red), H120A mutant (green), and XSophe simulations (blue) (Xie et al., 2008). Reprinted with permission from X. Xie, S.I. Gorelsky, R. Sarangi, D.K. Garner, H.J. Hwang, K.O. Hodgson, B. Hedman, Y. Lus, E.I. Solomon, *Journal of the American Chemical Society*, **130** (15), 5194–5205; © 2008, American Chemical Society.

Table 16.5 EPR parameters of the high- and low-pH forms of Cu_A azurin and its H120A mutant (Xie et al., 2008).

	g_x	g_y	g_z	A_x^{Cu1}, A_x^{Cu2} $(10^{-4}\,cm^{-1})$	A_y^{Cu1}, A_y^{Cu2} $(10^{-4}\,cm^{-1})$	A_z^{Cu1}, A_z^{Cu2} $(10^{-4}\,cm^{-1})$
High-pH form	2.022	2.022	2.173	25, 27	25, 26	53, 53
Low-pH form[a]	2.010	2.010	2.222	26, 21	21, 24	43, 12
H120A	2.010	2.010	2.225	24, 19	19, 24	46, 7

a) Note that the low-pH form still has ~10% of the high-pH form, as the pK_a is ~5. This is included in the EPR spectral simulation.

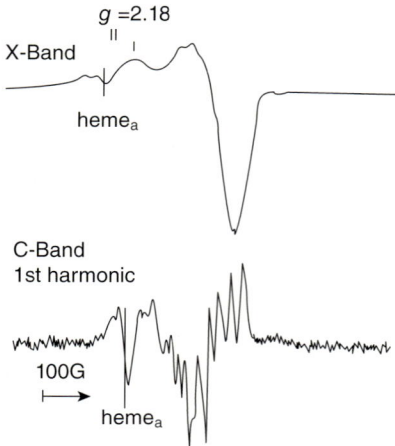

Figure 16.33 Upper spectrum: X-band spectrum of CcO with $g_\parallel$ indicated by the vertical marker. The middle line for the low-spin Fe$_A$ in heme$_a$, also indicated by the vertical marker, is superimposed on the [Cu$_A$(1.5+)Cu$_A$(1.5+)] signal. Lower spectrum: First harmonic of the C-band spectrum of CcO. Note the multiline pattern and improved resolution of the C-band spectrum. Sample below 20 K (Antholine, 1997). Reprinted with permission from A.C. Bush, *Advances in Biophysical Chemistry*, **6**, 217–246; © 1997, Elsevier.

Table 16.6 g-Values calculated from multifrequency EPR spectra assuming binuclear coppers, Cu-Cu (a seven-line pattern), or mononuclear copper, Cu (a four-line pattern) (Antholine, 1997).

Sample	Frequency (GHz)	$g_z^{Cu\text{-}Cu}$	g_z^{Cu}	$g_x^{Cu\text{-}Cu}$	g_x^{Cu}
CcO	9.130 (X-band)	–	–	2.00	–
CcO	4.530 (C-band)	–	–	2.00	1.96
CcO	2.760 (S-band)	–	–	2.00	1.94
N$_2$OR	9.130 (X-band)	2.18	2.22	2.02	1.98
N$_2$OR	4.670 (C-band)	2.18	2.26	2.02	1.97
N$_2$OR	3.480 (S-band)	2.18	2.29	2.02	1.95
N$_2$OR	2.397 (S-band)	2.18	2.34	2.02	1.92

for the Cu$_A$ site in both N$_2$OR (nitrous oxide reductase) and CcO, while the calculated value for the four-line monocupric site varies (Table 16.6). Here, the spectrum and calculated g-values for N$_2$OR are a template for the spectrum and calculated g-values for CcO, where overlap of a heme$_a$ signal makes an assignment at low field difficult.

16.4.3.4 Model Diamond Core Complexes, {Cu(LXL)}$_2^+$

{Cu(LXL)}$_2^+$ complexes are four-coordinate copper centers of highly distorted tetrahedral geometries (Figure 16.34). Two LXL^{-1} ligands with bridging amido or phosphido ligands link the coppers (Mankad et al., 2009b), which are mixed-valence complexes, [Cu$^{1.5}$Cu$^{1.5}$], where one unpaired electron is distributed over two coppers.

16.4.3.5 X-Band EPR Spectra of {Cu(PPP)}$_2^+$, {Cu(PNP)}$_2^+$, and {Cu(SNS)}$_2^+$

It is difficult to extract EPR parameters from the X-band spectra (Figure 16.35). The g-values in the spectrum for {Cu(SNS)}$_2^+$ are most apparent, with $g_\perp > g_\parallel$. Hyperfine lines are resolved, but a clear seven-line pattern is not easily assigned, as the pattern appears to have more than seven lines. A similar signal is observed for {Cu(PNP)}$_2^+$, but the lines are broadened due to superhyperfine lines from P, and it is difficult to assign g-values without a signal from {Cu(SNS)}$_2^+$ as a template. Only a crossover point suggesting a single g_{iso}-value is evident from the EPR signal from {Cu(PPP)}$_2^+$.

16.4.3.6 Q-Band EPR Spectra of {Cu(PPP)}$_2^+$, {Cu(PNP)}$_2^+$, and {Cu(SNS)}$_2^+$

A high-field line assigned to $g_\parallel$ is evident in the spectra for {Cu(PNP)}$_2^+$ and {Cu(SNS)}$_2^+$, but not for {Cu(PPP)}$_2^+$ (Figure 16.36). The spectra for {Cu(PNP)}$_2^+$ and {Cu(SNS)}$_2^+$ are cut off on the low field side. A longer scan reveals no new peaks for {Cu(PNP)}$_2^+$, but a signal for a Type 2 Cu is evident in the spectrum for {Cu(SNS)}$_2^+$. This complicates the assignment for $g_\perp$, and it is assumed only that $g_\perp > g_\parallel$.

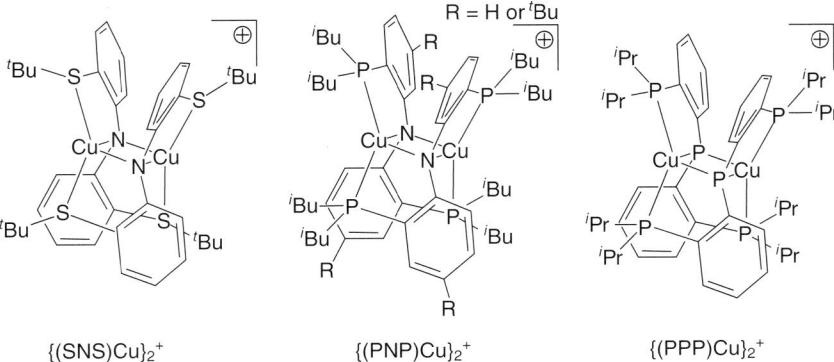

Figure 16.34 Cu$_2(\mu - XR_2)_2$ diamond core systems (Mankad et al., 2009b). Reprinted with permission from N.P. Mankad, S.B. Harkins, W.E. Antholine, J.S. Peters, *Inorganic Chemistry*, **48** (15), 7026–7032; © 2009, American Chemical Society.

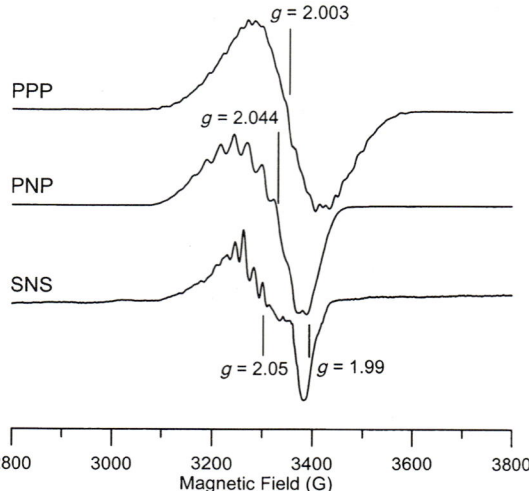

Figure 16.35 X-band EPR spectra of {Cu(PPP)}$_2^+$, {Cu(PNP)}$_2^+$, and {(SNS)Cu}$_2^+$ in 2-methyl-tetrahydrofuran at 120 K (Mankad et al., 2009b). Reprinted with permission from N.P. Mankad, S.B. Harkins, W.E. Antholine, J.S. Peters, *Inorganic Chemistry*, **48** (15), 7026–7032; © 2009, American Chemical Society.

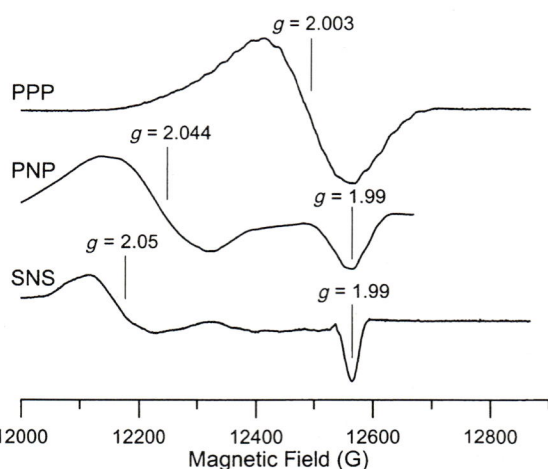

Figure 16.36 Q-band EPR spectra of {Cu(PPP)}$_2^+$, {Cu(PNP)}$_2^+$, and {Cu(SNS)}$_2^+$ in 2-methyl-tetrahydrofuran at 17 K (Mankad et al., 2009b). Reprinted with permission from N.P. Mankad, S.B. Harkins, W.E. Antholine, J.S. Peters, *Inorganic Chemistry*, **48** (15), 7026–7032; © 2009, American Chemical Society.

16.4.3.7 S-Band Spectra of $\{Cu(PPP)\}_2^+$, $\{Cu(PNP)\}_2^+$, and $\{Cu(SNS)\}_2^+$

The hyperfine structure is better resolved at low frequency (Figure 16.37). Additional lines in the spectrum for $\{Cu(PNP)\}_2^+$ are more evident on the high-field side of the S-band spectrum. The g-value closest to the crossover point in Figure 16.37 is assigned as g_{iso} for $\{Cu(PPP)\}_2^+$, and as g_{mid} for $\{Cu(PNP)\}_2^+$ and $\{Cu(SNS)\}_2^+$. The g-anisotropy is not as evident at S-band because g-values become closer as determined by the magnetic field. Equally-spaced lines about g_{mid} are hyperfine. The observed lines for $\{Cu(PNP)\}_2^+$ and $\{Cu(SNS)\}_2^+$ are consistent with a 1:4:10:18:25:28:25:10:4:1 pattern, where hyperfine couplings are about equal for copper and nitrogen (Figure 16.37). There appears to be too many lines for the expected seven lines, with a 1:2:3:4:3:2:1 pattern for only copper couplings from a dicopper mixed-valence complex.

16.4.3.8 EPR Parameters and Simulations for $\{Cu(SNS)\}_2^+$

Second-derivative EPR spectra emphasize sharp lines and deemphasize broad lines as obtained for $\{Cu(SNS)\}_2^+$ at X- and S-band (Figures 16.38 and 16.39). To reproduce experimental spectra, simulations were performed with the following EPR parameters: g_{max}, g_{mid}, and g_{min} equal to 2.069, 2.066, and 2.00, respectively; A^{Cu} equal to 44, 17, and 5 G; and A^N values equal to 12, 17, and 5 G, which fit well to experimental spectra and add confidence to obtaining EPR parameters.

16.4.3.9 First-Harmonic S-Band Spectrum for $\{Cu(PPP)\}_2^+$

The first-harmonic spectrum for $\{Cu(PPP)\}_2^+$ has more lines than expected from just copper couplings (Figure 16.40); in fact, the lines are so well resolved that the

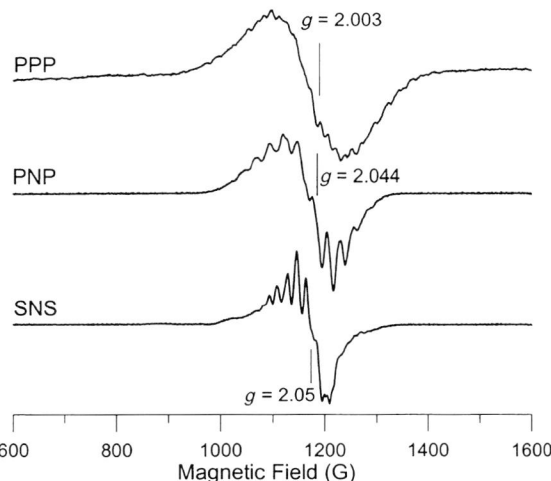

Figure 16.37 S-band EPR spectra of $\{Cu(PPP)\}_2^+$, $\{Cu(PNP)\}_2^+$, and $\{Cu(SNS)\}_2^+$ in 2-methyl-tetrahydrofuran at 133 K (Mankad et al., 2009b). Reprinted with permission from N.P. Mankad, S.B. Harkins, W.E. Antholine, J.S. Peters, Inorganic Chemistry, 48 (15), 7026–7032; © 2009, American Chemical Society.

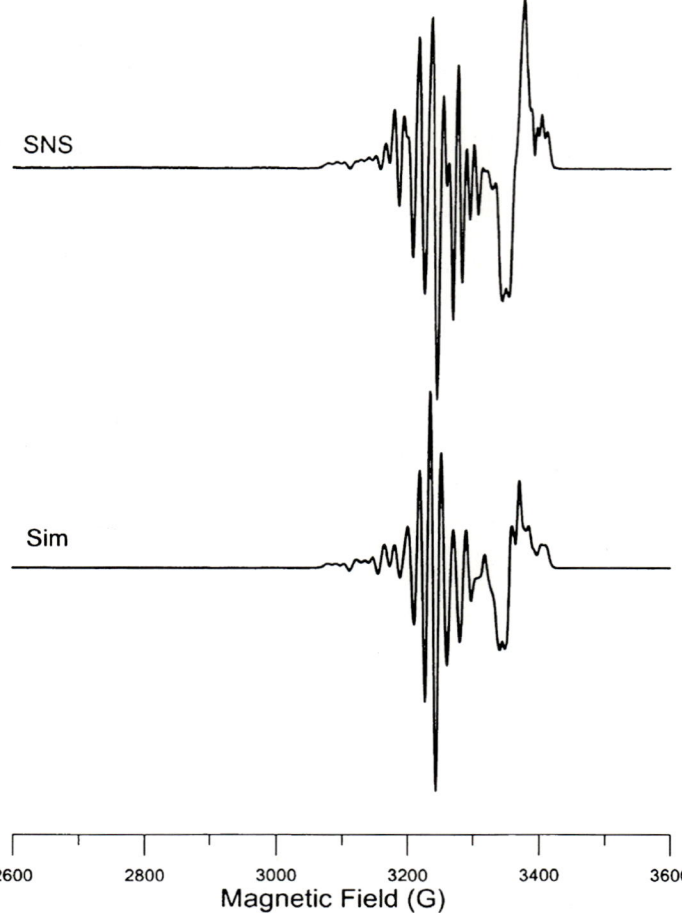

Figure 16.38 First harmonic of X-band spectrum for {Cu(SNS)}$_2^+$ and simulation. EPR parameters for simulation: g_{max}, g_{mid}, g_{min}, 2.069, 2.066, 2.00, respectively; A_{max}^{Cu}, A_{mid}^{Cu}, A_{min}^{Cu}, 44, 17, 5 G, respectively; A_{max}^{N}, A_{mid}^{N}, A_{min}^{N}, 12, 17, 5 G, respectively; line width, 7, 7, 4 G; microwave frequency, 9.434 GHz (Mankad et al., 2009b). Reprinted with permission from N.P. Mankad, S.B. Harkins, W.E. Antholine, J.S. Peters, *Inorganic Chemistry*, **48** (15), 7026–7032; © 2009, American Chemical Society.

relative intensity of the lines becomes more evident. On the low-field side of the spectrum, a box locates lines that give the minimum splitting in the spectrum. A value of 12.5 G is assigned to $A^{P(terminal)}$. It is concluded that the electron is delocalized over the two coppers, the bridging phosphorus, and the terminal phosphorus to account for all the lines in the spectrum for {Cu(PPP)}$_2^+$ and the quenched g-values.

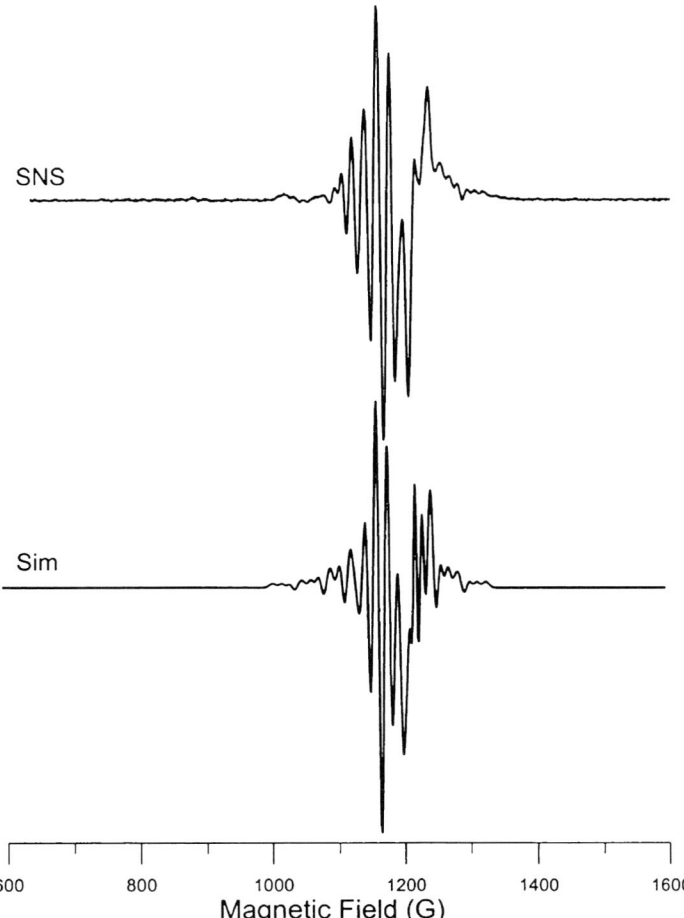

Figure 16.39 First harmonic S-band spectrum for {Cu(SNS)}$_2^+$ and simulation. EPR parameters for simulation given in the caption for Figure 16.38, except here the microwave frequency is 3.366 GHz (Mankad et al., 2009b). Reprinted with permission from N.P. Mankad, S.B. Harkins, W.E. Antholine, J.S. Peters, *Inorganic Chemistry*, **48** (15), 7026–7032; © 2009, American Chemical Society.

16.5
Structural Characterization of Copper(II) Cyclic Peptide Complexes Employing Multifrequency EPR and Computational Chemistry

Numerous cyclic peptides of variable size and shape have been isolated from the Ascidiacea class of tunicates. In particular, the patellamide family of cyclic octapeptides which have a 24-azacrown-8 macrocyclic structure, and the smaller 18-azacrown-6 hexapeptides, have been isolated from of *Lissoclinum patella* and

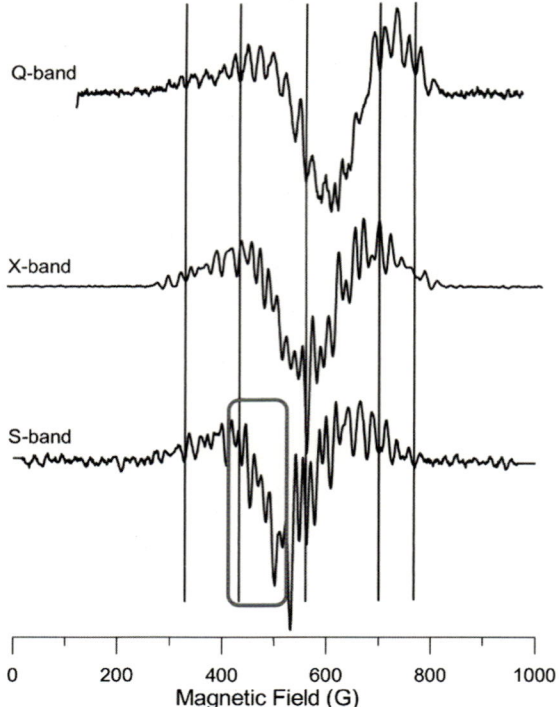

Figure 16.40 First harmonic of Q- (17 K), X- (120 K), and S-band (133 K) spectra for {(CuPPP)}$_2^+$ in 2-methyltetrahydrofuran (Mankad et al., 2009b). The box highlights a particularly well-resolved portion of the S-band spectrum. Reprinted with permission from N.P. Mankad, S.B. Harkins, W.E. Antholine, J.S. Peters, *Inorganic Chemistry*, 48 (15), 7026–7032; © 2009, American Chemical Society.

L. bistratum, respectively. Structural characterization of the cyclic peptides in both the solid and solution states (Ishida et al., 1987, 1988, and 1995; McDonald et al., 1992; Schmitz et al., 1989; Haberhauer and Rominger, 2002, 2003; Haberhauer et al., 2007), their biosynthesis (Schmidt et al., 2005), pharmaceutical activities (Degnan et al., 1989), and Cu(II) complexes have all been reported (Comba et al., 2008; Bernhardt et al., 2002; Comba et al., 1998; van den Brenk et al., 1994a, 1994b; van den Brenk et al., 2004; Grondahl et al., 1999). While *L. patella* is known to concentrate a range of metal ions, including Cu(II), very little information is available concerning their biological function in the ascidians metabolism, though carbon fixation has been suggested (van den Brenk et al., 1994a). Brief descriptions of the application of multifrequency EPR, in conjunction with circular dichroism (CD) and electronic absorption spectral titrations, mass spectrometry and quantum chemistry calculations to the elucidation of the geometric and electronic structures of a number of mono- and di-nuclear copper(II) cyclic peptide complexes, are given below.

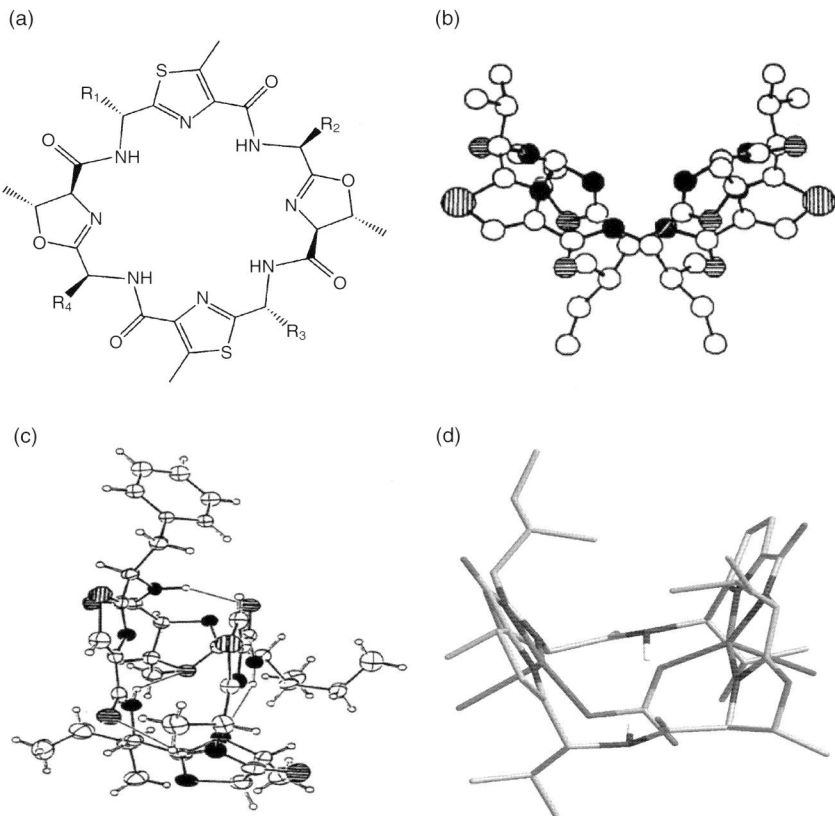

Figure 16.41 (a) Schematic of patellamides and X-ray crystal structures of (b) ascidiacyclamide, (c) patellamide D, and (d) [$Cu_2(ascH_2)(CO_3)$].

16.5.1
Copper(II) Complexes with Marine Cyclic Peptides

Ascidiacyclamide (ascH$_4$: $R_1 = R_3$ = D-Val; $R_2 = R_4$ = L-Ile) and patellamide D (patH$_4$: R_1 = D-Ala; R_3 = D-Phe; $R_2 = R_4$ = L-Ile) isolated from *L. patella* have a 24-azacrown-8 macrocyclic structure in which there are two oxazoline and thiazole rings, formed through condensation of threonine with an adjacent amino acid (Figure 16.41a) (van den Brenk *et al.*, 1994a, 1994b). The peptide chain of asidiacyclamide (Figure 16.41b) takes on a saddle-shaped conformation (Type II) (Ishida *et al.*, 1987, 1988, and 1995; Schmitz *et al.*, 1989), with the thiazole and oxazoline rings at each corner of a rectangular cavity and the eight nitrogen atoms directed towards the interior of the macrocycle, which is ideal for metal ion chelation. In contrast, the the X-ray structure of patellamide D (Figure 16.41c) has a twisted figure-of-eight conformation (Type III) (Schmitz *et al.*, 1989) that is stabilized by two transannular N-H. ... O = C- and two transannular N-H. ... O- (oxazoline ring) hydrogen bonds.

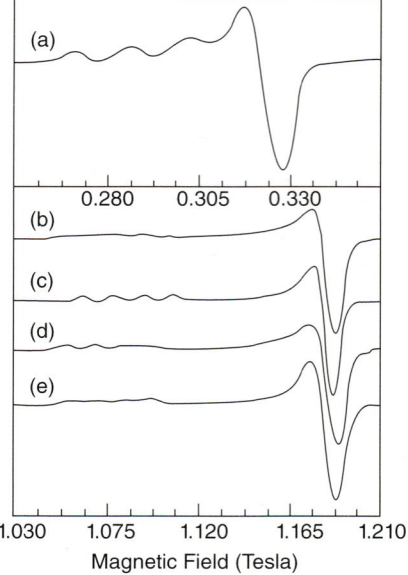
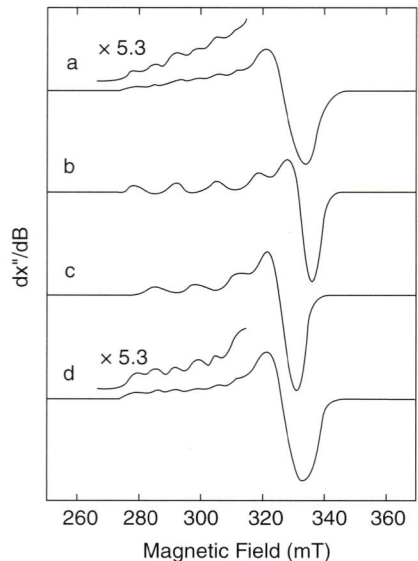

Figure 16.42 EPR spectra of [Cu(PatH₃)]⁺ in methanol (left panel) and acetonitrile:toluene (50:50) (right panel). Left panel: (a, b) X- and Q-band EPR spectra $v = 9.2357$ and $34.0194\,GHz$, respectively; (c) computer simulation of species 1; (d) Q-band spectrum of species 2 and (e) its computer simulation. Right panel: (a) Experimental spectrum, $v = 9.5028\,GHz$; (b) computer simulation of species 1; (c) computer simulation of species 2; (d) addition of (b) + 1.4 × (c).

The coordination chemistry of both patellamide D and ascidiacyclamide has been investigated thoroughly with electronic absorption, CD, mass spectrometry and EPR spectroscopy (van den Brenk et al., 1994a, 1994b; van den Brenk et al., 2004). The formation of both mono- and di-nuclear copper(II) complexes as a function of base have been found using these techniques. Formation of the mononuclear copper(II) complex [Cu(PatH₃)]⁺ requires the addition of one equivalent of base (deprotonation of the amide nitrogen). The EPR spectra of this mononuclear complex are shown in methanol and acetonitrile/toluene in Figure 16.42. While the X-band spectrum of [Cu(PatH₃)]⁺ in methanol is consistent with a single species with a tetragonally distorted geometry, the Q-band spectrum clearly reveals five parallel copper hyperfine resonances attributable to two species (van den Brenk et al., 1994a). This is even more apparent in the X-band EPR spectra measured in acetonitrile:toluene, where the acetonitrile stabilizes the coordination of chloride to the copper(II) ion (van den Brenk et al., 2004). A computer simulation of these spectra with the parameters in Table 16.7 yields the spectra shown in Figure 16.42. Since patH₄ is assymmetric, it is not surprising that copper(II) can form two complexes, namely [Cu(PatH₃)]⁺ and [Cu(Pat'H₃)]⁺. In contrast, the EPR spectra of [Cu(AscH₃)]⁺ revealed only a single species, as ascidiacyclamide is symmetric (van den Brenk et al., 1994b).

16.5 Structural Characterization of Copper(II) Cyclic Peptide Complexes

Table 16.7 Spin Hamiltonian parameters for Cu(II) complexes of patellamide D (van den Brenk et al., 1994a; van den Brenk et al., 2004).

Complex	$g_\parallel$	$g_\perp$	$A_\parallel(Cu)$[a]	$A_\perp(Cu)$[a]
$[Cu(MeOH)]^{2+}$ [b]	2.4235	2.0884	119.5	3.3
$[Cu(patH_3)(MeOH)]^{+}$ [b]				
Species 1b	2.2390	2.055	153.7	11.3
Species 2b	2.2601	2.053	133.0	11.3
$[Cu(patH_3)X]^{n+}$ in MeCN:toluene (1:1)[b]				
Species 1	2.2230	2.087	135.0	60.0
Species 2	2.2730	2.040	137.0	40.0

a) The units for hyperfine coupling constants are $10^{-4}\,cm^{-1}$.
b) X = Cl, n = 1, X = MeCN, n = 0

The formation equilibria can be described by:

$$patH_4 + [Cu(Sol)]^{2+} + NEt_3 \rightarrow$$
$$k[Cu(patH_3)(Sol)]^+ + l[Cu(patH'_3)(Sol)]^+ + NEt_3H^+$$
$$\downarrow \qquad\qquad \downarrow$$
$$m[Cu(patH_3)Cl] + n[Cu(patH'_3)Cl]$$
$$k + l + m + n = 1, \; Sol = Solvent \qquad (16.8)$$

Computer simulation of the $M_I = -1/2$ parallel copper hyperfine resonance measured at S-band frequencies reveals resonances attributable to three magnetically equivalent ^{14}N nuclei arising from a deprotonated peptide amide nitrogen and the nitrogens from the oxazoline and thiazole rings. The remaining coordination site is occupied by either solvent or chloride, depending on the solvent employed. Additional proton hyperfine coupling was also observed in HYSCORE and low-frequency CW-EPR spectra (van den Brenk et al., 1994a).

The titration of a solution containing $patH_4$ and two equivalents of $CuCl_2$ with base has been extensively studied using electronic absorption and EPR spectroscopy, CD, and mass spectrometry. The EPR spectra of the allowed and forbidden resonances as a function of base (triethylamine) are shown in Figure 16.43 (van den Brenk et al., 1994a).

In methanol, the EPR spectra reveal the formation of three dinuclear complexes which can be described by the equilibria (van den Brenk et al., 1994a):

$$[Cu(MeOH)]^{2+} + patH_4 + NEt_3 \rightarrow [Cu(patH_3)]^+ + NEt_3H^+$$
$$[Cu(MeOH)]^{2+} + [Cu(patH_3)]^+ + NEt_3 \rightarrow [Cu_2(patH_2)]^{2+} + NEt_3H^+$$
$$[Cu_2(patH_2)]^{2+} + H_2O + NEt_3 \rightarrow [Cu_2(patH_2)(OH)]^+ + NEt_3H^+$$
$$[Cu_2(patH_2)(OH)]^+ + H_2O + CO_2 + NEt_3 \rightarrow [Cu_2(patH_2)(CO_3)] + NEt_3H^+ \qquad (16.9)$$

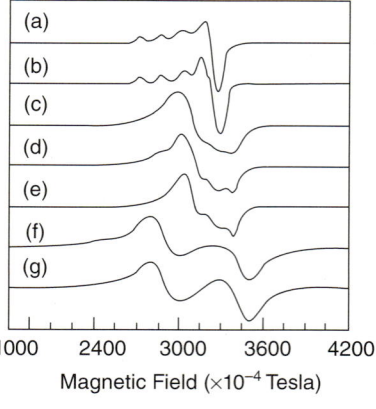 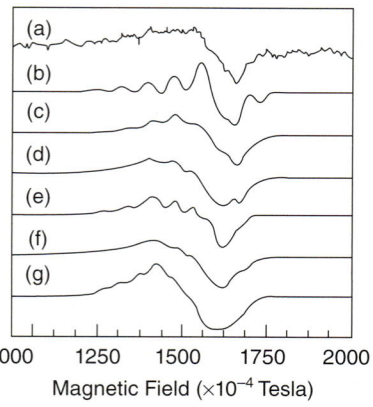

Figure 16.43 EPR spectra of dinuclear copper(II) patellamide D complexes in methanol showing resonances arising from allowed (left panel) and formally forbidden transitions. Left and right panels: Experimental spectra of (a) $[Cu_2(patH_2)]^{2+}$, $v = 9.2926\,GHz$; (c,d) $[Cu_2(patH_2)(OH)]^+$, $v = 9.2983$ and $9.2991\,GHz$; (f) $[Cu_2(patH_2)(CO_3)]$, $v = 9.2858\,GHz$; (b,e,g) Computer simulations of the spectra shown in (a,d,f) (van den Brenk et al., 1994a).

Interestingly, the EPR spectrum of the $[Cu_2(patH_2)]^{2+}$ complex (Figure 16.43, left panel) is typical of mononuclear copper(II) complexes containing a single unpaired electron rather than a dipole–dipole coupled dinuclear copper(II) complex with $S = 1$ (van den Brenk et al., 1994a). Examination of the second rank dipole–dipole coupling tensor ($S \cdot J \cdot S$, Equations 16.2 and 16.10):

$$J_{xx} = g_x^2(1 - 3\sin^2\beta\sin^2\alpha)\rho_r$$
$$J_{yy} = g_y^2(1 - 3\cos^2\beta\sin^2\alpha)\rho_r$$
$$J_{zz} = g_z^2(1 - 3\cos^2\alpha)\rho_r$$
$$J_{xy} = J_{yx} = -3g_xg_y\sin\alpha\sin\beta\cos\beta\,\rho_r$$
$$J_{xz} = J_{zx} = -3g_xg_y\sin\alpha\cos\beta\sin\beta\,\rho_r$$
$$J_{yz} = J_{zy} = -3g_yg_z\sin\alpha\cos\beta\cos\beta\,\rho_r$$
$$\rho_r = \beta_e/r^3 \tag{16.10}$$

reveals that all of the matrix elements are inversely proportional to the internuclear distance cubed; thus, increasing the distance will minimize the dipole–dipole interaction. Through computer simulation studies, a distance of greater than 10 Å was found to eliminate the dipole–dipole coupling; however, this distance is too large for the internal cavity of the patellamide D macrocycle (~7 Å). A closer inspection of the diagonal elements (Equation 16.10) shows that the dipole–dipole interaction can also be eliminated by setting α to 54.7° and β to 45.0°, which correspond to angles used in "magic angle" spinning in solid-state NMR to remove anisotropic and dipolar affects. A computer simulation of the spectrum with these angles and a distance of 6.8 Å (Table 16.8) satisfactorily reproduces the allowed resonances.

Table 16.8 Spin Hamiltonian parameters for the dinuclear copper(II) complexes of patellamide D (van den Brenk et al., 1994a).

Parameter	$[Cu_2(patH_2)]^{2+}$	$[Cu_2(patH_2)(OH)]^+$	$[Cu_2(patH_2)(CO_3)]$
g_x	2.060	2.150	2.075
g_y	2.060	2.050	2.075
g_z	2.242	2.300	2.267
A_x ($\times 10^{-4}$ cm^{-1})	11.00	3.00	70.00
A_y ($\times 10^{-4}$ cm^{-1})	11.00	3.00	70.00
A_z ($\times 10^{-4}$ cm^{-1})	160.00	150.00	111.00
$\alpha°$	54.74	15.00	140.00
$\beta°$	45.00	70.00	0.00
$\gamma°$	0.00	40.00	0.00
r (Å)	6.8	4.8	3.7

A simulation of the forbidden transitions (spectral line shape, resonant field positions and intensity) predicted the presence of extremely weak forbidden resonances, which were found experimentally after substantial signal averaging.

Computer simulation of the EPR spectra of $[Cu_2(patH_2)(OH)]^+$ and $[Cu_2(patH_2)(CO_3)]$ with the spin Hamiltonian parameters listed in Table 16.8 reveals that the cyclic peptide backbone becomes increasingly saddle-shaped (r decreases) as the base concentration increases. Similar results were found for the formation of dinuclear Cu(II) ascidiacyclamide complexes (van den Brenk et al., 1994b). The formation of a carbonate-bridged complex, confirmed by mass spectrometry and crystallographically characterized ($[Cu_2(ascH_2)(CO_3)]$; Figure 16.44d; van den Brenk et al. 1994b) is interesting in that $[Cu_2(patH_2)(OH)]^+$ appears to fix carbon dioxide from the air, as the complex is not formed when the reaction is carried out anaerobically. The factors governing this reaction and its biological relevance are currently being explored (P. Comba and G. Hanson, personal communication).

When the corresponding experiments were performed in dry acetonitrile, a chlorobridged ferromagnetically coupled dinuclear copper complex $[Cu_2(patH_2)Cl]^+$ was found (van den Brenk et al., 2004). When a small amount of water was added, species described by Equation 16.9 were found to be present in solution.

In the absence of X-ray crystal structures for binuclear copper(II) patellamide D complexes, molecular modeling was employed to elucidate plausible structures for the carbonate and chloro-bridged dinuclear Cu(II) complexes in both methanol and acetonitrile. A solvation sphere was modeled using the dielectric constant for the appropriate solvent. The X-ray crystal structure for $[Cu_2(ascH_2)(\mu\text{-}CO_3)]$ (Figure 16.44a) was used as a starting point for these studies, in which the amino acid side chains for ascidiacyclamide were replaced by those from patellamide D, and the structure optimized. The minimized structure of $[Cu_2(patH_2)(\mu\text{-}CO_3)]$ (dielectric constant, $\varepsilon = 1$) is shown in Figure 16.44c which is almost identical (RMS deviation for the backbone atoms is 0.05) to that of the minimized crystal structure

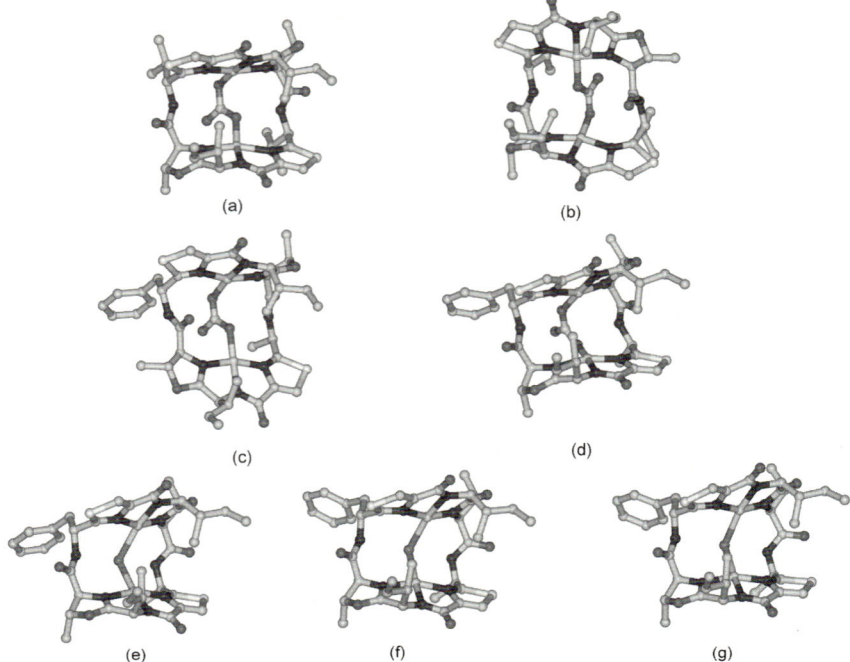

Figure 16.44 Minimized binuclear copper(II) patellamide D structures. (a) X-ray structure of [Cu$_2$(ascH$_2$)(μ-CO$_3$)]; (b) [Cu$_2$(ascH$_2$)(μ-CO3)], $\varepsilon = 1$; (c) [Cu$_2$(patH$_2$)(μ-CO$_3$)], $\varepsilon = 1$; (d) [Cu$_2$(patH$_2$)(μ-CO$_3$)], $\varepsilon = 33$ corresponding to methanol; (e) [Cu$_2$(patH$_2$)(μ-Cl)]$^+$, $\varepsilon = 1$; (f) [Cu$_2$(patH$_2$)(μ-Cl)]$^+$, $\varepsilon = 19$ corresponding to a 1:1 mixture of acetonitrile:toluene; (g) the same as (f) except $\varepsilon = 36$ corresponding to acetonitrile.

of [Cu$_2$(ascH$_2$)(μ-CO$_3$)] (Figure 16.44b). Substitution of the carbonate bridge in [Cu$_2$(patH$_2$)(μ-CO$_3$)] with a chloride bridge ([Cu$_2$(patH$_2$)(μ-Cl)]$^+$), and subsequent minimization, produced a structure (Figure 16.44e) in which the shorter Cu..Cu distance of 3.82 Å (reduced from 4.77 Å) tightened the saddle shape of the cyclic peptide. Increasing the dielectric constant from 1 to 19 (acetonitrile:toluene, 1:1) to 36 (acetonitrile) (Figure 16.44e, f, and g, respectively) had little effect on the peptide backbone (the Cu..Cu distance only varied by 0.03 Å). This was not surprising, as the chloride bridge constrains the macrocyclic structure. In contrast, increasing the dielectric constant from 1 to that for methanol ($\varepsilon = 33$) changed the structure of [Cu$_2$(patH$_2$)(μ-CO$_3$)] (Figure 16.44c) dramatically. The bridging mode of the carbonate anion changes from *syn, anti* for [Cu$_2$(patH$_2$)(μ-CO$_3$)] (Figure 16.44c, $\varepsilon = 1$) and [Cu$_2$(ascH$_2$)(μ-CO$_3$)] (Figure 16.44a) in the solid state to *syn, syn* for [Cu$_2$(patH$_2$)(μ-CO$_3$)] in methanol (Figure 16.44d, $\varepsilon = 33$) (van den Brenk et al., 2004). The consequence of this change in bridging mode was a shorter Cu..Cu distance of 3.86 Å, compared to 4.77 Å found in the X-ray structure of [Cu$_2$(ascH$_2$)(μ-CO$_3$)] (van den Brenk et al., 1994b). The shorter Cu..Cu distance (3.86 Å) was also consistent with the Cu..Cu distance (3.7 Å) obtained from the computer simulation of the EPR spectra of [Cu$_2$(patH$_2$)(μ-CO$_3$)] (see Figure 16.42c and d), measured in frozen methanolic solutions (van den Brenk et al., 1994a).

16.5.2
Copper(II) Complexes with Westiellamide and Synthetic Analogs

The copper(II) coordination chemistry of westiellamide (H$_3$L^{wa}), isolated and purified from the ascidian *L. bistratum*, and three synthetic analogs with an 18-azacrown-6 macrocyclic structure with three imidazole (H$_3$L^1), oxazole (H$_3$L^2) and thiazole (H$_3$L^3) rings instead of oxazoline (Figure 16.45), have been reported (Comba et al., 2008).

The N$_{heterocycle}$–N$_{peptide}$–N$_{heterocycle}$ binding site has been shown to be pre-organized for Cu(II) coordination, in a similar manner to the larger patellamides. The formation of mono- and di-nuclear Cu(II) cyclic peptide complexes was monitored by mass spectrometry, spectrophotometric titrations with base (^{n}Bu$_4$NOMe or NEt$_3$), EPR, and infrared (IR) spectroscopy (Comba et al., 2008).

EPR spectra of the mononuclear Cu(II) complexes with L = L^1, L^2 and L^{wa} in methanol at 50 K revealed signals of the pure mononuclear Cu(II) complexes (see Figure 16.46a–c). In contrast, EPR spectra of a solution of Cu(II) trifluormethansulfonate, H$_3$L^3 and base [(^{n}Bu$_4$N)(OCH$_3$)] in various ratios (x:2:y; x = 1,2; y = 1, 2, 3) show resonances attributable to both mono- and di-nuclear Cu(II) complexes.

An examination of the perpendicular region of the mononuclear complexes revealed nitrogen hyperfine coupling. Differentiation of the spectra and Fourier filtering procedures (see Section 16.2.6) produced well-resolved EPR spectra with nitrogen hyperfine coupling on the perpendicular resonances, and for [Cu(H$_2$L^{wa})(CH$_3$OH)$_n$]$^+$ n = 1, 2 also on the parallel M_I = 3/2 resonance (Figure 16.46). Computer simulations of both the first- and second-derivative EPR spectra, based on the spin Hamiltonian (Equation 16.11)

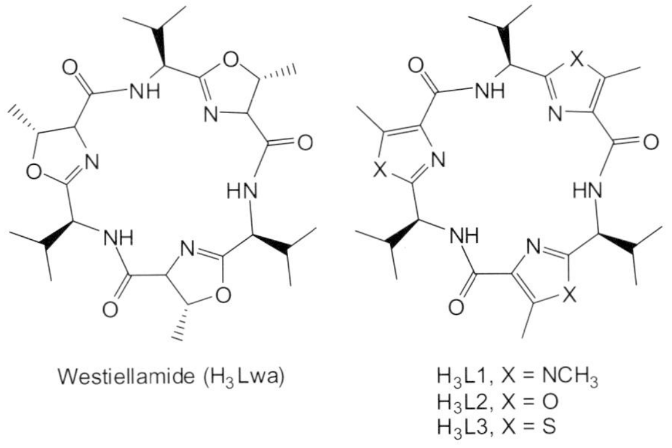

Westiellamide (H$_3$Lwa)

H$_3$L1, X = NCH$_3$
H$_3$L2, X = O
H$_3$L3, X = S

Figure 16.45 Schematics of Westiellamide and the synthetic analogues (Comba et al., 2008).

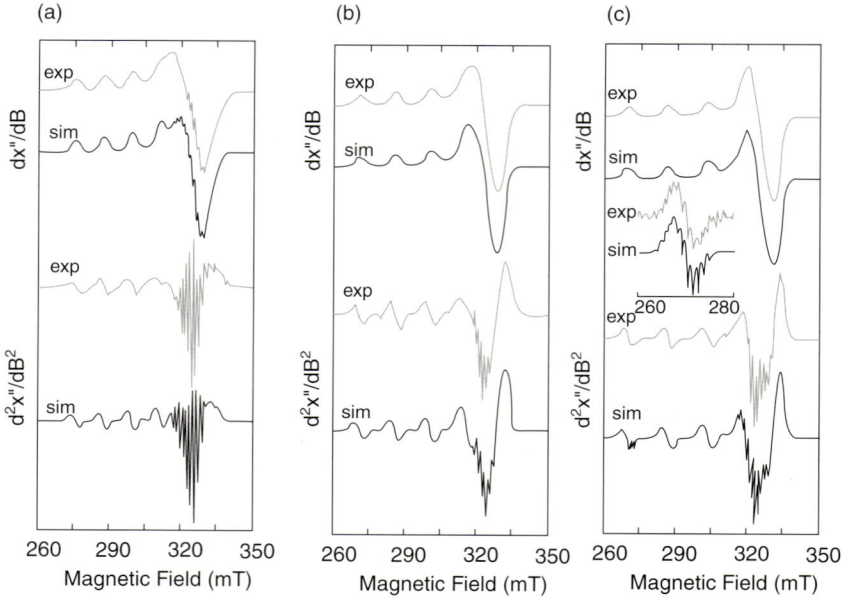

Figure 16.46 EPR spectra of the mononuclear copper(II) complexes with L = L¹, L² and L^wa. (a) [Cu (H₂L¹)]⁺, ν = 9.3571 GHz; (b) [Cu (H₂L²)]⁺, ν = 9.3597 GHz; (c) [Cu (H₂L^wa)]⁺, ν = 9.3588 GHz.

$$H = \sum_{i=x,y,z} (B_i \cdot g_i \cdot S_i + S_i \cdot A_i(Cu) \cdot I_i(Cu) - g_n \beta_n B_i \cdot I_i)$$
$$+ \sum_{j=1}^{3,4} (S_i \cdot A_i(N) \cdot I_i(N) - g_n \beta_n B_i \cdot I_i) \tag{16.11}$$

and with the spin Hamiltonian parameters (Table 16.5) yield the spectra shown above. The EPR spectra of [Cu(H₂L^{1-2})(CH₃OH)_n]⁺ (n = 1,2) (Figure 16.46a and b) were simulated, assuming ligand hyperfine coupling to two magnetically equivalent heterocyclic nitrogen nuclei and one peptide nitrogen nucleus. In contrast, for the simulation of the EPR spectrum of [Cu(H₂L^wa)(CH₃OH)_n]⁺ (n = 1, 2), ligand hyperfine coupling to two magnetically equivalent heterocyclic nitrogen donors and two peptide nitrogen donors were required for a satisfactory fit. The spin Hamiltonian parameters reveal a rhombically distorted square–pyramidal geometry for the Cu(II) center in these cyclic peptide complexes. While the g matrices for the four Cu(II) complexes were quite similar, A_z for [Cu(H₂L^wa)(CH₃OH)_n]⁺ was considerably larger (174 × 10⁻⁴ cm⁻¹ versus ~150 × 10⁻⁴ cm⁻¹), which was consistent with the coordination of an additional nitrogen donor (Peisach and Blumberg, 1974).

Previously, DFT calculations were used to model the structures of the Cu(II) complex on the basis of the EPR spectroscopic data and their simulations (Frisch

Table 16.9 Anisotropic spin Hamiltonian parameters of the mononuclear copper(II) species [Cu^{II}(H$_2$L^1)(CH$_3$OH)$_2$]$^+$, [Cu^{II}(H$_2$L^2)(CH$_3$OH)$_2$]$^+$, [Cu^{II}(H$_2$L^3)(CH$_3$OH)$_2$]$^+$ and [Cu^{II}(H$_2$L^{wa})(CH$_3$OH)]$^+$.[a] (Comba et al., 2008).

Parameter	[Cu^{II}(H$_2$L^1)(CH$_3$OH)$_2$]$^+$	[Cu^{II}(H$_2$L^2)(CH$_3$OH)$_2$]$^+$	[Cu^{II}(H$_2$L^{wa})(CH$_3$OH)]$^+$
g_x	2.088	2.083	2.083
g_y	2.051	2.034	2.051
g_z	2.278	2.279	2.267
A_x (^{63}Cu)	17.0	17.3	14.0
A_y (^{63}Cu)	15.4	17.2	16.2
A_z (^{63}Cu)	153.4	123.	175
A_x (^{14}N) – N$_{hetcyc}$	14.5	15.7	12.4
A_y (^{14}N) – N$_{hetcyc}$	7.1	7.1	6.2
A_z (^{14}N) – N$_{hetcyc}$	9.0	9.0	10.4
A_x (^{14}N) – N$_{pept}$(+ N$_{hetcyc}$)[b]	13.2	13.4	16.5
A_y (^{14}N) – N$_{pept}$(+ N$_{hetcyc}$)[b]	15.2	14.1	12.7
A_z (^{14}N) – N$_{pept}$(+ N$_{hetcyc}$)[b]	9.5	9.5	13.4

a) Units for hyperfine coupling constants are 10^{-4} cm^{-1}.
b) For [Cu^{II}(H$_2$L^{wa})]$^+$.

et al., 2004; Comba et al., 2008). The mononuclear Cu(II) complexes of H$_3$L (L = L^1, L^2, L^3) have a distorted square–pyramidal coordination geometry, with a peptide and two heterocyclic nitrogen atoms as well as a methanol oxygen donor in the basal plane, and a methanol oxygen donor with a significantly longer bond in the apical position (Figure 16.47). In agreement with the interpretation of the EPR spectra, structure optimization of the mononuclear Cu(II) complex of H$_3$L^{wa} yields a different coordination mode, with the Cu(II) center coordinated to all three oxazoline and one peptide nitrogen donor. In contrast to the macrocycles H$_3$L (L = L^1, L^2, L^3), the higher flexibility of H$_3$L^{wa} (due to the unsaturated heterocycles) leads to a conformation where the third oxazoline nitrogen is able to coordinate to the Cu(II) ion. A solvent molecule, methanol, coordinates axially to complete the coordination sphere.

The absence of EPR signals for dinuclear Cu(II) complexes with (H$_3$L^{wa}, H$_3$L^1 and H$_3$L^2) results from antiferromagnetic coupling (H$_3$L^1) and/or from low concentrations of the dinuclear Cu(II) complexes (H$_3$L^{wa}, H$_3$L^2), in agreement with the mass spectrometric data (Comba et al., 2008). An EPR spectrum was observed for the dinuclear Cu(II) complex of H$_3$L^3, and a computer simulation of the spectrum with a dipole–dipole-coupled spin Hamiltonian (Equation 16.2) produced the following parameters for site 1 ($g_\parallel = 2.2090, g_\perp = 2.090, A_\parallel = 152.8, A_\perp = 5.2 \times 10^{-4}$ cm^{-1}, $\beta = -66.2°$), and for site 2 ($g_\parallel = 2.2090, g_\perp = 2.090, A_\parallel = 35.1, A_\perp = 21.4 \times 10^{-4}$ cm^{-1}, $\beta = -66.2°$) with an internuclear distance $r = 5.0$ Å (Figure 16.48). While this

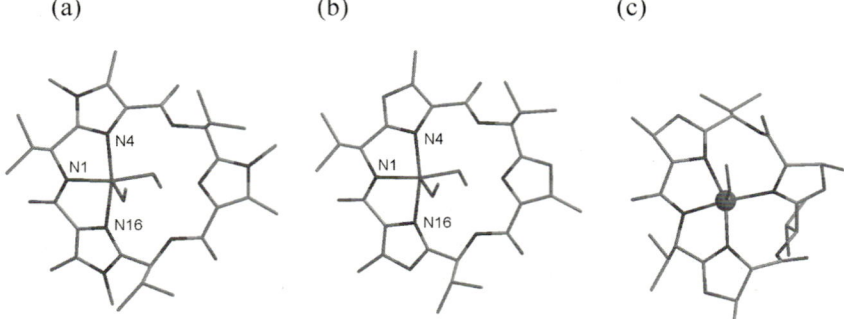

Figure 16.47 DFT calculations of the mononuclear copper(II) complexes. (a) [Cu (H$_2$L^1)]$^+$; (b) [Cu (H$_2$L^2)]$^+$; (c) [Cu (H$_2$L^{wa})]$^+$ (Comba et al., 2008).

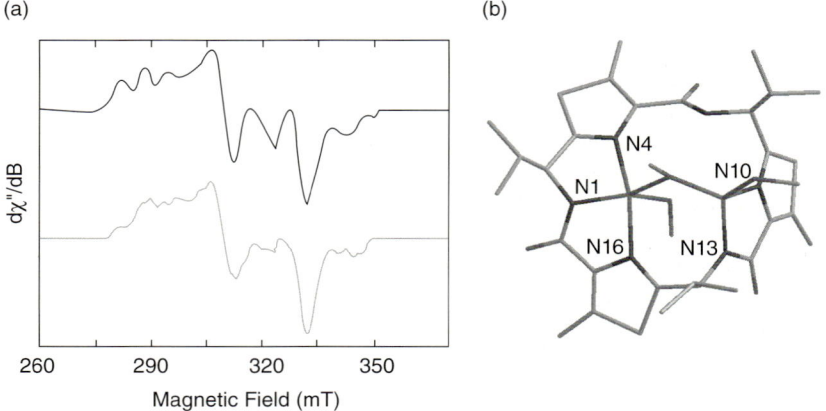

Figure 16.48 EPR spectra and DFT-calculated structure of [Cu$_2^{II}$(HL^3)(μ-OCH$_3$)(CH$_3$OH)$_n$]$^+$. (a) EPR spectrum, ν = 9.359 GHz (top) and computer simulation (bottom; see text for parameters); (b) DFT-optimized structure, showing the square–pyramidal and tetrahedrally coordinated Cu(II) ions.

distance appears to be too large for the 18-membered azacrown-6 macrocyclic cavity, it is known that an anisotropic term proportional to $(\Delta g/g)^2 J$ can contribute to the zero field splitting if the isotropic exchange coupling constant is large ($J > 30 \, \text{cm}^{-1}$) (Atanasov et al., 2006). Consequently, the internuclear distance is over estimated when the J-values are large.

The structures of dinuclear Cu(II) complexes with H$_3$L^1 and H$_3$L^3 refined using DFT, and the two dinuclear copper complexes [Cu$_2$(HL^1)(μ-OCH$_3$)(CH$_3$OH)$_n$]$^+$ and [Cu$_2^{II}$(HL^3)(μ-OCH$_3$)(CH$_3$OH)$_n$]$^+$ ($n = 0,2$) existed as stable minima on the potential energy surface and the coordination geometries of the two Cu(II) centers in [Cu$_2^{II}$(HL^3)(μ-OCH$_3$)(CH$_3$OH)$_n$]$^+$ were found to be square–pyramidal and distorted tetrahedral geometries, as predicted from EPR spectroscopy (Figure 16.48).

16.6 Summary

Within this chapter, an overview has been provided of the factors governing the choice of microwave frequency for a particular metal ion center, including: g-value resolution; magnitude of the microwave energy; state mixing of the various spin Hamiltonian interactions; angular anomalies or off-axis extrema; orientation selection; nuclear frequency separation for electron nuclear double (triple) resonance END(T)OR experiments; and distributions of spin Hamiltonian parameters. In general, g-values are best determined at higher microwave frequencies, while copper hyperfine lines are best resolved at low frequencies. In conjunction with multifrequency CW-EPR and high-resolution pulsed-EPR methods, computer simulation and quantum chemistry calculations provide a sophisticated approach to determining the geometric and electronic structures of Cu(II) centers in biological systems.

The use of multifrequency EPR simulations for the interrogation of square–planar-based Cu(II) species has multiple uses that include: (i) identifying frequencies at which specific features will be observable and interpretable in the experimental spectrum; (ii) obtaining more precise spin Hamiltonian parameters and a more complete spin Hamiltonian, including Euler angles, quadrupolar coupling, and forbidden transitions; and (iii) characterizing spectra due to more than one paramagnetic species. The use of simulations at more than one frequency to obtain spin Hamiltonian parameters has long been established, whereas the use of multifrequency simulations to design experiments and to characterize mixtures is an emerging methodology. One particular advantage is the application of global analyses in the elucidation of structural information.

The application of multifrequency CW- and pulsed-EPR spectroscopy, in conjunction with quantum chemistry calculations and/or molecular modeling techniques, has allowed the geometric and electronic structures of Type II Cu(II) centers, three-coordinate copper and mixed-valence copper metalloproteins and mono- and di-nuclear Cu(II) cyclic peptide complexes to be determined. It is concluded that new complex EPR signals would not have been understood without multifrequency data. The prime example is the signal assigned to one electron shared over two copper atoms (i.e., the Cu_A signal from CcO [see Section 16.4.3.3]). Although the signal is named Cu_A, it is really $Cu_A^{1.5+}Cu_A^{1.5+}$, as deduced from multifrequency spectra, which better describes the structure of the site. As for Cu_A, it is difficult to unravel the line patterns for spectra for which the unpaired electron is delocalized over more than a single copper atom, and for which hyperfine and superhyperfine couplings are about equal in magnitude (see Sections 16.4.2.2 and 16.4.3.5). Finally, a word of caution – the EPR assignment should fit data from other techniques, for example orientation-selected HYSCORE or END(T)OR.

Acknowledgments

These studies were supported by grant P41 EB001980 (National Biomedical EPR Center; PI: J. S. Hyde) from the National Institutes of Health. GRH would like to thank his collaborators, in particular Dr Christopher Noble, Prof. Peter Comba, Prof. Lawrence Gahan, Dr Kevin Gates, and the many PhD students (Dr Simon Benson, Dr Bjoern Seibold, Dr Nina Dovalil, Dr Mark Griffin) who were involved in the research described herein.

Pertinent Literature

Section 16.3

The following list of publications that are pertinent to this chapter is not intended to be exhaustive, but rather to provide "trailheads" for the reader (i.e., accessible and informative entry points into the literature on the field as a whole). The references from each of the other chapters – particularly Chapters 4.2, 5.2, 5.3, 6–9 and 16.2 – will also be useful. The references below are collected into areas of interest for the convenience of the reader. For that reason too, some references cited in this chapter are reproduced here.

General copper and copper EPR: B. J. Hathaway and D. E. Billing, *Coord. Chem. Rev.* **5**, 143 (1970); J. Peisach and W. E. Blumberg, *Arch. Biochem. Biophys.* **165**, 691 (1974); H. R. Gersmann and J. D. Swaten, *J. Chem. Phys.* **36**, 3221 (1962); J. Ammeter, G. H. Rist, and H. A. Günthard, *J. Chem. Phys.* **57**, 3852 (1972); B. G. Malmstrøm and T. Vänngård, *J. Mol. Biol.* **2**, 118 (1960); J. S. Hyde, W. E. Antholine, and R. Basosi, in *Biological and Inorganic Copper Chemistry* (eds. K. D. Karlin and J. Zubieta), Adenine Press, Schenectady, New York, pp. 239–246 (1985); G. H. Rist and J. S. Hyde, *J. Chem. Phys.* **52**, 4633 (1970).

Low-frequency EPR of Cu(II): Methodology and theory: J. S. Hyde and W. Froncisz, *Annu. Rev. Biophys. Bioeng.* **11**, 391 (1982); W. Froncisz and J. S. Hyde, *J. Chem. Phys.* **73**, 3123 (1980); W. Froncisz and J. S. Hyde, *J. Magn. Reson.* **47**, 515 (1982); R. Basosi, W. E. Antholine, and J. S. Hyde, *Biol. Magn. Reson.* **13**, 103 (1993); H. Hirata, T. Kuyama, M. Ono, and Y. Shimoyama, *J. Magn. Reson.* **164**, 233 (2003); G. A. Rinard, R. W. Quine, S. S. Eaton, and G. R. Eaton, *J. Magn. Reson.* **156**, 113 (2002).

Applications of low-frequency EPR to square-planar Cu(II) in biology: M. Chattopadhyay, E. D. Walter, D. J. Newell, P. J. Jackson, E. Aronoff-Spencer, J. Peisach, G. J. Gerfen, B. Bennett, W. E. Antholine, and G. L. Millhauser, *J. Am. Chem. Soc.* **127**, 12647 (2005); C. S. Burns, E. Aronoff-Spencer, G. Legname, S. B. Prusiner, W. E. Antholine, G. J. Gerfen, J. Peisach, and G. L. Millhauser, *Biochemistry* **42**, 6794 (2003); C. S. Burns, E. Aronoff-Spencer, C. M. Dunham, P. Lario, N. I. Avdievich, W. E. Antholine, M. M. Olmstead, A. Vrielink, G. J. Gerfen, J. Peisach, W. G. Scott, and G. L. Millhauser, *Biochemistry* **41**, 3991 (2002); G. Rakhit, W. E. Antholine, W. Froncisz, J. S. Hyde, J. R. Pilbrow, G. R. Sinclair, and B. Sarkar, *J. Inorg. Biochem.* **25**, 217 (1985); M. Pasenkiewicz-Gierula, W. Froncisz, R. Basosi, W. E. Antholine, and J. S. Hyde, *Inorg. Chem.* **26**, 801 (1987).

Applications of low-frequency EPR to other Cu(II) in biology: W. E. Antholine, D. H. W. Kastrau, G. C. M. Steffens, G. Buse, W. G. Zumft, and P. M. H. Kroneck, *Eur. J. Biochem.* **209**, 875 (1992); P. M. H. Kroneck, W. E. Antholine, J. Riester, and W. G. Zumft, *FEBS Lett.* **242**, 70 (1988); F. Neese, W. G. Zumft, W. E. Antholine, and P. M. H. Kroneck, *J. Am. Chem. Soc.* **118**, 8692 (1996); W. Froncisz, C. P. Scholes, J. S. Hyde, Y. H. Wei, T. E. King, R. W. Shaw, and H. Beinert, *J. Biol. Chem.* **254**, 7482 (1979); F. Neese, W. G. Zumft, W. E. Antholine, and P. M. H. Kroneck, *J. Am. Chem. Soc.* **118**, 8692 (1996); W. E. Antholine, P. M. Hanna, and D. McMillint, *Biophys. J.* **64**, 267 (1993).

EPR and Cu–Cu distances: S. S. Eaton and G. R. Eaton, in *Distance Measurements in Biological Systems by EPR* (eds. L. J. Berliner, S. S. Eaton, G. R. and Eaton), Academic/Plenum, New York, pp. 1–28 (2001); B. Bennett, W. E. Antholine, V. M. D'souza, G. Chen, L. Ustinyuk, and R. C. Holz, *J. Am. Chem. Soc.* **124**, 13025 (2002); Z. Yang, J. Becker, and S. Saxena, *J. Magn. Reson.* **188**, 337 (2007).

EPR and Cu(II) quadrupolar and forbidden transitions: R. S. Abrakhmanov and T. A. Ivanova, *J. Struct. Chem.* **19**, 145 (1978); R. L. Belford and D. C. Duan, *J. Magn. Reson.* **29**, 293 (1978); L. K. White and R. L. Belford, *J. Am. Chem. Soc.* **98**, 4428 (1976); K. S. Rothenberger, M. J. Nilges, T. E. Altman, K. Glab, R. L. Belford, W. Froncisz, and J. S. Hyde, *Chem. Phys. Lett.* **124**, 295 (1986); H. Soo and R. L. Belford, *J. Am. Chem. Soc.* **91**, 2392 (1969); L. D. Rollmann and S. I. Chan, *J. Chem. Phys.* **50**, 3416 (1969); D. L. Liczwek, R. L. Belford, J. R. Pilbrow, and J. S. Hyde, *J. Phys. Chem.* **87**, 2509 (1983).

Computer simulations: D. M. Wang and G. R. Hanson, *J. Magn. Reson. A* **117**, 1 (1995); G. R. Hanson, K. E. Gates, C. J. Noble, A. Mitchell, S. Benson, M. Griffin, and K. Burrage, in *EPR of Free Radicals in Solids: Trends in Methods and Applications* (eds. M. Shiotani and A. Lund), Kluwer, Dordrecht, Netherlands, pp. 197–237 (2003); S. Stoll and A. Schweiger, *Chem. Phys. Lett.* **380**, 464 (2003); S. Stoll and A. Schweiger, *J. Magn. Reson.* **178**, 42 (2006); S. Stoll and A. Schweiger, *Biol. Magn. Reson.* **27**, 299 (2007); G. R. Hanson, K. E. Gates, C. J. Noble, M. Griffin, A. Mitchell, and S. Benson, *J. Inorg. Biochem.* 98, 903 (2004); G. R. Hanson, C. J. Noble, and S. Benson, in High Resolution EPR: Applications to Metalloenzymes and Metals in Medicine, Biological Magnetic Resonance, vol. 28 (eds. G. R. Hanson and L. J. Berliner), Springer, New York, p. 105 (2009).

Section 16.4

The author has applied multifrequency EPR, primarily low-frequency EPR, for over 30 years after Hyde and Froncisz (1982) first utilized a loop-gap resonator to study cupric complexes at low frequency. These references, which were not cited in this chapter, are included so that the reader can choose a paper that is more relevant to his or her interests. The literature is as follows:

Cupric sites in prion proteins: J. S. Hyde, B. Bennett, E. D. Walter, G. L. Millhauser, J. W. Sidabras, W. E. Antholine, *Biophys. J.* **96**, 1 (2009); M.

Chattopadhyay, E. D. Walter, D. J. Newell, P. J. Jackson, E. Aronoff-Spencer, J. Peisach, G. J. Gerfen, B. Bennett, W. E. Antholine, G. L. Millhauser, *J. Am. Chem. Soc.* **127**, 12647 (2005); C. S. Burns, E. Aronoff-Spencer, G. Legname, S. B. Prusiner, W. E. Antholine, G. J. Gerfen, J. Peisach, G. L. Millhauser, *Biochemistry* **42**, 6794 (2003); C. S. Burns, E. Aronoff-Spencer, C. M. Dunham, P. Lario, N. I. Avdievich, W. E. Antholine, M. M. Olmstead, A. Vrielink, G. J. Gerfen, J. Peisach, W. G. Scott, G. L. Millhauser, *Biochemistry* **41**, 3991 (2002); E. Aronoff-Spencer, C. S. Burns, N. I. Avdievich, G. J. Gerfen, J. Peisach, W. E. Antholine, H. L. Ball, F. E. Cohen, S. B. Prusiner, G. L. Millhauser, *Biochemistry* **39**, 13760 (2000).

Type 2 site in particulate methane monooxygenase: S. S. Lemos, M. L. Perille Collins, S. S. Eaton, G. R. Eaton, W. E. Antholine, *Biophys. J.* **79**, 1085 (2000); H. Yuan, M. L. Collins, W. E. Antholine, *Biophys. J.* **76**, 2223 (1999); H. Yuan, M. L. P. Collins, and W. E. Antholine, *J. Am. Chem. Soc.* **119**, 5073 (1997).

Trinuclear copper cluster site in native laccase: S.-K. Lee, S. D. George, W. E. Antholine, B. Hedman, K. O. Hodgson, and E. I. Solomon, *J. Am. Chem. Soc.* **124**, 6180 (2002).

Additional studies on mixed-valence Cu(1.5+)Cu(1.5+) sites: Y. Zhen, B. Schmidt, U. G. Kang, W. E. Antholine, S. Ferguson-Miller, *Biochemistry* **41**, 2288 (2002); B. P. Hay, O. Clement, G. Sandrone, D. A. Dixon, *Inorg. Chem.* **37**, 5887 (1998); W. E. Antholine, D. H. Kastrau, G. C. Steffens, G. Buse, W. G. Zumft, P. M. Kroneck, *Eur. J. Biochem.* **209**, 875 (1992); Kroneck, P. M., Antholine, W. A., Riester, J., Zumft, W. G., *FEBS Lett.* **248**, 212 (1989); Kroneck, P. M., Antholine, W. A., Riester, J., Zumft, W. G., *FEBS Lett.* **242**, 70 (1988).

More broad-ranging studies: C. E. Ruggiero, S. M. Carrier, W. E. Antholine, J. W. Whittaker, C. J. Cramer, W. B. Tolman, *J. Am. Chem. Soc.* **115**, 11285 (1993); W. E. Antholine, P. M. Hanna, D. R. McMillin, *Biophys. J.* **64**, 267 (1993); J. B. Li, D. R. McMillin, W. E. Antholine, *J. Am. Chem. Soc.* **114**, 725 (1992); P. M. Hanna, D. R. McMillin, M. Pasenkiewicz-Gierula, W. E. Antholine, B. Reinhammar, *Biochem. J.* **253**, 561 (1988); S. O. Pember, S. J. Benkovic, J. J. Villafranca, M. Pasenkiewicz-Gierula, W. E. Antholine, *Biochemistry* **26**, 4477 (1987); R. Basosi, V. Kushnaryov, T. Panz, C. S. Lai, *Biochem. Biophys. Res. Commun.* **139**, 991 (1986); M. M. Morie-Bebel, D. R. McMillin, W. E. Antholine, *Biochem J.* **235**, 415 (1986); G. Rakhit, W. E. Antholine, W. Froncisz, J. S. Hyde, J. R. Pilbrow, G. R. Sinclair, B. Sarkar, *J. Inorg. Biochem.* **25**, 217 (1985).

References

Abragam, A. and Bleaney, B. (1970) *Electron Paramagnetic Resonance of Transition Ions*, Clarendon Press, Oxford, UK.

Abrakhmanov, R.S. and Ivanova, T.A. (1978) *J. Struct. Chem.*, **19**, 145.

te Velde, G., Bickelhaupt, F.M., Baerends, E.J., Fonseca Guerra, C., van Gisbergen, S.J.A., Snijders, J.G., and Ziegler, T. (2001) *J. Comp. Chem.*, **22**, 931.

Adman, E.T., Stenkamp, R.E., Sieker, L.C., and Jensen, L.H. (1978) *J. Mol. Biol.*, **125**, 35.

Ammeter, J., Rist, G., and Günthard, H.H. (1972) *J. Chem. Phys.*, **57**, 3852.

Antholine, W.E. (1997) Evolution of Mononuclear to Binuclear Cu$_A$: An EPR Study, in *Advances in Biophysical Chemistry*, vol. 6 (ed. A.C. Bush), JAI Press, Greenwich, CT, pp. 217–246.

Antholine, W.E. (2005) Low Frequency EPR of Cu^{2+} in Proteins, in *Biological Magnetic Resonance*, vol. 24 *Biomedical EPR* (eds S.S. Eaton, G.R. Eaton, and L.J. Berliner), Kluwer/Plenum, New York, pp. 417–450.

Antholine, W.E., Kastrau, D.H., Steffens, G.C., Buse, G., Zumft, W.G., and Kroneck, P.M. (1992) *Eur. J. Biochem.*, **209**, 875.

Antholine, W.E., Hanna, P.M., and McMillin, D.R. (1993) *Biophys. J.*, **64**, 267.

Aronoff-Spencer, E., Burns, C.S., Avdievich, N.I., Gerfen, G.J., Peisach, J., Antholine, W.E., Ball, H.L., Cohen, F.E., Prusiner, S.B., and Millhauser, G.L. (2000) *Biochemistry*, **39**, 13760.

Atanasov, M., Comba, P., Martin, B., Mueller, V., Rajaraman, G., Rohwer, H., and Wunderlich, S. (2006) *J. Comp. Chem.*, **27**, 1263.

Basosi, R., Antholine, W.E., Froncisz, W., and Hyde, J.S. (1984) *J. Chem. Phys.*, **81**, 4849.

Basosi, R., Valensin, G., Gaggelli, E., Froncisz, W., Pasenkiewicz-Gierula, M., Antholine, W.E., and Hyde, J.S. (1986) *Inorg. Chem.*, **25**, 3006.

Basosi, R., Antholine, W.E., and Hyde, J.S. (1993) Multifrequency ESR of Copper: Biophysical Applications, in *Biological Magnetic Resonance*, vol. 13 *EMR of Paramagnetic Molecules* (eds L.J. Berliner and J. Reuben), Plenum, New York, pp. 103–150.

Belford, R.L. and Duan, D.C. (1978) *J. Magn. Reson.*, **29**, 293.

Bencini, A. and Gatteschi, D. (1990) *EPR of Exchange Coupled Systems*, Springer-Verlag, Berlin, Germany.

Bernhardt, P.V., Comba, P., Hambley, T.W., Massoud, S.S., and Stebler, S. (1992) *Inorg. Chem.*, **31**, 2644.

Bernhardt, P.V., Comba, P., Fairlie, D.P., Gahan, L.R., Hanson, G.R., and Lötzbeyer, L. (2002) *Chem. Eur. J.*, **8**, 1527.

Burns, C.S., Aronoff-Spencer, E., Dunham, C.M., Lario, P., Avdievich, N.I., Antholine, W.E., Olmstead, M.M., Vrielink, A., Gerfen, G.J., Peisach, J., Scott, W.G., and Millhauser, G.L. (2002) *Biochemistry*, **41**, 3991.

Burns, C.S., Aronoff-Spencer, E., Legname, G., Prusiner, S.B., Antholine, W.E., Gerfen, G.J., Peisach, J., and Millhauser, G.L. (2003) *Biochemistry*, **42**, 6794.

Carrell, C.J., Ma, J.K., Antholine, W.E., Hosler, J.P., Mathews, F.S., and Davidson, V.L. (2007) *Biochemistry*, **46**, 1900.

Chattopadhyay, M., Walter, E.D., Newell, D.J., Jackson, P.J., Aronoff-Spencer, E., Peisach, J., Gerfen, G.J., Bennett, B., Antholine, W.E., and Millhauser, G.L. (2005) *J. Am. Chem. Soc.*, **127**, 12647.

Comba, P. and Hilfenhaus, P. (1995) *J. Chem. Soc., Dalton Trans.*, 3269.

Comba, P. and Martin, B. (2005) Modelling of Macrocyclic Ligand Complexes, in *Macrocyclic Chemistry, Current Trends and Future Perspectives*, vol. 303 (ed. K. Gloe), Springer, New York, USA, pp. 303–325.

Comba, P., Hambley, T.W., Hilfenhaus, P., and Richens, D.T. (1996) *J. Chem. Soc., Dalton Trans.*, 533.

Comba, P., Cusack, R., Fairlie, D.P., Gahan, L.R., Hanson, G.R., Kazmaier, U., and Ramlow, A. (1998) *Inorg. Chem.*, **37**, 6721.

Comba, P., Lampeka, Y.D., Prik'hoda, A., and Rajaraman, G. (2006) *Inorg. Chem.*, **45**, 3632.

Comba, P., Gahan, L.R., Haberhauer, G., Hanson, G.R., Noble, C.J., Seibold, S., and van den Brenk, A.L. (2008) *Chem. Eur. J.*, **14**, 4393.

Deeth, R.J. (2010) Molecular Modelling for Systems Containing Transition Metal Centres, in *Structure and Function* (ed. P. Comba), Springer, New York, pp. 21–51.

Degnan, B.M., Hawkins, C.J., Lavin, M.F., McCaffrey, E.J., Parry, D.L., and Watters, D.J. (1989) *J. Med. Chem.*, **32**, 1354.

Doonan, C.J., Wilson, H.L., Bennett, B., Prince, R.C., Rajagopalan, K.V., and George, G.N. (2008) *Inorg. Chem.*, **47**, 2033.

Drew, S.C., Hill, J.P., Lane, I., Hanson, G.R., Gable, R.W., and Young, C.G. (2007) *Inorg. Chem.*, **46**, 2373.

Drew, S.C., Young, C.G., and Hanson, G.R. (2007) *Inorg. Chem.*, **46**, 2388.

Drew, S.C., Noble, C.J., Hanson, G.R., Masters, C.L., and Barnham, K.J. (2009) *J. Am. Chem. Soc.*, **131**, 1195.

Frisch, M.J., Trucks, G.W., Schlegel, H.B., Scuseria, G.E., Robb, M.A., Cheeseman,

J.R., Montgomery, J.A., Jr, Vreven, T., Kudin, K.N., Burant, J.C., Millam, J.M., Iyengar, S.S., Tomasi, J., Barone, V., Mennucci, B., Cossi, M., Scalmani, G., Rega, N., Petersson, G.A., Nakatsuji, H., Hada, M., Ehara, M., Toyota, K., Fukuda, R., Hasegawa, J., Ishida, M., Nakajima, T., Honda, Y., Kitao, O., Nakai, H., Klene, M., Li, X., Knox, J.E., Hratchian, H.P., Cross, J.B., Bakken, V., Adamo, C., Jaramillo, J., Gomperts, R., Stratmann, R.E., Yazyev, O., Austin, A., Cammi, R., Pomelli, C., Ochterski, J.W., Ayala, P.Y., Morokuma, K., Voth, G.A., Salvador, P., Dannenberg, J.J., Zakrzewski, V.G., Dapprich, S., Daniels, A.D., Strain, M.C., Farkas, O., Malick, D.K., Rabuck, A.D., Raghavachari, K., Foresman, J.B., Ortiz, J.V., Cui, Q., Baboul, A.G., Clifford, S., Cioslowski, J., Stefanov, B.B., Liu, G., Liashenko, A., Piskorz, P., Komaromi, I., Martin, R.L., Fox, D.J., Keith, T., Al-Laham, M.A., Peng, C.Y., Nanayakkara, A., Challacombe, M., Gill, P.M.W., Johnson, B., Chen, W., Wong, M.W., Gonzalez, C., and Pople, J.A. (2004) *Gaussian 03, Revision B.03*, Gaussian Inc., Wallingford CT, USA.

Froncisz, W. and Hyde, J.S. (1980) *J. Chem. Phys.*, **73**, 3123.

Froncisz, W. and Hyde, J.S. (1982) *J. Magn. Reson.*, **47**, 515.

Froncisz, W., Scholes, C.P., Hyde, J.S., Wei, Y.H., King, T.E., Shaw, R.W., and Beiner, H. (1979) *J. Biol. Chem.*, **254**, 7482.

Froncisz, W., Sarna, T., and Hyde, J.S. (1980) *Arch. Biochem. Biophys.*, **202**, 289.

Gaffney, B. (2009) EPR of Mononuclear Non-Heme Iron Proteins, in *High Resolution EPR: Applications to Metalloenzymes and Metals in Medicine, Biological Magnetic Resonance*, vol. 28 (eds G.R. Hanson and L.J. Berliner), Springer, New York, USA, pp. 233–268.

Griffin, M., Muys, A., Noble, C., Wang, D., Eldershaw, C., Gates, K.E., Burrage, K., and Hanson, G.R. (1999) *Mol. Phys. Rep.*, **26**, 60.

Grinberg, O.Y. and Berliner, L.J. (eds) (2004) *Very High Frequency (VHF) ESR/EPR, Biological Magnetic Resonance*, vol. 22, Springer, New York, USA.

Grondahl, L., Sokolenko, N., Abberate, G., Fairlie, D.P., Hanson, G.R., and Gahan, L.R. (1999) *J. Chem. Soc., Dalton Trans.*, 1227.

Haberhauer, G. and Rominger, F. (2002) *Tetrahedron Lett.*, **43**, 6335.

Haberhauer, G. and Rominger, F. (2003) *Eur. J. Org. Chem.*, 3209.

Haberhauer, G., Pinter, A., Oeser, T., and Rominger, F. (2007) *Eur. J. Org. Chem.*, 1779.

Hanson, G.R. (2003) XSophe Release Notes, 1.1.3,1 Bruker Biospin, Germany.

Hanson, G.R. and Berliner, L.J. (2009) *High Resolution EPR: Applications to Metalloenzymes and Metals in Medicine, Biological Magnetic Resonance*, vol. 28, Springer, New York, USA.

Hanson, G.R. and Berliner, L.J. (2010) *Metals in Biology: Applications of High Resolution EPR to Metalloenzymes*, vol. 29, Springer, New York, USA.

Hanson, G.R., Wilson, G.L., Bailey, T.D., Pilbrow, J.R., and Wedd, A.G. (1987) *J. Am. Chem. Soc.*, **109**, 2609.

Hanson, G.R., Gates, K.E., Noble, C.J., Mitchell, A., Benson, S., Griffin, M., and Burrage, K. (2003) XSophe-Sophe-XeprView A Computer Simulation Software Suite for the Analysis of Continuous Wave EPR Spectra, in *EPR of Free Radicals in Solids: Trends in Methods and Applications* (eds A. Lund and M. Shiotani), Kluwer, Dordrecht, Netherlands, pp. 197–237.

Hanson, G.R., Gates, K.E., Noble, C.J., Griffin, M., Mitchell, A., and Benson, S. (2004) *J. Inorg. Biochem.*, **98**, 903.

Hanson, G.R., Noble, C.J., and Benson, S. (2009) Molecular Sophe, An Integrated Approach to the Structural Characterization of Metalloproteins, The Next Generation of Computer Simulation Software, in *High Resolution EPR: Applications to Metalloenzymes and Metals in Medicine, Biological Magnetic Resonance*, vol. 28 (eds G.R. Hanson and L.J. Berliner), Springer, New York, USA, pp. 105–173.

Harmer, J., Mitrikas, G., and Schweiger, A. (2009) Advance Pulse EPR Methods for the Characterization of Metalloproteins, in *High Resolution EPR: Applications to Metalloenzymes and Metals in Medicine, Biological Magnetic Resonance*, vol. 28 (eds G.R. Hanson and L.J. Berliner), Springer, New York, USA, pp. 13–61.

Hathaway, B.J. and Billing, D.E. (1970) *Coord. Chem. Rev.*, **5**, 143.

Heichel, M., Höfer, P., Kamlowski, A., Griffin, M., Muys, A., Noble, C., Wang, D., Hanson, G.R., Eldershaw, C., Gates, K.E., and Burrage, K. (2000) *Bruker Rep.*, **148**, 6.

Hirata, H., Kuyama, T., Ono, M., and Shimoyama, Y. (2003) *J. Magn. Reson.*, **164**, 233.

Holland, P.L. and Tolman, W.B. (1999) *J. Am. Chem. Soc.*, **121**, 7270.

Hori, H., Ikeda-Saito, M., Froncisz, W., and Yonetani, T. (1983) *J. Biol. Chem.*, **258**, 12368.

Hustedt, E.J., Smirnov, A.I., Laub, C.F., Cobb, C.E., and Beth, A.H. (1997) *Biophys. J.*, **74**, 1861.

Hyde, J.S. and Froncisz, W. (1982) *Annu. Rev. Biophys. Bioeng.*, **11**, 391.

Hyde, J.S. and Froncisz, W. (1986) *Spec. Period. Rep. R. Soc. Chem.*, **10**, 175.

Hyde, J.S., Antholine, W.E., and Basosi, R. (1985) Sensitivity Analysis in Multifrequency EPR Spectroscopy, in *Biological & Inorganic Copper Chemistry* (eds K.D. Karlin and J. Zubieta), Adenine Press, Schenectady, NY, pp. 239–246.

Hyde, J.S., Pasenkiewicz-Gierula, M., Basosi, R., Froncisz, W., and Antholine, W.E. (1989) *J. Magn. Reson.*, **82**, 63.

Hyde, J.S., Bennett, B., Walter, E.D., Millhauser, G.L., Sidabras, J.W., and Antholine, W.E. (2009) *Biophys. J.*, **96**, 3354.

Ishida, T., Inoue, M., Hamada, Y., Kato, S., and Shiori, T. (1987) *J. Chem. Soc., Chem. Commun.*, 370.

Ishida, T., Tanaka, M., Nabae, M., and Inoue, M. (1988) *J. Org. Chem.*, **53**, 107.

Ishida, T., In, Y., Shinozaki, F., Doi, M., Yamamoto, D., Hamada, Y., Shioiri, T., Kamigauchi, M., and Sugiwra, M. (1995) *J. Org. Chem.*, **60**, 3944.

Jazdzewski, B.A., Holland, P.L., Pink, M., Young, V.C., Jr, Spencer, D.J.E., and Tolman, W.B. (2001) *Inorg. Chem.*, **40**, 6097.

Kremer, S., Henke, W., and Reinen, D. (1982) *Inorg. Chem.*, **21**, 3013.

Kroneck, P.M.H. (2001) Binuclear Copper: CuA Copper, in *Handbook of Metalloproteins*, vol. 2 (eds A. Messerschmidt, R. Huber, T. Poulos, and K. Wieghardt), John Wiley & Sons, Inc., New York, pp. 1333–1341.

Kroneck, P.M.H., Antholine, W.A., Riester, J., and Zumft, W.G. (1988) *FEBS Lett.*, **242**, 70.

Lukoyanov, D., Lu, Y., Berry, S.M., Antholine, W.E., and Scholes, C.P. (2002) *Biophys. J.*, **82**, 2758.

Lunga, G.D., Pogni, R., and Basosi, R. (1995) *Magn. Res. A*, **114**, 174.

McDonald, L.A., Foster, M.P., Phillips, D.R., Ireland, C.M., Lee, A.Y., and Clardy, J. (1992) *J. Org. Chem.*, **57**, 4616.

Mabbs, F.E. and Collison, D.C. (1992) *Electron Paramagnetic Resonance of Transition Metal Compounds*, Elsevier, Amsterdam.

Mankad, N.P., Antholine, W.E., Szilagyi, R.K., and Peters, J.C. (2009a) *J. Am. Chem. Soc.*, **131**, 3878.

Mankad, N.P., Harkins, S.B., Antholine, W.E., and Peters, J.C. (2009b) *Inorg. Chem.*, **48**, 7026.

Mobius, K. and Savitsky, A.N. (2009) *High-Field EPR Spectroscopy on Proteins and Their Model Systems*, Royal Society of Chemistry, United Kingdom.

Nar, H., Messerschmidt, A., Huber, R., van de Kamp, M., and Canters, G.W. (1991) *J. Mol. Biol.*, **218**, 427.

Neese, F. (2009) Spin-Hamiltonian Parameters from First Principle Calculations: Theory and Application, in *High Resolution EPR: Applications to Metalloenzymes and Metals in Medicine, Biological Magnetic Resonance*, vol. 28 (eds G.R. Hanson and L.J. Berliner), Springer, New York, USA, pp. 175–229.

Neese, F., Zumft, W.G., Antholine, W.E., and Kroneck, P.M.H. (1996) *J. Am. Chem. Soc.*, **118**, 8692.

Nielsen, P., Toftlund, H., Bond, A.D., Boas, J.F., Pilbrow, J.R., Hanson, G.R., Noble, C.J., Riley, M.J., Neville, S.M., Moubaraki, B., and Murray, K.S. (2009) *Inorg. Chem.*, **48**, 7033.

Ovchinnikov, I.V. and Konstantinov, V.N. (1978) *J. Magn. Reson.*, **32**, 179.

Pasenkiewicz-Gierula, M., Froncisz, W., Basosi, R., Antholine, W.E., and Hyde, J.S. (1987) *Inorg. Chem.*, **26**, 801.

Peisach, J. and Blumberg, W.E. (1974) *Arch. Biochem. Biophys.*, **165**, 691.

Pilbrow, J.R. (1984) *J. Magn. Reson.*, **58**, 186.

Pilbrow, J.R. (1990) *Transition Ion Electron Paramagnetic Resonance*, Oxford University Press, Oxford, UK.

Pilbrow, J.R., Sinclair, G.R., Hutton, D.R., and Troup, G.R. (1983) *J. Mag. Reson.*, **52**, 386.

Rakhit, G., Antholine, W.E., Froncisz, W., Hyde, J.S., Pilbrow, J.R., Sinclair, G.R., and Sarkar, B. (1985) *J. Inorg. Biochem.*, **25**, 217.

Rinard, G.A., Quine, R.W., Eaton, S.S., and Eaton, G.R. (2002) *J. Magn. Reson.*, **156**, 113.

Rinard, G.A., Quine, R.W., Eaton, S.S., and Eaton, G.R. (2004) *Biol. Magn. Reson.*, **21**, 115.

Rist, G.H. and Hyde, J.S. (1970) *J. Chem. Phys.*, **52**, 4633.

Rothenberger, K.S., Nilges, M.J., Altman, T.E., Glab, K., Belford, R.L., Froncisz, W., and Hyde, J.S. (1986) *Chem. Phys. Lett.*, **124**, 295.

Schmidt, E.W., Nelson, J.T., Rasko, D.A., Sudek, S., Eisen, J.A., Haygood, M.G., and Ravel, J. (2005) *Proc. Natl Acad. Sci. USA*, **102**, 7315.

Schmitz, F.J., Ksebati, M.B., Chang, J.S., Wang, J.L., Hossain, M.B., and van der Helm, D. (1989) *J. Org. Chem.*, **54**, 3463.

Schweiger, A. and Jeschke, G. (2001) *Principles of Pulse Electron Paramagnetic Resonance*, Oxford University Press, Oxford, UK.

Sinclair, G.R. (1989) Modelling Strain Broadened Spectra. PhD Thesis. Monash University, Victoria, Australia.

Smith, T.D. and Pilbrow, J.R. (1974) *Coord. Chem. Rev.*, **13**, 173.

Stoll, S. and Schweiger, A. (2006) *J. Magn. Reson.*, **178**, 42.

Stoll, S. and Schweiger, A. (2007) *Biol. Magn. Reson.*, **27**, 299.

van den Brenk, A.L., Fairlie, D.P., Hanson, G.R., Gahan, L.R., Hawkins, C.J., and Jones, A. (1994a) *Inorg. Chem.*, **33**, 2280.

van den Brenk, A.L., Byriel, K.A., Fairlie, D.P., Gahan, L.R., Hanson, G.R., Hawkins, C.J., Jones, A., Kennard, C.H.L., Moubaraki, B., and Murray, K.S. (1994b) *Inorg. Chem.*, **33**, 3549.

van den Brenk, A.L., Tyndall, J.D.A., Cusack, R.M., Jones, A., Fairlie, D.P., Gahan, L.R., and Hanson, G.R. (2004) *J. Inorg. Biochem.*, **98**, 1857.

Vänngård, T. (1972) Copper Proteins, in *Biological Applications of Electron Spin Resonance* (eds H.M. Swartz, J.R. Bolton, and D.C. Borg), John Wiley & Sons, Inc., New York, pp. 411–447.

Wang, D.M. and Hanson, G.R. (1995) *J. Magn. Reson. A*, **117**, 1.

Weil, J.A., Bolton, J.R., and Wertz, J.E. (2007) *Electron Paramagnetic Resonance, Elementary Theory and Practical Applications*, Wiley Interscience, USA.

Werst, M.M., Davoust, C.E., and Hoffman, B.M. (1991) *J. Am. Chem. Soc.*, **113**, 1533.

White, L.K. and Belford, R.L. (1976) *J. Am. Chem. Soc.*, **98**, 4428.

Wilson, G.L., Greenwood, R.J., Pilbrow, J.R., Spence, J.T., and Wedd, A.G. (1991) *J. Am. Chem. Soc.*, **113**, 6803.

Wilson, G.L., Kony, M., Tiekink, E.R.T., Pilbrow, J.R., Spence, J.T., and Wedd, A.G. (1988) *J. Am. Chem. Soc.*, **110**, 6923.

Xie, X., Gorelsky, S.I., Sarangi, R., Garner, D.K., Hwang, H.J., Hodgson, K.O., Hedman, B., Lu, Y., and Solomon, E.I. (2008) *J. Am. Chem. Soc.*, **130**, 5194.

17
Multifrequency Electron Spin-Relaxation Times
Gareth R. Eaton and Sandra S. Eaton

17.1
Introduction and Scope of the Chapter

The background theories of EPR and electron spin relaxation are discussed in Chapters 3 and 10, respectively, of this book. In this chapter, attention is focused specifically on distinctions between the physical mechanisms of relaxation that are dependent on the microwave frequency (magnetic field), and those that are not. These principles are illustrated by experimental data that confirm the theoretical predictions, data that extend prior concepts, and other data that await convincing physical description. Because this chapter is tutorial and not an archival research review, the literature citations are illustrative, rather than comprehensive.

Historically, such a larger fraction of EPR experiments has been performed at X-band that there has been much less discussion of the frequency dependence of relaxation rates/times in EPR than in NMR. As most EPR spectrometers operate at a narrow range of frequencies, the majority of EPR measurements are made at constant frequency with a variable magnetic field. The impact on relaxation from orientation dependence due to anisotropy of the spin system must be considered (Bowman and Kevan, 1979; Brown, 1979; Gorchester, Millhauser, and Freed, 1990).

Effects on relaxation that could be magnetic field-dependent include: (i) relaxation at low temperature via the direct process, one type of Raman process, vibrational effects on zero-field splitting (ZFS) for $S > \frac{1}{2}$, vibrational effects on spin–orbit coupling (SOC), cross-relaxation, and spin–spin coupling; and (ii) relaxation at higher temperatures via the modulation of g or hyperfine anisotropy, electron–nuclear dipolar relaxation, spin–rotational coupling, collisions between unlike species, Heisenberg exchange, collisional effects on SOC or ZFS, cross-relaxation, and spin–spin coupling. However, the following effects could obscure field dependence: instantaneous diffusion; cross-relaxation, and B_1 being greater than $dB/d\Theta$ (the change in resonant magnetic field with change in orientation of the spin system) in pulse experiments. These effects will be discussed in greater detail in the following text.

Multifrequency Electron Paramagnetic Resonance, First Edition. Edited by Sushil K. Misra.
© 2011 Wiley-VCH Verlag GmbH & Co. KGaA. Published 2011 by Wiley-VCH Verlag GmbH & Co. KGaA.

Discussions of electron spin relaxation use the nomenclature of T_1 and T_2 from Bloembergen, Purcell, and Pound (1948), and the Bloch equations (see Section 10.3). Starting prior to the discovery of magnetic resonance, there has been a focus on the theory and experiment of physical mechanisms by which electron spins give up energy to the lattice. To a first approximation, it is surprising that electron spin–lattice relaxation occurs, because the electron spin angular momentum does not couple directly with thermal motions. The physical mechanisms involved in spin relaxation are explained in the following sections.

17.2
Spin–Spin Relaxation, T_2 and T_m

Experimental measurements of T_2 are usually not measurements of just electron–electron spin–spin (also called transverse in the Bloch equations) relaxation. Rather, what is measured is anything that takes spins off resonance or affects the phase of the electron spins in the rotating frame. Hence, the measured result is called the "phase memory time," T_m. (This topic is described in detail in Eaton and Eaton, 2000b.) Although T_2 in the Bloch equations is not formally dependent on the magnetic field, T_m can depend on magnetic field at constant microwave frequency because of orientation dependence. The EPR line width is a measure of T_2 only when it is relaxation-determined. More commonly, the line is inhomogeneously broadened by unresolved nuclear hyperfine. In such case, the line width parameter is called T_2^*. Similarly, for an inhomogenously broadened line a free induction decay (FID) rate constant after a microwave pulse is T_2^*, and not T_2.

Because of the high concentrations of nuclear spins in most samples, it is relatively rare that measurements are made of a pure electron spin–electron spin interaction that is well-described by the Bloch equations. The dilute nuclear spins in irradiated SiO_2 provide an approximation to pure electron spin–electron spin interaction, which could be described properly as a T_2 in the Bloch equation sense (Eaton and Eaton, 1993). Analogously, pure electron–electron dipole–dipole T_2 was observed for dilute spins in single-wall carbon nanotubes (SWNTs) (Corzilius, Dinse, and Hata, 2007).

In most EPR samples, the electron spins are magnetically dilute, but the nuclear spins are in high concentration. For example, proton spins in water are about 110 M. The proton concentration is almost as high in various organic solvents used to dissolve paramagnetic species for EPR study; for example, toluene is 75 M in protons. A protein has about the same proton concentration as organic solvents. Even in various ionic or covalent solids, there are many magnetic nuclei. A few matrices – such as CaO and SiO_2 – have few nuclear spins, but these are the exception rather than the rule. In rigid lattices, the dominant effect on T_m is the spin diffusion of the nuclear spins of the matrix, which results in T_m of a few microseconds (Zecevic et al., 1998). Lambe et al. (1961) described a relaxation mechanism in which the dynamic nuclear polarization of nuclear spins contributed to the ENDOR effect. To a first approximation, T_m in a rigid lattice is predicted to be independent of the magnetic field, because the physical model is a fluctuating magnetic dipole, and does not involve a Larmor energy difference for the pairs of spins.

The value of T_m is decreased if B_1 is large enough that the second pulse of a two-pulse sequence flips neighboring spins as well as the spin of interest – this process is called "instantaneous diffusion" (Brown, 1979). For a given g dispersion, and a given spin concentration (in units of spins per cm³ or moles per liter), a higher magnetic field means fewer spins per gauss. The fewer the spins per gauss, the less important is instantaneous diffusion to relaxation times measured by pulse methods. Therefore, instantaneous diffusion makes a smaller contribution to T_m at higher magnetic field.

Librational motion provides an additional phase memory decay mechanism. When the time constant for motions that modulate g and/or hyperfine splitting is comparable to the time scale of a spin-echo experiment, T_m is decreased. Since the enhancement of spin-echo dephasing increases as the anisotropy is increased, the contribution to dephasing is field-dependent. Dzuba and coworkers (Dzuba, Tsvetkov, and Marysov, 1992) presented a formula that included the adiabatic rate constant, T_2^a:

$$\frac{1}{T_2^a} = \tau_c \left(\begin{array}{c} \frac{\gamma^2}{4} H_0^2 \left\langle (g_{zz} - \langle g_{zz} \rangle)^2 \right\rangle + m_I^2 \frac{\left\langle [(A \cdot A_0) - A_0^2]^2 \right\rangle}{A_0^2} \\ + \gamma H_0 m_I \frac{\left\langle (g_{zz} - \langle g_{zz} \rangle)(A \cdot A_0 - A_0^2) \right\rangle}{A_0} \end{array} \right)$$

This formula is the basis for the statement that the phase memory rate constant depends on the magnetic-field squared. Dzuba, Tsvetkov, and Marysov (1992) stated that this formula was valid only if τ_c is between 10^{-8} and 10^{-6} s; if it is faster, then "… the model does not describe the experiment," but if it is slower then "… the Redfield approximation is no longer valid." These findings were subsequently quoted (Onischuk et al., 1998) as stating that $1/T_2$ depends on the magnetic field squared if the radicals have rapid librations with low amplitude, and suggested the application of this to α-Si:H particles. The T_2 data were not actually available at more than one frequency, so reliance was placed on line widths rather than a direct measure of T_2.

Barbon et al. (1999) demonstrated librational motion of nitroxyl radicals in single crystals, by using variable-temperature continuous-wave (CW) EPR, ENDOR, and spin-echo studies of oriented crystals at 95 GHz. Additional discussion of the librations of nitroxyl radicals includes the role of anharmonic motion in glassy media (Dzuba, Kirlina, and Salnikov, 2006).

In fluid solution, motional averaging of g and A anisotropy is a major contributor to T_2. g-anisotropy has a greater effect at higher microwave frequency, since small changes in molecular orientation have a greater effect on the resonant field and $dB/d\theta$ is larger. Thus, relaxation is more sensitive to faster motion at higher microwave frequencies and higher magnetic fields, and this contribution to spectral diffusion increases at a higher frequency. The greater sensitivity to faster motion, and to the specific model for rotational diffusion, is fundamental to the studies of molecular motion at 250 GHz by the Freed group (Budil, Earle, and Freed, 1993; Earle, Budil, and Freed, 1993; Earle et al., 1997; Barnes et al., 1999; Freed, 2005), who have compared CW line widths at 9.5 and 250 GHz (see Chapter 11).

Spin-echo envelope modulation (ESEEM), double quantum coherence (DQC), and double electron-electron resonance (DEER) measurements have been performed at 17 (Bennati et al., 2005; Chiang, Borbat, and Freed, 2005), 35 (Biglino et al., 2006), 94–95 (Fursman et al., 2001; Elsaesser et al., 2005), 130 (Kulik, Pashenko, and Dzuba, 2002), and 180 GHz (Denysenkov et al., 2005), without any obvious frequency-dependence of the echo dephasing, as would be expected if the nuclear spin diffusion were to dominate.

For Cr(V)-doped K_3NbO_8, T_2^{-1} was largely frequency-independent, but there was a strong temperature-dependence of the rate near 50 K at X-band (Nellutla et al., 2008).

There is a general observation that spin-echo dephasing rates at constant frequency are shorter for the lower-field portion of broad transition metal spectra. For example, in a series of Co(II) complexes, T_1 was largely independent of the position in the spectrum at 6 K, but T_m varied through the spectrum with longer values being observed at higher magnetic fields (Kang, Eaton, and Eaton, 1994). One consequence of the dependence of T_m on position in the spectrum is that a field-swept, echo-detected spectrum may not have the same field extent as a CW spectrum of the same sample. ESEEM may further change the appearance of the spectrum because of the dependence of the nuclear modulation frequency on the magnetic field.

Spin-echo dephasing rates for nitroxyl radicals depend on the position in the spectrum. Relaxation times are very sensitive to molecular motion. To first order, the effect of motion on T_m depends on the anisotropy that is averaged (Eaton and Eaton, 2000b). After an initial microwave pulse, motion of the molecular framework containing the unpaired spin moves a spin that was excited by the first pulse off resonance, so that it is not refocused by the second pulse of a two-pulse spin-echo sequence. Hence, the number of spins refocused is reduced, the echo amplitude is reduced, and T_m is reduced in a T_m measurement. Faster relaxation is observed for orientations at which the resonant field is more strongly dependent on molecular motion (Du et al., 1992; Du, Eaton, and Eaton, 1995b). These phenomena have been extensively investigated by Dzuba and coworkers (Muromtsev et al., 1975; Dzuba, Tsvetkov, and Marysov, 1992), and extensively interpreted in terms of the details of the motion by Freed and coworkers (Freed, 2005).

Prisner (2004) pointed out that high-frequency EPR (HFEPR) increases the orientation selection in powder (frozen solution) spectra, providing more detailed information regarding the behavior of relaxation times as a function of position in the spectrum. T_m varies across a nitroxyl spectrum, with sharp changes at the turning points (see Figure 4 in Prisner, 2004). The effect of motion on T_m depends on the details of the motion, but in general the effect is increased at higher microwave frequencies. For example, T_m was shorter at 180 GHz than at 95 GHz for a semiquinone radical in quinol oxidase. The dependence of T_m on position in the line differed because of different spectral dispersion at the two magnetic field strengths, but the ratio of the relaxation times was less than 2. This, together with temperature-dependence studies, showed that the radical was in the slow motion regime.

Kutter, Moll, and coworkers (Kutter et al., 1995; Moll et al., 1999) found a longer T_m at 604 GHz than at 9.5 GHz for a sample of tempone diluted 1:30 in polystyrene, with measurements being made in the range 4 to 13 K. The authors pointed out that, at 604 GHz, the upper Zeeman level was almost empty, making possible very few flip-flop processes, and hence resulting in slow decay rates (Kutter et al., 1995). It is likely that these numerical values reflect some instantaneous diffusion and spectral diffusion, due to the high spin concentrations in the samples.

The spin-echo dephasing for nitroxyl radicals above about 80 K, and also of other compounds in which there is hyperfine coupling to a methyl group, is dominated by rotation of the methyl group at a rate that is comparable to inequivalences in hyperfine coupling that are averaged by methyl rotation (Nakagawa et al., 1992; Du, Eaton, and Eaton, 1994). This contribution to dephasing is independent of the magnetic field.

At high concentrations (0.25 to 5.78 M) of di-*tert*-butylnitroxide in CCl_4, T_2^{-1} was larger at low frequency (51, 87, and 115 MHz) than at X-band (Davis, Mao, and Kreilick, 1975).

Many studies in the Freed laboratory illustrate the use of high frequency to increase the sensitivity of EPR to molecular motion of nitroxyl radicals and other species with anisotropic g factors. At 17.3 GHz, T_2 was measured for the spin label 16-PC in DPPC membrane as a function of temperature and added cholesterol (Chiang, Costa-Filho, and Freed, 2007). In the absence of cholesterol and at room temperature the T_2 was 62 ns, T_2 increased as the temperature and percentage of cholesterol were each increased.

When factors other than T_2 contribute to the lineshape, it can be useful to describe such phenomena in terms of the second moment of the line – effectively, the mean frequency deviation. The second moment is a strong function of the intensity in the wings of the spectrum (Abragam, 1961, p. 108; Poole and Farach, 1971, p. 49).

17.2.1
T_m for Fremy's Salt in Glassy Solvents

In glassy 1:1 water:glycerol at X-band, T_m for Fremy's salt shows little dependence on field position up to about 80 K, which is attributed to the relatively rigid hydrogen-bonded lattice (Eaton and Eaton, 2000b). Librational motion is much greater in glassy organic solvents such as ethanol. In a number of reports over the years, Dzuba (also transliterated Dzyuba) and coworkers have shown the effect of molecular motion on echo-detected field-swept EPR spectra (Dzuba et al., 1984; Dzuba, 1996, 2000; Kirilina, Grigoriev, and Dzuba, 2004; Kirilina et al., 2005). Molecular dynamics of Fremy's salt in glycerol glass, studied with multifrequency

EPR (3, 9.5, 95, and 180 GHz), helped to discriminate between the different relaxation mechanisms (Kirilina et al., 2005). By comparing W-band (95 GHz) and G-band (180 GHz), the dependence on magnetic field could be demonstrated more accurately than by comparing with lower-field spectra, because at high field the g_z components were well separated from the x and y components, whereas at X-band they overlapped in such a way that a clean comparison of relaxation at a g_z field position was not practicable. Kirilina et al. (2005) found that, for Fremy's salt at 185 K, the phase memory relaxation rate, $1/T_m$, was larger at higher fields. Specifically, these authors showed that the difference in relaxation rate at the g_{yy} position, and at the position half-way between g_{xx} and g_{yy}, was increased quadratically with field, as predicted by the Redfield theory. The simulation of spectra fit best for a model in which the librations are fast isotropic fluctuations of the radical in the solvent cage, and dynamic g-strain effects.

17.2.2
Exchange-Narrowed Species and the 10/3 Effect

In fluid solution, an organic free radical exhibits a multitude of EPR lines that extend over many gauss. The same radical in a magnetically concentrated solid has a narrow EPR spectrum, with no resolved hyperfine structure, due to spin exchange. Diphenylpicrylhydrazyl (DPPH), α,γ-bisdiphenylene-β-phenylallyl (BDPA), and lithium phthalocyanine (LiPc) are commonly encountered species, the concentrated solids of which exhibit EPR spectra that are dramatically narrower than the spectra of radicals in dilute fluid solution. For example, the spectrum of DPPH in fluid solution extends over more than 50 G (Haniotis and Guenthard, 1968), but the spectrum of the crystalline radical is about 1 G wide (Kolaczkowski, Cardin, and Budil, 1999). The exchange interaction is reduced if there are diamagnetic diluents, such as solvent molecules or impurities in the radical. This is why the line width of DPPH depends on the solvent from which it was crystallized (Kolaczkowski, Cardin, and Budil, 1999), and increases as the DPPH decomposes. The very good stacking of the planar LiPc molecules results in a very narrow line for this radical in the concentrated solid (Ilangovan, Zweier, and Kuppusamy, 2000). Although the dipolar broadening correlation function is Gaussian, in the limit of strong exchange the EPR line shapes in some systems are approximately Lorentzian (Henderson and Rogers, 1966; Scaringe, Shia, and Kokoszka, 1973; Thorpe and Hossain, 1980); consequently, these line widths are approximate measures of the relaxation times (Lloyd and Pake, 1953; Bloembergen and Pound, 1954).

The magnitude of the exchange relative to the Larmor energy and the hyperfine couplings determines the extent of narrowing. The result is a magnetic field-dependent line width that is called the "10/3 effect." This name derives from the prediction "… for Larmor frequencies well below the exchange interaction frequency the width of an exchange-narrowed resonance should be 10/3 as great as that for Larmor frequencies well above" (Rogers, Anderson, and Pake, 1959). Although the theory and initial experiment were described long ago by Anderson and Weiss (1953), relatively few examples of the magnetic field dependence have been reported.

The 10/3 effect has been demonstrated in copper salts. For example, Henderson and Rogers (1966) found that the line widths of $K_2CuCl_4 \cdot 2H_2O$ and $(NH_4)_2CuCl_4 \cdot 2H_2O$ narrowed as the frequency was increased from 3 to 60 GHz. Mauriello, Scaringe, and Kokoszka (1974) confirmed the measurements on $K_2CuCl_4 \cdot 2H_2O$. Rogers, Carboni, and Richards (1967) showed that the line widths of the linear chain material $Cu(NH_3)_4SO_4 \cdot H_2O$ narrowed from approximately 3 GHz to about 50 GHz (five frequencies), consistent with calculations for $J = 3.15$ K. Subsequently, Pleau and Kokoszka (1973) found that the line widths decreased from 9 GHz to 36 GHz for six Mn(II) complexes, Thorpe and Hossain (1980) showed the 10/3 effect to be operative within the liquid He temperature range. Scaringe, Shia, and Kokoszka (1973) showed the 10/3 effect to occur also for mixed-valence compounds. Ramakrishna and Manoharan (1984) observed the 10/3 effect in Ni(III) compounds between X-band and Q-band, although its angular dependence did not fit any existing theory.

Comparing spectra at 30 MHz, 6 GHz and 9 GHz, Rogers, Anderson, and Pake (1959) observed a narrowing of the EPR spectrum of concentrated solutions of DPPH at higher frequencies, consistent with the exchange $J \approx 6 \times 10^9 \, s^{-1}$ for a 0.4 M solution. For concentrated solid DPPH, Krzystek et al. (1997) showed that the line width decreased from X-band to Q-band, and then increased at higher microwave frequencies (magnetic fields). The increased linewidth at higher magnetic fields was not consistent with theory, and caused the authors to speculate that either the line width was dominated by g-factor anisotropy, or that there was frequency dependence of T_2. Anisotropy in g is the most likely explanation.

The EPR spectrum of the S_3^- radical in ultramarine blue is exchange-narrowed. Samples were studied from 258 MHz to 217 GHz, and the line width increased as the frequency was reduced below 2.7 GHz, as expected for the 10/3 effect (Eaton et al., 2001a).

In contrast to the above examples, the CW linewidth of $K_{0.97}B_6$ was found to be independent of magnetic field/frequency, from 1.38 to 186 GHz. The spectrum was simulated with a broad (318 G peak-to-peak) Lorentzian line, but the lack of dependence of the line width on frequency was attributed to pure dipolar interaction. This was stated to yield a Lorentzian line if a small percentage of lattice sites was occupied by paramagnetic ions (Ammar et al., 2004).

17.2.3
Conducting Systems

By using a multifrequency spectrometer with a high-pressure resonator (up to 1.6 GPa), Náfrádi et al. (2008) showed that the EPR line width of polycrystalline KC_{60} increased with microwave frequency and with pressure. For the radical cation salt (fluoranthenyl)$_2^+$ AsF_6^- at 180 MHz, $T_1 = T_2$ over the temperature range from approximately 150 to 300 K (Maresch et al., 1984).

Conduction electron spin resonance line widths in single-crystal Al increased in approximately linear fashion, from 1.27 GHz to 35 GHz (Lubzens and Schultz, 1976).

17.2.4
Metal Ions in Solution

Linewidths in fluid solution are functions of T_2, which depends on g and hyperfine anisotropy. The well-known Kivelson formula (Atkins and Kivelson, 1966) relates the EPR line width, ΔB, for each nuclear hyperfine line to the parameters a, b, and c, which depend on the anisotropic parameters in rigid lattice and the tumbling correlation time:

$$\Delta B = a + bm_I + cm_I^2,$$

where m_I is the nuclear hyperfine spin.

This equation describes many fluid solution spectra of paramagnetic metal ions, and is especially useful for $S = ½$ ions (see, for example, Chapter 10 in Weil, Bolton, and Wertz, 1994). Changing from X-band to Q-band, or to some other frequency, usually has dramatic effects on the CW-EPR line shape, because T_2 is field-dependent when anisotropies are not fully averaged. T_2^{-1} increases at a higher microwave frequency for $S = ½$ species. However, as will be discussed in more detail for Gd^{3+}, the line widths of $S > ½$ species tend to be narrower at a higher microwave frequency for a given correlation time (Bertini, Martini, and Luchinat, 1994a, 1994b). Bryant and coworkers have measured several ions in fluid solution by EPR to obtain relaxation time values relevant to NMR studies (Hernandez, Tweedle, and Bryant, 1990).

Poupko, Baram, and Luz (1974) pointed out that the prior development of the theory of dynamic frequency shifts led to the prediction that the superposition of several transitions in $S > ½$ species could result in non-Lorentzian lines shifted from the resonance frequency. The CW-EPR line widths tend to be narrower at Q-band than at X-band.

17.2.5
Pb^{3+} in Calcite

Hyperfine interactions for Pb^{3+} in calcite are so large (ca. 38 GHz) that the relaxation of this species has become a test case for multilevel spin systems (Martinelli et al., 2003). Line width measurements were made at X-, K-, Q-, and W-band, and at 190 and 285 GHz. Line widths due to the Raman process were independent of frequency. At high frequencies, the direct relaxation by modulation of the hyperfine interaction contributed significantly to the line width even at room temperature, and increased with increasing frequency.

17.3
Spin–lattice Relaxation, T_1

The name "spin–lattice" relaxation is a carry-over from the early studies of ionic solids at low temperature. In these samples, the lattice was the ionic environment in which the paramagnetic species was an impurity dopant. T_1 is also referred to as "longitudinal relaxation" in the context of the Bloch equations. The fundamental Bloch definition is in terms of a return to thermal equilibrium, without reference to the mechanisms by which this may occur. The most general view of T_1 is the conversion of spin angular momentum into thermal energy of the environment of the spins. There are many physical mechanisms by which this can occur. Building on the early phenomenological relaxation studies that gave the terms direct Raman, and Orbach, Bowman and Kevan distinguished processes and mechanisms (Bowman and Kevan, 1979). (Details of the direct, Raman and Orbach processes are provided in Section 10.6) The physical mechanism by which relaxation can occur via such processes is discussed by Eaton and Eaton (2000b), and also later in this section. The essence is that there must be a way for the electron spin to transfer energy to the lattice. Sometimes, there is an exact match of energies (see the discussion of spectral densities in the next section), in which case resonant energy transfer occurs. Sometimes, the mechanism requires more complicated descriptions as detailed in the next section and in some of the cited examples. The formulae in Chapter 10 provide the field/frequency dependence for several cases, and some examples are also provided in the following paragraphs.

The physical models raise the question of intramolecular versus intermolecular contributions. As a first approximation, it is reasonable to say that the relaxation times are characteristic of the radical and its environment, and that it is not meaningful to speak of the relaxation of a radical without considering its environment. A somewhat more refined view was suggested by the Stern–Gerlach-type measurements of Amirav and Navon (1981, 1983) who, by creating an almost collision-free environment, were able to measure the intramolecular relaxation times of nitroxyl radicals and other paramagnetic species. Under effusive conditions, the spin-flip relaxation time of tempo was about 0.5 µs, which was similar to the room temperature fluid solution value. However, in cooled supersonic beams, the relaxation time was increased to about 200 µs. It was observed that the more complex and flexible the molecule (e.g., di-tertbutylnitroxide), the faster the relaxation. The physical model is a modulation of spin interaction with the external field via spin–orbit coupling that depends on a sufficient density of vibrational and rotational states to satisfy the conservation of total angular momentum (Amirav and Navon, 1981, 1983; Knickelbein, 2004). Intramolecular rearrangements that modulate the g-tensor were identified as the cause of electron spin relaxation in metal complexes (Doddrell et al., 1978). These results are strongly supportive of the application of the model invoked in this chapter (and as described in Chapter 10) that includes Raman, thermally activated, and internal mode relaxation mechanisms.

17.3.1
Phonon Densities

Inherent in the discussions of relaxation, either by the direct process or by the Raman process, is the phonon density at the relevant frequencies. The plot in

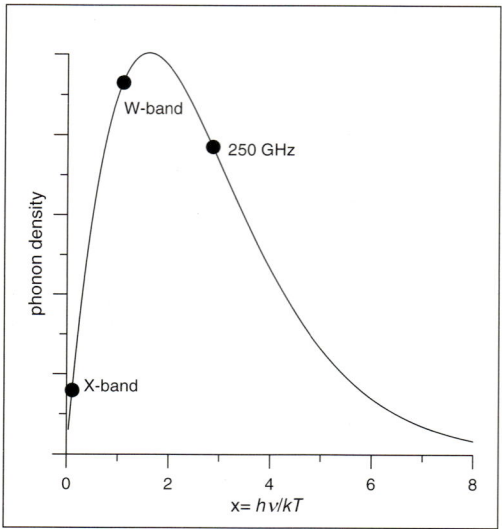

Figure 17.1 Relative phonon density as a function of the scaled phonon frequency. The points were calculated for $T = 4.2$ K.

Figure 17.1 is similar to those shown in Orton (1968, p. 154) and in Abragam and Bleaney (1970, p. 555). In this case, the *x*-axis is phonon energy (in units of kT, $x = h\nu/kT$). For X-band at $T = 4.2$ K, $x \sim 0.1$, and most of the phonon density is at higher energies. At 4.2 K, $x \sim 1.1$ for 95 GHz (W-band) and $x \sim 2.9$ at 250 GHz, which is well past the peak of the distribution curve. Since temperature is in the denominator, lower temperatures will correspond to larger values of x. The direct process–which dominates only at low temperatures–depends on the number of phonons at the relevant frequency, so that the relaxation rate would be expected to increase with an increase in microwave frequency, at a given low temperature for a given lattice. For values of x beyond the peak of the distribution curve, the higher the microwave frequency the smaller will be the number of phonons available at that frequency. Thus, there should be a frequency/temperature ratio above which the relaxation rates become slower, and this has been confirmed for cases in which the direct process dominates. For example, Witowski, Kutter, and Wyder (1997) investigated the effect of transverse acoustic phonons in the crystal $Hg_{1-x-y}Cd_xMn_yTe$ as a function of phonon energy in magnetic fields up to 23 T. At 2 K, $1/T_1$ increased strongly between 12 T and 19.5 T, but then dropped sharply. The spin–lattice relaxation depended on the direction of the wave vector of the phonons involved, on the density of phonon states, and also on the field-dependent cross-relaxation to fast-relaxing centers.

Detailed comparison with experimental data should take into consideration the fact that the Debye approximation for phonon distribution, while superior to the Einstein approximation, ignores the vibrational structure inherent in real solids;

see, for example, Ziman (1960), Zahlan (1968), Blakemore (1985), Srivastava (1990), and Pobell (2007).

17.3.2
Practical Interpretation of Relaxation Time Data as a Function of Temperature

Each relaxation process has a characteristic temperature dependence. For many samples one process will dominate the relaxation over a particular temperature range. When experimental spin–lattice relaxation rate data are available over a sufficiently wide temperature range, it is possible to model the data as the sum of contributions from the direct, Raman, local mode, Orbach, and thermally activated processes, as shown in the following equation (Zhou et al., 1999; Eaton and Eaton, 2000b; Sato et al., 2007, 2008a):

$$\frac{1}{T_1} = A_{dir}T + A_{Ram}\left(\frac{T}{\theta_D}\right)^9 J_8\left(\frac{\theta_D}{T}\right) + A_{loc}\left[\frac{e^{\Delta_{loc}/T}}{(e^{\Delta_{loc}/T}-1)^2}\right]$$
$$+ A_{orb}\frac{\Delta_{Orb}^3}{e^{\Delta_{Orb}/T}-1} + A_{therm}\left[\frac{2\tau_c}{1+\omega^2\tau_c^2}\right]$$

Care should be taken not to apply a model outside the region for which it has been shown to be applicable. For example, the direct process is known to be important at very low temperatures. A linear dependence on temperature over a region of temperature above about 10 K probably is not a direct process as described in terms of phonons in crystalline lattices. Other interpretations may be more physically meaningful. Furthermore, unless the relaxation rate has been measured over a large temperature range, multiple relaxation mechanisms may simulate the data equally well. Measurement at more than one microwave frequency may also be necessary to distinguish between relaxation mechanisms.

17.3.3
Glasses versus Crystals

Much of the theory of relaxation in low-temperature solids was derived using the models of vibrations of perfect single crystals. Studies of single crystals at low temperature led to the development of most modern concepts of electron spin relaxation (Standley and Vaughan, 1969). Most chemical and biochemical applications of EPR relaxation are in fluid, or glassy frozen, solutions. Discussion of electron spin relaxation continues to use the language of direct, Raman, and Orbach processes for spin systems other than the ionic lattices for which the expressions for these processes were originally derived. The central feature inherent in the concepts of the direct and Raman processes is the spectral density at the Larmor frequency. Although the calculations were made for phonons, any modulation of the environment of the spin by intermolecular jostling, or by intramolecular vibrations and rotations, can accomplish relaxation and may result in similar temperature dependence of the relaxation processes. Empirically, it is found that: (i) many species exhibit spin–lattice relaxation rates that increase linearly with temperature in a small temperature region at low temperatures; (ii) the spin–lattice relaxation for $S > ½$ metal complexes exhibit strong temperature dependence, as predicted for the Raman process at low temperature; and (iii) at a temperature near and above the Debye temperature, the spin–lattice relaxation rate increases approximately proportional to T^2, as predicted for the high-temperature limit of the Raman process (Zhou et al., 1999; Eaton and Eaton, 2000b). These behaviors are observed for many samples in dilute frozen solutions of aqueous and organic solvents, for organic radicals doped in diamagnetic molecular crystals, as well as for metal ions doped into ionic crystals. Thus, even though the assumption of phonons in ionic lattices is not fulfilled, the predicted trends apply, and hence the labels continue to be used.

The length scales in glasses are shorter and more irregular than those in crystals, so the phonon distribution in glasses is different from that in crystals. Measurements have been reported of phonon localization in glasses, and there is a universal temperature-independent region of the thermal conductivity in the range 1 to 30 K for glasses (Graebner, Golding, and Allen, 1986). The correlation length, and hence the phonon localization length scale, in glasses was judged to be of the order of 20–50 Å. In glasses, the T^2-dependence of relaxation by the Raman process is expected to begin at a lower temperature than in a crystalline system with a similar Debye temperature (Huber, 1982). Optical spectroscopy line widths in glasses display T^2-dependence at a temperature above about half the nominal Debye temperature. This was attributed to the increased density of low-frequency vibrational modes in glasses, relative to the Debye model (Huber, 1987). It is reasonable to speculate that similar considerations apply to electron spin relaxation, resulting in both a Raman-like temperature-dependence of T_1 and a T^2-dependence of T_1 at temperatures lower than would be predicted by the Debye temperature. This analogy also explains why the fitting of the Raman process to experimental T_1 data for glassy samples results in a lower-than-expected Debye temperature (Zhou et al., 1999; Eaton and Eaton, 2000a; Sato et al., 2007).

In general, it has been observed that relaxation rates are somewhat faster in glasses than in crystalline lattices. The slopes of the plots of $\log(1/T_1)$ versus temperature (ca. 10 to 160 K) were very similar for nitroxyl radicals dissolved in common solvents, spin-labeled proteins in water:glycerol, and a small spin label doped into a diamagnetic host (Zhou et al., 1999). Small differences in rates were interpreted in terms of hydrogen bonding and other restrictions on motion (Du, Eaton, and Eaton, 1995b). Over the temperature range of 20 to 80 K, $1/T_1$ for two nitroxyl radicals in glassy and crystalline ethanol increased as $T^{2.6}$ (Kveder et al., 2006). The rate was somewhat faster for the same nitroxyl in glassy than in crystalline ethanol, and the derived Debye temperatures were smaller for the glassy matrix than for the crystal, although the Raman model fitted both environments equally well. These observations were not restricted to nitroxyl radicals. The Cu(II) complex of diethyldithiocarbamate was studied in glassy 2:1 toluene:chloroform and doped into the diamagnetic Ni(II) analog (Du, Eaton, and Eaton, 1995a). Between 25 and 60 K, $1/T_1$ was systematically slightly faster in the glassy solution than in the doped solid, but the slopes of the rate versus temperature curves were almost the same until the solvent softened, at which point the relaxation rate increased more rapidly for the solution than for the doped solid (Du, Eaton, and Eaton, 1995a) because of tumbling-dependent processes. Similar patterns were observed for vanadyl, Cu(II) and Ag(II) porphyrins in glassy solvents and doped solids (Du, Eaton, and Eaton, 1996).

The T^2 dependence, characteristic of the Raman process, was observed for the relaxation of a triplet spin probe in glassy o-terphenyl, and it was suggested that the relaxation was due to motion of the probe within the glass (Kaiser and Friedrich, 1991).

Analogous results have been obtained in NMR studies of glasses and crystals. Rubinstein and Resing (1976) reported that values of ^{11}B T_1 were shorter in glassy B_2O_3 than in crystalline B_2O_3. T_1 was proportional to T^2 above the Debye temperature in both the glassy and crystalline samples, as expected for the high temperature limit of the Raman process.

In limiting cases, it is common to speak of "solid-state effects" and "solution-state effects" on relaxation. However, to the extent that intramolecular vibrations modulating spin–orbit coupling dominate the relaxation mechanism, one would expect little change in slope of T_1^{-1} versus T, as the solid softens to become a fluid. If the viscosity of the liquid phase is high, the radicals exhibit little or no change in the slope of T_1^{-1} versus T as the solvent melts (Owenius, Eaton, and Eaton, 2005; Sato et al., 2008a).

17.3.4
Spectral Diffusion and Cross-Relaxation

Beyond the cases in which the theory explicitly includes the magnetic field in the expression for spin relaxation, there can be a magnetic field effect because of the dispersion of the spectrum due to g anisotropy. For a given g dispersion, and a given spin concentration (in units of spins per cm^3 or moles per liter), a higher magnetic field will mean fewer spins per gauss. Relaxation mechanisms that rely on overlap of the spectral lines are diminished at higher magnetic fields where the spectral overlap is less. High magnetic fields and frequencies can bring other energy separations into resonance, such as transitions that are between states separated from the ground state by a ZFS.

When the relaxation rate that is actually measured is dominated by spectral diffusion, and not T_1^{-1} or T_2^{-1}, as often is the case in CW power saturation measurements, the increased spectral dispersion at higher field can result in a decrease in the effective relaxation rate because the spins per gauss, and, hence, this contribution to spectral diffusion, are decreased (Ghim et al., 1995).

There are special cases of apparent magnetic field dependence that involve resonant cross-relaxation for two species with different g and nuclear hyperfine values (Larson and Jeffries, 1966). The tradeoffs between g-factor dispersion of the spectrum and magnetic field dispersion of separate lines in the spectrum gives rise to the variety of effects that result in an optimum magnetic field for resolution of the EPR spectrum of any particular species.

17.3.5
Effect of Pairs and Clusters

During the early stages of these investigations, many of the studies of relaxation were focused on ions in ionic lattices. Following the statistics of Behringer (1958), at a 1% doping level of a simple cubic lattice, 5% of the dopant would exist as pairs, and 0.4% would be in triples. In real systems, charge compensation, lattice strain, and so on can cause deviations in either direction from the random distribution calculations. However, the likelihood remains that significant fractions of the paramagnetic ions would exist in pairs, and a nontrivial fraction would exist in triples. It should be noted also that closed triples (where each is a neighbor of the other two) would exhibit spin frustration; that is, the antiferromagnetic exchange interactions could not be satisfied over more than two of the three edges at the same time. Since, at the time, a 1% doping was considered to be a relatively low level, it was important to assess the role of pairs and clusters in relaxation. At the low temperatures used to conduct most of the studies, the direct process dominated. Under these conditions, pairs and/or triples made major contributions to the relaxation, because relaxation was faster in the pairs or triples, and cross-relaxation between isolated ions and the pairs/triples enhanced the relaxation for all spins in the sample. The following references provide an entrée to this literature: Van Vleck (1960); Van Vleck (1961); Abragam and Bleaney (1970, p. 522); Gill (1962); Atsarkin (1966); Harris and Yngvesson (1968); Dugdale and Thorpe (1969); Al'tshuler, Kirmse, and Solov'ev (1975). At the concentrations of paramagnetic species present in magnetically dilute solutions (mM or less) of molecular species, the concentrations of pairs and triples present in randomly oriented samples is much lower than was considered in the early studies. However, the earlier studies made the key point that when there is strong spin–spin interaction between paramagnetic ions, such as in dimers or trimers, the energy levels are different from the monomers, and this can result in enhanced relaxation.

The complicated temperature and field dependence of relaxation in $NiSiF_6 \cdot 6H_2O$ was attributed both to concentration effects and to phonon bottleneck effects (van der Bilt and van Duyneveldt, 1980).

17.3.6
Magnetic Field Dependence of Relaxation

In the following subsections, the term "field dependence" means that both the field and the resonant frequency were changed, such that measurements were performed at a constant g-value. The dependence of T_1 on magnetic field position within a spectrum obtained at a fixed microwave frequency is discussed in a separate section.

17.3.6.1 The Direct Process
The mechanism of the direct process is a resonant exchange of electron spin Larmor energy with vibrational modes of the lattice (phonons). This process is

characterized by a linear increase of the relaxation rate with an increase in temperature, and an increase in rate with an increase in the magnetic field. The transition moments for the direct process were originally derived for nonlocalized vibrations (phonons) of ionic crystals (Bowman and Kevan, 1979). However, spectral densities at the relevant energies are non-negligible even in covalent lattices. Consequently, the trends, if not the absolute magnitudes, carry over to spin environments other than those for which the formal derivations were made.

It was shown in Chapter 10.6.1 that the direct process results in a relaxation rate that is proportional to B^2 (non-Kramers' species, even number of spins) or to B^4 (Kramers' species, odd number of spins) (Standley and Vaughan, 1969). This predicted dependence on B^4 has been demonstrated for Fe^{3+} in $K_3Co(CN)_6$ from 3.9 to 12.4 GHz at temperatures between 2.16 and 1.38 K (Davids and Wagner, 1964). The same species exhibits T_1^{-1} proportional to T^9 (Rannestad and Wagner, 1963), which suggests that the Raman process dominates at higher temperatures. The faster relaxation rates at higher magnetic fields have been exploited to use a higher observe power to improve the signal-to-noise ratio (SNR) of otherwise too-slowly relaxing species (Muller et al., 1989). Honig and Stupp (1958) measured the relaxation rate of phosphorus-doped silicon at 1.27 K and 2.06 K in fields up to about 10 kG; in this case there was a near-quartic dependence on the field at high fields. Chang and Yang (1968) verified the expected B^4 dependence of the relaxation rate of Mo^{5+} in TiO_2 at approximately 1 K.

The relaxation times of F-centers in irradiated LiD were estimated to be about 1 s at 136 and 182 GHz, but about 2.8 s at 71 GHz (Bouffard et al., 1980). The exact temperature was not stated, but the context indicated that it was between 0.43 K and 1.9 K. Hence, the direct process was likely to dominate the electron spin–lattice relaxation, and the relative values were consistent with expectations for the direct process. Witowski and coworkers (Strutz, Witowski, and Wyder, 1992; Witowski and Bardyszewski, 1995; Witowski, Kutter, and Wyder, 1997) showed that, within the temperature range where the direct process dominated, the electron spin relaxation rate increases at higher magnetic fields were more accurately simulated by replacing the Debye approximation with a more realistic description of the phonon distribution. The rate in CdMnTe increased as $B^{1.5}$ to approximately 15 T and as B^5 at higher fields.

Caution There are cases in which relaxation rates increase linearly with temperature in temperature regimes at which the Raman or other process would be expected to dominate over the direct process. This occurs commonly when the spin concentration is too high to be considered magnetically dilute. A longstanding puzzle that the relaxation rate for DPPH in polystyrene increases linearly with temperature (Turkevich, Soria, and Che, 1972) is likely explained by the fact that the sample was 5% DPPH in polystyrene. The authors have seen this effect in many samples in which the bulk concentration is high, or in which local aggregation of radicals has occurred (for an example, see Sato et al., 2008b). The linear dependence of relaxation rate on temperature may be due to the partial thermal excitation or modulation of interactions between pairs. A linear dependence on temperature is also the high-temperature limit of relaxation via two-level systems (TLS; also called tunneling level systems) (Bowman and Kevan, 1977; Stutzmann and Biegelsen, 1983; Misra, 1998).

17.3.6.2 The Raman Process

For $S = ½$, the relaxation rate for the Raman process below the Debye temperature is proportional to T^9 (second-order Raman process) or $T^7 B^2$ (first-order Raman process). Abragam and Bleaney (1970; p. 565) stated that the first-order Raman process, which depends on $B^2 T^7$, is usually not very important except at very low temperatures, where the direct process is likely to dominate. Thus, as a practical matter, for most cases the relevant Raman process depends on T^9 and is independent of magnetic field. However, both the $B^2 T^7$ and T^9 terms have been observed in cobalt tutton salts (Van Duyneveldt and Pouw, 1973). A T^9 dependence was observed for rare earth ions in LaF_3 (Schulz and Jeffries, 1966). For even spin systems, the Raman process is predicted to be proportional to T^7 but independent of magnetic field (Standley and Vaughan, 1969; Pake and Estle, 1973). Above about half of the Debye temperature, where many phonons exist with energies greater than the resonance energy, the dependence decreases from T^7 to T^2. The higher the microwave frequency, the closer the EPR transition energy is to the Debye cut-off, and the relaxation rate will not increase as strongly with increasing temperature as at 9 GHz. None of the data yet published confirm these predictions. The limits of application of the two Raman processes are discussed by Rannestad and Wagner (1963). A T^2 dependence is also predicted for relaxation via TLS, although such relaxation is usually observed at very low temperatures, and is also predicted to be field-dependent (Stutzmann and Biegelsen, 1983; Misra, 1998).

The T_1 relaxation times of Cr in ruby at 34.6 GHz did not depend on microwave frequency (Pace, Sampson, and Thorp, 1960); rather, the results obtained at 7.2, 9.3, 24.2 and 34.6 GHz were judged to be "of the same order." At 1.4 K, T_1 was different for different transitions, so that the relaxation time was not the same for all levels. The relaxation times approached a common value at 77 K.

The relaxation of paramagnetic centers (P1 and P2) in diamond was studied at X-band and W-band (Terblanche and Reynhardt, 2000). The P1 centers are substitutional N impurities in the diamond, while the P2 centers are vacancies with three N in neighboring positions. These centers relax via a field-independent process at room temperature, with $T_1 = 2.2$ ms. The authors attributed the relaxation to a second-order Raman process.

The usual formulae apply to dilute spins. There is an extensive literature on magnetically concentrated samples, such as crystals or powders of transition metal salts. In these cases, the Raman process has a form that depends on the relative magnitudes of hyperfine, dipolar, and exchange interactions, and T_1 usually depends on the magnetic field (Nogatchewsky, Ablart, and Pescia, 1977).

17.3.6.3 The Orbach Process

The Orbach process is independent of the magnetic field, except that the field may alter the energy of the excited state. High magnetic fields and frequencies can bring other energy separations into resonance, such as transitions that are between states separated from the ground state by a ZFS.

A strong field dependence was observed by Aminov et al. (1997) for Nd^{3+} in $Y_3Al_5O_{12}$ crystals. When the frequency was increased from 9.25 to 36.4 GHz, the relaxation rate increased by a factor of 10. The effect on the rate was interpreted in terms of the degeneracy of the excited level. In other cases cited in the same report, a B^2 dependence was observed for Yb^{3+} in BaY_2F_8, whereas a B^{-2} dependence was observed for Tb^{3+} in $CaWO_4$ (see Aminov et al. 1997 for details of this relaxation process). At $T > 10$ K, two Orbach relaxation pathways were identified for Cr(V)-doped K_3NbO_8, and relaxation rates at higher temperatures were independent of the frequency (Nellutla et al., 2008).

17.3.6.4 The Thermally Activated Process

Electron spin interactions with the environment (lattice) may be thermally activated, such as the rotation of a methyl group (Harbridge, Eaton, and Eaton, 2002, 2003), motions of a hydrogen-bonded proton, or motion of a molecule as a glassy solvent softens (Bowman and Kevan, 1979, p. 85). The frequency dependence of relaxation rate for a thermally activated process depends on spectral density functions that have the form $\frac{\tau_c}{1+\omega^2\tau_c^2}$. Collisional modulation of spin–orbit coupling could also have this frequency dependence (Bowman and Kevan, 1977, 1979). These contributions to relaxation have a strong dependence on microwave frequency, with the maximum relaxation rate occurring when $\omega \sim \tau_c$; at X-band, $\omega \sim 6 \times 10^{10}$ rad s^{-1}. Bowman argued that in the solid state, usually $\tau_c \gg \omega^{-1}$ so that the frequency-dependent term would become $1/(\omega^2\tau_c)$ and increasing frequency would decrease the relaxation rate. The rate of change with temperature would then depend on the activation energy.

The rate of relaxation via modulation of tunneling modes in glassy matrices is predicted to depend on B_0^2 (Bowman and Kevan, 1977). In some systems, the relaxation of which was interpreted in terms of tunneling mechanisms, a slight decrease in relaxation rate was observed at 16.5 GHz relative to 9.5 GHz (Kurtz and Stapleton, 1980; Askew et al., 1984).

17.3.6.5 Local Modes

Somewhat analogous to an Orbach relaxation process, relaxation may occur due to a specific vibrational mode local to the unpaired electron (Murphy, 1966). Such a vibration would change the detailed electronic environment of the unpaired spin, and consequently change the orbital angular momentum of the spin. For species with a non-zero orbital angular momentum, the spin angular momentum couples to the orbital angular momentum (spin–orbit coupling), and thus to the thermal bath that excites the vibrational modes. The net result is a T_1 relaxation mechanism. Sometimes, the local mode required to fit the temperature dependence of the relaxation rate can be identified with a vibrational mode that is reasonable for the molecular species being studied. Relaxation via a local mode is frequency independent.

Between 1.1 and 9.2 GHz, the T_1 for the E' center created by the γ-irradiation of fused SiO_2 was approximately independent of frequency, and the relaxation process was assigned as a local mode (Ghim et al., 1995). There was an approximately 50% increase in T_1, from 9.2 GHz to 19.4 GHz, and preliminary results indicated a further increase to 94 GHz. It was pointed out that the energy of an EPR quantum at 94 GHz (ca. $3 cm^{-1}$) is of the order of some of the vibrational modes that had been inferred from temperature-dependence studies. Consequently, the assumption that energies are much larger than the EPR quantum may no longer have been well satisfied for those modes, and hence the relaxation would become slower (Ghim et al., 1995). Quantitative comparisons of T_1 as a function of frequency, even among this data set, must take into consideration that, at a given magnetic field and microwave frequency, the relative contribution of different orientations to the set of spins measured in the relaxation experiment is not constant, and since T_1 was shown to depend on the position in the line, it is not possible to measure T_1 of the same subset of spins at different frequencies.

Over narrow temperature ranges, and with only one microwave frequency, it is difficult (or even impossible) to distinguish between relaxation via a local mode and via a thermally activated process. A multifrequency and variable temperature study was performed of nitroxyl, copper and vanadyl species doped into diamagnetic hosts (Eaton et al., 2001b). Below about 50 K, relaxation of the vanadyl porphyrin was faster at W-band than at X-band, because of a much larger contribution of the direct process. For the copper diethyldithiocarbamate complex, relaxation was only slightly faster at W-band than at X-band, although the rates at both frequencies were faster for copper than for vanadyl. Above ca. 100 K for both vanadyl and copper, the relaxation was dominated by the Raman process and a local mode, and the rates were frequency-independent.

17.3.7
Dependence of Relaxation on Magnetic Field Position in a CW-EPR Spectrum

The dependence of T_1 on magnetic field position within a spectrum obtained at a fixed microwave frequency is discussed in this section.

T_1 depends on the mixing of spin and orbital angular momentum. When one examines the local environment of the spin, it is evident that there are more vibrational modes that can accomplish spin–orbit mixing in some directions than in others. For example, for a threefold symmetric site, there are more vibrational modes in the plane perpendicular to the principal axis of rotation than along the axis; hence, the relaxation rate is slower along the principal axis than perpendicular to it.

The many field-dependent terms, variously called g-strain and A-strain (Hyde and Froncisz, 1982), confound attempts to compare relaxation times measured at two different magnetic fields (microwave frequencies) for the same set of spins, because the same spectral components cannot be measured at two different microwave frequencies. Unless an isolated EPR line of an oriented single crystal is measured, there will be overlap of transitions in powder distributions of transitions. If, as is commonly the case, the relaxation times depend on the orientation of paramagnetic species in the magnetic field, the spins that are within the excitation and observation window at X-band will differ from those at some other frequency, such as Q-band or W-band. Another practical matter must also be considered, namely that it is much more difficult to remove oxygen for the geometry of samples used in some higher-field spectrometers, so that any residual paramagnetic O_2 in the sample could shorten the relaxation time. In some W-band spectrometers, the resonator and sample geometry renders gas exchange by diffusion fairly quick and effective (Froncisz et al., 2008).

Lee, Patyal and Freed (1993) found T_1 to be orientation-dependent for irradiated malonic acid. The T_1 for the E' signal created by γ-irradiation of fused SiO_2 depends on position in the line (Eaton and Eaton, 1993), which was interpreted in terms of coupling to local vibrational modes (Ghim et al., 1995). Kordas (1992) reported that the T_1 of the E' center at L-band and X-band "... reveals a dependence on the external magnetic field," but did not provide any specifics.

For nitroxyl radicals doped into solid diamagnetic hosts or in glassy solvents, T_1 depends on the position in the spectrum (Du, Eaton, and Eaton, 1995b). T_1 is longest along the parallel direction, and shortest in the perpendicular plane, for both ^{14}N and ^{15}N radicals. The same effect was observed for oriented single crystals and for orientations within a frozen solution spectrum. Spin–orbit coupling is believed to be the dominant spin–lattice relaxation mechanism for nitroxyl radicals in rigid media. The relaxation times were interpreted in terms of the larger number of vibrational modes that are available for modulating spin–orbit coupling perpendicular to the z-axis than parallel to it.

The spectral dispersion achievable at 140 GHz permitted orientational selectivity of the T_1 of the tyrosyl radical in ribonucleotide reductase, and discrimination between isotropic and anisotropic interactions between this radical and the diiron cofactor (Bar et al., 2001).

17.3.8
Case Studies of Experimental Data

Experimental assessments of relaxation mechanisms, and especially comparisons of relaxation as a function of microwave frequency, should ensure that concentration dependence, accuracy of temperature at the sample, position in the line, and method of measurement are all taken into account.

17.3.8.1 Nitroxyl Spin Labels

Currently, there is a strong focus on nitroxyl radical EPR to probe motion, and for protein structure determination using relaxation enhancement, DEER, and DQC. Chapter 11 describes in detail the use of multifrequency EPR to measure molecular motion.

Relaxation in Rigid Lattice The T_1 of tempol (4-hydroxyl-2,2,6,6-tetramethylpiperidinoloxy) doped into a diamagnetic host was studied at S-, X-, and W-band (95 GHz)

(Eaton et al., 2001b). Below about 130 K, the Raman process dominated and the relaxation was independent of frequency. At higher temperatures the relaxation time increased in the order S-band (2–4 GHz) < X-band < W-band (95 GHz). This trend was interpreted in terms of a model in which T_1 has a small contribution from a thermally activated process that was assigned to rotation of methyl groups on the ring.

In glassy sucrose octaactetate, spin–lattice relaxation rates at temperatures between 100 and 295 K for tempone and for a nitroxyl without ring methyl groups, as well as galvinoxyl radical and BDPA, were indistinguishable at X-band and Q-band, which is consistent with a local mode and not a thermally activated process (Sato et al., 2007).

For tempone in glassy 1 : 1 water : glycerol, or in glycerol below the glass transition temperature (T_g), deuteration of the radical decreased the relaxation rates by a factor of 1.3, which was approximately the ratio expected for the isotope effect on a vibration with a large C–H(D) contribution (Owenius et al., 2004; Sato et al., 2008a). In the same solvents, the replacement of ^{14}N by ^{15}N in tempone had a negligible impact on spin–lattice relaxation below T_g, which was consistent with the prediction that the modulation of hyperfine anisotropy is not a major contributor to relaxation in rigid lattices (Sato et al., 2008a).

Relaxation Processes in Fluid Solution Multiple processes have been identified that might contribute to relaxation in fluid solution, and their relative significance depends on the tumbling correlation time and resonance frequency. The relaxation rate for spin-rotation process is described (Atkins, 1972) as:

$$\frac{1}{T_1^{SR}} = \frac{\sum_{i=1}^{3}(g_i - g_e)^2}{9\tau_R}$$

where τ_R is the tumbling correlation time, which is frequency-independent. Mailer, Nielsen, and Robinson (2005) derived expressions for spin rotation involving anisotropic motion, which is frequency-dependent.

Molecular tumbling affects relaxation via a modulation of g anisotropy (sometimes called the "chemical shift anisotropy contribution"), and by the modulation of nuclear hyperfine coupling (sometimes called the "electron–nuclear dipolar contribution") (Robinson, Haas, and Mailer, 1994; Andreozzi et al., 1999; Robinson et al., 1999c; Owenius et al., 2004; Mailer, Nielsen, and Robinson, 2005):

$$\frac{1}{T_1^{g,A}} = (C_g + C_A)J(\omega) = C_{A,g}J(\omega)$$

$$C_g = \frac{2}{5\hbar^2}\left(\frac{\Delta g^2}{3} + \delta g^2\right)\mu_B^2 B^2$$

$$C_A = \frac{2}{9}I(I+1)\sum_i (A_i - \bar{A})^2$$

where $\Delta g = g_{zz} - 0.5(g_{xx} + g_{yy})$, $\delta g = 0.5(g_{xx} - g_{yy})$, A_i is a component of the nitrogen nuclear hyperfine coupling, $\bar{A}$ is the average nitrogen hyperfine coupling, I is the nitrogen nuclear spin, and $J(\omega)$ is the frequency-dependent spectral density function, typically:

$$J(\omega) = \frac{1}{1+\omega^2\tau^2}.$$

It has also been proposed that a generalized spin diffusion mechanism can contribute to nitroxyl spin–lattice relaxation (Mailer, Nielsen, and Robinson, 2005). This mechanism involves the modulation of through-space dipolar interaction of the electron spin with nuclei in the nitroxyl or the solvent. One contribution, due to rotation of nitroxyl methyl groups is represented (Mailer, Nielsen, and Robinson, 2005) by:

$$\frac{1}{T_1^{GSD-A}} \alpha \frac{\omega \tau_{therm}}{1+(\omega\tau_{therm})^2},$$

where τ_{therm} is the correlation time for the thermally activated motion. In the slow motion limit, this contribution to relaxation is proportional to $1/\omega$. Modulation of the interaction with solvent nuclei is represented (Mailer, Nielsen, and Robinson, 2005) by

$$\frac{1}{T_1^{GSD-B}} \alpha \left(\frac{2\omega_x \tau_d}{1+(\omega\tau_d)^{3/2}}\right)^{1/4},$$

where τ_d is the relative solvent-nitroxyl diffusion rate, ω_x is the reference frequency (2π 9.3 GHz), and ω is the angular frequency at which the experiment is performed. In the slow motion limit, the GSD-B contribution is proportional to $\omega^{-3/8}$.

Robinson, Mailer, and Reese (1999a) showed that apparent inconsistencies between approximate treatments of the m_I dependence of relaxation for nitroxyl radicals in the fast-motion limit are removed by extending the operator method of Abragam. In agreement with experiment, they predict that T_1 for nitroxyl radicals is independent of m_I (Robinson, Mailer, and Reese, 1999a, 1999b).

Collision with molecular oxygen increases the relaxation rate independent of the microwave frequency (Hyde et al., 2004).

It has been proposed that for tempone at X-band, solvent spin diffusion dominates for $\tau_c > 10^{-9}$ s (Mailer, Robiinson, and Haas, 1992; Robinson, Haas, and Mailer, 1994). However, if spin diffusion dominates relaxation, then solvent deuteration should cause a dramatic decrease in relaxation rate. For tempol in water:glycerol, the effect of solvent deuteration on relaxation at X-band and S-band was negligible (Owenius et al., 2004), which was not consistent with spin diffusion. Alternate proposals for the dominant contributions at longer tumbling correlation times are a thermally activated process (Owenius et al., 2004) or a local mode process analogous to what is observed in glassy solvents (Sato et al., 2008a).

For small water-soluble spin labels in water:glycerol mixtures at 20 °C, and for spin-labeled stearic acids in DMPC liposomes at 25–37 °C, the spin–lattice relaxation rates decrease with increasing microwave frequency between L-band (1.9 GHz) and Q-band (35 GHz) (Hyde et al., 1990, 2004; Owenius et al., 2004; Mailer, Nielsen, and Robinson, 2005). Relaxation rates are slower for ^{15}N-nitroxyls than for ^{14}N-nitroxyls (Hyde et al., 2004; Owenius et al., 2004; Sato et al., 2008a), consistent with significant contributions from modulation of hyperfine anisotropy by molecular tumbling (the END contribution) (Robinson, Mailer, and Reese, 1999a; Owenius et al., 2004; Mailer, Nielsen, and Robinson, 2005). Deuteration of tempol decreases relaxation at L-, S- and X-band in water:glycerol mixtures at room temperature (Owenius et al., 2004), consistent with a contribution for the GSD-A term. However, the deuteration of tempone had a negligible impact on spin–lattice relaxation at X-band in 1:1 water:glycerol near 295 K (Sato et al., 2008a).

Mailer, Nielsen, and Robinson (2005) further analyzed the multifrequency saturation recovery measurements of T_1 of nitroxyl radicals, including anisotropic motion, and fitted the theory to the L-, S-, X-, and K-band data from the Hyde laboratory for spin-labeled stearic acids in lipsomes (Hyde et al., 2004). For the range of data interpreted, the primary relaxation mechanism responsible for the microwave frequency dependence of the relaxation was the END, the interaction of the electron spin with the nuclear spin of the N and H in the spin label itself. At higher frequencies, analysis of the data indicated that spin rotation and generalized spin diffusion mechanisms dominated. At K-band, the chemical shift anisotropy (CSA) mechanism (modulation of g anisotropy) became important.

By using a 95 GHz pulse spectrometer, Hofbauer et al., (2004) found T_1 = 97 ± 6 ns for tempo in decane solvent at room temperature. This was described as "... significantly shorter than T_1 values typically obtained for similar nitroxides in solution at lower frequencies." Hyde and coworkers measured T_1 for nitroxyls at W-band by SR and pulsed ELDOR, and found them shorter than at Q-band (Froncisz et al., 2008). There appears to be a mechanism for coupling the spin angular momentum to the lattice that becomes more effective at higher frequency, and is also effective for both the small nitroxyls and for the spin-labeled cholestane and stearic acids, which presumably are in very different motional regimes, unless the motions of importance are within, or directly affect, the nitroxyl moiety itself. One such possibility would be collisions of solvent molecules (water) with the nitroxyl, affecting mixing of the spin angular momentum with the orbital angular momentum. If this is a valid model, then at frequencies much higher than W-band, or in much more viscous solvents (and not only aqueous glycerol), the relaxation time should become longer again.

The possibility of a vibrational modulation of spin–orbit coupling analogous to solid-state mechanisms does not appear to have been included in these analyses. An important judgment expressed by Mailer, Nielsen, and Robinson (2005) was that "... when a greater library of multifrequency time-domain data becomes available, we anticipate that more detailed models will become necessary."

Nitroxyl Spin Relaxation in Low Magnetic Fields Much interest has been shown in the relaxation of radicals at low fields, because of the use of low magnetic fields and low radiofrequency (RF)/microwave frequencies for *in vivo* EPR and dynamic nuclear polarization (DNP). Note that at very low frequencies, forbidden lines occur in addition to the familiar three lines observed for nitroxyl radicals at X-band (Lloyd and Pake, 1954; Lurie, Nicholson, and Mallard, 1991). For a discussion of the energy levels at low magnetic fields, see Atherton (1973, p. 114ff) and Pake and Estle (1973). Peroxylamine disulphonate (Fremy's salt) at resonance near 30 G exhibits six unequally spaced energy levels. CW saturation was used to monitor the relaxation behavior, and the relaxation mechanism was identified as spin–orbit coupling (Lloyd and Pake, 1954). The T_1 of tanol was longer at X-band than at low field, and at low field the relaxation rates were different for the five nitrogen hyperfine lines (Sünnetçioğlu and Bingöl, 1993; Horasan et al., 1997; Sert et al., 2000). In very low magnetic fields, the traditional high-field calculations for hyperfine-induced and spin-rotational interaction relaxation do not yield correct results, and new equations have been derived (Fedin, Purtov, and Bagryanskaya, 2001; Fedin et al., 2002, 2006; Fedin, Purtov, and Bagyanskaya, 2003). Hyperfine effects result in slower relaxation at lower fields, while spin rotation results in a faster relaxation at lower fields. A time-resolved EPR study conducted by Bagryanskaya et al. (2002) showed that the signals detected in high magnetic field decayed an order of magnitude more rapidly than did those at low magnetic field, due to hyperfine-induced spin relaxation.

Banci, Bertini, and Luchinat (1991, p. 37) asserted that the cross-relaxation rates for a system of $S = ½$, $I = ½$ due to dipolar coupling will be:

$$\frac{1}{T_{x1}} = \frac{1}{T_{1e}}(A_x + A_y)$$

$$\frac{1}{T_{x2}} = \frac{1}{T_{1e}}(A_x - A_y)$$

If, for example, the T_1 rate is 10^6s^{-1}, then any value of A_i ($I = x, y$) will make the T_{1x} and T_{2x} rates faster, unless $x = y$. These forbidden transitions become partially allowed by the mixing of the electron and nuclear spin wave functions. Based on the reports by Fedin, Lurie, and others cited above, one would expect this mixing to be more important at lower magnetic fields. As the g anisotropy term yields a rate that is proportional to B^2, the relaxation rate would increase with the magnetic field. The A anisotropy term is not directly dependent on the field, although it might be anticipated that the full theory would include field dependence due to mixing at lower field.

17.3.8.2 Semiquinones

Tetrachlorosemiquinone and tetrabromosemiquinone have g anisotropy that is similar to that of nitroxyls, but a much smaller hyperfine anisotropy, so the modulation of hyperfine anisotropy is a minor contributor to the relaxation of semiquinones. The dominant contributor for semiquinones in fluid solution is spin rotation (Prabhananda and Hyde, 1986), which is frequency-independent. The modulation of g anisotropy by molecular tumbling (see the discussion on nitroxyl radicals) causes T_1 to increase with increasing microwave frequency for $\omega\tau < 1$. However, the observation that T_1 was nearly equal at 1.15 GHz and 9 GHz indicated that this was not an important contribution to T_1 for these radicals, and that spin rotation dominates.

17.3.8.3 Triarylmethyl (Trityl) Radicals

Trityl radicals are used for *in vivo* oximetry, so their relaxation properties are of intense interest (Halpern et al., 1994). Is there an optimum frequency at which to use the relaxation of trityl radicals to measure collisions with oxygen? In a step towards answering this question, and towards designing optimum trityl molecules for oximetry, T_1 has been measured as a function of solvent viscosity and temperature, at 250 MHz, and 1.5, 3.1, and 9.2 GHz (Yong et al., 2001; Fielding et al., 2005; Owenius, Eaton, and Eaton, 2005). At room temperature in fluid solution, T_1 was longer at higher frequency; this result was the opposite of that observed at low temperature, where the direct process makes T_1 shorter at higher frequency. The dependence of T_1 on viscosity was much stronger at 250 MHz than that at higher microwave frequencies (Owenius, Eaton, and Eaton, 2005). The tumbling correlation times were of the order of $1/\omega$ at 250 MHz, which makes modulation of both intermolecular and intramolecular electron–proton dipolar interaction an efficient relaxation process. The long relaxation times of the trityl radicals – for example, $T_1 = 17\,\mu s$ for the deuterated symmetric trityl in H_2O solution at room temperature – is consistent with the interpretation that anisotropic interactions are significant contributors to relaxation in other species with larger anisotropies and shorter relaxation times.

17.3.8.4 DPPH

At 60 MHz Larmor frequency, the line width of solid DPPH and its power saturation was fitted with $T_1 = T_2 = 63\,ns$ (Lloyd and Pake, 1953). A wider line width was observed at X-band. This was interpreted as possibly indicating a shorter T_1 at higher frequency, but the effect of g anisotropy, or differences between samples in different laboratories, was not discussed.

Bloembergen and Wang (1954) used a saturation method to measure $T_1 = T_2 = 62\,ns$ at 77 and 300 K at X-band. Goldsborough, Mandel, and Pake (1960) found that the relaxation times depended on the solvent from which the DPPH was crystallized. At both 77 K and room temperature, and at 10 and 24 GHz, $T_1 = T_2 = 80\,ns$ when crystallized from chloroform; however, when crystallized from benzene $T_1 = 50\,ns$ and $T_2 = 24\,ns$. Mandel (1962) obtained $T_1 \approx 50\,ns$ at 300 K, also at X-band.

17.3.8.5 Conducting Spin Systems

For the radical cation salt (fluoranthenyl)$_2^+$ AsF$_6^-$, at 180 MHz $T_1 = T_2$ over the temperature range from ca. 150 to 300 K (Maresch et al., 1984). At 9.8 GHz down to about 180 K, $T_1 = T_2$ and were equal to the values at 180 MHz. However, below 180 K, T_1^{-1} at 9.8 GHz goes through a maximum and remains below the rate at 180 MHz. This was tentatively attributed to a structural phase transition at ca. 174 K

(Maresch et al., 1984). The formulae in this report were consistent with a lack of any frequency dependence between 180 MHz and 9.8 GHz, and with $\omega^{-1/2}$ dependence at higher frequencies. However, this would predict a 95 GHz value of 3.9, instead of 1.4 µs as observed by Prisner (2004).

Mizoguchi, Kume, and Shirakawa (1984) found that relaxation in polyacetylene as a function of frequency from 5 to 450 MHz fitted to $1/T_1 \propto 1/\sqrt{\omega}$. The data at X-band obtained by others also lies on their curve. Mizoguchi, Kume, and Shirakawa, (1984) stated that for "fixed" spins, as opposed to the diffusing spins in polyacetylene, $1/T_1 \propto \omega^{2n}$, where n is between 0 and 2 for direct, Raman, and Orbach–Aminov processes.

17.3.8.6 Metal ions in Fluid Solution

The relaxation rates of metal ions in fluid solution are usually too short to measure directly by time-domain EPR. Instead, line shape changes are interpreted in terms of relaxation time changes, or alternatively the electron relaxation time is estimated from the effect of the electron spin on nuclear relaxation times (e.g., see Bertini, Martini, and Luchinat, 1994b).

The magnetic field/frequency dependence of the electron spin relaxation of metal ions in solution involves several relaxation mechanisms. For ions with $S > \frac{1}{2}$, the principal relaxation mechanism in fluid solution is usually modulation of the ZFS due to collisions with solvent molecules (Rubinstein, Baram, and Luz, 1971; Poupko, Baram, and Luz, 1974). The contribution to relaxation is given by (Bertini and Luchinat, 1986):

$$\frac{1}{T_1} = \frac{1}{5\tau_{S0}}\left\{\frac{1}{1+\omega_s^2\tau_v^2} + \frac{4}{1+4\omega_s^2\tau_v^2}\right\},$$

where the electronic relaxation at zero field, τ_{S0}, is proportional to the quadratic ZFS, ω_s is the electron Larmor frequency, and τ_v is the correlation time for the collision process. At some value of ω_s the electron T_1^{-1} will no longer decrease as predicted by this equation, but will instead become dominated by the rotational correlation time or the lifetime of labile adducts of the metal (Bertini and Luchinat, 1986).

The case of Ti(III) in aqueous solution (Bertini, Luchinat, and Xia, 1992) highlighted the need to focus on the relevant relaxation mechanism. These authors found that the relaxation time for aqueous Ti(III) was decreased from 5.7×10^{-11} s to 3.2×10^{-11} s with increasing temperature (5–35 °C), whereas prior studies of VO^{2+}, Mn^{2+}, Cr^{3+}, and Fe^{3+} showed the opposite temperature-dependent effects. It was noted that a decrease in relaxation time with increasing temperature had also been observed for a Mn(III)porphyrin. The decrease of the Ti(III) T_1 with an increase in temperature, and independence of the magnetic field, was attributed to the dominance of an Orbach process for this species (Bertini, Luchinat, and Xia, 1992).

Since g anisotropy has increased importance at higher magnetic fields, spin rotation could have an increased contribution to relaxation at higher fields (Wilson and Kivelson, 1966; Atkins, 1972).

Cu(II) in Solution and in Proteins Gaber et al. (1972) stated that their NMR dispersion (NMRD) results for a copper protein could be interpreted only in terms of a dependence of the Cu(II) relaxation time on magnetic field, being 5×10^{-10} s at low field, and longer at >750 Oe. They attributed this to a general dependence of the form:

$$\frac{1}{\tau_S} = \sum_i D_i \tau_{Vi} \frac{1}{1+(\omega_S \tau_{Vi})^2} \quad \text{at low fields} \quad \omega_S \tau_{Vi} \ll 1$$

"... where τ_{Vi} are correlation times for the various unknown interactions of the Cu^{2+} spin with its immediate environment." This report cites the results of Rubinstein, Baram, and Luz, (1971).

Koenig and Brown (1985) argued that metal bound to proteins "... must be considered as being in the solid state (as contrasted to rapidly tumbling, highly isotropic aquo-complex)." The correlation time used in the calculation of T_1 for the nucleus includes the longitudinal and transverse relaxation times of the metal ion, denoted τ_{S1}, τ_{S2}. It was argued that, for homogeneous solution, these must become equal at zero field, although the usual formulae for relaxation times predict very different values at zero field. Hence, it was argued that the relevant values must be "... averages over the anisotropic relaxation tensor of the ion." There are "... in general $(2S + 1)/2$ different relaxation times" (Koenig and Brown, 1985):

$$\frac{1}{\tau_{S1}} = \frac{1}{5\tau_{S0}} \left[\frac{1}{1+\omega_S^2 \tau_V^2} + \frac{4}{1+4\omega_S^2 \tau_V^2} \right]$$

$$\frac{1}{\tau_{S2}} = \frac{1}{10\tau_{S0}} \left[3 + \frac{5}{1+\omega_S^2 \tau_V^2} + \frac{2}{1+4\omega_S^2 \tau_V^2} \right]$$

where "... τ_V is a correlation time that describes the interaction of the spin S with the protein responsible for relaxation; its source in protein complexes is unclear," τ_{S0} is the relaxation time in the zero-field limit, and ω_S is the Larmor frequency of the electron spin. Theoretical equations for electron spin relaxation are compared with experimental data for Fe^3, Mn^{2+} ($S = 5/2$), and Cr^{3+} ($S = 3/2$). Relaxation is "... controlled by modulation of the quadratic ZFS interaction." There is a decrease in $1/T_{1e}$ with increasing magnetic field, by about a factor of 5 from 30 to 100 MHz. The T_1 for Mn^{2+} increased with frequency (from 14 to 60 MHz proton NMR frequencies), and for V^{2+} from 2.7 to 60 MHz. For Ni^{2+}, Co^{2+}, and Fe^{2+} there was no frequency dependence up to 60 MHz (Rubinstein, Baram, and Luz, 1971).

Ma et al. (2000) used high-field NMR to determine the longitudinal relaxation rate of Cu(II) in plastocyanin. The method required measurement of the relaxation of two different nuclei at two magnetic field strengths. They found $T_1 = 0.18$ ns at 11.7 T, and 0.4 ns at 17.6 T.

Gadolinium The use of gadolinium (Gd) complexes as contrast agents in magnetic resonance imaging (MRI) has stimulated renewed interest in the field dependence of its electron spin relaxation. The relaxation of this $S = 7/2$ spin system is complicated, and only a brief summary of recent reports will be presented here. The relaxation rates of Gd^{3+} in solution at room temperature are too fast to measure directly by pulsed-EPR. Early EPR estimates of spin–lattice relaxation rates for Gd^{3+} were based on the linewidths of CW signals, and the assumption that $T_1 \sim T_2$, which is not valid in many cases. Most tabulations of relaxation times of the lanthanide ions are for the aquo ions in solution at room temperature, based on NMR measurements of proton relaxation times, as in Alsaadi, Rossotti, and Williams (1980). The relaxation of Gd^{3+} is much slower than that for other common lanthanide ions, and is more strongly dependent on the magnetic field because the rotation correlation time is more important for the more slowly relaxing Gd^{3+}. The relaxation of Eu^{2+} complexes is faster than that of the isoelectronic Gd^{3+} complexes (Seibig, Tóth, and Merbach, 2000). The higher the symmetry of the Gd^{3+} complex, the narrower is the CW-EPR line, and the longer is the electron spin T_1 (Sur and Bryant, 1995, 1996; Werner et al., 2007). Recently, effort has focused on complexes of the type used as MRI contrast agents. Multifrequency EPR studies, including X, Q, 2 mm (Powell et al., 1996), 9.5, 35, 94, and 249 GHz (Clarkson et al., 1998), 35, 94.3, and 249 GHz (Smirnova et al., 1998), 9.4, 75, 150, 225 GHz (Borel et al., 2000), 9.1 and 94.2 GHz (Borel et al., 2006), 9.4–325 GHz (Benmelouka et al., 2007), showed that Gd^{3+} CW-EPR lines become narrower at higher frequencies. The line widths have been interpreted in terms of the static and transient ZFS (Rast, Fries, and Belorizky, 1999).

The first direct measurement of the T_1 of Gd^{3+} in solution at room temperature was accomplished by modulating the microwave field amplitude at a rate of the order of T_{1e}^{-1} (Atsarkin et al., 2001). Amplitude modulation EPR with longitudinal detection at X-band was used to determine $T_1 = 1.2$–4.7 ns for four Gd^{3+} complexes related to MRI contrast agents at room temperature (Atsarkin et al., 2001; Borel et al., 2002).

Belorizky and Fries (2004) calculated that in the Redfield limit T_{1e} for aquo-Gd^{3+} is about 0.2 ns at X-band, increasing to about 1 ns at Q-band. A detailed study of the pressure, temperature, and magnetic field dependence of ^{17}O NMR spectra for Gd (diethylenetriamine-pentaacetate-bis(methylamide) (Gd(DTPA-BMA)) included modeling of the water-exchange kinetics, rotational motion of the complex, and the electron spin relaxation rates. The contribution to the Gd^{3+} spin–lattice relaxation depends on the microwave frequency and on a correlation time:

$$\frac{1}{T_{1e}} = \frac{1}{25}\Delta^2 \tau_v \{4S(S+1)-3\}\left[\frac{1}{1+\omega_s^2\tau_v^2} + \frac{4}{1+4\omega_s^2\tau_v^2}\right],$$

where ω_s is the electron Larmor frequency, Δ^2 is the sum of the square of the diagonal components of the ZFS tensor, and τ_V is the correlation time for modulation of the ZFS. It was noted that modulation of the ZFS may be due either to rotation or to the lifetime of transient distortions (Gonzalez et al., 1994). This expression, with the parameters for Gd(DTPA-BMA) ($\Delta^2 = 0.38 \times 10^{20}\,s^{-1}$, $\tau_V = 3.4 \times 10^{-11}\,s$), gives $T_{1e} = 0.8$ ns at X-band.

Belorizky et al. (2008) compared three different methods for computing the proton relaxation effect of Gd^{3+} and Ni^{2+} complexes. For slowly rotating molecules ($T_{1e} \ll \tau_r$), the electron relaxation is the dominant contributor to modulation of the electron–nuclear interaction. A distinction is made in the theory between "static" and "transient" ZFS. The transient component results from intermolecular collisions. When the molecular reorientation is fast the static ZFS also provides a mechanism of spin relaxation. Often, the motion of Gd complexes is slow enough that Redfield theory would not be strictly applicable, although in some cases simple theory gave good agreement with experiment. Theory and mechanisms of relaxation were reviewed (Helm, 2006). Fries and Belorizky (2007) showed how relaxation theory could be applied beyond the Redfield limit. The relative contribution of static and transient ZFS changes strongly with magnetic field, and at proton frequencies above about 60 MHz transient ZFS dominates for small molecules, and the Gd^{3+} $1/T_1$ values were $<3 \times 10^8 s^{-1}$ for the complexes studied (Benmelouka et al., 2007) . Borel et al. (2007) reported two Gd^{3+} complexes in which the static ZFS was nearly zero, and the CW-EPR lines were very narrow.

The reported values suggest that T_{1e} for Gd^{3+} complexes would be longer at Q-band than at X-band, and shorter at lower frequencies such as S-band. Values of T_2 are shorter than T_1, and also become longer at higher Larmor frequencies (Atsarkin et al., 2001; Borel et al., 2002).

The magnetic field dependence of electron spin relaxation in dry metalloproteins yielded $T_1 = 12$ ns for Mn^{2+} and 17 ns for Gd^{3+}, which are much longer than those in aqueous solutions. Apparently, in the lyophilized protein the motions are decreased and the metal relaxation time is longer (Korb, Diakova, and Bryant, 2006).

In summary, near room temperature, coordination compounds of Gd^{3+} have T_1 values in the range of several nanoseconds, and are strongly dependent on magnetic field (frequency). Although much shorter values are commonly given for T_2, many of these appear to be line width values, and not direct measures of T_2. In other cases it is assumed that $T_2 = T_1$. The relaxation times are generally interpreted in terms of static and transient ZFS. The static ZFS is attributed to a low symmetry of the molecule. The transient ZFS is attributed to vibrations and collisions with solvent molecules. At low NMR frequencies (magnetic fields), the rotation of the static ZFS is presumed to dominate relaxation, whereas, at high frequencies transient ZFS dominates.

17.3.8.7 Relaxation at 2 mm Wavelength (150 GHz)

Many EPR studies performed at 2 mm wavelength (150 GHz) were summarized in the monograph by Krinichnyi (1995). One interesting statement was that passage effects are more pronounced at 2 mm than at lower frequencies. This was attributed to a theory which assumes that T_1 and T_2 are dominated by electron–nuclear dipolar interactions. The relaxation rates are predicted to be proportional to I(I + 1), and to spectral density functions of rotational or translational diffusion, which are frequency-dependent. The relaxation rates were predicted to decrease at microwave frequencies higher than the inverse of the relevant correlation time. The discussion of relaxation of nitroxyl radicals and trityl radicals (see Sections 17.3.8.1) pointed out that there were relatively small effects of isotope substitutions on relaxation at lower frequencies. Consequently, it would be surprising for electron–nuclear couplings to be dominant in these higher-frequency spectra. One possible reason for greater passage effects at W-band and 2 mm, relative to X-band, was due to a decreased spectral diffusion, and not to field-dependent T_1 or T_2. In agreement with this proposal, Krinichnyi states that cross-relaxation decreases with increasing B_0.

17.3.9
Fullerenes

N@C_{60} was studied at 9.5 and 95 GHz by Dinse (1999). T_1 was about fourfold longer at 95 GHz than at 9.5 GHz. This was attributed to a solvent correlation time of 10 ps contributing to relaxation via collision-induced fluctuation of ZFS. The EPR of single-walled nanotubes from X-band to 319 GHz showed that the CW-EPR line was inhomogeneously broadened (Corzilius et al., 2008).

17.3.10
Summary

The recent availability of pulsed, time-domain, variable-temperature multifrequency EPR has facilitated the identification of relaxation processes in many spin systems. Because of spectral diffusion, different aspects of spin–lattice and spin–spin relaxation are measured using CW progressive saturation, pulsed saturation recovery, and pulsed inversion recovery methods for measuring T_1. The CW linewidth and pulsed spin-echo methods for measuring T_2 can yield different values, due to instantaneous diffusion and molecular dynamics.

A key statement by Bowman and Kevan (1979), relevant to many complicated spin systems, is that "The microwave frequency dependence of the relaxation rate reveals the frequency dependence of the integral involving the spin states, which, in turn, can reveal the mechanism responsible for the spin–lattice coupling for the direct process."

Among the commonly observed relaxation processes that are important in doped low-temperature crystalline solids, only the direct process has been predicted and observed to cause relaxation times to depend on microwave frequency to the fourth power. Nonetheless, many other mechanisms exist by which relaxation rates depend on frequency via a spectral density function such as $1/(1 + \omega^2 \tau^2)$. Relaxation rates generally are faster in glasses than in crystals, but analogous mechanisms may apply to both environments.

Relaxation rates may, in some cases, depend on the magnetic field in the sense that they depend on the position in the EPR spectrum, even if at the same position in the EPR spectrum the relaxation rate does not depend on the microwave frequency.

The study of relaxation rates, although preceding the discovery of EPR itself, is still "young," in that many of the complexities have only very recently been recognized. There is opportunity for many advances.

Acknowledgments

This chapter is based in part on research supported by NIH NIBIB grant EB000557 and EB02807 (GRE and SSE), and EB002034 (Howard Halpern, PI). Comments made by James S. Hyde helped to improve the chapter.

Pertinent Literature

The most extensive prior reviews of field dependence of spin relaxation are the books by Standley and Vaughan (1969) and Poole and Farach (1971), the reviews by Orbach and Stapleton (1972), by Bertini, Martini, and Luchinat (1994a, 1994b), and the high-field EPR review by Prisner (2004). Many of the plots of data that established the frequency dependence of the direct and Raman processes were reproduced in Standley and Vaughan (1969) and in Bertini, Martini, and Luchinat (1994a, 1994b). Krinichnyi (1995) discussed relaxation at 2 mm (D-band) EPR, especially that of conducting polymers. Eaton and Eaton (2000b) provided a tutorial review of the physical mechanisms of electron spin relaxation. References in these sources have stimulated the choice of some of the examples in this chapter.

References

Abragam, A. (1961) *The Principles of Nuclear Magnetism*, Oxford University Press, Oxford.

Abragam, A. and Bleaney, B. (1970) *Electron Paramagnetic Resonance of Transition Ions*, Oxford University Press, Oxford.

Al'tshuler, S.A., Kirmse, R., and Solov'ev, B.V. (1975) *J. Phys. C. Solid State Phys.*, **8**, 1907–1920.

Alsaadi, B.M., Rossotti, R.J.C., and Williams, R.J.P. (1980) *J. Chem. Soc. Dalton Trans.*, **1980**, 2147–2150.

Aminov, L.K., Kurkin, I.N., Lokoyanov, D.A., Salikhov, I.K., and Rakhmatullin, R.M. (1997) *J. Exp. Theor. Phys.*, **84**, 183–189.

Amirav, A. and Navon, G. (1981) *Phys. Rev. Lett.*, **47**, 906–909.

Amirav, A. and Navon, G. (1983) *J. Chem. Phys.*, **82**, 253–267.

Ammar, A., Menetrier, M., Villesuzanne, A., Matar, S., Chevalier, B., Etourneau, J., Villeneuve, G., Rodriguez-Carvajal, J., Koo, H.-J., Smirnov, A.I., and Whangbo, M.-H. (2004) *Inorg. Chem.*, **43**, 4974–4987.

Anderson, P.W. and Weiss, P.R. (1953) *Rev. Modern Phys.*, **25**, 269–276.

Andreozzi, L., Faetti, M., Giordano, M., and Leporini, D. (1999) *J. Phys. Chem. B*, **103**, 4097–4103.

Askew, T.R., Muench, P.J., Stapleton, H.J., and Brower, K.L. (1984) *Solid State Commun.*, **49**, 667–670.

Atherton, N.M. (1973) *Electron Spin Resonance: Theory and Applications*, John Wiley & Sons, Ltd.

Atkins, P.W. (1972) *Electron Spin Relaxation in Liquids* (eds L.T. Muus and P.W. Atkins), Plenum Press, New York, pp. 279–312.

Atkins, P.W. and Kivelson, D. (1966) *J. Chem. Phys.*, **44**, 169–174.

Atsarkin, V.A. (1966) *Soviet Phys. JETP*, **22**, 106–113.

Atsarkin, V.A., Demidov, V.V., Vasneva, G.A., Odintsov, B.M., Belford, R.L., Raduchel, B., and Clarkson, R.B. (2001) *J. Phys. Chem. A*, **105**, 9323–9327.

Bagryanskaya, E.G., Yashiro, H., Fedin, M.V., Purtov, P.A., and Forbes, M.D.E. (2002) *J. Phys. Chem. A*, **106**, 2820–2828.

Banci, L., Bertini, I., and Luchinat, C. (1991) *Nuclear and Electron Relaxation*, VCH, Weinheim.

Bar, G., Bennati, M., Nguyen, H.-H.T., Ge, J., Stubbe, J., and Griffin, R.G. (2001) *J. Am. Chem. Soc.*, **123**, 3569–3576.

Barbon, A., Brustolon, M., Maniero, A.L., Romanelli, M., and Brunel, L.C. (1999) *Phys. Chem. Chem. Phys.*, **1**, 4015–4023.

Barnes, J.P., Liang, Z., Mchaourab, H., Freed, J.H., and Hubbell, W.L. (1999) *Biophys. J.*, **76**, 3298–3306.

Behringer, R.E. (1958) *J. Chem. Phys.*, **29**, 537–539.

Belorizky, E. and Fries, P.H. (2004) *Phys. Chem. Chem. Phys.*, **6**, 2341–2351.

Belorizky, E., Fries, P.H., Helm, L., Kowalewski, J., Kruk, D., Sharp, R.R., and Westlund, P.-O. (2008) *J. Chem. Phys.*, **128**, 052315-1–052315-17.

Benmelouka, M., Borel, A., Moriggi, L., Helm, L., and Merbach, A.E. (2007) *J. Phys. Chem. B*, **111**, 832–840.

Bennati, M., Robblee, J.H., Mugnaini, V., Stubbe, J., Freed, J.H., and Borbat, P.P. (2005) *J. Am. Chem. Soc.*, **127**, 15014–15015.

Bertini, I. and Luchinat, C. (1986) *NMR of Paramagnetic Molecules in Biological Systems*, Benjamin Cummings Publ. Co., Menlo Park, CA, p. 77.

Bertini, I., Luchinat, C., and Xia, C. (1992) *Inorg. Chem.*, **31**, 3152–3154.

Bertini, I., Martini, G., and Luchinat, C. (1994a) *Handbook of Electron Spin Resonance: Data Sources, Computer Technology, Relaxation, and ENDOR* (eds C.P. Poole Jr and H.A. Farach), American Institute of Physics, New York, pp. 79–310.

Bertini, I., Martini, G., and Luchinat, C. (1994b) *Handbook of Electron Spin Resonance* (eds J.C.P. Poole, Jr., and H.A. Farach). American Institute of Physics, New York, pp. 51–77.

Biglino, D., Schmidt, P.P., Reijerse, E.J., and Lubitz, W. (2006) *Phys. Chem. Chem. Phys.*, **8**, 58–62.

van der Bilt, A. and van Duyneveldt, A.J. (1980) *Physica*, **100B**, 324–332.

Blakemore, J.S. (1985) *Solid State Physics*, Cambridge University Press, Cambridge.

Bloembergen, N. and Pound, R.V. (1954) *Phys. Rev.*, **95**, 8–12.

Bloembergen, N. and Wang, S. (1954) *Phys. Rev.*, **93**, 72–83.

Bloembergen, N., Purcell, E.M., and Pound, R.V. (1948) *Phys. Rev.*, **73**, 679–712.

Borel, A., Toth, E., Helm, L., Janossy, A., and Merbach, A.E. (2000) *Phys. Chem. Chem. Phys.*, **2**, 1311–1317.

Borel, A., Helm, L., Merbach, A.E., Atsarkin, V.A., Demikov, V.V., Odintsov, B.M., Belford, R.L., and Clarkson, R.B. (2002) *J. Phys. Chem.*, **A106**, 6229–6231.

Borel, A., Kang, H., Gateau, C., Mazzanti, M., Clarkson, R.B., and Belford, R.L. (2006) *J. Phys. Chem. A*, **110**, 12434–12438.

Borel, A., Laus, S., Ozarowski, A., Gateau, C., Nonat, A., Mazzanti, M., and Helm, L. (2007) *J. Phys. Chem. A*, **111**, 5399–5407.

Bouffard, V., Roinel, Y., Roubeau, P., and Abragam, A. (1980) *J. Phys.*, **41**, 1447–1451.

Bowman, M.K. and Kevan, L. (1977) *J. Phys. Chem.*, **81**, 456–461.

Bowman, M.K. and Kevan, L. (1979) *Time Domain Electron Spin Resonance* (eds L. Kevan and R.N. Schwartz), John Wiley & Sons, Inc., New York, pp. 68–105.

Brown, I.M. (1979) *Time Domain Electron Spin Resonance* (eds L. Kevan and R.N. Schwartz), John Wiley & Sons, Inc., New York, pp. 195–229.

Budil, D.E., Earle, K.A., and Freed, J.H. (1993) *J. Phys. Chem.*, **97**, 1294–1303.

Chang, T. and Yang, G.C. (1968) *J. Chem. Phys.*, **48**, 2546–2549.

Chiang, Y.-W., Borbat, P.P., and Freed, J.H. (2005) *J. Magn. Reson.*, **172**, 279–295.

Chiang, Y.-W., Costa-Filho, A.J., and Freed, J.H. (2007) *J. Phys. Chem.*, **111**, 11260–11270.

Clarkson, R.B., Smirnov, A.I., Smirnova, T., Kang, H., Belford, R.L., Earle, K.A., and Freed, J.H. (1998) *Mol. Phys.*, **95**, 1325–1332.

Corzilius, B., Dinse, K.-P., and Hata, K. (2007) *Phys. Chem. Chem. Phys.*, **9**, 6063–6072.

Corzilius, B., Dinse, K.-P., Hata, K., Haluška, M., Skákalová, V., and Roth, S. (2008) *Phys. Status Solidi B*, **245**, 2251–2254.

Davids, D.A. and Wagner, E.P. (1964) *Phys. Rev. Lett.*, **12**, 141–142.

Davis, M.S., Mao, C., and Kreilick, R.W. (1975) *Mol. Phys.*, **29**, 665–672.

Denysenkov, V.P., Prisner, T., Stubbe, J., and Bennati, M. (2005) *Appl. Magn. Reson.*, **29**, 375–384.

Dinse, K.-P. (1999) *EPR Newsletter*, **10** (2), 3–7.

Doddrell, D.M., Pegg, D.T., Bendall, M.R., and Gregson, A.K. (1978) *Aust. J. Chem.*, **31**, 2355–2365.

Du, J.L., More, K.M., Eaton, S.S., and Eaton, G.R. (1992) *Israel J. Chem.*, **32**, 351–355.

Du, J.L., Eaton, G.R., and Eaton, S.S. (1994) *Appl. Magn. Reson.*, **6**, 373–378.

Du, J.-L., Eaton, G.R., and Eaton, S.S. (1995a) *J. Magn. Reson. A*, **117**, 67–72.

Du, J.-L., Eaton, G.R., and Eaton, S.S. (1995b) *J. Magn. Reson. A*, **115**, 213–221.

Du, J.-L., Eaton, G.R., and Eaton, S.S. (1996) *J. Magn. Reson. A*, **119**, 240–246.

Dugdale, D.E. and Thorpe, J.S. (1969) *J. Phys. C Solid State Phys.*, **2**, 1376–1383.

Dzuba, S.A. (1996) *Physics Lett. A*, **213**, 77–84.

Dzuba, S.A. (2000) *Spectrochim. Acta A*, **56**, 227–234.

Dzuba, S.A., Maryasov, A.G., Salikhov, A.K., and Tsvetkov, Y.D. (1984) *J. Magn. Reson.*, **58**, 95–117.

Dzuba, S.A., Tsvetkov, Y.D., and Marysov, A.G. (1992) *Chem. Phys. Lett.*, **188**, 217–222.

Dzuba, S.A., Kirlina, E.P., and Salnikov, E.S. (2006) *J. Chem. Phys.*, **125**, 054502-1–054502-5.

Earle, K.A., Budil, D.E., and Freed, J.H. (1993) *J. Phys. Chem.*, **97**, 13289–13297.

Earle, K.A., Moscicki, J.K., Polimeno, A., and Freed, J.H. (1997) *J. Chem. Phys.*, **106**, 9996–10015.

Eaton, S.S. and Eaton, G.R. (1993) *J. Magn. Reson.*, **102**, 354–356.

Eaton, S.S. and Eaton, G.R. (2000a) *Biol. Magn. Reson.*, **19**, 347–381.

Eaton, S.S. and Eaton, G.R. (2000b) *Biol. Magn. Reson.*, **19**, 29–154.

Eaton, G.R., Eaton, S.S., Stoner, J.W., Quine, R.W., Rinard, G.A., Smirnov, A.I., Weber, R.T., Krzystek, J., Hassan, A.K., Brunel, L.C., and Demortier, A. (2001a) *Appl. Magn. Reson.*, **21**, 563–570.

Eaton, S.S., Harbridge, J., Rinard, G.A., Eaton, G.R., and Weber, R.T. (2001b) *Appl. Magn. Reson.*, **20**, 151–157.

Elsaesser, C., Monien, B., Haehnel, W., and Bittl, R. (2005) *Magn. Reson. Chem.*, **43**, 526–533.

Fedin, M.V., Purtov, P.A., and Bagryanskaya, E.G. (2001) *Chem. Phys. Lett.*, **339**, 395–404.

Fedin, M.V., Yashiro, H., Purtov, P.A., Bagyanskaya, E.G., and Forbes, M.D.E. (2002) *Mol. Phys.*, **100**, 1171–1180.

Fedin, M.V., Purtov, P.A., and Bagyanskaya, E.G. (2003) *J. Chem. Phys.*, **118**, 192–201.

Fedin, M.V., Shakirov, S.R., Purtov, P.A., and Bagryanskaya, E.G. (2006) *Russ. Chem. Bull. Intl. Ed.*, **55**, 1703–1716.

Fielding, A.J., Carl, P.J., Eaton, G.R., and Eaton, S.S. (2005) *Appl. Magn. Reson.*, **28**, 239–249.

Freed, J.H. (2005) *Biol. Magn. Reson.*, **24**, 239–268.

Fries, P.H. and Belorizky, E. (2007) *J. Chem. Phys.*, **126**, 204503/1–204503/13.

Froncisz, W., Camenisch, T.G., Ratke, J.J., Anderson, J.R., Subczynski, W.K., Strangeway, R.A., Sidabras, J.W., and Hyde, J.S. (2008) *J. Magn. Reson.*, **193**, 297–304.

Fursman, C.E., Bittl, R., Zech, S.G., and Hore, P.J. (2001) *Chem. Phys. Lett.*, **342**, 162–168.

Gaber, B.P., Brown, R.D., Koenig, S.H., and Fee, J.A. (1972) *Biochim. Biophys. Acta*, **271**, 1–5.

Ghim, B.T., Eaton, S.S., Eaton, G.R., Quine, R.W., Rinard, G.A., and Pfenninger, S. (1995) *J. Magn. Reson. A*, **115**, 230–235.

Gill, J. (1962) *Proc. Phys. Soc.*, **79**, 58–68.

Goldsborough, J.P., Mandel, M., and Pake, G.E. (1960) *Phys. Rev. Lett.*, **4**, 13–15.

Gonzalez, G., Powell, D.H., Tissieres, V., and Merbach, A.E. (1994) *J. Phys. Chem.*, **98**, 53–59.

Gorchester, J., Millhauser, G.L., and Freed, J.H. (1990) *Modern Pulsed and Continuous Wave Electron Spin Resonance* (eds L. Kevan and M.K. Bowman), John Wiley & Sons, Inc., New York, pp. 119–194.

Graebner, J.E., Golding, B., and Allen, L.C. (1986) *Phys. Rev. B*, **34**, 5696–5701.

Halpern, H.J., Yu, C., Peric, M., Barth, E., Grdina, D.J., and Teicher, B.A. (1994) *Proc. Natl Acad. Sci. USA*, **91**, 13047–13051.

Haniotis, Z. and Guenthard, H.H. (1968) *Helv. Chim. Acta*, **51**, 561–564.

Harbridge, J.R., Eaton, S.S., and Eaton, G.R. (2002) *J. Magn. Reson.*, **159**, 195–206.

Harbridge, J.R., Eaton, S.S., and Eaton, G.R. (2003) *J. Phys. Chem. A*, **107**, 598–610.

Harris, E.A. and Yngvesson, K.S. (1968) *J. Phys. C. Solid State Phys.*, **1**, 990–1010.

Helm, L. (2006) *Prog. NMR Spectrosc.*, **49**, 45–64.

Henderson, A.J., Jr and Rogers, R.N. (1966) *Phys. Rev.*, **152**, 218–222.

Hernandez, G., Tweedle, M.F., and Bryant, R.G. (1990) *Inorg. Chem.*, **29**, 5109–5113.

Hofbauer, W., Earle, K.A., Dunnam, C.R., Moscicki, J.K., and Freed, J.H. (2004) *Rev. Sci. Instrum.*, **75**, 1194–1208.

Honig, H. and Stupp, E. (1958) *Phys. Rev. Lett.*, **1**, 275–276.

Horasan, N., Sünnetçioğlu, M.M., Sungur, R., and Bingöl, G. (1997) *Z. Naturforsch.*, **52a**, 485–489.

Huber, D.L. (1982) *J. Noncryst. Solids*, **51**, 241–244.

Huber, D.L. (1987) *J. Luminescence*, **36**, 327–329.

Hyde, J.S. and Froncisz, W. (1982) *Annu. Rev. Biophys. Bioeng.*, **11**, 391–417.

Hyde, J.S., Yin, J.J., Feix, J.B., and Hubbell, W.L. (1990) *Pure Appl. Chem.*, **62**, 255–260.

Hyde, J.S., Yin, J.-J., Subczynski, W.K., Camenisch, T.G., Ratke, J.J., and Froncisz, W. (2004) *J. Phys. Chem. B*, **108**, 9524–9529.

Ilangovan, G., Zweier, J.L., and Kuppusamy, P. (2000) *J. Phys. Chem. B*, **104**, 4047–4059.

Kaiser, G. and Friedrich, J. (1991) *J. Phys. Chem.*, **95**, 1053–1057.

Kang, P.C., Eaton, G.R., and Eaton, S.S. (1994) *Inorg. Chem.*, **33**, 3660–3665.

Kirilina, E.P., Grigoriev, I.A., and Dzuba, S.A. (2004) *J. Chem. Phys.*, **121**, 12465–12471.

Kirilina, E.P., Prisner, T., Bennati, M., Endeward, B., Dzuba, S.A., Fuchs, M.R., Mobius, K., and Schnegg, A. (2005) *Magn. Reson. Chem.*, **43**, S119–S129.

Knickelbein, M.B. (2004) *J. Chem. Phys.*, **121**, 5281–5283.

Koenig, S.H. and Brown, R.D., III (1985) *J. Magn. Reson.*, **61**, 426–439.

Kolaczkowski, S.V., Cardin, J.T., and Budil, D.E. (1999) *Appl. Magn. Reson.*, **16**, 293–298.

Korb, J.-P., Diakova, G., and Bryant, R.G. (2006) *J. Chem. Phys.*, **124**, 134910–134911.

Kordas, G. (1992) *Phys. Chem. Glasses*, **33**, 143–148.

Krinichnyi, V.I. (1995) *2-mm Wave Band EPR Spectroscopy of Condense Systems*, Boca Raton, Fl. CRC Press.

Krzystek, J., Sienkiewicz, A., Pardi, L., and Brunel, L.C. (1997) *J. Magn. Reson.*, **125**, 207–211.

Kulik, L.V., Pashenko, S.V., and Dzuba, S.A. (2002) *J. Magn. Reson.*, **159**, 237–241.

Kurtz, S.R. and Stapleton, H.J. (1980) *Phys. Rev. B*, **22**, 2195–2205.

Kutter, C., Moll, H.P., Van Tol, J., Zuckerman, H., Maan, J.C., and Wyder, P. (1995) *Phys. Rev. Lett.*, **74**, 2925–2928.

Kveder, M., Merunka, D., Ilakovac, A., Makarevic, J., Jokic, M., and Rakvin, B. (2006) *Chem. Phys. Lett.*, **419**, 91–95.

Lambe, J., Laurance, N., McIrvine, E.C., and Terhune, R.W. (1961) *Phys. Rev.*, **122**, 1161–1170.

Larson, G.H. and Jeffries, C.D. (1966) *Phys. Rev.*, **145**, 311–324.

Lee, S., Patyal, B.R., and Freed, J.H. (1993) *J. Chem. Phys.*, **63**, 165–199.

Lloyd, J.P. and Pake, G.E. (1953) *Phys. Rev.*, **92**, 1576.

Lloyd, J.P. and Pake, G.E. (1954) *Phys. Rev.*, **94**, 579–591.

Lubzens, S. and Schultz, S. (1976) *Phys. Rev. Lett.*, **36**, 1104–1106.

Lurie, D.J., Nicholson, I., and Mallard, J.R. (1991) *J. Magn. Res.*, **95**, 405–409.

Ma, L., Jorgensen, A.M., Sorensen, G.O., Ulstrup, J., and Led, J.J. (2000) *J. Am. Chem. Soc.*, **122**, 9473–9485.

Mailer, C., Robiinson, B.H., and Haas, D.A. (1992) *Bull. Magn. Reson.*, **14**, 30–35.

Mailer, C., Nielsen, R.D., and Robinson, B.H. (2005) *J. Phys. Chem. A*, **109**, 4049–4061.

Mandel, M. (1962) *Rev. Sci. Instrum.*, **33**, 247–248.

Maresch, G.G., Mehring, M., von Schutz, J.U., and Wolf, H.C. (1984) *Chem. Phys. Lett.*, **85**, 333–340.

Martinelli, M., Massa, C.A., Pardi, L.A., Bercu, V., and Popescu, F.F. (2003) *Phys. Rev. B*, **67**, 014425.

Mauriello, L., Scaringe, R., and Kokoszka, G. (1974) *J. Inorg. Nucl. Chem.*, **36**, 1565–1568.

Misra, S.K. (1998) *Spectrochim. Acta A*, **54**, 2257–2267.

Mizoguchi, K., Kume, K., and Shirakawa, H. (1984) *Solid State Commun.*, **50**, 213–218.

Moll, H.P., Kutter, C., Tol, J.V., Zuckerman, H., and Wyder, P. (1999) *J. Magn. Reson.*, **137**, 46–58.

Muller, F., Hopkins, M.A., Coron, N., Grynberg, M., Brunel, L.C., and Martinez, G. (1989) *Rev. Sci. Instrum.*, **60**, 3681–3684.

Muromtsev, V.I., Shteinshneider, N.Y., Safronov, S.N., Golikov, V.P., Kuznetsov, A.I., and Zhidomirov, G.M. (1975) *Sov. Phys. Solid State*, **17**, 517.

Murphy, J. (1966) *Phys. Rev.*, **145**, 241–247.

Náfrádi, B., Gaál, R., Sienkiewicz, A., Fehér, T., and Forró, L. (2008) *J. Magn. Reson.*, **195**, 206–210.

Nakagawa, K., Candelaria, M.B., Chik, W.W.C., Eaton, S.S., and Eaton, G.R. (1992) *J. Magn. Reson.*, **98**, 81–91.

Nellutla, S., Morley, G.W., van Tol, J., Patti, M., and Dalal, N.S. (2008) *Phys. Rev. B*, **78**, 054426.

Nogatchewsky, M., Ablart, G., and Pescia, J. (1977) *Solid State Commun.*, **24**, 493–497.

Onischuk, A.A., Samoilova, R.I., Strunin, V.P., Chesnokov, E.N., Musin, R.N., Bashurova, V.S., Maryasov, A.G., and Panfilov, V.N. (1998) *Appl. Magn. Reson.*, **15**, 59–94.

Orbach, R. and Stapleton, H.J. (1972) Electron spin–lattice relaxation, in *Electron Paramagnetic Resonance* (ed. S. Geschwind), Plenum, New York, pp. 121–216.

Orton, J.W. (1968) *Electron Paramagnetic Resonance: An Introduction to Transition Group Ions in Crystals*, Gordon and Breach.

Owenius, R., Terry, G.E., Williams, M.J., Eaton, S.S., and Eaton, G.R. (2004) *J. Phys. Chem. B*, **108**, 9475–9481.

Owenius, R., Eaton, G.R., and Eaton, S.S. (2005) *J. Magn. Res.*, **172**, 168–175.

Pace, J.H., Sampson, D.F., and Thorp, J.S. (1960) *Phys. Rev. Lett.*, **4**, 18–19.

Pake, G.E. and Estle, T.L. (1973) *The Physical Principles of Electron Paramagnetic Resonance*, 2nd edn, W. A. Benjamin, Inc., Reading, Mass.

Pleau, E. and Kokoszka, G. (1973) *J. Chem. Soc., Faraday Trans.*, **2** (69), 355–382.

Pobell, F. (2007) *Matter and Methods at Low Temperature*, 3rd edn, Springer, New York.

Poole, C.P. and Farach, H. (1971) *Relaxation in Magnetic Resonance*, Academic Press, New York.

Poupko, P., Baram, A., and Luz, Z. (1974) *Mol. Phys.*, **27**, 1345–1357.

Powell, D.H., Dhubhghaill, O.M.N., Pubanz, D., Helm, L., Lebedev, Y.S., Schlaepfer, W., and Merbach, A.E. (1996) *J. Am. Chem. Soc.*, **118**, 9333–9346.

Prabhananda, B.S. and Hyde, J.S. (1986) *J. Chem. Phys.*, **85**, 6705–6712.

Prisner, T.F. (2004) *Biol. Magn. Reson.*, **22**, 249–276.

Ramakrishna, B.L. and Manoharan, P.T. (1984) *Mol. Phys.*, **52**, 65–80.

Rannestad, A. and Wagner, P.E. (1963) *Phys. Rev.*, **131**, 1953–1960.

Rast, S., Fries, P.H., and Belorizky, E. (1999) *J. Chem. Phys.*, **96**, 1543–1550.

Robinson, B.H., Haas, D.A., and Mailer, C. (1994) *Science*, **263**, 490–493.

Robinson, B.H., Mailer, C., and Reese, A.W. (1999a) *J. Magn. Reson.*, **138**, 199–209.

Robinson, B.H., Mailer, C., and Reese, A.W. (1999b) *J. Magn. Reson.*, **138**, 210–219.

Robinson, B.H., Reese, A.W., Gibbons, E., and Mailer, C. (1999c) *J. Phys. Chem. B*, **103**, 5881–5894.

Rogers, R.N., Anderson, M.E., and Pake, G.E. (1959) *Bull. Am. Phys. Soc. 2nd Ser.*, **4**, 261.

Rogers, R.N., Carboni, F., and Richards, P.M. (1967) *Phys. Rev. Lett.*, **19**, 1016–1018.

Rubinstein, M. and Resing, H.A. (1976) *Phys. Rev. B*, **13**, 959–968.

Rubinstein, M., Baram, A., and Luz, Z. (1971) *Mol. Phys.*, **20**, 67–80.

Sato, H., Kathirvelu, V., Fielding, A.J., Bottle, S.E., Blinco, J.P., Micallef, A.S., Eaton, S.S., and Eaton, G.R. (2007) *Mol. Phys.*, **105**, 2137–2151.

Sato, H., Bottle, S.E., Blinco, J.P., Micallef, A.S., Eaton, G.R., and Eaton, S.S. (2008a) *J. Magn. Reson.*, **191**, 66–77.

Sato, H., Kathirvelu, V., Spagnol, G., Rajca, S., Eaton, S.S., and Eaton, G.R. (2008b) *J. Phys. Chem. B.*, **112**, 2818–2828.

Scaringe, R., Shia, L., and Kokoszka, G. (1973) *Chem. Phys. Lett.*, **18**, 519–520.

Schulz, M.B. and Jeffries, C.D. (1966) *Phys. Rev.*, **149**, 270–288.

Seibig, S., Tóth, E., and Merbach, A.E. (2000) *J. Am. Chem. Soc.*, **122**, 5822–5830.

Sert, I., Sünnetçioğlu, M.M., Surgur, R., and Bingöl, G. (2000) *Z. Naturforsch.*, **55a**, 682–686.

Smirnova, T.I., Smirnov, A.I., Belford, R.L., and Clarkson, R.B. (1998) *J. Am. Chem. Soc.*, **120**, 5060–5072.

Srivastava, G.P. (1990) *The Physics of Phonons*, Adam Hilger, Bristol.

Standley, K.J. and Vaughan, R.A. (1969) *Electron Spin Relaxation Phenomena in Solids*, Plenum Press.

Strutz, T., Witowski, A.M., and Wyder, P. (1992) *Phys. Rev. Lett.*, **68**, 3912–3915.

Stutzmann, M. and Biegelsen, D.K. (1983) *Phys. Rev. B. Condens. Matt. Mater. Phys.*, **28**, 6256–6261.

Sünnetçioğlu, M. and Bingöl, G. (1993) *Phys. Chem. Liq.*, **26**, 47–58.

Sur, S.K. and Bryant, R.G. (1995) *J. Phys. Chem.*, **99**, 6301–6308.

Sur, S.K. and Bryant, R.G. (1996) *J. Magn. Reson. B*, **111**, 105–108.

Terblanche, C.J. and Reynhardt, E.C. (2000) *Chem. Phys. Lett.*, **322**, 273–279.

Thorpe, J.S. and Hossain, M.D. (1980) *J. Mater. Sci.*, **15**, 2647–2650.

Turkevich, J., Soria, J., and Che, M. (1972) *J. Chem. Phys.*, **56**, 1463–1466.

Van Duyneveldt, A.J. and Pouw, C.L.M. (1973) *XVII Congress Ampere* (ed. V. Hovi), North-Holland Publishing Co., pp. 404–406.

Van Vleck, J.H. (1960) *Quantum Electronics* (ed. C.H. Townes) Columbia University Press, New York, pp. 392–409.

Van Vleck, J.H. (1961) *Advances in Quantum Electronics* (ed. J.R. Singer), Columbia University Press, New York, pp. 388–398.

Weil, J.A., Bolton, J.R., and Wertz, J.E. (1994) *Electron Paramagnetic Resonance: Elementary Theory and Practical Applications*, John Wiley & Sons, Inc., New York.

Werner, E.J., Avedano, S., Botta, M., Hay, B.P., Moore, E.G., Aime, S., and Raymond, K.N. (2007) *J. Am. Chem. Soc.*, **129**, 1870–1871.

Wilson, R. and Kivelson, D. (1966) *J. Chem. Phys.*, **44**, 4445–4452.

Witowski, A.M. and Bardyszewski, W. (1995) *Solid State Commun.*, **94**, 9–12.

Witowski, A.M., Kutter, C., and Wyder, P. (1997) *Phys. Rev. Lett.*, **78**, 3951–3954.

Yong, L., Harbridge, J., Quine, R.W., Rinard, G.A., Eaton, S.S., Eaton, G.R., Mailer, C., Barth, E., and Halpern, H.J. (2001) *J. Magn. Reson.*, **152**, 156–161.

Zahlan, A.B. (1968) *Excitons, Magnons and Phonons in Molecular Crystals*, Cambridge University Press, Cambridge.

Zecevic, A., Eaton, G.R., Eaton, S.S., and Lindgren, M. (1998) *Mol. Phys.*, **95**, 1255–1263.

Zhou, Y., Bowler, B.E., Eaton, G.R., and Eaton, S.S. (1999) *J. Magn. Reson.*, **139**, 165–174.

Ziman, J.M. (1960) *Electrons and Phonons*, Oxford University Press, Oxford.

18
EPR Imaging: Theory and Instrumentation

Rizwan Ahmad and Periannan Kuppusamy

18.1
Introduction

Electron paramagnetic resonance (EPR), also called electron spin resonance, is a branch of spectroscopy in which electrons with unpaired spins absorb electromagnetic radiation which is typically in the microwave frequency range. EPR was first discovered by Zavoisky in 1944, and has since experienced a continuous growth. EPR is capable of detecting free radicals as well as transition metals and rare-earth ions, and has found numerous applications in biology, chemistry, physics, and medicine (Eaton, Eaton, and Berliner, 2005).

Over the past two decades, EPR-based imaging (EPRI) has been used to observe the distribution of free radicals in biological systems (Kuppusamy *et al.*, 1994). For a typical continuous-wave (CW) EPR imager, the images are reconstructed from data which are collected by sweeping the external magnetic field in the presence of a magnetic field gradient and monochromatic radio frequency (RF) radiation. Image reconstruction is generally performed using a filtered backprojection (FBP) method. Low-frequency (2 GHz or less) CW-EPR instrumentations with a capability for up to three-dimensional (3-D) spatial and up to four-dimensional (4-D) spectral–spatial imaging are commonly used for biological samples.

In principle, EPR is similar to NMR, where the transition of protons between two energy levels is observed. The practical challenges associated with EPR and EPRI, however, are distinct from those of NMR and MRI. Over the past few decades many advances – both in hardware and software – have been made to improve the overall performance of EPR imagers to the level where realization of many clinical applications seems plausible. Further progress, however, is still required to enable the widespread use of EPR and EPRI for biological applications.

In this chapter, the basic concepts of EPR imaging, data acquisition, CW instrumentation, and image reconstruction are discussed, along with some basic details of commonly used strategies to enhance image quality and resolution. A few examples that are representative of biological applications are also presented.

18 EPR Imaging: Theory and Instrumentation

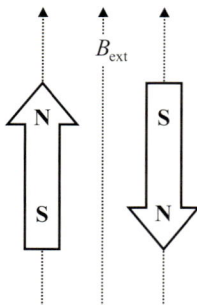

Figure 18.1 Lowest (left) and highest (right) energy orientations of the magnetic moment of an unpaired electron in the presence of an external magnetic field B_{ext}.

18.2
EPR Principle: Zeeman Effect

In EPR, the absorption of energy by an unpaired election to transition from a low-energy to a high-energy state in the presence of magnetic field is observed. These energy levels are created by the interaction between electronic magnetic moments and the external magnetic field (Griffiths, 1995), B_{ext}, usually called the "main magnetic field." Due to the magnetic moment, an unpaired election acts like a tiny bar magnet in the presence of an external magnetic field. It possesses the lowest energy if the magnetic moment is aligned with B_{ext} and the highest energy if it is aligned opposite to B_{ext} as shown in Figure 18.1. This effect is generally called the Zeeman effect.

Since an electron is a spin 1/2 particle, the low-energy and high-energy states can be designated as $M_s = -1/2$ and $M_s = +1/2$, respectively. The energy E of an unpaired electron in an external magnetic field B_{ext} can be defined as

$$E = g\mu_B M_s B_{ext} = \pm \frac{1}{2} g\mu_B B_{ext} \tag{18.1}$$

where dimensionless g, termed as g-factor, is the ratio of magnetic moment to the angular momentum expressed by means of Bohr magneton μ_B which is the unit of magnetic moment. Also, $\gamma = g\mu_B/\hbar$ is the gyromagnetic ratio, with $\hbar$ being the reduced Plank's constant. The resonance condition is satisfied if,

$$B_{ext} = B_0 \tag{18.2}$$

where $B_0 = 2\pi v_0/\gamma = \omega_0/\gamma$, with v_0 being the radiation frequency in Hertz.

It is apparent from Equation 18.1 and Figure 18.2 that the splitting between the energy states varies linearly with the external magnetic field B_{ext}, and is degenerate when $B_{ext} = 0$, which means that the two states have the same energy in the absence of an external magnetic field. In the presence of the external magnetic field, absorption of the electromagnetic radiation occurs whenever Equation 18.2

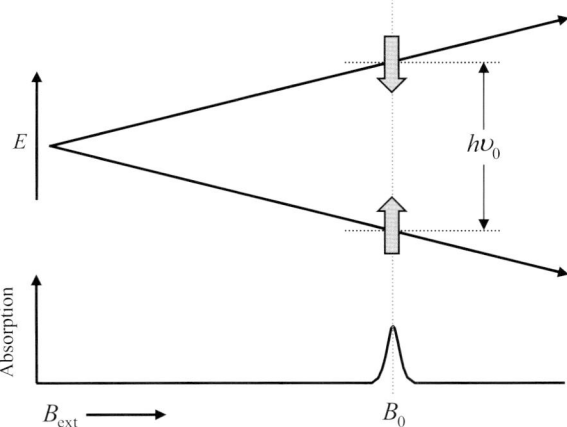

Figure 18.2 Zeeman splitting in the presence of an external magnetic field B_{ext} (neglecting second-order effects such as hyperfine splitting).

is satisfied. The condition can be satisfied either by fixing the radiation frequency and sweeping the main magnetic field B_{ext}, or by fixing B_{ext} and sweeping the radiation frequency. The limitation of RF (Laukemper-Ostendorf et al., 2002) hardware, however, makes the latter choice unattractive. At resonance, the unpaired electrons at the lower energy level absorb the RF radiation and jump to the higher-energy state; these excited electrons move back to the lower-energy state by releasing the excess energy. It is this transition between low- and high-energy states that is recorded in EPR.

18.2.1
Hyperfine Splitting

The unpaired electron is influenced by its local surrounding. For instance, if there are nuclei with a magnetic moment in the vicinity of an unpaired electron, the local magnetic field experienced by the unpaired electron may change according to the interaction between the two spins. Therefore, the external magnetic field value required to satisfy Equation 18.2 will also shift accordingly. If the magnetic effect, B_I, of the nucleus on the unpaired electron is in the direction of B_{ext}, the resonance will occur at $B_{ext} < B_0$, and if B_I opposes B_{ext} at the location of unpaired electron, the absorption will occur at $B_{ext} > B_0$ for a given v_0.

Since we are considering a system with large numbers of nuclei and unpaired electrons, at any given moment, one fraction of nuclei will enhance while the other fraction will suppress the net magnetic field on the nearby unpaired electrons. Therefore, for spin 1/2 nuclei ($M_I = \pm 1/2$), the observed EPR spectrum splits into two signals which are each B_I away from the original B_0 location, as shown in Figure 18.3. This interaction between the electron and nuclei–called the hyperfine

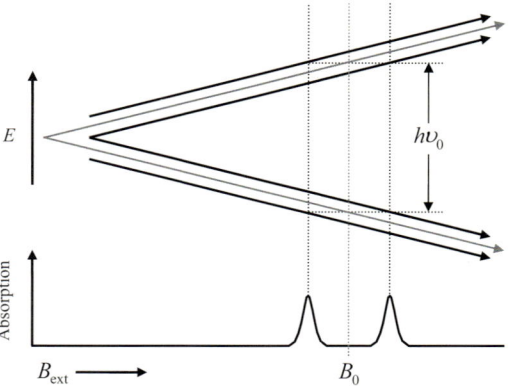

Figure 18.3 Hyperfine splitting of an unpaired electron in the presence of spin 1/2 nuclei.

interaction (Griffiths, 1987) – provides very useful information about the identity and structure of the molecule. If there is a second nucleus, each of the two signals will further subdivide into a pair. For n nuclei with spin M_I, there are $(2M_I+1)^n$ hyperfine components (Poole, 1997) if all coupling constants differ and there is no degeneracy. For k different nuclei groups, the total number of hyperfine components N_{hf} is given by

$$N_{hf} = \prod_k (2M_{I_k} + 1)^{n_k} \tag{18.3}$$

Several nitroxide spin labels are used to study the biological systems (Chumakov et al., 1972; Jiang et al., 1992). These spin labels, depending on the isotope of nitrogen, may give rise to two- or three-peak EPR spectrum. While the presence of hyperfine structures in an EPR spectrum can provide a wealth of information regarding the paramagnetic species under study, it may impose additional constraints on the data acquisition and processing for EPRI.

Most EPRI experiments using nitroxides have been performed by restricting the spectral (sweep) window to a single hyperfine line. The gradients in these experiments are limited to magnitudes that would not cause an overlap of the selected line with the other hyperfine lines. However, this constraint severely limits the magnitude of the gradient that can be applied, as well as the nature of paramagnetic species that can be imaged. Furthermore, as the resolution of an EPR image is dependent on the gradient magnitude, the presence of multiple lines limits the resolution of the image that can be obtained. On the other hand, the inclusion of a full hyperfine spectrum will increase the size of the spectral window, and consequently may require much larger gradient magnitudes to achieve resolutions comparable to those obtained with a single line spectrum. A few procedures (Eaton, Eaton, and Ohno, 1991) have been reported to either reduce or eliminate hyperfine artifacts in EPR spectroscopy and imaging.

18.2.2
Spin Relaxation

The EPR spectrum provides a wealth of information regarding the evolution of various terms in the spin Hamiltonian. Analysis of the EPR spectrum provides information not only about the molecular structure of the paramagnetic species but also about its local environment. The majority of biological EPR applications are built around quantifiable changes in the EPR spectrum as a function of a physiologically meaningful environmental variable. For example, hyperfine splitting of nitroxides predictably varies with local pH values, and hence has been used to measure pH in biological systems.

Many important EPR applications, however, are only concerned with the relaxation times of the unpaired electrons. In general, the relaxation time reflects how quickly the unpaired electrons that have been excited by the absorption of RF radiation can dissipate excess energy and return to the lower-energy state. The dynamics of these transitions between the two energy states are captured by the Bloch equations (Bloch, 1946). After applying an excitation pulse, the decay of magnetization, called free induction decay (FID), due to transition of unpaired electrons back to the lower-energy state, can be recorded. In situations where it is not feasible to directly record FID, due to extremely short relaxation times, an absorption line—which is the Fourier transform (FT) of FID—may instead be collected in CW mode by subjecting the sample to an RF excitation and sweeping the main magnetic field. A careful analysis of the width (called the linewidth) and shape (called the lineshape) of a resonant absorption line can provide a great deal of information about the paramagnetic species being studied. The most common lineshapes are Lorentzian and Gaussian, but other nonparametric lineshapes are not uncommon.

18.2.3
Comparison to NMR

Since EPR and NMR are based on the same principle, it is inevitable to compare the two techniques. From an applications view point, the abundance of unpaired nucleons in biological systems makes NMR suitable for broad use, while EPR relies on exogenously introduced spin labels (also called probes), which may limit its scope of *in vivo* applications. Some of the fundamental differences between the two resonance modalities are detailed in Table 18.1.

18.2.4
EPR Probes

Because the concentrations of endogenous free radicals in most biological systems are too low for detection with modern EPR spectrometers, the EPR-sensitive materials (probes) are exogenously introduced into the biological system. Generally, two types of probe are used: particulate (solid-state); and soluble. For purely spatial

Table 18.1 Fundamental differences between EPR/EPRI and NMR/MRI.

EPR/EPRI	NMR/MRI
Transition of unpaired electrons is measured	Transition of unpaired nucleons is observed
Due to higher gyromagnetic ratio of electrons, the measurements are made at higher frequencies (usually in GHz) and lower magnetic fields. Designing RF circuitry in EPR can be a challenge	Data are collected at lower frequencies and higher magnetic fields. Ensuring a homogeneous magnetic field can be a challenge in NMR and especially in MRI
Since the penetration depth is inversely related to the frequency, imaging large samples using EPR is not feasible	Both NMR and MRI are better suited for studying large biological systems
Due to the higher magnetic moment of electrons, EPR is inherently more sensitive, but sparsity of unpaired electrons in biological systems may result in poor signal-to-noise ratio (SNR)	Due to the abundance of unpaired nucleons such as hydrogen nuclei, SNR in biological systems is generally adequate for a majority of applications
In most applications, the EPR probes are exogenously introduced	Unpaired nucleons present in the biological system are adequate to generate a strong NMR signal in most applications
Since the relaxation times are short, observed absorption lineshapes are broader, which may result in loss of spatial resolution for EPRI	Lineshapes are narrower, resulting in high spatial resolution of MRI
Due to the hardware challenges associated with the measurement of short relaxation times, EPR is mostly conducted in CW domain. In CW-EPR, the data are collected in the form of projections	Due to all the benefits associated with digital firmware, pulsed-NMR and MRI are preferred for most of the applications. In pulsed domain, the data are collected in the Fourier domain

EPRI, the probes with narrow lineshapes are preferred as the broad linewidths usually result in a poor spatial resolution. For oximetry, a higher lineshape sensitivity to pO_2 is also desirable. While selecting a probe, other factors such as the stability of the probe in a biological system, reproducibility, the oxygen sensitivity, and contamination with insensitive isomorphs in the preparation should also be taken into account. Single-line triarylmethyl (TAM) (Ardenkjaer-Larsen *et al.*, 1998; Kuppusamy *et al.*, 1997) and lithium phthalocyanine (LiPc) (Liu *et al.*, 1993) are two of the most commonly used EPR probes.

18.3
CW-EPR Imager

A general layout of a CW-EPR spectrometer and imager is shown in Figure 18.4. A stable static magnetic field is provided by a main magnet, while the gradient

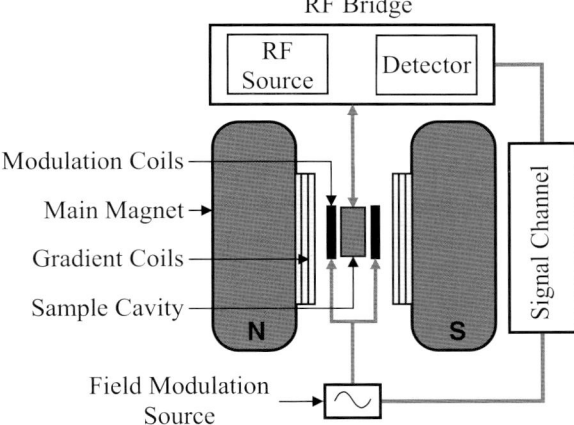

Figure 18.4 A basic layout of a CW-EPR imager.

coils provide a linear magnetic field gradient necessary for imaging. The sample to be studied is placed in the cavity (also called resonator), which helps in amplifying the weak EPR signals. The detector and electromagnetic radiation sources are in a box called RF bridge. The signal channel consists primarily of a phase-sensitive detector (Poole, 1997).

18.3.1
Magnets and Magnetic Field Control

In the existing EPR imaging designs, three types of magnet are commonly utilized. In the low-frequency/low-field systems with static fields up to 300–500 G (depending on the sample volume size), nonferrous electromagnets can be used. These magnets may utilize a Helmholtz coil pair design, or a solenoidal design, or some type of hybrid multicoil design for improved homogeneity of the field. These electromagnets provide the advantage of a simple current control of the field due to an absence of hysteresis effect associated with the iron core. However, the poor energy efficiency of such magnets limits the achievable field strength to several hundred Gauss. Permanent magnet-based systems have not gained a wide use due to the poor thermal stability of the field and, thus, a need for elaborate systems of compensation coils and electronics. Historically, the iron core electromagnets used in EPR spectrometers have found the most widespread use in EPR imaging systems, at frequencies from 300 MHz to X-band. Typically, such a system utilizes a large poleface electromagnet (≥40 cm) with custom-machined pole pieces to further increase the gap (Kuppusamy *et al.*, 1994).

To ensure the stability of the magnetic field, the use of Hall effect sensors is still desirable for direct magnetic field measurements. The use of a Hall probe-based field control loop in the iron core electromagnets is imperative due to the magnetic hysteresis effects in the core. In addition, the presence of strong 3-D field gradients influences the Hall probe signal, causing static field "pulling" and resulting in image corruption. In the early EPR imaging studies, the Hall probe was simply moved away from the magnet gap to decrease the influence of the gradient field. In this position, the probe would sense a reduced field intensity of the field, such that the field setting should be scaled accordingly. Unfortunately, this method does not provide a full compensation of the pulling effect, but merely decreases its magnitude. This procedure also assumes that the correction factor for the center field remains constant during field sweeps, but this is not true due to variations of the magnetic permeability of the yoke. Another known method of compensation employs coils placed in the close vicinity of the Hall probe, and circuitry to produce compensation currents to these coils. The major shortcomings of this method are that it is indirect in nature, while the compensation coils themselves may perturb the surrounding field pattern.

A method for discrete compensation (Kuppusamy, Chzhan, and Zweier, 1995a) has been reported. In this approach, a conventional Hall probe is mounted on the Z-axis of the gradient coil system inside the magnet gap on one of the pole pieces of the magnet, exactly on the Z-axis of the magnet, in such a way that it "feels" only the Z-gradient coils' field influence. This can be achieved due to the symmetry of the X and Y gradient coils around the Z-axis. Before each projection acquisition, the correction shift for the static field setting is calculated and introduced by the control computer, thus, compensating the "pulling" on the Hall probe caused by the Z-gradient. This correction method, though successfully implemented has the following shortcomings:

- It is discrete, and so cannot be used for continuous correction, as would be needed, for example, in a swept gradient method.
- Each field correction takes a few milliseconds to execute, so that the overall settling time could be as high as 1 to 2 s, which may slow down the acquisition.
- Since all the devices are digital, fine tuning is limited to the overall accumulated digital resolution.

18.3.2
Gradient Coil Assembly

The design of the magnetic field system for EPRI (magnet, gradients and field-controlling instrumentation) defines the size and geometry of samples that can be studied, as well as the quality and resolution of the resulting image, and temporal efficiency of the imaging apparatus. The free gap must be sufficient to accommodate the desired sample size resonator, and set of gradients. The system must provide the required field with homogeneity ideally better than 20 mG over the sample volume. EPRI imposes extremely stringent requirements for magnetic field gradient design as compared to that for MRI. This is due to the fact that strong linear gradients (>100-fold stronger than for MRI) are required to obtain

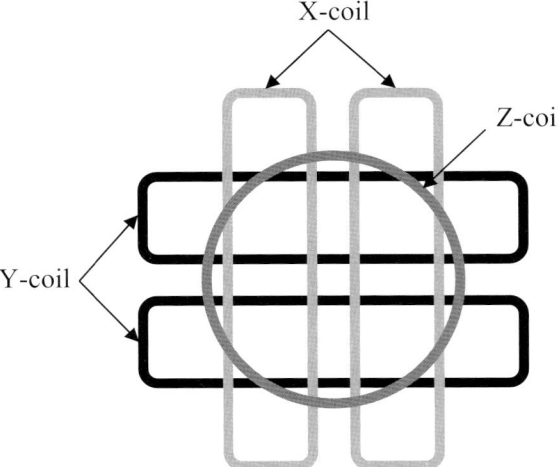

Figure 18.5 One side of a 3-D gradient coil assembly. The Z-gradient is generated using Maxwell coils, while X and Y gradients are generated using flat pair coils. Here, the x-axis is along the horizontal axis, the y-axis is along the vertical axis, and the z-axis is orthogonal to the x-y plane.

high spatial and spectral resolution, due to the much broader spectral linewidths of the EPR labels. Likewise, in CW-EPRI the gradients must be able to dissipate peak power during long periods of time. A transient response, however, is less important than that in MRI, so that active shielding is usually not necessary. As the sample size increases, problems imposed by the need for powerful field gradients escalate, since the power required to generate a given gradient increases with the cube of the gap distance between the coils. Dissipation of the power in the coils also becomes more difficult as the surface/volume ratio of the coil decreases.

For the two major geometries for homogeneous static magnetic field production—Helmholtz pair coils or a solenoidal coil (Thomas, Busse, and Scheuck, 1985)—distinct gradient coil designs are dictated. The gradients for Z in both designs are of a Maxwell pair coil configuration. For X and Y, the Helmholtz static field geometry calls for the use of flat square pairs (four coils per axis), while in the solenoid electromagnet a curved Golay gradient coil design is required to conform to the cylindrical geometry. The flat coil gradient design provides a factor of 1.8 higher gradient efficiency compared to the cylindrical Golay configuration (Golay, 1958). In addition, it offers a greatly simplified fabrication due to its planar structure. Sample and resonator access is also substantially improved. For these important reasons, a Helmholtz coil static field design with Maxwell coils for the Z-gradient and flat pair coils for the X- and Y-gradients have found the most use in EPRI systems with powerful field gradients. The geometry of such a gradient coils system is shown in Figure 18.5, where one half of the coil system is shown.

18 EPR Imaging: Theory and Instrumentation

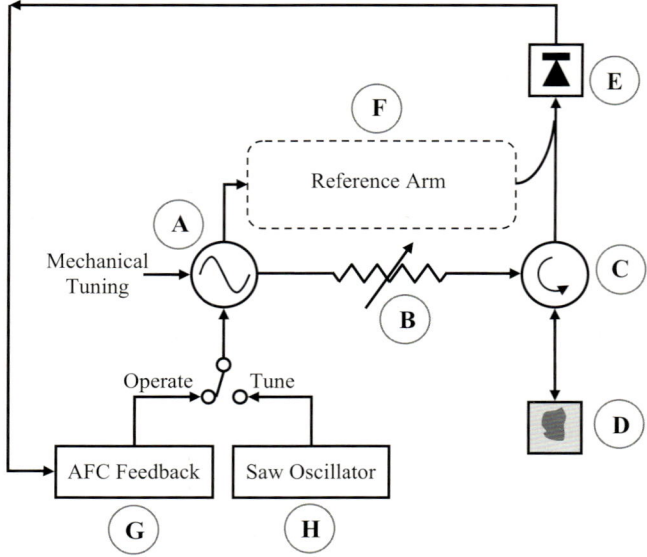

Figure 18.6 Layout of a typical EPR bridge. For details, see the text. Illustration courtesy of Eric Kesselring.

A type of cooling system must also be provided to achieve a desirable gradient strength; typically, a set of flat coils will be enclosed in a nonconductive enclosure with channels for coolant circulation.

A system of multiple computer-controlled power supplies must be utilized to drive the gradient coils as required by the projection acquisition sequence. Ideally, these power supplies are a bank of switching mode operational power amplifiers capable of four-quadrant operation with an IEEE-488 or other computer interface.

18.3.3
RF Bridge

Various parts of a typical RF EPR bridge, as labeled in Figure 18.6, are described as follows:

- **Oscillator (A):** This generates the RF energy used to irradiate the sample. The frequency of the RF energy is generally varied via mechanical and electrical means; while mechanical adjustments are used for coarse tuning, electronic adjustments are applied for fine tuning. The oscillator must have a very stable output frequency and amplitude, as even slight changes can introduce distortion in the data.

- **Attenuator (B):** This precisely controls the amount of RF energy delivered to the sample. It must be very stable over time and temperature.

- **Circulator (C):** This allows reflected energy from sample/resonator to reach the detector diode, while blocking high level excitation energy from the oscillator to reach the diode detector directly.

- **Resonator (D):** This amplifies the weak EPR signal. The changes in energy absorbed by the sample also change the impedance or quality factor, Q, of the resonator. The Q is the ratio of energy stored to energy lost. A higher Q allows larger detectable changes during absorption. The depth and width of the notch during tuning is a visual representation of Q.

- **Detector (E):** This down-converts the RF energy reflected from the resonator to the baseband signal. At RF power levels <1 µW, the detector operates in a nonoptimal lower-sensitivity region. A higher sensitivity can be achieved when the diode operates with a bias current above a minimum (typically >100 µA). The electrical output signal from a typical detector diode is 1500 mV per mW of RF input. As excessive RF power can easily damage the diode, an additional protection circuitry is included to monitor and limit the diode current

- **Reference arm (F):** This applies a small RF power to bias the detector diode into the more sensitive operating region. A phase shifter synchronizes the reference arm power and reflection from the resonator. Many homebuilt units do not have a reference arm, and require off-resonance coupling of the resonator for the bias.

- **AFC controller (G):** This processes the signal from the preamplifier to electronically match the RF oscillator and resonator frequencies.

- **SAW oscillator (H):** This generates ~400 Hz saw tooth waveform to provide a frequency sweep for tuning. It also allows visual feedback for tuning the oscillator frequency to the resonator frequency.

18.3.4
EPR Resonator

Sample resonator design is critical to achieve maximum sensitivity in a given application, and must be tailored to accommodate the sample with the highest possible [*filling factor* × Q] product. The resonator must be a mechanically stable structure, and should make the most efficient use of the space within the magnet. Space constraints present a major consideration in the choice of resonator design for EPRI. Innovations in resonator design, which enable automatic coupling adjustment and frequency tuning, can be used to suppress motion-induced noise that occurs in biological samples. In recent years, much effort has been focused on the development of lumped-parameter microwave sample cavities for L- and S-bands. Two major types of such resonator have been introduced and extensively discussed (Froncisz and Hyde, 1982; Sotgiu, 1985); these resonator designs have been utilized to perform *in vivo* EPR studies on whole animals and organs.

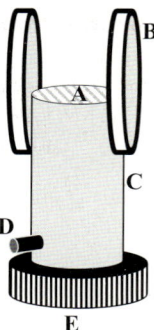

Figure 18.7 Layout of a typical surface bridged loop-gap resonator and modulation coils. A, sample area of the probe surface; B, modulation coils; C, microwave resonator shield; D, microwave input connector; E, coupling adjustment ring.

- Loop-gap resonators (LGRs) provide straightforward design and high filling factors compared to standard distributed parameters microwave cavities. However, due to the open structure of the inductive loop element, LGRs have significant radiation losses unless a shield is provided. The need for a shield leads to problems in achieving an optimum magnitude of modulation field, and also to at least a 20% increase in the overall resonator dimensions. For special applications, such as the topical imaging of skin (Kuppusamy et al., 1998c) or of externally localized tumors (Kuppusamy et al., 1998a), however, surface coil type LGRs provide an excellent approach. A sketch of a bridged loop-gap surface resonator design, originally developed by the Swartz group (Bacic et al., 1989) is shown in Figure 18.7. This design is better suited than the conventional reentrant resonator for topical studies or measurements on a small localized region, which extends only a limited distance from the body of the animal, for example, the head or a tumor. It can also be easily modified by scaling to fit a different than L-band frequency band from 500 to 4000 MHz. The mechanical design of the bridged LGR can also be modified to allow mechanical frequency tuning by variation of the capacitance of the bridged areas.

- The reentrant resonator (RER design) was introduced and described in great detail by Sotgiu (Sotgiu, 1985). Since the resonant structure of the RER forms a closed volume, it does not require any additional shield, thus providing substantial space savings as compared to LGRs. A number of RERs constructed of ceramic silver plated material were also reported (Chzhan et al., 1993), offering improved rigidity and stability.

In order to couple the microwave power from the bridge to the resonator, two techniques are used: inductive loop coupling; and capacitive coupling. Whilst providing a better decoupling from the modulation pickup, capacitive coupling designs require special resonator designs with access to the areas of electric field concentration. The high-stability/high-Q ceramic RER has a minimized thickness, which is about 3 mm wider than the sample. This enables a closer placement of the gradient coils, making it possible to achieve higher magnetic field gradients for a given coil's driving power. In addition, the dimensional stability and mechanical stiffness of the ceramics make this resonator design highly stable for imaging.

Another example of a resonator design which is well-suited for EPR spectroscopy and the imaging of larger samples at frequencies of 300 to 800 MHz has been recently described (Chzhan et al., 1999). This resonator design is optimized for large lossy biological samples at 750 MHz, and has a mechanical resonance frequency tuning and capacitive coupling.

The capacitive area of the resonator is formed by two plates protruding into the center of the resonator from the opposing sidewalls. The plane of the gap in the resulting flat capacitor is transverse to that in the conventional RER. At the frequencies of interest (RF frequency band), this structure can be well approximated by a lump circuit. For frequencies below 800 MHz, considering the structure as one turn toroidal inductance and lump capacitor, a simple equation predicting the resonant frequency f_0 of the structure can be used:

$$f_0 = \frac{1}{2\pi a} \sqrt{\frac{ld}{\varepsilon_0 \mu_0 s}} \tag{18.4}$$

where a is width of the square reentrance, d is the capacitor gap thickness, s is the area of interleaving capacitor plates, and l is the length of the toroidal path along the center line of the resonator. Due to the predominantly lumped circuit, the nature of the resonator lines of the electric field are concentrated in the gap between two opposing plates. The direction of electric field lines inside the capacitor is independent of the direction of magnetic field H that forms closed force lines around the capacitive region.

One important advantage of this design – besides its obvious simplicity – is that the geometry of the resonator offers a greatly improved accessibility to the capacitive region. This, in turn, enables convenient wide-range frequency tuning by the introduction of a flexible metal plate, controlled by a precision linear movement mechanism. This also allows the incorporation of a simple wide range capacitive coupling mechanism into the structure.

It seems promising to incorporate real-time electronic adjustments of both resonant frequency and coupling of the sample resonator, as this will suppress any noise arising from motion, volume changes, or other sources. An approach for obtaining fixed-frequency EPR measurements, utilizing a low-phase noise fixed-frequency oscillator as a microwave source and an electronically tunable L-band resonator, which is locked to the oscillator via a modified automatic frequency control (AFC) circuit, was reported (Chzhan, Kuppusamy, and Zweier, 1995). This approach is especially well suited to the low-frequency EPRI of *in vivo* samples, whereby the tunable resonator provides major advantages:

- It enables the use of low-phase noise fixed-frequency quartz oscillators (phase noise <−160 dBc Hz^{-1} at 100 kHz offset).

- It compensates for frequency changes due to motional and temporal instabilities of the sample during relatively long image acquisitions, preventing the violation of the isofrequency condition. This prevents possible artifacts/loss of resolution which would occur with conventional resonators.

- It enables the use of a fixed-frequency microwave bridge design using narrow band components, which provide better electrical characteristics, with lower losses.

In order to further suppress noise due to motion of the sample, a mechanism for automatic coupling control (ACC) via electronic and/or electromechanical adjustments in the cavity Q (loss) is necessary. This can be achieved through an additional coupling provision (either inductive or capacitive), which couples the resonant circuit to an external electronically controllable source of loss, such as a PIN diode, a LED-controlled photoresistor, or a similar technique. This loss must be electronically adjusted via a servo loop to hold the detector current constant in spite of changes in resonator loss caused by the sample. The AFC servo loop will compensate for changes in the reactance (frequency) of the resonator via the frequency tuning of the resonator.

The microwave or RF bridges for use with tunable resonators may be optimized by taking advantage of the single-frequency operation using narrow band microwave or RF components for excitation and detection, as well as circuitry for the implementation of frequency and coupling control of the resonator. Due to the use of single-frequency oscillators, such bridges must also incorporate a means for tuning visualizations, such as a built-in frequency sweeper. These innovations in the resonator and bridge design are currently under active development by several *in vivo* and imaging EPR groups around the world.

18.3.5
Signal Channel

To improve the system sensitivity, it is a common practice in CW-EPR to modulate B_{ext} by adding an oscillating magnetic field by using a pair of modulation coils (as shown in Figure 18.4), and to detect the signal using a phase-locked loop detector (also called a phase-sensitive detector or lock-in detector). The lock-in detector compares the EPR signal from the crystal with the reference signal coming from the same oscillator that generates the magnetic field. The lock-in detector only accepts the EPR signal that locks in to the reference signal (Poole, 1997). The advantages of lock-in detection include less $1/f$ noise from the detection diode, and an elimination of the baseline instabilities due to drift in DC electronics. It is important to note that, for certain data collection methods – for example, a rapid scan (Stoner *et al.*, 2004) – no field modulation is applied and the absorption signal is recorded directly.

18.4
Data Acquisition for CW-EPR and EPRI

Most EPR experiments are conducted in the CW domain, as the technical challenges associated with pulsed EPR (Alecci et al., 1998) limit its broad use. Based on these applications, EPR can be divided into spectroscopy, spatial EPRI, and spectral-spatial EPRI, with each having different requirements on data collection.

18.4.1
Spectroscopy

The spectrum is observed in the presence of RF excitation by sweeping the main magnetic field B_{ext}

$$B_{ext} = B_0 + B_\delta \tag{18.5}$$

where B_δ represents the field sweep; B_0, as defined by Equation 18.2, is the magnetic field strength at which magnetic resonance occurs; and B_{ext} represents the net external magnetic field. In spectroscopy, as the measurements take place in the absence of a magnetic field gradient, no information is captured regarding the spatial distribution of the spin probe. Rather, the collected information is solely limited to the shape of the composite spectrum. If more than one species is present in the sample, then the observed signal is the superposition of individual spectra.

Although, in principle, it is possible to sweep the RF frequency for a fixed B_{ext}, the related hardware challenges make it an unattractive option. Therefore, in EPR an equivalent method is adopted where the magnetic field is swept for a monochromatic RF excitation. For a number of EPR probes, the lineshape l belongs to family of parametric functions such as Lorentzian, Gaussian, or Voigt (Bales, Peric, and Lamy-Freund, 1998). A Lorentzian lineshape with full-width at half-maximum (FWHM) linewidth τ and spin density κ (which is the area under the absorption curve), is defined in Equation 18.6 and shown in Figure 18.8:

$$l(\tau,\kappa) = \frac{\kappa}{\pi} \frac{0.5\tau}{(B_\delta)^2 + (0.5\tau)^2} \tag{18.6}$$

As mentioned earlier, the signal-to-noise ratio (SNR) can be improved by modulating the magnetic field and observing the first harmonic which, for a smaller modulation amplitude B_m, represents the first derivative (Williams, Pan, and Halpern, 2005) of l, as shown in Figure 18.9.

In many applications, τ alone encodes important information about the species, or its environment. For example, τ for oxygen-sensitive probes increases linearly with the pO_2 that is in direct contact with the probe; hence, measuring τ for biological applications can provide a direct reading of pO_2. Further, measuring κ, which reflects the total concentration of spins in the system, over a period of time can also provide the concentration decay of volatile paramagnetic species.

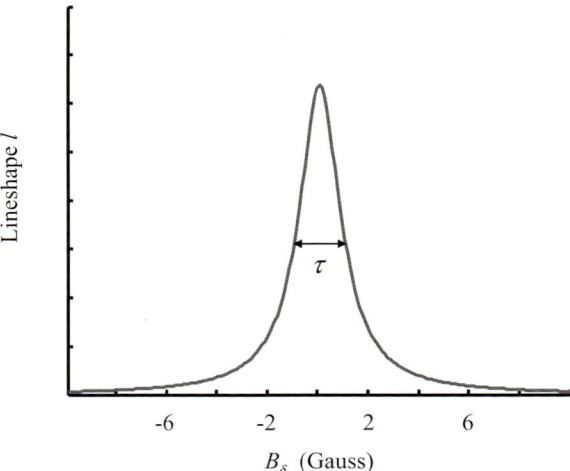

Figure 18.8 Lorentzian lineshape with FWHM linewidth $\tau = 1$ and spin density $\kappa = 1$.

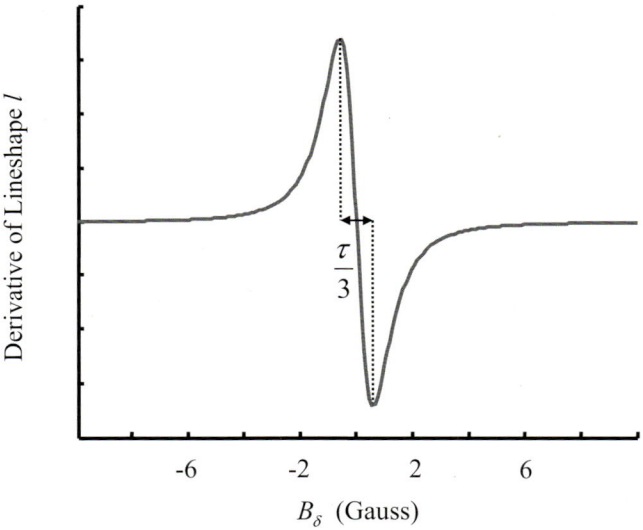

Figure 18.9 First derivative of the Lorentzian lineshape shown in Figure 18.8. Here, peak-to-peak linewidth $= \tau/\sqrt{3}$.

18.4.2
Spatial EPRI

Spatial EPRI is capable of providing the distribution of paramagnetic spins in one-demensional (1-D), 2-D, or 3-D spatial domain $u \in R^d$ (Kuppusamy, Wang, and Zweier, 1995b). The application of a magnetic field gradient, g_u, is used to resolve the spatial location of probes. In spatial imaging, it is assumed that the lineshape of the probe is spatially invariant. However, in cases where the lineshape changes with the location (possibly due to change in the environment), or where there are multiple probes with different lineshapes, it may not be possible to provide an accurate distribution of the probes using purely spatial EPRI.

The external magnetic field B_{ext} in the presence of gradient g_u is

$$B_{\text{ext}} = B_0 + B_\delta + \langle g_u, u \rangle$$
$$B_{\text{ext}} - B_0 = B_\delta + \langle g_u, u \rangle \quad (18.7)$$

where $\langle .,. \rangle$ represents the inner product. The resonance condition is only met where $B_\delta + \langle g_u, u \rangle = 0$. Each term in Equation 18.7 has units of magnetic field, but can be converted to spatial units by dividing both sides by $-|g_u|$. A projection p, the Radon transform (RT) of the object f, is the superposition of spin densities at the locations where resonance condition is satisfied. Therefore, p in spatial units can be written as

$$p(\rho, \hat{g}_u) = \int_u f(u) \delta(\rho - \langle \hat{g}_u, u \rangle) du \quad (18.8)$$

where $\rho = -B_\delta/|g_u|$ and $\hat{g}_u = g_u/|g_u|$. Here, $\langle \hat{g}_u, u \rangle$ is the plane to be integrated, and ρ is the distance of this plane from the center of image space, as shown in Figure 18.10.

For a probe with Lorentzian lineshape $l_p(\tau_p, \kappa)$ with linewidth τ_p defined in spatial units, the projection can be written as:

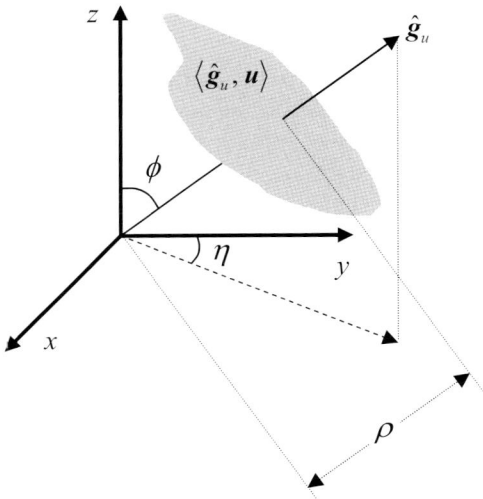

Figure 18.10 Parallel-ray projection acquisition for a 3-D object for an arbitrary gradient direction defined by angles η and ϕ.

$$p_l(\rho, \hat{g}_u) = \int_{\rho'} \int_u \frac{f(u)}{\pi} \frac{0.5\tau_p}{(\rho-\rho')^2 + (0.5\tau_p)^2} \delta(\rho' - \langle \hat{g}_u, u \rangle) du d\rho'$$
$$= \int_{\rho'} p(\rho', \hat{g}_u) \frac{1}{\pi} \frac{0.5\tau_p}{(\rho-\rho')^2 + (0.5\tau_p)^2} d\rho' \quad (18.9)$$
$$= p(\rho, \hat{g}_u) \otimes l_p(\tau_p, 1)$$

where $\otimes$ stands for convolution in first variable. It should be noted that $l_p(\tau_p, \kappa)$ is the lineshape in u (i.e., spatial domain), and $l(\tau, \kappa)$ is the lineshape in terms of magnetic field and is related to the former by

$$l(\tau, \kappa) = \frac{1}{|\hat{g}_u|} l_p(\tau_p, \kappa)$$
$$= \frac{1}{\pi |\hat{g}_u|} \frac{0.5\tau_p}{(\rho)^2 + (0.5\tau_p)^2} \quad (18.10)$$
$$= \frac{1}{\pi} \frac{0.5\tau}{(B_\delta)^2 + (0.5\tau)^2}$$

where $\tau = |g_u| \tau_p$. In other words, the gradient magnitude $|g_u|$ scales the lineshape l_p that gets convolved with the projection p in spatial domain. Hence, the spatial resolution is improved by increasing the applied gradient strength. A higher gradient strength, however, implies a larger sweepwidth in a given time, which results in a reduced signal amplitude. The effect of blurring can be partially suppressed by performing deconvolution (many techniques exist to perform deconvolution, during or before the image reconstruction).

18.4.3
Spectral–Spatial EPRI

For samples having variable linewidths or multiple radical species, it is not possible to obtain an accurate map of the spin distribution using purely spatial EPRI. Besides, the information obtained by purely spatial EPRI is limited to the spin density, and cannot resolve the nature of the spins at each spatial volume element (voxel). In order to overcome this limitation, an additional dimension – the spectral dimension – is required to capture the spectral shape function at each voxel. The imaging technique that includes a spectral dimension along with one or more spatial dimensions is termed as spectral–spatial imaging (Maltempo, Eaton, and Eaton, 1987). While the spatial information is captured by collecting projections along different orientations of the gradient vector $\hat{g}_u$, the spectral information is captured by varying the gradient strength in addition to the orientation. The spectral–spatial imaging can be performed in one, two, or three spatial dimensions, giving rise to 2-D, 3-D, or 4-D spectral–spatial images, respectively. While the information provided by the additional spectral dimension is immensely useful in many biological applications, it requires additional hardware capability, manageable experimental conditions, and additional acquisition time. The potential application of the spectral–spatial technique has been recognized in performing EPRI oximetry (Kuppusamy, Shankar, and Zweier, 1998b), that is based on the effect of an oxygen-induced broadening of the lineshape. Again, considering Equation 18.7

$$B_{ext} - B_0 = B_\delta + \langle g_u, u \rangle \tag{18.11}$$

A projection $\breve{p}$ can be written as

$$\begin{aligned}\breve{p}(B_\delta, g_u) &= \int_B \int_u f(B, u)\delta(B - (B_{ext} - B_0))dudB \\ &= \int_B \int_u f(B, u)\delta(B - B_\delta - \langle g_u, u \rangle)dudB\end{aligned} \tag{18.12}$$

A demonstration of 1-D spatial and 1-D spectral imaging is provided in Figure 18.11. Now, by introducing an angle θ such that $g_u = -k\tan\theta \hat{g}_u$, where $k = \Delta B/\Delta L$ with ΔB and ΔL being the spectral window and spatial field-of-view (FOV), we have

$$\begin{aligned}\breve{p}(B_\delta, g_u) &= \int_B \int_u f(B, u)\delta(B - B_\delta + k\tan\theta\langle \hat{g}_u, u\rangle)dudB \\ &= \cos\theta \int_B \int_u f(B, u)\delta(B\cos\theta + k\sin\theta\langle \hat{g}_u, u\rangle - B_\delta\cos\theta)dudB \\ &= \frac{\cos\theta}{k} \int_B \int_u f(B, u)\delta(\langle(\hat{g}_u\sin\theta, \cos\theta), (u, B/k)\rangle - B_\delta\cos\theta/k)dudB\end{aligned} \tag{18.13}$$

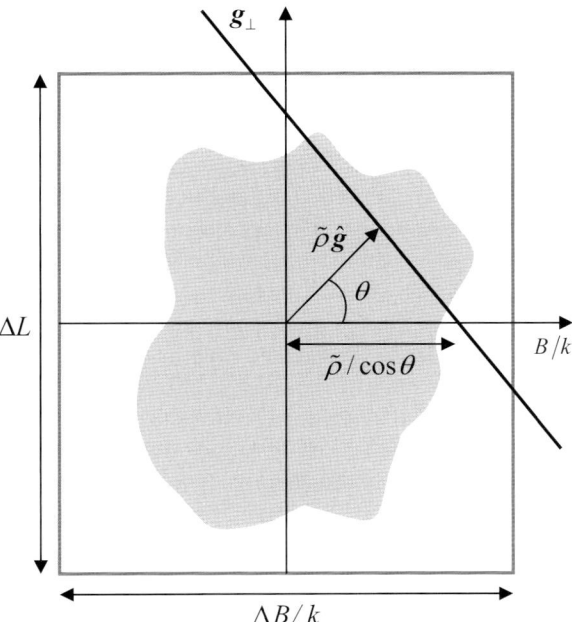

Figure 18.11 Data collection for 2-D spectral-spatial imaging with one spatial (x-axis).

Let $\hat{g} = (\hat{g}_u \sin\theta, \cos\theta)$, $r = (u, B/k)$, and $\tilde{\rho} = B_\delta \cos\theta/k$. The unit vector $\hat{g}$ can be thought of as a generalized gradient vector that also incorporates spectral direction.

$$\breve{p}(B_\delta, g_u) = \frac{\cos\theta}{k} \int_r f(r)\delta(\langle \hat{g}, r\rangle - \tilde{\rho})dr \qquad (18.14)$$

The term after $\cos\theta/k$ is the Radon transform of $(n+1)$-D object with n-D spatial and 1-D spectral dimensions. Therefore,

$$\breve{p}(B_\delta, g_u) = \frac{\cos\theta}{k} p(\tilde{\rho}, \hat{g}) \qquad (18.15)$$

Consequently, data acquisition for a 4-D spectral–spatial object, after rescaling with $1/\cos\theta$, is identical to the RT of an equivalent 4-D spatial object. In spectral–spatial imaging, as the SNR goes down with the $1/\cos\theta$, high gradient projections will suffer from a poor SNR. A similar expression for EPRI has been presented previously (Williams, Pan, and Halpern, 2005).

18.5
Important Imaging Parameters

18.5.1
Time Constant of Lock-In Amplifier

Usually, lowpass filtering follows the lock-in detector, as the EPR signal emerging from the detector is fairly noisy. The bandwidth of the EPR signal depends on the intrinsic linewidth of the species, the applied gradient strength, and the scan rate. A filter with an appropriate cut-off can be designed to find a reasonable trade-off between the SNR and resolution (Deng et al., 2003).

In general, CW-EPR data are comprised of absorption versus magnetic field sweep, although for a fixed magnetic field sweep rate the EPR data can be converted to absorption versus time. The Fourier analysis of this absorption versus time plot reveals an appropriate choice of the cut-off frequency for the lowpass filter. The cut-off frequency of the lowpass should be large enough to accommodate most of the EPR signal, and small enough to eliminate most of the noise. For the first derivative of the Lorentzian EPR signal, a rule of thumb (Poole, 1997) is to design a filter with a time-constant t_c (which is inversely related to the cut-off frequency of the filter) which is less than one-tenth of the peak-to-peak linewidth, τ_l, in the units of sweep time. It is very important not to apply a filter with a very large t_c as this may broaden the EPR signal. In some applications, it may also result in a loss of important features in the EPR spectra. For example, two fine absorption peaks occurring close to each other may become merged together if t_c is not significantly smaller than the separation between the two peaks in the units of sweep time.

18.5.2
Modulation Amplitude

Modulation amplitude is another important parameter that affects the data quality. In the presence of a modulating field $B_m \cos \omega_m t$, and without application of the gradient, Equation 18.5 can be written as:

$$B_{ext} - B_0 = B_\delta + B_m \cos \omega_m t \tag{18.16}$$

where B_m is the modulation amplitude. Therefore, Equation 18.6 becomes

$$l(\tau, \kappa, B_m) = \frac{\kappa}{\pi} \frac{0.5\tau}{(B_\delta + B_m \cos \omega_m t)^2 + (0.5\tau)^2}$$

$$= \frac{\kappa_0}{(2/\tau)^2 (B_\delta + B_m \cos \omega_m t)^2 + 1} \tag{18.17}$$

where $\kappa_0 = 2\kappa/\pi\tau$ is the amplitude of l at $B_\delta = 0$. In terms of a Fourier series

$$l(\tau, \kappa, B_m) = \kappa_0 \left(a_0 + \sum_{n=1}^{\infty} a_n(\tau, B_\delta, B_n) \cos n\omega_m t \right) \tag{18.18}$$

where

$$a_n = \frac{\omega_m}{\pi} \int_{-\pi/\omega_m}^{\pi/\omega_m} \frac{0.5\tau \cos n\omega_m t}{(B_\delta + B_m \cos \omega_m t)^2 + (0.5\tau)^2} \tag{18.19}$$

A closed form expression for a_n can be calculated by contour integration (Poole, 1997). In order to observe the first harmonic a_1, the signal at the output of the lock-in detector is multiplied with $\cos \omega_m t$ and lowpass-filtered. When B_m increases, so does the signal intensity, but the signal becomes distorted and no longer remains Lorentzian, as shown in Figure 18.12.

For B_m much smaller than τ, the first harmonic a_1 is approximately equivalent to the first derivative of the absorption signal. For larger B_m, a_1 signal is no longer Lorentzian, but fortunately this modulation induced distortion is well-characterized (Robinson, Mailer, and Reese, 1999). To improve the SNR, the use of B_m that is much larger than τ has been reported (Deng et al., 2006).

18.5.3
Gradient Strength

In EPRI, the magnetic field gradient is used to encode the spatial information. The magnitude of gradient vector $|\mathbf{g}_u|$, along with the intrinsic lineshape, determines the thickness of the physical plane of the sample that becomes excited by RF irradiation. Increasing the gradient magnitude increases the spatial resolution, since a thinner plane will become excited but at the same time the SNR will be reduced, for the same reason. Therefore, there is a trade-off between the spatial resolution and SNR. For spatial imaging, the SNR is proportional to $1/|\mathbf{g}_u|$, while for spectral–spatial imaging the SNR is proportional to $1/(\text{sweep width})$ for the

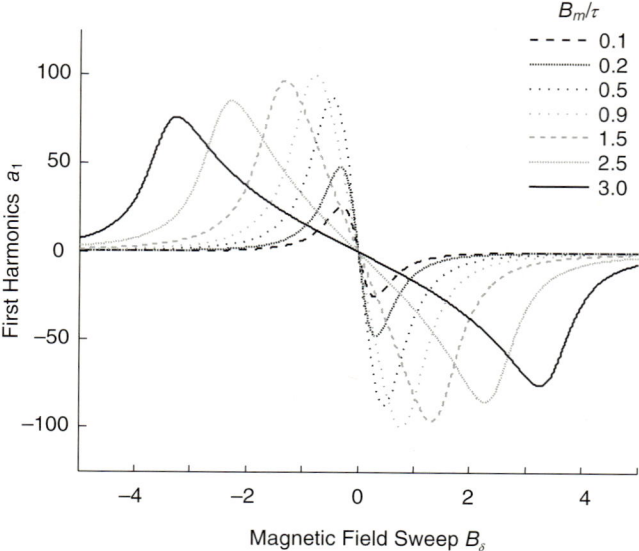

Figure 18.12 Effect of field modulation amplitude B_m on lineshape.

direct detection (Stoner et al., 2004), and to 1/(sweep width)^2 for lock-in detection (Williams, Pan, and Halpern, 2005).

18.6
Image Reconstruction

In EPRI, the reconstruction problem is to obtain an estimate of $f(u)$ given a finite number of projections which generally can be written as

$$q_i = \int_u f(u)\alpha_i(u)du \qquad (18.20)$$

which, for $\alpha_i(u) = \delta(\rho - \langle \hat{g}_i, u \rangle)$, becomes line integral given in Equation 18.8, although α_i can also take different forms to represent different physical measurements. Therefore, α_i can be thought of a blur function that generates q_i from $f(u)$. In terms of Radon operator $\mathcal{A}$

$$q = \mathcal{A}f \qquad (18.21)$$

In terms of vector notation

$$q_i = \langle a_i, f \rangle = a_i^T f$$
$$q = Af \qquad (18.22)$$

Here, vector q represents a collection of all measurements made, while the matrix A represents a linear operator that corresponds to the measurement process. Each measurement q_i can be thought of a projection of f on the vector a_i (Hanson,

1983). Therefore, the measurements only constitute those components of f that lie in the subspace spanned by a_i; this subspace is called the measurement space. Any component of f in the remaining orthogonal subspace constitutes the null space. Since the measurements carry no information about the null space, any content of f that lies in the null space cannot be recovered from the measurements alone. Therefore, whenever the null space exits, the inverse problem of recovering f from q becomes ill-posed (Katz, 1979), and f cannot be determined unambiguously without additional information. For parallel-ray projections, such a recovery problem is called limited angle tomography.

The methods employed to reconstruct images from the data q are numerous, with each method having advantages and disadvantages in terms of speed, reconstruction quality, and ability to incorporate any a priori information. A few of the most common reconstruction choices are discussed here.

18.6.1
Direct Methods

Direct methods, also called transform methods, are based on the direct inversion of the Radon operator $\mathcal{A}$.

18.6.1.1 Filtered Backprojection (FBP) Method

If A^* is the adjoint operator of A, from Equation 18.22

$$A^*q = A^*Af$$
$$f = (A^*A)^{-1} A^*q \tag{18.23}$$

which for full rank A can be written as

$$f = A^* (AA^*)^{-1} q \tag{18.24}$$

For parallel-ray projections, $(AA^*)^{-1}$ represents a Ram–Lak or an equivalent filter. Therefore, in the FBP each projection is first filtered, the shape of which depends on the dimensionality, and then projected back to generate the reconstructed image. Note that A is full-rank only when the null space associated with the measurement is empty, which is only true when there are large number projections. Therefore, the FBP works satisfactorily if there is a large number of equally spaced projections available. In cases of fewer projections, the FBP-based reconstruction gives rise to severe reconstruction artifacts called streak artifacts, as shown in Figure 18.13. Due to this streak artifact, the FBP-based reconstruction generates a strong background that may mask some important features of the object. Usually, thresholding is applied to clear the background, but this may result in the discarding of important information. A few modifications have been suggested to suppress the streak artifact; one such technique is termed projection–reprojection (Stillman and Levin, 1986), whereby additional projections are estimated iteratively in between the measured projections, and the reconstruction is carried out from all the projections.

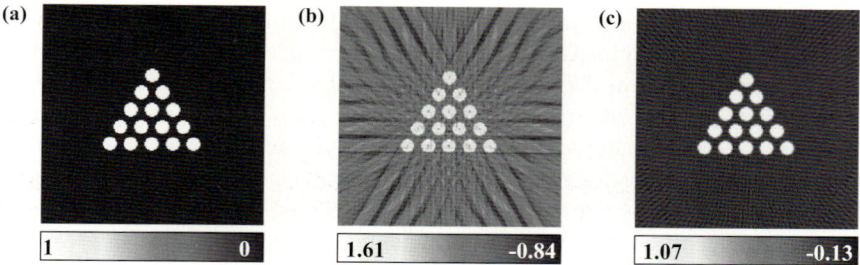

Figure 18.13 The FBP for two different numbers of projections. (a) Input digital phantom of size 192 × 192 pixels; (b) FBP with 18 equally spaced projections; (c) FBP with 90 equally spaced projections.

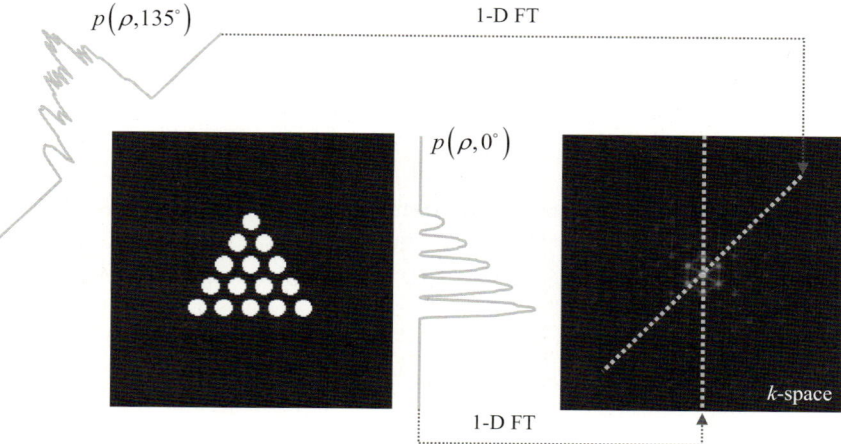

Figure 18.14 Pictorial representation of Fourier slice theorem.

18.6.1.2 Fourier-Based Reconstruction

For Fourier-based reconstruction (FBR), the processing is carried out in the Fourier domain. The Fourier slice theorem states that each parallel-ray projection corresponds to the 1-D inverse Fourier transform (IFT), of n-D FT of f, evaluated along the radial line with orientation defined by $\hat{g}$, as demonstrated in Figure 18.14. It is equivalent to knowing the Fourier amplitudes along the corresponding radial lines in n-D Fourier domain, as shown in Figure 18.15. Therefore, limited-angle tomography, equivalently, corresponds to determining the Fourier amplitudes at those locations that are not covered by the radial lines on which data are available (Cho, Jones, and Singh, 1993). One simple way is to use interpolation in k-space before taking n-D IFT. The interpolation error can further introduce artifacts into the reconstruction. For example, Figure 18.16 shows the reconstruction results from two different numbers of projections.

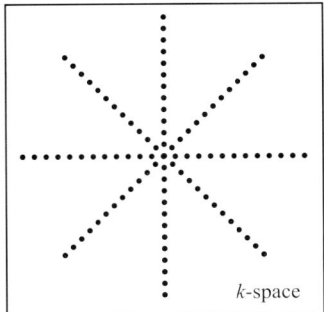

Figure 18.15 Acquired data in k-space. The dots indicate the locations at which data are known.

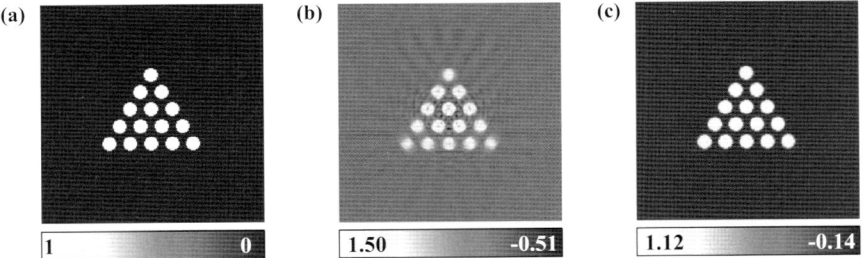

Figure 18.16 The FBR for two different numbers of projections. (a) Input digital phantom of size 192 × 192 pixels; (b) FBR from 18 equally spaced projections with bilinear interpolation; (c) FBR from 90 equally spaced projections with bilinear interpolation.

18.6.2
Iterative Methods

Iterative methods – which are also known as series expansion reconstruction methods or algebraic reconstruction methods – have been used for tomographic reconstruction for the past few decades (Herman and Lent, 1976). These methods are based on a discretization of the image domain prior to any mathematical analysis; this is in contrast to the transform methods (such as FBP), where the continuous problem is only discretized as the last step of the reconstruction process (Lewitt, 1983). A discretized version of a 2-D image is shown in Figure 18.17, where a Cartesian grid of square pixels covers the entire image domain. All of the pixels are numbered in some known deterministic pattern; for example, from top-left corner to the bottom-right corner. The intersection of ith ray q_i with the jth pixel is denoted by a_{ij}. Therefore,

$$q_i = \sum_j a_{ij} f_j = \boldsymbol{a}_i^T \boldsymbol{f} \tag{18.25}$$

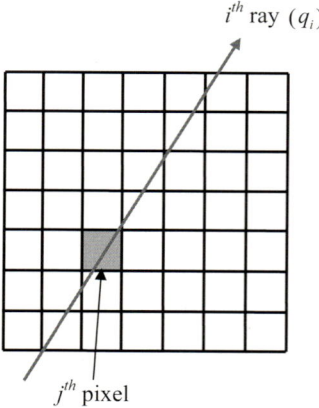

Figure 18.17 Discretized 2-D image. Each square represents a pixel.

The algebraic reconstruction technique (ART), which is one of the most commonly used interactive methods, was first introduced in 1970 (Gordon, Bender, and Herman, 1970). Instead of solving the entire $Af = q$, this method updates the image ray-by-ray to obtain the least-squares solution. The iterative process starts with some initialization $f^{(0)} \in \mathbb{R}^n$. In each iteration, the estimate $f^{(k)}$ is updated to $f^{(k+1)}$ by considering only one ray q_i and updating the image pixels along q_i. The discrepancy between the measurement q_i and the estimate $\sum_j a_{ij} f_j^{(k)}$ is redistributed among the pixels along q_i according to their weights a_{ij}. Therefore,

$$f^{(0)} \in \mathbb{R}^n$$
$$f^{(k+1)} = f^{(k)} + a_i \frac{q_i - \langle a_i, f^{(k)} \rangle}{\|a_i\|^2} \tag{18.26}$$

A geometric interpretation of ART is given in Figure 18.18, which shows that $f^{(k+1)}$ is just the orthogonal projection of $f^{(k)}$ onto the hyperplane that defines the set of solutions to $q_i - \sum_j a_{ij} f_j^{(k)} = 0$. The iterations are performed for all q_i cyclically until some convergence criterion is reached. In order to control the convergence process, a relaxation parameter λ is used, which is a real number usually confined to an interval.

$$f^{(k+1)} = f^{(k)} + \lambda^{(k)} a_i \frac{q_i - \langle a_i, f^{(k)} \rangle}{\|a_i\|^2} \tag{18.27}$$

Although ART is computationally intensive, the reconstruction is usually superior to the FBP or the FBR. The most attractive feature of ART is its ability to merge a variety of constraints into the iterative process. For example, a non-negativity constraint can be readily implemented by setting the negative portions of $f^{(k)}$ to zero in each iteration. Further, ART and other iterative methods do not require projections to be uniformly distributed, and hence the missing angle problem is

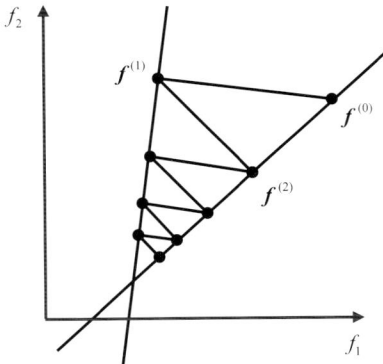

Figure 18.18 Geometric interpretation of ART for a two-pixel object.

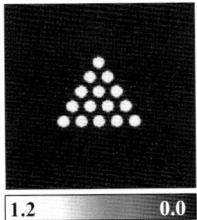

Figure 18.19 ART-based reconstruction from 12 noiseless projections.

handled seamlessly. An ART reconstruction from 12 noiseless projections after 100 iterations with non-negativity constraint is shown in Figure 18.19. Another major advantage of iterative methods is their ability to incorporate the deconvolution step into the interactive scheme, eliminating the need for a separate deconvolution, which generally yields poor results.

18.6.3
Spectral–Spatial Reconstructions

Intuitively, n-D spatial and 1-D spectral imaging is identical to $(n+1)$D spatial imaging. From this point of view, spectral–spatial imaging can be performed in the same way as spatial imaging, and all the iterative techniques discussed for spatial imaging are also applicable for the spectral–spatial imaging. Since the lineshapes usually belong to a parametric family of functions, the estimated image in each iteration can be curve-fitted to take that particular shape along the spectral dimension. This approach is extremely slow, however, and may not be feasible for large images. Recently, a parametric approach has been suggested (Som et al., 2007) to embed the spectral lineshape information in the forward model to write the projections as a function of local spin density and linewidth. In this way, the reconstruction problem becomes a parameter estimation problem. The parametric approach has the potential to reduce the acquisition time substantially.

18.6.4
Image Quality and Resolution

The accuracy of an image can be assessed in terms of its quality and resolution. The *image quality* is the fidelity of the reconstructed image to the original object, and mainly depends on the SNR and artifacts in the computed image. Image *resolution* is defined as the minimum significant image element (distance, area, or volume) that can be meaningfully interpreted from the image. The resolution depends on several factors, both instrumental and computational. Hoch and Ewert (Eaton, Eaton, and Ohno, 1991) have delineated the factors that affect the resolution of an EPR image, and discussed methods of estimating the resolution on an individual-factor basis. Of all the factors that control the overall accuracy of an image, the gradient magnitude is the most important factor that must be optimized to the maximum value for a given experiment. Under circumstances when the resolution is not already limited by other factors, the absolute resolution based on the gradient magnitude, g, is given as τ/g. If ΔL is the maximum size of the object space (spatial window) and ΔB is the field sweep corresponding to the gradient g, then the resolution can be defined as follows:

$$\sigma = \frac{\tau}{g\Delta L} = \frac{\tau}{\Delta B} \tag{18.28}$$

where σ is expressed as a ratio of the resolution to the size of the object. In order to exclude overlap of the next hyperfine line, the sweep width must satisfy the following condition:

$$\Delta B + \tau < \beta \tag{18.29}$$

where β is the hyperfine coupling constant. This imposes a lower resolution limit given by:

$$\sigma_{min} = \frac{\tau}{\beta - \tau} \tag{18.30}$$

Thus, the maximum obtainable resolution of the image is limited by the spectral parameters β and τ. Consequently, if a study demands image resolutions better than those achievable under these conditions, this hyperfine-based limitation must be overcome to obviate the adventitial effects of extending the spectral window into regions of other hyperfine lines. The maximum obtainable resolution of a spectral–spatial image depends on the number of projections (N) that are acquired and used in the reconstruction. As the number of projections increases, the sweep width of some of the high-gradient projections may now contain other hyperfine lines of the multiple-line paramagnetic sample causing the hyperfine contamination. Depending on the number of hyperfine lines and the magnitude of the contamination, the resulting backprojected image will contain streak artifacts, which will decrease both the resolution and quality of the image.

Spatial imaging, however, requires the use of gradients of constant magnitude, and fixed field sweep, ΔB. If the sweep width is larger than the hyperfine coupling constant, then the hyperfine contamination may occur in all of the projections, thus causing artifacts in the image. The backprojection reconstruction method may cancel this contamination for a theoretically required or large number of projections. Howwever, in practice artifacts will be observed where the number measured projections is less than this limit.

18.7
Other Data Collection Modalities

It is not necessary to acquire the data in the form of parallel-ray projections, although this is the most commonly employed method for data collection in EPRI. Some other modalities of acquiring data, which may have advantages over the conventional parallel-ray projection modality, are discussed in the following subsections.

18.7.1
Pulsed-EPR

In CW-EPR, the data are collected by providing the RF excitation and sweeping the magnetic field gradually. As it may take a couple of seconds to acquire one projection, the acquisition of data for higher dimensions, where hundreds and even thousands of projections may be required, becomes impractical for many applications where the conditions may change rapidly. Whilst pulsed-EPR has the potential to reduce the acquisition time substantially (Bourg et al., 1993), there are numerous technical challenges still to be addressed before it overtakes CW-EPR for broad EPR applications (Giuseppe et al., 1999). In pulsed-EPR, the excitation is provided by a train of short RF pulses, after which the emitted signals from the spins are digitized and recorded; this is termed free induction decay (FID). After applying a Fourier transform, the EPR spectrum is obtained in the frequency domain. Although the basic principle of the pulsed-EPR spectrometer is similar to that of modern NMR spectrometers, the extremely short relaxation times used for EPR cause the instrumentation to be much more challenging. At low frequencies, the dead time of an EPR resonator (defined as the time required by the resonator to dissipate the RF energy) can approach the relaxation time of many commonly used EPR spin probes, which makes it extremely difficult to collect the data. Therefore, probes with longer relaxation times are highly preferred to exploit the advantages of the pulsed system.

18.7.2
Single Point Imaging

Single point imaging (SPI) or constant-time imaging represents a special means of collecting data in pulsed-EPR. In SPI, for every pulsed RF excitation, a single datum point of the FID after a fixed delay in the presence of static magnetic field gradients is acquired (Subramanian et al., 2002a). The process is analogous to performing pure phase-encoding in all dimensions to fill the n-D Fourier domain, called k-space, point-by-point. As the phase-encoding time (the fixed delay after excitation) remains constant for a given image data set, the spectral information (lineshape) is automatically deconvolved, providing well-resolved pure spatial images. Therefore, SPI has the potential to provide high-resolution, artifact-free images that may be useful for many biological applications. Since only one datum point is recorded per excitation, the acquisition times for SPI can be longer than those for CW-EPRI. For spectral–spatial imaging, the spectral information can also be extracted from a series of SPI images, each of which corresponds to a different delay from the excitation pulse.

18.7.3
Rapid Scan

In traditional CW-EPR, a slow linear scan is made to collect a projection. The magnetic field modulation and lock-in detection are applied to increase the SNR, which is inversely proportional to the square of the sweep width (Williams, Pan, and Halpern, 2005). In contrast, in direct detection – where the absorption rather than its derivative is directly observed – the decrease in SNR is proportional to the sweep width. However, if the scan speed is comparable to the relaxation time, this introduces distortion, which fortunately can be well-characterized by using Bloch equations and easily accounted for (Dadok and Sprecher, 1974). When the RF power is adjusted for maximum signal amplitude as a function of the scan rate, the signal intensity for a given number of scans is enhanced by up to a factor of three relative to conventional slow scans (Stoner et al., 2004). In addition, the data in a projection of traditional CW-EPRI are highly correlated. By using rapid scanning, the redundancy can be reduced by increasing the number of projections. The performance of a rapid scan has been evaluated for EPR recently (Joshi et al., 2005). High scan rates (usually in kHz) have led to rapid scan becoming the method of choice for imaging moving objects, such as the beating heart.

18.7.4
Spinning Gradient

To acquire a projection, the main magnetic field is swept for a given gradient $\hat{g}$, and the process is repeated for a number of different gradient directions. The adjacent data points in each projection are highly correlated, and hence reduce the acquisition efficiency. Therefore, rapidly rotating the gradient $\hat{g}$ for a given magnetic field B_δ, and repeating it for different B_δ to cover the entire object, reduces the data correlation and hence improves the reconstructed results. The performance of a spinning gradient has been explored for EPR (Deng et al., 2004). Usually, special low-inductance gradient coils are used to spin the gradients rapidly. If the spinning frequency is low (<100 Hz), the existing systems with phase-sensitive detection can be used without any major hardware modification.

18.8
Constraints for Biological Applications

The penetration and distribution of the RF fields into lossy biological samples is the fundamental determining factor in the selection of an EPR frequency for a given size of sample. Although the most common operating frequency for EPR spectrometers is X-band (~9 GHz), X-band instrumentation is limited to nonaqueous samples of a few millimeters in size. Biological samples are aqueous in nature, and hence cause serious heating problems due to the dielectric loss. A further limitation is the problem associated with penetration of the excitation field at higher microwave frequencies. It has been estimated, for example, for a water-containing tissue, that the penetration depth at X-band is ~1 mm (Bottomley and Andrew, 1978), which severely restricts the use of conventional X-band techniques for studying larger biological samples. Hence, low-frequency spectrometers operating in the range of (200 MHz to 3 GHz) have been developed to enable EPR measurements on large aqueous samples (Berliner and Fujii, 1985; Halpern et al., 1989; Quaresima et al., 1992; Zweier and Kuppusamy, 1988). The main feature of these low-frequency spectrometers is their resonators, which are capable of accommodating large aqueous samples with minimal dielectric dissipation. There is, however, a major disadvantage in the use of lower frequencies, as the sensitivity of the measurement is directly related to the square of the operating frequency. The decrease in sensitivity at lower frequencies is, to some extent, compensated by the higher filling factor for the sample. For example, while X-band EPR measurements on lossy samples are limited to a volume of a few microliters, L-band (1–2 GHz) measurements are possible on volumes of a few milliliters, while RF resonators (200–300 MHz) are used for samples of up to 200 ml. Over the past decade, a variety of low-frequency instrumentation suitable for lossy biological samples has been developed, and this has enabled spectroscopic measurements to be made on biological samples of sizes up to a whole rat (Alecci et al., 1990; Halpern et al., 1994; He et al., 1999; Kuppusamy et al., 1998a; Kuppusamy, Chzhan, and Zweier, 1995a; Kuppusamy, Shankar, and Zweier, 1998b; Zweier and Kuppusamy, 1988).

In addition to the penetration depth, the selection of an EPR probe can also be an important variable in determining the reconstruction quality. Probe selection is made based on its spin density, biostability, toxicity, lineshape, and oxygen sensitivity. A high spin density will improve the SNR which, in turn, will improve the spatial resolution and reduce the acquisition time, which may be important for biological applications where conditions can change over time. Likewise, the use of probes with narrow linewidths also translates to a higher spatial resolution. A higher oxygen sensitivity may also be desired to improve the spectral resolution.

18.9
Special Imaging Applications

18.9.1
EPR Oximetry Mapping

Currently, numerous applications of EPR spectroscopy and imaging are available, including *in vivo* oximetry (Halpern *et al.*, 1994; Ilangovan *et al.*, 2004; Subczynski, Lukiewicz, and Hyde, 1986; Zweier, Thompson-Gorman, and Kuppusamy, 1991). This involves measurement of the partial pressure of oxygen (pO_2) by observing an oxygen-induced broadening in the lineshape of the introduced probe. EPR oximetry, as first reported by Hyde (Popp and Hyde, 1981) and later used extensively by Swartz and colleagues (Swartz and Glockner, 1991; Swartz and Walczak, 1998), enables repeated measurements to be made of oxygen concentrations within living tissues. In cancer, the oxygen concentration within a tumor is considered an important factor for determining the response of the tumor to different treatment options (O'Hara *et al.*, 1997). Likewise, the presence of oxygen plays a critical role in the pathophysiology of myocardial injury during both ischemia and subsequent reperfusion. Therefore, the ability of EPR oximetry to make repeated minimally invasive measurements of oxygen over time can provide vital information to characterize the progression of a disease state, and determine the efficacy of different treatment options. Figure 18.20 (Bratasz *et al.*, 2007) shows the ability of EPRI to map both spin density and pO_2 in an RIF tumor in a mouse.

The principle of EPR oximetry is based on the effect of molecular oxygen on the lineshape (mostly linewidth) (Liu *et al.*, 1993) of a known paramagnetic spin probe. Molecular oxygen exists in the triplet ground state, and hence it is paramagnetic. On interaction with a spin probe, oxygen can increase the relaxation rate of the probe via a Heisenberg exchange, and this leads to an oxygen concentration-dependent increase in the linewidth of the probe. The changes in lineshape as a function of pO_2 have been well characterized for a number of EPR probes. Hence, for any known probe the change in EPR linewidth can be regarded as a direct measure of pO_2.

In EPR oximetry, a probe is exogenously introduced at the site of interest. The two most important factors to be considered when evaluating an oximetry probe are:

- *EPR sensitivity*: this is the detection sensitivity of the probe; as a good SNR is essential to keep the acquisition times reasonable, probes with high

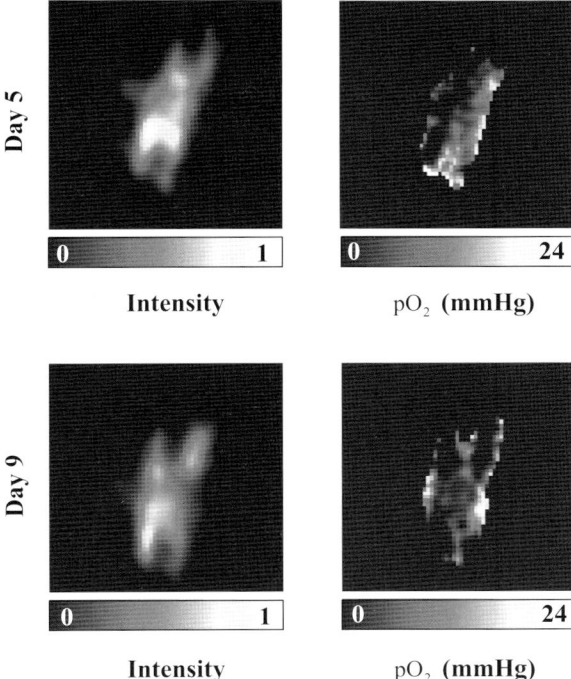

Figure 18.20 Representative images (10 × 10 mm) of probe and oxygen distribution in a growing RIF-1 tumor. The images were obtained on day 5 (volume = 86 mm³) and day 9 (volume = 113 mm³) after the mouse was inoculated with RIF-1 cells internalized with the nanoparticulate spin probes (LiNc-BuO). The probe is seen to be distributed in the core of the tumor, which is about 62 ± 9% of the tumor volume on day 5, and 42 ± 6% on day 9. Note that the oxygen information is obtained only from the regions where the particulates are present. The images show the presence of significantly hypoxic pockets in the region of the tumor under examination.

densities of unpaired spins and a single narrow EPR lineshape are generally preferred.

- *Oxygen sensitivity*: this is the change in EPR linewidth as a function of pO_2. Oxygen sensitivity depends on the combination of a variety of factors, including electron–electron interaction, medium, viscosity, and the nature of the probe's molecular structure.

EPR oximetry offers several unique advantages. For example, methods based on EPR oximetry are direct, minimally invasive, highly sensitive, and capable of providing high spatial (submillimeter) and temporal resolution (<1 s), without consuming oxygen. Most importantly, EPR oximetry enables measurements of oxygen concentrations ranging from normal physiological levels down to millitorr values. In the case of a single point implant, the oximetry is performed by spectroscopy as there is not much spatial variation of spin response, whereas for larger probes or multiple implants distributed over a volume of varying pO_2, spectral–spatial imaging is generally required to capture the complete information.

18.9.2
Imaging Redox Metabolism in Tissues

EPRI has the capability to provide unique functional imaging of physiological information, noninvasively. Stable nitroxide free radicals are valuable biophysical probes that have been used in EPRI for many years, but more recently they have also been identified as biological antioxidants (Krishna et al., 1996; Samuni et al., 1991). The nitroxide free radical can be detected by EPRI. The one-electron reduction product, namely the hydroxylamine, is EPR-silent and can be converted back to the EPR-detectable nitroxide radical by oxidation (Swartz, 1990). When the spin probe is administered to an animal, cellular redox processes convert the nitroxide to the corresponding hydroxylamine, which is then reoxidized back to the nitroxide. As the ratio of nitroxide to hydroxylamine depends on several physiological parameters (such as tissue oxygenation and redox status), the nitroxide level can be correlated with the tissue metabolism, and localized EPR spectroscopy and imaging can be performed noninvasively on a tissue to monitor its metabolism. In addition, some spin probes used in EPR (namely, nitroxide free radicals) are susceptible to differential bioreduction, depending on the oxygen status of the tissue. The chemical and biological nature of these probes make them ideal candidates to explore their potential in functional imaging in at least two areas: (i) differential redox metabolism between normal and tumor tissues; and (ii) the determination of pO_2 levels in tissues. Examples of these applications are presented below, based on research conducted in the present authors' laboratories.

18.9.2.1 Differential Distribution of Nitroxide Probes in Normal versus Tumor Tissue

Recently, experiments have been conducted to determine whether: (i) nitroxide could be detected in tumor tissue; and (ii) differential reduction rates of nitroxide probes could be discerned between tumor versus normal tissues (Kuppusamy et al., 1998a). For this, EPRI experiments were conducted using an EPR spectrometer operating at 1.2 GHz, corresponding to a magnetic field of 420 G. A specially built bridged-loop surface resonator was used for the imaging experiments. The open structure of this resonator was ideal for the localized study of metabolic activity in large objects; when using this imaging system a cylindrical volume of only 10 mm diameter and 5 mm depth could be probed. Mice bearing tumors of ~1 cm diameter were anesthetized, and the tail vein was cannulated with a heparin-filled 30-gauge catheter for nitroxide (3-carbamoyl-2,2,5,5-tetramethylpyrrolidine-N-oxyl) infusion. Either, the right leg with a tumor or the left leg with normal tissue (muscle, skin) was utilized for the imaging studies. The presence of nitroxide in normal and tumor tissue was detected readily by using EPRI (see Figure 18.21). It is clear from these spatial images that the nitroxide distribution differs

Normal Tissue

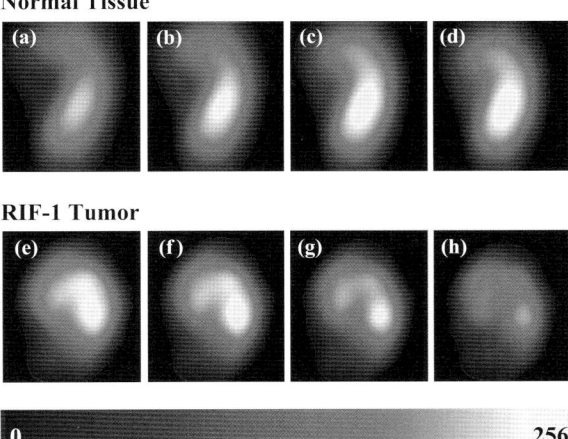

RIF-1 Tumor

Figure 18.21 Visualization of tumor heterogeneity monitored by nitroxide uptake. Following a tail vein infusion of 160 mg kg^{-1} of 3-carbamoyl-proxyl (a commonly used soluble EPR probe), 3-D EPR images of the nitroxide in normal muscle and tumor were measured. A few selected adjacent slices of 0.3 mm thickness obtained from the 3-D image of normal (a–d) and tumor (e–h) tissue are shown. The images show differences between the two tissues in terms of anatomy and physical architecture. The tumor tissue shows a significant heterogeneity of nitroxide uptake compared to the normal tissue. Image acquisition parameters: projections, 100; gradient, 15 G cm^{-1}; acquisition time, 10 min.

between the normal and tumor tissue, which may reflect differences in the vasculature of the microenvironment associated with tumors.

18.9.2.2 Differential Metabolism of Nitroxide Probes in Normal versus Tumor Tissue

Whilst EPR spectroscopic studies indicate the total amount of nitroxide in the volume under study, imaging experiments provide information regarding the spatial distribution of the nitroxides. Such imaging experiments, when carried out as a function of time, provide details of the decay of nitroxides in tissues under study. A 2-D image of the concentration of nitroxide in the normal tissue and RIF-1 tumor also was compared as a function of time after infusion, as shown in Figure 18.22. The normal tissue had a significant amount of nitroxide present compared to that of the tumor, while the rate of clearance of nitroxide from the RIF-1 tumor was shown to be significantly higher than from normal tissue. This was consistent with previous observations that tumors would provide a strong reducing environment compared to normal tissue, resulting in a faster reduction of the nitroxide. Such observations were also in agreement with earlier studies, which suggested that hypoxic cells within tumors would reduce nitroxides more efficiently than well-oxygenated normal tissues (Hahn et al., 1997). Such properties

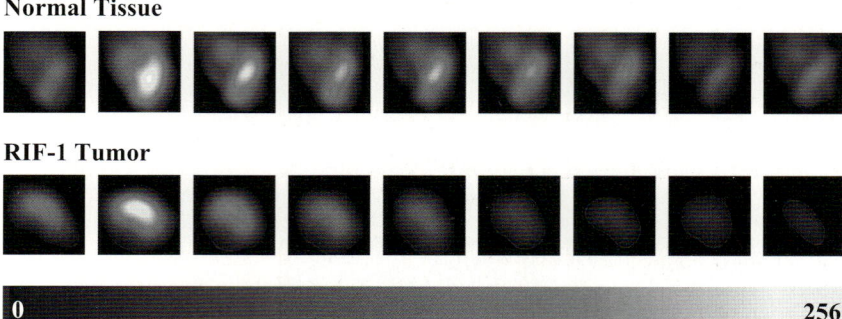

Figure 18.22 Spatially resolved clearance of nitroxide in normal and tumor tissue. Following a tail vein infusion of 160 mg kg^{-1} of 3-carbamoyl-proxyl, a series of 2-D images of the nitroxide from normal muscle (top) and tumor (bottom) were measured using L-band EPR imaging instrumentation. The first image from left was taken at 3 min after injection; each subsequent image was taken after a delay of 90 s. It is evident that the nitroxide in normal tissue persisted for longer than 16 min, while in tumor it was cleared within 10 min of infusion. Image acquisition parameters: projections, 16; gradient, 15 G cm^{-1}; acquisition time, 1.5 min.

enable the free radical probes to provide information on spatial and metabolic differences in tissues noninvasively, by using EPRI.

18.10
Scope and Limitations

In recent years, considerable progress has been made in both the development and application of EPRI to important biological applications, with the number of laboratories investigating EPRI having increased extensively. Moreover, such increased interest is driven almost in parallel by innovations in low-frequency instrumentation suitable for biological objects, and the importance of unique application problems in free radical biology/medicine that can be resolved using this technique. Despite all the progress that has been made during the past two decades, the acquisition of high-quality images of biological samples has been limited by several technical factors, including a lack of resolution, poor SNRs, long acquisition times, and a limit to the number of available probes with desired linewidths and oxygen sensitivities. Although alternative approaches such as pulsed techniques may provide certain advantages over CW methods, they will eventually hit the sensitivity bottleneck in the case of clinically useful applications. In the meantime, "crossbreed" techniques such as proton–electron double resonance imaging (PEDRI) or Overhauser-enhanced magnetic resonance imaging (OMRI), which utilize the advantages of NMR detection, also deserve attention.

Acknowledgments

The development of the CW-EPRI systems described in this chapter was carried out at the Davis Heart and Lung Research Institute, School of Medicine, The Ohio State University, Columbus. The authors gratefully acknowledge the contributions made by their colleagues and collaborators.

Pertinent Literature

The elementary theory of EPR is adequately covered by Weil, Bolton, and John (1994), while the instrumentation of EPR spectrometers and imagers is extensively discussed by Poole (1997). The principle of EPR imaging and its practical aspects are detailed in Chapter 6 of Berliner (1993) and in Chapter 11 of Eaton, Eaton, and Berliner (2005). The biological applications of EPR and EPRI are extensively covered by Berliner (2003); Gallez and Swartz (2004); Swartz and Dunn (2003); and Swartz *et al.* (2004).

References

Alecci, M., Colacicchi, S., Indovina, P.L., Momo, F., Pavone, P., and Sotgiu, A. (1990) *J. Magn. Reson. Imag.*, **8**, 59.

Alecci, M., Brivati, J.A., Placidi, G., and Sotgiu, A. (1998) *J. Magn. Reson.*, **130**, 272.

Ardenkjaer-Larsen, J.H., Laursen, I., Leunbach, I., Ehnholm, G., Wistrand, L.G., Petersson, J.S., and Golman, K. (1998) *J. Magn. Reson.*, **133**, 1.

Bacic, G., Nilges, M.J., Magin, R.L., Walczak, T., and Swartz, H.M. (1989) *Magn. Reson. Med.*, **10**, 266.

Bales, B.L., Peric, M., and Lamy-Freund, M.T. (1998) *J. Magn. Reson.*, **132**, 279.

Berliner, J.L. (1993) *In Vivo EPR (ESR): Theory and Applications*, Klewer Academic/Pleanum Publishers, New York.

Berliner, J.L. (2003) *In Vivo EPR (ESR): Theory and Applications*, Kluwer Academic/Plenum Publishers, New York.

Berliner, J.L. and Fujii, H. (1985) *Science*, **227**, 517.

Bloch, F. (1946) *Phys. Rev. B*, **70**, 460.

Bottomley, P.A. and Andrew, E.R. (1978) *Phys. Med. Biol.*, **23**, 630.

Bourg, J., Krishna, M.C., Mitchell, J.B., Tschudin, R.G., Pohida, T.J., Friauf, W.S., Smith, P.D., Metcalfe, J., Harrington, F., and Subramanian, S. (1993) *J. Magn. Reson. B*, **102**, 112.

Bratasz, A., Pandian, R.P., Deng, Y., Petryakov, S., Grecula, J.C., Gupta, N., and Kuppusamy, P. (2007) *Magn. Reson. Med.*, **57**, 950.

Cho, Z.H., Jones, J.P., and Singh, M. (1993) *Foundations of Medical Imaging*, John Wiley & Sons, Inc., New York.

Chumakov, V.M., Ivanov, V.P., Yaguzhinskii, L.S., Rozantsev, E.G., and Kalmanson, A.E. (1972) *Mol. Biol.*, **6**, 188.

Chzhan, M., Shtenbuk, M., Kuppusamy, P., and Zweier, J.L. (1993) *J. Magn. Reson. A*, **105**, 49–53.

Chzhan, M., Kuppusamy, P., and Zweier, J.L. (1995) *J. Magn. Reson. B*, **108**, 67.

Chzhan, M., Kuppusamy, P., Samouilov, A., He, G., and Zweier, J.L. (1999) *J. Magn. Reson.*, **137**, 373.

Dadok, J. and Sprecher, R.F. (1974) *J. Magn. Reson.*, **13**, 243.

Deng, Y., He, G., Kuppusamy, P., and Zweier, J.L. (2003) *Magn. Reson. Med.*, **50**, 444.

Deng, Y., He, G., Petryakov, S., Kuppusamy, P., and Zweier, J.L. (2004) *J. Magn. Reson.*, **168**, 220.

Deng, Y., Pandian, R.P., Ahmad, R., Kuppusamy, P., and Zweier, J.L. (2006) *J. Magn. Reson.*, **181**, 254.

Eaton, G.R., Eaton, S.S., and Ohno, K. (1991) *EPR Imaging and in Vivo EPR*, CRC Press, Inc, Boca Raton, FL.

Eaton, S.S., Eaton, G.R., and Berliner, L.J. (2005) *Biomedical EPR–Part A: Free Radicals, Metals, Medicine and Physiology*, Kluwer Academic, New York.

Froncisz, W. and Hyde, J.S. (1982) *J. Magn. Reson.*, **47**, 515.

Gallez, B. and Swartz, H.M. (2004) *NMR Biomed.*, **17**, 223.

Giuseppe, S.D., Placidi, G., Brivati, J.A., Alecci, M., and Sotgiu, A. (1999) *Phys. Med. Biol.*, **44**, N137.

Golay, M.J. (1958) *Rev. Sci. Instrum.*, **29**, 313.

Gordon, R., Bender, R., and Herman, G.T. (1970) *J. Theor. Biol.*, **29**, 471.

Griffiths, D.J. (1987) *Introduction to Elementary Particles*, John Wiley & Sons, Inc., New York.

Griffiths, D.J. (1995) *Introduction to Quantum Mechanics*, Prentice Hall, New York.

Hahn, S.M., Sullivan, F.J., DeLuca, A.M., Krishna, C.M., and Wersto, N. (1997) *Free Radic. Biol. Med.*, **22**, 1211.

Halpern, H.J., Spencer, D.P., Polen, J.V., Bowman, M.K., Nelson, A.C., Dowey, E.M., and Teicher, E.A. (1989) *Rev. Sci. Instrum.*, **60**, 1040.

Halpern, H.J., Yu, C., Peric, M., Barth, E., Grdina, D.J., and Teicher, B.A. (1994) *Proc. Natl Acad. Sci. USA*, **91**, 13047.

Hanson, K.M. (1983) *J. Opt. Soc. Am.*, **73**, 1501.

He, G., Shankar, R.A., Chzhan, M., Samouilov, A., Kuppusamy, P., and Zweier, J.L. (1999) *Proc. Natl Acad. Sci. USA*, **96**, 4586.

Herman, G.T. and Lent, A. (1976) *Comput. Biol. Med.*, **6**, 273.

Ilangovan, G., Bratasz, A., Li, H., Schmalbrock, P., Zweier, J.L., and Kuppusamy, P. (2004) *Magn. Reson. Med.*, **52**, 650.

Jiang, J.J., Bank, J.F., Zhao, W.W., and Scholes, C.P. (1992) *Biochemistry*, **31**, 1331.

Joshi, J.P., Ballard, J.R., Rinard, G.A., Quine, R.W., Eaton, S.S., and Eaton, G.R. (2005) *J. Magn. Reson.*, **175**, 44.

Katz, M.B. (1979) *Questions of Uniqueness and Resolution in Reconstruction from Projections*, Springer, Berlin.

Krishna, M.C., Samuni, A., Taira, J., Goldstein, S., Mitchell, J.B., and Russo, A. (1996) *J. Biol. Chem.*, **271**, 26018.

Kuppusamy, P., Chzhan, M., Vij, K., Shteynbuk, M., Lefer, D.J., Giannella, E., and Zweier, J.L. (1994) *Proc. Natl Acad. Sci. USA*, **91**, 3388.

Kuppusamy, P., Chzhan, M., and Zweier, J.L. (1995a) *J. Magn. Reson. B*, **106**, 122.

Kuppusamy, P., Wang, P., and Zweier, J.L. (1995b) *Magn. Reson. Med.*, **34**, 99.

Kuppusamy, P., Wang, P., Chzhan, M., and Zweier, J.L. (1997) *Magn. Reson. Med.*, **37**, 479.

Kuppusamy, P., Afeworki, M., Shankar, R.A., Coffin, D., Krishna, M.C., Hahn, S.M., Mitchell, J.B., and Zweier, J.L. (1998a) *Cancer Res.*, **58**, 1562.

Kuppusamy, P., Shankar, R.A., and Zweier, J.L. (1998b) *Phys. Med. Biol.*, **43**, 1837.

Kuppusamy, P., Wang, P., Shankar, R.A., Ma, L., Trimble, C.E., Hsia, C.J., and Zweier, J.L. (1998c) *Magn. Reson. Med.*, **40**, 806.

Laukemper-Ostendorf, S., Scholz, A., Burger, K., Heussel, C.P., Schmittner, M., Weiler, N., Markstaller, K., Eberle, B., Kauczor, H.U., Quintel, M., Thelen, M., and Schreiber, W.G. (2002) *Magn. Reson. Med.*, **47**, 82.

Lewitt, R.M. (1983) *IEEE Proc.*, **71**, 390.

Liu, K.J., Gast, P., Moussavi, M., Norby, S.W., Vahidi, N., Walczak, T., Wu, M., and Swartz, H.M. (1993) *Proc. Natl Acad. Sci. USA*, **90**, 5438.

Maltempo, M.M., Eaton, S.S., and Eaton, G.R. (1987) *J. Magn. Reson.*, **72**, 449.

O'Hara, J.A., Goda F., Dunn, J.F., and Swartz, H.M. (1997) *Adv. Exp. Med. Biol.*, **411**, 233.

Poole, C.P.J. (1997) *Electron Spin Resonance: A Comprehensive Treatise on Experimental Techniques*, Dover Publications.

Popp, C.A. and Hyde, J.S. (1981) *J. Magn. Reson.*, **43**, 249.

Quaresima, V., Alecci, M., Ferrari, M., and Sotgiu, A. (1992) *Biochem. Biophys. Res. Commun.*, **183**, 829.

Robinson, B.H., Mailer, C., and Reese, A.W. (1999) *J. Magn. Reson.*, **128**, 199.

Samuni, A., Winkelsberg, D., Pinson, A., Hahn, S.M., Mitchell, J.B., and Russo, A. (1991) *J. Clin. Invest.*, **87**, 1526.

Som, S., Potter, L.C., Ahmad, R., and Kuppusamy, P. (2007) *J. Magn. Reson.*, **186**, 1–10.

Sotgiu, A. (1985) *J. Magn. Reson.*, **65**, 206.

Stillman, A.E. and Levin, D.N. (1986) *J. Magn. Reson.*, **69**, 168.

Stoner, J.W., Szymanski, D., Eaton, S.S., Quine, R.W., Rinard, G.A., and Eaton, G.R. (2004) *J. Magn. Reson.*, **170**, 127.

Subczynski, W.K., Lukiewicz, S., and Hyde, J.S. (1986) *Magn. Reson. Med.*, **3**, 747.

Subramanian, S., Devasahayam, N., Murugesan, R., Yamada, K., Cook, J., Taube, A., Mitchell, J.B., Lohman, J.A., and Krishna, M.C. (2002a) *Magn. Reson. Med.*, **48**, 370.

Swartz, H.M. (1990) *Free Radic. Res. Commun.*, **9**, 399.

Swartz, H.M. and Dunn, J.F. (2003) *Adv. Exp. Med. Biol.*, **530**, 1.

Swartz, H.M. and Glockner, J.F. (1991) *Measurements of Oxygen by EPRI and EPRS*, CRC Press, Inc., Boca Raton, FL.

Swartz, H.M. and Walczak, T. (1998) *Adv. Exp. Med. Biol.*, **454**, 243.

Swartz, H.M., Khan, N., Buckey, J., Comi, R., Gould, L., Grinberg, O., Hartford, A., Hopf, H., Hou, H., Hug, E., Iwasaki, A., Lesniewski, P., Salikhov, I., and Walczak, T. (2004) *NMR Biomed.*, **17**, 335.

Thomas, S.R., Busse, L.J., and Scheuck, V.F. (1985) *AAPM Summer School in Medical Physics*, Seattle, WA.

Weil, J.A., Bolton, J.R., and John, E.W. (1994) *Electron Paramagnetic Resonance: Elementry Theory and Practical Applications*, John Wiley & Sons, Inc., New York.

Williams, B.B., Pan, X., and Halpern, H.J. (2005) *J. Magn. Reson.*, **174**, 88.

Zweier, J.L. and Kuppusamy, P. (1988) *Proc. Natl Acad. Sci. USA*, **85**, 5703.

Zweier, J.L., Thompson-Gorman, S., and Kuppusamy, P. (1991) *J. Bioenerg. Biomembr.*, **23**, 855.

19
Multifrequency EPR Microscopy: Experimental and Theoretical Aspects

Aharon Blank

19.1
General

This chapter describes some experimental and theoretical aspects of multifrequency EPR microscopy (EPRM).[1] A general formal definition of the field of EPR microscopy, an historical overview, and a brief comparison of various detection techniques are provided in the Introduction, and this is followed by an outline of some experimental aspects of continuous-wave (CW) and pulsed-EPR microscopy. Attention will be focused on instrumentation issues, and a theoretical analysis of signal-to-noise ratio (SNR) and resolution in the context of EPRM will also be provided. Further sections will describe specific aspects of multifrequency EPRM, with an emphasis on the relations between SNR and resolution and frequency/temperature. Following some experimental results demonstrating the current capabilities of EPRM, some general conclusions and future prospects for this field of research are provided. (Note: The information provided in this chapter relates to EPRM methodology and applications conducted at the author's laboratory.)

19.2
Introduction

19.2.1
Definition

In the broad context of EPR imaging, it is commonplace to define EPR microscopy as related to images acquired with a resolution better than ~100 μm. This

---

1) Note: To conform with this book, the notation "EPR" is used in this chapter rather than "ESR"; most reports on this subject use notations of "ESR" and "ESRM".

Multifrequency Electron Paramagnetic Resonance, First Edition. Edited by Sushil K. Misra.
© 2011 Wiley-VCH Verlag GmbH & Co. KGaA. Published 2011 by Wiley-VCH Verlag GmbH & Co. KGaA.

definition is compatible with what is customary in the field of NMR imaging (Callaghan, 1991).

19.2.2
Historical Overview

The field of EPR microscopy has closely followed the developments of EPR imaging in general. The first published account of EPR imaging, performed in 1978, introduced the capabilities of EPR microscopy (Karthe and Wehrsdorfer, 1979) when the authors employed a simple static magnetic field gradient to resolve, in one dimension, two grains of 2,2-diphenyl-1-picrylhydrazyl (DPPH), separated by 200 μm. The resolution in this CW-based method, which employed a conventional TE_{102} rectangular resonator, was calculated as ~12 μm. Closely following this, other groups expanded the field of CW-EPR microscopy to two dimensions by acquiring several projections with several directions of static gradients of a DPPH sample and a diamond with paramagnetic defects (Hoch and Day, 1979; Hoch, 1981). The two-dimensional (2-D) resolution achieved in these studies was ~100 μm². In addition to acquiring images with the projection reconstruction (PR) method, early CW-EPR microimaging studies also employed the modulated field gradient (MFG) method (Herrling et al., 1982).

These early developments were closely followed by more advanced experimental set-ups, based on CW and pulsed systems, which further improved the resolution. Significant past resolution achievements in the field, excluding those by the present author and coworkers (which are described below) have been as follows:

- In CW-EPRM, the modulated-field-gradient method achieved a resolution better than ~10 μm in one dimension (Herrling et al., 1982; Herrling, Fuchs, and Groth, 2002), whereas a resolution of ~100 μm² was achieved in 2-D images (DDR, 1987).

- The CW technique with static magnetic field gradients achieved a resolution of ~12 μm for one-dimensional (1-D) experiments (Karthe and Wehrsdorfer, 1979), while the 2-D resolution was of the order of ~100 μm², achieved by employing CW with projection reconstruction (Eaton, Eaton, and Ohno, 1991).

- A commercial CW imaging system developed by Bruker (E540) can obtain images with three-dimensional (3-D) resolution down to ~50 μm³ in special samples (Drescher and Dormann, 2004).

- Pulsed EPR imaging techniques requires a fast gradient switching and spin probes with a sufficiently long T_2. Key studies with pulsed high-resolution EPR microscopy were conducted with a $(FA)_2X$ (FA = fluoranthene; X = PF_6, AsF_6) crystal, which is a unique organic conductor, characterized by the spin–spin relaxation time (T_2) of ~6 μs at room temperature. Using this material, 1-D with ~10 μm resolution was reported for pulsed X-band (Maresch, Mehring, and Emid, 1986).

- Low-frequency pulse radiofrequency (RF) EPR, employing standard NMR microscopy gradients, achieved 2-D and 3-D images with a resolution of ~20–30 µm³ (Coy, Kaplan, and Callaghan, 1996; Feintuch et al., 2000) after ~10 h of data acquisition. It should be noted that the unique crystals used in these experiments, and the long acquisition time required for 2-D/3-D images, are not attractive for biophysical, botanical, and many other potential applications.

- An early 1-D imaging study at high field (5 T, 140 GHz) achieved a resolution of ~200 µm, but this was limited by the gradient system (Smirnov, Poluectov, and Lebedev, 1992).

- Early efforts in pulsed X-band spin-echo imaging (Milov et al., 1985) should also be noted.

19.2.3
"Induction Detection" versus Other Detection Methods

Magnetic resonance is a very effective but, nevertheless, very insensitive spectroscopic tool that requires a large amount of material for evaluation (relative to methods such as mass and fluorescence spectroscopy). When confined to measurements at ambient temperature and pressure – as would be required, for example, for live specimens – by using state-of-the-art NMR, with a conventional coil – the so-called "induction" or "Faraday" detection – it is possible to measure $\sim 3 \times 10^{12}$ hydrogen nuclei (Ciobanu, Seeber, and Pennington, 2002). Under similar ambient conditions, with conventional EPR and employing a highly sensitive resonator detection (analogous to the NMR coil), it is possible to measure a signal of $\sim 2 \times 10^{7}$ electron spins (Blank et al., 2004b). By lifting the constraint of ambient conditions, better results can be obtained with other new techniques. In recent years, there has been an explosion in the field of new, sensitive magnetic resonance detection techniques, which have included: scanning tunneling microscopy EPR (STM-EPR) (Manassen et al., 1989; Durkan and Welland, 2002); magnetic resonance force microscopy (MRFM) (Rugar et al., 2004); Hall detection (Boero, Besse, and Popovic, 2001); superconducting quantum interference device (SQUID) detection (McDermott et al., 2002); optically detected MR (Wrachtrup et al., 1993); quantum dot spin detection (Elzerman et al., 2004); electrically detected magnetic resonance (Xiao et al., 2004); and indirect detection via diamond nitrogen-vacancy (NV) centers (Balasubramanian et al., 2008; Maze et al., 2008; Taylor et al., 2008).

These methods have all approached or surpassed the sensitivity and resolution obtained with the conventional induction-detection method. Unfortunately, this sensitivity comes at a price, and each of the new detection methods has its own significant limitations:

- The sensitivity of Hall probes and SQUIDs degrades substantially as the frequency is elevated, depending heavily on the operating temperature. In addition, in order to make use of the high sensitivity in imaging applications, these detection methods require sequential point-to-point measurements, which are very inconvenient and limited, especially for 3-D imaging.

- STM-EPR functions only under high vacuum and ultra-low temperatures, and requires that the samples be deposited on a conductive surface. Moreover, it cannot provide 3-D images, and it also has to scan the surface point by point.

- The MRFM technique [which recently demonstrated single electron spin detection (Rugar et al., 2004) and 2-D imaging with 90 nm resolution (Mamin et al., 2007)] operates efficiently only at low cryogenic temperatures and under high vacuum. It has very limited spectroscopic capabilities due to the extreme gradients generated by the magnetic tip used for detection, and cannot be applied in conjunction with modern pulse techniques. Furthermore, MRFM offers only limited 3-D imaging capability as its sensitivity degrades quickly as the distance between the magnetic tip and the sample increases, and it enables only a sequential, single-point, readout of information, as the tip scans the sample.

Similar limitations exist for the other techniques mentioned above, which function only in very special cases, and thus are not of general use.

It can be concluded that, despite the many new ideas and vast activity in the field, induction-detection remains the only general-purpose approach available today for both spectroscopy and imaging applications. In view of this, a route was chosen that calls for concentrated effort on improving the capability of induction-detection to provide enhanced sensitivity and resolution. This effort complements the development of these new and more "exotic" techniques. With these objectives in mind, it was decided that the first line of action would be to significantly enhance the sensitivity and resolution of induction-detection in EPR, down to the micron scale and below. Thus, hereafter, the discussion of EPRM will be given uniquely in the context of induction-detection.

19.3
General Experimental Aspects of EPR Microscopy

EPR microscopy can be based both on CW- and/or pulsed-EPR spectrometers. CW systems are more favorable for measurements of radicals with short T_2 (commonly <100–200 ns), while pulsed-EPR is often better to use for systems with long T_2 (commonly >200 ns). Generally speaking, unless T_2 is too short, it is preferable to use pulsed-EPRM due to its high versatility and good capability to image parameters such as T_1, T_2 and diffusion coefficient in a relatively short time. The main experimental aspects of typical CW- and pulsed-EPRM systems, and an analysis of their expected performance in terms of SNR and resolution are outlined in the following subsections.

19.3.1
CW-EPR Microscopy

19.3.1.1 System Configuration
A typical CW-EPR microscope has the following components (see Figure 19.1):

19.3 General Experimental Aspects of EPR Microscopy

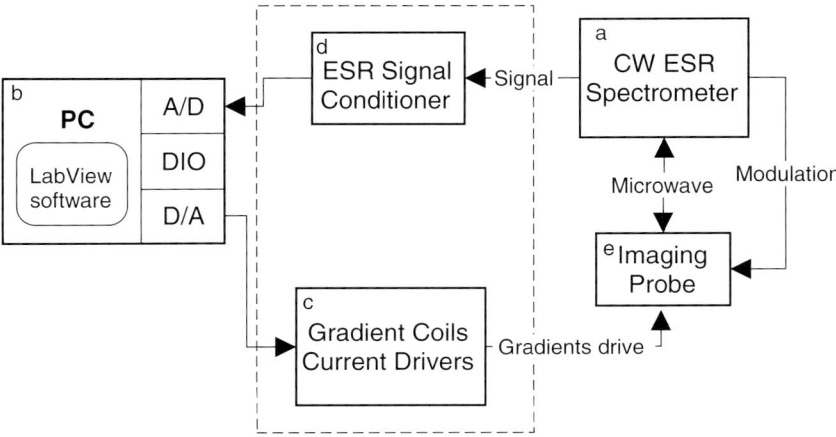

Figure 19.1 Block diagram of a typical CW-EPRM system.

- A conventional CW-EPR spectrometer;
- A dedicated computer which controls the imaging process and acquires the EPR signal;
- Current drivers for the gradient coils;
- A baseband (up to ~250 kHz) amplifier and filter unit (signal conditioner);
- An imaging probe that includes the microwave resonator, a mechanical fixture for holding the sample, and gradient coils.

In order to image a sample, it should be placed in the imaging probe and the EPR spectrometer set to acquire the signal due to the sample with the desired conditions of microwave (MW) power, static magnetic field, and field modulation amplitude. After fixing the spectrometer on the maximum of the EPR signal, the computer-controlled imaging procedure can be initiated, after which the gradient coils are activated to obtain the image via one of the known imaging procedures, for example, projection reconstruction or the modulated field gradient method (Eaton, Eaton, and Ohno, 1991). During the imaging period, the EPR microscope incorporates a field frequency lock (FFL) system that adjusts the static magnetic field by biasing one of the gradient coil pairs on the probe and maintains the on-resonance condition throughout the experiment. Thus, in this mode of operation the basic CW-EPR spectrometer is completely unaware of the imaging procedure, with the interfaces between the spectrometer and the imaging probe/system kept to a minimum. This enables the incorporation of either a home-built, or a commercial, CW-EPR spectrometer as part of the microscope, with minimal adaptations. During and following the acquisition, the image is displayed on the control computer in real time, and can be saved and/or manipulated as necessary.

Following this introductory description, additional details of the individual components of a typical 3-D CW-EPRM system are described:

CW-EPR spectrometer: As mentioned above, most commercial (e.g., Bruker, Varian, or JEOL) or home-built CW-EPR spectrometers can be used for constructing the imaging system. The spectrometer serves as a good stable, amplitude-controlled, MW source that is frequency-locked independently to the resonance frequency of the imaging probe, by means of the spectrometer's automatic frequency control (AFC). The spectrometer can also provide a current drive for the field modulation coils of the imaging probe, and should allow for the operator to set the external static magnetic field sufficiently close to the resonance field of the sample to be imaged. The MW EPR signal returning from the imaging probe is detected and preamplified by the spectrometer's MW bridge. Most spectrometers have this signal available in one of the bridge output connectors. The diode-detected baseband signal is directly fed from the bridge preamplifier, similar to the case of time-resolved EPR measurements (Gonen and Levanon, 1984) to a signal conditioning unit, and then sent to a PC for sampling and further analysis (see below).

Control computer and imaging software: The entire imaging process is controlled by a standard PC equipped with analog input and output cards. At the author's laboratory, National Instruments cards (NI 6023E and NI 6713) are used, although many other PC cards have similar functionalities. The analog output card should enable the generation of arbitrary waveforms in the kHz region, while the analog input card should have sampling capability for signals ranging from DC up to several hundred kHz. In the case of using a lock-in amplifier with the diode detected signal, where the modulation signal is used as a reference signal for the lock-in, the sampling rate of the analog input card can be significantly reduced, down to the ~kHz range. The PC cards and the entire imaging procedure are controlled by an appropriate software. Prior to acquisition, the control software (based on LabView™ in the author's systems) obtains the imaging parameters from the user. These parameters include, for example, the number of pixels in the image, the type of image to be acquired (i.e., 2-D spatial, 3-D spatial, 3-D spectral–spatial, or 4-D spectral–spatial) the current amplitude in the gradient coils, the image extent (in mm), and parameters related to the functionality of the FFL system. On completion of the imaging process the data can be saved and/or further processed with the aid of Matlab™ software. Such post-processing includes, for example, deconvolution of the projections and applying the inverse Radon transformation (Poularikas, 2000) to obtain the image.

Current drivers for the gradient coils: The gradient coils are driven by six programmable current sources, one for each gradient coil in the probe (as described below), each capable of supplying up to 3 A of arbitrary waveform current, in the DC-10 kHz range. This provides enough flexibility in fast image acquisition modes with constantly varying gradient magnitudes. The maximum gradient current requirements of the system depend on the efficiency of the gradient coils in the imaging probe, the type of sample measured, and the required resolution (as described below). In practice, the coils in the present system at the author's laboratory do not require more than 1–2 A to generate high enough gradient fields. The relatively low current consumption of the system greatly simplifies the design and space requirements for the driver unit. This issue is significantly different in EPRM than conventional EPR imaging, where the imaging probes are much larger and thus the current consumption – and system complexity – of the gradient drivers are relatively large.

Baseband amplifier and filter unit (signal conditioner): In most EPR spectrometers, the EPR signal, as detected by the MW diode, passes through a baseband preamplifier that is part of the commercial bridge. The signal level after this preamplification is not large enough to be sampled directly by the A/D card in the computer. To facilitate proper A/D sampling, a signal-conditioning unit comprised of band-pass filters and a high-gain amplifier is employed. The dual band-pass filter transfers the signal only at the regular modulation frequency generated by the CW spectrometer, and its second harmonic. For example, in the present system there is a filter that transfers 25 kHz and 50 kHz, and the spectrometer is operated at 25 kHz modulation frequency. The high-gain amplifier (e.g., Tektronix AM502) that follows the dual band-pass filter produces a variable gain, manually controlled, in the range of 40–100 dB. As mentioned above, if a lock-in amplifier is available it can be used instead of the signal conditioning unit.

Imaging probe: The imaging probe is the heart of the CW imaging system. A schematic drawing and a photograph of the imaging probe are shown in Figure 19.2. The probe is based on either a single- or double-stacked ring resonator (Jaworski, Sienkiewicz, and Scholes, 1997; Blank et al., 2003a, 2003b), machined from high-permittivity material such as $SrTiO_3$, or TiO_2, single crystal. These types of crystals have a permittivity of ~100–300 and $\tan\delta$ ~5×10^{-4} at room temperature. The dimensions of the resonator depend on the type of crystal employed, and the frequency of measurement. Typically, at X-band, the ring diameter would be about 2.5 mm, but much smaller at higher frequencies, being inversely proportional to the frequency. The resonator is excited by a loop at the end of a thin (~0.4 mm) semi-rigid coaxial transmission line. The excitation geometry and the calculated fields of such resonator at the resonance frequency (carried out with finite elements software CST Microwave Studio) are shown in Figure 19.3.

The effective volume of the resonator (Blank et al., 2003a), that is roughly the volume from which most of the signal originates, ranges from a few mm^3 at X-band to much less than $1\,mm^3$ at higher frequencies (e.g., 17, 35, and 60 GHz). The resonator ring(s) and the sample are held by a Rexolite piece at an exact position with respect to each other and at the center of the structure of the gradient coils. Variable coupling is achieved by changing the distance between the center of the resonator ring(s) and the loop at the end of the coaxial line using two linear 1-D nonmagnetic stages (e.g., model MDE 255 from Elliot Scientific, GB). In addition, the vertical position of the rings with respect to the coaxial line feed

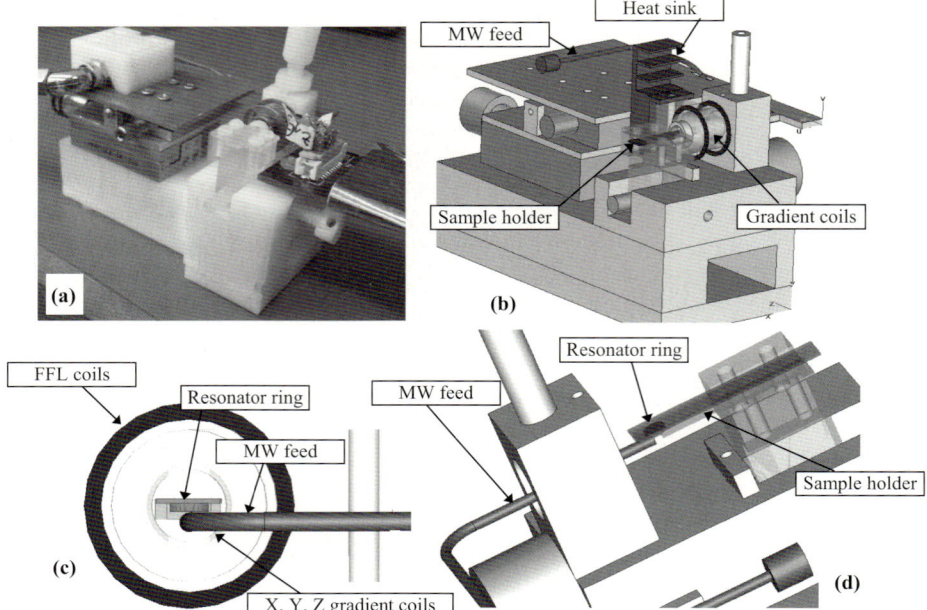

Figure 19.2 The structure of the CW and pulsed-EPRM probe. (a) Photographic image of the probe; (b) A schematic drawing of the probe, showing an isometric view; (c) Lateral view of the probe, showing the microwave feed, the resonator, and the gradient coils located on a cylindrical shield around the resonator; (d) An isometric view of the resonator, sample and microwave feed, without the gradient coils.

(a)

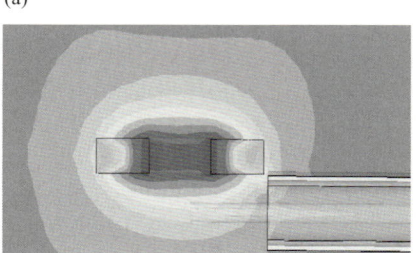

(b)

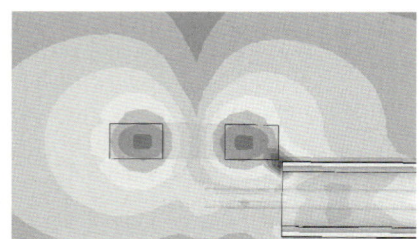

Figure 19.3 Calculated magnetic (a) and electric (b) field distribution around the single-ring dielectric resonator inside the cylindrical shield at 17 GHz. The resonator is made from Rutile, and is 2.4 mm in diameter, 0.5 mm high, and has an inner diameter of 0.9 mm. The magnetic field is maximal at the center of the resonator, while the electrical field has a node at the center of the resonator.

can be varied slightly by adjusting the vertical position of the coaxial line. This variability enables optimal control of the rings' coupling for a wide variety of samples.

The resonator is surrounded by a thin (4.2 mm i.d, 4.6 mm o.d) quartz tube with 1 μm of gold deposited on the outer part of the tube by means of a standard physical vapor deposition technique. The gold shields the resonator from microwave frequencies, but is transparent to low frequencies, which are mostly static-magnetic-field gradients and the field modulation frequencies used in CW spectrometers, which are commonly in the 25–100 kHz range. This type of structure enables a relatively high Q-factor (typically ~1000) to be maintained for the resonator, even at close proximity to the gradient coils. The gradient X, Y, and Z coils are arranged around the cylindrical shield, along with the regular modulation coils (see Figure 19.2b, d). A variety of combinations are possible for the exact arrangement and characteristics of the gradient coils (Blank and Freed, 2004a, 2004b, 2006). In general, for X-band CW-EPR the typical resistance of the coils would be a few ohms, and the typical inductance would be ~10 µH. The typical gradient efficiency is a few T/(m × A). At higher frequencies, smaller probes can be employed with gradient coils having a smaller resistance and inductance and a larger gradient efficiency (see Section 19.4).

One of the major problems of CW-EPRM is the excessive heat generated in the gradient coils. In order to avoid this problem, the gradient coils in some of the author's designs are embedded in a heat-conductive adhesive that is in contact with a heat sink. In addition, a continuous air flow is applied to the heat sink and into the resonator shield to maximize heat dissipation and to maintain a constant temperature of operation. This cooling is necessary because high-permittivity single-crystal materials, such as $SrTiO_3$, are highly sensitive to temperature changes; for example, at X-band the drift may be as high as ~20 MHz per °K (Krupka et al., 1994). The imaging probe can accommodate flat samples with dimensions of ~1.5 × 1.5 mm, corresponding to the active area/volume of the probe (as shown in Figure 19.3), and a height of up to ~1 mm. In practice, for liquid samples the sample should be contained in a glass structure, such as those produced via a photolithography technique (see Figure 19.4) (Halevy, Talmon, and Blank, 2007). This enables the optimal positioning of samples inside the resonator for maximum sample size, and without significant loss in resonator Q. The samples can be sealed, if necessary, under an argon atmosphere, by using an ultraviolet light-curable glue.

19.3.1.2 Signal-to-Noise Ratio

Whilst, in principle, both the SNR and gradient strength can limit the resolution achievable in an EPRM experiment, in most cases the former is the fundamental limiting factor. Therefore, a quantitative analysis of the sensitivity in CW mode EPR experiments is now provided. The SNR in a CW experiment is given by (Blank et al., 2003a):

$$SNR_{CW} \approx \frac{\chi_0'' V_v Q_u \sqrt{P}}{8V_c \sqrt{k_b T \Delta f}} \cdot \frac{1}{1+(Q_u \mu_0 / 2V_c \omega_0) P \gamma^2 T_1 T_2} \qquad (19.1)$$

where χ_0'' is the resonant magnetic susceptibility of the sample per unit of volume, Q_u is the unloaded quality factor of the resonator, μ_0 is the permeability of the free space, and P is the incident MW power. The sample volume is given by V_v, and

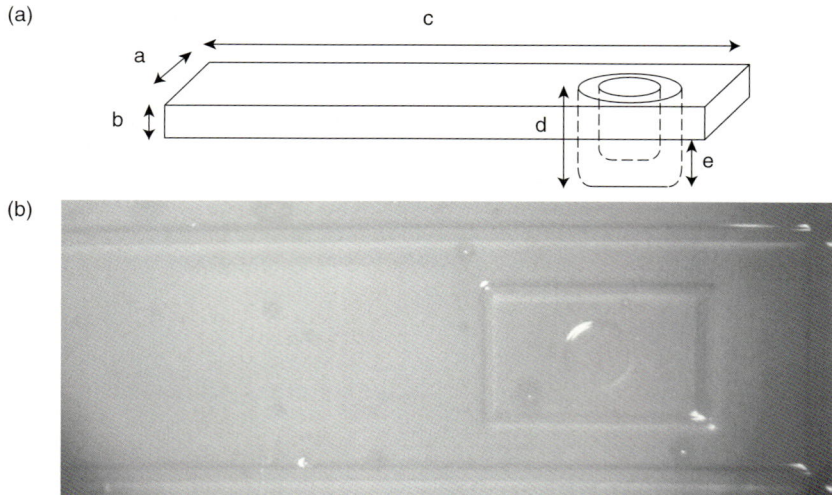

Figure 19.4 (a) Schematic drawing of the "cup-like" sample holder, prepared by a photolithography technique. a = 3.7 mm, b = 0.17 mm, c = 21.9 mm, d = 0.2 mm, e = 0.03 mm; (b) Photographic image of the manufactured glass sample holder.

the resonator effective volume (Blank et al., 2003a) is V_c. The other parameters in Equation 19.1 are Boltzmann's constant (k_b), the electron gyromagnetic ratio (γ), the temperature (T), the detection bandwidth (Δf), the MW frequency (ω_0), and the relaxation times of the spins (T_1, T_2). The maximum SNR is achieved for an MW power level that obeys the relation (Poole, 1983):

$$(Q_u \mu_0 / 2 V_c \omega_0) P \gamma^2 T_1 T_2 = 1/2 \tag{19.2}$$

after which Equation 19.1 can be written as (Blank et al., 2003a):

$$\begin{aligned} SNR_{CW}^{optimal\ P} &\approx \frac{\chi_0'' V_v \sqrt{2Q_u} \sqrt{\omega_0}}{8\sqrt{2}\sqrt{\mu_0}\gamma\sqrt{T_1 T_2}\sqrt{V_c}\sqrt{k_b T \Delta f}} \frac{2}{3} \\ &\approx \frac{\mu_0 H_0 \chi''(\omega_0/\Delta\omega) V_v \sqrt{2Q_u} \sqrt{\omega_0}}{8\sqrt{2}\sqrt{\mu_0 \mu_0 H_0 \gamma T_2}\sqrt{V_c}\sqrt{k_b T \Delta f}} \frac{2}{3} \\ &\approx \frac{\sqrt{2}\sqrt{2\mu_0} M \omega_0 V_v}{24\sqrt{V_c}\sqrt{k_b T \Delta f}} \sqrt{\frac{Q_u}{\omega_0}} = \frac{\sqrt{\mu_0} M \omega_0 V_v}{12\sqrt{V_c}\sqrt{k_b T \Delta f}} \sqrt{\frac{Q_u}{\omega_0}} \end{aligned} \tag{19.3}$$

Here, M is the specific magnetization of the sample (units of [JT^{-1}m^{-3}]), as given by the Curie law (Rinard et al., 1999), and H_0 is the static magnetic field (in H/m^{-1}). The derivation here used the optimal P from Equation 19.2. Furthermore, both the numerator and the denominator were multiplied by $\mu_0 H_0$ while considering the fact that the resonant susceptibility, χ_0'', is related to the static susceptibility, χ'', through $\chi_0'' = \chi''(\omega_0/\Delta\omega)$ (where $\Delta\omega$ is the linewidth, which is equal to $2/T_2$). In addition, it was also assumed, for simplicity, that $T_1 = T_2$. Equation 19.3 will be analyzed in detail in Section 19.4, where the multifrequency aspects of EPRM will be considered.

19.3.1.3 Resolution

In case of sufficient SNR, the resolution in a CW-EPRM experiment is determined by the strength of the applied gradients. In order to obtain high gradient values suitable for EPRM, the intuitive approach is to try and minimize the size of the gradient coils, and thereby to increase the magnitude of the gradients. With CW detection, it is possible to employ either the MFG or the PR imaging method (as described above). In the MFG method, the image resolution is given by (Herrling et al., 1982):

$$\Delta z \approx 2\Delta B_{1/2}/G_z \tag{19.4}$$

Thus, for a radical with linewidth of $\Delta B_{1/2} = 0.01$ mT (which is a typical value for a trityl radical; Yong et al., 2001), a gradient of $G_z \sim 20\,\text{T}\,\text{m}^{-1}$ corresponds to 1 μm resolution. Similar gradient values are required for the PR method (Eaton, Eaton, and Ohno, 1991), although a deconvolution procedure may improve the resolution by a factor of between 2 and 5, assuming an identical lineshape throughout the sample. For small-scale imaging probes, a gradient efficiency of a few T/(m·A) is available (e.g., in the probe described in Section 19.3.1.1), and therefore achieving such gradients requires currents of approximately 5–8 A (such current drives are easily obtained from relatively simple electronic drivers). However, as the resistance of the coils in the micro-imaging probe described above is a few ohms, currents of a few amperes would result in a relatively large heat dissipation of up to 100 W in the small coils structure. In order to prevent coil damage, cooling methods that are used in computers to cool central processing units with a similar heat-generation capacity can be adapted. For this, toils are embedded in an adhesive that is characterized by a high (>4 W mK^{-1}) heat conductivity, connected to a copper heat sink and subjected to a compressed air flow. Further aspects related to gradient strength and SNR, along with their effect on resolution limits, will be considered in Section 19.4.

19.3.2
Pulsed-EPR Microscopy

19.3.2.1 System Configuration

As in the case of CW micro-imaging, pulsed EPRM systems can also be based on either commercial or home-built spectrometers. In general, the latter option is preferred because it provides much better performances and enables system flexibility with lower overall costs. A block diagram of a typical pulsed EPRM system is shown in Figure 19.5. Here, the system is constructed from the following main components:

- A PC which controls the image acquisition process via LabView software (National Instruments). The software controls four PCI cards installed on the PC with the following functionality.

19 Multifrequency EPR Microscopy: Experimental and Theoretical Aspects

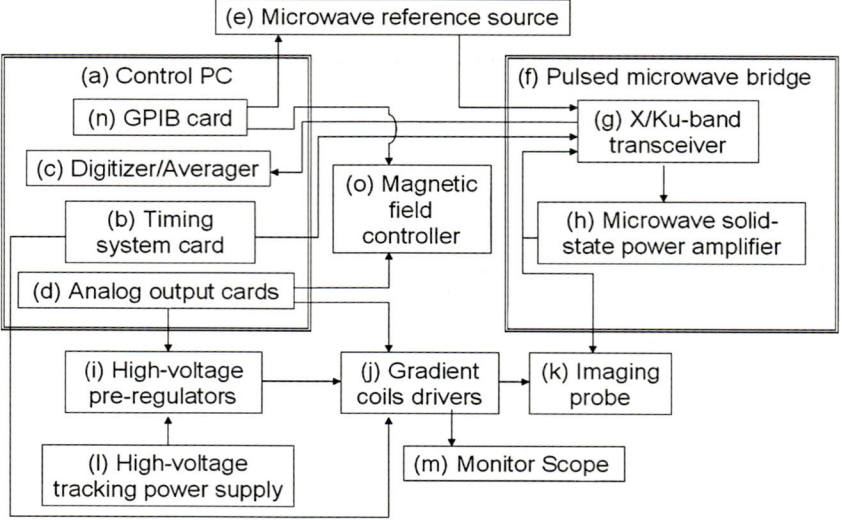

Figure 19.5 A block diagram of a typical pulsed-EPR micro-imaging system.

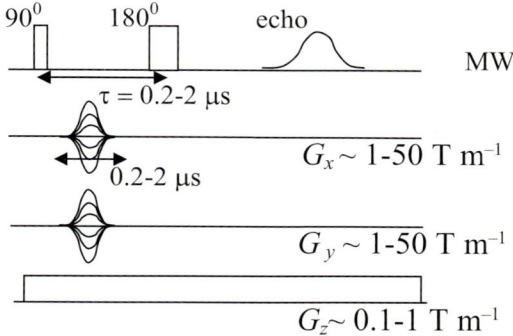

Figure 19.6 Typical EPR pulse sequence used for micro-imaging. A simple Hahn echo with pulse separation of τ, two phase gradients in the X and Y axes, and a constant read gradient along Z.

- A timing card (PulseBlasterEPR-Pro from SpinCore) which has 21 TTL outputs, a time resolution of 2.5 ns and a minimum pulse length of 2.5 ns. This card generates the control signals for the MW and gradient pulses of the required imaging sequence (see Figure 19.6), as well as the control for the phases of the transmitted pulses.

- An 8-bit two channels PCI-format digitizer card for raw data acquisition and averaging having a sampling rate of 500 MHz and averaging capability of up to 0.7 M waveforms s^{-1} (AP-235, Acqiris). This card acquires the raw data EPR signal, digitizes and averages it according to the user requirements.

- A PCI analog output card having eight outputs with 16-bit resolution (PCI-6733; National Instruments). This card generates analog voltages that determine the magnitude of the various gradients employed, and controls the static magnetic field via the magnetic field controller.
- A GPIB card that controls the functionality of the MW source and the field controller.

In addition to the PC-control software and its attached PCI cards, the system includes the following components.

- A MW reference source (HP8672A) with a power output of 10 dBm in the 2–18 GHz range.
- A home-built pulsed microwave bridge containing a 6–18 GHz low-power transceiver, and a solid-state power amplifier with 1 W output, 35 dB gain (home made). The phases of the transmitted pulses from this bridge are digitally controlled by the timing card, and can be varied in less than 20 ns.
- High-voltage pre-regulators power supply for the gradient coil drivers, which provides a current drive to the pulsed gradient coils in the imaging probe.
- The final – and probably the most important component in the system – is the micro-imaging probe, which facilitates its high sensitivity and resolution.

The imaging probe for the pulsed-EPRM is very similar to that employed in the CW system (see Section 19.3.1.1). The resonator part of the probe is the same as that used for CW, with a relatively high loaded Q of ~1000 at critical coupling, as compared to the common pulsed-EPR resonators which have Q ~100. This is because most of the species employed for pulsed-EPRM have a relatively long T_2, so that the measurement dead time – which decreases with decreasing Q – is not a problem. In addition to that, the excitation bandwidth, which is $\sim f_0/Q$ (where f_0 is the MW frequency) is sufficient to excite the entire sample for species with long T_2, even when a constant Z gradient is applied (see Figure 19.6). Nevertheless, if in some experiments Q is found to be too high, it can be easily reduced by overcoupling the resonator using the linear stage that controls the resonator–coaxial feed distance, and thus affects the coupling. The gradient coils are based on the same geometric structure described for the CW probe, but in contrast to the CW case they are characterized by an approximate fourfold lower inductance and resistance. This is achieved by connecting the coil pairs that constitute the gradient coils in parallel rather than in serial. Thus, for example, a typical probe that operates at 17 GHz, which includes a set of X-, Y-, and Z-gradient coils, has the following coil characteristics:

- The structure of the X-gradient coil is a simple Maxwell pair, the coils of the pair are connected in parallel and have total inductance of $1.1\,\mu H$, a resistance of $0.5\,\Omega$, and produce a magnetic gradient of $1.37\,T/(m \cdot A)$.
- The Y-gradient coil is based on Golay geometry (Jin, 1999), has total inductance of $2.09\,\mu H$, a resistance of $0.55\,\Omega$, and produces magnetic gradient of $1.25\,T/m \cdot A$.

Both, the X- and Y- gradient coils are driven by the pulse current drivers. The Z-gradient coil is also based on Golay geometry, but its coil pairs are connected in serial. It has magnetic field gradient efficiency of $1.31\,\text{T/m}\cdot\text{A}$, a total inductance of $8.9\,\mu\text{H}$, and a resistance of $1.8\,\Omega$, making it more suitable for static gradient rather than pulsed operation. The maximum magnetic-field gradient achieved by this system for the X and Y coils with short (0.5–1 µs) current pulse of 40 A (out of a 620 V source) is $\sim 55\,\text{T}\,\text{m}^{-1}$.

19.3.2.2 SNR

In this subsection, a rigorous derivation of the SNR in pulsed-EPRM experiments is provided (Blank and Freed, 2003a, 2006). In general, for imaging applications the SNR per voxel can be determined to a good approximation by dividing the SNR of the entire sample by the number of voxels. More accurate calculations take into account the details of the exact acquisition technique, the distribution of the magnetic fields in the resonator, and the data-processing techniques used to acquire the image (Callaghan, 1991). However, these higher-order issues are not considered here. While the methodology for calculating the free induction decay (FID) or echo signal in NMR is well established and validated (Hoult and Richards, 1976; Hoult and Ginsberg, 2001; Jeener and Henin, 2002), for pulsed-EPR the situation requires some slight adjustments of the existing formulas. The usual formula for the signal in the NMR literature (Hoult and Richards, 1976) gives the voltage produced in a coil characterized by field efficiency – that is, the field produced by unit current, $\varepsilon = B_1/I$, due to a precessing specific magnetization M:

$$S_{pulse} = \omega \int_r \mathbf{M} \cdot \boldsymbol{\varepsilon} \tag{19.5}$$

The voltage in the detection coil, along with the equivalent circuit which describes the resonator's coupling to the detector, results in the voltage in the detector, S_{pulse}^{β} (Rinard et al., 1994, 1999):

$$S_{pulse}^{\beta} = \frac{\beta}{1+\beta}\sqrt{\frac{R_0}{\beta R}} S_{pulse} \tag{19.6}$$

Here, β is the coupling to the resonator, R_0 is the transmission line impedance, and R is the resonator's equivalent resistance. (This resistance will also be used later for noise calculations.) Such formalism has been found useful for the estimation of the SNR in EPR experiments in the case of a loop–gap resonator that is essentially a coil with one loop, for which B_1/I and R are well-defined (Rinard et al., 1999). Consider now the more general case, which applies to any EPR resonator (e.g., rectangular, cylindrical, dielectric), where it is unknown what B_1/I corresponds to, and what is the equivalent R of the resonator.

Every resonator can be represented by an equivalent RLC circuit, coupled to a transmission line. Following a 90° RF/MW pulse, the signal ζ in the equivalent receiving coil, induced by the magnetization of a point sample, m, is given by (Hoult and Bhakar, 1997):

$$\zeta = \frac{\partial}{\partial t}(\boldsymbol{\varepsilon} \cdot \boldsymbol{m}) \tag{19.7}$$

On the other hand, the noise voltage is given by:

$$N = \sqrt{4k_b TR\Delta f} \tag{19.8}$$

where k_b is Boltzmann's constant, T is the temperature, and Δf is the bandwidth of detection. Thus, the SNR is:

$$SNR_{pulse} = \frac{\frac{\partial}{\partial t}(\boldsymbol{\varepsilon} \cdot \boldsymbol{m})}{\sqrt{4k_b TR\Delta f}} = \frac{1}{\sqrt{4k_b T\Delta f}} \frac{\partial}{\partial t}\left(\frac{\boldsymbol{\varepsilon}}{\sqrt{R}} \cdot \boldsymbol{m}\right) \tag{19.9}$$

As noted above, the field efficiency, ε, corresponds to the magnetic field in the resonator due to 1 A of induced current. Whilst this quantity is difficult to measure for an arbitrary resonator, it is known that that 1 W of power entering the resonator, under matched conditions, would result in an effective RMS current of $I = 1/\sqrt{R}$, produced in the equivalent RLC circuit representing the resonator. Thus, it is clear that the *amplitude* of the B_1 field in the *laboratory frame* produced by this 1 W of power is $\sqrt{2/R} \cdot \varepsilon$. Therefore, the field produced in the *rotating frame* is (avoiding hereafter, for simplicity the vector notation of ε, and assuming that it is directed along one of the transverse axes of the laboratory frame of reference):

$$B_1 \text{ for 1 W of induced power} = \frac{1}{\sqrt{2}}\left(\frac{\varepsilon}{\sqrt{R}}\right) \equiv C_p \tag{19.10}$$

The value of C_p, can be measured by using standard pulse techniques, or calculated using the expression (Poole, 1983):

$$C_p \approx \sqrt{Q_L \mu_0 / V_c \omega_0} \tag{19.11}$$

By substituting Equations 19.11 and 19.10 in Equation 19.9, and integrating over the entire active volume of the resonator, the following approximation for the FID/echo SNR is obtained:

$$SNR_{pulse} \approx \frac{M\omega_0}{4\sqrt{4kT\Delta f}} V_c \sqrt{2} \, C_p \approx \frac{\sqrt{2\mu_0 V_c} \, \omega_0 M}{8\sqrt{kT\Delta f}} \sqrt{\frac{Q_L}{\omega_0}} \tag{19.12}$$

Here, a noise level that is fourfold the theoretical thermal noise is considered, to account for unavoidable losses and impedance mismatches. Furthermore, in the derivation of Equation 19.12 it is taken into account that the magnetization m is only for a point sample, and that slight differences between the filling factor calculation for CW and pulsed experiments that result in slightly different active volumes have been ignored (Blank and Levanon, 2002). Here, M can be considered as the average specific magnetization, which takes into consideration the field inhomogeneity within the resonator. Based on this expression, which is valid for a resonator that is completely filled with the sample, the SNR of a sample with a volume V_v in a pulsed-EPR experiment is given by a similar expression as that derived for the CW case (see Equation 19.3), with equivalent approximations up to a constant:

$$SNR_{pulse} \approx \frac{\sqrt{2\mu_0} M \omega_0 V_V}{8\sqrt{V_c} \sqrt{k_b T \Delta f}} \sqrt{\frac{Q_u}{\omega_0}} \quad (19.13)$$

From Equations 19.3 and 19.13 it is clear that, since M is linear with frequency, for the Curie law, the voxel SNR (both in the CW and the pulse cases) is proportional to $\omega_0^{1.5}/\sqrt{V_c}$. Therefore, in order to improve the voxel SNR, the resonator size must be decreased as much as possible and/or a higher frequency should be employed.

From Equation 19.13, it can be seen that for 1 s of acquisition time:

$$SNR_{pulse} \approx \frac{\sqrt{2\mu_0} M \omega_0 V_V}{8\sqrt{V_c} \sqrt{k_b T (1/\pi T_2)}} \sqrt{\frac{Q_u}{\omega_0}} \sqrt{\frac{1}{T_1}}, \quad (19.14)$$

where an averaging with a $1/T_1$ repetition rate for SNR improvement has been assumed.

As a numerical example, it is possible to consider a sample of 1 mM trityl radical ($T_1 = 17\,\mu s$ and $T_2 = 11\,\mu s$; Owenius, Eaton, and Eaton, 2005) in an X-band resonator with active volume of $[1\,mm]^3$ and $Q_u = 1000$, which yields a single-shot SNR of ~5000 when the resonator is completely filled with the sample. This corresponds to a single-shot SNR of $\sim 5 \times 10^{-6}$ for a $[1\,\mu m]^3$ voxel. In 1 s of acquisition time, this seemingly low figure can be improved by a factor close to ~1000. This is achieved by ~100 K–1 M averages per second, which is supported by today's state-of-the-art acquisition cards. Such an averaging scheme may be based on a CPMG sequence (Meiboom and Gill, 1958) with approximately 10 echoes separated by ~0.2 μs. This sequence can be repeated every T_1 (~10 μs, depending on the O_2 concentration), resulting in a total of $\sim 10 \times 100\,K = 1\,M$ echoes in 1 s. More detailed numerical examples for SNR and resolution will be provided in Section 19.4.

19.3.2.3 Resolution

In pulsed-EPR systems, which employ pulsed-field gradients, the effectiveness of the gradients should be discussed in the context of the imaging sequence employed. When choosing an imaging sequence suitable for pulsed-EPR, many constraints are encountered which do not exist for NMR imaging. For example, shaped excitation pulses used for slice selection are difficult to realize at MW frequencies and on the nanosecond time-scale. In addition, short rectangular gradient pulses with sharp edges are also difficult to produce (Ewert et al., 1991a, 1991b). The simple sequence shown in Figure 19.6 avoids most of these difficulties (Callaghan, 1991; Feintuch et al., 2000; Blank et al., 2004b), and was used in recent pulse-EPRM experiments carried out at the author's laboratory. Considering this sequence, it is possible to deduce that, for a radical characterized by a Lorentzian lineshape with a width of $1/(\pi T_2)$, the constant gradient (G_z in Figure 19.6) is related to the resolution, Δz, through the expression (Maresch, Mehring, and Emid, 1986):

$$G_z = 2/\gamma \Delta z T_2 \quad (19.15)$$

As a numerical example, it can be calculated from Equation 19.15 that, in order to obtain a spatial resolution of $\Delta z \sim 5\,\mu m$ along the Z-axis of the sample, the gradient G_z should be ~0.2 T m^{-1} for $T_2 = 11\,\mu$s, which is readily achieved (Blank and Freed, 2004b, 2006). It should be noted that diffusion of the spins in the solute in the presence of such a constant gradient, does not significantly affect the echo amplitude; this is in contrast to NMR, where the maximum applied gradient is limited by this phenomenon. This advantage of EPRM is apparent after reviewing the following expression (Callaghan 1991) that describes the echo decay due to diffusion in the presence of a static gradient:

$$\text{Echo Amplitude} = e^{(-\frac{1}{3}\gamma^2 G_z^2 D \tau^3)} \tag{19.16}$$

Employing Equation 19.16, even with relatively fast diffusion of $D = 10^{-9}\,\text{m}^2\,\text{s}^{-1}$, a very high G_z of 40 T m^{-1} and $\tau = 1\,\mu$s (Figure 19.6), results in an echo decay of only ~2%. Thus, clearly even with gradients that correspond to submicron image resolution, there is no significant prohibiting decay due to diffusion. Nevertheless, some applications do require the measurement of diffusion or motion in short time scales (1–100 μs), and this can be performed in EPRM systems by means of a stimulated echo sequence, taking advantage of spin system having a relatively long T_1 with respect to T_2 (Norwood, 1993). Such capability was recently demonstrated in the author's laboratory, where direct measurements of the diffusion of paramagnetic species, such as deuterated trityl and N@C$_{60}$, were performed in liquid solutions (Blank et al., 2008).

In addition to a constant Z-gradient, X and Y- phase gradients are now considered (Figure 19.6). These gradients are related to the resolution, Δx, through the expression (Callaghan 1991):

$$\int_t G_x dt = \frac{1}{(\gamma/2\pi)\Delta x} \tag{19.17}$$

It is clear from this expression that the shape of the phase gradient is not important, thereby relaxing the technical constraints for its generation. The parameter T_2 does not appear explicitly in Equation 19.17, but since the duration of the gradient pulse should be shorter than τ (Figure 19.6) it is clear that, in practice, having a longer T_2 enables one to employ a longer gradient pulse leading to a better resolution. As a numerical example, a half-sine-shaped gradient pulse with a duration of 1 μs and a peak amplitude of 50 T m^{-1} can be considered; in this case, the resolution will be ~1.13 μm. As detailed above, the typical gradient efficiency of the X and Y coils is ~1.5 T/m·A, which means that a peak current of ~35 A is required to reach a resolution of 1 μm. Such currents are readily available from the drivers employed in the author's laboratory, based on the charge of a capacitor and its subsequent discharge into the gradient coil (see Conradi et al., 1991).

19.4
Specific Aspects of Multifrequency EPR Microscopy at Various Temperatures

Based on the theoretical analysis presented above, and supported by many experimental results (Blank et al., 2003a, 2004a, 2004b, 2006; see also Section 19.5), a prediction for the SNR and resolution when carrying out EPRM experiments at varying set of frequencies and temperatures can now be made. The following discussion assumes the use of a pulsed-EPRM technique.

19.4.1
SNR in a Multifrequency Context

Although the SNR can, in principle, be calculated using Equation 19.14, these predictions must be based on the knowledge of parameters such as resonator's effective volume, V_c, the quality factor, Q_u, and the radical T_1 and T_2. Assuming that dielectric ring resonators of the type presented here are employed (see Figures 19.2 and 19.3), the "effective" volume for a given frequency can be calculated based on the crystal permittivity. A typical dielectric resonator has an outer radius, a, approximately 2.5-fold larger than its height. Thus, for the crystal relative permittivity, ε_r, the resonator radius can be calculated by the expression (Kajfez and Guillon, 1986):

$$a \approx \frac{202}{f\sqrt{\varepsilon_r}}, \tag{19.18}$$

where a is expressed in millimeters and f in GHz. As most of the RF magnetic field is concentrated in a circle of radius $a/2$ (see Figure 19.3), the effective volume of the resonator can be approximated by the expression:

$$V_c \approx \pi \left(\frac{a}{2}\right)^2 \left(\frac{a}{2.5}\right) \approx \frac{7.8 \times 10^6}{f^3 \varepsilon_r^{3/2}} \tag{19.19}$$

Whereas, in Equation 19.19, it was taken into account that although most of the magnetic field was confined to a circular cross-section of radius $a/2$ and a height of $a/2.5$, there was still a considerable amount of magnetic energy outside this volume, which effectively reduced the filling factor and thus increased the "effective volume" by a factor of approximately 3 (Blank and Freed, 2006).

Equations 19.14 and 19.19 can now be used, along with the dielectric properties of the resonator's crystals and the relaxation times of the measured spins, to provide an estimate for the spin sensitivity as a function of temperature and frequency. Figure 19.7 shows the crystal dielectric properties as functions of temperature and frequency. These data were used to calculate the minimum number of detectable spins for 60 min of acquisition time for a typical sample of LiPc, assuming fixed $T_1 = 3.5\,\mu s$ and $T_2 = 2.5\,\mu s$. The calculations are displayed in Figure 19.8. The LiPc sample has properties that can be similar to common paramagnetic defects encountered in the fields of solid-state and semiconductors (Blank and Freed, 2003a, 2006). Clearly, T_1 and T_2 vary with field and temperature; however, for the LiPc sample the variations are not expected to significantly affect the results of these calculations. It is evident from Figure 19.8 that at high fields and low temperatures, single electron spin sensitivity could be approached and even surpassed. Above 60 GHz and below 4 K the available information is not reliable enough to provide an extrapolation of the SNR calculations with a reasonable

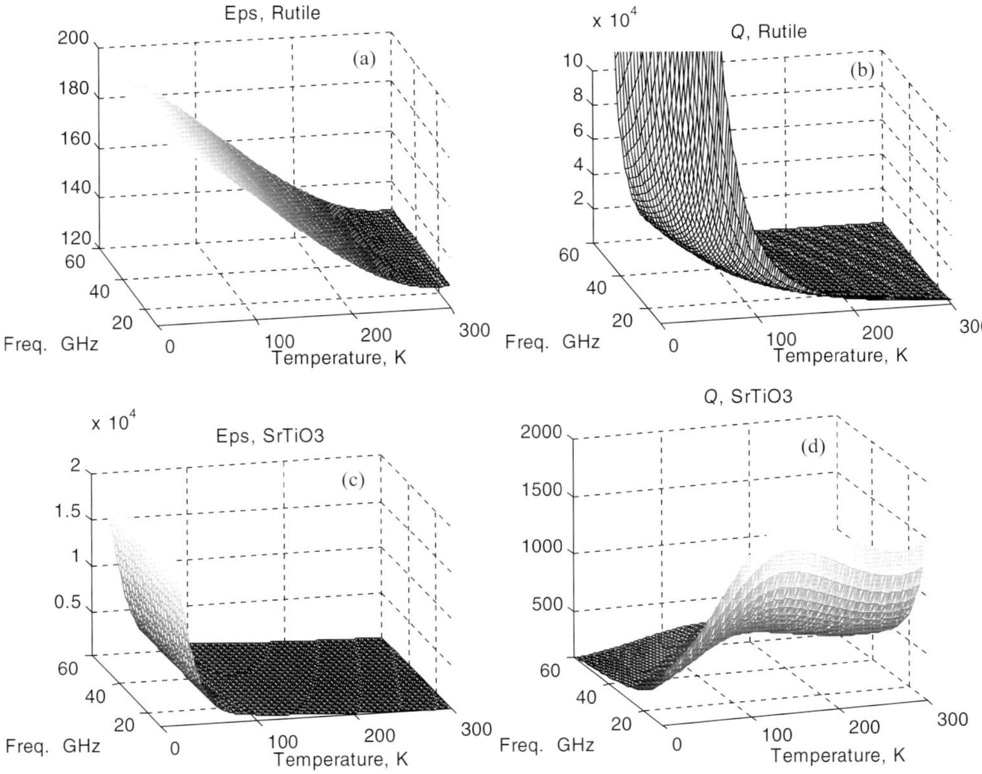

19.4 Specific Aspects of Multifrequency EPR Microscopy at Various Temperatures | 813

Figure 19.7 Single crystal dielectric properties as a function of frequency and temperature. (a) Rutile crystal, ε; (b) Rutile crystal, Q; (c) SrTiO$_3$ crystal ε; (d) SrTiO$_3$ crystal Q.

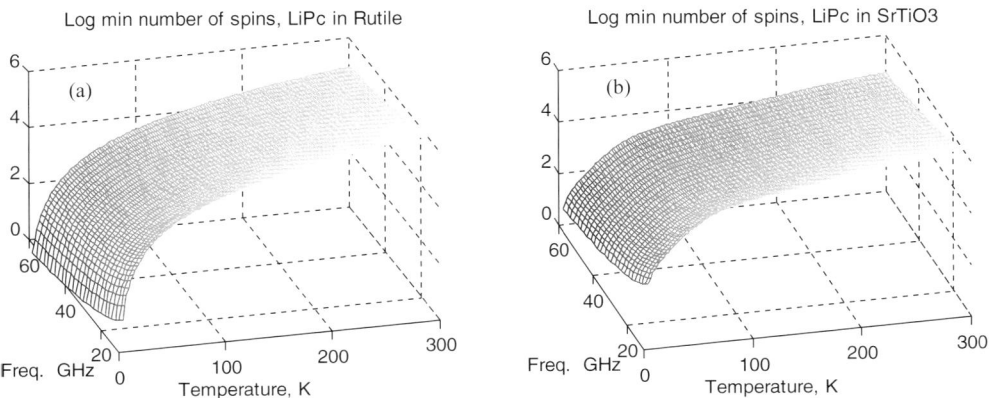

Figure 19.8 The minimum number of detectable spins for LiPc sample ($\log_{10}(N_{min})$), 1 h of acquisition time, as a function of frequency and temperature for Rutile (a) and SrTiO$_3$ (b) resonators.

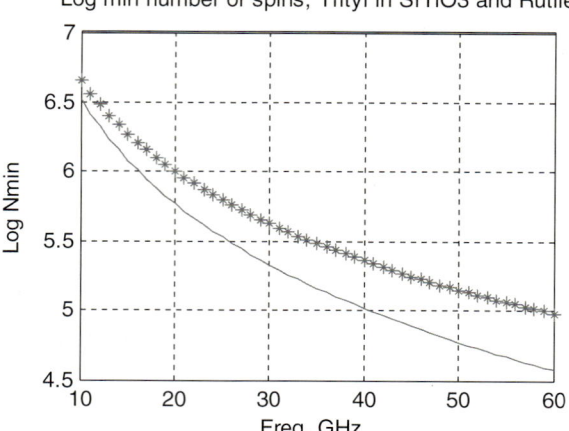

Figure 19.9 The minimum number of detectable spins for 1 mM trityl water solution ($\log_{10}(N_{min})$), 1 h of acquisition time, as a function of frequency and temperature for Rutile (*) and SrTiO$_3$ (solid line) resonators.

degree of certainty. A similar calculation is given in Figure 19.9 for a 1 mM liquid solution of trityl radical. This type of sample is appropriate for use in conjunction with biologically related applications at room temperature. Therefore, the sensitivity data for trityl is given only for 300 K as a function of frequency, assuming frequency invariant $T_1 = 17\,\mu s$ and $T_2 = 11\,\mu s$ (Owenius, Eaton, Eaton 2005). Based on this calculation, at 60 GHz it would be expected to achieve a spin sensitivity better than 10^5 spins.

19.4.2
Resolution in a Multifrequency Context

In principle, the SNR calculations that were provided above can be used to readily obtain the image resolution based on the sample spin concentration. However, the possibilities should also be considered of generating gradients that are large enough to support SNR-limited image resolution. The multifrequency aspects of EPRM resolution, using the examples of the representative LiPc and trityl samples, will now be discussed.

LiPc crystals typically have 10^8 spins per $[1\,\mu m]^3$ (Blank et al., 2003a). Thus, with single spin sensitivity it would be possible, in principle, to approach the 1 nm-scale resolution. On the other hand, for 1 mM of trityl the spin concentration is 6×10^5 spins per $[1\,\mu m]^3$, which means that at room temperature it would be expected to approach a resolution of ~1 μm for such samples. EPRM past experiments had already provided evidence for generating gradients that are sufficient to achieve a resolution of ~1 μm. Thus, for the trityl solution there is no question that the resolution is determined by the SNR, rather than being gradient-limited. In samples such as LiPc measured at the nanometer scale. Such high resolution also requires the use of very powerful gradients, the achievement of which is beyond the current state of the art. For example, the current 17 GHz probe at the author's laboratory has a gradient coil with an efficiency of ~1.4 T/m·A, and the current drivers can produce up to 40 A peak current for a duration of 1.5 μs. By plugging these values into Equation 19.17, an available theoretical resolution of ~700 nm is obtained. In Section 19.5 below, some experimental results carried out with the 17 GHz probe at room temperature, that indeed approach these figures, are discussed. These experiments stretched to the limits the current system capabilities in terms of SNR, gradient strength, and heat dissipation in the gradient coils, yet they were still a long way from the 1 nm-scale resolution mentioned above. In order to estimate the extent of these effects resolution–wise, the reasoning of Section 19.4.1 can be followed. The effects on the attainable resolution as a function of increasing the magnetic field and/or decreasing the temperature are considered.

The gradient coils efficiency, in T/m·A, keeping the number of windings in the coil constant, scales as $1/a^2$, where a is a typical radius of the gradient coil (Jin 1999), and can be estimated using Equation 19.18. Thus, if the 17 GHz probe based on a Rutile single crystal is considered as a reference with $a \sim 2.5$ mm at room temperature, it is found that at 60 GHz the expected resonator radius would be $a \sim 0.6$ mm at 4 K. For the $SrTiO_3$ probe, the resonator size would be even smaller, and probably too small to produce or handle with the current machining capabilities. Such a decrease in resonator size immediately translates to a factor of $(2.5/0.6)^2 \approx 17$ in gradient coil efficiency. Furthermore, by maintaining the coil inductance and resistance proportional to a^2 and a, respectively, and similar to those of the 17 GHz probe, the number of windings can be increased by a factor of $2.5/0.6 \approx 4$, implying an overall increase of a factor of ~65 in the gradient coil efficiency. This means that, with the same current pulse of 40 A/1.5 μs described above, a theoretical resolution of $700/65 \approx 10.8$ nm would be obtained. Additional upgrades in the gradient current drivers, such as increasing the drive voltage and the circuit components, can raise the available current to ~120 A, and with longer gradient pulses of up to 3 μs, for species having a long enough T_2, the theoretical resolution would reach ~1.8 nm. An additional limiting factor relevant to these extreme cases is the heat dissipated in the gradient coils. As the coils become smaller, heating becomes an increasingly greater problem. The ways to overcome this include either decreasing the repetition rate of the acquisition, which means a longer acquisition to achieve a sufficient SNR, or employing superconducting wires as gradient coils. However, the latter approach is relevant only when the measurements are carried out at cryogenic temperatures.

19.5
Illustrative Examples

A few representative practical examples highlighting the current state-of-the-art in EPR microscopy are now described.

19.5.1
Pulsed-EPR Microscopy of Solid Samples at Room Temperature

The best resolution in EPRM is obtained for solid samples with a high spin concentration and a long relaxation time, T_2. LiPc crystals, under de-oxygenized conditions, are suitable for these types of measurement, as their spin concentration is ~10^8 spins per $[1\,\mu m]^3$ and their T_2 is ~$2.5\,\mu s$ for crystals that are larger than ~$100\,\mu m$ (Stoner et al. 2004). Furthermore, T_1 is not much larger than T_2 (~$3.5\,\mu s$), which enables the pulse repetition rate (i.e., the number of averages per given measurement time) to be relatively high. Figure 19.10 provides examples of a typical LiPc crystal 2-D EPR micro-images with in-plane resolution varying from $4\,\mu m$ to less than $1\,\mu m$. The images were acquired without Z-gradient, so the voxel thickness was roughly equal to the crystal thickness (~$10\,\mu m$). In all cases, whilst the net acquisition time was no more than a few minutes, the *actual* acquisition time was longer, due mainly to the heating effects in the current coil set-up that forced a reduction in the repetition rate to $20\,kHz$ from an optimal repetition rate of ~$100\,kHz$. At the highest resolution, each voxel contains approximately 10^9 spins and the image SNR was ~100, conforming well with the theoretical predictions made in Section 19.4.1.

19.5.2
Pulsed-EPR Microscopy of Liquid Samples at Room Temperature

In liquid samples, the spins concentration is approximately two orders of magnitude smaller than that in solid samples. In addition, samples of interest–for example, live cells or other biologically related samples–must be measured anywhere within a few minutes to up to an hour, so that they can be maintained alive in the small sample containers employed in EPRM (see Figure 19.4). All of this means that, whereas for the solid samples a resolution better than $1\,\mu m$ was achieved for a liquid sample, under the same conditions the 2-D resolution could not be better than 5–$10\,\mu m$, with an acceptable SNR. Figure 19.11 shows an example of an image that demonstrates the available resolution in a typical liquid solution of a stable radical (trityl) (Blank et al., 2004b). This type of sample has a well-defined 3-D geometry and a known spin concentration. The 3-D image of $256 \times 256 \times 50$ voxels was acquired after $40\,min$, and the theoretically expected image resolution, based on the amplitude of the field gradients, was $8.8 \times 10.5 \times 19\,\mu m$. The actual resolution could be estimated from the unique features of the sample as they were resolved in the EPR image. By using the Rayleigh resolution criterion, it should be possible to resolve two ~$10\,\mu m$ voxels with signal in the XY-plane, separated by a ~$10\,\mu m$ voxel without signal. Although the current sample was not fine enough to obtain such information directly, it readily showed an excellent separation between the $50\,\mu m$ voxels separated by the $39\,\mu m$ wires. The Z-axis showed a good separation along the sample (~$90\,\mu m$ high) where, in the lower part of the image, only the solution was observed, and on the upper part where the ~$39\,\mu m$ high mesh was visible. Thus, in terms of resolution,

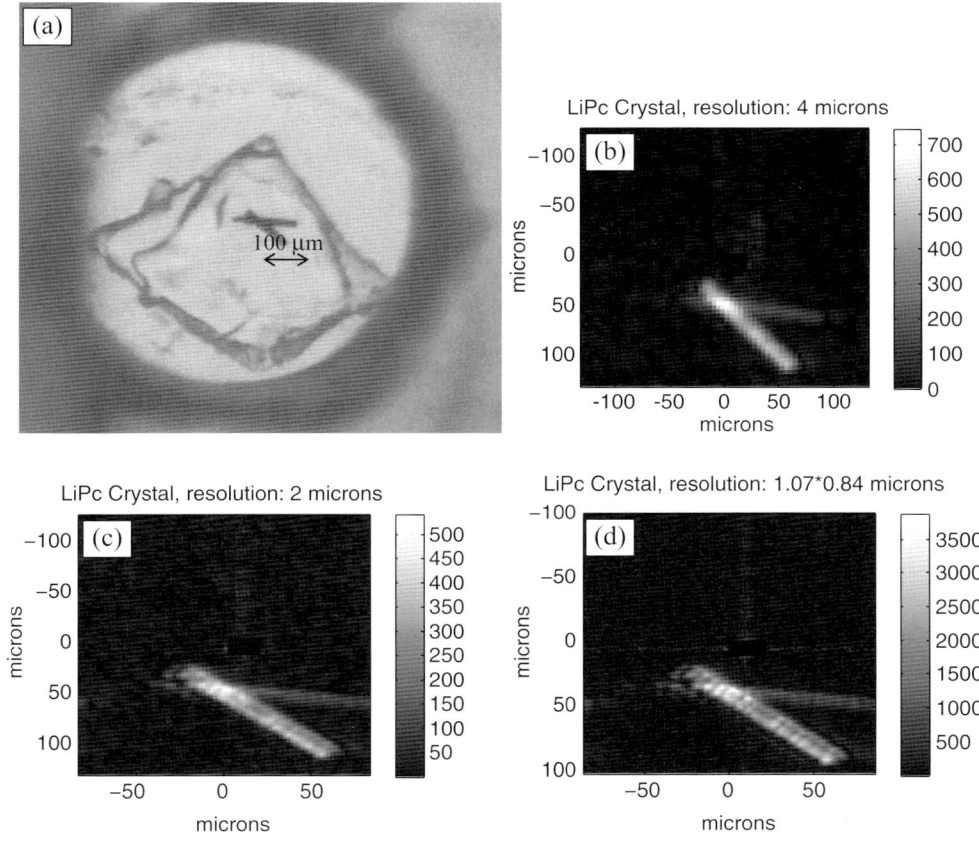

Figure 19.10 Optical (a) and EPR (b–d) images of two LiPc crystals; (b) A 64 × 64 pixels EPR image acquired with 4 × 4 µm resolution; (c) A 80 × 128 pixels EPR image acquired with 2 × 2 µm resolution; (d) A 160 × 240 pixels EPR image acquired with 1.07 × 0.84 µm resolution. This high-resolution image was obtained after 6 h of acquisition time at a repetition rate of 20 kHz, and 14% duty cycle (i.e., leaving 86% of the time periods of rest to avoid excessive gradient heating). The theoretical net acquisition time for this case is 10.3 min, assuming 100 kHz sequence reputation rate and 100% acquisition duty cycle. The EPR images were acquired with the 17 GHz pulsed probe. In all cases the imaging sequence of Figure 19.6 was used with $\tau = 1500$ ns and gradient half-sine pulse of 1 µs. The vertical scale in the EPR images shows the magnitude (in arbitrary units).

it can be concluded that for both samples the theoretical image resolution – as obtained from the measured field gradients – may very well be the actual measured image resolution, though direct evidence could only be obtained with specially prepared 3-D test samples. In terms of SNR, the single-voxel SNR of ~40 achieved in the trityl solution image was in good agreement with the predicted/measured results, taking into account the fact that there were ~6×10^5 spins per $[1\,\mu m]^3$, or ~10^9 spins per imaged voxel.

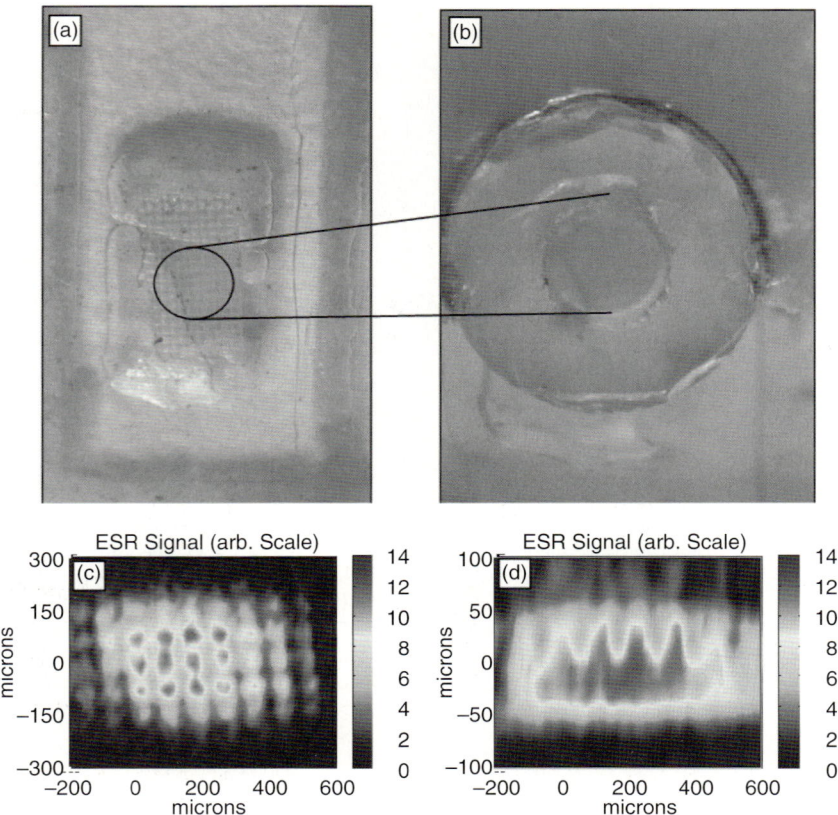

Figure 19.11 Optical and EPR images of trityl solution embedded in a fine woven Nylon mesh with a mesh aperture of 50 × 50 μm and a wire diameter of 39 μm. (a, b) Optical images showing the circular area inside the imaging probe (white ring); (c) A single Z slice of the 3-D EPR image showing the mesh; (d) A vertical cut through the sample (ZY plane).

One of the most important potential applications of EPRM in liquids is the mapping of oxygen, based on the measurement of T_2 (Swartz, 2002). Measurements of T_2 can be made by acquiring images with several different values of τ (Figure 19.6). For each voxel in the image the EPR signal can be fitted to an exponential decay curve with two fitting parameters, T_2 and the amplitude of the signal extrapolated to $\tau = 0$. An example of such an experiment, carried out with a test sample, is shown in Figure 19.12, where the measured spatially resolved T_2 of a special sample of trytil solution prepared with four different viscosities (different glycerol/water mixtures) are shown. It can be seen from this micro-imaging example that the viscosity has a significant effect on the T_2 of trityl solution (Owenius, Eaton, Eaton 2005),

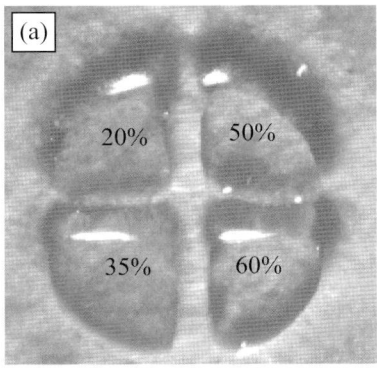 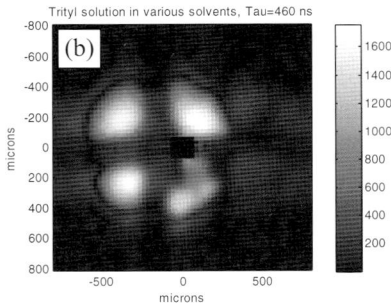
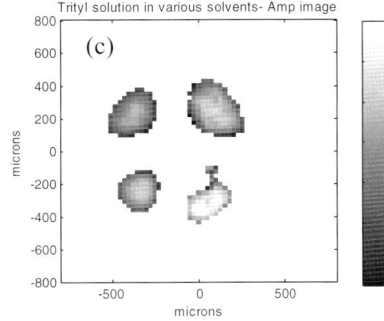 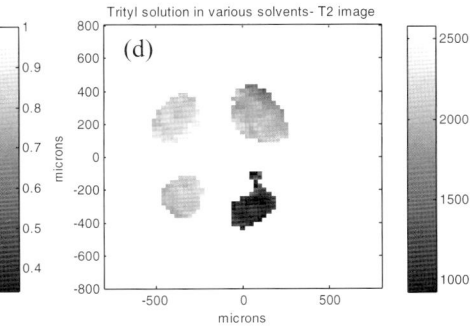

Figure 19.12 Optical and EPR images of 1 mM trityl solution in four different viscosities. (a) Optical image showing the four quadrants, where each contains different solution of trityl (the % of glycerol in water is marked on the photograph). The quadrants were prepared in the glass by photolithography; (b) A 64 × 64 pixels EPR image acquired with ~25 × 25 µm resolution. This image was obtained after 12 min of acquisition time at a repetition rate of 35 kHz, and 100% duty cycle. The theoretical net acquisition time for this case is ~4 min, assuming a 100 kHz sequence reputation rate. The EPR images were acquired with the 17 GHz pulsed probe with $\tau = 460$ ns and gradient half-sine pulse of 300 ns. The vertical scale in the EPR images shows the magnitude (in arbitray units). Similar images were acquired for τ values ranging up to 2060 ns, and processed to obtain the normalized amplitude image (c) and the T_2 (in ns) map (d).

19.5.3
CW-EPR Microscopy of Solid and Liquid Samples at Room Temperature

As mentioned above, only CW-EPRM can be employed for radicals having a short T_2. Furthermore, this is the method of choice for obtaining the spatially resolved complex EPR lineshapes in heterogeneous samples by spectral–spatial imaging. An example of the latter case is shown in Figure 19.13 for a sample containing both solid LiNc-BuO (Pandian et al. 2003) and liquid trityl solution (Blank et al., 2006). The different EPR lineshapes of these two paramagnetic species is resolved here with a resolution of approximately 40 µm².

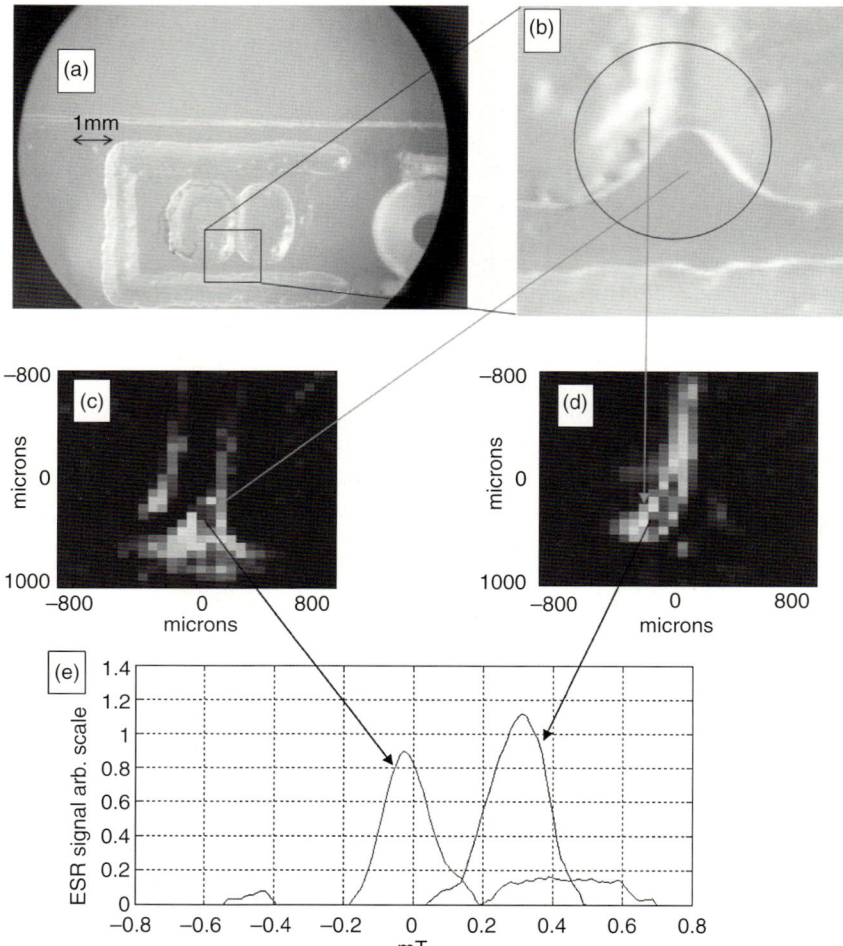

Figure 19.13 Example of spectral-spatial image obtained with the 9 GHz CW-EPR microscope. (a) Optical image of the sample made from trityl solution and solid LiNc-BuO (dark traces left after evaporation); (b) A closer examination of the area imaged inside the resonator effective volume (superimposed circle); (c) EPR image of the trityl obtained by taking only the part in the spectrum where the trityl is dominant; (d) The same as panel (c), but for the LiNc-BuO; (e) EPR spectrum obtained with the spectral-spatial algorithm at two points as marked by the arrows, corresponding to the spectrum of the trityl solution (left peak) and the solid LiNc-BuO.

19.6
Conclusions and Future Prospects

The field of induction-detection-based EPRM has progressed significantly during the past few years, improving sensitivity by approximately four orders of magnitude, and resolution by almost two orders of magnitude, while still operating at room temperature and using moderate static fields. In view of the past and present capabilities it has been found that, by going to higher fields and/or lower temperatures, it would be possible to further improve sensitivity by at least four to six orders of magnitude, and resolution by another about two orders of magnitude, reaching the deep nanoscale level and approaching single electron spin sensitivity. In addition to the SNR and resolution aspects of multifrequency EPRM, it should also be noted that different samples/applications may be sought at different EPRM frequencies. For example, the examination of liquid samples with sizes of several hundreds of microns (e.g., large cells and tissue extracts) may be best pursued at relatively low frequencies, in order to accommodate the entire sample within the resonator and enable its good MW penetration. On the other hand, the examination of solid-state samples—such as discrete micron-sized semiconductor devices—may be pursued at the highest possible frequency with the smallest possible resonators, to achieve the best possible resolution.

Clearly, although many further investigations are required in this field, major advances may very well be achieved in the near future as there are no fundamental physical laws to contradict such progress. These new capabilities may lead to many applications of interest, both in the biological and materials sciences, and these would complement other sensitive EPR imaging methods such as STM-EPR, MRFM, and indirect detection via diamond NV centers.

Acknowledgments

The author wishes to thank Prof. Jack Freed, Curt Dunnam, and Dr Peter Borbat from Cornell University, and Dr Lazar Shtirberg, Dr Ygal Twig, Michael Shklyar and the other group members at Technion for their contribution to the development of EPRM, and to many discussions. These activities have been supported along the years through the NIH/NCRR (Grant P41-RR016292), the Israeli Science Foundation (grants 169/05 and 1143/05), the US-Israel Bi-national Science Foundation (grant 2005258), the European research Council (ERC grant 201665), and by the Russell Berrie Nanotechnology Institute at the Technion.

Pertinent Literature

The following is a list of recommended literature for further reading on the subject of this chapter:

Callaghan, P. (1991) *Principles of Nuclear Magnetic Resonance Microscopy*. Oxford University Press, Oxford.

Eaton, G.R., Eaton, S.S., and Ohno, K. (1991) *EPR imaging and in vivo EPR*. CRC Press, Boca Raton.

Jin, J.M. (1999) *Electromagnetic Analysis and Design in Magnetic Resonance imaging*. CRC Press, Boca Raton.

Feintuch, A., Alexandrowicz, G., Tashma, T., Boasson, Y., Grayevsky, A., and Kaplan, N. (2000) *J. Magn. Reson.*, **142**, 382.

Blank, A., Dunnam, C.R., Borbat, P.P., and Freed, J.H. (2003) *J. Magn. Reson.*, **165**, 116.

Blank, A. and Freed, J. H. (2006) *Isr. J. Chem.*, **46**, 423.

References

Balasubramanian, G., Chan, I.Y., Kolesov, R., Al-Hmoud, M., Tisler, J., Shin, C., Kim, C., Wojcik, A., Hemmer, P.R., Krueger, A., Hanke, T., Leitenstorfer, A., Bratschitsch, R., Jelezko, F., and Wrachtrup, J. (2008) *Nature*, **455**, 648.

Blank, A. and Freed, J.H. (2006) *Isr. J. Chem.*, **46**, 423.

Blank, A. and Levanon, H. (2002) *Spectrochim Acta [A]*, **58**, 1329.

Blank, A., Dunnam, C.R., Borbat, P.P., and Freed, J.H. (2003a) *J. Magn. Reson.*, **165**, 116.

Blank, A., Stavitski, E., Levanon, H., and Gubaydullin, F. (2003b) *Rev. Sci. Instrum.*, **74**, 2853.

Blank, A., Dunnam, C.R., Borbat, P.P., and Freed, J.H. (2004a) *Rev. Sci. Instrum.*, **75**, 3050.

Blank, A., Dunnam, C.R., Borbat, P.P., and Freed, J.H. (2004b) *Appl. Phys. Lett.*, **85**, 5430.

Blank, A., Freed, J.H., Kumar, N.P., and Wang, C.H. (2006) *J. Control. Release*, **111**, 174.

Blank, A., Talmon, Y., Shklyar, M., Shtirberg, L., and Harneit, W. (2008) *Chem. Phys. Lett.*, **465**, 147.

Boero, G., Besse, P.A., and Popovic, R. (2001) *Appl. Phys. Lett.*, **79**, 1498.

Callaghan, P.T. (1991) *Principles of Nuclear Magnetic Resonance Microscopy*, Oxford University Press, Oxford.

Ciobanu, L., Seeber, D.A., and Pennington, C.H. (2002) *J. Magn. Reson.*, **158**, 178.

Conradi, M.S., Garroway, A.N., Cory, D.G., and Miller, J.B. (1991) *J. Magn. Reson.*, **94**, 370.

Coy, A., Kaplan, N., and Callaghan, P.T. (1996) *J. Mag. Reson. Ser. A*, **121**. 201.

Akademie der Wissenschaften der DDR (1987) ZZG1 User manual.

Drescher, M. and Dormann, E. (2004) *Europhys. Lett.*, **67**, 847.

Durkan, C. and Welland, M.E. (2002) *Appl. Phys. Lett.*, **80**, 458.

Eaton, G.R., Eaton, S.S., and Ohno, K. (1991) *EPR Imaging and in Vivo EPR*, CRC Press, Boca Raton.

Elzerman, J.M., Hanson, R., van Beveren, L.H.W., Witkamp, B., Vandersypen, L.M.K., and Kouwenhoven, L.P. (2004) *Nature*, **430**, 431.

Ewert, U., Crepeau, R.H., Dunnam, C.R., Xu, D.J., Lee, S.Y., and Freed, J.H. (1991a) *Chem. Phys. Lett.*, **184**, 25.

Ewert, U., Crepeau, R.H., Lee, S.Y., Dunnam, C.R., Xu, D.J., and Freed, J.H. (1991b) *Chem. Phys. Lett.*, **184**, 34.

Feintuch, A., Alexandrowicz, G., Tashma, T., Boasson, Y., Grayevsky, A., and Kaplan, N. (2000) *J. Magn. Reson.*, **142**, 382.

Gonen, O. and Levanon, H. (1984) *J. Phys. Chem.*, **88**, 4223.

Halevy, R., Talmon, Y., and Blank, A. (2007) *Appl. Magn. Reson.*, **31**, 591.

Herrling, T., Klimes, N., Karthe, W., Ewert, U., and Ebert, B. (1982) *J. Magn. Reson.*, **49**, 203.

Herrling, T., Fuchs, J., and Groth, N. (2002) *J. Magn. Reson.*, **154**, 6.

Hoch, M.J.R. (1981) *J. Phys. C Solid State Phys.*, **14**, 5659.

Hoch, M.J.R. and Day, A.R. (1979) *Solid State Commun.*, **30**, 211.

Hoult, D.I. and Bhakar, B. (1997) *Concept. Magnetic Res.*, **9**, 277.

Hoult, D.I. and Ginsberg, N.S. (2001) *J. Magn. Reson.*, **148**, 182.

Hoult, D.I. and Richards, R.E. (1976) *J. Magn. Reson.*, **24**, 71.

Jaworski, M., Sienkiewicz, A., and Scholes, C.P. (1997) *J. Magn. Reson.*, **124**, 87.

Jeener, J. and Henin, F. (2002) *J. Chem. Phys.*, **116**, 8036.

Jin, J.M. (1999) *Electromagnetic Analysis and Design in Magnetic Resonance Imaging*, CRC Press, Boca Raton.

Kajfez, D. and Guillon, P. (1986) *Dielectric Resonators*, Artech House, Dedham, MA.

Karthe, W. and Wehrsdorfer, E. (1979) *J. Magn. Reson.*, **33**, 107.

Krupka, J., Geyer, R.G., Kuhn, M., and Hinken, J.H. (1994) *IEEE Trans. Microw. Theory Tech.*, **42**, 1886.

McDermott, R., Trabesinger, A.H., Muck, M., Hahn, E.L., Pines, A., and Clarke, J. (2002) *Science*, **295**, 2247.

Mamin, H.J., Poggio, M., Degen, C.L., and Rugar, D. (2007) *Nat. Nanotechnol.*, **2**, 301.

Manassen, Y., Hamers, R.J., Demuth, J.E., and Castellano, A.J. (1989) *Phys. Rev. Lett.*, **62**, 2531.

Maresch, G.G., Mehring, M., and Emid, S. (1986) *Physica B C*, **138**, 261.

Maze, J.R., Stanwix, P.L., Hodges, J.S., Hong, S., Taylor, J.M., Cappellaro, P., Jiang, L., Dutt, M.V.G., Togan, E., Zibrov, A.S., Yacoby, A., Walsworth, R.L., and Lukin, M.D. (2008) *Nature*, **455**, 644.

Meiboom, S. and Gill, D. (1958) *Rev. Sci. Instrum.*, **29**, 688.

Milov, A.D., Pusep, A.Y., Dzuba, S.A., and Tsvetkov, Y.D. (1985) *Chem. Phys. Lett.*, **119**, 421.

Norwood, T.J. (1993) *J. Magn. Reson. A*, **103**, 258.

Owenius, R., Eaton, G.R., and Eaton, S.S. (2005) *J. Magn. Reson.*, **172**, 168.

Pandian, R.P., Parinandi, N.L., Ilangovan, G., Zweier, J.L., and Kuppusamy, P. (2003) *Free Radic. Biol. Med.*, **35**, 1138.

Poole, C.P. (1983) *Electron Spin Resonance: A Comprehensive Treatise on Experimental Techniques*, John Wiley & Sons, Inc., New York.

Poularikas, A.D. (2000) *The Transforms and Applications Handbook*, CRC Press, Boca Raton.

Rinard, G.A., Quine, R.W., Eaton, S.S., Eaton, G.R., and Froncisz, W. (1994) *J. Magn. Reson. A*, **108**, 71.

Rinard, G.A., Quine, R.W., Song, R.T., Eaton, G.R., and Eaton, S.S. (1999) *J. Magn. Reson.*, **140**, 69.

Rugar, D., Budakian, R., Mamin, H.J., and Chui, B.W. (2004) *Nature*, **430**, 329.

Smirnov, A.I., Poluectov, O.G., and Lebedev, Y.S. (1992) *J. Magn. Reson.*, **97**, 1.

Stoner, J.W., Szymanski, D., Eaton, S.S., Quine, R.W., Rinard, G.A., and Eaton, G.R. (2004) *J. Magn. Reson.*, **170**, 127.

Swartz, H.M. (2002) *Biochem. Soc. Trans.*, **30**, 248.

Taylor, J.M., Cappellaro, P., Childress, L., Jiang, L., Budker, D., Hemmer, P.R., Yacoby, A., Walsworth, R., and Lukin, M.D. (2008) *Nat. Phys.*, **4**, 810.

Wrachtrup, J., Vonborczyskowski, C., Bernard, J., Orrit, M., and Brown, R. (1993) *Nature*, **363**, 244.

Xiao, M., Martin, I., Yablonovitch, E., and Jiang, H.W. (2004) *Nature*, **430**, 435.

Yong, L., Harbridge, J., Quine, R.W., Rinard, G.A., Eaton, S.S., Eaton, G.R., Mailer, C., Barth, E., and Halpern, H.J. (2001) *J. Magn. Reson.*, **152**, 156.

20
EPR Studies of Nanomaterials
Alex Smirnov

20.1
Introduction

Nanotechnology is today emerging as one of the most rapidly evolving technological areas, with the current "nanorevolution" being fueled by massive research and development efforts aimed at designing and characterizing devices, structures, and materials with ca. 100 nm feature size, and below. The reasons beyond such a major undertaking are the principal advantages achieved by both a further miniaturization of the conventional devices and a continuing exploration of the entirely new phenomena dominating at the nanoscale. Examples of the former benefit include the evolution of the semiconductor industry and computer hardware, proceeding in pace with Moore's law. Currently, the feature sizes of many computer devices are already at or below ca. 100 nm. Semiconductor structures at such a scale exhibit a plethora of new properties, including "quantum size/quantum confinement" effects that alter the electronic properties of solids. Some of the most notable manifestations of quantum confinement are the properties of quantum dots (QD), which are nanocrystals of semiconductors that are just 2–10 nm in diameter. The QDs provide three-dimensional (3-D) quantum confinement for excitons, the discrete energy levels of which can be tuned by adjusting not only the bulk semiconductor properties but also the QD size, a phenomenon encountered only at the nanoscale!

The inherently small feature scales of nanomaterials open the prospects of developing many new technologies with capabilities that far surpass the current state of the art. One such example is magnetic nanoparticles, which could be employed for magnetic field-assisted drug delivery or diagnostic purposes. Yet, biomedicine is not the only potential application field for magnetic nanostructures. With the decreasing size of magnetic particles beyond a critical value – typically approximately 30 nm diameter, when the reduction in magnetostatic energy is no longer balanced by the wall energy – such particles become single-domain ferromagnets. The latter are envisioned to revolutionize data storage technology; indeed, by recording one bit of data per nanoparticle arranged in an array with a 25 nm periodicity, a genuine technological breakthrough of a 1 Tbit in^{-2} (0.155 Tbit cm^2) storage density in a single plane could be achieved. Such a data density would provide about a 100-fold improvement over the current hard disk technology, which is based on multilayer magnetic films.

Another essential aspect of nanomaterials is the vastly increased surface-to-volume ratio. As a consequence, the interfacial and surface properties will become the dominant factors determining the chemistry of nanomaterials, and their interactions with biological systems. The surface chemistry could also be significantly altered; for example, an element such as gold, although chemically inert at the macroscopic scale, may become a potent catalyst. The properties of other known catalysts may also be altered significantly, by converting them to nanoparticles and/or nanopattering.

Nanotechnology also provides unique opportunities for a rational design of interfaces between solid man-made structures and delicate biological molecules and systems. Such hybrid nanomaterials are thought to pave the way for the next generation of protein biochips, bioinspired materials and biomimetic systems capable of combining the robustness of inorganic structures with sophisticated functionality and, perhaps, even the regeneration ability of living organisms.

As transdisciplinary research in nanomaterials and related technologies continues to grow, so too do the needs for spectroscopic techniques suitable for characterizing nanoscale objects and their interfacial properties – preferably, in a nondestructive way. Some of the needs for gathering detailed images of nanoscale objects, and assessing their physical properties, might be fulfilled by various types of scanning microscopy, including scanning electron microscopy (SEM), transmission electron microscopy (TEM), atomic force microscopy – (AFM), or its variation, magnetic force microscopy (MFM). Unfortunately, these otherwise informative methods are incapable of providing data at the molecular level on the structure and dynamics of interfacial layers that are responsible for the reactivity and biorecognition properties of nanoparticles and hybrid nanomaterials. Research on spin-dependent electron transport phenomena in solid-state devices – that is, *spintronics* – would also benefit from ascertaining the electronic spin properties of magnetic semiconductor and ferromagnetic materials at the nanoscale, these data being seldom obtained from the above-mentioned microscopic techniques alone.

Fortunately, some of the missing experimental capabilities for characterizing nanostructures and hybrid nanomaterials may be filled by applying electron paramagnetic resonance (EPR) methods that have already been developed for other applications. In brief, EPR provides spectroscopic data on unpaired electronic spin states at both the atomic (conventional EPR) and the nanoscale (ferromagnetic resonance; FMR). The latter, for example, could be achieved by directly identifyng spin-waves in FMR spectra and relating spin-wave parameters to the properties of the microscopic object (Berger *et al.*, 1998). Moreover, EPR appears to be uniquely positioned to make rapid progress in understanding the molecular structure and dynamics of solid-to-liquid interface layers in nanoparticles and hybrid materials. Indeed, the surface area of many hybrid nanostructures, such as ligand-protected nanoparticles and nanoporous substrates, is sufficiently high to be studied by spin-labeling EPR methods, even at a monolayer (ML) surface coverage and doping levels of ca. 1 mol% of the spin label per ligand. The recent progress in site-directed spin labeling (SDSL) EPR methods for studying membranes and proteins has opened avenues for further extensions of the same approaches to the field of hybrid nanomaterials. Specifically, there are reasons to expect that spin-labeling EPR, in combination with time-domain and multifrequency/high magnetic field methods, would be eminently successful in studying such interfacial properties as local structure and dynamics, as well as electrostatics, hydrogen bonding, and hydrophobicity. The latter local properties play a critical role in the protein–protein and protein–nucleic acid interactions that form the basis for cell targeting by nanoparticle-based drug-delivery vectors.

The aim of this chapter is to cover the wide spectrum of recent applications of multifrequency EPR in the fields of nanoscience and hybrid biomaterials. From this perspective, the use of multifrequency methods is especially important, as the agility to choose the resonance conditions allows research groups to emphasize specific interactions or the properties of a particular interest, while suppressing any undesirable phenomena. The use of EPR to study collective electronic spin phenomena in magnetic nanoparticles from FMR spectra will be reviewed first. This will be followed by a discussion of the use of more conventional EPR methods for characterizing the nanostructured oxide semiconductors that are currently being developed for photoactivated catalysis and solar energy conversion. Many such catalytic processes invoke free radical mechanisms and unpaired electron species that could be characterized by EPR. While the unusually high surface reactivity of nanostructured materials is highly beneficial for catalysis, it is also a source of free radicals and consequent oxidative DNA damage when introduced into cells. These free radical-mediated processes can be successfully studied by the well-developed spin trapping EPR methods. Finally, the recent developments of spin-labeling methods for studying hybrid nanostructures and nanomaterials will be reviewed.

20.2
EPR Studies of Magnetic Nanostructures

To date, several types of magnetic nanostructure have undergone extensive investigation by means of magnetic resonance. One of the first experimental demonstrations of physical effects specific for magnetic structures with small feature sizes was provided by Seavey and Tannenwald (1958) more than 50 years ago. These authors reported on the direct observation of spin waves in a 560 nm-thick 80–20% permalloy upon absorption of 8.89 GHz microwave radiation. When the static magnetic field was directed perpendicular to the film and irradiated with microwave, several resonance peaks emerged. These resonances were interpreted as having arisen from excitation of the spin waves at the angular frequency:

$$\omega = \frac{2J\gamma}{M}k^2 + \omega_0 \quad (20.1)$$

where J is the exchange integral, M is the magnetization, and $k = n\pi/L$, where L is the film thickness and n an integer. Such spin waves have been observed in a number of magnetic films, including a recent report for a Ge:Mn magnetic semiconductor (Morgunov et al., 2008). Such data allow for evaluating the exchange integral J, which otherwise is difficult to determine. Other examples include – but are not limited to – studies of polyimide foils implanted with 49 keV Fe^+ and Co^+ ions (Rameev et al., 2003).

Another class of magnetic nanostructures that has been successfully examined with magnetic resonances is represented by magnetic nanoparticles and clusters. Although, typically, the clusters are sufficiently small, the individual spins can form a single ferromagnetic domain along the directions of the net magnetic moments of the individual particles, such that they are still susceptible to thermal fluctuations. Thus, the whole sample can become magnetized only in the presence of an external magnetic field. This phenomenon is termed superparamagnetism, while the associated microwave absorption by such a system in the external magnetic field is known as superparamagnetic resonance (SPR).

Usually, an analysis of experimental SPR and FMR spectra is based on a Landau–Lifshitz model (Landau and Lifshitz, 1935), which combines the classical equation for the precession of the magnetization vector $\vec{M}$ in the external magnetic field $\vec{B}$ with a phenomenological damping of the magnetic moment:

$$\dot{\vec{M}} = \gamma \vec{M} \times \vec{B} - \frac{\lambda}{|\vec{M}|^2} \vec{M} \times (\vec{M} \times \vec{B}) \qquad (20.2)$$

In this equation the damping parameter $\lambda > 0$ accounts for the finite width of the experimental SPR spectra. Previously, Berger, Bissey, and Kliava, (2000) have shown that when the magnetization becomes saturated at sufficiently high magnetic fields, the Landau–Lifshitz model yields the following normalized absorption line:

$$f(B, B_0, \Delta_B) = \frac{B_0^2 \Delta_B \left[(B_0^2 + \Delta_B^2) B^2 + B_0^4 \right]}{\pi \left[B_0^2 (B - B_0)^2 + \Delta_B^2 B^2 \right] \left[B_0^2 (B + B_0)^2 + \Delta_B^2 B^2 \right]} \qquad (20.3)$$

where the line width parameter Δ_B is proportional to the damping parameter λ as:

$$\Delta_B = \lambda B_0 / |\gamma \vec{M}| \qquad (20.4)$$

When the line width parameter Δ_B is small compared with the resonance field B_0, the line-shape given by Equation 20.3 is well approximated by a Lorentzian function, especially when the first derivative signal is displayed (Figure 20.1, spectra A–C). With an increase in Δ_B, the deviation from the Lorentzian shape and, especially, asymmetry become apparent, although the differences are still rather modest when a linear baseline is included in the fit (Figure 20.1, spectrum D, dashed line).

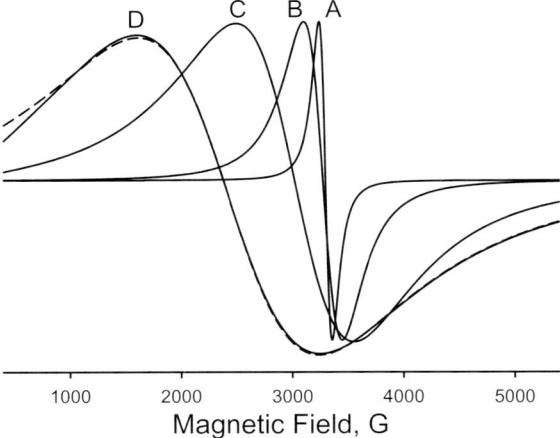

Figure 20.1 Representative peak-to-peak amplitude normalized SPR line shapes simulated according to Equation 20.3 with $B_0 = 3300\,G$ and the following line width parameters Δ_B: A = 100 G, B = 300 G, C = 1000 G, D = 2000 G. The dashed line shows a least-squares fit of spectrum D to a Lorentzian shape, including an adjustable linear baseline.

One of the important properties of Equation 20.3 is the dependence of the absorption maximum or the zero cross-point for the corresponding first-derivative form on the line width parameter Δ_B. With an increase in Δ_B, the zero cross-point shifts progressively to lower fields (see Figure 20.1). For experimental samples of ferromagnetic nanoparticles with a distribution of diameters, the individual line width parameters Δ_B are expected to be different. For such a sample, the individual SPR spectra with a larger Δ_B will be shifted to lower fields, yielding the overall spectra characterized by broad low-field shoulders. Such features are typically observed in experimental spectra of ferromagnetic nanoparticles (Figure 20.2).

As the temperature increases, so too do the magnitude and frequencies of thermal fluctuations. These fluctuations effectively reduce the magnetic anisotropy and the spread in magnetic field experienced by the nanoparticles, leading to a line-width narrowing with temperature. In order to understand such a line-width behavior, Berger et al. (2001) considered several mechanisms that would lead to magnetization fluctuations. In addition, these authors carried out detailed simulations of experimental SPR spectra, assuming a log-normal form for distribution of the volume fraction of the nanoparticles with a specific diameter, D. Such an assumption was verified earlier (Berger et al., 1998). The simulated line shapes and the temperature dependence of parameters associated with the experimental spectra, as exhibited by the apparent resonance field, peak-to-peak width, and double-integrated intensity, reproduced the experimental spectra rather well and verified the theory developed by Berger et al. (2001). Other experimental parameters, such as sample blocking temperature, could also be deduced from such data

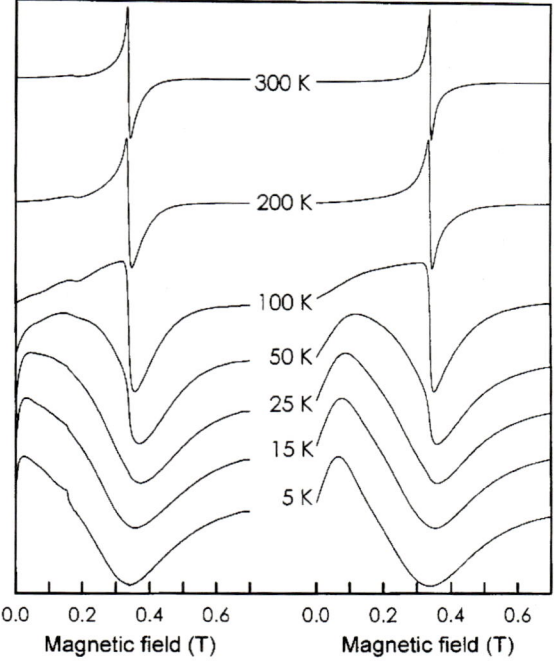

Figure 20.2 Experimental (left) and computer-generated (right) variable-temperature SPR spectra of maghemite (γ-Fe$_2$O$_3$) nanoparticles formed in a sol–gel silicate glass with a molar ratio Fe/Si of 2%. The spectra were normalized to the peak-to-peak amplitude. Reproduced with permission from Berger et al. (2001).

(Berger et al., 2001) An application of this theory to the analysis of SPR spectra from Ni nanoparticles placed in bundles of single-wall carbon nanotubes (SWNTs) has been provided by Konchits et al. (2006).

While the theory developed by Berger et al. (2001) had already produced satisfactory results for the analysis of experimental spectra from iron oxide nanoparticles formed in sol–gel silicate glasses, it would further benefit from a critical examination of other nanoparticles classes. Specifically, it was deemed beneficial to examine the SPR spectra of monodispersed ferromagnetic nanoparticles of variable diameter sizes, in order to determine whether SPR could be used for the sizing of magnetic nanoparticles and evaluating their size distribution from SPR spectra. Ferromagnetic nanoparticles also await examination with available multifrequency and high-field EPR instrumentation. The latter may prove to be advantageous for achieving a regime of complete magnetization saturation that would simplify spectral analysis.

For nanoparticles of smaller sizes (ca. 5 nm diameter), an alternative model based on a giant spin approximation has been suggested (Noginova et al., 2007). In this model, the magnetic moment of a nanoparticle is considered as a single gigantic spin, and the observed EPR spectrum is computed by summation over all the quantum transitions between the associated energy levels quantized in the external magnetic field. Noginov *et al.* (2008) evoked this model to explain the so-called half-field ($g \approx 4.0$) transitions observed in the EPR spectrum of a commercially available magnetite Fe_3O_4 ferrofluid containing particles with an average diameter of ca. 9 nm. The same model also explained satisfactorily the $g \approx 4.0$ and an additional $g \approx 6.0$ transition reported earlier for γ-Fe_2O_3 nanoparticles (Noginova *et al.*, 2008).

Both, SPR and FMR can also be used successfully to probe magnetic anisotropy in various magnetic nanomaterials. Ebels *et al.* (2001) examined arrays of macroscopically ordered Ni nanowires formed by an electrodeposition inside the pores of polycarbonate membranes. These membranes have randomly distributed pores that are aligned parallel to each other at densities not exceeding $1.5 \times 10^9 \, cm^{-2}$ for the smallest 35 nm pores. For such pore densities, the dipolar interactions between the individual wires were considered to be negligible. In having a length of ca. 22 µm, these nanowires could be modeled well as infinite cylinders. The pumping radiofrequency field of 23.6 GHz (K-band) or 34.4 GHz (Q-band) was applied perpendicularly to both the wire axis and the direction of the static magnetic field B_0. The FMR spectra were recorded as a function of the orientation of the nanowire array with respect to B_0 and the effective anisotropy field. The distribution of the wire orientations were evaluated from the observed shifts in the resonance spectra (Ebels *et al.*, 2001).

Dipolar interactions between the individual nanostructures can contribute significantly to the observed FMR spectra, as demonstrated by Jung *et al.* (2002), of a sample of arrays of submicron-sized particles composed of a permalloy produced by electron-beam lithography (EBL). The FMR spectra of such arrays show several additional resonances when compared to those of ferromagnetic films. The positions of these resonances were found to depend on the array orientation with respect to B_0, and also on the interparticle distances (Jung *et al.*, 2002). Micromagnetic time-dependent simulations of the a.c. response confirmed that these resonances were associated with dipolar waves propagating within such nanoparticle arrays. Further insight into the role of the dimensionality and the dipolar coupling on the dynamic response of magnetic systems due to microwave excitation in an external magnetic field has been provided by Zakeri *et al.* (2006), who examined the FMR spectra of: (i) ultrathin magnetic 5–20 ML-thick Fe films grown epitaxially on GaAs substrates (a sample two-dimensional; 2-D system); (ii) magnetic MnAs stripes (one-dimensional, 1-D system); and (iii) arrays of monodisperse FePt nanoparticles (a quasi-1-D system).

Further advances in nanofabrication technology are expected to lead to the development of nanostructured magnetic materials with well-defined shapes and anisotropy axes via, for example, epitaxial growth and/or electron-beam lithography, as well as achieving a narrow size distribution (e.g., by employing self-assembly principles). It is anticipated that the latter materials would stimulate both FMR and SPR studies, as the distribution of shapes and sizes would no longer obscure the individual spectral features.

Parallel efforts in advancing the underlying theory and least-squares modeling of the experimental spectra would provide essential data on magnetic anisotropy, size and shape distribution and interparticle coupling. In this way they would assist in the even further development of magnetic nanoparticle technology.

20.3
Characterization of Nanostructured Oxide Semiconductors for Photoactivated Catalysis and Solar Energy Conversion

Nanostrictured oxide semiconductors are envisioned as the likely architectural elements for the next generation of solar energy convertors, as well as for photoactivated catalysis. In order to perform these tasks efficiently, such structures should be capable of capturing the photons by forming spatially separated charge carriers, and then allowing for further interfacial energy transfer steps. These requirements could be satisfied by: (i) employing wide band-gap semiconductors with suitable electronic structure for charge separation; and (ii) building nanostructures from such materials in order to enhance the role of interfacial phenomena. In order to achieve these research objectives, many research groups have focused on various nanomaterials based on TiO_2. The principal advantages of TiO_2 are its wide availability, chemical stability, and low toxicity. The main disadvantages include a relatively high band gap of $E_g > 3.2\,eV$ (for the anatase crystalline phase) that require excitation by ultraviolet (UV) light, and a high rate of the electron-hole recombination.

As the search for suitable photoactivated catalysts continues, a better understanding of the coordination geometry of the dopant atoms, as well as locations of the trapping and recombination sites, is becoming ever more desirable. EPR spectroscopy appears to be an experimental tool that could fulfill these needs. For example, the localization of a charge in a TiO_2 matrix typically yields paramagnetic Ti^{3+} (d^1) ions and O^- trapped holes that could be stabilized and studied by EPR at 77 K, and below (Hurum, Agrios, and Gray, 2003). The EPR lines of Ti^{3+} are relatively narrow, allowing for differentiation between two types of Ti^{3+} species and following the evolution of these and O^- species upon, for example, UV illumination and room-temperature annealing (Hurum, Agrios, and Gray, 2003; Berger et al., 2007).

The efficiency of TiO_2 as a photocatalyst could be further improved with assistance from several approaches, including the deposition or doping of titania nanoparticles with noble metals (Sclafani and Herrmann, 1998), transition metal molybdates (Ghorai et al., 2007), and even carbon (Lee et al., 2008). For many of these systems, EPR can provide valuable information on the formation of active species, and even follow the charge localization upon sample illumination. For example, the observation of a characteristic Ni^{3+} EPR signal upon illuminating TiO_2 nanoparticles with visible light at 77 K provided an indication of an efficient transformation of the photogenerated hole initially trapped in Ni^{2+} into one trapped in Ni^{3+} (Ghorai et al., 2007). Furthermore, paramagnetic metal ions such as Mn^{2+} could be successfully used as sensitive reporters of the coordination environment of the host lattice of the anatase TiO_2 nanoparticles (Saponjic *et al.*, 2006). Specifically, light excitation EPR data suggested that the majority of the photogenerated electrons (Ti^{3+}) are formed via either fast hole trapping at the Mn substitutional sites, or via a direct excitation of the Mn dopant atoms. In addition, Mn ions incorporated into the surface layers were found to participate in charge separation, while those located deeply in the nanoparticle core were fully inactive. These authors argued that doping of the core of nanoparticles with Mn^{2+} ions would enable the fabrication of optically transparent films with superparamagnetic behavior at room temperatures, and a saturation magnetic moment of $1.23\,\mu_B$ per Mn atom (Saponjic *et al.*, 2006). Analysis of the Mn^{2+} coordination in the host lattice has been facilitated by the availability of multifrequency EPR data at X- (9.5 GHz) and D-band (130 GHz) (Saponjic *et al.*, 2006).

20.4
Surface Radicals, Catalytic Activity, Cytotoxicity, and Radical-Scavenging Properties of Nanomaterials

20.4.1
Cayalytic Activity

While the charge separation that arises upon the absorption of a photon by a wide-gap semiconductor nanoparticle (as described in the preceding section) can, in principle, occur anywhere within the oxide volume, catalytic processes such as the oxidation of organic compounds or the splitting of water to produce hydrogen must take place at the surface. Many such catalytic processes invoke free-radical mechanisms and unpaired electron species that could be characterized by EPR.

Recently, many aspects of free radicals on reactive surfaces have been reviewed. For example, Murphy and Giamello (2002) described the progress in EPR studies of the nature and reactivity of a series of oxide surfaces, while Rhodes (2005) provided a comprehensive overview of the reactive radicals on the reactive surfaces, with emphasis placed on the use of EPR to characterize free-radical species involved in heterogeneous catalysis and environmental pollution control. While the latter review did not focus specifically on nanomaterials, it largely described nanostructured surfaces, as the latter possess the exceptionally large surface areas required for heterogeneous catalytic processes. Such high-surface-area materials are exemplified by zeolites – the most widely studied and commercially important family of nanostructured catalysts.

Zeolites are aluminosilicates in which the pores (of ca. 1.3 nm diameter) are arranged in a honeycomb fashion. While most of the zeolites used in the manufacture of concrete are sourced from natural deposits, catalytic zeolites are typically of synthetic origin because of a need to tailor the surface properties to specific processes. Various aspects of zeolite surface properties can be deduced from the EPR of organic free radicals, either introduced as probes or occurring spontaneously (see Rhodes, 2005 and references cited therein).

20.4.2
Cytotoxicity

The wide industrial use of zeolites has raised much environmental concern, as some of these materials are suspected to be carcinogenic in nature. For example, in aqueous solution the zeolite erionite has been shown to generate hydroxyl radicals, HO·, most likely during the course of a Fenton reaction involving Fe^{3+} and Fe^{2+} ions trapped at the surface (Ruda and Dutta, 2005).

As increasing numbers of nanostructured and nanoparticle-based materials continue to be developed and introduced into the human environment, further research will be required to monitor them for possible toxic and/or carcinogenic effects, as well as to ascertain the mechanisms of their biotoxicity. Even materials previously considered to be biologically inert should be retested when their particulate size approaches the nanoscale. A recent example was provided by Reeves *et al.* (2008), who described the cytotoxicity and oxidative DNA damage in fish cells induced by TiO_2 nanoparticles. Conventional (>100 nm) TiO_2 particles are frequently added to white paint, or used as food colorants, sunscreens and cosmetic products, based on their assumed biological safety in human and animals. In contrast, anatase TiO_2, with an average particle size of 5 nm, was shown to cause a significant decrease in cell viability that was dependent on both the concentration of TiO_2 and exposure to UVA radiation. Although, even in the absence of UVA, an elevated level of oxidative DNA damage was detected, the UVA irradiation of TiO_2-treated cells resulted in a further increase in DNA damage. Subsequent spin-trapping EPR experiments indicated that the observed biotoxicity of TiO_2 nanoparticulates may be attributed to the production of hydroxyl radicals, HO·, at the nanoparticle surface. Additional spin-trapping EPR studies have been conducted to elucidate the kinetics and possible mechanisms of hydroxyl radical formation for photocatalytic nanocrystalline TiO_2 (Sroirayaa *et al.*, 2008). The latter study also evaluated the effects of particle size distribution, concentration, and agglomeration. It should be noted that other catalytically inert materials – including gold, which is regarded as one of the most chemically inert elements – can become catalytically active when employed in a nanoparticulate form. Ionita *et al.* (2007a) reported the air-oxidation of organic substrates containing active hydrogen atoms (such as amines and phosphine oxides) by phosphine- and amine-protected gold nanoparticles, whereas nanoparticles protected by more strongly bound ligands (e.g., thiols) were inactive in these reactions. Overall, it was concluded that the mechanisms of such reactions could be best studied with EPR spectroscopy and available spin-trapping methods (Ionita *et al.*, 2007a).

20.4.3
Radical-Scavenging Properties

In contrast to the examples given above, some nanoparticles possess radical-scavenging properties, and may be effective in protecting cells and tissues against oxidative stress. One such material is nanoparticulate cerium oxide (ceria), which today is widely used as a component of exhaust catalytic systems in automobiles. Nanoceria has already been shown to protect against oxidative stress in both cell culture and animal models. Recently, Heckert et al. (2008) employed a spin-trap (5-diethoxyphosphoryl-5-methyl-1-pyrroline N-oxide; DEPMPO), and EPR to demonstrate that nanoceria is effective at scavenging superoxide anions and, thus, could be considered a superoxide dismutase (SOD) mimic. The nature of nanoceria's antioxidant properties could be attributed to specific ion-oxidation states that exist at the surface, specifically, cerium(III) surface concentrations (Heckert et al., 2008).

While the current rapid development of nanotechnologies is expected to revolutionize the field of heterogeneous catalysis, and lead also to many other beneficial applications, the potential cytotoxic and carcinogenic effects of these new nanomaterials on humans and animals should not be overlooked. Clearly, such potential adverse consequences must be addressed by developing efficient screening protocols, and attention paid to studying the fundamental mechanisms responsible for radical production or scavenging at nanostructured surfaces. As the presently available spin-trapping EPR methods are exceptionally suited to such purposes, a rapid development of this research area is to be expected in the near future.

20.5
Spin-Labeling EPR Studies of Ligand-Protected Nanoparticles and Hybrid Nanostructures

Monolayer-protected clusters (MPCs) exhibit many unusual optical, electrical, catalytic, and other properties that today are finding numerous new applications in biodiagnostics, different varieties of sensors, and drug delivery, to name but a few. The properties of MPCs could be further manipulated through rational ligand synthesis that allows for the positioning of specific functional groups at the monolayer interface. Indeed, the decoration of the surface of a nanoparticle with different ligands having specific functionalities that, to some extent, would resemble protein surfaces or, perhaps, the surfaces of even larger macromolecular assemblies such as protein complexes and viruses, can be envisioned.

Several properties of nanoparticle surfaces – such as electrostatic potential, the presence and state of the diffusion-ion layer, as well as hydrogen bonding between the ligands – are deemed crucial not only for stabilizing the nanoparticle in a solution but also for controlling molecular interactions and tailoring the nanoparticle's functional properties. Notably, many of the methods currently employed for MPC studies are focused on the "bulk" properties of supramolecular assemblies, without providing detailed information on the structure and dynamics of the diffuse layer formed at the interface, and the associated electrostatics. Yet, this essential information could be obtained from molecular probes – specifically, nitroxide-based EPR probes – that could be positioned at specific sites with respect to the monolayer, affording a direct spectral readout from the probe location. Although spin-labeling methods are well established in biophysics, their applications to nanostructures have only recently begun to emerge.

Spin-labeling EPR methods can be employed to study both the mechanism of formation and exchange of the ligand-protecting layer on the nanoparticle surface, and the local properties of the assemblies formed. A primary example of the former was provided by Ionita et al. (2004), who described the mechanism of a place-exchange reaction of ligand-protected Au nanoparticles. In this case, EPR of the diradical disulfide spin labels was employed to determine the local concentration profile of the label on Au surfaces, and to propose a kinetic model for the ligand-exchange reaction. As continuous-wave (CW) EPR allows a straightforward resolution between the mono- and biradical nitroxides species, it could be used directly to monitor the ligand-exchange reactions, even at small nanoparticle concentrations. Hence, by carrying out a series of EPR experiments, Ionita et al. (2002, 2004) showed that only one branch of the disulfide ligand would be adsorbed onto the nanoparticle surface during the exchange, while the other branch formed a mixed disulfide with the outgoing ligand. These data also suggested the presence of different binding sites (most likely to be surface defects), with different reactivities in the exchange reaction (Ionita et al., 2002, 2004). More recently, disulfide nitroxide spin labels have been employed to study ligand dynamics and the dissociation on a surface of semiconductor nanoparticle QDs formed from CdSe and capped by trioctylphospine oxide (Billone et al., 2007). The significance of the latter study, which was focused on the molecular mechanisms of the interactions of RS–H and RS–SR bonds with the QD surface, should be viewed from a perspective of the potential applications of QDs for biodiagnostic purposes. Because the disulfide linkage has proven to be useful for the robust attachment of many ligands to nanoparticle surfaces, there are many reasons to believe that spin labeling EPR of the corresponding mono- and biradical species can be widely used to monitor ligand-exchange processes on the nanoparticle surfaces.

Magnetic spin–spin interactions between the nitroxides were used many years ago to study lateral diffusion in lipid bilayer membranes, using CW-EPR. However, recent developments in pulsed-EPR – and particularly in pulsed dipolar electron–electron resonance (DEER) and double quantum coherence (DQC) – have led to improvements in the quality of long-distance constraints that could be obtained from spin-labeling EPR experiments. Moreover, the EPR data could be analyzed not only in terms of average spin–spin distances, but also of distance distributions. Subsequently, Ionita et al. (2008) took advantage of these developments to study the lateral diffusion of thiol ligands on a gold nanoparticle surface. In this case, the average distance between the spin-labeled ligand branches on the Au nanoparticle surface was shown to increase only marginally on heat treatment at 90 °C;

20.5 Spin-Labeling EPR Studies of Ligand-Protected Nanoparticles and Hybrid Nanostructures

this implied that, even at these elevated temperatures, the lateral diffusion of the thiolate ligands was very slow. Thus, the likely ligand redistribution mechanism appeared to be "ligand hopping" between the nanoparticles, rather than diffusion on the particle surface (Ionita et al., 2008). Spin–spin interactions may also be used to investigate the exchange of solutes between the ligand shell of monolayers of protected Au nanoparticles and the bulk aqueous solutions (Lucarini et al., 2004).

Further detailed information on the local properties of nanoparticle ligand shells could be obtained by analyzing the EPR spectra of spin-labeled ligands for dynamic effects. Bertholon et al. (2006) reported on the surface characteristics of a series of poly(isobutylcyanoacrylate) nanoparticles prepared by redox radical emulsion polymerization with polysaccharides of different molecular weight and nature. These studies were carried out by grafting a nitroxide spin label onto the polysaccharide chains following synthesis of the nanoparticles, and then analyzing the CW-EPR line shapes. These experiments yielded a rather unexpected result – that the percentage of the fast moving nitroxide was highest when linked to the shortest chitosan, the most rigid polysaccharide tested. Thus, it was concluded that in these nanostructures the polysaccharide chains are likely to fold, so as to permit hydrophobic interactions between the label and the nanoparticle core (Bertholon et al., 2006).

A more detailed least-squares CW-EPR line shape simulation analysis has been reported of multifrequency EPR data at X- (9.5 GHz) and Q-band (35 GHz) to deduce the rotational diffusion tensors of several spin-labeled ligands attached to Au nanoparticles (Ionita et al., 2007b). This report was, perhaps, the first example of multifrequency EPR being used to study ligand-protected Au nanoclusters. There is much reason to believe that future studies of MPCs would benefit greatly from the high-field spin-labeling EPR methods currently being developed for broad applications in biophysics (Smirnov, 2008).

Spin-labeling EPR studies of the nanoparticle ligand layer could also benefit from employing pH-sensitive nitroxides, as has been demonstrated by Khlestkin et al. (2008). These authors described a series of ligands with variable length of the hydrocarbon bridge between the anchoring sulfur and the pH-sensitive imidazoline nitroxide, bearing an amidine functionality (Figure 20.3). The protonation state of this nitroxide can be directly observed by X-band EPR spectroscopy from changes in the isotropic nitrogen hyperfine coupling constant. The results of a series of EPR titration experiments for tiopronin-protected Au nanoparticles with an average diameter of 1.8 nm enabled the first-ever acquisition of data on the radial profile of the electrostatic and heterogeneous dielectric environment of MPCs. For the carboxyl-terminated monolayer the surface potential, Ψ, was negative, yielding a positive shift, ΔpK_a^{el}, in the pK_a value of the probe that decayed progressively with the distance between the probe and the surface layer. The probe pK_a was also affected by the effective local dielectric constant: notably, there was a reduction in the Gibbs free energy for the protonation of a probe experiencing a somewhat lower effective dielectric constant, ε, than in bulk water. The presence of organic ligands decreased the effective ε, thus resulting in a negative ΔpK_a^{pol} that also vanished with the distance as the effective ε approaches that of the bulk

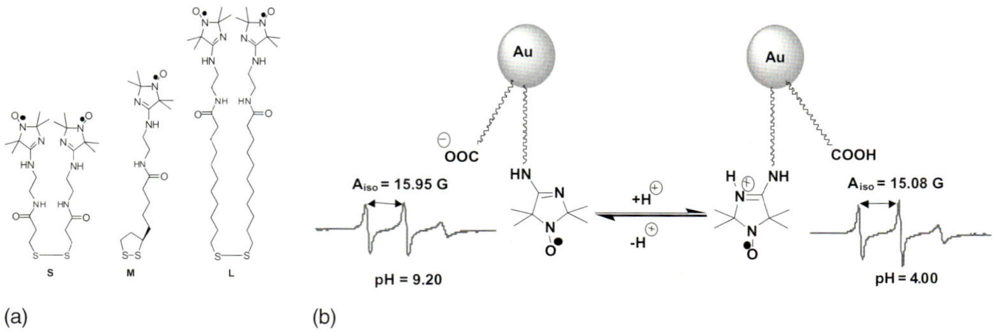

Figure 20.3 (a) Structures of ligands designed for anchoring pH-sensitive nitroxide spin-labels to the gold surface; (b) A cartoon illustrating an EPR titration experiment with spin-labeled nanoparticles. Further details of these experiments may be found in Khlestkin et al. (2008).

water. This interplay between the positive ΔpK_a^{el} and the negative ΔpK_a^{pol} due to varying ligands yielded a pK_a profile with a maximum at a certain distance from the nanoparticle surface that did not necessarily coincide with the length of the thiopronin ligand. The nitroxide pK_a data obtained for three progressively longer tethers (see Figure 20.3) indicated that the dielectric interface of the thiopronin monolayer extended further from the nanoparticle surface than the ligand shell (Khlestkin et al., 2008). Overall, these studies demonstrated the utility of combining EPR spectroscopy with the synthesis of tailored, pH-sensitive nitroxide ligands for probing electrostatic phenomena in the polar interface of ligand-protected Au nanoclusters. This approach, of employing pH-sensitive nitroxide ligands, is expected to stimulate further research on the complex electrostatics of the organized monolayer interfaces of hybrid nanomaterials and the phenomena of monolayer charge compensation on DNA and/or protein binding.

Spin-labeling EPR methods can also be of valuable assistance in developing and characterizing the structural and dynamical organization of hybrid nanoscale materials and devices including, potentially, protein biochips. These hybrid nanostructures are typically constructed from man-made substrates and various biomolecules such as proteins and/or oligonucleotides, or even molecular assemblies (e.g., lipid bilayers). The role of the substrate is to afford the addressability for the individual classes of biomolecules and/or to provide the means for an analytical readout of the events involving specific biospecies. One of the main challenges of biochip technology is to ensure that the biospecies under study retain their conformations as close as possible to those found in functional biological systems. While DNA biochip technology has already matured, its success has been attributed to the relatively simple structure and high stability of the oligonucleotides. In contrast, many proteins are subject to irreversible denaturation when they encounter substrate surfaces or even mild changes in environmental conditions. These problems become even more significant for membrane proteins that would fold and function only when embedded into lipid bilayers with the composition close to the native cellular membranes. For protein biochips, it is often the case that if a binding event and/or a catalytic activity is detected for a protein immobilized onto a substrate, then the same biochemical processes would also be observed for the same protein in a solution. Unfortunately, however, this simple general rule does not work in reverse, as the biological function could be altered by the substrate. Consequently, further research is required to understand the structural conformational changes that occur upon the immobilization of proteins and other biospecies on solid surfaces, as well as to elucidate the magnitude of the dynamic effects.

It appears that spin-labeling EPR is uniquely positioned to assist in elaborating these issues of developing biochip technology, due to it possessing a sufficiently high sensitivity, being suitable for even opaque samples, for the small molecular volume of the nitroxide tags employed, and its well-developed methodology for studying the structure and dynamics of biological systems. One of the first steps in this direction has been taken by Smirnov and Poluektov (2003), who reported on the use of spin-labeling EPR at 95 GHz (W-band) to elucidate the macroscopic organization of lipid bilayers self-assembled inside nanoporous anodic aluminum oxide (AAO). These structures represent a new type of substrate-supported lipid bilayer that has many attractive features for membrane biochip technology and for biophysical studies of lipid bilayers and membrane proteins. Specifically, these structures – which are also lipid nanotube arrays – have a high density of the nanoporous channels, thus providing at least a 600-fold gain in the bilayer surface area for a similar-sized planar substrate chip. The lipid bilayers are also adequately protected from any environmental contamination by the rigid structure of the AAO substrate. A cartoon of the nanotubular bilayer, and a scanning electron microscopy image of the AAO, are shown in Figure 20.4.

Smirnov and Poluektov (2003) investigated the self-assembly of synthetic DMPC (1,2-dimyristoyl-sn-glycero-3-phosphocholine) lipids inside the pores of a commercial AAO (Whatman, United Kingdom) with an average pore size of ca. 175 nm.

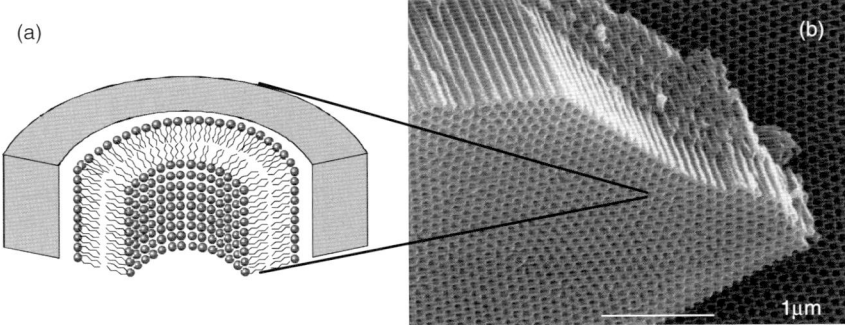

Figure 20.4 (a) A cartoon of a nanotubular lipid bilayer confined inside a cylindrical nanopore; (b) SEM image of a nanoporous AAO substrate fabricated at North Carolina State University, Raleigh, NC, USA.

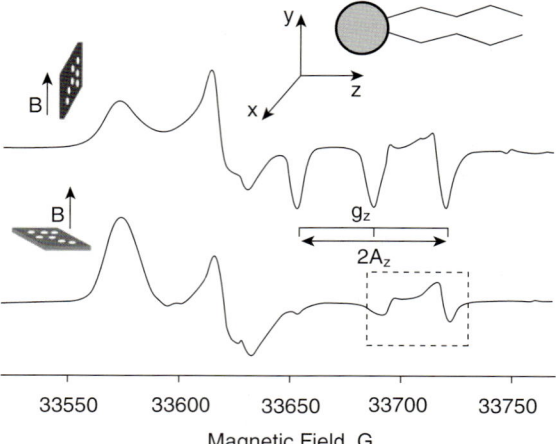

Figure 20.5 Experimental rigid-limit ($T = 150$ K) high-resolution 94.4 GHz (W-band) EPR spectra of AAO substrate with deposited DMPC:5PC (100:1) at two orientations of the substrate surface in the magnetic field. The cartoon on top shows orientations of the magnetic axes with respect to the phospholipid. Note that the bottom EPR spectrum has a low intensity in the g_z-region (the feature in the dashed box is due to a paramagnetic AAO impurity), indicating that at this substrate orientation the lipid chains are perpendicular to the magnetic field. Reproduced with permission from Smirnov and Poluektov (2003).

Before preparing multilamellar vesicles and depositing these into the AAO substrates, the lipids were doped at 1 mol% with an EPR-active 1-palmitoyl-2-stearoyl-(5-doxyl)-sn-glycero-3-phosphocholine (5PC). Here, 5PC contains the doxyl nitroxide moiety at the position 5 of the acyl chain where the local order parameter, S, of the lipid bilayer is still relatively high ($S \approx 0.7$). The dynamic lipid disorder and partial averaging of spectral anisotropies were further reduced by taking W-band EPR spectra at a temperature of $T = 150$ K, at which the dynamics of the phospholipids is approaching the rigid limit. An enhanced angular resolution of high-field (HF) 95 GHz (W-band) EPR for nitroxide spin probes facilitated interpretation of the experimental EPR spectra and verified the formation of lipid nanotubes. It was observed that the relative intensities of the characteristic peaks of the experimental W-band EPR spectra, measured for two different orientations of the AAO substrate with respect to the external magnetic field, B_0, were clearly different (Figure 20.5). Particularly noticeable were the intensity changes in the g_z-region that spread from ca. 3.3850 to 3.3940 T. When the surface of the AAO substrate was perpendicular to B_0 (lower spectrum), the three nitrogen hyperfine coupling z-components almost completely disappeared when compared to the upper spectrum in Figure 20.5, taken at a different orientation. (Note: The signals inside the dashed box were mainly due to paramagnetic impurities in the AAO substrate.) This means that, at this substrate orientation, only a very small fraction of molecules has the z-axis of the N–O molecular frame aligned with the external magnetic field. Thus, it was concluded that the majority of the phospholipids inside the nanopores were assembled with their long axis perpendicular to the magnetic field and, therefore, perpendicular to the direction of pores, as the surface phospholipids were mechanically removed during sample preparation. Subsequent solid-state ^{31}P NMR studies of nanotubular bilayers confirmed the results of these spin-labeling EPR experiments (Chekmenev et al., 2005).

Follow-up studies using spin-labeling EPR, solid-state NMR, differential scanning calorimetry and other methods, showed that the lipid nanotube arrays have several important advantages over conventional, mechanically aligned planar bilayers. Among these advantages are included a very high surface area, a long-term stability of the aligned lipid assemblies over an exceptionally wide range of temperatures, pH-values and salt concentrations, a high hydration level of lipid bilayers, and protection from surface contaminations. Another feature is the excellent accessibility of the bilayer surface for exposure to solute molecules, which makes it possible for the membrane protein samples to be exposed to different solution media, and for this procedure to be repeated multiple times with the same sample. In addition, fully hydrated membrane proteins can be exposed, for example, to buffers at various pH-values and to ion and drug concentrations, so as to facilitate a wider range of structural and functional studies in the native-like environment by means of both NMR and EPR. The lipid nanotube arrays also facilitate a spin-labeling EPR study to derive the helical tilt of transmembrane peptides containing the rigid unnatural amino acid, 2,2,6,6-tetramethylpiperidine-1-oxyl-4-amino-4-carboxylic acid (TOAC) (Karp et al., 2006).

20.6
Summary and Future Perspectives

In this chapter, an attempt has been made to review one of the emerging fields of the application and development of EPR methods for studying nanoscale objects. In particular, it should be noted that nanotechnology itself continues to evolve synergetically at a rather rapid pace and that, although still in its early stages, EPR-mediated research into nanomaterials has already demonstrated the wealth of information that it is capable of gathering. It is anticipated by the present author that this area of EPR is poised to expand rapidly during the coming years, and that the original studies highlighted throughout the chapter should inspire the readers to new discoveries. Specifically, detailed EPR studies of the surface catalytic properties of nanoparticles and nanostructural materials are expected to emerge rapidly, because of the close involvement of free-radical species in the most essential steps of catalytic cycles. The availability and advanced state of the development of spin-labeling methods makes this technology readily employable in elucidating the detailed structure and dynamics of the nanomaterials' interfacial layers. The latter studies would further benefit from mutifrequency spin-labeling EPR methods, as well as from advances in the chemistry of spin-labeled ligands and other custom-made spin-probes with new functionalities. A combination of nitroxide synthetic organic chemistry and future progress in multifrequency EPR could pave the way to new hybrid nanomaterials capable of taking advantage of free radical redox chemistry. The identification of possible biotoxicity mechanisms, by using available spin-trapping methods, is also expected to play a major role in evaluating the many new nanostructural materials currently being introduced. Finally – but not lastly – EPR may emerge as a competing technology for the "sizing" of magnetic nanoparticles and their arrays, as well as a simple – but informative – tool to maintain the quality control of these important magnetic nanomaterials.

Acknowledgments

The author thanks NIH 1R01GM072897 and NSF ECS 0420775 for providing partial financial support to this work. The author was also partially supported by U.S. DOE Contract DE-FG02-02ER15354 (to A.I.S.) and as a part of The Center for LignoCellulose Structure and Formation, an Energy Frontier Research Center funded by the U.S. DOE, Office of Science, Office of Basic Energy Sciences under Award Number DE-SC0001090. Gratitude is also expressed to Prof. T. Smirnova and all members of the EPR group at NCSU. Special thanks are due to Drs Maxim Voinov and Saritha Nellutla (NCSU) for many useful discussions.

Pertinent Literature

Berger, R., Bissey, J.-C., Kliava, J., and Daubric, H. (2001) *J. Magn. Magn. Mater.*, **234**, 535–544.

Ionita, P., Volkov, A., Jeschke, G., and Chechik, V. (2008) *Anal. Chem.*, **80**, 95–106.

Khlestkin, V.K., Polienko, J.F., Voynov, M., Smirnov, A.I., and Chechik, V. (2008) *Langmuir*, **24**, 609–612.

Rhodes, C.J. (2005) *Prog. React. Kinet. Mech.*, **30**, 145–213.

Seavey, M.H., Jr and Tannenwald, P.E. (1958) *Phys. Rev. Lett.*, **1**, 168–169.

References

Berger, R., Kliava, J., Bissey, J.-C., and Baietto, V. (1998) *J. Phys. Condens. Matter*, **10**, 8559–8572.

Berger, R., Bissey, J.-C., and Kliava, J. (2000) *J. Phys. Condens. Matter*, **12**, 9347–9360.

Berger, T., Diwald, O., Knözinger, E., Napoli, F., Chiesa, M., and Giamello, E. (2007) *Chem. Phys.*, **339**, 138–145.

Bertholon, I., Hommel, H., Labarre, D., and Vauthier, C. (2006) *Langmuir*, **22**, 5485–5490.

Billone, P.S., Maretti, L., Maurel, V., and Scaiano, J.C. (2007) *J. Am. Chem. Soc.*, **129**, 14150–14151.

Chekmenev, E.Y., Hu, J., Gor'kov, P.L., Brey, W.W., Cross, T.A., Ruuge, A., and Smirnov, A.I. (2005) *J. Magn. Reson.*, **173**, 322–327.

Ebels, U., Duvail, J.-L., Wigen, P.E., Piraux, L., Buda, L.D., and Ounadjela, K. (2001) *Phys. Rev. B*, **64**, 144421.

Ghorai, T.K., Dhak, D., Biswas, S.K., Dalai, S., and Pramanik, P. (2007) *J. Mol. Catal. A: Chem.*, **273**, 224–229.

Heckert, E.G., Karakoti, A.S., Seal, S., and Self, W.T. (2008) *Biomaterials*, **29**, 2705–2709.

Hurum, D.C., Agrios, A.G., and Gray, K.A. (2003) *J. Phys. Chem. B*, **107**, 4545–4549.

Ionita, P., Caragheorgheopol, A., Gilbert, B.C., and Chechik, V. (2002) *J. Am. Chem. Soc.*, **124**, 9048–9049.

Ionita, P., Caragheorgheopol, A., Gilbert, B.C., and Chechik, V. (2004) *Langmuir*, **20**, 11536–11544.

Ionita, P., Conte, M., Gilbert, B.C., and Chechik, V. (2007a) *Org. Biomol. Chem.*, **5**, 3504–3509.

Ionita, P., Wolowska, J., Chechik, V., and Caragheorgheopol, A. (2007b) *J. Phys. Chem. C*, **111**, 16717–16723.

Jung, S., Watkins, B., DeLong, L., Ketterson, J.B., and Chandrasekhar, V. (2002) *Phys. Rev. B*, **66**, 132401.

Karp, E.S., Inbaraj, J.J., Laryukhin, M., and Lorigan, G.A. (2006) *J. Am. Chem. Soc.*, **128**, 12070–12071.

Konchits, A.A., Motsnyi, F.V., Petrov, Y.N., Kolesnik, S.P., and Yefanov, V.S. (2006) *J. Appl. Phys.*, **100**, 124315.

Landau, L.D. and Lifshitz, E.M. (1935) *Phys. Z. Soviet Union*, **8**, 153–159.

Lee, S., Yun, C.Y., Hahn, M.S., Lee, J., and Yi, J. (2008) *Korean J. Chem. Eng.*, **25**, 892–896.

Lucarini, M., Franchi, P., Pedulli, G.F., Pengo, P., Scrimin, P., and Pasquato, L. (2004) *J. Am. Chem. Soc.*, **126**, 9326–9329.

Morgunov, R., Farle, M., Passacantando, M., Ottaviano, L., and Kazakova, O. (2008) *Phys. Rev. B*, **78**, 045206.

Murphy, D.M. and Giamello, E. (2002) *Specialist Periodical Reports: Electron Paramagnetic Resonance*, vol. 18, The Royal Society of Chemistry, Cambridge. p. 183.

Noginov, M.M., Noginova, N., Amponsah, O., Bah, R., Rakhimov, R., and Atsarkin, V.A. (2008) *J. Magn. Magn. Mater.*, **320**, 2228–2232.

Noginova, N., Chen, F., Weaver, T., Giannelis, E.P., Bourlinos, A.B., and Atsarkin, V.A. (2007) *J. Phys. Condens. Matter*, **19**, 246208.

Noginova, N., Weaver, T., Giannelis, E.P., Bourlinos, A.B., Atsarkin, V.A., and Demidov, V.V. (2008) *Phys. Rev. B*, **77**, 014403.

Rameev, B.Z., Yıldız, F., Aktas, B., Okay, C., Khaibullin, R.I., Zheglov, E.P., Pivin, J.C., and Tagirov, L.R. (2003) *Microlectron. Eng.*, **69**, 330–335.

Reeves, J.F., Davies, S.J., Dodd, N.J.F., and Jha, A.N. (2008) *Mutat. Res.*, **640**, 113–122.

Ruda, T.A. and Dutta, P.K. (2005) *Environ. Sci. Technol.*, **39**, 6147–6152.

Saponjic, Z.V., Dimitrijevic, N.M., Poluektov, O.G., Chen, L.X., Wasinger, E., Welp, U., Tiede, D.M., Zuo, X., and Rajh, T. (2006) *J. Phys. Chem. B*, **110**, 25441.

Sclafani, A. and Herrmann, J.-M. (1998) *J. Photochem. Photobiol. A Chem.*, **113**, 181–188.

Smirnov, A.I. (2008) *Nitroxides. Applications in Chemistry, Biomedicine, and Materials Science*, (eds G.I. Likhtenshtein, J. Yamauchi, S. Nakatsuji, A.I. Smirnov, and R. Tamura), Wiley-VCH Verlag GmbH, Berlin, Germany, pp. 121–159.

Smirnov, A.I. and Poluektov, O.G. (2003) *J. Am. Chem. Soc.*, **125**, 8434–8435.

Sroirayaa, S., Triampob, W., Moralesd, N.P., and Triampo, D. (2008) *J. Ceram. Process. Res.*, **9**, 146–154.

Zakeri, K.H., Kebe, T., Lindner, J., Antkowiak, C., Farle, M., Lenz, K., Tolinski, T., and Baberschke, K. (2006) *Phase Transit.*, **79**, 793–813.

21
Single-Molecule Magnets and Magnetic Quantum Tunneling
Sushil K. Misra

21.1
Introduction

Apart from a general introduction to single-molecule magnets (SMMs), including descriptions of intramolecular coupling, a list of the various SMMs reported in the literature, and the applications of SMMs, this chapter will discuss in detail the multifrequency (MF) EPR studies of SMMs, as well as magnetic quantum tunneling (MQT) in SMMs.

The discovery that individual molecules can function as nanoscale magnets – referred to as SMMs – was made during the 1990s. This represented a significant development, as many specialized applications of magnets require monodisperse, nanoscale magnetic particles, wherein each molecule functions as a nanoscale, single-domain magnetic particle that exhibits the classical macroscale property of a magnet. SMMs become magnetized in a magnetic field; in addition, they show slow relaxation when the magnetic field is removed.

The magnetization in SMMs occurs due to the existence of a large energy barrier between the spin-up and spin-down states. The magnetic order exists within the molecules, which can be considered as single domains. The requirements to exhibit SMM behavior are: (i) a very high spin state; and (ii) a large magnetic anisotropy. SMMs have many important advantages over conventional nanoscale magnetic particles composed of metals, metal alloys, or metal oxides. Such benefits include, among others, a uniformity of size, solubility in organic solvents, and readily alterable peripheral ligands. SMMs also offer the advantages of being handled by molecular chemistry, such as room-temperature synthesis, a true solubility in many solvents, a well-defined protective shell of diamagnetic organic groups, and a crystalline ensemble of monodisperse units, as well as displaying the superparamagnetism of a much larger, classical magnetic particle. SMMs can also be true mesoscale materials, straddling the interface between classical and quantum behavior by exhibiting quantum tunneling of magnetization (MQT).

SMMs exhibit magnetic hysteresis of purely molecular in origin. In particular, they exhibit magnetization hysteresis below its blocking temperature (T_B), defined as the temperature below which the relaxation of the magnetization becomes slow compared to the time scale of a particular investigation technique (Gatteschi, Sessoli, and Cornia, 2000). For example, a molecule magnetized at 2 K will retain 40% of its magnetization after 2 months, but by lowering the temperature to 1.5 K it would take 40 years (Gatteschi, Sessoli, and Cornia, 2000) to reach the same state. Contrary to conventional bulk magnets and molecule-based magnets, the collective long-range magnetic ordering of magnetic moments is not necessary here. SMMs are characterized by a combination of a large ground-state spin (S) value and an easy-axis (Ising-type) anisotropy, which produces a significant barrier versus $k_B T$ to magnetization relaxation. SMMs thus represent a molecular (or bottom-up) approach to new nanoscale magnetic materials, in contrast to classical nanomagnets which are obtained via a top-down approach.

21.1.1
Intramolecular Coupling

SMMs contain a finite number of interacting spin centers – for example, paramagnetic ions – and thus provide an ideal opportunity to study the basic concepts of magnetism. The magnetic coupling between the spins of the metal ions is mediated via superexchange interactions, and can be described by the following isotropic Heisenberg Hamiltonian: $H_{HB} = \sum_{i<j} J_{ij} S_i \cdot S_j$, where $J_{i,j}$ is the coupling constant between spin S_i and spin S_j. [For positive $J_{i,j}$, the coupling is called *ferromagnetic* (parallel alignment of spins), while for negative $J_{i,j}$ the coupling is called *antiferromagnetic* (antiparallel alignment of spins).] This coupling leads to: (i) a high spin ground state; (ii) a high zero-field splitting (ZFS) (due to high magnetic anisotropy); and (iii) negligible magnetic interaction between the molecules. The combination of these properties – and, in particular, the spin anisotropy – manifests itself as an energy barrier that the spins must overcome when they switch from a parallel alignment to an antiparallel alignment so that, at low temperatures, the system can be trapped in one of the high-spin energy wells. This barrier (U) is defined as $U = S^2 |D|$, where S is the dimensionless total spin state and D the ZFS parameter. (Note: D can be negative, but only its absolute value is considered in the equation.) The higher the barrier, the longer a material remains magnetized; a high barrier is obtained when the molecule contains many unpaired electrons and when its ZFS value is large. For example, in the case of the Mn_{12}-ac SMM cluster the spin state is 10 (involving 20 unpaired electrons) and $D = -0.5\,cm^{-1}$, resulting in a barrier of $50\,cm^{-1}$ (equivalent to 60 Kelvin).

SMMs possess magnetic ground states, and give rise to hysteresis effects and metastable magnetic phases. They also show MQT, which is related to coherent dynamics in such systems. The effect is also observed by hysteresis experienced when the magnetization is measured in a magnetic field sweep; this implies that, on lowering the magnetic field again after reaching the maximum magnetization, the magnetization remains at high levels and requires a reversed field to bring it back to zero.

21.1.2
Examples of SMMs Reported in the Literature

The first SMM reported dates back to 1991 (Caneschi et al., 1991), although the term "single-molecule magnet" was first employed by David Hendrickson, a chemist at the University of California, San Diego and George Christou (Indiana University) in 1996 (Aubin et al., 1996), The following is a description of the various well-known SMMs reported so far.

$Mn_{12}Ac_{16}$: The European researchers discovered a $Mn_{12}O_{12}(MeCO_2)_{16}(H_2O)_4$ complex ($Mn_{12}Ac_{16}$), first synthesized in 1980 (Lis, 1980). This exhibits a slow relaxation of the magnetization at low temperatures. This manganese oxide compound is composed of a central $Mn(IV)_4O_4$ cube surrounded by a ring of eight Mn(III) units connected through bridging oxo ligands (Figure 21.1). In addition, it has 16 acetate and four water ligands (Yang, 2003). It was discovered in 2006 that the deliberate structural distortion of a Mn_6 compound via the use of a bulky salicylaldoxime derivative switches the intra-triangular magnetic exchange from antiferromagnetic to ferromagnetic, resulting in an $S = 12$ ground state (Milios et al., 2007a). Mn_{12}-acetate consists of ferro- or ferrimagnetically coupled magnetic ions, wherein the organic ligands isolate the molecules, characterized by a giant spin ($S = 10$) at low temperatures ($T < 10$ K), with a magnetic moment of up to $26 \mu_B$. The archetype of single-molecule

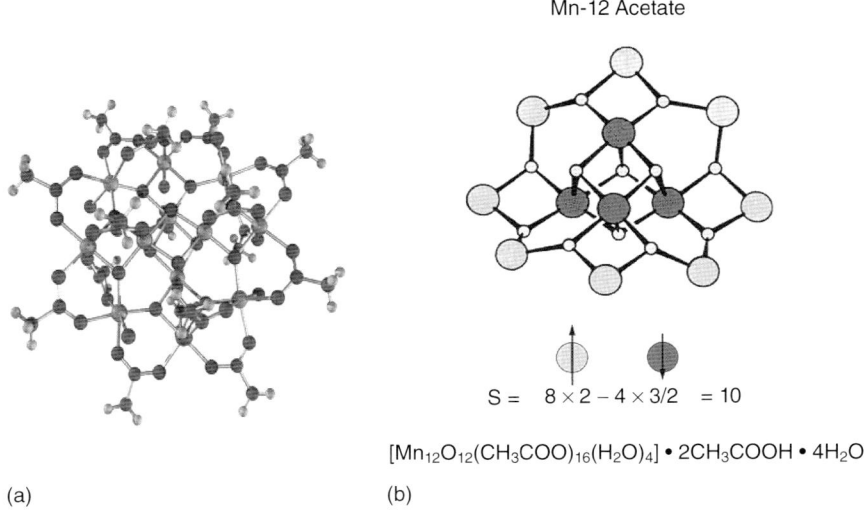

Figure 21.1 (a) Diagram showing the structure of Mn_{12}-acetate, consisting of ferro- or ferri-magnetically coupled magnetic ions. The organic ligands isolate the molecules; (b) An isolated Mn_{12}-acetate molecule, showing how the various ions couple to produce a resultant spin of $S = 10$.

magnets is called "Mn$_{12}$"; this is a polymetallic manganese (Mn) complex having the formula [Mn$_{12}$O$_{12}$(OAc)$_{16}$(H$_2$O)$_4$], as shown in Figure 21.1. It has the remarkable property of showing an extremely slow relaxation of its magnetization below the blocking temperature (Terazzi et al., 2008). [Mn$_{12}$O$_{12}$(OAc)$_{16}$(H$_2$O)$_4$]·4H$_2$O·2AcOH, which is called "Mn$_{12}$-acetate" (hereafter Mn$_{12}$-ac), is a common form of this used in research. Mn$_{12}$-acetate, which has a spin $S = 10$, characterized by $D \sim -0.5\,\text{cm}^{-1}$, is a very well-studied molecular magnet. It is a disc-shaped organic molecule in which 12 Mn ions are embedded; eight of these ions form a ring, each having a charge of +3 and a spin $S = 2$. The other four ions form a tetrahedron, each having a charge of +4 and a spin $S = 3/2$. The exchange interactions within the molecule are such that the spins of the ring align themselves in opposition to the spins of the tetrahedron, giving the molecule a total net spin $S = 10$.

Mn(III)$_6$: A record magnetization was reported in 2007 (Milios et al., 2007b) for the compound [Mn(III)$_6$O$_2$(sao)$_6$(O$_2$CPh)$_2$(EtOH)$_4$], related to MnAc$_{12}$, with $S = 12$, $D = -0.43\,\text{cm}^{-1}$, and hence $U = 62\,\text{cm}^{-1}$ or 86 K, characterized by a blocking temperature of 4.3 K. This was accomplished by replacing acetate ligands by the bulkier salicylaldoxime, thus distorting the manganese ligand sphere. It is prepared by mixing the perchlorate of manganese, the sodium salt of benzoic acid, a salicylaldoxime derivate and tetramethylammonium hydroxide in water and collecting the filtrate.

The SMM "Mn$_{12}$": [Mn$_{12}$O$_{12}$(O$_2$CR)$_{16}$(H$_2$O)$_4$], with liquid-crystalline phases was synthesized at the Institut de Physique et Chimie des matériaux de Strasbourg (IPCMS) in 2008 (Terazzi et al., 2008). The approach here consists of functionalizing the Mn$_{12}$ by mesogenic ligands in order to favor the self-organization into liquid-crystalline phases. Depending on the characteristics of the mesogenic ligands (size and position of the substituting groups), clusters possessing fluid mesophases with three-dimensional (3-D; cubic) or one-dimensional (1-D; smectic) positional order have been obtained. Although short-ranged, the intralayer ordering of the molecular cores in the smectic phase is square-like. The magnetic properties are preserved upon functionalization, and the molecules are thermally stable up to 150 °C. Another very interesting result is the fact that the main molecular axes of the magnetic cores share a common alignment direction, in a nematic sense. It seems reasonable to expect that these remarkable features may facilitate the two-dimensional (2-D) ordering of these clusters on surfaces, which is the ultimate goal. In order to tailor more precisely the mesomorphic behavior of these SMMs, the modification of the ligand structure, as well as the regioselective substitution of two different ligands (at the axial and equatorial positions), are currently in progress at IPCMS.

Cubic Ni$_4$-complexes: An example is [Ni(hmp)(t-BuEtOH)Cl]$_4$, the structure of which consists of four O^s in the hmp ligands bridging four $S = 1$ Ni ions to form a cube (hmp$^-$ is the monoanion of 2-hydroxymethylpyridine); the Ni ions couple ferromagnetically to give the total spin $S = 4$. The geometry of the molecular

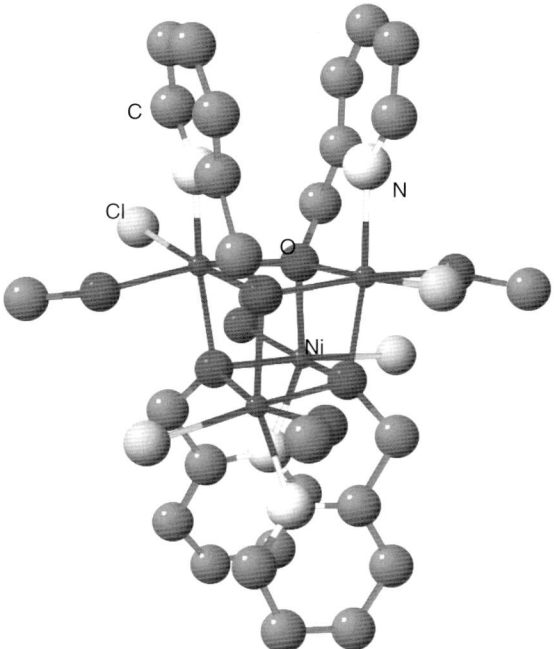

Figure 21.2 Geometry of the molecular magnet [Ni(hmp)(MeOH)Cl]$_4$. For simplicity, the hydrogen atoms are not shown.

magnet [Ni(hmp)(MeOH)Cl]$_4$ is shown in Figure 21.2; for simplicity, the hydrogen atoms are not shown.

Mn$_4$ Complexes: (a) [Mn$_4$O$_3$X]R, are characterized by a significant $S = 9/2$ value in the ground state, due to coupling between the three Mn^{3+} and one Mn^{4+} ions (Wang et al., 1991), characterized by an easy-axis, Ising-type, anisotropy, as shown in Figure 21.3. The advantages of these are: (i) variation in the core X group, which bridges the three Mn^{3+} ions, with constant ligand groups (X = MeCO$_2^-$, Cl$^-$, Br$^-$, N$_3^-$, NCO$^-$, OH$^-$, MeO$^-$); (ii) variation in the ligand R groups, for a constant Mn$_4$O$_3$X core; (iii) solubility in organic solvents; and (iv) sub-nanoscale dimensions (core volume ~ 0.01 nm^3). (b) The other three of these Mn$_4$ SMMs are (Yang et al., 2003a): [Mn$_4$(hmp)$_6$(NO$_3$)$_2$(MeCN)$_2$](ClO$_4$)$_2$·2MeCN, [Mn$_4$(hmp)$_6$(NO$_3$)$_4$]·(MeCN), and [Mn$_4$(hmp)$_4$(acac)$_2$(MeO)$_2$](ClO$_4$)$_2$·2MeOH. In each of these Mn$_4$ complexes, there is a planar diamond core of MnIII 2MnII ions. (hmp$^-$ is the monoanion of 2-hydroxymethylpyridine.) As revealed by an analysis of the variable-temperature and variable-field magnetization data, all three molecules have intramolecular ferromagnetic coupling and an $S = 9$ ground state. The observation that the alternating-current susceptibility signal is frequency-dependent indicates a significant energy barrier between the spin-up and spin-down states for each of these three MnIII 2MnII

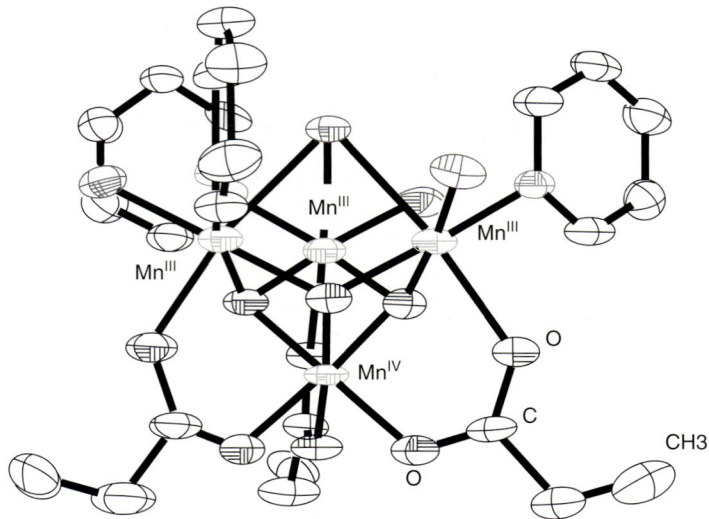

Figure 21.3 Structure of $Mn_4O_3X(OSiMe_3)(O_2CMe_3)(dbm)_3$, a distorted cubane, where Mn^{3+} ($S = 2$) and Mn^{4+} ($S = 3/2$) ions couple ferrimagnetically to produce a total ground-state spin of $S = 9/2$.

ions (Yang et al., 2003a). The magnetization orientation energy is $\Delta E = (S^2 - 1/4)|D| = 20|D|$, where the value of the ZFS parameter (D) ranges from −0.65 to −0.75 K.

Fe$_8$, Iron-clusters-based SMMs (Gatteschi, Sessoli, and Cornia, 2000): These iron clusters potentially have large spin states. (In addition, the biomolecule ferritin is considered a nanomagnet.) In the cluster Fe_8Br, Fe_8Br $[Fe_8O_2(OH)_{12}(tacn)_6]$ $Br_8 \cdot 9H_2O$, the cation Fe_8 stands for $[Fe_8O_2(OH)_{12}(tacn)_6]^{8+}$, with tacn representing 1,4,7-triazacyclononane; this was originally reported by Wieghardt et al. (1984) (see Figure 21.4). The eight Fe^{3+} ions, each with spin 5/2, are antiferromagnetically coupled to give an $S = 10$ ground state. The value of $D = -0.21$ cm^{-1}, with $E/D = 0.2$, was determined by polycrystalline powder EPR spectra. In accordance with Mn_{12}-ac, the negative D parameter results in an Ising-type magnetic anisotropy, leading to an anisotropy barrier, which slows down the relaxation of the magnetization. However, in contrast to Mn_{12}-ac, the anisotropy barrier is smaller and the presence of a sizeable value of the non-axial parameter E provides a transverse field, which is responsible for enhancing the relaxation of magnetization to become faster (Sangrgorio et al., 1997).

Trinuclear heterobimetalllic compounds: More recently there have been reported examples of 3d-4f complexes that show SMM behavior (Chandrasekhar et al., 2007, 2008). These exploit a phosphorus-supported multisite coordination ligand to assemble trinuclear heterobimetalllic compounds $L_2Co_2Ln][X]\}$ and $\{[L_2Ni_2Ln][X]\}$; Ln = lanthanide, X = e.g. $[NO_3] \times 2CHCl_3$ ($LH_3 =$ (S)P[N(Me)

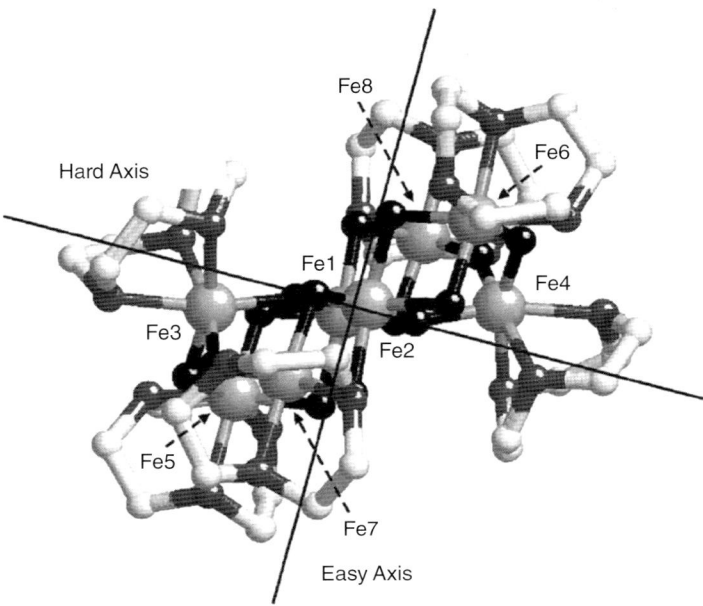

Figure 21.4 Diagram showing the molecular structure of Fe$_8$Br cluster. The orientations of the hard and easy magnetic anisotropy axes as determined by single crystal HF-EPR spectra are indicated. The iron atoms are represented by large gray balls. Oxygen, carbon and nitrogen atoms are represented by small black, white and gray spheres, respectively. Adapted from Barra, Gatteschi, and Sessoli (2000).

N=CH–C$_6$H$_3$-2-OH-3-OMe]$_3$. The molecular structures of this family of compounds reveal that they are isostructural, where all three metal ions are arranged in a perfectly linear manner, and are held together by two trianionic ligands, L^{3+}. The two transition metal ions are present in the terminal positions, and are bridged to the central lanthanide ion by phenolate oxygen ligands. Some members of these new 3d-4f assemblies, for example, linear trinuclear mixed-metal Co(II)–Gd(III)–Co(II) single-molecule magnet: [L$_2$Co$_2$Gd][NO$_3$] × 2CHCl$_3$ (LH$_3$ = (S)P[N(Me)N=CH–C$_6$H$_3$-2-OH-3-OMe]$_3$, show SMM behavior at low temperatures.

21.1.3
Applications

SMMs are potential candidates for technological applications that require highly controlled thin films and patterns (Cavallini et al., 2008). As of 2008, there were many discovered types and potential uses. Research on the ability of a single molecule to behave like a tiny magnet has grown rapidly over the past few years such that, today, SMMs represent the smallest possible magnetic devices and represent a controllable, bottom-up approach to nanoscale magnetism. Potential applications of SMMs include quantum computing, high-density information storage, magnetic refrigeration (Milios, Piligkos, and Brechin, 2008), and biomedical applications, such as magnetic resonance imaging (MRI) contrast agents.

SMMs can be used for information storage at the molecular level, and also as *qubits* in quantum computing. A SMM is an example of a macroscopic quantum system. If spin flips can be detected in a single atom or molecule, then the spin can be used to store information, thus enabling an increase in the storage capacity of computer hard disks. To this end, a good starting point is to identify a molecule with a spin of several Bohr magnetons. Due to the typically large, bistable spin anisotropy, SMMs have the potential to perhaps become the smallest practical unit for magnetic memory, and thus represent possible building blocks for a quantum computer. As a result, great efforts have been expended into the synthesis of additional SMMs, although the $Mn_{12}O_{12}$ complex and analogous complexes remain the canonical SMM, with a $50\,cm^{-1}$ spin anisotropy. In order for SMMs to be used for information storage, it is necessary that their properties are both well understood and better controlled, particularly the MQT. In this context, the $[Mn_{12}O_{12}(O_2CR)16(H_2O)_4]$ and $[Mn_4O_3Cl_4(O_2CR)_3(py)_3]$ (R = various, e.g. Et, py = pyridine) compounds (or simply Mn_{12} and Mn_4), with $S = 10$ and 9/2, respectively, are among the best currently understood. It has been discovered that the magnetic properties – including MQT – of these and other SMMs can be significantly altered in a controllable manner by using standard chemistry techniques. For example, this can be accomplished by varying the peripheral organic groups, which can affect the intermolecule separations and the local molecular symmetry within the crystal or, alternatively, by the addition of one or more extra electrons to each molecule. For Mn_4 SMMs with $S = 9/2$, their crystallization as $[Mn_4]_2$ supramolecular dimers within the crystal has been used as a possible modification. These have fascinating MQT properties due to the exchange interactions between the two halves of each dimer, thus establishing the feasibility of tuning the MQT in SMMs.

21.2
Multifrequency EPR of SMMs: Magnetic Hysteresis and MQT

Multifrequency (MF) EPR is a valuable tool for studying SMMs – especially as single-crystal samples, when they are monodisperse. This means that each molecule in the crystal has the same spin, orientation, magnetic anisotropy, and structure. As a consequence, fundamental studies of the properties intrinsic to the magnetic nanostructures of SMMs – which so far have been inaccessible – can be carried out. The following discussion is centered on MF-EPR studies with SMMs in an effort to understand their properties, and is based mainly on the reviews by Hill (2004) and by Gatteschi *et al.* (2006b).

21.2.1
The Effective Spin Hamiltonian

The effective spin Hamiltonian applicable to SMMs is (Christou et al., 2000; Gatteschi and Sessoli, 2003; Barra, Gatteschi, and Sessoli, 1997; Hill et al., 2002c, 2003a; Park et al., 2002b, 2002c; Mirebeau et al., 1999):

$$H = \mu_B B \cdot \tilde{g} \cdot S + D S_z^2 + H';$$
$$H' = B_2^2 O_2^2 + B_4^0 O_4^0 + B_4^2 O_4^2 + B_4^4 O_4^4 \quad (21.1)$$
$$+ \ldots + \text{intermolecular dipolar and exchange interactions}$$

(For a definition of spin operators O_l^m, see Chapter 7). In particular,

$$O_2^2 = \frac{1}{2}(S_+^2 + S_-^2); \quad O_4^0 = \left[35 S_z^4 - \{30 S(S+1) - 25\} S_z^2 + 3 S^2 (S+1)^2 - 6 S(S+1)\right];$$
$$O_4^4 = \frac{1}{2}(S_+^4 + S_-^4)$$

In Equation 21.1, D (< 0) is the zero-field uniaxial anisotropy constant, and S_z is the z-component of the spin operator S. The energy eigenstates can be labeled by the electron magnetic quantum number M_S ($-S \leq M_S \leq S$) for the strictly axial situation, obtained for $B \parallel z$ and by substituting $H' = 0$ in Equation 21.1. The eigenvalues of Equation 21.1 in zero field are then expressed as $\varepsilon(M_S) = -|D| M_S^2$. This results in an energy barrier, with separated doubly degenerate ($M_S = \pm i$; i = integer $\leq S$) "up" and "down" spin states, as shown in Figure 21.5, which also shows the energy level diagram as a function of the magnetic field strength, with B being parallel to the easy axis of the crystal, along with the EPR transitions which can be excited up to 5 T.

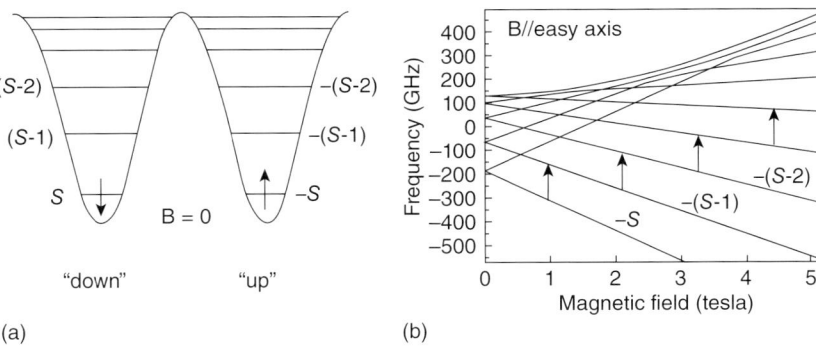

Figure 21.5 (a) Potential wells for "spin-up" and "spin-down" for a SMM in the absence of a magnetic field, as described by Equation 21.1; labeling of the magnetic sub-levels is in accordance with their M_S values ($M_S = \pm i$, i = integer $\leq S$); (b) Energy levels plotted as functions of magnetic field strength parallel to the easy axis. The allowed EPR transitions ($\Delta M_S = \pm 1$) that can be excited between the different M_S sub-levels are shown by arrows. Adapted from Hill (2004).

21.2.2
Magnetic Quantum-Mechanical Tunneling (MQT) and MF-EPR

There is exhibited a continuous slow magnetic relaxation by all SMMs, even when the temperature approaches 0 K. This relaxation is due to quantum mechanical tunneling, which is effected by the admixture of states of opposite spins brought about by the non-axial terms in H' in Equation 21.1 (Christou et al., 2000; Chudnovsky and Tajeda, 1998; Gatteschi and Sessoli, 2003). The tunneling rates depend on the "degree of symmetry breaking" due to H', determined rather precisely by single-crystal EPR (Hill et al., 1998a, 2002c, 2003a, 2003b; Park et al., 2002b, 2002c). This, however, requires EPR measurements to be made over a wide range of frequencies, as the molecular spin S is large, resulting in an extended $(2S+1)$-fold quantum energy level structure. In addition, the large ZFS due to the large value of magnetocrystalline anisotropy ($|D|$), associated with the large value of S necessitates the use of considerably higher frequencies, 50–500 GHz, than are commonly used (as seen from Figure 21.5b). This resulted in the development of high-field (HF) EPR spectrometers with MF capabilities (Edwards et al., 2003a, 2003b, 2003c; Hill, Brooks, and Dalal, 1999a; Hill, Dalal, and Brooks, 1999b; Hill et al., 1998a, 1998b, 2002a, 2003a, 2003c; Maccagnano et al., 2001; Mola et al., 2000; Park et al., 2002a, 2002b, 2002c; Yang et al., 2003b).

21.2.3
Zero-Field EPR with Variable Frequency

Zero-field EPR (ZFR) has been used to study SMMs (Dressel et al., 2001; Mukhin et al., 1998, 2001; Parks et al., 2001, 2002; Park et al., 2002a, 2002b, 2002c). In this case, it is possible to investigate large pressed-powder pellets in order to obtain an intense signal for the EPR lineshape analysis, as sample alignment is not required (see Chapter 4, Section 4.1 of this book for more details on ZFR). The use of ZFR leads to an over-parameterization of any data fitting. In particular, although the quantum energy-level spacings are found to be in agreement with those obtained with HF-EPR, the conclusions drawn from analyses of inhomogeneous linewidths of zero-field EPR transitions (Dressel et al., 2001; Mukhin et al., 1998, 2001; Parks et al., 2001, 2002; Park et al., 2002a, 2002b, 2002c) are quite different from those obtained from HF-EPR data analyses.

21.2.4
Low-Field (X-band) EPR

Low-frequency measurements also provide limited information on SMMs. X-band measurements on a molecular cluster nanomagnet (Blinc et al., 2001) enabled only the observation of EPR transitions due to thermally activated populations close to the top of the barrier, with no new information otherwise available from HF-EPR being obtained. Furthermore, the T_1 and T_2 relaxation times could not be determined from these data due to the broad inhomogeneous linewidths.

21.2.5
MF High-Frequency EPR

It appears that it is really MF high-frequency EPR, and the use of single-crystal samples, that is vital in obtaining in-depth information on SMMs.

21.2.5.1 EPR Spectrometers with MF Cavity (40–350 and Extended Range 18–350 GHz), and up to 650 GHz Without a Cavity

An important instrumental breakthrough was made at the National High Magnetic Field Laboratory, Tallahassee, Florida, USA (NHMFL) by the development of a sensitive cavity-perturbation device for EPR measurements at variable frequencies in the range of 40 to 350 GHz on tiny single crystals (~$1 \times 0.2 \times 0.2$ mm^3) (Hill, Brooks, and Dalal, 1999a; Hill, Dalal, and Brooks, 1999b; Hill *et al.*, 1998a, 1998b, 2002a, 2002b, 2003a, 2003b, 2003c; Maccagnano *et al.*, 2001; Mola *et al.*, 2000; Park *et al.*, 2002a, 2002b, 2002c). This technique enables a better than three orders-of-magnitude improvement in detection sensitivity, as compared to conventional single-pass techniques (Hill, 2004; Barra, Gatteschi, and Sessoli, 1997; Cornia *et al.*, 2002b; Krzystek *et al.*, 2001; Yoo *et al.*, 2000, 2001), thus opening up the possibility to probe highly anisotropic SMMs in many different ways. Improvements to the spectrometer developed at the NHMFL have resulted in a new facility at the University of Florida, enabling single-crystal studies to be conducted in a cavity in the frequency range 18 to 350 GHz (Hill *et al.*, 2002a). In addition, it is possible to measure single crystals up to 650 GHz – albeit at a somewhat reduced sensitivity – without cavities, in a broad-band quasi-optical spectrometer (Mola *et al.*, 2000).

21.2.5.2 Polycrystalline Powder EPR Spectrum

As an example, the powder spectrum of Cu_6 at 9 and 245 GHz is shown in Figure 21.6 (Rentschler *et al.*, 1996). Only three partially structured transitions are observed at 9 GHz, along with a transition close to zero field, implying that the ZFS of some levels is close to the microwave frequency. These transitions are in agreement with those observed at the high-frequency 245 GHz, which shows a regular fine structure of the EPR lines. The fact that the intensities of the low-field lines are increased relative to those at the high field, implies that the absolute sign of the ZFS parameter D is negative. Another example of a powder spectrum is shown in Figure 21.7 for Mn_{12}-ac at 525 GHz at 30 K (Gatteschi *et al.*, 2006a). By exploiting the tetragonal symmetry of the single crystal, the value of the ZFS parameter $D_4 = D + [30S(S+1) - 25]B_4^0$; $S = 10$ was determined to be -0.65 K. The assignments of the low-field transitions were: $M_S = -10 \rightarrow -9$, $-9 \rightarrow -8$, $-8 \rightarrow -7$, and $-7 \rightarrow -6$.

21.2.5.3 The Virtues of Single-Crystal Measurements

The single-crystal measurements on Mn_{12}-ac made by Hill *et al.* (1998a), when compared to those on a powder sample by Barra, Gatteschi, and Sessoli (1997), indicated several inadequacies of the simple giant-spin ($S = 10$) model as described by Equation 21.1. This was further confirmed by additional measurements. In the

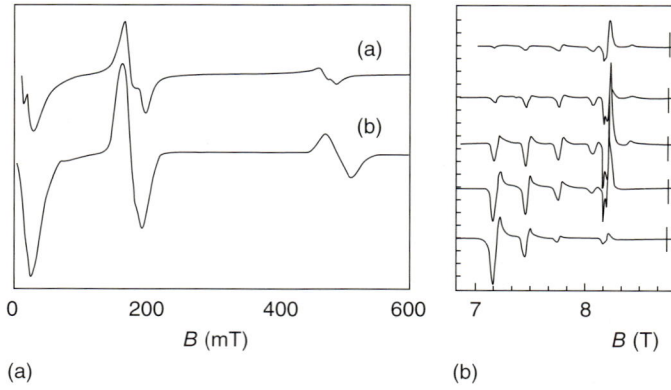

Figure 21.6 (a) Experimental (spectrum a) and simulated (spectrum b) X-band EPR spectra of Cu_6 at 4 K; (b) Temperature evolution of the experimental HF-EPR spectra (245 GHz) of Cu_6. The sharp signal at 8.8 T is the DPPH signal used as reference for $g = 2.00$. Adapted from Rentschler et al. (1996).

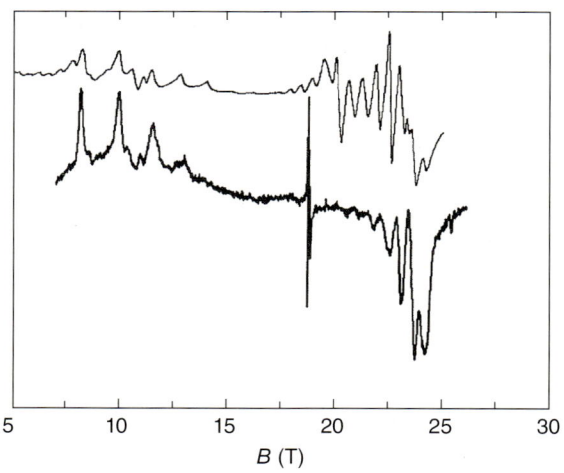

Figure 21.7 Experimental (lower) and simulated (upper) 525 GHz EPR powder spectra of Mn_{12}-ac at 30 K. Adapted from Gatteschi et al. (2006a).

past few years, single-crystal EPR studies conducted at the University of Florida have focused, among others, on a number of Mn-, Fe-, and Ni-based SMMs of varying spin, anisotropy, and crystal structures, covering a broad range of frequencies, fields, orientations, and temperatures. These extensive measurements have provided considerable new information on the physics of SMMs, which have huge implications on the MQT phenomenon. In particular, in essentially all SMMs, it has been found that there are: (i) significant crystal-field distributions (D, E, and g strains); (ii) significant intermolecular exchange interactions; and (iii) a breakdown of the giant-spin model, as described by Equation 21.1, for the most widely studied Mn_{12}-ac and Fe_8Br SMMs. These are described in detail below.

21.2.5.4 A Typical SMM Spectrum

Figure 21.8 shows a part of the spectrum for Fe_8 at 89.035 GHz at 10 K for the orientation of the crystal at ~8°, with the easy axis recorded by using the cavity perturbation technique (Takahashi *et al.*, 2004), displaying typical features of the EPR spectrum for a SMM. The lines fit best to a Gaussian lineshape. The broadening of the lines with increasing $|M_S|$ is due to a D-strain effect, which scales almost linearly with $|M_S|$, as discussed in the next section. (The slight decrease of the linewidth with increasing frequency, not shown here, is due to decrease of the inter-cluster dipolar broadening.)

21.2.5.5 EPR Linewidth Measurements: Effect of D-Strain, g-Strain, Dipolar and Exchange Interactions

EPR linewidths for Fe_8Br, measured for various frequencies for the field parallel to the easy axis of magnetization, versus M_S, the level from which the transition was excited, are shown in Figure 21.9. The D-strain is responsible for the pronounced – almost linear – increase in linewidth as a function of M_S. This is because the energy difference of levels, or the EPR transition frequency, is proportional to M_S, since the energy of a level is proportional to M_S^2, which translates

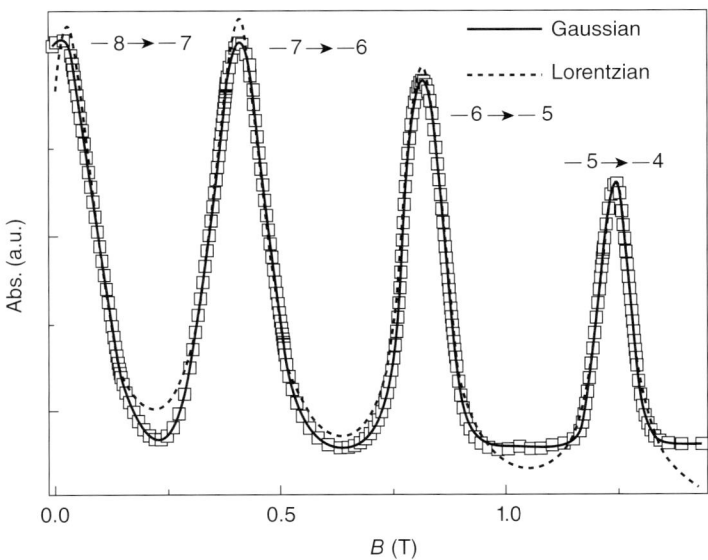

Figure 21.8 Single-crystal EPR spectrum ($v = 89.035$ GHz) of Fe_8 in absorption mode at 10 K and magnetic field oriented at ca. 8° from the easy axis. The continuous and dashed lines represent the best fit obtained by using Gaussian and Lorentzian line shape, respectively. Adapted from Takahashi *et al.* (2004).

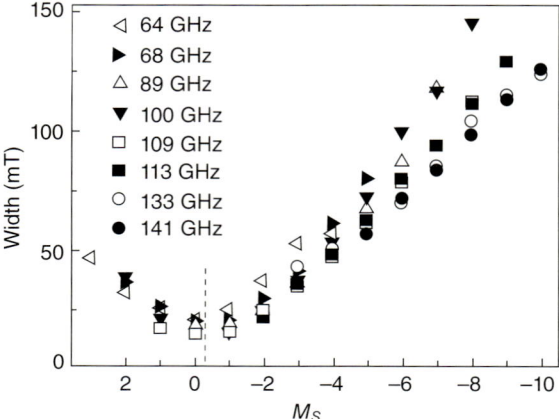

Figure 21.9 MF-EPR linewidth data at 10 K for the SMM Fe$_8$Br for the magnetic-field orientation along the easy axis as a function of the M_S value from which the transition was excited. The various frequencies are listed in the figure. Adapted from Hill (2004).

into the linewidth that depends linearly on M_S. The bottoming in linewidth close to $M_S = 0$ in Figure 21.9 arises due to a convolution of the intrinsic lifetime broadening and the M_S-dependent contribution. The observed slight narrowing in linewidth as a function of frequency, and the weak symmetry about $M_S = -1/2$ in Figure 21.9 is due to the fact that the higher frequency (lower $|M_S|$) transitions occur at higher magnetic fields, where the inter-molecular dipolar broadening is weaker. No measureable g-strain is observed for Fe$_8$Br. As for Mn$_{12}$-ac, qualitatively similar D-strain effects are observed, together with the effect of a slight g-strain, which produces the opposing dependence on M_S as compared to that due to the D-strain, since higher M_S transitions are observed at lower fields. A theoretical treatment of these effects and a summary of main findings for Mn$_{12}$-ac and Fe$_8$Br is given by Hill *et al.* (2002b, 2002c), and Park *et al.* (2002b, 2002c). Separating the comparable contributions due to, for example, D-strain, g-strain, dipolar and exchange interactions, becomes possible thanks to the MF high-frequency capabilities, in addition to the ability to measure over a wide field and temperature range. The origins of the D-strain observed in Mn$_{12}$-ac and Fe$_8$Br are to be found in studies on a series of cubic Ni$_4$ complexes, [Ni(hmp)(ROH)Cl]$_4$, where R is CH$_3$ (complex 1), CH$_2$CH$_3$ (complex 2), or CH$_2$CH$_2$ C(CH$_2$)$_2$ (complex 3), and hmp$^-$ is the monoanion of 2-hydroxymethylpyridine (Edwards *et al.*, 2003c; Yang *et al.*, 2003b). The EPR spectra at ~190 GHz at 10 K for each of these three complexes are shown in Figure 21.10, for the sweep of the magnetic field over the 0–5 T range, and for the orientation of the field roughly aligned with the easy axis. Three sets of more-or-less evenly spaced double peaks are observed in each case, with the intensity of the highest-field doublet being less than those at lower fields. Each pair of split peaks is attributed to one of the transitions indicated by arrows

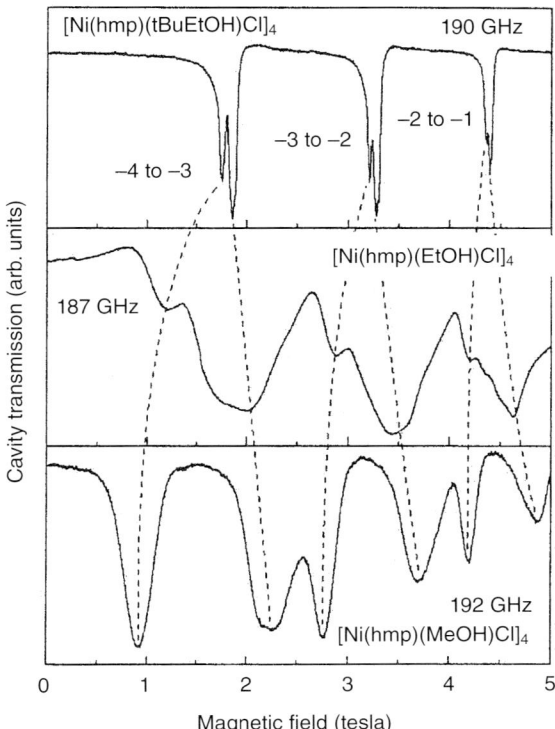

Figure 21.10 Transmission-mode EPR spectra for three Ni_4 SMMs for the field orientation nearly parallel to the easy axis at 10 K at frequencies (~190 GHz), as indicated in the figure. The resonances are labeled in the top panel according to the quantum numbers M_S involved in the transitions. The splittings of the various peaks increase for the complex in the top panel to the complex in the bottom panel, with their trends of increase indicated by dashed lines. Adapted from Hill (2004).

in Figure 21.5b. The reduction in the intensity of the peaks with increasing magnetic field is due to a Boltzmann distribution of populations which favors the levels at lower fields. The splitting of each peak into a doublet is, presumably, due to the existence of two species of molecule within the crystal, characterized by slightly different sets of crystal-field parameters. The spectra for complexes 1 and 2 show considerably broader absorptions and dramatically enhanced splittings, as compared to that for complex 3. Figure 21.11 shows separate fits to the $S = 4$ spin Hamiltonian, as given by Equation 21.1, for the lower (L) and upper (U) split peaks for complex 3. (Twinning and sample misalignment as causes for splitting were ruled out by careful angle-dependent studies.) The assignments of the peaks were made by their temperature dependences, and by fits similar to those shown in Figure 21.11. The evolution of these peaks for the three complexes on moving from complex 3 to 2 to 1 is shown by dashed lines in Figure 21.10, while

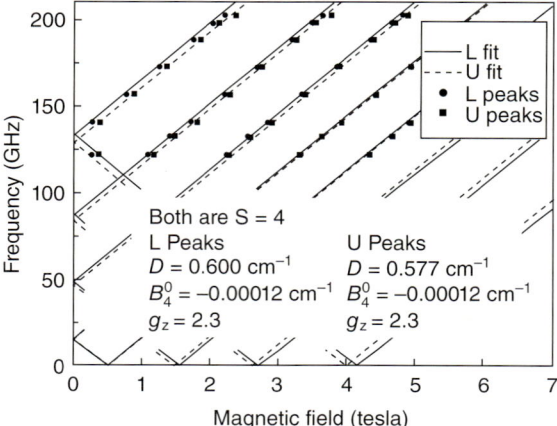

Figure 21.11 Individual fits to Equation 21.1 for the lower (L) and upper (U) peaks within each split doublet for Ni_4 complex #3: $[Ni(hmp)(ROH)Cl]_4$, where $R = CH_2CH_2C(CH_3)_2$. These fits confirm that the ground state is $S = 4$, and provide values for the diagonal components of the effective spin Hamiltonian. Similar fits (not shown here) were obtained for complexes #1 and #2. Adapted from Hill (2004).

assignments of the various transitions in terms of the M_S quantum numbers are given in the upper part of Figure 21.10. The fits to parameters are excellent, yielding the following values of the parameters: Complex 1: $D_L = -0.715\,cm^{-1}$, $D_U = -0.499\,cm^{-1}$, $B_4^0 = -2 \times 10^{-4}\,cm^{-1}$, and $g_z = 2.24$; Complex 2: $D_L = -0.673\,cm^{-1}$, $D_U = -0.609\,cm^{-1}$, $B_4^0 = -1.2 \times 10^{-4}\,cm^{-1}$, and $g_z = 2.20$. The presence of two crystallographically inequivalent Ni_4 species is confirmed by X-ray structures, so that the different D-values arise due to slightly different ligand environments, and may also be due to the different inter-SMM exchange pathways for these two species. In the case of complex 3, for which all the Ni_4 SMMs are crystallographically identical, the picture is not as clear. However, the slight splitting seen in the spectra for complex 3, which may explain the two apparent D-values, may be due to a weak disorder associated with the ligands.

21.2.5.6 Study of Intermolecular Exchange Interactions and Dipolar Interactions

The temperature-dependent shifts in the EPR line positions for Fe_8Br for easy-axis alignment of the magnetic field are shown in the upper part of Figure 21.12. It is seen that below 20 K, there is a clear cross-over in the temperature dependence for all the transitions from levels with $|M_S| < 10$, which may be due to a polarization of the spin system. In addition, there is observed opposite temperature dependence for the $M_S = -10$ to -9 transition, as seen in Figure 21.12, which cannot be explained on the basis of dipolar spin–spin interaction alone. In fact, this opposite trend is considered due to competition between the local intermolecular exchange interactions and longer range dipolar interactions (Park et al., 2002c). As shown in the lower part of Figure 21.12, indeed, a quantitative

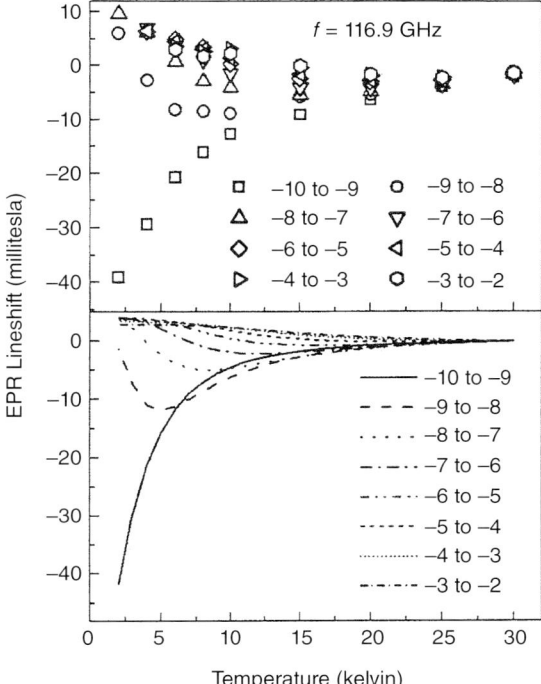

Figure 21.12 Upper panel: Temperature-dependence of the EPR line positions at 116.931 GHz for the SMM Fe$_8$Br for the magnetic-field orientation along the easy axis, observed for the various spin transitions indicated in the figure. The line positions [B(T)] have been normalized to their positions at 30 K [B(30 K)]; Lower panel: A plot of calculated line shifts versus temperature for Fe$_8$Br at a frequency of 116.9 GHz. For details, see Park et al. (2002c). Adapted from Hill (2004).

agreement is obtained based on these considerations. Of the many competing interactions considered, it was only the combination of a weak effective intermolecular ferromagnetic exchange interaction ($J = -7$ Gauss) and a longer-range antiferromagnetic dipolar coupling (~20 Gauss) which accounted for the line shifts (Park et al., 2002c). Similar intermolecular exchange interactions have been observed in various other SMMs, pointing to a host of novel low-temperature quantum phenomena, such as the exchange biasing of hysteresis loops (Edwards et al., 2003a; Hill et al., 2002a; Wernsdorfer et al., 2002).

21.2.5.7 EPR Spectra for Mn$_4$ Family

This type of cluster is not only interesting as a SMM, but also because clusters of this type were first synthesized to model the manganese cluster present in the water-oxidizing complex of Photosystem II (Bashkin et al., 1987). Sharp signals were obtained at 327.9 GHz by Aubin et al. (1998) from oriented samples of [Mn$_4$O$_3$Cl(O$_2$CMe)$_3$(dbm)$_3$], where Hdbm is dibenzoylmethane. An analysis of the

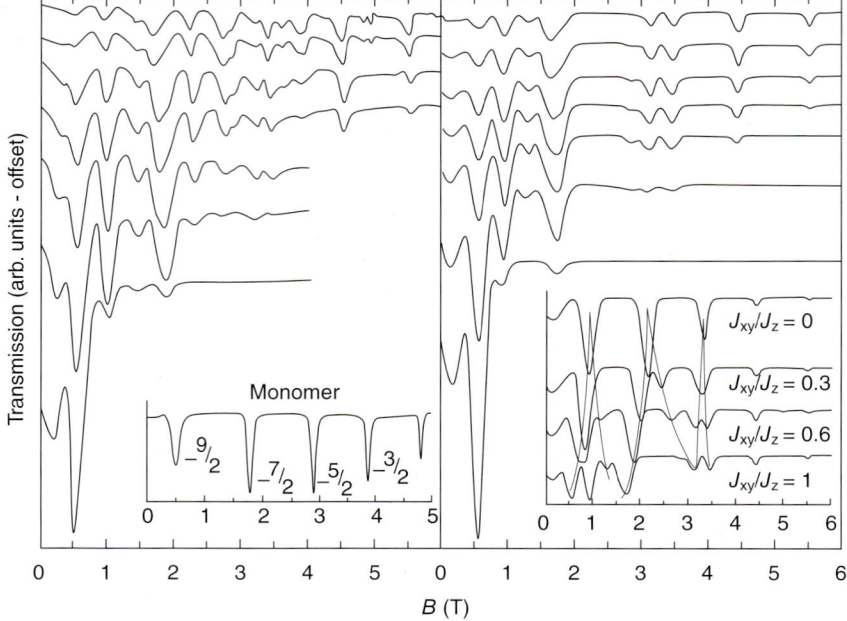

Figure 21.13 Left panel: Transmission EPR spectra (145 GHz) obtained for [Mn$_4$O$_3$Cl$_4$(O$_2$CEt)$_3$(py)$_3$]$_2$, [Mn$_4$]$_2$ with static magnetic field applied along the easy axis at variable temperatures (from bottom to top: 2, 4, 6, 8, 10, 15, and 18 K). The inset shows a single 6 K, 140 GHz spectrum obtained for a monomeric Mn$_4$ complex; Right panel: Simulations of the dimer data obtained by using $D/k = -0.750$ K, $B_4^0/k = -5 \times 10^{-5}$ K, $g_z = 2$, and $J_{zz}/k = J_{xy}/k = J/k = 0.12$ K. The inset illustrates the effect of the transverse part of the exchange (J_{xy}) for four values of J_{xy}/J_z ($T = 8$ K). Adapted from Hill et al. (2003d).

data yielded the crystal-field parameters $D/k_B = -0.76$ K, $B_4^0/k_B = -1.1 \times 10^{-1}$ K, which were confirmed subsequently by INS data (Andres et al., 2000). EPR spectra for [Mn$_4$O$_3$Cl$_4$(O$_2$CEt)$_3$(py)$_3$] at variable frequencies (136–144 GHz) and at various temperatures (Edwards et al., 2003a), and a simulation at 145 GHz (Hill et al., 2003d) are shown in Figure 21.13. These members of the Mn$_4$ family show the presence of significant intermolecular interactions, wherein Mn$_4$ clusters are connected pairwise through C—H···Cl and Cl···Cl interactions (Wernsdorfer et al., 2002). The ground state of the individual clusters is typical $S = 9/2$. In the Hamiltonian, the interaction between them is accounted for by the term:

$$H = J_z S_{1z} \cdot S_{2z} + J_{xy}(S_{1x} \cdot S_{2x} + S_{1y} \cdot S_{2y}), \qquad (21.2)$$

where the subscripts 1 and 2 refer to the interacting clusters. The states of the dimmer are described by the kets $|M_{S1}\ M_{S2}\rangle$. Consideration of the term given by Equation 21.2 results in the splitting of the resonance lines due to the allowed transitions: $|M_{S1}\ M_{S2}\rangle \to |M_{S1}+1\ M_{S2}\rangle$ and $|M_{S1}\ M_{S2}\rangle \to |M_{S1}\ M_{S2}+1\rangle$. These splittings can be measured from the spectra shown in Figure 21.13. An analysis of the HF-EPR spectra revealed that the decoherence rates of the coupled states, or the electronic transverse relaxation rates T_2, are considerably smaller than the characteristic quantum splitting (=Δ/h, where Δ is the energy splitting and h is Planck's constant) as induced by the exchange couplings between the dimmers.

21.2.6
Effect of Molecular Site Symmetry on Tunneling Phenomenon (MQT) as Revealed by EPR

The studies described above show that slight ligand variations and/or disorder can lead to profound changes in the SMM environmental couplings which, in turn, can lead to dramatic changes in the quantum dynamics of SMMs (Edwards et al., 2003a, 2003c; Yang et al., 2003b).This is because the tunneling phenomenon is very sensitive to the molecular site symmetry. In particular, angle-dependent EPR studies shed light on this effect, as shown in Figures 21.14 and 21.15 for Mn_{12}-ac (Hill et al., 2003a, 2003c). The upper part of Figure 21.14 shows two transmission-mode EPR spectra obtained for the field orientation along two different directions ($\phi = -15°$ and $30°$), measured with respect to one of the edges of the approximately square cross-section of the sample in the hard magnetic plane of the sample, as illustrated in the inset in the lower part of the figure. In the figure, α_m resonances correspond to transitions within tunnel-split $M_S = \pm m$ zero-field doublets, where it is the applied field that contributes mainly to this tunnel splitting. It is clear from the figure that the spectra change dramatically with the orientation of the magnetic field, ϕ, for example, there are shoulders on the HF sides of the $\phi = -15°$ peaks, which are absent in the $\phi = 30°$ spectrum. This is exhibited by the appreciable shift in the peak positions with ϕ, as illustrated by the contour plots in the lower part of Figure 21.14, which shows a pronounced fourfold pattern in the central positions in the field of the main EPR absorptions. The fourfold pattern arises due to the fourth-order transverse zero-field interaction (O_4^4) in the Hamiltonian H' in Equation 21.1. Fits to spectra yield the value of the parameter $B_4^4 = 3.2 \pm 0.1 \times 10^{-5}$ cm^{-1}, which is in excellent agreement with that determined by neutron scattering (Mirebeau et al., 1999). It is noted that, unlike neutron-diffraction and powder-EPR studies, the B_4^4 value obtained here is completely independent of all other parameters in Equation 21.1, and thus represents the most direct confirmation for the existence of this contribution to the transverse anisotropy in Mn_{12}-ac. The angular (ϕ) dependence of the peak positions for the hard-plane spectra (Figure 21.14) is shown in Figure 21.15 for the first four EPR peaks (α_6–α_9) in Figure 21.14. In Figure 21.15a are plotted the full width at half maximum (FWHM) for each of the resonances as determined graphically, whereas in Figure 21.15b are plotted the splitting of the individual peaks as obtained by fitting two Gaussians to each peak. It is seen that both sets of plots in Figure 21.15 exhibit the same fourfold angular dependence. However, the origin of this effect is due to a disorder-induced twofold rhombic distortion, in accordance with that proposed on the basis of X-ray measurements by Cornia et al. (2002b). The rhombic

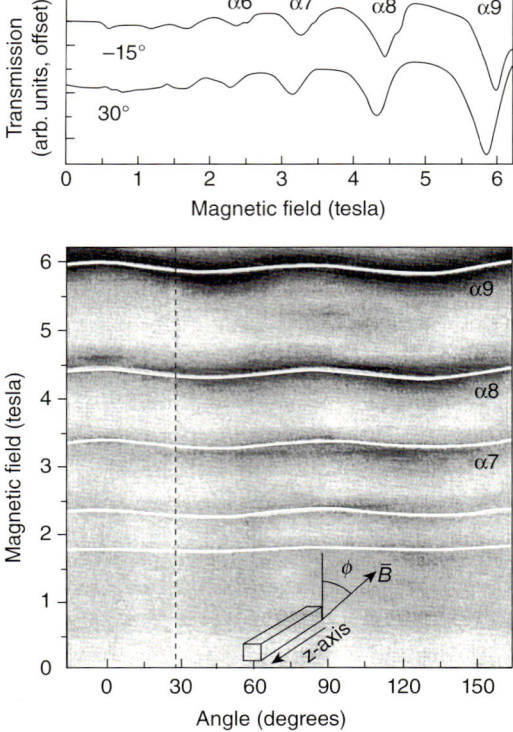

Figure 21.14 Upper panel: EPR spectra for Mn$_{12}$-ac at 49.86 GHz at 15 K. The field orientations, as indicated in the figure, are along two different directions in the hard magnetic plane of the sample. The angle ϕ is measured relative to one of the square edges of the sample. The resonances are labeled in accordance with those described by Blinc et al. (2001); Lower panel: Contour plot showing the angular variation of the peak positions in the upper panel. The data were recorded at 15° intervals (greater darkness indicates stronger absorption). The dashed line corresponds to the 30° curve in the upper panel, while the y-axis corresponds to the 15° curve. The geometry of the experiment is depicted in the inset. Adapted from Hill (2004).

term (O_2^2) in the Hamiltonian H' produces a twofold pattern for field rotations in the hard plane. The angular dependence in Figure 21.15 is in excellent accord with Cornia's solvent disorder picture (Cornia et al., 2002b). In addition, the magnitude of the zero-field interaction rhombic term (O_2^2) appears to be in excellent agreement with that determined by the low-temperature magnetic measurements as noted by Hill (2004). These experiments, therefore, suggest a direct connection between the symmetry breaking caused by the solvent disorder, and the previously unexplained MQT in Mn$_{12}$-ac, the first system to exhibit this behavior (Christou et al., 2000; Gatteschi and Sessoli, 2003). As a matter of fact, it has been explicitly discussed in the theoretical reports by Chudnovsky and Garanin (2001) and by Garanin and Chudnovsky (2002), that many previously unexplained aspects of the

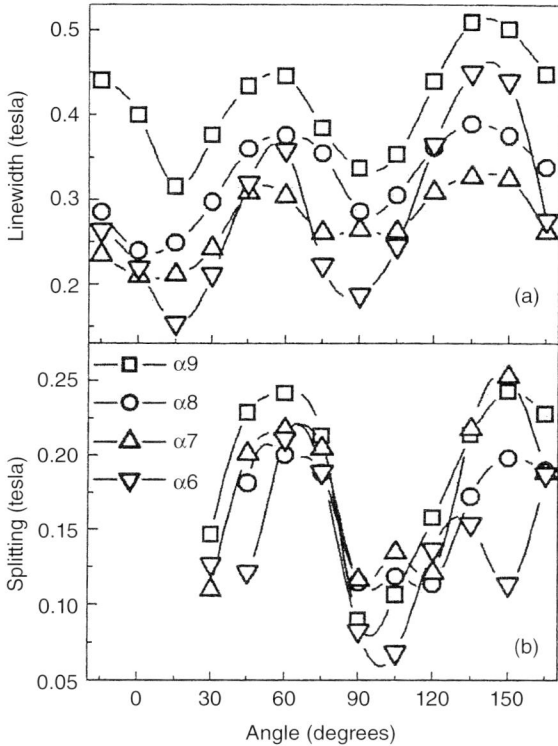

Figure 21.15 Angular variations of (a) the linewidths, and (b) the line splitting for the four strongest EPR peaks in Figure 21.14 (α_6–α_9). Adapted from Hill (2004).

MQT phenomenon are naturally explained when disorder is taken into account. It should be emphasized that this disorder is discrete in nature – that is, it is associated with the multiple, yet finite, number of configurations involving ligand and solvent coordination to the SMMs, which can be controlled chemically. In particular, this disorder is less serious than the disorder associated with, for example, crystal imperfections and dislocations. Further study of the hard-axis EPR data, establishing the virtues of single-crystal EPR, involves attempts to fit hard-axis data points showing the line positions versus the field, as shown in Figure 21.16, obtained for Mn_{12}-ac over a wide frequency range (48–180 GHz). The open circles in Figure 21.16 fit perfectly with the simulated line positions for spin $S = 10$ for axial symmetry using the widely accepted SH parameters for Mn_{12}-ac, as listed in Figure 21.16. The α-resonances are the same as those shown in Figure 21.14, whereas the black squares correspond to the β-resonances originally reported by Hill et al. (1998a). These latter data points lie on or close to the $S = 10$ curve for frequencies above 100 GHz. However, below 100 GHz, there is found a dramatic departure from the expected $S = 10$ model, such that even a forced fit is not

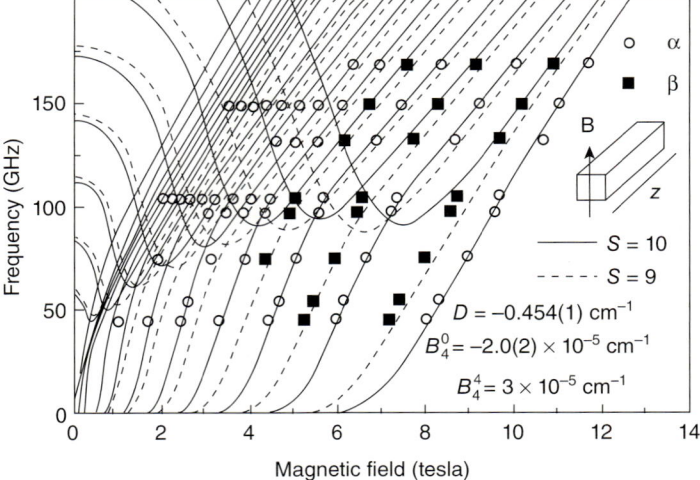

Figure 21.16 Fits of the frequency dependence of hard-axis spectra obtained for Mn$_{12}$-ac. The α-resonances (open circles) fit the $S = 10$ model (solid lines) exceptionally well. On the other hand, the β-resonances (solid squares) only fit the $S = 10$ picture at higher frequencies. The $S = 10$ fits cannot be forced through the β-resonance data points observed below 95 GHz. It is, however, found that these low-frequency points lie quite close to the behavior expected for $S = 9$ (dashed curves). The orientation of the applied field relative to the sample is depicted in the inset. The obtained quadratic and quartic SH parameters are listed in the figure. Adapted from Hill (2004).

feasible. It is seen from Figure 21.16, that the solid curves which fit the high-frequency β-resonances each go through a minimum, located at about 95 GHz for the highest field resonances. The minimum cut-off frequency depends only on the spin S and the ZFS parameter D to a fairly good approximation. As these values are so widely accepted for Mn$_{12}$-ac, one is led to the unequivocal conclusion that the low-frequency behavior of the β-resonances is certainly beyond the predictions of the $S = 10$ single-spin behavior. Furthermore, this low-frequency behavior of β-resonances agrees very well with an $S = 9$ model, shown by the dashed curves in Figure 21.16, and found using the same SH parameters as for the $S = 10$ case. As for the easy-axis spectra, the parameter values would be required to be very similar in order to explain any evidence for an $S = 9$ state or, in other words, in this case the coexistence of $S = 9$ and $S = 10$ states is difficult to distinguish with the same SH parameters. The $S = 9$ state lies only 10–15 k_B above the ground state of the $S = 10$ multiplet, as deduced from the temperature-dependence of the low-frequency $S = 9$ peaks. This is considerably lower than was previously believed. The existence of such a low-lying state is supported, for example, by neutron-scattering data (Hennion *et al.*, 1997), though it is not clear at present what effect this has on the dynamics and MQT in Mn$_{12}$-ac.

21.3
Magnetic Quantum Tunneling (MQT): Pure and Thermally Assisted Tunneling

SMMs provide useful systems for the study of quantum mechanics. MQT was first observed in $Mn_{12}O_{12}$, characterized by evenly spaced steps in the hysteresis curve. The periodic quenching of this tunneling rate in the compound Fe_8 has been observed and explained with geometric phases. The effective Hamiltonian to describe MQT is:

$$H \cong DS_z^2 + E\left(S_x^2 - S_y^2\right) + \hat{O}(4) + H'.$$

This Hamiltonian exhibits transverse anisotropy, as it contain the terms $E\left(S_x^2 - S_y^2\right) + \hat{O}(4) + H'$, that do not commute with $\hat{S}_z$. They mix states in the two wells, leading to quantum tunneling. Pure tunneling occurs between states with opposite values of M_S; that is, between $+M_S$ and $-M_S$. In thermally assisted tunneling a transition occurs first to a higher-lying state $+M_S$, from the lowest-lying $+M_S$ state, The lowest-lying $+M_S$ state then undergoes a transition by pure tunneling to the $-M_S$ state, which then makes a transition to the lowest-lying $-M_S$ state. Other sources of transverse anisotropy are higher-order crystal-field terms in $\hat{O}(4)$, disorder, strain, inter-molecular exchange, and dipolar interactions, as well as hyperfine fields in H'.

21.3.1
Relaxation of Magnetization for SMMs

The relaxation of magnetization is typically very slow for SMMs, and is almost independent of temperature at low temperatures. Furthermore, highly axially symmetric axial systems, such as Mn_{12}-ac, exhibit the slowest relaxation. As for the relaxation time (τ), in some systems it may exceed the time of the measurement—that is, days, weeks, or years—as shown in Figure 21.17. As for the mechanism of the anomalously slow relaxation of Mn_{12}-ac clusters, the following is noted. Normally, it is explained on the basis of Figure 21.18. Using the analogy with superparamagnets (Morrish, 1966), the energies of the split state of the ground state with $S = 10$ are plotted in a double well, wherein the energies of the negative M_S states are shown in the left well, and those of the positive M_S states in the right well. The discreteness of the levels indicates the quantum nature of the system. Only the $M_S = \pm 10$ level will be populated at the lowest temperature, with the magnetization saturated under the action of a magnetic field. If the field is turned off, thermal equilibrium will be re-established when the $M_S = -10$ and 10 states acquire the same population. In order to accomplish this, the whole ladder of levels will have to be climbed, with an activation energy equal to the separation in energy (Δ_T) of the $M_S = \pm 10$ and $M_S = 0$ level. By applying the Orbach mechanism of relaxation to this process, one obtains for the temperature-dependent relaxation time: $\tau = \tau_0 \exp(\Delta/kT)$. On the other hand, under the action of "thermally activated tunneling" (Thomas *et al.*, 1996; Friedman *et al.*, 1996), a spin can simply "tunnel" through the barrier, instead of climbing the whole ladder. A transition from an

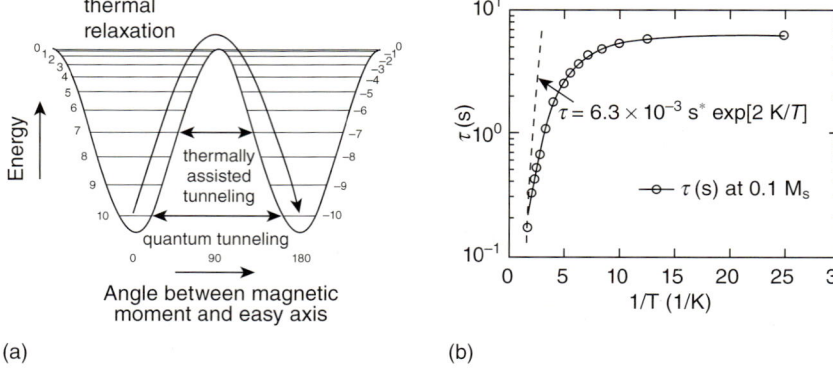

Figure 21.17 (a) Diagram showing the quantum tunneling of magnetization. In the static picture the system is located both on the left and right of the energy barrier, whereas in the dynamic picture the system oscillates coherently between the two sides with frequency determined by the tunnel splitting (Δ_T), which is very small in a SMM. The magnetization can tunnel, but only incoherently; (b) Magnetization relaxation in the pure tunneling regime. In Mn_{12}-ac the quantum tunneling is always thermally assisted because Δ_T is very small. In other systems, MQT from the ground state is the only relaxation process; the relaxation time (τ) is independent of temperature. Tunneling rates up to $2 \times 10^{-1} s^{-1}$ have been observed.

excited $-M_S$ level to the corresponding $+M_S$ level is effected by a transverse field, provided that the value of $2M_S$ is small. As for Mn_{12}-ac, characterized by tetragonal symmetry, an important issue is the existence of transverse fields which effect the tunneling. For tetragonal symmetry, the O_4^4 term in the Hamiltoian admixes only the states with difference in M_S by ± 4, implying that the tunneling should occur only for the pair of levels whose M_S values differ by ± 4. In contrast, the experimental data show that the relaxation due to tunneling occurs for all pairs of levels, necessitating the existence of additional transverse fields. Chudnovsky and Garanin (2001) suggested that dislocations play a crucial role in this mechanism. Another suggestion was made by Cornia et al. (2002b), who reanalyzed the X-ray crystal structure of Mn_{12}-ac reported earlier (Cornia et al., 2002a). The latter authors found that the acetic acid molecule of salvation is disordered on a binary axis, such that the crystal can be described as a mixture of clusters with n = 4, 3, 2, 1, and 0 hydrogen-bound acetic acid molecules. Thus, a rigorous fourfold symmetry is seen only by clusters with n = 4 and 0.

21.3.2
Magnetic Hysteresis, Resonant Magnetization Tunneling in High-Spin Molecules and Thermally Assisted Resonant Tunneling Between Quantum States

The barrier shown in Figure 21.5 is responsible for magnetic bistability and magnetic hysteresis at low temperature (Christou et al., 2000; Chudnovsky and Tajeda,

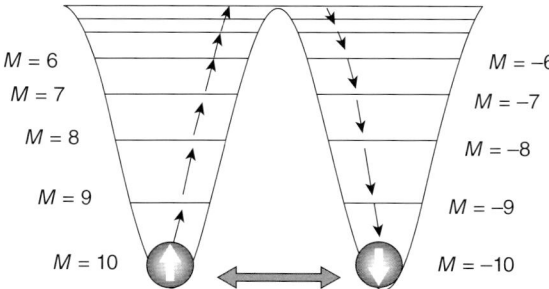

Figure 21.18 Pattern of energy for M_S sublevels for a spin state $S = 10$ and easy-axis anisotropy in zero field, exhibiting the two possible mechanisms for magnetization reversal through a thermally activated process shown by black arrows or by quantum tunneling shown by the gray arrow, short-cutting the energy barrier. Adapted from Gatteschi et al. (2006a).

1988; Gatteschi and Sessoli, 2003). This hysteresis sets in below a characteristic temperature, called the blocking temperature, T_B ($\ll DS^2/k_BT$). It is intrinsic to each individual molecule, unlike that in bulk magnets, and this is why the term SMM is used.

Friedman et al. (1996) reported the observation of steps at regular magnetic field intervals in the hysteresis loop of a microscopic sample of oriented Mn_{12}-ac crystals. Subsequently, the magnetic relaxation rate was found to increase substantially when the field was tuned to a step. It was proposed that these steps were manifestations of thermally assisted, field-tuned, resonant tunneling between quantum spin states, and the observation of quantum-mechanical phenomena on a macroscopic scale was attributed to tunneling in a large (Avogadro's) number of magnetically identical molecules. The hysteresis loops taken with the magnetic field applied along the easy axis of the oriented sample at six temperatures between 1.7 and 2.8 K are shown in Figure 21.19. As the field is increased steps are seen, but when it is reduced back to zero no noticeable steps occur. No steps are exhibited by an orientationally disordered control sample. It is shown in the inset of Figure 21.19 that the steps occur only at specific values of the magnetic field; the field at which a step occurs is plotted as a function of step number, with the one at zero field being labeled 0. Seven steps, at equal intervals of ~0.46 T, are observed, and more steps are expected to be observed at lower temperatures. When the temperature is lowered, new steps emerge out of the saturation curve, while others observed at higher temperatures disappear. These "frozen" steps are recovered when the magnetic field is swept more slowly. Furthermore, the magnetic relaxation was found to be more rapid when the magnetic field was in the neighborhood of a step. These observations were attributed (Friedman et al., 1996) to "thermally assisted resonant tunneling between quantum states" in Mn_{12}-ac. The spin of the molecule has two degenerate ground states separated by an anisotropy barrier in zero magnetic field, corresponding to spin parallel ($m = S$) or antiparallel ($m = -S$) to the c-axis. When a magnetic field is applied, it breaks the symmetry, rendering one state the true ground state (as shown in Figure 21.20), where the system is

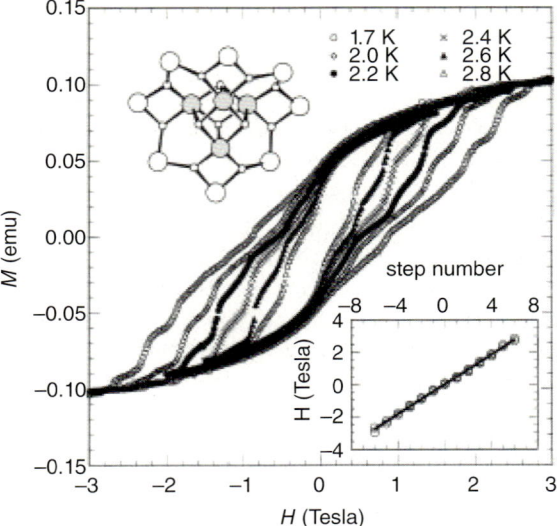

Figure 21.19 Magnetization of Mn$_{12}$ as a function of magnetic field at six different temperatures, as shown (field sweep rate of 67 mT min^{-1}). The inset shows the fields at which steps occur versus step number (with step 0 at zero field). The straight line is a least-squares fit, yielding a slope of 0.46 T per step. The structure of the Mn$_{12}$ molecule (Sessoli et al., 1993a,b) is represented at the top. Only the Mn^{4+} (large shaded circles), Mn^{3+} (large open circles), and oxygen (small circles) ions are shown. Adapted from Friedman et al. (1996).

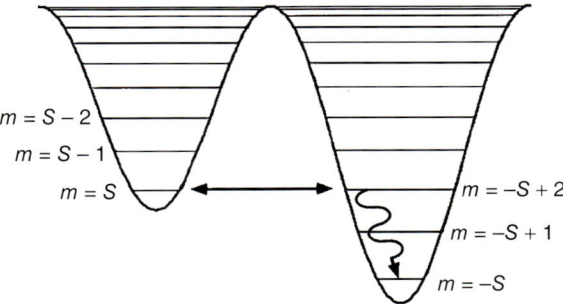

Figure 21.20 Schematic diagram of the resonant tunneling model. Tunneling from the metastable state $m = S$ to an excited state $m = -S + n$ is followed by a rapid spontaneous decay into the ground state. Adapted from Friedman et al. (1996).

initially populated in the metastable ground state, $m = S$, in the left-hand well. When an applied field brings this state in resonance with an excited level in the right-hand well, tunneling across the barrier is induced. Such tunneling is followed by a rapid spontaneous decay from the excited state to the ground state, with each step in the magnetization corresponding to such a resonance. Using the first two terms of the spin Hamiltonian (as expressed by Equation 21.1), and denoting the eigenstates of the system as $|S, M_S\rangle$, when the field is applied along the easy axis a simple calculation shows that the field at which the initial state $|S, M_S = S\rangle$ of the system coincides in energy with the state $|S, -S + n\rangle$, is:

$$H_{S,-S+n} = -Dn/g\mu_B, \qquad (21.3)$$

which shows that steps occur at even intervals of field, in accordance with the experimental data.

The tunneling must arise due to a perturbation, since the effective Hamiltonian given by the first two terms of Equation 21.1 commutes with S_z. This perturbation can be, for example, a transverse anisotropy or a small off-axis component of the external magnetic field. The proposed model implies that whenever the field is tuned to a step, each state in the left well coincides with a state in the right well, and this sets up multiple resonances. Using the fact that a step occurs every 0.46 T, one obtains $D/g = 0.21\,\text{cm}^{-1}$, which is consistent with the published value of $D \sim 0.5\,\text{cm}^{-1}$ and $g \sim 1.9$ obtained from HF-EPR (Caneschi et al., 1991; Sessoli et al., 1993b). This was further checked by the value of the anisotropy barrier at zero field, which was estimated to be $g(D/g)S^2 = 41\,\text{cm}^{-1}$, consistent with the value of $49\,\text{cm}^{-1}$ obtained from the blocking temperature. There are an expected 21 steps for an $S = 10$ system ($n = 0$–20), with the last one corresponding to the end of the barrier. Now, assuming that the blocking temperature on resonance scales as $T_B \sim (1 - H/H_c)^2 \sim (1 - n/n_c)^2$, a zero-temperature intercept of $n_c = 21.6$ is found, indicating roughly the number of steps predicted by this model. The observation of higher-numbered steps requires going to lower temperatures; it is estimated that the step 19 should be observed at ~10 mK and 8.74 T.

21.4
Concluding Remarks

The low-lying levels of SMMs can be investigated very effectively by EPR, and in particular at very high frequencies, by the use of single crystals. It is also possible to use zero-field EPR and neutron scattering to study SMMs, which have direct access to the energy-level patterns of the various spin multiplets, both being zero-field techniques. However, the advantage of high-frequency, high-field EPR is that the field-dependent and field-independent contributions can be separated rather precisely. It is also possible to use X-band EPR to study the so-called "EPR-silent" species by using parallel-mode EPR, wherein the static magnetic field (**B**) is parallel to the excitation magnetic field (**B₁**), as reported, for example, by Piligkos et al. (2004), who investigated the low-lying levels of a dodeca-nuclear Cr^{3+} cluster with a ground state of $S = 6$.

Recent research has provided much information on the various couplings with the environment experienced by SMMs, for example, crystal-field distributions, intermolecular dipolar and exchange interactions, and spin–lattice interactions, the fluctuations of which produce quantum decoherence. However, much information remains to be acquired, and in particular the factors that affect quantum coherence and MQT, as pointed out by Stamp and Tupitsyn (2003), are not fully understood. If there exists no decoherence, then coherence can be achieved by the superposition of "spin-up" and "spin-down" states as produced by MQT, which is the basis of quantum computation, or quantum information processing, as discussed by Leuenberger and Loss (2001).

SMMs are amenable to alterations of their properties in a systematic manner, so that inter-SMM exchange and dipolar interactions can be controlled by varying the peripheral ligands. Furthermore, crystal-field parameters and tunneling anisotropy can be altered by changing the core of the SMM. Hyperfine interactions can also be varied, for example, by deuteration, but parameters such as solvent disorder and low-lying excited states do not pose any serious problems for many SMMs. Nevertheless, Mn_{12}-ac, which has been studied so extensively, does not appear to be the best candidate as regards its quantum properties, as research has shown.

More sophisticated techniques, such as pulsed-EPR operating at frequencies well above 100 GHz, are needed to acquire more detailed information on SMMs.

Acknowledgments

The author is grateful to Prof. C. P. Poole, Jr for providing helpful comments on this chapter.

Pertinent Literature

Of the references cited in this chapter, much information is provided in the review by Hill (2004) on SMMs investigated by EPR. The review by Gatteschi et al. (2006b) also deals with the EPR studies on molecular nanomagnets, while MQT and related phenomena are discussed in detail by Gatteschi and Sessoli (2003). Some interesting conclusions are drawn by Brunel (2004).

References

Andres, H., Basler, R., Gudel, H.U., Aromi, G., Christou, G., Buttner, H., and Ruffle, B. (2000) *J. Am. Chem. Soc.*, **122**, 12469.

Aubin, S.M.J., Wemple, M.W., Adams, D.M., Tsai, H.-L., Christou, G., and Hendrickson, D.N. (1996) *J. Am. Chem. Soc.*, **118**, 7746.

Aubin, S.M.J., Diley, N.R., Pardi, L., Krzystek, J., Wemple, M.W., Brunel, L.-C., Maple, B., Christou, G., and Hendrickson, D.N. (1998) *J. Am. Chem. Soc.*, **120**, 4991.

Barra, A.L., Gatteschi, D., and Sessoli, R. (1997) *Phys. Rev.*, **B56**, 8192.

Barra, A.L., Gatteschi, D., and Sessoli, R. (2000) *Chem. Eur. J*, **6**, 1608.

Bashkin, J.S., Chang, H.-R., Streib, W.E., Huffman, J.C., Hendrickson, D.N., and Christou, G. (1987) *J. Am. Chem. Soc.*, **109**, 6502.

Blinc, R., Cevc, P., Arcon, D., Dalal, N.S., and Achey, R.M. (2001) *Phys. Rev.*, **B63**, 21240.

Brunel, L.-C. (2004) Single molecule nanomagnets, in *Biological Magnetic Resonance*, vol. 24 (eds O.G. Grinberg and L.A. Berliner), Kluwer Academic/Plenum Publishers, New York, pp. 534–535.

Caneschi, A., Gatteschi, D., Sessoli, R., Barra, A.L., Brunel, L.-C., and Guillot, M. (1991) *J. Am. Chem. Soc.*, **113**, 5873.

Cavallini, M., Facchini, M., Albonetti, C., and Biscarini, F. (2008) *Phys. Chem. Chem. Phys.*, **10**, 784.

Chandrasekhar, V., Murugesapandian, B., Azhakar, R., Vittal, J.J., and Clérac, R. (2007) *Inorg. Chem.*, **46**, 5140.

Chandrasekhar, V., Murugesapandian, B., Boomishankar, R., Steiner, A., Vittal, J.J., Houri, A., and Clérac, R. (2008) *Inorg. Chem.*, **47**, 4918.

Christou, G., Gatteschi, D., Hendrickson, D.N., and Sessoli, R. (2000) *MRS Bull.*, **25**, 66.

Chudnovsky, E.M. and Tajeda, J. (1998) *Macroscopic Tunneling of the Magnetic Moment.* Cambridge University Press, Cambridge.

Chudnovsky, E.M. and Garanin, D.A. (2001) *Phys. Rev. Lett.*, **87**, 187203.

Cornia, A., Fabretti, A.C., Sessoli, R., Sorace, L., Gatteschi, D., Barra, A.L., Daiguebonne, C., and Roisnel, T. (2002a) *Acta Crystallogr. C*, **58**, m371.

Cornia, A., Sessoli, R., Sorace, L., Gatteschi, D., Barra, A.L., and Daiguebonne, C. (2002b) *Phys. Rev. Lett.*, **89**, 257201.

Dressel, M., Gorshunov, B., Rajagopal, K., Vongtragool, S., and Mukhin, A.A. (2001) *Phys. Rev.*, **B67**, 060405.

Edwards, R.S., Hill, S., Bhaduri, S., Aiaga-Alcade, N., Bolin, E., Maccagnano, S., Christou, G., and Hendrickson, D.N. (2003a) *Polyhedron*, **22**, 1911.

Edwards, R.S., Maccagnano, S., Hill, S., North, J.M., and Dalal, N.S. (2003b) *Condens. Matter*, 0302052.

Edwards, R.S., Maccagnano, S., Yang, E.-C., Hill, S., Wernsdorfer, W., Hendrickson, D.N., and Christou, G. (2003c) *J. Appl. Phys.*, **93**, 7807.

Friedman, J.R., Sarachik, M.P., Tajeda, J., and Ziolo, R. (1996) *Phys. Rev. Lett.*, **76**, 3830.

Garanin, D.A. and Chudnovsky, E.M. (2002) *Phys. Rev.*, **B65**, 094423.

Gatteschi, D. and Sessoli, R. (2003) *Agnew. Chem. Int. Ed.*, **4**, 268.

Gatteschi, D., Sessoli, R., and Cornia, A. (2000) *Chem. Commun.*, 725–732.

Gatteschi, D., Sessoli, R., and Villain, J. (2006a) *Molecular Nanomagnets*, Oxford University Press, Oxford.

Gatteschi, D., Barra, A.L., Caneschi, A., Cornia, A., Sessoli, R., and Sorace, L. (2006b) *Coord. Chem. Rev.*, **250**, 1514.

Hennion, M., Pardi, L., Mirebeau, I., Suard, E., Sessoli, R., and Caneschi, A. (1997) *Phys. Rev.*, **B56**, 8819.

Hill, S. (2004) Single molecule nanomagnets, in *Biological Magnetic Resonance*, vol. 24 (eds O.G. Grinberg and L.A. Berliner), Kluwer Academic/Plenum Publishers, New York, pp. 506–524.

Hill, S., Perenboom, J.A.A.J., Dalal, N.S., Hathaway, T., Stalcup, T., and Brooks, J.S. (1998a) *Phys. Rev. Lett.*, **80**, 2453.

Hill, S., Perenboom, J.A.A.J., Stalcup, T., Dalal, N.S., Hathaway, T., and Brooks, J.S. (1998b) *Physica B*, **549**, 246.

Hill, S., Brooks, J.S., and Dalal, N.S. (1999a) Dipole-dipole coupling in Mn_{12}-acetate, in *Physical Phenomena in High Magnetic Fields -III* (eds Z. Fisk, L.P. Gor'kov, and J.R. Schrieffer), World Scientific, Singapore, p. 469.

Hill, S., Dalal, N.S., and Brooks, J.S. (1999b) *Appl. Magn. Reson.*, **16**, 237.

Hill, S., Edwards, R.S., Jones, S.I., Maccagnano, S., North, J.M., Aliaga, N., Yang, E.-C., Dalal, N.S., Christou, G., and Hendrickson, D.N. (2002a) *MRS Proc.*, **746**, Q1.1.1.

Hill, S., Maccagnano, S., Dalal, N., Achey, R.M., and Park, K. (2002b) *Int. J. Mod. Phys B*, **16**, 3326.

Hill, S., Maccagnano, S., Park, K., Achey, R.M., North, J.M., and Dalal, N.S. (2002c) *Phys. Rev.*, **B65**, 224410.

Hill, S., Edwards, R.S., North, J.M., and Dalal, N.S. (2003a) *Phys. Rev. Lett.*, **90**, 217204.

Hill, S., Edwards, R.S., North, J.M., Park, K., and Dalal, N.S. (2003b) *Polyhedron*, **22**, 1897.

Hill, S., Edwards, R.S., North, J.M., Maccagnano, S., and Dalal, N.S. (2003c) *Polyhedron*, **22**, 1889.

Hill, S., Edwards, R.S., Aliaga-Alcade, N., Hendrickson, D.N., and Christou, G. (2003d) *Science*, **302**, 1015.

Krzystek, J., Tesler, J., Knapp, M.J., Hendrickson, D.N., Aromi, G., Christou, G., Angerhofer, A., and Brunel, L.-C. (2001) *Appl. Mag. Res.*, **21**, 571.

Leuenberger, M. and Loss, D. (2001) *Nature*, **410**, 789–793.

Lis, T. (1980) *Acta Crystallogr. B*, **36**, 2042.

Maccagnano, S., Achey, R., Negusse, E., Lussier, A., Mola, M.M., Hill, S., and Dalal, N.S. (2001) *Polyhedron*, **20**, 1441.

Milios, C.J., Vinslava, A., Wernsdorfer, W., Moggach, S., Parsons, S., Perlepes, S.P., Christou, G., and Brechin, E.K. (2007a) *J. Am. Chem. Soc.*, **129**, 2754.

Milios, C.J., Vinslava, A., Wood, P.A., Parsons, S., Wernsdorfer, W., Christou, G., Perlepes, S.P., and Brechin, E.K. (2007b) *J. Am. Chem. Soc.*, **129**, 8.

Milios, C.J., Piligkos, S., and Brechin, E.K. (2008) *Dalton Trans.*, 1809.

Mirebeau, I., Hennion, M., Casalta, H., Andres, H., Gudel, H.U., Irodova, A.V., and Caneschi, A. (1999) *Phys. Rev. Lett.*, **83**, 628.

Mola, M., Hill, S., Goy, P., and Gross, M. (2000) *Rev. Sci. Instrum.*, **71**, 186.

Morrish, A.H. (1966) *The Physical Principles of Magnetism*, John Wiley & Sons, Inc., New York.

Mukhin, A., Gorshunov, B., Dressel, M., Sangrgorio, C., Gatteschi, D., and Gatteschi, D. (1998) *Europhys. Lett.*, **44**, 778.

Mukhin, A.A., Travkin, V.D., Zvezdin, A.K., Lebdev, S.P., Caneschi, A., Gorshunov, B., Dressel, M., Sangrgorio, C., and Gatteschi, D. (2001) *Phys. Rev.*, **B63**, 214411.

Park, K., Novotny, M.A., Dalal, N.S., Hill, S., and Rikvold, P.A. (2002a) *J. Appl. Phys.*, **91**, 7167.

Park, K., Novotny, M.A., Dalal, N.S., Hill, S., and Rikvold, P.A. (2002b) *Phys. Rev.*, **B65**, 14426.

Park, K., Novotny, M.A., Dalal, N.S., Hill, S., and Rikvold, P.A. (2002c) *Phys. Rev.*, **B66**, 144409.

Parks, B., Loomis, J., Rumberger, E., Hendrickson, D.N., and Christou, G. (2001) *Phys. Rev.*, **B64**, 184426.

Parks, B., Loomis, J., Rumberger, E., Yang, E.-C., Hendrickson, D.N., and Christou, G. (2002) *J. Appl. Phys.*, **91**, 7170.

Piligkos, S., Collision, D., Oganesyan, V.S., Rajaraman, G., Timco, G.A., Thomson, A.J., Winpenny, R.E.P., and Mc Innes, E.J.L. (2004) *Phys. Rev. B*, **69**, 134424.

Rentschler, E., Gatteschi, D., Cornia, A., Fabretti, A.C., Barra, A.L., Schegolikhina, O.I., and Zhdanov, A.A. (1996) *Inorg. Chem.*, **35**, 4427.

Sangrgorio, C., Ohm, T., Paulsen, C., Sessoli, R., and Gatteschi, D. (1997) *Phys. Rev. Lett.*, **78**, 4645.

Sessoli, R., Gatteschi, D., Caneschi, A., and Novak, M.A. (1993a) *Nature*, **365**, 141.

Sessoli, R., Tsai, H.-L., Schake, A.R., Wang, S., Vincent, J.B., Folting, K., Gattesci, D., Christou, G., and Hendrickson, D.N. (1993b) *J. Am. Chem. Soc.*, **115**, 1804.

Stamp, P.C.E. and Tupitsyn, I.S. (2003) *Condens. Matter*, **5**, 5371.

Takahashi, S., Edwards, R.S., North, J.M., Hill, S., and Dalal, N.S. (2004) *Phys. Rev. B*, **70**, 094429.

Terazzi, E., Bourgogne, C., Welter, R., Gallani, J.-L., Guillon, D., Rogez, G., and Donnio, B. (2008) *Angew. Chem. Int. Ed.*, **47**, 490.

Thomas, L., Lionti, F., Ballou, R., Gatteschi, D., Sessoli, R., and Barbara, B. (1996) *Nature*, **383**, 145.

Wernsdorfer, W., Aliaga-Alcalde, N., Hendrickson, D.N., and Christou, G. (2002) *Nature*, **416**, 406.

Wieghardt, K., Pohl, K., Jibril, I., and Huttner, G. (1984) *Agnew. Chem. Int. Ed. Engl.*, **23**, 77.

Wang, S.Y., Folting, K., Streib, W.E., Schmitt, E.A., McCusker, J.K., Hendrickson, D.N., and Christou, G. (1991) *Angew. Chem. Int. Ed. Engl.*, **307**, 305.

Yang, P. (ed.) (2003) *Chemistry of Nanostructured Materials*, World Scientific Publ., Hong Kong.

Yang, E.-C., Harden, N., Wernsdorfer, W., Zakharov, L.N., Brechin, E.K., Rheingold, A.L., Christou, G., and Hendrickson, D.N. (2003a) *Polyhedron*, **22**, 1857.

Yang, E.-C., Wernsdorfer, W., Hill, S., Edwards, R.S., Nakano, M., Maccagnano, S., Zakharov, L.N., Rheingold, A.L., Christou, G., and Hendrickson, D.N. (2003b) *Polyhedron*, **22**, 1727.

Yoo, J., Brechin, E.K., Yamaguchi, A., Nakano, M., Huffman, J.C., Maniero, A.L., Brunel, L.-C., Awaga, K., Ishimoto, H., Christou, G., and Hendrickson, D.N. (2000) *Inorg. Chem.*, **39**, 3615.

Yoo, J., Yamaguchi, A., Nakano, M., Krzystek, J., Streib, W.E., Brunel, L.-C., Ishimoto, H., Christou, G., and Hendrickson, D.N. (2001) *Inorg. Chem.*, **40**, 4604.

22
Multifrequency EPR on Photosynthetic Systems

Sushil K. Misra, Klaus Möbius, and Anton Savitsky

22.1
Introduction

This chapter focuses on a selection of photosynthetic systems that were previously crystallized and for which X-ray structures of high or, at least medium resolution, are available by now: these include bacterial "reaction centers" (RCs) from non-oxygenic as well as photosystem I (PS I) and photosystem II (PS II) from oxygenic photosynthetic organisms executing light-induced electron transfer across the membrane. These proteins have been characterized in detail using powerful spectroscopic techniques, that include ultra-fast laser spectroscopy (Michel-Beyerle, 1996), FT-IR (Mäntele, 1993; Rödig et al., 1999; Gerwert, 2002; Remy and Gerwert, 2003), solid-state NMR (de Groot, 2000; Lansing et al., 2003; Mason et al., 2004; Hong, 2006) and, last but not least, multifrequency EPR (Shin et al., 1993; Kay et al., 1999, 2002, 2005a, 2005b; Richter et al., 1999; Salwinski and Hubbell, 1999; Möbius, 2000; Weber, 2000, 2005; Un, Dorlet, and Rutherford, 2001; Weber et al., 2001a, 2001b, 2002; Fuchs et al., 2002; Riedi and Smith, 2002; Smirnov, 2002; Steinhoff, 2002; Plato et al., 2003; Lubitz, 2004; Savitsky et al., 2004; Möbius et al., 2005; Schnegg et al., 2006; Weber and Bittl, 2007; Möbius and Goldfarb, 2008). In addition, sophisticated theoretical studies have been performed to elucidate the light-induced electron- and proton-transfer characteristics of proteins (Jortner and Bixon, 1999; Crystal and Friesner, 2000; Parson and Warshel, 2004), emphasizing the importance of electrostatic energies (Parson, Chu, and Warshel, 1990; Warshel and Parson, 1991; Gunner, Nicholls, and Honig, 1996; Warshel, Kato, and Pisliakov, 2007; Parson and Warshel, 2008) and electronic couplings (Michel-Beyerle et al., 1988; Plato et al., 1988; Plato and Winscom, 1988; Warshel, Creighton, and Parson, 1988; Scherer and Fischer, 1989; Bixon, Jortner, and Michel-Beyerle, 1991; Scherer, Scharnagl, and Fischer, 1995; Ivashin et al., 1998; Zhang and Friesner, 1998; Kolbasov and Scherz, 2000; Petrenko and Redding, 2004). Hence, these proteins represent paradigm systems of general interest. They are well suited for new multifrequency high-field EPR experiments to study functionally important transient states at time windows relevant for biological action. These states are often directly detectable by EPR via their transient paramagnetism as, for instance,

Multifrequency Electron Paramagnetic Resonance, First Edition. Edited by Sushil K. Misra.
© 2011 Wiley-VCH Verlag GmbH & Co. KGaA. Published 2011 by Wiley-VCH Verlag GmbH & Co. KGaA.

the radical-ion and radical-pair states of cofactors involved in photosynthetic electron transfer in RCs.

High-field EPR experiments allow specific protein regions to be identified and characterized as molecular control subunits for vectorial transfer processes across membranes. Such subunits function by optimizing the electronic structure of cofactors, for example by invoking weak cofactor–protein interactions via H-bonds, by polarity gradients, or even by substantial cofactor and/or helix displacements.

The time scale of photosynthetic processes extends over a huge range of 20 orders of magnitude, ranging from less than 100 femtoseconds to more than 100 megaseconds. Photosynthesis, therefore, is a research target in vastly different fields of science, ranging from ultra-short radiation physics, condensed matter physics and chemistry, photochemistry, biophysics and biochemistry, physiology, and botany (Feher, 1992).

Life on Earth is enabled by photosynthetic processes, which convert the energy of sunlight into electrochemical energy. They can be performed by certain archaea, bacteria, algae and green plants, and provide the energetic basis needed by all higher organisms for synthesis, growth, and replication. These organisms utilize sunlight to power cellular processes and ultimately derive their biomass through secondary – that is, dark – chemical reactions. They are driven by the light-initiated primary reactions of energy and electron transfer between pigment molecules in a protein scaffold. The solar energy is first collected by an "antenna" system, consisting of highly organized protein–pigment complexes, and then is rapidly transferred, most likely via efficient Förster-type energy-transfer processes, to a reaction-center complex where primary electron transfer takes place. Thereafter, secondary reactions lead to the reduction of carbon dioxide (CO_2) with or without oxidation of water, depending on the class of organisms. Antenna complexes permit an organism to increase greatly the absorption cross-section for light absorption needed to drive photosynthetic electron transfer. Thus, antenna systems are also called "light-harvesting complexes"; these regulate aggregates optimized by evolution for maximum energy-collection efficiency, while avoiding photodamage of the pigments. Such a pigment can be (bacterio)chlorophyll, carotenoid or bilin (open chain tetrapyrrole) molecules, depending on the type of organism. A wide variety of different antenna complexes are found in different photosynthetic systems.

Light-induced electron-transfer (ET) reactions between protein-bound donor (D) and acceptor (A) pigments across the cytoplasmic membrane are the primary processes in the reaction-center protein complexes of photosynthetic organisms. The cascade of charge-separating ET steps of primary photosynthesis competes extremely favorably with wasteful charge-recombination ET steps, thereby providing almost 100% quantum yield.

Green plants, algae, and cyanobacteria provide the largest impact of photosynthesis on life on Earth. In their two interconnected reaction centers, PS I and PS II, a reversible ET photocycle occurs in which water serves as the electron donor. CO_2 is fixed in the form of carbohydrates, and oxygen gas (O_2) is released as a byproduct, thereby stabilizing the CO_2/O_2 composition of the Earth's atmosphere. Thus, oxygenic photosynthesis reaction can be expressed as

$$n \cdot CO_2 + n \cdot H_2O \xrightarrow{h\nu, Chl} C_n(H_2O)_n + n \cdot O_2 \quad (22.1)$$

In order for this photo-reaction to proceed, chlorophylls, and other cofactors, are required as biocatalysts.

As for RCs, there exist two different types. These are related, but differ in their abilities to reduce the terminal electron acceptors, either iron–sulfur complexes (type I) or quinones (type II), whereas the purple nonsulfur photosynthetic bacteria contain only type II RCs. Cyanobacteria, algae, and green plants contain both type I and type II RCs. PS I and PS II are connected in series to split water, and to produce the transmembrane electrochemical potential for NADPH and ATP synthesis, as well as for CO_2 assimilation.

The PS I and PS II RCs are used by the organisms capable of oxygenic photosynthesis in two light-induced ET reactions associated with a sequential electron transport scheme, known as the Z-scheme (Moser *et al.*, 1997) (see Figure 22.1), in which PS I and PS II operate in tandem. After light excitation of the primary donors (chlorophyll dimers known as P_{700} and P_{680}, which absorb predominantly at 700 and 680 nm, respectively), ET proceeds from the water-splitting Mn complex (left) via the cytochrome b_6f, plastocyanine and PS I complexes to $NADP^+$ (right), where the energetic coupling of the light reactions to the dark reactions for CO_2 fixation occurs. The ellipses in Figure 22.1 represent the various protein–cofactor complexes, with the vertical position of each cofactor indicating its redox potential. Light excitation of the primary donors to their first excited singlet states, P*, is symbolized by wavy arrows $h\nu$, whereas the other arrows signify the ET pathways.

The energy-rich molecules NADPH and ATP are the final products of the chloroplasts' light reactions. They provide, respectively, reducing power and free energy for the subsequent light-independent reactions to drive the Calvin cycle, in which CO_2 is reduced and incorporated into carbohydrates. (Note: M. Calvin received the Nobel Prize in Chemistry 1961 for this perception.) If, for example, the CO_2 molecules are incorporated into hexose sugars, such as glucose or fructose, the overall reaction of the Calvin cycle can be formulated as:

$$6CO_2 + 18ATP + 12NADPH + 12H_2O \rightarrow$$
$$C_6(H_2O)_6 + 18ADP + 18P_i + 12NADP^+ + 6H^+ \quad (22.2)$$

Three billion years before green plants evolved, the Earth and its atmosphere were very different from what they are today. Nevertheless, photosynthetic energy conversion could be performed in those early times by certain bacteria, for instance the purple bacterium *Rhodobacter (Rb.) sphaeroides*, which still exists in ecological niches.

Among the numerous spectroscopic techniques utilized in photosynthesis research, high-field/high-frequency EPR (HF-EPR) spectroscopy – and, in particular, its pulse and double-resonance options – play an important role in the endeavor to understand, on the basis of structure and dynamics data, the dominant factors

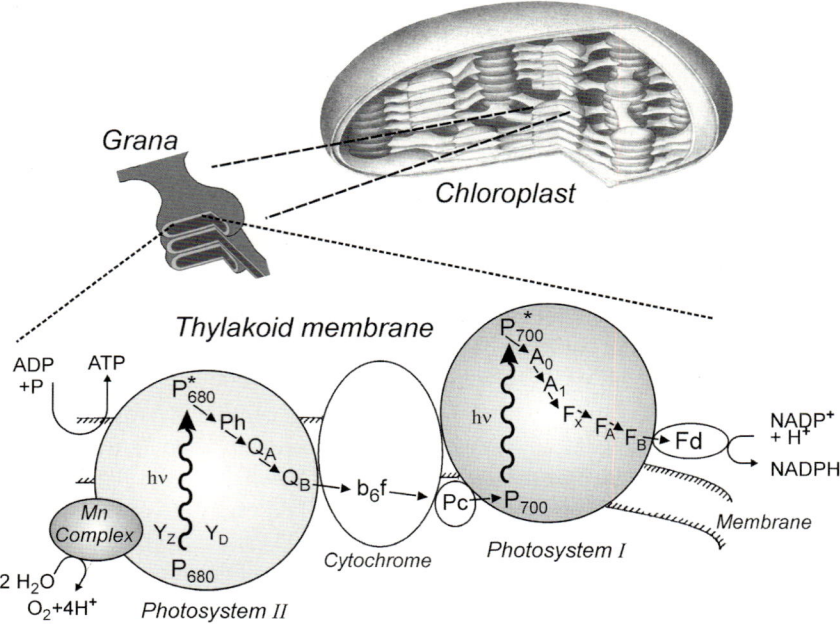

Figure 22.1 The sequential electron transport scheme (Z-scheme). The thylakoid phospholipid membrane in the chloroplast organelles of plant cells, where photosynthesis is conducted by absorbing light energy, and storing it in the form of sugars. The chloroplast contains stacks of thylakoid membranes. Each stack of thylakoid disks represents one granum. The light reactions of photosynthesis occur in the grana. The area between the grana is called the stroma. This is where the dark reactions of photosynthesis occur. The light reactions follow the "Z-scheme" according to which electrons from excited chlorophyll molecules flow through a cytochrome transport system along the thylakoid membranes. During this electron-transport process, ATP and NADPH are generated; these are required to fuel the dark reactions in the stroma, by which CO_2 is converted into glucose through a series of reactions called the Calvin cycle. In the Z-scheme of light-induced electron-transfer pathways in oxygenic photosynthetic systems the photosystems I and II (PS I, PS II) are interconnected by the cytochrome b_6f and plastocyanin (Pc) complexes. Abbreviations: P_{680} and P_{700}, primary electron donors; Y_Z and Y_D, tyrosines; Mn complex, oxygen-evolving complex; Ph, pheophytin; Q_A and Q_B, quinones; A_0, monomeric chlorophyll; A_1, phylloquinone; F_X, F_B, F_A, iron-sulfur centers; Fd, ferredoxin/flavodoxin complexes.

that control the specificity and efficiency of light-induced electron- and proton-transfer processes in primary photosynthesis. Short-lived transient intermediates of the photocycle can be characterized by high-field EPR techniques, and detailed structural information can be obtained even from disordered sample preparations. This chapter describes how multifrequency EPR methodology, in conjunction with mutation strategies for site-specific isotope or spin labeling and with the support of modern quantum-chemical computation methods for data interpretation, is capable of providing new insights into the photosynthetic transfer processes. The information obtained is complementary to that of protein crystallography, solid-state NMR and laser spectroscopy. Thereby, our understanding of the relation between structure, dynamics and function is considerably improved. This is especially true with respect to the fine-tuning of electronic properties of donors and acceptors by means of weak interactions with their protein environment, such as hydrogen bonding to specific amino acid residues.

The question "what do we learn from high-field EPR" can be answered in terms of a few summarizing statements:

- Most organic cofactors in photosynthetic RCs have only small g-anisotropies and, therefore, require much higher fields than in X-band EPR to resolve the canonical g-tensor orientations in their powder spectra and, thereby, to trace orientation-selective hydrogen bonding in the binding sites. This is important complementary information to what is available from X-ray diffraction.

- During the course of the ET processes, several radical species are often generated. To distinguish them by the small differences in their g-factor and hyperfine interactions, high fields are required.

- This argument also holds for separating different sites of different cofactor orientations in the RC unit cell.

- Often, high-purity protein samples can be prepared only in minute quantities. This is the rule for protein crystals and RCs of site-directed mutants. Accordingly, to study them by EPR, very high sensitivity is needed. This can be accomplished only with dedicated high-field/high-frequency spectrometers.

- High-field/high-frequency continuous wave (CW)-EPR generally provides, by lineshape analysis, shorter time windows down into the picosecond range for studying the correlation times of important dynamic processes, such as protein motion and folding over wide temperature ranges.

- Pulsed high-field/high-frequency EPR in the form of field-swept electron-spin echo (ESE) spectroscopy, provides real-time access to specific cofactor–protein slow motions on the nanosecond time-scale, and even to their spatial anisotropy that is generated by anisotropic weak interactions such as hydrogen bonds within the binding site.

- In metalloprotein high-spin systems, such as the Mn^{2+} oxygen-evolving complex in PS II, the EPR spectrum analysis can be drastically simplified at high fields due to suppression of second-order effects. Some high-spin metalloproteins with large zero-field splittings cannot be observed at all at X-band frequencies.

EPR in photosynthesis has been recently reviewed in a comprehensive manner by Lubitz (Lubitz, 2004). In what follows, HF-EPR applications to the transient intermediates of nonoxygenic photosynthesis will be discussed first, followed by a discussion of oxygenic photosynthesis.

22.2
Nonoxygenic Photosynthesis

Nonoxygenic photosynthetic energy conversion is achieved by certain bacteria, for example, the purple bacteria *Rhodobacter (Rb.) sphaeroides* and *Rhodopseudomonas (R.) viridis*. These photosynthetic organisms are simple, unicellular protein-bound donor–acceptor complexes that contain only one RC for light-induced charge separation. They cannot split water, but rather use hydrogen sulfide or organic compounds as electron donors to reduce CO_2 to carbohydrates with the help of sunlight and bacteriochlorophylls and quinones as biocatalysts. Thus, they do not produce oxygen. The net reaction for nonoxygenic bacterial photosynthesis is

$$2H_2S + CO_2 \xrightarrow{h\nu, BChl} C(H_2O) + H_2O + 2S \tag{22.3}$$

The incoming light quanta are harvested by "antenna" pigment/protein complexes in primary photosynthesis, and channeled to the actual RC complexes by ultra-fast energy transfer to excite the primary donor—a "special pair" of bacteriochlorophylls—to its excited electronic singlet state. A sequence of ET steps is thus initiated, going downhill energetically from the protein-bound donor to the quinone acceptors situated in their binding sites at the opposite side of the cytoplasmic membrane. A unique feature of the RC is a quantum yield of 100%, when converting the sun's energy into electrochemical potential.

Molecular models of the light-harvesting complexes, LH1 and LH2, from photosynthetic bacteria are shown in Figure 22.2; these are based on crystal structures and molecular modeling calculations by various research groups (Kühlbrandt, 1995; McDermott *et al.*, 1995; Cogdell *et al.*, 1996; Freer *et al.*, 1996; Hu and Schulten, 1997; Papiz *et al.*, 2003; Roszak *et al.*, 2003). The light-harvesting complexes are integral membrane proteins that form ring-like structures, oligomers of alpha beta-heterodimers, in the photosynthetic membranes of purple bacteria. They contain a large number of chromophores organized in ring structures, that are optimal for light absorption and rapid light-energy migration. One ring comprises 16 B850 BChl molecules (absorbing at 850 nm) that are oriented perpendicularly to the membrane plane, nd the other ring eight B800 BChl molecules (absorbing at 850 nm) that are nearly parallel to the membrane plane. The transition dipole moments of neighboring B850 and B800 BChl molecules are nearly parallel to each other and, thus, are optimally aligned for Förster exciton transfer. Specifically, LH2 consists of a circular array of heterodimers, each comprising an α- and β-apoprotein. Together, they bind three BChl molecules and one carotenoid molecule, such that the full structure is an $\alpha_9\beta_9$ nonamer. The pigment molecules are enclosed by two concentric rings of transmembrane α-helices. The BChl molecules are arranged in two distinct groups absorbing light at somewhat different wavelengths. Nine monomeric BChl molecules absorb at 800 nm (B800), and 18 tightly coupled BChl molecules absorb at 850 nm (B850). The RC-containing LH1 complex has a very similar, but larger ring structure; in the LH1 case it is a larger $\alpha_{16}\beta_{16}$ oligomer. Thereby, a central pocket is provided which is large enough to accommodate the RC. The LH1 complex only contains a single ring of 32 tightly

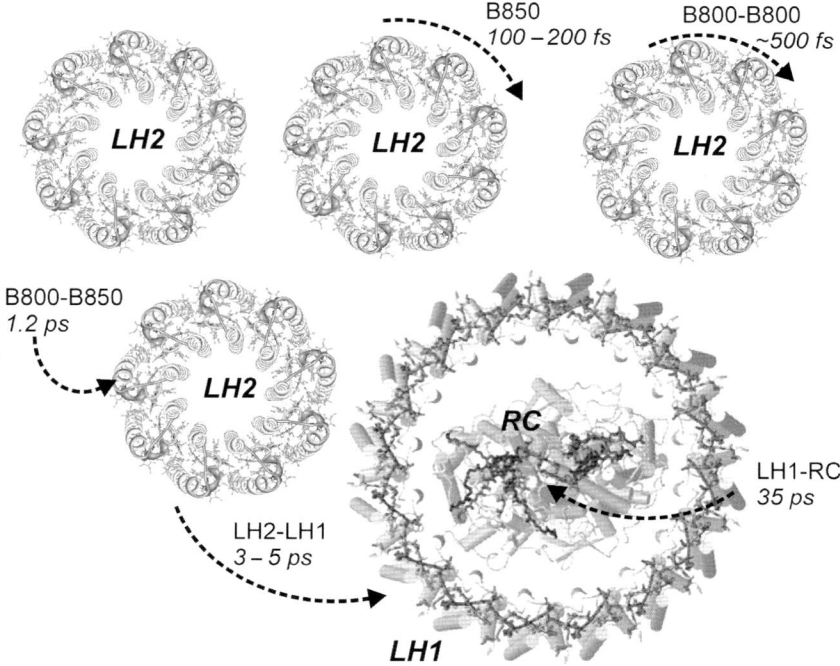

Figure 22.2 Details of the light-harvesting complexes, LH1 and LH2, of the antenna domain of purple bacteria (Cogdell *et al.*, 1999). The time constants of the energy-transfer processes between the various BChl domains are marked next to the arrows in the figure.

coupled BChl molecules that absorb at 875 nm (B875). Thus, the protein cage provides, in addition to an optimal cofactor orientation, a subtle cofactor–protein interaction mechanism for tuning the absorption wavelength of the BChl aggregates involved in energy transfer into energetically low-lying exciton states. Thereby, an energy funnel is created which is needed for initiating the primary ET processes in the RC. For details and references, see Papiz *et al.* (2003) and Roszak *et al.* (2003).

No paramagnetic species are created during light harvesting, since the antenna complexes LH1 and LH2 operate via singlet exciton transfer between the pigment molecules. EPR spectroscopy cannot, therefore, be used a priori for characterizing transient intermediates of the energy transfer pocesses. Antenna pigments can be, however, turned into paramagnetic doublet or triplet states by manipulating the antenna systems, for example by the chemical oxidation of LH1 sub-complexes or intact LH1 domains (Gingras and Picorel, 1990; Srivatsan and Norris, 2001; Kolbasoy *et al.*, 2003; Srivatsan *et al.*, 2003a, 2003b), or by 532 nm pulsed laser excitation of carotinoidless preparations of LH1 and LH2 complexes (Zhang *et al.*, 2001). Lubitz (Lubitz, 2004) has extensively reviewed up to 2003 X-band EPR experiments on such manipulated antenna complexes, and their results in comparison to the cation radical of the primary donor in the electron-transfer cascade of illuminated RCs. Remarkable progress has been made to understand primary photosynthesis by the application of X-ray crystallography, ultra-fast laser spectroscopy in the visible and infrared (IR) regions and, last not least, with modern EPR and NMR techniques (Deisenhofer and Norris, 1993; Hoff and Deisenhofer, 1997; Page et al., 1999; Noy, Moser, and Dutton, 2006).

It is now well established (see, e.g., Onuchic, Luthey-Schulten, and Wolynes, 1997; Balabin and Onuchic, 2000; Fenimore et al., 2002; Finkelstein and Ptitsyn, 2002; Frauenfelder, 2002; Frauenfelder, McMahon, and Fenimore, 2003; Kriegl, Forster, and Nienhaus, 2003; Kriegl and Nienhaus, 2004) that the intrinsic flexibility of proteins, which are thermally fluctuating between different conformations of cofactors in their binding sites ("energy landscape") at wide time scales including nanosecond to millisecond, can strongly affect functional processes with corresponding time constants. Since this ns–ms time window is open to EPR experiments, both CW and pulsed HF-EPR can be used to resolve T_2 anisotropies orientationally. These are caused by anisotropic librational fluctuations of donor and acceptor radical ions in the H-bonding network of their binding sites. As described below, such dynamics have been studied, for example, for the quinone electron acceptors in RCs, by using pulsed high-field EPR (Rohrer et al., 1996; Schnegg et al., 2002; Möbius et al., 2005).

22.3
Multifrequency EPR on Bacterial Photosynthetic Reaction Centers (RCs)

22.3.1
X-band EPR Experiments

Primary photosynthesis provides a "Garden of Eden" for the EPR spectroscopist (Hoff, 1987), because in each ET step a transient paramagnetic intermediate is formed. In fact, EPR spectroscopy has led to the identification and characterization of the primary ion radicals of P_{865} and Q_A. A striking difference in the linewidth of the EPR spectra was revealed when comparing the frozen-solution EPR spectra of monomeric BChl$^+$ in an organic solvent and of $P_{865}^{\bullet+}$ in the RC: for $P_{865}^{\bullet+}$ the linewidth is $1/\sqrt{2}$ narrower than for BChl$^{\bullet+}$. Norris, et al. (Norris et al., 1971) explained this observation by the "special pair" hypothesis; that is, the unpaired electron on $P_{865}^{\bullet+}$ is equally shared between a (BChl)$_2$ dimer cation radical, which they assumed to be a symmetric dimer. Feher et al. (Feher, Hoff et al., 1973; Feher et al., 1975) confirmed the existence of the special pair by frozen-solution electron-nuclear double resonance (ENDOR) experiments. In order to contribute to a solution of the "unidirectionality enigma," the electronic structure of the primary donor cation radicals, the dimers $P_{865}^{\bullet+}$ in Rb. sphaeroides and $P_{960}^{\bullet+}$ in R. viridis, as well as their monomeric constituents, BChl a$^{\bullet+}$ and BChl b$^{\bullet+}$, respectively, have been studied in great detail by using EPR, ENDOR and electron-nuclear-nuclear triple resonance (TRIPLE). These investigations were conducted in liquid and frozen solutions, as well as in single crystals of RCs. (For a chronological account, see Feher, 1992; for reviews, see Lubitz, 1991; Plato, Möbius, and Lubitz, 1991.)

Whilst the good feature of fluid-solution EPR and ENDOR is the narrow linewidth with a concomitant increase of spectral resolution, the bad feature is that the information of anisotropic hyperfine interactions is lost – that is, of the molecular orientation in the laboratory coordinate system. This information is retained by studying RC single crystals. For non-metalloproteins, such as the bacterial RC, the g-anisotropy is very small ($<10^{-3}$). hence, a full resolution of the g-tensor components of $P_{865}^{\bullet+}$ definitely requires high-field EPR on RC single crystals (by 95 GHz EPR) or on RC frozen solutions (by 360 GHz EPR).

A few introductory remarks are made here concerning the quinone acceptors in bacterial photosynthesis, Q_A and Q_B. In the light-driven ET processes of *Rb. sphaeroides*, the primary and secondary quinones – Q_A and Q_B – are the same ubiquinones-10, acting as one- and two-electron gates, respectively. Their different function in the ET processes is clearly induced by different interactions with the amino-acid environment in their binding sites. To learn about these interactions within the binding pocket – for example the specific H-bonding patterns – EPR and ENDOR on quinone anion radicals in bacterial RCs (with Fe^{2+} replaced by Zn^{2+} to avoid fast spin relaxation) and in organic solvents have been performed at several microwave (MW) frequencies by various groups, both in fluid and frozen solution. These multifrequency experiments on the primary donor and acceptor radical ions will be reviewed in the next subsections.

22.3.2
95-GHz EPR on Primary Donor Cations P$^{\bullet+}$ in Single-Crystal RCs

95 GHz high-field EPR on illuminated single-crystal RCs of *Rb. sphaeroides* was performed at 12 °C (i.e., at physiological temperature; Klette *et al.*, 1993), in order to determine the characteristic symmetry properties of the electronic structure of the primary donor. At a Zeeman field of 3.4 T even the magnetically inequivalent sites in the unit cell of the RC crystal were resolved, and the angular dependence of their g-factors in the three symmetry planes of the crystal was measured and analyzed. The RCs were excited predominantly in the BChl dimer band at 865 nm. The space group of the orthorhombic crystals of the RC from *Rb. sphaeroides* is $P2_12_12_1$ and, thus, a complication is encountered in that there are four RCs ("sites") per unit cell, pairwise related by a twofold symmetry axis (Allen *et al.*, 1986). In an arbitrary direction of the Zeeman field, B_0, in the crystal axes system $\{a, b, c\}$ there is a fourfold site-splitting of the $P_{865}^{\bullet+}$ EPR signal. Fortunately, for B_0 lying in the symmetry planes *ab, ac, bc*, there remain only two magnetically inequivalent sites, A and B, each of which consists of two magnetically equivalent RCs; and for $B_0 \parallel a$, $B_0 \parallel b$, $B_0 \parallel c$ all four RCs are magnetically equivalent (Budil *et al.*, 1988). The rotation patterns $g(\theta)$ of the g-values of $P_{865}^{\bullet+}$ in the three crystallographic symmetry planes were recorded (Klette *et al.*, 1993), and the site splitting (A, B) could be clearly resolved in the *ac* and *ab* planes, but not in the *bc* plane. Diagonalization of the g-matrix yields the principal g-values ($g_{\alpha\alpha} = 2.00329$,

$g_{\beta\beta} = 2.00239$, $g_{\gamma\gamma} = 2.00203$) in the g-tensor axes system (α, β, γ), together with the principal axes directions of the g-tensor. (See Table 5.1 in Möbius and Savitsky, 2009; see also Atherton, 1993; Klette et al., 1993; Prisner et al., 1995). It was found that the principal directions of the g-tensor are tilted in the molecular axes system, to reveal a breaking of the local C_2 symmetry of the electronic structure of $P_{865}^{\bullet+}$, consistent with the ENDOR/TRIPLE results for the hyperfine structure. The g-tensor of $P_{865}^{\bullet+}$ in the heterodimer mutant HL(M202) of *Rb. sphaeroides* (Huber and Törring, 1995) was determined by HF-EPR (95 GHz) on RC single crystals. Since, in this mutant, the unpaired electron of $P_{865}^{\bullet+}$ is localized on one of the bacteriochlorophylls, the g-tensor reflects the monomer properties. The directions of the principal axes of the g-tensor were found to be similar for the mutant and the wild-type. In recent years, very promising g-tensor calculations for cofactor radical ions in photosynthesis have been performed on the basis of density functional theory (DFT) methods (Fuchs et al., 2003; Möbius et al., 2005; Möbius and Savitsky, 2009).

22.3.3
360-GHz EPR on Primary Donor Cations P$^{\bullet+}$ in Mutant RCs

The effect on the electronic structure caused by mutations can be measured via the oxidation-potential changes of the primary donor (Ivancich et al., 1998; Schulz et al., 1998; Lubitz, Lendzian, and Bittl, 2002). An alternative is to measure the mutation effect via characteristic shifts of g-tensor and hyperfine-tensor components, which can be resolved by using high-field EPR (Klette et al., 1993; Huber and Törring, 1995; Fuchs et al., 2003) and ENDOR (Rautter et al., 1995; Huber et al., 1996; Artz et al., 1997; Müh et al., 1998, 2002), respectively. To gain further insight into the origins and consequences of the asymmetry in the electronic structure of the special-pair cation, an attempt was made to investigate various the tailor-made site-directed mutants of the RC from *Rb. spaeroides* by 360 GHz/12.9 T high-field EPR. To this end, at first the heterodimer mutants HE(M202) and HL(M202) were studied in which the ligands to the magnesium of the bacteriochlorophylls were altered (Fuchs et al., 2003; Möbius et al., 2005). The asymmetry in the electron spin density (and charge) distribution in P$^{\bullet+}$ is not a priori the major contributor to the unidirectionality of the ET in bacterial RCs, but its effect is enhanced by specific cofactor–protein interactions in addition to those affecting predominantly the spin distribution over the P$^{\bullet+}$ supermolecule. As shown by previous W-band measurements on similar mutants (Huber and Törring, 1995; Huber et al., 1995), the expected *shifts* of the g-tensor components upon mutation are very small – that is, on the order of 10^{-4}. A further increase of EPR frequency and field by a factor of 4 to 360 GHz and 12.9 T, however, provided the spectral resolution necessary to both fully resolve all three principal g-tensor components of P$^{\bullet+}$ randomly oriented in frozen solution (Bratt et al., 1999), and to measure their mutation-induced shifts with high precision (Fuchs et al., 2003). A small g-strain-induced increase of the intrinsic linewidth compared to W-band EPR measurements (Huber and Törring, 1995) from 0.89(6) mT to 1.4(2) mT was noticed

(Fuchs et al., 2003), but owing to the concurrent increase in spectral resolution by a factor of 4, this did not pose any resolution problems for the bacterial RCs.

22.3.4
Results of g-tensor Computations of P$^{\bullet+}$

Experimental results on the M202 mutants from non-EPR studies strongly indicate that the primary donors of the HL(M202) and the HE(M202) mutants are heterodimers of bacteriochlorophyll *a* and bacteriopheophytin *a* (abbreviated BChl:BPhe), in contrast to the homodimeric special-pair species (BChl)$_2$ of wild-type and R26 RCs. Both, EPR studies and energetic considerations have led to the conclusion that this structural modification of the primary donor in its oxidized paramagnetic state is characterized by an almost complete localization of the spin density on the BChl(L) half-bound to the L protein subunit (Huber et al., 1996; Schulz et al., 1998). The g-component shifts observed in a 360 GHz EPR experiment, indicate that for P$^{\bullet+}$ of HL(M202), the overall g-tensor anisotropy $\Delta g = (g_{xx} - g_{zz})$ becomes smaller (more like that of the monomer), while for HE(M202) Δg increases to a value considerably larger than that for R26 (Möbius and Savitsky, 2009). This unexpected behavior of the g-tensor strongly suggests a local structural change of BChl(L) as a consequence of the M202 ligand mutations. This conclusion is also supported by the observation of considerable rearrangements in the spin density distributions on BChl(L) in the two mutants (Käss et al., 1994; Rautter et al., 1995; Müh et al., 1998; Müh, Jones, and Lubitz, 1999). DFT calculations of the g-tensors of the three P$^{\bullet+}$ species as a function of different torsional angles $\theta_{ac}(L)$ of the acetyl group at position 3 with respect to the plane of the adjacent aromatic ring were performed (Fuchs et al., 2003; Möbius et al., 2005; Möbius and Savitsky, 2009). The essential conclusion from the computational results is that this model predicts a range of increasing angles $\theta_{ac}(L)$ between 0° and 35° in which both R_L (the ratio of methyl hyperfine couplings of P$^{\bullet+}$ in *Rb. sphaeroides* mutants) and Δg increase parallel to each other, in accordance with the experimental results. The increase in Δg in this series is due entirely to a decrease of g_{zz}. To model the remaining experimentally observed strong shift of the g_{xx} component, one needs to take into account H-bonding interactions with the environment. The high sensitivity of very-high-frquency EPR to subtle changes in the structure and microenvironment of the ET chromophores can provide valuable insight into the influence of structural features and energetics on the effectiveness of the photosynthetic ET steps, which cannot be obtained from X-ray crystallography because of insufficient precision of the extracted atomic positions so obtained.

22.3.5
95-GHz EPR and ENDOR on the Acceptors $Q_A^{\bullet-}$ and $Q_B^{\bullet-}$

The light-driven ET processes of photosynthesis are prominent examples of how quinones play important roles in many biological systems. For example, in the photosynthetic bacterium *Rb. sphaeroides*, the primary and secondary quinones,

Q_A and Q_B, act as one- and two-electron gates, respectively: $Q_A^{\bullet-}$ simply passes the extra electron to Q_B which, in a second photoinitiated ET step, becomes doubly reduced, binds two protons, dissociates from the RC, and releases protons on the periplasmic side of the membrane. In *Rb. sphaeroides*, Q_A and Q_B are the same ubiquinones-10. It appears that their different functions in the ET processes are induced by different interactions with the protein environment at their binding sites.

In the conventional g-tensor frame of quinones, the g_{xx} component lies along the line connecting the two oxygen atoms carrying most of the spin density; g_{zz} is perpendicular to the molecular plane, and g_{yy} is perpendicular to both (Isaacson *et al.*, 1995b; Möbius and Savitsky, 2009). EPR and ENDOR on quinone anion radicals in bacterial RCs (with Fe^{2+} replaced by Zn^{2+} to avoid fast spin relaxation) and in organic solvents have been performed at several MW frequencies, both in fluid and frozen solution, to learn about such cofactor–protein interactions within the quinone $Q_A^{\bullet-}$ or $Q_B^{\bullet-}$ binding pockets exhibiting, for example, specific H-bonding networks. For example, W-band high-field EPR and ENDOR experiments on a series of quinones related to photosynthesis were performed, from which their intramolecular and intermolecular proton hyperfine interactions were discerned (Burghaus *et al.*, 1993; Rohrer *et al.*, 1995). The aim of the W-band high-field EPR and ENDOR experiments on quinone radical anions in frozen solutions was to measure their anisotropic interactions with the organic solvent matrix ("*in vitro*") or protein microenvironment ("*in vivo*"), by means of the cofactors' g- and hyperfine-tensor components as well as T_2 relaxation times. The aim was to learn more about the anisotropic hydrogen bonding of the quinones to specific amino acid residues, and about the motional dynamics of the quinones at their binding sites. Thus, powder-type high-field EPR spectra were recorded on more than a dozen quinone anion radicals, both natural and model systems. Due to the high Zeeman magneto-selection capability of W-band EPR, a high degree of orientational selectivity was achieved that is unaccessible by X-band EPR (cf. Figure 22.3a and b) (Rohrer *et al.*, 1995, 1998). The measured g-tensor components exhibit the general pattern $g_{xx} > g_{yy} > g_{zz}$, where x is along the >C=O bond direction and z is perpendicular to the quinone plane (Burghaus *et al.*, 1993). The g-tensor values of anion radicals of Q_A in RCs from *Rb. sphaeroides* (*in vivo*), and of ubiquinones-10 in the organic solvent isopropyl alcohol (*in vitro*) are collected in Möbius and Savitsky (2009).

The orientational distribution of molecules for which the Zeeman field is lined up along the g_{yy} direction is considerably broader than for the g_{xx} and g_{zz} directions (see Figure 22.3c). This reflects the still rather poor Zeeman resolution of the quinone g-anisotropy by W-band EPR. Nevertheless, when varying the solvent (protic and aprotic, with and without perdeuteration), characteristic changes of hyperfine-tensor components (predominantly along the y-direction) and g-tensor components (predominantly along the x-direction) could be discerned. These were attributed to hydrogen-bond formation at the lone-pair orbitals on the oxygens: dipolar hyperfine interactions with the solvent protons will result in line broadening along the oxygen lone-pair direction – that is, a broadening of the g_{yy} part of

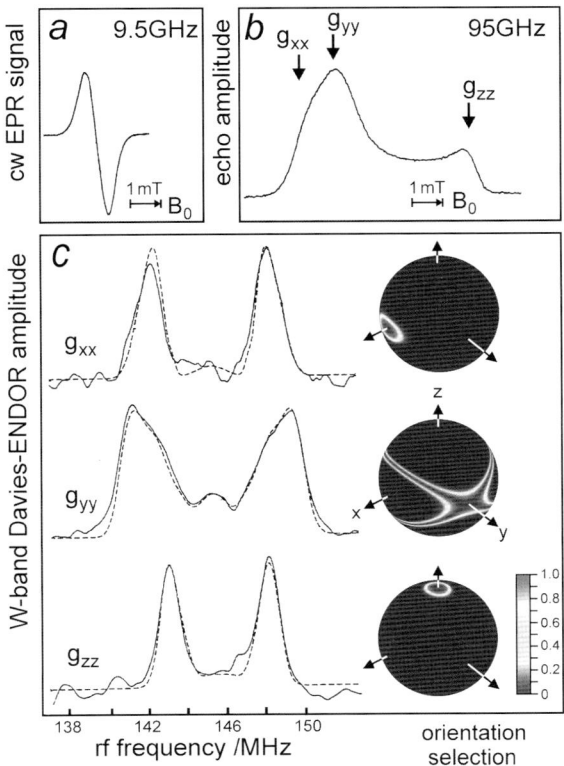

Figure 22.3 (a) X-band CW-detected and (b) W-band ESE-detected EPR spectra of ubiquinone-10 anion radicals in frozen perdeuterated isopropanol solution ($T = 115$ K); (c) Davies-type ENDOR spectra taken at the three B_0 positions marked in (b) by the principal g-tensor components g_{xx}, g_{yy}, g_{zz}. The dotted lines show the simulated ENDOR spectra from which the degrees of orientation selection of contributing molecules (right) were derived (Burghaus et al., 1993; Rohrer et al., 1995).

the EPR spectrum, while changes in the lone-pair excitation energy $\Delta E_{n\pi^*}$ and/or spin density ρ_π^O at the oxygen due to H-bonding will predominantly shift the g_{xx} component of the g-tensor (Rohrer et al., 1995). This explanation conforms to the simplified approach to g-factor theory, as suggested by Stone (Stone, 1963), to approximate the state energies E_0, E_i in the energy denominator of the g-tensor components by the corresponding molecular orbital energies. Since the spin–orbit coupling parameter for p-electrons at the oxygen in the C=O bond is much larger than that of the carbon atoms ($\zeta(C) = 28\,\text{cm}^{-1}$, $\zeta(O) = 151\,\text{cm}^{-1}$; Carrington and McLachlan, 1969), the dominant contribution to the g-tensor components comes from the π spin density at the oxygen. In first-order approximation, g_{rt} is given by (Stone, 1963; Burghaus et al., 1993):

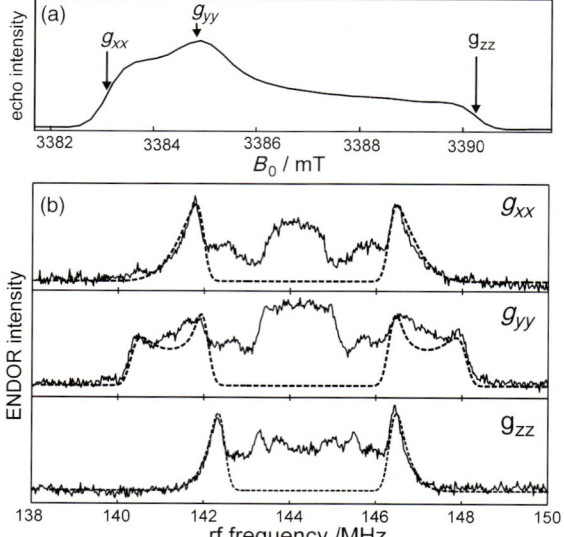

Figure 22.4 (a) Echo-detected W-band powder EPR spectrum of $Q_B^{\bullet-}$ in frozen-solution Zn-substituted RCs from *Rb. sphaeroides* ($T = 120\,\text{K}$). The resonance-field positions of the principal g-tensor values g_{xx}, g_{yy} and g_{zz} are indicated by arrows; (b) W-band Davies-ENDOR spectra (solid lines) recorded at the g_{xx}, g_{yy} and g_{zz} positions of $Q_B^{\bullet-}$ in RCs from *Rb. sphaeroides* ($T = 120\,\text{K}$). The dashed lines indicate the respective contribution of the methyl hfc to the ENDOR spectra. Note that the Larmor frequency and, thereby, the center of the ENDOR spectra increases proportional to the external magnetic field from g_{xx} to g_{zz}. For details, see Schnegg *et al.* (2007).

$$\begin{aligned}
g_{xx} &\approx g_e + 2 \cdot \zeta(O) \cdot \rho_\pi^O \cdot c_{ny}^2/\Delta E_{n\to\pi} \\
g_{yy} &\approx g_e + 2 \cdot \zeta(O) \cdot \rho_\pi^O \cdot c_{nx}^2/\Delta E_{n\to\pi} \\
g_{zz} &\approx g_e \\
g_{rt} &= 0 \text{ for } r \neq t, \, r, t = x, y, z,
\end{aligned} \qquad (22.4)$$

where ρ_π^O is the π spin density c_{SOMO}^2 on the oxygen $2p_z$ atomic orbital; c_{nx}, c_{ny} are the MO coefficients of the $2p_x$ and $2p_y$ atomic orbitals contributing to the oxygen lone-pair orbital ψ_n (briefly n), and $\Delta E_{n\to\pi}$ is the $n \to \pi$ excitation energy. Equations 22.4 can be justified by the fact that the lone-pair orbital n lies energetically very close to the lowest half-filled π orbital (ground state SOMO).

The orientation-selection benefit of EPR and ENDOR at high B_0 field is highlighted in Figure 22.4, which shows the echo-detected W-band EPR spectrum (a) and Davies-type ENDOR spectra (b) of $Q_B^{\bullet-}$ in RCs in frozen solution at $T = 120\,\text{K}$ (Schnegg *et al.*, 2007). The EPR lineshape exhibits the typical powder pattern of a rather well-resolved anisotropic g-tensor. The respective resonance-field positions of the canonical g-tensor values are indicated by arrows. At these field positions the respective ENDOR measurements were performed.

The most elaborate EPR and ENDOR studies on the trapped radical anion states $Q_A^{\bullet-}$ and $Q_B^{\bullet-}$ of the ubiquinone acceptors in Zn-substituted RCs from *Rb. sphaeroides* are described in detailed review articles (Möbius, 1994; Lubitz and Feher, 1999; Weber, 2000; Lubitz, 2004). More recently, the hydrogen-bonding network of $Q_A^{\bullet-}$ (Flores et al., 2007a) and that of $Q_B^{\bullet-}$ (Paddock et al., 2007) were studied by means of 35 GHz (Q-band) EPR and ENDOR. From the frozen solutions of deuterated Q_A in H_2O buffer and protonated Q_A in D_2O buffer, the proton hyperfine and deuteron quadrupole coupling tensors of $Q_A^{\bullet-}$ in its binding site were obtained. The results provide a precise picture of the three-dimensional (3-D) geometry of the Q_A binding site in the charge-separated state of the RC. Nonequivalent hydrogen bonds from the carbonyl oxygens of the quinone to close-by histidine and alanine residues dominate the bonding network of the protein pocket. This information is important for understanding the competition between charge-separation and recombination kinetics in bacterial photosynthesis.

In conclusion, one can say that the multifrequency EPR and ENDOR results show that both $Q_A^{\bullet-}$ and $Q_B^{\bullet-}$ have two different H-bonds to the two carbonyl oxygens; their strength, however, is different in $Q_A^{\bullet-}$ and $Q_B^{\bullet-}$. The EPR and ENDOR results of the primary donor and acceptor ion radicals nicely complement the geometric structure information from X-ray crystallography of RCs. In addition, they provide detailed information of the electronic structure of the transient intermediate states of primary photosynthesis which is not directly obtained by X-ray crystallography. Such information is, however, crucial for understanding the light-induced reactions on the molecular level.

Magnetic resonance can be used to probe the details of both the *static* structure of a molecule and its *dynamic* properties (Thomann, Dalton, and Dalton, 1990; Saxena and Freed, 1997; Kirilina et al., 2005). If the molecular motion is on the time scale of the EPR experiment, spin relaxation and, thereby, line broadening can be observed in the CW EPR spectrum. In many cases, the analysis of this effect is obscured by static ("inhomogeneous") broadening effects from unresolved hyperfine interactions or g-strain. Therefore, pulsed spin–echo techniques – which can separate dynamic and static contributions to the spectrum – are used to study molecular motion (Millhauser and Freed, 1984; Dzuba, Tsvetkov, and Maryasov, 1992). The molecular fluctuations that contribute to the spin relaxation in biomolecules may be vibrational, librational, and rotational motion, intra- and intermolecular conformational dynamics, spin flips of surrounding nuclei or fluctuating coupling to solvent phonons. However, by choosing the correct pulse sequences and resonance conditions, these effects may be discriminated against each other.

High-field/high-frequency (3.4 T/95 GHz) two-pulse echo experiments were performed to investigate the molecular motion of the ubiquinone-10 (UQ-10) cofactor anions, $Q_A^{\bullet-}$ and $Q_B^{\bullet-}$, in RCs of *Rb. sphaeroides* R26. In the investigated RCs, the paramagnetic Fe^{2+} was replaced by diamagnetic Zn^{2+} to reduce the linewidth of the EPR spectra (Debus, Feher, and Okamura, 1986). Thereby, the g-tensor anisotropy is left as the linewidth-determining relaxation contribution to the W-band spectrum. W-band EPR is required to spread the spectrum sufficiently (over almost 10 mT) and to separate the x, y, and z components of the g-tensor (Burghaus et al., 1993) because the g-tensor anisotropies of quinone radical anions are about 4×10^{-3} and the inhomogeneous linewidth is about 0.5 mT. One can, thus, spectroscopically select those molecules in the sample that are oriented with one of these principal g-tensor axes along B_0 to measure their phase-memory time T_{mem} (Rohrer et al., 1996). T_{mem} is the time constant of the echo decay and, since this is monoexponential, T_{mem} can be identified as the transverse relaxation time, T_2. At high magnetic fields, the dominant anisotropic T_2 relaxation is induced by molecular motion modulating the effective g-value of the tumbling radical. At W-band Zeeman fields, this leads to an orientation-dependent modulation of the resonance condition. The anisotropic T_2 contributions have minima (longest T_2) along the canonical orientations of the g-tensor. Thus, the determination of T_2 as a function of the resonance position in the high-field EPR spectrum provides information about the directions and amplitudes of molecular motions and their correlation times.

In order to learn more about the slow motion of the quinone cofactors in photosynthesis, the anisotropic stochastic oscillatory motion of $Q_A^{\bullet-}$ in frozen RC solutions of *Rb. sphaeroides* were studied by pulsed ESE-detected EPR at high-field by Rohrer et al. (1996). The two-dimensional (2-D) field-swept electron spin–echo technique directly reveals the homogeneous linewidth parameter T_2 and, due to the high Zeeman field, resolves its variation over the powder spectrum. The 2-D W-band ESE spectrum of $Q_A^{\bullet-}$ at 115 K in frozen-solution RCs of the *Rb. sphaeroides* mutant HC(M266), in which Fe^{2+} is replaced by Zn^{2+}, shows that the canonical orientations of the g-tensor are rather well resolved at this Zeeman field. The monoexponential echo decay curves $S(2\tau, B_0)$ at the g_{xx}, g_{yy} and g_{zz} orientations are described by Rohrer et al. (1996); Weber et al. (1998): $S(2\tau, B_0) = S_0 \cdot \exp[-2\tau/T_2(B_0)]$. The decays have different time constants T_2 in different directions with respect to the B_0 field. Since T_2 relates solely to the dynamic process, the resolved anisotropy of T_2 directly provides information about the axes of torsional fluctuations (librations) of $Q_A^{\bullet-}$ at a given temperature. At high B_0 fields, the dominant contribution to anisotropic T_2 relaxation stems from the wobbling motion of $Q_A^{\bullet-}$, and depends on the orientation of the g-tensor with respect to the direction of B_0. The magnitudes of the T_2 contributions are determined by random walk on the surface of the g-tensor ellipsoid. This leads to time- and angular-dependent fluctuations δg that translate to fluctuations of the Larmor frequency of the electron spins (Rohrer et al., 1996). As stated above, in high fields, T_2 is dominated by fluctuations of the g-values of the radical when tumbling in the matrix. In the fast-motion limit, changes in $1/T_2$ due to changes in fluctuation frequency and/or motional correlation time are proportional to B_0^2, making high-field ESE decays particularly sensitive for probing matrix effects on the cofactor dynamics. In the case of $Q_A^{\bullet-}$, at $T = 115$ K, the magnitude of T_2 varies strongly over the powder EPR spectrum and clearly peaks at the g_{xx} orientation. For $Q_A^{\bullet-}$, this result is compatible with the X-ray structure of the RC in its neutral ground state Q_A (Ermler et al., 1994). The x-axis of the quinone, which is along the C=O bonds, points to the nearby His(M219) histidine residue and to the more distant Ala(M260) alanine, allowing a strong and a weak H-bond to be formed between the imidazole and peptide nitrogens and the two respective carbonyl oxygens. This asymmetric H-bond pattern is apparently preserved in the anion state $Q_A^{\bullet-}$, that is probed by EPR. This conclusion is in accordance with the H-bond situation observed for ^{13}C-labeled Q_A by NMR (Breton et al., 1994; van Liemt et al., 1995) and FT-IR (Brudler, de Groot et al., 1994) and for ^{13}C-labeled $Q_A^{\bullet-}$ by X- and Q-band EPR (van den Brink et al., 1994; Bosch et al., 1995; Isaacson et al., 1995a; Lubitz and Feher, 1999), but is in contrast to the $Q_B^{\bullet-}$ situation (see below).

The next system to consider is the secondary quinone, Q_B, with its pronounced differences in amino-acid environment as compared to the Q_A site. An intriguing question is how the orientation dependence of T_2 changes the 95 GHz echo decays of $Q_B^{\bullet-}$ in comparison to $Q_A^{\bullet-}$ in RCs from *Rb. sphaeroides* R26 (Schnegg et al., 2002; Möbius et al., 2005).

The anisotropy of the relaxation times T_2 of $Q_A^{\bullet-}$ and $Q_B^{\bullet-}$ for different temperatures has been discussed in detail (Schnegg et al., 2002; Möbius et al., 2005; Möbius and Savitsky, 2009). The T_2 values of $Q_A^{\bullet-}$ and $Q_B^{\bullet-}$ at the resonance-field positions for g_{xx}, g_{yy} and g_{zz} are clearly different, but the orientation dependence is strikingly similar for both quinone anions: The observed field dependence of T_2 of $Q_A^{\bullet-}$ and $Q_B^{\bullet-}$ is very similar, and at 120 K exhibits a pronounced T_2 maximum along the g_{xx} axis – despite their different binding pockets in terms of amino-acid composition and spatial constraints (Stowell et al., 1997). The T_2 values in the various directions mirror the local restrictions to the motion of the quinone in its binding environment. In their protein pockets, at 120 K both $Q_A^{\bullet-}$ and $Q_B^{\bullet-}$ fluctuate uniaxially about a dominant binding axis, that lies along a strong H-bond to an amino acid of the protein pocket. For $Q_A^{\bullet-}$ and $Q_B^{\bullet-}$ the H-bond directions are approximately parallel to g_{xx}, as is reflected by their similar relaxation pattern. As for the Q_B binding site it is found that in the ground-state X-ray structure (Ermler et al., 1994) the C=O bonds of Q_B do not point to a nearby H-bonding amino-acid candidate! Hence, the high-field spin-echo result can only be understood by assuming structural changes of the Q_B binding site upon charge separation, shifting an H-bond candidate like the His(L190) histidine close to the C=O bond. Such structural changes of the Q_B site have, indeed, been unveiled in a more recent high-resolution (1.9 Å) X-ray crystallography experiment conducted by Stowell et al. (Stowell et al., 1997). The structural change of the Q_B site under illumination provides a molecular switch for vectorial ET steps in bacterial photosynthesis. The present results show that low-temperature, high-field spin–echo measurements of the relaxation times at different spectral positions can be used to reveal important structural information, such as H-bond directions in functional proteins at work.

Another important work related to quinone radical anions embedded in hydrogen bond networks (Sinnecker et al., 2004) is reviewed as follows. In a combined Q-band pulse EPR/ENDOR and DFT study of quinone-solvent interactions, the impact of hydrogen bonding on the *p*-benzosemiquinone radical anion BQ$^{\bullet-}$ in coordination with heavy water, D_2O, or alcohol molecules was investigated. After complete geometry optimizations, ^{1}H, ^{13}C and ^{17}O hyperfine, as well as ^{2}H nuclear quadrupole couplings and g-tensor components, were computed. It is noteworthy to comment on the DFT result that there exists a linear correlation between $1/R^2_{O\cdots H}$ and the g_{xx} tensor component. On the other hand, the strong dependence of the g-shifts on the detailed structure and polarity of the surroundings of the radicals makes it difficult – if not impossible – to derive a quantitative relationship between the g-shift and the H-bond length that is sufficiently general to be used in practice.

22.3.6
95-GHz ESE-Detected EPR on the Spin-correlated Radical Pair $P^{\bullet+}Q_A^{\bullet-}$

W-band high-field, field-swept two-pulse ESE experiments on the laser-pulse-generated short-lived radical pairs $P^{\bullet+}Q_A^{\bullet-}$ in frozen RC solution of *Rb. sphaeroides* (Prisner *et al.*, 1995) are reviewed here. These were performed to determine the 3-D structure of the charge-separated donor–acceptor system via spin–polarization effects. Principally, this charge-separated-state structure may differ from the neutral ground-state structure. Indeed, upon the illumination of RC crystals of *Rb. sphaeroides*, drastic changes have been observed in the X-ray structure of the secondary quinone, Q_B, binding site in comparison with the dark-adapted X-ray structure (Stowell *et al.*, 1997). In order to avoid fast spin relaxation of $Q_A^{\bullet-}$, the non-heme Fe^{2+} ion was replaced by Zn^{2+}, and the charge-separated radical pairs were generated by 10 ns laser flashes. Their time-resolved EPR spectrum is strongly electron-spin polarized because the transient RPs are suddenly born in a spin-correlated non-eigenstate of the spin Hamiltonian with pure singlet character. Such spin-polarized spectra with lines in enhanced absorption and emission originate from the CCRP (spin-correlated coupled radical-pair) mechanism (Prisner *et al.*, 1995). They contain important structural information on the magnitude and orientation of the g-tensors of the two radical partners, $P^{\bullet+}$ and $Q_A^{\bullet-}$ with respect to each other and to the dipolar axis r_{QP} connecting the two radicals (Prisner *et al.*, 1995). Several parameters critically determine the lineshape of the CCRP polarization pattern, such as the principal values and orientations of the g- and electron dipolar-coupling tensors, the exchange coupling *J*, and the inhomogeneous linewidths of both radicals (Prisner *et al.*, 1995).

In the pulsed W-band ESE experiments, unlike those at X-band (9.5 GHz), K-band (24 GHz) and Q-band (35 GHz) (Stehlik and Möbius, 1997), the Zeeman field is strong enough to sufficiently separate the spectral contributions from $P^{\bullet+}$ and $Q_A^{\bullet-}$. Thus, the overall spectrum is dominated by the characteristics of the two g-tensors, and its interpretation is simplified. This allows for an unambiguous analysis of the tensor orientations. The most important result of this high-field ESE study is that, within an estimated error margin of ±0.3 Å, no detectable light-induced structural changes of the quinone site occur, as compared to the ground-state configuration $P_{865}Q_A$. This is in accordance with recent results from various other studies, including X-ray crystallography (Stowell *et al.*, 1997), and contrasts with the $Q_B^{\bullet-}$ situation. However, as has been shown recently (Savitsky *et al.*, 2007a), there exist small – but significant – conformational changes of $Q_A^{\bullet-}$ with respect to Q_A, which require the enhanced orientational resolution of high-field PELDOR for their detection (see below).

22.3.7
95-GHz RIDME and PELDOR on the Spin-Correlated Radical Pair $P^{\bullet+}Q_A^{\bullet-}$

The light-induced transient radical pairs $P^{\bullet+}Q_A^{\bullet-}$ of the primary electron donor and quinone acceptors in bacterial and plant photosynthetic reactions centers have been characterized by a variety of time-resolved EPR methods (Weber, 2000; Thurnauer, Poluektov, and Kothe, 2004; Möbius and Savitsky, 2009). After pulsed laser excitation of the primary donor, $P^{\bullet+}Q_A^{\bullet-}$ appears in the CCRP state, which is characterized by a weak electron spin–spin coupling in a fixed geometry of the radicals in the pair, and an initial singlet state of the system. The EPR responses of the CCRP state display a number of interesting and useful spectroscopic features: spin polarization, quantum beats, transient nutations, as well as echo-envelope modulation and out-of-phase echo effects. The magnetic-interaction parameters and the geometry of the system can be obtained from the analysis of these features. The 3-D geometric information about the radical pair is of particular importance, as it allows one to: (i) extract structural information about the transient charge-separated states in photosynthetic RCs for which detailed X-ray data are only rarely available (Stowell *et al.*, 1997); and (ii) recognize and characterize the structural changes occurring in RCs upon charge-separation and charge-recombination processes. An additional EPR methodology was developed which allows one to obtain the orientation information directly, and this with predictable accuracy. This methodology is known as pulsed electron–electron double resonance (PELDOR) or double electron–electron resonance (DEER) dipolar spectroscopy (Milov, Salikhov, and Shirov, 1981; Milov, Ponomarev, and Tsvetkov, 1984), in which two MW fields of different frequencies are applied. This experiment is favorably performed on the CCRP state at high magnetic fields in conjunction with the one-frequency RIDME (relaxation-induced dipolar modulation enhancement) dipolar spectroscopy (Kulik, Paschenko, and Dzuba, 2002). Recent studies of orientation-resolving high-field dipolar spectroscopy on $P_{865}^{\bullet+}Q_A^{\bullet-}$ via PELDOR and RIDME at 95 GHz/3.4 T (Savitsky *et al.*, 2007a; Schnegg *et al.*, 2007) has been reviewed by Möbius and Savitsky (Möbius and Savitsky, 2009), summarizing the principles of these dipolar spectroscopies on CCRP and thermalized radical-pair states as an example.

The geometry parameters characterizing the relative positioning of the paired radicals $Q_A^{\bullet-}$ and $P_{865}^{\bullet+}$, as determined by high-field dipolar EPR methods (Savitsky *et al.*, 2007a), are compared with those derived for a model pair from the X-ray structures of RCs from *Rb. sphaeroides* (these are listed in Möbius and Savitsky (2009). Small changes were thus revealed with respect to the torsional angle around the C=O axis (x-axis) of the quinone. The significance of the PELDOR results in relation to the drastic conformational changes upon charge separation observed for the Q_B binding site have been discussed (Savitsky *et al.*, 2007a; Möbius and Savitsky, 2009).

RCs from *Rb. sphaeroides* R-26 exhibit rather large changes in the recombination kinetics of the charge-separated radical-pair state, $P^{\bullet+}_{865}Q^{\bullet-}_A$, depending on whether the RCs are cooled to cryogenic temperatures in the dark or under continuous illumination (Kleinfeld, Okamura, and Feher, 1984). Structural changes upon charge separation had been suggested as the cause for this observation (Kleinfeld, Okamura, and Feher, 1984). To explore the nature of the proposed structural changes, W-band EPR and PELDOR was employed to obtain the necessary orientation selectivity and spectral resolution of the radical-pair spectra (Flores et al., 2007b; Savitsky et al., 2007a). High-field PELDOR spectroscopy (Savitsky et al., 2007a) was used to monitor the relative orientation of the cofactor ions in the pair $P^{\bullet+}_{865}Q^{\bullet-}_A$, the orientational distribution of $Q^{\bullet-}_A$, and the distance between the donor and acceptor ions in deuterated RCs frozen in the dark or frozen under illumination. Details of the data analysis are given in Savitsky et al. (2007a). Finally, it was found that all the structural parameters of the $P^{\bullet+}_{865}Q^{\bullet-}_A$ radical pair are the same for both the dark-adapted and light-adapted RC states, which implies that the Q_A site does not experience significant conformational changes upon light-induced ET. Apparently, other factors are responsible for the retardation of charge separation in light-adapted RCs, possibly unbound water molecules in the protein that reorient under the electric-field changes upon charge separation (Iwata et al., 2009).

Reference is also made to the recent 180 GHz high-field PELDOR experiments on the tyrosyl radicals in R2 from mouse ribonucleotide reductase (Denysenkov et al., 2008). The pronounced orientational selectivity was employed to determine the biradical structure, and to elucidate the relative orientation of radical sites in a protein complex. It is clear that high-field PELDOR is an important step towards the full structure determination of proteins by EPR methodologies, provided that suitable paramagnetic probes are available for creating weakly coupled radical pairs. For the frequently occurring case of proteins with no paramagnetic states involved in their biological action, the construction of site-directed mutants for selective nitroxide spin labeling has been developed as a widely used technique in molecular biology. Radical-pair states can be generated by double-spin labeling, or by taking advantage of intrinsic paramagnetic transition-metal ions that either weakly interact among each other or with site-specifically attached nitroxide spin labels.

This section is concluded by drawing attention to recent work on bacterial photosynthetic reaction centers, and applying such a strategy of combining site-directed spin labeling with natural cofactor radicals to generate two-spin systems in RC proteins (Borovykh et al., 2006; Gajula et al., 2007). Although, so far only X-band EPR techniques have been applied to this end, this strategy shows much promise also for high-field EPR investigations. The RC of *Rb. sphaeroides* contains five native cysteines necessary for site-directed spin labeling. The EPR experiments, in conjunction with molecular dynamics (MD) simulations, show that only one cysteine, Cys(H156) located on the H protein subunit, is accessible for spin labeling. Using the two-frequency pulsed DEER method, a distance of 3.05 nm between the paramagnetic semiquinone anion state of the primary acceptor (Q_A) and the spin label at the native cysteine at position 156 in the H-subunit is found.

MD simulations are performed to interpret the distance. For more details, see Borovykh *et al.* (2006) and Gajula *et al.* (2007).

22.3.8
Multifrequency EPR on Primary Donor Triplet States in RCs

Doublet state ($S = 1/2$) cofactor ions occuring as paramagnetic intermediates in the ET cascade initiated by light excitation of the primary donor, P, in the photosynthetic reaction center have been the focus of most EPR and ENDOR studies at various MW frequencies. Pulsed EPR experiments at Q-band (34 GHz) on the triplet state of the primary donor P_{865} in wild-type and mutant RCs from *Rb. spaeroides* at 10 K have been performed recently (Marchanka *et al.*, 2007) to investigate the relative activities of the A- and B-branch for charge separation in the RC protein complex. On the other hand, very few studies on $^3P^*$ employing high-field EPR have yet appeared (Labahn and Huber, 2001; Paschenko, Gast, and Hoff, 2001; Pachtchenko, 2002; Zeng *et al.*, 2003). There exist excellent review articles dealing with triplet states in photosynthesis (e.g., Lubitz, 2002; Lubitz, Lendzian, and Bittl, 2002; Lubitz, 2004), focusing on the primary donor in bacterial RCs as well as PS I and PS II of oxygenic photosynthesis. Zeng *et al.* (2003) have reported an interesting application of 240 GHz high-field EPR on the primary donor triplet state in randomly oriented RCs from *Rb. sphaeroides*, embedded in dried plastic films. These authors studied the temperature dependence of the $^3P^*$ g-tensor; due to the high Zeeman resolution of 240 GHz EPR, the measured g-tensor data allowed determination of both the principal values and the principal-axes directions with respect to the axes of the zero-field splitting tensor. In contrast to the doublet-state cation $P^{\bullet+}$, the triplet state $^3P^*$ exhibited a significant temperature dependence of its g-tensor, particularly in the directions of the principal axes. Over the temperature range studied (10–230 K), the g-tensor principal axes system rotated by about 30° around the x-direction. The only previous observation of temperature-dependent principal g-values of $^3P^*$ was reported by Hoff and Proskuryakov (Hoff and Proskuryakov, 1985). Apparently, the much lower resolution available at 9 GHz X-band EPR (Hoff and Proskuryakov, 1985) greatly limits the reliability of the reported temperature effects. Among discussed reasons for the observed temperature dependence of the g-tensor, the most probable include: (i) temperature-dependent conformational changes of the P molecule (Scherer *et al.*, 1985; Reddy, Kolaczkowski, and Small, 1993); (ii) temperature-dependent delocalization of triplet excitation onto an accessory bacteriochlorophyll (Hoff and Proskuryakov, 1985; Aust *et al.*, 1990; Angerhofer *et al.*, 1998); and (iii) spin–orbit coupling (SOC) between ^3P and an electronic state with temperature-dependent energy, such as the first excited state of P. It is suggested that SOC of the acetyl oxygen atom may be quite significant in determining the rotation of the g-tensor in ^3P as a function of temperature.

Measurement of the g-tensor of $^3P^*$ offers a unique opportunity to compare the g-tensors of two different states of the same molecule. For $^3P^*$, the principal g-values – particularly g_{xx} and g_{yy}, – are considerably larger than the corresponding values of P•+. The g-axis orientations in the two states are also significantly different, and well beyond experimental uncertainty. The observed differences between the g-tensors of ^{3}P* and P•+ may be rationalized in terms of the electronic structures of the two states. Whereas, the unpaired spin in P•+ occupies only the HOMO of P, the second unpaired electron in ^{3}P* occupies the LUMO; thus, the g-tensor of ^{3}P probes the properties of both HOMO and LUMO. Because the LUMO is singly occupied in the radical-anion states, one might expect, to a first approximation, that the g-tensor of ^{3}P* reflects the "average" properties of the radical cation and anion states, P•+ and P•−.

This section is concluded by reviewing a most recent Q- and W-band study on doublet and triplet states of a quadruple mutant of the purple bacterium *Rb. sphaeroides*, which was performed with the aim of affecting the directionality of light-induced ET along the A- or B-branch of the RC protein (Marchanka *et al.*, 2010). This directionality of light-induced charge transfer in bacterial RCs is still poorly understood on the electronic level. Site-directed mutants with specific alterations of the cofactor binding sites with respect to the native system can reveal useful information towards a better understanding of the directionality enigma. This has been demonstrated by EPR studies of the quadruple mutant, LDHW, which is derived from H*L*(M182)/G*D*(M203)/L*H*(M214)/A*W*(M260) and contains crucial mutations in the ET pathway. The directionality of the charge separation process was studied under light- or dark-freezing conditions, both *directly* by W-band (95 GHz) high-field EPR spectroscopy examining the charge-separated radical-pair state, $P_{865}^{•+}Q_B^{•-}$, of the primary donor P_{865} and terminal acceptor Q_B, and *indirectly* by Q-band (34 GHz) EPR examining the triplet state of the primary donor, $^3P_{865}$, that occurs as byproducts of the photo-reaction. The triplet state $^3P_{865}$ in the LDHW mutant has been investigated in a wide range of temperatures and at different excitation wavelengths. At 10 K, the triplet state has been found to derive mainly from an intersystem-crossing mechanism, indicating the absence of any charge-separated radical-pair states with a lifetime longer than 10 ns. B-branch charge separation and formation of the triplet state $^3P_{865}$ via a radical-pair mechanism can be induced with low yield at 10 K by direct excitation of the bacteriopheophytins in the B-branch at 537 nm. At this wavelength, charge separation most probably proceeds via hole transfer from bacteriopheophytin to the primary donor. The triplet state of the primary donor is found to be quenched by the carotenoid cofactor present in the RC. The light-induced radical-pair state of the charge-separated primary donor and B-branch quinone, $P_{865}^{•+}Q_B^{•-}$, in RCs from the LDHW mutant has been characterized using pulsed-EPR spectroscopy at 95 GHz, following similar cooling/excitation protocols as in the triplet-state measurements. The RC samples were cooled to 90 K either in the dark (dark-adapted ground-state protein conformation) or under continuous illumination (light-adapted charge-separated state protein conformation). About 70% of the RCs illuminated upon freezing are trapped in the long-lived ($\tau > 10^4$ s) charge-separated state $P_{865}^{•+}Q_B^{•-}$. The temperature behavior of the EPR signals from $P_{865}^{•+}Q_B^{•-}$ points to two factors responsible for the forward ET to the terminal acceptor Q_B and for the charge-recombination reaction. The first factor involves a significant protein conformational change to initiate $P^{\bullet+}_{865}Q^{\bullet-}_B$ charge separation, presumably by moving the quinone from the distal to the proximal position relative to the non-heme iron. The second factor includes a protein relaxation process governing the charge-recombination process along the B-branch pathway of the LDHW mutant.

22.4
Oxygenic Photosynthesis

Detailed reviews are available on doublet-state radicals occurring in PS I and PS II during light-induced ET reactions, as characterized using EPR in general (Lubitz, 2004), and high-field EPR in particular (Un, Dorlet, and Rutherford, 2001). These authors provide precise g-tensor data of tyrosine-, quinone-, pheophytin- and chlorophyll-based radicals, from which valuable information about radical–protein interactions can be derived. Likewise, EPR results concerning spin pairs in PS I and PS II have been thoroughly reviewed. The present understanding of the structure–function relationship for the primary processes in oxygenic photosynthesis of green plants is still considerably less developed as compared to that in bacterial photosynthesis, owing to the much greater complexity of the charge-transfer photomachine of the interconnected photosystems PS I and PS II. Moreover, it is only recently that the 3-D structures of PS I and PS II have become known with sufficient accuracy. Whilst this situation is in contrast to that of RCs from the purple photosynthetic bacteria, it has recently dramatically improved due to progress in high-resolution X-ray crystallography of PS I and PS II (Jordan *et al.*, 2001; Zouni *et al.*, 2001).

22.4.1
Multifrequency EPR on Doublet States in Photosystem I (PS I)

Whilst the 2001 X-ray structure of PS I has reached 2.5 Å resolution (Jordan *et al.*, 2001), before then it was impossible to even locate the quinone acceptor A_1, a phylloquinone. It was, therefore, a major challenge to apply the whole arsenal of modern multifrequency time-resolved EPR and ENDOR techniques to PS I with the goal of determining the $A_1^{\bullet-}$ location and orientation; for reviews, see Füchsle *et al.* (1993); Bittl and Zech (2001); Teutloff *et al.* (2001); Lubitz (2002), and Lubitz (2004). For instance, time-resolved EPR studies, at X-, K- and W-band frequency/field settings, either with direct detection or with ESE detection, were performed on the transient spin-correlated radical pair $P^{\bullet+}_{700}A_1^{\bullet-}$ of PS I, and the results compared with those of $P^{\bullet+}_{865}Q^{\bullet-}_A$ in RCs from *Rb. spaeroides* (van der Est *et al.*, 1997).

From the spin-polarized EPR spectra of the radical pair $P^{\bullet+}_{700}A_1^{\bullet-}$ in highly purified PS I particles, the values of both the magnetic parameters of the radical pair and the relative orientation of the two radical species were obtained. It was found that $A_1^{\bullet-}$ was oriented such that the carbonyl bonds are parallel to the vector joining the centers of $P^{\bullet+}_{700}$ and $A_1^{\bullet-}$. The anisotropy of the g-tensor was found to be considerably larger than that obtained for chemically reduced phylloquinone in frozen

2-propanol solution, most likely due to specific cofactor–protein interactions in the $A_1^{\bullet-}$ binding site. The relative orientation of $P_{700}^{\bullet+}$ and $A_1^{\bullet-}$ was compared with that measured earlier by W-band ESE-detected EPR for the radical pair $P_{865}^{\bullet+}Q_A^{\bullet-}$ in Zn-substituted bacterial RCs from *Rb. sphaeroides* R-26, in which the non-heme iron was replaced by zinc (Prisner *et al.*, 1995) (see Figure 22.5). This information was used to compare the structural and magnetic properties of the charge-separated state in the two systems. The low-field parts of the two W-band spectra (Figure 22.5) are very different from each other, an effect which results from differences in the orientations of $A_1^{\bullet-}$ and $Q_A^{\bullet-}$ with respect to their inter-spin distance vectors as a consequence of their different binding sites. This finding was fully confirmed later by the 2.5 Å high-resolution X-ray structure of PS I (Jordan *et al.*, 2001).

Additional variations on the PS I theme have been published, with special emphasis on the advantages of high-field EPR. For example, W-band EPR was

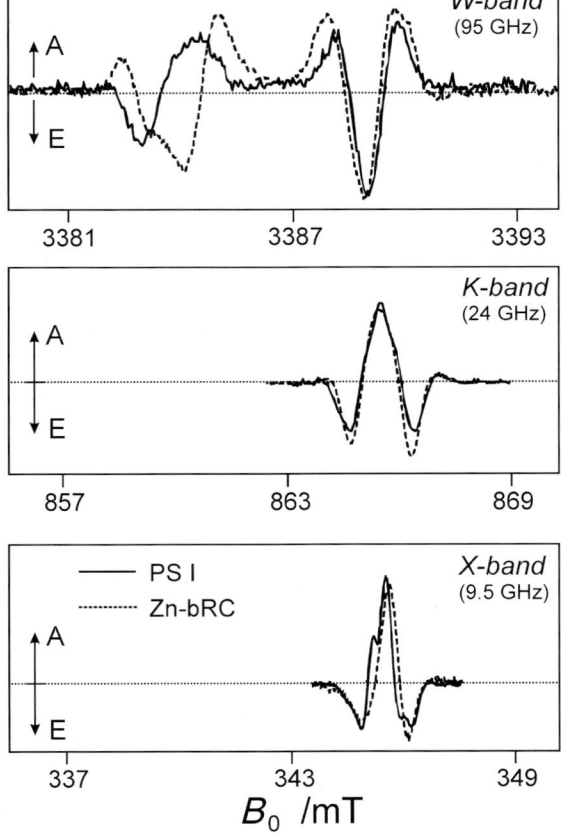

Figure 22.5 Comparison of spin-polarized EPR spectra of $P_{700}^{\bullet+}A_1^{\bullet-}$ in PSI and $P_{865}^{\bullet+}Q_A^{\bullet-}$ in Zn-bRCs at three different microwave frequencies. For details, see Prisner *et al.* (1995).

used to resolve the g-tensor components of the primary chlorophyll donor cation $P_{700}^{\bullet+}$ and quinone acceptor $A_1^{\bullet-}$ in frozen-solution and single-crystal preparations of PS I (Teutloff et al., 2001). The measured principal values of the g-tensor of $P_{700}^{\bullet+}$ were compared to those of the isolated pigment radicals in organic solvents. Information concerning cofactor–protein interactions in the binding sites was obtained from the observed differences. The measured g-tensor principal axes of $P_{700}^{\bullet+}$ were assigned to the molecular structure of the primary donor by means of an analysis of the spin-polarized spectra of the photoinduced radical pair $P_{700}^{\bullet+}A_1^{\bullet-}$. DFT calculations on a structural model of the A_1 binding pocket, as derived from the 2.5 Å X-ray structure (Jordan et al., 2001), were used to correlate the EPR parameters with structural elements of the protein. Another detailed g-tensor analysis has been made for the cation radicals and triplet states of the primary donor P_{700} in PS I and in vitro chlorophyll a (Poluektov et al., 2002).

To summarize these multifrequency studies on PS I, it can be said that by combining all pieces of information from the various state-of-the-art EPR and ENDOR experiments, the goal of localizing A_1 in PS I was finally achieved during the time before crystallization ("BC time"). The results obtained with EPR were in agreement with the structures determined later with high-resolution X-ray crystallography (Jordan et al., 2001). Multifrequency time-resolved EPR was found to be particularly suited to the characterization of radical and radical-pair intermediates in the RCs of different photosynthetic organisms, both wild-type and mutants. Subsequently, these aspects of mutation-induced changes of the PS I complex and their characterization led to several pulsed high-field EPR and ENDOR studies being conducted (Zybailov et al., 2000; Xu et al., 2003).

The technique of orientation selection by high-field EPR was exploited to study the transient radical pair $P_{700}^{\bullet+}A_1^{\bullet-}$ in photosynthetic PS I multilayers, by applying time-resolved W-band EPR (TREPR) with direct detection (Fuhs et al., 2002). The width of the orientation distribution of the transient radical pairs $P_{700}^{\bullet+}A_1^{\bullet-}$ was determined as $(30 \pm 10)°$. Simulations of the 1-D-oriented spectra show that subtle structural changes in the radical-pair complex in membrane fragments (e.g., by point mutations) can be detected more easily in oriented multilayers of PS I than in disordered frozen-solution samples. The combination of multifrequency TREPR on disordered samples and on oriented multilayers was found to represent a very appealing strategy for the structural analysis of transient radical-pair systems, and their mutation-induced changes, in photo-induced ET processes.

By combining X-band and W-band experiments on PS I complexes of intact cyanobacterial cells Synechocystis sp. PCC 6803, a significant step forward was taken recently in identifying the complexity of electron transport processes in whole cells (Savitsky et al., 2007b). It should be noted that the energy of light quanta absorbed by the LHCs is delivered to the pigment–protein complexes, to PS I and PS II, and is then converted via an electron-transfer chain (ETC) into the energy of macroenergetic chemical compounds, NADPH and ATP (see Figure 22.1). The ETC contains three large protein complexes located in the thylakoid membrane: PS I, PS II, and the cytochrome b_6f-complex (bf-complex). Light-induced electron transport from the water-splitting complex of PS II to the terminal electron acceptor of PS I (NADP$^+$) is mediated by the bf-complex and three mobile electron carriers, plastoquinone (Q), plastocyanine (Pc), and ferredoxin (Fd), that provide the electron transport between PS II, bf-complex, PS I and the ferredoxin–NADP-reductase complex. It was found that the kinetic behavior of the cation-radical $P_{700}^{\bullet+}$, generated by illumination with continuous light, and the EPR intensity of the radical pair $P_{700}^{\bullet+}A_1^{\bullet-}$, generated by laser pulse illumination, depended heavily on the illumination prehistory (either the sample was frozen in the dark or during illumination). Both processes were sensitive to the presence of electron-transport inhibitors, which block the electron flow between PS I and PS II in the cell. In line with the X-band EPR data on the kinetics of light-induced redox transients of P_{700}, the high-field W-band EPR study of the spin-correlated radical-pair state $P_{700}^{\bullet+}A_1^{\bullet-}$ showed that photosynthetic electron flow through PS I is controlled both on the donor and on the acceptor sides of PS I. That is, it is controlled at two check-points: (i) at the plastoquinone segment of the ETC; and (ii) at the acceptor side of PS I. In conclusion, it is emphasized that the high spectral and time resolution, as well as excellent detection sensitivity of high-field W-band EPR, has allowed the elucidation of the details of the complex electron-transfer pathways in the interconnected ETCs of both intact and chemically treated cells of photosynthetic organisms.

22.4.2
Multifrequency EPR on Doublet States in Photosystem II (PS II)

There are many similarities between PS II and the simpler bacterial RC with regards to the primary electron-transfer cofactors embedded in their protein subunits. Both of these RCs exhibit a pseudo-C_2 symmetry axis running through the non-heme iron and the dimeric primary donor sitting in the symmetry-related D1 and D2 protein subunits. As for the complete PS II protein complex, however, its increased complexity is attributed to the evolutionary invention of photosynthetic water splitting by the oxygen-evolving complex (OEC), with its protein-bound tetranuclear manganese cluster, and to protect the fragile photosynthetic machinery against the destructive interaction with molecular oxygen produced in the OEC of PS II (Raymond and Blankenship, 2004). The following discussion is focused on a small selection of multifrequency EPR studies on PS II cofactors. The comprehensive reviews mentioned above (Un, Dorlet, and Rutherford, 2001; Lubitz, 2004) can be consulted for a broader coverage of the field.

Currently, there is a growing interest in advanced EPR studies of the tyrosyl radicals in D1 and D2 polypeptides. In particular, the determination of individual g-tensor components in frozen PS II preparations by high-field/high-frequency EPR has proved to be very informative in probing the different hydrogen-bonding interactions of the Y_Z and Y_D tyrosyl and the Q_A and Q_B quinone radicals. Their frozen-solution (Un et al., 1995; Stehlik and Möbius, 1997) and single-crystal (Hofbauer et al., 2001) spectra exhibit resolved Zeeman and, in some cases, hyperfine structure. Similar CW high-field EPR studies have been performed to characterize the pheophytin anion radical in the wild-type and D1-E130 mutants of PS II from *Chlamydomonas reinhardtii* (Dorlet et al., 2001). Both, EPR and X-ray absorption spectroscopies, with their complementary information contents, have provided much of what is now known about the structure of the Mn complex. Recently, the combined efforts of laboratories using X-ray diffraction, extended X-ray absorption fine structure (EXAFS), X-ray absorption near edge structure (XANES) and multifrequency EPR and ENDOR techniques have led to a significant improvement in spatial resolution. For the Mn_4O_xCa complex, more than 10 different structural models have been suggested, considering two or three 2.7 Å Mn–Mn distances and one or two 3.3 Å Mn–Mn distances (DeRose et al., 1994; Yachandra, Sauer, and Klein, 1996; Peloquin et al., 2000; Messinger et al., 2001; Peloquin and Britt, 2001; Junge et al., 2002; Yachandra, 2002; Britt et al., 2004; Glatzel et al., 2004, 2005; Haumann et al., 2005a, 2005b; Kulik et al., 2005a, 2005b; Kulik, Lubitz, and Messinger, 2005).

Although high-field EPR (at 95 GHz and above) should be suitable for studying the Mn cluster that constitutes the OEC, such high-field studies on the Mn cluster in PS II have not yet been carried out, although the details of multifrequency (9, 95, and 285 GHz) high-field EPR studies of binuclear Mn(III)–Mn(IV) complexes as relevant model systems were reported some ten years ago (Policar et al., 1998). This excursion to the "hidden secrets" of the electronic structure of the Mn_4O_xCa cluster in the OEC of PS II is concluded by reviewing a very recent breakthrough in the theoretical interpretation of pulse Q-band EPR and ^{55}Mn-ENDOR results (Kulik et al., 2005b). The recently obtained ^{55}Mn hyperfine coupling constants of the S_0 and S_2 states of the OEC were analyzed on the basis of Y-shaped spin-coupling schemes with up to four non-zero exchange-coupling constants, J. This analysis ruled out the presence of one or more Mn(II) ions in S_0, establishing that the oxidation states of the manganese ions in S_0 and S_2 are, at 4 K, Mn_4(III, III, III, IV) and Mn_4(III, IV, IV, IV), respectively. From this analysis a new structural model is favored that is fully consistent with the EPR and ^{55}Mn-ENDOR data. Furthermore, Mn oxidation states were assigned to the individual Mn ions. It was proposed that the known structural changes of the Mn_4O_xCa cluster, when passing through the Kok cycle, namely the shortening of one 2.85 Å Mn–Mn distance in S_0 to 2.75 Å in S_1, corresponded to a deprotonation of a μ-hydroxo bridge between Mn_A and Mn_B – that is between the outer Mn and its neighboring Mn of the $μ_3$-oxo bridged moiety of the cluster. The exchange coupling J_{AB} for the suggested models of the Mn_4O_xCa cluster is significantly smaller in the S_0 state as compared to the S_2 state, while the other J couplings between the four Mn ions in the possible coupling schemes (dimer of dimers, trimer-monomers, tetramers) are hardly changed. This was the first time that an assignment of a structural change within the Mn_4O_xCa cluster to a specific Mn–Mn bridge was made, namely that between Mn_A and Mn_B. Apparently, the Mn_4O_xCa cluster actively takes part in the photosynthetic water-splitting chemistry via structural changes within the cluster. These results are expected to be useful for the synthesis of artificial catalysts for solar water-splitting, which is certainly a most challenging project for renewable energy resources.

22.5
Concluding Remarks

The field of photosynthesis research is currently concerned with understanding the dominant factors that control the specificity and efficiency of electron- and ion-transfer processes in membrane proteins, on the basis of structure and dynamics data. Modern CW and pulsed EPR, operated at high magnetic fields and MW frequencies and extended by multiple-resonance capabilities, is currently playing an important role in this endeavor, particularly in view of the fact that no single-crystal protein preparations are required here to obtain any detailed structural information at atomic resolution. During the past decade, biologists, chemists and physicists have all attempted to develop high-field EPR techniques [see Möbius and Savitsky (2009) for a comprehensive account on the instrumentation developments]. Such techniques were applied to functional proteins, thus demonstrating that this type of spectroscopy is particularly powerful when characterizing the structure and dynamics of the transient states of proteins in action on biologically relevant time scales. The information thus obtained is unique in its specifity, and is complementary to that provided by protein crystallography, solid-state NMR, and laser spectroscopy.

The salient features of the studies on high-field EPR spectroscopy on photosynthesis in this chapter are as follows:

1) Many organic cofactors in proteins possess only small g-anisotropies. Thus, to resolve their canonical g-tensor orientations in their powder spectra from disordered samples ("Zeeman magnetoselection"), much higher magnetic fields B_0 are required than those applied in X-band EPR. Thus, even in disordered samples, orientation-selective hydrogen bonding and polar interactions in the protein binding sites can be traced. This information is important, and complementary to that available from the high-resolution X-ray diffraction of protein crystals.

2) Often, several paramagnetic organic species are generated in photochemical reactions and ET processes as transient intermediates with overlapping EPR spectra. In order to distinguish them by the notoriously small differences in their g-factors and hyperfine interactions, high Zeeman fields are required, so that high-frequency EPR becomes the method of choice.

3) High-purity protein samples can often be prepared only in minute quantities; examples include the RCs of site-directed mutants, or with isotopically labeled cofactors. The problem with a small concentration of paramagnetic molecules is the rule for single crystals of membrane proteins, which are often tiny in all dimensions. To study these using EPR, a very high absolute detection sensitivity is thus needed, which can be accomplished only with dedicated high-field/high-frequency spectrometers.

4) High-field/high-frequency CW EPR generally provides shorter time windows by lineshape analysis down to the picosecond range for studying correlation

times and fluctuating local fields over a wide temperature range. They are associated with characteristic dynamic processes, such as protein motion and refolding or cofactor libration and reorientation in the binding sites of photosynthetic RCs.

5) 2-D field-swept ESE spectroscopy, as a high-field/high-frequency pulsed-EPR technique, provides real-time access to specific cofactor/protein slow motions in the nanosecond time-scale. With this technique, even their motional anisotropy, generated by anisotropic interactions (e.g., hydrogen bonding within the binding site), can be traced, and may lead to temperature-dependent anisotropic relaxation.

6) High magnetic-field ENDOR takes advantage of the additional orientation selection of molecular sub-ensembles in powder or frozen-solution samples, as obtained by double orientation selection by the Zeeman field and ENDOR frequency. Thus, ENDOR can provide single-crystal-like information about hyperfine interactions, including anisotropic hydrogen bonding of the cofactors to the protein, even in the case of small g-anisotropies.

7) Analysis of the EPR spectra of metalloprotein high-spin systems can be drastically simplified at high Zeeman fields, owing to the suppression of second-order effects in metalloprotein high-spin systems. A prominent example is the Mn^{2+} oxygen-evolving complex in the photosynthetic RC, PS II. EPR transitions of certain high-spin metalloproteins with large zero-field splittings cannot be observed at all at X-band frequencies, but they become accessible at the higher quantum energies of millimeter or sub-millimeter microwave fields.

8) Information on the electronic structure of the ET redox partners, and on the 3-D structure of radical-pair systems with large inter-spin distances (up to ca. 8 nm), even in disordered frozen solutions, can be obtained with high precision by the powerful tools of pulsed high-field electron–electron dipolar EPR spectroscopy, such as the PELDOR and RIDME techniques. In photosynthetic ET proteins, the distance and relative orientation of functional cofactors within the protein domains, and their conformational changes during the photocycle, determine the selectivity and efficiency of the biological processes. Not only interspin distances, but the full 3-D structure of laser-flash-induced transient charge-separated radical pairs in frozen-solution RCs from photosynthetic organisms can be obtained by orientation-resolving 95 GHz high-field PELDOR, and related techniques.

In conclusion, it should be noted that CW and pulse multifrequency EPR at high magnetic fields has matured recently to add substantially to the capabilities of "classical" spectroscopic and diffraction techniques for determining the structure–dynamics–function relations of biosystems, since transient intermediates can be observed in real time in their working states at biologically relevant time scales. Today, the role of high-field EPR in biology, chemistry, and physics is growing rapidly and, as a consequence, the scientific literature on high-field EPR also continues to grow rapidly. It was for these reasons that only a small part of the relevant research could be included in this chapter.

Acknowledgments

S.K.M. acknowledges partial financial support from the Natural Sciences and Research Council of Canada (NSERC). K.M. and A.S. gratefully acknowledge financial support by the Deutsche Forschungsgemeinschaft in the frame of the Priority Program SPP 1051 ("High-field EPR in Biology, Chemistry and Physics"), the Collaborative Research Center SFB 498 ("Protein-Cofactor Interactions in Biological Processes") and the Group Project MO 132/19-2 ("Protein Action Observed by Advanced EPR").

Pertinent Literature

The material covered in this chapter is largely extracted from Chapter 5 of the monograph by K. Möbius and A. Savitsky, *High-Field EPR Spectroscopy on Proteins and their Model Systems: Characterization of Transient Paramagnetic States*; RSC Publishing, London, (2009). This book describes how multifrequency EPR, and in particular high-field EPR methodology, in conjunction with mutation strategies for site-specific spin labeling and support of modern quantum-chemical computation methods for data interpretation, provides new insights into biological processes. Specifically, the theoretical and instrumental background of CW and pulsed high-field EPR, and its multiple-resonance extensions ENDOR, TRIPLE and PELDOR, as well as high-field RIDME and ESEEM, are discussed.

Much of the excitement of modern research on photosynthesis can be shared when reading the seminal overview by A.J. Hoff and J. Deisenhofer (1997) "Photophysics of Photosynthesis," *Phys. Rep.*, **287**, 1–247. This starts with the fundamental photochemical reaction of photosynthetic energy conversion. Second, the high-resolution X-ray diffraction analysis of two bacterial RCs is highlighted. Third, a detailed account of investigations on the functioning of the RC with a number of optical and magnetic resonance spectroscopic techniques is presented, and the results are discussed in the framework of current theories of photosynthetic electron transport.

Comprehensive updates of the pertinent literature in terms of techniques and applications are presented in the following references:

T. J. Aartsma and J. Matysik (eds) (2008) *Biophysical Techniques in Photosynthesis*, Vol. II, Springer, Dordrecht. Since the first volume of this series (published in 1996), new experimental techniques and methods have been devised at a rapid pace. The present volume is a sequel which complements the first volume by providing a comprehensive overview of the most important new techniques developed over the past ten years, especially those that are relevant for research on the mechanism and fundamental aspects of photosynthesis. In five chapters, expert authors describe imaging, structure, laser spectroscopy, magnetic resonance and theory, emphasizing the basic concepts, practical applications and scientific results.

J. Messinger, A. Alia and Govindjee (eds) (2009) Special Educational Issue on "Basics and Application of Biophysical Techniques in Photosynthesis and Related Processes"–Part B, *Photosynth. Res.*, **102**, 311–333. This special issue stresses its educational intention to interest newcomers in this fascinating field of science. Part A of this issue focuses on optical techniques, and Part B on the following categories: Imaging techniques; methods for determining the structures of proteins and cofactors; magnetic resonance techniques for elucidating the electronic structures of proteins and cofactors; and theory and modeling methods. Special attention is given to a broad variety of advanced magnetic-resonance methods, which are convincingly considered as the driving force to access photosynthesis at the molecular level. The multifrequency EPR techniques covered comprise ESE, ENDOR, ESEEM, and PELDOR, applied to photosynthetic cofactor radicals, radical pairs and triplet states; optically detected magnetic resonance (ODMR) is also discussed. Magic angle spinning (MAS) NMR is shown to have created its own niche in studies, at atomic resolution, on photosynthetic membrane proteins. A novel and promising application of MAS NMR in photosynthesis research, exploiting photochemically induced dynamic nuclear polarization (photo-CIDNP) effects, is treated in detail; strong NMR signals can be obtained directly from the active site in native RCs, even without isotopic enrichment.

Finally, attention is focused on a recent, rather unconventional textbook on EPR: M. Brustolon and E. Giamello (eds) (2009) *Electron Paramagnetic Resonance: A Practioner's Toolkit*, Wiley, Hoboken, New Jersey. This book offers a pragmatic guide to navigating through the complex maze of EPR spectroscopy fundamentals, techniques, and applications. The first part presents basic fundamentals and advantages of EPR spectroscopy, while the second part explores several application areas, including chemistry, biology, medicine, materials, and geology.

References

Allen, J.P., Feher, G., Yeates, T.O., Rees, D.C., Eisenberg, D.S., Deisenhofer, J., Michel, H., and Huber, R. (1986) *Biophys. J.*, **49**, A583.

Angerhofer, A., Bornhauser, F., Aust, V., Hartwich, G., and Scheer, H. (1998) *Biochim. Biophys. Acta*, **1365**, 404.

Artz, K., Williams, J.C., Allen, J.P., Lendzian, F., Rautter, J., and Lubitz, W. (1997) *Proc. Natl Acad. Sci. USA*, **94**, 13582.

Atherton, N.M. (1993) *Principles of Electron Spin Resonance*, Ellis Horwood, New York.

Aust, V., Angerhofer, A., Parot, P.H., Violette, C.A., and Frank, H.A. (1990) *Chem. Phys. Lett.*, **173**, 439.

Balabin, I.A. and Onuchic, J.N. (2000) *Science*, **290**, 114.

Bittl, R. and Zech, S.G. (2001) *Biochim. Biophys. Acta*, **1507**, 194.

Bixon, M., Jortner, J., and Michel-Beyerle, M.E. (1991) *Biochim. Biophys. Acta*, **1056**, 301.

Borovykh, I.V., Ceola, S., Gajula, P., Gast, P., Steinhoff, H.J., and Huber, M. (2006) *J. Magn. Reson.*, **180**, 178.

Bosch, M., Gast, P., Hoff, A.J., Spoyalov, A.P., and Tsvetkov, Y.D. (1995) *Chem. Phys. Lett.*, **239**, 306.

Bratt, P.J., Ringus, E., Hassan, A., Tol, H.V., Maniero, A.-L., Brunel, L.-C., Rohrer, M., Bubenzer-Hange, C., Scheer, H., and Angerhofer, A. (1999) *J. Phys. Chem. B*, **103**, 10973.

Breton, J., Boullais, C., Burie, J.-R., Nabedryk, E., and Mioskowski, C. (1994) *Biochemistry*, **33**, 14378.

van den Brink, J.S., Spoyalov, A.P., Gast, P., van Liemt, W.B.S., Raap, J., Lugtenburg, J., and Hoff, A.J. (1994) *FEBS Lett.*, **353**, 273.

Britt, R.D., Campbell, K.A., Peloquin, J.M., Gilchrist, M.L., Aznar, C.P., Dicus, M.M., Robblee, J., and Messinger, J. (2004) *Biochim. Biophys. Acta*, **1655**, 158.

Brudler, R., de Groot, H.J.M., van Liemt, W.B.S., Steggerda, W.F., Esmeijer, R., Gast, P., Hoff, A.J., Lugtenburg, J., and Gerwert, K. (1994) *EMBO J.*, **13**, 5523.

Budil, D.E., Taremi, S.S., Gast, P., Norris, J.R., and Frank, H.A. (1988) *Israel J. Chem.*, **28**, 59.

Burghaus, O., Plato, M., Rohrer, M., Möbius, K., MacMillan F., and Lubitz, W. (1993) *J. Phys. Chem.*, **97**, 7639.

Carrington, A. and McLachlan, A.D. (1969) *Introduction to Magnetic Resonance*, Harper and Row, New York.

Cogdell, R.J., Fyfe, P.K., Barrett, S.J., Prince, S.M., Freer, A.A., Isaacs, N.W., McGlynn, P., and Hunter, C.N. (1996) *Photosynth. Res.*, **48**, 55.

Cogdell, R.J., Isaacs, N.W., Howard, T.D., McLuskey, K., Fraser, N.J., and Prince, S.M. (1999) *J. Bacteriol.*, **181**, 3869.

Crystal, J. and Friesner, R.A. (2000) *J. Phys. Chem. A*, **104**, 2362.

Debus, R.J., Feher, G., and Okamura, M.Y. (1986) *Biochemistry*, **25**, 2276.

Deisenhofer, J. and Norris, J.R. (eds) (1993) *The Photosynthetic Reaction Center*, vol. I and II, Academic Press, San Diego.

Denysenkov, V.P., Biglino, D., Lubitz, W., Prisner, T.F., and Bennati, M. (2008) *Angew. Chem. Int. Ed.*, **47**, 1224.

DeRose, V.J., Mukerji, I., Latimer, M.J., Yachandra, V.K., Sauer, K., and Klein, M.P. (1994) *J. Am. Chem. Soc.*, **116**, 5239.

Dorlet, P., Xiong, L., Sayre, R.T., and Un, S. (2001) *J. Biol. Chem.*, **276**, 22313.

Dzuba, S.A., Tsvetkov, Y.D., and Maryasov, A.G. (1992) *Chem. Phys. Lett.*, **188**, 217.

Ermler, U., Fritzsch, G., Buchanan, S.K., and Michel, H. (1994) *Structure*, **2**, 925.

van der Est, A., Prisner, T., Bittl, R., Fromme, P., Lubitz, W., Möbius, K., and Stehlik, D. (1997) *J. Phys. Chem. B*, **101**, 1437.

Feher, G. (1992) *J. Chem. Soc. Perkin Trans.* 2, 1861.

Feher, G., Hoff, A.J., Isaacson, R.A., and McElroy, J.D. (1973) *Abstr. Biophys. Soc.*, **17**, 611a.

Feher, G., Hoff, A.J., Isaacson, R.A., and Ackerson, L.C. (1975) *Ann. N. Y. Acad. Sci.*, **244**, 239.

Fenimore, P.W., Frauenfelder, H., McMahon, B.H., and Parak, F.G. (2002) *Proc. Natl Acad. Sci. USA*, **99**, 16047.

Finkelstein, A.V. and Ptitsyn, O. (2002) *Protein Physics: A Course for Lectures*, Academic Press, San Diego.

Flores, M., Isaacson, R., Abresch, E., Calvo, R., Lubitz, W., and Feher, G. (2007a) *Biophys. J.*, **92**, 671.

Flores, M., Savitsky, A., Abresch, E., Lubitz, W., and Möbius, K. (2007b) *Photosynth. Res.*, **91**, 155.

Frauenfelder, H. (2002) *Proc. Natl Acad. Sci. USA*, **99**, 2479.

Frauenfelder, H., McMahon, B.H., and Fenimore, P.W. (2003) *Proc. Natl Acad. Sci. USA*, **100**, 8615.

Freer, A., Prince, S., Sauer, K., Papiz, M., Hawthornthwaite-Lawless, A., McDermott, G., Cogdell, R., and Isaacs, N.W. (1996) *Structure*, **4**, 449.

Fuchs, M.R., Schleicher, E., Schnegg, A., Kay, C.W.M., Törring, J.T., Bittl, R., Bacher, A., Richter, G., Möbius, K., and Weber, S. (2002) *J. Phys. Chem. B*, **106**, 8885.

Fuchs, M.R., Schnegg, A., Plato, M., Schulz, C., Müh, F., Lubitz, W., and Möbius, K. (2003) *Chem. Phys.*, **294**, 371.

Füchsle, G., Bittl, R., van der Est, A., Lubitz, W., and Stehlik, D. (1993) *Biochim. Biophys. Acta*, **1142**, 23.

Fuhs, M., Schnegg, A., Prisner, T., Köhne, I., Hanley, J., Rutherford, A.W., and Möbius, K. (2002) *Biochim. Biophys. Acta*, **1556**, 81.

Gajula, P., Borovykh, I.V., Beier, C., Shkuropatova, T., Gast, P., and Steinhoff, H.J. (2007) *Appl. Magn. Reson.*, **31**, 167.

Gerwert, K. (2002) Molecular reaction mechanisms of proteins monitored by time-resolved FT-IR spectroscopy, in *Handbook of Vibrational Spectroscopy* (eds J.M. Chalmers and P.R. Griffiths), John Wiley & Sons, Ltd, Chichester, p. 3536.

Gingras, G. and Picorel, R. (1990) *Proc. Natl Acad. Sci. USA*, **87**, 3405.

Glatzel, P., Bergmann, U., Yano, J., Visser, H., Robblee, J.H., Gu, W.W., de Groot, F.M.F., Christou, G., Pecoraro, V.L., Cramer, S.P., and Yachandra, V.K. (2004) *J. Am. Chem. Soc.*, **126**, 9946.

Glatzel, P., Yano, J., Bergmann, U., Visser, H., Robblee, J.H., Gu, W.W., de Groot, F.M.F., Cramer, S.P., and Yachandra, V.K. (2005) *J. Phys. Chem. Solids*, **66**, 2163.

de Groot, H.J.M. (2000) *Curr. Opin. Struct. Biol.*, **10**, 593.

Gunner, M.R., Nicholls, A., and Honig, B. (1996) *J. Phys. Chem.*, **100**, 4277.

Haumann, M., Liebisch, P., Müller, C., Barra, M., Grabolle, M., and Dau, H. (2005a) *Science*, **310**, 1019.

Haumann, M., Müller, C., Liebisch, P., Iuzzolino, L., Dittmer, J., Grabolle, M., Neisius, T., Meyer-Klaucke, W., and Dau, H. (2005b) *Biochemistry*, **44**, 1894.

Hofbauer, W., Zouni, A., Bittl, R., Kern, J., Orth, P., Lendzian, F., Fromme, P., Witt, H.T., and Lubitz, W. (2001) *Proc. Natl Acad. Sci. USA*, **98**, 6623.

Hoff, A.J. (1987) Electron paramagnetic resonance in photosynthesis, in *Photosynthesis* (ed. J. Amesz), Elsevier, Amsterdam, p. 97.

Hoff, A.J. and Deisenhofer, J. (1997) *Phys. Rep.*, **287**, 2.

Hoff, A.J. and Proskuryakov, I. (1985) *Chem. Phys. Lett.*, **115**, 303.

Hong, M. (2006) *Acc. Chem. Res.*, **39**, 176.

Hu, X.C. and Schulten, K. (1997) *Phys. Today*, **50**, 28.

Huber, M. and Törring, J.T. (1995) *Chem. Phys.*, **194**, 379.

Huber, M., Törring, J.T., Plato, M., Möbius, K., Finck, U., Lubitz, W., Feick, R., and Schenck, C.C. (1995) *Sol. Energy Mater. Sol. Cells*, 38, 119.

Huber, M., Isaacson, R.A., Abresh, E.C., Gaul, D., Schenck, C.C., and Feher, G. (1996) *Biochim. Biophys. Acta*, **1273**, 108.

Isaacson, R.A., Abresch, E.C., Lendzian, F., Boullais, C., Paddock, M.L., Mioskowski, C., Lubitz, W., and Feher, G. (1995a) Asymmetry of the binding sites of Q_A^- and Q_B^- in reaction centers of Rb. sphaeroides probed by Q-Band EPR with ^{13}C-Labeled quinones, in *The Reaction Center of Photosynthetic Bacteria* (ed. M.E. Michel-Beyerle), Springer, Berlin, p. 353.

Isaacson, R.A., Lendzian, F., Abresch, E.C., Lubitz, W., and Feher, G. (1995b) *Biophys. J.*, **69**, 311.

Ivancich, A., Artz, K., Williams, J.C., Allen, J.P., and Mattioli, T.-A. (1998) *Biochemistry*, **37**, 11812.

Ivashin, N., Kallebring, B., Larsson, S., and Hansson, O. (1998) *J. Phys. Chem. B*, **102**, 5017.

Iwata, T., Paddock, M.L., Okamura, M.Y., and Kandori, H. (2009) *Biochemistry*, **48**, 1220.

Jordan, P., Fromme, P., Klukas, O., Witt, H.T., Saenger, W., and Krauß, N. (2001) *Nature*, **411**, 909.

Jortner, J. and Bixon, M. (eds) (1999) *Electron Transfer – From Isolated Molecules to Biomolecules. Part 1*; Advances in Chemistry and Physics, vol. 106, John Wiley & Sons, Inc., New York.

Junge, W., Yachandra, V.K., Dismukes, C., and Hammarstrom, L. (2002) *Philos. Trans. R. Soc. Lond. B Biol. Sci.*, **357**, 1357.

Käss, H., Rautter, J., Zweygart, W., Struck, A., Scheer, H., and Lubitz, W. (1994) *J. Phys. Chem.*, **98**, 354.

Kay, C.W.M., Feicht, R., Schulz, K., Sadewater, P., Sancar, A., Bacher, A., Möbius, K., Richter, G., and Weber, S. (1999) *Biochemistry*, **38**, 16740.

Kay, C.W.M., Mögling, H., Schleicher, E., Hitomi, K., Möbius, K., Todo, T., Bacher, A., Richter, G., and Weber, S. (2002) A comparative time-resolved electron paramagnetic resonance study of the flavin cofactor photoreduction in *Escherichia coli* cyclobutane pyrimidine dimer photolyase and *Xenopus laevis* (6-4) photolyase, in *Flavins and Flavoproteins 2002* (eds. S. Chapman, R. Perham and N. Scrutton), Rudolf Weber Agency for Scientific Publishing, Berlin, p. 713.

Kay, C.W.M., Bittl, R., Bacher, A., Richter, G., and Weber, S. (2005a) *J. Am. Chem. Soc.*, **127**, 10780.

Kay, C.W.M., Schleicher, E., Hitomi, K., Todo, T., Bittl, R., and Weber, S. (2005b) *Magn. Reson. Chem.*, **43**, S96.

Kirilina, E.P., Prisner, T.F., Bennati, M., Endeward, B., Dzuba, S.A., Fuchs, M.R., Möbius, K., and Schnegg, A. (2005) *Magn. Reson. Chem.*, **43**, S119.

Kleinfeld, D., Okamura, M.Y., and Feher, G. (1984) *Biochemistry*, **23**, 5780.

Klette, R., Törring, J.T., Plato, M., Möbius, K., Bönigk, B., and Lubitz, W. (1993) *J. Phys. Chem.*, **97**, 2051.

Kolbasov, D. and Scherz, A. (2000) *J. Phys. Chem. B*, **104**, 1802.

Kolbasov, D., Srivatsan, N., Ponomarenko, N., Jager, M., and Norris, J.R. (2003) *J. Phys. Chem. B*, **107**, 2386.

Kriegl, J.M. and Nienhaus, G.U. (2004) *Proc. Natl Acad. Sci. USA*, **101**, 123.

Kriegl, J.M., Forster, F.K., and Nienhaus, G.U. (2003) *Biophys. J.*, **85**, 1851.

Kühlbrandt, W. (1995) *Nature*, **374**, 497.

Kulik, L.V., Paschenko, S.V., and Dzuba, S.A. (2002) *J. Magn. Reson.*, **159**, 237.

Kulik, L.V., Lubitz, W., and Messinger, J. (2005) *Biochemistry*, **44**, 9368.

Kulik, L.V., Epel, B., Lubitz, W., and Messinger, J. (2005a) *J. Am. Chem. Soc.*, **127**, 2392.

Kulik, L., Epel, B., Messinger, J., and Lubitz, W. (2005b) *Photosynth. Res.*, **84**, 347.

Labahn, A. and Huber, M. (2001) *Appl. Magn. Reson.*, **21**, 381.

Lansing, J.C., Hu, J.G., Belenky, M., Griffin, R.G., and Herzfeld, J. (2003) *Biochemistry*, **42**, 3586.

van Liemt, W.B.S., Boender, G.J., Gast, P., Hoff, A.J., Lugtenburg, J., and de Groot, H.J.M. (1995) *Biochemistry*, **34**, 10229.

Lubitz, W. (1991) EPR and ENDOR Studies of chlorophyll cation and anion radicals, in *Chlorophylls* (ed. H. Scheer), CRC Press, Boca Raton, Florida, p. 903.

Lubitz, W. (2002) *Phys. Chem. Chem. Phys.*, **4**, 5539.

Lubitz, W. (2004) EPR in Photosynthesis, in *Electron Paramagnetic Resonance*, vol. 19 (eds B.C. Gilbert, M.J. Davies, and D.M. Murphy), Royal Society of Chemistry, Cambridge, p. 174.

Lubitz, W. and Feher, G. (1999) *Appl. Magn. Reson.*, **17**, 1.

Lubitz, W., Lendzian, F., and Bittl, R. (2002) *Acc. Chem. Res.*, **35**, 313.

McDermott, G., Prince, S.M., Freer, A.A., Isaacs, N.W., Papiz, M.Z., Hawthornthwaite-Lawless, A.M., and Cogdell, R.J. (1995) *Protein Eng.*, **8**, 43.

Mäntele, W. (1993) Infrared vibrational spectroscopy of photosynthetic. Reaction centers, in *The Photosynthetic Reaction Center*, vol. 2 (eds J. Deisenhofer and J.R. Norris), Academic Press, New York, p. 239.

Marchanka, A., Paddock, M., Lubitz, W., and van Gastel, M. (2007) *Biochemistry*, **46**, 14782.

Marchanka, A., Savitsky, A., Lubitz, W., Möbius, K., and van Gastel, M. (2010) *J. Phys. Chem. B*, doi: 10.1021/jp1003424.

Mason, A.J., Grage, S.L., Straus, S.K., Glaubitz, C., and Watts, A. (2004) *Biophys. J.*, **86**, 1610.

Messinger, J., Robblee, J.H., Bergmann, U., Fernandez, C., Glatzel, P., Visser, H., Cinco, R.M., McFarlane, K.L., Bellacchio, E., Pizarro, S.A., Cramer, S.P., Sauer, K., Klein, M.P., and Yachandra, V.K. (2001) *J. Am. Chem. Soc.*, **123**, 7804.

Michel-Beyerle, M.-E. (ed.) (1996) *The Reaction Center of Photosynthetic Bacteria, Structure and Dynamics*, Springer, Berlin.

Michel-Beyerle, M.E., Plato, M., Deisenhofer, J., Michel, H., Bixon, M., and Jortner, J. (1988) *Biochim. Biophys. Acta*, **932**, 52.

Millhauser, G.L. and Freed, J.H. (1984) *J. Chem. Phys.*, **81**, 37.

Milov, A.D., Salikhov, K.M., and Shirov, M.D. (1981) *Fiz. Tverd. Tela*, **23**, 975.

Milov, A.D., Ponomarev, A.B., and Tsvetkov, Y.D. (1984) *Chem. Phys. Lett.*, **110**, 67.

Möbius, K. (1994) *Electron Spin Resonance*, vol. 14 (eds N.M. Atherton, E.R. Davies, and B.C. Gilbert), Royal Society of Chemistry, Cambridge, p. 203.

Möbius, K. (2000) *Chem. Soc. Rev.*, **29**, 129.

Möbius, K. and Goldfarb, D. (2008) High-field/high-frequency electron paramagnetic resonance involving single- and multiple-transition schemes, in *Biophysical Techniques in Photosynthesis*, vol. II (eds T.J. Aartsma and J. Matysik), Springer, Dordrecht, p. 267.

Möbius, K. and Savitsky, A. (2009) *High-Field EPR Spectroscopy on Proteins and Their Model Systems: Characterization of Transient Paramagnetic States*, RSC Publishing, London.

Möbius, K., Savitsky, A., Schnegg, A., Plato, M., and Fuchs, M. (2005) *Phys. Chem. Chem. Phys.*, **7**, 19.

Moser, C.C., Page, C.C., Chen, X., and Dutton, P.L. (1997) *J. Biol. Inorg. Chem.*, **2**, 393.

Müh, F., Bibikova, M., Lendzian, F., Oesterhelt, D., and Lubitz, W. (1998) Pigment-protein interactions in reaction centers of Rhodopseudomonas viridis: ENDOR-study of the oxidized primary donor in site-directed mutants, in *Photosynthesis: Mechanisms and Effects*, vol. 2 (ed. G. Garab), Kluwer, Dordrecht, p. 763.

Müh, F., Jones, M.R., and Lubitz, W. (1999) *Biospectroscopy*, **5**, 35.

Müh, F., Lendzian, F., Roy, M., Williams, J.C., Allen, J.P., and Lubitz, W. (2002) *J. Phys. Chem. B*, **106**, 3226.

Norris, J.R., Uphaus, R.A., Crespi, H.L., and Katz, J.J. (1971) *Proc. Natl Acad. Sci. USA*, **68**, 625.

Noy, D., Moser, C.C., and Dutton, P.L. (2006) *Biochim. Biophys. Acta*, **1757**, 90.

Onuchic, J.N., Luthey-Schulten, Z., and Wolynes, P.G. (1997) *Annu. Rev. Phys. Chem.*, **48**, 545.

Pachtchenko, S.V. (2002) Primary electron donor triplet states in photosynthetic reaction centers as studied by high-field EPR, PhD Thesis, Leiden University.

Paddock, M.L., Flores, M., Isaacson, R., Chang, C., Abresch, E.C., and Okamura, M.Y. (2007) *Biochemistry*, **46**, 8234.

Page, C.C., Moser, C.C., Chen, X.X., and Dutton, P.L. (1999) *Nature*, **402**, 47.

Papiz, M.Z., Prince, S.M., Howard, T., Cogdell, R.J., and Isaacs, N.W. (2003) *J. Mol. Biol.*, **326**, 1523.

Parson, W.W. and Warshel, A. (2004) *J. Phys. Chem. B*, **108**, 10474.

Parson, W.W. and Warshel, A. (2008) Calculations of electrostatic energies in proteins: using microscopic, semimicroscopic and macroscopic models and free energy perturbation approaches, in *Biophysical Techniques in Photosynthesis*, vol. II (eds T.J. Aartsma and J. Matysik), Springer, Dordrecht, p. 401.

Parson, W.W., Chu, Z.T., and Warshel, A. (1990) *Biochim. Biophys. Acta*, **1017**, 251.

Paschenko, S.V., Gast, P., and Hoff, A.J. (2001) *Appl. Magn. Reson.*, **21**, 325.

Peloquin, J.M. and Britt, R.D. (2001) *Biochim. Biophys. Acta*, **1503**, 96.

Peloquin, J.M., Campbell, K.A., Randall, D.W., Evanchik, M.A., Pecoraro, V.L., Armstrong, W.H., and Britt, R.D. (2000) *J. Am. Chem. Soc.*, **122**, 10926.

Petrenko, A. and Redding, K. (2004) *Chem. Phys. Lett.*, **400**, 98.

Plato, M. and Winscom, C.J. (1988) A configuration interaction (CI) description of vectorial electron transfer in bacterial reaction centers, in *The Photosynthetic Bacterial Reaction Center* (eds J. Breton and A. Verméglio), Plenum, New York, p. 421.

Plato, M., Möbius, K., Michel-Beyerle, M.E., Bixon, M., and Jortner, J. (1988) *J. Am. Chem. Soc.*, **110**, 7279.

Plato, M., Möbius, K., and Lubitz, W. (1991) Molecular orbital calculations on chlorophyll radical ions, in *Chlorophylls* (ed. H. Scheer), CRC Press, Boca Raton, Florida, p. 1015.

Plato, M., Krauss, N., Fromme, P., and Lubitz, W. (2003) *Chem. Phys.*, **294**, 483.

Policar, C., Knüpling, M., Frapart, Y.M., and Un, S. (1998) *J. Phys. Chem. B*, **102**, 10391.

Poluektov, O.G., Utschig, L.M., Schlesselman, S.L., Lakshmi, K.V., Brudvig, G.W., Kothe, G., and Thurnauer, M.C. (2002) *J. Phys. Chem. B*, **106**, 8911.

Prisner, T.F., van der Est, A., Bittl, R., Lubitz, W., Stehlik, D., and Möbius, K. (1995) *Chem. Phys.*, **194**, 361.

Rautter, J., Lendzian, F., Schulz, C., Fetsch, A., Kuhn, M., Lin, X., Williams, J.C., Allen, J.P., and Lubitz, W. (1995) *Biochemistry*, **34**, 8130.

Raymond, J. and Blankenship, R.E. (2004) *Biochim. Biophys. Acta*, **1655**, 133.

Reddy, N.R.S., Kolaczkowski, S.V., and Small, G.J. (1993) *Science*, **260**, 68.

Remy, A. and Gerwert, K. (2003) *Nat. Struct. Biol.*, **10**, 637.

Richter, G., Kay, C.W.M., Struck, K., Sadewater, P., Möbius, K., and Weber, S. (1999) Radical intermediates in Escherichia coli DNA photolyase: EPR and ENDOR studies, in *Flavins and Flavoproteins* (eds S. Ghisla, P. Kroneck, P. Macheroux, and H. Sund), Rudolf Weber Berlin, p. 91.

Riedi, P.C. and Smith, G.M. (2002) Progress in High-field EPR, in *Electron Paramagnetic Resonance*, vol. 18 (eds B.C. Gilbert, M.J. Davies, and D.M. Murphy), Royal Society of Chemistry, Cambridge, p. 254.

Rödig, C., Chizhov, I., Weidlich, O., and Siebert, F. (1999) *Biophys. J.*, **76**, 2687.

Rohrer, M., Plato, M., MacMillan, F., Grishin, Y., Lubitz, W., and Möbius, K. (1995) *J. Magn. Reson. A*, **116**, 59.

Rohrer, M., Gast, P., Möbius, K., and Prisner, T.F. (1996) *Chem. Phys. Lett.*, **259**, 523.

Rohrer, M., MacMillan, F., Prisner, T.F., Gardiner, A.T., Möbius, K., and Lubitz, W. (1998) *J. Phys. Chem. B*, **102**, 4648.

Roszak, A.W., Howard, T.D., Southall, J., Gardiner, A.T., Law, C.J., Isaacs, N.W., and Cogdell, R.J. (2003) *Science*, **302**, 1969.

Salwinski, L. and Hubbell, W.L. (1999) *Protein Sci.*, **8**, 562.

Savitsky, A., Kühn, M., Duche, D., Möbius, K., and Steinhoff, H.J. (2004) *J. Phys. Chem. B*, **108**, 9541.

Savitsky, A., Dubinskii, A.A., Flores, M., Lubitz, W., and Möbius, K. (2007a) *J. Phys. Chem. B*, **111**, 6245.

Savitsky, A., Trubitsin, B.V., Möbius, K., Semenov, A.Y., and Tikhonov, A.N. (2007b) *Appl. Magn. Reson.*, **31**, 221.

Saxena, S. and Freed, J.H. (1997) *J. Phys. Chem. A*, **101**, 7998.

Scherer, P.O.J. and Fischer, S.F. (1989) *Chem. Phys.*, **131**, 115.

Scherer, P.O.J., Fischer, S.F., Hörber, J.K.H., and Michel-Beyerle, M.E. (1985) On the temperature dependence of the long wavelength fluorescence and absorption of rhodopseudomonas viridis reaction centers, in *Antennas and Reaction Centers of Photosynthetic Bacteria* (ed. M.E. Michel-Beyerle), Springer, Berlin.

Scherer, P.O.J., Scharnagl, C., and Fischer, S.F. (1995) *Chem. Phys.*, **197**, 333.

Schnegg, A., Fuhs, M., Rohrer, M., Lubitz, W., Prisner, T.F., and Möbius, K. (2002) *J. Phys. Chem. B*, **106**, 9454.

Schnegg, A., Kay, C.W.M., Schleicher, E., Hitomi, K., Todo, T., Möbius, K., and Weber, S. (2006) *Mol. Phys.*, **104**, 1627.

Schnegg, A., Dubinskii, A.A., Fuchs, M.R., Grishin, Y.A., Kirilina, E.P., Lubitz, W., Plato, M., Savitsky, A., and Möbius, K. (2007) *Appl. Magn. Reson.*, **31**, 59.

Schulz, C., Müh, F., Beyer, A., Jordan, R., Schlodder, E., and Lubitz, W. (1998) Investigation of *Rhodobacter sphaeroides* reaction center mutants with changed ligands to the primary donor, in *Photosynthesis: Mechanisms and Effects*, vol. 2 (ed. G. Garab), Kluwer Academic Publishers, Dordrecht, p. 767.

Shin, Y.K., Levinthal, C., Levinthal, F., and Hubbell, W.L. (1993) *Science*, **259**, 960.

Sinnecker, S., Reijerse, E., Neese, F., and Lubitz, W. (2004) *J. Am. Chem. Soc.*, **126**, 3280.

Smirnov, A.I. (2002) Spin-labeling in High-field EPR, in *Electron Paramagnetic Resonance*, vol. 18 (eds B.C. Gilbert, M.J. Davies, and D.M. Murphy), Royal Society of Chemistry, Cambridge, p. 109.

Srivatsan, N. and Norris, J.R. (2001) *J. Phys. Chem. B*, **105**, 12391.

Srivatsan, N., Kolbasov, D., Ponomarenko, N., Weber, S., Ostafin, A.E., and Norris, J.R. (2003a) *J. Phys. Chem. B*, **107**, 7867.

Srivatsan, N., Weber, S., Kolbasov, D., and Norris, J.R. (2003b) *J. Phys. Chem. B*, **107**, 2127.

Stehlik, D. and Möbius, K. (1997) *Annu. Rev. Phys. Chem.*, **48**, 745.

Steinhoff, H.-J. (2002) *Front. Biosci.*, **7**, 97.

Stone, A.J. (1963) *Proc. R. Soc. Lond. A*, **271**, 424.

Stowell, M.H.B., McPhillips, T.M., Rees, D.C., Soltis, S.M., Abresch, E., and Feher, G. (1997) *Science*, **276**, 812.

Teutloff, C., Hofbauer, W., Zech, S.G., Stein, M., Bittl, R., and Lubitz, W. (2001) *Appl. Magn. Reson.*, **21**, 363.

Thomann, H., Dalton, L.R., and Dalton, L.A. (1990) Biological applications of time domain ESR, in *Biological Magnetic Resonance*, (ed. L.J. Berliner), vol. 6, Plenum Press, New York.

Thurnauer, M.C., Poluektov, O.G., and Kothe, G. (2004) Time-resolved high-frequency and multifrequency EPR studies of spin-correlated radical pairs in photosynthetic reaction center proteins, in *Very High Frequency (VHF) ESR/EPR; Biological Magnetic Resonance*, vol. 22 (eds O. Grinberg and L.J. Berliner), Kluwer/Plenum Publishers, New York, p. 166.

Un, S., Atta, M., Fontecave, M., and Rutherford, A.W. (1995) *J. Am. Chem. Soc.*, **117**, 10713.

Un, S., Dorlet, P., and Rutherford, A.W. (2001) *Appl. Magn. Reson.*, **21**, 341.

Warshel, A. and Parson, W.W. (1991) *Annu. Rev. Phys. Chem.*, **42**, 279.

Warshel, A., Creighton, S., and Parson, W.W. (1988) *J. Phys. Chem.*, **92**, 2696.

Warshel, A., Kato, M., and Pisliakov, A.V. (2007) *J. Chem. Theory Comput.*, **3**, 2034.

Weber, S. (2000) Recent EPR studies on the bacterial photosynthetic reaction centre, in *Electron Paramagnetic Resonance*, vol. 17 (eds N.M. Atherton, M.J. Davies, and B.C. Gilbert), Royal Society of Chemistry, Cambridge, p. 43.

Weber, S. (2005) *Biochim. Biophys. Acta*, **1707**, 1.

Weber, S. and Bittl, R. (2007) *Bull. Chem. Soc. Jpn.*, **80**, 2270.

Weber, S., Möbius, K., Richter, G., and Kay, C.W.M. (2001a) *J. Am. Chem. Soc.*, **123**, 3790.

Weber, S., Richter, G., Schleicher, E., Bacher, A., Möbius, K., and Kay, C.W.M. (2001b) *Biophys. J.*, **81**, 1195.

Weber, S., Kay, C.W.M., Mögling, H., Möbius, K., Hitomi, K., and Todo, T. (2002) *Proc. Natl Acad. Sci. USA*, **99**, 1319.

Yachandra, V.K. (2002) *Philos. Trans. R. Soc. Lond. B Biol. Sci.*, **357**, 1347.

Yachandra, V.K., Sauer, K., and Klein, M.P. (1996) *Chem. Rev.*, **96**, 2927.

Zhang, L.Y. and Friesner, R.A. (1998) *Proc. Natl Acad. Sci. USA*, **95**, 13603.

Zhang, J.P., Nagae, H., Qian, P., Limantara, L., Fujii, R., Watanabe, Y., and Koyama, Y. (2001) *J. Phys. Chem. B*, **105**, 7312.

Zouni, A., Witt, H.T., Kern, J., Fromme, P., Krauß, N., Saenger, W., and Orth, P. (2001) *Nature*, **409**, 739.

Zybailov, B., van der Est, A., Zech, S.G., Teutloff, C., Johnson, T.W., Shen, G.Z., Bittl, R., Stehlik, D., Chitnis, P.R., and Golbeck, J.H. (2000) *J. Biol. Chem.*, **275**, 8531.

Xu, W., Chitnis, P., Valieva, A., van der Est, A., Pushkar, Y.N., Krzystyniak, M., Teutloff, C., Zech, S.G., Bittl, R., Stehlik, D., Zybailov, B., Shen, G.Z., and Golbeck, J.H. (2003) *J. Biol. Chem.*, **278**, 27864.

Zeng, R.H., van Tol, J., Deal, A., Frank, H.A., and Budil, D.E. (2003) *J. Phys. Chem. B*, **107**, 4624.

23
Measurement of Superconducting Gaps

Sushil K. Misra

23.1
Introduction

Consideration of superconducting gaps (SCGs) is of relevance in the context of multifrequency EPR, in that today, microwave sources covering frequencies up to ~1000 GHz are available to measure the values of SCGs. This chapter will review the data showing the presence of an SCG, and how to measure it. The discussion presented here is based on books by Owens and Poole (1999, 2008), and by Poole, Farach, and Creswick (1988). Before proceeding any further, the details of the SCG are described.

23.2
The Superconducting Gap

In the superconducting state, below the superconducting transition temperature, T_c, the electrons at the Fermi surface of the metal, which carry the current, are bound together in pairs called Cooper pairs. This is based on the fact that experiments show that the carriers of current in the superconducting state have a charge 2e. These bound pairs of electrons modify the energy band-gap structure of the metal in the superconducting state. The top-occupied conduction band in a metal is not full. The energy of the uppermost filled state in the band is defined by the Fermi level. The presence of bound-electron pairs in the superconducting state in the band implies that there exists an energy gap, $E_g = 2\Delta$, at the Fermi level, the magnitude of which corresponds to the binding energy of the Cooper pairs; this is equal to the energy difference between the normal electrons and the bound electron pairs at the Fermi level. The Bardeen–Cooper–Schrieffer (BCS) theory predicts that, in the weak-coupling limit, the band gap, $E_g(0)$, at absolute zero is related to the transition temperature, T_c, as follows: $E_g(0) = 3.528 k_B T_c$. Experimental data show that $\dfrac{T}{T_c} \propto (T - T_c)^{1/2}$, consistent with the well-known mean-field result for the order parameter of a second-order phase transition. Figure

Multifrequency Electron Paramagnetic Resonance, First Edition. Edited by Sushil K. Misra.
© 2011 Wiley-VCH Verlag GmbH & Co. KGaA. Published 2011 by Wiley-VCH Verlag GmbH & Co. KGaA.

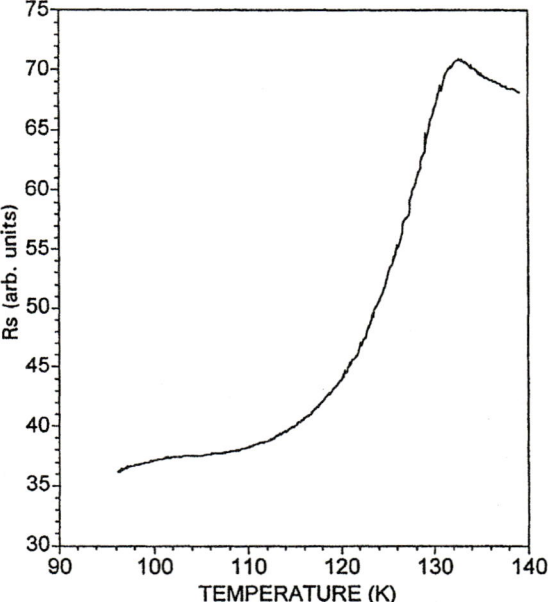

Figure 23.1 Plot of the temperature variation of the surface resistance through the transition to the superconducting state in the granular pellet of Hg–Pb–Ba–Cu–O at 9.2 GHz, showing a pronounced reduction in the surface resistance, and thus in the absorption of microwave energy, above $T_c \sim 132$ K. Adapted from Owens and Poole (1999).

23.1 shows the temperature variation of surface resistance through the transition to the superconducting state in the granular pellet of the high-T_c superconductor Hg–Pb–Ba–Cu–O at 9.2 GHz, which shows a pronounced reduction in the surface resistance, and thus in the absorption microwave energy above $T_c \sim 132$ K.

23.3
Measurement of SCG

Two techniques used to measure SCG are described as follows (Owens and Poole, 1999):

Absorption of electromagnetic radiation method: An energy gap, increasing with the temperature dependence of $(T - T_c)^{1/2}$, significantly enhances the absorption of electromagnetic radiation near the transition temperature, because when the gap equals the energy of the electromagnetic radiation at some temperature near T_c, the absorption of radiation will occur because of excitation across the gap. This manifests in a frequency dependence of the surface resistance at constant temperature, and provides an example of a technique to measure the SCG.

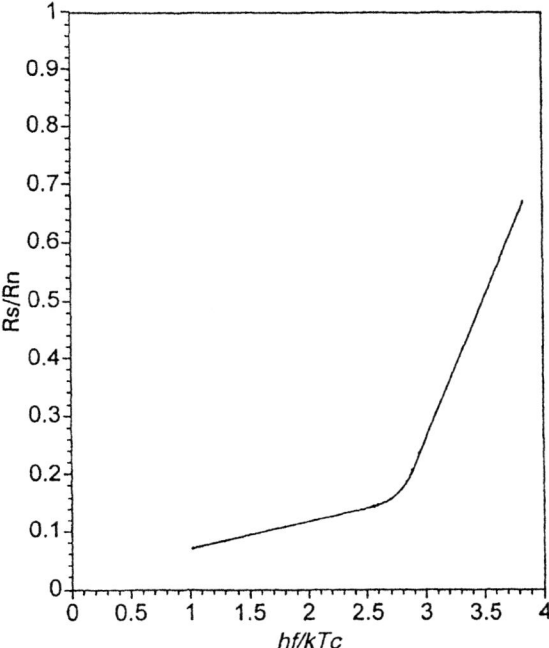

Figure 23.2 Frequency dependence of the ratio of the microwave absorption-determined surface resistance in the superconducting state to that in the normal state in aluminum at the constant temperature of 0.83 K, where $T_c = 1.18$ K. (Adapted from Owens and Poole, 1999).

Experimental data exhibiting this behavior are presented in Figure 23.2, which shows the frequency dependence of the ratio of the microwave absorption-determined surface resistance in the superconducting state to that in the normal state in aluminum for which $T_c = 1.18$ K at the constant temperature of 0.83 K. It is seen clearly from this figure that there occurs a sharp increase in the surface resistance just above $\dfrac{hf}{k_B T_c} = 2.5$, because the photons of the incident radiation cause transitions across the gap due to the breaking-up of Cooper pairs to form normal carriers, which are capable of absorbing electromagnetic energy.

Infrared method of measuring SCG: Vibrational spectroscopy involving infrared radiation has been used to determine the SCG. At absolute zero, electromagnetic radiation with frequencies lower than $\dfrac{E_g}{h}$ will be reflected by a superconductor to a greater extent than that with frequencies greater than $\dfrac{E_g}{h}$, the same as that for a metal in the normal state (here, h is Planck's constant). When the temperature is above absolute zero, the latter frequencies excite quasi-particles, and induce a photoconductive response. Low-temperature experimental data showing

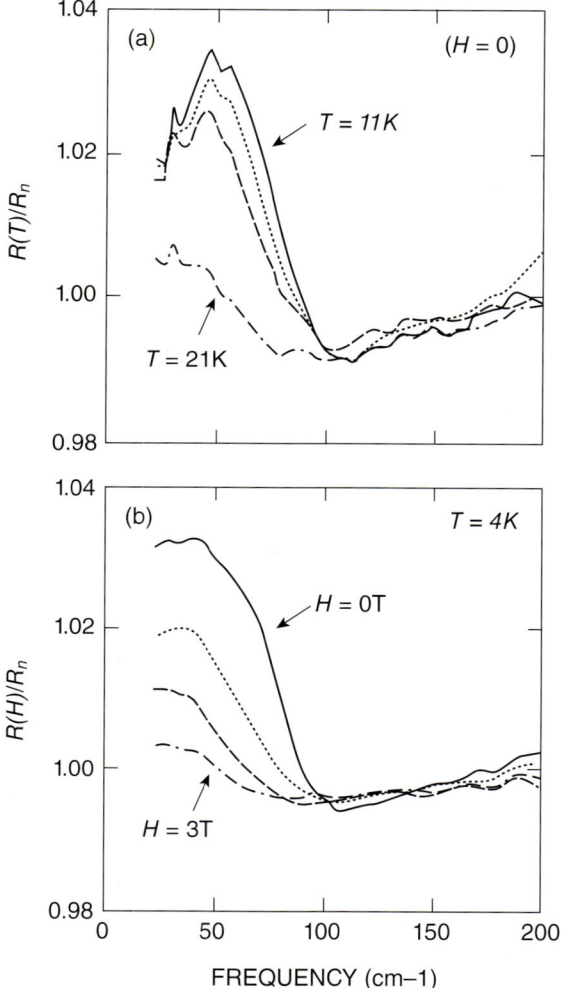

Figure 23.3 Data on infrared reflectance spectra, which show the abrupt change in transmission at the energy gap for the superconductor $Ba_{0.6}K_{0.4}BiO_3$. (a) Data are plotted for the temperatures $T = 11$, 14, 17, and 21 K; (b) Suppression of the low-frequency reflectivity enhancement by successively increasing applied magnetic-field values $B_{app} = 0$, 1, 2, and 3 T. Adapted from Owens and Poole (1999).

$R(T)/R_n$ for the reflection of infrared radiation at frequencies below the gap value $E_g \approx 70 \, cm^{-1}$, and the drop in reflectivity for frequencies above this value for the cubic perovskite superconductor $Ba_{0.6}K_{0.4}BiO_3$, are shown in Figure 23.3. The data in Figure 23.3a show that an increase in the temperature causes a decrease in the frequency at which the reflectivity undergoes a sharp drop in value. This can be explained by the temperature dependence of the energy gap, which has the form:

$$E_g(T) = E_g[1 - T/T_c]^{1/2},$$

where $E_g = E_g(0)$, with the BCS value given by $E_g = 3.52 k_B T_c$. On the other hand, Figure 23.3b shows that increasing the applied magnetic field has the same effect as increasing the temperature, due to the dependence of the critical magnetic field as follows:

$$B_c(T) = B_c\left[1 - (T/T_c)^2\right]^{1/2},$$

where $B_c = B_c(0)$. The above equation can be inverted by treating T as the critical temperature $T_c(B)$ as a function of the applied field:

$$T_c(B) = T_c\left[1 - (B/B_c)^2\right]^{1/2},$$

where B stands for $B_c(T)$, and T_c is the critical temperature for $B = 0$. With these expressions, it is seen that $T_c(B)$ decreases as the applied field increases. In the presence of a magnetic field, the gap equation becomes:

$$E_g(T, B) = E_g\left[1 - \frac{T}{T_c(B)}\right]^{1/2};$$

This formula is only valid for $T < T_c(B)$. This expression shows that the energy gap decreases when either the temperature or the magnetic field increases. Thus, increasing T or B also decreases the concentration, n_s, of the Cooper pairs, and this in turn causes the reflectivity peak below the gap to decrease in magnitude, as shown in Figure 23.3a,b. The infrared method can also be applied to measuring energy gaps at much lower frequencies to elemental and other classical superconductors characterized by a much lower T_c. An example of this is shown in Figure 23.4, which illustrates the temperature dependence of the normalized microwave resistivity $\rho(T)/\rho_n$ for aluminum for five frequencies in the range from 12 to 80 GHz. Each curve is labeled by its microwave photon energy, $h\upsilon$, expressed in units of $k_B T_c$, where $T_c = 1.2$ K for aluminum, bearing in mind that a temperature of 1 K is equivalent to 20.84 GHz (here k_B is Boltzmann's constant). In this figure, the three lowest curves extrapolate to zero resistivity, indicating that the super electrons are not excited above the gap; in contrast, the two upper curves extrapolate to a finite resistivity, indicating the presence of excited quasi-particles. The microwave resistivity of each frequency is plotted versus the energy in the lower figure at the temperature of $T = 0.7\, T_c$. Here, it is seen that the slope of the curve is small up to the energy of $2.6 k_B T_c$, with a larger value of the slope beyond this point, indicating a gap energy of $E_g \approx 2.6 k_B T_c$, which is lower than the BCS prediction, $E_g \approx 3.53 k_B T_c$. The faster rise in resistivity beyond this point is caused by the presence of super electrons that have been excited to the quasi-particle state.

23.4 Concluding Remarks

This short chapter discusses the SCG, and how it can be measured. This is relevant in the multifrequency context, since with the present availability of microwave

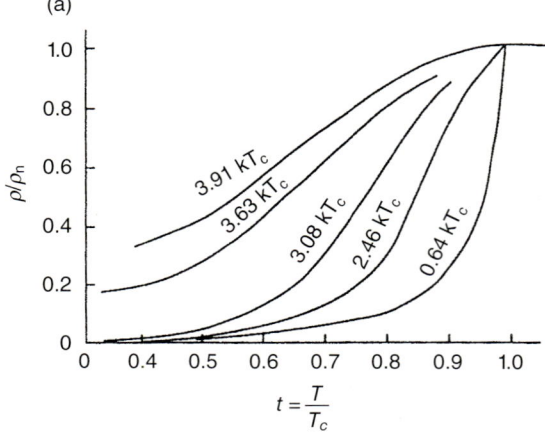
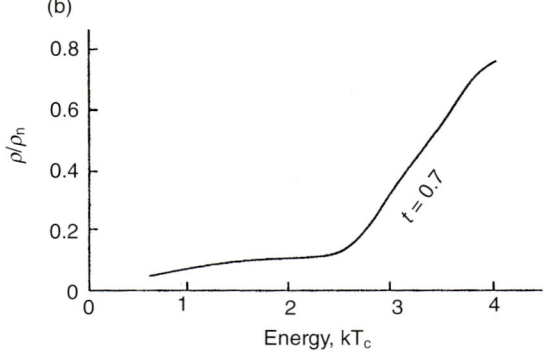

Figure 23.4 Plot of the temperature dependence of the normalized microwave surface resistivity ρ/ρ_n of aluminum (upper figure) for microwave frequencies in the range 12 to 80 GHz. Here, ρ_n is the normal-state surface resistivity. Each curve is labeled with its equivalent $k_B T_c$ value. There is a break shown in the plot of the normalized resistivity ρ/ρ_n versus energy at the reduced temperature $T/T_c = 0.7$ (lower figure) at the energy 2.6 $k_B T_c$, corresponding to the energy gap $E_g = 2.6\, k_B T_c$. Adapted from Owens and Poole (1999).

radiation over a large range of frequencies, the SCGs can be measured using the techniques described here. Additional details on this topic are available in the books by Owens and Poole (1999, 2008) and by Poole et al. (1988).

Acknowledgments

The author is grateful to Prof. C.P. Poole for his critical reading of this chapter, and his useful comments.

References

Owens, F.J. and Poole, C.P. (1999) *Electromagnetic Absorption in the Copper Oxide Superconductors*, Kluwer Academic/Plenum Publishers, New York.

Owens, F.J. and Poole, C.P. (2008) *The Physics and Chemistry of Nanosolids*, John Wiley & Sons, Inc., Hoboken, NJ.

Poole, C.P., Farach, H.A., and Creswick, R.J. (1988) *Superconductivity*, Academic Press, New York.

24
Dynamic Nuclear Polarization (DNP) at High Magnetic Fields
Thomas Prisner and Mark J. Prandolini

24.1
Introduction

Dynamic nuclear polarization (DNP) refers to various mechanisms which attempt to cross-polarize coupled electronic and nuclear levels, and thereby increase the NMR signal intensity. The maximum DNP enhancement of the NMR signal is generally given by the electron to nuclear gyromagetic ratios (γ_e / γ_n: for protons $\gamma_e / \gamma_p \sim 660$). Improving the sensitivity is a key issue in NMR spectroscopy and magnetic resonance imaging (MRI). In the past, NMR signal and resolution were improved by increasing the external magnetic field; however, this method will reach practical and technical limits at around 23.5 T (proton frequency 1000 MHz), not to mention the costs and space required for supporting large high-resolution NMR magnet systems. Therefore, every effort should be made to increase the NMR signal-to-noise ratio (SNR) in existing NMR magnets. The main source of noise is thermal fluctuations in the induction coil, called *Johnson noise*. Cryogenic cooling of the coil, while keeping the sample at room temperature, can reduce the SNR by a factor of 3 to 4, thereby reducing acquisition time by a factor of 3^2 to 4^2. On the other hand, improving the NMR signal, by means other than increasing the external magnetic field, has been an ongoing development. A significant advancement came with the introduction of pulsed Fourier transform NMR (Ernst and Anderson, 1966). Other improvements were made using nuclear–nuclear cross-polarization schemes, usually between the larger gyromagnetic ratio of the proton to the smaller gyromagnetic ratios of carbon-13 ($\gamma_p/\gamma_c \sim 4$) or nitrogen-15 ($\gamma_p/\gamma_n \sim 10$) (Ernst, Bodenhausen, and Wokaun, 1987). Therefore the cross-polarizing of electrons to nuclei using DNP has the potential to greatly increase the sensitivity of NMR beyond nuclear–nuclear cross-polarization schemes, and would have profound effects for the NMR study of biological systems and MRI applications.

24.2
Historical Aspects (Metals, Solids and Liquids) at Lower Magnetic Fields

In 1953, Albert W. Overhasuer described a double-resonance scheme to polarize nuclei in metals by applying a saturation pulse to the electron resonance (Overhauser, 1953). Although the proposal was first met with skepticism, the first published experimental proof also came in the same year; polarizing Li nuclei in Li metal at fields below 50 Gauss (Carver and Slichter, 1953). It should be noted that this experiment was the first double-resonance, cross-polarization, between different spin species (i.e., between conduction electrons and Li nuclei), while the first double-resonance and polarization transfer experiment was carried out by Pound on quadrupolar split ^{23}Na nuclear levels (Pound, 1950). It was not long before the method was extended to proton polarization using free radicals (Beljers, van der Kint, and van Wieringen, 1954). This method, which is referred to as the Overhauser effect (OE), found its main application in liquids and metals (Carver and Slichter, 1956; Hausser and Stehlik, 1968; Müller-Warmuth and Meise-Gresch, 1983). The polarization transfer is mediated via relaxation mechanisms, such as scalar or dipolar coupling between electrons and nuclei.

In insulating solids, spin–spin relaxation processes become ineffective as a cross-polarization mechanism. However, it was recognized by Jeffries (Jeffries, 1957, 1960) and independently by Abragam and coworkers (Abragam, Combrisson, and Solomon, 1958) that the forbidden electron transition in solids involves spin–flip terms that can be used for nuclear polarization. (The effect was also experimentally discovered by Erb and colleagues; Erb, Montchane, and Uebersfeld, 1958.) In the literature, this method was later termed the solid effect (SE), and has been defined when the EPR homogeneous line width δ and the corresponding inhomogeneous line width Δ is less than the nuclear Larmor frequency ω_n; that is, when $\delta, \Delta < \omega_n$ (Maly et al., 2008). Shortly afterwards, another related effect was identified, called the cross effect (CE) (Kessenikh et al., 1963; Kessenikh, Manenkov, and Pyatnitskii, 1964; Hwang and Hill, 1967a, 1967b; Wollan, 1976), which covers the case when the inhomoegnenous EPR line is broader than the nuclear Larmor frequency and, concurrently, the homogeneous EPR line width remains smaller than the nuclear Larmor frequency ($\delta < \omega_n < \Delta$), which is often the case with high magnetic fields.

A final "classical" DNP mechanism using continuous microwave pumping covers the case where the homogeneous EPR line width is larger than the nuclear Larmor frequency ($\omega_n < \delta$); this is referred to as thermal mixing (TM). It was first recognized by Provotorov (Provotorov, 1962) that, in a strongly coupled spin system, the Zeeman and dipolar energies could be assigned separate spin temperatures. The application of an off-resonant microwave field ($\Delta\omega = \omega - \omega_e$) can "cool" ($\Delta\omega < 0$) or "heat" ($\Delta\omega > 0$) the dipolar bath of a spin system. If the dipolar bath is in thermal contact with the nuclear Zeeman system, then either "cooling" or "heating" of the nuclear spin system can be achieved. Provotorov's equations can be elegantly re-derived using the concept of spin temperature (Abragam and

Goldman, 1982), and were extended to low temperatures (below the high temperature limit) by Borghini (Borghini, 1968).

Shortly after the discovery of the OE, it was demonstrated that all continuous microwave DNP mechanisms rapidly become inefficient at high magnetic fields. In the case of the OE, the DNP enhancement was expected to be negligibly small for magnetic fields $B > 1/(\gamma_e \tau)$, where τ is the correlation time between unpaired electrons and nuclei (Hausser and Stehlik, 1968; Müller-Warmuth and Meise-Gresch, 1983). For the SE, involving the excitation of the forbidden transitions, the DNP enhancement depends on the magnetic field as B^{-2} (Abragam and Goldman, 1978), and for the CE, involving allowed EPR transitions and since the EPR line width scales the magnetic field, the DNP enhancement therefore scales as B^{-1} (Maly et al., 2008). Thus, with the development of high-field NMR spectrometers, DNP was not considered a viable option and, as a result, DNP has been largely ignored by the magnetic resonance community, except in some specialized areas, such as polarizing nuclear targets (Abragam and Goldman, 1982; Goertz, Meyer, and Reicherz, 2002), studying magnetic order in the μK region (Abragam and Goldman, 1982; Sprenkels, Wenckebach, and Poulis, 1983), or other solid-state NMR applications (Wind et al., 1985; Singel et al., 1989; Afeworki, McKay, and Schaefer, 1992). Another reason for this stunted development of high-field DNP was the underdevelopment of microwave technology at submillimeter wavelengths. The development at high magnetic fields was largely performed in Griffin's group at MIT, who constructed high-power gyrotrons for solid-state magic angle spinning (MAS) DNP, with applications to large biomolecules (Maly et al., 2008).

Another approach to high-field DNP arose from the Amersham Health Research Laboratory, in Sweden, where carbon-13 and nitrogen-15 were DNP-polarized at low temperatures. The same group then developed a cleaver dissolution process that allowed the transfer of highly polarized carbon-13 and nitrogen-15 nuclei (over >10 000) from a low-field, low-temperature environment into a room temperature, high-field NMR spectrometer (Ardenkjaer-Larsen et al., 2003). These extremely large enhancements were a result of the DNP enhancement at low temperatures, multiplied by a Boltzmann factor resulting from the transfer of spins from low to high temperatures with minimal loss though spin-lattice relaxation. This process works well for small molecules or small contrast agents for MRI.

Finally, for completion, the microwave pumping of unpaired electrons is not the only method of polarizing nuclei for NMR; other methods include laser-polarized noble gases (Leawoods et al., 2001), chemical-induced dynamic nuclear polarization (CIDNP) (Hore and Broadhurst, 1993), as well as para-hydrogen-induced polarization (PHIP) (Bowers and Weitekamp, 1986). In the cases of CIDNP and PHIP, the polarized states are generated by spin-sensitive chemical reactions and, while they are very successful, they are generally system-specific. In contrast, microwave-driven DNP experiments are evolving as a broadly applicable approach to sensitivity enhancement in solid-state and solution NMR.

24.3
Theory

Before starting to describe the different DNP mechanisms, it is important to review nuclear polarization and its relation to thermal equilibrium (Abragam and Goldman, 1982). For an ensemble of N particles with nuclear spin I, the polarization is defined as:

$$P = \frac{Tr(\rho I_z)}{I} = \frac{\langle I_z \rangle}{I}, \quad (24.1)$$

where ρ is the normalized density matrix and $\langle I_z \rangle$ is the ensemble average. For a spin-1/2 system, Equation 24.1 simplifies to

$$P = \frac{N_+ - N_-}{N_+ + N_-}, \quad (24.2)$$

where N_+ and N_- are corresponding populations for spin-up and spin-down, respectively. In thermal equilibrium, a Boltzmann's distribution is established and the polarization for a spin-1/2 is given by:

$$P_0 = \tanh\frac{\hbar\gamma_n B}{2kT} \approx \frac{\hbar\gamma_n B}{2kT}, \quad (24.3)$$

where the approximation given in Equation 24.3 is called the "high-temperature limit". Thus, thermal polarization increases linearly with magnetic field in the high-temperature limit. Finally, DNP enhancement (E) is defined as $E = P/P_0$.

We now proceed to discuss the various "classical" DNP mechanisms – OE, SE, TM, and CE – all of which involve continuous microwave excitation. The labeling of the different DNP mechanisms, as described initially in Section 24.2, is not uniform in the literature. Whilst the OE and TM are fairly easy to define, this is not the case with SE and CE. Some authors do not have a separate classification for the CE; for example, the SE can take into account all process involving flip-flop transitions between electronic spins and nuclear spins induced by a microwave field at low unpaired electron concentration. In this model, the SE can be further divided into a well-resolved SE ($\omega_n > \Delta$) and a differential SE ($\omega_n < \Delta$) (Wenckebach, 2008). Other authors have termed the CE as indirect thermal mixing (Wind et al., 1985). For pedagogic and historic reasons, we have used the distinction between the SE and CE following the reference of Maly et al. (2008).

24.3.1
The Overhauser Effect (OE)

In the classical OE, continuous microwave pumping of allowed EPR transitions brings the unpaired electron spins away from thermal equilibrium; in this new microwave-driven equilibrium state, cross-relaxation (flip-flop terms) between electronic and nuclear spins can significantly enhance the polarization of the nuclear spins, provided that other non-electronic relaxation processes are comparatively smaller. This method works well in metals, and in liquids where significant electron spin saturation is possible and electron–nuclear cross-relaxation rates dominate other relaxation processes. Restricting our discussion to liquids – as this has potential application to enhance the sensitivity of high-resolution NMR – the essential interactions between an electron spin S and a nuclear spin I are represented by the spin Hamiltonian:

$$H = -\gamma_e B S_z - \gamma_n B I_z + a S \cdot I + \frac{\mu_0 \gamma_e \gamma_n \hbar}{4\pi}\left[\frac{S \cdot I}{r^3} - \frac{3(I \cdot \mathbf{r})(S \cdot \mathbf{r})}{r^5}\right] + H_{nL}(t), \quad (24.4)$$

where the first two terms are the static electronic and nuclear Zeeman terms, the third and fourth terms represent time-dependent scalar and dipolar coupling, respectively, and the final term represents the remaining time-dependent non-electronic nuclear relaxation processes. The time-dependent terms of Equation 24.4 can be used to calculate relaxation rates between the nuclear and electronic levels (Figure 24.1a).

The relaxation rate equation of a nuclear spin coupled to an electron spin can be described by the Solomon equation (Hausser and Stehlik, 1968):

$$\frac{d\langle I_z \rangle}{dt} = -(\rho_I + W^0)(\langle I_z \rangle - I_0) - \sigma_{IS}(\langle S_z \rangle - S_0), \quad (24.5)$$

where I_0 and S_0 are the nuclear and electronic spin Boltzmann equilibrium expectation values, respectively, ρ_I and σ_{IS} are the relaxation rates derived from nuclear–electron scalar and dipolar interactions, and W^0 is the relaxation rate of other non-electronic relaxation processes. Applying continuous microwave pumping (i.e., $\langle S_z \rangle \to 0$), the DNP enhancement of the nuclear spins can be derived from the steady-state solution of Equation 24.5:

$$E = \frac{\langle I_z \rangle}{I_0} = 1 + \xi f s \frac{\gamma_e}{\gamma_n}, \quad (24.6)$$

where the ratio of the Boltzmann equilibrium spin expectations S_0/I_0 is replaced by the ratio of the gyromagnetic ratios of the electron and nucleus (γ_e / γ_n). The other parameters, ξ, f and s are defined below (Hausser and Stehlik, 1968):

$$\xi = \frac{\sigma_{IS}}{\rho_I} = \frac{W_2 - W_0 - W_0^S}{W_0^S + W_0 + 2W_I + W_2}, \quad (24.7)$$

$$f = \frac{\rho_I}{\rho_I + W^0} = \frac{W_0^S + W_0 + 2W_I + W_2}{W_0^S + W_0 + 2W_I + W_2 + W^0} = 1 - \frac{T_{1n}}{T_{1n}^0}, \quad (24.8)$$

$$s = \frac{S_0 - \langle S_z \rangle}{S_0} \approx \frac{\gamma_e^2 B_1^2 T_{1e} T_{2e}}{1 + \gamma_e^2 B_1^2 T_{1e} T_{2e}}. \quad (24.9)$$

The relaxation rates W_i are defined in Figure 24.1a, and T_{1n} and T_{1n}^0 are the nuclear relaxation rates with and without paramagnetic species, respectively. Historically, the term $\xi f s \gamma_e/\gamma_n$ has been called the "signal enhancement factor" (Hausser and Stehlik, 1968), where ξ is the coupling factor, f the leakage factor, and s the elec-

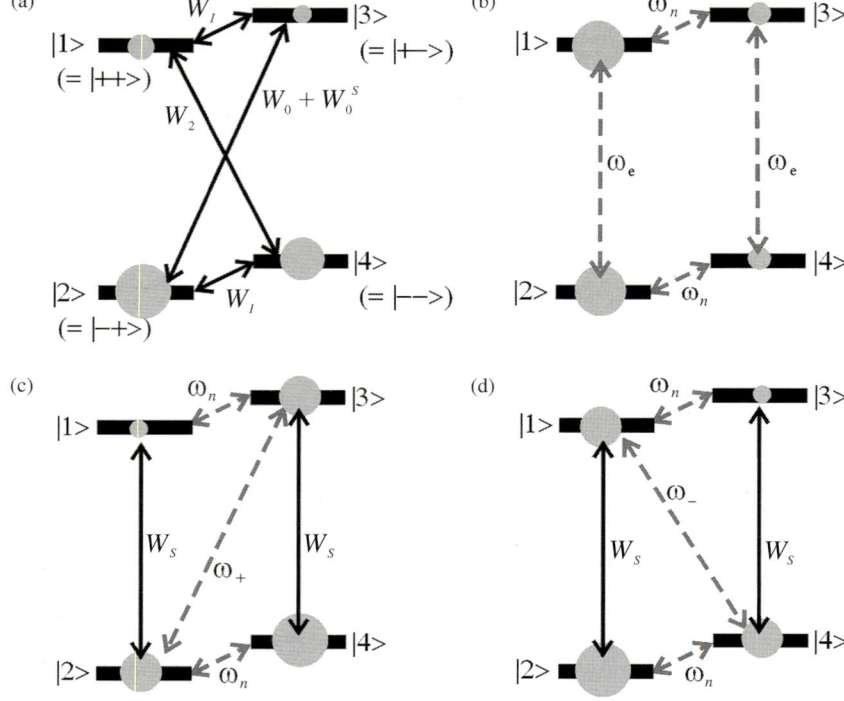

Figure 24.1 Energy-level diagrams ($|m_S\ m_I\rangle$), transitions (ω) and relaxation rates W_i associated with a SI-spin system. (a) The relaxation rates are given for dipolar coupling [W_I (nuclear), W_0 (zero quantum) and W_2 (double quantum)]; and scalar coupling [W_0^S (zero quantum)]; (b) The OE is achieved by applying saturating microwaves to the allowed EPR transitions (ω_e), while observing the NMR transitions (ω_n). In this example, it is assumed that pure scalar relaxation (W_0^S) is the dominant relaxation mechanism; (c, d) These represent the solid effect, where W_S is the electron spin relaxation rate and $\omega_\pm$ are the forbidden transitions. The gray spheres symbolize the population probabilities (p_1, p_2, p_3, p_4) in the steady state without the application of microwaves (panel a) and with microwaves (panels b, c, and d).

tronic saturation which can be approximated using the Bloch equations (Equation 24.9) in the case of a single homogenous line width.

Scientifically, the aim is usually to determine the coupling factor ξ by measuring the parameters of Equation 24.6. E and f are easily determined experimentally, the latter by measuring the nuclear relaxation time with and without paramagnetic centers, and is dependent on the paramagnetic concentration, through $\rho_I = 1/T_{1n} - 1/T_{1n}^0 = kC$, where k is the relaxivity and C the concentration. In the past, however, the accurate experimental determination of s has been more difficult to obtain (Hausser and Stehlik, 1968), because the electron saturation is complicated by Heisenberg exchange (Bates and Drozdoski, 1977) and nuclear relaxation (Robinson, Haas, and Mailer, 1994). Additionally, direct experimental

Table 24.1 The main interactions and parameters for the spectral density functions (SDF) for inner- and outer-sphere contributions are tabulated. Time-dependent functions, given in the Hamiltonian Equation 24.4, are listed in row 3. The various correlations times and total correlation times are given in rows 4, 5, and 6: τ_e is the chemical exchange time, T_{1e} and T_{2e} are the electronic relaxation times, and τ_r and τ_t are the rotational and translational correlation times, respectively. The SDFs all take the same Lorentzian form $J_x(\omega) = \tau_x/(1+\tau_x^2\omega^2)$, except for the case of outer-sphere relaxation (see text, Equation 24.12).

Interaction	Inner-sphere		Outer-sphere
	Scalar 1st and 2nd kind	Molecular motion (dipolar)	
Function	$a(t), S(t)$	$\mathbf{r} = (r, \theta(t), \phi(t))$	$\mathbf{r} = (r(t), \theta(t), \phi(t))$
Correlation times	τ_e, T_{1e}, T_{2e}	$\tau_e, T_{1e}, T_{2e}, \tau_r$–rotational	T_{1e}, τ_t–translational
	$1/\tau_{s1} = 1/\tau_e + 1/T_{1e}$	$1/\tau_{d1} = 1/\tau_r + 1/\tau_e + 1/T_{1e}$	(see Equation 24.12)
	$1/\tau_{s2} = 1/\tau_e + 1/T_{2e}$	$1/\tau_{d2} = 1/\tau_r + 1/\tau_e + 1/T_{2e}$	
SDF	$J_{s1}(\omega), J_{s2}(\omega)$	$J_{d1}(\omega), J_{d2}(\omega)$	$J_d(\omega) = J_d(z(\omega, \tau_t, T_{1e}))$

determination of the electronic relaxation times at high frequencies (above ~100 GHz) by EPR for liquids is difficult. Therefore, in the literature, the determination of the coupling factor ξ at high magnetic fields has often been made using nuclear magnetic relaxation dispersion (NMRD) experiments, which effectively measure ρ_I (see Equation 24.5) as a function of magnetic field (see Equation 63 in Hausser and Stehlik, 1968). To overcome this problem, two different approaches have been recently implemented to directly determine s (Armstrong and Han, 2007; Sezer et al., 2009b). In particular, in Sezer et al. (2009b), a rigorous semiclassical relaxation theory including electron coherences has been applied to a nitroxide radical.

Experimentally, at large paramagnetic concentrations and optimal microwave electron saturation, both s and f can approach 1. However, the coupling factor ξ can show a large variation depending on the mixture of dipolar to scalar coupling, and on the magnetic field. The effects of a paramagnetic center on the nuclear relaxation process can be divided up into inner-sphere and outer-sphere contributions (Bertini, Luchinat, and Parigi, 2001). Outer-sphere relaxation is caused by the dipolar interaction between paramagnetic centers and protons freely diffusing. Inner-sphere relaxation is caused by dipolar and scalar interactions of nuclei bound to a paramagnetic center or in a chemical exchange with a paramagnetic center, with an exchange rate of τ_e. Here, we will restrict ourselves to paramagnetic centers within the Solomon–Bloembergen–Morgen (SBM) theory, assuming that zero-field splitting (ZFS) effects can be neglected and a point-dipole approximation is assumed. A comprehensive review of the NMR relaxation processes between paramagnetic molecules in solution is given elsewhere (Bertini, Luchinat, and Parigi, 2001).

The relaxation rates W_i shown in Figure 24.1a can be expressed in terms of spectral density functions ($J_x(\omega)$; see Table 24.1), which results from frequency-dependent fluctuations of the scalar and dipolar interactions. Assuming isotropic motion within the Redfield limit (Hausser and Stehlik, 1968), the coupling constant takes the form

$$\xi = \frac{6J_{d2}(\omega_e + \omega_n) - J_{d2}(\omega_e - \omega_n) - 5MJ_{s2}(\omega_e - \omega_n)}{6J_{d2}(\omega_e + \omega_n) + J_{d2}(\omega_e - \omega_n) + 3J_{d1}(\omega_n) + 5M(J_{s2}(\omega_e - \omega_n) + \beta J_{s1}(\omega_n))}, \quad (24.10)$$

where M is the ratio of dipolar to scalar coupling,

$$M = \frac{a^2}{(\mu_0/4\pi)^2 \gamma_e^2 \gamma_n^2 \hbar^2 / r^6}. \quad (24.11)$$

In addition to the scalar relaxation rate W_0^S at frequencies $\omega_e - \omega_n$ (as shown in Figure 24.1a), electron–nuclear relaxation can also be induced between NMR nuclear levels if there is mixing of the wavefunctions between the terms $|+-\rangle$ and $|-+\rangle$ (Abragam, 1961), which results in an additional term $\beta J_{s1}(\omega_n)$. Here, β can be considered as a further leakage term that can take on values from 0 ($\tau_e \ll T_{1e}, T_{2e}$) to 1 ($\tau_e \gg T_{1e}, T_{2e}$). Depending on the type of interaction between the paramagnetic center and the nuclei (inner-sphere or outer-sphere), and on the various correlation and relaxation times, the spectral density function can be read from Table 24.1 and substituted into Equation 24.10.

In the case of a freely diffusing solution consisting of radical and solvent molecules, solvent-protons generally show no scalar coupling (Hausser and Stehlik, 1968; Müller-Warmuth and Meise-Gresch, 1983), and can be described by outer-sphere relaxation (pure translational motion). A typically used analytical SDF to describe this motion is given by:

$$J_d(z) = \frac{1 + 5z/8 + z^2/8}{1 + z + z^2/2 + z^3/6 + 4z^4/81 + z^5/81 + z^6/648}, \quad (24.12)$$

where $z = \sqrt{(2(\omega \tau_t + \tau_t/T_{1e}))}$ (Bertini, Luchinat, and Parigi, 2001). The translational diffusion time τ_t depends on the free diffusion coefficients of the paramagnetic molecule D_P and the solvent molecules D_n, and on the distance of closest approach d:

$$\tau_t = \frac{d^2}{D_P + D_n}. \quad (24.13)$$

In the case of bound nuclei at a fixed distance to a paramagnetic center (i.e., $\tau_e = \infty$), where the distance is long enough so that there is no scalar coupling, the relaxation is classified as inner-sphere (purely rotational motion). If we further assume $\tau_t, \tau_r \ll T_{1e}, T_{2e}$, a plot of the coupling factors for both cases – pure rotational and pure translational – against $\omega_e \tau$ is given in Figure 24.2, using Equation 24.10

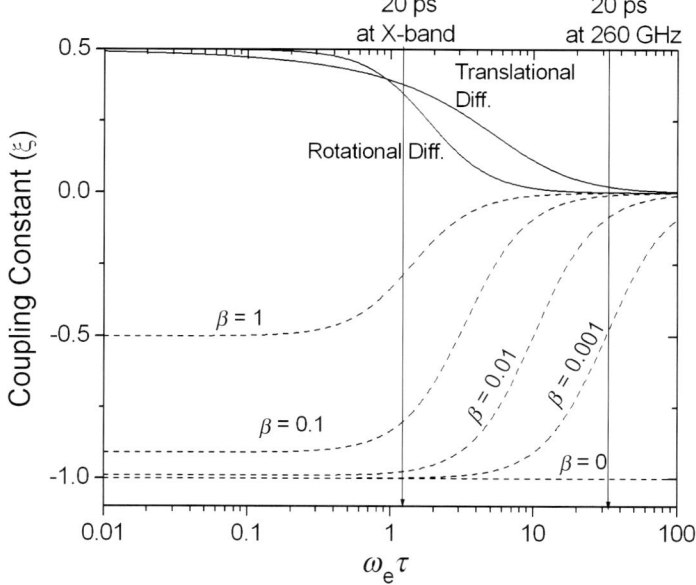

Figure 24.2 The dependence of the coupling factor ξ on the product of electron Larmor frequency and correlation time. Using Equation 24.10 and Table 24.1, plots of pure rotational and translational diffusion of molecular motions, and of scalar coupling for various β values are plotted. The vertical lines show two cases corresponding to X-band (9.8 GHz) and 260 GHz, both with a correlation time of 20 ps.

and the SDFs from Table 24.1. At higher magnetic fields for similar correlation times, rotational motion is ineffective compared to translational motion (Figure 24.2). Also plotted are several cases of pure scalar coupling at β values of 0, 0.001, 0.01, 0.1, and 1 (Figure 24.2).

While the dipolar coupling decays as $1/r^3$, the scalar interaction has a shorter range and is proportional to the unpaired electron density at the nucleus. Assuming $\beta = 0$ and $\tau_e \ll \tau_r$, the ratio of dipolar to scalar coupling (M) can be derived from measuring the coupling factor in the extreme narrowing limit ($\omega_n \tau_e \ll 1 \rightarrow J(\omega) \approx \tau_e$):

$$\xi = \frac{1-M}{2+M}. \tag{24.14}$$

In the case of pure scalar coupling ($M \rightarrow \infty$), $\xi = -1$, while in the case of pure dipolar coupling ($M \rightarrow 0$), $\xi = 1/2$. Experimental evidence suggests that protons exhibit in most cases a pure dipolar character, especially using organic radicals (Hausser and Stehlik, 1968; Müller-Warmuth and Meise-Gresch, 1983); other nuclei, such as ^{19}F, ^{13}C, ^{31}P, and ^{15}N, have shown quite high scalar-coupled DNP enhancements, even at high magnetic fields (Loening et al., 2002).

Finally, the above-described analytical approaches often simplify the geometry and describe the complex relative motions between unpaired electrons and nuclei by a single correlation time. Recently, molecular dynamic (MD) simulations have offered a better method to model the complex relative motion between the radical and solvent molecules. For example, coupling factors were calculated from MD and compared well with experimental data over a wide range of magnetic fields for the radical TEMPOL in water (Sezer, Prandolini, and Prisner, 2009a).

24.3.2
Two-Spin Cross-Polarization: Solid Effect (SE)

The SE can be found in solids containing fixed paramagnetic centers, where the inhomogeneous EPR line width Δ is less that the nuclear Larmor frequency ($\omega_n > \Delta$) (well-resolved SE). In the SE, the electron–nuclear flip-flop transitions are induced by applying saturating microwave power to the so-called "forbidden transitions" at microwave frequencies ω_+ ($|2\rangle \leftrightarrow |3\rangle$; Figure 24.1c) or ω_- ($|1\rangle \leftrightarrow |4\rangle$; Figure 24.1d). A fast electron relaxation (W_S) keeps the population ratio of the allowed transitions ($p_1/p_2 = p_3/p_4 = \exp(-\hbar\gamma_e B/kT)$) in thermal equilibrium. With the additional condition of normalized population probabilities $\Sigma_i\, p_i = 1$ and defining nuclear polarization as $P = (p_1 + p_2 - p_3 - p_4)/2$, it is easy to show that the application of saturating microwave pumping (i.e., $\omega_+ \rightarrow p_2 = p_3$ or $\omega_- \rightarrow p_1 = p_4$) induces a nuclear polarization of $P \approx -/+\hbar\gamma_e B/2kT$. Therefore, compared to thermal equilibrium P_0 (see Equation 24.3), the maximum possible DNP enhancement (P/P_0) for the SE is $-/+\gamma_e/\gamma_n$ using applied microwave pumping at $\omega_\pm$ (Figure 24.1c, d).

In practice, the complete saturation of forbidden transitions is not possible, and other relaxation processes will short-circuit the DNP enhancement. For simplicity, we consider an isolated electron–nuclear pair, again described by Equation 24.4; however, this time the nuclear–electron relative positions **r** become time-independent in an insulating solid. In the case of a dipolar interaction, the pure states $|i\rangle = |m_S\, m_I\rangle$ (Figure 24.1a) become mixed $|1'\rangle = \kappa|1\rangle - q^*|3\rangle$, $|2'\rangle = \kappa|2\rangle + q^*|4\rangle$, $|3'\rangle = \kappa|3\rangle + q|1\rangle$ and $|4'\rangle = \kappa|4\rangle - q|2\rangle$, where

$$|q| = \frac{\mu_0}{4\pi}\frac{3}{4}\frac{\gamma_e\gamma_n\hbar}{r^3\omega_n}\sin\theta\cos\theta, \tag{24.15}$$

and $|\kappa| = (1 - |q|^2)^{1/2} \approx 1$ and θ is the polar coordinate describing the electron–nuclear vector to the magnetic field (Wind et al., 1985). This mixing allows the "forbidden transition" at $\omega_\pm = \omega_e \pm \omega_n$ frequencies to be excited. In the fast diffusion limit, the transition probability $\omega_\pm$ is dependent on the isotopic average of $\langle |q|^2 \rangle$, and given by

$$\omega_\pm = 2\pi\langle |q|^2 \rangle \gamma_e^2 B_1^2 g(\omega_e - \omega \pm \omega_n), \tag{24.16}$$

where g is the normalized EPR line shape. Here, we note that $\omega_\pm$ falls off as the inverse square of the applied field B^{-2}, because it depends on q^2. Finally, the DNP enhancement can be derived after solving the rate equations, including electron–nuclear relaxation rates W_I (Figure 24.1a) and other non-electronic rates W^0. For

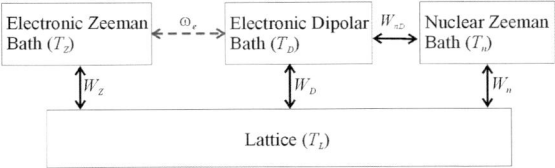

Figure 24.3 A thermodynamic model to describe thermal mixing. The spin system can be broken up into separate heat baths with corresponding assigned temperatures. The relaxation rates, W_i, are given between the baths; resonant microwave radiation (ω_e) brings the Zeeman and dipolar baths into thermal contact.

example, by applying continuous microwave radiation at $\omega_e + \omega_n$ ($=\omega_+$), the DNP enhancements is given by (Wind et al., 1985):

$$E = P/P_0 = 1 + \frac{\omega_+}{\omega_+ + W_I + W^0}\left(\frac{\gamma_e}{\gamma_n} - 1\right). \tag{24.17}$$

Similarly, positive enhancement can be derived at $\omega_e - \omega_n$. Note that for an electron and a proton ($\gamma_e < 0$ and $\gamma_p > 0$).

24.3.3
Many-Spin Cross-Polarization: Thermal Mixing (TM)

The TM effect can be found in solids containing fixed paramagnetic centers, where the homogeneous linewidth δ is greater than the nuclear Larmor frequency ($\omega_n < \delta$). Let us initially ignore the nuclear spin system and concentrate on an electronic, dipolar-coupled, spin-1/2 system in a magnetic field. It is easy to show that the Zeeman terms, $H_Z = \sum_i g_e\mu_B B S_z^i$, and the secular dipolar terms, $H_D = \sum_{i,j} D_{i,j}(3S_z^i S_z^j - \mathbf{S}^i\cdot\mathbf{S}^j)$, commute ($[H_Z, H_D] = 0$). If we further assume that this spin system is isolated from the external world—that is, $T_{1e} \gg T_{2e}$—then it is possible to assign separate spin temperatures to the Zeeman (T_Z) and dipoalar (T_D) "baths" (Figure 24.3). In thermal equilibrium, the temperature of each bath is equal to the lattice temperature (T_L) (Figure 24.4a). A peculiarity about such "isolated" spin systems, with an upper and lower energy bound to the energy spectrum, is that it is possible to generate positive and negative spin temperatures (see Figure 24.4b, c). These above ideas briefly summarize the concept of spin temperature (Goldman, 1970).

Now the application of resonant microwaves brings both electronic baths into thermal contact in the rotating frame (Figure 24.3): for a given microwave field at frequency ω, $\hbar\omega_e$ is absorbed by the Zeeman bath, and the remaining energy $\hbar(\omega - \omega_e)$ is either absorbed ($\omega > \omega_e$) or emitted ($\omega < \omega_e$) by the electronic dipolar system (Figure 24.4). Formally, this process can be described by the Provotorov equations (Provotorov, 1962):

$$\frac{d\alpha}{dt} = -W(\alpha - \beta) - W_Z(\alpha - \alpha_L), \tag{24.18}$$

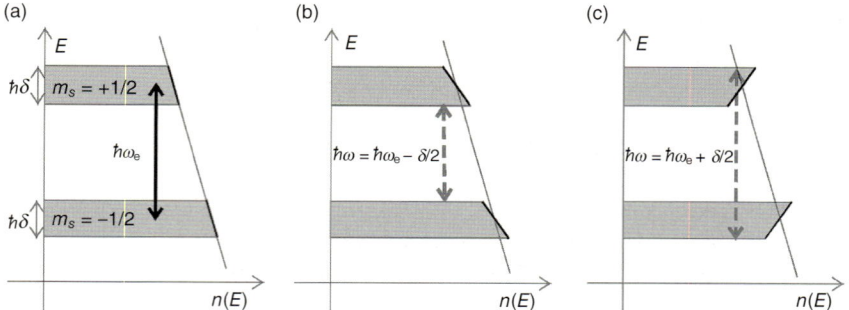

Figure 24.4 Energy distribution of populations ($n(E)$) between electronic spin sate manifolds $m_z = 1/2$ and $m_z = -1/2$, separated by energy $\hbar\omega_e$. The width of each state ($\hbar\delta$) is determined by electron–electron dipolar broadening; ω represents off-resonant application of microwaves. (a) Thermal equilibrium $T_L = T_D = T_Z$; (b) The application of off-resonant microwave radiation at $\omega < \omega_e$ "cools" the dipolar bath $0 < T_D < T_Z$; (c) The application of off-resonant microwave radiation at $\omega > \omega_e$ "heats" the dipolar bath to a negative spin temperature $0 < |T_D| < T_Z$, $T_D < 0$.

$$\frac{d\beta}{dt} = -W\frac{(\omega_e - \omega)^2}{\delta^2}(\alpha - \beta) - W_D(\beta - \beta_L), \qquad (24.19)$$

where W is the allowed microwave driven transition rate, and $\alpha = \hbar/(kT_Z)$ and $\beta = \hbar/(kT_D)$ are effective inverse spin temperatures for the electronic Zeeman and dipolar thermal baths, respectively. From the stationary solutions of the Provotorov equations, it is possible to derive the maximum inverse spin temperature assuming a fully saturating microwave field:

$$|\beta_{max}| = \beta_L \frac{\omega_e}{2\delta}\sqrt{\frac{W_Z}{W_D}}. \qquad (24.20)$$

Hence, maximum cooling (heating) takes place when the EPR line width δ is narrow and the dipolar relaxation rate W_D is small.

In the final step, we add a nuclear Zeeman bath (Figure 24.3). The relaxation rate between the electronic dipolar and nuclear Zeeman baths (W_{nD}) is governed by energy-conserving, three-spin electron–electron–nuclear exchange processes (Buishvili, 1966; Wenckebach et al., 1969). The maximum nuclear polarization that can be achieved is proportional to the maximum inverse spin temperature (Abragam and Goldman, 1978; Goertz, Meyer, and Reicherz, 2002):

$$P_{I,max} = \frac{I+1}{3}\frac{\beta_{max}\omega_n}{\sqrt{1+f}}, \qquad (24.21)$$

where a leakage factor $\sqrt{(1+f)}$ is simply added to account for all the processes that do not proceed via the electronic-dipolar reservoir.

As an experimental example from the literature, we consider the system of the radical 1,3-bisdiphenylene-2-phenylallyl (BDPA) doped in polystyrene at 1.4 T, corresponding to a proton frequency of 60 MHz, a carbon-13 frequency of 15 MHz, and an EPR frequency of 40 GHz (Q-band) at room temperature. The half-width of the EPR line was $\delta/(2\pi) = 13$ MHz. The maximum proton DNP enhancement was measured to be ~30, and the mechanism was dominated by the SE, since $\omega_p > \delta$; however, the carbon-13 DNP enhancement was found to be ~160, and dominated by TM, since $\omega_c \sim \delta$ (Wind et al., 1985).

The above description is unsuitable at low temperatures, since the Provotorov equations are only valid in the high-temperature limit. Extensions to low temperatures were performed by Borghini (Borghini, 1968). Additionally, Borghini included the effects of inhomogeneous EPR broadening, which becomes important at high magnetic fields, potentially reducing the DNP enhancement. Another consideration of TM at high fields is the desire to choose a paramagnetic center with a narrow EPR line width, thereby maximizing β_{max} (see Equation 24.20), and fulfilling the condition $\omega_n < \delta$ at the same time. At high magnetic fields this is not always possible, however. In the above example, this was already the case for protons at 1.4 T using in the system of BDPA doped in polystyrene.

24.3.4
Three-Spin Cross-Polarization: Cross Effect (CE)

The CE covers the case of solids with fixed paramagnetic centers, where the nuclear Larmor frequency ω_n is less than the inhomogeneous EPR line width Δ but remains greater than the homogeneous line with δ, ($\Delta > \omega_n > \delta$) (also differential SE). The basic concept involves a three-spin process, where two dipolar-coupled electrons, with EPR frequencies at ω_{e1} and ω_{e2}, are separated in energy by the nuclear Zeeman energy:

$$\omega_n = \omega_{e2} - \omega_{e1}. \tag{24.22}$$

The frequency separation between the electron spins can be generated either through strong hyperfine couplings, or through differences in the electronic g-factor. Frequency matching using hyperfine coupling can be applied only at a single matched external magnetic field, while matching using the g-factor scales with the external magnetic field.

The DNP mechanism can be qualitatively described using an energy level diagram of a three-spin system (Figure 24.5). If the separation between the two electronic spins satisfies Equation 24.22, then two of the levels become degenerate (|4> and |5> or |3> and |6>); for example, in Figure 24.5a levels |4> and |5> are chosen to be degenerate. In thermal equilibrium, the initial polarizations are described by the Boltzmann distribution (Figure 24.5a). The application of saturating microwave pumping of electron resonance ω_{e1}, concurrently with the CE transition (ω_{CE}) to a second electron and energy-conserving three-spin cross-relaxation (|4> ↔ |5>), ensures a positive nuclear polarization (Figure 24.5b).

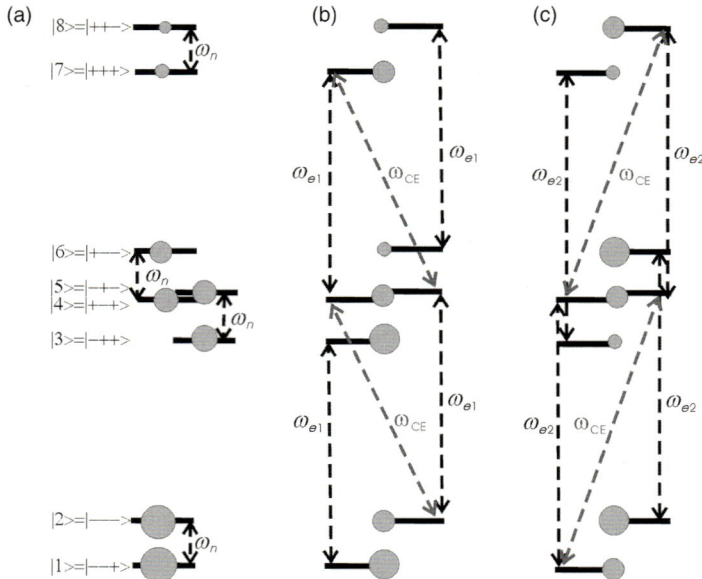

Figure 24.5 Energy-level diagrams ($|m_{S1}\ m_{S2}\ m_I\rangle$) and transitions ω_i associated with a three-spin (1/2) *SSI*-spin system. (a) Thermal equilibrium, where ω_n is the nuclear Larmor frequency; (b, c) Equilibrium established after applying a saturating microwave power to electron spin 1 (ω_{e1}) (panel b) or spin 2 (ω_{e2}) (panel c), and cross-relaxation involving three spins *SSI*. The gray spheres symbolize the population probabilities in the steady state without the application of microwaves (panel a) and with microwaves (panels b and c).

A similar argument can be used to describe the negative enhancement by pumping ω_{e2} (Figure 24.5c).

In practice, at low magnetic fields ($\omega_e/(2\pi) \le 40\,\text{GHz}$), with small inhomogeneous broadening by g-tensor anisotropy, it is impossible to separate TM and CE mechanisms (Wind et al., 1985). At higher fields, the concept of a single spin temperature can break down and microwave pumping will result in hole-burning. However, the inhomogeneous EPR line can become effectively homogeneous at low temperatures, where DNP processes are much faster than T_{1e} and at high paramagnetic concentrations (Atsarkin, 1978). For example, DNP and ELDOR experiments on 40 mM 4-amino-TEMPO at 11 K and at a magnetic field of 5 T, showed an effective homogeneous EPR line via strong electron–electron relaxation (Farrar et al., 2001). In this case, DNP enhancement was calculated using a TM/CE thermodynamic model (Farrar et al., 2001). At lower concentrations, the DNP enhancement using single radicals is dominated by the CE, and the efficiency is usually very low because the CE relies on pairs of radicals having the correct orientation (satisfying the condition Equation 24.22) and being close enough to have sufficient dipole–dipole coupling. However, this is seldom the case at low concentrations in a random distribution of single radicals. Therefore, at high magnetic fields an efficient CE requires not only the correct frequency matching and low

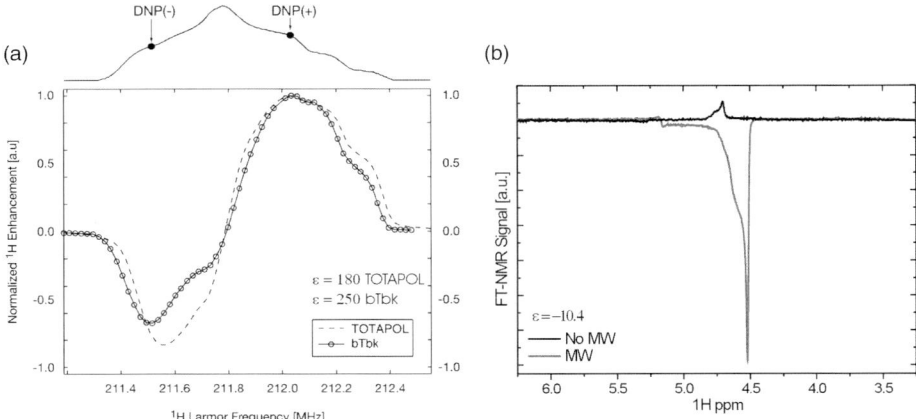

Figure 24.6 (a) Field-dependent normalized DNP enhancement profiles (bottom) and EPR spectrum of the biradical bTbk (top) at 95 K and at a field of 5 T, corresponding to 140 GHz electron and 210 MHz proton frequencies. The actually measured enhancements ε for different radicals in DMSO/H$_2$O/D$_2$O are given in the insert (for details, see Matsuki *et al.*, 2009); (b) Water proton NMR spectra with microwaves (dashed line) and without microwaves (solid line) for Fremy's Salt in an aqueous solution at room temperature at a field of 9.2 T, corresponding to 260 GHz electron and 400 MHz proton frequencies (for details, see Prandolini *et al.*, 2009).

temperatures, but the electron–electron–nuclear cross-relaxation, which is a function of the average dipolar coupling between the spins, must be strong (Wollan, 1976; Farrar *et al.*, 2000, 2001).

Thus, in order to improve the efficiency of the CE a number of paramagnetic systems have been proposed. Mixtures of two radicals with the correct *g*-factor separation, for example the g_{yy} of TEMPO and the *g*-value of trityl, have almost the correct separation for protons (Hu *et al.*, 2007). A further improvement has been made by tethering together two radicals (biradicals), ensuring a reliable and strong dipole–dipole coupling between the unpaired electrons (Hu *et al.*, 2004). By careful optimization of the biradical distance and orientation, large DNP enhancements have been observed at high magnetic fields (Matsuki *et al.*, 2009).

24.3.5
Beyond Classical DNP Methods: Coherent Polarization Transfer

Presently, the classical CE/TM mechanisms using optimized biradicals and high-power microwave sources (gyrotrons) at temperatures below <100 K appear to offer the largest DNP enhancements on bimolecules in solid-state MAS NMR experiments at high magnetic fields (Matsuki *et al.*, 2009) (Figure 24.6a). In liquids, large polarizations have been achieved using the OE on ^{15}N, ^{13}C, ^{31}P via scalar coupling at 5 T (Loening *et al.*, 2002). In the case of liquid water protons, enhancements of over 10 have been achieved via dipolar relaxation at 9.2 T (Prandolini *et al.*, 2009)

(Figure 24.6b). These results have been achieved despite the predications that the classical DNP mechanisms have reduced efficiency at higher magnetic fields. Coherent cross-polarization using pulsed DNP should not show magnetic field dependence, and promises large polarization transfers at high fields if the technical problems can be solved. For example, polarization transfers using Hartmann–Hahn (HH) condition (i.e., the Rabi frequencies of the nuclear and electronic oscillating fields must be equal, $\omega_I B_{1I} = \omega_e B_{1S}$; Hartmann and Hahn, 1962) can be used at any applied magnetic field.

The above-described classical DNP mechanisms of OE, SE, TM, and CE all involve continuous microwave pumping and non-conserving processes: that is, the absorption or emission of energy with the lattice (OE) or the electron-broadening system (TM and CE), or the irradiating field (SE). Additionally, the polarization transfer rate is proportional to the nuclear spin-lattice relaxation rate (T_{1n}^{-1}). If we could apply the HH condition, polarization transfer would be much faster via energy-conserving processes but unfortunately in practice, because of the differences in electronic to proton magnetogyric ratios, this condition is impossible to fulfill. For example, if $B_{1S} = 0.1\,\text{mT}$, then the proton Rabi frequency would need to be $B_{1I} = 0.066\,\text{T}$.

A number of methods involving coherent transfer have shown promise for applications. Nuclear orientation via electron spin locking (NOVEL) is a DNP method that achieves the HH condition between the electrons in the rotating frame and the nuclear spins in the laboratory frame (Brunner, Fritsch, and Hausser, 1987; Henstra et al., 1988a; Henstra and Wenckebach, 2008). A variant of the NOVEL technique is the integrated solid effect (ISE) (Henstra, Dirksen, and Wenckebach, 1988b, Henstra et al., 1990), which uses continuous microwave irradiation and a fast field sweep through the EPR line, thereby rotating all spins though the HH condition. Unfortunately, these methods at high magnetic fields require extremely large microwave power. Another method, called "rotating frame DNP," was first demonstrated by Bloembergen and Sorokin on a single crystal of CsBr (Bloembergen and Sorokin, 1958); In this method, the CE and TM are performed in the rotating frame (Wind et al., 1988; Wind and Lock, 1990; Farrar et al., 2000). Finally, a promising method is the dressed-state solid effect (DSSE) (Weis et al., 2000; Weis and Griffin, 2006; Pomplun et al., 2008), which does not rely on nonsecular hyperfine couplings and therefore should also function in liquids.

24.4
Hardware (High-Frequency Microwave Equipment, SS-MAS DNP, HF-Liquid DNP, Dissolution DNP, Shuttle-DNP)

24.4.1
High-Frequency Microwave Sources

Currently, semiconductor microwave technology (Gunn and IMPATT diodes) reaches its limit at frequencies of ~100 GHz, corresponding to a magnetic field of 3.5 T (150 MHz, ^{1}H NMR). Higher frequencies can be attained most conveniently by generating higher harmonics, but with significant losses in power. Alternatives are vacuum electron devices, where an accelerated electron beam is modulated by suitable slow-wave structures or magnetic fields. A number of different designs exist for continuous-wave (CW) or pulsed operation, variable or fixed frequency. Devices such as backward-wave oscillators, orotrons, and carcinotrons can be used at high frequencies. Because of the presence of a slow-wave structure, which has a size comparable to the microwave wavelength, the magnitude of the electron beam density close to this structure is limited, and this leads to maximum deliverable microwave powers in the range of 0.1 to 1 W.

Gyrotrons, which are referred to as "fast-wave devices," circumvent this problem by replacing the slow-wave structure with a high-mode cavity immersed in a magnetic field. In this configuration, CW output powers in the Watt range are achieved in devices designed specifically for DNP at MIT (Bajaj *et al.*, 2003, 2007), and more recently at Fukui University (Saito *et al.*, 2007; Idehara *et al.*, 2008), CPI/Bruker and GYCOM. In a gyrotron, an electron beam is launched from an annular cathode and accelerated through the field of a superconducting magnet. The field profile is designed to compress the beam as it moves through the vacuum tube to a resonant cavity that converts the transverse kinetic energy from the helical motion of the electrons into microwaves. A quasi-optical mode converter finally couples the radiation to the output window of the device and into a transmission line to the sample. Details of the physics and engineering of gyrotrons have been described (Bajaj *et al.*, 2003, 2007; Saito *et al.*, 2007; Idehara *et al.*, 2008), their great virtue for DNP/NMR experiments being that they are scalable to experiments at high magnetic fields. Specifically, because they are fast-wave devices, they can generate tens of Watts of microwave power at frequencies up to 800 GHz, corresponding to ^{1}H NMR frequencies of 1.2 GHz. Furthermore, they operate in true CW mode for periods of days, enabling multidimensional NMR experiments that are commonly used.

24.4.2
Transmission Lines

Transmitting the microwaves to the sample in the probe with minimal loss, and then monitoring the microwave power output, is important experimentally. Fundamental mode waveguides have unacceptable insertion losses, and do not couple to a free-space propagation of a Gaussian beam, which is typically used for quasi-optical transmission outside of the probe. Corrugated overmoded or metallodielectric waveguides can be used inside the DNP probe for transmission (Woskov *et al.*, 2005; Denysenkov *et al.*, 2008). These differ from classical fundamental waveguides, in that the total losses in such DNP probes are typically less than 1–2 dB. Detection of the EPR signal requires quasioptical duplexing devices to prohibit the strong excitation power entering into the microwave detector.

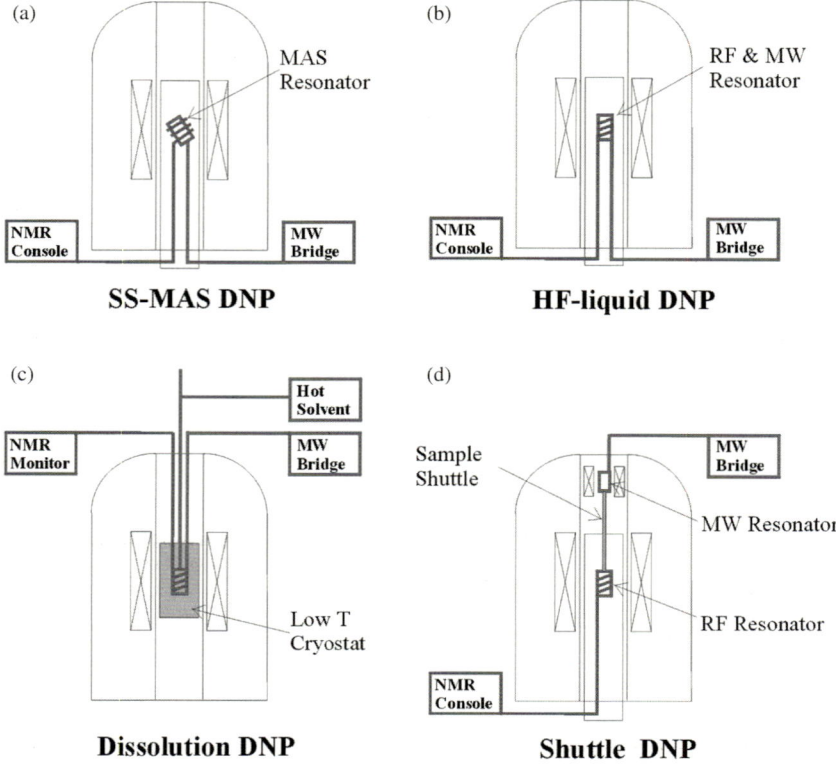

Figure 24.7 Typical experimental approaches for DNP spectrometers. (a) *In situ* solid-state MAS DNP spectrometer; (b) *In situ* high-field (HF) liquid-DNP spectrometer; (c) Dissolution DNP spectrometer: high-resolution NMR is performed in another magnet system (not shown); (d) Shuttle DNP spectrometer.

24.4.3
Spectrometer Types

Figure 24.7 illustrates some typical DNP/NMR spectrometer configurations. In the first two examples (Figure 24.7a, b), the sample can be simultaneously irradiated *in situ* at both the electronic and nuclear frequencies, which is ideal for multidimensional NMR. The other two spectrometer designs (Figure 24.7c, d) require that the sample is first DNP-polarized in a lower magnetic field, and then transferred to a high-field NMR spectrometer or MRI. Some of the features of the design and future developments of these spectrometers are discussed below.

24.4.3.1 Solid-State Magic Angle Spinning (MAS) DNP

At lower fields (1.4 T, Q-band), MAS DNP spectrometers operating at room temperature were developed to study polymers and carbonaceous materials (Wind et al., 1985; Afeworki, McKay, and Schaefer, 1992). Later on, high-field MAS DNP spectrometers suitable for experiments on biological systems were developed by the Griffin group at MIT. These operated at 5 T (Becerra et al., 1995) and 9 T (Bajaj et al., 2003) at low temperatures (<100 K), and gyrotrons were used for the first time as a microwave source for DNP (Figure 24.7a). The generation of strong B_1 fields at the nuclear Larmor frequencies is usually accomplished with solenoid radiofrequency (RF) coils. The coil and MAS apparatus complicate the design of a microwave-resonance structure. Typically, the inside of the stator is coated with a thin layer of silver, which reflects the microwaves and increases the magnetic microwave field slightly. Initially, the microwave radiation was introduced parallel to the MAS axis, but more recent probes employ irradiation perpendicular to the rotor axis. As long as the spacing between the turns of the coil is $>\lambda/2$, the microwaves will penetrate into the sample. Using a gyrotron microwave source, the investigator can satisfy, simultaneously, the constraints of MAS at kHz frequencies at cryogenic temperatures, high NMR sensitivity, and efficient microwave pumping of the electrons. The MAS-DNP probe has been used down to 85 K and with spinning frequencies of 10–15 kHz.

Enhancements of 400 for the proton NMR signals have been obtained using the TM effect with 30 mM concentrations of radicals at 12 K, with static samples (Weis et al., 1999). In MAS experiments, enhancements of up to ~300 have been observed with 5 mM biradicals (10 mM unpaired electrons) at ~95 K, using the CE (Hu et al., 2004, 2008; Matsuki et al., 2009) (Figure 24.6a). Details of the polarization transfer mechanism in mono- and biradicals are the subject of ongoing research investigations.

Recently, solid-state MAS DNP instrumentation at 260 GHz – including gyrotrons, transmission lines, and low-temperature probes – has become commercially available and should accelerate propagation of the technique to many laboratories. In Griffin's laboratory at MIT, MAS DNP experiments are currently operating at 140 and 250 GHz.

24.4.3.2 Low-Temperature Dissolution Polarizer

This method utilizes DNP in the solid state at very low temperatures (1.2 K) and at magnetic fields of 3–5 T. The DNP polarization step is followed by rapid dissolution, with a suitable warm solvent. Immediately thereafter, the sample is transferred either to a high-resolution NMR spectrometer or to a MR imager (Figure 24.7c). Very high polarizations for ^{13}C and ^{15}N (over >10 000) can be achieved within the dissolution and transfer process (Ardenkjaer-Larsen et al., 2003). In this case, the signal enhancement is the product of the DNP enhancement (up to 2600 for ^{13}C) multiplied by the large Boltzmann enhancement (about 250) because of the change in temperature from 1.2 K (where the DNP process occurs) to room temperature (where the NMR measurement occurs).

At a magnetic field of 3.35 T, all the necessary microwave components for the excitation of the electron spin system are commercially available (Wolber et al., 2004; Granwehr, Leggett, and Köckenberger, 2007). Microwave excitation at 95 GHz (W-band, 3.35 T) is realized by solid-state or vacuum tube devices with a typical output power of more than 200 mW. The solid sample containing a high concentration of a suitable radical (usually trityl, but recently TEMPO) is excited by continuous microwave radiation for a period of 1000 to 10 000 s, and the progress of the polarization transfer can be monitored using ^{13}C NMR detection (Ardenkjaer-Larsen *et al.*, 2003; Reynolds and Patel, 2008; Comment *et al.*, 2007). When the saturating DNP polarization has been obtained, the dissolution and transfer times for subsequent NMR or MRI measurements are on the order of a few seconds.

Complete dissolution-DNP spectrometers are also commercially available, spreading the usage of this method. The limitations of the approach are the fact that the polarization times are rather long, that the sample is diluted irreversibly by a factor of ~70 in the dissolution process, and that the sample must be shuttled between magnets. Some of these problems are partially circumvented by integrating single-scan 2D-NMR methods into the technique (Frydman and Blazina, 2007). These experiments have been used to boost NMR sensitivity, time resolution, and contrast for applications in spectroscopy and imaging, and will undoubtedly be developed further in the future (Comment *et al.*, 2007; Reynolds and Patel, 2008; Emwas *et al.*, 2008).

24.4.3.3 *In-Situ* Temperature-Jump DNP (Laser Melting)

Another approach to enhance the NMR spectra of liquids is to polarize the sample in the solid state, by using the same approaches that were developed for MAS DNP experiments, followed by *in situ* rapid laser melting and subsequently recording a liquid-state NMR spectrum. The melting is accomplished with an infrared laser and an optic fiber, and is highly reproducible (Joo, Bryant, and Griffin, 2006). The advantage of this approach for analytical NMR is that the sample is not diluted as with the dissolution experiment; neither is it necessary to physically move the sample between two different magnetic fields. Further, the existing repertoire of multidimensional solution state NMR experiments can be inserted as mixing periods into this scheme. Finally, as the scheme includes a ^{1}H–^{13}C cross-polarization step, the long polarization times associated with directly polarizing ^{13}C or ^{15}N are shortened to that of the ^{1}H T_1.

It is likely that this method will become most useful for small molecules that can be repeatedly frozen and thawed. It may also be quite useful for analytical NMR, when combined with multiple receivers, sparse sampling, and other improvements.

24.4.3.4 High-Field (HF) Liquid-DNP Spectrometers

Encouraged by the experimental DNP enhancements obtained in liquid solution at 5 T (Loening *et al.*, 2002), a dedicated HF liquid-DNP spectrometer was designed at Frankfurt University, Germany, operating at 9.2 T, corresponding to 260 GHz EPR and 400 MHz proton NMR frequencies (Denysenkov *et al.*, 2008) (Figure 24.7b). The key problem here is to avoid excessive microwave heating of the liquid–water solution, which is mandatory for high-resolution NMR studies of proteins in their native state. Therefore, a microwave resonance structure which spatially separates the magnetic fields from the electrical fields of the microwave must be used. The microwave cavity has two important features: first, it drastically reduces the microwave electrical field strength at the sample position, thus avoiding excessive heating of the liquid sample; and second it strongly enhances the microwave magnetic field strength at the sample position, which allows significant DNP enhancements already with a very low incident microwave power. For the first prototype HF liquid-DNP spectrometer, a helical structure developed for ENDOR experiments was used, where a cylindrical TE_{011} mode microwave resonator functioned simultaneously as a multiresonance NMR coil (Weis et al., 1999). The main disadvantages of this fundamental mode design are the sample size, which is only 3–4 nl, and the rather poor NMR coil-filling factor.

Unexpected high DNP enhancements of more than 10 were achieved in liquid water samples at room temperature, with a very low incident microwave power of only 45 mW (Prandolini et al., 2009) (Figure 24.6b). These liquid-state DNP enhancements suggest water/radical translational correlation times similar to that found by nuclear magnetic relaxation dispersion measurements (Höfer et al., 2008). Even larger enhancements can be achieved with higher microwave powers, and at slightly higher temperatures. Pulsed microwaves at spacing similar to T_{1e} of the radical might even double the achieved enhancement (incoherently) on water (Un et al., 1992).

Currently, more elaborate resonance structures are under development, which could potentially increase the sample volume up to few microliters and achieve a higher NMR filling factor. Such extended structures would require microwave power levels of about 5 W to saturate the electronic transitions of nitroxide radicals. Thus, if the technical problems related to sample size, electric field heating and field homogeneity can be solved, this method may become very useful for *in situ* multidimensional NMR measurements typically encountered in structural studies on biological macromolecules. Additionally, the possibility of exciting electron and nuclear spins simultaneously without any time delays is unique for this approach, and offers versatile prospects for coherent spin manipulations which might lead to improved polarization transfer pathways and new types of experiments.

24.4.3.5 Shuttle DNP

This approach exploits the fact that polarization transfer processes in liquids can be more efficient and technically less demanding at magnetic field values below 1 T (Hausser and Stehlik, 1968; Müller-Warmuth and Meise-Gresch, 1983). In this case, the microwave irradiation is typically performed at X-band frequencies (9 GHz/0.3 T) in either an electromagnet (Dorn et al., 1989; Reese et al., 2009) or in a permanent Halbach magnet (Armstrong et al., 2008; Münnemann et al., 2008). In this case, the liquid sample must reside inside a microwave cavity in order to achieve high microwave magnetic field strengths at the sample; this condition is required for saturation of the electron spin system of the radicals, which have very short relaxation times in liquid solution at room temperature. Following the polarization process (which takes only a few seconds), the sample is shuttled to a high magnetic field for NMR detection. This experimental approach is used in flow systems (Dorn et al., 1989), fast shuttling of the sample (Reese et al., 2009), or by a rapid transfer of the whole DNP probe (Miesel et al., 2008). Typical transfer times are on the order of several 100 ms. Because the polarization is transferred from a very low magnetic field to a higher field, a "Boltzmann penalty" is present; this is the ratio of the DNP polarization magnetic field (0.3 T) over the NMR detection magnetic field (2–14 T). Depending on the application, which vary from MRI, analytical chemistry or high-resolution spectroscopy, this factor might range from 1/5 to 1/60. A second problem related to this approach concerns the magnetic field profile during the transfer from the DNP polarization field to the NMR detection field. However, care must be taken so that the polarization enhancement of the nuclear spins does not become reduced by passage through low magnetic fields (Miesel et al., 2008).

Experimentally, a DNP enhancement factor of +15 for carbon-13 in chloroform for a DNP pump field of 0.3 T and a NMR detection field of 14 T was achieved (Reese et al., 2009). The limitations of the present apparatus, which shuttles the sample between two magnets through a very low-field region, result in no observed DNP enhancement on molecules larger than urea. However, if the transfer times are reduced and the magnetic field profile between the two magnetic centers can be increased, thus reducing relaxation losses, DNP enhancements on large biomolecules should be possible (Reese et al., 2009). Additionally, it might be possible to use non-Boltzmann-polarized electron spin systems (e.g., optical excited triplet states or radical pairs) in the DNP polarizer.

24.5
First Applications and Outlook

24.5.1
Application Areas of High-Field DNP

Some recent applications of high-field DNP highlight the potential of the method:

Biomolecular structure studies on membrane proteins: MAS DNP was used to investigate the intermediate states in the photocycle of bacteriorhodopsin. The enhanced sensitivity of DNP permitted, for the first time, the characterization of the retinal conformation in the K, L, and M states (Bajaj et al., 2007; Mak-Jurkauskas et al., 2008). In addition, DNP-enhanced MAS spectra of amyloid nanocrystals illustrated the potential of polarizing structures with dimensions of 100–200 nm via ^{1}H spin diffusion from the solvent.

Ultrafast 2-D-NMR spectroscopy: A combination of gradient-encoded ultrafast 2D-NMR methods (Frydman and Blazina, 2007) with dissolution DNP permit multidimensional NMR spectra of polarized small molecules to be obtained, with very high sensitivity (Mishkovsky and Frydman, 2008).

MRI with hyperpolarized metabolites: ^{13}C-pyruvate was polarized with a dissolution spectrometer and used for cardiac MR imaging (Golman et al., 2006; Kohler et al., 2007).

Time-resolved DNP spectroscopy: Reactions such as enzyme catalysis can be followed in real time with hyperpolarized NMR spectroscopy, by following the kinetics of the substrate and product resonances (Bowen and Hilty, 2008).

Hydration and local water dynamics in lipids and proteins: Nitroxide spin labels, covalently attached along the hydrophobic tail of stearic acid molecules or on proteins, have permitted the study of local hydration dynamics via DNP in micelles and vesicle assemblies, and on protein surfaces (McCarney *et al.*, 2008; Armstrong and Han, 2009).

24.5.2
Outlook

All of the approaches to high-field DNP are currently undergoing extensive instrumental development and simultaneous refinement of the experimental methods. The successful development of these approaches will have an enormous impact in the fields of biology, chemistry, physics, and medicine. Recently, a number of academic and industrial research groups have initiated research efforts to overcome the current limitations of the techniques. Technical developments of high-frequency microwave sources, components and of the various DNP spectrometers will be of vital importance for the further development of this method. Other areas of active research include the optimization of polarizing agents, the development of new types of polarization transfer methods, and the design of new experiments concentrating on selectivity, contrast, and additional structural restraints. Thus, collaborative efforts between scientists from chemistry, physics, and biology will be required to optimize DNP for applications to high-field NMR and MRI.

Acknowledgments

The authors would like to thank Prof. R. G. Griffin at the Massachusetts Institute of Technology for support and scientific advice. They also thank their colleagues of the European Design Study Bio-DNP for their hard work and support, and the EU for financial support.

Pertinent Literature

Abragam, A. and Goldman, M. (1982) *Nuclear Magnetism: Order and Disorder*, Clarendon Press, Oxford.

Ardenkjaer-Larsen, J.H., Fridlund, B., Gram, A., Hansson, G., Hansson, L., Lerche, M.H., Servin, R., Thaning, M., and Golman, K. (2003) *Proc. Natl Acad. Sci. USA*, **100**, 10158–10163.

Atsarkin, V.A. (1978) *Sov. Phys. Usp.*, **21**, 725–745.

Goertz, S., Meyer, W., and Reicherz, G. (2002) *Prog. Part. Nucl. Phys.*, **49**, 403–489.

Hausser, K.H. and Stehlik, D. (1968) *Adv. Magn. Reson.*, **3**, 79–139.

Maly, T., Debelouchina, G.T., Bajaj, V.S., Hu, K.-N., Joo, C.-G., Mak-Jurkauskas,

M.L., Sirigiri, J.R., van der Wel, P.C.A., Herzfeld, J., Temkin, R.J., and Griffin, R.G. (2008) *J. Chem. Phys.*, **128**, 052211.

Müller-Warmuth, W. and Meise-Gresch, K. (1983) *Adv. Magn. Reson.*, **11**, 1–45.

Wind, R.A., Duijvestijn, M.J., van der Lugt, C., Manenschijn, A., and Vriend, J. (1985) *Prog. NMR Spectrosc.*, **17**, 33–67.

References

Abragam, A. (1961) *Principles of Nuclear Magnetism*, Oxford University Press, Oxford.

Abragam, A. and Goldman, M. (1978) *Rep. Prog. Phys.*, **41**, 395–467.

Abragam, A. and Goldman, M. (1982) *Nuclear Magnetism: Order and Disorder*, Clarendon Press, Oxford.

Abragam, A., Combrisson, J., and Solomon, I. (1958) *C. R. Acad. Sci.*, **247**, 2337–2340.

Afeworki, M., McKay, R.A., and Schaefer, J. (1992) *Macromolecules*, **25**, 4084–4091.

Ardenkjaer-Larsen, J.H., Fridlund, B., Gram, A., Hansson, G., Hansson, L., Lerche, M.H., Servin, R., Thaning, M., and Golman, K. (2003) *Proc. Natl Acad. Sci. USA*, **100**, 10158–10163.

Armstrong, B.D. and Han, S. (2007) *J. Chem. Phys.*, **127**, 104508.

Armstrong, B.D. and Han, S. (2009) *J. Am. Chem. Soc.*, **131**, 4641–4647.

Armstrong, B.D., Lingwood, M.D., McCarney, E.R., Brown, E.R., P. Blümler, and Han S. (2008) *J. Magn. Reson.*, **191**, 273–281.

Atsarkin, V.A. (1978) *Sov. Phys. Usp.*, **21**, 725–745.

Bajaj, V.S., Farrar, C.T., Hornstein, M.K., Mastovsky, I., Vieregg, J., Bryant, J., Eléna, B., Kreischer, K.E., Temkin, R.J., and Griffin, R.G. (2003) *J. Magn. Reson.*, **160**, 85–90.

Bajaj, V.S., Hornstein, M.K., Kreischer, K.E., Sirigiri, J.R., Woskov, P.P., Mak-Jurkauskas, M.L., Herzfeld, J., Temkin, R.J., and Griffin, R.G. (2007) *J. Magn. Reson.*, **189**, 251–279.

Bates, R.D. and Drozdoski, W.S. (1977) *J. Chem. Phys.*, **67**, 4038–4044.

Becerra, L.R., Gerfen, G.J., Bellew, B.F., Bryant, J.A., Hall, D.A., Inati, S.J., Weber, R.T., Un, S., Prisner, T.F., McDermott, A.E., Fishbein, K.W., Kreischer, K.E., Temkin, R.J., Singel, D.J., and Griffin, R.G. (1995) *J. Magn. Reson. Ser. A*, **117**, 28–40.

Beljers, H.G., van der Kint, L., and van Wieringen, J.S. (1954) *Phys. Rev.*, **95**, 1683.

Bertini, I., Luchinat, C., and Parigi, G. (2001) *Solution NMR of Paramagnetic Molecules*, Elsevier Press, Amsterdam.

Bleombergen, N. and Sorokin, P.P. (1958) *Phys. Rev.*, **110**, 865–875.

Borghini, M. (1968) *Phys. Rev. Lett.*, **20**, 419–421.

Bowen, S. and Hilty, C. (2008) *Angew. Chem. Int. Ed.*, **47**, 5235–5237.

Bowers, C.R. and Weitekamp, D.P. (1986) *Phys. Rev. Lett.*, **57**, 2645–2648.

Brunner, H., Fritsch, R.H., and Hausser, K.H. (1987) *Z. Naturforsch. A*, **42**, 1456–1457.

Buishvili, L.L. (1966) *Sov. Phys. JETP*, **22**, 1277–1281.

Carver, T.R. and Slichter, C.P. (1953) *Phys. Rev.*, **92**, 212–213.

Carver, T.R. and Slichter, C.P. (1956) *Phys. Rev.*, **102**, 975–980.

Comment, A., van den Brandt, B., Uffmann, K., Kurdzesau, F., Jannin, S., Konter, J.A., Hautle, P., Wenckebach, W.T., Gruetter, R., and van der Klink, J.J. (2007) *Concepts Magn. Reson. B*, **31**, 255–269.

Denysenkov, V.P., Prandolini, M.J., Krahn, A., Gafurov, M., Endeward, B., and Prisner, T.F. (2008) *Appl. Magn. Reson.*, **34**, 289–299.

Dorn, H.C., Gitti, R., Tsai, K.H., and Glass, T.E. (1989) *Chem. Phys. Lett.*, **155**, 227–232.

Emwas, A.-H., Saunders, M., Ludwig, C., and Günther, U.L. (2008) *Appl. Magn. Reson.*, **34**, 483–494.

Erb, E., Montchane, J.-L., and Uebersfeld, J. (1958) *C. R. Acad. Sci.*, **246**, 2121–2123.

Ernst, R.R. and Anderson, W.A. (1966) *Rev. Sci. Instrum.*, **37**, 93–102.

Ernst, R.R., Bodenhausen, G., and Wokaun, A. (1987) *Principles of Nuclear Magnetic Resonance in One and Two Dimensions*, Clarendon Press, Oxford.

Farrar, C.T., Hall, D.A., Gerfen, G.J., Rosay, M., Ardenkjaer-Larsen, J.-H., and Griffin, R.G. (2000) *J. Magn. Reson.*, **144**, 134–141.

Farrar, C.T., Hall, D.A., Gerfen, G.J., Inati, S.J., and Griffin, R.G. (2001) *J. Chem. Phys.*, **114**, 4922–4933.

Frydman, L. and Blazina, D. (2007) *Nat. Phys.*, **3**, 415–419.

Goertz, St., Meyer, W., and Reicherz, G. (2002) *Prog. Part. Nucl. Phys.*, **49**, 403–489.

Goldman, M. (1970) *Spin Temperature and Nuclear Magnetic Resonance in Solids*, Oxford University Press, Oxford.

Golman, K., in't Zandt, R., Lerche, M., Pehrson, R., and Ardenkjaer-Larson, J.H. (2006) *Cancer Res.*, **66**, 10855–10860.

Granwehr, J., Leggett, J., and Köckenberger, W. (2007) *J. Magn. Reson.*, **187**, 266–276.

Hartmann, S.R. and Hahn, E.L. (1962) *Phys. Rev.*, **128**, 2042–2053.

Hausser, K.H. and Stehlik, D. (1968) *Adv. Magn. Reson.*, **3**, 79–139.

Henstra, A. and Wenckebach, W.Th. (2008) *Mol. Phys.*, **106**, 859–871.

Henstra, A., Dirksen, P., Schmidt, J., and Wenckebach, W.Th. (1988a) *J. Magn. Reson.*, **77**, 389–393.

Henstra, A., Dirksen, P., and Wenckebach, W.Th. (1988b) *Phys. Lett. A*, **134**, 134–136.

Henstra, A., Lin, T.-S., Schmidt, J., and Wenckebach, W.Th. (1990) *Chem. Phys. Lett.*, **165**, 6–10.

Höfer, P., Parigi, G., Luchinat, C., Carl, P., Guthausen, G., Reese, M., Carlomagno, T., Griesinger, C., and Bennati, M. (2008) *J. Am. Chem. Soc.*, **130**, 3254–3255.

Hore, P.J. and Broadhurst, R.W. (1993) *Prog. NMR Spectr.*, **25**, 345–402.

Hu, K.-N., Yu, H.-H., Swager, T.M., and Griffin, R.G. (2004) *J. Am. Chem. Soc.*, **126**, 10844–10845.

Hu, K.-N., Bajaj, V.S., Rosay, M., and Griffin, R.G. (2007) *J. Chem. Phys.*, **126**, 044512.

Hu, K.-N., Song, C., Yu, H., Swager, T.M., and Griffin, R.G. (2008) *J. Chem. Phys.*, **128**, 052302.

Hwang, C.F. and Hill, D.A. (1967a) *Phys. Rev. Lett.*, **18**, 110–112.

Hwang, C.F. and Hill, D.A. (1967b) *Phys. Rev. Lett.*, **19**, 1011–1014.

Idehara, T., Saito, T., Ogawa, I., Mitsudo, S., Tatematsu, Y., Agusu, La., Mori, H., and Kobayashi, S. (2008) *Appl. Magn. Reson.*, **34**, 265–275.

Jeffries, C.D. (1957) *Phys. Rev.*, **106**, 164–165.

Jeffries, C.D. (1960) *Phys. Rev.*, **117**, 1056–1069.

Joo, C.-G., Hu, K.-N., Bryant, J.A., and Griffin, R.G. (2006) *J. Am. Chem. Soc.*, **128**, 9428–9432.

Kessenikh, A.V., Lushchikov, V.I., Manenkov, A.A., and Taran, Y.V. (1963) *Sov. Phys. Solid State*, **5**, 321–329.

Kessenikh, A.V., Manenkov, A.A., and Pyatnitskii, G.I. (1964) *Sov. Phys. Solid State*, **6**, 641–643.

Kohler, S.J., Yen, Y., Wolber, J., Chen, A.P., Albers, M.J., Bok, R., Zhang, V., Tropp, J., Nelson, S., Vigneron, D.B., Kurhanewicz, J., and Hurd, R.E. (2007) *Magn. Reson. Med.*, **58**, 65–69.

Leawoods, J.C., Yablonskiy, D.A., Saam, B., Gierada, D.S., and Conradi, M.S. (2001) *Concepts Magn. Reson.*, **13**, 277–293.

Loening, N.M., Rosay, M., Weis, V., and Griffin, R.G. (2002) *J. Am. Chem. Soc.*, **124**, 8808–8809.

McCarney, E.R., Armstrong, B.D., Kausik, R., and Han, S. (2008) *Langmuir*, **24**, 10062–10072.

Mak-Jurkauskas, M.L., Bajaj, V.S., Hornstein, M.K., Belenky, M., Griffin, R.G., and Herzfeld, J. (2008) *Proc. Natl Acad. Sci. USA*, **105**, 883–888.

Maly, T., Debelouchina, G.T., Bajaj, V.S., Hu, K.-N., Joo, C.-G., Mak-Jurkauskas, M.L., Sirigiri, J.R., van der Wel, P.C.A., Herzfeld, J., Temkin, R.J., and Griffin, R.G. (2008) *J. Chem. Phys.*, **128**, 052211.

Matsuki, Y., Maly, T., Ouari, O., Karoui, H., Le Moigne, F., Rizzato, E., Lyubenova, S., Herzfeld, J., Prisner, T., Tordo, P., and Griffin, R.G. (2009) *Angew. Chem. Int. Ed.*, **121**, 5096–5100.

Miesel, K., Ivanov, K.L., Köchling, T., Yurkovskaya, A.V., and Vieth, H.-M. (2008) *Appl. Magn. Reson.*, **34**, 423–437.

Mishkovsky, M. and Frydman, L. (2008) *ChemPhysChem*, **9**, 2340–2348.

Müller-Warmuth, W. and Meise-Gresch, K. (1983) *Adv. Magn. Reson.*, **11**, 1–45.

Münnemann, K., Bauer, C., Schmiedeskamp, J., Spiess, H.W., Schreiber, W.G., and Hinderberger, D. (2008) *Appl. Magn. Reson.*, **34**, 321–330.

Overhauser, A.W. (1953) *Phys. Rev.*, **92**, 411–415.

Pomplun, N., Heitmann, B., Khaneja, N., and Glaser, S.J. (2008) *Appl. Magn. Reson.*, **34**, 331–346.

Pound, R.V. (1950) *Phys. Rev.*, **79**, 685–702.

Prandolini, M.J., Denysenkov, V.P., Gafurov, M., Endeward, B., and Prisner, T.F. (2009) *J. Am. Chem. Soc.*, **131**, 6090–6092.

Provotorov, B.N. (1962) *Sov. Phys. JETP*, **14**, 1126–1131.

Reese, M., Türke, M.-T., Tkach, I., Parigi, G., Luchinat, C., Marquardsen, T., Tavernier, A., Höfer, P., Engelke, F., Griesinger, C., and Bennati, M. (2009) *J. Am. Chem. Soc.*, **131**, 15086–15087.

Reynolds, S. and Patel, H. (2008) *Appl. Magn. Reson.*, **34**, 495–508.

Robinson, B.H., Haas, D.A., and Mailer, C. (1994) *Science*, **263**, 490–493.

Saito, T., Nakano, T., Hoshizuki, H., Sakai, K., Tatematsu, Y., Mitsudo, S., Ogawa, I., Idehara, T., and Zapevalov, V.E. (2007) *Int. J. Infrared Millimeter Waves*, **28**, 1063–1078.

Sezer, D., Prandolini, M.J., and Prisner, T.F. (2009a) *Phys. Chem. Chem. Phys.*, **11**, 6626–6637.

Sezer, D., Gafurov, M., Prandolini, M.J., Denysenkov, V.P., and Prisner, T.F. (2009b) *Phys. Chem. Chem. Phys.*, **11**, 6638–6653.

Singel, D.J., Seidel, H., Kendrick, R.D., and Yannoni, C.S. (1989) *J. Magn. Reson.*, **81**, 145–161.

Sprenkels, J.C.M., Wenckebach, W.T., and Poulis, N.J. (1983) *J. Phys. C*, **16**, 4425–4445.

Un, S., Prisner, T., Weber, R.T., Seaman, M.J., Fishbein, K.W., McDermott, A.E., Singel, D.J., and Griffin, R.G. (1992) *Chem. Phys. Lett.*, **189**, 54–59.

Weis, V. and Griffin, R.G. (2006) *Solid State Nucl. Magn. Reson.*, **29**, 105–117.

Weis, V., Bennati, M., Rosay, M., Bryant, J.A., and Griffin, R.G. (1999) *J. Magn. Reson.*, **140**, 293–299.

Weis, V., Bennati, M., Rosay, M., and Griffin, R.G. (2000) *J. Chem. Phys.*, **113**, 6795–6802.

Wenckebach, W.T. (2008) *Appl. Magn. Reson.*, **34**, 227–235.

Wenckebach, W.T., van den Heuvel, G.M., Hoogstraate, H., Swanenburg, T.J.B., and Poulis, N.J. (1969) *Phys. Rev. Lett.*, **22**, 581–583.

Wind, R.A. and Lock, H. (1990) *Adv. Magn. Opt. Reson.*, **15**, 51–77.

Wind, R.A., Duijvestijn, M.J., van der Lugt, C., Manenschijn, A., and Vriend, J. (1985) *Prog. NMR Spectr.*, **17**, 33–67.

Wind, R.A., Li, L., Lock, H., and Maciel, G.E. (1988) *J. Magn. Reson.*, **79**, 577–582.

Wolber, J., Ellner, F., Fridlund, B., Gram, A., Jóhannesson, H., Hansson, G., Hansson, L.H., Lerche, M.H., Månsson, S., Servin, R., Thaning, M., Golman, K., and Ardenkjaer-Larsen, J.H. (2004) *Nucl. Instr. Meth. Phys. Res. A*, **526**, 173–181.

Wollan, D.S. (1976) *Phys. Rev. B*, **13**, 3671–3685.

Woskov, P.P., Bajaj, V.S., Hornstein, M.K., Temkin, R.J., and Griffin, R.G. (2005) *IEEE Trans. Microw. Theory Tech.*, **53**, 1863–1869.

25
Chemically Induced Electron and Nuclear Polarization

Lawrence J. Berliner and Elena Bagryanskaya

25.1
Introduction

The phenomenon of chemically induced dynamic nuclear polarization (CIDNP) was discovered almost accidentally by scientists at the Technical University of Darmstadt, Germany, and at Brown University, Providence, Rhode Island, USA, while studying novel organic chemistry reactions. The research groups observed not only remarkable intensity enhancements, but more strangely also emission lines that gave negative peaks! In particular, Bargon, Fischer, and Johansen (1967) were studying the decomposition of dibenzoyl peroxide, diproprionyl peroxide and diacetyl peroxide, where they observed products that showed both emission and enhanced absorption lines. Ward and Lawler (1967) observed essentially similar phenomena during the reactions of alkyl halides with alkyl lithium compounds. Subsequently, the two groups shared their results and interpretations and coined the phenomenon of CIDNP.

An understanding of the CIDNP phenomenon led to development of the radical pair theory by Closs (1969), Closs and Trifunac (1970) and Kaptein and Oosterhoff (1969a, 1969b) and, as a consequence, to the discovery of numerous related techniques: chemically induced dynamic electron polarization (CIDEP) (Fessenden and Schuler, 1963) and the magnetic field effect (Sagdeev *et al.*, 1972), magnetic isotope effect (Molin *et al.*, 1976; Buchachenko *et al.* 1976) and reaction yield-detected magnetic resonance (RYDMR) (Frankevich and Kubarev, 1982) (Muus *et al.*, 1977; Nagakura, Hayashi, and Azumi, 1998; Salikhov *et al.*, 1984).

Both, CIDNP and CIDEP effects arise from the spin selection process that occurs during a recombination reaction of radical pairs. It leads to significant nonequilibrium electron and nuclear spin state populations that result in enhanced absorption and emission spectral lines, the signs of which follow a set of well-defined rules. While this chapter will cover the theory behind the technique in general, the real power for biochemists is the ability to apply these methods to proteins for probing molecular details on the surface. This also leads to interesting studies of protein–protein interactions. When attention is focused on the biological applications of CIDNP, the technique requires laser irradiation for the key photochemical reactions necessary to induce the phenomenon; it is then referred to as photo-CIDNP.

25.2
History of the CIDNP Phenomenon

Although CIDNP was discovered by Bargon, Fischer, and Johansen (1967) and Ward and Lawler (1967), as noted above it was not totally understood until Closs (1969) and Kaptein and Oosterhoff (1969a, 1969b) realized that the effect originated from the interactions of radical pairs (RPs), and not from isolated free radicals. This led to the introduction of the "radical pair theory," that was shown to explain a quite large number of the results obtained from previous experiments and publications. In addition, the triplet mechanism operates frequently in photochemical reactions: in this case, the electron spin polarization is transferred from a photoexcited triplet molecule to free radicals and, following electron–nuclear cross-relaxation, leads to nuclear spin polarization. However, the latter occurs only under a very special set of conditions, and is rather rare.

25.3
The Radical Pair Mechanism

The collision of two reactive radicals in solution may, or may not, lead to a recombination reaction; the alternative situation is termed "escape." The recombination reaction assumes any RP reaction: coupling; disproportionation; back electron transfer and others. The reaction pathway depends on the electron spin states of the radicals, as shown schematically in Figure 25.1. Here, a photoexcited molecule dissociates into two spin-correlated radicals with the formation of a so-called "geminate" RP. The spin state of the geminate RP remains the same as that of the photoexcited precursor molecule. For a diffusive collision between two radicals, one would expect a statistical distribution of spin states, with one-fourth in the singlet state (S = 0), and three-fourths in the triplet state.

Since the radicals R_1 and R_2 are surrounded by solvent molecules, a fast separation is hindered. The radicals are located in a "cage" of solvent molecules, and have a certain probability of re-encounter; the probability of re-encounters at time $t > \tau$ is described by the following expression $f(t) \sim (1/t^{3/2})\exp(-t/\tau)$, where τ is the lifetime of the RP. In the model of continuous diffusion the characteristic lifetime of RPs is equal to $\tau = R^2/D$, where R is the recombination radius which is equal to the van der Waals radius for the uncharged radical, and D is the diffusion coefficient of the radicals. The typical lifetime of RPs in solution is about 1 ns, whereas ion–radical pairs in nonpolar solution, biradicals and radical pairs in micelles, cyclodextrins, and so on, are characterized by much longer lifetimes of 10 ns to 1 μs.

In most cases, recombination is allowed only for singlet RPs; however, in some cases the RPs formed from triplet excited molecules recombine with formation of

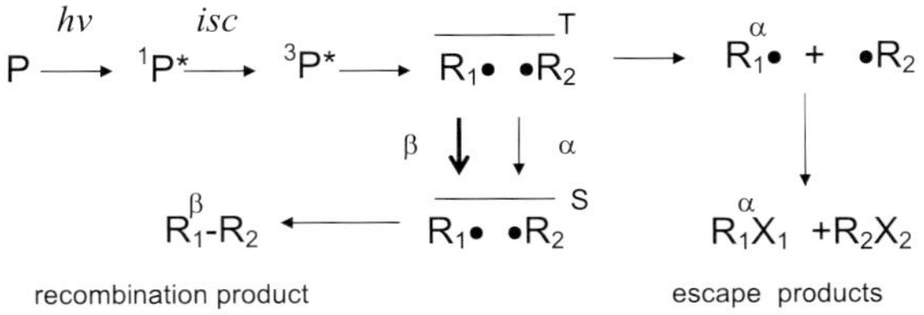

$$P \xrightarrow{h\nu} {}^1P^* \xrightarrow{isc} {}^3P^* \xrightarrow{T} \overline{R_1\bullet\ \bullet R_2} \xrightarrow{\alpha} R_1\bullet\ +\ \bullet R_2$$

$$\beta \downarrow\ \downarrow \alpha$$

$$R_1^\beta\text{-}R_2 \xleftarrow{\beta} \overline{R_1\bullet\ \bullet R_2}^S \xrightarrow{\alpha} R_1X_1\ +R_2X_2$$

recombination product   escape products

Figure 25.1 Schematic representation of the nuclear spin sorting (selection) process in a photoinduced radical reaction. P is photoexcited to the $^1P^*$ singlet state, after which intersystem crossing (isc) converts it to the $^3P^*$ triplet state. Chemical dissociation occurs to a triplet radical pair state; the bar indicates that the two radical "partners" are spin-correlated. Subsequently, a random walk dissociation occurs, during which intersystem crossing T → S transitions can occur. The rate of T → S conversion depends on the nuclear spin state; in the example here it is faster for β nuclei than for α nuclei. Singlet pairs form faster with β than for α, and the recombination product is rich in excess β spins, since only singlet pairs recombine. The recombination product is overpopulated in β spins, leading to an emissive NMR signal. Radicals with an excess of α nuclear spins remain in solution and lead to what are termed "escape" products, where the NMR spectrum shows enhanced absorption signals. Reproduced with permission from Kaptein (1982).

the product in the excited triplet state. Magnetic interactions (Zeeman, electron–nuclear hyperfine interaction) induce singlet–triplet (S–T) conversion and, as a result, change the yield of the recombination product. Thus, the ratio of geminate and escape diamagnetic products depends on the probability of S–T conversion; in turn the rate of S–T conversion depends on the magnetic field, leading to the magnetic effect, the magnetic isotope effect, CIDNP, and CIDEP. It should be noted that the typical energy of the magnetic interaction is several orders of magnitude smaller than the thermal energy kT (0.025 eV) or chemical reaction energy (1–10 eV), but the influence of magnetic field on the S–T conversion is still capable of changing the reaction pathway.

25.3.1
The Mechanism of Singlet–Triplet Conversion in RPs

For a qualitative consideration of S–T conversion, one must consider the vector model shown in Figure 25.2. The unpaired electron spins of RP, S_1 and S_2, are shown as vectors rotating in a magnetic field B_0. In the case of equal precession frequencies of S_1 and S_2, the spin state of RPs does not change in time; however, in the case of different precession rates the RP spin state alternates between S and T_0. In addition, $T_0 - T_-$ and $T_0 - T_+$ transitions can be induced by electron spin relaxation or the resonance microwave magnetic field.

The Schrödinger equation governs the time dependence of the RP wave function, $\psi(t)$:

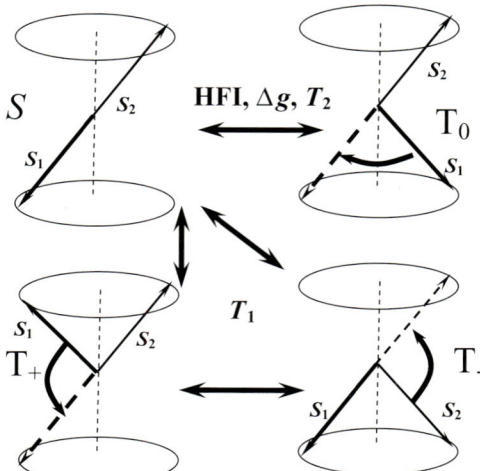

Figure 25.2 Vector model of singlet–triplet conversion in radical pairs. S = singlet state; T_0, T_-, T_+ = triplet states; T_1 and T_2 are electron spin relaxation times; s_1 and s_2 correspond to spins of radical 1 and radical 2. Relative dephasing induced by $\Delta\omega$ of the spin precessions is sufficient for transitions between the S and T_0 states. Electron spin relaxation T_1 induced transitions between the T_0 and T_-, T_- states.

$$i\frac{d\psi(t)}{dt} = \hat{H}\psi(t), \qquad (25.1)$$

In the general case the Hamiltonian of a RP in a magnetic field B_0 can be written as:

$$\hat{H} = g_1\mu_B\hat{B}_0\hat{S}_{1z} + g_2\mu_B\hat{B}_0\hat{S}_{2z} + \sum_i^a A_i\hat{S}_1\hat{I}_i + \sum_k^b A_k\hat{S}_2\hat{I}_k - J(1/2 + 2\hat{S}_1\hat{S}_2) \qquad (25.2)$$

where J is the exchange interaction of spins S_1 and S_2, μ_B is the Bohr magneton, g_1 and g_2 are the electron g-factors of radicals 1 and 2, and A_i, A_k are the hyperfine (HFI) constants of electrons with nuclei. The eigenfunctions of the RP can be expressed by the products of their electron (S, T_0, T_+, T_-) and nuclear $|\chi_N\rangle$ spin states:

$$S = (|\alpha_1\beta_2\rangle - |\beta_1\alpha_2\rangle)/\sqrt{2};\ T_0 = (|\alpha_1\beta_2\rangle + |\beta_1\alpha_2\rangle)/\sqrt{2};\ T_+ = |\alpha_1\alpha_2\rangle;\ T_- = |\beta_1\beta_2\rangle,$$

where α_1, α_2, β_1, and β_2 are the common designations for spin states.

The energies of the singlet and triplet eigenstates are:

$$\langle S\chi_N|H_{RP}|S\chi_N\rangle = J;\quad \langle T_0\chi_N|H_{RP}|T_0\chi_N\rangle = -J \qquad (25.3)$$

$$\langle T_+\chi_N|H_{RP}|T_+\chi_N\rangle = -J + (g_1+g_2)\mu_B B_0 + \left(\sum_i^a A_iM_i + \sum_k^b A_kM_k\right)/2 \qquad (25.4)$$

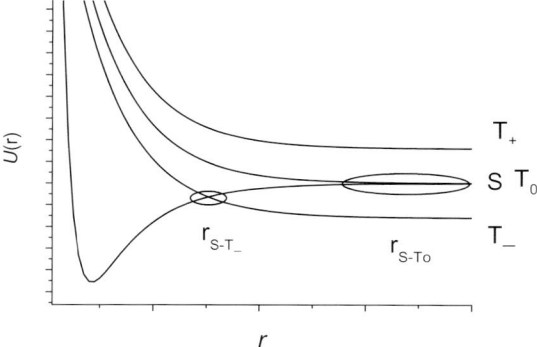

Figure 25.3 The dependence of radical pair energy $U(r)$ on distance r between two radicals for negative exchange interaction in the presence of a magnetic field. $r_{S\text{-}T_0}$ and $r_{S\text{-}T_-}$ denote the regions where effective S–T_0 and S–T_- transitions occur.

$$\langle T_-\chi_N|H_{RP}|T_-\chi_N\rangle = -J - (g_1+g_2)\mu_B B_0 - \left(\sum_i^a A_i M_i + \sum_k^b A_k M_k\right)/2 \quad (25.5)$$

where the sums must be evaluated with the nuclear spin quantum numbers M_i and M_k pertaining to the state $|\chi_N\rangle$.

The exchange interaction J is frequently assumed to be exponentially dependent on the distance between the radicals r: $J(r) = J_0 \exp(-r/r_d)$, where r_d is the parameter characterizing the exchange interaction decay. Figure 25.3 shows the energies of spin levels of a RP in a magnetic field as a function of the inter-radical distance. At large distances $r > r_d$, the S and T_0 levels become degenerate and the magnetic interaction effectively induces transitions between the S and T_0 states. In the reaction zone where $r < r_d$, the S and T_0 levels are split by the exchange interaction and the transitions between them are suppressed. The crossing of S and T_- levels at short distances leads to an effective S–T_- conversion in RPs.

The time dependence of the wave function of a RP is given by:

$$\psi(t) = C_{SN}(t)|S\chi_N\rangle + C_{TN}(t)|T_0\chi_N\rangle \quad (25.6)$$

In Equation 25.6, the S and T_0 states are not stationary states because they are coupled by the Zeeman and HFI terms of the Hamiltonian. Equation 25.1 can be solved using Equation 25.6 for different initial conditions $C_S(0)$ and $C_T(0)$. The probabilities of transition from the T_0 and T_- triplet states to the singlet S state are proportional to the squares of the matrix elements, specifically

$$P_{S-T_0} \sim (\langle T_0\chi_N|H_{RP}|S\chi_N\rangle)^2 = \left((g_1-g_2)\mu_B B_0/2 + \left(\sum_i^a A_i M_i - \sum_k^b A_k M_k\right)/4\right)^2 \quad (25.7)$$

$$P_{S-T_-} \sim (\langle S\chi_N|H_{RP}|T_-\chi_N\rangle)^2 = \left(\left(\sum_i^a A_i M_i + \sum_k^b A_k M_k\right)/4\right)^2 \quad (25.8)$$

The S–T conversion in RPs can be induced by the following mechanisms: (i) differences in Zeeman frequencies (Δg-mechanism); (ii) isotropic HFI of electrons with nuclei (HFI-mechanism); and (iii) by electron spin relaxation. For organic radicals, in most of these cases the Δg- and HFI- mechanisms make major contributions. The dependence of the S–T conversion rate on the intramolecular spin interactions of electrons and nuclei leads to a different probability of RP recombination, depending on the nuclear spin state. The resulting nonequilibrium population of nuclear spin states in the recombination products of a RP is referred to as CIDNP. The recombination probability for a RP with nuclear spin state $|\chi_N\rangle$ is given by:

$$P_N^S = \lambda \int_0^\infty |C_{SN}(t)|^2 f(t) dt \tag{25.9}$$

where λ is called the "steric factor."

The spin dynamics in RPs or precursor triplet molecules also results in the formation of nonequilibrium populations of electron spin levels in free radicals or triplet molecules in a phenomenon that is called "CIDEP."

The microwave (mw) magnetic field can also affect spin polarization by inducing S–T conversion in RPs. If it is assumed that the mw field is applied at frequencies corresponding to the energy splitting between T_- and T_0 and T_+ and T_0 states, then as the T_0 and S states experience mutual interconversion (due either to the HFI- or to Δg-mechanisms), the mw field-induced $T_- - T_0$ and $T_+ - T_0$ transitions produce changes in the singlet state population. The influence of the mw field on reaction parameters [such as yield, isotopic ratio, nuclear magnetic resonance (NMR), fluorescence, etc.] as a function of magnetic field provides information regarding resonance transitions in the RP (Salikhov et al., 1984). This approach is used in numerous reaction yield-detected magnetic resonance (RYDMR) techniques (Frankevich and Kubarev, 1982; Yu et al., 1984), in optically detected electron paramagnetic resonance (ODEPR) (Molin et al., 1980), and stimulated nuclear polarization (Bagryanskaya and Sagdeev, 1993; Sagdeev and Bagryanskaya, 1990).

25.4
Chemically Induced Dynamic Nuclear Polarization

Once created, the RP may follow two possible scenarios: (i) recombination to form geminate products; or (ii) diffusive separation and recombination to form escape products. If the radicals are created directly from the singlet state they may recombine immediately after formation; however, if they are created from the triplet state then recombination is forbidden and they must diffuse away to form escape products, unless the triplet–singlet conversion has enough time to occur. As mentioned above, the Δg- and HFI- mechanisms can promote either geminate recombination or the formation of escape products, the consequence being that both the geminate recombination and escape products have nonequilibrium populations of nuclear Zeeman levels, leading to emissive or enhanced absorptive peaks in the NMR spectra of the products.

There are two types of CIDNP effect: net and multiplet (Figure 25.4). In the case of net CIDNP, the nuclear spins in the reaction product have predominant orientations either along or opposite to the external field direction. Multiplet CIDNP corresponds to the case of mutual spin orientations in radicals and their recombination products (Figure 25.4a). Figure 25.4 shows the spin energy levels for two nonequivalent nuclei I_a and I_b, possessing chemical shifts δ_a, δ_b and coupled by spin–spin interaction j_{ab}. In the case of the multiplet effect, the polarization observed in the EPR lines of one multiplet is opposite to that of the other (Figure 25.4b). In the general case, a combination of net and multiplet effect is observed.

Based on qualitative predictions, Kaptein (1971) formulated the CIDNP sign rules in high- and low-magnetic fields. The CIDNP sign depends on the multiplicity of the RP precursor (μ), the pathway of product formation (ε), the radical g-factor differences (Δg), and the sign of HFI constant (A). The signs of net Γ_n and multiplet Γ_m effects are determined by:

$\Gamma_n = \mu \cdot \varepsilon \cdot \Delta g \cdot A$, where $\mu = +1$ for triplet RP, and -1 for singlet RP, $\varepsilon = +1$ for in-cage and -1 for escape products.

$$\Gamma_m = \mu \cdot \varepsilon \cdot A_i \cdot A_j \cdot j_{ij} \cdot \sigma_{ij}$$

where A_i, A_j are the HFI constants of i and j-nuclei, j_{ij} is the nuclear spin–spin exchange interaction, $\sigma_{ij} = +1$, if both nuclear spins reside in the same radical, and $\sigma_{ij} = -1$ if they reside in different radicals.

Biradicals and RPs in micelles belong to a special case, as the lifetime of these RPs is longer than that typical of singlet–triplet conversion times. In this case, the S–T$_0$ transition proceeds with the same efficiency for RPs, independent of their nuclear spin orientations. Thus, CIDNP due to the S–T$_0$ transitions in these RPs

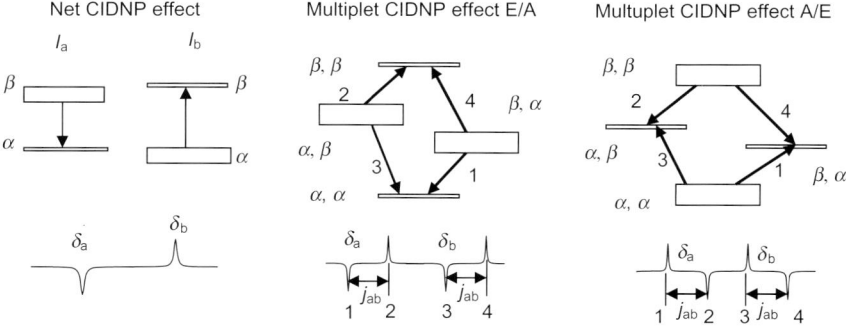

Figure 25.4 Schematic diagram of nuclear spin level populations for net CIDNP effect for two nuclear spins I_a and I_b with chemical shifts δ_a, δ_b, and multiplet CIDNP effect for two nonequivalent nuclear spins I_a and I_b, belonging to the same molecule and with spin–spin coupling j_{ab}. The corresponding simulated NMR spectra are shown below.

should be equal to zero and can only be formed due to the S–T_ transitions proceeding in the zone of S and T_ level crossing for $J < 0$. The HFI interaction induces the flip–flop transitions, so that the T_$\alpha \rightarrow$Sβ transition becomes allowed, whereas the T_$\beta \rightarrow$Sα transition becomes forbidden. In the case of positive J, CIDNP is formed in a similar way, but induced by S and T_+ level crossing and allowed T_+$\beta \rightarrow$Sα transitions. Thus, in the case of long-lived RPs, the signs of CIDNP are the same for in-cage and escape products, and do not depend on the sign of the HFI constant. The Kaptein (1971) rule in this case can be written as $\Gamma = \mu \cdot J_0$, where J_0 is the sign of the exchange interaction. The CIDNP intensity maximum due to this mechanism is expected in magnetic fields at which S-T_ levels crossing occurs. Thus, an analysis of the CIDNP represents a good approach for obtaining the value and sign of the exchange interaction in biradicals (de Kanter et al., 1977; Closs, Miller, and Redwine, 1985; Morozova et al., 1998; Maeda, Terazima, and Azumi, 1991) and micellized RPs (Lehr and Turro, 1981; Bagryanskaya et al., 1992).

The basic principles of RP theory were proposed by Kaptein and Oosterhoff (1969a, 1969b), Kaptein, Dijkstra, and Nicolay (1978), Closs (1969) and Closs and Trifunac (1970). During the past three decades, many new theoretical considerations of magnetic and spin effects (e.g., CIDNP, CIDEP, isotope effect, optical detected EPR, stimulated nuclear polarization, magnetic effects) have been developed and applied experimentally (Muus et al., 1977; Nagakura, Hayashi, and Azumi, 1998; Salikhov et al., 1984). These techniques took into account the different exchange–interaction models; for example, stepwise, distance-dependent, isotropic and anisotropic; and different models of molecular motion of partners (two-state jump model, continuous diffusion, etc.) including electron spin relaxation induced by various mechanisms. CIDNP theory in geminate RPs, for the case of viscous liquids in high magnetic fields, was developed for RPs confined in micelles and evaluated both numerically (Shkrob, Tarasov and Bagryanskaya, 1991) and analytically (Shushin et al., 1994; Shushin, 1991a). Low-field CIDNP theory for RPs with large HFI interactions was developed by Ananchenko, Purtov, and Bagryanskaya (1997); CIDNP and CIDEP theory employing Green's function was also proposed (Purtov and Doktorov, 1993). Recently, time-resolved CIDNP in stereospecific photoreactions of proteins with excited dye molecules was studied from a theoretical basis by Ivanov, Yurkovskaya, and Vieth (2008), using the integral encounter theory. CIDNP theory in the solid state, caused by the electron–electron-nuclear three-spin mixing (TSM) mechanism (where a net nuclear polarization is created in the spin-correlated radical pair (SCRP) due to the presence of both anisotropic hyperfine interaction and coupling between two electron spins) was proposed by Jeschke (1998) and later successfully applied to explain a CIDNP mechanism in the bacterial photosynthetic center (Jeschke and Matysic, 2003). Goez and Heun (2001, 2002) carried out Monte Carlo simulations of radical ion pair diffusion in homogeneous solution and in micelles, respectively. Another approach to CIDNP simulations in micelles was based on a numerical solution of the diffusion equation in a "microreactor model" (Shkrob, Tarasov, and Bagryanskaya, 1991).

Nowadays, CIDNP is applied widely to the study of the mechanisms of photochemical and thermal radical reactions. It was shown that the maximum nonequilibrium nuclear spin level populations is 0.1–0.3, whereas in a magnetic field of 1T, the corresponding equilibrium value is only 10^{-5}. Thus, typical CIDNP enhancement factors are 10^3–10^5, and this sensitivity allows for the detection of very low steady-state concentrations of nuclear spins. CIDNP is used widely, for example, in determining the multiplicity of intermediate RPs and their precursor excited molecules, the identification of radical intermediates, determination of the signs of the HFI constants, and studies of reversible electron-transfer reactions. A high sensitivity of CIDNP is achieved due to the "chemical accumulation" of polarization in the diamagnetic products. The lifetime of RPs lies on a nanosecond time scale, whereas nuclear relaxation times of reaction products are typically several seconds.

Since some reactions that produce CIDNP also proceed by nonradical pathways, one should employ other techniques (e.g., laser flash-photolysis, analysis of product yields, time-resolved EPR) in combination with the existence of predominant radical processes, or not.

25.4.1
The CIDNP Experiment

The basic requirements for the CIDNP experiment are a high-resolution NMR instrument, a light source (usually a laser, although some powerful diodes are now available), and an optical device to guide the light into the NMR tube without causing sample heating or excessive photodamage. Figure 25.5 depicts one of the earlier layouts, where light was guided by mirrors onto the side of the NMR tube.

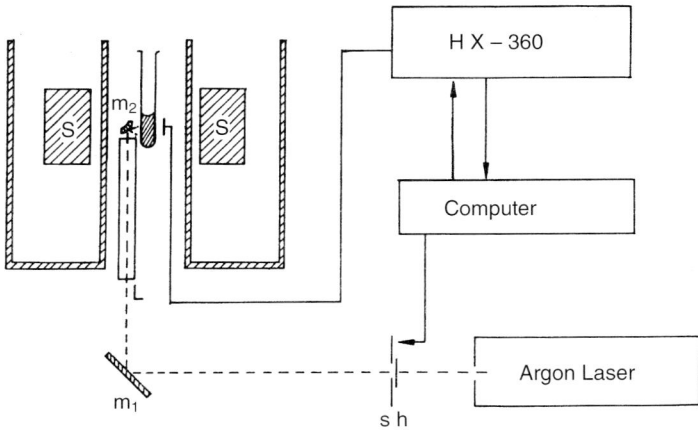

Figure 25.5 An experimental set-up for photo-CIDNP experiments on a high-resolution, narrow-bore magnet. S = superconducting magnet; sh = shutter; m_1 = flat mirror; m_2 cylindrical mirror; L = quartz light guide. Reproduced with permission from Kaptein (1982).

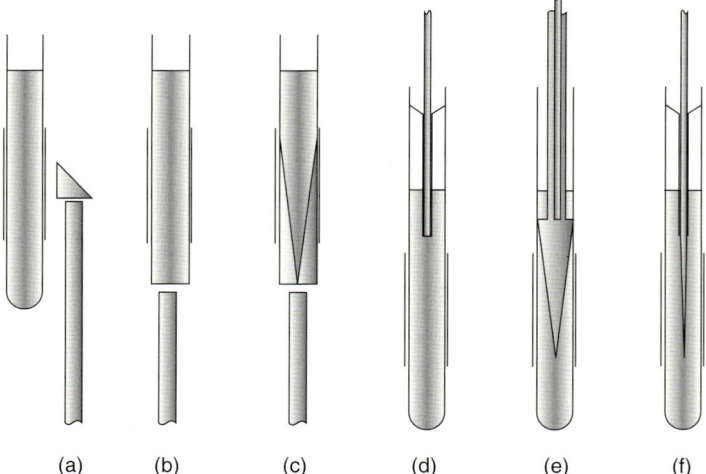

Figure 25.6 Schematic drawings of various NMR sample illumination methods. (a) Illumination from the side through the receiver coil via a cylindrical quartz rod installed inside the probe body surmounted by a prism or mirror; (b) Illumination from below with a quartz light guide and a flat-bottomed NMR tube; (c) A variant of (b), using a "V-cone" NMR tube to permit a more homogeneous irradiation of optically dense samples; (d) Illumination from above, using an optical fiber held inside a coaxial glass insert; (e) A variant of (d), in which light is distributed by means of a "pencil tip" insert; (f). The arrangement demonstrated here with a stepwise tapered optical fiber. Reproduced with permission from Kuprov and Hore (2004).

This was later "refined," for both economic and other reasons, into a fiber-optic cable system where the end of the cable was essentially "inserted into" the NMR tube (Scheffler, Cottrell, and Berliner, 1985; Kuprov and Hore, 2004). (Figure 25.6). Although sample spinning with this later approach is difficult, high-resolution NMR magnets are sufficiently homogeneous to provide decent spectra without spinning. For a steady-state CIDNP experiment, the pulse sequences needed are fairly straightforward (Figure 25.7a); this simply involves a "dark" cycle followed by a "light" cycle, after which the two NMR spectra are subtracted from one another in order to obtain the net CIDNP signal from the reaction products.

25.4.2
Time-Resolved CIDNP

The time-resolved (TR) CIDNP experiment, as proposed in 1974 by R. Ernst (Schaublin, Honener, and Ernst, 1974), has been developed and improved substantially during the past few years as more powerful lasers have become commercially available (Closs and Miller, 1979; Miller and Closs, 1981). A diagram of a typical TR CIDNP experiment is shown in Figure 25.7b, where a laser pulse is followed by a radiofrequency (RF) pulse after a variable delay (τ). The CIDNP

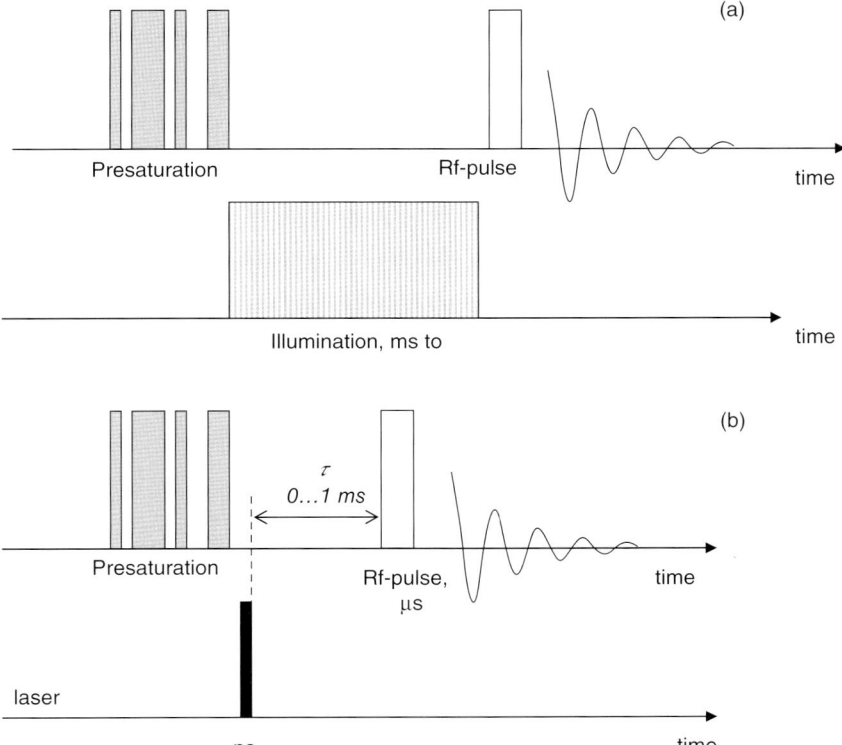

Figure 25.7 Time diagrams for photo-CIDNP experiments. (a) Without time resolution; (b) For time-resolved CIDNP (τ = variable delay).

intensity dependence on this time delay is determined by the radical reactions kinetics, as well as the nuclear spin relaxation times. It is possible to measure CIDNP formation on microsecond and submicrosecond time scales by monitoring the NMR lines of different functional groups from the reaction products. The signal of in-cage products of a geminate RP does not depend on the time delay τ; rather, escape products are formed on a microsecond time scale and their CIDNP kinetics reflect their accumulation. In addition, the nuclear relaxation time of intermediate radicals can affect the CIDNP kinetics of the escape products.

The advantage of TR CIDNP exists in the possibility of completely saturating the diamagnetic NMR signal. For saturation, a series of Rf pulses are used with different duration and uncorrelated phases (Goez, Mok, and Hore, 2005). A presaturation pulse sequence precedes the laser pulse initiating the TR CIDNP sequence; the time resolution of TR CIDNP is then determined by the duration of the Rf pulse, after which the free induction decay (FID) is detected. The typical duration of a 90° Rf pulse in modern NMR spectrometers is about 3–10 μs, although for a higher time-resolution it is possible to use a resonance circuit with a switchable Q-factor and a high-power transmitter (Miller and Closs, 1981).

The time-resolution of the method can be improved by the use of staggered observation pulses and an iterative reconvolution of pulse and CIDNP magnetization (Laufer, 1986; Goez, 1990). When this approach was applied to systems with high CIDNP enhancement factors, an increase was achieved in time-resolution of up to 100 ns, while the Rf pulse duration was 0.3–0.5 μs, which is typically achieved in commercial NMR spectrometers.

The separation of net and multiplet effects, as well as higher multiplet effects, can be achieved by applying Rf pulses of different lengths (flip angle) (Hany, Vollenweider, and Fischer, 1988). An alternative approach for analyzing the low magnetic field spin polarization of a nonequilibrium system of N coupled nuclei is based on a Fourier analysis of the measured intensity dependence of each line in the CIDNP spectrum on the Rf excitation pulse length. In this case, the authors demonstrated a relationship between the spectral components at various harmonic order and the alignment in the spin multiplet (Ivanov et al., 2003, 2006).

Sensitivity optimization is a key goal in time-resolved photo-CIDNP experiments. Recently, Goez et al. (2006) showed that a significant improvement could be be achieved by using multiple laser flashes per acquisition, and by storing the polarizations temporarily in the spin system. These authors developed a new pulse sequence that sums the time-resolved photo-CIDNP signals from n successive laser flashes, not in the acquisition computer of the NMR spectrometer but rather in the experiment itself, and this resulted in a greatly improved signal-to-noise ratio (SNR). For this experiment, the CIDNP is stored first in the transverse plane, then along the z axis, and then finally superimposed with the CIDNP produced by the next flash. These storage cycles result in very efficient background suppression. Because only one free induction decay is acquired for n flashes, the noise is digitized only once. As the main advantage, compared to signal averaging in the usual way, these sequences yield the same SNR with fewer laser flashes; the theoretical improvement is a factor of $\sqrt{n}$. From an analysis of the transfer pathways, it was shown that multiplet signals and CIDNP multiplet effects can also be investigated this way, even for strongly coupled spin systems.

The detection of CIDNP in the solid state requires special experimental modifications, in particular magic-angle spinning (MAS) NMR spectroscopy for different nuclei (^{15}N and ^{13}C) (Zysmilich and McDermott, 1994; Schulten et al., 2002). A presaturation pulse sequence, based on three ($\pi/2$) ^{13}C pulses, optimized timing and phase cycling, allows the repetition time of the experiment to be increased up to 4 Hz, and an efficient polarization extinction to be achieved (Daviso et al., 2008).

25.4.3
Low Magnetic Field CIDNP

CIDNP is usually detected in the NMR probe in a strong magnetic field (>2 T). As a rule, this field is not optimum for CIDNP observation, and consequently it is often necessary to study CIDNP in variable magnetic fields (0 to 7 T). It is known that the magnetic field dependence of CIDNP can provide information concerning the exchange interaction in long-lived RPs and biradicals, electron exchange reactions, and so on (Muus *et al.*, 1977; Nagakura, Hayashi, and Azumi, 1998; Salikhov *et al.*, 1984). For measurements of the magnetic field dependence of CIDNP, the photolysis step is carried out in a separate magnetic field; the sample is then transferred into the NMR spectrometer probe and the spectrum recorded. Different approaches for this sample transfer can be performed using either continuous (Lawler and Halfon, 1974; Bagryanskaya *et al.*, 1985; Meng *et al.*, 1990) or stop-flow systems (Lyon *et al.*, 2002b), or even pneumatic devices (Redfield, 2003; Grosse *et al.*, 2001). Lawler and Halfon (1974) described the construction of a flow system that allowed the use of normal NMR sample spinning while maintaining a high spectral resolution. Lyon *et al.* (2002a) also developed a new method for measuring and exploiting magnetic field CIDNP dependence. In this case, a sample is irradiated by laser light at a position in the bore of a superconducting NMR magnet, where the field magnitude is between 0.1 T and 7.0 T. The polarized sample is then transferred by rapid injection into an NMR tube at the center of the magnet (at 9.4 T), and the spectrum subsequently recorded. This technique was applied to the study of real-time protein folding and structure determination of partially denatured states of proteins (Mok *et al.*, 2003). Modern solvent suppression techniques use "excitation sculpting" such as DPFGSE (double pulsed-field gradient spin-echo; Hwang and Shaka, 1994), WATERGATE (WATER suppression through GrAdient Tailored Excitation; Piotto, Saudek, and Sklenar, 1992; Liu *et al.*, 1998) and CHESS (CHEmical Shift Selective) excitation (Haase *et al.*, 1985). Both, the DPFGSE and WATERGATE techniques rely on a gradient echo to selectively dephase the solvent magnetization, leaving the desired signals intact. In contrast, CHESS consists of a selective 90° pulse on the solvent resonance, followed by a dephasing gradient; this is repeated three times using orthogonal dephasing gradients prior to the acquisition pulse.

Grosse *et al.* (1999) developed a novel field-cycling unit with fast digital positioning of a high-resolution NMR probe in a spatially varying magnetic field between 0 and 7 T. Redfield (2003) described an apparatus that can pneumatically move an aqueous NMR liquid sample sealed in a 5- or 8-mm standard tube from the center of a standard Varian 500 MHz spectrometer, and place it at any position in its fringe field and back. This device allowed the study of the low-magnetic field relaxation of ^{31}P, ^{15}N and ^{13}C, using proton detection employing the full power of the commercial system for preparation before, and detection after, the mixing interval at low magnetic field. For this, a pneumatic glass shuttle tube, held in an aluminum support tube, is temporarily inserted in place of the upper tube. A standard thin-walled NMR tube is then connected to a plastic piston shuttle that is moved up or down inside the shuttle tube by a low vacuum or pressure applied from the top. It should be noted that sample transfer leads to polarization transfer and its redistribution between nuclear spins. De Kanter and Kaptein (1979) considered this effect and formulated the conditions for adiabatic and nonadiabatic polarization transfer, and its dependence on the transfer time of the sample. It was shown that, at a sufficiently low field – especially in the region of level-crossings for strongly coupled spins – the magnetic relaxation dispersion exhibited characteristic features and the polarization was transferred among the spins. Later,

Ivanov and Sagdeev (2006) and Miesel *et al.* (2006, 2008) discussed the manipulation and possible utilization of polarization transfer in field-cycling experiments. In this case, the low magnetic field CIDNP measurements were performed out of the NMR probe, such that the CIDNP kinetics could not be monitored. However, this problem was overcome using additional pulse switching of the external magnetic field (SEMF) during photolysis. This new time-resolved technique for the study of low magnetic field CIDNP kinetics was developed by Bagryanskaya, Gorelik, and Sagdeev (1997), and by Bagryanskaya and Sagdeev (2000). SEMF CIDNP involves a fast switching of the magnetic field during the lifetime of the radical intermediates. Information concerning the spin dynamics and chemical kinetics is then obtained from the effect of the magnetic field switch on nuclear polarization of the diamagnetic products of the radical reactions. This method has been used to measure the rate constants of a degenerate electron-exchange reaction (Bagryanskaya, Gorelik, and Sagdeev, 1997), the rate constants of stable nitroxides reacting with alkyl radicals (Lebedeva *et al.*, 2004), the electron spin relaxation times of short-lived radicals in low magnetic fields (Fedin *et al.*, 2002a), the rates of radical escape from micelles (Fedin, Bagryanskaya, and Purtov, 1999), and to obtain information on the electron polarization formed during chemical reactions in weak magnetic fields.

25.4.4
The Application of CIDNP to Biological Systems

Numerous applications of CIDNP to study the pathways of chemical reactions and identify radical intermediates have been summarized in various books (Muus *et al.*, 1977; Nagakura, Hayashi, and Azumi, 1998; Salikhov *et al.*, 1984) and reviews (Goez, 2009; Steiner and Ulrich, 1989; Kuhn and Bargon, 2007). For example, CIDNP was used to study the radical mechanism of singlet carbenes, the mechanism of acylperoxide photolysis, the photochemical decay of several ketones, the kinetics and dynamics of flexible biradicals formed during the photolysis of cyclic ketones, and also to establish radical mechanisms in numerous reactions of electron (Roth, 2001, 2003, 2008; Bargon, 2006) and proton transfer, and fragmentation. CIDNP has also been detected in several different nuclear spins and their isotopes (^{1}H, ^{2}D, ^{13}C, ^{15}N, ^{19}F, ^{29}Sb, etc.). Some selected examples of CIDNP applications to biological systems are described in the following subsections.

During the past 10 years CIDNP has been applied successfully to a range of biological systems, and in particular to the study of light-induced structural changes in bacterial photosynthetic centers, to understanding the mechanisms of photochemical reactions of amino acids and peptides, and the mechanism and kinetics of protein folding. For biological systems, the CIDNP method relies entirely on cyclic photoreactions, which might suggest that the phenomenon is not observable. As a cyclic reaction has no net chemical change, no polarization would be expected to be observed, since a geminate recombination of the radical pair would lead to one polarization and the escaping radicals would carry the opposite polarization, thus canceling out the entire effect. Fortunately, there are some competing processes – particularly electron-induced nuclear relaxation and electron exchange – that tend to diminish escape and recombination product polarization.

25.4.5
Photo-CIDNP in the Study of Protein Folding

Photo-CIDNP in the study of protein folding employs a reversible chemical reaction between the protein and a photoexcited dye to generate nuclear spin polarization in the side chains of certain amino acid residues (Stob and Kaptein, 1998; Kaptein, 1982). The polarization, which is detected as enhancements in the NMR spectrum, is observed only for tryptophan, tyrosine, and histidine residues that are physically accessible to the excited dye, which usually is a flavin mononucleotide (Kaptein, Dijkstra, and Nicolay, 1978) (Figure 25.8). For the native state of a protein, the method provides a "picture" of the extent of exposure of these residues (Kaptein, Dijkstra, and Nicolay, 1978). The CIDNP of several proteins in their unfolded or partially folded states has been applied in order to shed light on both the structure and coformational changes of the native state (Garssen *et al.*, 1978; Canioni, Cozzone, and Kaptein, 1960; Akasaka, Fujii, and Kaptein, 1981; Hincke, Sykes, and Kay, 1981; Muszkat, Khait, and Weinstein, 1984; Vogel and Sykes, 1984; Weiss *et al.*, 1989). Broadhunt *et al.* (1991) were the first to use the CIDNP pulse labeling technique to follow the unfolding transition of the protein hen egg-white

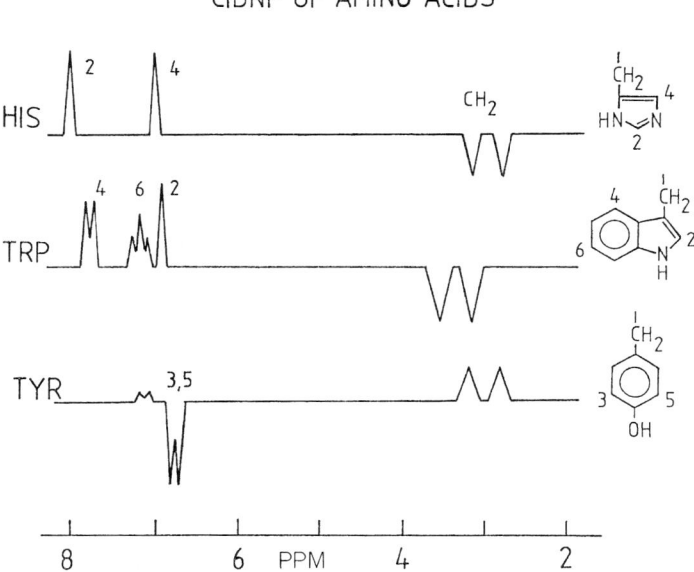

Figure 25.8 Schematic representation of the flavin-induced photo-CIDNP for the three amino acids histidine, tryptophan and tyrosine. Reproduced with permission from Kaptein (1982).

lysozyme, and to characterize its denatured state. This technique (Lyon et al., 2002a) utilizes the high-resolution spectrum of the native state in order to obtain information regarding the denatured state, the spectra of which are often comparatively poorly resolved (Dobson and Hore, 1998; Balbach et al., 1997). In order for the nuclear polarization to be detected with adequate sensitivity, this experiment requires the folding to be faster than the spin-lattice relaxation of the residues involved. The results of CIDNP pulse labeling experiments for partially denatured hen egg white lysozyme in 1,1,1-trifluoroethanol (TFE) are shown in Figure 25.9. In the presence of 45% TFE (v/v), lysozyme forms a partially structured state with a high degree of helical structure. In the CIDNP pulse-labeling procedure, nuclear polarization is induced in a solution of the unfolded state by

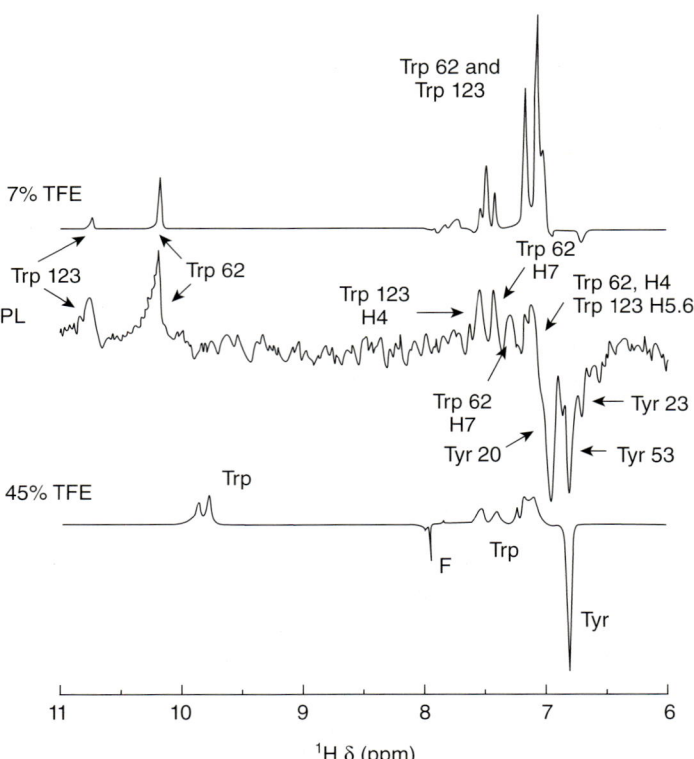

Figure 25.9 CIDNP pulse-labeled spectrum of hen egg white lysozyme (HEWL) diluted from 45% (v/v) to 7% in 2,2,2-trifluoroethanol (TFE). The spectrum was recorded at 600 MHz, averaged over four light and dark pairs. Laser illumination for 500 ms was used, with an injection time of 50 ms and post-injection delay of 100 ms. Upper: Equilibrium CIDNP spectrum of lysozyme in 7% TFE; Lower: Equilibrium CIDNP spectrum of lysozyme in 45% TFE. Center: Pulse-labeled CIDNP spectrum. The peaks corresponding to the solvent-accessible tyrosine (Tyr) and tryptophan (Trp) residues are shown. F denotes a signal arising from a polarized flavin (Mok et al., 2003).

means of a photochemical reaction between flavin mononucleotide and solvent-accessible tyrosine, tryptophan, and histidine residues. The protein is then rapidly folded back to the native state by its injection into a refolding medium, after which the nuclear polarization is detected.

Real-time, rapid-injection protein refolding experiments permit the observation of changes in the accessibility of specific residues during the folding process. Heteronuclear two-dimensional (2-D) ^{15}N–^{1}H CIDNP techniques allow the identification of surface-accessible residues with improved resolution and sensitivity (Mok and Hore, 2004). Recently, by using a "Trp-cage" miniprotein, Mok et al. (2007) the results of reported CIDNP pulse-labeling experiments that involved rapid in situ protein refolding. These authors found that there was residual structure due to hydrophobic collapse in the unfolded state of this small protein, with strong inter-residue contacts between side chains that were relatively distant from one another in the native state. A new and very promising approach to enhancing the sensitivity of heteronuclear nuclear Overhauser effects (NOEs) in ^{19}F-labeled proteins was proposed by Hore and coworkers (Kuprov et al., 2007; Khan et al., 2006). In this case, CIDNP was explored as a source of nuclear hyperpolarization in heteronuclear Overhauser effect experiments. For this, a cyclic photochemical reaction, proceeding through a radical pair intermediate, was used to enhance the ^{19}F nuclear magnetization in 3-fluorotyrosine by more than an order of magnitude, with a corresponding increase in the amplitudes of ^{19}F–^{1}H cross-relaxation and cross-correlation effects.

The importance of using only geminate CIDNP to interpret CIDNP data quantitatively in terms of residue accessibility was shown by Morozova et al. (2002). CIDNP in proteins depends not only on the accessibility of the amino acid residues and the mechanism of the photochemical reaction, but also on the reactivity of the radical intermediates and the efficiency of the intramolecular reactions. Only the initial polarization is directly related to the accessibility of the aromatic side chains to the photoexcited dye, and this occurs within a few nanoseconds by geminate radical recombination. On a time scale of hundreds of microseconds, rapid degenerate electron hopping and polarization contributions of opposite sign from the recombination products of escaped radicals may affect the polarization. Subsequently, TR CIDNP kinetics were used to measure the intermediate nuclear relaxation time, T_1, and the rate constants of the electron-transfer reaction (Morozova et al., 2004). The correlation times for side chain motion, as determined from the T_1 of the radicals, were shown to correlate with the accessibility of the side chains in the intact protein.

25.4.6
CIDNP Application to Study Primary Processes in the Bacterial Photosynthetic Center

Observations of the effect of photo-CIDNP on photosynthetic reaction centers (RCs) by solid-state NMR opened a new area for applications of this technique. The first observation of ^{15}N CIDNP in primary radical pairs in the quinine-blocked frozen bacterial reaction center was demonstrated by Zysmilich and McDermott (1994). Unfortunately, these studies were conducted mainly on the purple bacterium *Rhodobacter sphaeroides* WT and R-26, followed by photosystem II (PS II) from plants, which belong to the group of type II RCs. In a recent series of reports from the Matysic/Jeschke groups (Prakash *et al.*, 2005, 2006; Diller *et al.*, 2007), the solid-state photo-CIDNP effect was shown to be a general feature of primary RPs in natural photosynthesis. Due to the minimal Zeeman splitting in radical intermediates formed in RCs, and a resultant unfavorable Boltzmann distribution, all magnetic resonance methods have an intrinsically low sensitivity. Photo-CIDNP magic-angle spinning (MAS) NMR has been demonstrated as the method of choice to overcome this limitation, particularly by producing non-Boltzmann nuclear spin distributions via photochemical reactions in solids that, in turn, allow for the detailed study of the photochemical machinery of RCs at nanomolar concentrations. Photo-CIDNP also provides information at the atomic scale of the electronic structure of the photochemically active regions of RCs in the ground state. Prakash *et al.* (2007) observed photo-CIDNP signals from plant photosystem I (PS I), and RCs from green sulfur bacteria and heliobacteria, with the major representative groups of organisms having type I RCs. The RCs labeled by ^{13}C were first measured with 2-D ultrahigh field cross-polarization/MAS (CP/MAS) solid-state dipolar correlation spectroscopy without illumination, followed by ^{13}C photo-CIDNP MAS NMR. For this, a Bruker DSX-750 NMR spectrometer equipped with a double-resonance MAS probe operating at 750 MHz for ^{1}H and at 188 MHz for ^{13}C, was used. The combination of partial ^{13}C enrichment and photo-CIDNP yielded a large enhancement of the NMR intensity, and reduced the number of signals. Subsequent 2-D ^{13}C–^{13}C photo-CIDNP solid-state MAS NMR experiments enabled chemical shift assignments for the labeled carbons that participated in the photo-CIDNP process.

The origin of net CIDNP in RCs has been under intense discussion over the past few years. Initially, it was assumed to be due to significant differential relaxation (DR) between nuclear spins in the special triplet, ^{3}P, and nuclear spins in the singlet ground state, ^{1}P. Currently, two other mechanisms are under discussion, however: (i) the electron–electron-nuclear three spin mixing (TSM) mechanism (Jeschke, 1998); and (ii) the differential decay (DD) mechanism. In the DD mechanism, a net photo-CIDNP effect is caused by anisotropic hyperfine coupling, without any explicit requirement for electron–electron coupling if the spin-correlated pairs have different lifetimes in their singlet and triplet states (Figure 25.10). In order to differentiate between these two mechanisms, photo-CIDNP was performed at different NMR spectrometer frequencies (200, 400, and 750 MHz), because the TSM and DD mechanisms are predicted to have different magnetic field dependencies. It was shown that, for the quinine-depleted RC of wild-type purple bacterium *Rh. sphaeroides*, the highest CIDNP intensity was observed at 4.7 T, and that both CIDNP mechanisms contributed with the TSM mechanism being dominant.

In contrast, for the RC from the caratenoid-less strain R26 of the purple bacterium *Rh. sphaeroides*, the observed CIDNP was due to a contribution of all three

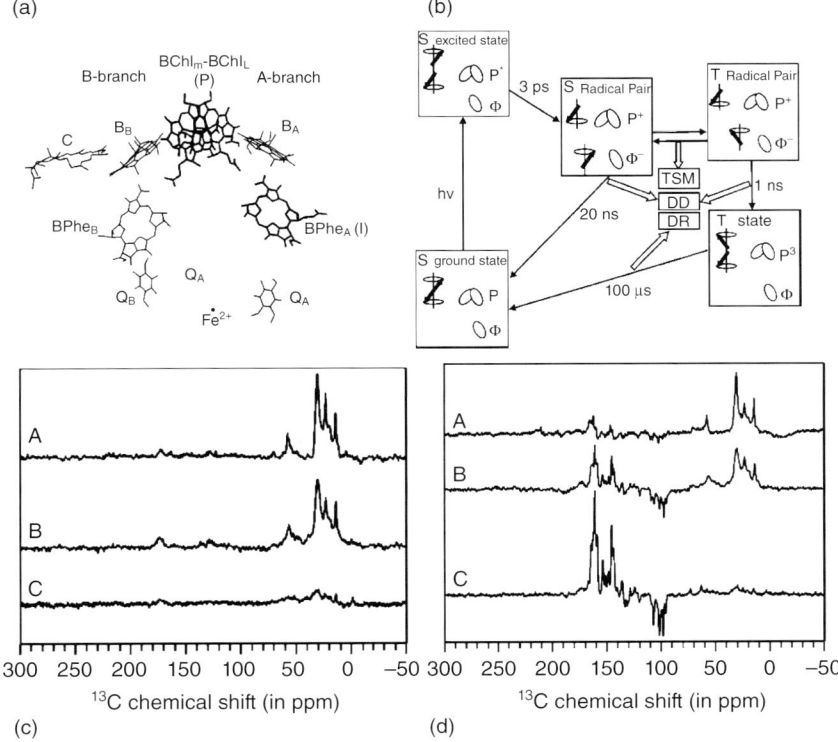

Figure 25.10 (a) Arrangement of cofactors in RC of *Rhodobacter sphaeroides* WT. The aliphatic chains of BChl, BPhe, and Q have been omitted for clarity. The cofactors that participate in the photo-CIDNP experiment (P and I) are depicted in bold (Schulten et al., 2002); (b) Reaction cycle in quinone blocked bacterial RCs. After light-induced electron transfer from the primary donor (P) to the bacteriopheophytin (Φ), an electron-polarized singlet radical pair is formed. The electron polarization is transferred to nuclei via three-spin mixing (TSM) within the radical pair and via differential decay (DD), owing to the difference in lifetime of the two radical pair states. Cancellation of incomplete nuclear spin polarization during long-lived donor triplet is by differential relaxation (DR); (c) ^{13}C MAS NMR spectra of quinine-depleted RCs of *Rb. sphaeroides* obtained at 223 K in the dark at different magnetic field strengths of 17.6 T (spectrum A), 9.4 T (spectrum B), and 4.7 T (spectrum C); (d) ^{13}C photo-CIDNP MAS NMR spectra of quinine-depleted RCs of *Rb. sphaeroides* obtained at 223 K in continuous illumination with white light at different magnetic field strengths of 17.6 T (spectrum A), 9.4 T (spectrum B), and 4.7 T (spectrum C) (Prakash et al., 2006).

mechanisms. Enhancement factors of about 1500 have been observed in several RCs, although when combined with ^{13}C-isotope labeling the signal was shown to increase by a factor of 100, up to a total enhancement of one million. The application of photo-CIDNP yields an almost complete set of assignments of all aromatic carbon atoms in the macrocycles of BChl and Bphe; this allowed, for the first time, a comprehensive map to be constructed of the ground-state electronic structure of the photochemical active cofactors (Prakash et al., 2007; Schulten et al., 2002).

25.4.7
CIDNP Applications to Electron Transfer in Peptide and Amino Acids

The first basic, detailed study of the photo-CIDNP of amino acids was performed by Stob and Kaptein (1998). In the recent reports of Morozova et al. (2007, 2008, 2009), TR CIDNP was used to study the mechanism and rate constants of electron-transfer (ET) reactions in peptides and amino acids, electron transfer from the aromatic amino acid tyrosine to the purine base radical guanosine monophosphate (Morozova et al., 2007), intramolecular electron transfer in tryptophan–tyrosine dipeptides (Morozova, Kiryutin, and Yurkovskaya, 2008), and the photooxidation of glycylmethionine and methionylglycine (Morozova et al., 2009). A quantitative analysis of CIDNP kinetics, obtained at different concentrations of amino acids at different pH-values, was used to determine the rate constant of the reductive ET reaction for different reactant RPs with different protonation states. The values of the ET rate constants are of particular importance for understanding the mechanism of chemical repair of nucleic acid damage by proteins. It should be noted that the amino acids methionine, cysteine, histidine, and tyrosine were each found to be involved in DNA repair (Milligan et al., 2003).

CIDNP can be applied to study ET in proteins of moderate size with somewhat more distant redox partners, in order to unravel the ET pathways and to acquire information concerning protein reaction mechanisms, as demonstrated recently by Eisenreich et al. (2008, 2009) and Richter et al. (2007). ^{13}C photo-CIDNP signals were observed upon the photoexcitation of flavin mononucleotide-binding domains from the blue-light receptor, phototropin, involved in the phototropic response of higher plants. The origin of the CIDNP was concluded to originate from hyperfine-selective branching into singlet and triplet products of different lifetimes. Spin-polarized ^{13}C NMR signals in the emission, arising from ^{13}C nuclei at natural abundance in the apoprotein, were assigned to a tryptophan residue that is located about 14 Å from the cofactor undergoing photoinduced ET to the flavin. It should be noted that the RP proposed in these studies was not detected directly by either optical spectroscopy nor TR-EPR, despite intense efforts. In fact, TR-EPR can be used to probe radical intermediates on a time scale as short as 10 ns, while the RP may be either too short-lived (<10 ns) or its electron-spin polarization relaxes too rapidly for direct detection.

A recent review by Verhoeven (2006) considered examples of the application of magnetic field-dependent CIDNP to study artificial multichromophoric ET systems. Over the past decades, much effort has been expended to produce systems in which photoexcitation leads to one or more intramolecular ET events, ultimately resulting in a charge-transfer excited state with a relatively long lifetime. This process is generally considered to be a mimic of natural photosynthesis; although it is not relevant to solar energy conversion, it is relevant to molecular information storage, molecular electronics, and molecular photonics.

25.5
Chemically Induced Dynamic Electron Polarization

Short-lived reactive radicals in solution often exhibit a nonequilibrium distribution of spin energy level populations; this phenomenon is caused by a chemical reaction termed chemically induced dynamic electron polarization (CIDEP). Since many reviews covering CIDEP have been produced in recent years (McLauchlan, 1989; van Willigen et al., 1993; Freed, 2005; McLauchlan and Yeung, 1994; Murai, Tero-Kubota, and Yamauchi, 2000; Savitsky and Moebius, 2006, 2009), only a brief overview of the mechanism of CIDEP, and some applications to biological systems, will be presented in this chapter. CIDEP provides information for characterizing radicals from their g-factors and hyperfine interactions. In addition to these parameters, the sign, absorptive (A) or emissive (E), lineshape and linewidth of the EPR transitions are also be affected by CIDEP processes. These processes alter the population of each Zeeman level, and result in different spectral patterns and intensities. The manifestations of CIDEP vary depending on the identity of the RP involved, as well as the environment in which the RP was formed. For example, the solvent viscosity, spin state of the radical precursors (singlet, triplet, etc.) and limits to diffusion (solvent boundaries such as the lamellar phase of a surfactant) will affect the intensity, phase, and lineshape of CIDEP spectra. The theoretical background behind CIDEP has been reported (Adrian, 1974); the most important CIDEP mechanisms are the triplet mechanism (TM), the radical pair mechanism (RPM), the SCRP mechanism, and the radical-triplet pair mechanism (RTPM).

25.5.1
Triplet Mechanism of CIDEP

Polarization due to the TM originates from the electron spin polarization of a triplet state precursor generated by an electron-spin-selective intersystem crossing (ISC) from a photoexcited singlet, as shown in Figure 25.11. The triplet CIDEP mechanism was first proposed by Wong, Hutchinson, and Wan (1973). The triplet sublevel splittings are induced by spin–spin, dipole–dipole, or spin–orbit interactions of unpaired electrons or by their interaction with the external field:

$$\hat{H} = g\mu_B \vec{B}\hat{S} + DS_z^2 + E(S_x^2 - S_y^2)$$

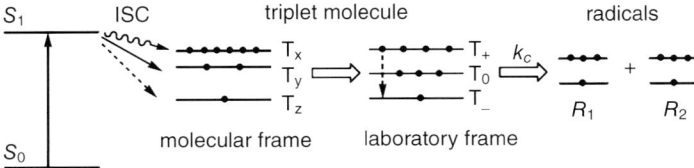

Figure 25.11 Polarization due to the triplet mechanism originates from the electron spin polarization of a triplet state precursor generated by an electron spin-selective intersystem crossing from a photoexcited singlet.

Here, D and E are the zero-field-splitting constants, x, y, and z are the principal axes of the zero field tensor fixed in the molecular frame, and k_x, k_y, and k_z are the rate constants of intersystem conversion from the singlet state to triplet states. In many molecules, the rates of ISC are different (van der Waals and Groot, 1967). The initial polarization of a randomly oriented triplet state is given by

$$P_{TM}^0 = -\frac{4}{15} \times \frac{D}{B_0}(P_x + P_y - 2P_z)$$

where P_x, P_y, and P_z are the populations of the three T_x, T_y, and T_z sublevels. In a magnetic field intramolecular electron polarization, the nonequilibrium population of T_x, T_y, and T_z levels is partially transferred to the electron Zeeman levels T_{-1}, T_0, and T_1. TM CIDEP is commonly observed in excited triplet state molecules in molecular crystals or in frozen liquid crystals (Yamauchi and Pratt, 1979). The spectral analysis allows the zero-field-splitting parameters caused by the electron dipole–dipole interaction to be obtained.

The electron spin relaxation of a triplet state molecule in liquids is very short (10^{-8} to 10^{-10} s) (van Willigen et al., 1993; Freed, 2005; McLauchlan and Yeung, 1994; Clancy and Forbes, 1999; Savitsky and Moebius, 2006), due to modulation of the strong anisotropic dipole–dipole interaction by fast molecular tumbling; consequently, CIDEP in triplet-state molecules can be detected in only a few cases (Closs et al., 1993; Regev et al., 1993; Zhang et al., 1993; Steren, Willigen, and Dinse, 1994). The TR-EPR spectra of the excited triplet state of fullerene C_{60} were first reported by several groups (van Willigen et al., 1993; Freed, 2005; McLauchlan and Yeung, 1994) while later, in 1996, the excited triplet state of porphyrins polarized by the TM was observed in fluid solution at room temperature by Fujisawa, Ohba, and Yamauchi (1997) and others (Freed, 2005; Savitsky and Moebius, 2006).

If the lifetime of the triplet molecule is shorter than the electron spin relaxation time (e.g., due to chemical reactions), then the nonequilibrium populations of the T_{-1}, T_0 and T_1 levels are transferred to the electron spin levels of the RP formed either by decay or by the reaction of the triplet with a suitable substrate. The ensuing radicals are formed with increased spin density in the higher or lower Zeeman levels, which results in net E or A transitions, respectively. The TM produces a net initial electron spin polarization of the same sign in both radicals.

Triplet mechanism CIDEP theory in solution has been developed by Atkins and Evans (1974a, 1974b) and Pedersen and Freed (1975), using a density matrix formalism. For a rapidly rotating triplet molecule reacting with rate constant k_T to form an RP, the equation for polarization is given by:

$$P_{TM}^0 = \tfrac{2}{15}\left(\frac{k_T T_1^T}{1+k_T T_1^T}\right)\left[D\hat{K} + 3E\hat{I}\left\{\frac{\omega_0 \tau_R}{1+\omega_0^2 \tau_R^2} + \frac{4\omega_0 \tau_R}{1+4\omega_0^2 \tau_R^2}\right\}\right]$$

$$\hat{K} = \frac{1}{2}(W_x + W_y) - W_z$$

$$\hat{I} = \frac{1}{2}(W_y - W_x)$$

$$W_i = \frac{k_i}{k}$$

where ω_0 is the microwave frequency, τ_R is the rotational correlation time which is assumed to be isotropic, T_1^T is the spin-lattice relaxation time of the precursor triplet state, and W_i are the relative populating rates of the sublevels. The magnitude of polarization depends on the triplet molecule parameters D and E, the rotational correlation time, and k_T. In most of these systems, $D \gg E$ and the sign of TM polarization is determined by the sign of the expression:

$$\Gamma = -D(k_x + k_y - 2k_z).$$

25.5.2
Radical-Pair Mechanism of CIDEP

25.5.2.1 CIDEP Due to S–T₀ Transitions

As noted above, the RPM assumes that the cage reaction of the RP can be affected by magnetic interactions that produce nuclear spin-dependent mixing of the reactive singlet and unreactive triplet electron spins states of the RP. The CIDEP process due to RPM is more complicated than CIDNP, and involves the combined effects of S–T$_0$ mixing by magnetic interactions and S–T$_0$ splitting by the exchange interaction. A detailed description of CIDEP due to RPM requires application of the stochastic Liouville equation to describe the spin evolution of RP, and to also take into account radical diffusion in the RP. A simple quasi-adiabatic model, as proposed by Adrian (1986), provides a qualitative picture of CIDEP due to RPM, and allows prediction of the sign rules for CIDEP. At short inter-radical distances, the singlet and triplet states are split by the exchange interaction; thus, the eigenfunctions of RP are $T_0\alpha_n$ and $T_0\beta_n$. At a distance r, where the exchange interaction is zero, the eigenfunctions of RP are $\beta_1\beta_n$, $\alpha_1\alpha_n$, $\alpha_1\beta_n$, $\beta_1\alpha_n$ for R_1, and β_2 and α_2 for R_2. The separation of the radicals can be either adiabatic or nonadiabatic, depending on the radical size, the solution viscosity, and the HFI constants. The criterion of adiabacity can be written as $\lambda^2/D \sim a^{-1}$, where a is the HFI constant, D is the mutual diffusion coefficient, and λ is the characteristic scale for exponential decay of the exchange interaction in its usual form: $J(r) = J_0 \cdot \exp(-(r - R)/\lambda)$.

In the case of adiabatic separation, the population of each eigenstate remains the same at any moment in time, although the eigenfunctions themselves change significantly. In the case of a triplet precursor, during the separation the $T_0\alpha_n$ state transforms to the $\alpha_1\alpha_n\beta_2$, $\beta_1\alpha_n\alpha_2$ states, whereas $T_0\beta_n$ transforms to the $\beta_1\beta_n\alpha_2$, $\alpha_1\beta_n\beta_2$ states. Thus, in the case of a triplet RP precursor the $\alpha_1\alpha_n$ and $\beta_1\beta_n$ states of R_1 will be overpopulated. As a consequence, for radicals with equal g-values, A/E multiplet polarization arises. However, whereas for radicals with HFI constants of 1–2 mT in nonviscous solutions the adiabatic contribution is not significant, it can be important for RPs with large HFI constants in solutions of high viscosity, for example, in micelles.

Adrian (1971) has performed a model calculation for the effect of the re-encounter sequences of SCRP on the basis of the Noyes diffusion model, and obtained the following result for the unpaired electron spin density $<\rho_1>$ on R_1.

$$<\rho_1> = <\psi(t)|S_{z1} - S_{z2}|\psi(t)> = 0.085(Q_{ab}J)/|Q_{ab}J|(Q_{ab}\tau_D)^{1/2}\{|C_S(0)|^2 - |C_T(0)|^2\}$$

where τ_D is the time of diffusion $\tau_D = d^2/3D$, and d is the distance of closest approach.

The magnitude of CIDEP is proportional to $Q_{ab}^{1/2}$, which means that the polarization increases with increasing difference in resonance frequencies and with decreasing solvent viscosity. The sign of the polarization is determined by the sign of Q_{ab}, which means that the differences in polarization correspond to different nuclear spin states. Thus, the sign of the polarization (A/E or E/A) is determined by the sign of the exchange interaction and the multiplicity of the precursor RP, and is given by:

$\Gamma = \mu \cdot \text{sign}(J)$, where $\mu = +1$ for triplet precursor, $\mu = -1$ for singlet precursor. $\Gamma > 0$ corresponds to a E/A pattern, while $\Gamma < 0$ – corresponds to a A/E pattern.

After pulsed radical generation, the geminate RPM polarization decays to thermal equilibrium with electron spin relaxation time T_1. The additional source for RPM CIDEP in reacting radical systems is a continuous formation of F-pairs; that is, random encounters of independently formed spin-uncorrelated radicals. The RPs in T_+ and T_- spin states do not contribute to CIDEP formation; the probability of RP encounters in the S and T_0 states is the same. Selective recombination in the S-state leaves more radicals in the T_0 state, and the polarization mechanism proceeds as for geminate T_0-pairs. This leads to F-pairs CIDEP which, in Adrians model, is described by:

$$<\rho_1^F> = <\rho_1>(1 - 2\varphi|C_S(0)|^2$$

where 2φ is the singlet reaction probability. In the case of g-value differences in radicals in RP, the net CIDEP is formed with a sign determined by: $\Gamma = \mu \cdot \Delta g \, \text{sign}(J)$, where $\Gamma > 0$ corresponds to enhanced absorption, and $\Gamma < 0$ corresponds to emission. The sign of the polarization will be opposite for radical partners.

25.5.3
CIDEP Due to S–T_ and S–T_+ Transitions

In cases when the HFI cannot be neglected in comparison to the Zeeman interaction, or at high solvent viscosity for example, short-lived biradicals (Maeda, Terazima, and Azumi, 1991; Closs, Forbes, and Piotrowiak, 1992; Forbes and Ruberu, 1993), micellized radical pairs, RPs in low magnetic field or at low temperatures, the S–T_ transition becomes important and CIDEP is generated in the region of the S–T_ level crossing (see Figure 25.3). S–T_ transitions proceed via simultaneous flips of both electron and nuclear spins. The magnitude of the polarization is given by Adrian and Monchik (1979):

$$P_{ST\_M} = \frac{\pi A^2 r_c}{4g\mu_\beta B_0 \lambda D},$$

where $r_c = \lambda^{-1}\ln(J_0/B_0)$, with $J = J_0\exp(\lambda r)$.

The net polarization due to the S–T_ mechanism is proportional to A^2, inversely proportional to the diffusion coefficient D, and is more sensitive to HFI and diffusion than CIDEP due to S–T$_0$ transitions. The CIDEP sign is given by $\Gamma = \mu \cdot \text{sign}(J)$. For a typical negative J, polarization due to the S–T_ mechanism is predicted to be *emissive*. In the case of a positive exchange interaction, transitions occur in the region of S–T$_+$ crossing terms; thus an overpopulation of β-spins of the radical is predicted and a positive polarization is expected.

The simple analytical solutions for CIDEP due to S–T_ and S–T$_+$ transitions in the level crossing region induced by isotropic and anisotropic HFI, as well as electron– electron, dipole–dipole interactions at high magnetic and low magnetic fields, were obtained by Shushin (1995a, 1991c).

25.5.4
CIDEP Due to the Radical-Triplet Pair Mechanism

The RTPM is responsible for electron polarization generated in F-pairs formed in the encounters of radicals and excited triplet-state molecules (Blattler, Jent, and Paul, 1990; Kawai, Okutsu, and Obi, 1991; Kawai and Obi, 1993). Triplet quenching by radicals is a well-known process that can proceed via several mechanisms, such as enhanced ISC or energy transfer (Gizeman, Kaufman, and Porter, 1973). Upon encounter, a triplet molecule and a radical may form either via a quartet, Q, or a doublet, D. During the encounter of a radical and a triplet molecule, the pair spin states are split by the exchange interaction, as shown in Figure 25.12. Polarization can be produced in two regions of the level crossing; the net polarization is generated in the region of level crossing r_n, r'_n, where the pair states are mixed by the zero-field-splitting in the triplet. As this interaction is often one or two orders of magnitude larger than the HFI, the polarization is generated very effectively, and even on a short time scale. Multiplet polarization is created in the region where the exchange interaction is approximately equal to the difference in HFI, and proceeds in the same way as in RPM. Recently, it was found that a variety of observed spin polarizations could be explained successfully by the RTPM (Blattler, Jent, and Paul, 1990; Kandrashkin, Asano, and Van der Est, 2006).

The analytical formulae for the probability and rate of net and multiplet CIDEP generation and triplet radical quenching in the case of relatively strong quenching, was derived by Shushin (1993, 2002). The analysis considers fairly strong nonadiabatic transitions between states of the triplet radical spin Hamiltonian in two limits of fast and slow molecular rotation. It was shown that the net contribution is due to nonadiabatic transitions between the quartet and doublet terms of the triplet–doublet pair, and that the value of this contribution was independent of the details of the triplet–doublet relative motion in the level crossing region. Another approach for the calculation of RTPM CIDEP is based on the numerical solutions of the stochastic Liouville equation for a diffusively rotating system. In this model, the magnetic dipolar isotropic Heisenberg exchange and anisotropic Zeeman electron spin interactions were taken fully into account, whereas ISC processes

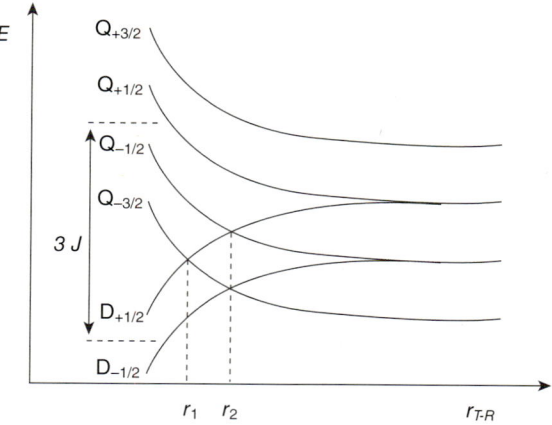

Figure 25.12 Spin sublevels of RTP in a magnetic field B_0 as a function of the radical–triplet relative distance, r_{TR}. The quartet–doublet energy separation $3J$ is assumed to change exponentially with r_{T-R}. The figure refers to the case $J < 0$, usually found for the interaction between neutral systems. At the distances r_1 and r_2 the $Q_{\pm 3/2}$ and $D_{\pm 1/2}$ spin sublevels do cross each other. Actually, level crossing is avoided by the electron spin dipolar interaction, which mixes Q and D states. In the level crossing regions an effective mixing takes place and spin population is transferred from the excited quartet state Q to the excited doublet state D, which decays rapidly to the corresponding ground state sublevels of the free radical. For large r_{TR} values, the $Q_{\pm 1/2}$ and $D_{\pm 1/2}$ sublevels become nearly degenerate. (Corvaja et al., 2000b).

between the singlet and triplet states were considered in terms of kinetic equations for their relevant spin density matrices (Tarasov et al., 2006a).

The electron spin polarization associated with electronic relaxation in molecules with triplet–quartet and triplet–doublet excited states were calculated by Kandrashkin and Est (2004). Such molecules typically relax to the lowest triplet–quartet state via ISC from the triplet–doublet, and it is known that, when spin–orbit coupling provides the main mechanism for this relaxation pathway, this leads to a spin polarization of the triplet quartet. Analytical expressions for this polarization were derived using first- and second-order perturbation theory in order to calculate powder spectra for typical sets of magnetic parameters (Adrian, 1986).

25.5.5
CIDEP Due to the SCRP Mechanism

The SCRP mechanism is a special CIDEP process that affects transitions attributed to radicals that are confined to microenvironments such as micelles (Buckley et al., 1987; Closs, Forbes, and Norris, 1987), vesicles, or as biradicals. Figure 25.13 shows the energy-level diagram for a SCRP at high magnetic fields. At the instant after RP generation, the populations of the three triplet states are equal to 1/3 and zero for the singlet state. After S–T_0 conversion induced by HFI, the population

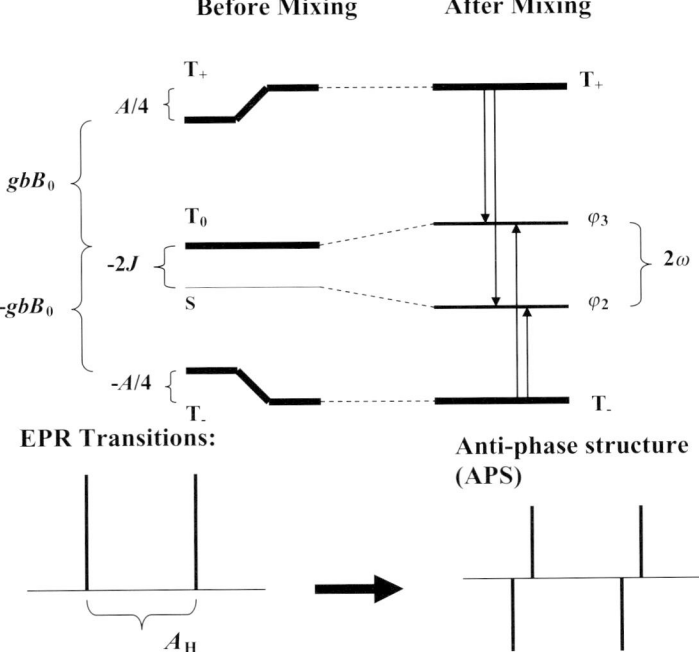

Figure 25.13 The energy level diagram of spin-correlated radical pair (SCRP)-demonstrated formation of the SCRP CIDEP mechanism.

of the T_0 and S states become equal to 1/6, leading to a large difference in spin populations between the T_-, T_+, and T_0 states. In the absence of the exchange interaction the frequencies of the EPR transitions $T_+\chi_N \rightarrow \alpha\beta\chi_N$; $T_- \rightarrow \beta\alpha\chi_N$, and $T_+ \rightarrow \beta\alpha\chi_N$; $T_- \rightarrow \alpha\beta\chi_N$ become equivalent to one another. Thus, the EPR lines of opposite phase overlap, leading to an absence of electron polarization. The exchange interaction shifts the resonance frequencies and splits normal monoradical EPR transitions into doublets of opposite phase (E/A or A/E); this is often referred to as the anti-phase structure (APS). Closs, Forbes, and Norris (1987) and Buckley et al. (1987) were the first to explain APS as the spectroscopic manifestation of the spin exchange interaction. In this very simple explanation, the exchange interaction was assumed to be constant. In a RP confined in micelles or flexible biradicals, the exchange is modulated by diffusion or intermolecular motion, which should be taken into account in CIDEP calculations (Tarasov and Forbes, 2000; Salikhov, 1997; Shushin, 1991b; Tarasov et al., 1996; Avdievich and Forbes, 1995; Maeda, Terazima, and Azumi, 1991). APS has been observed in a variety of different micelles (Tarasov et al., 2006b) and biradicals (Maeda, Terazima, and Azumi, 1991; Closs, Forbes, and Piotrowiak, 1992), and it stems from the fact that the radicals cannot diffuse far enough away from each other to be considered as independent doublet states. Radicals within a certain distance of a partner are linked by J, and are therefore not completely independent. In reference to the above-mentioned model, radical diffusion is assumed to be relatively fast with respect to spin exchange, and an averaged single J_{avg} can be obtained by measuring the distance between the APS spectral lines. The theoretical description of CIDEP due to the SCRP mechanism, as well as different qualitative explanations of this phenomenon, have been the subject of numerous publications (Shushin, 1995a, 1995b; Neufeld and Pedersen, 2000; Shushin, Pedersen, and Lolle, 1994).

Shushin (1995a, 1995b) obtained a simple formula for the APS of the CIDEP of SCRP diffusing in a small volume, by using the sudden perturbation approach. In the theoretical consideration by Neufeld and Pedersen (2000) and Shushin, Pedersen, and Lolle (1994), the use of a simple one re-encounter model calculation allowed it to be shown that the reference EPR spectra of RP could be written formally as a superposition of Lorentzian-like lines and the corresponding dispersive lines. The coefficients of these spectral components were shown to depend heavily on the frequency and amplitude of the mw field, and consequently a variety of spectral forms may result. The coefficients – and thus the spectral form – are determined by the interradical interaction induced rate of change of the longitudinal and transverse electron spin polarization. The longitudinal polarization is responsible for the CIDEP, which changes the intensity of the lines without affecting the lineshape, whereas the transverse polarization is responsible for the APS. The principal role of spin exchange relaxation on the lineshape of APS in micellized pairs was shown recently by Tarasov et al. (1996).

25.5.6
CIDEP Kinetics

25.5.6.1 Modified Bloch Equations
The result of TR-EPR experiments is a three-dimensional (3-D) data set containing time profiles at different field positions. The detailed analysis of these data, by using modified Bloch equations, allows information to be obtained concerning many of the processes that determine CIDEP kinetics, namely electron spin relaxation T_1 and T_2, reaction kinetics, the generation of electron spin polarization in diffusion RPs (Verma and Fessenden, 1976) due to RPM and RTMP (Blattler, Jent, and Paul, 1990), and processes of electron–nuclear cross-relaxation (Bagryanskaya et al., 1999).

25.5.7
Time-Resolved EPR Spectroscopy

For radicals in low-viscosity liquids, CIDEP typically relaxes to thermal equilibrium within a few microseconds. As the conventional continuous-wave (CW) EPR detection technique measures a time-integrated EPR intensity, it is rather insensitive towards CIDEP. Hence, in order to improve the time-resolution of CW EPR, field modulation frequencies of up to 2 MHz (Smaller, Remko, and Avery, 1968; Atkins, McLauchlan, and Simpson, 1970) were employed; however, the time-resolution proved to be insufficient and the penetration depth of the modulation field too small. Consequently, investigations and detailed analyses of the CIDEP phenomenon have received major attention only when fast time-resolved EPR spectrometers became available. Today, many research groups continue to study CIDEP effects by using a combination of fast time-resolved EPR detection and a strong laser for pulsed photolytic radical generation. In the so-called "direct detection" method known as TR-EPR, the signals of the transient radicals are detected using continuous microwave excitation, without magnetic field modulation. The equipment for the TR-EPR experiments comprises a CW EPR spectrometer with a fast time response, a transient signal recorder, field and frequency measuring equipment, a sample thermostat, and a computer-controlled data acquisition system. A time scheme for a TR-EPR experiment is shown in Figure 25.14; here, the rise and decay of the EPR signals are determined by the chemical reaction kinetics of the transients and the dynamics of their spin ensemble under the influence of the mw field and relaxation processes. Electron spin relaxation times of radicals in low-viscosity solutions are typically in the range between 0.1 and 10 µs; thus, a time resolution of at least 100 ns is required. The EPR signal of the radical species is detected by using an integrator or digital oscilloscope with a time-window, Δt, and time-delay τ after the laser pulse. The measured signal consists of 2-D arrays (time delay over magnetic field). The time resolution of a TR-EPR spectrometer is limited by two factors: (i) the rise and decay time of the preamplifier; and (ii) the ringing time of the cavity, given by $t_r = 2Q_L/\omega$, where Q_L is the cavity Q-factor and ω is the resonance frequency. The sensitivity of TR-EPR

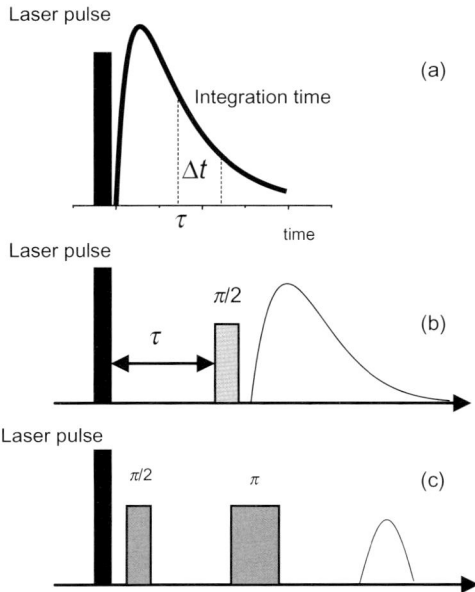

Figure 25.14 Time schemes of (a) TR-EPR, (b) FT-EPR, and (c) ESE-EPR experiments.

is substantially less than that of CW EPR, due to the absence of any magnetic field modulation and the use of broad-frequency detectors and amplifiers. The SNR is enhanced by averaging the time profiles at the same field position. During recent years, pulsed EPR methods such as Fourier-transform (FT) EPR and electron-spin echo (ESE) spectroscopy have been developed and can be used for CIDEP detection. However, the applicability of FT-EPR is limited to radical systems with an overall EPR linewidth less than the dead time of the spectrometer (Prakash *et al.*, 2006). Likewise, ESE-EPR spectroscopy requires inhomogeneously broadened EPR lines, and thus is only applicable for radicals in solid or frozen solution.

25.5.8
CIDEP Applications

Recently, TR-EPR has been developed for high-magnetic field applications (Forbes, 1992; Savitsky and Moebius, 2006), which leads to substantial increases in both time-resolution and sensitivity. The application of multifrequency CIDEP can help in understanding the polarization mechanisms in radical systems (Tarasov *et al.*, 2006; Fujisawa *et al.*, 1999; Fedin *et al.*, 2002b; Forbes, 1992), as well as confirming predictions from CIDEP theory (Shushin, 1995b; Goudsmit *et al.*, 1993c; Jent *et al.*, 1987; Makarov *et al.*, 2005). Another very important advantage of high-field TR-EPR is the ability to observe radicals in their thermal equilibrium state, as this opens up the possibility of directly determining the absolute value of the initial polarization by scaling the initial signal to the thermal equilibrium signal. Moreover, the time dependence of the EPR time profile provides direct kinetic information concerning the radical reaction.

Quantitative proof of TM CIDEP theory was provided by using multifrequency CIDEP following the photolysis of phosphine oxide photoinitiators in different magnetic fields (Makarov *et al.*, 2005). Radical CIDEP was measured using TR-EPR at different microwave frequencies and magnetic fields: S-band (2.8 GHz, 0.1 T), X-band (9.7 GHz, 0.34 T), Q-band (34.8 GHz, 1.2 T), and W-band (95 GHz, 3.4 T) (Figure 25.15). CIDEP was found to originate from a TM superimposed on a RP mechanism comprising both $S-T_0$ and $S-T_-$ mixing. The contributions of the different CIDEP mechanisms were separated, and the dependence of the TM polarization on mw frequency was determined; this agreed well with the numerical solution of the relevant stochastic Liouville equations (Shushin, 1995a). In a number of TR-EPR spectroscopic investigations of radicals created by laser flash photolysis, the spectra exhibited an E/A polarization pattern at early times after the laser flash, as would be expected for radicals created from triplet precursors, but produces an A/E pattern typically 15–40 μs later (McLauchlan, 1990; Carmichael and Paul, 1979; Basu, Grant, and McLauchlan, 1983; McLauchlan and Stevens, 1985b, 1986; Jent *et al.*, 1987; Valyaev *et al.*, 1988; Muus, 1989; Borbat, Milov, and Molin, 1989; Jent and Paul, 1989; Goudsmit, Jent, and Paul, 1993). In fact, this behavior has been reported for a number of radicals. Although it transpired that, in many cases, the phase inversions originated from instrumental

25.5 Chemically Induced Dynamic Electron Polarization

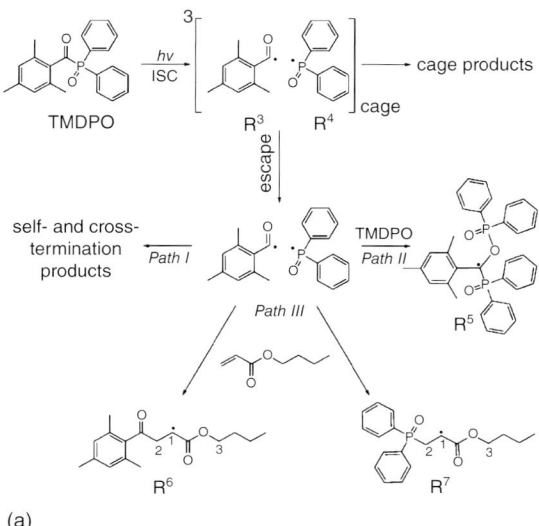

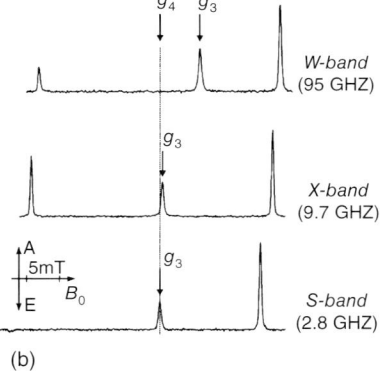

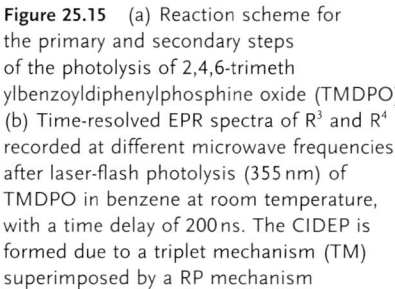

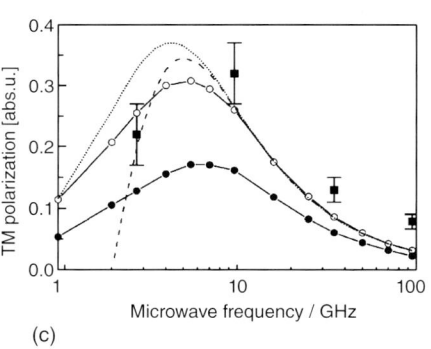

Figure 25.15 (a) Reaction scheme for the primary and secondary steps of the photolysis of 2,4,6-trimethylbenzoyldiphenylphosphine oxide (TMDPO); (b) Time-resolved EPR spectra of R^3 and R^4 recorded at different microwave frequencies after laser-flash photolysis (355 nm) of TMDPO in benzene at room temperature, with a time delay of 200 ns. The CIDEP is formed due to a triplet mechanism (TM) superimposed by a RP mechanism comprising S–T_0, as well as S–T_- mixing; (c) The experimental TM polarization of phosphinoyl radical R^4 after photolysis of TMDPO in comparison with the theoretical calculations invoking the numerical solution of the stochastic Liouville equation (SLE). The dashed and dotted lines show the polarization calculated by analytical expressions of Atkins–Evans and Pedersen–Freed. Figure adapted from Savitsky and Moebius (2006).

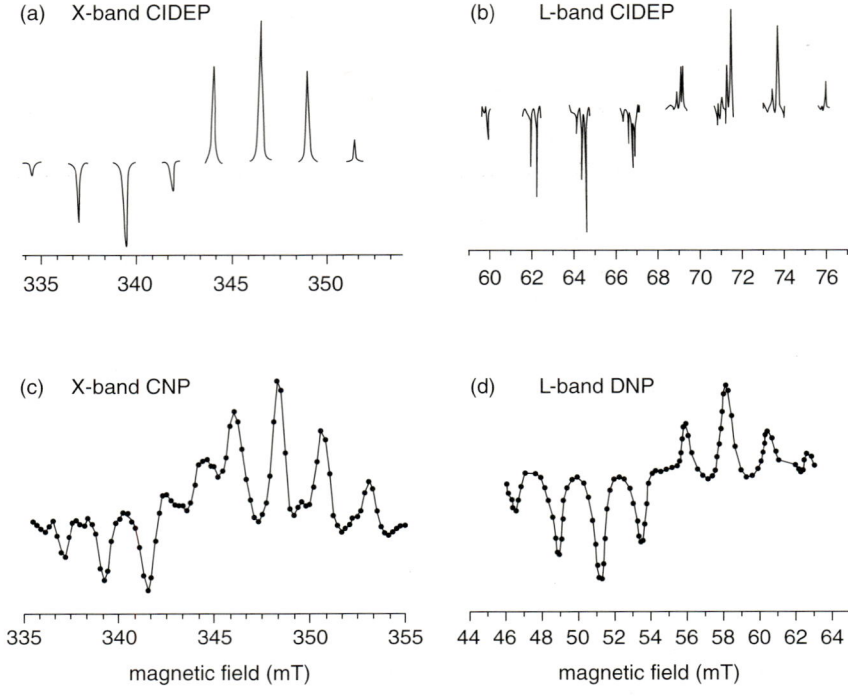

Figure 25.16 TR-ESR and DNP spectra detected after photolysis of $((CH_3)_3C)_2CO$ in benzene. (a) X-band TRESR spectrum integrated from 0.5 to 1.5 µs; (b) L-band TR-ESR spectrum of $CH_3)_3C\bullet$ radical integrated from 1.5 to 2 µs; (c) X-band and (d) L-band DNP spectra detected on the proton resonance of $(CH_3)_3CC(CH_3)_3$ for $B_1 = 0.2$ mT. The same ratio of signal intensity of CIDEP and DNP lines proves that an efficient flip-flop electron–nuclear cross-relaxation has occurred (Bagryanskaya et al., 1999).

causes, the phenomenon has been found to be genuine for $(CH_3)_2(OH)C\bullet$, $(CH_3)_2$–HC•, and $(CH_3)_3C\bullet$ radicals. After a long discourse, the CIDEP phase change observed for these radicals has been attributed to an efficient electron–nuclear flip-flop, cross-relaxation process. This problem was finally resolved by the application of multifrequency CIDEP and dynamic nuclear polarization (DNP). Subsequently, CIDEP and DNP of the radical products in solution were measured at the L- and X-bands to investigate their electron–nuclear cross-relaxation. It was shown experimentally, and confirmed theoretically, that a strong multiplet-type electron spin polarization leads to a multiplet-type DNP spectrum of the products with the same relative line intensity as observed in the CIDEP spectrum. The measurements of CIDEP and DNP spectra for the radicals $(CH_3)_2XC\bullet$ with X) CH_3, OH, D, and $C(O)CH_3$ showed the same signal intensity of CIDEP and DNP lines as shown in Figure 25.16. The observation of DNP spectra proves that an efficient flip-flop electron-nuclear cross-relaxation has occurred for t-butyl and 2-hydroxy-

2-propyl radicals. These results indicated that phase inversions, observed in time-resolved CIDEP spectra of *t*-butyl and 2-hydroxy-2-propyl radicals, were attributable to an efficient flip-flop electron-nuclear cross-relaxation.

The theory of CIDEP due to the RTPM was tested experimentally by the photolysis of benzophenone in the presence of TEMPO between 193 and 298 K. With decreasing temperature, the relative diffusion coefficient of the particles decreased from 5.9×10^{-5} to $0.9 \times 10^{-5}\,\mathrm{cm^2\,s^{-1}}$, and the absolute values of net and multiplet CIDEP increased from 0.6 to 9.0 and from 0.07 to 0.26, respectively, in units dictated by the Boltzmann population distribution. The CIDEP was attributed to the mixing and splitting of doublet and quartet spin states in radical–triplet pairs. It was shown that CIDEP is generated predominantly in regions where the exchange interaction is smaller than the Zeeman energy, and that the multiplet polarization is diminished effectively by a fast T_2 relaxation of the triplet spin (Goudsmit *et al.*, 1993b). The magnetic field dependence of CIDEP due to RTPM was measured experimentally and analyzed theoretically by Stavitski, Wagnert, and Levanon (2005). In this case, the TR-EPR measurements were performed using a mw set-up that consisted of low-loss dielectric ring resonators that were tunable at mw frequencies corresponding to the specific magnetic field. CIDEP due to the RTPM of the radical was observed in the 170–370 mT range, while the results of calculations based on the numerical solution of the stochastic Liouville equation were found to be in line with the experimental data; these showed that the CIDEP was decreased when the magnetic field was increased (Stavitski, Wagnert, and Levanon, 2005).

CIDEP in short-lived flexible and rigid biradicals at different temperatures and viscosity was studied intensively, both experimentally and theoretically (Burns, Rochelle, and Forbes, 2001; Mizuochi, Ohba, and Yamauchi, 1999; Maeda *et al.*, 1992; Forbes, 1993; Forbes *et al.*, 1994). CIDEP in long-lived SCRPs in micelles was studied at both high (Ohara *et al.*, 2003) and low magnetic fields (Bagryanskaya, Fedin, and Forbes, 2005). The analysis of multifrequency CIDEP provided detailed information regarding the value of the exchange interaction and the mobility of radical centers for SCRPs in micelles (Ohara *et al.*, 1997). TM CIDEP was observed during the photolysis of C_{60} fullerenes in solution and in liquid crystals (Goudsmit and Paul, 1993). CIDEP due to RPTM was experimentally observed for both single (Mizuochi, Ohba, and Yamauchi, 1999) and double (Corvaja, Franco, and Mazzoni, 2001) spin-labeled C_{60} and C_{70} derivatives in solution (Conti *et al.*, 2009). Spectral simulation allowed determination of the magnitude of the exchange coupling constant between the triplet C_{60} molecule and nitroxide radicals, and the sign of the exchange interaction was inferred from the sign of the spin polarization (Mazzoni *et al.*, 2006).

TR-EPR and FT-EPR represent two of the major techniques used for studying triplet porphyrins, and their role as donors in intramolecular electron transfer as well as their intermolecular interactions with free radicals. Two applications of TR-EPR to processes in which the photoexcited singlet and/or triplet are involved are: (i) intramolecular ET in photoexcited donor–acceptor systems embedded in liquid crystals, where the porphyrins and the electron donors are attached to different types of acceptor; and (ii) intermolecular magnetic interactions between photoexcited porphyrin triplets and free radicals (Blank, Galili, and Levanon, 2001). An FT-EPR study of energy and electron transfer from the triplet excited state of porphyrins to C_{60} in toluene and benzonitrile was reported by Martino and Willigen (2000). Based on the spin polarization and time profile of the signal from the anion radical, it was deduced that the primary route of ET was oxidative quenching of the porphyrin triplets (Martino and Willigen, 2000).

TR-EPR may also serve as a good method for measuring the electron relaxation time of short-lived radicals (Tsentalovich and Forbes, 2002; Makarov, Bagryanskaya, and Paul, 2004), when the relaxation determines the decay rate of the polarization. In various studies of spin dynamics in RPs, benzoyl-type radicals have been one species in the paramagnetic pair, their electron spin relaxation having been assumed to be slow enough to be neglected in the data analysis. This assumption was checked by measuring the electron spin relaxation in a sequence of three acyl radicals (benzoyl, 2,4,6-trimethylbenzoyl, and hexahydrobenzoyl) by using TR-EPR. In contrast to the slow relaxation assumption, rather short spin-lattice relaxation times (100–400 ns) were found for benzoyl and 2,4,6-trimethylbenzoyl radicals from the decay of the initial electron polarization to thermal equilibrium at different temperatures and viscosities. The relaxation is induced by spin–rotation coupling arising from two different radical motion types: overall rotation of the whole radical; and hindered internal rotation of the CO group. The results obtained were explained with the framework of Bull's theory, using a modified rotational correlation time (Burns, Rochelle, and Forbes, 2001).

TR-EPR spectra and SCRP polarization decay kinetics in an acyl-benzyl biradical were measured over a wide temperature range (180–274 K) (Stavitski, Wagnert, and Levanon, 2005). The major mechanism – that is, ISC in the biradical – was explained by the spin rotation-induced relaxation of the acyl moiety, which is associated with a rotation of the carbonyl group about the neighboring C–C bond axis. This relaxation time was viscosity-independent, and changed by a factor of less than 2 when going from room temperature (60 ns) to 180 K (110 ns) in 2-propanol.

25.5.9
Applications of CIDEP to Biological Systems

The most important applications include studies of:

- The photochemical RCs of plants and bacteria
- Energy, charge, and spin transport in molecules and self-assembled nanostructures involved in photosynthesis (Wasielewski, 2006)
- The reaction mechanism and structure of peptides
- Protein dynamics, protein–surface interactions, and amino acid photooxidation.

Some examples of these applications are reviewed in the following subsections.

25.5.10
Applications of CIDEP to Study Photochemical Reaction Centers

CIDEP an important technique for studying the photochemical RCs of plants and bacteria, allowing details to be obtained regarding the RCs' structure and function. Over recent decades, various spectroscopic studies have focused on the light-induced transient RPs formed in RCs. It was shown that, after pulsed laser excitation of the primary donor in plant PS I and PS II, the SCRP appears and is characterized by a weak electron spin–spin coupling between a fixed geometry of the radicals in the pair and an initial singlet state of the system. In many cases, it was shown that the CIDEP analysis, quantum beats, transient nutations, as well as echo-envelope modulation and out-of-phase echo effects, provided the geometry of the system and magnetic interaction parameters. The CIDEP of SCRPs allows structural information to be extracted concerning the transient charge-separated states in photosynthetic RCs for which detailed X-ray data are sparingly available. A detailed consideration of multifrequency TR-EPR, FT-EPR, and ESE-EPR applications to the study of photochemical RCs, as well as to the study of ET in a system of photosynthetic centers, is provided in Chapter 22 and Savitsky and Moebius (2009).

25.5.11
RTPM CIDEP in Spin-Labeled Peptides

The first example of *intra*molecular radical–triplet interaction in peptides was shown in the cyclic dipeptide **1** (Figure 25.17), obtained by condensing the 2,2,6,6-tetramethylpiperidine-1-oxyl-4-amino-4-carboxylic acid (TOAC) radical amino acid with triplet precursor amino acids (Corvaja *et al.*, 2000a, b). The TR-EPR spectrum of this compound in 10 mM toluene, when photoexcited by a 308 nm light pulse from an excimer laser, consists of three strong emission lines at field positions corresponding to the ^{14}N hyperfine components of the TOAC nitroxide radical. In the case of the linear dipeptide **2**, which has an unfolded structure, where the radical center and the triplet precursor are separated by the same number of bonds as the cyclic dipeptide **1**, no signal was recorded; this indicates that radical–triplet interaction takes place through space, and that the through bond interaction is not operative.

The CIDEP effects of intramolecular quenching of singlet and triplet excited states by nitroxide radicals in oligopeptides represent a potentially useful new method for investigating peptide secondary structures in solution, as developed by Sartori *et al.* (2003). CIDEP was observed in the photolysis of two hexapeptides, each bearing one photoactive alpha-amino acid (Bin or Bpa) and one nitroxide-containing TOAC residue. Two amino acid units separate the photoactive residue from TOAC in the peptide sequence; the two moieties face each other at a distance of about 6 Å after one complete turn of the ternary helix. Irradiation from an excimer laser light pulse populates the excited states localized on the chromophores. An intramolecular interaction between the singlet (Bin) or triplet (Bin and Bpa) excited states and the doublet state of the TOAC nitroxide makes a

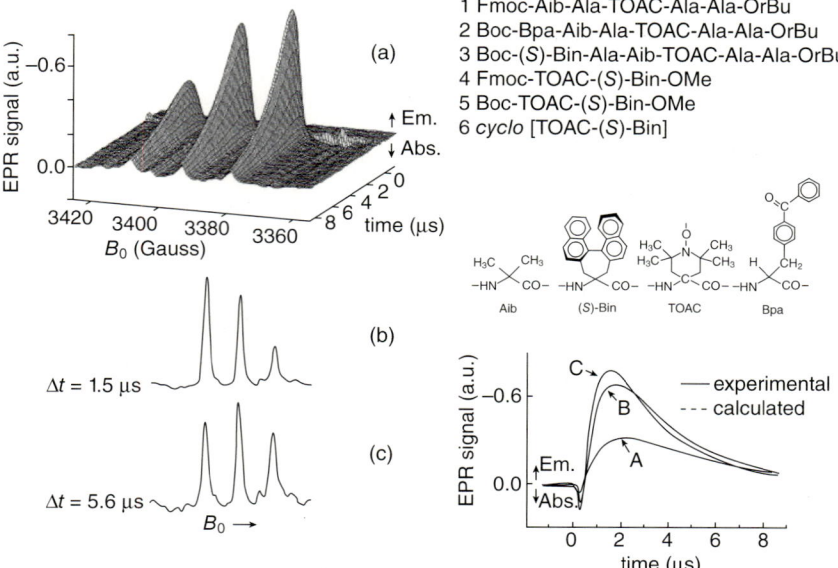

Figure 25.17 Chemical formulae of amino acids. (a) 2-D TR-ESR spectrum of various peptides 3 in CH$_3$CN solution at 260 K; (b, c) Sections of the 2-D spectrum at time delay 1.5 μs and 5.6 μs after the laser pulse. Section of the surface parallel to the time axis at (a) and their fittings: A = high-field line, B = central line, C = low-field line. The spectrum shows a strong three-line spin polarization signal in which the intensity ratios among the lines are due to a contribution of hyperfine interaction to the polarization mechanism (Corvaja et al., 2000a).

spin-selective decay pathway possible, producing CIDEP. In order to determine whether the intramolecular exchange interaction occurs through-bond or through-space, a CIDEP study of linear and cyclic TOAC-Bin dipeptide units was performed, the results of which revealed that a through-space intramolecular interaction is operative. The observation of spin polarization makes the two helical hexapeptides suitable models for applying this novel technique to conformational studies of peptides in solution (Corvaja et al., 2000a).

25.5.12
Applications of CIDEP to Studies of Biological Function: Protein Dynamics and Protein–Surface Interactions

TR-EPR spectroscopy may be used to explore the local surface characteristics of a protein, and to study dynamic biological functions mediated by reactions with specific ligands and their subsequent release. The possibility of using CIDEP to explore protein–surface reactions to investigate drug delivery and protein–substrate catalysis was shown recently by Kobori and Norris (2006). In this case, TR-EPR

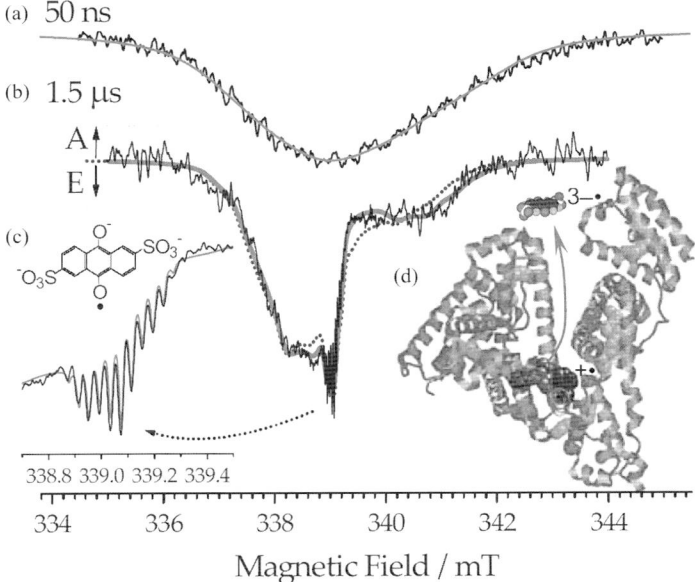

Figure 25.18 Time-resolved EPR spectra obtained for the HAS–AQDS$_2$-system at (a) 50 ns and (b) 1.5 μs after the 355 nm laser irradiation; (c) Expansion of the 339.1 mT region of (b); (d) Ribbon structure of the HAS–warfarin complex. The structure was obtained from the Protein Data Bank (ID code 1ha2). The ligand and W214 are shown by the space-filling structures, respectively. The CIDEP spectra were fitted, as shown by lines in (a–c), by considering a 1-D diffusion model of radicals, whereas the conventional 3-D diffusion model deviates from the experimental result, as denoted by the dotted line in (b) (Kobori and Norris, 2006).

spectroscopy was used to observe a photoinduced, proton-coupled ET between 9,10-anthraquinone-2,6-disulfonate (AQDS^{2-}) and a tryptophan residue in human serum albumin (HSA), a major protein in blood plasma that plays a significant role in transporting hydrophobic ligands (fatty acids, hormones, and drugs) (He and Carter, 1992). The CIDEP spectra showed that photoinduced ET takes place specifically from the tryptophan residue (W214) to the excited triplet state of AQDS^{2-} bound in a cleft in HAS. Polarized EPR signals clearly demonstrated that the anion radical of the ligand could escape out of the pocket towards the bulk water, undergoing a one-dimensional (1-D) motion towards the protein outer surface (Figure 25.18).

25.5.13
CIDEP Study of Amino Acid Photooxidation

The damage of amino acids, peptides, and proteins caused by oxidative processes – and especially modifications of biological molecules induced by various reactive oxidative species – are of intense interest. During recent years, CIDEP has been applied successfully to study the photooxidation mechanisms of amino acids, and the structure and kinetics of the related intermediate species.

In a series of reports from Beckert, FT-EPR was used to investigate the oxidation mechanism of a series of nucleosides and amino acid derivatives, including cytosine, 1-methylcytosine (Geimer et al., 2000), alpha-glycine, l-alpha-alanine, alpha-aminoisobutyric acid (Lu et al., 2001), glycine, alpha-alanine, alpha-aminoisobutyric acid, beta-alanine (Tarabek, Bonifacic, and Beckert, 2006), cyclic dipeptides glycine (Tarabek et al., 2004a), alanine and sarcosine anhydrides (Tarabek, Bonifacic, and Beckert, 2004b), and glycine methyl and ethyl esters (Tarabek, Bonifacic, and Beckert, 2007), caused by excited triplets of anthraquinone-2,6-disulfonate dianion over a wide pH range. Anthraquinone radical trianions with strong emissive CIDEP were formed, indicating that a fast ET from the quenchers to the spin-polarized quinone triplet was the primary step (Figure 25.19). None of the primary radicals formed upon one-electron oxidation of the quenchers could be detected on the FT-EPR nanosecond time scale, due to their very rapid transformation into secondary products. The secondary products were attributed to decarboxylated alpha-aminoalkyl radicals for the alpha-amino acids anions and zwitterions, to beta-aminoalkyl radicals for the beta-alanine zwitterions, and to methyl radicals for the acetate anions. The corresponding aminyl radicals were the first EPR-detectable products from beta-alanine anions and methylamine. The absolute triplet quenching rate constants, as well as the rate constants of the decarboxylation reaction, were determined from CIDEP kinetic analysis. A typical CIDEP spectrum is shown in Figure 25.19.

Subsequently, Forbes and coworkers investigated the direct photoionization mechanism for a series of amino acids at basic pH using TR-EPR at X- and Q-band. The photoionization of deprotonated tyrosine leads to CIDEP, due to RPM with the tyrosyl radical as emission and the solvated electron as absorption, implying a triplet precursor and negative exchange interaction J (Clancy and Forbes, 1999). Peroxyl radicals formed from the addition of oxygen to carbon radicals of N-acetyl glycine, serine, and diglycine (White and Forbes, 2006) were observed directly at room temperature (Figure 25.20), while isotopic labeling confirmed the identity of the N-acetyl glycine and diglycine peroxyl analogues. The peroxyl radicals showed an unusually strong chemically induced electron spin polarization, which was discussed in terms of the RP mechanism and spin polarization transfer processes.

25.6
Conclusion

In this short overview, it has been shown that both CIDNP and CIDEP can provide detailed information regarding the structure and dynamics of transient radicals, as well as of RPs, occurring in photochemical reactions in solution. CIDNP and CIDEP contain important information about the structure and dynamics of

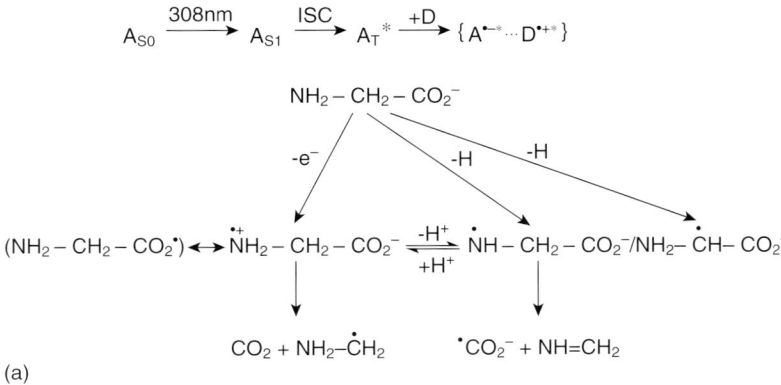

(a)

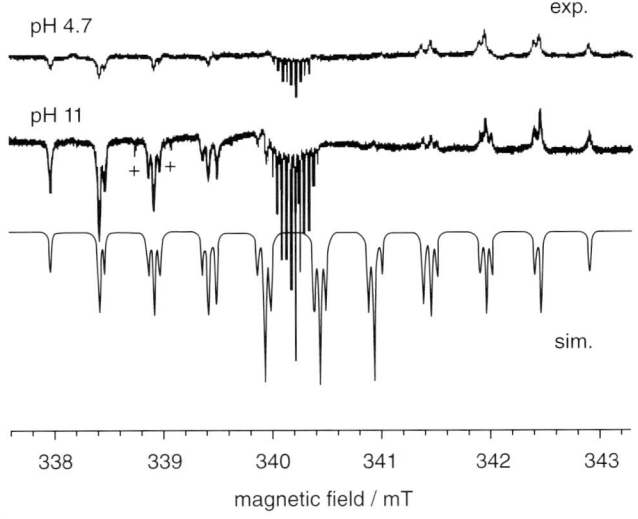

(b)

Figure 25.19 (a) The first step is excitation of the photosensitizer (A, anthraquinone-2,6-disulfonate in this study) by the laser pulse with photons of 308 nm. ISC denotes the intersystem crossing to the spin-polarized triplet state AT*, and the last step is the electron transfer from the donor D to the acceptor triplet A_T^* (spin-polarized states are marked by *); (b) The main reactions and radical products of amino acid anions (for glycine) are shown, with experimental and simulated FT-EPR spectra of the R-aminomethyl radical $NH_2-•CH_2$. Samples: 0.3 mM 2,6-AQDS and 1 M glycine (pH 4.7) or 0.1 M glycine (pH 11) in aqueous solution; delay 96 ns. Lines denoted by + belong to the radical $NH_2-•CHCO_2^-$ with a yield of ca. 5%. The spectra of 2,6-AQDS•- (central line groups) are multiplied by 0.25 (Tarabek et al., 2007).

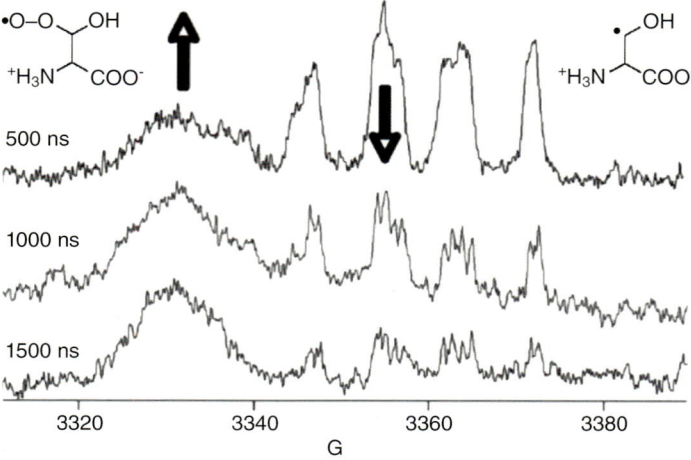

Figure 25.20 X-band TR-EPR spectra acquired at the delay times indicated after 248 nm photolysis of an aqueous pH 7 solution of the amino acid serine with H_2O_2 (upper scheme). The broad low-field signal is assigned to the peroxyl radical shown Clancy and Forbes (1999).

transient reaction intermediates which can also be exploited for signal enhancement. Applications of CIDEP and CIDNP include the study of photoinduced electron and hydrogen transfer, fragmentation reactions, *cis–trans* isomerization and cycloaddition, reactions proceeding through formation of short-lived biradicals, and inorganic and metal organic substrates (Goez, 2009). Some of the most exciting results obtained with CIDNP and CIDEP in biological systems have included the photoreduction of animo acids, folding processes in proteins, and photosynthetic RCs.

Pertinent Literature

The material covering CIDNP and CIDEP and their applications to biological systems are based on the following references:

- L.T. Muus, P.W. Atkins, K.A. McLauchlan, and J.B. Pedersen (1977) *Chemically Induced Magnetic Polarization*, NATO Advanced Study Institutes Series, Dordrecht-Holland, Boston, USA, vol. 34. This book was written 10 years after the RP theory was proposed, summarized the basic principles of CIDEP and CIDEP, and provided an overview of the first results based on these phenomena, obtained during the period 1967–1977.

- K.M. Salikhov, Yu.N. Molin, R.Z. Sagdeev, and A.L. Buchachenko (1984), *Spin Polarization and Magnetic Effects in Radical Reactions* (ed. Yu.N. Molin), Elsevier, Amsterdam. The theory of radical recombination, theory magnetic effects in radical reactions, theory of CIDEP and CIDNP and detailed review of experimental results obtained using these techniques to chemical reaction covering time period from the discovery of CIDNP 1967 until 1984.

- A.J. Hoff (ed.) (1989) *Advanced EPR. Applications in Biology and Biochemistry*. Elsevier, Amsterdam, Oxford, New York, Tokyo. Chapter 10, "Time-Resolved EPR," by K.A. McLauchlan reviews CIDEP theory, technique and its application to chemical reactions.

- S. Nagakura *et al.* (eds) (1998) Dynamic Spin Chemistry. Magnetic Control and Spin Dynamics of Chemical Reactions, in *Chemically Induced Dynamic Electron Polarization (CIDEP) Studies in Photochemical Reactions*, John Wiley & Sons. Chapter 7. This chapter, by N. Hirota, overviews the basic CIDEP mechanism, its theory, and application to study photochemical reactions.

- A. Savitsky and K. Mobius (2006) Review: Photochemical reactions and photoinduced electron-transfer processes in liquids, frozen solutions, and proteins as studied by multifrequency time-resolved EPR spectroscopy, *Helvetica Chimica Acta* **89**, 2544; and A. Savitsky and K. Mobius (2009) *High-Field EPR Spectroscopy on Proteins and their Model Systems*, Royal Society of Chemistry, Cambridge, UK. In this review and book, modern multifrequency EPR spectroscopy, in particular at high magnetic fields, is shown to provide detailed information about the structure, motional dynamics, and CIDEP of transient radicals occurring in photochemical reactions. They cover the theoretical background as well as state-of-art research in term of instrumentation and application to biological systems.

- Reviews of E.G. Bagryanskaya and R.Z. Sagdeev, "Stimulated and dynamic nuclear polarization in photochemical radical reactions," *Russian Chemical Reviews* 2000, **69**, 925; and "Kinetic and mechanistic aspects of stimulated nuclear polarization," *Progress in Reaction Kinetics* 1993, **18**, 63. In these reviews, the applications of highly sensitive TR magnetic resonance techniques, based on the influence of radiofrequency and switching magnetic fields on the nuclear polarization diamagnetic radical products to the investigation of short-lived radical intermediates (radical pairs, biradicals, free radicals and radical-ions) formed in photochemical reactions in homogeneous and molecular-organized media, are considered.

- Reviews of M. Goez, "Photo-CIDNP Spectroscopy," *Annual Reports on NMR Spectroscopy* 2009, **66**, 77; and *Advances in Photochemistry* 1997, **23**, 63. These reviews provide an introduction to the theory, explain the instrumentation and techniques with emphasis on new developments, and overview the recent applications of photo-CIDNP spectroscopy to chemical and biochemical problems. The latter review covers the literature from the discovery of CIDNP until mid 1996, and the former from mid 1996 to mid 2008.

- Review of U.E. Steiner and T. Ulrich, *Chemical Reviews* 1989, **89**, 51, which covered the background and application of CIDNP up until 1988.

References

Adrian, F.J. (1971) *J. Chem. Phys.*, **54**, 3918.
Adrian, F.J. (1974) *J. Phys. Chem.*, **61**, 4875.
Adrian, F.J. (1986) *Rev. Chem. Interm*, **7**, 173.
Adrian, F.J. and Monchik, L. (1979) *J. Chem. Phys.*, **71**, 2600.
Akasaka, K., Fujii, S., and Kaptein, R. (1981) *J. Biochem.*, **89**, 1945.
Ananchenko, G.S., Purtov, P.A., and Bagryanskaya, E.G. (1997) *J. Phys. Chem. A*, **101**, 3848.
Atkins, P.W. and Evans, G.T. (1974a) *Chem. Phys. Lett.*, **25**, 108.
Atkins, P.W. and Evans, G.T. (1974b) *Mol. Phys.*, **27**, 1633.
Atkins, P.W., McLauchlan, K.A., and Simpson, A.F. (1970) *J. Phys. E*, **3**, 547.
Avdievich, N.I. and Forbes, M.D.E. (1995) *J. Phys. Chem.*, **99**, 9660.
Bagryanskaya, E.G. and Sagdeev, R.Z. (1993) *Progr. React. Kinet.*, **18**, 63.
Bagryanskaya, E.G. and Sagdeev, R.Z. (2000) *Russ. Chem. Rev.*, **69**, 925.
Bagryanskaya, E.G., Grishin, Y.A., Sagdeev, R.Z., Leshina, T.V., Polyakov, N.E., and Molin, Y.N. (1985) *Chem. Phys. Lett.*, **117**, 220.
Bagryanskaya, E.G., Tarasov, V.F., Avdievich, N.I., and Schrob, I.A. (1992) *Chem. Phys.*, **162**, 213.
Bagryanskaya, E.G., Gorelik, V.R., and Sagdeev, R.Z. (1997) *Chem. Phys. Lett.*, **264**, 655.
Bagryanskaya, E.G., Ananchenko, G.S., Nagashima, T., Maeda, K., Milikisyants, S., and Paul, H. (1999) *J. Phys. Chem.*, **103**, 11271.
Bagryanskaya, E., Fedin, M., and Forbes, M.D.E. (2005) *J. Phys. Chem.*, **109**, 5064.
Balbach, J., Forge, V., Lau, W.S., Jones, J.A., van Nuland, N.A.J., and Dobson, C.M. (1997) *Proc. Natl Acad. Sci. USA*, **94**, 7182.
Bargon, J. (2006) *Photochem. Photobiol. Sci.*, **5**, 970.
Bargon, J., Fischer, H., and Johansen, U. (1967) *Z. Naturforsch.*, **22a**, 1551.
Basu, S., Grant, A.I., and McLauchlan, K.A. (1983) *Chem. Phys. Lett.*, **94**, 517.
Blank, A., Galili, T., and Levanon, H. (2001) *J. Porphy. Phthalocya.*, **5**, 58.
Blattler, C., Jent, F., and Paul, H. (1990) *Chem. Phys. Lett.*, **166**, 375.

Borbat, P.P., Milov, A.D., and Molin, Y.N. (1989) *Chem. Phys. Lett.*, **164**. 330.
Broadhunt, R.W., Dobson, C.M., Hore, P.J., Radford, S.E., and Reed, M.L. (1991) *Biochemistry*, **30**, 405.
Buchachenko, A.L., Galimov, E.M., Yersgov, V.V., Nikiforov, G.A., and Pershin, A.D. (1976) *Dokl. A. N. S.S.S.R.*, **228**, 379.
Buckley, C.D., Hunter, D.A., Hore, P.J., and McLauchlan, K.A. (1987) *Chem. Phys. Lett.*, **135**, 307.
Burns, C.S., Rochelle, L., and Forbes, M.D.E. (2001) *Org. Lett.*, **3**, 2197.
Canioni, P., Cozzone, P.J., and Kaptein, R. (1960) *FEBS Lett.*, **111**, 219.
Carmichael, J. and Paul, H. (1979) *Chem. Phys. Lett.*, **67**, 519.
Clancy, C.M.R. and Forbes, M.D.E. (1999) *Photochem. Photobiol.*, **69**, 16.
Closs, G.L. (1969) *J. Am. Chem. Soc.*, **91**, 4552.
Closs, G.L. and Miller, R.J. (1979) *J. Am. Chem. Soc.*, **101**, 1639.
Closs, G.L. and Trifunac, A.D. (1970) *J. Am. Chem. Soc.*, **92**, 2183.
Closs, G.L., Miller, R.J., and Redwine, O.D. (1985) *Acc. Chem. Res.*, **18**, 196.
Closs, G.L., Forbes, M.D.E., and Norris, J.R. (1987) *J. Phys. Chem.*, **91**, 3592.
Closs, G.L., Forbes, M.D.E., and Piotrowiak, P. (1992) *J. Am. Chem. Soc.*, **114**, 3285.
Closs, G.L., Gautam, P., Zhang, D., Krusic, P.J., Hill, S., and Wasserman, E. (1993) *J. Phys. Chem.*, **96**, 3671.
Conti, F., Corvaja, C., Busolo, F., Zordan, G., Maggini, M., and Weber, S. (2009) *Phys. Chem.*, **11**, 495.
Corvaja, C., Sartori, E., Toffoletti, A., Formaggio, F., Crisma, M., Toniolo, C., Mazaleyrat, J.P., and Wakselman, M. (2000a) *Chem. Eur. J*, **6**, 2775.
Corvaja, C., Sartori, E., Toffoletti, A., Formaggio, F., Crisma, M., and Toniolo, C. (2000b) *Biopolymers*, **55**, 486.
Corvaja, C., Franco, L., and Mazzoni, M. (2001) *Appl. Magn. Res.*, **20**, 71.
Daviso, E., Diller, A., Alia, A., Matysik, J., and Jeschke, G.J. (2008) *J. Magn. Reson.*, **190**, 43.
De Kanter, F.J.J. and Kaptein, R. (1979) *Chem. Phys. Lett.*, **62**, 421.

Diller, A., Roy, E., Gast, P., van Gorkom, H.J., de Groot, H.J.M., Glaubitz, C., Jeschke, G., Matysic, J., and Alia, A. (2007) *Proc. Natl Acad. Sci. USA*, **104**, 12967.

Dobson, C.M. and Hore, P.J. (1998) *Nat. Struct. Biol.*, **5**, 504.

Eisenreich, W., Joshi, M., Weber, S., Bacher, A., and Fischer, M. (2008) *J. Am. Chem. Soc.*, **130**, 13544.

Eisenreich, W., Fischer, M., Roemisch-Margl, W., Joshi, M., Richter, G., Bacher, A., and Weber, S. (2009) *Biochem. Soc. Trans.*, **37**, 382.

Fedin, M.V., Bagryanskaya, E.G., and Purtov, P.A. (1999) *J. Phys. Chem.*, **111**, 5491.

Fedin, M.V., Bagryanskaya, E.G., Purtov, P.A., Makarov, T.N., and Paul, H. (2002a) *J. Phys. Chem.*, **117**, 6148.

Fedin, M.V., Yashiro, H., Purtov, P.A., Forbes, M.D.E., and Bagryanskaya, E.G. (2002b) *Mol. Phys.*, **100**, 1171.

Fessenden, R.W. and Schuler, R.H. (1963) *J. Chem. Phys.*, **39**, 2147.

Forbes, M.D.E. (1992) *J. Phys. Chem.*, **96**, 7836.

Forbes, M.D.E. (1993) *Z. Phys. Chem. Intern. J. Res. Phys. Chem. Chem. Phys.*, **182**, 63.

Forbes, M.D.E. and Ruberu, S.R. (1993) *J. Phys. Chem.*, **97**, 13223.

Forbes, M.D.E., Barborak, J.C., Dukes, K.E., and Ruderu, S.R. (1994) *Macromolecules*, **27**, 1020.

Frankevich, E.L. and Kubarev, S.I. (1982) *Triplet State Optical Detection Magnetic Resonance Spectroscopy* (ed. R.H. Clarke), John Wiley & Sons, Inc., New York.

Freed, J.H. (2005) *Annu. Rev. Phys. Chem.*, **1**, 655.

Fujisawa, J., Ohba, Y., and Yamauchi, S. (1997) *J. Am. Chem. Soc.*, **119**, 8736.

Fujisawa, J., Ishii, K., Ohba, Y., Yamauchi, S., Fuhs, M., and Mobius, K. (1999) *J. Phys. Chem. A*, **103**, 213.

Garssen, G.J., Kaptein, R., Schoenmakers, J.G.G., and Hilbers, C.W. (1978) *Proc. Natl Acad. Sci. USA*, **75**, 5281.

Geimer, J., Hildenbrand, K., S. Naumov, and Beckert D. (2000) *Phys. Chem. Chem. Phys.*, **2**, 4199.

Gizeman, O.L., Kaufman, F., and Porter, G. (1973) *J. Chem. Faraday Trans. II*, **69**, 727.

Goez, M. (1990) *Chem. Phys. Lett.*, **165**, 11.

Goez, M. (2009) *Annu. Rep. NMR Spectrosc.*, **66**, 77.

Goez, M. and Heun, R. (2001) *J. Phys. Chem. A*, **105**, 10446.

Goez, M. and Heun, R. (2002) *Phys. Chem. Chem. Phys.*, **4**, 5531.

Goez, M., Mok, K.H., and Hore, P.J. (2005) *J. Magn. Res.*, **177**, 236.

Goez, M., Kuprov, I., Mok, K.H., and Hore, P.J. (2006) *Mol. Phys.*, **104**, 1675.

Goudsmit, G.H. and Paul, H. (1993) *Chem. Phys. Lett.*, **208**, 73.

Goudsmit, G.H., Jent, F., and Paul, H. (1993a) *Z. Phys. Chem.*, **180**, 51.

Goudsmit, G.H., Paul, H., and Shushin, A.I. (1993b) *J. Phys. Chem.*, **97**, 50.

Goudsmit, G.H., Paul, H., and Shushin, A.I. (1993c) *J. Phys. Chem.*, **97**, 13243.

Grosse, S., Gubaydullin, F., Scheelken, H., Vieth, H.-M., and Yurkovskaya, A.V. (1999) *Appl. Magn. Reson.*, **17**, 211.

Grosse, S., Yurkovskaya, A., Lopez, J., and Vieth, H.M. (2001) *J. Phys. Chem. A*, **105**, 6311.

Haase, A., Frahm, J., Haenicke, W., and Matthaei, D. (1985) *Phys. Med. Biol.*, **30**, 341.

Hany, R., Vollenweider, J.K., and Fischer, H. (1988) *Chem. Phys.*, **120**, 169–175.

He, X.M. and Carter, D.C. (1992) *Nature*, **358**, 209.

Hincke, M.T., Sykes, B.D., and Kay, C.M. (1981) *Biochemistry*, **20**, 4185.

Hwang, T.-L. and Shaka, A.J. (1994) *J. Magn. Reson. A*, **112**, 275.

Ivanov, K.L. and Sagdeev, R.Z. (2006) *Dokl. Phys. Chem.*, **409**, 221–223.

Ivanov, K.L. and Lukzen, N.N. (2008) *J. Chem. Phys.*, **128**, 155105.

Ivanov, K.L., Miesel, K., Vieth, H.M., Yurkovskaya, A.V., and Sagdeev, R.Z. (2003) *Z. Phys. Chem. Int. J. Res. Phys. Chem. Chem. Phys.*, **217**, 1641–1659.

Ivanov, K.L., Miesel, K., Yurkovskaya, A.V., Korchak, S.E., Kityutin, A., and Vieth, H.-M. (2006a) *Appl. Magn. Res.*, **30**, 513.

Ivanov, K.L., Yurkovskaya, A.V., Hore, P.J., and Lukzen, N.N. (2006b) *Mol. Phys.*, **104**, 1687.

Ivanov, K.L., Yurkovskaya, A., and Vieth, H.-M. (2008) *J. Phys. Chem.*, **128**, 154701.

Jent, F. and Paul, H. (1989) *Chem. Phys. Lett.*, **160**, 632.

Jent, F., Paul, H., McLauchlan, K.A., and Stevens, D.G. (1987) *Chem. Phys. Lett.*, **141**, 443.

Jeschke, G. (1998) *J. Am. Chem. Soc.*, **120**, 4425.

Jeschke, G. and Matysic, J. (2003) *Chem. Phys.*, **294**, 239.

Kandrashkin, Y. and van der Est, A. (2004) *J. Chem.Phys.*, **120**, 4790.

Kandrashkin, Y., Asano, M.S., and van der Est, A. (2006) *Phys. Chem. Chem. Phys.*, **8**, 2129.

de Kanter, F.J.J., den Hollander, J.A., Huizer, A.H., and Kaptein, R. (1977) *Mol. Phys.*, **34**, 857.

Kaptein, R. (1971) *Chem. Commun.*, 132.

Kaptein, R. (1982) *Biol. Magn. Reson.*, **4**, 145.

Kaptein, R. and Oosterhoff, L.J. (1969a) *Chem. Phys. Lett.*, **4**, 195.

Kaptein, R. and Oosterhoff, L.J. (1969b) *Chem. Phys. Lett.*, **4**, 214.

Kaptein, R., Dijkstra, K., and Nicolay, K. (1978) *Nature*, **274**, 293.

Kawai, A. and Obi, K. (1993) *Res. Chem. Intermed.*, **19**, 865.

Kawai, A., Okutsu, T., and Obi, K. (1991) *J. Phys. Chem.*, **95**, 9130.

Khan, F., Kuprov, I., Craggs, T.D., Hore, P.J., and Jackson, S.E. (2006) *J. Am. Chem. Soc.*, **128**, 10729.

Kobori, Y. and Norris, J.R. (2006) *J. Am. Chem. Soc.*, **128**, 4.

Kuhn, L.T. and Bargon J. (2007) *Top. Curr. Chem.*, **276**, 125.

Kuprov, I.V. and Hore, P.J. (2004) *J. Magn. Reson.*, **171**, 171.

Kuprov, I.V., Craggs, T.D., Jackson, S.E., and Hore, P.J. (2007) *J. Am. Chem. Soc.*, **129**, 9004.

Laufer, M. (1986) *Chem. Phys. Lett.*, **127**, 136.

Lawler, R.G. and Halfon, M. (1974) *Rev. Sci. Instrum.*, **45**, 84.

Lebedeva, N.V., Zubenko, D.P., Bagryanskaya, E.G., Sagdeev, R.Z., Ananchenko, G.S., Marque, S., Bertin, D., and Tordo, P. (2004) *Phys. Chem. Chem. Phys.*, **6**, 2254.

Lehr, G.F. and Turro, N.J. (1981) *Tetrahedron*, **37**, 4211.

Liu, M., Ye, X.-A., Mao, C., Huang, H., Nicholson, J.K., and Lindon, J.C. (1998) *J. Magn. Reson.*, **132**, 125.

Lu, J.M., Wu, L.M., Geimer J., and Beckert, D. (2001) *Phys. Chem. Chem. Phys.*, **3**, 2053.

Lyon, C.E., Suh, E.S., Dobson, C.M., and Hore, P.J. (2002a) *J. Am. Chem. Soc.*, **124**, 13018.

Lyon, C.E., Lopez, J.J., Cho, B.M., and Hore, P.J. (2002b) *Mol. Phys.*, **100**, 1261.

McLauchlan, K.A. (1989) Time-resolved EPR, in *Advanced EPR, Application in Biology and Biochemistry* (ed. A.J. Hoff), Elsevier, Amsterdam, pp. 345–370.

McLauchlan, K.A. (1990) *Modern Pulsed and Continuous Wave Electron Spin Resonance* (eds L. Kevan and M.K. Bowman), John Wiley & Sons, Inc., New York, pp. 285–363.

McLauchlan, K.A. and Stevens, D.G. (1985a) *J. Chem. Soc. Faraday Trans.*, **83**, 473.

McLauchlan, K.A. and Stevens, D.G. (1985b) *J. Magn. Reson.*, **63**, 473.

McLauchlan, K.A. and Stevens, D.G. (1986) *Mol. Phys.*, **57**, 223.

McLauchlan, K.A. and Yeung, M.T. (1994) *Electron Spin Resonance*, vol. 14, Royal Society of Chemistry, Cambridge, p. 32.

Maeda, K., Terazima, M., and Azumi, T. (1991) *J. Phys. Chem.*, **95**, 197.

Maeda, K., Meng, Q.X., Aizawa, T., Terazima, M., Azumi, T., and Tanimoto, Y. (1992) *J. Phys. Chem.*, **96**, 4884.

Makarov, T.N., Bagryanskaya, E.G., and Paul, H. (2004) *Appl. Magn. Reson.*, **26**, 197.

Makarov, T.N., Savitsky, A.N., Mobius, K., Beckert, D., and Paul, H. (2005) *J. Phys. Chem.*, **109**, 2254.

Martino, D.M. and van Willigen, H. (2000) *J. Phys. Chem.*, **104**, 1070.

Mazzoni, M., Corvaja, C., Gubskaya, V.P., Berezhnaya, L.S., and Nuretdinov, I.A. (2006) *Mol. Phys.*, **104**, 1543.

Meng, Q., Suzuki, K., Terazima, M., and Azumi, T. (1990) *Chem. Phys. Lett.*, **175**, 364.

Miesel, K., Ivanov, K.L., Yurkovskaya, A.V., and Vieth, H.M. (2006) *Chem. Phys. Lett.*, **425**, 71.

Miesel, K., Ivanov, K.L., Köchling, T., Yurkovskaya, A.V., and Vieth, H.-M. (2008) *Appl. Magn. Res.*, **34**, 423.

Miller, R.J. and Closs, G.L. (1981) *Rev. Sci. Instrum.*, **52**, 1876.

Milligan, J.R., Aguilera, J.A., Ly, A., Hoang, O., Tran, N.Q., and Ward, J.F. (2003) *Nucleic Acids Res.*, **31**, 6258.

Mizuochi, N., Ohba, Y., and Yamauchi, S. (1999) *J. Phys. Chem. A*, **103**, 7749.

Mok, K.H. and Hore, P.J. (2004) *Methods*, **34**, 75.

Mok, K.H., Nagashima, T., Day, I.J., Jones J.A., Dobson, C.M., and Hore, P.J. (2003) *J. Am. Chem. Soc.*, **125**, 12484.

Mok, K.H., Kuhn, L.T., Goez, M., Day, I.J., Lin, J.C., Andersen, N.H., and Hore, P.J. (2007) *Nature*, **447**, 106.

Molin, Y.N., Anisimov, O.A., Melekhov, V.I., and Smirnov, S.N. (1984) *Faraday Discuss. Chem. Soc.*, **78**, 1.

Molin, Yu.N. and Sagdeev, R.Z. (1976) Lecture at the All-Union Xonference in Chemical Kinetics, dedicated to the 80th Birthday of Academician N.N. Semenov, Moscow.

Molin, Yu.N., Anisimov, O.A., Grigoryants, V.M., Molchanov, V.K., and Salikhov, K.M. (1980) *J. Phys. Chem.*, **84**, 1853.

Morozova, O.B., Tsentalovich, Y.P., Yurkovskaya, A.V., and Sagdeev, R.Z. (1998) *J. Phys. Chem. A*, **102**, 3492.

Morozova, O.B., Yurkovskaya, A.V., Tsentalovich, Y.P., Forbes, M.D.E., Hore, P.J., and Sagdeev, R.Z. (2002) *Mol. Phys.*, **100**, 1187.

Morozova, O.B., Yurkovskaya, A.V., Sagdeev, R.Z., Mok, K.H., and Hore, P.J. (2004) *J. Phys. Chem.*, **39**, 15355.

Morozova, O.B., Kiryutin, A.S., Sagdeev, R.Z., and Yurkovskaya, A.V. (2007) *J. Phys. Chem. B*, **111**, 7439.

Morozova, O.B., Kiryutin, A.S., and Yurkovskaya, A.V. (2008) *J. Phys. Chem. B*, **112**, 2747.

Morozova, O.B., Korchak, S.E., Vieth, H.M., and Yurkovskaya, A.V. (2009) *J. Phys. Chem. B*, **113**, 7398.

Murai, H., Tero-Kubota, S., and Yamauchi, S. (2000) Pulsed and time-resolved EPR studies of transient radicals, radical pairs and excited states in photochemical systems, in *Electron Spin Resonance* (ed. B.C. Gilbert, M.J. Davies, and K.A. McLauchlan), Royal Society of Chemistry, Cambridge, pp. 130–163.

Muszkat, K.A., Khait, I., and Weinstein, S. (1984) *Biochemistry*, **23**, 5.

Muus, L.T. (1989) *Chem. Phys. Lett.*, **160**, 17.

Muus, L.T., Atkins, P.W., McLauchlan, K.A., and Pedersen, J.B. (1977) Chemically Induced Magnetic Polarization, NATO Advanced Study Institutes Series. Dordrecht-Holland/Boston-U.S.A. vol. 34.

Nagakura, S., Hayashi, H., and Azumi, T. (1998) *Dynamic Spin Chemistry*, Kodansha Ltd and John Wiley & Sons, Inc., Tokyo, New York.

Neufeld, A.A. and Pedersen, J.B. (2000) *J. Phys. Chem.*, **113**, 1595.

Ohara, K., Miura, Y., Terazima, M., and Hirota, N. (1997) *J. Phys. Chem.*, **101**, 605.

Ohara, K., Watanabe, R., Mizuta, Y., Nagaoka, S., and Mukai, K. (2003) *J. Phys. Chem.*, **107**, 11527.

Pedersen, J.B. and Freed, J.H. (1975) *J. Chem. Phys.*, **62**, 1706.

Piotto, M., Saudek, V., and Sklenar, V. (1992) *J. Biomol. NMR*, **2-6**, 661.

Prakash, S., Alia, Gast, P., de Groot, H.J.M., Jeschke, G., and Matysic, J. (2005) *J. Am. Chem. Soc.*, **127**, 14290.

Prakash, S., Alia, Gast, P., de Groot, H.J.M., Matysic, J., and Jeschke, G. (2006) *J. Am. Chem. Soc.*, **128**, 12794.

Prakash, S., Alia, Gast, P., de Groot, H.J.M., Jeschke, G., and Matysic, J. (2007) *Biochemistry*, **46**, 8953.

Purtov, P.A. and Doktorov, A.B. (1993) *Chem. Phys.*, **178**, 47.

Redfield, A.G. (2003) *Magn. Res. Chem*, **41**, 753.

Regev, A., Gamliel, D., Meiklyar, V., Micheali, S., and Levanon, H. (1993) *J. Phys. Chem.*, **97**, 3671.

Richter, G., Weber, S., Romisch, W., Bacher, A., Fischer, M., and Eisenreich, W. (2007) *J. Phys. Chem. B*, **111**, 7439.

Roth, H.D. (2001) *J. Photochem. Photobiol. C*, **2**, 93.

Roth, H.D. (2003) *J. Phys. Chem. A*, **107**, 3432.

Roth, H.D. (2008) *Photochem. Photobiol. Sci. A.*, **7**, 540.

Sagdeev, R.Z. and Bagryanskaya, E.G. (1990) *Pure Appl. Chem.*, **62**, 1547.

Sagdeev, R.Z., Salikhov, K.M., Leshina, T.V., Kamkha, M.A., Shein, S.M., and Molin, Y.N. (1972) *Pis'ma Zh. Eksp. Teor. Fiz.*, **16**, 599. [*JETP Lett.*, **16** 422 (1972)].

Salikhov, K.M. (1997) *Appl. Magn. Res.*, **13**, 415.

Salikhov, K.M., Molin, Y.N., Sagdeev, R.Z., and Buchachenko, A.L. (1984) Theory of radical recombination. The theory of chemically induced dynamic nuclear and electron spin polarization, in *Spin Polarization and Magnetic Effects in Radical*

Reactions (ed. Y.N. Molin), [translated by G.Z. Ribina and L.Ya. Yuzina] Elsevier, Amsterdam.

Sartori, E., Toffoletti, A., Rastrelli, F., Corvaja, C., Bettio, A., Formaggio, F., Oancea, S., and Toniolo, C. (2003) *J. Phys. Chem. A*, **107**, 6905.

Savitsky, A. and Moebius, K. (2006) *Helv. Chim. Acta*, **89**, 2544.

Savitsky, A. and Mobius, K. (2009) *High-Field EPR Spectroscopy on Proteins and their Model Systems*. Royal Society of Chemistry, Cambridge, UK.

Schaublin, S., Honener, A., and Ernst, R.R. (1974) *J. Magn. Reson.*, **13**, 196.

Scheffler, J.E., Cottrell, C.E., and Berliner, L.J. (1985) *J. Magn. Reson.*, **63**, 199.

Schulten, E.A.M., Matysik, J., Alia, A., Kiihne, S., Raap, J., Lugtenburg, J., Gast, P., Hoff, A.J., and de Groot, J.M. (2002) *Biochemistry*, **41**, 8708.

Shkrob, I.A., Tarasov, V.R., and Bagryanskaya, E.G. (1991) *Chem. Phys.*, **153**, 427.

Shushin, A.I. (1991a) *Chem. Phys.*, **152**, 133.

Shushin, A.I. (1991b) *Chem. Phys. Lett.*, **177**, 338.

Shushin, A.I. (1991c) *Chem. Phys. Lett.*, **183**, 321.

Shushin, A.I. (1993) *Z. Phys. Chem. Int. J. Res. Phys. Chem.*, **182**, 9.

Shushin, A.I. (1995a) *Chem. Phys. Lett.*, **237**, 177.

Shushin, A.I. (1995b) *Chem. Phys. Lett.*, **245**, 183.

Shushin, A.I. (1995c) *J. Phys. Chem.*, **101**, 8747.

Shushin, A.I. (1996) *Chem. Phys. Lett.*, **260**, 261.

Shushin, A.I. (2002) *Mol. Phys.*, **100**, 1303.

Shushin, A.I., Pedersen, J.B., and Lolle, L.I. (1994) *Chem. Phys.*, **188**, 1.

Smaller, B.J., Remko, J.R., and Avery, E.C. (1968) *J. Chem. Phys.*, **48**, 5174.

Stavitski, E., Wagnert, L., and Levanon, H. (2005) *J. Phys. Chem.*, **109**, 976.

Steiner, U.E. and Ulrich, T. (1989) *Chem. Phys. Rev.*, **89**, 51.

Steren, C.A., van Willigen, H., and Dinse, K.-P. (1994) *J. Phys. Chem.*, **98**, 7464.

Stob, S. and Kaptein, R. (1998) *Photochem. Photobiol.*, **49**, 565.

Tarabek, P., Bonifacic, M., Naumov, S., and Beckert, D. (2004a) *J. Phys. Chem. A*, **108**, 929.

Tarabek, P., Bonifacic, M., and Beckert, M.D. (2004b) *J. Phys. Chem. A*, **108**, 3467.

Tarabek, P., Bonifacic, M., and Beckert, M.D. (2006) *J. Phys. Chem. A*, **110**, 7293.

Tarabek, P., Bonifacic, M., and Beckert, M.D. (2007) *J. Phys. Chem. A*, **111**, 4958.

Tarasov, V.F. and Forbes, M.D.E. (2000) *Spectrochim. Acta A*, **56**, 245.

Tarasov, V.F., Yashiro, H., K. Maeda, Azumi T., and Shkrob, I.A. (1996) *Chem. Phys.*, **212**, 353.

Tarasov, V.F., Saiful, I.S.M., Iwasaki, I., Ohba, Y., Savitsky, A., Mobius, K., and Yamauchi, S. (2006a) *Appl. Magn. Reson.*, **30**, 619.

Tarasov, V.F., White, R.C., and Forbes, M.D.E. (2006b) *Spectrochim. Acta. Part A: Mol. Biomol. Spectrosc.*, **63**, 776.

Tsentalovich, Y.P. and Forbes, M.D.E. (2002) *Mol. Phys.*, **100**, 1209.

Valyaev, V.I., Molin, Y.N., Sagdeev, R.Z., Hore, P.J., McLauchlan, K.A., and Simpson, N.J.K. (1988) *Mol. Phys.*, **63**, 891.

Van der Waals, J.H. and de Groot, M.S. (1967) *The Triplet State* (ed. A.B. Zahlan), Cambridge University Press, Cambridge.

Verhoeven, J.W. (2006) *J. Phoochem. Photobiol. C Photochem. Rev.*, **7**, 40.

Verma, N.C. and Fessenden, R.W. (1976) *J. Chem. Phys.*, **65**, 2139.

Vogel, H.J. and Sykes, B.D. (1984) *J. Magn. Reson.*, **59**, 197.

Ward, H.R. and Lawler, R.G. (1967) *J. Am. Chem. Soc.*, **89**. 5518.

Wasielewski, M.R. (2006) *J. Org. Chem.*, **71**, 5051.

Weiss, M.A., Nguyen, D.T., Khait, I., Inouye, K., Frank, B.H., Beckage, M., O'Shea, E., Shoelson, S.E., Karplus, M., and Neuringer, L.J. (1989) *J. Biochem.*, **28**, 9855.

White, R. and Forbes, M.D.E. (2006) *Org. Lett.*, **8**, 6027.

van Willigen, H., Levstein, P.R., and Ebersole, M.H. (1993) *Chem. Rev.*, **93**, 173.

Wong, S.K., Hutchinson, D.A., and Wan, J.K.S. (1973) *J. Chem. Phys.*, **58**, 985.

Yamauchi, S. and Pratt, D.W. (1979) *Mol. Phys.*, **37**, 541.

Zhang, D., Norris, J.R., Krusic, P.J., Wasserman, E., Chen, C.C., and Lieber, C.M. (1993) *J. Phys. Chem.*, **97**, 5886.

Zysmilich, M.G. and McDermott, A.E. (1994) *J. Am. Chem. Soc.*, **116**, 8362.

**Part Four
Future Perspectives**

26
Future Perspectives

Sushil K. Misra

EPR has recently gone through a *renaissance*, and the future appears to be very bright. However, in order to outline the future perspectives, it is first important to review current EPR spectroscopic techniques, as well as cutting-edge topics and desirable advancements in EPR instrumentation. Historically, applications have driven instrumentation development and, to some degree, the converse is also true. Whilst the forecasting of applications that are driven by instrumental developments is feasible, the forecasting of applications that are *not* driven by instrumentation progress must be based on existing knowledge that lies outside the field of EPR. The following description is based not only on the author's opinion, but also relies equally on input from prominent investigators in the field (see below, Acknowledgments). Detailed descriptions of many of the topics mentioned below in this chapter are included in this book.

26.1
Spectroscopic Techniques Currently Available in EPR

These can be listed as multifrequency continuous wave (CW) and pulse EPR, augmented by time-resolved (TR) and rapid-scan capabilities, EPR operating at high magnetic fields and high microwave frequencies (HFHF-EPR) up to the sub-millimeter region. These are further extended by multiresonance and multipulse capabilities, such as electron-nuclear double resonance (ENDOR), electron spin echo envelope modulation (ESEEM), hyperfine sublevel correlation spectroscopy (HYSCORE), double quantum coherence (DQC), pulsed electron double resonance (PELDOR) – also known as double electron-electron resonance (DEER) – which play important roles, and will continue to do so. (For example, Figure 26.1 shows the various ongoing research projects in operation at ACERT [National Biomedical Center for Advanced ESR Technology] in the Department of Chemistry and Chemical Biology, Cornell University, Ithaca [USA].) At the high-frequency end, for commercially available spectrometers, Bruker now has a W-band (95 GHz) spectrometer available and a 243 GHz machine will be on the market shortly. Further improvements of new pulse schemes, such as those developed by the late Arthur

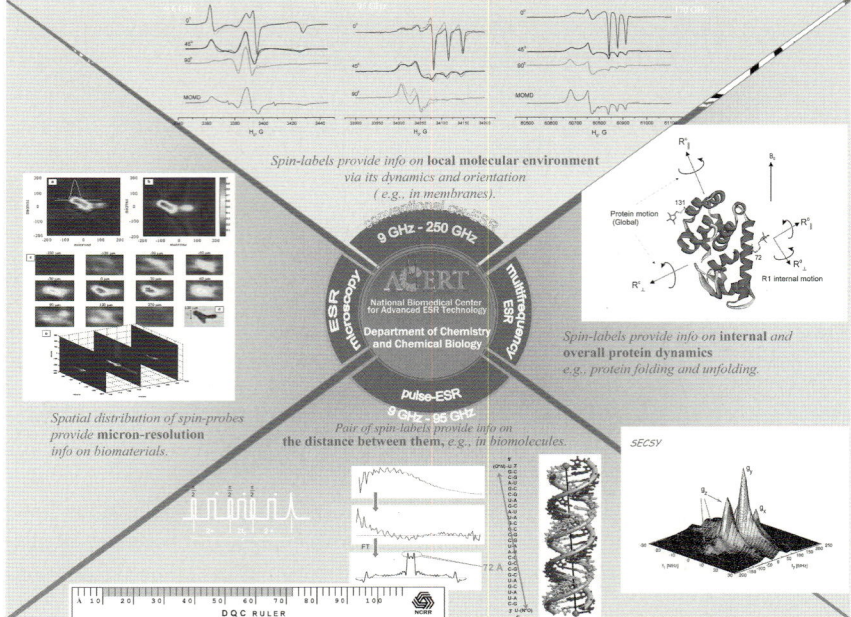

Figure 26.1 The various research projects in operation at ACERT (National Biomedical Center for Advanced ESR Technology) in the Department of Chemistry and Chemical Biology, Cornell University, Ithaca (USA). Illustration from the lecture given by Prof. Freed at the EUROMAR-2005 conference in Eindhoven, The Netherlands.

Schweiger and his group (Figure 26.2), is very important, and is still in progress. Examples include modified versions of HYSCORE, introduced originally by Höffer and Mehring, for measuring small couplings of metal nuclei and heteronuclei in organic systems. The development of dedicated multifrequency/multiple-resonance EPR instrumentation has enabled several outstanding applications of state-of-the-art EPR spectroscopy. It is anticipated that the advancement of EPR will depend largely on the development of novel and sophisticated instrumentation involving the latest innovations in microwave and magnet technologies. The trend to higher and higher magnetic fields will continue. In NMR, the "magic" 1 GHz limit has already been passed, and in EPR, although the THz domain has been reached, it has not yet exceeded the g-strain limit of the Zeeman field for many systems, beyond which no more resolution improvement is expected. In fact, enhanced resolution at increased field is very species-dependent. For some species, the trade-off between hyperfine, Zeeman, and g-strain favors lower frequencies, for example, for the Cu^{2+} ion. The choice depends on the information sought. The following examples of recent instrumental developments demonstrate how modern EPR and NMR complement each other to the benefit of future magnetic resonance spectroscopy: (i) Dynamic nuclear polarization (DNP) at high magnetic

26.1 Spectroscopic Techniques Currently Available in EPR | 997

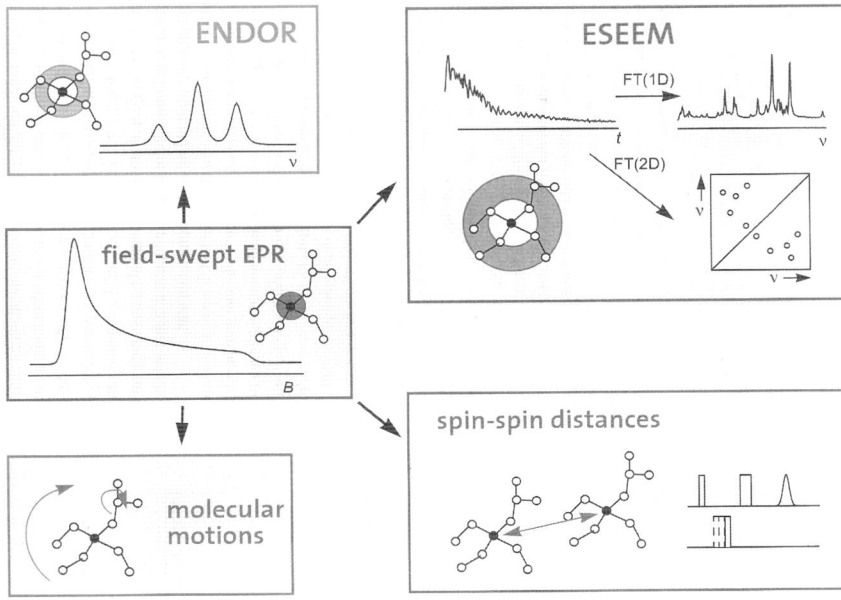

Figure 26.2 Some of the pulsed-EPR projects in operation in the research laboratory of the late Arthur Schweiger in Zurich. Illustration from the lecture given by the late Arthur Schweiger, at the Asia-Pacific Conference in 2004.

fields, as well as photochemically induced DNP in magic-angle-spinning NMR (DNP is a very 'hot topic' that brings HFHF-EPR and NMR together); (ii) pulsed THz EPR with synchrotron radiation and magnetic fields; and (iii) Fourier-transform high-field EPR with broadband stochastic microwave excitation.

26.1.1
Future Perspectives in EPR Instrumentation

Based on natural evolutionary steps in improved computers and digital devices, the following appear feasible:

1) The exploitation of software to accomplish the processes of phase shifting, mixing, quadrature detection, and filtering, or narrow banding such as that achieved with time-constant selection in CW-EPR. Analog-to-digital (A/D) converters are now used routinely for direct digitization at 3 GHz, with 10 GHz not far away.

2) As for resonators, digital design and automated fabrication will be the rule rather than the exception (see Chapter 5). There will be an increased use of resonators, specifically optimized for a particular experiment.

3) With respect to sample excitation, hard pulses will generally remain beyond reach; however, hybrid methods, such as those initially used in NMR under the rubric of correlation spectroscopy, will be used. There will be an increased use of very fast field sweeps, very fast microwave frequency sweeps, and the use of multiple irradiation arms, with each arm under its own temporal control. The distinction between pulsed and CW methods will blur. In addition, improved computers will enable advanced solutions of Bloch equations, spin Hamiltonians, and Liouville equation, required to interpret EPR data.

4) There will be continued interest in multifrequency EPR, including zero-field EPR, especially because the high-field approximation is no longer a limitation with modern simulation programs.

5) In view of points 1 and 2 above, the cost of EPR spectrometers will be sharply reduced. The long-sought goal of a really cheap EPR spectrometer that can be used in mass routine applications may be within reach. EPR-based assays may become "routine."

6) Digital design will vastly improve stability of EPR spectrometers. This will "open the door" to a much-improved sensitivity by being able to scan over long periods of time, as in NMR. This increased sensitivity will become the principal driver of new applications. There will be further developments in sensitivity and resolution with induction detection and also with new methods, such as magnetic resonance force microscopy (MRFM), nitrogen vacancy (NV) centers, and Hall probes.

26.1.2
Desirable Advancements in EPR Instrumentation

First, the combination of optical (laser) excitation with pulsed-EPR has a much larger application field than is currently realized. This technique should be developed to a point where it is no longer the specialty of a few groups, but can be applied with relative ease as a routine technique.

Second, the exploitation of high-field pulsed-EPR with ultra-short pulses and ultra-short dead times is already underway, to increase sensitivity and time resolution significantly. This would allow for new Fourier-transform EPR experiments and a broader application of many of Jack Freed's techniques that have been confined almost exclusively to his own group in the past. Although financial constraints on both the users and the manufacturers direct us towards multipurpose spectrometers, single-purpose instruments can be designed to yield better signal-to-noise ratios) (SNRs) and more nearly optimized information for cutting-edge research. Improvements in instrumentation, and primarily improvements in SNR, have made possible entirely new areas of research throughout the history of EPR.

Thus, the main advances that will facilitate new applications of HFHF-EPR are improvements in low-noise sources, low-noise amplifiers, and detectors.

26.2 Cutting-Edge Topics

These involve applications of HFHF EPR, and its multiple-resonance extensions, representing near-future challenges in EPR. They can be exploited to investigate challenging problems in the studies of structure and dynamics, notwithstanding the fact that, currently, sensitivity issues exclude the application to several interesting classes of (biological) samples. The following examples are noteworthy:

Resolution of small g-anisotropies: (a) The canonical g-tensor orientations in their powder spectra can be resolved for cofactor radicals, even in disordered samples. Thereby, orientation-selective spin interactions within the molecule and with its microenvironment can be detected. (b) Study of transient intermediates of several radical species with overlapping EPR spectra. These are often generated in chemical reactions, and distinguished by clearly detecting small differences in their g-factors and hyperfine interactions using high Zeeman fields in HFHF-EPR.

Study of biological samples: (a) High-purity samples. These are often produced only in minute quantities, and a very high detection sensitivity of HFHF spectrometers is quite suitable in this context. However, many HFHF spectrometers are not designed to facilitate control of the environment of the biological sample, such as the exclusion of oxygen, and keeping the sample cold or in the dark at all times. (b) Study of mixed samples. The high spectral range allows the discernment between, for example, different carbon-based radical nitrone adducts that could never be resolved at lower frequency.

Determination of the absolute sign of the zero-field splitting (ZFS) **parameter** (D): ZFS characterizes, for example, a two-spin system, such as a biradical or triplet state, or a single spin with $S \geq 1$. Considerable Boltzmann thermal spin polarization to accomplish this can be easily achieved well above liquid-helium temperature at high fields. On the other hand, as for the signs of HF couplings, simulations yield the relative signs of the principal values; initial steps have been made to determine the absolute signs, but a robust method for reliably measuring the absolute signs of HF couplings is needed.

Study of dynamic processes: HFHF CW-EPR generally provides shorter time windows down into the picoseconds (p) range for determining correlation times and fluctuating local fields over a wide temperature range.

Real-time access to specific molecular slow motions in the nanosecond (ns) time-scale: This is achieved by pulsed HFHF-EPR in the form of 2-D field-swept electron spin echo (ESE) spectroscopy. Here, even motional anisotropy, generated by anisotropic interactions, for example, hydrogen bonding, can be traced.

Orientation selection of molecular sub-ensembles in powder or frozen-solution samples: This is achieved by ENDOR at high magnetic fields, so that even in the case of small g-anisotropies, ENDOR can provide single-crystal like information about hyperfine interactions, including, for example, anisotropic hydrogen bonding to the protein.

Differentiation of strongly and weakly hyperfine-coupled nuclei in protein systems: This can be accomplished by properly adjusting the Zeeman field in multifrequency pulsed EPR experiments to record ENDOR and ESEEM spectra, for example, those of remote and coordinated nitrogens in histidines of metalloproteins.

Revealing subtle changes in polarity and proticity profiles along segments of proteins or their microenvironment of the plasma or membrane: This is well accomplished by a combination of CW- and pulsed high-field EPR spectroscopy with site-directed spin labeling (SDSL) techniques employing nitroxide radicals. This information can be obtained from the g_{xx} and A_{zz} (hyperfine) components of the nitroxide spin label, but also from the quadrupole-tensor component P_{yy} (also denoted as Q_{yy}) of the ^{14}N nucleus, measured by high-field ESEEM, which can be exploited for probing subtle matrix effects.

Characterization of the environment of the probe in more detail: This can be achieved by advanced instrumentation in conjunction with SDSL. The complex motional information so acquired may lead to new insights on biomolecular dynamics. As for choices of SDSL, it is desirable to use probes other than nitroxides.

Other studies of proteins using spin labels: For example, multifrequency EPR of nitroxide spin labels for increased resolution in the analysis of protein rotational dynamics; bifunctional spin labels attached to two Cys residues, used to detect global protein orientation and microsecond rotational dynamics (using saturation transfer [ST] EPR); unnatural nitroxide-bearing amino acids such as TOAC that could be rigidly inserted into the protein backbone to report on the rotational dynamics and inter-spin distances; Pulsed EPR used to resolve multiple structural states of motor proteins;

Redox imaging: EPR is used in conjunction with magnetic resonance imaging (MRI) to study blood flow and metabolism;

High-field EPR applications to inorganic materials, containing various transition-metal ions: This has not yet achieved its full potential. The main unresolved issue here is improving the sensitivity of EPR spectrometers operating at very high magnetic fields/frequencies, as these lack suitable resonator cavities currently.

Determination of 3-D structure (distance and relative orientation) of stable or transient radical-pair systems with large inter-spin distances (up to ca. 8 nm),

even in frozen solutions: The possibility to measure not only distances up to about 10 nm, but also to obtain orientational information (at high field) using DQC, PELDOR (DEER) would provide much-needed data for biophysicists. These data would be complementary to solid-state NMR for short-distance NOE constraints, and are envisaged to become extremely important for the structure determination of large proteins and/or protein complexes. Here, the development of new spin labels, in addition to nitroxides, and labeling techniques is desirable to increase the sensitivity and the distance range.

Study of molecular magnets and other high-spin systems: For this, multifrequency HFHF high-spin EPR is very well suited.

Study of disordered systems: This can be achieved with the use of currently available EPR techniques, without requiring single-crystal preparations (e.g., that of a protein) to obtain detailed structural information with atomic resolution. Likewise, the structure and dynamics of transient states of proteins in action on biologically relevant time scales are readily characterized. The combination of EPR with labeling techniques opens the possibility to probe diamagnetic compounds that are otherwise EPR-silent. Here, an important research objective is to understand, on the basis of structural and dynamics data, the dominant factors that control the specificity and efficiency of electron- and ion-transfer processes in proteins. Parallel to high-resolution X-ray crystallography, theory-assisted EPR spectroscopy, in all its facets, is being used currently in this endeavor. The structure and dynamics of biomacromolecules and biomacromolecular complexes – in particular of membrane proteins and their complexes – are fields where EPR can potentially yield unique insights.

Study of transient paramagnetic states: When detected immediately after (photo) initiation of the reaction, characteristic electron spin-polarization (CIDEP, chemically induced dynamic electron polarization) effects are exhibited. These can be exploited for signal enhancement in EPR and ENDOR experiments to provide valuable information on the structure, dynamics, and reaction pathways of short-lived intermediates.

Study of catalysis: EPR is a very sensitive tool to study catalytic processes. It is unique for monitoring catalytic processes in which species with unpaired electrons are involved. For example, one can study supported and unsupported transition metal oxide catalysts and/or radical intermediates. Some applications include the oxidation and ammoxidation of aliphatic and aromatic hydrocarbons over metal oxide catalysts, the catalytic reduction of nitrogen oxides over supported metal oxide catalysts and zeolites modified with transition metal ions, and the dehydrogenation and aromatization of paraffins

Study of organic systems: Here, important developments are emerging thanks to the enhanced sensitivity and time resolution of EPR. For example, triplet states – characterized by a few microseconds lifetime – are not only easily detected by time-resolved (TR) EPR, but also by pulsed ENDOR, for example, at Q-band. As a consequence, the spin-density distribution in this $S = 1$ state can be investigated via the resolved hyperfine structure. Other examples of interesting applications to organic radicals are: amino-acid radicals that occur in many reactions, for example, in water-splitting complex of photosystem II and in ribonucleotide reductase of "radical enzymes." Flavine radicals and triplet states have seen a renaissance, for example, in the study of cryptochromes.

Study of transition-metal ions in bioinorganic chemistry: EPR and related methods are particularly useful to study transition metals in chemical and biological systems. Together with methods to trap intermediates, EPR has been instrumental to the set-up reaction mechanisms for many enzymes, and to study the action of other proteins, for example, nitrogenase, hydrogenase, and water oxidase – a topic related to solve future energy problems.

Some other cutting-edge topics are listed as follows. [Additional information on the latest research can be found in recent conference proceedings, for example, that of the Asia-Pacific conference (APES08) held in Cairns, Australia in 2008, with proceedings published in *Applied Magnetic Resonance*, **36** (2009).]

- Measurements using the torque method with cantilever.
- Organic molecular magnetism, with applications to molecular spin quantum computing and quantum information processing.
- Electron-spin quantum information processing using the techniques of high-field pulse EPR, ENDOR, and EDMR.
- HFHF EPR using accelerator-based THz light sources.
- Single-spin detection.
- EPR microscopy and nanoscopy.
- Application of EPR to quantum-information processing.
- *In-vivo* EPR (300 MHz–9 GHz).
- Low-temperature tissue EPR, for example, that of myocardial tissue.

26.2.1
Topics Related to the Theoretical Interpretation of EPR Data

The objective here is the deduction of structure and dynamics from EPR parameters.

Spin-Hamiltonian parameters: The following parameters are not yet well predicted: ZFS, hyperfine couplings (HFS) of transition metal ions, as opposed to the corresponding ones of ligand nuclei, which can be predicted reasonably well, and exchange couplings (*J*). However, quantum chemistry is on the verge of predicting these.

***Ab initio* methods to calculate EPR parameters:** With a focus for future research on systems with non-dynamic correlation untreatable with single-determinant methods (Hartree–Fock, Kohn–Sham, density-functional theory: DFT).

Dynamics of spin-labeled macromolecules: Despite much research having been conducted during the past four decades, detailed, precise information cannot yet be extracted from the CW-EPR spectra of nitroxides. Jack Freed's new rotamer approach to the problem bridges the gap between more abstract, fast simulations and more detailed, slow MD simulations that are fraught with problems of full sampling of the conformational space.

Interpretation of distance measurements: EPR is currently the most precise method for measuring distance distributions in the range from 1.8 to 5 nm (in favorable cases, to 8 nm) in disordered systems. This allows for insights into macromolecular conformation that no other method can obtain. The relation between statistical mechanics descriptions of macromolecule conformation and the measured data needs to be further investigated to make full use of this potential.

Pulsed-EPR theory: Most pulse sequences, with good application potential, are well understood for systems consisting of a single electron spin $S = 1/2$ and a single nuclear spin $I = 1/2$. They are also reasonably well understood for a single electron spin and a single nuclear spin $I > 1/2$. However, the response of systems with more than one electron spin, with electron group spins $S > 1/2$, and with multiple nuclear spins is generally not very well understood, and requires further development for its application to detailed data analysis of many systems of interest. Multifrequency EPR is crucial to understanding electron-spin relaxation mechanisms. One advantage of HF-EPR is that, for the same resonator Q, the dead time due to ring-down following a pulse is shorter, allowing shorter relaxation times to be measured.

26.3
Desirable Applications of EPR

Systematic strategies that apply a whole arsenal of EPR techniques and well-defined measurement protocols need to be developed. Current molecular modeling approaches are not very well suited to the information that EPR can provide, nor to combining information from diverse techniques. The EPR-restraint modeling of structures and structural transitions needs to be developed. Analogous techniques and approaches are well-suited for materials based on supramolecular assemblies. This field is almost unexplored, due partly to non-health-related funding being scarce in the USA.

26.4
Future of EPR

The future of EPR appears very promising, particularly in anticipation of the availability of cheaper spectrometers in the near future, and with continuous progress being made in both experimental and theoretical research. Today, there are many EPR centers and specialized research laboratories in existence worldwide, and in particular in Australia, France, Germany, Israel, Italy, Japan, Russia, UK, and the

Figure 26.3 A vision for the future: a 2 THz pulsed-EPR spectrometer. Illustration from the lecture given by the late Arthur Schweiger at the Asia-Pacific Conference in 2004.

USA. In view of its applications to science and industry (as described above), to biology and medicine, quantum computing, photosystems, and magnetism, and the effective role played in understanding the environment of a paramagnetic ion, EPR offers excellent opportunities to attract strategic funding and commercial development.

A vision for the future, a 2 THz pulsed-EPR spectrometer, is shown in Figure 26.3.

Acknowledgments

The author is grateful, in particular, to Aharon Blank, Larry Berliner, Peter Dinse, Gareth R. Eaton, Sandra S. Eaton, Daniella Goldfarb, Steve Hill, Brian Hoffman, Jim Hyde, Gunnar Jeschke, Wolfgang Lubitz, Klaus Moebius, Thomas Prisner, Alex Smirnov, and Hans van Tol for detailed specific input on the topics covered here. In addition, thanks are expressed to the following for their inputs: Chris Chang, Jack Freed, Balaraman Kalyanraman, Gavin Morley, Hitoshi Ohta, Hal Swartz, David Thomas, and Dmitry Tipikin. The author apologizes to anyone whose name might have been inadvertently omitted. Finally, Charlie Poole is thanked for his critical reading of this chapter.

Appendix A1 Fundamental Constants and Conversion Factors used in EPR

Adapted from *Electron Paramagnetic Resonance: A Practitioner's Toolkit*, Eds. M. Brustolon and E. Giamello, John Wiley, New Jersey, USA (2009), pp. 520–522

Constant	Symbol	SI Value
Speed of light (vacuum)	c_0	299 792 458 m s^{-1} (defined)
Permeability of vacuum	μ_0	$4\pi \times 10^{-7}$ H m^{-1} or N A^{-2} (defined)
Permittivity of vacuum	$\varepsilon_0 = 1/(\mu_0 c_0^2)$	$8.854\,187\,817 \times 10^{-12}$ F m^{-1} (defined)
Planck's constatnt	h $\hbar = h/2\pi$ (au)	$6.626\,0693\,(11) \times 10^{-34}$ J s $1.054\,571\,68\,(18) \times 10^{-34}$ J s
Elementary charge	$|e|$	$1.602\,176\,53\,(14) \times 10^{-19}$ C
Electron's rest mass	m_e	$9.109\,3826\,(16) \times 10^{-31}$ kg
Proton's rest mass	m_p	$1.672\,621\,71\,(29) \times 10^{-27}$ kg
Proton/electron mass ratio	m_p/m_e	1836.152 672 61 (85)
Neutron's rest mass	m_n	$1.674\,927\,28\,(29) \times 10^{-27}$ kg
Deuteron's rest mass	m_d	$3.343\,583\,35\,(57) \times 10^{-27}$ kg
Atomic mass unit (^{12}C/12)	$m_u = 1\,u = 1$ Da	$1.660\,538\,86\,(28) \times 10^{-27}$ kg
Avogadro's number	N_A	$6.022\,1415\,(10) \times 10^{23}$ mol^{-1}
Boltzmann's constant	k_B	$1.380\,6505\,(24) \times 10^{-23}$ J K^{-1}
Fine structure constant Inverse Fine structure constant	$\alpha = \mu_0 e^2 c_0/2h$ $1/\alpha$	$7.297\,352\,568\,(24) \times 10^{-3}$ 137.035 999 11 (46)
Bohr radius (au)	$a_0 = 4\pi\varepsilon_0 \hbar^2/m_e e^2$	$0.529\,177\,2108\,(18) \times 10^{-10}$ m
Compton wavelength (electron)	$\lambda_c = h/m_e c_0$	$2.426\,310\,238\,(16) \times 10^{-12}$ m
Bohr magneton (β, β_e)	$\mu_B = e\hbar/2m_e$	$9.274\,009\,49\,(80) \times 10^{-24}$ J T^{-1}

Appendix A1 Fundamental Constants and Conversion Factors used in EPR

Constant	Symbol	SI Value
Electron magnetic moment	μ_e	$-9.284\,764\,12\,(80) \times 10^{-24}$ J T^{-1}
Electron magnetogyric ratio	$\gamma_e = 2\,\mu_0/\hbar$	$1.760\,859\,74\,(15) \times 10^{11}$ s^{-1} T^{-1}
	$\gamma_e/2\pi$	$28.024\,9532\,(24) \times$ GHz T^{-1}
Free electron Landé g factor	$g_e = 2\,\mu_e/\mu_B$	$-2.002\,319\,304\,3718\,(75)$
Nuclear magneton (β_N)	$\mu_N = (m_e/m_p)\,\mu_B$	$5.050\,783\,43\,(43) \times 10^{-27}$ J T^{-1}
Proton magnetic moment (free) (Shielded H$_2$O sphere, 25 °C – correction for diamagnetism)	μ_p μ_p'	$1.410\,606\,71\,(12) \times 10^{-26}$ J T^{-1} $1.410\,570\,47\,(12) \times 10^{-26}$ J T^{-1}
Proton magnetogyric ratio (free) (Shielded H$_2$O sphere, 25 °C – correction for diamagnetism)	γ_p γ_p'	$2.675\,222\,05\,(23) \times 10^8$ s^{-1} T^{-1} $2.675\,153\,33\,(23) \times 10^8$ s^{-1} T^{-1}
Proton MR frequency in H$_2$O	$\gamma_p'/2\pi$	$42.576\,3875\,(37)$ MHz T^{-1}
Electron/Proton mag. mom. ratio	μ_e/μ_p	$-658.210\,6862\,(66)$
Deuteron magnetic moment	μ_d	$0.433\,073\,482\,(38) \times 10^{-26}$ J T^{-1}
$\pi = 3.141\,592\,653\,59$	e (exponential) = $2.718\,281\,828\,46$	$\ln 10 = 2.302\,585\,092\,99$

N.B. au = atomic units; uncertainty of last digit is shown in ( ).

Conversion factors for energy equivalents

From\to	Joule	Hertz	m^{-1}	Kelvin	eV
Joule	1	1.5091904×10^{33}	5.0341172×10^{24}	7.2429626×10^{22}	6.2415095×10^{18}
Hertz	$6.6260693 \times 10^{-34}$	1	3.3356410×10^{-9}	$4.7992372 \times 10^{-11}$	$4.1356674 \times 10^{-15}$
m^{-1}	$1.9864456 \times 10^{-25}$	2.9979246×10^8	1	1.4387751×10^{-2}	1.2398419×10^{-6}
Kelvin	$1.3806505 \times 10^{-23}$	2.0836644×10^{10}	6.9503564×10^1	1	8.6173432×10^{-5}
eV	$1.6021765 \times 10^{-19}$	2.4179894×10^{14}	8.0655444×10^5	1.1604505×10^4	1

Useful relations in EPR

Magnetic moment of the electron
$\mu_e = -g_e\,\mu_B\,S = -g_e\,\mu_B/2$ (g_e is the magnitude)

Magnetic moment for nucleus n with spin I_n
$\mu_n = g_n\,\mu_N I_n = \gamma_n\,\hbar\,I_n$

EPR resonance condition
$\nu_e = |g|\mu_B B_0/h$

$\nu_e(GHz) = 13.996246|g|B_0(T)$

$B_0(T) = 0.071447730\,\nu_e(GHz)/|g|$

$|g| = 0.071447730\,\nu_e(GHz)/B_0(T)$

$|g| = 3.04198626\,\nu_e(GHz)/\nu_{H_2O}(MHz)$

NMR resonance condition
$\nu_n = |g_n|\mu_N B_0/h = |\gamma_n|B_0/2\pi$

for ^1H:

$\nu_{H_2O}(MHz) = 42.5763875\,B_0(T)$

$B_0(T) = 0.0234871970\,\nu_{H_2O}(MHz)$

Hyperfine coupling

$A(MHz) = 2.99792458 \times 10^4\,A(cm^{-1})$
$\qquad = 13.996246|g|A(mT)$
$\qquad = 1.3996246|g|A(G)$

$A(cm^{-1}) = 0.333564095 \times 10^{-4}\,A(MHz)$
$\qquad = 4.6686451 \times 10^{-4}|g|A(mT)$
$\qquad = 0.46686451 \times 10^{-4}|g|A(G)$

Magnetic field (flux density) = B_0 (Tesla, T)

$1\,T = 10^4\,G = 10\,kG;\ 1\,mT = 10\,G;\ 1\,G = 0.1\,mT$

Index

2,2,6,6-tetramethylpiperidine-1-oxyl, *see* TEMPO
2,2-diphenyl-1-picrylhydrazyl, *see* DPPH
2-D-ELDOR, *see* two-dimensional electron-electron double resonance

a

absorption spectrum   100, 203, 428, 449, 536, 666, 668, 669
actinide ions   354
AFC, *see* automatic frequency control
AFM, *see* atomic force microscopy
amorphous systems   474, 475
analog-to-digital converter (ADC)   196, 203
analytic derivative approaches   311
angular frequency   97, 200, 205, 405, 526, 562, 739, 827
angular overlap model (AOM)   57, 58, 88, 89, 91, 92
anisotropic interactions   737, 741–742, 886, 903, 1000
anisotropic tumbling   505
antibonding orbital   85, 87
antiferromagnetic exchange interaction   479, 732
AOM, *see* angular overlap model
aqueous samples   33, 193, 229, 246, 252, 282, 785
arbitrary waveform generator (AWG)   195, 196, 234
asymmetry parameter   104, 319, 393
atomic force microscopy (AFM)   826
attenuator   5, 6, 162, 173, 194, 195, 764
automatic frequency control (AFC)   5, 9, 176, 238, 764, 765, 767, 768, 800
AWG, *see* arbitrary waveform generator

b

backward diode   8, 164, 165
backward-wave oscillators   12, 937

baseline   575, 655, 768, 828, 829
BC time   899
bimodal resonator   239
biological systems   29, 192, 197, 541, 619, 620, 622, 624, 626–628, 630, 632, 634, 636, 638, 640, 660, 661, 675, 711, 755, 758–760, 826, 838, 839, 885, 921, 939, 960, 967, 980, 1002
– free radicals   18, 19, 197, 755, 759, 788
– membranes   619, 632
Bloch decay   203
Bloch's phenomenological equations   185, 459, 720, 727, 759, 784, 926, 974, 998
– modified   974
blocking temperature   829, 845, 848, 869, 871
Bohr magneton   118, 126, 146, 185, 299, 335, 551, 620, 647, 649, 756, 852, 950
Boltzmann distribution   859, 933, 964
Boltzmann factor   30, 118, 250, 350, 535, 923
Boltzmann population   979
Born-Oppenheimer approximation   95
Born-Oppenheimer (BO) Hamiltonian operator   300
box-car integrator   194, 196
Breit-Pauli approximation   305
broadening, *see* electron paramagnetic resonance (EPR), line broadening
Brownian motion   499, 514

c

carbenes   960
catalytic activity   833, 839
CEF, *see* crystalline electric field
CFT, *see* crystal-field theory
chemical exchange   518, 927
chemical kinetics   960
chemical oxidation   881
chemical shift anisotropy (CSA)   738, 740

chemically induced dynamic electron polarization (CIDEP) 499, 947, 949, 952, 954, 967–984, 1001
– mechanism of 967, 969, 984
chemically induced dynamic nuclear polarization (CIDNP) 16, 499, 923, 947–949, 952–966, 969, 984
CIDEP, see chemically induced dynamic electron polarization
CIDNP, see chemically induced dynamic nuclear polarization
circulator 4, 5, 7, 138, 141, 173, 177, 182, 193–195, 765
– four-port 173, 174
Clebsch-Gordan coefficients 69, 96
coaxial components 6, 10, 119, 127–129, 131, 154, 157, 159, 161, 162, 164, 168, 173–175, 194, 200, 258, 262, 265, 281, 801, 802, 807, 956
coaxial transmission line 119, 161, 262, 801
cofactors 619, 639, 876, 877, 879, 882, 886, 890, 900, 902, 903, 965, 966
collisions 482, 485, 719, 727, 739, 740, 742, 743, 746, 747, 948
computer simulation of EPR spectrum, see electron paramagnetic resonance (EPR)
conduction electrons 495, 922
copper complexes 677, 679, 681, 686, 710
correlated ab initio approaches 303
correlation energy 303
correlation spectroscopy (COSY) 2, 215, 466, 518
correlation time, see rotational correlation time
COSY, see correlation spectroscopy
coulomb interaction 60, 61, 71, 73, 92
coupled-perturbed self-consistent field (CP-SCF) method 311, 320
coupled representation 81
coupling parameter 339, 887
Covalency 105
covalency effects 352, 354
CP-SCF method, see coupled-perturbed self-consistent field (CP-SCF) method
Crystal-field potential 65–67, 386, 469
crystal-field theory (CFT) 57–83, 85, 87, 102, 338
– Critique of 82
– d-orbital 63–65
– p-orbital 63
crystalline electric field (CEF) 59, 65, 66, 347, 348
crystalline materials 117

crystals
– coordinate system 883
– types of 504, 816
CSA mechanism, see chemical shift anisotropy (CSA) mechanism
Curie law 804, 810
cutting-edge topics in EPR 20, 999–1002
CW-EPR, electron paramagnetic resonance (EPR)
CYCLOPS 199
cyclotron resonance 12
cytotoxicity 833, 834

d
data collection 200, 285, 572, 768, 769, 773, 783
Dead time 191, 192, 200
DEER, see double electron-electron resonance
density functional theory (DFT) 298, 304, 305, 310, 311, 317, 324, 449, 648, 659, 672, 673, 675, 708, 710, 884, 885, 891, 892, 899, 1002
dephasing 456, 457, 483, 484, 549, 556, 721–723, 950, 959
descent of symmetry 329, 332, 381, 382
detection system 4, 5, 8, 14, 132, 146, 151, 180, 186, 187, 192, 272–278, 286, 405, 466
detector current 8, 134, 768
deuteron 167, 889
DFT, see density functional theory
diamond core 695, 849
dielectric constant 7, 192, 246, 248, 252, 254, 257, 266, 268, 625, 705, 706, 837
– complex 248, 252, 268, 705
dielectric resonator oscillator (DRO) 180, 182
dipolar interactions 16, 548–550, 554, 556, 719, 720, 724, 726, 734, 738, 739, 741, 742, 746, 831, 860, 867, 872, 925, 928
– electron-electron 548, 549
– electron-nuclear 549
dipole moment 880
disordered state (glassy) spectrum 437, 438
dispersion 148, 149, 187, 203, 217, 277, 288, 289, 469, 721, 723, 731, 737, 744, 927, 941, 959
distance measurement by EPR 545–575
DNP, see dynamic nuclear polarization
double electron-electron resonance (DEER), same as pulsed electron double resonance (PELDOR) 2, 20, 218, 219, 288, 291, 545, 546, 549–559, 570–575, 577, 579, 581, 583, 722, 737, 836, 893, 894, 903, 995, 1001
– thermal noise 187, 191, 271, 272, 275, 278, 280, 290, 809

double-perturbation theory (DPT) 311
double quantum coherence (DQC) 2, 17,
  20, 220–223, 288, 291, 545, 546, 549, 559–
  576, 583, 584–588, 722, 737, 836, 995, 1001
DONUT-HYSCORE 213, 217, 218
DPPH (2,2-diphenyl-1-picrylhydrazyl) 47,
  139, 140, 238, 281, 724, 725, 733, 742, 796,
  856
DPT, see double-perturbation theory
DQC, see double quantum coherence
dressed-state solid effect (DSSE) 936
DRO, see dielectric resonator oscillator
DSSE, see dressed-state solid effect
dynamic nuclear polarization (DNP) 188,
  253, 720, 741, 921–926, 928–943, 978, 996,
  997
– high-field DNP 923, 942, 943

e
echo amplitude 215, 216, 465, 550, 553,
  554, 557, 560, 570, 581, 588, 722, 811
EDM, see electrical discharge machining
effective spin Hamiltonian 110, 853, 860
EFG, see electric field gradient
EHMO, see extended Hückel molecular
  orbital theory
eigenvalue equation 450, 453, 454
eigenvalues of spin Hamiltonian 102, 333,
  411
ELDOR, see electron-electron double
  resonance
electric field gradient (EFG) 307, 319
electrical discharge machining (EDM) 251
electromagnetic field 239, 247, 264, 268
electromagnetic radiation 755, 756, 914
– excitation 914
electron
– charge on 339
– mass of 95, 104
electron density 302, 304, 319, 683, 685, 929
– at the nucleus 929
electron distribution 16
electron-electron double resonance (ELDOR)
  2, 16, 17, 20, 129, 180, 218, 223, 229, 235,
  238–242, 288, 466, 497, 498, 516–522, 540,
  549, 556, 740, 934
electron magnetic resonance (EMR) 1
electron-nuclear double resonance (ENDOR)
  129, 180, 181, 208, 209, 211–217, 223, 281,
  289, 386, 404, 412, 415, 429, 437, 498, 599,
  648, 650, 656, 658, 690, 691, 720, 721,
  882–889, 891, 895, 897, 899, 901, 903, 941,
  995, 1000–1002
– cavity 181, 281, 941

– CW 129, 180, 211, 212, 281, 289, 648,
  656, 658, 721, 889, 995, 1000
– double 180, 209, 217, 404, 429, 498, 648,
  882, 895, 903, 941, 995
– energy levels 129, 209, 386, 404,
  599
– enhancement 211, 941, 1001
– frequency 129, 180, 181, 211, 212, 216,
  281, 289, 386, 437, 599, 648, 650, 656, 658,
  720, 721, 757, 883, 884, 888, 889, 897, 899,
  901, 903, 995, 1000, 1001
– lines 217
– pulsed 129, 180, 208, 211, 212, 214, 223,
  289, 648, 650, 656, 658, 889, 895, 899, 903,
  941, 995, 1000, 1001
electron paramagnetic resonance (EPR)
– Abragam-Pryce spin Hamiltonian (iron
  group), see electron paramagnetic
  resonance (EPR) 102
– absorption signal 4, 9, 134, 177, 185, 248,
  428, 449, 768, 775, 863, 949
– – first derivative of 9, 428, 449, 775
– applications of EPR 15–17, 19, 20, 623,
  729, 786, 818, 1003
– continuous-wave (CW) EPR, see electron
  paramagnetic resonance (EPR) 1, 128–
  173, 231, 448, 461, 500–513, 545–548, 558,
  648, 690, 721, 736, 755, 760–774, 798–805,
  836, 879, 974, 995
– – spectrometer 3, 9, 128, 129, 135, 136,
  140, 141, 143, 145, 147–149, 153, 159,
  162, 168, 169, 173, 176, 177, 760, 799,
  800, 975
– ENDOR-induced 212
– EPR data, theoretical interpretation
  1002
– experimental considerations 229–290,
  663
– frequency range of 9, 230, 242
– future 20, 1003
– high-field 19, 176, 282, 283, 291,
  500–514, 830, 837, 854, 871, 875–879,
  882–884, 886, 890, 893–904, 976, 997,
  998, 1000
– imaging (EPRI) 2, 19, 191, 197, 199–201,
  689, 755–790, 795–801, 803, 805–808, 810,
  817–819, 821, 852, 1000
– in vivo 19, 163, 197, 741, 759, 765, 767,
  768, 786, 886
– line broadening 35
– lineshapes 148, 499, 506, 508, 511, 515,
  726, 787, 819, 854, 888
– low-frequency EPR 34, 35, 37, 262, 497,
  661–663, 677, 767

– microscopy   795, 796, 798, 799, 801–803, 805, 807, 808, 810–812, 815, 816, 818, 819, 821
– pulsed   2, 7, 15–17, 20, 129, 156, 180, 182, 190–224, 235, 279, 280, 284–286, 288, 290, 463, 466, 497, 514–522, 541, 545–586, 623, 639, 648–650, 655, 656, 658, 711, 745, 769, 783, 784, 795, 796, 798, 805–810, 816, 836, 872, 882, 895, 896, 902, 903, 976, 997–1000, 1003, 1004
– scope of   139, 222, 276, 305, 759
– sensitivity   53, 132, 270–292, 786
– signals   19, 199, 229, 241, 248, 341, 463, 624, 709, 711, 761, 896, 975, 983
– spectra
– – computer simulation of   36, 437, 447, 648, 655–659, 673, 702–705, 707, 709, 711
– – extent of   664, 724
– – Fourier-transformed   655
– – polycrystalline (powder) (EPR) spectrum   42, 47, 117, 153, 409, 410, 417, 421, 423–427, 429–433, 435–437, 449, 451, 591, 593, 594, 628, 725, 850, 855
– – single-crystal EPR spectrum   417, 419
– – turning points in EPR spectra   37, 592–594, 722
– spectrometer   3–5, 9–11, 24, 128–133, 135–141, 143, 145–149, 151, 153, 155–157, 159, 161–169, 171–177, 179, 181, 183, 185, 187, 192–198, 202, 270–292, 591, 632, 633, 719, 759–761, 783, 785, 788, 798–801, 854, 855, 975, 998, 1000, 1004
– – block diagram of   10, 11, 133, 141, 145, 159, 162, 171, 172
– – cavity   5, 6, 8, 9, 12, 130, 137–139, 141, 143, 153, 238
– – CW   3, 9, 128–133, 135–141, 143, 145, 147–149, 151, 153, 155–157, 159, 161–169, 171, 173, 176, 177, 197, 202, 270–292, 760, 783, 798–801, 975, 998, 1000
– – frequency of   130, 132, 135, 141, 143, 149, 151, 157, 164, 171, 194, 800, 801
– – pulsed   129, 151, 177, 192–194, 197, 202, 279, 280, 284, 286, 290, 783, 798, 975, 998, 1000, 1004
– – sample tubes (extruded)   255
– tetrahedral field   77, 338, 340, 341, 344
– theory   57–111, 300, 498, 1003
– transitions   28, 43, 48, 49, 100–102, 110, 209–212, 361, 591, 650, 853, 854, 903, 923, 924, 926, 967, 973

– W-band EPR   12, 27, 28, 41, 52, 176, 177, 179, 185, 210, 229, 230, 235, 236, 240–242, 257, 263–265, 282, 285, 290, 429, 623, 625, 626, 641, 653, 724, 726, 728, 734, 736–738, 740, 747, 839, 840, 884, 886–890, 892, 894, 896–900, 939, 976, 977, 995
– X-band EPR   5, 23, 49, 281, 285, 423, 637, 677, 680, 681, 684, 691, 693, 695, 696, 702, 785, 837, 856, 871, 879, 881, 882, 886, 894, 895, 900, 902
– zero-field EPR (ZFR)   16, 28, 117–128, 159, 167, 168, 209, 314, 341, 392, 417, 441, 591, 592, 620, 650, 719, 853, 854, 863, 871, 879, 895, 903, 927, 968, 998, 999
electron spin
– angular momentum   209, 720, 727, 740
– effective   95, 216, 235, 273, 298, 361, 521, 638, 740, 797, 812, 890, 971, 972
– energy levels of   105, 209
electron spin echo (ESE)   2, 219, 429, 437, 463–465, 514–517, 641, 879, 887, 890, 892, 897, 898, 903, 975, 976, 981, 999
electron spin echo envelope modulation (ESEEM)   2, 40–42, 208, 209, 211, 214, 215, 223, 291, 466, 575, 648, 655, 658, 659, 722, 995, 1000
electron transfer   639, 875, 876, 965, 966, 979, 980, 985
electron-transfer reaction   17, 955, 963
electronic states   85, 95, 96, 103
electronic transitions   941
EMR, same as EPR, see electron paramagnetic resonance (EPR)
ENDOR, see electron-nuclear double resonance
energy barrier   845, 846, 849, 853, 868, 869
EPR, desirable applications of   1003
EPR, see electron paramagnetic resonance
EPR instrumentation, desirable advances in   995, 998
EPR toolkit   649–659
EPRI, see electron paramagnetic resonance (EPR), imaging
EPRM, see electron paramagnetic resonance (EPR), microscopy
ESE, see electron spin echo
ESEEM, see electron spin echo envelope modulation
ESR, same as EPR, see electron paramagnetic resonance (EPR)
EXAFS, see extended X-ray absorption fine structure
exchange integral   828
exchange narrowing   724–725

excitation magnetic field   871
excitation pulse   190, 572, 587, 759, 784, 810, 958
expectation value   319, 424, 529, 531, 533, 554, 556, 579, 925
extended Hückel molecular orbital theory (EHMO)   88
extended time excitation (ETE)   215
extended X-ray absorption fine structure (EXAFS)   901

f
Fabry-Pérot resonator   177, 187, 201
fast dynamics   522, 528
fast internal motion (FIM) model   630, 631
Felgett advantage   204
Fermi-contact interaction   339, 344, 360
ferromagnetic materials   826
ferromagnetic resonance (FMR)   826–828, 831
ferromagnets   20, 825
FID, *see* free induction decay
field modulation   11, 14, 119–121, 127, 130–136, 141, 143, 145, 148, 159, 162, 165, 172, 178, 241, 246, 247, 263, 268, 281, 282, 663, 678, 761, 768, 776, 784, 799, 800, 803, 974–976
filling factor   13, 118, 146, 154, 156, 157, 164, 168, 183, 201, 248, 254, 255, 272–276, 278, 281, 289, 290, 572, 663, 765, 766, 785, 809, 812, 941
FIM model, *see* fast internal motion model
fine structure   30, 35, 103, 123, 124, 208, 305, 339, 343, 353, 357–360, 362, 387, 394, 402, 429, 437, 442, 454, 467, 652, 670, 674, 697, 758, 855, 884, 900, 901, 1001
first-derivative   428, 449
flash-photolysis   955
flat cell   33, 246, 255–257
fluorescence   546, 797, 952
fluorescence resonance energy transmission (FRET)   546, 558, 559
FMR, *see* ferromagnetic resonance
forbidden transitions   44, 211, 289, 432, 438, 651, 662, 704, 705, 711, 741, 923, 926, 930
Fourier transform EPR (FTEPR)   2, 19, 198, 517
free electron   47, 79, 238, 459
free induction decay (FID)   190–192, 197, 202–207, 238, 241, 242, 280, 281, 284, 517, 518, 526, 530, 658, 720, 759, 783, 784, 808, 809, 957
Fremy's salt   723, 724, 741

frequency bands   23, 24, 35
frequency domain   203, 231, 536, 552, 557, 783
frequency multiplication   231, 232, 235–237
frequency translation   231, 235, 237
FRET, *see* fluorescence resonance energy transmission
Friis equation   278
FTEPR, *see* Fourier transform EPR
full-width at half-maximum (FWHM)   203, 420, 569, 769, 770, 863
fullerenes   747
FWHM, *see* full-width at half-maximum

g
gadolinium   726, 744–746
Gaussian lineshape   148, 420, 428, 434, 857
Gunn diodes   10

h
Hahn echo, two pulse   207, 214, 215, 553
Hall-effect   8
Hamiltonian, *see also* spin Hamiltonian
Hamiltonian operator   300, 509
hard pulses   998
Helmholtz coils   7, 8, 167
HFEPR, *see* high-frequency and high-field (HFHF) EPR   16, 18, 175–177, 179, 181, 183, 185, 187, 188, 276, 281, 284, 285, 287, 291, 497, 503, 525, 591, 619, 722, 830, 855, 871, 875, 876, 877, 878, 879, 882–884, 886, 890, 894–904, 995, 997, 999, 1000, 1001, 1002
high-resolution NMR   218, 921, 938–940, 955, 956, 959
highest occupied molecular orbital (HOMO)   671, 896
hindered rotation   101
HOMO, *see* highest occupied molecular orbital
homogeneous broadening   9, 41, 100, 456, 461, 467, 497, 499, 514, 518, 521, 541, 548, 934
homotopy   402, 409, 417, 421, 423–427, 429, 430, 432
Hund's rule   60, 335, 347, 349
hybrid functionals   305, 321, 322
hydrocarbon   268, 623, 626, 837, 1001
hyperfine interaction   16, 30, 33, 35, 37, 39, 41, 92, 98, 100, 103, 104, 127, 208, 289, 361, 386, 392, 457, 475, 478, 638, 640, 641, 648, 650, 652, 655, 661, 670, 674, 721, 726, 738, 741, 872, 879, 883, 886, 889, 902, 903, 949, 954, 967, 982, 999, 1000

– anisotropic   16, 100, 648, 655, 879, 883, 886, 903, 954, 1000
– conformers   522
– coupling constant   703, 709, 782, 783, 837, 901
– dipolar   16, 208, 475, 872, 886, 903
– isotropic   16, 41, 100, 103, 127, 361, 386, 638, 648, 655, 879, 883, 886, 903, 954, 1000
– – theoretical considerations   954
– second-order effects   638, 879
hyperfine lines   204, 238, 242, 353, 360, 430, 431, 607, 671, 678, 683, 686, 691, 695, 711, 741, 758, 782, 783
hyperfine splittings   16, 191, 212, 217, 238, 289, 352, 360, 467, 481, 498, 621, 648, 664, 683, 721, 757–759
– isotropic   498, 648
hyperfine structure   35, 103, 123, 124, 339, 343, 353, 357–360, 362, 387, 394, 402, 429, 437, 454, 467, 670, 674, 697, 724, 758, 884, 900, 1001
hyperfine sublevel correlation experiment (HYSCORE)   2, 208, 214, 215, 217, 218, 648, 655, 656, 658, 703, 711, 995, 996
hyperpolarization   963
HYSCORE, see hyperfine sublevel correlation experiment

i

imaging, see electron paramagnetic resonance (EPR), imaging (EPRI)
impedance matching   201
induction-mode detection   177, 181, 187
inelastic neutron scattering (INS)   597, 862
inhomogeneous broadening   9, 100, 456, 461, 467, 497, 499, 514, 518, 521, 541, 548, 934
inorganic radicals   191, 204
INS, see inelastic neutron scattering
integrated solid effect (ISE)   936
intensity, see electron paramagnetic resonance (EPR), line intensity
Inversion Recovery   204, 206
iris   5, 7, 154, 155, 157, 201, 247, 251, 261–265
irreducible spherical tensor operator (ISTO)   583
ISE, see integrated solid effect
isotropic tumbling   507
ISTO, see irreducible spherical tensor operator

j

Jahn-Teller (JT)
– distortion   92, 93, 342–344
– effect   92–95, 97, 99, 101, 342–344, 346, 347, 350
– – dynamic JT effect   342, 346, 347
– – static JT effect   347
– Mexican-hat type adiabatic potential   94
– perturbation with the vibronic ground state   98–100
– theorem   92
– theory of JT effect   95–98
– three-state model   100
– transition from dynamic to static JT effect   101
JT effect, see Jahn-Teller (JT) effect
jump model   954

k

kinetic energy   59, 96, 300, 305, 307, 937
kinetic-energy operator   318
klystron   4–6, 8, 9, 12, 119, 130, 137–139, 141, 143–145, 151, 153, 162, 238, 397, 401, 422, 429, 432
– frequency   5, 6, 8, 130, 141, 143, 151, 397, 422, 432
– local oscillator   8, 139, 141, 143, 151
– mode   4, 5, 8, 12, 130, 138, 238
– temperature of   130
Kohn-Sham (KS) procedure   304
Kramers degeneracy   92, 339, 340, 345, 350, 354, 360
Kramers doublet   100, 101, 122, 334, 340–343, 345, 350–352, 354, 356–358, 360–362, 449, 470, 486, 637, 650
Kramers ions   29, 30, 40, 42, 81, 82, 92, 100, 101, 122, 148, 334, 339–343, 345, 350–352, 354, 356–358, 360–362, 449, 470–473, 486–489, 592, 593, 595, 620, 637, 650, 733
Kramers-Kronig relation   148
Kramers theorem   82, 92

l

Lanczos algorithm (LA)   499, 508
Landé interval rule   61
Larmor frequency   211, 212, 348, 459, 460, 463, 526, 527, 529, 563, 564, 587, 724, 729, 742–746, 888, 890, 922, 929–931, 933, 934, 939
Larmor precession   2, 525–528
LCAO, see linear combination of atomic orbitals

least-squares fitting (LSF) method 32, 41, 324, 385, 395–405, 407–410, 416, 421, 422, 430, 432–436, 450, 591, 604, 606–609, 611, 613–615, 617
LFT, *see* ligand-field theory
LGR, *see* loop-gap resonator librations 721
lifetime broadening 467, 515, 858
ligand-field theory (LFT) 57, 58, 85, 637
– Atomic-orbital model (AOM) 88–92
line broadening, *see* electron paramagnetic resonance (EPR), line broadening
line intensity, *see* electron paramagnetic resonance (EPR), line intensity
linear combination of atomic orbitals (LCAO) 58, 85, 86, 88, 89, 105
linear response theory (LRT) 310, 311
linear scaling 303
lineshapes, *see* electron paramagnetic resonance (EPR), lineshapes
lithium phthalocyanine 724, 760
LNA, *see* low-noise amplifier
local magnetic field, *see* magnetic field, local
local polarity 619, 626, 628
lock-in amplifier 11, 172, 281, 774, 800, 801
lock-in detector 137, 139, 768, 774, 775
LODEPR, *see* longitudinally detected paramagnetic resonance
LODPEPR, *see* longitudinally detected pulsed EPR
longitudinally detected paramagnetic resonance (LODEPR) 463
longitudinally detected pulsed EPR (LODPEPR) 463, 466
loop-gap resonator (LGR) 7, 10, 52, 121, 163, 200, 201, 230, 231, 238–240, 244–247, 251, 258–262, 268, 270, 274–277, 575, 663, 664, 675, 692, 766, 808
Lorentzian lineshape 148, 420, 428, 434, 435, 442–446, 461, 769–771, 810
low-noise amplifier (LNA) 179, 180, 235, 237, 241
lowest unoccupied molecular orbital (LUMO) 671–673, 896
LRT, *see* linear response theory
LSF method, *see* least-squares fitting method
LUMO, *see* lowest unoccupied molecular orbital

m

magic angle spinning (MAS) 95, 97, 104, 141, 223, 307, 336, 468, 485, 700, 702, 703, 705, 707, 709, 797, 923, 935, 936, 938–940, 942, 958, 964, 965, 998

magnet system 5, 921, 938
magnetic and spin effects 954
magnetic circular dichroism (MCD) 597, 691
magnetic dipole 314, 345, 564, 565, 585, 721
– electron 314, 564
– – spin 314, 564
– nuclear 345, 585
magnetic energy 156, 812, 915
magnetic field
– applied 25, 105, 154, 248, 250, 264, 647, 917, 936
– direction 551, 632
– gradients 755, 761, 762, 767, 769, 770, 775, 784, 796, 808
– local 550, 757
– modulation 14, 120, 127, 130–136, 141, 143, 145, 148, 159, 162, 165, 172, 178, 263, 278, 281, 282, 784, 975, 976
– orientation of 2, 16, 31, 33, 44, 105, 119, 133, 221, 250, 385, 386, 394, 402, 404, 406, 409, 417, 418, 425, 430, 510, 520, 565, 607, 631, 632, 652, 719, 737, 831, 858, 863, 890
– oscillating 663, 768, 936
– resonant 386, 397, 412, 414, 450
– stability of 7, 8, 18, 138, 166, 167, 761, 841
– static 7, 8, 117, 133, 247–250, 273, 307, 520, 745, 746, 760, 761, 763, 784, 796, 799, 800, 804, 807, 808, 825, 827, 831, 862, 871
magnetic force microscopy (MFM) 826
magnetic moment 2, 15–17, 80, 103, 104, 223, 270, 311, 336, 346, 348, 349, 405, 424, 458, 475, 481, 547, 756, 757, 760, 828, 830, 833, 846, 847, 868
– classical 828, 846
– electron 2, 15–17, 103, 104, 223, 270, 311, 405, 424, 475, 481, 547, 756, 757, 760, 833
– nuclear 15, 16, 103, 104, 311, 336, 346, 348, 475, 481
– proton 223
magnetic quantum tunneling (MQT) 845, 846, 852–857, 859, 861, 863–869, 872
magnetic resonance force microscopy (MRFM) 797, 798, 821, 998
magnetic resonance imaging (MRI) 191, 193, 199, 262, 277, 455, 744, 745, 755, 760, 852, 921, 923, 938, 940, 942, 943, 1000
magnetic susceptibility 92, 148, 156, 184, 185, 206, 248, 272, 273, 350, 635, 803
– complex 148, 156, 248
– static 273

magnetization   2, 203–207, 216, 218, 220, 221, 248–250, 273–275, 282, 456–460, 463, 465, 466, 515–517, 524, 527–528, 531, 536, 579–581, 759, 804, 808, 809, 828–830, 845–850, 857, 867–870, 958, 959, 963
– longitudinal   458, 517, 528
– time dependence of   524, 719–748
– transverse   203, 205–207, 216, 220, 221, 456, 458, 463, 524, 527–529, 579, 719–748, 850, 958
magnetogyric ratio   9, 479, 936
– electron   9, 936
– nuclear   479, 936
magnetrons   151, 457
magnets
– pulsed   17, 921, 1001
– superconducting   14, 23, 51, 176
many-body perturbation theory (MBPT)   303
marine cyclic peptides   701
MAS, see magic angle spinning
maximum-entropy method (MEM)   574
MBPT, see many-body perturbation theory
MCD, see magnetic circular dichroism
MEM, see maximum-entropy method
metalloenzymes   208, 217, 648, 654
metalloproteins   29, 429, 489, 593, 620, 636–639, 649, 650, 658, 675, 690, 711, 746, 879, 883, 903, 1000
methyl radical   280, 984, 985
MFG method, see modulated field gradient method
MFM, see magnetic force microscopy
micro-imaging   805–807, 818
microscopic order-macroscopic disorder (MOMD) model   498, 504, 505, 509–513, 520, 521, 525, 531, 535–536, 539, 630–632, 635
microwave detector   129–131, 136, 164, 937
microwave magnetic field   7, 117, 133, 274, 282, 353, 424, 439, 563, 941, 949
microwave power   6, 8, 129, 130, 132, 136, 139–141, 143, 146, 147, 149, 151, 153, 165, 168, 171, 173–175, 231, 235, 245, 246, 248–250, 271, 272, 275, 276, 428, 461, 462, 678, 692, 766, 930, 934, 936, 937, 941
microwave radiation   4, 5, 7, 10, 146, 164, 827, 931, 932, 939, 940
microwave rectifier   8
mineralogical systems   20
mixed-valence dinuclear copper   677, 686
mixer   10, 14, 15, 136, 138–140, 142, 147, 176–178, 180–183, 186, 195, 199, 231–233, 235, 278

mixing period   216, 940
modulated field gradient (MFG) method   796, 805
modulation amplitude   9, 133, 148, 668, 691, 692, 769, 775, 776, 799
modulation field   121, 132, 264, 766, 975
modulation frequency   4, 8, 119, 130, 132–136, 139–141, 162, 165, 172, 175, 241, 722, 801
modulation of saturation (MOS)   241, 242
modulation sidebands   229, 242
modulation system   5, 133, 147
molecular dynamics and stochastic trajectories   522, 524, 539–541
molecular motion   497–540
molecular motion studied by EPR   497, 540
Molecular Orbital (MO) approach   85–91
molecular-orbital (MO) approach   57, 58, 85–91, 301, 338, 415
molecular Sophe   656, 658, 659, 675
molecular tumbling   484, 520, 738–741, 968
– anisotropic   484, 738, 740, 968
– isotropic   484, 738, 740, 968
MOMD model, see microscopic order-macroscopic disorder (MOMD) model
MOS, see modulation of saturation
MQEPR, see multi-quantum EPR
MQT, see magnetic quantum tunneling
MRFM, see magnetic resonance force microscopy
MRI, see magnetic resonance imaging
multi-quantum EPR (MQEPR)   466
multiarm EPR spectroscopy   229–239, 241, 242
multifrequency EPR   18, 20, 23–53, 229–291, 522–539, 649–710, 811–814, 852–863, 882–901
– advantages   24
– data   385–416, 498, 591, 833, 837
multiquantum ELDOR   218
multiquantum (MQ) EPR   229

n

nanoparticles   720, 825–839, 841, 842
nanoscale magnets   845
narrow-band detection, see phase-sensitive detection (PSD)
nitrogen atom   701, 709, 851
nitroxide spin labels   17, 19, 503, 514, 521, 524, 541, 548, 619, 620, 628, 669, 737–741, 758, 836, 894, 943, 1000
NMR, see nuclear magnetic resonance
noise factor   146, 147, 185, 278
noise, source   278

noninteracting reference system   304
NQR, *see* nuclear quadrupole resonance
nuclear hyperfine interaction, *see* hyperfine interaction
nuclear magnetic resonance (NMR)
– chemical shift   318, 953, 959, 964
– signal   760, 921, 939, 949, 957, 966
– spectroscopy   156, 159, 164, 168, 202, 223, 745, 783, 921, 923, 935, 938–940, 942, 943, 949, 953, 956–959, 961, 964, 965
– transition   211, 926
nuclear magneton   104, 387
nuclear quadrupole resonance (NQR)   119, 429, 437
nuclear spin
– eigenfunctions   969
– quantum numbers   37, 386, 951
nuclear Zeeman energy, *see* Zeeman energy, nuclear
nuclear Zeeman terms, *see* Zeeman terms, nuclear
nuclear-quadrupole   387
– interaction   387
– spin Hamiltonian   387

o
observable   40, 140, 221, 301, 499, 596, 665, 711, 960
octahedral
– field   336, 338–345, 362
– symmetry   64, 69, 77, 96, 98, 333–335, 342, 344–346, 354, 357, 361
ODEPR, *see* optically detected electron paramagnetic resonance
operators
– Equivalent   69–71
– linear   58, 301, 469
– product   58, 70, 328, 586
– summation   328
optical spectrometer   855
optically detected electron paramagnetic resonance (ODEPR)   952
optimum sensitivity   212
orbital degeneracy   72, 73, 78, 92, 333, 343
organic radical   191, 202, 204, 324, 468, 479, 485, 498, 619, 620, 622, 730, 929, 952, 1002
orthorhombic symmetry, *see* rhombic symmetry
oscillator stability   139, 141, 153, 171
Overhauser effect (DNP)   218, 495, 790, 922, 924, 963
oximetry   19, 742, 760, 773, 786, 787
oxygen atom   339, 886, 895

p
parahydrogen-induced polarization (PHIP)   923
Pauli spin matrices   350, 532
PCM, *see* point-charge model
PDS, *see* pulsed dipolar spectroscopy
PELDOR, same as DEER, *see* double electron-electron resonance (DEER)
performance network analyzer (PNA)   251
permeability   263, 273, 762, 803
perturbation theory   62, 102, 303, 308, 311, 412, 425, 486, 600, 637, 656, 657, 661, 972
– second-order   303, 972
phase coherence   457, 556
phase-cycling modality   221
phase-locked loop (PLL)   131, 176, 179, 233
phase memory   207, 214, 216, 720, 721, 724
phase noise   9, 179, 180, 186, 187, 234, 236, 237, 280, 281, 282, 663, 767, 768
phase-sensitive detection (PSD)   8, 119, 134, 279, 785
phase-sensitive detector   9, 172, 278, 761, 768
– output polarity of   9
phase shifter   8, 137, 162, 174, 194, 195, 765
PHIP, *see* parahydrogen-induced polarization
phonon density   470, 727, 728
phonons   467–475, 477, 478, 485–488, 494, 495, 727–730, 732–734, 889
photo-CIDNP, *see* photochemically induced dynamic nuclear polarization
photoactivated catalysis   827, 832
photochemically induced dynamic nuclear polarization (photo-CIDNP)   948, 955, 957, 958, 961, 963–966
photolysis   955, 959, 960, 976–979, 981
photon   153, 219, 832, 833, 915, 917, 966, 985
– absorption of   833
– energy of   917
– microwave   153, 915, 917
– types of   832
photosynthesis   222, 639, 876–886, 889–891, 895, 897, 899, 901, 902, 964, 966, 980
photosynthetic reaction center   18, 46, 882, 883, 885, 887, 889, 891, 893–895, 963
PLL, *see* phase-locked loop
PMR, same as EPR, *see* electron paramagnetic resonance (EPR)

PNA, *see* performance network analyzer
point-charge model (PCM)   57, 67, 83
polarization   14, 16, 103, 118–120, 177–179, 188, 211–213, 249, 250, 253, 255, 256, 258, 273, 322, 467–469, 499, 720, 860, 892, 893, 921–924, 926, 928, 930–936, 938–943, 947, 948, 950, 952–984, 996, 999, 1001
polymers   19, 20, 221, 518, 938
population difference   212, 489
power saturation   202, 249, 250, 273, 457, 731, 742
principal axes   41, 95, 338, 341, 401, 421, 438, 510, 565, 599, 606–611, 614–616, 631, 633, 672, 884, 895, 899, 968
principal-axes system   566
principal directions   388, 390, 415, 605, 608, 884
principal values   27, 41, 103, 344, 449, 607, 609, 611, 615–617, 619, 621, 622, 640, 673, 892, 895, 899, 999
proton transfer   960
PSD, *see* phase-sensitive detection
pulse sequence   204, 209, 213–215, 218, 221, 288, 464, 466, 555–557, 560, 571, 578, 582, 584, 655, 656, 721, 806, 889, 956–958, 1003
pulsed dipolar EPR spectroscopy (PDS)   219, 220, 288, 545–586
pulsed EPR   2, 7, 15–17, 20, 129, 156, 180, 192–199, 201–203, 205, 208, 209, 220, 222, 223, 235, 279, 280, 284–286, 288, 290, 463, 466, 497, 514, 515, 517, 519, 521, 540, 545, 556, 648–650, 656, 769, 796, 805, 895, 902, 976, 1000

q

quadrature detection   199, 997
quadrupole interaction   35, 98, 104, 105, 110, 307, 386, 602, 650
– electronic   98, 307, 386
quantum chemistry   297, 298, 300, 305, 311, 639, 648, 649, 656–659, 700, 711, 1002
quasi-optics   13, 177, 180
Quenching of angular momentum   78–80

r

Rabi frequency   936
radiation source   192, 761
radical pair mechanism (RPM)   896, 967, 969–971, 974, 984
radical pairs   18, 892–894, 899, 903, 942, 947, 948, 950, 963, 970
radical scavenging   833, 835

rare-earth ions $4f^n$ 15, 17, 59, 80, 81, 83, 347–352, 359, 360, 362, 448, 449, 489, 755
– ground states of rare-earth ions   347, 348, 351, 360, 362
reaction centers   18, 875, 876, 882, 883, 885, 887, 889, 891, 893–895, 963, 981
reaction yield-detected magnetic resonance (RYDMR)   947, 952
receiver   130, 132, 138, 191, 193–197, 199–201, 230, 235, 237, 239, 571, 940, 956
redox   19, 197, 639, 788, 837, 842, 877, 900, 903, 966
refocusing pulse   206, 215
relativistic corrections   305, 307, 354
relaxation in amorphous materials   474
relaxation in dilute liquids   468, 479
relaxation-induced dipolar modulation enhancement (RIDME)   893, 903
relaxation mechanisms   46, 53, 205, 468, 469, 471, 473, 475, 477–479, 481, 483, 485, 487, 489, 495, 536, 720–748, 922, 1003
relaxation times
– electron spin-lattice   720–748
– measurement of   191, 242, 463, 744, 745, 760
resonance condition   42, 386, 393, 426, 437, 603, 607, 649, 756, 771, 799, 827, 889, 890
resonant frequency   4–6, 132, 135, 147, 149, 154–157, 159, 160, 163, 164, 166, 231, 268, 275, 515, 657, 732, 767
resonant mode   247
resonant structure   129, 136, 154, 163, 164, 168, 171, 632, 766
resonator
– helical   164, 941
– loop-gap (LGR)   7, 10, 52, 121, 154, 163, 200, 201, 230, 244, 245, 258, 259, 270, 274, 276, 575, 663, 692, 766, 808
resonator efficiency parameter   231, 246, 248, 258
resonator frequency   233, 765
resonator quality factor (Q)   118, 146, 185, 199–201, 204, 273, 275, 464, 765, 803, 812
restricted open-shell HF (ROHF) method   302
rhombic symmetry   108, 343, 361, 386, 413
RIDME, *see* relaxation-induced dipolar modulation enhancement
ROHF method, *see* restricted open-shell HF method
rotating frame   202, 204, 205, 219, 241, 246, 248, 249, 459, 527, 532, 578, 720, 809, 931, 936

rotation matrix   509, 532
rotational correlation time   25, 35, 36, 481, 482, 484, 500, 743, 969, 980
RPM, *see* radical pair mechanism
ruby   20, 84, 85, 119–121, 734
RYDMR, *see* reaction yield-detected magnetic resonance

s

S-state ions   84, 333, 341, 346, 350–352, 358–360, 408
sample access stacks   265
sample size   13, 132, 167, 168, 271, 274, 275, 277, 279, 575, 762, 763, 803, 941
samples, reference   281–290
saturation
– partial   741
saturation recovery   229, 463, 464, 740, 747
saturation transfer   456, 1000
SBM theory, *see* Solomon-Bloembergen-Morgen theory
scanning electron microscopy (SEM)   826, 839
SCGs, *see* superconducting gaps
SCRP, *see* spin-correlated radical pair
SDF, *see* spectral density functions
SDSL, *see* site-directed spin labeling
SECSY, *see* spin-echo correlated spectroscopy
secular determinant   79, 86, 361
secular equation   88, 89
SEM, *see* scanning electron microscopy
SEMF, *see* switching of the external magnetic field
sensitivity, optimum   212
SHF, *see* super-hyperfine interaction
SHP, *see* spin-Hamiltonian parameters (SHP)   404, 416
sidebands   139, 143, 150, 151, 165, 229, 231, 232, 234, 235, 241, 242
signal-to-noise ratio (SNR)   10, 52, 53, 139–141, 143, 149, 191, 196, 201, 271–285, 287, 288, 290, 291, 522, 570–572, 574, 655, 663, 675, 733, 760, 769, 774, 775, 782, 784, 786, 790, 795, 798, 803–805, 808–812, 814–817, 821, 921, 958, 976, 998
silicon crystal   6, 8
similarity transformation   530
single-molecule magnets (SMMs)   16, 20, 845–872
single pole double throw (SPDT) switch   195
single pole-single throw (SPST) switch   195
single-sideband modulator (SSM)   232

single-wall carbon nanotubes (SWNTs)   720, 747, 830
singlet state   96, 102, 343, 877, 880, 893, 948–950, 952, 968, 972, 981
singly occupied molecular orbital (SOMO)   672, 888
site-directed spin labeling (SDSL)   219, 245, 497, 545, 827, 1000
site symmetries   327, 328, 330, 332, 334, 336, 338, 340, 342, 344, 346, 348, 350, 352, 354, 356, 358, 360, 362, 366, 368, 370, 372, 374, 376, 378, 380, 382–384, 437
six-pulse DQC   559–563, 565, 567, 584, 586, 587
Slater determinant   57, 60, 304
SLE, *see* stochastic Liouville equation
slide-screw tuner   7
slow-motion EPR spectrum, simulation of   499, 505, 506, 508, 511, 515
slowly relaxing local structure (SRLS) model   498, 504, 505, 511, 512–514, 521, 525, 531, 536, 539, 620, 630
SLR, *see* spin-lattice relaxation
SMART HYSCORE   217
SMMs, *see* single-molecule magnets
SNR, *see* signal-to-noise ratio
SOC, *see* spin-orbit coupling (SOC)
solar energy conversion   827, 832, 966
solenoid   10, 11, 14, 199, 259, 761, 763, 939
Solomon-Bloembergen-Morgen (SBM) theory   927
SOMO, *see* singly occupied molecular orbital
SPDT switch, *see* single pole double throw switch
spectral density   480, 489, 514, 632, 729, 735, 739, 746, 747, 927, 928
spectral density functions (SDF)   927–929
spectral extent   191, 621
spectral purity   138, 143, 146, 147, 149, 153, 164, 171
spectral-spatial imaging   773, 819
spectroscopic notation   60
spherical harmonics   64–66, 70, 539
spin concentrations   457, 723
spin-correlated radical pair (SCRP)   954, 967, 969, 972–974, 979–981
spin density   103, 315, 319, 505, 529, 562, 639, 769, 770, 772, 782, 785, 786, 884–888, 968, 969, 972
– at the nucleus   103
spin diffusion   207, 208, 456, 457, 483, 515, 720, 722, 739, 740, 942
spin dynamics   526, 528, 529, 952, 960, 980

spin echo   2, 41, 208, 211, 214, 219, 280, 284, 456, 463, 464, 483, 514, 515, 556, 564, 575, 641, 648, 879, 976, 995, 999
spin-echo correlated spectroscopy (SECSY)   2, 466, 517, 518, 658
spin echo, dephasing   720–724
spin flip   209, 456, 478, 528, 852, 889
spin-fracton relaxation   468, 485, 488, 489
spin Hamiltonian   102, 298, 328, 358, 385, 387, 417, 429, 599
– dipolar   315, 566, 892, 971
– effective   110, 853, 860
– eigenvalues of   361, 386, 411, 425
– electron Zeeman   35, 414, 449, 647
– ENDOR   211, 289, 386, 648, 656, 658
– field-dependent terms   331
– generalized   110
– isotropic hyperfine interaction   386
– nuclear quadrupole   35, 386
– nuclear Zeeman   35, 104, 108, 289, 386, 414, 526, 647
– operator   509
– permittivity   252, 253, 801, 803, 812
– perturbation expressions for eigenvalues of spin Hamiltonian   398, 437
– phenomenological   102, 106
– Zeeman   35, 103, 104, 106, 108, 209, 289, 329, 338, 350, 386, 387, 414, 425, 449, 504, 507, 526, 584, 591, 637, 647, 664, 892, 971
– zero-field terms   330
spin-Hamiltonian matrix   359, 418, 439
– computer diagonalization of SH matrix   439
spin-Hamiltonian parameters (SHP)   16, 20, 30, 32, 33, 41, 59, 84, 88, 297–324, 327, 330, 333, 360, 363, 385, 386, 388, 390–392, 394–396, 398, 400, 402, 404–406, 408–410, 412–417, 421, 430, 437, 438, 441, 447, 448, 591, 604, 606, 607, 609, 611, 613, 614, 616, 617, 1002
– $\tilde{g}^2$, $\tilde{A}^2$, $\tilde{D}$, and $\tilde{P}$ anisotropic tensors   599–616
– evaluation   385, 386, 395, 398, 404, 405, 408, 415, 606–613, 616
– signs of   30, 441
spin labels   17–19, 52, 204, 221, 238, 244, 245, 247, 249–251, 253, 255, 257, 259, 261, 263, 265, 267, 484, 499, 503, 514, 515, 521, 524, 539, 541, 547, 548, 619, 620, 625, 628, 629, 669, 737–741, 758, 759, 836, 894, 943, 1000, 1001
spin-lattice relaxation (SLR)   3, 46, 206, 207, 242, 335, 336, 341, 362, 455, 457, 460, 468, 473–479, 485–489, 494, 495, 505, 514, 570, 571, 720, 725–747, 923, 936, 962, 969, 980, *see also* "spin relaxation"
– techniques to measure relaxation times   455, 460, 461, 463, 465, 467
spin operators   20, 81, 102, 104, 106, 109, 111, 305, 327, 328, 358, 361, 365, 367, 369, 371, 373, 375, 377, 379, 387, 398, 408, 409, 507, 509, 527, 584, 586, 617, 647, 853
– spin matrices   350, 530, 532
– – matrix elements of spin operators   111, 327, 328, 365–379
spin-orbit coupling (SOC)   44, 59, 62–64, 72, 74, 78, 79, 81, 82, 103, 244, 305, 306, 309, 310, 315–319, 333–335, 339–341, 343, 347, 348, 354–358, 362, 412, 437, 468, 660, 673, 719, 731, 735, 737, 740, 741, 887, 895, 972
spin-orbit coupling parameter   887
spin packet   463, 467
spin polarization   211, 467, 893, 948, 952, 958, 961, 965–968, 971, 972, 974, 978–980, 982, 984, 999
spin population   315, 972, 973
spin relaxation   47, 191, 199, 206, 284, 456–458, 461, 466, 482, 498, 517, 520, 521, 549, 556, 654, 719–748, 759, 796, 883, 886, 889, 892, 922, 926, 949, 950, 952, 954, 957, 960, 968, 970, 974, 975, 980, 1003
spin-rotational coupling   719, 740
spin-rotational interaction   480, 482, 741
spin sensitivity   185, 291, 571, 812, 814, 821
spin-spin relaxation   47, 191, 199, 206, 456–458, 461, 466, 654, 720, 721, 723, 725, 796, 922
spin system   41, 148, 191, 202, 203, 205, 206, 209, 216, 218, 273, 454, 457–459, 465, 470, 495, 568, 638, 650, 652, 656, 719, 726, 729, 734, 742, 744, 747, 811, 860, 879, 894, 903, 922, 926, 931, 933, 934, 939, 941, 942, 958, 999, 1001
spin temperature   922, 931, 932, 934
– negative   931, 932, 934
spin trapping   18–20, 827
spin vector   551, 575
spins, minimum detectable   270
SPM, *see* superposition model
SPST switch, see single pole-single throw switch
square-planar copper   660, 662, 664, 670
SRLS, *see* slowly relaxing local structure
SSM, *see* single-sideband modulator
standing waves   7, 177, 290

stochastic Liouville equation (SLE) 498–500, 505–507, 511, 513, 514, 523–526, 528, 536–539, 977
sucking-in effect 256
super-hyperfine interaction (SHF) 103, 104
superconducting gaps (SCGs) 913–918
superparamagnetic resonance 828
superparamagnetism 17, 828, 845
superposition model (SPM) 57, 58, 83–85, 88, 89
surface radicals 833
surroundings 16, 59, 85, 438, 483, 892
susceptibility, *see* magnetic susceptibility
switching of the external magnetic field (SEMF) 960
SWNTs, *see* single-wall carbon nanotubes
Symmetry 107–108
Symmetry, descent of 329
– Cubic 62–64, 67, 73, 96, 108, 110, 330, 343, 351–357, 637, 657
– low symmetry 320, 361, 413, 414, 449, 599, 746
– Monoclinic 108, 330, 347, 359, 657
– Orthorhombic 108, 361, 386, 413
– Spherical 59, 108, 110
– Tetragonal 64, 65, 78, 334, 343, 351, 855, 868
– Triclinic 107, 448, 618, 657
symmetry axis 121, 123, 126, 330, 350, 357, 430, 883, 900
symmetry, local 110, 852, 884
synchronous demodulation 139, 141, 143–145, 149, 159, 162, 172, 175

t

Tanabe-Sugano diagrams 73
TEM, *see* transmission electron microscopy
temperature
– effective 46, 95, 201, 273, 278, 341, 348, 362, 455, 470, 473, 474, 483, 495, 521, 628, 797, 801, 804, 812, 829, 837, 861, 871, 890, 922, 932, 934, 979
– liquid helium 14, 23, 404, 405, 463, 999
terminating load 5
tetragonal field 64, 336, 340, 342, 354
thermal equilibrium 460, 561, 727, 867, 924, 930, 931, 933, 934, 970, 974, 976, 980
three-coordinate copper 677, 678, 681, 711
three-/four-pulse DEER/PELDOR 546, 549, 555, 557, 575

three spin mixing (TSM) 954, 964, 965
Tikhonov regularization 221, 223, 547, 573, 574
time constant 9, 199, 200, 203, 205, 208, 275, 285, 455, 456, 462, 464–467, 479, 536, 678, 692, 721, 774, 881, 882, 890
time domain 203, 231, 524, 536, 554, 557, 568
time resolution 19, 193, 196, 806, 900, 940, 957, 975, 998, 1001
toggling frame 219
tomography 777, 778
transition metal ions 57, 59, 80, 85, 327, 328, 330, 332–363, 366, 368, 370, 372, 374, 376, 378, 380, 382, 384, 412, 596, 620, 637, 655, 743, 851, 1001, 1002
– First transition series ($3d^n$) ions
– raising operator 529
– Second and Third transition series ($4d^n$, $5d^n$) ions
transition moment 499, 733
transition probability 413, 414, 418, 424, 441, 450, 472, 657, 658, 930
transmission electron microscopy (TEM) 826
transmission line 119, 154, 155, 161, 194, 230, 262, 272, 273, 291, 801, 808, 937, 939
transmitter 130, 190, 193–197, 199, 200, 957
trapped atoms and molecules 20
trapped electrons 477
traveling-wave tube amplifier (TWTA) 195, 196
traveling-wave tubes 12
trigonal field 336, 339, 342
triplet state 105, 338, 344, 881, 895, 896, 899, 942, 948–952, 964, 967–969, 972, 983, 985, 999, 1001, 1002
TSM, *see* three spin mixing
two-dimensional electron-electron double resonance (2D-ELDOR) 497, 517–522, 541
TWTA, *see* traveling-wave tube amplifier

u

uniform field cavities 249, 258–260

v

VCO, *see* voltage-controlled oscillator
vibronic coupling 94, 95, 99–101
visible light 832
voltage standing wave ratio (VSWR) 184
voltage-controlled oscillator (VCO) 134, 180, 182
VSWR, *see* voltage standing wave ratio

W

W-band EPR, see electron paramagnetic resonance (EPR)
wavefunctions   42, 57–60, 62, 63, 74, 75, 80, 82, 86, 92, 95–100, 103, 301, 335, 347, 430, 672, 928
– antisymmetrized   301
– electronic   59, 74, 95, 96, 99
– orthogonal   58
westiellamide complexes   707
Wigner-Eckart theorem   64, 69, 77, 96, 308

X

X-band EPR, see electron paramagnetic resonance (EPR)
X-ray absorption near edge structure (XANES)   901
XANES, see X-ray absorption near edge structure
Xsophe   658, 665, 673, 689, 693

Z

Zeeman energy   50, 59, 105, 307, 393, 933, 979
– dominant   59, 979
– electronic   59, 105
– nuclear   59, 393, 933, 979
Zeeman field   42, 201, 204–206, 211, 212, 218, 219, 329, 340, 356, 358, 609, 611, 883, 886, 890, 892, 902, 903, 996, 999, 1000
Zeeman interaction   25, 28, 30, 35, 38, 39, 42, 59, 78, 101, 103, 104, 108, 209, 338, 414, 425, 606, 607, 647, 650, 652, 664, 970
– electron and nuclear   647, 970
– second-order   28, 38, 414, 652
Zeeman terms   106, 414, 416, 507, 579, 664, 925, 931
– electronic   925, 931
– high-spin   416
– nuclear   106, 414, 925, 931
zero-field EPR (ZFR)   16, 28, 117–128, 159, 167, 168, 209, 314, 341, 392, 417, 441, 591, 592, 620, 650, 853, 854, 863, 871, 879, 895, 903, 927, 968, 998, 999
– examples   119, 121
– spectra   119, 124
– spectrometer   117, 119, 120
– zero-field energy levels   122
zero-field splitting (ZFS)   16, 28, 105, 106, 117, 167, 298, 308, 314, 341, 392, 417, 473, 591–596, 599, 620, 650, 719, 731, 735, 743–747, 846, 879, 895, 903, 927, 999
zero-order regular approximation (ZORA)   673
zero-quantum coherence (ZQC)   220, 221
ZFR, see zero-field EPR
ZFS, see zero-field splitting
ZORA, see zero-order regular approximation
ZQC, see zero-quantum coherence